Lecture Notes in Computer Science 5369

Commenced Publication in 1973
Founding and Former Series Editors:
Gerhard Goos, Juris Hartmanis, and Jan van Leeuwen

Seok-Hee Hong   Hiroshi Nagamochi
Takuro Fukunaga (Eds.)

Algorithms and Computation

19th International Symposium, ISAAC 2008
Gold Coast, Australia, December 15-17, 2008
Proceedings

 Springer

Volume Editors

Seok-Hee Hong
School of Information Technologies
University of Sydney, Australia
E-mail: shhong@it.usyd.edu.au

Hiroshi Nagamochi
Graduate School of Informatics
Kyoto University, Kyoto, Japan
E-mail: nag@amp.i.kyoto-u.ac.jp

Takuro Fukunaga
Graduate School of Informatics
Kyoto University, Kyoto, Japan
E-mail: takuro@amp.i.kyoto-u.ac.jp

Library of Congress Control Number: 2008940692

CR Subject Classification (1998): F.2, C.2, G.2, I.3.5, C.2.4

LNCS Sublibrary: SL 1 – Theoretical Computer Science and General Issues

ISSN   0302-9743
ISBN-10   3-540-92181-8 Springer Berlin Heidelberg New York
ISBN-13   978-3-540-92181-3 Springer Berlin Heidelberg New York

Springer is a part of Springer Science+Business Media

springer.com

© Springer-Verlag Berlin Heidelberg 2008
Printed in Germany

Typesetting: Camera-ready by author, data conversion by Scientific Publishing Services, Chennai, India
Printed on acid-free paper   SPIN: 12578184   06/3180   5 4 3 2 1 0

Preface

This volume contains the proceedings of the 19th International Symposium on Algorithms and Computation (ISAAC 2008), held on the Gold Coast, Australia, December 15–17, 2008. In the past, it was held in Tokyo (1990), Taipei (1991), Nagoya (1992), Hong Kong (1993), Beijing (1994), Cairns (1995), Osaka (1996), Singapore (1997), Daejeon (1998), Chennai (1999), Taipei (2000), Christchurch (2001), Vancouver (2002), Kyoto (2003), Hong Kong (2004), Hainan (2005), Kolkata (2006), and Sendai (2007).

ISAAC is an annual international symposium that covers the very wide range of topics in the field of algorithms and computation. The main purpose of the symposium is to provide a forum for researchers working in algorithms and theory of computation from all over the world. In response to our call for papers, we received 229 submissions from 40 countries. The task of selecting the papers in this volume was done by our Program Committee and many other external reviewers. After an extremely rigorous review process and extensive discussion, the Committee selected 78 papers. We hope all accepted papers will eventually appear in scientific journals in a more polished form. Two special issues, one of *Algorithmica* and one of the *International Journal on Computational Geometry and Applications*, with selected papers from ISAAC 2008 are in preparation.

The best paper award was given to Takehiro Ito, Takeaki Uno, Xiao Zhou and Takao Nishizeki for "Partitioning a Weighted Tree to Subtrees of Almost Uniform Size." Selected from seven submissions authored by students only, the best student paper award was given to Ludmila Scharf and Marc Scherfenberg for "Inducing Polygons of Line Arrangements." Three prominent invited speakers, Tetsuo Asano, JAIST, Japan, Peter Eades, University of Sydney, Australia, and Robert Tarjan, Princeton University, HP, USA, also contributed to the program.

We would like to thank all the Program Committee members and external reviewers for their excellent work, especially given the time constraints. We also thank all those who submitted papers for consideration, thereby contributing to the high quality of the conference. We would like to thank our supporting organizations for their assistance and support. Finally, we are deeply indebted to the Organizing Committee members whose excellent effort and professional service to the community made the conference an unparalleled success.

December 2008

Seok-Hee Hong
Hiroshi Nagamochi
Takuro Fukunaga

Organization

Program Committee

Kazuyuki Amano	Gunma University, Japan
Lars Arge	University of Aarhus, Denmark
Cristina Bazgan	Universite Paris-Dauphine, France
Christoph Buchheim	University of Cologne, Germany
Francis Chin	University of Hong Kong, Hong Kong
Kyung-Yong Chwa	KAIST, Korea
Ding-Zhu Du	University of Texas at Dallas, USA
Thomas Erlebach	University of Leicester, UK
Sandor Fekete	Braunschweig University of Technology, Germany
Andrew Goldberg	Microsoft Research, USA
Refael Hassin	Tel Aviv University, Israel
Seok-Hee Hong (Co-chair)	University of Sydney, Australia
Costas Iliopoulos	University of London, UK
Hiro Ito	Kyoto University, Japan
Ming-Yang Kao	Northwestern University, USA
Michael Kaufmann	University of Tübingen, Germany
Giuseppe Liotta	University of Perugia, Italy
Hsueh-I Lu	National Taiwan University, Taiwan
Anna Lubiw	University of Waterloo, Canada
Tomomi Matsui	Chuo University, Japan
Brendan McKay	Australian National University, Australia
Pat Morin	Carleton University, Canada
Hiroshi Nagamochi (Co-chair)	Kyoto University, Japan
Kunsoo Park	Seoul National University, Korea
Frank Ruskey	University of Victoria, Canada
Ileana Streinu	Smith College, USA
Takeshi Tokuyama	Tohoku University, Japan
Anastasios Viglas	University of Sydney, Australia
Koichi Wada	Nagoya Institute of Technology, Japan
Lusheng Wang	City University of Hong Kong, Hong Kong
Osamu Watanabe	Tokyo Institute of Technology, Japan
Koichi Yamazaki	Gunma University, Japan
Hsu-Chun Yen	National Taiwan University, Taiwan

Organizing Committee

Sharon Chambers	University of Sydney, Australia
Takuro Fukunaga (Co-chair)	Kyoto University, Japan

Seok-Hee Hong (Co-chair)   University of Sydney, Australia
Tony (WeiDong) Huang   University of Sydney, Australia
Takashi Imamichi   Kyoto University, Japan
Ehab Morsy   Kyoto University, Japan
Wu Quan   University of Sydney, Australia
Jiexun Wang   Kyoto University, Japan

ISAAC Advisory Committee

Francis Chin   University of Hong Kong, China
Kyung-Yong Chwa   KAIST, Korea
Ding-Zhu Du   University of Texas at Dallas, USA
Peter Eades   University of Sydney, Australia
Wen-Lian Hsu   Academia Sinica, Taiwan
Der-Tsai Lee   Academia Sinica, Taiwan
Takao Nishizeki   Tohoku University, Japan, Chair
Takeshi Tokuyama   Tohoku University, Japan

Sponsors

The University of Sydney
The IEICE Information and Systems Society, Technical Committee on COMP
Information Processing Society of Japan, Special Interest Groups on Algorithms

External Reviewers

Mohammad Abam   Hans Bodlaender
Umut Acar   Janina Brenner
Mustaq Ahmed   Tianmin Bu
Khaled Almiáni   Hans-Joachim Böckenhauer
Andris Ambainis   Valentina Cacchiani
Patrizio Angelini   Joseph Chan
Spyros Angelopoulos   Kun-Mao Chao
Pavlos Antoniou   Chandra Chekuri
Toru Araki   Kuan-Ling Chen
Marta Arias   Ho-Lin Chen
Yuichi Asahiro   Siu-Wing Cheng
Yasuhito Asano   Victor Chepoi
James Aspnes   Taenam Cho
Yossi Azar   Sunghee Choi
Maxim Babenko   Jinhee Chun
Sang Won Bae   Tom Coleman
Holger Bast   Sébastien Collette
Carla Binucci   Colin Cooper

Camil Demetrescu
Der-Jiunn Deng
Walter Didimo
Bojan Djordjevic
Stephan Eidenbenz
Leah Epstein
Bruno Escoffier
Piotr Faliszewski
Chun-I Fan
Mike Fellows
Henning Fernau
Simon Fischer
Alan Frieze
Bernhard Fuchs
Toshihiro Fujito
Akihiro Fujiwara
Hiroshi Fujiwara
Takuro Fukunaga
Peter Gacs
Markus Geyer
Emilio Di Giacomo
Xavier Goaoc
Laurent Gourvès
Chris Gray
Peter Hachenberger
Mohammad Taghi Hajiaghayi
Xin Han
Sariel Har-Peled
Toru Hasunuma
Meng He
Keld Helsgaun
Hiroshi Hirai
Markus Holzer
Takashi Horiyama
Kuen-Bang Hou
Sheng-Ying Hsiao
Dai-Sin Hsu
John Iacono
Katsunobu Imai
Toshimasa Ishii
Takeshiro Ito
Toshiya Itoh
Kazuo Iwama
Chuzo Iwamoto
Krishnam Raju Jampani

Inuka Jayasekera
Hirotsugu Kakugawa
Sayaka Kamei
Tom Kamphans
George Karakostas
Yoshiyuki Karuno
Naoki Katoh
Akinori Kawachi
Ken-ichi Kawarabayashi
Mark Keil
Shuji Kijima
David Kirkpatrick
Masashi Kiyomi
Rolf Klein
Kojiro Kobayashi
Stephen Kobourov
Stavros Kolliopoulos
Matias Korman
Takeshi Koshiba
Stephan Kottler
Marc van Kreveld
Alexander Kroeller
Vincent Kuo
Benjamin Lafreniere
Rob LeGrand
Mun-Kyu Lee
Jon Lenchner
Janice Leung
Asaf Levin
Weifa Liang
Xuemin Lin
Chun-Cheng Lin
Tsung-Hao Liu
Zvi Lotker
Chi-Jen Lu
Marco Lübbecke
Gary MacGillivray
Meena Mahajan
Michael Mahoney
Toshiya Mashima
Toshimitsu Masuzawa
Jiri Matousek
Akihiro Matsuura
Alexander Meduna
Henk Meijer

Spiros Michalakopoulos
Michael Miller
Ilya Mironov
Joseph Mitchell
Hiroyoshi Miwa
Eiji Miyano
Shuichi Miyazaki
Manal Mohamed
Jerome Monnot
Ehab Morsy
Gabriel Moruz
Mitsuo Motoki
Laurent Mouchard
Ian Munro
S. Muthukrishnan
Wendy Myrvold
Koji Nakano
Shin-ichi Nakayama
Rolf Niedermeier
Mitsunori Ogihara
Yoshio Okamoto
Seiya Okubo
Hirotaka Ono
Fukuhito Ooshita
Aris Pagourtzis
Amit Patel
David Peleg
Ugo Pietropaoli
Valentin Polishchuk
Sheung-Hung Poon
Tomasz Radzik
Mina Razaghpour
Joachim Reichel
Michal Ren
Fran Rosamond
Srinivasa Rao Satti
Joe Sawada
Guido Schaefer
Christiane Schmidt
Michael Schulz
Danny Segev
Jeffrey Shallit
Ting Shen
Hung-Yi Shih
Akiyoshi Shioura

Takayoshi Shoudai
Martin Skutella
Michiel Smid
Shakhar Smorodinsky
Sagi Snir
Venkatesh Srinivasan
Grzegorz Stachowiak
Kathleen Steinhofel
Lorna Stewart
Svetlana Stolpner
Kokichi Sugihara
Yien-Pong Sung
Wing-Kin Sung
Siamak Taati
Hisao Tamaki
Arie Tamir
Takeyuki Tamura
Keisuke Tanaka
Yuuki Tanaka
Seiichi Tani
Dimitrios Thilikos
Alex Thomo
Ioan Todinca
Iannis Tourlakis
Shi-Chun Tsai
Ryuhei Uehara
Marc Uetz
Alper Ungor
Takeaki Uno
Jan Vahrenhold
Jenn Vandenbussche
Yann Vaxès
Santosh Vempala
Chang-Yi Wang
Bow-Yaw Wang
Ian Wanless
John Watrous
Renato Werneck
Aaron Williams
Hans-Christoph Wirth
Ronald de Wolf
Alexander Wolff
Damien Woods
Kevin Wortman
Takuya Yoshihiro

Masaki Yamamoto
Hiroki Yanagisawa
Makoto Yokoo
Norbert Zeh
Zhao Zhang

Yong Zhang
Zhang Zhao
Liang Zhao
Feng Zou
Katharina Zweig

Table of Contents

Invited Talk

1A Approximation Algorithm I

1B Online Algorithm

2A Data Structure and Algorithm

2B Game Theory

3A Graph Algorithm I

3B Fixed Parameter Tractability

4A Distributed Algorithm

4B Database

5A Approximation Algorithm II

5B Computational Biology

6A Computational Geometry I

6B Complexity I

7A Computational Geometry II

7B Network

8A Optimization

8B Routing

9A Graph Algorithm II

9B Complexity II

Constant-Working-Space Algorithms:
How Fast Can We Solve Problems without
Using Any Extra Array?

Tetsuo Asano

School of Information Science, Jaist,
1-1 Asahidai, Nomi, Ishikawa, 923-1292, Japan

In this talk I will present a new direction of algorithms which do not use any extra working array. More formally, we want to design efficient algorithms which require no extra array of size depending on input size n but use only constant working storage cells (variables), each having $O(\log n)$ bits. As an example, consider a problem of finding the median among n given numbers. A linear-time algorithm for the problem is well known. An ordinary implementation of the algorithm requires another array of the same size. It is not very hard to implement the algorithm without using any additional array, in other words, to design an in-place algorithm. Unfortunately, it is *not* a constant working space algorithm in our model since it requires some space, say $O(\log n)$ space, for maintaining recursive calls. A good news is that an efficient algorithm is known which finds the median in $O(n^{1+\epsilon})$ time using $O(1/\epsilon)$ working space for any small positive constant ϵ. The algorithm finds the median without altering any element of an input array. In other words, the input array is considered as a read-only array.

A main interest in this talk is how to design algorithms using only constant working space in addition to input arrays. There are two different situations depending on whether input arrays are read-only or not. If input data are stored in a read-only array and only constant working space is allowed in an algorithm, it is called a *constant working space algorithm with a read-only array*. If we can read any array element and write any information of $\log n$ bits into any array element in constant time, it is called a *constant working space algorithm with a read-write array*. The latter one is usually called as an in-place algorithm.

In this talk I will introduce several constant working space algorithms with read-write input arrays or with read-only input arrays. Such problems have been investigated in the community of complexity theory under the name of *log-space* computation. The *log-space* implies the working space of $O(\log n)$ bits for an input size n. There is no difference but their names. In the complexity theory a main concern is whether a problem belongs to *log-space*, that is, whether it is solvable in polynomial time using only small working space of $O(\log n)$ bits in total. My concern is not only polynomial-time solvability but also computational performance of the algorithm.

S.-H. Hong, H. Nagamochi, and T. Fukunaga (Eds.): ISAAC 2008, LNCS 5369, p. 1, 2008.
© Springer-Verlag Berlin Heidelberg 2008

Some Constrained Notions of Planarity

Peter Eades

School of Information Technologies,
University of Sydney
peter@it.usyd.edu.au

Graph Drawing is the art and science of making pictures of graphs. Planarity has always played a central role in graph drawing research. Effective, efficient and elegant methods for drawing planar graphs were developed over the course of the last century by Wagner, Hopcroft and Tarjan, Read, de Frassieux, Pach and Pollack, amongst others.

In the past 30 years, diverse sections of the information industry have motived a shift in graph drawing, from pure mathematics to industrial research. This has come from the need to make large and complex data sets comprehensible to humans. In the mid 1990s, Purchase carried out a set of human experiments that justified algorithmic planarity research, by showing that edge crossings inhibit human understanding of graphs.

The shift to a focus on applications brought a number of constraints, and a number of constrained notions of planarity. These include:

- Upward planarity, for directed acyclic graphs, where edges are drawn monotonically upward;
- Hierarchical planarity, a variation on upward planarity where vertices are constrained to a set of horizontal lines;
- Clustered planarity, where the vertices have a cluster hierarchy and each cluster is drawn as a region in the plane;
- Orthogonal planarity, where the edges must consist of horizontal and vertical line segments;
- Symmetric planarity, where the drawing must display a given automorphism.

In this talk we discuss algorithmic research on some such constrained notions of planarity.

S.-H. Hong, H. Nagamochi, and T. Fukunaga (Eds.): ISAAC 2008, LNCS 5369, p. 2, 2008.

Reachability Problems on Directed Graphs

Robert E. Tarjan[1,2][*]

[1] Princeton University, Princeton NJ 08544
[2] HP Laboratories, Palo Alto CA 94304
robert.tarjan@hp.com

Abstract. I will present recent work and open problems on two directed graph reachability problems, one dynamic, one static. The dynamic problem is to detect the creation of a cycle in a directed graph as arcs are added. Much progress has been made recently on this problem, but intriguing questions remain. The static problem is to compute dominators and related information on a flowgraph. This problem has been solved, but the solution is complicated, and there are related problems that are not so well understood. The work to be discussed is by colleagues, other researchers, and the speaker.

[*] Research at Princeton University partially supported by NSF Grants CCF-0830676 and CCF-0832797. The information contained herein does not necessarily reflect the opinion or policy of the federal government and no official endorsement should be inferred.

S.-H. Hong, H. Nagamochi, and T. Fukunaga (Eds.): ISAAC 2008, LNCS 5369, p. 3, 2008.

Greedy Construction of 2-Approximation Minimum Manhattan Network[*]

Zeyu Guo[1], He Sun[1,**], and Hong Zhu[2]

[1] Fudan University, China
[2] East China Normal University, China

Abstract. Given a set T of n points in $\mathbb{R}^2$, a Manhattan Network G is a network with all its edges horizontal or vertical segments, such that for all $p, q \in T$, in G there exists a path (named a Manhattan path) of the length exactly the Manhattan distance between p and q. The Minimum Manhattan Network problem is to find a Manhattan network of the minimum length, *i.e.*, the total length of the segments of the network is to be minimized. In this paper we present a 2-approximation algorithm with time complexity $O(n \log n)$, which improves the 2-approximation algorithm with time complexity $O(n^2)$. Moreover, compared with other 2-approximation algorithms employing linear programming or dynamic programming technique, it was first discovered that only greedy strategy suffices to get 2-approximation network.

Keywords: Minimum Manhattan Network, approximation algorithm, greedy strategy.

1   Introduction

A *rectilinear path* between two points $p, q \in \mathbb{R}^2$ is a path connecting p and q with all its edges horizontal or vertical segments. Furthermore, a *Manhattan path* between p and q is a rectilinear path with its length exactly $\mathrm{dist}(p, q) :=$ $|p.x - q.x| + |p.y - q.y|$, *i.e.*, the Manhattan distance between p and q.

Given a set T of n points in $\mathbb{R}^2$, a network G is said to be a *Manhattan network* on T, if for all $p, q \in T$ there exists a Manhattan path between p and q with all its segments in G. For the given network G, let the length of G, denoted by $L(G)$, be the total length of all segments of G. For the given point set T, the *Minimum Manhattan Network* (MMN) Problem is to find a Manhattan network G on T with minimum $L(G)$.

From the problem description, it is easy to show that there is a close relationship between the MMN problem and planar t-spanners. For $t \geq 1$, if there exists a planar graph G such that for all $p, q \in T$, there exists a path in G connecting p and q of length at most t times the distance between p and q, G is said to be

---

[*] This work is supported by Shanghai Leading Academic Discipline Project(Project Number:B412), National Natural Science Fund (grant #60496321), and the ChunTsung Undergraduate Research Endowment.
[**] Correspondence author.

S.-H. Hong, H. Nagamochi, and T. Fukunaga (Eds.): ISAAC 2008, LNCS 5369, pp. 4–15, 2008.
© Springer-Verlag Berlin Heidelberg 2008

a *t-spanner* of T. The MMN Problem for T is exactly the problem to compute the 1-spanner of T under the L_1-norm.

Related works: Due to the numerous applications in city planning, network layout, distributed algorithms, and VLSI circuit design, the MMN problem was first introduced by Gudmundsson, Levcopoulos *et al.* [5], and until now, it is open whether this problem belongs to the complexity class P. Gudmundsson *et al.* [5] proposed an $O(n^3)$-time 4-approximation algorithm, and an $O(n \log n)$-time 8-approximation algorithm. Kato, Imai *et al.* [7] presented an $O(n^3)$-time 2-approximation algorithm. However, the proof of their algorithm correctness is incomplete [3]. In spite of that, their paper still provided a valuable idea, that it suffices for G to be a Manhattan network if for each of $O(n)$ certain pairs there exists a Manhattan path connecting its two points. Thus it is not necessary to enumerate all the pairs in $T \times T$. Following this idea, Benkert, Wolff *et al.* [1,2] proposed an $O(n \log n)$-time 3-approximation algorithm. They also described a mixed-integer programming (MIP) formulation of the MMN problem. After that, Chepoi, Nouioua *et al.* [3] proposed a 2-approximation rounding algorithm by solving the linear programming relaxation of the MIP. In this paper, the notions Pareto Envelope and a nice strip-staircase decomposition has been proposed first of all. In K. Nouioua's Ph.D thesis [8], the primal-dual based algorithm with 2-approximation and running time $O(n \log n)$ has been presented. After these works, it was Z. Guo *et al.* [6] who observed that the same approximation ratio can also be achieved using combinatorial construction. In their paper, the dynamic programming speed-up technique of quadrangle inequality was first used in this problem and, therefore the time complexity $O(n^2)$ has been achieved. In [9], S. Seibert and W. Unger proposed a 1.5-approximation algorithm. However, their proof is incorrect and 2-approximation is, to our best knowledge, the lowest ratio for this problem.

Our contributions: In this paper, we present a very simple 2-approximation algorithm for constructing Manhattan network with running time $O(n \log n)$. Compared with the simple 3-approximation algorithm with running time $O(n \log n)$ proposed recently [4] and the previous 2-approximation result [6] relying on dynamic programming speed-up technique, a highlight in our paper is that, except Pareto Envelope which is widely used in the previous literatures, it is proven simply greedy strategy is enough for constructing 2-approximation Minimum Manhattan Network.

Outline of our approach: From a high-level overview, our algorithm is as follows: partition the input into several blocks (ortho-convex regions) that can be solved independently of each other. For the blocks, some can be trivially solved optimally, whereas only one type of blocks is difficult to solve. For such a non-trivial block there are some horizontal and vertical strips which can be solved by horizontal and vertical nice covers plus switch segments to connect neighboring points in the same strip. In such manner, we divide each block into several staircases. In order to connect the points in each staircase, simple greedy strategy has been used.

2 Preliminaries

Basic notations: For $p = (p.x, p.y) \in \mathbb{R}^2$, let $\mathcal{Q}_k(p)$ denote the k-th closed quadrant with respect to the origin p, e.g., $\mathcal{Q}_1(p) := \{q \in \mathbb{R}^2 \mid p.x \leq q.x, p.y \leq q.y\}$.

Define $R(p, q)$ as a closed rectangle (possibly degenerate) where $p, q \in \mathbb{R}^2$ are its two opposite corners. $B_V(p, q)$ is defined as the vertical closed band bounded by p, q, whereas $B_H(p, q)$ denotes the horizontal closed band bounded by p, q.

For the given point set T, let Γ be the union of vertical and horizontal lines which pass through some point in T. In addition, we use $[c, d]$ to represent the vertical or horizontal segment with endpoints c and d, as Fig. 1 shows.

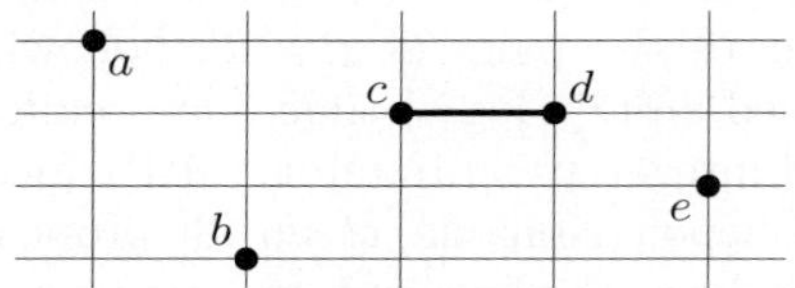

Fig. 1. $T = \{a, b, c, d, e\}$. The vertical and horizontal lines compose Γ.

Pareto envelope: The Pareto envelope, originally proposed by Chepoi *et al.* [3], plays an important role in our algorithm and we give a brief introduction.

Given the set of points T, a point p is said to be *dominated* by q if $\big(\forall t \in T : \mathrm{dist}(q, t) \leq \mathrm{dist}(p, t)\big) \wedge \big(\exists t \in T : \mathrm{dist}(q, t) < \mathrm{dist}(p, t)\big)$. A point is said to be an *efficient point* if it is not dominated by any point in the plane. The *Pareto envelope* of T is the set of all efficient points, denoted by $\mathcal{P}(T)$. Fig. 2 shows an example of $\mathcal{P}(T)$. It is not hard to prove that $\mathcal{P}(T) = \bigcap_{u \in T} \bigcup_{v \in T} R(u, v)$. For $|T| = n$, $\mathcal{P}(T)$ can be built in $O(n \log n)$ time. [3] also presented some other properties of $\mathcal{P}(T)$. In particular, $\mathcal{P}(T)$ is *ortho-convex*, *i.e.*, the intersection of $\mathcal{P}(T)$ with any vertical or horizontal line is continuous, which is equivalent to the fact that for any two points $p, q \in \mathcal{P}(T)$, there exists a Manhattan path in $\mathcal{P}(T)$ between p and q.

In [3] Chepoi *et al.* also showed that the Pareto envelope is the union of some ortho-convex (possibly degenerate) rectilinear polygons (called *blocks*). Two blocks can overlap at only one point which is called a *cut vertex*. We denote by C the set of cut vertices, and let $T^+ := T \cup C$. For a block B, denote by H_B and W_B its height and width respectively. Let $T_B := T^+ \cap B$. We say B is *trivial* if B is a rectangle (or degenerate to a segment) such that $|T_B| = 2$. It is known that the two points in T_B must be two opposite corners of B when it is trivial. In Fig. 2, $C = \{a, b, c, d\}$ and only the block between c and d is non-trivial.

Chepoi *et al.* [3] proved that an MMN on T^+ is also an MMN on T, and to obtain an MMN on T^+, it suffices to build an MMN on T_B for each $B \subseteq \mathcal{P}(T)$. The MMN in any trivial block B can be built by simply connecting the two points in T_B using a Manhattan path. So we have reduced the MMN problem on T to MMN on non-trivial blocks.

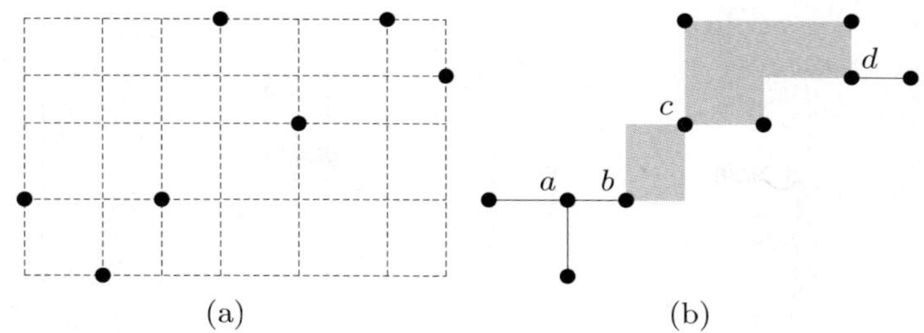

(a)   (b)

Fig. 2. An example of a Pareto envelope. The black points in (a) are the set T. The two separate grey regions in (b) are non-degenerate blocks, whereas the black lines are degenerate blocks. All these blocks form the Pareto envelope $\mathcal{P}(T)$.

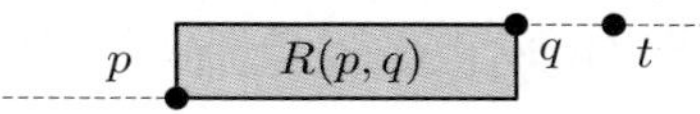

Fig. 3. The rectangle is a horizontal strip. Any point in T_B within $B_H(p,q)$ can only be placed on the dashed lines, *e.g.*, the point t.

For a non-trivial block B denote its border by ∂B and let $\Gamma_B := \Gamma \cap B$. We call a corner p in ∂B a *convex corner* if the interior angle at p equals to $\pi/2$, otherwise p is called a *concave corner*.

Lemma 1. *[3] For any non-trivial block B and any convex corner p in ∂B, it holds that $p \in T_B$.*

Lemma 2. *[3] For any non-trivial block B, there exists an MMN G_B on T_B such that $G_B \subseteq \Gamma_B$. Furthermore, any MMN $G_B \subseteq \Gamma_B$ on T_B contains ∂B.*

Strips and staircase components: Informally, for $p, q \in T_B, p.y < q.y$, we call $R(p,q)$ a *vertical strip* if it does not contain any point of T_B in the region $B_V(p,q)$ except the vertical lines $\{(x,y)|x = p.x, y \le p.y\}$ and $\{(x,y)|x = q.x, y \ge q.y\}$. Similarly, for the points $p, q \in T_B, p.x < q.x$, we call $R(p,q)$ a *horizontal strip* if it does not contain any point in the region $B_H(p,q)$ except the horizontal lines $\{(x,y)|x \le p.x, y = p.y\}$ and $\{(x,y)|x \ge q.x, y = q.y\}$. Especially, we say a vertical or horizontal strip $R(p,q)$ is *degenerate* if $p.x = q.x$ or $p.y = q.y$. Fig. 3 gives an example of a horizontal strip.

The other notion which plays a critical role in our algorithm is the staircase component. There are four kinds of staircase components specified by four quadrants, and without loss of generality we only describe the one with respect to the third quadrant. Suppose $R(p,q)$ is a vertical strip and $R(p',q')$ is a horizontal strip, such that $q \in \mathcal{Q}_1(p)$, $q' \in \mathcal{Q}_1(p')$, $p, q \in B_V(p',q')$, $p', q' \in B_H(p,q)$, *i.e.*, they cross in the way as Fig. 4 shows. Denote by $T_{pp'|qq'}$ the set of any point $v \in T_B$ such that $v.x > q.x, v.y > q'.y$, where p is the leftmost point and p' is the topmost point in $\mathcal{Q}_3(v)$ besides v. A non-empty $T_{pp'|qq'}$ is said to be a *staircase component* (see Fig. 4). In this figure, no point in T_B is located in the dark grey area and the two light grey unbounded regions except those in $T_{pp'|qq'}$.

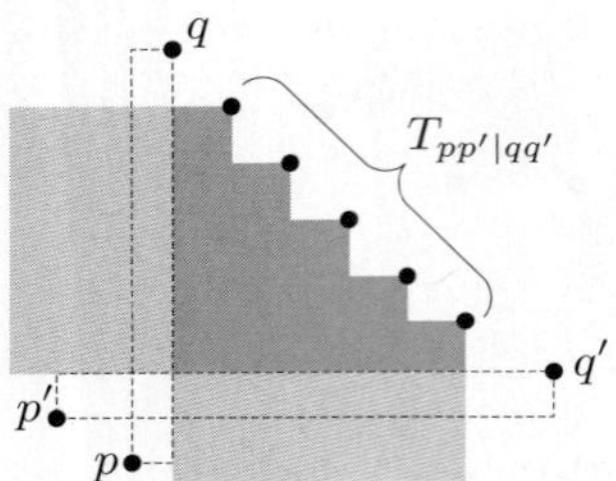

Fig. 4. A staircase component

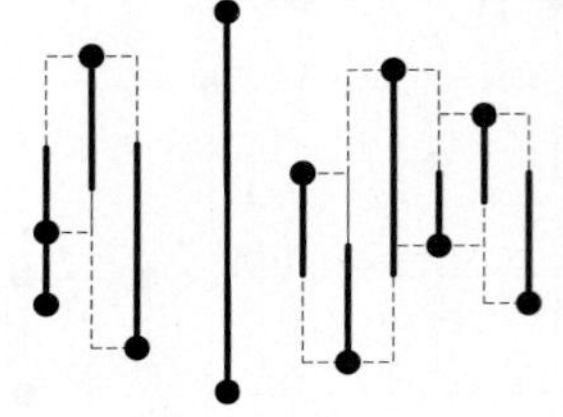

Fig. 5. An NVC consisting of black lines

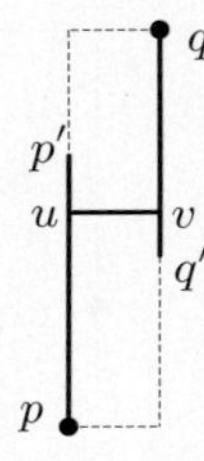

Fig. 6. A switch segment

For a strip $R(p, q)$, (p, q) is called a *strip pair*. For each staircase component $T_{pp'|qq'}$ and each point v in $T_{pp'|qq'}$, (v, p) (also (v, p')) is called a *staircase pair*.

Theorem 1. *[3] A network G_B is a Manhattan network on T_B if and only if for any strip pair or staircase pair (p, q), $p, q \in T_B$, there exists a Manhattan path in G_B connecting p and q.*

3   Algorithm Description

Following the approach of [1], a union of vertical segments C_V is said to be *a vertical cover* if for any horizontal line ℓ and any vertical strip R that ℓ intersects, it holds that $\ell \cap R \cap C_V \neq \emptyset$. Similarly, a union of horizontal segments C_H is said to be *a horizontal cover* if for any vertical line ℓ and any horizontal strip R that ℓ intersects, it holds that $\ell \cap R \cap C_H \neq \emptyset$. Furthermore, a *nice vertical cover* (NVC) is a vertical cover such that any of its segments contains at least one point of T_B. A *nice horizontal cover* (NHC) is defined symmetrically. Fig. 5 shows an NVC.

For an NVC C_V, obviously $[p, q] \subseteq C_V$ for every degenerate vertical strip $R(p, q)$. Assume $R(p, q)$ is a non-degenerate vertical strip where $p.y < q.y$, then there exists vertical segments $[p, p']$ and $[q, q']$ in C_V where $p'.y \geq q'.y$ (it is possible that $p = p'$ or $q = q'$), as Fig. 6 shows. Obviously, a Manhattan path connecting p and q can be built by adding a horizontal segment $[u, v]$ where $u.x = p.x, v.x = q.x, q'.y \leq u.y = v.y \leq p'.y$. Such a segment $[u, v]$ is said to be a *switch segment* of $R(p, q)$. The same concept for NHC is defined symmetrically.

Now we present an iterative algorithm `CreateNVC` to construct an NVC. In the initialization step, let C_V be the union of segments $[p, q]$ for each degenerate vertical strip $R(p, q)$, whereas N is the set of non-degenerate ones. In addition, let the set X be T_B. The main part of the procedure consists of two loops. Regarding the first loop, a vertical segment of ∂B in some $R(p, q) \in N$ is chosen in each round. Lemma 1 and the definition of strips guarantee that such segments must be connected to some point in X. Let the segment lying in the non-degenerate strip $R(p, q)$ be $[p, p']$, as Fig. 7 shows. Then p' is added to X, and $[p, p']$ is added to C_V. And by invoking `Update(p')`, N is updated to be the set of non-degenerate strips when the new set X is considered as the input point set. Define

$\bigcup N := \bigcup_{R(p,q)\in N} R(p,q)$. It is easy to see that the part of $\bigcup N$ adjacent to $[p, p']$ is eliminated in each round, which turns out that $\bigcup N$ becomes smaller. It can be demonstrated that $[p, p']$ is the unique vertical segment excluded from $\bigcup N$ in ∂B. We repeat the operations above until all the vertical segments initially falling in $\partial B \cap \bigcup N$ are excluded from $\bigcup N$ and added to C_V.

In the second loop, we choose $R(p, q) \in N$ arbitrarily and both its left and right edges are added to C_V. Two points $(p.x, q.y), (q.x, p.y)$ are added to X. And N is updated in the similar manner as Fig. 8 shows. The formal description is as follows.

Input: T_B
1 $C_V \leftarrow \bigcup [p, q]$ where $R(p, q)$ is a degenerate vertical strip;
2 $X \leftarrow T_B$;
3 $N \leftarrow \{R(p, q) \mid R(p, q)$ is a non-degenerate vertical strip$\}$;
4   **while** *there exists a vertical segment* $[p, p'] \subseteq \partial B \cap R(p, q)$, *where* $R(p, q) \in N$ **do**
5 $X \leftarrow X \cup \{p'\}$;
6 $C_V \leftarrow C_V \cup [p, p']$;
7 $Update(p')$;
8   **end**
9   **while** $N \neq \emptyset$ **do**
10   Let $R(p, q)$ be an arbitrary vertical strip in N;
11 $p' \leftarrow (p.x, q.y)$; $q' \leftarrow (q.x, p.y)$;
12 $X \leftarrow X \cup \{p', q'\}$;
13 $C_V \leftarrow C_V \cup [p, p'] \cup [q, q']$;
14 $Update(p')$; $Update(q')$;
15 **end**

Algorithm 1. CreateNVC

Lemma 3. CreateNVC *takes* $O(n)$ *time to output an NVC* C_V.

Proof. Since C_V initially contains $[p, q]$ for any degenerate vertical strip $R(p, q)$, a horizontal line ℓ that crosses $R(p, q)$ always intersects C_V. Therefore we only need to consider non-degenerate vertical strips.

We prove the following invariant maintains: let $R(p, q)$ be a vertical strip in the original N and ℓ be a horizontal line that intersects $R(p, q)$, then at any stage of the algorithm, either $\ell \cap R(p, q) \cap C_V \neq \emptyset$ or $\ell \cap R(p, q) \subseteq \bigcup N$ holds.

Input: v
1   **for** *each* $R(p, q) \in N$ *such that* $v.x = p.x, [p, v] \cap R(p, q) \neq \{p\}$ **do**
2 $N \leftarrow N \backslash \{R(p, q)\}$;
3   **if** $v \in R(p, q)$ *and* $v.y \neq q.y$ **then** $N \leftarrow N \cup \{R(v, q)\}$;
4   **end**

Algorithm 2. Update

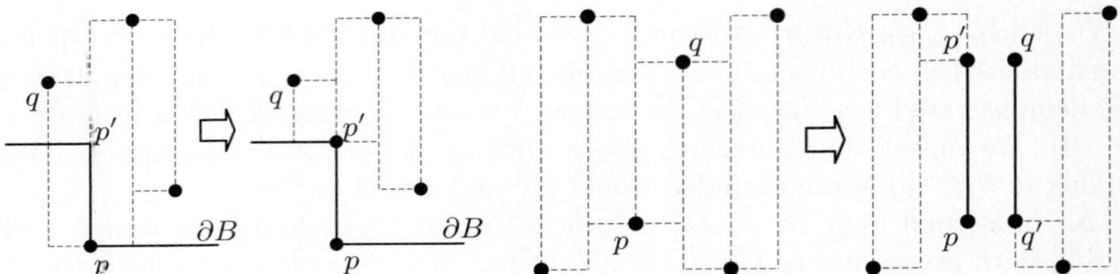

Fig. 7. The change in the first loop   **Fig. 8.** The change in the second loop

At the beginning obviously $\ell \cap R(p,q) \subseteq \bigcup N$ holds. Each time when N is updated, the part of $R(p,q)$ eliminated from $\bigcup N$ (if existing) must be adjacent to some segment which is added to C_V, so either $\ell \cap R(p,q) \cap C_V \neq \emptyset$ or $\ell \cap R(p,q) \subseteq \bigcup N$ still holds for the updated N. The set N will be updated iteratively until $\bigcup N = N = \emptyset$, which implies $\ell \cap R(p,q) \cap C_V \neq \emptyset$.

Secondly, we consider the running time of the procedure.

Line 1 takes $O(n)$ time since $O(n)$ degenerate vertical strips exist. Initially N contains $O(n)$ non-degenerate vertical strips and $\bigcup N$ contains $O(n)$ vertical segments of ∂B. The first loop reduces one such vertical segment in each round, whereas the second loop eliminates at least one strip in N in each round. Moreover, each invoking of the procedure **Update** takes $O(1)$ time since when a point is added to X, $O(1)$ strips need to be removed or replaced. Therefore the overall time complexity is $O(n)$.   □

After invoking **CreateNVC**, we add the topmost and bottommost switch segments for each non-degenerate vertical strip, as Fig. 9 shows. Then for each vertical strip $R(p,q)$, at least one Manhattan path between p and q is built. Symmetrically, we can use the algorithm **CreateNHC** to compute an NHC. Furthermore, for each horizontal strip, the leftmost and the rightmost switch segments are added. All these procedures guarantee that the Manhattan paths for all the strip pairs have been constructed.

Now we turn to the discussion of staircases. For simplicity, we only describe the definition of the staircase with respect to the third quadrant. The other cases are symmetric.

Definition 1 (staircase). *For a staircase component $T_{pp'|qq'}$ with respect to the third quadrant, assume $R(p,q)$ is a vertical strip and $R(p',q')$ is a horizontal strip. Let M_{pq} be the Manhattan path between p and q which passes through the bottommost switch segment. Let $M_{p'q'}$ be the Manhattan path between p' and q' which passes through the leftmost switch segment. The part of $\bigcup_{v \in T_{pp'|qq'}} Q_3(v)$ bounded by M_{pq} and $M_{p'q'}$, excluding $M_{pq}, M_{p'q'}, C_V, C_H$ is said to be a staircase, denoted by $S_{pp'|qq'}$.*

Fig. 10 gives an example of staircase.

Lemma 4. *There exists a procedure **CreateStaircasePath** such that for the given staircase $S_{pp'|qq'}$ with the staircase component $T_{pp'|qq'}$, $|T_{pp'|qq'}| = n$, the*

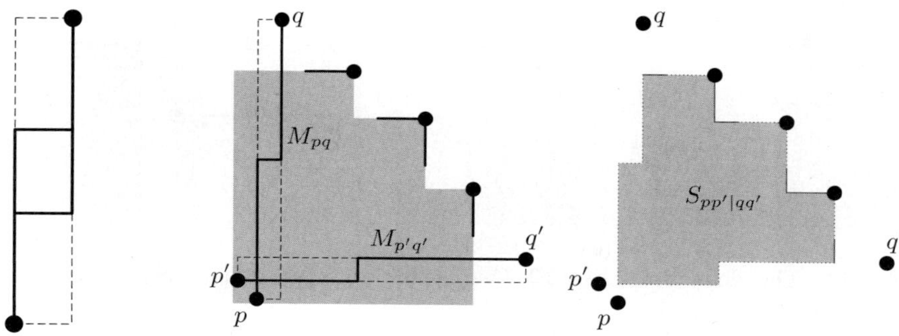

Fig. 9.   Adding switch segments

Fig. 10. The definition of the staircase $S_{pp'|qq'}$. The dotted lines in the right picture is not included in $S_{pp'|qq'}$.

procedure takes $O(n \log n)$ time to construct a network $G_{pp'|qq'} \subseteq S_{pp'|qq'}$ such that $G_{pp'|qq'} \cup C_H \cup C_V$ connects each point in $T_{pp'|qq'}$ to either M_{pq} or $M_{p'q'}$.

Proof. Without loss of generality, assume $R(p, q)$ is a vertical strip and $R(p', q')$ is a horizontal strip where $q \in \mathcal{Q}_1(p)$, $q' \in \mathcal{Q}_1(p')$, as Fig. 10 shows. Let $t_0 := q, t_{n+1} := q'$. Express the points in $T_{pp'|qq'}$ as $t_1, t_2, \cdots, t_n$ in the order from the topmost and leftmost one to the bottommost and rightmost one.

For $1 \leq i \leq n$, define the horizontal segment $h_i := \{(x, y) \mid y = t_i.y\} \cap S_{pp'|qq'}$ and the vertical segment $v_i := \{(x, y) \mid x = t_i.x\} \cap S_{pp'|qq'}$, as Fig 11 shows. We use $\mathrm{Right}_i(S_{pp'|qq'})$ to represent the staircase polygon on the right of v_i whereas $\mathrm{Top}_i(S_{pp'|qq'})$ represents the one on the top of h_i. Note that $\mathrm{Right}_i(S_{pp'|qq'})$ and $\mathrm{Top}_i(S_{pp'|qq'})$ are all smaller staircase polygons. Assume S is a general staircase polygon in $S_{pp'|qq'}$. Let $h_i^S := h_i \cap S, v_i^S := v_i \cap S$. It can be observed that $\langle h_i^S \rangle$ is ascending whereas $\langle v_i^S \rangle$ is descending. Define $\mathrm{Right}_i(S)$ and $\mathrm{Top}_i(S)$ in a similar way. The partial network $G_{pp'|qq'} \cap S$ is constructed in a recursive manner.

Input: S
1 **if** $S = \emptyset$ **then** return $\emptyset$;
2 **else if** $L\big(h_1^S\big) \geq L\big(v_1^S\big)$ **then** return $v_1^S \cup$ CreateStaircasePath($\mathrm{Right}_1(S)$);
3 **else if** $L\big(h_n^S\big) \leq L\big(v_n^S\big)$ **then** return $h_n^S \cup$ CreateStaircasePath($\mathrm{Top}_n(S)$);
4 **else**
5   Choose k such that $L\big(h_k^S\big) \leq L\big(v_k^S\big)$ and $L\big(h_{k+1}^S\big) \geq L\big(v_{k+1}^S\big)$;
6   return $h_k^S \cup v_{k+1}^S \cup$ CreateStaircasePath($\mathrm{Top}_k(S)$) $\cup$
        CreateStaircasePath($\mathrm{Right}_{k+1}(S)$);
7 **end**

Algorithm 3. CreateStaircasePath

Initially we invoke CreateStaircasePath($S_{pp'|qq'}$). For any non-empty S one of the three branches is chosen. In the third case, binary search guarantees the

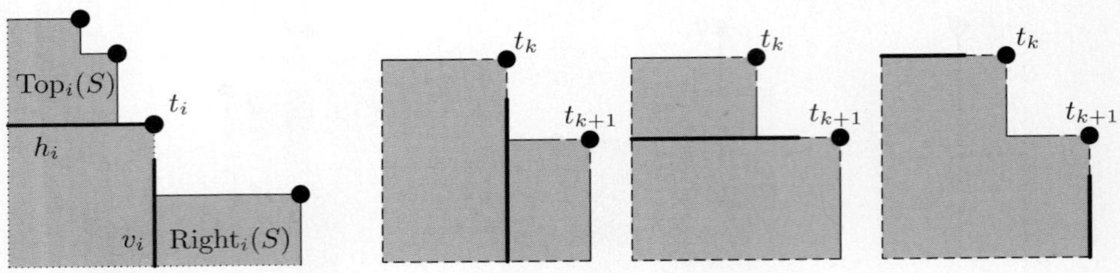

Fig. 11. The definition of h_i and v_i

Fig. 12. One of the three connections for t_k and t_{k+1} is optimal

proper k can be obtained with running time $O(\log n)$ whereas the procedure is invoked recursively at most $O(n)$ times, which results in the total running time $O(n \log n)$.

The correctness proof simply follows from the induction method. $\square$

In the following, we present the global algorithm `CreateMMN`.

Input: T
1 Compute $\mathcal{P}(T)$.
2 **for** *each trivial block $B \subseteq \mathcal{P}(T)$* **do**
3   connect the two points in T_B with a Manhattan path.
4 **for** *each non-trivial block $B \subseteq \mathcal{P}(T)$* **do**
5   `CreateNVC`;
6   **for** *each vertical strip $R(p,q)$* **do**
7   add the topmost and bottommost switch segments of $R(p,q)$;
8   `CreateNHC`;
9   **for** *each horizontal strip $R(p,q)$* **do**
10   add the leftmost and rightmost switch segments of $R(p,q)$;
11   **for** *each staircase $S_{pp'|qq'}$* **do** `CreateStaircasePath`$(S_{pp'|qq'})$;
12 **end**

Algorithm 4. `CreateMMN`

Theorem 2. *For the given point set T of size n, `CreateMMN` takes $O(n \log n)$ time to compute a Manhattan network G on T.*

Proof. For any non-trivial block, NVC, NHC and switch segments form the Manhattan paths for strip pairs, whereas some segments are added in staircases such that there exist Manhattan paths for staircase pairs. By Theorem 1, the final network is a Manhattan network.

Regarding the running time, it is well-known that computing the Pareto envelope and constructing the networks in staircases can be implemented in $O(n \log n)$ time, and the time required for decomposing each block into staircases and strips is also $O(n \log n)$ using the method similar to [1]. The other steps, including computing NVC, NHC and adding switch segments, can be implemented in linear time. Thus the overall time complexity is $O(n \log n)$. $\quad\square$

4 Approximation Analysis

The rest of this paper is devoted to the approximation analysis of this problem. Let G denote the Manhattan network constructed by our algorithm, whereas $G^\star$ is the optimal one demonstrated by Lemma 2 with the property that $\partial B \subseteq G^\star \cap B \subseteq \Gamma_B$ for every non-trivial block B. For any block B, let $G_B := G \cap B$ and $G_B^\star := G^\star \cap B$.

Let B be a non-trivial block. Denote by G_S the switch segments our algorithm adds when computing G_B. Let $\mathcal{S} := \bigcup S_{pp'|qq'}, G_U := G_B \cap \mathcal{S}$. From the algorithm description obviously $G_B := C_V \cup C_H \cup G_S \cup G_U$.

Let $G_C^\star := G_B^\star \cap (C_V \cup C_H)$, whereas $G_U^\star := G_B^\star \cap \mathcal{S}$.

Lemma 5. $L(C_V \cup C_H) \leq 2L(G_C^\star) - 2H_B - 2W_B$.

Proof. We divide $C_V \cup C_H$ into two parts: let C_1 be the set of segments for each degenerate vertical and horizontal strip, as well as the segments added in the first loop of procedures `CreateNVC` and `CreateNHC`. Let C_2 represent the union of the segments added in the second loop. In addition, denote $C_1^\star := G_C^\star \cap C_1$, and $C_2^\star := G_C^\star \cap C_2$.

Observing that C_1 is the union of the segments in degenerate strips and ∂B, it is easy to show that $\partial B \subseteq C_1 = C_1^\star$. Therefore $L(C_1) \leq 2L(C_1^\star) - L(\partial B) = 2L(C_1^\star) - 2H_B - 2W_B$.

On the other hand, let us consider the second loop of the procedure `CreateNVC` and `CreateNHC`. By symmetric property, we only analyze the procedure `CreateNVC`. In a round, two segments $[p, p']$, $[q, q']$ of length ℓ are added into C_V. By our algorithm, $R(p, q)$ is contained in some vertical strip $R(s, t)$. Since $G_B^\star$ is a Manhattan network, $C_2^\star \cap ([p, p'] \cup [q, q'])$ contains segments of length at least ℓ to connect s and t. Since the relation holds for each round and also the procedure `CreateNHC`, we obtain $L(C_2) \leq 2L(C_2^\star)$.

Combining the two inequalities above, we obtain the lemma. $\quad\square$

For any staircase $S_{pp'|qq'}$, let $G_{pp'|qq'}^\star := G_U^\star \cap S_{pp'|qq'}$.

Lemma 6. *For any staircase $S_{pp'|qq'}$, it holds $L(G_{pp'|qq'}) \leq 2L(G_{pp'|qq'}^\star)$.*

Proof. Without loss of generality, let $S_{pp'|qq'}$ be a staircase with respect to the third quadrant, as Fig. 10 shows. Let S be a staircase polygon in $S_{pp'|qq'}$ such that `CreateStaircasePath`(S) is invoked. We will prove $L\left(G_{pp'|qq'} \cap S\right) \leq 2L\left(G_{pp'|qq'}^\star \cap S\right)$ using induction.

The inequality obviously holds in the trivial case $S = \emptyset$. Assume the relation holds for smaller staircase polygons in S. For the case $L\left(h_1^S\right) \geq L\left(v_1^S\right)$, t_1 is connected down and the original problem is reduced to the small one with region $\mathrm{Right}_1(S)$. Let $S_R := S \backslash \mathrm{Right}_1(S)$. By assumption, we only need to prove

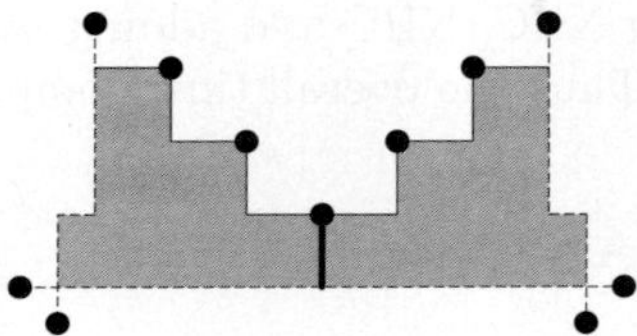

Fig. 13. The segment lying in two different staircases

$L\left(G_{pp'|qq'} \cap S_R\right) \le 2L\left(G^\star_{pp'|qq'} \cap S_R\right)$. Note that $L\left(G_{pp'|qq'} \cap S_R\right) = L\left(v_1^S\right)$, and in $G_{pp'|qq'} \cap S_R$ segments of length $\min\left\{L\left(h_1^S\right), L\left(v_1^S\right)\right\} = L\left(v_1^S\right)$ is necessary to connect t_1 to either the left or the bottom boundary of S. Thus the relation holds. The analysis for the case $L\left(h_n^S\right) \le L\left(v_n^S\right)$ is analogous.

Regarding the last case, let $S_R := S\backslash\left(\mathrm{Top}_k(S)\cup\mathrm{Right}_{k+1}(S)\right)$. We only need to prove $L\left(G_{pp'|qq'} \cap S_R\right) \le 2L\left(G^\star_{pp'|qq'} \cap S_R\right)$. As Fig. 12 shows, segments of length at least $\min\left\{L\left(v_k^S\right), L\left(h_{k+1}^S\right), L\left(h_k^S\right) + L\left(v_{k+1}^S\right)\right\}$ are necessary to connect t_k and t_{k+1} to either the left or the bottom boundary. By monotonicity, $L\left(v_{k+1}^S\right) \le L\left(v_k^S\right)$ and $L\left(h_k^S\right) \le L\left(h_{k+1}^S\right)$. By the choice of k, we obtain $L\left(v_{k+1}^S\right) \le L\left(h_k^S\right)$ and $L\left(h_k^S\right) \le L\left(v_{k+1}^S\right)$. Therefore $L\left(G_{pp'|qq'} \cap S_R\right) = L\left(h_k^S\right) + L\left(v_{k+1}^S\right) \le 2L\left(G^\star_{pp'|qq'} \cap S_R\right)$. $\qquad\square$

Now we estimate $L(G_U)$. Note that it is possible that some segments of $G_U^\star$ lie in two different staircases. Let $G_D^\star$ denote the union of these segments. Fig. 13 illustrates this special condition.

Lemma 7. $L(G_U) \le 2L(G_U^\star) + 2L(G_D^\star)$.

Proof. Since the segments of $G_D^\star$ are counted twice, we obtain that $L(G_U) \le \sum L(G_{pp'|qq'}) \le 2\sum L(G^\star_{pp'|qq'}) \le 2L(G_U^\star) + 2L(G_D^\star)$. $\qquad\square$

Lemma 8. $2L(G_D^\star) + L(G_S) \le 2H_B + 2W_B$.

Proof. The lemma can be obtained by the following fact: let ℓ be a vertical or horizontal line such that $\ell \not\subseteq \Gamma$, then ℓ may cross at most one segment in $G_D^\star$, and at most two segments in G_S. Furthermore, due to the definitions of strips and staircase components, ℓ cannot intersect both of $G_D^\star$ and G_S. We omit the details here. $\qquad\square$

Theorem 3. *For any block B, $L(G_B) \le 2L(G_B^\star)$.*

Proof. For any trivial block B, the relation obviously holds since $L(G_B) = H_B + W_B$. Let B be a non-trivial block, $L(G_B) \le L(C_V \cap C_H) + L(G_S) + L(G_U) \le 2L(G_C^\star) + 2L(G_U^\star) + 2L(G_D^\star) + L(G_S) - 2H_B - 2W_B \le 2L(G_C^\star) + 2L(G_U^\star)$.

Recall that $G_C^\star = G_B^\star \cap (C_V \cup C_H), G_U^\star = G_B^\star \cap S_U$. By the definition of staircases, it holds that $(C_V \cup C_H) \cap S_U = \emptyset$. This means $G_C^\star$ and $G_U^\star$ are disjoint parts of $G_B^\star$. Therefore $L(G_B) \le 2L(G_C^\star) + 2L(G_U^\star) \le 2L(G_B^\star)$. $\qquad\square$

Corollary 1. $L(G) \le 2L(G^\star)$. $\qquad\square$

References

1. Benkert, M., Wolff, A., Widmann, F.: The minimum Manhattan network problem: a fast factor-3 approximation. Technical Report 2004-16, Fakultät für Informatik, Universität Karlsruhe. In: Proceedings of the 8th Japanese Conference on Discrete and Computational Geometry, pp. 16–28 (2005) (short version)
2. Benkert, M., Shirabe, T., Wolff, A.: The minimum Manhattan network problem: approximations and exact solution. In: Proceedings of the 20th European Workshop on Computational Geometry, pp. 209–212 (2004)
3. Chepoi, V., Nouioua, K., Vaxès, Y.: A rounding algorithm for approximating minimum Manhattan networks. Theoretical Computer Science 390, 56–69 (2008); In: Proceedings of the 8th International Workshop on Approximation Algorithms for Combinatorial Optimization, pp. 40–51 (2005)
4. Fuchs, B., Schulze, A.: A simple 3-approximation of minimum Manhattan networks. Technical Report (2008),
   `http://www.zaik.uni-koeln.de/~paper/unzip.html?file=zaik2008-570.pdf`
5. Gudmundsson, J., Levcopoulos, C., Narasimhan, G.: Approximating a minimum Manhattan network. Nordic Journal of Computing 8, 219–232 (2001); In: Proceedings of the 2nd International Workshop on Approximation Algorithms for Combinatorial Optimization, pp. 28–37 (1999) (preliminary version)
6. Guo, Z., Sun, H., Zhu, H.: A fast 2-approximation algorithm for the minimum Manhattan network Problem. In: Proceedings of 4th International Conference on Algorithmic Aspect in Information Management, pp. 212–223 (2008)
7. Kato, R., Imai, K., Asano, T.: An improved algorithm for the minimum Manhattan network problem. In: Proceedings of the 13th International Symposium on Algorithms and Computation, pp. 344–356 (2002)
8. Nouioua, K.: Enveloppes de Pareto et Réseaux de Manhattan: Caractérisations et algorithmes, Ph.D. thesis, Université de la Méditerranée (2005)
9. Seibert, S., Unger, W.: A 1.5-approximation of the minimal Manhattan network problem. In: Proceedings of the 16th International Symposium on Algorithms and Computation, pp. 246–255 (2005)

The Complexity of Minimum Convex Coloring

Frank Kammer and Torsten Tholey

Institut für Informatik, Universität Augsburg, D-86135 Augsburg, Germany
{kammer,tholey}@informatik.uni-augsburg.de

Abstract. A coloring of the vertices of a graph is called convex if each subgraph induced by all vertices of the same color is connected. We consider three variants of recoloring a colored graph with minimal cost such that the resulting coloring is convex. Two variants of the problem are shown to be $\mathcal{NP}$-hard on trees even if in the initial coloring each color is used to color only a bounded number of vertices. For graphs of bounded treewidth, we present a polynomial-time $(2+\epsilon)$-approximation algorithm for these two variants and a polynomial-time algorithm for the third variant. Our results also show that, unless $\mathcal{NP} \subseteq DTIME(n^{O(\log\log n)})$, there is no polynomial-time approximation algorithm with a ratio of size $(1 - o(1))\ln\ln n$ for the following problem: Given pairs of vertices in an undirected graph of bounded treewidth, determine the minimal possible number l for which all except l pairs can be connected by disjoint paths.

Keywords: Convex Coloring, Maximum Disjoint Paths Problem.

1   Introduction

A colored graph (G, C) is a tuple consisting of a graph G and a *coloring* C of G, i.e., a function assigning each vertex v a color that is either 0 or a so-called *real color*. A vertex colored with 0 is also called *uncolored*. A coloring is an (a, b)-coloring if the color set used for coloring the vertices contains at most a real colors and if each real color is used to color at most b vertices. Two equal-colored vertices v and w in a colored graph (G, C) are *C-connected* if there is a path from v to w whose vertices are all colored with the color of u and v. A coloring C is called *convex* if all pairs of vertices colored with the same real color are C-connected. For a colored graph (G, C_1), another arbitrary coloring C_2 of G is also called a *recoloring* of (G, C_1). We then say that C_1 is the *initial coloring* of G and that C_2 *recolors* or *uncolors* a vertex v of G if $C_2(v) \neq C_1(v)$ and $C_2(v) = 0$, respectively. The *cost* of a recoloring C_2 of a colored graph (G, C_1) with $G = (V, E)$ is $\sum_{v \in V : 0 \neq C_1(v) \neq C_2(v)} w(v)$, where $w(v)$ denotes the weight of v with $w(v) = 1$ in the case of an unweighted graph. This means that we have to pay for recoloring or uncoloring a real-colored vertex, but not for recoloring an uncolored vertex. In the *minimum convex recoloring problem* (MCRP) we are given a colored graph and search for a convex recoloring with minimal cost.

The MCRP describes a fundamental problem in graph theory with different applications in practice: a first systematic study of the MCRP on trees is from

S.-H. Hong, H. Nagamochi, and T. Fukunaga (Eds.): ISAAC 2008, LNCS 5369, pp. 16–27, 2008.

Moran and Snir [10] and was motivated by applications in biology. Further applications are so-called multicast communications in optical wavelength division multiplexing networks; see, e.g., [6] for a short discussion of these applications. Here we focus on the MCRP as a special kind of routing problem. Suppose we are given a telecommunication or transportation network modeled by a graph whose vertices represent routers. Moreover, assume that each router can establish a connection between itself and an arbitrary set of adjacent routers. Then routers of the same initial color could represent clients that want to be connected by the other routers to communicate with each other or to exchange data or commodities. More precisely, connecting clients of the same color means finding a connected subgraph of the network containing all the clients, where the subgraphs for clients of different colors should be disjoint. If we cannot establish a connection between all the clients, we want to give up connecting as few clients to the other clients of the same color as possible in the unweighted case ($w(v) = 1$ for all $v \in V$) or to give up a set of clients with minimal total weight in the weighted case. Hence, our problem reduces to the MCRP, where a recoloring colors all those vertices with color c that represent routers used to connect clients of color c. The case in which routers can connect a constant number of disjoint sets of adjacent routers can be handled by copying vertices representing a router.

We also introduce a new relaxed version of the problem that we call the *minimum restricted convex recoloring problem* (MRRP). In this problem we ask for a convex recoloring C' that does not recolor any real-colored vertex with a different real color. In practice clients often cannot be used for routing connections for other clients so that a clear distinction between clients and routers should be made. This can be modeled by the MRRP, where a client that cannot be connected to the other clients of the same color may only be uncolored.

Finally, we consider a variant of the MCRP where we search for a convex recoloring, but assign costs to each color c. We have to pay the cost for color c if at least one vertex of color c is recolored. We call this coloring problem the *minimum block recoloring problem* or MBRP. In an unweighted version we assign cost 1 to each color. The MBRP is useful if in an application it is not useful to connect only a proper subset of clients that want to be connected.

The MCRP, the MRRP, and the MBRP can also be considered as generalizations of the *maximum disjoint paths problem* (MDPP) and the *disjoint paths problem* (DPP), where in the first case a maximum number and in the second case all pairs of given pairs of vertices of a graph are to be connected by vertex-disjoint paths, if possible. Indeed any algorithm solving one of our recoloring problems on $(\infty, 2)$-colorings to optimality also can solve the DPP. Given an algorithm for the MBRP one can also solve the problem of connecting a subset of a given set of weighted node pairs $(s_1, t_1), \ldots, (s_l, t_l)$ by disjoint paths such that the sum of the weights of the connected pairs is maximized.

Previous results. The $\mathcal{NP}$-hardness of the unweighted MCRP, MRRP, and MBRP follows directly from the $\mathcal{NP}$-hardness of the MDPP [8,9]. However, non-approximability results for our recoloring problems do not follow from corresponding results for the MDPP since the latter problem is a maximization and not a minimization problem. Moran and Snir [10] showed that the MCRP on (∞, ∞)-colorings remains $\mathcal{NP}$-hard on trees, and the same is true for the MRRP, as follows implicitly from the results in [10] concerning leaf colored trees.

Snir [13] presented a polynomial-time 2-approximation algorithm for the weighted MCRP on strings and a polynomial-time 3-approximation algorithm for the weighted MCRP on trees also published in [11]. Bar-Yehuda, Feldman, and Rawitz [3] could improve the approximation ratio on trees to $2 + \epsilon$.

New results. In contrast to the work of Bar-Yehuda et al. and of Snir here we consider initial (a, b)-colorings with a and b different from ∞. In addition, we also consider graphs of bounded treewidth instead of only trees.

Surprisingly, the variants of the three coloring problems all have different complexities on graphs of bounded treewidth, as we prove in Section 2 and 3. We show that the MCRP is NP-hard even on trees initially colored with $(\infty, 2)$-colorings whereas the MRRP can be solved in polynomial time for the more general $(\infty, 3)$-colorings as input colorings even on weighted graphs of bounded treewidth. We also observe the NP-hardness of the MRRP on trees colored with $(\infty, 4)$-colorings. Moreover, we present a polynomial-time algorithm for the MBRP on weighted graphs of bounded treewidth for general colorings.

Extending the result of Bar-Yehuda et al., we present a polynomial-time $(2 + \epsilon)$-approximation algorithm for the MCRP and the MRRP on weighted graphs of bounded treewidth. However, if we follow their approach in a straight forward way, we would have to store too much information at each node of a so-called tree decomposition tree. Therefore, we would obtain a running time of size $\Omega(n^k)$ with k being the treewidth of the graph considered. Additional ideas allows us to guarantee a quadratic running time.

Beside our results on graphs of bounded treewidth we show that the unweighted versions of our recoloring problems cannot be approximated within an approximation ratio of $(1 - o(1)) \ln \ln n$ in polynomial time unless $\mathcal{NP} \subseteq DTIME(n^{O(\log \log n)})$ even if the initial coloring is restricted to be an $(\infty, 2)$-coloring. As a consequence of this result, if we are given pairs of vertices, there is no good approximation possible for approximating in polynomial-time the minimal l such that all except l pairs are connected by disjoint paths, unless $\mathcal{NP} \subseteq DTIME(n^{O(\log \log n)})$. Determining l can be considered in some kind as the inverse of the MDPP problem. Due to space limitations some proofs in this article are omitted. They can be found in the full version of this paper.

2   Hardness Results

Theorem 1. *Given an unweighted n-vertex graph with an $(\infty, 2)$-coloring, no polynomial-time algorithm for the MCRP, the MRRP or the MBRP has an approximation ratio of $(1 - o(1)) \ln \ln n$ unless $\mathcal{NP} \subseteq DTIME(n^{O(\log \log n)})$.*

Theorem 2. *The MCRP on unweighted graphs is $\mathcal{NP}$-hard even if the problem is restricted to trees colored by an initial $(\infty, 2)$-coloring.*

Proof. The theorem can be proven by a reduction from 3-SAT. Let F be an instance of 3-SAT, i.e., F is a Boolean formula in 3-CNF. W.l.o.g. we assume that each clause in F has exactly three literals. Let n be the number of literals in F, let m be the number of clauses of F and let r be the minimal number such that each literal in F appears at most r times in F. For the time being let us construct a forest G which later can be easily connected to a tree. We construct G by introducing for each variable x a so-called *gadget* G_x consisting of an uncolored vertex v_x, leaves $v_x^{L,i}, v_x^{R,i}$ colored with a color c_x^i and an edge $\{v_x, v_x^{L,i}\}$ for each $i = \{1, 2\}$, and two internally disjoint paths of length $r+1$, one from v_x to $v_x^{R,1}$, and the other from v_x to $v_x^{R,2}$. Let us call the internal vertices of the path connecting v_x and $v_x^{R,i}$ for $i = 1$ the *positive* and for $i = 2$ the *negative* vertices in the gadget G_x. For each clause K, we introduce a similar *gadget* G_K consisting of an uncolored vertex v_K, leaves $v_K^{L,j}, v_K^{R,j}$ colored with a color c_K^j and an edge $\{v_K, v_K^{L,j}\}$ for each $j \in \{1, 2, 3\}$, and three internally disjoint paths of length 2, all starting in v_K but ending in different endpoints, $v_K^{R,1}, v_K^{R,2}$, and $v_K^{R,3}$, respectively. In addition, we also introduce $2nr$ extra vertices without any incident edges called the *free* vertices of G. From this forest we obtain a tree T if we simply connect all gadgets and all free vertices by the following two steps. First, add two adjacent vertices v^1 and v^2 into G that both are colored with the same new color. Second, for each variable x, connect v_x to v^1, for each clause K, connect v_K to v^2 and finally also all free vertices to v^2.

Concerning the coloring C of T, we want to color further vertices of T. For each literal x or $\overline{x}$ part of clause K, color in the gadget for x one positive vertex (in case of literal x) or one negative vertex (in case of literal $\overline{x}$) as well as one of the non-leaves adjacent to v_K with a new color $c_{x,K}$. If after these colorings there is at least one uncolored positive or negative vertex, we take for each such vertex y a new color c_y and assign it to y as well as to exactly one uncolored free vertex. One can show that F is satisfiable if and only if (T, C) has a convex recoloring C' with cost $\leq 2nr + (n + 2m)$. The proof of this equivalence is omitted. $\square$

Although the MCRP is $\mathcal{NP}$-complete when being restricted to initial $(\infty, 2)$-colorings, this is not the fact for the MRRP as we show in Theorem 6. However, a slight modification of the reduction above shows that the MRRP on weighted graphs with an initial $(\infty, 4)$-coloring is also $\mathcal{NP}$-hard even for trees. The idea is, for each colored non-leaf x, to add two new vertices x_1, x_2, and edges (x, x_1), (x, x_2), to color x_1, x_2 with the color of x, and finally to uncolor x.

3   Exact Algorithms

In this section we present algorithms on graphs with bounded treewidth. For defining graphs of bounded treewidth we have to define tree decompositions. Tree decompositions and treewidth were introduced by Robertson and Seymour [12] and a survey for both is given by Bodlaender [4].

Definition 3. *A* tree decomposition of treewidth k *for a graph $G = (V, E)$ is a pair (T, B), where $T = (V_T, E_T)$ is a tree and B is a mapping that maps each* node w of V_T to a subset $B(w)$ of V such that (1) $\bigcup_{w \in V_T} B(w) = V$, (2) for each edge $(u,v) \in E$, there exists a node $w \in V_T$ such that $\{u,v\} \subseteq B(w)$, (3) $B(x) \cap B(y) \subseteq B(w)$ for all $w, x, y \in V_T$ with w being a vertex on the path from x to y in T, (4) $|B(w)| \leq k+1$ for all $w \in V_T$. Moreover, a tree decomposition is called nice if (5) T is a rooted and binary tree, (6) $B(\overline{w}) = B(\overline{w}_1) = B(\overline{w}_2)$ holds for each node $\overline{w}$ of T with two children $\overline{w}_1$ and $\overline{w}_2$, (7) either $|B(w) \setminus B(w_1)| = 1$ and $B(w) \supset B(w_1)$ or $|B(w_1) \setminus B(w)| = 1$ and $B(w_1) \supset B(w)$ holds for all nodes w of T with exactly one child w_1.

The treewidth of a graph G is the smallest number k for which a tree decomposition of G with treewidth k exists. If $k = O(1)$, G has *bounded treewidth*. For an n-vertex graph of constant treewidth k, one can determine a nice tree decomposition (T, B) with T consisting of $O(n)$ nodes in linear time [5].

In this section we therefore will assume that we are given an n-vertex graph $G = (V, E)$ and a nice tree decomposition (T, B) of G of treewidth $k - 1$ ($k \in \mathbb{N}$) with T having $O(n)$ nodes. Before presenting our algorithm we introduce some further notations and definitions. For clarity, overlined vertices—as for example $\overline{v}$—should always denote nodes of T. Moreover, we will refer to nodes and arcs instead of vertices and edges if we mean the vertices or edges of T. By $\overline{v}_l$ and $\overline{v}_r$ we denote the left and the right child of $\overline{v}$ in T, respectively. If $\overline{v}$ has only one child, we define it to be a left child. We also introduce a new set consisting of k gray colors—in the following always denoted by Y—and we allow for each recoloring additionally to use the gray colors. A gray colored vertex w intuitively means that w is uncolored and will later be colored with a real color. We therefore define the cost for recoloring a gray colored vertex to be 0 and do not consider the gray colors as real colors. A convex coloring from now on should denote a coloring C where all pairs of vertices of the same gray or real color are C-connected. For each node $\overline{v}$ in T, each subset S of vertices of G, each subgraph H of G, and each coloring C of G we let

- $G(\overline{v})$ be the subgraph of G induced by all vertices contained in at least one set $B(\overline{w})$ of a node $\overline{w}$ contained in the subtree of T rooted in $\overline{v}$.
- $C(S), C(H)$ be the set of colors used by a coloring C for coloring the vertices of S and of H, respectively.
- $\mathrm{SEP}(C, \overline{v})$ be the set of real colors used to color vertices except from $B(\overline{v})$ in more than one of the subgraphs $G(\overline{v}_l), G(\overline{v}_r)$ and $G - G(\overline{v})$.

Finally, for each subgraph H of G, a *legal recoloring* of (H, C) is a recoloring C' of (H, C) such that for each real color c assigned by C' there is a vertex u of H with $c = C(u) = C'(u)$. Observe that, if there is a convex recoloring C'' of a colored graph (H, C) of cost k, there is also a legal convex recoloring C' of (H, C) with cost k. C' can be obtained from C'' without increasing the cost by uncoloring all vertices colored with a color c for which no vertex u with $C''(u) = C(u) = c$ exists. Hence for solving the MCRP, the MRRP, and the MBRP we only need to search for legal recolorings solving the problem.

For the rest of this section we assume that our given graph G is colored by an initial coloring C not using gray colors. We first present an algorithm for the MCRP. This algorithm considers the nodes of T in a bottom-up strategy and computes for each node $\overline{v}$ a set of so-called *characteristics*. Intuitively, each characteristic represents a recoloring C' of $G(\overline{v})$ that will be stepwise extended to a convex recoloring of the whole graph G. *Extending a (re-)coloring C_1 for a graph H_1* means replacing C_1 by a new color function C_2 for a graph $H_2 \supseteq H_1$ with $C_2(w) = C_1(w)$ for all vertices w of H_1 with the following exception: A vertex colored with a gray color c_1 may be recolored with a real color c_2 if all vertices of color c_1 are recolored with c_2. We next define a characteristic for a node $\overline{v}$ precisely as a tuple $(P, \{P_S \mid S \in P\}, \{c_S \mid S \in P\}, Z)$, where

- P is a partition of $B(\overline{v})$, i.e., a family of nonempty pairwise disjoint sets $S_1, \ldots, S_j$ with $\bigcup_{1 \leq i \leq j} S_i = B(\overline{v})$. These sets are called *macro sets*.
- P_S is a partition of the macro set S, where the subsets of S contained in P_S are called *micro sets*.
- c_S for each macro set S is a value in $\mathrm{SEP}(C, \overline{v}) \cup Y \cup \{0, \mathrm{b}\}$, where b is an extra value different from 0 and the real and gray colors.
- $Z \subseteq \mathrm{SEP}(C, \overline{v})$. The colors in Z are called the *forbidden colors*.

In the following for a characteristic Q and a macro set S of Q we denote the values P, P_S, c_S and Z above by P^Q, P_S^Q, c_S^Q and Z^Q. We next describe a first intuitive approach of solving the MCRP extending the ideas of Bar-Yehuda et al. [3] from trees to graphs of bounded treewidth by introducing macro and micro sets but not using gray colors or the extra value b.

A characteristic Q for a node $\overline{v}$ should represent a coloring C' of $G(\overline{v})$ such that the following holds: A macro set S of Q denotes a maximal subset of vertices in $B(\overline{v})$ that are colored by C' with the same unique color equal to the value c_S^Q stored with the macro set—maximal means that there is no further vertex in $B(\overline{v}) \setminus S$ colored with c_S^Q. A micro set is a maximal subset of a macro set that is C'-connected in $G(\overline{v})$. When later extending the recoloring C' we need to know which of the colors not in $C'(B(\overline{v}))$ are used by C' to color vertices of $G(\overline{v})$ since these colors may not be used any more to color a vertex outside $G(\overline{v})$. These colors are exactly the forbidden colors of the characteristic. Note that there can be more than one recoloring of $G(\overline{v})$ leading to the same characteristic for $\overline{v}$. Hence, a characteristic does not really represent one recoloring, but an equivalence class of recolorings. The main idea of our algorithm is the following:

Given all characteristics for the children of a node $\overline{v}$ and, for each equivalence class $\mathcal{E}$ described by one of these characteristics, the minimal cost among all costs of recolorings in $\mathcal{E}$, our algorithm uses a bottom-up strategy to compute the same information also for $\overline{v}$ and its ancestors. Since we only want to compute convex recolorings, at the root of T we have to remove all characteristics having a real colored macro set that consists of at least two micro sets. The cost of an optimal convex recoloring is the minimal cost among all costs computed for the remaining characteristics. An additional top-down traversal of T can also determine a recoloring having optimal cost. Unfortunately, the number of characteristics to be considered by the approach above would be too high for an efficient algorithm. The problem is that for graphs of bounded treewidth, in contrary to what is the case for trees, a path connecting two vertices outside $G(\overline{v})$

may use vertices in $G(\overline{v})$ and vice versa. Hence, as a further change compared to the algorithm of Bar Yehuda et al., we use gray colors and the extra value b. These colors are intuitively used as follows:

If a color c is used by C only to color vertices outside $G(\overline{v})$, a recoloring C' of G may possibly also want to recolor a set S of vertices in $G(\overline{v})$ with c in order to C'-connect some vertices with color c. The cost for recoloring vertices of $G(\overline{v})$ with c are independent from the exact value of c, and can be computed as the costs of uncoloring all vertices of S (since $C(\overline{w}) \neq c$ for all $\overline{w}$ in $G(\overline{v})$) and of recoloring it (without any cost) with color c. Therefore, when considering recolorings of the graph $G(\overline{v})$, we do not allow to color it with a real color $c \notin C(G(\overline{v}))$. Instead of c we use a gray color, since coloring a vertex gray has the same costs as of uncoloring the vertex but allows us to distinguish the vertex from vertices colored with another gray color or being uncolored. Note that our definition of extending a recoloring allows us with zero costs to recolor gray and uncolored vertices in a later step with a real color, whereas recoloring real-colored vertices is forbidden when extending a recoloring.

If a recoloring C' of $G(\overline{v})$ colors a macro set S with a color c that is only used by C to color vertices of $G(\overline{v})$, then for extending the recoloring C' to a recoloring C'', we do not need to know the exact color of S. The reason for this is that, for any vertex w outside $G(\overline{v})$, the cost for setting $C''(w) = c$ can be computed again independently from the color of S: We have to pay the weight of w as costs if w is real-colored by C and zero costs otherwise. Therefore, we use the extra value b to denote that a macro set S is real-colored with a color c that with respect to C only appears in $G(\overline{v})$ and, in this case, we will set $c_S^{\mathcal{Q}} = $ b instead of setting $c_S^{\mathcal{Q}} = c$.

Following the ideas described above we let our algorithm consider only a restricted class of characteristics. For a node $\overline{v}$ of T, we define $C_{|G(\overline{v})}$ to be the coloring C restricted to $G(\overline{v})$. We call a characteristic $\mathcal{Q}$ a *good characteristic* if there exists a legal recoloring C' of $(G(\overline{v}), C_{|G(\overline{v})})$ with the properties (A1)-(A7). C' is then said to be *consistent* with $\mathcal{Q}$.

(A1) $C'(G(\overline{v})) \subseteq C(G(\overline{v})) \cup Y \cup \{0\}$.

(A2)   For each macro set S of $\mathcal{Q}$, C' colors all vertices of S with one color, namely with $c_S^{\mathcal{Q}}$ if $c_S^{\mathcal{Q}} \neq $ b, and with a real color not in $C(G - G(\overline{v}))$ if $c_S^{\mathcal{Q}} = $ b.

(A3) C' colors two different macro sets of $\mathcal{Q}$ with different colors.

(A4)   A micro set is a maximal subset of $B(\overline{v})$ that is C'-connected in $G(\overline{v})$.

(A5) C' is a convex recoloring for the graph obtained from $G(\overline{v})$ by adding, for each macro set S, edges of an arbitrary simple path visiting exactly one vertex of each micro set of S.

(A6)   Every gray colored vertex in $G(\overline{v})$ is C'-connected to a vertex in $B(\overline{v})$.

(A7) $Z^{\mathcal{Q}} = \mathrm{SEP}(C, \overline{v}) \cap (C'(G(\overline{v})) \setminus C'(B(\overline{v})))$.

Note that each convex legal recoloring C' of the initial colored graph (G, C) is consistent with a good characteristic $\mathcal{Q}$ for the root $\overline{r}$ of T. More explicitly, we obtain $\mathcal{Q}$ by dividing $B(\overline{r})$ into macro sets each consisting of all vertices of one color with respect to C', by defining the partition of each macro set to consist of only one micro set, by setting $Z^{\mathcal{Q}} = \emptyset$ and by defining, for each macro set S, $c_S^{\mathcal{Q}} = \mathrm{b}$ if $C'(S)$ is a real color, or $c_S^{\mathcal{Q}} = 0$ otherwise.

Our algorithm computes in a bottom-up process for each node $\overline{v}$ of T all good characteristics of $\overline{v}$ from the good characteristics of the children of $\overline{v}$. However not all pairs of good characteristics of the children can be combined to good characteristics of $\overline{v}$. Therefore we call a characteristic $\mathcal{Q}_{\mathrm{l}}$ of $\overline{v}_{\mathrm{l}}$ and a characteristic $\mathcal{Q}_{\mathrm{r}}$ of $\overline{v}_{\mathrm{r}}$ *compatible* if they satisfy the following three conditions:

- Two vertices $v_1, v_2 \in B(\overline{v}_{\mathrm{l}}) = B(\overline{v}_{\mathrm{r}})$ belong to the same macro set in $\mathcal{Q}_{\mathrm{l}}$ if and only if this is true for $\mathcal{Q}_{\mathrm{r}}$.
- Let S be a macro set of $\mathcal{Q}$ and hence also of $\mathcal{Q}_{\mathrm{l}}$ and $\mathcal{Q}_{\mathrm{r}}$. Then either $c_S^{\mathcal{Q}_{\mathrm{l}}} = c_S^{\mathcal{Q}_{\mathrm{r}}} \neq \mathrm{b}$ or exactly one of $c_S^{\mathcal{Q}_{\mathrm{l}}}$ and $c_S^{\mathcal{Q}_{\mathrm{r}}}$ is a gray color.
- The sets of forbidden colors of $\mathcal{Q}_{\mathrm{r}}$ and of $\mathcal{Q}_{\mathrm{l}}$ are disjoint.

The following algorithm computes for each node $\overline{v}$ of T a set $M_{\overline{v}}$ of characteristics from which we will show in the full version of this paper that it is exactly the set of good characteristics of $\overline{v}$. First of all, in a preprocessing phase compute by a bottom-up and a top-down traversal of T, for each node $\overline{v}$ of T, the set $\mathrm{SEP}(C, \overline{v})$ as well as the set of colors that are used by C to color vertices in $G(\overline{v})$ but no vertex outside $G(\overline{v})$. The latter set is in the following denoted by $U(C, \overline{v})$. Next, for each leaf $\overline{v}$ of T, $M_{\overline{v}}$ is obtained by taking into account all possible divisions of the vertices of $B(\overline{v})$ into macro sets and all possible colorings of the macro sets with different colors of $C(B(\overline{v})) \cup Y \cup \{0\}$. More precisely, for each choice, a characteristic $\mathcal{Q}$ is obtained and added to $M_{\overline{v}}$ by defining, for each macro set S colored with c, the micro sets of S to be the connected components of the subgraph of G induced by the vertices of S, and by setting $c_S^{\mathcal{Q}} = \mathrm{b}$ if c is a real color in $U(C, \overline{v})$, and $c_S^{\mathcal{Q}} = c$ otherwise. The set $Z^{\mathcal{Q}}$ of forbidden colors is set to $\emptyset$.

Next start a bottom-up traversal of T. At a non-leaf $\overline{v}$ all already computed characteristics of the children are considered. In detail, for each characteristic $\mathcal{Q}_{\mathrm{l}}$ of $M_{\overline{v}_{\mathrm{l}}}$ and—if $\overline{v}$ has two children—for each compatible good characteristic $\mathcal{Q}_{\mathrm{r}}$ of $M_{\overline{v}_{\mathrm{r}}}$, we add to $M_{\overline{v}}$ the set of characteristics that also could be obtained as output by the following non-deterministic algorithm:

- Take for $\mathcal{Q}$ and the vertices in $B(\overline{v}) \cap B(\overline{v}_{\mathrm{l}})$ the same division into macro sets as for $\mathcal{Q}_{\mathrm{l}}$. If $\overline{v}$ has only one child and there is also a vertex $w \in B(\overline{v}) \setminus B(\overline{v}_{\mathrm{l}})$, choose one of the $\leq k$ possibilities of assigning w to one of the macro sets of $B(\overline{v}) \cap B(\overline{v}_{\mathrm{l}})$ or choose $\{w\}$ to be its own new macro set.
- For dividing the vertices of $B(\overline{v})$ into micro sets, construct the graph H consisting of the vertices in $B(\overline{v})$ and having an edge between two vertices if and only if both vertices belong to the same macro set and either this edge exists in G or both vertices belong to the same micro set in $\mathcal{Q}_{\mathrm{l}}$ or $\mathcal{Q}_{\mathrm{r}}$. Define the vertices of each connected component in H to be a micro set of $\mathcal{Q}$.
- For each macro set S obtained by the construction above, distinguish between three cases.
    - $S \subseteq S'$ for a macro set S' of $\mathcal{Q}_{\mathrm{l}}$: If $\overline{v}_{\mathrm{r}}$ does not exist or if $c_{S'}^{\mathcal{Q}_{\mathrm{l}}} = c_S^{\mathcal{Q}_{\mathrm{r}}}$, set $c_S^{\mathcal{Q}} = c_{S'}^{\mathcal{Q}_{\mathrm{l}}}$. Otherwise, set $c_S^{\mathcal{Q}}$ to the non-gray value in $\{c_{S'}^{\mathcal{Q}_{\mathrm{l}}}, c_S^{\mathcal{Q}_{\mathrm{r}}}\}$.

- S has a vertex $w \in B(\overline{v}) \setminus B(\overline{v}_1)$ and $|S| > 1$: Choose $c_S^Q \in \{c_{S \setminus \{w\}}^{Q_1}, C(w)\}$ if $c_{S \setminus \{w\}}^{Q_1}$ is a gray color, otherwise, $c_S^Q = c_{S \setminus \{w\}}^{Q_1}$.
  - $S = \{w\}$ with $w \in B(\overline{v}) \setminus B(\overline{v}_1)$: Choose for c_S^Q a value of $Y \cup \{0, C(w)\}$. After defining c_S^Q as described above, if c_S^Q is a real color and $c_S^Q \in U(c, \overline{v})$, redefine $c_S^Q = \mathrm{b}$.
- Reject the computation if there is a micro set S' part of a macro set S in $\mathcal{Q}_1$ with $S' \cap B(\overline{v}) = \emptyset$ and either $c_S^{Q_1}$ is a gray color or $S \setminus S' \neq \emptyset$.
- If there is macro set $S = B(\overline{v}_1) \setminus B(\overline{v})$ of $\mathcal{Q}_1$ and if $c_S^{Q_l}$ is a real color, set $Z' = \{c_S^{Q_l}\}$ and $Z' = \emptyset$ otherwise. Finally, set $Z^Q = \mathrm{SEP}(C, \overline{v}) \cap (Z' \cup Z^{Q_1} \cup Z^{Q_r})$.

As mentioned before, one can show that our algorithm correctly computes for each node $\overline{v}$ the set of all good characteristics of $\overline{v}$ and that our algorithm can be extended such that it computes with each good characteristic $\mathcal{Q}$ the costs of a recoloring consistent with $\mathcal{Q}$ that among all such recolorings has minimal costs. One can also show that our algorithm has a running time of $O(n^2 + 4^s(k + s + 2)^{6k+1}(k^2 + s)n)$, where $s = \max_{\overline{v} \text{ node of } T} |\mathrm{SEP}(C, \overline{v})|$.

After the removal of all characteristics having a real colored macro set that consists of at least two micro sets or having a gray colored macro set we obtain the cost of an optimal legal convex recoloring as the minimal costs among all costs stored with the remaining characteristics constructed for the root of T. Finally by an additional top-down traversal our algorithm can—beside the minimal costs of a legal recoloring—also determine the coloring itself within the same time bound. We obtain the following theorem.

Theorem 4. *Given a colored graph (G, C) and a nice tree decomposition (T, B) of width $k - 1$ as input the MCRP can be solved in $O(n^2 + 4^s(k + s + 2)^{6k+1}(k^2 + s)n)$ time, where $s = \max_{\overline{v} \text{ node of } T} |\mathrm{SEP}(C, \overline{v})|$.*

It is easy to modify the algorithm above such that it solves the MRRP within the same time bound. In each bottom-up step we only have to exclude recolorings that recolor a real colored vertex with a gray or another real color.

Unfortunately, the algorithms above for the MCRP and the MRRP are exponential in s since there are 2^s different possible lists of forbidden colors. The good news concerning the MBRP on general initial colorings and the MRRP with its initial coloring being an $(\infty, 3)$-coloring is that we can omit to store the forbidden colors explicitly. We next describe the necessary modifications.

For the MBRP we use the same basic algorithm as for the MCRP. However, we compute as a solution for the MBRP w.l.o.g. only recolorings that, for each real color c, either recolor all or none of the vertices initially colored with c. Following this approach, a characteristic of a node $\overline{v}$ should only represent recolorings that, for each real color c, either recolor all or none of the vertices u in $G(\overline{v})$ for which $C(u) = c$ holds. If in the latter case there is a vertex in $G(\overline{v})$ and also a vertex outside $G(\overline{v})$ initially colored with c, we therefore claim that a vertex of $B(\overline{v})$ is also colored with c since otherwise the recoloring can not be extended to a legal convex recoloring not recoloring any vertex of c. This implies an additional rule for constructing characteristics:

Assume that—as in our basic algorithm—we want to construct a characteristic Q of a non-leaf $\overline{v}$ from a characteristic $Q_{\overline{u}}$ of a child $\overline{u}$ of $\overline{v}$. Then we are only allowed to color a macro set S of Q with c if (1) all vertices in $B(\overline{v})$ that are initially colored with c are contained in S and (2) either $Q_{\overline{u}}$ also contains a macro set S' with $c_{S'}^{Q_{\overline{u}}} = c$ or $c \notin C(G(\overline{u}))$.

For efficiently testing condition (2), we construct in a preprocessing phase for each node $\overline{v}$ an array $A_{\overline{v}}$, with the following entries: For each color $c \in C(G)$, $A_{\overline{v}}[c] = 1$ if $G(\overline{v})$ contains a vertex u with $C(u) = c$. Otherwise $A_{\overline{v}}[c]$ is defined to be 0. If the array is computed by a bottom-up traversal of T, the preprocessing phase takes $O(n^2)$ time. After the preprocessing phase we can test for each characteristic Q of a node $\overline{v}$ and each color c in $O(k)$ time whether (1) and (2) hold. Hence, the asymptotic running time of our algorithm does not increase. Moreover, our additional rules enables us to find out, for each color $c \in \mathrm{SEP}(C, \overline{v})$, whether a vertex of $G(\overline{v}_l)$ or $G(\overline{v}_r)$ is colored with c by considering $A_v[c]$ and by testing whether a macro set S of Q_l or Q_r, respectively, is colored with c. Hence, there is no need to store the forbidden colors.

Theorem 5. *On graphs of bounded treewidth the MBRP is solvable in polynomial time.*

More complicated modifications are necessary for the MRRP. We assume w.l.o.g. that, for each color c, there are either no or at least two vertices colored with c by C. The main idea of our algorithm is the following: For improving the running time at a node $\overline{v}$ of T we only want to consider recolorings C' of $G(\overline{v})$ such that for each color $c \in C(G)$ the following condition (D,c) holds. The correctness of this step will be discussed later.

(D,c) If u is a vertex in $G(\overline{v})$ with $C'(u) = C(u) = c$, either there exists a vertex w outside $G(\overline{v})$ with $C(w) = c$ and a vertex $w' \in B(\overline{v})$ with $C'(w') = c$, or there exists another vertex $w \in G(\overline{v})$ with $C'(w) = C(w) = c$.

This property guarantees that, for a node $\overline{v}$ of highest depths with $G(\overline{v}_l)$ containing a vertex u_l initially colored with c and $G(\overline{v}_r)$ containing a vertex u_r initially colored with c, a recoloring C' with property (D,c) colors u_l or u_r with c if and only if $B(\overline{v}_l)$ and $B(\overline{v}_r)$, respectively, also contains a vertex colored with c. Therefore, there is no need to store c explicitly as a forbidden color in a characteristic of $\overline{v}_l$ and of $\overline{v}_r$ any more. With similar arguments one can show that for no node its characteristic has to store c explicitly as a forbidden color.

The problem is that some legal recolorings are permitted by (D,c). However, each convex recoloring C_{opt} of optimal cost either is a recoloring with property (D,c) or it colors w.l.o.g. exactly one vertex u with c. In the latter case a coloring with the same cost as C_{opt} can be obtained from a recoloring with property (D,c) not coloring any vertex with c by undoing the recoloring of the vertex originally colored with c that among all such vertices has a maximal weight. Therefore, for computing the costs of an optimal convex recoloring, we only have to consider the costs of recolorings with property (D,c) and eventually to subtract the maximal weight over all vertices originally colored with c. Let us call such a subtraction a *c-cost adaption*. Our goal now is to describe an algorithm that runs the

c-cost-adaption during the bottom-up traversal of T at a certain node $\bar{v}$—called the c-decision node—having the following property:

For each characteristic Q of $\bar{v}$, either each recoloring C' extending a recoloring consistent with Q and having property (D,c) C'-connects at least two vertices initially colored with c (and we therefore must not run a c-cost-adaption) or all such recolorings uncolor all vertices initially colored with c (and therefore we have to run a c-cost-adaption).

If, for each color c, we know the c-decision node, our algorithm runs as follows: For each node $\bar{v}$ (also above the c-decision node), we only compute characteristics representing recolorings for which property (D,c) holds for each color c. If we reach the c-decision node $\bar{v}$, for each characteristic Q of $\bar{v}$, we test whether all recolorings extending Q do not use color c and if so, we run a c-cost adaption for Q. One can show that, for all colors c, a c-decision node exists and that one can efficiently determine the characteristics representing the recoloring with property (D,c).

Theorem 6. *On graphs of bounded treewidth the MRRP restricted to initial* $(\infty, 3)$*-colorings is solvable in polynomial time.*

Note that the running times for the MCRP and the MRRP on arbitrary initial colorings are also polynomial if s—defined as in Theorem 4—is of size $O(\log n)$. This is the case if an (a, b)-coloring with $a = O(\log n)$ is given.

Theorem 7. *On graphs of bounded treewidth the MCRP and the MRRP, both restricted to initial* (a, b)*-colorings with* $a = O(\log n)$*, are solvable in polynomial time.*

4 Approximation Algorithms

Since the MCRP is $\mathcal{NP}$-hard even on trees, we can not hope for a polynomial-time algorithm that solves the problem to optimality—even if we consider graphs of bounded treewidth. Using the algorithm of the last section we now present for graphs of bounded treewidth a $(2 + \epsilon)$-approximation algorithm for the MCRP and the MRRP given an arbitrary (∞, ∞)-coloring. The following algorithm is inspired by the algorithm of Bar-Yehuda et al. [3]. We extend the algorithm from trees to graphs of bounded treewidth and present a slightly different description for proving the correctness of the algorithm.

Given a graph G with a coloring C and a nice tree decomposition (T, B) of width $k-1$ for G our results can be obtained by iteratively modifying the coloring C and the weights of the vertices such that finally $|\text{SEP}(C, \bar{v})| < s$ for all nodes $\bar{v}$ of T and a fixed $s \in \mathbb{N}$ with $s > k$. Let $\bar{v}$ be a node of T such that there is a set $R' \subseteq \text{SEP}(C, \bar{v})$ containing exactly s colors and let V' be a set consisting of two vertices of color c for all $c \in R'$ such that for each pair of vertices $x, y \in V'$ of the same color the vertices x and y are in different components in $G - B(\bar{v})$. Moreover, let α be the minimal weight of a vertex in V'. The size of $\text{SEP}(C, \bar{v})$ is decremented by reducing the weight of all vertices in V' by α and subsequently uncoloring the vertices of zero-weight.

On the one hand, this weight reduction decreases the cost of an optimal convex (re-)coloring C' of G by at least $(s-k)\alpha$ since the at most k vertices in $B(\overline{v})$ allow to color connect only k of the s colors in R', i.e., $s-k$ vertices in V' can not be C'-connected. On the other hand, if we have a solution for the MCRP (or the MRRP) with the reduced weight function, we can simply take this solution as a solution for the MCRP (or the MRRP) with the original weights and our costs increase by at most $2s\alpha$. Thus, in each iteration our costs decrease at most by a factor of $2s/(s-k)$ more than the decrease of the costs of an optimal solution. If at the end no further reduction is possible, we can use the exact algorithms from the previous section, i.e., we can solve the instance obtained by this weight reduction as good as an optimal algorithm. Altogether, we have only recoloring costs that are a factor of $2s/(s-k)$ bigger than the costs of an optimal solution. Choosing s large enough, we obtain the following.

Corollary 8. *For graphs of bounded treewidth a $(2+\epsilon)$-approximation algorithms exist for the MCRP and the MRRP with quadratic running time.*

References

1. Alon, N., Moshkovitz, D., Safra, S.: Algorithmic construction of sets for k-restrictions. ACM Transactions on Algorithms 2, 153–177 (2006)
2. Andrews, M., Zhang, L.: Hardness of the undirected edge-disjoint paths problem. In: Proc. 37th Annual ACM Symposium on Theory of Computing (STOC 2005), pp. 276–283 (2005)
3. Bar-Yehuda, R., Feldman, I., Rawitz, D.: Improved approximation algorithm for convex recoloring of trees. In: Erlebach, T., Persinao, G. (eds.) WAOA 2005. LNCS, vol. 3879, pp. 55–68. Springer, Heidelberg (2006)
4. Bodlaender, H.L.: A partial k-arboretum of graphs with bounded tree width. Theoret. Comput. Sci. 209, 1–45 (1998)
5. Bodlaender, H.L., Kloks, T.: Efficient and constructive algorithms for the pathwidth and treewidth of graphs. J. Algorithms 21, 358–402 (1996)
6. Chen, X., Hu, X., Shuai, T.: Inapproximability and approximability of maximal tree routing and coloring. J. Comb. Optim. 11, 219–229 (2006)
7. Feige, U.: A threshold of $\ln n$ for approximating set cover. J. ACM 45, 634–652 (1998)
8. Karp, R.M.: On the computational complexity of combinatorial problems. Networks 5, 45–68 (1975)
9. Lynch, J.F.: The equivalence of theorem proving and the interconnection problem. (ACM) SIGDA Newsletter 5, 31–36 (1975)
10. Moran, S., Snir, S.: Convex recolorings of strings and trees: definitions, hardness results and algorithms. In: Dehne, F., López-Ortiz, A., Sack, J.-R. (eds.) WADS 2005. LNCS, vol. 3608, pp. 218–232. Springer, Heidelberg (2005)
11. Moran, S., Snir, S.: Efficient approximation of convex recoloring. J. Comput. System Sci. 73, 1078–1089 (2007)
12. Robertson, N., Seymour, P.D.: Graph minors. I. Excluding a forest. J. Comb. Theory Ser. B 35, 39–61 (1983)
13. Snir, S.: Computational Issues in Phylogenetic Reconstruction: Analytic Maximum Likelihood Solutions, and Convex Recoloring. Ph.D. Thesis, Department of Computer Science, Technion, Haifa, Israel (2004)

On the Complexity of Reconfiguration Problems

Takehiro Ito[1], Erik D. Demaine[2], Nicholas J.A. Harvey[2],
Christos H. Papadimitriou[3], Martha Sideri[4], Ryuhei Uehara[5], and Yushi Uno[6]

[1] Graduate School of Information Sciences, Tohoku University,
Aoba-yama 6-6-05, Sendai, 980-8579, Japan
[2] MIT Computer Science and Artificial Intelligence Laboratory,
32 Vassar St., Cambridge, MA 02139, USA
[3] Computer Science Division, University of California at Berkeley,
Soda Hall 689, EECS Department, Berkeley, CA 94720, USA
[4] Department of Computer Science,
Athens University of Economics and Business,
Patision 76, Athens 10434, Greece
[5] School of Information Science, JAIST,
Asahidai 1-1, Nomi, Ishikawa 923-1292, Japan
[6] Graduate School of Science, Osaka Prefecture University,
1-1 Gakuen-cho, Naka-ku, Sakai 599-8531, Japan
`takehiro@ecei.tohoku.ac.jp`, `edemaine@mit.edu`, `nickh@mit.edu`,
`christos@cs.berkeley.edu`, `sideri@aueb.gr`, `uehara@jaist.ac.jp`,
`uno@mi.s.osakafu-u.ac.jp`

Abstract. Reconfiguration problems arise when we wish to find a step-by-step transformation between two feasible solutions of a problem such that all intermediate results are also feasible. We demonstrate that a host of reconfiguration problems derived from NP-complete problems are PSPACE-complete, while some are also NP-hard to approximate. In contrast, several reconfiguration versions of problems in P are solvable in polynomial time.

1 Introduction

Consider the bipartite graph with weighted vertices in Fig.1(a) (both solid and dotted edges). It models a situation in which power stations with fixed capacity (the square vertices) provide power to customers with fixed demand (the round vertices). It can be seen as a feasible solution of a particular instance of a search problem which we may call the POWER SUPPLY problem [10,11]: Given a bipartite graph $G = (U, V, E)$ with weights on the vertices, is there a forest covering all vertices in G, and with exactly one vertex from U in each component, such that the sum of the demands of the V vertices (customers) in each component is no more than the capacity of the U vertex (power station) in it?

But suppose now that we are given *two* feasible solutions of this instance (the leftmost and rightmost ones in Fig.1), and we are asked: Can the solution on the left be transformed into the solution on the right *by moving only one customer at a time, and always remaining feasible?* This problem, which we call the POWER

S.-H. Hong, H. Nagamochi, and T. Fukunaga (Eds.): ISAAC 2008, LNCS 5369, pp. 28–39, 2008.

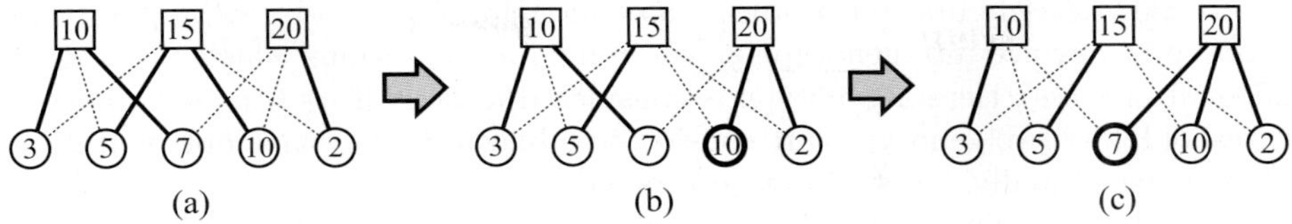

Fig. 1. A sequence of feasible solutions for the POWER SUPPLY problem

SUPPLY RECONFIGURATION problem, is an exemplar of the kind of problems we discuss in this paper. (In this particular instance, it turns out that the answer is "yes"; see Fig.1.) As one may have expected, the most basic reconfiguration problem is the SATISFIABILITY RECONFIGURATION problem: Given a CNF formula and two satisfying truth assignments s_0 and s_t, are these connected in the subgraph of the hypercube induced by the satisfying truth assignments? This problem has been shown PSPACE-complete in [6].

In more generality, *reconfiguration problems* have the following structure: Fix a search problem $\mathcal{S}$ (a polynomial-time algorithm which, on instance I and candidate solution y of length polynomial in that of I, determines whether y is a feasible solution of I); and fix a polynomially-testable symmetric *adjacency relation* A on the set of feasible solutions, that is, a polynomial-time algorithm such that, given an instance I of $\mathcal{S}$ and two feasible solutions y' and y'' of I, it determines whether y' and y'' are adjacent. (In almost all problems discussed in this paper, the feasible solutions can be considered as sets of elements, and two solutions are adjacent if their symmetric difference has size 1 — or, in some cases such as POWER SUPPLY RECONFIGURATION, 2.) The RECONFIGURATION PROBLEM FOR $\mathcal{S}$ AND A is the following computational problem: Given instance I of $\mathcal{S}$ and two feasible solutions y_0 and y_t of I, is there a sequence of feasible solutions $y_0, y_1, \ldots, y_t$ of I such that y_{i-1} and y_i are adjacent for $i = 1, 2, \ldots, t$?

Reconfiguration problems can also arise from optimization problems, if one turns the optimization problem into a search problem by giving a threshold. For example, the CLIQUE RECONFIGURATION problem is the following: Given a graph G, an integer k, and two cliques C_0 and C_t of G, both of size at least k, is there a way to transform C_0 into C_t via cliques, each of which results from the previous one by adding or subtracting one node of G, without ever going through a clique of size less than $k - 1$?

Reconfiguration problems are useful and entertaining, have been coming up in recent literatures [1,6,9], and are interesting for a variety of reasons. First, they may reflect, as in the POWER SUPPLY RECONFIGURATION problem above, a situation where we actually seek to implement such a sequence of elementary changes in order to transform the current configuration to a more desirable one, in a context in which intermediate steps must also be fully feasible, and only restricted changes can occur — in our example, no two customers can change providers simultaneously, and we certainly do not wish customers to be without power. In a complex, dynamic environment in which changing circumstances affect the feasible solution of choice, determining whether such adaptation is possible may be crucial. Reconfiguration problems also model questions of *evolvability*: Can genotype y_0 evolve into genotype y_t via individual mutations which are each of adequate fitness? Here a genotype is considered feasible if its fitness is above a threshold, and two genotypes are considered adjacent if one is a simple mutation of the other. Finally, reconfiguration versions of constraint satisfaction problems (the first kind studied in the literature [6]) yield insights into the structure of the solution space, and heuristics, such as survey propagation, whose performance depends crucially on connectivity and other properties of the solution space.

In this paper we embark on a systematic investigation of the complexity of reconfiguration problems. Our main focus is showing that a host of reconfiguration problems (including all those mentioned above and many more) are PSPACE-complete. The proof for the POWER SUPPLY RECONFIGURATION problem and those for certain other problems are explained in Section 2. In Section 3 we point out that certain reconfiguration problems arising from problems in P (such as the MINIMUM SPANNING TREE and MATCHING problems) can be solved in polynomial time, and in Section 4 we show certain approximability and inapproximability results for reconfiguration problems.

2   PSPACE-Completeness

In this section we show that a host of reconfiguration problems are PSPACE-complete. We first give a proof for the POWER SUPPLY RECONFIGURATION problem in Subsection 2.1, and then give proof sketches for certain other reconfiguration problems in Subsection 2.2.

2.1   POWER SUPPLY RECONFIGURATION

The POWER SUPPLY RECONFIGURATION problem was defined informally in the Introduction. An instance is given in terms of a bipartite graph $G = (U, V, E)$, where each vertex in U is called a *supply vertex* and each vertex in V is called a *demand vertex*. Each supply vertex $u \in U$ is assigned a positive integer $\mathrm{sup}(u)$, called the *supply of u*, while each demand vertex $v \in V$ is assigned a positive integer $\mathrm{dem}(v)$, called the *demand of v*. We wish to find a forest which covers all vertices in G such that each tree T in the forest has exactly one supply vertex whose supply is at least the sum of demands of all demand vertices in T. We call an assignment $f : V \to U$ a *configuration of G* if there is an edge $(v, f(v)) \in E$ for each demand vertex $v \in V$. A configuration f of G is called *feasible* if the following condition holds: for each supply vertex $u \in U$,

$$\mathrm{sup}(u) \geq \sum \{\mathrm{dem}(v) \mid v \in V \text{ such that } f(v) = u\}.$$

The *adjacency relation* on the set of feasible configurations is defined as follows: two feasible configurations f and f' are *adjacent* if $\left|\{v \in V : f(v) \neq f'(v)\}\right| = 1$, that is, f' can be obtained from f by changing the assignment of a single demand vertex. Then, for given a bipartite graph $G = (U, V, E)$ and two feasible configurations f_0 and f_t of G, the POWER SUPPLY RECONFIGURATION problem is to determine whether there is a sequence of feasible configurations $f_0, f_1, \ldots, f_t$ of G such that f_{i-1} and f_i are adjacent for $i = 1, 2, \ldots, t$.

Fig.1 illustrates three feasible configurations of a bipartite graph G, where each supply vertex is drawn as a square, each demand vertex as a round, and the supply or demand is written inside. Fig.1 also illustrates an example of a transformation from the feasible configuration in Fig.1(a) to one in Fig.1(c), where the demand vertex whose assignment was changed from the previous one is depicted by a thick round. The optimization problem for finding a certain configuration of a given graph has been studied in [10,11].

Theorem 1. POWER SUPPLY RECONFIGURATION *is PSPACE-complete.*

Proof. It is easy to see that this problem, as well as any reconfiguration version of a problem in NP, can be solved in (most conveniently, nondeterministic [13]) polynomial space.

We give a reduction to this problem from the SATISFIABILITY RECONFIGURA-TION problem, which was recently shown to be PSPACE-complete [6]. In that problem we are given a Boolean formula ϕ in conjunctive normal form, say with n variables $x_1, x_2, \ldots, x_n$ and m clauses $C_1, C_2, \ldots, C_m$, and two satisfying truth assignments $\mathbf{s}_0$ and $\mathbf{s}_t$ of ϕ. Then, we are asked whether there is a sequence of satisfying truth assignments, starting with $\mathbf{s}_0$ and ending in $\mathbf{s}_t$, and each differing from the previous one in only one variable. Let c be the maximum number of clauses in which a literal occurs, and hence no literal appears in more than c clauses in ϕ.

Given such an instance of SATISFIABILITY RECONFIGURATION, we construct an instance of POWER SUPPLY RECONFIGURATION as follows. We first make a *variable gadget* G_{x_i} for each variable x_i, $1 \leq i \leq n$; G_{x_i} is a binary tree with three vertices as illustrated in Fig.2(a); the root F_i is a demand vertex of demand c, and the two leaves x_i and $\bar{x}_i$ are supply vertices of supply c. Then the corresponding bipartite graph G_ϕ is constructed as follows. For each variable x_i, $1 \leq i \leq n$,

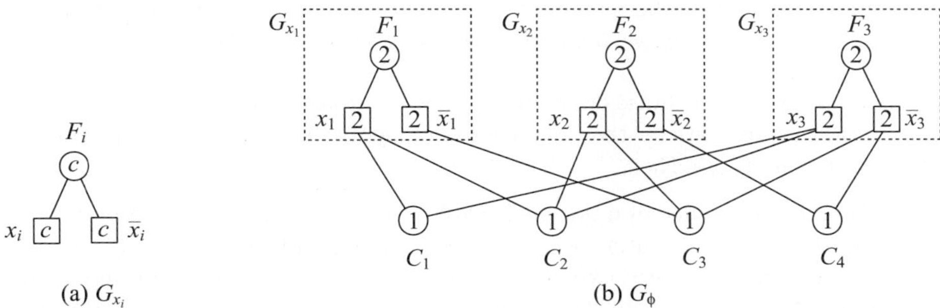

Fig. 2. (a) Variable gadget G_{x_i}, and (b) bipartite graph G_ϕ corresponding to a Boolean formula ϕ with four clauses $C_1 = (x_1 \vee x_3)$, $C_2 = (x_1 \vee x_2 \vee x_3)$, $C_3 = (\bar{x}_1 \vee x_2 \vee \bar{x}_3)$ and $C_4 = (\bar{x}_2 \vee \bar{x}_3)$, and hence $c = 2$

put the variable gadget G_{x_i} to the graph, and for each clause C_j, $1 \le j \le m$, put a demand vertex C_j of demand 1 to the graph. Finally, for each clause C_j, $1 \le j \le m$, join a supply vertex x_i (or $\bar{x}_i$) in G_{x_i}, $1 \le i \le n$, with the clause demand vertex C_j if and only if the literal x_i (respectively, $\bar{x}_i$) is in the clause C_j. (See Fig.2(b) as an example.) Clearly, G_ϕ is a bipartite graph.

Consider a feasible configuration of G_ϕ. Then each demand vertex F_i, $1 \le i \le n$, must be assigned to one of x_i and $\bar{x}_i$; *a literal is considered false if F_i is assigned to the corresponding supply vertex.* Notice that, since supply vertices have supply c and the F_i's have demand c, a false-literal supply vertex cannot provide power to any of the other demand vertices. Hence, all clause demand vertices C_j, $1 \le j \le m$, must be assigned to true-literal supply vertices that occur in them. Since each literal x_i (or $\bar{x}_i$), $1 \le i \le n$, appears in at most c clauses in ϕ, the corresponding supply vertex x_i (respectively, $\bar{x}_i$) in G_{x_i} can provide power to all clause demand vertices C_j whose corresponding clauses have x_i (respectively, $\bar{x}_i$).

To complete the reduction, we now create two feasible configurations f_0 and f_t of G_ϕ corresponding to the satisfying truth assignments $\mathbf{s}_0$ and $\mathbf{s}_t$ of ϕ, respectively. Each demand vertex F_i, $1 \le i \le n$, is assigned to the supply vertex whose corresponding literal is false, while each clause demand vertex C_j, $1 \le j \le m$, is assigned to an arbitrary true-literal supply vertex adjacent to C_j. Clearly, f_0 and f_t are feasible configurations of G_ϕ. This completes the construction of the corresponding instance of the POWER SUPPLY RECONFIGURATION problem.

We know that a feasible configuration of G_ϕ corresponds to a satisfying truth assignment of ϕ *plus* an assignment of each clause to a true literal. It is easy to see that this correspondence goes backwards: every satisfying truth assignment of ϕ can be mapped to at least one (in general, to exponentially many) feasible configurations of G_ϕ.

How about adjacent configurations — defined to be configurations differing in the assignment of just one demand vertex? One can easily observe that there are only two types of reassignments to go from a feasible configuration of G_ϕ to an adjacent one, as follows:

(1) One could change the assignment of a demand vertex F_i from x_i to $\bar{x}_i$, or vice versa, *if any clause demand vertex is currently assigned to neither supply vertices x_i nor $\bar{x}_i$.*

(2) Alternatively, if a clause demand vertex C_j is adjacent to more than one true-literal supply vertices, then one could change the assignment of C_j from the current one to another.

Therefore, any sequence of adjacent feasible configurations of G_ϕ can be broken down to subsequences, intermittently with a reassignment of type (1) above; in each subsequence, every two adjacent configurations can go from one to another via a reassignment of type (2) above. Therefore, all feasible configurations in each subsequence correspond to the same satisfying truth assignment of ϕ, while any two consecutive subsequences correspond to adjacent satisfying truth assignments (namely, differing in only one variable). Conversely, for given any sequence of adjacent satisfying truth assignments of ϕ, there is a corresponding sequence of adjacent feasible configurations of G_ϕ, obtained as follows: Consider a flip of a variable x_i from true to false. (A flip of x_i from false to true is similar.) Then we wish to change the assignment of the demand vertex F_i from the supply vertex $\bar{x}_i$ to x_i. (Remember that the literal to which F_i is assigned is considered false.) We first change the assignments of all clause demand vertices, which are currently assigned to x_i, to another true-literal supply vertex: since we are about to flip the variable x_i and we know that the truth assignment of ϕ after the flip will be also satisfying, there must be a second true-literal supply vertex for every clause demand vertex currently assigned to x_i. After all such reassignments, we finally change the assignment of F_i from $\bar{x}_i$ to x_i.

It is easy now to see that there is a sequence of adjacent satisfying truth assignments of ϕ from $\mathbf{s}_0$ to $\mathbf{s}_t$ if and only if there is a sequence of adjacent feasible configurations of G_ϕ from f_0 to f_t. This completes a proof of Theorem 1. $\square$

2.2   Other Intractable Reconfiguration Problems

There is a wealth of reconfiguration versions of NP-complete problems which can be shown PSPACE-complete via extensions, often quite sophisticated, of the original NP-completeness proofs; in this subsection we only sample the realm of possibilities.

We have already defined the CLIQUE RECONFIGURATION problem in the Introduction as an example of a general scheme whereby any optimization problem can be transformed into a reconfiguration problem by giving a threshold (upper bound for minimization problems, lower bound for maximization problems) for the allowed values of the objective function of the intermediate feasible solutions; the INDEPENDENT SET RECONFIGURATION and VERTEX COVER RECONFIGURA-TION problems are defined similarly. In the INTEGER PROGRAMMING RECON-FIGURATION problem, we are given a 0-1 linear program seeking to maximize cx subject to $Ax \leq b$, and we consider two solutions adjacent if they only differ in one variable.

Theorem 2. *The following problems are PSPACE-complete*: INDEPENDENT SET RECONFIGURATION, CLIQUE RECONFIGURATION, VERTEX COVER RECONFIGU-RATION, SET COVER RECONFIGURATION, INTEGER PROGRAMMING RECONFIG-URATION.

Proof sketch. We sketch a proof for the INDEPENDENT SET RECONFIGURATION problem. The reduction can be obtained by extending the well-known reduction from the 3SAT problem to the INDEPENDENT SET problem [12]. We construct a graph $\rho(\phi)$ from a given 3SAT formula ϕ with n variables and m clauses, as follows. For each variable x in ϕ, we put an edge to the graph; the two endpoints are labeled x and $\bar{x}$. Then, for each clause C in ϕ, we put a clique of size $|C|$ to the graph; each node in the clique corresponds to a literal in the clause C. Finally, we add an edge between two nodes in different components if and only if the nodes correspond to opposite literals. Then, any maximum independent set in $\rho(\phi)$ contains at least n nodes; the n nodes are chosen from the endpoints of edges corresponding to the variables; *a literal is considered true if the corresponding endpoint is chosen.* Clearly, $\rho(\phi)$ has a maximum independent set of size $k = n + m$ if and only if ϕ is satisfiable. Consider all independent sets of size k in $\rho(\phi)$; they can be partitioned into subclasses of the form $\rho(\mathbf{s})$ corresponding to the satisfying truth assignments $\mathbf{s}$ of ϕ (the various independent sets in the subclass $\rho(\mathbf{s})$ correspond to the different possible ways to satisfy each clause by $\mathbf{s}$). It is easy to see that all independent sets in $\rho(\mathbf{s})$ are connected via intermediate independent sets of size at least $k - 1$. Therefore, by similar arguments in the proof of Theorem 1, one can easily observe that telling whether two independent sets of size k in $\rho(\phi)$ can be transformed into one another via intermediate independent sets of size at least $k - 1$ is PSPACE-complete.

Similarly as the NP-completeness proofs [5, §3.1.3], the result for INDEPENDENT SET RECONFIGURATION yields those for CLIQUE RECONFIGURATION and VERTEX COVER RECONFIGURATION. Then, the result for SET COVER RECONFIGURATION is immediate since it is a generalization of VERTEX COVER RECONFIGURATION. INTEGER PROGRAMMING RECONFIGURATION generalizes CLIQUE RECONFIGURATION via the well-known integer program for CLIQUE.   □

3   Reconfiguration Problems in P

Reconfiguration problems arise in relation to polynomially solvable problems as well. For example, in the MINIMUM SPANNING TREE RECONFIGURATION problem, we are given an edge-weighted graph G, a threshold k, and two spanning trees of G, both of weight at most k, and wish to transform one tree into another via edge exchanges, without ever getting into a tree with weight $> k$. The MATCHING RECONFIGURATION problem is defined similarly (the formal definition will be given later). We show in this section that both problems can be solved in polynomial time.

The result for the MINIMUM SPANNING TREE RECONFIGURATION problem can be obtained from the following more general proposition.

Proposition 1. *Given a weighted matroid* $\mathbf{M}$ *and two bases* B_0 *and* B_t *of* $\mathbf{M}$, *both of weight at most* k, *there always exists a sequence of* $|B_0 \setminus B_t|$ *exchanges that transforms one into the other without ever exceeding weight* k.

Proof sketch. For an unweighted matroid, this result follows trivially from the properties of a base family [15, §39.5]. For a weighted matroid $\mathbf{M}$, we outline a proof for the case when B_0 and B_t are both of maximum weight. Then, the result follows from the fact that the set of *maximum weight bases* of $\mathbf{M}$ also form the base family of another matroid [2, p. 287] [3, p. 130]. By generalizing this proof appropriately, one can obtain the full result. (Due to the page limitation, we omit the details.)   □

In the MATCHING RECONFIGURATION problem, we are given an unweighted graph G, a threshold k, and two matchings M_0 and M_t of G, both of size at least k, and we are asked whether there is a sequence of matchings of G, starting with M_0

and ending in M_t, and each resulting from the previous one by either addition or deletion of an edge in G, without ever going through a matching of size less than $k - 1$.

Proposition 2. MATCHING RECONFIGURATION *can be solved in polynomial time.*

Proof sketch. Since the adjacency relation is symmetric, we may assume without loss of generality that $|M_0| \leq |M_t|$. Consider the subgraph H of G induced by all edges in $(M_0 \setminus M_t) \cup (M_t \setminus M_0)$. Then, H consists of single edges, and alternating paths and cycles with respect to M_0 and M_t. The *greedy algorithm* for transforming M_0 into M_t is the following. Divide the components of H into the following four categories: (1) single edges of $M_t \setminus M_0$; (2) alternating paths starting with an edge of $M_t \setminus M_0$; (3) alternating cycles; and (4) all the rest. In this category order, transform M_0 into M_t by repeatedly adding edges of $M_t \setminus M_0$ and deleting edges of $M_0 \setminus M_t$ along each component of H. Notice that, after exchanging the edges in Categories (1) and (2), the obtained matching M has size at least $|M_t|$ $(\geq |M_0|)$. Therefore, one can easily observe that intermediate matchings have size at least $|M_0| - 1$ for exchanging edges in Category (2), and have size at least $|M_t| - 2$ for exchanging edges in Categories (3) and (4).

For the case $|M_t| \geq k + 1$, the greedy algorithm always transforms M_0 into M_t without ever going through a matching of size less than $k - 1$. For the case $|M_0| = |M_t| = k$, there does not always exist a desired sequence of matchings if H has components of Category (3). Nonetheless, existence can be determined in polynomial time, as follows. If M_0 and M_t are *not* maximum matchings of G, we first transform M_t into a matching M_t' of size $k+1$ along an arbitrary augmenting path with respect to M_t; then, the greedy algorithm works for transforming M_0 into M_t'. Therefore, a desired sequence always exists for this subcase. If M_0 and M_t are maximum matchings of G and H contains alternating cycles, we have the following lemma, whose proof is omitted due to the page limitation.

Lemma 1. *There is a sequence of adjacent matchings from M_0 to M_t such that all intermediate matchings have size at least $k - 1$ if and only if every cycle in H contains a vertex that begins an even-length alternating path in G with respect to M_0 ending at an unmatched vertex by M_0.*

By Lemma 1 one can easily determine whether there exists a desired sequence for this subcase in polynomial time. □

We note in passing that the MATCHING RECONFIGURATION problem for edge-weighted graphs seems quite a bit more complicated; however, we conjecture that it also can be solved in polynomial time.

Besides MINIMUM SPANNING TREE RECONFIGURATION and MATCHING RE-CONFIGURATION, it turns out that all polynomial-time solvable special cases of SATISFIABILITY, as characterized by Schaefer [14], give rise to polynomially solvable reconfiguration problems:

Theorem 3 ([6]). SATISFIABILITY RECONFIGURATION *for linear, Horn, dual Horn and 2-literal clauses are all in* P.

4 Approximation

We have seen that an optimization problem gives rise to a reconfiguration problem by bounding the objective of intermediate configurations. In turn, we can get a natural optimization problem if we try to *optimize the worst objective among all configurations* in the reconfiguration path. For example, in the problem that we call the MAXMIN CLIQUE RECONFIGURATION problem, we are given a graph and two cliques C_0 and C_t, and we are asked to transform C_0 into C_t by a sequence of additions and removals of nodes so that the minimum size of any clique in the sequence is as large as possible.

Theorem 4. MAXMIN CLIQUE RECONFIGURATION *cannot be approximated within any constant factor unless* P = NP.

Proof. We give a reduction in an approximation-preserving manner from the CLIQUE problem to this problem. For a given graph G with n nodes, we construct a new graph G' with $3n$ nodes as a corresponding instance of MAXMIN CLIQUE RECONFIGURATION: a set of n nodes is connected as G, while two new sets of n nodes are connected each as a clique (these two cliques of G' are called C_0 and C_t); finally, there are edges in G' between each new node and each node in G.

Consider any sequence of cliques of G', each resulting from the previous one by insertion or deletion of a node, starting from C_0 and ending in C_t. We claim that one of them will be a clique of G — this follows directly from the absence of any edges from C_0 to C_t. Conversely, for any clique C of G, there exists a sequence from C_0 to C_t via C (add the nodes of C to the clique C_0, then remove those of C_0, then add those of C_t). Therefore, the minimum clique size in the sequence is the size of C, and hence solving (or approximating) this instance of MAXMIN CLIQUE RECONFIGURATION is the same as solving (respectively, approximating) the CLIQUE problem for G. Since it is known that CLIQUE cannot be approximated within any constant factor unless P = NP [7], the result follows. □

A similar argument establishes the following:

Theorem 5. MAXMIN MAXSAT RECONFIGURATION *cannot be approximated within a factor better than* $\frac{15}{16}$ *unless* P = NP.

Proof. We reduce in an approximation-preserving manner the MAXSAT problem to this problem. Suppose that we are given an instance ϕ of MAXSAT with n variables $x_1, x_2, \ldots, x_n$ and m clauses $C_1, C_2, \ldots, C_m$. We construct a new instance ϕ' in which each clause C_j, $1 \le j \le m$, is replaced by $(C_j \vee y \vee z)$ where y and z are new variables, and the additional clause $(\bar{y} \vee \bar{z})$ *with weight m*. Note that the truth assignments $s_0 : z = 1, y = 0, x_1 = x_2 = \cdots = x_n = 1$ and $s_t : z = 0, y = 1, x_1 = x_2 = \cdots = x_n = 0$ are both satisfying all $2m$ clauses.

Consider now an optimal path in the hypercube between s_0 and s_t. Since at $s_0 : z = 1, y = 0$ and at $s_t : z = 0, y = 1$, there must exist a truth assignment on this path such that $y = z$. Since the clause $(\bar{y} \vee \bar{z})$ has weight m and the path is assumed optimal, it must be that $y = z = 0$. Thus, the remaining variables must spell an optimum satisfying truth assignment of the original formula ϕ. Hence, from an optimum path for the corresponding instance of MAXMIN MAXSAT RECONFIGURATION, we can obtain an optimum truth assignment for the original instance of MAXSAT. Similarly, from an α-approximation for MAXMIN MAXSAT RECONFIGURATION, it is easy to see that we get a $(2\alpha - 1)$-approximation of the MAXSAT instance. Since it is known that MAXSAT cannot be approximated within a factor better than $\frac{7}{8}$ unless P $=$ NP [8], the result follows. $\square$

By a similar maneuver, it can be shown that the MINMAX SET COVER RECONFIG-URATION problem cannot be approximated within a factor better than $o(\log n)$ unless NP is contained in DTIME$\left(n^{O(\log \log n)}\right)$ [4].

Returning to the POWER SUPPLY problem, there is a natural optimization version of the problem, in which the constraint that the total demand of all demand vertices in each tree T be within the supply of the supply vertex in T is replaced by a "soft" criterion: we allow that the total demand in T exceeds the supply in T, but wish to minimize the sum of the "deficient power" of all supply vertices in the graph.

We now define the MINMAX POWER SUPPLY RECONFIGURATION problem. For a configuration f of a bipartite graph $G = (U, V, E)$ and a supply vertex $u \in U$, the *deficient power* $d(f, u)$ *of* u *on* f is defined as follows:

$$d(f, u) = \sum \{\text{dem}(v) \mid v \in V \text{ such that } f(v) = u\} - \sup(u).$$

If f is infeasible, then there is at least one supply vertex u such that $d(f, u) > 0$. On the other hand, if f is feasible, then $d(f, u) \leq 0$ for all supply vertices $u \in U$; in fact, a nonpositive deficient power $d(f, u)$ represents the *marginal power* of u on f. The *cost* $c(f)$ *of a configuration* f is defined to be $c(f) = \sum_{u \in U} |d(f, u)|$. Clearly, $c(f) = \sum_{u \in U} \sup(u) - \sum_{v \in V} \text{dem}(v)$ for every feasible configuration f of G. In the problem that we call the MINMAX POWER SUPPLY RECONFIGURATION problem, we are given a bipartite graph $G = (U, V, E)$ and two feasible configurations f_0 and f_t of G, and we are asked to transform f_0 into f_t by a sequence of reassignments of single demand vertices so that the maximum cost of any configuration in the sequence is as small as possible. It is easy to see that a sequence $f_0, f_1, \ldots, f_t$ which consists of only feasible configurations is optimum, and the optimum value is $\sum_{u \in U} \sup(u) - \sum_{v \in V} \text{dem}(v)$.

One can observe that the MINMAX POWER SUPPLY RECONFIGURATION problem is strongly NP-hard (by a reduction from the 3-PARTITION problem [5], for example). However, the problem can be solved in linear time for the following special case. Suppose in the remainder of this section that we are given a bipartite graph $G = (U, V, E)$ having exactly two supply vertices. For a configuration f of G, let $W(f) = \{v \in V \mid f(v) \neq f_t(v)\}$, that is, $W(f)$ is the set of demand vertices which are assigned to "wrong" supply vertices on f. Note that all (demand) vertices in $W(f)$ are adjacent to both the two supply vertices. For a given initial configuration f_0 of G, let v^* be a demand vertex in $W(f_0)$ having the maximum demand, that is, $\text{dem}(v^*) = \max\{\text{dem}(v) \mid v \in W(f_0)\}$. Then, we have the following lemma.

Lemma 2. *If $c(f_0) \geq 2 \cdot \mathrm{dem}(v^*)$, then the optimum sequence for* MINMAX POWER SUPPLY RECONFIGURATION *consists of only feasible configurations, and it can be found in linear time.*

Proof. Suppose without loss of generality that $W(f_0) \neq \emptyset$. If all demand vertices in $W(f_0)$ are assigned to the same supply vertex, then we just change the assignments of all demand vertices in $W(f_0)$ from the current supply vertex to the other. Since both f_0 and f_t are feasible, all intermediate configurations are also feasible. Therefore, we assume in the following that each of the two supply vertices has at least one demand vertex in $W(f_0)$.

Since f_0 is feasible, the cost $c(f_0)$ denotes the sum of marginal powers of the two supply vertices. Moreover, since the sum is at least $2 \cdot \mathrm{dem}(v^*)$, one of the two supply vertices has marginal power of at least $\mathrm{dem}(v^*)$. Therefore, we can change the assignment of at least one demand vertex $v \in W(f_0)$ from the "wrong" supply vertex to the "correct" one, since $\mathrm{dem}(v) \leq \mathrm{dem}(v^*)$. Clearly, the resulting configuration f_1 is also feasible, and satisfies $c(f_1) \geq 2 \cdot \mathrm{dem}(v^*)$. By repeatedly executing such a reassignment, we can obtain a desired sequence $f_0, f_1, \ldots, f_t$ which consists of only feasible configurations. Therefore, the sequence is an optimum solution. The length of the sequence is $|W(f_0)|$ $(\leq |V|)$ since each demand vertex in $W(f_0)$ moves exactly once and any of the other demand vertices does not move in the sequence. We can thus find an optimum solution in linear time. $\qquad\square$

Theorem 6. *There is a linear-time 2-approximation algorithm for* MINMAX POWER SUPPLY RECONFIGURATION *having exactly two supply vertices.*

Proof. Let OPT be the optimum value for an instance of MINMAX POWER SUPPLY RECONFIGURATION. Since we have to change the assignment of the demand vertex v^* for obtaining the target configuration f_t, we have $\mathrm{OPT} \geq \mathrm{dem}(v^*)$.

By Lemma 2 it suffices to consider the case $c(f_0) < 2 \cdot \mathrm{dem}(v^*)$. In this case, consider a slightly modified instance in which the supplies of the two supply vertices are increased so that the total supply is equal to $2 \cdot \mathrm{dem}(v^*)$. In the modified instance, both the configurations f_0 and f_t remain feasible and $c(f_0) = 2 \cdot \mathrm{dem}(v^*)$. Therefore, by Lemma 2 we can find in linear time an optimum sequence which consists of only feasible configurations for the modified instance; the optimum value is thus $2 \cdot \mathrm{dem}(v^*)$. Note that some configurations in the sequence may be infeasible for the original instance. We take the sequence as our approximation solution for the original instance, and hence our approximate value A is $A = 2 \cdot \mathrm{dem}(v^*) \leq 2 \cdot \mathrm{OPT}$. $\qquad\square$

5   Open Problems

There are many open problems raised by this work, and we mention some of these below:

- Do all problems in P give rise, in a natural way, to polynomially solvable reconfiguration problems? We conjecture that the answer is negative, but we have yet to identify a counterexample (even a conjectured one).

- Is the TRAVELING SALESMAN RECONFIGURATION problem (where two tours are adjacent if they differ in two edges) PSPACE-complete?
- Are there better approximation algorithms for the MINMAX POWER SUPPLY RECONFIGURATION problem? Lower bounds?
- Are the problems in Section 4 PSPACE-complete to approximate (not just NP-hard)?

References

1. Bonsma, P., Cereceda, L.: Finding paths between graph colourings: PSPACE-completeness and superpolynomial distances. In: Kučera, L., Kučera, A. (eds.) MFCS 2007. LNCS, vol. 4708, pp. 738–749. Springer, Heidelberg (2007)
2. Cook, W.J., Cunningham, W.H., Pulleyblank, W.R., Schrijver, A.: Combinatorial Optimization. Wiley, Chichester (1997)
3. Edmonds, J.: Matroids and the greedy algorithm. Math. Programming 1, 127–136 (1971)
4. Feige, U.: A threshold of $\ln n$ for approximating set cover. J. ACM 45, 634–652 (1998)
5. Garey, M.R., Johnson, D.S.: Computers and Intractability: A Guide to the Theory of NP-Completeness. Freeman, San Francisco (1979)
6. Gopalan, P., Kolaitis, P.G., Maneva, E.N., Papadimitriou, C.H.: The connectivity of Boolean satisfiability: computational and structural dichotomies. In: Bugliesi, M., Preneel, B., Sassone, V., Wegener, I. (eds.) ICALP 2006. LNCS, vol. 4051, pp. 346–357. Springer, Heidelberg (2006)
7. Håstad, J.: Clique is hard to approximate within $n^{1-\varepsilon}$. Acta Mathematica 182, 105–142 (1999)
8. Håstad, J.: Some optimal inapproximability results. J. ACM 48, 798–859 (2001)
9. Hearn, R.A., Demaine, E.D.: PSPACE-completeness of sliding-block puzzles and other problems through the nondeterministic constraint logic model of computation. Theoretical Computer Science 343, 72–96 (2005)
10. Ito, T., Zhou, X., Nishizeki, T.: Partitioning trees of supply and demand. International J. Foundations of Computer Science 16, 803–827 (2005)
11. Ito, T., Demaine, E.D., Zhou, X., Nishizeki, T.: Approximability of partitioning graphs with supply and demand. In: Asano, T. (ed.) ISAAC 2006. LNCS, vol. 4288, pp. 121–130. Springer, Heidelberg (2006)
12. Papadimitriou, C.H.: Computational Complexity. Addison-Wesley, Reading (1994)
13. Savitch, W.J.: Relationships between nondeterministic and deterministic tape complexities. J. of Computer and System Sciences 4, 177–192 (1970)
14. Schaefer, T.J.: The complexity of satisfiability problems. In: Proc. of 10th ACM Symposium on Theory of Computing, pp. 216–226 (1978)
15. Schrijver, A.: Combinatorial Optimization: Polyhedra and Efficiency. Springer, Heidelberg (2003)

Multiobjective Disk Cover Admits a PTAS

Christian Glaßer[1], Christian Reitwießner[1], and Heinz Schmitz[2]

[1] Julius-Maximilians-Universität Würzburg, Germany
{glasser,reitwiessner}@informatik.uni-wuerzburg.de
[2] Fachhochschule Trier, Germany
schmitz@informatik.fh-trier.de

Abstract. We introduce multiobjective disk cover problems and study their approximability. We construct a polynomial-time approximation scheme (PTAS) for the multiobjective problem where k types of points (customers) in the plane have to be covered by disks (base stations) such that the number of disks is minimized and for each type of points, the number of covered points is maximized. Our approximation scheme can be extended so that it works with the following additional features: interferences, different services for different types of customers, different shapes of supply areas, weighted customers, individual costs for base stations, and payoff for the quality of the obtained service.

Furthermore, we show that it is crucial to solve this problem in a multiobjective way, where all objectives are optimized at the same time. The constrained approach (i.e., the restriction of a multiobjective problem to a single objective) often used for such problems can significantly degrade their approximability. We can show non-approximability results for several single-objective restrictions of multiobjective disk cover problems. For example, if there are 2 types of customers, then maximizing the supplied customers of one type is not even approximable within a constant factor, unless P = NP.

1   Introduction

Geometric cover problems have received much attention in recent years, mostly due to their applicability to wireless networks. Typically, a service provider aims to deliver various kinds of services to customers and therefore has to choose base station locations such that customer locations can be covered. Various optimization problems arise in this context in a natural way. For example, for a given set P of customer locations and a set D of possible base station locations in the Euclidean plane, the Unit Disk Cover Problem tries to find a minimal subset of D such that all customers in P are covered by unit disks whose centers belong to D (hence assuming equivalent base stations and ignoring obstacles to the signal propagation) [10,5,19]. In another version of the problem one tries to maximize the number of supplied customers with a given budget of base stations, e.g. [17]. So far these problems have been studied only in terms of single-objective optimization where either disk locations or customer supply have been optimized.

In contrast, here we are interested in the complete trade-offs when both objectives are considered *at once*. These trade-offs give not only a better insight in

S.-H. Hong, H. Nagamochi, and T. Fukunaga (Eds.): ISAAC 2008, LNCS 5369, pp. 40–51, 2008.
© Springer-Verlag Berlin Heidelberg 2008

the nature of the problem, but also allow a human decision-maker to choose an appropriate solution according to aspects that are perhaps not quantifiable or that differ from instance to instance.

This paper introduces *multiobjective* disk cover problems and presents the first study of their approximability. We want to minimize the number of base stations and simultaneously maximize the number of supplied customer locations. More than that, we allow *different types* of customers such that for each type, the number of supplied customers has to be maximized. For example, for $k = 2$ types of customers this captures the scenario where a service provider wants to optimize a wireless network with customers having subscribed to two different services. More generally, an instance of the Disc Cover Problem with k types of customers (k-DC) has k sets $P_1, \ldots, P_k$ of customer locations, a set of potential base station locations D and a disk radius r describing the range of action of a base station. We seek a valid subset of base station locations (i.e., respecting a minimum-distance constraint) such that the number of base stations is minimized, and for each type of customers, the numbers of covered customers is maximized. In practice, several additional aspects can be taken into account to obtain more realistic models. For instance, if a customer receives signals from two base stations, then interferences may have a negative effect on the quality of service. So actually, here one wants to maximize the number of points that are covered by *exactly* one disk. To capture the aspect of interference, we also investigate exact versions of k-DC, i.e., where the numbers of *uniquely* covered customers of the different types are considered (k-EDC).

Trade-offs in multiobjective optimization are captured by the notion of the so-called *Pareto curve* which is the set of all solutions whose vector of objective values is not dominated by any other solution (for an introduction see, e.g., [13]). In most interesting cases however, Pareto curves are computationally hard in the sense that we do not know polynomial-time algorithms computing them. The reason for this is that the Pareto curve may have exponential size (which is not the case in our setting), or because it comprises optimal solutions of NP-hard single-objective optimization problems (which is the case here). A reasonable approach to avoid these difficulties is to approximate the set of non-dominated solutions using the concept of the ϵ-approximate Pareto curve. Informally, for every solution S of the Pareto curve there is a solution S' in the ϵ-approximate Pareto curve that is within a factor $(1 + \epsilon)$ of S, or better, in all objectives. The question whether there exist fast approximation schemes for Pareto curves has been addressed for several multiobjective optimization problems [21,22,20]. The systematic study of the theory of multiobjective approximation was initiated by Papadimitriou and Yannakakis [20], see also [23,11].

Our contribution. We introduce *multiobjective* disk cover problems and study their approximability. We construct polynomial-time approximation schemes for the multiobjective problems k-DC and k-EDC where $k \geq 1$. So for each of the problems there exists an algorithm, which, given a problem instance I and some $\epsilon > 0$, outputs an ϵ-approximate Pareto curve in time polynomial in $|I|$

(Theorems 1 and 2). On the methodological side we extend the *shifting strategy* introduced by Hochbaum and Maass [18] to the multiobjective case.

We also discuss the possibility to extend our algorithms so that they work with the following features: different services for different types of customers, different shapes of supply areas, weighted customers, individual costs for base stations, and payoff for the quality of the obtained service. Although we mention only problems where the number of covered points of different types have to be maximized, one can also think of an application where some types of points have to be maximized, while others are to be minimized. Only minor modifications of our algorithms are needed to take this into account as well.

Our paper also shows that we should be careful when looking for an appropriate model for a given practical problem. The choice of the right model can be crucial for a successful algorithmic solution. We see this at two places:

1. Our models contain a minimum-distance constraint for disk locations. On one hand, this constraint plays an important role in the construction of the PTAS. Without this assumption, the problem becomes more difficult such that the shifting strategy does not yield a PTAS. On the other hand, this minimum-distance constraint is actually present in practical settings: It usually makes no sense to build base stations, fire departments, drugstores, etc. *arbitrarily* close to each other, and a small constant specifying their minimum distance can always be identified. So we may add the constraint to our model and exploit it to achieve better approximation algorithms. This shows that a *too general choice* of the model can complicate the solution of the underlying practical problem.

2. The approximability of a multiobjective problem does not necessarily imply that the restriction to a single objective is approximable. The reason for this apparent contradiction is that an optimization algorithm can exploit trade-offs between the single objectives if it optimizes all objectives at the same time. We show non-approximability results for several restrictions of k-DC and k-EDC (Theorems 4 and 5). For example, for $k \geq 2$, no restriction of k-EDC to a single criterion is approximable within a constant factor (unless $\mathrm{P} = \mathrm{NP}$), while the general (multiobjective) version of k-EDC admits even a PTAS. This shows that the frequently used constrained approach (i.e., the restriction of a multiobjective problem to a single objective) can considerably degrade the approximability. In other words, also a *too restricted choice* of the model can complicate the solution of the underlying problem.

Related work. The single-objective disk cover problem was initially examined in the continuous version (i.e., no given disk centers) by Hochbaum and Maass [18] who construct a PTAS for this problem. The version with given disk centers has been studied by several authors [5,19,7,4], but in general only constant-factor approximation results are known (which shows again the influence of the minimum-distance constraint). Calinescu, Mandoiu, Wan and Zelikovsky [5] investigated a variant of the disk cover problem where the number of disks needed to cover all points is to be minimized. Some of their results were improved by

Narayanappa and Vojtechovsky [19]. Very recently, the Unique Cover Problem on unit disks (the number of points covered by exactly one disk have to be maximized) and its approximability has been studied by Erlebach and van Leeuwen [15]. Several other variations of the disk cover problem where a solution includes specifying radii for the individual disks were analyzed in [8,14,1]. Cannon and Cowen [6] studied the single-objective problem of minimizing the number of disks where one type of customers must be covered while the other one has to be avoided. Various partial covering problems were investigated by Gandhi, Kuller and Srinivasan [16]. These problems are concerned with covering a given amount of elements while minimizing the cost of such a covering. In contrast to most of the afore mentioned papers, in [16] the multiobjective version of some special covering problem on graphs is also examined.

2   Definitions

We recall some standard notations, see e.g., [20,23]. A *multiobjective optimization problem* Π has a set of valid instances $\mathcal{I}$, and for every instance $I \in \mathcal{I}$ there is a set $\mathcal{S}(I)$ of feasible and polynomially length-bounded solutions for I. As usual, we assume that $\mathcal{I}$ is decidable in polynomial time, and that there is a polynomial-time algorithm that decides on input (I, S) whether $S \in \mathcal{S}(I)$. Moreover, we have $K \geq 1$ polynomial-time computable objective functions f_i that map every $I \in \mathcal{I}$ and $S \in \mathcal{S}(I)$ to some value $f_i(I, S) \in \mathbb{N}$. Note that every optimization problem with objective functions that have values in $\mathbb{Q}$ can be transformed into an equivalent problem satisfying the previous definition. A vector *goal* $\in \{\min, \max\}^K$ specifies whether the i-th objective has to be minimized or maximized, respectively. So for an instance I we can evaluate every $S \in \mathcal{S}(I)$ to the K-vector $f(I, S) = (f_1(I, S), \ldots, f_K(I, S))$ of values with respect to the given objective functions.

We say a solution $S \in \mathcal{S}(I)$ *dominates* a solution $S' \in \mathcal{S}(I)$ if for all $1 \leq i \leq K$ it holds that $f_i(I, S) \leq f_i(I, S')$ if f_i is to be minimized (and $f_i(I, S) \geq f_i(I, S')$ if f_i is to be maximized), with at least one strict inequality. Denote by $P^{sol}(I) \subseteq \mathcal{S}(I)$ the *Pareto-solution* set for I, i.e., the set of all non-dominated solutions for I. The *Pareto-value* set for I is $P^{val}(I) = \{f(I, S) \mid S \in P^{sol}(I)\}$.

Let $\epsilon = (\epsilon_1, \ldots, \epsilon_K)$ be a K-vector of numbers $\epsilon_i \geq 0$. A solution $S \in \mathcal{S}(I)$ ϵ-*covers* a solution $S' \in \mathcal{S}(I)$ if for all $1 \leq i \leq K$ it holds that $f_i(I, S) \leq (1 + \epsilon_i)f_i(I, S')$ if f_i is to be minimized (and $(1 + \epsilon_i)f_i(I, S) \geq f_i(I, S')$ if f_i is to be maximized).

A set $P^{sol}_\epsilon(I) \subseteq \mathcal{S}(I)$ is an ϵ-*approximate Pareto-solution* set for I if for all $S' \in P^{sol}(I)$ there is some $S \in P^{sol}_\epsilon(I)$ that ϵ-covers S'. (So an ϵ-approximate Pareto-solution set can contain dominated points.) We call $P^{val}_\epsilon(I) \subseteq \mathbb{N}^K$ an ϵ-*approximate Pareto-value* set for I if $P^{val}_\epsilon(I) = \{f(I, S) \mid S \in P^{sol}_\epsilon(I)\}$ for some set $P^{sol}_\epsilon(I)$. Note that for fixed ϵ there may be more than one ϵ-approximate Pareto-solution set for I. Moreover, if $\mathcal{S}(I) \neq \emptyset$ then $P^{sol}(I) \neq \emptyset$ and $P^{sol}_\epsilon(I) \neq \emptyset$. If $\epsilon = (\delta, \ldots, \delta)$ for some $\delta > 0$ we simply write $P^{sol}_\delta(I)$ and the like.

A multiobjective optimization problem Π is *ϵ-approximable in polynomial time* if there is a polynomial-time algorithm, which on input $I \in \mathcal{I}$ outputs an ϵ-approximate Pareto-solution set $P_\epsilon^{sol}(I)$. Problem Π has a *polynomial-time approximation scheme* (PTAS) if there is an algorithm, which, given $I \in \mathcal{I}$ and $\delta > 0$, outputs an δ-approximate Pareto-solution set $P_\delta^{sol}(I)$ in time polynomial in $|I|$. APX is the class of all single-objective optimization problems that are δ-approximable for some $\delta > 0$. For some vector x denote by $|x|$ its Euclidean norm. If S is a finite set, then $|S|$ gives the cardinality of S. Both cases will be distinguishable from the context without confusion. Moreover, we use $[a, b]$ as an abbreviation for $\{a, a + 1, \ldots, b\}$.

Next we define $(k + 1)$-objective disk-cover problems. As is standard for such problems, we always want to minimize the number of disks which is the first objective in all of the following problems. The parameter $k \geq 1$ denotes the number of different types of points we want to cover. Moreover, $\varrho \in (0, 2]$ is a fixed rational constant that determines the minimal distance $\varrho \cdot r$ between different disks of radius r. This *minimum-distance constraint* plays an important role in our model, since it is crucial for the polynomial running time of the approximation algorithm we construct in section 3. With our method we cannot well approximate instances that essentially depend on coverings where the disks are very close to each other. This insight has an important consequences for the choice of an appropriate model: If such degenerated instances can be excluded by practical reasons (e.g., because it makes no sense to build base stations, fire departments, drugstores, etc. arbitrarily close to each other), then we should add the minimum-distance constraint to our model and exploit it to achieve a better approximability.

k-Objective Disk Cover (k-DC$_\varrho$)
Instance: k finite sets of points $P_1, \ldots, P_k \subseteq \mathbb{Z} \times \mathbb{Z}$, disk radius $r \in \mathbb{N}$, finite set
         of disk positions $D \subseteq \mathbb{Z} \times \mathbb{Z}$
Solution: a selection $S \subseteq D$ such that for all different $x, y \in S$, $|x - y| \geq \varrho \cdot r$
Goals: $(\min|S|, \max|C_1|, \ldots, \max|C_k|)$ where $C_i = \{x \in P_i \mid \exists y \in S, |x - y| \leq r\}$

k-Objective Exact Disk Cover (k-EDC$_\varrho$)
Instance: k finite sets of points $P_1, \ldots, P_k \subseteq \mathbb{Z} \times \mathbb{Z}$, disk radius $r \in \mathbb{N}$, finite set
         of disk positions $D \subseteq \mathbb{Z} \times \mathbb{Z}$
Solution: a selection $S \subseteq D$ such that for all different $x, y \in S$, $|x - y| \geq \varrho \cdot r$
Goals: $(\min|S|, \max|C_1|, \ldots, \max|C_k|)$ where $C_i = \{x \in P_i \mid \exists! y \in S, |x - y| \leq r\}$

The value of ϱ will be always clear from the context. So for simplicity we write k-DC and k-EDC instead of k-DC$_\varrho$ and k-EDC$_\varrho$.

We also discuss single-objective versions of these problems. Following Diakonikolas and Yannakakis [11] we define the restricted versions of multiobjective problems (also known as the ε-constraint problem [13]). Let Π be a K-objective optimization problem with objectives $(f_1, \ldots, f_K)$ and goals $(g_1, \ldots, g_K)$. The restriction to the i-th objective is the following single-objective problem.

Restriction of Π to the i-th objective ($\text{Restricted}_i\text{-}\Pi$)
Instance: an instance I of Π and numbers $B_1, \ldots, B_{i-1}, B_{i+1}, \ldots, B_K \in \mathbb{N}$
Solution: a solution S for I such that for $j \in [1, K] - \{i\}$ it holds that
$$(g_j = \max \Rightarrow f_j(I, S) \geq B_j) \text{ and } (g_j = \min \Rightarrow f_j(I, S) \leq B_j)$$
Goal: $\max f_i(I, S)$ if $g_i = \max$, $\min f_i(I, S)$ otherwise

3   PTAS for Multiobjective Disk Cover

In this section we construct polynomial-time approximation schemes for the multiobjective problems k-DC where $k \geq 1$. To keep the exposition simple, we concentrate on the 3-objective problem 2-DC and explain a polynomial-time algorithm that computes ϵ-approximate Pareto-solution sets for this problem. Our algorithm extends the shifting strategy introduced by Hochbaum and Maass [18] to the multiobjective case. For this we need a combinatorial argument showing that this strategy works for multiple objectives. Moreover, we use dynamic programming for efficiently combining the solutions of sub-problems. At the end of the section we discuss the possibility to extend our algorithms so that they work with the following additional features: interferences, different services for different types of customers, different shapes of supply areas, weighted customers, individual costs for base stations, and payoff for the quality of the obtained service. In particular, an appropriate modification provides a PTAS for k-EDC.

We start with the description of the algorithm. Fix some shifting parameter $l \in \mathbb{N} \setminus \{0\}$. The larger l is, the better the approximation will be. The input to the algorithm are two finite sets of points $B, G \subseteq \mathbb{Z} \times \mathbb{Z}$ (blue and green points), a disk radius $r \in \mathbb{N}$ and a finite set of disk positions $D \subseteq \mathbb{Z} \times \mathbb{Z}$. For finite $P, S \subseteq \mathbb{Z} \times \mathbb{Z}$, where S is a valid solution (it respects the minimum distance constraint), define $c(P, S) \stackrel{df}{=} |\{p \in P \mid \exists x \in S, |p - x| \leq r\}|$ as the number of points from P covered by solution S.

In the algorithm, some functions $p \colon \mathbb{N} \times \mathbb{N} \to \mathcal{P}(D) \times \mathbb{N}$ with different indices will be defined. For a given number of disks and blue points to cover, such a function provides a partial solution for this sub-problem together with the number of green points covered in this solution. For any of these functions we address their components as $(p^{sol}(k, b), p^{val}(k, b)) \stackrel{df}{=} p(k, b)$ for $k, b \in \mathbb{N}$.

```
2-DC-APPROX(B,G,r,D):
```

1. let $\{a_1, a_2, \ldots, a_m\} \stackrel{df}{=} \{a \in rl \cdot (\mathbb{Z} \times \mathbb{Z}) \mid (B \cup G \cup D) \cap (a + [0, 2rl]^2) \neq \emptyset\}$
2. for every $s \in r \cdot [0, 1)^2$ do
3.   for every $i \in [1, m]$ do
4. $D_i \stackrel{df}{=} D \cap (s + a_i + [r, rl - r)^2)$
5. $B_i \stackrel{df}{=} B \cap (s + a_i + [2r, rl - 2r)^2)$
6. $G_i \stackrel{df}{=} G \cap (s + a_i + [2r, rl - 2r)^2)$
7.   for every $k \in [0, |D_i|]$ and every $b \in [0, |B_i|]$ do
8. $V_{k,b} \stackrel{df}{=} \{S \subseteq D_i \mid S \text{ is valid solution}, |S| \leq k, c(B_i, S) \geq b\}$
9.   if $V_{k,b} = \emptyset$ then $p_{s,i}(k, b) \stackrel{df}{=} (\bot, \bot)$
10.   else $p_{s,i}(k, b) \stackrel{df}{=} (S, g)$ for $S \in V_{k,b}$ such that
$$g = c(G_i, S) = \max\{c(G_i, S') \mid S' \in V_{k,b}\}$$

```
11.            done
12.        done
13.        for every k ∈ [0, |D|] and every b ∈ [0, |B|] do
14.            by dynamic programming choose k₁,...,kₘ,b₁,...,bₘ ∈ ℕ
```

$$\text{such that } \sum_{i=1}^{m} \mathsf{p}_{s,i}^{val}(k_i,b_i) \text{ is maximal, } \forall_{i\in[1,m]}\mathsf{p}_{s,i}^{val}(k_i,b_i) \neq \perp,$$

$$\sum_{i=1}^{m} k_i \leq k, \text{ and } \sum_{i=1}^{m} b_i \geq b$$

```
15.            if this succeeded,
```

$$\text{then let } \mathsf{p}_s(k,b) \overset{df}{=} \left(\bigcup_{i=1}^{m} \mathsf{p}_{s,i}^{sol}(k_i,b_i), \sum_{i=1}^{m} \mathsf{p}_{s,i}^{val}(k_i,b_i)\right)$$

```
16.        done
17. done
18. P ≝ {pₛˢᵒˡ(k,b) | s ∈ r · [0,1)², k,b ∈ ℕ}
19. remove all dominated solutions from P
20. return P
```

Explanation of the algorithm. First, we want to give an overview of the algorithm. The plane is divided into a grid of squares of side length rl. In each of these squares, the problem is solved independently (i.e., a small Pareto curve is calculated). By not considering the points at the border of width r of the squares, we obtain that an optimal solution needs at least as much disks as our calculated solution to cover the points in the square. Then, these solutions are combined. This is repeated for l^2 different positions (*shifts*) of the grid and the best solution is chosen.

The algorithm starts by partitioning the plane into squares of side length rl. Of course, there are infinitely many such squares, but many of them are empty and only some are of interest. The points $a_1, a_2, \ldots, a_m$ are the lower-left corner points of squares we need to consider. Because we will shift these squares later, we also have to include squares that contain a point for some, but possibly not all shifts, and thus we look for points in a square of side length $2rl$. These points a_i are all points such that there is at least one blue point, green point or one disk position in the square of side length $2rl$ which has a_i as its lower-left corner point.

Next, the algorithm loops over all l^2 (l in each dimension) shifts s of hop size r (the radius of a disk). In line 3, we loop over every index of the rl-grid points a_i which were found worth considering at the beginning.

In lines 4 to 6, we prepare a spatially restricted sub-problem of the general problem. The expression $[0, rl)^2$ denotes the set of points in a square of side length rl. Modified to $[r, rl - r)^2$ it denotes the set of points in such a square where a border of width r is removed from every edge. We only retain those points D_i from the set of disk positions D which lie in the restricted square that is positioned at the grid point a_i and shifted by s. In this way, disk positions from different sub-problems are guaranteed to have a minimal distance of ϱr (recall that $\varrho \in (0, 2]$). We also restrict the blue and green points, but here we use a larger border of width $2r$. The points on the $2r$-border are completely ignored in every sub-problem. By this method, as we will argue later, we get an optimal solution for each square restricted in this way. We can combine the solutions of these sub-problems to obtain a global solution.

We now calculate the whole Pareto curve of this sub-problem starting in line 7. To this end, we loop over every possible number of disks k and covered blue points b and calculate the solution $S \subseteq D_i$ that maximizes the number of covered green points g using at most k disks in positions from D_i such that at least b blue points are also covered. This can be done by exhaustive search in polynomial time, as we will explain next. We first argue that there are only polynomially many valid solutions $|V_{k,b}|$. Because all disk positions in a valid solution $S \in V_{k,b}$ must have a mutual distance of at least ϱr, virtual circles of diameter ϱr around these positions can touch each other but must not overlap. The area covered by these virtual circles is $|S| \left(\frac{\varrho r}{2}\right)^2 \pi$. Since $\frac{\varrho r}{2} \leq r$ and $S \subseteq D_i$, these virtual circles are all located in a square of side length rl, and we get $|S| \left(\frac{\varrho r}{2}\right)^2 \pi \leq (rl)^2$. Solved for the number of disks we obtain $|S| \leq \frac{4l^2}{\pi \varrho^2} \stackrel{df}{=} c$, which is a constant. Since there are only polynomially many ways to choose at most c elements from the polynomially sized set D_i, we see that $|V_{k,b}|$ is polynomial in the input size. Since c can be calculated effectively, $V_{k,b}$ can be searched exhaustively for a solution that maximizes the number of covered green points in polynomial time. If such a solution does not exist (because $V_{k,b} = \emptyset$) then both components of $p_{s,i}(k,b)$ are set to a special undefined value $\perp$, which we will need later.

After the i-loop, we have small Pareto curves $p_{s,i}$ for each sub-problem given by shift s and point a_i. In the loop starting in line 13, we combine them into a larger Pareto curve p_s for the current shift s. To this end, we try to find a solution that maximizes the number of covered green points for a given number of disks k and blue points b using the solutions $p_{s,i}$ in line 14. We distribute the k disks and b blue points over all squares in any possible way. The number of disks available to square i is called k_i and the number of blue points that must be covered in square i is called b_i. The distribution that maximizes the total number of covered green points is chosen and the combination of the individual solutions $p_{s,i}(k_i, b_i)$ for this distribution is stored in $p_s(k,b)$.

In general, there are exponentially many ways to distribute the numbers k and b, but the search for an optimal distribution can be done efficiently by dynamic programming, as can be seen from the following algorithm. We need to consider the iterations of the k, b-loop starting in line 15 of the 2-DC-APPROX-algorithm all at once, so the following code can be used as a replacement of the lines 13–16 of the 2-DC-APPROX-algorithm. The return value of CombinePartialSolutions is the function p_s.

```
CombinePartialSolutions(p_{s,1}, p_{s,2}, ..., p_{s,m}):
1.  p'_1 =df p_{s,1}
2.  for t := 2 to m do
3.      for every k ∈ [0, |D|] and every b ∈ [0, |B|] do
4.          find maximal p'val_t(k, b) =df p'val_{t-1}(k̄_1, b̄_1) + p^val_{s,t}(k̄_2, b̄_2) for
                k̄_1, b̄_1, k̄_2, b̄_2 ∈ ℕ such that k̄_1 + k̄_2 ≤ k, b̄_1 + b̄_2 ≥ b,
                p'val_{t-1}(k̄_1, b̄_1) ≠ ⊥ and p^val_{s,t}(k̄_2, b̄_2)) ≠ ⊥
```

```
5.              if this is not possible then let p'ₜ(k, b) ≝ ⊥
6.          else let p'ₜˢᵒˡ(k, b) ≝ p'ₜ₋₁ˢᵒˡ(k̄₁, b̄₁) ∪ pₛ,ₜˢᵒˡ(k̄₂, b̄₂)
7.      done
8. done
9. return p'ₘ
```

In every iteration of the t-loop of CombinePartialSolutions, another square is incorporated into the Pareto curve, each time solving some kind of knapsack problem. Since there are only polynomially many combinations of $\bar{k}_1, \bar{b}_1, \bar{k}_2, \bar{b}_2$ such that the constraints in line 4 are met, p'_t can be computed in polynomial time. The correctness of this dynamic programming method follows by induction. Since m is polynomial in the input length, the computation of CombinePartial-Solutions and thus also the computation of lines 13–16 in 2-DC-APPROX can be done in polynomial time.

Back at the 2-DC-APPROX-algorithm, we have an approximate Pareto curve p_s for every of the shift values after the end of the second loop over s. In Line 18, we simply put all the previously obtained solutions $p_s^{sol}(k, b)$ for all s, k, b in one set P, remove the dominated solutions in line 19 and return that set in line 20.

Correctness of the algorithm. We now argue for the correctness of the algorithm by showing that for fixed l it runs in polynomial time and that the relative error becomes arbitrarily small if l is increased.

Lemma 1. *For every fixed $l \geq 5$, the algorithm 2-DC-APPROX works in polynomial time.*

We argue that by choosing l large enough, the algorithm 2-DC-APPROX has an arbitrarily small relative error.

Lemma 2. *Fix an $l \geq 5$ and let $\epsilon \overset{df}{=} (0, \frac{16}{l}, \frac{16}{l})$. On input of a 2-DC instance $I = (B, G, r, D)$ the algorithm 2-DC-APPROX computes an ϵ-approximate Pareto-solution set P for I.*

Theorem 1. *Fix $k \geq 1$ and $\varrho \in (0, 2]$. For all $\delta > 0$, k-DC is $(0, \delta, \ldots, \delta)$-approximable in polynomial time (and hence has a PTAS).*

The multiobjective shifting strategy used in 2-DC-APPROX is a very general method that can be applied to several other multiobjective covering problems. It is easy to see that the algorithm 2-DC-APPROX can be adapted such that it takes interferences into account.

Theorem 2. *Fix $k \geq 1$ and $\varrho \in (0, 2]$. For all $\delta > 0$, k-EDC is $(0, \delta, \ldots, \delta)$-approximable in polynomial time (and hence has a PTAS).*

Besides interferences, also other parameters can be added to the problem. For instance it might be the case that the single services have different operating distances. This brings us to the version of 2-DC where we have to place simultaneously two disks of different radii on the selected locations (one disk for each type of customers). 2-DC-APPROX can be easily adapted such that it gives a

PTAS also for this variant of the problem. In general, we can allow even more complicated rules that determine whether or not a customer is supplied by a base station. Here the different services can have supply areas of very general shape as long as

- we can efficiently test whether a point belongs to such an area and
- the minimum distance constraint is satisfied (i.e., the distance of two base stations is at least $\varrho \cdot r$ where $2r$ is the maximal diameter of the area and ϱ is a fixed constant).

Further generalization could handle weights for the customers, individual costs for the base stations, and payoffs that depend on the quality of the service obtained by the single customers. For these scenarios, appropriate versions of 2-DC-APPROX provide polynomial-time approximation schemes.

4   Non-approximability of the Restricted Version

The approximability of a multiobjective problem does not necessarily imply that the single-objective restrictions of this problem are approximable. The reason for this apparent contradiction is that all solutions for a restricted version of the problem must *strictly* satisfy the additional constraints on the values of the objectives that are not optimized any more (i.e., the constraints $f_j(I, S) \geq B_j$ or $f_j(I, S) \leq B_j$ in the definition of the restricted problem). An approximation algorithm has more freedom if it can optimize all objectives at the same time, since here the algorithm can *exploit trade-offs* between the single objectives. In fact, the problems k-DC and k-EDC are examples where such trade-offs yield a significantly better approximability. In this section we will show that several restrictions of k-DC and k-EDC are not approximable within a constant factor, unless P = NP. For instance, for $k \geq 2$, no restriction of k-EDC is in APX (unless P = NP), while the general (multiobjective) version of k-EDC even admits a PTAS. Angel, Bampis, and Kononov [2,3], Cheng, Janiak, and Kovalyov [9], and Dongarra et al. [12] discovered similar phenomena for multiobjective scheduling problems.

For our results in this section we need the NP-completeness of the following versions of geometric disk cover problems.

Disk Cover
DC = $\{(P, D, k) \mid P, D \subseteq \mathbb{Z} \times \mathbb{Z}$ are finite sets, $k \in \mathbb{N}$, and there exists an $S \subseteq D$ such that $|S| \leq k$ and $\forall x \in P \, \exists y \in S, |x - y| \leq 2\}$

Exact Disk Cover
EDC = $\{(P, D) \mid P, D \subseteq \mathbb{Z} \times \mathbb{Z}$ are finite sets and there exists an $S \subseteq D$ such that $\forall x \in P \, \exists! y \in S, |x - y| \leq 2\}$

Note that there is a minimum-distance constraint for disk locations implicitly given in these definitions since we consider points from $\mathbb{Z} \times \mathbb{Z}$ and a fixed radius $r = 2$.

Theorem 3. DC *and* EDC *are* NP-*complete.*

As we have seen in Section 3, the minimum-distance constraint helps us to find a PTAS for k-DC. Nevertheless, the problem remains difficult. Although the Pareto curve is only polynomial in size, we cannot hope to discover an algorithm that computes it exactly: For $\varrho \le \frac{1}{2}$, an algorithm that exactly determines the Pareto curve for k-DC$_\varrho$ in polynomial time would also solve the problem DC in polynomial time. Since DC is NP-complete, this would imply P = NP.

By Theorem 2, the multiobjective problem k-EDC has good approximation properties (a PTAS). Now we will see (Theorem 4) that in contrast, the restricted versions of k-EDC vary with respect to their approximation behavior. While 1-EDC restricted to the first component is not approximable (i.e., not in APX unless P = NP), the restriction to the second component has good approximation properties (a PTAS). Even more surprisingly, for $k \ge 2$ the Pareto curve of k-EDC is approximable, but no restriction of k-EDC is approximable (i.e., not in APX unless P = NP). Theorem 5 states similar results for k-DC. For instance, while the Pareto curve of 2-EDC is approximable, the restriction of 2-EDC to the second component is not (i.e., not in APX unless P = NP).

Theorem 4. *Fix some* $\varrho \in (0, \frac{1}{2}]$.

1. *If* P $\ne$ NP, *then for all* $k \ge 2$ *and all* $i \in [1, k+1]$, *Restricted$_i$-k-EDC is not in* APX.
2. *If* P $\ne$ NP, *then Restricted$_1$-1-EDC is not in* APX.
3. *Restricted$_2$-1-EDC admits a PTAS.*

Theorem 5. *Fix some* $\varrho \in (0, \frac{1}{2}]$.

1. *Restricted$_2$-1-DC has a PTAS.*
2. *If* P $\ne$ NP, *then for all* $k \ge 2$ *and all* $i \in [2, k+1]$, *Restricted$_i$-k-DC is not in* APX.

References

1. Alt, H., Arkin, E.M., Brönnimann, H., Erickson, J., Fekete, S.P., Knauer, C., Lenchner, J., Mitchell, J.S.B., Whittlesey, K.: Minimum-cost coverage of point sets by disks. In: Symposium on Computational Geometry, pp. 449–458 (2006)
2. Angel, E., Bampis, E., Kononov, A.: A FPTAS for approximating the unrelated parallel machines scheduling problem with costs. In: Meyer auf der Heide, F. (ed.) ESA 2001. LNCS, vol. 2161, pp. 194–205. Springer, Heidelberg (2001)
3. Angel, E., Bampis, E., Kononov, A.: On the approximate tradeoff for bicriteria batching and parallel machine scheduling problems. Theoretical Computer Science 306(1-3), 319–338 (2003)
4. Ben-Moshe, B., Carmi, P., Katz, M.J.: Approximating the visible region of a point on a terrain. GeoInformatica 12(1), 21–36 (2008)
5. Calinescu, G., Mandoiu, I.I., Wan, P.-J., Zelikovsky, A.: Selecting forwarding neighbors in wireless ad hoc networks. MONET 9(2), 101–111 (2004)
6. Cannon, A.H., Cowen, L.J.: Approximation algorithms for the class cover problem. Annals of Mathematics and Artificial Intelligence 40(3-4), 215–223 (2004)

7. Carmi, P., Katz, M.J., Lev-Tov, N.: Covering points by unit disks of fixed location. In: Tokuyama, T. (ed.) ISAAC 2007. LNCS, vol. 4835, pp. 644–655. Springer, Heidelberg (2007)
8. Chan, T.M.: Polynomial-time approximation schemes for packing and piercing fat objects. Journal of Algorithms 46(2), 178–189 (2003)
9. Cheng, T.C.E., Janiak, A., Kovalyov, M.Y.: Bicriterion single machine scheduling with resource dependent processing times. SIAM Journal on Optimization 8(2), 617–630 (1998)
10. Clark, B.N., Colbourn, C.J., Johnson, D.S.: Unit disk graphs. Discrete Mathematics 86(1-3), 165–177 (1990)
11. Diakonikolas, I., Yannakakis, M.: Small approximate pareto sets for bi-objective shortest paths and other problems. In: Charikar, M., Jansen, K., Reingold, O., Rolim, J.D.P. (eds.) RANDOM 2007 and APPROX 2007. LNCS, vol. 4627, pp. 74–88. Springer, Heidelberg (2007)
12. Dongarra, J., Jeannot, E., Saule, E., Shi, Z.: Bi-objective scheduling algorithms for optimizing makespan and reliability on heterogeneous systems. In: Proceedings 19th Annual ACM Symposium on Parallel Algorithms and Architectures, pp. 280–288. ACM, New York (2007)
13. Ehrgott, M.: Multicriteria Optimization. Springer, Heidelberg (2005)
14. Erlebach, T., Jansen, K., Seidel, E.: Polynomial-time approximation schemes for geometric intersection graphs. SIAM Journal on Computing 34(6), 1302–1323 (2005)
15. Erlebach, T., van Leeuwen, E.J.: Approximating geometric coverage problems. In: Proceedings of 19th Annual Aymposium on Discrete Algorithms, pp. 1267–1276 (2008)
16. Gandhi, R., Khuller, S., Srinivasan, A.: Approximation algorithms for partial covering problems. J. Algorithms 53(1), 55–84 (2004)
17. Glaßer, C., Reith, S., Vollmer, H.: The complexity of base station positioning in cellular networks. Discrete Applied Mathematics 148(1), 1–12 (2005)
18. Hochbaum, D., Maass, W.: Approximation schemes for covering and packing problems in image processing and VLSI. Journal of the ACM 32, 130–136 (1985)
19. Narayanappa, S., Vojtechovsky, P.: An improved approximation factor for the unit disk covering problem. In: CCCG (2006)
20. Papadimitriou, C.H., Yannakakis, M.: On the approximability of trade-offs and optimal access of web sources. In: FOCS, pp. 86–92 (2000)
21. Safer, H.M., Orlin, J.B.: Fast approximation schemes for multi-criteria combinatorial optimization. Working papers 3756-95, Massachusetts Institute of Technology, Sloan School of Management (1995)
22. Safer, H.M., Orlin, J.B.: Fast approximation schemes for multi-criteria flow, knapsack, and scheduling problems. Working papers 3757-95, Massachusetts Institute of Technology, Sloan School of Management (1995)
23. Vassilvitskii, S., Yannakakis, M.: Efficiently computing succinct trade-off curves. Theoretical Computer Science 348(2-3), 334–356 (2005)

Data Stream Algorithms via Expander Graphs

Sumit Ganguly

IIT Kanpur, India
`sganguly@cse.iitk.ac.in`

Abstract. We present a simple way of designing deterministic algorithms for problems in the data stream model via lossless expander graphs. We illustrate this by considering two problems, namely, k-sparsity testing and estimating frequency of items.

1 Introduction

We say that an n-dimensional vector f from $\mathbb{Z}^n$ is k-*sparse* if it has at most k non-zero entries. The problem is to test whether f is k-sparse or no after it has been subject to a sequence of coordinate wise updates in arbitrary order, that is, f is the frequency vector of a data stream. More formally, a data stream over the domain $[n] = \{1, 2, \ldots, n\}$ is a sequence σ of records of the form $(index, i, v)$, where, $index$ is the position of the record in the sequence, $i \in [n]$ and $v \in \mathbb{Z}$. Associated with each data stream σ is an n-dimensional *frequency vector* $f(\sigma)$, such that $f_i(\sigma)$ is the frequency of i, or the cumulative sum of the updates to $f_i(\sigma)$, made by the sequence σ. That is,

$$f_i(\sigma) = \sum_{(index, i, v) \in \sigma} v, \quad i \in [n] \ .$$

The k-sparsity testing problem is as follows: design a data structure, referred to as a k-*sparsity tester*, that (a) processes any stream σ of updates over the domain $[n]$, and, (b) provides a test to check whether $f(\sigma)$ is k-sparse, that is, whether, f has at most k non-zero entries. The problem is to obtain solutions whose space requirement is $o(n)$.

We first review work on the following well-studied and closely related problem, namely, k-*sparse vector reconstruction* problem, where it is required to design a structure that can process a data stream σ and can retrieve the frequency vector $f(\sigma)$ provided $f(\sigma)$ is k-sparse. However, the structure is not required to actually test whether $f(\sigma)$ is k-sparse or not and may present an incorrect answer if $f(\sigma)$ is not k-sparse. Let m denote $\max_{i=1}^n |f_i|$. It is easy to show [15] that the k-sparse reconstruction problem requires $\Omega(k \log(mn/k))$ bits. Minsky et. al. [22] study a constrained version of the k-sparse vector reconstruction problem where $f(\sigma) \in \{-1, 0, 1\}^n$ and present a space-optimal algorithm for this scenario. Eppstein and Goodrich [11] present a space-optimal solution for the case when $f(\sigma) \in \{0, 1\}^n$. The k-set structure [15] presents a k-sparse vector reconstruction technique for the general case when $f \in \{-m, \ldots, m\}^n$. We reproduce their theorem since we will refer to it later.

S.-H. Hong, H. Nagamochi, and T. Fukunaga (Eds.): ISAAC 2008, LNCS 5369, pp. 52–63, 2008.

Theorem 1 ([15]). *For vectors $f \in \{-m, \ldots, m\}^n$, there exists a data structure for the k-sparse reconstruction problem that requires space $O(k \log(mn) \log(n/k))$ bits. The time taken to process any coordinate-wise update to f is $O(k(\log(n/k)))$ elementary arithmetic operations over a finite field of size $O(mn)$ and characteristic at least $mn + 1$.* $\qquad\qquad\square$

The work on compressed sensing [3,10] has independently considered the problem of k-sparse vector reconstruction. Based on previous work in [3,10,18], Indyk [20] presented the first deterministic algorithm in the compressed sensing framework for k-sparse vector reconstruction using space $O(k2^{(\log \log n)^E} \log(mn))$ bits, where, $E > 1$ is a constant that depends on the best known construction of a class of extractors.

We now review prior work on k-sparsity testing. It is known that k-sparsity testing for vectors in $\{-m, \ldots, m\}^n$ requires $\Omega(n)$ space, for any $m \geq 1$ and $k \geq 0$ [11,16]. In view of this negative result, in this paper, we will restrict our attention to non-negative frequency vectors from $\{0, \ldots, m\}^n$, that is, $0 \leq f_i(\sigma) \leq m$, for each $i \in [n]$. A space-optimal 1-sparsity tester was presented in [12] that requires $O(\log(mn))$ bits. A k-sparsity tester can be constructed by using *strongly selective families* [6,8] as follows. An (n, k) strongly selective family is a family of sets $\{S_i\}_{1 \leq i \leq t}$ such that for any $A \subset \{1, 2, \ldots, n\}$ such that $|A| \leq k$ and for any $x \in A$, there exists a member S_j of the family such that $S_j \cap A = \{x\}$. In other words, each member of the set A is *selected via intersection* by some member of the family. Constructive solutions for (n, k)-strongly selective families are known for which the size of the family $t = O(k^2 \cdot \text{polylog}(n))$. A k-sparsity test can be designed by keeping a 1-sparsity tester [12] for each of the sets $\{S_i\}_{1 \leq i \leq t}$. The space requirement is $O(k^2 \cdot \text{polylog}(n) \log(mn))$ bits. This line of work cannot be used to obtain significantly more space efficient k-sparsity tests, since, there is a space lower bound of $\Omega(k^2(\log(n/k))/(\log k))$ for the size of (n, k)-strongly selective family [6]. The k-set structure [15] presents a technique that can be used to test k-sparsity of vectors in $\{0, \ldots, m\}^n$ using space $O(k^2 \log n + k \log(mn))$. Finite fields based solutions to k-sparse vector construction of Minsky et.al. [22], Eppstein and Goodrich [11], our previous work in [15], and, the compressed sensing approach of Indyk [20] are not known to directly extend to deterministic k-sparsity testing.

Deterministic estimation of data stream frequency. We consider deterministic solution to the APPROXFREQ(ϵ) problem, namely, to design a low-space data structure that can (a) process any stream σ over the domain $[n]$, and, (b) given any $i \in [n]$, it returns a deterministic estimate $\hat{f}_i$ for $f_i(\sigma)$ satisfying $|\hat{f}_i - f_i(\sigma)| \leq \epsilon \|f(\sigma)\|_1$, where, $\|f(\sigma)\|_1$ is the ℓ_1 norm of the frequency vector $f(\sigma)$. The APPROXFREQ(ϵ) problem is a well-studied and basic problem in data stream processing. Deterministic algorithms requiring $O(\epsilon^{-1} \log m)$ space are known for insert-only (i.e., no deletions) streams [9,23,21]. For streams with arbitrary insertions and deletions, the CR-PRECIS algorithm solves the problem APPROXFREQ(ϵ) [14] using space $\tilde{O}(\epsilon^{-2}(\log^{-2}(1/\epsilon))(\log^2 n)(\log mn))$. A space lower bound of $\Omega(\epsilon^{-2} \log m)$ for deterministic algorithms is known for solving APPROXFREQ(ϵ) over streams with frequency vector in $\{-m, \ldots, m\}^n$ [13].

Contributions. We present a simple paradigm for designing deterministic algorithms over data streams by using appropriately chosen lossless expander graphs. The paradigm consists of two steps, namely, identifying the expansion properties needed to solve the problem at hand, and, a low space representation of the expander graph (or an object that closely resembles it). We illustrate our paradigm by designing algorithms for k-sparsity testing and estimating item frequencies.

We first present a novel solution for deterministic k-sparsity tester for the frequency vector $f(\sigma)$ of a data stream σ, when, $f(\sigma) \in \{0,\ldots,m\}^n$. This technique, based on lossless expander graphs, requires space $O(k \cdot D(1/4, n, k))$, where, $D(\epsilon, n, r) = o(n)$ is the smallest known degree function in the construction of (k, ϵ)-lossless expanders. We subsequently improve upon this algorithm to present a space upper bound of $O(k(\log(n/k))(\log mn))$ bits. This improves the current upper bound of $O(k^2 \log n + k \log(mn))$ space [15] and nearly matches the space lower bound up to a logarithmic factor, which we show to be $\Omega(k \log(mn/k))$. We also use the expander graphs based approach to design a family of deterministic algorithms for APPROXFREQ(ϵ), of which the CR-PRECIS algorithm [14] is a special case. The algorithms derived in this manner are slightly more efficient in space and update time than the CR-PRECIS algorithm.

Organization. The remainder of the paper is organized as follows. In Section 2, we consider the k-sparsity testing problem over data stream and and in Section 3 we consider the APPROXFREQ(ϵ) problem.

2   Testing k-Sparsity

In this section, we design a deterministic k-sparsity tester for frequency vectors in $\{0,\ldots,m\}^n$ based on lossless expander graphs. We first present a space lower bound for k-sparsity testing of vectors over $\{0,\ldots,m\}^n$ in the data stream model.

Lemma 1. *For each value of $1 \leq k < n/2$, a deterministic k-sparsity tester for vectors over $\{0,\ldots,m\}^n$ requires $\Omega(k \log(mn/k))$ bits.*

Proof (Of Lemma 1). Let $k < n/2$. Suppose f, g are distinct k-sparse vectors such that $\|f\|_\infty \leq m/2$ and $\|g\|_\infty \leq m/2$. We first show that there exists $h \in \{-m,\ldots,m\}^n$ such that both $f + h$ and $g + h$ are non-negative and one of them is k-sparse and the other is not k-sparse. This would imply that any k-sparsity tester must map distinct k-sparse vectors in $\{0,\ldots,\lfloor m/2 \rfloor\}^n$ to distinct summaries. Since, the number of k-sparse vectors in $\{0,\ldots,\lfloor m/2 \rfloor\}^n$ is $\sum_{r=0}^{k} \binom{n}{k}(\lfloor m/2 \rfloor + 1)^r$, the space requirement would be at least the logarithm of this quantity, which is $O(k \log(mn/k))$.

Let S_f and S_g denote the set of coordinates of f and g respectively with non-zero entries. Let T be any set of size $k - |S_g|$ such that $T \cap (S_f \cup S_g) = \phi$.

Such a T exists since, $k < n/2$. Denote by 1_T the characteristic vector of T. Let $h = g + 1_T$. Then $g + h = 2g + 1_T$ and is k-sparse. Further, $f + h = f + g + 1_T$ and therefore,

$$|S_{f+g+1_T}| = |(S_f \cup S_g)| + |T| > |S_g| + 1 + |T| = k + 1$$

and so, $f + h$ is not k-sparse. $\qquad\square$

2.1   Sparsity Separator Structure

We first define the (k, l)-*sparsity separator structure* that will be used later to test k-sparsity in sub-linear space.

Definition 1. *A (k, l)-sparsity separator structure, where, $k \leq l$, is a data structure that (a) supports updates corresponding to any stream σ over $[n]$, and, (b) supports a deterministic operation called* SEPARATESPARSITY *that returns* TRUE *if the sparsity of $f(\sigma)$ is at most k and returns* FALSE *if the sparsity of $f(\sigma)$ is at least l.* $\qquad\square$

There is an indeterminate region, namely, if the sparsity of $f(\sigma)$ is between $k+1$ and $l - 1$, then the function SEPARATESPARSITY(f) is allowed to return either TRUE or FALSE.

Lossless expander graphs. We design a $(k, 2k)$ sparsity separator structure using lossless expander graphs. We first recall some standard concepts from expander graphs [19]. Let $G = (V_L, V_R, E, d)$ denote a left-regular bipartite graph where, $V_L = \{v_1, \ldots, v_n\}$ is the set of vertices in the left partition, $V_R = \{u_1, \ldots, u_r\}$ is the set of vertices in the right partition, E is the set of edges of G and d is the degree of each vertex in the left partition.

Definition 2 (Lossless Expanders [19].). *A left-regular bipartite graph $G = (V_L, V_R, E, d)$ is said to be be a $(K_{\max}, \epsilon)$-lossless expander if every set of $K \leq K_{\max}$ vertices from the left partition has at least $(1 - \epsilon)dK$ neighbors in V_R.* [1]

The work in [4] presents non-trivial, explicit constructions of lossless expanders using the zig-zag product of expanders.

Theorem 2 ([4]). *For any $\epsilon > 0$ and $r \leq n$, there is an explicit family of left-regular bipartite graphs with $|V_L| = n$, $|V_R| = r$ that is an $(c'\epsilon r/d, \epsilon)$-lossless expander with left degree $D(\epsilon, n, r) \leq (n/\epsilon r)^c$ for some constants $c, c' > 0$. The neighbors of any left vertex may be found in time $O(d \cdot \log^{O(1)}(n))$.* $\qquad\square$

Denote by $R(\epsilon, n, k)$ the smallest value of r for which there is a known efficiently constructible (k, ϵ)-lossless expander with n left vertices and r right vertices. Theorem 2 is optimized for the case $r = \Theta(n)$, in which case, the degree $d = d(n)$ is constant. Our approach will be the following. We will be interested in (K, ϵ)-lossless expander graphs with r as small as possible. We use Theorem 2 to obtain

---

[1] More accurately, an expander is a family of bipartite graphs $\{G_n\}_{n \geq n_0}$, for some n_0, where, $K_{\max} = K_{\max}(n), \epsilon = \epsilon(n)$ and $d = d(n)$.

expander graphs with the desired lossless expansion property to give us the wire-frame of an algorithm for the problem. We then replace the expander by a more explicit and low-space construction of a bipartite graph G with a smaller value of r and that has the desired lossless expansion property.

A $(k, 2k)$-sparsity separator using lossless expander graphs. Given n and using Theorem 2, we consider a left-regular bipartite graph $G = (V_L, V_R, E, d)$ that is a $(2k, \epsilon = 1/4)$-lossless expander such that

$$|V_L| = n, |V_R| = r \text{ and left-degree } d = D(\epsilon, n, r) \ .$$

Keep $r = |V_R|$ integer counters denoted as an r-dimensional vector $g = [g_1, \ldots, g_r]$, where, g_s is the counter associated with vertex $u_s \in V_R$. All counters are initialized to 0. The counter g_s maintains the following sum over the data stream.

$$g_s = \sum_{(v_i, u_s) \in E} f_i(\sigma), \quad s = 1, 2, \ldots, r \ .$$

Alternatively, if we let B be the $r \times n$ matrix such that $A_{s,i} = 1$ if v_i is adjacent to u_s and $A_{s,i} = 0$ otherwise, then, $g = A(f(\sigma))$. The counters are easily updated corresponding to a stream update (pos, i, v) as follows:

$$g_s := g_s + v, \quad \forall s \in [r] \text{ such that } u_s \text{ is adjacent to } v_i \ .$$

Equivalently, in matrix notation, $g := g + A_i$, where, A_i is the column corresponding to vertex v_i. Since, the neighbors of any left vertex v_i can be computed in $d \cdot \text{polylog}(n)$ time and $d = D(\epsilon, n, r)$, the update can be performed in time $D(\epsilon, n, 4k) \cdot \text{polylog}(n)$.

A procedure for SEPARATESPARSITY$(k, 2k)$ can be designed as follows. It first checks if g is not $(1 - 1/4)2dk = \frac{3dk}{2}$-sparse in which case it returns FALSE. Otherwise, the procedure returns TRUE.

procedure SEPARATESPARSITY$(k, 2k)$
Data Structure: A $(k, 2k)$-sparsity separator structure.
    if g is not $\left(\frac{3}{2}dk - 1\right)$-sparse **return** FALSE **else return** TRUE

We now show that the algorithm SEPARATESPARSITY$(k, 2k)$ correctly solves the approximate sparsity problem with parameter $k, 2k$.

Lemma 2. *Algorithm* SEPARATESPARSITY$(k, 2k)$ *correctly solves the problem* SEPARATESPARSITY$(k, 2k)$.

Proof. Suppose that f is not $2k - 1$-sparse. Then, it has at least $2k$ non-zero entries. Let $S_f = \{v_i \in V_L \mid f_i > 0\}$. Then, $|S_f| \geq 2k$. Choose any subset $T \subset S_f$ such that $|T| = 2k$. Let $\Gamma(S)$ denote the neighbors of any set $T \subset V_L$. By property of the $(2k, 1/4)$-lossless expander graph, $(1 - 1/4)d(2k) \leq |\Gamma(T)| \leq 2dk$, that is, $1.5dk \leq |\Gamma(T)| \leq 2dk$. For each $s \in \Gamma(T)$, $g_s > 0$, since, g_s is the sum of the (positive) f_i's of those i's such that v_i is adjacent to u_s. Therefore, g

procedure SPARSITYTEST(k)
Input: Data Stream σ with frequency vector $f(\sigma) \in \{0,\ldots,m\}^n$.
Output: Returns TRUE if f is k-sparse and FALSE otherwise.
Data Structure: (a) A $(k,2k)$-sparsity separator structure over $\{0,\ldots,m\}^n$, and,
(b) a $2k$-set structure over $\{-m,\ldots,m\}^n$ that supports operation RETRIEVEVECTOR($2k$) [15].
begin

    1.   **if** SEPARATESPARSITY($k,2k$) = FALSE **return** FALSE
    2.   **else** **if** SPARSITY(RETRIEVEVECTOR($2k$)) $\leq k$ **return** TRUE
    3.   **else** **return** FALSE

end.

Fig. 1. Procedure for testing k-sparsity

has at least $1.5dk$ non-zero entries and the algorithm returns FALSE. On the other hand, if f is k-sparse, $|\Gamma(S_f)| \leq kd$ and therefore g is kd-sparse and the algorithm returns TRUE. Hence the algorithm satisfies the properties of testing SEPARATESPARSITY($k,2k$). $\qquad\square$

The space requirement of Algorithm SEPARATESPARSITY consists of the r-dimensional vector g, each of whose entries is an integer between 0 and mn. Then, the space requirement is $O(R(\epsilon,n,2k)\log(mn))$. The time requirement to process each stream update is $D(\epsilon,n,R) \cdot O(\log^{O(1)}(n))$.

2.2   Algorithm for Testing k-Sparsity

We now use the sparsity separator $(k,2k)$ of Section 2.1 together with the k-set reconstruction procedure of [15] to design an algorithm for k-sparsity test.

We keep two data structures, namely, a $2k$-set structure for $2k$-set reconstruction as presented in [15] and a $(k,2k)$-sparsity separator structure, presented in Section 2.1. Both structures are maintained independently and in parallel in the face of stream updates. The procedure k-SPARSITYTEST is presented in Figure 1 and is described as follows. It first uses the $(k,2k)$ sparsity separator test on the r-dimensional vector g. If the approximate sparsity test returns FALSE, then, we know that f cannot be k-sparse. (Otherwise, the $(k,2k)$-sparsity separator test would have returned TRUE, by definition.) However, if the sparsity separator procedure returns TRUE, then, f is $2k$-sparse (otherwise, the sparsity-separator($k,2k$) would have returned FALSE). The reconstruction procedure of the $2k$-set structure [15] is then invoked to obtain f, and from f, we obtain its sparsity. If the sparsity is at most k, then the procedure returns TRUE, and otherwise returns FALSE. We summarize these properties and the space and time bounds in the following theorem.

Theorem 3. *There exists a k-sparsity tester for frequency vector of data stream in $\{0,\ldots,m\}^n$ using space $O(R(1/4,n,2k)\log mn))$, where, $R = R(1/4,n,2k)$ is the smallest value of r for which a $(2k,1/4)$-lossless expander can be efficiently constructed with n vertices in the left vertex partition V_L. The time required to process each stream update is $O(D(1/4,n,R) \cdot \log^{O(1)}(n))$.* $\qquad\square$

Improving the expander based sparsity test. The space requirement of the
the expander based k-sparsity separator presented in Section 2 can be improved
by using a different construction of an (approximate) expander graph than the
one given by Theorem 2. The set of vertices adjacent to a given subset of vertices
S in a graph is denoted as $\Gamma(S)$.

Lemma 3. *For any $n \geq 2$, $d > 3\log(ne/4k)$ and $r \geq 8kd$, there exist bipartite
graphs $G = (V_L, V_R, E)$ with $|V_L| = n$, $|V_R| = r$ satisfying the following property:
for any subset $S \subset V_L$ such that $|S| \leq k$, $|\Gamma(S)| \leq k$ and for any subset $S \subset V_L$
such that $|S| \geq 4k$, $|\Gamma(S)| > k$. Moreover, the bipartite graph can be succinctly
represented by a string of size kd^2 bits. The adjacency of a vertex in the left
partition may be computed in time $O(kd^2)$.*

Proof (Of Lemma 3). Let $V_L = \{1, 2, \ldots, n\}$ and $V_R = \{1, 2, \ldots, r\}$. Define d
independently chosen random hash functions $h_1, \ldots, h_d$ each mapping $[n] \to [r]$.
For $i \in V_L$ and $s \in V_R$, we say that there is an edge (i, s) provided there exists
$j \in \{1, \ldots, d\}$ such that $h_j(i) = s$.

By construction, for any $S \subset V_L$ be a set of left vertices of size k, $|\Gamma(S)| \leq kd$.
Now suppose $S \subset V_L$ and $|S| = 4k$. For $s \in V_R$, define an indicator variable y_s
that is 1 if $u_s \in \Gamma(S)$ and 0 otherwise.

$$\Pr\{y_s = 1\} = 1 - (1 - |S|/r)^d = p \text{ (say)} .$$

Thus,

$$\frac{|S|d}{r} - \frac{d^2|S|^2}{2r^2} \leq p \leq \frac{kd}{r} .$$

Let

$$W_S = \sum_{s=1}^{r} y_s .$$

Therefore,

$$\mathsf{E}[W_S] = rp \geq kd - \frac{d^2 k^2}{r} .$$

Further,

$$\mathsf{E}\left[W_S^2\right] = \left(\sum_{s=1}^{r} y_s\right)^2 = \sum_{s=1}^{r} y_s + 2 \sum_{1 \leq s_1 < s_2 \leq r} y_{s_1} y_{s_2}$$

$$= rp + 2\binom{r}{2}p^2 .$$

Thus,

$$\sigma^2(W_S) = \mathbf{Var}[W_S] = \mathsf{E}\left[W_S^2\right] - (\mathsf{E}[W_S])^2 = rp + 2\binom{r}{2}p^2 - (rp)^2$$

$$= rp - rp^2 .$$

Therefore,

$$\sigma^2(W_S) \leq rp \leq kd .$$

If the hash functions $h_1, \ldots, h_d$ are each t-wise independent, then, the y_s's are at least t-wise independent. By Chernoff's bound for t-wise independent variables [24], we have,

$$\Pr\left\{|W_S - \mathsf{E}\left[W_S\right]| > T\right\} \le \left(\frac{t \max(t, \sigma^2(W_S))}{e^{2/3}T^2}\right)^{t/2} \quad [24] \ .$$

Choose the degree of independence as $t = kd$ and let the deviation from the expectation be $T = \mathsf{E}\left[W_S\right] - (kd + 1)$. Then,

$$T \ge rp - (kd + 1) \ge 4kd - \frac{16k^2d^2}{r} \ge 2kd$$

by choosing $r \ge 8kd$. Substituting, we have

$$\Pr\left\{W_S \le k\right\} \le \Pr\left\{|W_S - \mathsf{E}\left[W_S\right]| > T\right\} \le \left(\frac{(kd)(4kd)}{e^{2/3}(2kd)^2}\right)^{kd/2} = e^{-kd/3} \ .$$

Therefore,

$$\Pr\left\{W_S \le k, \text{ for some } S \text{ s.t. } |S| = 4k\right\} \le \binom{n}{4k} e^{-kd/3} \le e^{k \ln(ne/4k) - kd/3} < 1$$

provided, $d \ge 3 \ln(ne/4k)$ Thus,

$$\Pr\left\{\forall S, |S| = 4k, W_S > 4k\right\} > 0$$

This proves the existence of a bipartite graph with the properties as stated in the Lemma.

Such a bipartite graph may be generated as follows. The random seed length is $kd^2 \log n$ bits, since, each of the hash functions may be implemented as a degree $kd - 1$ polynomial over a field of size $O(n)$. We iterate over the space of $kd^2 \log n$ bit strings, generate the corresponding bipartite graph and check for the property. If the property holds, then, the $kd^2 \log n$ bit seed is stored as the generator for the bipartite graph. The above proof assures us of the existence of such a seed. $\qquad\qquad\square$

$(k, 4k)$-*Sparsity separator.* A $(k, 4k)$-sparsity separator can be designed based on a succinctly representable bipartite graph $G = (V_L, V_R, E)$ obtained using Lemma 3 such that $|V_L| = n, d = 4\log(ne/k)$ and $|V_R| = r = 16kd$. By Lemma 3, for any subset $S \subset V_L$, if $|S| \ge 4k$, then, $|\Gamma(S)| > 2k$ and if $|S| \le k$, then, $|\Gamma(S)| \le k$. A $(k, 4k)$-sparsity separator is designed as follows. Keep r counters, $g_1, \ldots, g_r$, one each corresponding to each right vertex $u_s \in V_R$; all initialized to 0. The counter g_s maintains the following sum: $g_s = \sum_{i:(v_i, u_s) \in E} f_i$. Corresponding to update (pos, i, v) on the stream, the counters are updated as follows.

$$\textsc{Update}(pos, i, v): \quad g_s = g_s + v, \quad \forall s \text{ such that } (i, s) \in E.$$

The space requirement is $O(r) = O(k \log(n/k))$ counters of size $O(\log(mn))$ bits plus the succinct description length $O(kd^2) = O(k \log^2(n/k))$ bits. The time required to process each stream update of the form (pos, i, v) is to evaluate d polynomials of degree kd each to obtain the adjacency of vertex v_i; this requires time $O(kd^2) = O(k \log^2(n/k))$. The $(k, 4k)$-SEPARATESPARSITY test is as follows.

procedure BIPARTITE-SEPARATESPARSITY$(k, 4k)$
1.   **if** g is not k-sparse **then**   **return**   FALSE **else**   **return**   TRUE.

Rephrasing Lemma 3, if f is k-sparse, then, g is k-sparse, and, if f is not $4k$-sparse, then, g is not k sparse. The problem of k-sparsity testing can now be readily solved as before. Keep a $(k, 4k)$-sparsity separator for vectors in $\{0, \ldots, m\}^n$ based on succinct bipartite graphs and a $4k$-set structure from [15]. The algorithm for k-sparsity testing is identical to that presented in Figure 1, with the only change being that the use of the $2k$-set structure is replaced by a $4k$-set structure. We therefore have the following theorem.

Theorem 4. *There exists a structure for testing k-sparsity of vectors in $\{0, \ldots, m\}^n$ updated coordinate-wise as a data stream using space $O(k \log(n/k) \log(mn))$. The time taken to process each coordinate-wise update is $O(k \log^2 (n/k))$.* $\square$

The space requirement of the succinct bipartite graph based k-sparsity tester is within a logarithmic factor of the space lower bound of $\Omega(k \log(mn/k))$ proved in Lemma 1.

3   Deterministic Estimation of Frequency Vector

In this section, we present a novel, deterministic algorithm for approximating the frequency vector of a data stream based on the use of lossless expander graphs. Consider a $(2, \epsilon/2)$-lossless expander graph $G = (V_L, V_R, E, d)$, where, $V_L = \{v_1, \ldots, v_n\}$, $V_R = \{u_1, \ldots, u_r\}$. By Theorem 2, a $(2, \epsilon/2)$-lossless expander has $r = O(D(\epsilon, n, O(1))/\epsilon)$ and $d = D(\epsilon, n, r)$, where, $D(\epsilon, n, O(1))$ is the current best known degree function for an $(O(1), \epsilon)$-lossless expander given by Theorem 2. As before, we keep an integer counter g_s corresponding to each vertex $u_s \in V_R$. The counters are initialized to 0 and are updated corresponding to stream update exactly in the same manner as discussed in Section 2.1. The estimate $\hat{f}_i$ is the following.

$$\hat{f}_i = \frac{1}{d} \sum_{s:(v_i, u_s) \in E} g_s, \quad i \in [n].$$

Lemma 4. $|\hat{f}_i - f_i(\sigma)| \leq \epsilon \|f(\sigma)\|_1.$

Proof. For simplicity, fix the input stream σ and let f denote $f(\sigma)$. Fix i. By property of $(2, \epsilon)$-lossless expander, for any $i, j \in [n]$, $i \neq j$,

$$|\Gamma(v_i) \cap \Gamma(v_j)| = 2d - |\Gamma(v_i) \cup \Gamma(v_j)| \leq 2d - (1 - \epsilon/2)(2d) \leq \epsilon d \ . \qquad (1)$$

Therefore,

$$\sum_{s:(v_i,u_s)\in E} g_s = \sum_{s:(v_i,u_s)\in E} \sum_{j:(v_j,u_s)\in E} f_j = \sum_{j=1}^{n} f_j |\Gamma(v_j) \cap \Gamma(v_i)|$$

$$= df_i + \sum_{\substack{1 \leq j \leq n \\ j \neq i}} f_j |\Gamma(v_j) \cap \Gamma(v_i)| = df_i + \sum_{\substack{1 \leq j \leq n \\ j \neq i}} f_j(\epsilon d), \quad \text{by } (1) \ .$$

Dividing by d, transposing and taking absolute values, we have,

$$\left| \frac{1}{d} \sum_{s:(v_i,u_s)\in E} g_s - f_i \right| \leq \left| \sum_{j \neq i} \epsilon f_j \right| \leq \epsilon \|f\|_1 - |f_i| \ .$$

Since, $\frac{1}{d} \sum_{s:(v_i,u_s)\in E} g_s = \hat{f}_i$, this proves the lemma. $\qquad \square$

Theorem 2 can be applied by setting $r = O(\frac{1}{\epsilon} D(\epsilon, n, 1))$ and $d = D(\epsilon, n, r)$, thereby obtaining a (K, ϵ)-lossless expander, for some $K \geq 2$. We summarize this in the following theorem.

Theorem 5. *There exists a deterministic algorithm for solving* APPROXFREQ(ϵ) *over a data stream using space* $O(R(\epsilon, n, 2) \log(mn))$. *The time taken to process each stream update is* $O(D(\epsilon, n, R) \log^{O(1)} n)$. $\qquad \square$

An exercise along the lines of producing a succinctly representable bipartite graph using the probabilistic method instead of using the lossless expander family of Theorem 2 can be carried out (and has a slightly simpler argument). We state this in the following lemma.

Lemma 5. *There exists a bipartite graph* $G(n, \epsilon) = (V_L, V_R, E, d)$ *such that* $|V_L| = n$, $|V_R| = O((1/\epsilon^2) \log(n/\epsilon))$, $d = O(\log n)$ *such that the degree of every vertex in* V_L *is between* $(1 - \epsilon)d$ *and* d *and for any* $v_i, v_j \in V_L$, $i \neq j$, *the number of common neighbors of* v_i *and* v_j *do not exceed* ϵd. *Moreover, such a bipartite graph can be succinctly represented using* $O((\log(n/\epsilon))(\log n))$ *bits. The neighbors of any vertex in* V_L *can be computed in time* $O(\log(n/\epsilon)(\log n))$. $\qquad \square$

Note that the bipartite graph of Lemma 5 is not left-regular, but rather almost left-regular: $(1 - \epsilon)d \leq \deg(v_i) \leq d$. The analysis of Lemma 4 goes through with a slight modification, since the division by d gives rise to a factor that lies between 1 and $1/(1 - \epsilon)$, thereby, increasing the error factor by $O(\epsilon)$. Replacing the $(2, \epsilon/2)$-expander graph by the succinct bipartite graph $G(n, \epsilon/4)$ from Lemma 5 yields an algorithm for APPROXFREQ(ϵ). This is summarized in the following theorem.

Theorem 6. *There exists a deterministic algorithm for solving* APPROXFREQ(ϵ) *over a data stream using space $O(\epsilon^{-2} \log(n/\epsilon) \log(mn))$ bits. The time taken to process each stream update is $O(\log(n/\epsilon)(\log n))$.* $\square$

The only known previous algorithm for deterministic estimation of frequency is the CR-PRECIS algorithm [14] which requires space $O(\epsilon^{-2}(\log^{-2}(1/\epsilon))(\log^2 n) \cdot (\log mn))$. The algorithm of Theorem 6 is slightly better in its space requirement than the CR-PRECIS algorithm [14] by a small poly-logarithmic factor.

It is interesting to note that the primes residue based structure used by the CR-PRECIS structure [15] is an explicit construction of a $(2, \epsilon/2)$-lossless expander as follows. For t distinct primes $p_1, \ldots, p_t$, we define the bipartite graph $G_{CR}(p_1, \ldots, p_t) = (V_L, V_R, E, d)$ where, $V_L = \{v_1, \ldots, v_n\}$ and $V_R = \{u_{j,l} \mid 1 \le j \le t \text{ and } 0 \le l \le p_j - 1\}$. There is an edge between left vertex v_i and right vertex $u_{j,l}$ if and only if $l = i \mod p_j$. The degree of each left vertex is by construction t, since, each number has exactly one residue respectively for $p_1, \ldots, p_t$. For any $1 \le i < j \le n$, each common neighbor $u_{j,l}$ of v_i and v_j means that $p_l | j - i$. If there are s distinct common neighbors, then, there are s distinct primes that divides $j - i$. Since, $j - i \le n - 1$, and each $p_l \ge 2$, $s < \log n$. This shows that the graph is a $(2, \epsilon/2)$-lossless expander for $t = 2(\log n)/\epsilon$ and for any choice of primes $p_1, \ldots, p_t$. Since, $r = |V_R| = p_1 + \ldots + p_t$, r is minimized by choosing the first $2 \log n/\epsilon$ primes as $p_1, \ldots, p_t$. The well-known prime number theorem then guarantees that $p_1 + \ldots + p_t = O(p_t t) = O(t^2 \ln t) = O((\log^2 n/\epsilon^2) \log((\log n)/\epsilon))$.

References

1. Alon, N.: Perturbed identity matrices have high rank: proof and applications (2006), http://www.math.tau.ac.il/~nogaa/identity.pdf
2. Bose, P., Kranakis, E., Morin, P., Tang, Y.: Bounds for Frequency Estimation of Packet Streams. In: Proc. of SIROCCO, pp. 33–42 (2003)
3. Candès, E., Romberg, J., Tao, T.: Robust uncertainty principles: Exact signal reconstruction from highly incomplete frequency information. IEEE Trans. Inf. Theory 52(2), 489–509 (2006)
4. Capalbo, M., Reingold, O., Vadhan, S., Wigderson, A.: Randomness conductors and constant degree lossless expanders. In: Proc. of ACM STOC (2002)
5. Charikar, M., Chen, K., Farach-Colton, M.: Finding frequent items in data streams. In: Widmayer, P., Triguero, F., Morales, R., Hennessy, M., Eidenbenz, S., Conejo, R. (eds.) ICALP 2002. LNCS, vol. 2380, pp. 693–703. Springer, Heidelberg (2002)
6. Cheblus, B., Kowalski, D.R.: Almost Optimal Explicit Selectors. In: Liśkiewicz, M., Reischuk, R. (eds.) FCT 2005. LNCS, vol. 3623, pp. 270–280. Springer, Heidelberg (2005)
7. Cormode, G., Muthukrishnan, S.: An Improved Data Stream Summary: The Count-Min Sketch and its Applications. J. Algo. 55(1)
8. De Bonis, A., Gàsieniec, L., Vaccaro, U.: Generalized framework for selectors with applications in optimal group testing. In: Baeten, J.C.M., Lenstra, J.K., Parrow, J., Woeginger, G.J. (eds.) ICALP 2003. LNCS, vol. 2719, pp. 81–96. Springer, Heidelberg (2003)
9. Demaine, E.D., López-Ortiz, A., Munro, J.I.: Frequency estimation of internet packet streams with limited space. In: Möhring, R.H., Raman, R. (eds.) ESA 2002. LNCS, vol. 2461, pp. 348–360. Springer, Heidelberg (2002)

10. Donoho, D.: Compressed sensing. IEEE Trans. Inf. Theory 52(4), 1289–1306 (2006)
11. Eppstein, D., Goodrich, M.T.: Space-Efficient Straggler Identification in Round-Trip Data Streams. In: Dehne, F., Sack, J.-R., Zeh, N. (eds.) WADS 2007. LNCS, vol. 4619, pp. 637–648. Springer, Heidelberg (2007)
12. Ganguly, S.: Counting distinct items over update streams. In: Deng, X., Du, D.-Z. (eds.) ISAAC 2005. LNCS, vol. 3827, pp. 505–514. Springer, Heidelberg (2005)
13. Ganguly, S.: Lower bounds for frequency estimation over data streams. In: Hirsch, E.A., Razborov, A.A., Semenov, A., Slissenko, A. (eds.) Computer Science – Theory and Applications. LNCS, vol. 5010, pp. 204–215. Springer, Heidelberg (2008)
14. Ganguly, S., Majumder, A.: CR-precis: A Deterministic Summary Structure for Update Streams. In: Chen, B., Paterson, M., Zhang, G. (eds.) ESCAPE 2007. LNCS, vol. 4614, pp. 48–59. Springer, Heidelberg (2007)
15. Ganguly, S., Majumder, A.: Deterministic K-set Structure. In: Proc. ACM PODS, pp. 280–289 (2006)
16. Ganguly, S., Majumder, A.: Deterministic K-set Structure. Manuscript under review (July 2006), http://www.cse.iitk.ac.in/users/sganguly
17. Gilbert, A., Guha, S., Indyk, P., Kotidis, Y., Muthukrishnan, S., Strauss, M.: Fast Small-space Algorithms for Approximate Histogram Maintenance. In: Proc. ACM STOC, pp. 152–161 (2002)
18. Gilbert, A., Strauss, M., Tropp, J., Vershynin, R.: One sketch for all: Fast algorithms for Compressed Sensing. In: Proc. ACM STOC (2007)
19. Hoory, S., Linial, N., Wigderson, A.: Expander graphs and their applications. Draft of book (2006)
20. Indyk, P.: Explicit Constructions for Compressed Sensing of Sparse Signals. In: Proc. ACM SODA (2008)
21. Karp, R.M., Shenker, S., Papadimitriou, C.H.: A Simple Algorithm for Finding Frequent Elements in Streams and Bags. ACM TODS 28(1), 51–55 (2003)
22. Minsky, Y., Trachtenberg, A., Zippel, R.: Set Reconciliation with Nearly Optimal Communication Complexity. IEEE Trans. Inf. Theory 49(9), 2213–2218
23. Misra, J., Gries, D.: Finding repeated elements. Sci. Comput. Programm. 2, 143–152 (1982)
24. Schmidt, J., Siegel, A., Srinivasan, A.: Chernoff-Hoeffding Bounds with Applications for Limited Independence. In: Proc. ACM SODA, pp. 331–340 (1992)

Improving the Competitive Ratio of the Online OVSF Code Assignment Problem[*]

Shuichi Miyazaki[1] and Kazuya Okamoto[2]

[1] Academic Center for Computing and Media Studies, Kyoto University
shuichi@media.kyoto-u.ac.jp
[2] Graduate School of Informatics, Kyoto University
okia@kuis.kyoto-u.ac.jp

Abstract. Online OVSF code assignment has an important application to wireless communications. Recently, this problem was formally modeled as an online problem, and performances of online algorithms have been analyzed by the competitive analysis. The previous best upper and lower bounds on the competitive ratio were 10 and 5/3, respectively. In this paper, we improve them to 7 and 2, respectively. We also show that our analysis for the upper bound is tight by giving an input sequence for which the competitive ratio of our algorithm is $7 - \varepsilon$ for arbitrary $\varepsilon > 0$.

1   Introduction

Universal Mobile Telecommunication System (UMTS) is one of the third generation (3G) technologies, which is a mobile communication standard. UMTS uses a high-speed transmission protocol Wideband Code Division Multiple Access (W-CDMA) as the primary mobile air interface. W-CDMA was implemented based on Direct Sequence CDMA (DS-CDMA), which allows several users to communicate simultaneously over a single communication channel. DS-CDMA utilizes Orthogonal Variable Spreading Factor (OVSF) code to separate communications [1].

OVSF code is based on an *OVSF code tree*, which is a complete binary tree of height h. The leaves of the OVSF code tree are of level 0 and parents of vertices of level ℓ ($\ell = 0, \ldots, h - 1$) are of level $\ell + 1$. Therefore the level of the root is h. Fig. 1 shows an OVSF code tree of height 4. Each vertex of level ℓ corresponds to a code of level ℓ. In DS-CDMA, each communication uses a code of the specific level. To avoid interference, we need to assign codes (vertices of an OVSF code tree) to communications so that they are *mutually orthogonal*, namely, in any path from the root to a leaf of an OVSF code tree, there is at most one assigned vertex. However, it is not so easy to serve requests *efficiently* as we will see later.

Erlebach et al. [4] first modeled this problem as an online problem, called the online OVSF code assignment problem, and verified the efficiency of algorithms using the competitive analysis: An input σ consists of a sequence of *a-requests (assignment requests)* and *r-requests (release requests)*. An a-request a_i specifies

---

[*] This work was supported by KAKENHI 17700015, 19200001, 19·4017, and 20300028.

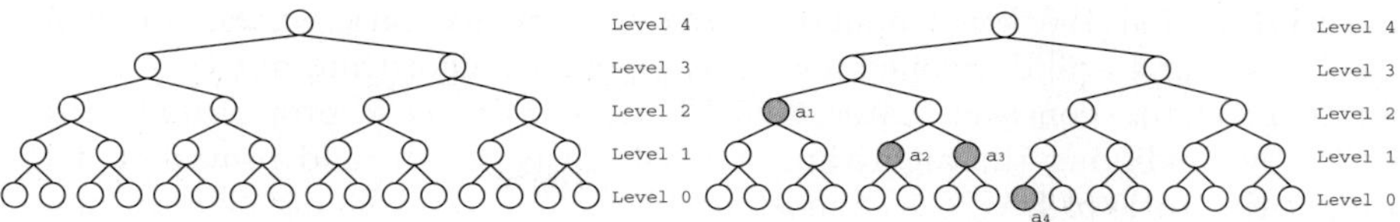

Fig. 1. An OVSF code tree of hight 4   **Fig. 2.** An example of an assignment

a required level, denoted $\ell(a_i)$. Upon receiving an a-request a_i, the task of an online algorithm is to assign a_i to one of the vertices of level $\ell(a_i)$ of OVSF code tree, so that the orthogonality condition is not broken. It may also reassign other requests (already existing in the tree). An r-request r_i specifies an a-request, denoted $f(r_i)$, which was previously assigned and is still assigned to the current OVSF code tree. When an r-request r_i arrives, the task of an online algorithm is to merely remove $f(r_i)$ from the tree (and similarly, it may reassign other requests in the tree). Each assignment and reassignment causes a cost of 1, but removing a request causes no cost. Without loss of generality, we may assume that σ does not include an a-request that cannot be assigned by any reassignment of the existing requests (in other words, the total bandwidth of requests at any point never exceeds the capacity).

For example, consider the OVSF code tree given in Fig. 1 and the input $\sigma = (a_1, a_2, a_3, a_4, r_1, a_5)$, where $\ell(a_1) = 2, \ell(a_2) = \ell(a_3) = 1, \ell(a_4) = 0, \ell(a_5) = 3$ and $f(r_1) = a_2$. Suppose that, for the first four requests, an online algorithm assigns a_1 through a_4 as depicted in Fig. 2. Then, r_1 arrives and a_2 is released. Next, a_5 arrives, but an algorithm cannot assign a_5 unless it reassigns other a-requests. If it reassigns a_4 to a child of the vertex to which a_2 was assigned, it can assign a_5 to the right vertex of level 3. In this case, it costs 6 (5 assignments and 1 reassignment). However, if it has assigned a_2 to a vertex in the right subtree of the root, and a_4 to a vertex in the left subtree, then the cost is 5, which is clearly optimal.

This problem also has an application in assigning subnets to users in computer network managements. An IP address space can be divided into subnets, each of which is a fragment of the whole IP address space consisting of a set of continuous IP addresses of size power of 2. This structure can be represented as a complete binary tree, in exactly the same way as our problem. Usually, the sizes of subnets requested by users depend on the number of computers they want to connect to the subnet, and the task of managers is to assign subnets to users so that no two assigned subnets overlap. Apparently, we want to minimize the number of reassignments because a reassignment causes a large cost for updating configurations of computers.

Online algorithms are evaluated by the competitive analysis. The *competitive ratio* of an online algorithm ALG is defined as $\max\{\frac{C_{ALG}(\sigma)}{C_{OPT}(\sigma)}\}$, where $C_{ALG}(\sigma)$ and $C_{OPT}(\sigma)$ are the costs of ALG and an optimal offline algorithm, respectively, for an input σ, and max is taken over all σ. If the competitive ratio of ALG is at most c, we say that ALG is c-competitive.

For the online OVSF code assignment problem, Erlebach et al. [4] developed a $\Theta(h)$-competitive algorithm (recall that h is the hight of the OVSF code tree), and proved that the lower bound on the competitive ratio of the problem is 1.5. Forišek et al. [6] developed a $\Theta(1)$-competitive algorithm, but they did not obtain a concrete constant. Later, Chin, Ting, and Zhang [2] proposed algorithm LAZY by modifying the algorithm of Erlebach et al. [4], and proved that the competitive ratio of LAZY is at most 10. Chin, Ting, and Zhang [2] also showed that no online algorithm can be better than 5/3-competitive.

Our Contribution. In this paper, we improve both upper and lower bounds, namely, we give a 7-competitive algorithm EXTENDED-LAZY, and show that no online algorithm can be better than 2-competitive. We further show that our upper bound analysis is tight by giving a sequence of requests for which the competitive ratio of EXTENDED-LAZY is $7 - \varepsilon$ for an arbitrary constant $\varepsilon > 0$.

Let us briefly explain an idea of improving an upper bound. Erlebach et al. [4] defined the "compactness" of the assignment, and their algorithm keeps compactness at any time. They proved that serving a request, namely assigning (or releasing) a request and modifying the tree to make compact, will cause at most one reassignment at each level, which leads to $\Theta(h)$-competitiveness. Chin, Ting, and Zhang [2] pointed out that always keeping the tree compact is too costly. Their algorithm LAZY does not always keep the compactness but makes the tree compact when it is necessary. To achieve this relaxation, they defined "tanks". By exploiting the idea of tanks, they proved that the cost of serving each request is at most 5, which provides 10-competitiveness. Our algorithm follows this line. We further relax the compactness by defining "semi-compactness". We also use amortized cost analysis. We prove that serving one a-request (r-request, respectively) and keeping semi-compactness costs at most 4 (3, respectively) and obtained 7-competitiveness of EXTENDED-LAZY.

Related Results. For the online OVSF code assignment problem, there are a couple of resource augmentations, namely, online algorithms are allowed to use more bandwidth than an optimal offline algorithm: Erlebach et al. [4] developed a 4-competitive algorithm in which an online algorithm can use a double-sized OVSF code tree. Chin, Zhang, and Zhu [3] developed a 5-competitive algorithm that uses 1/8 extra bandwidth.

Also, there are several *offline* models. One is the problem of finding a minimum number of reassignments to modify the current assignment so that the new request can be assigned, given a current assignment configuration of the tree and a new request. For this problem, Minn and Siu [7] developed a greedy algorithm. Moreover, Erlebach et al. [4] proved that this problem is NP-hard and developed a $\Theta(h)$-approximation algorithm. Another example is an offline version of our problem, namely, given a whole sequence of requests, we are asked to find a sequence of operations that minimizes the number of reassignments. Erlebach, Jacob, and Tomamichel [5] proved that this is NP-hard and gave an exponential-time algorithm.

2   Preliminaries

In this section, we define terminologies needed to give our algorithm, most of which are taken from [2]. Given an assignment configuration, we say that vertex

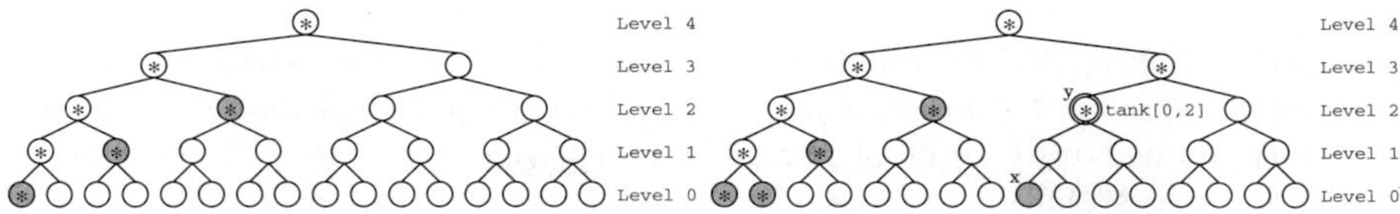

Fig. 3. A compact assignment   **Fig. 4.** A semi-compact assignment

v is *dead* if v or one of its descendants is assigned. In the example of Fig. 3, shaded vertices are assigned, and vertices with stars ($*$) are dead. If, at any level, all the left vertices (on the same level) to the rightmost dead vertex are dead, and all the assigned vertices are mutually orthogonal, then the assignment is called *compact*. For example, the assignment in Fig. 3 is compact.

Next, let us define a status of levels. Level ℓ is said to be *rich* if an a-request of level ℓ can be assigned to the leftmost non-dead vertex v at ℓ without reassigning other requests. In other words, none of descendants, ancestors, and v itself is assigned. Otherwise, the level ℓ is said to be *poor*. For example, in the assignment of Fig. 3, levels 0, 2, and 3 are rich and levels 1 and 4 are poor. A level ℓ is said to be *locally rich* if the rightmost assigned vertex is the left child of its parent. For example, in Fig. 3, only level 0 is locally rich. Note that in a compact assignment, locally rich levels are always rich.

Then, we define a *tank*. Consider an a-request a of level b, and suppose that it is assigned to a vertex x of level b. We sometimes consider as if a were assigned to a vertex y of a higher level t ($t > b$) if x is the only assigned descendant of y (see Fig. 4). In this case, the vertex y is called tank$[b, t]$. Levels b and t are called the *bottom* and the *top* of tank$[b, t]$, respectively. We say that level ℓ ($b \leq \ell \leq t$) *belongs to* tank$[b, t]$. Note that we consider that the vertex y is assigned and the vertex x is unassigned.

Finally, let us define the *semi-compactness*. An assignment is said to be *semi-compact* if the following five conditions are satisfied: (i) All the assigned vertices are mutually orthogonal; (ii) All left vertices of the rightmost dead vertex are dead at each level; (iii) Each level belongs to at most 1 tank; (iv) Suppose that there is a tank $v(=\text{tank}[b, t])$ at level t. Then level t contains at least one assigned vertex other than v, and there is no dead vertex to the right of v in t; (v) Levels belonging to tanks are all poor except for the top levels of tanks. Fig. 4 shows an example of a semi-compact assignment.

3   Algorithm EXTENDED-LAZY

To give a complete description of EXTENDED-LAZY, we first define the following four functions [2]. Note that a single application of each function keeps the orthogonality, but may break the semi-compactness. However, EXTENDED-LAZY combines functions so that the combination keeps the semi-compactness.

- AppendRich(ℓ, a): This function is available if the level of a-request a is less than or equal to ℓ (namely $\ell(a) \leq \ell$), and level ℓ is rich. It assigns a to the vertex immediately right of the rightmost dead vertex at ℓ. If there is no dead vertices at ℓ, it assigns a to the leftmost vertex v at ℓ. Note that if $\ell(a) \neq \ell$, this function creates tank$[\ell(a), \ell]$.
- AppendPoor(ℓ, a): This function is available if $\ell(a) \leq \ell$ and level ℓ is poor. It assigns a-request a to the vertex v immediately right of the rightmost dead vertex at ℓ. If there is no dead vertices at ℓ, it assigns a to the leftmost vertex v at ℓ. (If there is no such v, abort.) Note that if $\ell(a) \neq \ell$, tank$[\ell(a), \ell]$ is created. Then, it releases an a-request assigned to a vertex in the path from v to the root and returns it. (Such a request exists because ℓ was poor and v was non-dead. This request is unique because of the orthogonality.)
- FreeTail(ℓ): Release the a-request assigned to the rightmost assigned vertex at level ℓ, and return it.
- AppendLeft(ℓ, a): This function is available if $\ell(a) = \ell$. Assign a-request a to the leftmost non-assigned vertex at level ℓ that has no assigned ancestors or descendants.

Each of AppendRich, AppendPoor, and AppendLeft yields a cost of 1, and FreeTail does not yield a cost.

Now, we are ready to describe EXTENDED-LAZY. Its behaviors on an a-request and an r-request are given in Sects. 3.1 and 3.2, respectively. Executions of EXTENDED-LAZY is divided into several cases. In the description of each case, we explain the behavior of EXTENDED-LAZY, and in addition, for the later analysis, we will calculate the cost incurred and an upper bound on the increase in the number of locally rich levels due to the operations.

3.1   Executions of EXTENDED-LAZY for a-Requests

As summarized in Fig. 5, the behavior of EXTENDED-LAZY for an a-request a is divided into six cases based on the status of the level $\ell(a)$ (recall that $\ell(a)$ is the level to which a has to be assigned).

Case (1): The case that $\ell(a)$ does not belong to a tank and is rich. Execute AppendRich($\ell(a), a$). The execution of this case costs 1, and the number of locally rich levels increases by at most 1 because only $\ell(a)$ changes status.

Case (2): The case that $\ell(a)$ does not belong to a tank and $\ell(a)$ is poor. Furthermore, if we look at the higher levels from $\ell(a)$ to the root, namely in the

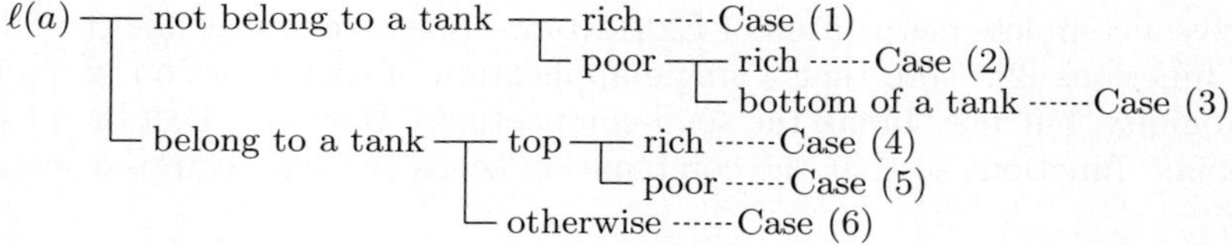

Fig. 5. Execution of EXTENDED-LAZY for an a-request a

order of $\ell(a)+1, \ell(a)+2, \ldots, h$ until we encounter a level that is rich or a bottom of a tank, we encounter a rich level (say, the level t) before a bottom of a tank. In this case, execute AppendRich(t, a). Note that the new tank$[\ell(a), t]$ is created. This case costs 1 and the number of locally rich levels increases by at most 1 since only level t changes status.

Case (3): The same as Case (2), but when looking at higher levels, we encounter a bottom b of tank$[b, t]$ before we encounter a rich level. This case is a little bit complicated. First, execute FreeTail(t) and receive the a-request a' of level b (because a level-b request was assigned to level t by exploiting a tank). Then execute AppendPoor(b, a') (note that b is poor by the condition (v) of semi-compactness), and receive another a-request a'' of level s. Note that there is an assigned vertex at t because of the condition (iv), and hence $b < s \le t$. Next, using AppendRich(t, a''), assign a'' to the vertex which was tank$[b, t]$, which creates the new tank$[s, t]$ if $s \ne t$. Now, recall that the level b was poor, and hence a' was assigned to a left child. So, the level b is currently locally rich. We execute AppendRich(b, a). Note that tank$[\ell(a), b]$ is newly created. In this case, the cost incurred is 3, and the number of locally rich levels does not change.

Case (4): The case that $\ell(a)$ is the top of a tank and is rich. First, execute FreeTail($\ell(a)$) and receive the a-request a' from the bottom of the tank. Then, execute AppendRich($\ell(a), a$) and AppendRich($\ell(a), a'$) in this order. Intuitively speaking, we shift the top of the tank to the right, and assign a to the vertex which was a tank. In this case, it costs 2 and similarly as Case (1), the number of locally rich levels increases by at most 1.

Case (5): The case that $\ell(a)$ is the top of a tank, say tank$[b, \ell(a)]$, and is poor. Execute FreeTail($\ell(a)$) and receive the a-request a' of level b from tank$[b, \ell(a)]$, and execute AppendRich($\ell(a), a$) to process a. Note that $\ell(a')$ $(= b)$ is poor because $\ell(a')$ was the bottom of tank$[b, \ell(a)]$. Also, note that b currently does not belong to a tank. Hence our current task is to process a' of level b where b does not belong to a tank and is poor. So, according to the status of levels higher than b, we execute one of Cases (2) or (3). Before going to Cases (2) or (3), the cost incurred is 1 and there is no change in the number of locally rich levels. Hence, the total cost of the whole execution of this case can be obtained by adding one to the cost incurred by the subsequently executed case (Cases (2) or (3)), and the change in the locally rich levels is the same as that of the subsequently executed case.

Case (6): The case that $\ell(a)$ belongs to tank$[b, t]$ and is not the top of tank$[b, t]$. Execute FreeTail(t) and receive the a-request a' of level b from tank$[b, t]$. Then, execute AppendPoor($\ell(a), a$) and receive an a-request a'' of level s. By a similar observation as Case (3), we can see that $\ell(a) < s \le t$. Then, assign a'' to the vertex which was tank$[b, t]$ using AppendRich(t, a''). (Note that tank$[s, t]$ is created if $s \ne t$.) Since $\ell(a)$ was poor, a was assigned to a left child. Hence $\ell(a)$ is currently locally rich. Execute AppendRich($\ell(a), a'$), which creates tank$[b, \ell(a)]$ if $\ell(a) \ne b$. The execution of this case costs 3, and the number of locally rich levels does not change.

Here we give one remark. Suppose that $\ell(a)$ does not belong to a tank and is poor. Then, we look at higher levels to see which of Cases (2) and (3) is executed. It may happen that we encounter neither a rich level nor a bottom of a tank, namely there is no tank in upper levels, and all upper levels are poor. However, we can claim that this does not happen since such a case happens only when the total bandwidth of all a-requests exceeds the capacity. As we have mentioned previously, we excluded this case from inputs. (Actually, we do not have to exclude this case because our algorithm detects this situation, and in such a case, we may simply reject the a-request.) Because of the space restriction, we omit the proof of this claim.

The following lemma proves the correctness of EXTENDED-LAZY on a-requests.

Lemma 1. EXTENDED-LAZY *preserves the semi-compactness on a-requests.*

Proof. In order to prove this lemma, we have to show that the five conditions (i) through (v) of semi-compactness are preserved after serving an a-request. It is relatively easy to show that (i) is preserved because EXTENDED-LAZY uses only AppendRich, AppendPoor, and FreeTail, each of which preserves the orthogonality even by a single application. So, let us show that conditions (ii) through (v) are preserved after the execution of each of Cases (1) through (6), provided that (i) through (v) are satisfied before the execution. Because of the space restriction, however, we treat here only Case (1) and omit all other cases.
Case (1): Let v be the vertex (of level $\ell(a)$) to which the request a is assigned. Note that by AppendRich($\ell(a), a$), some of ancestors of v may turn from non-dead to dead. It is easy to see that the condition (ii) is preserved at level $\ell(a)$ since we only appended the request to the right of the rightmost dead vertex. Now, suppose that vertex v_s of level s ($s > \ell(a)$) turned from non-dead to dead. Then, by the above observation, v_s is an ancestor of v. Next, let v' be the vertex which is immediately left of v. Then, since v' was dead, v_s is not an ancestor of v'. As a result, the ancestor of v' at level s is the vertex, say v'_s, immediately left of v_s, which implies that v'_s was dead. Thus, condition (ii) is satisfied at any level. It is not hard to see that other conditions, (iii) through (v), are also preserved because Case (1) does not create or remove tanks.   □

3.2   Executions of EXTENDED-LAZY for r-Requests

Next, we describe executions of EXTENDED-LAZY for r-requests. Similarly as Sec. 3.1, there are eight cases depending on the status of level $\ell(f(r))$ as summarized in Fig. 6, each of which will be explained in the following. Recall that $f(r)$ is the request that r asks to release.

Case (I): The case that $\ell(f(r))$ does not belong to a tank and is locally rich. Release $f(r)$. If $f(r)$ is the rightmost dead vertex at $\ell(f(r))$, do nothing. Otherwise, use FreeTail($\ell(f(r))$) and receive an a-request a of level $\ell(f(r))$. Then, using AppendLeft($\ell(f(r)), a$), assign a to the vertex to which $f(r)$ was assigned. Note that the vertex v which was the rightmost dead vertex of level $\ell(f(r))$ becomes non-dead after the above operations, which may turn some vertices in the path from v to the root non-dead from dead. As a result, an assignment may

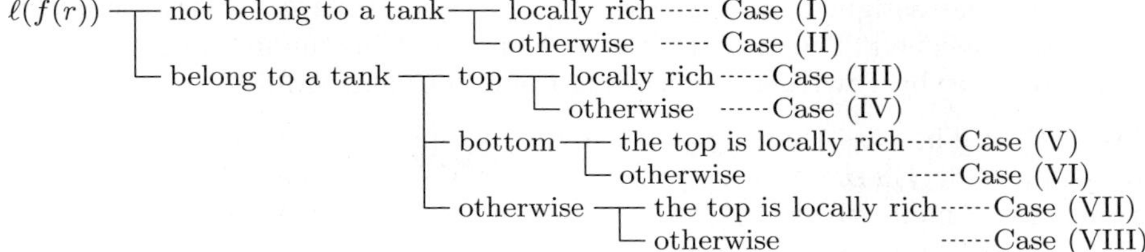

Fig. 6. Execution of EXTENDED-LAZY for an r-request r

become non-semi-compact. If the semi-compactness is broken, we use the operation REPAIR, which will be explained later, to retrieve the semi-compactness. The cost of this case is either 1 or 0, and the number of locally rich levels decreases by 1 without considering the effect of REPAIR. (We later estimate these quantities considering the effect of REPAIR.)

Case (II): The case that $\ell(f(r))$ does not belong to a tank and is not locally rich. EXTENDED-LAZY behaves exactly the same way as Case (I). Note that vertex v which was the rightmost dead vertex at level $\ell(f(r))$ becomes non-dead after the above operations, but v is a right child because $\ell(f(r))$ was not locally rich. Since the semi-compactness was satisfied before the execution, the vertex immediately left of v was (and is) dead, which implies that the parent and hence all ancestors of v are still dead. Thus, we do not need REPAIR in this case. It costs either 1 or 0, and the number of locally rich levels increases by 1 because $\ell(f(r))$ becomes locally rich.

Case (III): The case that $\ell(f(r))$ belongs to tank$[b, t]$, $\ell(f(r)) = t$, and t is locally rich. First, release $f(r)$. Next, execute FreeTail(t) and receive the a-request a of level b from tank$[b, t]$. If $f(r)$ was assigned to the vertex immediately left of tank$[b, t]$ at t, do nothing. Otherwise, using FreeTail(t), receive an a-request a' of level t, and using AppendLeft(t, a'), assign a' to the vertex to which $f(r)$ was assigned. We then find a level to which we assign the request a. Starting from level t, we see if the level contains at least one a-request, until we reach level $b + 1$. Let ℓ be the first such level. Then execute AppendRich(ℓ, a), which creates tank$[b, \ell]$. If there is no such level ℓ between t and $b + 1$, execute AppendRich(b, a). In this case, we may need REPAIR. Without considering the effect of REPAIR, it costs either 1 or 2. If it costs 1, the number of locally rich levels stays unchanged or decreases by 1, and if it costs 2, the number of locally rich levels decreases by 1.

Case (IV): The case that $\ell(f(r))$ belongs to tank$[b, t]$, $\ell(f(r)) = t$, and t is not locally rich. EXTENDED-LAZY behaves exactly the same way as Case (III). In this case, we do not need REPAIR by a similar observation as Case (II). It costs either 1 or 2, and the number of locally rich levels increases by 1.

Case (V): The case that $\ell(f(r))$ belongs to tank$[b, t]$, $\ell(f(r)) = b$, and t is locally rich. First, release $f(r)$. If $f(r)$ was the request assigned to tank$[b, t]$, stop here; otherwise, do the following: Execute FreeTail(t) and receive the a-request a of level b from tank$[b, t]$. Then, using AppendLeft(b, a), assign a to the vertex to which $f(r)$ was assigned. In this case, we may need REPAIR because tank$[b, t]$ becomes unassigned. The incurred cost is 1 or 0, and the number of locally rich levels decreases by 1 without considering the effect of REPAIR.

Case (VI): The case that $\ell(f(r))$ belongs to tank$[b, t]$, $\ell(f(r)) = b$, and t is not locally rich. EXTENDED-LAZY behaves exactly the same way as Case (V). In this case, we do not need REPAIR for the same reason as Case (II). The cost is 1 or 0, and the number of locally rich levels increases by 1.

Case (VII): The case that $\ell(f(r))$ belongs to tank$[b, t]$, $b < \ell(f(r)) < t$, and t is locally rich. First, release $f(r)$. Next, execute FreeTail(t) and receive the a-request a of level b from tank$[b, t]$. If $f(r)$ was assigned to the rightmost assigned vertex at $\ell(f(r))$, do nothing. Otherwise, using FreeTail($\ell(f(r))$), receive an a-request a' of level $\ell(f(r))$, and using AppendLeft($\ell(f(r)), a'$), assign a' to the vertex to which $f(r)$ was assigned. We then find a level to which we assign the request a in the same way as Case (III). Starting from level $\ell(f(r))$, we see if the level contains at least one a-request, until we reach level $b + 1$. Let ℓ be the first such level. Then execute AppendRich(ℓ, a), which creates tank$[b, \ell]$. If there is no such level ℓ between $\ell(f(r))$ and $b+1$, execute AppendRich(b, a). In this case, we may need REPAIR. Without considering the effect of REPAIR, it costs either 1 or 2. If it costs 1, the number of locally rich levels is unchanged or decreases by 1, and if it costs 2, the number of locally rich levels decreases by 1.

Case (VIII): The case that $\ell(f(r))$ belongs to tank$[b, t]$, $b < \ell(f(r)) < t$, and t is not locally rich. EXTENDED-LAZY behaves exactly the same way as Case (VII). In this case, we do not need REPAIR for the same reason as Case (IV). The cost is 1 or 2. The number of locally rich levels increases by 1 or 2 when the cost is 1, and by 1 when the cost is 2.

Recall that after executing Cases (I), (III), (V), or (VII), the OVSF code tree may not satisfy semi-compactness. In such a case, however, there is only one level that breaks the conditions of semi-compactness, and furthermore, there is only one broken condition, namely (ii) or (v) (again, the proof is omitted). If (ii) is broken at level ℓ, level ℓ consists of, from left to right, a sequence of dead vertices up to some point, then one non-dead vertex v, and then again a sequence of (at least one) dead vertices. This non-dead vertex was called a "hole" in [2]. We also use the same terminology here, and call level ℓ a *hole-level*. If (v) is broken at level ℓ, ℓ is a bottom of a tank tank$[\ell, t]$ and is rich. Furthermore, level ℓ consists of, from the leftmost vertex, a sequence of 0 or more dead vertices, a sequence of 1 or more non-dead vertices, and then the leftmost level-ℓ descendant of tank$[\ell, t]$ (which is non-dead by definition). We call level ℓ a *rich-bottom-level*. A level is called a *critical-level* if it is a hole-level or a rich-bottom-level.

The idea of REPAIR is to resolve a critical-level one by one. When we remove a critical-level ℓ by REPAIR, it may create another critical-level. However, we can prove that there arises at most one new critical level, and its level is higher than ℓ. Hence we can obtain a semi-compact assignment by applying REPAIR at most h times.

Let us explain the operation REPAIR (again because of a space restriction, we will give only a rough idea and omit detailed descriptions). If ℓ is a hole-level and ℓ is not a bottom of a tank, then we release the a-request a assigned to the rightmost assigned vertex at level ℓ, and reassign it to the hole to fill the hole by AppendLeft(ℓ, a). If ℓ is a hole-level and ℓ is a bottom of a tank $v(=\text{tank}[\ell, t])$, then there is a vertex u that is the leftmost level-ℓ descendant of v. Recall that the a-request virtually assigned to v is actually a request for level ℓ and is assigned to u. We release this request a using FreeTail(t) and perform AppendLeft(ℓ, a). Finally, if ℓ is a rich-bottom-level, then we will do the same operation, namely, release the a-request a from the tank, and execute AppendLeft(ℓ, a).

Lemma 2. EXTENDED-LAZY *preserves the semi-compactness on r-requests.*

Proof. Similarly as Lemma 1, we will check that five conditions (i) through (v) of semi-compactness are preserved for each application of Cases (I) through (VIII) (followed by appropriate number of applications of REPAIR). Because of the space restriction, it is omitted. $\square$

4 Competitive Analyses of EXTENDED-LAZY

First, we estimate the cost and the increase in the number of locally rich levels incurred by applications of REPAIR. By a single application of REPAIR, the cost of 1 is incurred and the number of locally rich levels increases or decreases by 1. In case that the number of locally rich levels increases by 1, the resulting OVSF code tree is semi-compact. On the other hand, if the number of locally rich levels decreases by 1, we may need one more application of REPAIR. Hence, if REPAIR is executed k times, then the total cost of k is incurred, and the number of locally rich levels decreases by $k - 2$ or k. (In the case of $k = 1$, "decreases by $k - 2$" means "increases by 1".)

Then, let us estimate the cost and the increase in the number of locally rich levels for each of the cases (1) through (6) and (I) through (VIII) of EXTENDED-LAZY. From the observations of Sects. 3.1 and 3.2, and the above observation on REPAIR, these quantities can be calculated as in Table 1. There are two values in Case (5): Left and right values correspond to the cases where Cases (2) and (3), respectively, are executed after Case (5). There are also two values in Cases (III), (VII), and (VIII), which correspond to behaviors of EXTENDED-LAZY. In the lower table, k denotes the number of applications of REPAIR. One can see that, from the upper table, the sum of the cost and the

Table 1. The costs and increases in the number of locally rich levels for each execution of EXTENDED-LAZY

Case	(1)	(2)	(3)	(4)	(5)		(6)
Cost	1	1	3	2	2	4	3
Increase	≤ 1	≤ 1	0	≤ 1	≤ 1	0	0

Case	(I)	(II)	(III)		(IV)	(V)	(VI)	(VII)		(VIII)	
Cost	$\leq k + 1$	≤ 1	$k + 1$	$k + 2$	≤ 2	$\leq k + 1$	≤ 1	$k + 1$	$k + 2$	1	2
Increase	$\leq -k + 1$	1	$\leq -k + 2$	$\leq -k + 1$	1	$\leq -k + 1$	1	$\leq -k + 2$	$\leq -k + 1$	≤ 2	1

increase in the number of locally rich levels is at most 4 for serving an a-request. This happens when EXTENDED-LAZY executes Case (5) followed by Case (3). Similarly, by the lower table, the sum of the cost and the increase in the number of locally rich levels for serving one r-request is at most 3, which happens in Cases (III), (IV), (VII), and (VIII).

Now, we are ready to calculate the competitive ratio of EXTENDED-LAZY. For an arbitrary input sequence σ, let A and R be the set of a-requests and the set of r-requests in σ, respectively. It is easy to see that the cost of an optimal offline algorithm is at least $|A|$ because each a-request incurs a cost of 1 in any algorithm. We then estimate the cost of EXTENDED-LAZY. For $a \in A$ and $r \in R$, let c_a and c_r be the costs of EXTENDED-LAZY for serving a and r, respectively. The cost of EXTENDED-LAZY for σ is then $\sum_{a \in A} c_a + \sum_{r \in R} c_r$. Also, for $a \in A$ and $r \in R$, let p_a and p_r be the increases in the number of locally rich levels caused by EXTENDED-LAZY in serving a and r, respectively. Define P to be the number of locally rich levels in the OVSF code tree at the end of the input σ. Then, $P = \sum_{a \in A} p_a + \sum_{r \in R} p_r$ since there is no locally rich level at the beginning. The cost of EXTENDED-LAZY for σ is

$$\sum_{a \in A} c_a + \sum_{r \in R} c_r \leq \sum_{a \in A} c_a + \sum_{r \in R} c_r + P$$

$$= \sum_{a \in A} (c_a + p_a) + \sum_{r \in R} (c_r + p_r)$$

$$\leq \sum_{a \in A} 4 + \sum_{r \in R} 3 \qquad (1)$$

$$= 4|A| + 3|R|$$

$$\leq 7|A|. \qquad (2)$$

(1) is due to the above analysis, and (2) is due to the fact that $|R| \leq |A|$ since for each r-request, there must be a preceding a-request corresponding to it. Now, the following theorem is immediate from the above inequality.

Theorem 1. *The competitive ratio of* EXTENDED-LAZY *is at most 7.*

Next, we give a lower bound on the competitive ratio of EXTENDED-LAZY.

Theorem 2. *The competitive ratio of* EXTENDED-LAZY *is at least* $7 - \epsilon$ *for any positive constant* $\epsilon > 0$.

Proof. As one can see in the upper bound analysis, the most costly operations are Case (5) followed by Case (3) for a-requests and Cases (III), (IV), (VII), and (VIII) for r-requests. We first give a short sequence which leads an OVSF code tree of EXTENDED-LAZY to some special configuration, and after that, we give a-requests and r-requests repeatedly, for which EXTENDED-LAZY executes Case (5) followed by Case (3), or Case (VIII) almost every time. Because of the space restriction, we omit the complete proof.   □

5 A Lower Bound

Theorem 3. *For any positive constant $\epsilon > 0$, there is no $(2 - \epsilon)$-competitive online algorithm for the online OVSF code assignment problem.*

Proof. Consider an OVSF code tree of height h (where h is even), namely, the number of leaves are $n = 2^h$. First, an adversary gives n a-requests of level 0 so that the vertices of level 0 are fully assigned, by which, an arbitrary online algorithm incurs the cost of n. Then, depending on the assignment of the online algorithm, the adversary requires to release one a-request from each subtree rooted at a vertex of level $h/2$. (Hereafter, we simply say "subtree" to mean a subtree of this size.) Since there are $\sqrt{n}$ such subtrees, the adversary gives $\sqrt{n}$ r-requests in total. Next, the adversary gives an a-request a_1 of level $h/2$. To assign a_1, the online algorithm has to make one of subtrees empty by reassignments, for which the cost of at least $\sqrt{n} - 1$ is required.

Again, depending on the behavior of the online algorithm, the adversary requires to release $\sqrt{n}$ a-requests of level 0 *uniformly* from each subtree except for the subtree to which a_1 is assigned. Here, "uniformly" means that the numbers of r-requests for any pair of subtrees differ by at most 1; in the current case, the adversary requires to release two a-requests from one subtree, and one a-request from each of the other $\sqrt{n} - 2$ subtrees. Subsequently, the adversary gives an a-request a_2 of level $h/2$. Similarly as above, the online algorithm needs $\sqrt{n} - 2$ reassignments to assign a_2.

The adversary repeats the same operation $\sqrt{n}$ rounds, where one round consists of $\sqrt{n}$ r-requests to release a-requests of level 0 uniformly from subtrees, and one a-request of level $h/2$. Eventually, all initial a-requests of level 0 are removed, and the final OVSF code tree contains $\sqrt{n}$ a-requests of level $h/2$.

The total cost of the online algorithm is $n + \sqrt{n} + (\sqrt{n} - 1) + (\sqrt{n} - 2) + (\sqrt{n} - 2) + \cdots = n + \sqrt{n} + \sum_{i=1}^{\sqrt{n}}(\sqrt{n} - \lceil \frac{\sqrt{n}}{\sqrt{n}+1-i} \rceil) > 2n - \sqrt{n}(\log \sqrt{n} + (\sum_{i=1}^{\sqrt{n}} \frac{1}{i} - \log \sqrt{n}))$. On the other hand, the cost of an optimal offline algorithm is $n + \sqrt{n}$ since it does not need reassignment. Hence, the competitive ratio is at least

$$\frac{2n - \sqrt{n}(\log \sqrt{n} + (\sum_{i=1}^{\sqrt{n}} \frac{1}{i} - \log \sqrt{n}))}{n + \sqrt{n}} = 2 - \frac{\sqrt{n}(\log \sqrt{n} + 2 + (\sum_{i=1}^{\sqrt{n}} \frac{1}{i} - \log \sqrt{n}))}{n + \sqrt{n}}.$$

Since $\lim_{n \to \infty}(\sum_{i=1}^{\sqrt{n}} \frac{1}{i} - \log \sqrt{n}) = \gamma$ ($\gamma \simeq 0.577$) is the Euler's constant, the term $\frac{\sqrt{n}(\log \sqrt{n} + 2 + (\sum_{i=1}^{\sqrt{n}} \frac{1}{i} - \log \sqrt{n}))}{n + \sqrt{n}}$ becomes arbitrarily small as n goes infinity. $\square$

References

1. Adachi, F., Sawahashi, M., Okawa, K.: Tree-structured generation of orthogonal spreading codes with different lengths for forward link of DS-CDMA mobile radio. Electronics Letters 33(1), 27–28 (1997)
2. Chin, F.Y.L., Ting, H.F., Zhang, Y.: A constant-competitive algorithm for online OVSF code assignment. In: Tokuyama, T. (ed.) ISAAC 2007. LNCS, vol. 4835, pp. 452–463. Springer, Heidelberg (2007)

3. Chin, F.Y.L., Zhang, Y., Zhu, H.: Online OVSF code assignment with resource augmentation. In: Kao, M.-Y., Li, X.-Y. (eds.) AAIM 2007. LNCS, vol. 4508, pp. 191–200. Springer, Heidelberg (2007)
4. Erlebach, T., Jacob, R., Mihaľák, M., Nunkesser, M., Szabó, G., Widmayer, P.: An algorithmic view on OVSF code assignment. Algorithmica 47(3), 269–298 (2007)
5. Erlebach, T., Jacob, R., Tomamichel, M.: Algorithmische aspekte von OVSF code assignment mit schwerpunkt auf offline code assignment. Student thesis as ETH Zürich
6. Forišek, M., Katreniak, B., Katreniaková, J., Královič, R., Koutný, V., Pardubská, D., Plachetka, T., Rovan, B.: Online bandwidth allocation. In: Arge, L., Hoffmann, M., Welzl, E. (eds.) ESA 2007. LNCS, vol. 4698, pp. 546–557. Springer, Heidelberg (2007)
7. Minn, T., Siu, K.Y.: Dynamic assignment of orthogonal variable-spreading-factor codes in W-CDMA. IEEE Journal on Selected Areas in Communications 18(8), 1429–1440 (2000)

Optimal Key Tree Structure for Deleting Two or More Leaves[*]

Weiwei Wu[1], Minming Li[2], and Enhong Chen[3]

[1] Department of Computer Science, University of Science and Technology of China
`wweiwei2@cityu.edu.hk`
[2] Department of Computer Science, City University of Hong Kong
`minmli@cs.cityu.edu.hk`
[3] Department of Computer Science, University of Science and Technology of China
`cheneh@ustc.edu.cn`

Abstract. We study the optimal tree structure for the key management problem. In the key tree, when two or more leaves are deleted or replaced, the updating cost is defined to be the number of encryptions needed to securely update the remaining keys. Our objective is to find the optimal tree structure where the worst case updating cost is minimum. We first prove the degree upper bound $(k + 1)^2 - 1$ when k leaves are deleted from the tree. Then we focus on the 2-deletion problem and prove that the optimal tree is a balanced tree with certain root degree $5 \leq d \leq 7$ where the number of leaves in the subtrees differs by at most one and each subtree is a 2-3 tree.

1 Introduction

In the applications that require content security, encryption technology is widely used. Asymmetric encryption is usually used in a system requiring stronger security, while symmetric encryption technology is also widely used because of the easy implementation and other advantages. In the applications such as teleconferencing and online TV, the most important security problem is to ensure that only the authorized users can enjoy the service. Centralized key management technology can achieve efficiency and satisfy the security requirement of the system. Hence, several models based on the key tree management are proposed to safely multicast the content. Two kinds of securities should be guaranteed in these applications: one is Future Security which prevents a deleted user from accessing the future content; the other is Past Security which prevents a newly joined user from accessing the past content. Key tree model, which was proposed by Wong el al. [8] is widely studied in recent years. In this model, the Group

[*] This work was supported in part by the National Basic Research Program of China Grant 2007CB807900, 2007CB807901, a grant from the Research Grants Council of the Hong Kong Special Administrative Region, China [Project No. CityU 116907], Program for New Century Excellent Talents in University (No.NCET-05-0549) and National Natural Science Foundation of China (No.60775037).

S.-H. Hong, H. Nagamochi, and T. Fukunaga (Eds.): ISAAC 2008, LNCS 5369, pp. 77–88, 2008.

Controller (GC) maintains a tree structure for the whole group. The root of the tree stores a Traffic Encryption Key (TEK) which is used to encrypt the content that should be broadcast to the authorized users. To update the TEK securely, some auxiliary keys are maintained. Whenever a user leaves or joins, the GC would update keys accordingly to satisfy Future Security and Past Security. Because updating keys for each user change is too frequent in some applications, [5] proposed Batch Rekeying Model where keys are only updated after a certain period of time. [10] studied the scenario of popular services with limited resources which always has the same number of joins and leaves during the batch period (because there are always users on the waiting list who will be assigned to empty positions whenever some authorized users leave). A recent survey for key tree management can be found in [2].

An important research problem in the key tree model is to find an optimal structure for a certain pattern of user behaviors so that the total number of encryptions involved in updating the keys is minimized. Graham et al. [3] studied the optimal structure in Batch Rekeying Model where every user has a probability p to be replaced in the batch period. They showed that the optimal tree for n users is an n-star when $p > 1 - 3^{-1/3} \approx 0.307$, and when $p \le 1 - 3^{-1/3}$, the optimal tree can be computed in $O(n)$ time. Specially when p tends to 0, the optimal tree resembles a balanced ternary tree to varying degrees depending on certain number-theoretical properties of n. Feng et al. [1] studied the optimal structure in Key Tree Model under the assumption that users in the group are all deleted one by one. Their result shows that the optimal tree is a tree where every internal node has degree at most 5 and the children of nodes which have degree $d \neq 3$ are all leaves. [9] improved the result of [1] and showed that a balanced tree where every subtree has nearly the same number of leaves can achieve the optimal cost. They then investigate the optimal structure when the insertion cost in the initial setup period is also considered and showed that the optimal tree is a tree where every internal node has degree at most 7 and children of nodes which have degree $d \neq 2$ and $d \neq 3$ are all leaves.

More related to this paper, Soneyink et al. [6] proved that any distribution scheme has a worst-case cost of $\Omega(\log n)$ for deleting a user. They also found an optimal structure when only one user is deleted from the tree. In this paper, we further investigate the problem when two or more users are deleted from a tree. We first prove a degree upper bound $(k + 1)^2 - 1$ for the problem of deleting k users. Then we give a tighter degree bound for the problem of deleting two users. After that, we investigate the maximum number of leaves that can be placed on the tree given a fixed worst case deletion cost. Based on this, we prove that a balanced tree with certain root degree $5 \le d \le 7$ where the number of leaves in the subtrees differs by at most one and each subtree is a 2-3 tree can always achieve the minimum worst case 2-deletion cost.

The rest of this paper is organized as follows. We review the key tree model in Section 2 and then prove the degree bound of the optimal tree for the k-deletion problem in Section 3. From Section 4 on, we focus on the 2-deletion problem and remove the possibility of root degree 2 and 3. In Section 5, we study the maximum number of leaves that can be placed on a tree given a fixed deletion cost and use the result to prove that the optimal tree for the 2-deletion problem is a tree where each subtree of the root is a 2-3 tree and has the number of leaves differed by at most 1. Finally, we conclude our work and propose a conjecture on the optimal tree cost for the general k-deletion problem in Section 6.

2 Preliminaries

We first review the Key Tree Model [8] which is also referred to in the literature as LKH(logical key hierarchy) [7].

In the Key Tree Model, there is a Group Controller maintaining a key tree for the group. A leaf on the key tree represents a user and stores an individual key that is only known by this user. An internal node stores a key that is shared by all its leaf descendants. Thus a user always knows the keys stored in the path from the leaf to the root. To guarantee content security, the GC encrypts the content by the Traffic Encryption Key (TEK) which is stored in the root and then broadcast it to the users. Only the authorized users knowing the TEK can decrypt the content. When a user joins or leaves, the GC will update the keys in a bottom-up fashion. As shown in Figure 1(a), there are 7 users in the group. We take the deletion of user u_4 as an example, since u_4 knows k_4, k_9 and k_{10}, the GC need to update the keys k_9 and k_{10} (the node that stores k_4 disappears because u_4 is already deleted from the group). GC will encrypt the new k_9 with k_5 and broadcast it to notify u_5. Note that only u_5 can decrypt the message. Then GC encrypts the new k_{10} with k_6, k_7, k_8 and the new k_9 respectively, and then broadcast the encrypted messages to notify the users. Since all the users and only the users in the group can decrypt one of these messages, the GC can safely notify the users except user u_4 about the new TEK. The deletion cost measured as the number of encryptions is 5 in this example.

In the following, we say that a node u has degree d if it has d children in the tree. Note that the worst case deletion cost of the tree shown in Figure 1(a) is 6 where one of the users u_1, u_2, u_3 is deleted. In [6], the authors investigate the optimal tree structure with n leaves where the worst case single deletion cost is minimum. Their result shows that the optimal tree is a special kind of 2-3 tree defined as follows.

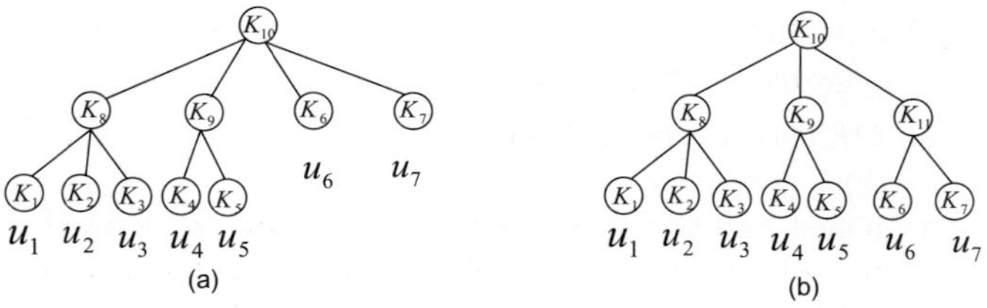

Fig. 1. Two structures of a group with 7 users

Definition 1. *In the whole paper, we use 2-3 tree to denote a tree with n leaves constructed in the following way.*

(1)When $n \geq 5$, the root degree is 3 and the number of leaves in three subtrees of the root differs by at most 1. When $n = 4$, the tree is a complete binary tree. When $n = 2$ or $n = 3$, the tree has root degree 2 and 3 respectively. When $n = 1$, the tree only consists of a root.

(2)Each subtree of the root is a 2-3 tree defined recursively.

In fact, [6] showed that 2-3 tree is a tree where the maximum ancestor weight (summation of degrees of all ancestors of a node) of the leaves on the tree is minimum among all the trees having the same number of leaves. As shown in Figure 1(b), the optimal tree for a single deletion for à group with 7 users has a worst case deletion cost 5.

In this paper, we study the scenario where two or more users leave the group during a period and we update keys at the end of this period. There are two versions of this problem to be considered here.

We denote the problem to find the optimal tree when k users are deleted as *pure k-deletion problem*. For example, deleting two users u_1 and u_4 in Figure 1(a) will incur cost 7 because we need to update k_8 and k_9 with 2 and 1 encryptions respectively and then to update k_{10} with 4 encryptions. This is also the worst case deletion. The objective is to find the optimal structure where the worst case cost is minimum.

In popular applications, there is a fixed number of positions and new users are always waiting to join. In such a scenario the number of joins and the number of leaves during the period are the same, which means that the newly joined k users will take the k positions which are vacant due to the leave of k users. In this setting, when two users u_1 and u_4 are replaced on Figure 1 (a), the updating cost is 9 which equals the summation of the ancestors' degrees of these two leaves where the common ancestors' degrees are only computed once. We denote the problem to find the optimal tree when k users are replaced as *k-deletion problem*.

We first define the k-deletion problem formally as follows.

Definition 2. *Given a tree T, we denote the number of encryptions incurred by replacing $u_{i_1}, \ldots, u_{i_k}$ with k new users as $C_T(u_{i_1}, \ldots, u_{i_k}) = \sum_{v \in (\bigcup_{1 \leq j \leq k} ANC(u_{i_j}))} d_v$ where $ANC(u)$ is the set of u's ancestor nodes and d_v is v's degree. We use k-deletion cost to denote the maximum cost among all possible combinations and write it as $C_k(T, n) = \max_{i_1, i_2, \ldots, i_k} C_T(u_{i_1}, u_{i_2}, \ldots, u_{i_k})$.*

We further define an optimal tree $T_{n,k,opt}$ (abbreviated as $T_{n,opt}$ if the context is clear) for k-deletion problem as a tree which has the minimum k-deletion cost over all trees with n leaves, i.e. $C_k(T_{n,opt}, n) = \min_{T} C_k(T, n)$. We also denote this optimal cost as $OPT_k(n)$. The k-deletion problem is to find the $OPT_k(n)$ and $T_{n,k,opt}$.

The pure k-deletion cost and the pure k-deletion problem are defined similarly by using the cost incurred by permanently deleting the leaves instead of the cost by updating the leaves (Some keys need not be updated if all its leaf descendants are deleted and the number of encryptions needed to update that key is also reduced if some branches of that node totally disappear after deletion). We will show the relationship between these two problems in the following.

Definition 3. *We say a node v is a pseudo-leaf node if its children are all leaves. In the following two lemmas, we use t to denote the number of pseudo-leaf nodes in a tree T.*

Lemma 1. *If $t \leq k$, then the pure k-deletion cost of T is at least $n - k$.*

Proof. When $t \leq k$, we claim that in order to achieve pure k-deletion cost, we need to delete at least one leaf from each pseudo-leaf node. Suppose on the contrary there exists one pseudo-leaf node v where none of its children belongs to the k leaves we delete. We divide the discussion into two cases.

First, if each of the k leaves is a child of the remaining $t-1$ pseudo-leaf nodes, then there exists one pseudo-leaf node u with at least two children deleted. In this case, a larger pure deletion cost can be achieved if we delete one child of v while keeping one more child of u undeleted.

Second, if some of the k leaves are not from the remaining $t - 1$ pseudo-leaf nodes, then we assume u is one of them whose parent is not a pseudo-leaf. Then there exists one of u's sibling w that contains at least one pseudo-leaf w' (w' can be w itself). If no children of w' belong to the k leaves, then deleting a child of w' while keeping u undeleted incurs larger pure deletion cost. If at least one child of w' belongs to the k leaves, then deleting a child of v while keeping u undeleted incurs larger pure deletion cost.

We see that in the worse case deletion, each pseudo-leaf node has at least one child deleted, which implies that all the keys in the remaining $n - k$ leaves should be used once as the encryption key in the updating process. Hence the pure k-deletion cost of T is at least $n - k$.

Lemma 2. *If $t > k$, then the pure k-deletion cost is $C_k(T, n) - k$ where $C_k(T, n)$ is the k-deletion cost.*

Proof. Using similar arguments as in the proof of Lemma 1, we can prove that when $t > k$, the pure k-deletion cost can only be achieved when the k deleted leaves are from k different pseudo-leaf nodes. Then it is easy to see that the pure k-deletion cost is $C_k(T, n) - k$ where $C_k(T, n)$ is the k-deletion cost.

Theorem 1. *When considering trees with n leaves, the optimal pure k-deletion cost is $OPT_k(n) - k$ where $OPT_k(n)$ is the optimal k-deletion cost.*

Proof. Note that in the tree where all n leaves have the same parent (denoted as one-level tree), the pure k-deletion cost is $n - k$. By Lemma 1, any tree with the number of pseudo-leaf nodes at most k has the pure k-deletion cost at least $n - k$. Hence we only need to search the optimal tree among the one-level tree and the trees with the number of pseudo-leaf nodes larger than k. Moreover, in the one-level tree T, the pure k-deletion cost is $n - k = C_k(T, n) - k$ where $C_k(T, n)$ is the k-deletion cost. Further by Lemma 2, all the trees in the scope for searching the optimal tree have pure k-deletion cost $C_k(T, n) - k$, which implies that the optimal pure k-deletion cost is $OPT_k(n) - k$ where $OPT_k(n)$ is the optimal k-deletion cost (The structure of the optimal trees in both problems are also the same).

The above theorem implies that the optimal tree for the pure k-deletion problem and the k-deletion problem are in fact the same. Therefore, we only focus on the k-deletion problem in the following and when we use "deleting", we in fact mean "updating".

3   Degree Bound for the k-Deletion Problem

In this section, we try to deduce the degree bound for the k-deletion problem. In the following proofs, we will often choose a template tree T and then construct a tree T' by deleting from T some leaves together with the exclusive part of leaf-root paths of those leaves. Here, the exclusive part of a leaf-root path includes those edges that are not on the leaf-root path of any of the remaining leaves. We also say that T is a template tree of T'. By the definition of the k-deletion cost, we have the following fact.

Fact 1. *If T is a template tree of T', then the k-deletion cost of T' is no larger than that of T.*

Lemma 3. *$OPT_k(n)$ is non-decreasing when n increases.*

Proof. Suppose on the contrary $OPT_k(n_1) > OPT_k(n_2)$ when $n_1 \leq n_2$, then there exist two trees T_1 and T_2 satisfying $C_k(T_1, n_1) = OPT_k(n_1)$ and $C_k(T_2, n_2) = OPT_k(n_2)$. We can take T_2 as a template tree and delete the leaves until the number of leaves decreases to n_1. The resulting tree T_1' satisfies $C_k(T_1', n_1) \leq C_k(T_2, n_2) < OPT_k(n_1)$ by Fact 1, which contradicts the definition of $OPT_k(n_1)$. The lemma is then proved.

Lemma 4. *$T_{n,opt}$ has root degree upper bounded by $(k + 1)^2 - 1$.*

Proof. We divide the value of root degree $d \geq (k + 1)^2$ into two sets, $\{d \mid (k + t)^2 \leq d < (k + t)(k + t + 1), d, k, t \in N, t \geq 1\}$ and $\{d \mid (k + t - 1)(k + t) \leq d < (k + t)^2, d, k, t \in N, t \geq 2\}$. Take $k = 2$ for instance, the first set is $\{9, 10, 11, 16, 17, 18, 19, 25, ..\}$ while the other is $\{12, 13, 14, 15, 20, 21, 22, 23, ...\}$.

Case 1: $(k + t)^2 \leq d < (k + t)(k + t + 1)$ $(t \geq 1)$.
We write d as $(k + t)^2 + r$ where $0 \leq r < k + t$. Given a tree T, we can transform it into a tree with root degree $k + t$ as Figure 2 shows. In the resulting tree T', subtrees $T_{u_1}, \ldots, T_{u_{k+t}}$ are $k + t$ subtrees where the root $u_1, \ldots, u_{k+t}$ are on level one. Among the $k + t$ subtrees, there are r subtrees with root degree $k + t + 1$ and $k + t - r$ subtrees with root degree $k + t$. Suppose that the k-deletion cost of T' is incurred by deleting $k_1, k_2, \ldots, k_s$ users from subtree $T_{i_1}, T_{i_2}, \ldots, T_{i_s}$ respectively where $k_1 + k_2 + \ldots + k_s = k$ and $s \leq k$. The corresponding cost is $C_k(T', n) = \sum_{j=1}^{s} C_{k_j}(T_{i_j}, n_{i_j}) + D_0$ where n_{i_j} is the number of leaves in T_{i_j} and D_0 is the cost incurred in the first two levels.

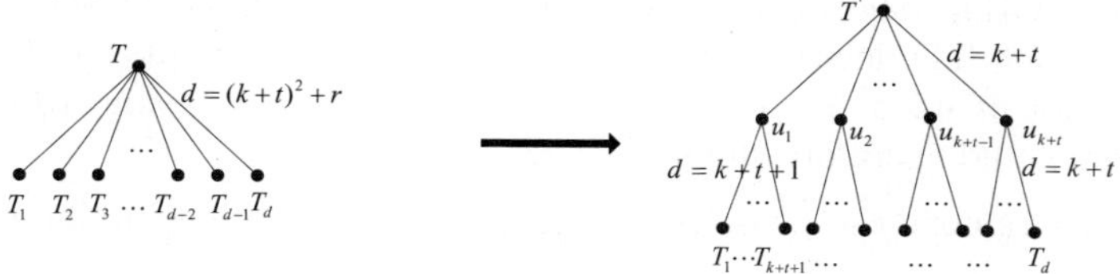

Fig. 2. Transformation of the tree which has root degree $d = (k + t)^2 + r$

In the original tree T, the corresponding cost for deleting those leaves is $C_k(T, n) = \sum_{j=1}^{s} C_{k_j}(T_{i_j}, n_{i_j}) + d = \sum_{j=1}^{s} C_{k_j}(T_{i_j}, n_{i_j}) + (k + t)^2 + r$. We will prove that when $t \geq 1$ we always have $C_k(T, n) \geq C_k(T', n)$, i.e. $D_0 \leq (k+t)^2 + r$.

Firstly, if $r \leq k$, the cost D_0 is at most $r(k + t + 1) + (k - r)(k + t) + k + t$ where there are r users coming from r subtrees with root degree $k + t + 1$ and $k - r$ users coming from $k - r$ subtrees with root degree $k + t$. Therefore, we have $D_0 \leq (k + t + 1)r + (k + t)(k - r) + k + t = (k + t)(k + 1) + r \leq (k + t)^2 + r$.

Secondly, if $r > k$, the cost D_0 is at most $(k + t) + (k + t + 1)k$ where the k users are all from k subtrees which have root degree $k + t + 1$. Therefore, we have $D_0 \leq (k + t) + (k + t + 1)k \leq (k + t)(k + 1) + k \leq (k + t)^2 + r$.

Hence, in both situations, the condition $t \geq 1$ ensures that the transformation from T to T' does not increase the k-deletion cost.

Case 2 can be proved similarly.

Lemma 4 suggests that we can find an optimal tree for k-deletion cost among trees whose root degree is at most $(k + t)^2 - 1$. Note that our degree bound in Lemma 4 is only for the root. We can also extend this property to all the internal nodes (proof is also omitted).

Lemma 5. *Any internal node in $T_{n,opt}$ has degree upper bounded by $(k+1)^2 - 1$.*

4   Degree Bound for 2-Deletion Problem

From this section on, we focus on the 2-deletion problem.

Definition 4. *We denote the maximum cost to delete a single leaf in a tree T as S_T and the maximum cost to delete two leaves as D_T, i.e. $S_T = C(T, 1)$ and $D_T = C(T, 2)$.*

According to Lemma 5, for 2-deletion, $T_{n,opt}$ has degree upper bounded by 8. Furthermore, in a tree T with root degree 1, the two deleted users in any combination are from the only subtree T_1. Therefore, the tree T_1 is better than T because the 2-deletion cost of T_1 is one less than that of T. Thus we need not consider root degree $d = 1$ when we are searching for the optimal tree.

Fact 2. *For 2-deletion problem, suppose that a tree T has root degree d where $d \geq 2$ and the d subtrees are $T_1, T_2, \ldots, T_d$. We have*

$$D_T = \max_{1 \leq i,j \leq d} \{D_{T_i} + d, S_{T_i} + S_{T_j} + d\}.$$

Proof. We know that deleting any two leaves from a subtree T_i will incur a cost at most $D_{T_i} + d$, while deleting two leaves from two different subtrees T_i and T_j will incur a cost at most $S_{T_i} + S_{T_j} + d$. The 2-deletion cost comes from one of the above cases and therefore the fact holds.

We can further remove the possibility of degree 8 by the following lemma.

Lemma 6. *For 2-deletion problem, we can find an optimal tree among the trees with node degrees bounded between 2 and 7.*

In the following, we show two important properties of the optimal tree (monotone property and 2-3 tree property) and then further remove the possibility of root degree 2 and 3 to reduce the scope of trees within which we search for the optimal tree. Due to space limit, most of the proofs are omitted in this version.

Lemma 7. *(monotone property) For 2-deletion problem, suppose a tree T has root degree d where $d \geq 2$ and d subtrees are $T_1, T_2, \ldots, T_d$. Without loss of generality, we assume that T has a non-increasing leaf descendant vector $(n_1, n_2, \ldots, n_d)$, where n_i is the number of leaves in subtree T_i. Then, there exists an optimal tree where $S_{T_1} \geq S_{T_2} \geq \ldots \geq S_{T_d}$ and T_i is a template of T_{i+1} for $2 \leq i \leq d - 1$.*

Fact 3. *For trees satisfying Lemma 7, we have*

$$D_T = \max\{D_{T_1} + d, S_{T_1} + S_{T_2} + d\}.$$

Proof. By Fact 2 we have $D_T = \max_{1 \leq i,j \leq d}\{D_{T_i} + d, S_{T_i} + S_{T_j} + d\}$. Lemma 7 further ensures that $\max_{1 \leq i,j \leq d}\{D_{T_i} + d, S_{T_i} + S_{T_j} + d\} = \max\{D_{T_1} + d, D_{T_2} + d, S_{T_1} + S_{T_2} + d\}$. Since $D_{T_2} < 2S_{T_2} \leq S_{T_1} + S_{T_2}$, we have $D_T = \max\{D_{T_1} + d, S_{T_1} + S_{T_2} + d\}$.

In the following, we further reduce the scope for searching the optimal tree by proving the following lemma.

Lemma 8. *(2-3 tree property) For a tree T satisfying Lemma 7, we can transform subtrees $T_2, \ldots, T_d$ into 2-3 trees without increasing the 2-deletion cost.*

Proof. Given a tree T satisfying Lemma 7 and Fact 3, we transform subtrees $T_2, T_3, \ldots, T_d$ into 2-3 trees $T_2', T_3', \ldots, T_d'$ to get a new tree T'. For $2 \leq i \leq d$, since $S_{T_i'} = OPT_1(n_i)$, we have $S_{T_d'} \leq \ldots \leq S_{T_3'} \leq S_{T_2'} \leq S_{T_2}$ (Lemma 3) and $D_{T_i'} \leq 2S_{T_i'} \leq 2S_{T_2'} \leq S_{T_1} + S_{T_2'}$ ($2 \leq i \leq d$). Thus $D_{T'} = \max\{D_{T_1} + d, D_{T_2'} + d, S_{T_1} + S_{T_2'} + d\} \leq \max\{D_{T_1} + d, S_{T_1} + S_{T_2} + d\} = D_T$, which implies the transformation does not increase 2-deletion cost. The lemma is then proved.

We denote the trees satisfying Lemma 8 as *candidate-trees*. By Lemma 8, we can find an optimal tree among all the candidate trees. For a candidate tree T with root degree d, we define branch B_i to be the union of T_i and the edge connecting the root of T with the root of T_i. We say the branch B_1 is the *dominating branch* and other branches $B_2, \ldots B_d$ are *ordinary branches*. We then prove the following theorem to further remove the possibility of root degree 2 and 3 in the optimal tree (details are omitted in this version).

Theorem 2. *For 2-deletion problem, a tree T with root degree 2 or 3 can be transformed into a tree with root degree 4 without increasing the 2-deletion cost.*

5   Optimal Structure of 2-Deletion Problem

Although we have removed the possibility of the root degree 2 and 3 in Section 4 and have fixed the structure of the ordinary branches, we still do not have an effective algorithm to exactly compute the optimal structure because we need to enumerate all the possible structures of the dominating branch. In this section, we will prove that among the candidate trees with n leaves, a balanced structure can achieve 2-deletion cost $OPT_2(n)$. The basic idea is to first investigate the capacity $g(R)$ for candidate trees with 2-deletion cost R (Theorem 3). Note that the optimal tree has the minimum 2-deletion cost with n leaves, which reversely implies that if we want to find a tree with 2-deletion cost R and at the same time has the maximum possible number of leaves, then computing the optimal tree for increasing n until $OPT_2(n) > R$ will produce one such solution. We then analyze and calculate the exact value for the capacity (maximum number of leaves) given a fixed 2-deletion cost R (Theorem 4). Finally, we prove that certain balanced structure can always be the optimal structure that minimizes the 2-deletion cost (Theorem 5).

Definition 5. *We use capacity to denote the maximum number of leaves that can be placed in a certain type of trees given a fixed deletion cost. According to [6], function $f(r)$ defined below is the capacity for 1-deletion cost r (among all the possible trees). We use function $g(R)$ to denote the capacity for 2-deletion cost R (among all the possible trees). In other words, when $g(R-1) < n \leq g(R)$, we have $OPT_2(n) = R$.*

$$f(r) = \begin{cases} 3 \cdot 3^{i-1} & \text{if} & r = 3i \\ 4 \cdot 3^{i-1} & \text{if} & r = 3i+1 \\ 6 \cdot 3^{i-1} & \text{if} & r = 3i+2 \end{cases}$$

To facilitate the discussion, according to Fact 3, we can divide the candidate trees with 2-deletion cost R and root degree d into two categories as summarized in the following definition.

Definition 6. *Candidate trees of category 1: The two leaves whose deletion cost achieves 2-deletion cost are from different branches, i.e. $D_T = S_{T_1} + S_{T_2} + d$, which implies $S_{T_1} + S_{T_2} \geq D_{T_1}$.*

Candidate trees of category 2: The two leaves whose deletion cost achieves 2-deletion cost are both from the dominating branch B_1, i.e. $D_T = D_{T_1} + d$, which implies $S_{T_1} + S_{T_2} < D_{T_1}$.

Correspondingly, we denote the capacity of the candidate trees belonging to category 1 with 2-deletion cost R as $g_1(R)$ and denote the capacity of the candidate trees belonging to category 2 with 2-deletion cost R as $g_2(R)$. Note that we can find the optimal tree among the candidate trees according to Lemma 8, which implies that with the same 2-deletion cost R, the best candidate tree can always have equal or larger number of leaves than the general trees. That is, we have $g(R) = max\{g_1(R), g_2(R)\}$. Thus in the following discussions, we only focus on the candidate trees. On the other hand, because we are finding trees with the maximum number of leaves, it is easy to see that we can assume the number of leaves in ordinary branches are all the same (Otherwise, we can make the tree bigger without affecting the 2-deletion cost).

In all candidate trees with 2-deletion cost R, by Fact 3, we only need to consider the case where at most one of the two leaves whose deletion cost achieves 2-deletion cost are from the ordinary branches. Suppose each ordinary branch has 1-deletion cost r^-, and correspondingly T_1 has 1-deletion cost r^+ where $r^+ \leq R - d - r^-$ (otherwise we have $D_T \geq r^+ + r^- + d > R$, a contradiction). For fixed cost R, Lemma 7 (monotonous property) implies that $r^+ \geq r^-$. We first prove the following capacity bound (details are omitted in this version).

Theorem 3. *We have* $g_i(R) \leq (R - 2r^-) \cdot f(r^-)(i = 1, 2)$.

In the following theorem, among these candidate trees we study the optimal structure which achieves the maximum capacity for different values of R.

Theorem 4. *For 2-deletion cost R, the maximum capacity is*

$$
g(R) = \begin{cases}
6 \cdot 3^{i-1} & \text{if} & R = 6i \\
7 \cdot 3^{i-1} & \text{if} & R = 6i + 1 \\
8 \cdot 3^{i-1} & \text{if} & R = 6i + 2 \\
10 \cdot 3^{i-1} & \text{if} & R = 6i + 3 \\
12 \cdot 3^{i-1} & \text{if} & R = 6i + 4 \\
15 \cdot 3^{i-1} & \text{if} & R = 6i + 5
\end{cases}
$$

After we have obtained the capacity for the 2-deletion cost R, we finally prove that among the candidate trees with n leaves, the optimal cost can be achieved by some balanced structure as shown below.

Definition 7. *We use the balanced tree to denote a tree with root degree d where each subtree is 2-3 tree and has number of leaves differed by at most 1.*

Theorem 5. *Among trees with n leaves,*

(1)when $n \in (15 \cdot 3^{i-1}, 18 \cdot 3^{i-1}]$, the optimal tree is a balanced tree which has root degree 6 and 2-deletion cost $6i+6$.
(2)when $n \in (12 \cdot 3^{i-1}, 15 \cdot 3^{i-1}]$, the optimal tree is a balanced tree which has root degree 5 and 2-deletion cost $6i+5$.
(3)when $n \in (10 \cdot 3^{i-1}, 12 \cdot 3^{i-1}]$, the optimal tree is a balanced tree which has root degree 6 and 2-deletion cost $6i+4$.
(4)when $n \in (8 \cdot 3^{i-1}, 10 \cdot 3^{i-1}]$, the optimal tree is a balanced tree which has root degree 5 and 2-deletion cost $6i+3$.
(5)when $n \in (7 \cdot 3^{i-1}, 8 \cdot 3^{i-1}]$, the optimal tree is a balanced tree which has root degree 6 and 2-deletion cost $6i+2$.
(6)when $n \in (6 \cdot 3^{i-1}, 7 \cdot 3^{i-1}]$, the optimal tree is a balanced tree which has root degree 7 and 2-deletion cost $6i+1$.

Proof. When $n \in (15 \cdot 3^{i-1}, 18 \cdot 3^{i-1}]$, we have $OPT_2(n) = 6i + 6$. We will prove that the balanced tree with root degree 6 can always achieve this optimal cost. In the balanced tree, each subtree T_j $(1 \leq j \leq 6)$ has the number of leaves $n_j = \lceil \frac{n-j+1}{6} \rceil \in [\lfloor \frac{5}{2} \cdot 3^{i-1} \rfloor, 3 \cdot 3^{i-1}]$. By function $f(\cdot)$, we have $S_{T_i} \leq 3i$. Thus any two leaves from the tree will incur a deletion cost at most $2 \cdot 3i + 6 = 6i + 6$.

When $n \in (12 \cdot 3^{i-1}, 15 \cdot 3^{i-1}]$ we have $OPT_2(n) = 6i + 5$. Then we will prove that the balanced tree with root degree 5 can always achieve the optimal cost. In the balanced tree, each subtree T_j $(1 \leq j \leq 5)$ has the number of leaves $n_j = \lceil \frac{n-j+1}{5} \rceil \in [\lfloor \frac{12}{5} \cdot 3^{i-1} \rfloor, 3 \cdot 3^{i-1}]$. By function $f(\cdot)$, we have $S_{T_i} \leq 3i$. Thus any two leaves from the tree will incur a deletion cost at most $2 \cdot 3i + 5 = 6i + 5$.

When $n \in (10 \cdot 3^{i-1}, 12 \cdot 3^{i-1}]$ we have $OPT_2(n) = 6i + 4$ and $n_j = \lceil \frac{n-j+1}{6} \rceil \in [\lfloor \frac{5}{3} \cdot 3^{i-1} \rfloor, 2 \cdot 3^{i-1}]$. By function $f(\cdot)$, we have $S_{T_i} \leq 3i - 1$. Thus any two leaves from the tree will incur a deletion cost at most $2 \cdot (3i - 1) + 6 = 6i + 4$.

When $n \in (8 \cdot 3^{i-1}, 10 \cdot 3^{i-1}]$, we have $OPT_2(n) = 6i + 3$ and $n_j = \lceil \frac{n-j+1}{5} \rceil \in [\lfloor \frac{8}{5} \cdot 3^{i-1} \rfloor, 2 \cdot 3^{i-1}]$. By function $f(\cdot)$, we have $S_{T_i} \leq 3i - 1$. Thus any two leaves from the tree will incur a deletion cost at most $2 \cdot (3i - 1) + 5 = 6i + 3$.

When $n \in (7 \cdot 3^{i-1}, 8 \cdot 3^{i-1}]$, we have $OPT_2(n) = 6i + 2$ and $n_j = \lceil \frac{n-j+1}{6} \rceil \in [\lfloor \frac{7}{6} \cdot 3^{i-1} \rfloor, \frac{4}{3} \cdot 3^{i-1}]$. By function $f(\cdot)$, we have $S_{T_i} \leq 3i - 2$. Thus any two leaves from the tree will incur a deletion cost at most $2 \cdot (3i - 2) + 6 = 6i + 2$.

When $n \in (6 \cdot 3^{i-1}, 7 \cdot 3^{i-1}]$, we have $OPT_2(n) = 6i + 1$. The balanced tree with root degree 7 where $n_j = \lceil \frac{n-j+1}{7} \rceil \in [\lfloor \frac{6}{7} \cdot 3^{i-1} \rfloor, 3^{i-1}]$ can always achieve the optimal cost. By function $f(\cdot)$, we have $S_{T_i} \leq 3i - 3$. Thus any two leaves from the tree will incur a deletion cost at most $2 \cdot (3i - 3) + 7 = 6i + 1$.

Note that in some cases a balanced tree with degree 4 can also be an optimal tree, but it is not necessary to consider this possibility because we do not need to find all the possible structures of an optimal tree with n leaves.

Finally we have fixed the structure of the dominating branch and obtained the optimal tree structure for the 2-deletion problem. We conjecture the general result for the k-deletion problem in the next section.

6   Conclusion

In this paper, we study the optimal structure for the key tree problem. We consider the scenario where two or more users are deleted from the key tree and aim to find an optimal tree in this situation. We first prove a degree upper bound $(k+1)^2 - 1$ for the k-deletion problem. Then we focus on the 2-deletion problem by firstly removing the possibility of root degree 2 and 3 to reduce the scope for searching the optimal tree. Then, we investigate the capacity of the key tree. Based on this, we prove that the optimal tree for the 2-deletion problem is a balanced tree with certain root degree $5 \leq d \leq 7$ where the number of leaves in each subtree differs by at most 1 and each subtree is a 2-3 tree.

The capacity $f(\cdot)$ where $k = 1$ and $g(\cdot)$ where $k = 2$ stimulates us to conjecture the general form of capacity $G_k(R)$ which denotes the maximum number of leaves that can be placed in a tree given the k-deletion cost R in the k-deletion problem. Based on the form of $f(\cdot)$ and $g(\cdot)$, we conjecture the capacity $G_k(R)$ to be of the form shown in Equation (1). Furthermore, if the conjecture is proved to be correct, it is also possible to obtain the optimal structure in a similar way as in the proof of Theorem 5.

$$G_k(R) = \begin{cases} (3k + \alpha) \cdot 3^{i-1} & \text{if} & R = 3k \cdot i + \alpha, \alpha \in [0, k) \\ (4k + 2(\alpha - k)) \cdot 3^{i-1} & \text{if} & R = 3k \cdot i + \alpha, \alpha \in [k, 2k) \\ (6k + 3(\alpha - 2k)) \cdot 3^{i-1} & \text{if} & R = 3k \cdot i + \alpha, \alpha \in [2k, 3k) \end{cases} \quad (1)$$

One of the possible future work is therefore to investigate the capacity and optimal structure for the general k-deletion problem. We believe that the concept of capacity will also be very important to this problem.

References

1. Chen, Z.Z., Feng, Z., Li, M., Yao, F.F.: Optimizing Deletion Cost for Secure Multicast Key Management. Theoretical Computer Science 401, 52–61 (2008)
2. Goodrich, M.T., Sun, J.Z., Tamassia, R.: Efficient Tree-Based Revocation in Groups of Low-State Devices. In: Franklin, M. (ed.) CRYPTO 2004. LNCS, vol. 3152, pp. 511–527. Springer, Heidelberg (2004)
3. Graham, R.L., Li, M., Yao, F.F.: Optimal Tree Structures for Group Key Management with Batch Updates. SIAM Journal on Discrete Mathematics 21(2), 532–547 (2007)
4. Li, M., Feng, Z., Graham, R.L., Yao, F.F.: Approximately Optimal Trees for Group Key Management with Batch Updates. In: Cai, J.-Y., Cooper, S.B., Zhu, H. (eds.) TAMC 2007. LNCS, vol. 4484, pp. 284–295. Springer, Heidelberg (2007)
5. Li, X.S., Yang, Y.R., Gouda, M.G., Lam, S.S.: Batch Re-keying for Secure Group Communications. In: Proceedings of the Tenth International Conference on World Wide Web, pp. 525–534 (2001)
6. Snoeyink, J., Suri, S., Varghese, G.: A lower bound for multicast key distribution. In: Proceedings of the Twentieth Annual IEEE Conference on Computer Communications, pp. 422–431 (2001)
7. Wallner, D., Harder, E., Agee, R.C.: Key Management for Multicast: Issues and Architectures, RFC 2627 (June 1999)
8. Wong, C.K., Gouda, M.G., Lam, S.S.: Secure Group Communications Using Key Graphs. IEEE/ACM Transactions on Networking 8(1), 16–30 (2003)
9. Wu, W., Li, M., Chen, E.: Optimal Tree Structures for Group Key Tree Management Considering Insertion and Deletion Cost. In: Proceedings of the 14th Annual International Computing and Combinatorics Conference, pp. 521–530 (2008)
10. Zhu, F., Chan, A., Noubir, G.: Optimal Tree Structure for Key Management of Simultaneous Join/Leave in Secure Multicast. In: Proceedings of Military Communications Conference, pp. 773–778 (2003)

Comparing First-Fit and Next-Fit for Online Edge Coloring

Martin R. Ehmsen[1,*], Lene M. Favrholdt[1,*], Jens S. Kohrt[1,*],
and Rodica Mihai[2]

[1] Department of Mathematics and Computer Science,
University of Southern Denmark
{ehmsen,lenem,svalle}@imada.sdu.dk
[2] Department of Informatics, University of Bergen
rodica.mihai@ii.uib.no

Abstract. We study the performance of the algorithms First-Fit and Next-Fit for two online edge coloring problems. In the min-coloring problem, all edges must be colored using as few colors as possible. In the max-coloring problem, a fixed number of colors is given, and as many edges as possible should be colored. Previous analysis using the competitive ratio has not separated the performance of First-Fit and Next-Fit, but intuition suggests that First-Fit should be better than Next-Fit. We compare First-Fit and Next-Fit using the relative worst order ratio, and show that First-Fit is better than Next-Fit for the min-coloring problem. For the max-coloring problem, we show that First-Fit and Next-Fit are not strictly comparable, i.e., there are graphs for which First-Fit is better than Next-Fit and graphs where Next-Fit is slightly better than First-Fit.

1 Introduction

In edge coloring, the edges of a graph must be colored such that no two adjacent edges receive the same color. This paper studies two variants of online edge coloring, min-coloring and max-coloring. For both problems, the algorithm is given the edges of a graph one by one, each one specified by its endpoints.

In the *min-coloring* problem, each edge must be colored before the next edge is received, and once an edge has been colored, its color cannot be changed. The aim is to color all edges *using as few colors as possible.*

For the *max-coloring* problem, a limited number k of colors is given. Each edge must be either colored or rejected before the next edge arrives. Once an edge has been colored, its color cannot be changed and it cannot be rejected. Similarly, once an edge has been rejected, it cannot be colored. In this problem, the aim is to *color as many edges as possible.*

For both problems we study the following two algorithms. *First-Fit* is the natural greedy algorithm which colors each edge using the lowest possible color.

---

[*] Supported in part by the Danish Agency for Science, Technology and Innovation (FNU).

S.-H. Hong, H. Nagamochi, and T. Fukunaga (Eds.): ISAAC 2008, LNCS 5369, pp. 89–99, 2008.

Next-Fit uses the colors in a cyclic order. It colors the first edge with the color 1 and keeps track of the last used color c_{last}. For the max-coloring problem, when coloring an edge (u, v), it uses the first color in the sequence $\langle c_{last} + 1, c_{last} + 2, \ldots, k, 1, 2, \ldots, c_{last} \rangle$ that is not yet used on any edge incident to u or v. For the min-coloring problem it only cycles through the set of colors that it has been forced to use so far.

Both algorithms are members of more general families of algorithms. For the max-coloring problem, we define the class of *fair* algorithms that never reject an edge, unless all k colors are already represented at adjacent edges. For the min-coloring problem, we define the class of *parsimonious* algorithms that do not take a new color into use, unless necessary.

The min-problem has previously been studied in [1], where the main result implies that all parsimonious algorithms have the same competitive ratio of approximately 2.

The max-problem was studied in [8]. For k-colorable graphs, First-Fit and Next-Fit have very similar competitive ratios of $1/2$ and $k/(2k - 1)$. For general graphs, there is an upper bound on the competitive ratio of First-Fit of $\frac{2}{9}(\sqrt{10} - 1) \approx 0.48$, and the competitive ratio of Next-Fit exactly matches the general lower bound for fair algorithms of $2\sqrt{3} - 3 \approx 0.46$. No fair algorithm can be better than 0.5-competitive, so the competitive ratio cannot vary much between fair algorithms. Moreover, there is a general upper bound (even for randomized algorithms) of $4/7 \approx 0.57$.

General intuition suggests that First-Fit should be better than Next-Fit, and thus comes the motivation to study the performance of the two algorithms using some other measure than the competitive ratio. There are previous problems, such as paging [5,3], bin packing [4], scheduling [7], and seat reservation [6] where the relative worst-order ratio was successfully applied and separated algorithms that the competitive ratio could not. The relative worst-order ratio is a quality measure that compares two online algorithms directly, without an indirect comparison via an optimal offline algorithm. Thus, the relative worst-order ratio in many cases give more detailed information than the competitive ratio.

For the min-problem, we prove that the two algorithms are comparable, and First-Fit is 1.5 times better than Next-Fit. For the max-problem, surprisingly, we conclude that First-Fit and Next-Fit are not comparable using the relative worst-order ratio.

2   Quality Measures

The standard quality measure for online algorithms is the competitive ratio. Roughly speaking, the competitive ratio of an online algorithm A is the worst-case ratio of the performance of A to the performance of an optimal offline algorithm over all possible request sequences [14,10].

In this paper, we use the competitive ratio only for the min-coloring problem. For that problem, the measure is defined in the following way. Let A be an edge coloring algorithm and let E be a sequence of edges. Then, $A(E)$ denotes the number of colors used by A. OPT denotes an optimal offline. The competitive ratio of algorithm A is

$$\mathrm{CR}_A = \inf\{c \mid \exists b\colon \forall E\colon\ A(E) \le c \cdot \mathrm{OPT}(E) + b\}$$

The relative worst-order ratio was first introduced in [4] in an effort to combine the desirable properties of the max/max ratio [2] and the random-order ratio [11]. The measure was later refined in [5]. We describe the measure using the terminology of the coloring problems. Let E be a sequence of n edges. If σ is a permutation on n elements, then $\sigma(E)$ denotes E permuted by σ.

For the max-coloring problem, $A(E)$ is the number of edges colored by algorithm A, and

$$A_W(E) = \min_{\sigma}\{A(\sigma(E))\}.$$

For the min-coloring problem, $A(E)$ is the number of colors used by A, and

$$A_W(E) = \max_{\sigma}\{A(\sigma(E))\}.$$

Thus, in both cases, $A_W(E)$ is the performance of A on a worst possible permutation of E.

Definition 1. *For any pair of algorithms A and B, we define*

$$c_u(A, B) = \inf\{c \mid \exists b\colon \forall E\colon\ A_W(E) \le cB_W(E) + b\}\ \text{and}$$
$$c_l(A, B) = \sup\{c \mid \exists b\colon \forall E\colon\ A_W(E) \ge cB_W(E) - b\}\,.$$

If $c_l(A, B) \ge 1$ or $c_u(A, B) \le 1$, the algorithms are said to be comparable *and the relative worst-order ratio* $\mathrm{WR}_{A,B}$ *of algorithm A to algorithm B is defined. Otherwise,* $\mathrm{WR}_{A,B}$ *is undefined.*

$$\text{If } c_u(A, B) \le 1,\ \text{ then } \mathrm{WR}_{A,B} = c_l(A, B),\ \text{ and}$$
$$\text{if } c_l(A, B) \ge 1,\ \text{ then } \mathrm{WR}_{A,B} = c_u(A, B)\,.$$

Intuitively, c_l and c_u can be thought of as tight lower and upper bounds, respectively, on the performance of A relative to B.

3 Min-coloring Problem

We first study the min-coloring problem, where all edges of a graph must be colored using as few colors as possible. The first result is an immediate consequence of a result in [1].

Theorem 1. *Any parsimonious algorithm has a competitive ratio of $2 - 1/\Delta$, where Δ is the maximum vertex degree.*

Proof. In [1], it is proven that, for any online algorithm A, there is a graph G with maximum vertex degree Δ, such that G can be Δ-colored, but A uses $2\Delta - 1$ colors. On the other hand, since no edge is adjacent to more than $2\Delta - 2$ other edges, no parsimonious algorithm will use more than $2\Delta - 1$ colors.   □

Thus, the competitive ratio does not distinguish First-Fit and Next-Fit. However, with the relative worst-order ratio, we get the result that First-Fit is better than Next-Fit:

Theorem 2. *The relative worst-order ratio of* Next-Fit *to* First-Fit *is at least* $\frac{3}{2}$.

The theorem follows directly from Lemmas 1 and 2 below.

Lemma 1. *Given any graph G with edges E, $\mathrm{NF_W}(E) \geq \mathrm{FF_W}(E)$.*

Proof. For any First-Fit coloring, we construct an ordering of the edges so that Next-Fit does the same coloring as First-Fit. Assume that First-Fit uses k colors and let C_i denote the set of edges that First-Fit colors with color i. The ordering of the edges given to Next-Fit consists of all the edges from C_1, then from C_2 and further till C_k. The edges in each set is given in an arbitrary order. By the First-Fit policy, each edge in C_i is adjacent to edges of $C_1, \ldots, C_{i-1}$. Thus, since Next-Fit only cycles through the colors that it used so far, it will color the edges the same way as First-Fit. This means that, for any First-Fit coloring, we can construct an ordering of the edges such that Next-Fit uses the same number of colors. The result follows.   □

Lemma 2. *There exists a graph with edges E such that $\mathrm{NF_W}(E) \geq \frac{3}{2}\mathrm{FF_W}(E)$.*

Proof. For any even k, consider the graph S consisting of $k/2+2$ stars $S_0, S_1, \ldots,$ $S_{k/2}$ and S_{center} as depicted in Figure 1. The center vertices of $S_1, \ldots, S_{k/2}$ have

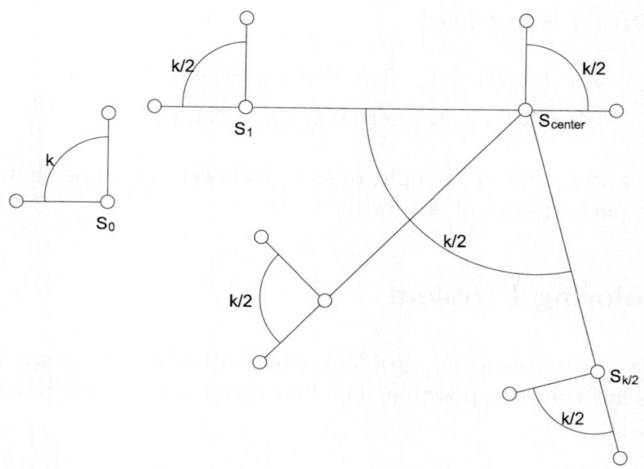

Fig. 1. The graph S used in the proof of Lemma 2

degree $k/2 + 1$, and the centers of S_0 and S_{center} have degree k. Edges for which one end-vertex has degree one are called outer edges. The remaining edges are called inner edges.

It is not difficult to see that, for any ordering of the edges, First-Fit uses exactly k colors: At most one outer edge in each star S_i, $i = 1, \ldots, k/2$, is colored with a color larger than $k/2$, and if this happens, the edge connecting S_i to S_{center} has already been colored.

Next-Fit will use $3k/2$ colors, if the edges are given in the following order: First the edges of S_0, forcing Next-Fit to use the first k colors. Then the outer edges of S_1 followed by the outer edges of S_{center}. Then the inner edge of S_1, which is colored with the color $k + 1$. Finally, for $i = 2, \ldots, k/2$, the outer edges of S_i followed by the inner edge of S_i, which is colored with the color $k + i$. This way, the inner edges will be colored with $k + 1, \ldots, 3k/2$. $\square$

4   Max-coloring Problem

In this section, we study the max-coloring problem, where a limited number k of colors are given, and as many edges as possible should be colored. We first describe a bipartite graph with edges E, such that $\text{FF}_\text{W}(E) \geq 9/8 \cdot \text{NF}_\text{W}(E)$. Then, we describe a family of graphs with edge set E_n such that $\text{NF}_\text{W}(E_n) = (1 + \Omega(\frac{1}{k^2})) \cdot \text{FF}_\text{W}(E_n)$. Thus, the two algorithms are not comparable.

4.1   First-Fit Can Be Better Than Next-Fit

Let $B_{k,k} = (X, Y, E)$ be a complete bipartite graph with $|X| = |Y| = k$. For simplicity, we assume that 4 divides k. For other values of k, we get similar results, but the calculations are a bit more messy. We denote by C_i the edges that First-Fit colors with color i.

Proposition 1. *In any* First-Fit *coloring of* $B_{k,k}$, $|C_i| \geq k - i + 1$, $i = 1, \ldots, k$.

Proof. Assume that color i has been used $j \leq k - i$ times. The induced subgraph containing all vertices *not* adjacent to an edge colored with color i is the complete bipartite graph $B_{k-j,k-j}$, where $k - j \geq i$. This subgraph cannot be colored with the colors $1, \ldots, i - 1$ only, and since this is a First-Fit coloring, the color i is going to be used. Thus, at least one more edge will be colored with color i. $\square$

Proposition 2. *If* First-Fit *colors at most* $\frac{9}{16}k^2$ *edges of* $B_{k,k}$, *then*

$$|C_i| \geq \frac{7k^2}{16(2k - 1 - i)}, \, i = 1, \ldots, k.$$

Proof. If First-Fit colors at most $9k^2/16$ edges, then it rejects at least $7k^2/16$ edges. Each rejected edge is adjacent to at least one edge of each color $i = 1, \ldots, k$. Each edge colored with color i has $2k - 2$ neighbor edges. Among those, at least $i - 1$ edges are already colored, since each edge colored with i is adjacent to all lower colors $1, 2, .., i-1$. Thus, at most $2k-1-i$ edges can be rejected for each edge colored with i. Hence, for First-Fit to reject $7k^2/16$ edges, it has to use color i at least $7k^2/(16(2k-1-i))$ times. $\qquad\square$

Lemma 3. *Given any ordering of the edges of $B_{k,k}$, First-Fit colors more than $\frac{9}{16}k^2$ edges.*

Proof. The number of edges colored by First-Fit is $\sum_{i=1}^{k} |C_i|$. We assume for the sake of contradiction that First-Fit colors at most $9k^2/16$ edges of $B_{k,k}$. Using Propositions 1 and 2 we get,

$$
\begin{aligned}
\sum_{i=1}^{k} |C_i| &\geq \sum_{i=1}^{3k/4} (k-i+1) + \sum_{i=3k/4+1}^{k} \frac{7k^2}{16(2k-1-i)} \\
&= \sum_{i=k/4+1}^{k} i + \frac{7}{16}k^2 \sum_{i=k-1}^{5k/4-2} \frac{1}{i} \\
&> \frac{15}{32}k^2 + \frac{7}{16}k^2 \ln\left(\frac{k+k/4-2}{k-2}\right) \quad \frac{15}{32}k^2 + \frac{7}{16}k^2 \ln(1+1/4) \\
&> \frac{15}{32}k^2 + \frac{7}{16}k^2 \frac{3}{14} = \frac{9}{16}k^2,
\end{aligned}
$$

which is a contradiction. Thus, First-Fit colors more than $\frac{9k^2}{16}$ edges. $\qquad\square$

Lemma 4. *Given the worst ordering of the edges of $B_{k,k}$, Next-Fit colors at most $k^2/2$ edges.*

Proof. We partition the vertex sets X, Y into equal-sized sets X_1, X_2, Y_1, Y_2. The induced subgraphs H_1 and H_2 with vertex sets X_1, Y_1 and X_2, Y_2, respectively, are complete bipartite graphs. We give the edges of H_1 and H_2 alternately such that Next-Fit colors the edges of H_1 using colors $1, 2, ..., k/2$ and the edges of H_2 with colors $k/2+1, ..., k$. After that, Next-Fit cannot color any of the $k^2/2$ edges between H_1 and H_2. Thus, Next-Fit colors at most $k^2/2$ edges of $B_{k,k}$. $\qquad\square$

Combining Lemmas 3 and 4, we arrive at:

Corollary 1. *Given the graph $B_{k,k} = (X, Y, E)$, $\mathrm{FF_W}(E) \geq \frac{9}{8} \mathrm{NF_W}(E)$.*

4.2 Next-Fit Can Be Slightly Better Than First-Fit

In this section, we prove that there exists a family of graphs where Next-Fit is $1 + \Omega(\frac{1}{k^2})$ times better than First-Fit. We first define the building blocks of the graph family.

Definition 2. *For any given number $k \geq 25$ of colors, a* superstar S_k *is a graph consisting of an* inner *star with k edges, each incident to the center of an* outer *star with $k-2$ edges of its own.*

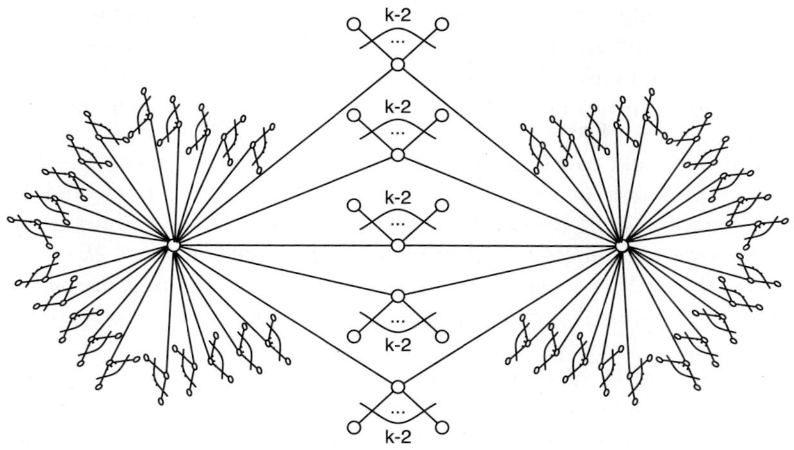

Fig. 2. Two superstars, for $k = 25$, connected through a link of five outer stars

A superstar graph is a graph consisting of superstars. Each pair of superstars in the graph may share a number of outer stars. The set of outer stars shared by a pair of superstars is called the link *between them. All outer stars are contained in a link. Each link contains at least five outer stars, and each superstar has links to between five and seven other superstars. See Figure 2 for an incomplete example.*

Clearly, fair algorithms never reject outer star edges. However, if all outer stars are colored using the same $k - 2$ colors, at least $k - 2$ edges of each inner star will be rejected. This leads to the following lemma.

Lemma 5. *Let $G_{n,k}$ be a superstar graph with n superstars. Then, on its worst ordering of the edges, First-Fit rejects at least $n(k - 2)$ edges.*

What remains to be shown is that there exists a family $G_{n,k}$ of superstar graphs, such that on a worst ordering of the edges of $G_{n,k}$, Next-Fit rejects only $n(k - 2) - \Omega(n)$ edges.

Proposition 3. *Consider a superstar graph G colored by a fair algorithm. Any superstar in G has at most $k - 1$ edges rejected. If some superstar S in G has $k - 1$ edges rejected, then each of its neighbor superstars has at most $k - 4$ edges rejected.*

Proof. Clearly, outer star edges are not rejected, so we only need to consider the inner star edges. At least one inner star edge will be colored in each superstar, since each inner star edge is only adjacent to $k - 1$ edges that are not inner star edges in the same superstar. Thus, at most $k - 1$ edges are rejected from any superstar in the graph.

Assume that some superstar S has $k - 1$ inner star edges rejected. Each of these edges must be adjacent to k colored edges. However, at most $k - 1$ of these colored neighbor edges belong to S ($k-2$ from the outer star, and the one colored inner edge of S). Hence, the kth colored neighbor edge must be an inner star edge in a neighboring superstar. Since each link contains at least five inner edges of S and at most one of them is colored, this completes the proof.   □

By Proposition 3, any pair of neighboring superstars have at most $2k-4$ rejected edges in total. A pair of neighboring superstars with $2k-4$ rejected edges in total is called a *bad pair*. Note that in a bad pair, exactly $k-2$ edges are rejected in each superstar. A pair of neighboring superstars with at most $2k-5$ rejected edges in total is called a *good pair*. A superstar contained only in bad pairs is called a *bad superstar*. A superstar contained in at least one good pair is called a *good superstar*.

Counting the good superstars, the extra colored edge from a good pair is counted at most eight times: once for the superstar S containing it and once for each of the at most seven neighbors of S. Thus, the following lemma follows directly from Proposition 3.

Lemma 6. *Consider a fair coloring of a superstar graph with n superstars. If there are m good superstars, then at most $n(k-2) - \frac{m}{8}$ edges are rejected.*

Thus, we just need to show that we can connect our building blocks, the superstars, such that, for any Next-Fit coloring, there will be $\Omega(n)$ good superstars. Such a construction is described in the proof of Lemma 8. The proof of Lemma 8 uses Proposition 4 and Lemma 7 below.

The *majority coloring* of a superstar is the set of colors used on the majority of its outer stars, breaking ties arbitrarily. An outer star is *isolated*, if it is not adjacent to at least one colored inner star edge.

Proposition 4. *If two neighboring superstars have different majority colorings, one of them is a good superstar.*

Proof. We prove the proposition by contraposition. Assume that two superstars S and S' are both bad superstars. Then, by Proposition 3, S, S', and their neighbors each have exactly $k-2$ edges rejected. Let c_1 and c_2 be the two colors used on inner star edges in S.

If S has m neighbors, the outer stars of S are adjacent to at most $2m+2$ colored inner star edges. Thus S has at least $k-2m-2$ isolated outer stars. Each of these outer stars must be colored with the $k-2$ colors different from c_1 and c_2. Hence, the isolated outer stars in S are colored the same, and since $m \geq 5$ and $k \geq 5m$ that coloring is the majority coloring of S. The same is true for S'. Since S and S' have exactly two colored edges and there are at least five edges in the link between them, they share at least one isolated outer star. This means that S and S' have the same majority coloring.   □

Lemma 7. *Assume that $k \geq 101$. Consider a Next-Fit coloring of a superstar graph $G_{n,k}$, $n \geq 6$. Among the bad superstars, there are at most $\frac{2}{3}n$ superstars with the same majority coloring.*

Proof. Any subgraph of $G_{n,k}$ containing x superstars has at least $x\frac{k}{2}$ outer stars. Thus, in any subgraph H of $G_{n,k}$ consisting of x bad superstars with the same majority coloring $\mathcal{M}$, there are at least $x\frac{k-16}{2} = x\left(\frac{k}{2} - 8\right)$ isolated outer stars colored with $\mathcal{M}$. Each time Next-Fit has used the colors in $\mathcal{M}$ once, the two colors $c_1, c_2 \notin \mathcal{M}$ must be used once, before it will use the colors in $\mathcal{M}$ on isolated outer stars again. Thus, an upper bound on the number of times c_1 and c_2 are used in $G_{n,k}$ gives an upper bound on x.

Clearly, c_1 and c_2 are each used at most once on inner star edges in each superstar. Inside H, c_1 and c_2 are not used on isolated outer stars. Thus, since each bad superstar has at least $k - 16$ isolated outer stars, c_1 and c_2 are used at most $17x$ times on superstars in H.

Outside H, c_1 and c_2 can each be used at most once per outer star, since using c_1 (c_2) on an inner star edge would prohibit the algorithm from using c_1 (c_2) on the adjacent outer star. Hence, since each superstar outside H share each outer star with another superstar, the superstars outside H can only contribute $(n - x)\frac{k}{2}$.

Thus, to create x bad superstars with majority coloring $\mathcal{M}$, we must have

$$x\left(\frac{k}{2} - 8\right) - 1 \le 17x + (n - x)\frac{k}{2}.$$

Solving for x, we obtain $x \le \frac{2}{3}n$, since $k \ge 101$ and $n \ge 6$. $\qquad\square$

Lemma 8. *For $k \ge 101$, there exists a family of superstar graphs $G_{n,k}$ where any Next-Fit coloring results in $\Omega(n)$ good superstars.*

Proof. We use a result from expander graphs [12,9]. Using the notation from [13], for any positive integer m, there exists an $\left(n = 2m^2, 7, \frac{2-\sqrt{3}}{2}\right)$-expander, i.e., a 7-regular bipartite multigraph $G(X \cup Y, E)$ with $|X| = |Y| = \frac{n}{2}$, such that for any $S \subseteq X$,

$$|\Gamma(S)| \ge \left(1 + \frac{2 - \sqrt{3}}{2}\left(1 - \frac{2|S|}{n}\right)\right)|S|,$$

where $\Gamma(S)$ is the set of edges between S and $\overline{S}$. The result also holds for any $S \subseteq Y$. The graph contains parallel edges, but each vertex has at least five neighbors. Replacing each set of parallel edges by one edge, we obtain a simple graph with the same Γ-function.

Now, we connect the superstars as in the simple expander graph. For any suitable n, let each vertex in the expander graph correspond to a superstar. Each edge in the expander graph corresponds to a link between the corresponding superstars. Thus, we obtain a superstar graph where each superstar has links to 5 to 7 other superstars.

Consider any Next-Fit coloring of this graph with n superstars. If there are at least $\frac{1}{3}n$ good superstars, the result follows immediately. Thus, we consider the case where there are at least $\frac{2}{3}n$ bad superstars. By Lemma 7, no majority coloring occurs on more than $\frac{2}{3}n$ bad superstars. Among the bad superstars, let S be the superstars with the most frequently occurring majority coloring. If

$|S| < \frac{1}{3}n$, add the bad superstars with the most frequently occurring majority coloring among the superstars not in S. Continue doing this until S reaches a size between $\frac{1}{3}n$ and $\frac{2}{3}n$. This is possible, since we consider the case where there are at least $\frac{2}{3}n$ bad superstars.

Define $S_X = S \cap X$ (similarly for S_Y), and assume without loss of generality that $|S_X| \geq |S_Y|$. Note that $|S_X| \geq \frac{1}{2}|S| \geq \frac{1}{6}n$. We can bound the size of $\Gamma(S)$ from below by the following

$$\begin{aligned}
|\Gamma(S)| &\geq |\Gamma(S_X)| - |S_Y| \\
&\geq \frac{2-\sqrt{3}}{2}\left(1 - \frac{2|S_X|}{n}\right)|S_X| + (|S_X| - |S_Y|).
\end{aligned} \tag{1}$$

We now have two cases depending on the size of S_X.

- $\frac{5}{12}n \leq |S_X| \leq \frac{1}{2}n$. Since $|S_X| + |S_Y| \leq \frac{2}{3}n$, we must have $|S_Y| \leq \frac{3}{12}n$. Thus, inequality (1) immediately yields a lower bound of $\frac{5}{12}n - \frac{3}{12}n = \frac{1}{6}n$, since $\frac{2-\sqrt{3}}{2}\left(1 - \frac{2|S_X|}{n}\right)$ is nonnegative.
- $|S_X| < \frac{5}{12}n$. Since $|S_X| - |S_Y| \geq 0$, inequality (1) gives a lower bound of $\left(\frac{2-\sqrt{3}}{2}\right)\frac{1}{6}|S_X| \geq \frac{2-\sqrt{3}}{144}n$.

Hence, in the coloring done by Next-Fit, we in both cases have $\Omega(n)$ links between S and $\overline{S}$. By the construction of S, each superstar in $\overline{S}$ linked to a superstar s in S is a good superstar or has a different majority coloring than s. Thus, by Proposition 4, there are $\Omega(n)$ good superstars. $\square$

This immediately yields the following theorem.

Theorem 3. First-Fit *and* Next-Fit *are not comparable by the relative worst-order ratio.*

Proof. By Lemma 5, there is an ordering of the edges in any superstar graph with n superstars, such that First-Fit rejects at least $n(k-2)$ edges.

By Lemmas 6 and 8, there are superstar graphs $G_{n,k}$ with n superstars such that, for any ordering of the edges, Next-Fit rejects only $n(k-2) - \Omega(n)$ edges. Hence, since First-Fit colors $\Theta(nk^2)$ edges,

$$\mathrm{NF_W}(G_{n,k}) = \left(1 + \Omega\left(\frac{1}{k^2}\right)\right)\mathrm{FF_W}(G_{n,k}).$$

On the other hand, by Corollary 1, there exists a graph S, such that

$$\mathrm{FF_W}(S) \geq \frac{9}{8}\mathrm{NF_W}(S). \qquad \square$$

5   Conclusion

We have proven that, with the relative worst-order ratio, First-Fit is strictly better than Next-Fit for the min-coloring problem. This is in contrast to the competitive ratio which is the same for all parsimonious algorithms, a class of algorithms to which First-Fit and Next-Fit belongs.

For the max-coloring problem, the answer is not as clear: With the relative worst-order ratio, there are graphs where First-Fit does significantly better than Next-Fit and graphs where Next-Fit does slightly better than First-Fit. This is somewhat in keeping with an earlier result saying that the two algorithms can hardly be distinguished by their competitive ratios.

Note that, for the max-coloring problem, the two algorithms may be *asymptotically* comparable [5]. Roughly speaking, it means that, as k tends to infinity, the algorithms "become comparable". This is left as an open problem. Note that if one were to prove that the algorithms are *not* asymptotically comparable, another construction than the superstar graphs would be required: even if Next-Fit colored all edges of a superstar graph, it would color only $1+\Theta(\frac{1}{k})$ times as many edges as First-Fit.

References

1. Bar-Noy, A., Motwani, R., Naor, J.: The greedy algorithm is optimal for on-line edge coloring. Information Processing Letters 44(5), 251–253 (1992)
2. Ben-David, S., Borodin, A.: A new measure for the study of on-line algorithms. Algorithmica 11(1), 73–91 (1994)
3. Boyar, J., Ehmsen, M.R., Larsen, K.S.: Theoretical evidence for the superiority of LRU-2 over LRU for the paging problem. In: Approximation and Online Algorithms, pp. 95–107 (2006)
4. Boyar, J., Favrholdt, L.M.: The relative worst order ratio for on-line algorithms. ACM Transactions on Algorithms 3(22) (2007)
5. Boyar, J., Favrholdt, L.M., Larsen, K.S.: The relative worst-order ratio applied to paging. Journal of Computer and System Sciences 73, 818–843 (2007)
6. Boyar, J., Medvedev, P.: The relative worst order ratio applied to seat reservation. ACM Transactions on Algorithms 4(4), article 48, 22 pages (2008)
7. Epstein, L., Favrholdt, L.M., Kohrt, J.: Separating online scheduling algorithms with the relative worst order ratio. Journal of Combinatorial Optimization 12(4), 362–385 (2006)
8. Favrholdt, L.M., Nielsen, M.N.: On-line edge coloring with a fixed number of colors. Algorithmica 35(2), 176–191 (2003)
9. Gabber, O., Galil, Z.: Explicit constructions of linear-sized superconcentrators. Journal of Computer and System Sciences 22, 407–420 (1981)
10. Karlin, A.R., Manasse, M.S., Rudolph, L., Sleator, D.D.: Competitive snoopy caching. Algorithmica 3, 79–119 (1988)
11. Kenyon, C.: Best-fit bin-packing with random order. In: 7th Annual ACM-SIAM Symposium on Discrete Algorithms, pp. 359–364 (1996)
12. Margulis, G.A.: Explicit constructions of concentrators. Problems of Information Transmission 9(4), 325–332 (1973)
13. Motwani, R., Raghavan, P.: Randomized Algorithms. Cambridge University Press, Cambridge (1995)
14. Sleator, D.D., Tarjan, R.E.: Amortized efficiency of list update and paging rules. Communications of the ACM 28(2), 202–208 (1985)

Selecting Sums in Arrays

Gerth Stølting Brodal and Allan Grønlund Jørgensen[*]

BRICS[**], MADALGO[***], Department of Computer Science,
University of Aarhus, Denmark
{gerth,jallan}@daimi.au.dk

Abstract. In an array of n numbers each of the $\binom{n}{2} + n$ contiguous sub-arrays define a sum. In this paper we focus on algorithms for selecting and reporting maximal sums from an array of numbers. First, we consider the problem of reporting k subarrays inducing the k largest sums among all subarrays of length at least l and at most u. For this problem we design an optimal $O(n + k)$ time algorithm. Secondly, we consider the problem of selecting a subarray storing the k'th largest sum. For this problem we prove a time bound of $\Theta(n \cdot \max\{1, \log(k/n)\})$ by describing an algorithm with this running time and by proving a matching lower bound. Finally, we combine the ideas and obtain an $O(n \cdot \max\{1, \log(k/n)\})$ time algorithm that selects a subarray storing the k'th largest sum among all subarrays of length at least l and at most u.

1   Introduction

In an array, $A[1, \ldots, n]$, of numbers each subarray, $A[i, \ldots, j]$ for $1 \leq i \leq j \leq n$, defines a sum, $\sum_{t=i}^{j} A[t]$. There are $\binom{n}{2} + n$ different subarrays each inducing a sum. Locating a subarray $A[i, \ldots, j]$ maximizing $\sum_{t=i}^{j} A[t]$ is known as the *maximum sum* problem, and it was formulated by Ulf Grenander in a pattern matching context. Algorithms solving the problem also have applications in areas such as Data Mining [12,13] and Bioinformatics [1]. In [5] Bentley describes the problem and an optimal linear time algorithm.

The problem can be extended to any number of dimensions. In the two dimensional version of the problem the input is an $m \times n$ matrix of numbers, and the task is to locate the connected submatrix storing the largest aggregate. This problem can be solved by a reduction to $\binom{m}{2} + m$ one-dimensional instances of size n, or a single one-dimensional instance in one array of length $O(m^2 n)$ created by separating each of the $\binom{m}{2} + m$ instances mentioned before by dummy $-\infty$ elements. However, this solution is not optimal since faster algorithms are known [22,21]. The currently fastest algorithm is due to Takaoka who

---

  [*] Supported in part by an Ole Roemer Scholarship from the Danish National Science Research Council.

 [**] Basic Research in Computer Science, research school.

[***] Center for Massive Data Algorithmics, a Center of the Danish National Research Foundation.

S.-H. Hong, H. Nagamochi, and T. Fukunaga (Eds.): ISAAC 2008, LNCS 5369, pp. 100–111, 2008.

describes an $O(m^2 n \sqrt{\log \log m / \log m})$ time algorithm in [21].[1] The only known lower bound for the problem is the trivial $\Omega(mn)$ bound. The two-dimensional version was the first problem studied, introduced as a method for maximum likelihood estimations of patterns in digitized images [5].

A natural extension of the maximum sum problem, introduced in [4], is to compute the k largest sums for $1 \le k \le \binom{n}{2} + n$. The subarrays are allowed to overlap, and the output is k triples of the form $(i, j, \sum_{t=i}^{j} A[t])$. An $O(nk)$ time algorithm is given in [4]. An algorithm solving the problem in optimal $O(n + k)$ time using $O(k)$ additional space is described in [7].

Another generalization of the maximal sum problem is to restrict the length of the subarrays considered. This generalization is considered in [15,18,9] mainly motivated by applications in Bioinformatics such as finding tandem repeats [23], locating GC-rich regions [14], and constructing low complexity filters for sequence database search [2]. In [15] Huang describes an $O(n)$ time algorithm locating the largest sum of length at least l, while in [18] an $O(n)$ time algorithm locating the largest sum of length at most u is described. The algorithms can be combined into at linear time algorithm finding the largest sum of length at least l and at most u [18]. In [9] it is shown how to solve the problem in $O(n)$ time when the input elements are given online one by one.

The *length constrained k maximal sums* problem is defined as follows. Given an array A of length n, find the k largest sums consisting of at least l and at most u numbers. The k maximal sums problem is the special case of this problem where $l = 1$ and $u = n$. Lin and Lee solved the problem using a randomized algorithm with an expected running time of $O(n \log(u - l) + k)$ [17]. Their algorithm is based on a randomized algorithm that selects the k'th largest length constrained sum from an array in $O(n \log(u - l))$ expected time. The authors state as an open problem whether this is optimal. Furthermore, in [16] Lin and Lee describe a deterministic $O(n \log n)$ time algorithm that selects the k'th largest sum in an array of size n. They propose as an open problem whether this bound is tight. This problem is known as the *sum selection* problem.

Our Contribution. In this paper we settle the time complexity for the sum selection problem and the length constrained k maximal sums problem. First, we describe an optimal $O(n + k)$ time algorithm for the length constrained k maximal sums problem in Section 2. This algorithm is an extension of our optimal algorithm solving the k maximal sums problem from [7]. Secondly, we prove a time bound of $\Theta(n \log(k/n))$ for the sum selection problem in Section 3. This is the main result of the paper. An $O(n \log(k/n))$ time algorithm that selects the k'th largest sum is described in Section 3.1, and in Section 3.2 we prove a matching lower bound using a reduction from the cartesian sum problem [11]. Finally, in Section 4 we combine the ideas from the two algorithms we have designed and obtain an $O(n \log(k/n))$ time algorithm that selects the k'th largest sum among all sums consisting of at least l and at most u numbers. This bound

---

[1] For simplicity of exposition by $\log x$ we denote the value $\max\{1, \log_2 x\}$.

Table 1. Overview of results on reporting and selecting sums in arrays

Problem	Previous Work	This Paper
Length Const. k Maximal Sums	$O(n \log(u - l) + k)$ exp. [17]	$O(n + k)$
Sum Selection	$O(n \log n)$ [16]	$\Theta(n \log(k/n))$
Length Const. Sum Selection	$O(n \log(u - l))$ exp. [17]	$O(n \log(k/n))$

is always as good as the previous randomized bound of $O(n \log(u - l))$ by Lin and Lee [17], since there are $\sum_{t=l}^{u} n - t + 1 \leq n(u - l + 1)$ subarrays of length between l and u in an array of size n and thus $k/n \leq u - l + 1$. Due to lack of space the details are deferred to the full version which will combine the results of this paper and the results in [7]. The results are summarized in Table 1.

2 The Length Constrained k Maximal Sums Problem

In this section we present an optimal $O(n + k)$ time algorithm that reports the k largest sums of an array A of length n with the restriction that each sum is an aggregate of at least l and at most u numbers. We reuse the idea from the k maximal sums algorithm in [7], and construct a heap[2] that implicitly represents all $\sum_{t=l}^{u} n - t + 1 = O(n(u - l))$ valid sums from the input array using only $O(n)$ time and space. The k largest sums are then retrieved from the heap using Fredericksons heap selection algorithm [10] that extracts the k largest elements from a heap in $O(k)$ time. We note that the k maximal sums algorithm from [7] can be altered to use a heap supporting deletions to obtain an $O(n \log(u - l) + k)$ algorithm solving the problem without randomization. The difference between our new $O(n + k)$ time algorithm and the algorithm solving the k maximal sums problem [7] is in the way the sums are grouped in heaps. This change enables us to solve the problem without deleting elements from a heap. In the following we assume that $l < u$. If $l = u$ the problem can be solved in $O(n)$ time using a linear time selection algorithm [6].

2.1 A Linear Time Algorithm

For each array index j, for $j = 1, \ldots, n - l + 1$, we build data structures representing all sums of length between l and u ending at index $j + l - 1$. This is achieved by constructing all sums ending at $A[j]$ with length between 1 and $u - l + 1$, and then adding the sum of the $l - 1$ elements, $A[j + 1, \ldots, j + l - 1]$, following $A[j]$ in the input array to each sum. To construct these data structures efficiently, the input array is divided into *slabs* of $w = u - l$ consecutive elements, and the sums are grouped in disjoint sets, $\hat{Q}_j$ and $\bar{Q}_j$ for $j = 1, \ldots, n$, depending on the slab boundaries.

---

[2] For simplicity of exposition, by heap we denote a heap ordered binary tree where the largest element is placed at the root.

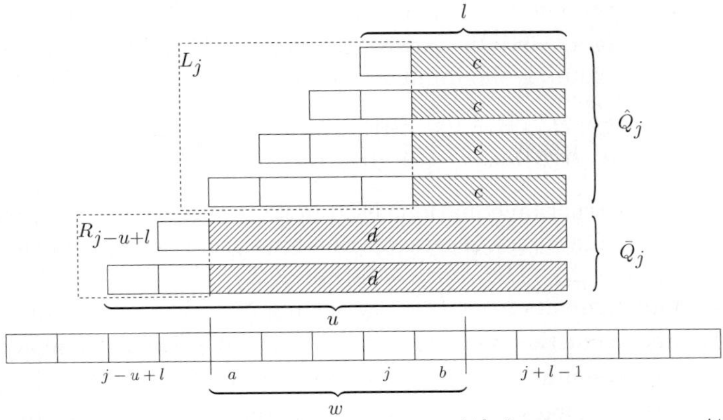

Fig. 1. Overview of the sets, $l = 4, u = 9, c = \sum_{t=j+1}^{j+l-1} A[t]$ and $d = \sum_{t=a}^{j+l-1} A[t]$. A slab is starting at index a and ending at index b.

Let a be the first index in the slab containing index j, i.e. $a = 1 + \lfloor \frac{j-1}{w} \rfloor w$. The set $\hat{Q}_j$ contains all sums of length between l and u ending at index $j + l - 1$ that start in the slab containing index j and is defined as follows:

$$\hat{Q}_j = \{(i, j + l - 1, sum) \mid a \leq i \leq j, \ sum = c + \textstyle\sum_{t=i}^{j} A[t]\} \ ,$$

where $c = \sum_{t=j+1}^{j+l-1} A[t]$ is the sum of $l - 1$ numbers in $A[j+1, \ldots, j+l-1]$. The set $\bar{Q}_j$ contains the $(u - l + 1) - (j - a + 1) = u - l - j + a$ valid sums ending at index $j + l - 1$ that start to the left of index a, thus:

$$\bar{Q}_j = \{(i, j + l - 1, sum) \mid j - u + l \leq i < a, \ sum = d + \textstyle\sum_{t=i}^{a-1} A[t]\} \ ,$$

where $d = \sum_{t=a}^{j+l-1} A[t]$ is the result of summing the $j - a + l$ numbers in $A[a, \ldots, j + l - 1]$. The sets are illustrated in Figure 1. By construction, the sets $\hat{Q}_j$ and $\bar{Q}_j$ are disjoint and their union is the $u - l + 1$ sums of length between l and u ending at index $j + l - 1$.

With the sets of sums defined we continue with the representation of these. The sets $\hat{Q}_j$ and $\bar{Q}_j$ are represented by pairs $\langle \hat{\delta}_j, \hat{H}_j \rangle$ and $\langle \bar{\delta}_j, \bar{H}_j \rangle$ where $\hat{H}_j$ and $\bar{H}_j$ are partially persistent heaps and $\hat{\delta}_j$ and $\bar{\delta}_j$ are constants that must be added to all elements in $\hat{H}_j$ and $\bar{H}_j$ respectively to obtain the correct sums. For the heaps we use the Iheap from [7] which supports insertions in amortized constant time. Partial persistence is implemented using the node copying technique [8].

We construct representations of two sequences of sets, L_j and R_j for $j = 1, \ldots, n$, that depend on the slab boundaries. Consider the slab $A[a, \ldots, j, \ldots, b]$ containing index j. The set L_j contains the $j - a + 1$ sums ending at $A[j]$ that start between index a and j. The set R_j contains the $b - j + 1$ sums ending at $A[b]$ starting between index j and b, see Figure 1.

Each set L_j is represented as a pair $\langle \delta_j^L, H_j^L \rangle$ where δ_j^L is an additive constant as above and H_j^L is a partially persistent Iheap. The pairs are incrementally constructed while scanning the input array from left to right as follows:

$$\begin{aligned} \langle \delta_a^L, H_a^L \rangle &= \langle A[a], \{0\} \rangle \wedge \\ \langle \delta_j^L, H_j^L \rangle &= \langle \delta_{j-1}^L + A[j], H_{j-1}^L \cup \{-\delta_{j-1}^L\} \rangle \,. \end{aligned}$$

This is also the construction equations used in [7]. Constructing a representation of L_a is simple, and creating a representation for L_j can be done efficiently given a representation of L_{j-1}. The representation of L_j is constructed by implicitly adding $A[j]$ to all elements from L_{j-1} by setting $\delta_j^L = \delta_{j-1}^L + A[j]$ and inserting an element to represent the sum $A[j]$. Since $\delta_{j-1}^L + A[j]$ needs to be added to all elements in the representation of L_j, an element with $-\delta_{j-1}^L$ as key is inserted into H_{j-1}^L, yielding H_j^L ending the construction. Partial persistence ensures that the Iheap H_{j-1}^L used to represent L_{j-1} is not destroyed. By the above description and the cost of applying the node copying technique [8] the amortized time needed to construct a pair is $O(1)$.

The R_j sets are represented by partially persistent Iheaps H_j^R, and these representations are built by scanning the input array from right to left. We get the following incremental construction equations:

$$\begin{aligned} H_b^R &= \{A[b]\} \wedge \\ H_j^R &= H_{j+1}^R \cup \{\textstyle\sum_{t=j}^b A[t]\} \,. \end{aligned}$$

Similar to the $\langle \delta_j^L, H_j^L \rangle$ pairs, constructing a partial persistent Iheap H_j^R also takes $O(1)$ time amortized. Therefore, the time needed to build the representation of the $2n$ sets L_j and R_j for $j = 1, \ldots, n$ is $O(n)$.

We represent the sets $\hat{Q}_j$ and $\bar{Q}_j$ using the representations of the sets L_j and R_{j-u+l}. Figure 1 illustrates the correspondence between $\hat{Q}_j$ and L_j and $\bar{Q}_j$ and R_{j-u+l}. Consider any index $j \in \{1, \ldots, n-l+1\}$, and let $A[a, \ldots, j, \ldots, b]$ be the current slab containing j. The set $\hat{Q}_j$ contains the sums of length between l and u that start in the current slab and end at index $j+l-1$. The set L_j contains the $j - a + 1$ sums that start in the current slab and end at $A[j]$. Therefore, adding the sum of the $l - 1$ numbers in $A[j+1, \ldots, j+l-1]$ to each element in L_j gives $\hat{Q}_j$ and thus:

$$\hat{Q}_j = \langle c + \delta_j^L, H_j^L \rangle \,,$$

where $c = \sum_{t=j+1}^{j+l-1} A[t]$.

Similarly, the set $\bar{Q}_j$ contains the $u - l + 1 - (j - a + 1) = u - l - j + a$ sums of length between l and u ending at $A[j+l-1]$ starting in the previous slab. The set R_{j-u+l} contains the $u - l - j - a$ shortest sums ending at the last index in the previous slab. Therefore, adding the sum of the $j + l - a$ numbers in $A[a, \ldots, j+l-1]$ to each element in R_{j-u+l} gives $\bar{Q}_j$ and thus:

$$\bar{Q}_j = \langle d, H_{j-u+l}^R \rangle \,,$$

where $d = \sum_{t=a}^{j+l-1} A[t]$.

Lemma 1. *Constructing the $2(n - l + 1)$ pairs that represent $\hat{Q}_j$ and $\bar{Q}_j$ for $j = 1, \ldots, n - l + 1$ takes $O(n)$ time.*

Proof. Constructing all $\langle \delta_j^L, H_j^L \rangle$ pairs and all H_j^R partial persistent Iheaps takes $O(n)$ time, and calculating sums c and d takes constant time using a prefix array. Constructing the prefix array takes $O(n)$ time. Therefore, constructing $\hat{Q}_j$ and $\bar{Q}_j$ for $j = 1, \ldots, n - l + 1$ takes $O(n)$ time. $\qquad\square$

After constructing the $2(n - l + 1)$ pairs, they are assembled into one large heap using $2(n - l + 1) - 1$ dummy ∞ keys as in [7]. The largest $2(n - l + 1) - 1 + k$ elements are then extracted from the assembled heap in $O(n + k)$ time using Fredericksons heap selection algorithm. The implicit sums given by adding δ values are explicitly computed while Fredericksons algorithm explores the final heap top down in the way described in [7]. The $2(n - l + 1) - 1$ dummy elements are discarded.

Theorem 1. *The algorithm described reports the k largest sums with length between l and u in an array of length n in $O(n + k)$ time.*

3   Sum Selection Problem

In this section we prove a $\Theta(n \log(k/n))$ time bound for the sum selection problem by designing an $O(n \log(k/n))$ time algorithm that selects the k'th largest sum in an array of size n and by proving a matching lower bound.

The idea of the algorithm is to reduce the problem to selection in a collection of sorted arrays and weight balanced search trees [19,3]. The trees and the sorted arrays are constructed using the ideas from Section 2 and [7]. Selecting the k'th largest element from a set of trees and sorted arrays is done using an essential part of the sorted column matrix selection algorithm of Frederickson and Johnson [11]. The part of Frederickson and Johnsons algorithm that we use is an iterative procedure named *Reduce*. In a round of the Reduce algorithm each array, A, is represented by the $1 + \lfloor \alpha|A| \rfloor$ largest element stored in the array, and a constant fraction of the elements in each array may be eliminated. This can be approximated in weight balanced search trees and the complexity analysis from [11] remains valid.

The lower bound is proved using a reduction from the $X + Y$ cartesian sum selection problem [11].

We note that if $k \leq n$ then the k maximal sums algorithm from [7] can be used to solve the problem optimally in $O(n)$ time.

To construct the sorted arrays efficiently, we use a heap data structure, that is a generalization of the Iheap, which we name *Bheap*. The Bheap is a heap ordered binary tree where each node of the tree contains a sorted array of size B. By heap order, we mean that all elements in a child of a node v must be smaller than the smallest element stored in v. Sorted arrays of B elements are required to be inserted in $O(B)$ time amortized. Our Bheap implementation is based on ideas from the functional random access lists in [20] and simple bubble up/down procedures based on merging sorted arrays.

3.1   An $O(n \log(k/n))$ Time Algorithm

In this section we reduce the sum selection problem to selection in a set of trees and sorted arrays. We use the weight balanced B-trees of Arge and Vitter [3] with degree $B = O(1)$. Similar to the grouping of sums in Section 2, each index j, for $j = 1, \ldots, n$, is associated with data structures representing all possible sums ending at $A[j]$. The set representing all sums ending at index j is defined as follows:

$$Q_j = \left\{ (i, j, sum) \mid 1 \leq i \leq j,\ sum = \sum_{t=i}^{j} A[t] \right\} .$$

The input array is divided into slabs of size $w = \lceil k/n \rceil$, and the set Q_j is represented by two disjoint sets WB_j and BH_j that depend on the slab boundaries. The set WB_j contains the sums ending at index j beginning in the current slab, and BH_j contains the sums ending at index j not beginning in the current slab. Let $a = 1 + \lfloor \frac{j-1}{w} \rfloor w$, i.e. the first index in the slab containing index j, then:

$$WB_j = \left\{ (i, j, sum) \mid a \leq i \leq j,\ sum = \sum_{t=i}^{j} A[t] \right\} \wedge$$
$$BH_j = \left\{ (i, j, sum) \mid 1 \leq i < a,\ sum = c + \sum_{t=i}^{a-1} A[t] \right\} ,$$

where $c = \sum_{t=a}^{j} A[t]$ is the sum of the $j - a + 1$ numbers in $A[a, \ldots, j]$. The sets WB_j and BH_j are disjoint, and $WB_j \cup BH_j = Q_j$ by construction. The sets are illustrated in Figure 2.

The set WB_j is represented as a pair $\langle \tau_j, T_j \rangle$ where T_j is a partial persistent weight balanced B-tree and τ_j is an additive constant that must be added to all elements in T_j to obtain the correct sums. The set BH_j is represented as a pair $\langle \delta_j, H_j \rangle$ where δ_j is an additive constant and H_j is a partial persistent Bheap with $B = w$.

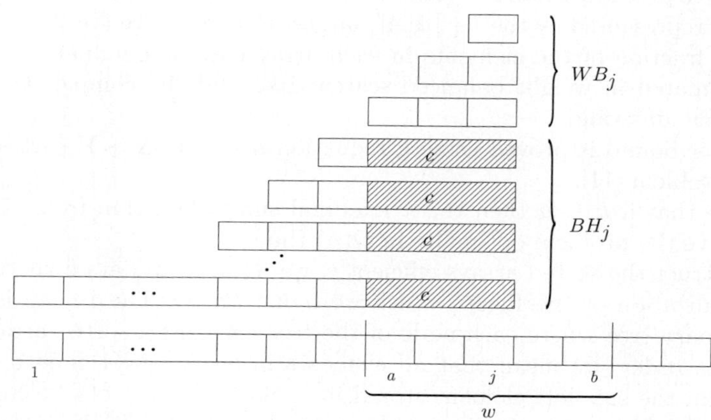

Fig. 2. Overview of the sets. Slab size $w = 5$, and $A[a, \ldots, b]$ is the slab containing index j.

The pairs $\langle \tau_j, T_j \rangle$ are constructed as follows. If j is the first index of a slab, i.e. $j = 1 + tw$ for some natural number t, then:

$$\langle \tau_j, T_j \rangle = \langle A[j], \{0\} \rangle \ .$$

This is the start of a new slab, and a new partial persistent weight balanced B-tree representing $A[j]$, the first element in the slab, is created. If j is not the first index in a slab then:

$$\langle \tau_j, T_j \rangle = \langle \tau_{j-1} + A[j], T_{j-1} \cup \{-\tau_{j-1}\} \rangle \ ,$$

i.e. we change the additive constant and insert $-\tau_{j-1}$ into the weight balanced tree T_{j-1}. These construction equations are identical to the construction equations from Section 2, and partial persistence ensures that T_{j-1} is not destroyed by constructing T_j.

For the $\langle \delta_j, H_j \rangle$ pairs representing the sets $\hat{Q}_j$, we observe that if $j \leq w$ then $BH_j = \emptyset$, thus:

$$\langle \delta_j, H_j \rangle = \langle 0, \emptyset \rangle \ .$$

If $j > w$ and j is not the first index in a slab, then adding $A[j]$ to all elements from the previous set yields the new set, thus:

$$\langle \delta_j, H_j \rangle = \langle \delta_{j-1} + A[j], H_{j-1} \rangle \ .$$

If j is the first index of a slab, i.e. $j = 1 + tw$ for some integer $t \geq 1$, all w sums represented in $\langle \tau_{j-1}, T_{j-1} \rangle$ are inserted into a sorted array S and each sum explicitly calculated. This sorted array then contains all sums starting in the previous slab ending at index $j - 1$. For each element in S the additive constant δ_{j-1} is subtracted and S is inserted into the Bheap H_{j-1}. The construction equation becomes:

$$\langle \delta_j, H_j \rangle = \langle \delta_{j-1} + A[j], H_{j-1} \cup S \rangle \ ,$$

where

$$S = \left\{ (i, j, s - \delta_{j-1}) \mid j - w \leq i < j, s = \textstyle\sum_{t=i}^{j-1} A[t] \right\} \ .$$

Again, partial persistence ensures that the previous version of the Bheap, H_{j-1}, is not destroyed.

Lemma 2. *Constructing the pairs $\langle \delta_j, H_j \rangle$ and $\langle \tau_j, T_j \rangle$ for $j = 1, \ldots, n$ takes $O(n \log(k/n))$ time.*

Proof. The Bheap and the weight balanced B-trees have constant in and out-degree. Therefore, partial persistence can be implemented for both using the node copying technique [8].

For the Bheap, amortized $O(1)$ pointers and arrays are changed per insertion. The extra cost for applying the node copying technique is $O(B) = O(w)$ time amortized per insert operation. Constructing the sorted array S from a weight balanced B-tree takes $O(w)$ time. An insert in a Bheap is only performed every w'th step, and calculating additive constants in each step takes constant time. Therefore, the time used for constructing all $\langle \delta_j, H_j \rangle$ pairs is $O(n + \frac{n}{w} w) = O(n)$.

Each insert in a weight balanced B-tree is performed on a tree containing at most w elements using $O(\log w)$ time. Therefore, the extra cost of using the node copying technique is $O(\log w)$ time amortized per insert operation. Calculating an additive constant τ_j takes constant time, thus, constructing all $\langle \tau_j, T_j \rangle$ pairs takes $O(n \log(k/n))$ time. $\qquad\square$

After the n pairs, $\langle \delta_i, H_i \rangle$, storing Bheaps are constructed, they are assembled into one large heap in the same way as in Section 2. That is, we construct a complete heap on top of the pairs using $n - 1$ dummy nodes storing the same array of w dummy ∞ elements. We then use Fredericksons heap selection algorithm in the same way as in Section 2 where the representative for each node is the smallest element in the sorted array stored in it. Using Fredericksons heap selection algorithm the $2n - 1$ nodes with the maximal smallest element and their $2n$ children are extracted. This takes $O(n)$ time and the nodes extracted from the Bheap gives $3n$ sorted arrays by discarding the $n - 1$ dummy nodes.

Lemma 3. *The $3n$ nodes found as described above contain the k largest sums contained in the n pairs $\langle \delta_i, H_i \rangle$.*

Proof. The $4n - 1$ nodes found by the heap selection algorithm forms a connected subtree T of the heap rooted at the root of the heap. Any element e stored in a node $v_e \notin T$ is smaller than all elements stored in any internal node $v \in T$ since, by heap order, e is smaller than the smallest element in the leaf of T that is on the path from v_e to the root. The smallest element in a leaf is smaller than the smallest element in any internal node since the leaf was not picked by the heap selection algorithm. There are $2n - 1$ internal nodes in T and n of these does not store dummy elements. Therefore, for each element not residing in T there at least $nw = n\lceil \frac{k}{n} \rceil \geq k$ larger elements in the $3n$ found nodes. $\qquad\square$

These $3n$ sorted arrays of size w and the n pairs $\langle \tau_i, T_i \rangle$ storing weight balanced B-trees of size at most w contain at most $4nw = 4n\lceil \frac{k}{n} \rceil \leq 4(k + n)$ sums. The $3n$ arrays and the n weight balanced B-trees are given as input to the adapted sorted column matrix selection algorithm, which extracts the k'th largest element from these in $O(n \log(k/n))$ time. The fact that the weight balanced B-trees are partially persistent versions of the same tree and contain additive constants is handled by expanding the trees and computing the sums explicitly during the top down traversals performed by the selection algorithm as in Section 2 and [7].

Theorem 2. *The algorithm described selects the k'th largest sum in an array of size n in $O(n \log(k/n))$ time.*

3.2   Lower Bound

In this section we prove a matching lower bound of $\Omega(n \log(k/n))$ time for the sum selection problem via a reduction from the $X + Y$ cartesian sum selection problem. In the $X + Y$ cartesian sum selection problem as defined in [11], the input is two unsorted arrays X and Y and an integer k, and the task is to select the k'th largest element in the cartesian sum $\{x + y \mid x \in X, y \in Y\}$.

Given an instance of the $X + Y$ cartesian sum selection problem, $X = \{x_1, \ldots, x_n\}$, $Y = \{y_1, \ldots, y_m\}$, and k, construct the following array A:

$x_1 - x_2$	$\cdots$	$x_i - x_{i+1}$	$\cdots$	$x_{n-1} - x_n$	$x_n + \infty + y_1$	$y_2 - y_1$	$\cdots$	$y_m - y_{m-1}$

where ∞ is a number larger than $(n + m) \cdot \max\{|x| \mid x \in X\} \cup \{|y| \mid y \in Y\}$. The sums in A have the following properties:

- A sum ranging from i to j where $i \leq n \leq j$ represents the sum $(\sum_{t=i}^{n-1} A[t]) + x_n + \infty + y_1 + (\sum_{t=n+1}^{j} A[t]) = x_i + y_{j-n+1} + \infty$.
- A sum including $A[n] = x_n + \infty + y_1$ is larger than any sum that does not

There are more sums in the sum selection instance than there are in the $X+Y$ cartesian sum instance since any sum not containing $A[n]$ does not correspond to an element in the cartesian sum. However, the k'th largest sum does contain $A[n]$ and corresponds to the k'th largest sum in the cartesian sum instance. Therefore, any algorithm that selects the k'th largest sum in an array can be used to select the k'th largest element in the cartesian sum.

The lower bound for selecting the k'th largest element in the cartesian sum $(X + Y)$ is $\Omega(m + p \log(k/p))$ comparisons where $|X| = n, |Y| = m$ with $n \leq m$ and $p = \min\{k, m\}$ [11]. In the reduction the size of the array A is $n + m - 1$, which is $\Theta(n + m) = \Theta(m)$, and it can be built in $O(m)$ time.

Theorem 3. *Any algorithm that selects the k'th largest sum in an array of size n uses $\Omega(n \log(k/n))$ comparisons.*

4   Length Constrained Sum Selection

In this section we sketch how to select the k'th largest sum consisting of at least l and at most u numbers from an array of size n in $O(n \log(k/n))$ time. The algorithm combines the ideas from Section 2 and Section 3. Similar to Section 3 the algorithm works by reducing the problem to selection in a collection of weight balanced search trees and sorted arrays. It should be noted that a deterministic algorithm with running time $O(n \log(u - l))$ can be achieved by using weight balanced B-trees instead of Iheaps in the algorithm from Section 2, and using these as input to the adapted sorted column matrix selection algorithm instead of the heap selection algorithm.

To achieve $O(n \log(k/n))$ time, we constrain the lengths of the sums considered and divide the input array into slabs of size $u - l$ as in Section 2. Subsequently, we efficiently construct representations of the sets $\hat{Q}_j$ and $\bar{Q}_j$ defined in Section 2 using weight balanced trees and Bheaps by subdividing each slab into sub-slabs of size $\lceil \frac{k}{n} \rceil$ as in Section 3, recall $k/n \leq u - l + 1$. Weight balanced B-trees are used to represent sums residing inside a sub-slab, and Bheaps are used to represent sums covering multiple sub-slabs. The sums are illustrated in Figure 3. The Bheaps and the weight balanced B-trees are constructed efficiently as in Section 3 using partial persistence.

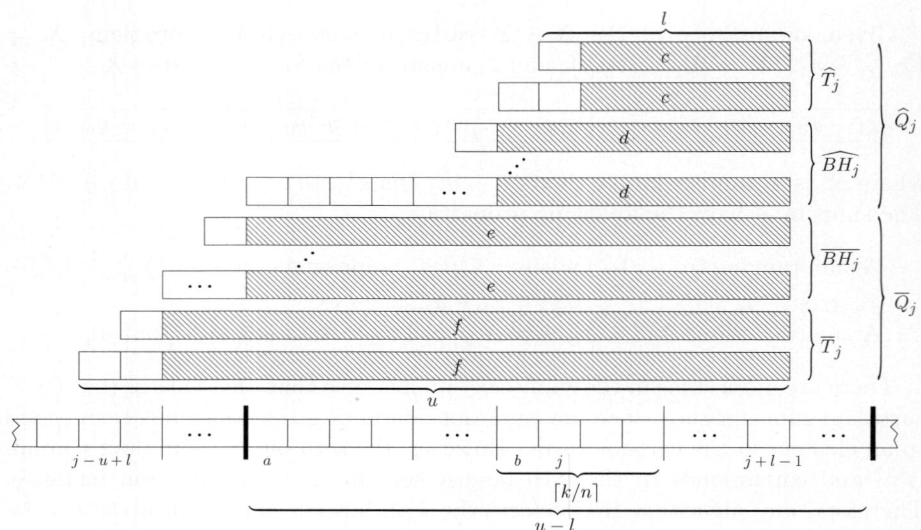

Fig. 3. Combining ideas - The sums associated with index j. A new slab of length $u - l$ starts at index a and a new subslab of length $\lceil k/n \rceil = 4$ starts at index b. $c = \sum_{t=j+1}^{j+l-1} A[t]$, $d = \sum_{t=b}^{j+l-1} A[t]$, $e = \sum_{t=a}^{j+l-1} A[t]$ and $f = \sum_{t=x}^{j+l-1} A[t]$ where x is the first index in the subslab following the subslab containing index $j - u + l$. The set $\hat{Q}_j$ is split into $\hat{T}_j$, represented by a weight balanced tree, and $\widehat{BH_j}$, represented by a Bheap. The set $\bar{Q}_j$ is split similarly.

After the representations of the sets $\hat{Q}_j$ and $\bar{Q}_j$ are constructed, the algorithm continues as in Section 3. The sorted arrays storing the k largest sums stored in the Bheaps are extracted using Fredericksons heap selection algorithm. The sorted arrays and the weight balanced B-trees are then given as input to the adapted sorted column matrix selection algorithm that selects the k'th largest sum.

Theorem 4. *The k'th largest sum of length between l and u in an array of size n can be selected in $O(n \log(k/n))$ time.*

References

1. Allison, L.: Longest biased interval and longest non-negative sum interval. Bioinformatics 19(10), 1294–1295 (2003)
2. Altschul, S.F., Gish, W., Miller, W., Myers, E.W., Lipman, D.J.: Basic local alignment search tool. Journal of Molecular Biology 215(3), 403–410 (1990)
3. Arge, L., Vitter, J.S.: Optimal external memory interval management. SIAM Journal on Computing 32(6), 1488–1508 (2003)
4. Bae, S.E., Takaoka, T.: Algorithms for the problem of k maximum sums and a vlsi algorithm for the k maximum subarrays problem. In: Proc. 7th International Symposium on Parallel Architectures, Algorithms, and Networks, pp. 247–253. IEEE Computer Society, Los Alamitos (2004)
5. Bentley, J.: Programming pearls: algorithm design techniques. Commun. ACM 27(9), 865–873 (1984)

6. Blum, M., Floyd, R.W., Pratt, V.R., Rivest, R.L., Tarjan, R.E.: Time bounds for selection. J. Comput. Syst. Sci. 7(4), 448–461 (1973)
7. Brodal, G.S., Jørgensen, A.G.: A linear time algorithm for the k maximal sums problem. In: Kučera, L., Kučera, A. (eds.) MFCS 2007. LNCS, vol. 4708, pp. 442–453. Springer, Heidelberg (2007)
8. Driscoll, J.R., Sarnak, N., Sleator, D.D., Tarjan, R.E.: Making data structures persistent. Journal of Computer and System Sciences 38(1), 86–124 (1989)
9. Fan, T.-H., Lee, S., Lu, H.-I., Tsou, T.-S., Wang, T.-C., Yao, A.: An optimal algorithm for maximum-sum segment and its application in bioinformatics. In: Proc. 8th International Conference on Implementation and Application of Automata. LNCS, pp. 46–66. Springer, Heidelberg (2003)
10. Frederickson, G.N.: An optimal algorithm for selection in a min-heap. Inf. Comput. 104(2), 197–214 (1993)
11. Frederickson, G.N., Johnson, D.B.: The complexity of selection and ranking in X+Y and matrices with sorted columns. J. Comput. Syst. Sci. 24(2), 197–208 (1982)
12. Fukuda, T., Morimoto, Y., Morishita, S., Tokuyama, T.: Data mining with optimized two-dimensional association rules. ACM Trans. Database Syst. 26(2), 179–213 (2001)
13. Fukuda, T., Morimoto, Y., Morishta, S., Tokuyama, T.: Interval finding and its application to data mining. IEICE Transactions on Fundamentals of Electronics, Communications and Computer Sciences E80-A(4), 620–626 (1997)
14. Hannenhalli, S., Levy, S.: Promoter prediction in the human genome. Bioinformatics 17, S90–S96 (2001)
15. Huang, X.: An algorithm for identifying regions of a DNA sequence that satisfy a content requirement. Computer Applications in the Biosciences 10(3), 219–225 (1994)
16. Lin, T.-C., Lee, D.T.: Efficient algorithms for the sum selection problem and k maximum sums problem. In: The 17th International Symposium on Algorithms and Computation. LNCS, pp. 460–473. Springer, Heidelberg (2006)
17. Lin, T.-C., Lee, D.T.: Randomized algorithm for the sum selection problem. Theor. Comput. Sci. 377(1-3), 151–156 (2007)
18. Lin, Y.-L., Jiang, T., Chao, K.-M.: Efficient algorithms for locating the length-constrained heaviest segments, with applications to biomolecular sequence analysis. In: Proc. 27th International Symposium of Mathematical Foundations of Computer Science 2002. LNCS, pp. 459–470. Springer, Heidelberg (2002)
19. Nievergelt, J., Reingold, E.M.: Binary search trees of bounded balance. In: STOC 1972: Proceedings of the fourth annual ACM symposium on Theory of computing, pp. 137–142. ACM, New York (1972)
20. Okasaki, C.: Purely functional random-access lists. In: Functional Programming Languages and Computer Architecture, pp. 86–95 (1995)
21. Takaoka, T.: Efficient algorithms for the maximum subarray problem by distance matrix multiplication. Electr. Notes Theor. Comput. Sci. 61 (2002)
22. Tamaki, H., Tokuyama, T.: Algorithms for the maximum subarray problem based on matrix multiplication. In: Proceedings of the ninth annual ACM-SIAM symposium on Discrete algorithms, pp. 446–452. Society for Industrial and Applied Mathematics, Philadelphia (1998)
23. Walder, R.Y., Garrett, M.R., McClain, A.M., Beck, G.E., Brennan, T.M., Kramer, N.A., Kanis, A.B., Mark, A.L., Rapp, J.P., Sheffield, V.C.: Short tandem repeat polymorphic markers for the rat genome from marker-selected libraries. Mammalian Genome 9(12), 1013–1021 (1998)

Succinct and I/O Efficient Data Structures for Traversal in Trees[*]

Craig Dillabaugh, Meng He, and Anil Maheshwari

School of Computer Science, Carleton University, Ottawa, Ontario, Canada

Abstract. We present two results for path traversal in trees, where the traversal is performed in an asymptotically optimal number of I/Os and the tree structure is represented succinctly. Our first result is for bottom-up traversal that starts with a node in the tree T and traverses a path to the root. For blocks of size B, a tree on N nodes, and for a path of length K, we design data structures that permit traversal of the bottom-up path in $O(K/B)$ I/Os using only $2N + \frac{\epsilon N}{\log_B N} + o(N)$ bits, for an arbitrarily selected constant, ϵ, where $0 < \epsilon < 1$. Our second result is for top-down traversal in binary trees. We store T using $(3 + q)N + o(N)$ bits, where q is the number of bits required to store a key, while top-down traversal can still be performed in an asymptotically optimal number of I/Os.

1 Introduction

Many operations on graphs and trees can be viewed as the traversal of a path. Queries on trees, for example, typically involve traversing a path from the root to some node, or from some node to the root. Often the datasets represented in graphs and trees are too large to fit in internal memory and traversal must be performed efficiently in external memory (EM). Efficient EM traversal in trees is important for structures such as suffix trees, and as a building block to graph searching and shortest path algorithms.

Succinct data structures were first proposed by Jacobson [1]. The idea is to represent data structures using space as near the information-theoretical lower bound as possible, while allowing efficient navigation. Succinct data structures, which have been studied largely outside the external memory model, also have natural application to large data sets.

In this paper, we present data structures for traversal in trees that are both efficient in the EM setting, and that encode the trees succinctly. We are aware of only the work by Chien *et al.* [2] on succinct full-text indices supporting efficient substring search in EM, that follows the same track. Our contribution is the first such work on general trees that bridges these two techniques.

Previous Work: The I/O-model [3] splits memory into two levels; the fast, but finite, internal memory; and slow but infinite EM. Data is transferred between these levels by an input-output operation (I/O). Algorithms are analyzed

---

[*] Research supported by NSERC.

S.-H. Hong, H. Nagamochi, and T. Fukunaga (Eds.): ISAAC 2008, LNCS 5369, pp. 112–123, 2008.

in terms of the number of I/O operations. The unit of memory that may be transferred in a single I/O is referred to as a *disk block*. The parameters B, M, and N denote the size (in terms of the number of data elements) of a block, internal memory, and the problem instance. *Blocking* of data structures in the I/O model has reference to the partitioning of the data into individual blocks that can subsequently be transferred with a single I/O.

Nodine *et al.* [4] studied the problem of blocking graphs and trees, for efficient traversal, in the I/O model. Among their main results they presented a bound of $\Theta(K/\log_d B)$ for d-ary trees where on average each vertex may be represented twice. Blocking of bounded degree planar graphs, such as Triangular Irregular Networks (TINs), was examined in Aggarwal *et al.* [5]. The authors show how to store a planar graph of size N, and of bounded degree d, in $O(N/B)$ blocks so that any path of length K can be traversed using $O(K/\log_d B)$ I/Os.

Hutchinson *et al.* [6] examined the case of bottom-up traversal, where the path begins with some node in T and proceeds to the root. They gave a blocking which supports bottom-up traversal in $O(K/B)$ I/Os when the tree is stored in $O(N/B)$ blocks. The case of top down traversal has been more extensively studied. Clark and Munro [7] describe a blocking layout that yields a logarithmic bound for root-to-leaf traversal in suffix trees. Given a fixed independent probability on the leaves, Gil and Itai [8], presented a blocking layout that yields the minimum expected number of I/Os on a root to leaf path. In the cache oblivious model, Alstrup *et al.* [9] gave a layout that yields a minimum worst case, or expected number of I/Os, along a root-to-leaf path, up to constant factors. Demaine *et al.* [10] presented an optimal blocking strategy to support top down traversals. They showed that for a binary tree T, a traversal from the root to a node of depth K requires the following number of I/Os: (1) $\Theta(K/\lg(1+B))$, when $K = O(\lg N)$, (2) $\Theta(\lg N/(\lg(1 + B \lg N/K)))$, when $K = \Omega(\lg N)$ and $K = O(B \lg N)$, and (3) $\Theta(K/B)$, when $K = \Omega(B \lg N)$. Finally, Brodal and Fagerberg [11] describe the giraffe-tree, which likewise permits a $O(K/B)$ root-to-leaf tree traversal with $O(N)$ space in the cache-oblivious model.

Our Results: Throughout this paper we assume that $B = \Omega(\lg N)$ (i.e. the disk block is of reasonable size). Our paper presents two main results:

1. In Section 3, we show how a tree T can be blocked in a succinct fashion such that a bottom-up traversal requires $O(K/B)$ I/Os using only $2N + \frac{\epsilon N}{\log_B N} + o(N)$ bits to store T, where K is the path length and $0 < \epsilon < 1$. This technique is based on [6], and achieves an improvement on the space bound by a factor of $\lg N$.

2. In Section 4, we show that a binary tree, with keys of size $q = O(\lg N)$ bits, can be stored using $(3+q)N+o(N)$ bits so that a root-to-node path of length K can be reported with: (a) $O\left(\frac{K}{\lg(1+(B \lg N)/q)}\right)$ I/Os, when $K = O(\lg N)$; (b) $O\left(\frac{\lg N}{\lg(1+\frac{B \lg^2 N}{qK})}\right)$ I/Os, when $K = \Omega(\lg N)$ and $K = O\left(\frac{B \lg^2 N}{q}\right)$; and (c) $O\left(\frac{qK}{B \lg N}\right)$ I/Os, when $K = \Omega\left(\frac{B \lg^2 N}{q}\right)$. This result achieves a $\lg N$ factor improvement on the previous space cost in [10] for the tree structure.

2   Preliminaries

Bit Vectors: A key data structure used in our research is a bit vector $B[1..N]$ that supports the operations *rank* and *select*. The operations $\mathtt{rank}_1(B,i)$ and $\mathtt{rank}_0(B,i)$ return the number of 1s and 0s in $B[1..i]$, respectively. The operations $\mathtt{select}_1(B,r)$ and $\mathtt{select}_0(B,r)$ return the position of the rth occurrences of 1 and 0, respectively. Several researchers [1,7,12] considered the problem of representing a bit vector succinctly to support $\mathtt{rank}$ and $\mathtt{select}$ in constant time under the word RAM model with word size $\Theta(\lg n)$ bits, and their results can be directly applied to the external memory model. The following lemma summarizes some of these results, in which part (a) is from Jacobson [1] and Clark and Munro [7], while part (b) is from Raman *et al.* [12]:

Lemma 1. *A bit vector B of length N can be represented using either: (a) $N + o(N)$ bits, or (b) $\lceil \lg \binom{N}{R} \rceil + O(N \lg \lg N / \lg N) = o(N)$ bits, where R is the number of 1s in B, to support the access to each bit,* $\mathtt{rank}$ *and* $\mathtt{select}$ *in $O(1)$ time (or $O(1)$ I/Os in external memory).*

Succinct Representations of Trees: As there are $\binom{2n}{n}/(n+1)$ different binary trees (or ordinal trees) on N nodes, various approaches [1,13,14] have been proposed to represent a binary tree (or ordinal tree) in $2N + o(N)$ bits, while supporting efficient navigation. Jacobson [1] first presented the *level-order binary marked* (LOBM) structure for binary trees, which can be used to encode a binary tree as a bit vector of $2N$ bits. He further showed that operations such as retrieving the left child, the right child and the parent of a node in the tree can be performed using $\mathtt{rank}$ and $\mathtt{select}$ operations on bit vectors. We make use of his approach to encode tree structures in Section 4. Another approach we use in this paper is based on the isomorphism between *balanced parenthesis sequences* and ordinal trees, proposed by Munro and Raman [13]. The balanced parenthesis sequence of a given tree can be obtained by performing a depth-first traversal, and outputting an opening parenthesis each time we visit a node, and a closing parenthesis immediately after all the descendants of this node are visited. Munro and Raman [13] designed a succinct representation of an ordinal tree of N nodes in $2N + o(N)$ bits based on the balanced parenthesis sequence, which supports the computation of the parent, the depth and the number of descendants of a node in constant time, and the ith child of a node in $O(i)$ time.

3   Bottom Up Traversal

In this section, we present a set of data structures that encode a tree T succinctly so that the I/Os performed in traversing a path from a given node to the root is optimal. Let A denote the maximum number of nodes that can be stored in a single block, and let K denote the length of the path. Given the bottom up nature of the queries, there is no need to encode a node's key value, since the path always proceeds to the current node's parent.

Data Structures: We begin with a brief overview of our technique. We partition T into layers of height τB where $0 < \tau < 1$. The top layer and the bottom layer can contain less than τB levels. We then group the nodes of each layer into blocks, and store with each block a *duplicate path*. To reduce the space required by these paths, we further group blocks into *superblocks* which also store a duplicate path. By loading at most the block containing a node, along with its associated duplicate path, and the superblock duplicate path we demonstrate that a layer can be traversed with at most $O(1)$ I/Os. A set of bit vectors[1] that map the nodes at the top of one layer to their parents in the layer above are used to navigate between layers.

Layers are numbered starting at 1 for the topmost layer. Let L_i be the ith layer in T. The layer is composed of a forest of subtrees whose roots are all at the top level of L_i. We now describe how the blocks and superblocks are created within L_i. We number L_i's nodes in preorder starting from 1 for the leftmost subtree and numbering the remaining subtrees from left to right. Once the nodes of L_i are numbered they are grouped into blocks of consequtive preorder number. Each block stores a portion of T along with the representation of its duplicate path, or the superblock duplicate path, if it is the first block in a superblock. We refer to the space used to store the duplicate path as *redundancy* which we denote W. In our succinct tree representation we require two bits to represent each node in the subtrees of L_i, so for blocks of $B \lg N$ bits we have $A = \left\lfloor \frac{B \lg N - W}{2} \right\rfloor$.

We term the first block in a layer the *leading block*. Layers are blocked in such a manner that the leading block is the only block permitted to be non-full (may contain less than A nodes). The leading block requires no duplicate structure and thus $W = 0$ for leading blocks. All superblocks except possibly the first, which we term the *leading superblock*, contain exactly $\lfloor \lg B \rfloor$ blocks.

For each block we store as the duplicate path, the path from the node with the minimum preorder number in the block to the layer's top level. Similar to Property 3 of Lemma 2 in [6] we have the following property:

Property 1. Given a block (or superblock) Y, for any node x in Y there exists a path from x to either the top of its layer, or to the duplicate path of Y, which consists entirely of nodes in Y.

Each block is encoded by three data structures:

1. An encoding of the tree structure, denoted B_e. The subtree(s) contained, or partially contained, within the block are encoded as a sequence of balanced parentheses (see Section 2). Note that in this representation, the ith opening parenthesis corresponds to the ith node in the preorder in this block.
2. The duplicate path array, $D_p[j]$, for $1 < j \leq \tau B$. Let v be the node with the smallest preorder number in the block. Entry $D_p[j]$ stores the preorder number of the node at the jth level on the path from v to the top level of the layer. The number recorded is the preorder number with respect to the block's superblock. It may be the case that v is not at the τB^{th} level of

---

[1] All bit vectors are represented using the structures of Lemma 1a or Lemma 1b.

the layer. In this case the entries below v are set to 0. Recall that preorder numbers begin at 1, so the 0 value effectively flags an entry as invalid.

3. The root-to-path array, $R_p[j]$, for $1 < j \leq \tau B$. A block may include many subtrees rooted at nodes on the duplicate path. Consider the set of roots of these subtrees. The entry at $R_p[j]$ stores the number of subtrees rooted at nodes on D_p from level τB up to level j. The number of subtrees rooted at node $D_p[j]$ can be calculated by evaluating $R_p[j] - R_p[j+1]$ when $j < \tau B$, or $R_p[j]$ when $j = \tau B$.

Now consider the first block in a superblock. The duplicate path of this block is the superblock's duplicate path. Unlike the duplicate path of a regular block, which stores the preorder numbers with respect to the superblock, the superblock's duplicate path stores the preorder numbers with respect to the preorder numbering in the layer. Furthermore, consider the duplicate paths of blocks within a superblock. These paths may share a common subpath with the superblock duplicate path. Each entry on a block duplicate path that is shared with the superblock duplicate path is set to -1.

For an arbitrary node $v \in T$, let v's layer number be ℓ_v and its preorder number within the layer be p_v. Each node in T is uniquely represented by the pair (ℓ_v, p_v). Let π define the lexicographic order on these pairs. Given a node's ℓ_v and p_v values, we can locate the node and navigate within the corresponding layer. The challenge is how to map between the roots of one layer and their parents in the layer above. Consider the set of N nodes in T. We define the following data structures, that will facilitate mapping between layers:

1. Bit vector $\mathcal{V}_{first}[1..N]$, where $\mathcal{V}_{first}[i] = 1$ iff the ith node in π is the first node within its layer.
2. Bit vector $\mathcal{V}_{parent}[1..N]$, where $\mathcal{V}_{parent}[i] = 1$ iff the the ith node in π is the parent of some node at the top level of the layer below.
3. Bit vector $\mathcal{V}_{first\_child}[1..N]$, where $\mathcal{V}_{first\_child}[i] = 1$ iff the ith node in π is a root in its layer and its parent in the preceeding layer differs from that of the previous root in this layer.

All leading blocks are packed together on disk, separate from the full blocks. Note that leading blocks do not require a duplicate path or root-to-path array, so only the tree structure need be stored for these blocks. Due to the packing, the leading block may overrun the boundary of a block on disk. We use the first $\lg B$ bits of each disk block to store an offset which indicates the position of the starting bit of the first leading block starting in the disk block. This allows us to skip any overrun bits from a leading block stored in the previous disk block.

We store two bit arrays to aid in locating blocks to index the partially full leading blocks and the full blocks. Let x be the number of layers on T, and let z be the total number of full blocks. The bit vectors are:

1. Bit vector $\mathcal{B}_l[1..x]$, where $\mathcal{B}_l[i] = 1$ iff the ith leading block resides in a different disk block than the $(i-1)^{\text{th}}$ leading block.
2. Bit vector $\mathcal{B}_f[1..(x+z)]$ that encodes the number of full blocks in each layer in unary. More precisely $\mathcal{B}_f[1..(x+z)] = 0^{l_1} 1 0^{l_2} 1 0^{l_3} 1 \ldots$ where l_i is the number of full blocks in layer i.

To analyze the space costs of our data structures we have the following lemma.

Lemma 2. *The data structures described above occupy* $2N + \frac{12\tau N}{\log_B N} + o(N)$ *bits.*

Proof (sketch). The number of bits used to store the actual tree structure of T is $2N$, as the structure is encoded using the balanced parentheses encoding. We must also account for the space required to store the duplicate paths and their associated root-to-path arrays. The space required for block and superblock duplicate paths differs: $\lceil \lg \left((B\lceil \lg B \rceil \lceil \lg N \rceil)/2 \right) \rceil$ bits are sufficient to store a node on a block duplicate path (as there are at most $(B\lceil \lg B \rceil \lceil \lg N \rceil)/2$ nodes in a superblock), while an entry of a superblock duplicate path require $\lceil \lg N \rceil$ bits.

We store the array R_p for each block. As a block may have as many as $(B\lceil \lg N \rceil)/2$ nodes, each entry in R_p requires $\lceil \lg B \rceil + \lceil \lg \lceil \lg N \rceil \rceil$ bits. Thus, for a regular block, the number of bits used to store both D_p and R_p is $\tau B(2\lceil \lg B \rceil + 2\lceil \lg \lceil \lg N \rceil \rceil + \lceil \lg \lceil \lg B \rceil \rceil)$.

Now consider the total space required for all duplicate paths and root-to-path arrays within a superblock. The superblock duplicate path requires $\tau B\lceil \lg N \rceil$ bits. The space cost of each of the $(\lceil \lg B \rceil - 1)$ remaining blocks is given in the previous paragraph. Thus the total redundancy per superblock is $W = \tau B\lceil \lg N \rceil + \tau B(\lceil \lg B \rceil + \lceil \lg \lceil \lg N \rceil \rceil) + (\lceil \lg B \rceil - 1)\tau B(2\lceil \lg B \rceil + 2\lceil \lg \lceil \lg N \rceil \rceil + \lceil \lg \lceil \lg B \rceil \rceil)$.

The average redundancy per block is then:

$$
\begin{aligned}
W_b &= \frac{\tau B\lceil \lg N \rceil}{\lceil \lg B \rceil} + \frac{\tau B(\lceil \lg B \rceil + \lceil \lg \lceil \lg N \rceil \rceil)}{\lceil \lg B \rceil} \\
&\quad + \frac{(\lceil \lg B \rceil - 1)\tau B(2\lceil \lg B \rceil + 2\lceil \lg \lceil \lg N \rceil \rceil + \lceil \lg \lceil \lg B \rceil \rceil)}{\lceil \lg B \rceil} \\
&\leq \tau B\lceil \log_B N \rceil + \tau B(3\lceil \lg B \rceil + 3\lceil \lg \lceil \lg N \rceil \rceil + \lceil \lg \lceil \lg B \rceil \rceil) \qquad (1)
\end{aligned}
$$

The value for the average redundancy, W_b, represents the worst case per block redundancy, as the redundancy for leading blocks is $\lceil \lg B \rceil / (B\lceil \lg N \rceil) < W_b$ bits. The total number of blocks required to store T is $\frac{2N}{B\lceil \lg N \rceil - W_b}$. The the total size of the redundancy for T is $W_t = \frac{2N}{B\lceil \lg N \rceil - W_b} \cdot W_b$, which is at most $W_t = \frac{2N \cdot 2W_b}{B\lceil \lg N \rceil}$ when $W_b < \frac{1}{2}B\lceil \lg N \rceil$. It is easy to show that when $\tau \leq \frac{1}{16}$, this condition is true. Finally, substituting the value for W_b from Eq. 1, to obtain $W_t = \frac{4N\tau\lceil \log_B N \rceil}{\lceil \lg N \rceil} + \frac{12N\tau\lceil \lg B \rceil}{\lceil \lg N \rceil} + \frac{12N\tau\lceil \lg \lceil \lg N \rceil \rceil}{\lceil \lg N \rceil} + \frac{4N\tau\lceil \lg \lceil \lg B \rceil \rceil}{\lceil \lg N \rceil} = \frac{12\tau N}{\lceil \log_B N \rceil} + o(N)$. We arrive at out final bound because the first, third, and fourth terms are each asymptotically $o(N)$ (recall that we assume $B = \Omega(\lg N)$).

The bit vectors $\mathcal{V}_{first}$, $\mathcal{V}_{parent}$, $\mathcal{V}_{first\_child}$, $\mathcal{B}_l$, and $\mathcal{B}_f$ can be stored in $o(N)$ bits using Lemma 1b, as they are spare bit vectors. $\qquad \square$

Navigation: The algorithm for reporting a node-to-root path is given by algorithms *ReportPath* (see Fig. 1) and *ReportLayerPath* (see Fig. 2). Algorithm *ReportPath*(T, v) is called with v being the number of a node in T given by π. *ReportPath* handles navigation between layers and calls *ReportLayerPath* to perform the traversal within each layer. The parameters ℓ_v and p_v are the layer number and the preorder value of node v within the layer, as previously

---

Algorithm *ReportPath(T, v)*
1. Find ℓ_v, the layer containing v. $\ell_v = \text{rank}_1(\mathcal{V}_{first}, v)$.
2. Find α_{ℓ_v}, the position in π of ℓ_v's first node. $\alpha_{\ell_v} = \text{select}_1(\mathcal{V}_{first}, \ell_v)$.
3. Find p_v, v's preorder number within ℓ_v. $p_v = v - \alpha_{\ell_v}$.
4. Repeat the following steps until the top layer has been reported.
    (a) Let $r = ReportLayerPath(\ell_v, p_v)$ be the preorder number of the root of the path in layer ℓ_v (This step also reports the path within the layer).
    (b) Find $\alpha_{(\ell_v - 1)}$, the position in π of the first node at the next higher layer. $\alpha_{(\ell_v - 1)} = \text{select}_1(\mathcal{V}_{first}, \ell_v - 1)$.
    (c) Find λ, the rank of r's parent among all the nodes in the layer above that have children in ℓ_v. $\lambda = (\text{rank}_1(\mathcal{V}_{first\_child}, \alpha_{\ell_v} + r)) - (\text{rank}_1(\mathcal{V}_{first\_child}, \alpha_{\ell_v} - 1))$.
    (d) Find which leaf δ, at the next higher layer corresponds to λ. $\delta = \text{select}_1(\mathcal{V}_{parent}, \text{rank}_1(\mathcal{V}_{parent}, \alpha_{(\ell_v - 1)}) - 1 + \lambda)$.
    (e) Update $\alpha_{\ell_v} = \alpha_{(\ell_v - 1)}$; $p_v = \delta - \alpha_{(\ell_v - 1)}$, and; $\ell_v = \ell_v - 1$.

Fig. 1. Algorithm for reporting the path from node v to the root of T

Algorithm *ReportLayerPath(ℓ_v, p_v)*
1. Load block b_v containing p_v. Scan B_e (the tree's representation) to locate p_v. If b_v is stored in a superblock, SB_v, then load SB_v's first block if b_v is not the first block in SB_v. Let $\min(D_p)$ be the minimum valid preorder number of b_v's duplicate path (let $\min(D_p) = 1$ if b_v is a leading block), and let $\min(SB_{D_p})$ be the minimum valid preorder number of the superblock duplicate path (if b_v is the first block in SB_v then let $\min(SB_{D_p}) = 0$).
2. Traverse the path from p_v to a root in B_e. If r is the preorder number (within B_e) of a node on this path report $(r - 1) + \min(D_p) + \min(SB_{D_p})$. This step terminates at a root in B_e. Let r_k be the rank of this root in the set of roots of B_e.
3. Scan the root-to-path array, R_p from $\tau B...1$ to find the smallest i such that $R_p[i] \geq r_k$. If $r_k \geq R_p[1]$ then r is on the top level in the layer so return $(r - 1) + \min(D_p) + \min(SB_{D_p})$ and terminate.
4. Set $j = i - 1$.
5. $\texttt{while}(j \geq 1$ and $D_p[j] \neq 1)$ report $D_p[j] + \min(SB_{D_p})$ and set $j = j - 1$.
6. If $j \geq 1$ then report $SB_{D_p}[j]$ and set $j = j - 1$ $\texttt{until}(j < 1)$.

Fig. 2. Steps to execute traversal within a layer, ℓ_v, starting at the node with preorder number p_v. This algorithm reports the nodes visited and returns the layer preorder number of the root at which it terminates.

described. *ReportLayerPath* returns the preorder number, within layer ℓ_v of the root of path reported from that layer. In *ReportLayerPath* we find the block b_v containing node v using the algorithm *FindBlock(ℓ_v, p_v)* described in Fig. 3. It is straightforward that this algorithm performs $O(1)$ I/Os per layer when performing path traversals, so a path of length K in T can be traversed in $O(K/\tau B)$ I/Os. Combing this with Lemma 2, we have the following theorem (to simplify our space result we define one additional term $\epsilon = 12\tau$):

Algorithm *FindBlock*(ℓ_v, p_v)

1. Find σ, the disk block containing ℓ_v's leading block. $\sigma = \text{rank}_1(\mathcal{B}_l, \ell_v)$.
2. Find α, the rank of ℓ_v's leading block within σ, by performing **rank/select** operations on $\mathcal{B}_l$ to find $j \leq \ell_v$ such that $\mathcal{B}_l[j] = 1$. $\alpha = p_v - j$.
3. Scan σ to find, and load, the data for ℓ_v's leading block (may required loading the next disk block). Note the size δ of the leading block.
4. If $p_v \leq \delta$ then p_v is in the already loaded leading block, terminate.
5. Calculate ω, the rank of the block containing p_v within the $\text{select}_1(\mathcal{B}_f, \ell_v + 1) - \text{select}_1(\mathcal{B}_f, \ell_v)$ full blocks for this level.
6. Load full block $\text{rank}_0(\mathcal{B}_f, \ell_v) + \omega$ and terminate.

Fig. 3. *FindBlock* algorithm

Theorem 1. *A tree T on N nodes can be represented in $2N + \frac{\epsilon N}{\log_B N} + o(N)$ bits such that given a node-to-root path of length K, the path can be reported in $O(K/B)$ I/Os, when $0 < \epsilon < 1$.*

For the case in which we wish to maintain a key with each node, we store each key in the same block that contains its corresponding node, and we also store each duplicate path and keys associated to the nodes in the path in the same block. This yields the following corollary:

Corollary 1. *A tree T on N nodes with q-bit keys, where $q = O(\lg N)$, can be represented in $(2 + q)N + q \cdot \left[\frac{6\tau N}{\lceil \log_B N \rceil} + \frac{2\tau q N}{\lceil \lg N \rceil} + o(N) \right]$ bits such that given a node-to-root path of length K, that path can be reported in $O(\tau K/B)$ I/Os, when $0 < \tau < 1$.*

In Corollary 1 it is obvious that the first and third terms inside the square brackets are small so we will consider the size of the the second term inside the brackets $((2\tau q N)/\lceil \lg N \rceil)$. When $q = o(\lg N)$ this term becomes $o(N)$. When $q = \Theta(\lg N)$ we can select τ such that this term becomes (ηN) for $0 < \eta < 1$.

4   Top Down Traversal

Given a binary tree T, in which every node is associated with a key, we wish to traverse a top-down path of length K starting at the root of T and terminating at some node $v \in T$. We follow our previous notation by letting A be the maximum number of nodes that can be stored in a single block. Let $q = O(\lg N)$ be the number of bits required to encode a single key. Keys are included in the top-down case because it is assumed that the path followed during the traversal is selected based on the key values in T.

Data Structures: We begin with a brief sketch of our data structures. A tree T is partitioned into subtrees, where each subtree T_i is laid out into a *tree block*. Each block contains a succinct representation of T_i and the set of keys associated with the nodes in T_i. The edges in T that span a block boundary are not explicitly stored within the tree blocks. Instead, they are encoded through a set of bit vectors that enable navigation between blocks.

To introduce our data structures, we give some definitions. If the root node of a block is the child of a node in another block, then the first block is a *child* of the second. There are two types of blocks: *internal* blocks that have one or more *child* blocks, and terminal blocks that have no *child* blocks. The *block level* of a block is the number of blocks along a path from the root of this block to the root of T.

We number the internal blocks in the following manner. First number the block containing the root of T as 1, and number its child blocks consecutively from left to right. We then consecutively number the internal blocks at each successive block level. The internal blocks are stored on the disk in an array I, such that the block numbered j is stored in entry $I[j]$.

Terminal blocks are numbered and stored separately. Starting again at 1, they are numbered from left to right. Terminal blocks are stored in the array Z. As terminal blocks may vary in size, there is no one-to-one correspondence between disk and tree blocks in Z; rather, the tree blocks are packed into Z to minimize wasted space. At the start of each disk block j, a $\lg B$ bit *block offset* is stored which indicates the position of the starting bit of the first terminal block stored in $Z[j]$. Subsequent terminal blocks are stored immediately following the last bits of the previous terminal blocks. If there is insufficient space to record a terminal block within disk block $Z[j]$, the remaining bits are stored in $Z[j+1]$.

We now describe how an individual internal tree block is encoded. Consider the block of subtree T_j; it is encoded using the following structures:

1. The block keys, B_k, is an A-element array which encodes the keys of T_j.
2. The tree structure, B_s, is an encoding of T_j using the LOBM sequence of Jacobson [1]. More specifically, we define each node of T_j as a *real* node. T_j is then augmented by adding *dummy* nodes as the left and/or right child of any real node that does not have a corresponding real child node in $T_j{}^2$. We then perform a level order traversal of T_j and output a 1 each time we visit a real node, and a 0 each time we visit a dummy node. If T_j has A nodes the resulting bit vector has A 1s for real nodes and $A+1$ 0s for dummy nodes. As the first bit is always 1, and the last two bits are always 0s, we do not store them explicitly. Thus, B_s can be represented with $2A-2$ bits.
3. The *dummy offset*, B_d. Let Γ be a total order over the set of all dummy nodes in internal blocks. In Γ the order of dummy node d is determined first by its block number, and second by its position within B_s. The dummy offset records the position in Γ of the first dummy node in B_s.

The encoding for terminal blocks is identical to internal blocks except: the dummy offset is omitted, and the last two 0s of B_s are encoded explicitly.

We now define a *dummy root*. Let T_j and T_k be two tree blocks where T_k is a child block of T_j. Let r be the root of T_k, and v be r's parent in T. When T_j is

---

^{2} The node may have a child in T, but if that node is not part of T_j, it is replaced by a dummy node.

encoded a dummy node is added as a child of v which corresponds to r. Such a dummy node is termed a dummy root.

Let ℓ be the number of dummy nodes over all internal blocks. We create:

1. $X[1..\ell]$ stores a bit for each dummy node in internal blocks. Set $X[i] = 1$ iff dummy node i is the dummy root of an internal block.
2. $S[1..\ell]$ stores a bit for each dummy node in internal blocks. Set $S[i] = 1$ iff dummy node i is the dummy root of a terminal block.
3. $S_B[1..\ell']$, where ℓ' is the number of 1s in S. Each bit in this array corresponds to a terminal block. Set $S_B[j] = 1$ iff the corresponding terminal block is stored starting in a disk block of Z that differs from that in which terminal block $j-1$ starts.

Block Layout: We have yet to describe how T is split up into *tree* blocks. This is achieved using the two-phase blocking strategy of Demaine *et al.* [10]. Phase one blocks the first $c \lg N$ levels of T, where $0 < c < 1$. Starting at the root of T the first $\lfloor \lg (A+1) \rfloor$ levels are placed in a block. Conceptually, if this first block is removed we are left with a forest of $O(A)$ subtrees. The process is repeated recursively until $c \lg N$ levels of T have thus been blocked.

In the second phase we block the rest of the subtrees by the following recursive procedure. The root, r, of a subtree is stored in an empty block. The remaining $A - 1$ capacity of this block is then subdivided, proportional to the size of the subtrees, between the subtrees rooted at r's children. During this process, if at a node the capacity of the current block is less than 1, a new block is created. To analyze the space costs of our structures, we have:

Lemma 3. *The data structures described above occupy* $(3+q)N + o(N)$ *bits.*

Proof. We first determine the maximum block size A. The encoding of subtree T_j requires $2A$ bits. We need Aq bits to store the keys, and $\lceil \lg N \rceil$ bits to store the dummy offset. Therefore, $2A + Aq + \lceil \lg N \rceil = \lfloor B \lg N \rfloor$. Thus, $A = \Theta\left(\frac{B \lg N}{q}\right)$.

During the first phase of the layout, non-full internal blocks may be created. However, the height of the phase 1 tree is bounded by $c \lg N$ levels, so the total number of wasted bits in such blocks is bounded by $o(N)$.

The arrays of blocks I and Z store the structure of T as LOBM which requires $2N$ bits. The dummy roots are duplicated as the roots of child blocks, but as the first bit in each block need not be explicitly stored, the entire tree structure still requires only $2N$ bits. The keys occupy $N \cdot q$ bits. Each of the $O(N/A)$ blocks in I stores a block offset of size $\lg(N/A)$ bits. The total space required for the offsets is $N/A \cdot \lg (N/A)$, which is $o(N)$ bits since $q = O(\lg N)$. The bit vectors X and S_B have size N, but in both cases the number of 1 bits is bounded by N/A. By Lemma 1b, we can store them in $o(N)$ bits. S can be encoded in $N + o(N)$ bits using Lemma 1a. The total space is thus $(3+q)N + o(N)$ bits. $\square$

Navigation: Navigation in T is summarized in Figures 4 and 5 which present the algorithms $Traverse(key, i)$ and $TraverseTerminalBlock(key, i)$ respectively. During the traversal the function `compare(key)` compares the value *key* to the

---

Algorithm $Traverse(key, i)$

1. Load block $I[i]$ to main memory. Let T_i denote the subtree stored in $I[i]$.
2. Scan B_s to navigate within T_i. At each node x use $\texttt{compare}(key, B_k[x])$ to determine which branch to follow until a dummy node d with parent p is reached.
3. Scan B_s to determine $j = \texttt{rank}_0(B_s, d)$.
4. Determine the position of j with respect to Γ by adding the *dummy offset* to calculate $\lambda = B_d + j$.
5. If $X[\lambda] = 1$, then set $i = \texttt{rank}_1(X, \lambda)$ and call $Traverse(key, i)$.
6. If $X[\lambda] = 0$ and $S[\lambda] = 1$, then set $i = \texttt{rank}_1(S, \lambda)$ and call $TraverseTerminalBlock(key, i)$.
7. If $X[\lambda] = 0$ and $S[\lambda] = 0$, then p is the final node on the traversal, so the algorithm terminates.

Fig. 4. Top down searching algorithm for a blocked tree

Algorithm $TraverseTerminalBlock(key, i)$

1. Load disk block $Z[\lambda]$ containing terminal block i, where $\lambda = \texttt{rank}_1(S_B, i)$.
2. Let B_d be the offset of disk block $Z[\lambda]$.
3. Find α, the rank of terminal block i within $Z[\lambda]$ by scanning from $S_B[i]$ backwards to find $j \le i$ such that $S_B[j] = 1$. Then $\alpha = i - j$.
4. Starting at B_d scan $Z[\lambda]$ to find the start of the αth terminal block. Recall that each block stores a bit vector B_s in the LOBM encoding, so we can determine when we have reached the end of one terminal block as follows:
   (a) Set two counters $\mu = \beta = 1$.
   (b) Scan B_s. When a 1 bit is encountered increment μ and β. When a 0 bit is encountered decrement β. Terminate the scan when $\beta < 0$ as the end of B_s has been reached.
   (c) Now μ records the number of nodes in the terminal block so calculate the length of the array B_k needed to store the keys and jump ahead this many bits. This will place the scan at the start of the next terminal block.
5. Once the αth block has been reached, the terminal block can be read in (process is the same as scanning the previous blocks). It may be the case the this terminal block overruns the *disk* block $Z[\lambda]$ into $Z[\lambda + 1]$. In this case skip the first $\lceil \lg B \rceil$ bits of $Z[\lambda + 1]$ and continue reading in the terminal block.
6. With the terminal block in memory, the search can be concluded in a manner analogous to that for internal blocks except that once a dummy node is reached, the search terminates.

Fig. 5. Performing search for a terminal block

key of a node to determine which branch of the tree to traverse. The parameter i is the number of a *disk* block. Traversal is initiated by calling $Traverse(key, 1)$.

It is easy to observe that a call to $TraverseTerminalBlock$ can be performed in $O(1)$ I/Os, while $Traverse$ can be executed in $O(1)$ I/Os per recursive call. Thus, the I/O bounds are then obtained directly by substituting our succinct block size A for the standard block size B in the result of Demaine *et al.* [10] (see Section 1). Combining this with Lemmas 3, we have the following result:

Theorem 2. *A rooted binary tree, T, of size N, with keys of size $q = O(\lg N)$ bits, can be stored using $(3+q)N + o(n)$ bits so that a root to node path of length K can be reported with:*

1. $O\left(\frac{K}{\lg(1+(B\lg N)/q)}\right)$ *I/Os, when $K = O(\lg N)$*

2. $O\left(\frac{\lg N}{\lg(1+\frac{B\lg^2 N}{qK})}\right)$ *I/Os, when $K = \Omega(\lg N)$ and $K = O\left(\frac{B\lg^2 N}{q}\right)$, and*

3. $O\left(\frac{qK}{B\lg N}\right)$ *I/Os, when $K = \Omega\left(\frac{B\lg^2 N}{q}\right)$.*

Acknowledgements

We wish to thank our anonymous reviewers for their helpful comments and corrections.

References

1. Jacobson, G.: Space-efficient static trees and graphs. FOCS 42, 549–554 (1989)
2. Chien, Y.F., Hon, W.K., Shah, R., Vitter, J.S.: Geometric Burrows-Wheeler transform: Linking range searching and text indexing. In: DCC, pp. 252–261 (2008)
3. Aggarwal, A., Jeffrey, S.V.: The input/output complexity of sorting and related problems. Commun. ACM 31(9), 1116–1127 (1988)
4. Nodine, M.H., Goodrich, M.T., Vitter, J.S.: Blocking for external graph searching. Algorithmica 16(2), 181–214 (1996)
5. Agarwal, P.K., Arge, L., Murali, T.M., Varadarajan, K.R., Vitter, J.S.: I/O-efficient algorithms for contour-line extraction and planar graph blocking (extended abstract). In: SODA, pp. 117–126 (1998)
6. Hutchinson, D.A., Maheshwari, A., Zeh, N.: An external memory data structure for shortest path queries. Discrete Applied Mathematics 126, 55–82 (2003)
7. Clark, D.R., Munro, J.I.: Efficient suffix trees on secondary storage. In: SODA, pp. 383–391 (1996)
8. Gil, J., Itai, A.: How to pack trees. J. Algorithms 32(2), 108–132 (1999)
9. Alstrup, S., Bender, M.A., Demaine, E.D., Farach-Colton, M., Rauhe, T., Thorup, M.: Efficient tree layout in a multilevel memory hierarchy. arXiv:cs.DS/0211010 [cs:DS] (2004)
10. Demaine, E.D., Iacono, J., Langerman, S.: Worst-case optimal tree layout in a memory hierarchy. arXiv:cs/0410048v1 [cs:DS] (2004)
11. Brodal, G.S., Fagerberg, R.: Cache-oblivious string dictionaries. In: SODA, pp. 581–590 (2006)
12. Raman, R., Raman, V., Rao, S.S.: Succinct indexable dictionaries with applications to encoding k-ary trees and multisets. In: SODA, pp. 233–242 (2002)
13. Munro, J.I., Raman, V.: Succinct representation of balanced parentheses and static trees. SIAM J. Comput. 31(3), 762–776 (2001)
14. Benoit, D., Demaine, E.D., Munro, J., Raman, R., Raman, V., Rao, S.S.: Representing trees of higher degree. Algorithmica 43(4), 275–292 (2005)

Space-Time Tradeoffs for Longest-Common-Prefix Array Computation*

Simon J. Puglisi and Andrew Turpin

School of Computer Science and Information Technology, RMIT University,
Melbourne, Australia
{sjp,aht}@cs.rmit.edu.au

Abstract. The suffix array, a space efficient alternative to the suffix tree, is an important data structure for string processing, enabling efficient and often optimal algorithms for pattern matching, data compression, repeat finding and many problems arising in computational biology. An essential augmentation to the suffix array for many of these tasks is the Longest Common Prefix (LCP) array. In particular the LCP array allows one to simulate bottom-up and top-down traversals of the suffix tree with significantly less memory overhead (but in the same time bounds). Since 2001 the LCP array has been computable in $\Theta(n)$ time, but the algorithm (even after subsequent refinements) requires relatively large working memory. In this paper we describe a new algorithm that provides a continuous space-time tradeoff for LCP array construction, running in $O(nv)$ time and requiring $n+O(n/\sqrt{v}+v)$ bytes of working space, where v can be chosen to suit the available memory. Furthermore, the algorithm processes the suffix array, and outputs the LCP, strictly left-to-right, making it suitable for use with external memory. We show experimentally that for many naturally occurring strings our algorithm is faster than the linear time algorithms, while using significantly less working memory.

1   Introduction

The suffix array SA_x of a string x is an array containing all the suffixes of x sorted into lexicographical order [10]. The suffix array can be *enhanced* [1] with the longest-common-prefix (LCP) array LCP_x which contains the lengths of the longest common prefixes of adjacent elements in SA_x. When preprocessed in this way the suffix array becomes equivalent to, though much more compact than, the suffix tree [13,16,17], a data structure with "myriad virtues" [2]. Conceptually LCP_x defines the "shape" of the suffix tree and thus allows top-down and bottom-up traversals to be simulated using SA_x. Such traversals are at the heart of many efficient string processing algorithms [1,2,8,15].

Despite its utility, the problem of constructing LCP_x efficiently has been little studied. The three algorithms that do exist [9,12] consume significant amounts of memory and, though asymptotically linear in their runtime, have poor locality

---

* This work is supported by the Australian Research Council.

S.-H. Hong, H. Nagamochi, and T. Fukunaga (Eds.): ISAAC 2008, LNCS 5369, pp. 124–135, 2008.

of memory reference and so tend to be slow in practice, certainly relative to the fastest algorithms for constructing SA_x. In this paper we describe a new algorithm for LCP array construction that provides a continuous space-time tradeoff, running in $O(nv)$ time and requiring $n + O(n/\sqrt{v} + v)$ bytes of working space, where v can be chosen to suit the available memory. We show experimentally that for many naturally occurring strings our algorithm is faster than the linear time algorithms, while always using a fraction of the working memory. Our basic strategy is to compute the LCP values for a special sample of the suffixes and then use these values to derive the rest with only a fixed extra cost per value. Importantly we require only sequential access to the SA and LCP array, and this locality of memory reference makes our algorithm suitable for low memory environments when these arrays must reside on disk, and non sequential access would cripple runtime.

The remainder of this paper is organized in the following manner. In the remainder of this section we set some notation, formalise basic ideas and then tour existing LCP array construction algorithms. Then, in Section 2 we describe two peices of algorithmic machinery that are essential to our methods. Our new algorithm is detailed in Section 3. Section 4 reports on our experiments with an efficient implementation of the algorithm, with reflections offered in Section 5.

1.1   Basic Definitions and Notation

Throughout we consider a string $x = x[0..n] = x[0]x[1] \ldots x[n]$ of $n + 1$ symbols. The first n symbols of x are drawn from a constant ordered alphabet, Σ, consisting of symbols σ_j, $j = 1, 2, \ldots, |\Sigma|$ ordered $\sigma_1 < \sigma_2 < \cdots < \sigma_{|\Sigma|}$. The final character $x[n]$ is a special "end of string" character, \$, lexicographically smaller than all the other characters in Σ, so \$ $< \sigma_1$. For purposes of accounting we assume the common case that $|\Sigma| \in 0..255$, where each symbol requires one byte of storage and that $n < 2^{32}$ so the length of x and any pointers into it require four bytes each. These are not inherant limits on our algorithm, which will work for any alphabet size and any string length.

For $i = 0, \ldots, n$ we write $x[i..n]$ to denote the **suffix** of x of length $n - i + 1$, that is $x[i..n] = x[i]x[i + 1] \cdots x[n]$. For simplicity we will frequently refer to suffix $x[i..n]$ simply as "suffix i". Similarly, we write $x[0..i]$ to denote the **prefix** of x of length $i + 1$. We write $x[i..j]$ to represent the **substring** $x[i]x[i+1] \cdots x[j]$ of x that starts at position i and ends at position j.

The **suffix array** of x, denoted SA_x or just SA, when the context is clear, is an array $SA[0..n]$ which contains a permutation of the integers $0..n$ such that $x[SA[0]..n] < x[SA[1]..n] < \cdots < x[SA[n]..n]$. In other words, $SA[j] = i$ iff $x[i..n]$ is the jth suffix of x in ascending lexicographical order.

Our focus in this paper is the computation of the an array derived from SA_x, the **lcp array** $LCP = LCP[0..n]$. Define $lcp(y, z)$ as the length of the longest common prefix of strings y and z. For every $j \in 1..n$,

$$LCP[j] = lcp(x[SA[j-1]..n], x[SA[j]..n]),$$

that is the length of the longest common prefix of suffixes $SA[j-1]$ and $SA[j]$. $LCP[0]$ is undefined. The following example, which we will return to later, illustrates these data structures.

	0	1	2	3	4	5	6	7	8	9	10
x	a	t	a	t	g	t	t	t	g	t	$\$$
SA	10	0	2	8	4	9	1	7	3	6	5
LCP	-	0	2	0	2	0	1	1	3	1	2

Thus the longest common prefix of suffixes 0 and 2 is at, of length 2, while that of suffixes 9 and 1 is t, of length 1.

1.2 Prior Art

When the average value in LCP is low it is feasible to compute each value by brute force. However, we do not know in advance the longest LCP value for a gien suffix array; in the worst case (a string of n identical symbols) the brute force approach requires $O(n^2)$ time. Kasai et al [9] give an elegant algorithm to construct LCP given SA and x in $\Theta(n)$ time. Including these arrays and another for working, their algorithm requires $13n$ bytes. Manzini [12] describes a refinement to Kasai et al's approach that maintains the linear running time but reduces space usage to $9n$ bytes. Manzini presents a further variant – again with linear runtime – that overwrites SA with LCP and uses $4H_k(x) + 6n$ bytes of space where H_k is the kth order emperical entropy [11], a measure that decreases the more compressible x is. Thus, for very regular strings this algorithm uses little more than $6n$ bytes, but in the worst case may use $10n$.

To gain an idea of the problematic nature of the space bounds for these algorithms, consider the task of construction LCP for the entire human genome; some 3Gb of data. DNA is notoriously uncompressible, so Manzini's second algorithm will have no advantage, leaving his $9n$ byte algorithm as the most space efficient option. To avoid the deleterious effect of using secondary memory would require 27Gb of main memory, much more than the capacity of most workstations.

2 Tools

In this section we introduce two of the important components of our algorithm: the concept of a Difference Cover, and efficient algorithms for solving Range Minimum Queries.

2.1 Difference Covers

Essential to our methods is the concept of a **difference cover**, which was recently used by Burkhardt and Kärkkäinen for the purposes of suffix array construction [4]. As they define it, a difference cover D modulo v is a set of integers in the range $0..v-1$ with the property that for all $i \in 0..v-1$ there exists $j, k \in D$ such that $i \equiv k-j (\bmod v)$. A difference cover must have size $\geq \sqrt{v}$, though small ones (with size $\leq \sqrt{1.5v}+6$) can be generated relatively easily with a method described by Colburn and Ling [5]. In any case, once computed a difference cover can be stored and reused – around a dozen of varying sizes are provided in the source code of Burkhardt and Kärkkäinen [4].

An example of a difference cover for $v = 7$ is $D = \{1, 2, 4\}$. Observe that for $i = 0..6$ it is always possible to find two elements of D (perhaps the same element twice) so that i equals the difference of the numbers, modulo v.

We also borrow from Burkhardt and Kärkkäinen a function $\delta(i, j)$ over a difference cover D defined as follows. For any integers i, j, let $\delta(i, j)$ be an integer $k \in 0..v-1$ such that $(i+k) \bmod v$ and $(j+k) \bmod v$ are both in D. They show that a lookup table of size $O(v)$ computable in time $O(v)$ is sufficient for $\delta(i, j)$ to be evaluted in constant time for any i, j.

2.2   Range Minimum Queries

Our algorithms also rely on efficient data structures for solving **range minimum queries (RMQs)**. Given an array of n integers $A[0..n-1]$, function $\mathrm{RMQ}_A(i, j)$, $i, j \in 0..n-1$, $i < j$, returns $k \in i..j$ such that $A[k]$ is the minimum (or equal minimum) of all values in $A[i..j]$. Remarkably, it is possible to preprocess A in $O(n)$ time to build a data structure requiring $O(n)$ space that subsequently allows arbitrary RMQs to be answered in constant time [3,6]. The most space efficient of these preprocessing algorithms requires just $2n + o(n)$ bits for the final data structure and working space during its construction [6,7].

3   Computing the LCP Array

Our strategy for saving space is to compute the lcp values for a sample of suffixes and then use those values to generate the rest. The particular sample we use is defined by a difference cover with density selected such that overall space requirements fit into the available memory.

At a high level the algorithm is comprised of the following steps:

1. Choose $v \geq 2$, a difference cover D modulo v with $|D| = \Theta(\sqrt{v})$ and compute the function δ.
2. Fill an array $S[0..n/|D|]$ with the suffixes whose starting position modulo v is in D. Build a data structure on S so that we can compute the lcp of an arbitrary pair of sample suffixes in constant time.
3. Scan SA left-to-right and compute all the LCP values. To compute $\mathrm{LCP}[i]$ we compare the first v symbols of $\mathrm{SA}[i]$ and $\mathrm{SA}[i-1]$. If we find a mismatch at $\mathrm{SA}[i]+j$ we know the lcp is $j-1$ and we are done. If they are equal after v symbols we compute the lcp using the fact that suffixes $\mathrm{SA}[i]+\delta(\mathrm{SA}[i-1], \mathrm{SA}[i])$ and $\mathrm{SA}[i-1]+\delta(\mathrm{SA}[i-1], \mathrm{SA}[i])$ are in the sample and we can compute their lcp in constant time using the data structure alluded to in the previous step.

As mentioned in Section 2.1, Step 1, the selection of a suitable difference cover, can be acheived in $O(\sqrt{v})$ time and anyway can be done offline. We now delve further into Step 2 and Step 3 in turn.

3.1   Step 2: Collecting and Preprocessing the Sample

Let $m = n|D|/v$ denote the number of sample suffixes in string $x[0..n]$. As mentioned above, we collect the sample suffixes in an left-to-right scan of $SA[0..n]$, filling an array $S[0..m-1]$ with the suffixes whose starting position modulo v is in D. This can be acheived in $O(n)$ if we first preprocess D and build a small table $\hat{D}[0..v-1]$. That is, $\hat{D}[i], i \in D$, is the number of elements less than $i \in D$ (ie. it's rank in D), or -1 if $i \notin D$. After the scan of SA all the sample suffixes appear in S in the order in which they appear in the SA.

We then preprocess S and build a data structure so that we can compute the lcp of an arbitrary pair of sample suffixes in constant time. The first part of this is a mapping that gives us the position in S of a given sample suffix in constant time. Because the values in S are periodic according to the difference cover, the mapping can be implemented with an array $\hat{S}[0..m-1]$ and the $\hat{D}$ table: the position of sample suffix i in S is simply $\hat{S}[|D|\lfloor i/v \rfloor + \hat{D}[i \bmod v]]$. Figure 1 summarises the process.

> — *Compute* $\hat{D}[0..v-1]$ *(assume set D is sorted).*
> 1: $j \leftarrow 0$
> 2: **for** $i \leftarrow 0$ **to** $v-1$ **do**
> 3: $\hat{D}[i] = -1$
> 4:   **if** $D[j] = i$ **then**
> 5: $\hat{D}[i] \leftarrow j$
> 6: $j \leftarrow j + 1$
>   — *Put sample suffixes in $S[0..m-1]$ and their ranks in $\hat{S}$.*
> 7: $j \leftarrow 0$
> 8: **for** $i \leftarrow 0$ **to** n **do**
> 9:   **if** $\hat{D}[SA[i] \bmod v] \neq -1$ **then**
> 10: $S[j] \leftarrow SA[i]$
> 11: $\hat{S}[|D|\lfloor S[j]/v \rfloor + \hat{D}[S[j] \bmod v]] \leftarrow j$
> 12: $j \leftarrow j + 1$

Fig. 1. Compute arrays $S[0..m-1]$, $\hat{S}[0..m-1]$ and $\hat{D}[0..v-1]$

Using the example string from Section 1.1, and the difference cover $D = \{0, 1\}$ and $v = 3$, we get arrays shown in Figure 2.

Having collected S, for each $S[i], i = 1..m$ we compute $L[i] = \text{lcp}(S[i-1], S[i])$ using the algorithm in Figure 3. The algorithm is an adaption of the original $O(n)$ time LCP algorithm from Kasai et al. [9]. The outer loop at Line 1 iterates over elements of our chosen difference cover D, s, and the while loop on Line 4 fills in the values of $L[\hat{S}[s]]$, $L[\hat{S}[s + |D|]]$, $L[\hat{S}[s + 2|D|]]$, and so on. Note we assume $S[-1] = n$ to finess the boundary case in Line 5 when $\hat{S}[k] = 0$.

We exploit the following lemma from Kasai et al. [9]

Lemma 1. *If $\ell(i)$ is lcp value for suffix i then $\ell(i + v) \geq \ell(i) - v$,*

which allows us to maintain ℓ never reducing by more than v with each iteration. The analysis of execution time for Figure 3 is similar to that of Kasai et al. [9].

	0	1	2	3	4	5	6	7	8	9	10
i	0	1	2	3	4	5	6	7	8	9	10
x	a	t	a	t	g	t	t	t	g	t	\$
$D[i]$	0	1									
$\hat{D}[i]$	0	1	-1								
SA	10	0	2	8	4	9	1	7	3	6	5
$SA[i] \bmod v$	1	0	2	2	1	0	1	1	0	0	2
j	0	1			2	3	4	5	6	7	
S	10	0			4	9	1	7	3	6	
$|D|\lfloor S[j]/v\rfloor$	6	0			2	6	0	4	2	4	
$\hat{D}[S[j] \bmod v]$	1	0			1	0	1	1	0	0	
$\hat{S}$	1	4			6	2	7	5	3	0	
L after $s = 0$		0			0				3	1	
L after $s = 1$	-	0			0	0	1	1	3	1	

Fig. 2. An example of the derivation of S, $\hat{S}$ and L

> — Using $S[0..m-1]$, $\hat{S}$ and $\hat{D}$ compute $L[0..m-1]$.

```
1:  for s ← 0 to |D|−1 do
2:      k ← s
3:      ℓ ← 0
4:      while k < m do
5:          s₀ ← S[Ŝ[k]−1]
6:          s₁ ← S[Ŝ[k]]
7:          while x[s₀ + ℓ] = x[s₁ + ℓ] do
8:              ℓ ← ℓ + 1
9:          L[Ŝ[k]] ← ℓ
10:         ℓ ← max(0, ℓ − v)
11:         k ← k + |D|
```

Fig. 3. Compute the values of $L[0..m-1]$

Consider the loop starting on Line 4, which computes the values in L corresponding to sample suffixes $k = D[s](\bmod v)$. The execution time of this loop is propotional to the number of times Line 8 is executed, as it is the innermost loop. The value of ℓ increases by 1 on Line 8 and is always less than n, the string length. Since ℓ is initially 0 and is decreased by at most v each time around the outer while loop (at Line 10), which is executed at most $m/|D|$ times, ℓ increases at most $2n$ times. The $O(n)$ time bound follows. In order to compute all the values in L we simply run the algorithm once for each $k \in D$. Since $|D| = O(\sqrt{v})$ it follows that we can compute L in $O(\sqrt{v}n)$ time. With L in hand we preprocess it for constant time RMQs, allowing us to take advantage of the following well known result (see, eg. [10]).

Lemma 2. *Let $A[0..n]$ be an array of lexicographically sorted strings and let $P[1..n]$ be an array where $P[i] = \mathrm{lcp}(A[i-1], A[i])$. Then for $i, j \in 0..n, i < j, \mathrm{lcp}(A[i], A[j]) = \mathrm{RMQ}_P(i, j)$.*

We now have a data structure, which can be built in $O(n\sqrt{v})$ time, that lets us find the lcp of two arbitrary sample suffixes in constant time. In the next section we use it to efficiently compute the entire LCP array.

3.2   Step 3: Computing the Entire LCP Array

We make a second scan of SA left-to-right to compute all the LCP values. For convenience, let s_0 denote suffix $\mathrm{SA}[i-1]$ and s_1 denote suffix $\mathrm{SA}[i]$. To compute $\mathrm{LCP}[i]$ we first compare $x[s_0..s_0+v]$ to $x[s_1..s_1+v]$. If they are not equal, the offset of the first mismatching character is the lcp value and we can move on. On the other hand, if the first v characters of these suffixes are equal, then $\mathrm{LCP}[i]$ can be computed by finding the beginning of the sample suffixes in $x[s_0..s_0+v]$ and $x[s_1..s_1+v]$, a_0 and a_1 respectively in Figure 4, and then using their precomputed lcp value to derive $\mathrm{LCP}[i]$. More specifically, because of the properties of the difference cover, $\delta(s_0, s_1)$ will return an offset k from s_0 and s_1 so that both suffixes beginning at s_0+k and s_1+k are in the sample set S. Using $\hat{S}$ we can locate these in constant time, giving their positions in S, as r_0 and r_1 as shown in Figure 4. Finally, $\mathrm{LCP}[i]$ will be equal to k plus the lcp of the two sample suffixes located at r_0 and r_1, which is computed as the $\mathrm{RMQ}_L(r_0, r_1)$.

Because we access SA strictly left-to-right it is possible to overwrite SA with LCP as the algorithms proceeds. This excellent memory access pattern also allows the possibility of leaving the SA on disk, overwriting it there with the LCP information.

```
1: for i ← 1 to n do
2:        s₀ ← SA[i−1]
3:        s₁ ← SA[i]
          — Check if lcp(s₀, s₁) < v.
4:        j ← 0
5:        while x[s₀ + j] = x[s₁ + j] and j < v do
6:             j ← j + 1
7:        if j < v then
8:             LCP[i] ← j
9:        else
               — Compute lcp(s₀, s₁) using L.
10:            a₀ ← s₀ + δ(s₀, s₁) ; r₀ ← Ŝ[|D|⌊a₀/v⌋ + D̂[a₀ mod v]]
11:            a₁ ← s₁ + δ(s₀, s₁) ; r₁ ← Ŝ[|D|⌊a₁/v⌋ + D̂[a₁ mod v]]
12:            LCP[i] ← δ(s₀, s₁) + RMQ_L(r₀, r₁)
```

Fig. 4. Compute $\mathrm{LCP}[1..n]$ ($\mathrm{LCP}[0]$ is undefined) given $\hat{S}$ and the RMQ_L data structure in $O(nv)$ time

4   Implementation and Experiments

In this section we report on experiments with the new algorithm. The purpose of the experiments was to compare the new algorithm to existing approaches and also to gauge the effect of parameter v, the size of the difference cover on performance. We implemented two different versions of the new algorithm, and three existing approaches as follows.

PT-mem. Overwrites a memory resident SA with the LCP information, requiring around $5n+13n/\sqrt{v}$ bytes of primary memory.

PT-disk. Streams and overwrites a disk resident SA with the LCP information, requiring around $n+13n/\sqrt{v}$ bytes of primary memory.

L-13. The original $O(n)$ algorithm of Kasai et al [9], requiring $13n$ bytes.

L-9. Manzini's first variant, requiring $9n$ bytes.

L-6. Manzini's most space efficient version requiring $4H_k(x) + 6n$ bytes.

The memory requirements as stated here include the cost of the text, the SA and the LCP array when they are required to be held in RAM. Note that while the PT-MEM and PT-DISK programs overwrite SA, if the SA is required it can be copied in $\Theta(n)$ time before the algorithm commences.

All tests were conducted on a 3.0 GHz Intel Xeon CPU with 4Gb main memory and 1024K L2 Cache and a 320 GB Seagate Barracuda 7200.10 disk. The machine was under light load, that is, no other significant I/O or CPU tasks were running. The operating system was Fedora Linux running kernel 2.6.9. The compiler was g++ (gcc version 3.4.4) executed with the -O3 option. Times were recorded with the standard Unix `getrusage` function. All running times given are the average of three runs.

The five approaches were tested on the four different data sets shown in Table 1 obtained from the Pizza-Chili Corpus[1]. The large lcp values for the ENGLISH corpus are due to duplicate documents in the text. Also note that we tested our algorithms on intermediate file sizes (100 Mb), but the resources required fell between those of the 50 Mb and 200 Mb data sets as expected, and so are not reported here.

Figure 5 shows plots of the running time versus the memory usage on the four data sets. Several points are immediately obvious. Firstly, comparing PT-MEM (squares) with PT-DISK (circles) horizontally in each panel, we see that using the disk based version requires very little extra time over and above the memory based version. That is, the circles are about level vertically with the squares for the same v value. This confirms the left-to-right nature of our implementations; because we process the data structures on disk left-to-right, the buffering and pre-fetching strategies employed by the operating system hide the cost of disk reads and seeks.

Secondly, examining the methods at the memory levels given by the L-6 method (leftmost triangle in curve in each panel), we see that PT-MEM (squares) is faster on all four data sets when $v \geq 64$. Moving the SA and LCP to disk, PT-DISK still remains faster than L-6, and uses about 60% less memory.

---

[1] `pizzachili.dcc.uchile.cl`

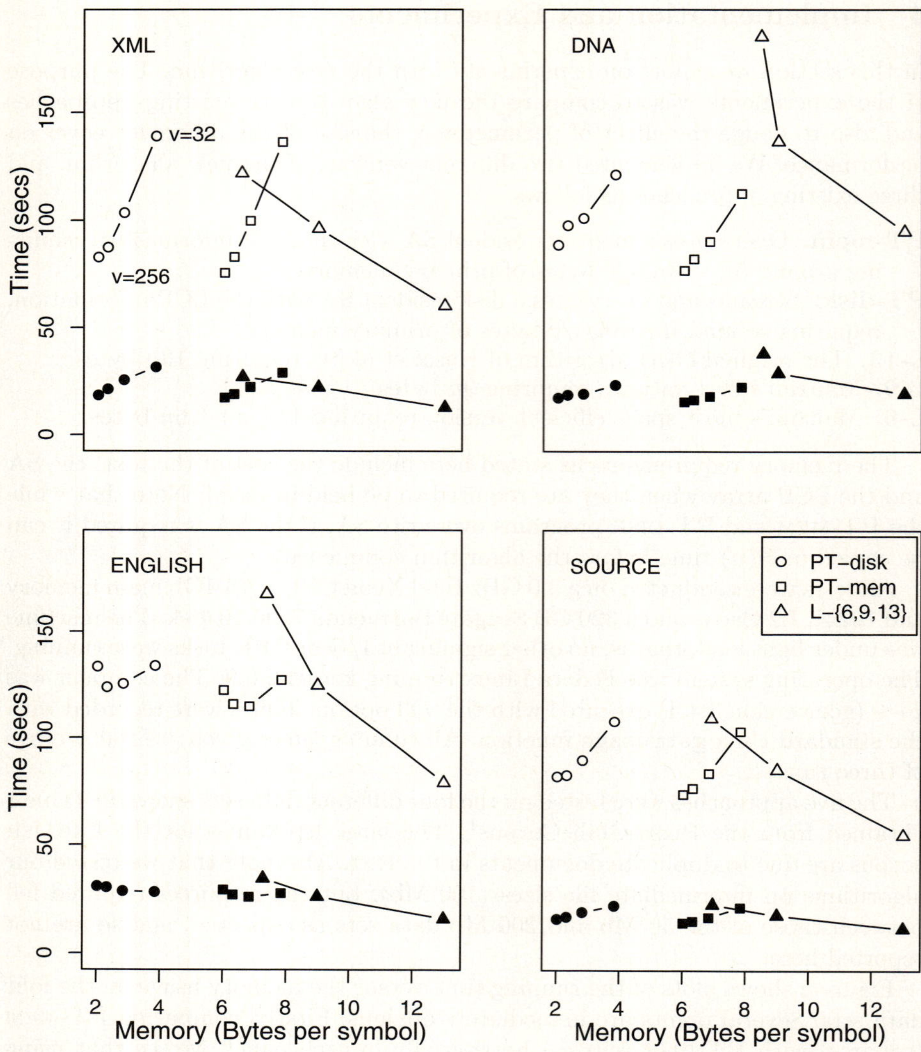

Fig. 5. Time-memory tradeoff for the three approaches. Each panel represents a different collection, with open symbols are for the 200Mb variants, and filled symbols for the 50Mb. Lines linking data points are for display only, and do not represent data points. Distinguishing the L methods (triangles) is obvious from the memory usage of each. For the PT methods, the four data points for each cuve represent $v = 32, 64, 128, 256$ respectively, with memory usage decreasing as v increases, as indicated for the PT-DISK method in the XML panel.

Thirdly, we note that on XML, DNA and SOURCES, the run-time of the new PT approaches decreases with increasing v. This seems counterintuitive, since we showed that the asymptotic runtime of the algorithm was directly proportional to

Table 1. Data sets used for empirical tests

| | | | | lcp | |
Data set name	Size (Mb)	$|\Sigma|$	H_5	mean	max
DNA-50	50	4	1.903	31	14,836
DNA-200	200	4	1.901	59	97,979
XML-50	50	97	0.735	42	1,005
XML-200	200	97	0.817	44	1,084
ENGLISH-50	50	239	1.764	2,221	109,394
ENGLISH-200	200	239	1.839	9,390	987,770
SOURCES-50	50	230	1.372	168	71,651
SOURCES-200	200	230	1.518	373	307,871

v, and so time should increase as v increases. The asymptotic analysis, however, hides the internal tradeoff between the time to establish the S, $\hat{S}$ and L arrays and the actual time to compute the LCP array. For these three data sets, the average lcp value is quite low, and so the time invested in establishing the data structures is not recouped during the LCP construction phase. As v decreases, the initialisation time increases, but the extra knowledge of sampled lcp values is not used as many lcp values are still less than v.

On the ENGLISH data set, which has longer lcp in general, there is a reward for decreasing v from 256 to 128: the increase in the time required to produce a denser sampling in S, $\hat{S}$ and L is recovered by faster production of LCP. Again, however, when v is too low, the setup time begins to dominate again; hence the "u-turn" in the curves for the PT methods in the bottom left panel of Figure 5.

Finally we can observe that if you have memory to spare, then L-13 is the algorithm of choice, consistently coming in faster on all but the DNA data set. The DNA data set has the property that it has very short lcp values, and so the algorithm has particularly poor locality of memory reference in comparison to the PT approaches. Note also that the memory use by L-6 is about 8.5 bytes per input symbol for the DNA data set because of the high value of H_k, or the poor compressibility, for that data. Similarly, Table 1 shows that the ENGLISH-200 collection has a high H_5 value, and likewise the memory required by L-6 is about 8 bytes per symbol.

5   Discussion

The LCP array is a vital component in many algorithms that exploit suffix arrays for efficient string processing. In this paper we have introduced a new algorithm for computing the LCP array from the Suffix Array which allows a controlled tradeoff between speed and memory. It requires $O(nv)$ time and $5n + O(n/\sqrt{v})$ bytes of primary memory. Choosing v to be larger than 32 allowed our method to run faster and in less memory than existing approaches for LCP construction on the data sets used in our experiments. Moreover, if the SA and LCP are stored on disk, the memory of our algorithm falls by a further $4n$ bytes, and in practice is still faster than both L-6 and L-9 while using about 70% less primary memory.

We remark that one could just keep the data structures representing the sample of suffixes as a surrogate LCP array and compute actual LCP values on demand, rather than explicitly computing the entire LCP array. Each "access" to an element in LCP would cost $O(v)$ time. On the other hand, instead of computing the full LCP values, one could choose to store only values in the range $0..v$ and compute the remainder of the length as needed using the surrogate LCP array. Each of these smaller entries can be stored in $O(\log v)$ bits, and constant time access to each full LCP value is maintained. We are currently investigating the efficacy of such succinct LCP representations.

There are several other avenues future work might take. Firstly, Manzini's algorithms are all refinements (particularly to space overheads) of the original algorithm by Kasai et. al. As the method we employ here for producing the L array (Figure 3) is a modification of Kasai et. al's approach, perhaps Manzini's tricks can be adapted to reduce the space requirements for this step in our algorithm. Another interesting question is whether one of the many fast algorithms for SA construction [14] can be modified to also output the LCP information in an efficient manner. Finally, our current algorithm makes the tacit assumption that the input text can reside in main memory so that random accesses to it are not too costly. Developing variants that make only batches of sequential accesses to the text to allow the algorithm to scale to massive, disk resident data is an important open problem.

References

1. Abouelhoda, M.I., Kurtz, S., Ohlebusch, E.: Replacing suffix trees with enhanced suffix arrays. Journal of Discrete Algorithms 2(1), 53–86 (2004)
2. Apostolico, A.: The myriad virtues of subword trees. In: Apostolico, A., Galil, Z. (eds.) Combinatorial Algorithms on Words. NATO ASI Series F12, pp. 85–96. Springer, Heidelberg (1985)
3. Bender, M.A., Farach-Colton, M.: The LCA problem revisited. In: Gonnet, G., Panario, D., Viola, A. (eds.) LATIN 2000. LNCS, vol. 1776, pp. 88–94. Springer, Heidelberg (2000)
4. Burkhardt, S., Kärkkäinen, J.: Fast lightweight suffix array construction and checking. In: Baeza-Yates, R., Chávez, E., Crochemore, M. (eds.) CPM 2003. LNCS, vol. 2676, pp. 55–69. Springer, Heidelberg (2003)
5. Colbourn, C.J., Ling, A.C.H.: Quorums from difference covers. Information Processing Letters 75(1–2), 9–12 (2000)
6. Fischer, J., Heun, V.: Theoretical and practical improvements on the RMQ-problem, with applications to LCA and LCE. In: Lewenstein, M., Valiente, G. (eds.) CPM 2006. LNCS, vol. 4009, pp. 36–48. Springer, Heidelberg (2006)
7. Fischer, J., Heun, V.: A new succinct representation of RMQ-information and improvements in the enhanced suffix array. In: Chen, B., Paterson, M., Zhang, G. (eds.) ESCAPE 2007. LNCS, vol. 4614, pp. 459–470. Springer, Heidelberg (2007)

8. Gusfield, D.: Algorithms on Strings, Trees, and Sequences: Computer Science and Computational Biology. Cambridge University Press, Cambridge (1997)
9. Kasai, T., Lee, G., Arimura, H., Arikawa, S., Park, K.: Linear-time longest-common-prefix computation in suffix arrays and its applications. In: Amir, A., Landau, G.M. (eds.) CPM 2001. LNCS, vol. 2089, pp. 181–192. Springer, Heidelberg (2001)
10. Manber, U., Myers, G.W.: Suffix arrays: a new method for on-line string searches. SIAM Journal of Computing 22(5), 935–948 (1993)
11. Manzini, G.: An analysis of the Burrows-Wheeler transform. Journal of the ACM 48(3), 407–430 (2001)
12. Manzini, G.: Two space saving tricks for linear time LCP computation. In: Hagerup, T., Katajainen, J. (eds.) SWAT 2004. LNCS, vol. 3111, pp. 372–383. Springer, Heidelberg (2004)
13. McCreight, E.M.: A space-economical suffix tree construction algroithm. Journal of the ACM 23(2), 262–272 (1976)
14. Puglisi, S.J., Smyth, W.F., Turpin, A.: A taxonomy of suffix array construction algorithms. ACM Computing Surveys 39(2), 1–31 (2007)
15. Smyth, B.: Computing Patterns in Strings. Pearson Addison-Wesley, Essex (2003)
16. Ukkonen, E.: Online construction of suffix trees. Algorithmica 14(3), 249–260 (1995)
17. Weiner, P.: Linear pattern matching algorithms. In: Proceedings of the 14th annual Symposium on Foundations of Computer Science, pp. 1–11 (1973)

Power Domination in $\mathcal{O}^*(1.7548^n)$ Using Reference Search Trees

Daniel Raible and Henning Fernau

Univ. Trier, FB 4—Abteilung Informatik, 54286 Trier, Germany
`{raible,fernau}@informatik.uni-trier.de`

Abstract. The POWER DOMINATING SET problem is an extension of
the well-known domination problem on graphs in a way that we enrich
it by a second propagation rule: Given a graph $G(V, E)$ a set $P \subseteq V$ is
a power dominating set if every vertex is observed after we have applied
the next two rules exhaustively. First, a vertex is observed if $v \in P$
or it has a neighbor in P. Secondly, if an observed vertex has exactly
one unobserved neighbor u, then also u will be observed as well. We
show that POWER DOMINATING SET remains $\mathcal{NP}$-hard on cubic graphs.
We designed an algorithm solving this problem in time $\mathcal{O}^*(1.7548^n)$ on
general graphs. To achieve this we have used a new notion of search trees
called reference search trees providing non-local pointers.

1   Introduction

We study an extension of DOMINATING SET. The extension originates not from
an additional required property for the solution set (e.g., CONNECTED DOMI-
NATING SET) but by adding a second rule. To be precise we look for a vertex
set, called power dominating set, such that every vertex is observed according
to the next two rules:

Observation Rule 1 (OR1): A vertex in the power domination set observes
itself and all of its neighbors.

Observation Rule 2 (OR2): If an observed vertex v of degree $d \geq 2$ is adjacent to $d - 1$ observed vertices, then the remaining unobserved neighbor becomes
observed as well.

By skipping the second rule we would exactly arrive at DOMINATING SET. The
second rule is responsible for the non-local character of the problem as it im-
plements a kind of propagation. Due to this propagation mechanism a vertex
can observe another vertex at arbitrary distance. Also the sequence of OR2 applications can be arbitrary but leading to the same set of observed vertices.
Indeed, many arguments relying on the locality of DOMINATING SET fail. There
is no transformation to SET COVER and thus the algorithm of [3] cannot simply
be modified. The problem occurs in the context of monitoring electric power
networks. One wishes to place a minimum number of measurement devices at
certain points in the network to measure the state variables (for example the

S.-H. Hong, H. Nagamochi, and T. Fukunaga (Eds.): ISAAC 2008, LNCS 5369, pp. 136–147, 2008.
© Springer-Verlag Berlin Heidelberg 2008

voltage magnitude and the current phase). In that sense OR2 stands for Kirchhoff's law. In this way, we arrive at the definition of the central problem:

POWER DOMINATING SET (PDS)
Given: An undirected graph $G = (V, E)$, and the parameter k.
We ask: Is there a set $P \subseteq V$ with $|P| \leq k$ which observes all vertices in V with respect to the two observation rules OR1 and OR2?

Discussion of Related Results. The study of PDS was initiated by Haynes et *al.* [7] where they showed $\mathcal{NP}$-hardness and gave a first polynomial time algorithm for trees. Guo et *al.* [6] and Kneis et *al.* [8] studied this problem independently with respect to parameterized complexity. They proved $W[2]$-hardness if the parameter is the size of the solution by reducing DOMINATING SET to PDS. As a by-product it turns out that PDS is still $\mathcal{NP}$-hard on graphs with maximum degree four and that there is a lower bound for any approximation ratio of $\Omega(\log n)$ modulo standard complexity assumptions. Additionally, they showed fixed-parameter tractability of PDS with respect to tree-width, where [6] also give a concrete algorithm. [6] achieve this by transforming PDS into a orientation problem on undirected graphs. The problem was also studied in the context of special graph classes like interval graphs [9] and block graphs [12] where linear time algorithms where obtained. Aazami and Stilp [1] improved the approximation lower bound to $\Omega(2^{\log^{1-\epsilon} n})$ and gave an $\mathcal{O}(\sqrt{n})$-approximation for planar graphs. On the other hand also domination problems have been studied in exact algorithmics. Fomin, Gradoni and Kratsch [3] gave a $\mathcal{O}^*(1.5137^n)$-algorithm for DOMINATING SET where they use the power of the measure and conquer approach. The currently fastest $\mathcal{O}^*(1.5134^n)$-algorithm by Rooj and Bodlaender [11] achieves this slight improvement by a new reduction rule. Fomin, Gradoni and Kratsch [4] showed that the variant CONNECTED DOMINATING SET can be solved in time $\mathcal{O}^*(1.9407^n)$.

New Results. First, we show that PDS remains $\mathcal{NP}$-hard for planar cubic graphs. As PDS is polynomial time solvable for max-degree-two graphs and $\mathcal{NP}$-hardness was shown for max-degree-four graphs [6,8], this result closes the gap inbetween. Furthermore, this justifies to follow a branching strategy even in the case of cubic graphs. Note that it is not always true that generally $\mathcal{NP}$-hard graph problems remain $\mathcal{NP}$-hard for cubic graphs. FEEDBACK VERTEX SET is a problem where as with PDS cycles play a role (see [6]). But in contrast to general graphs, it is solvable in polynomial time on cubic graphs [10]. Secondly, we present an algorithm solving PDS in time $\mathcal{O}^*(1.7548^n)$, which breaks the trivial 2^n-barrier. The run time analysis proceeds in an amortized fashion using the measure and conquer approach (see [3]). Furthermore, we introduce the concept of a reference search tree. In an ordinary search tree we usuallay cut off branches due to local structural conditions. In a reference search tree we also will cut off branches if we can point to another node of the search tree where we can find no worse solutions. This node must not be a neighbor of the current node but can be anywhere in the search tree, as long as the overall search structure remains acyclic.

Terminology and Notation. The *(open) neighborhood* of $v \in V$ is $N(v) = \{w \mid \{w, v\} \in E\}$ and the *closed neighborhood* $N[v] := N(v) \cup \{v\}$. A possible solution set will be denoted P. We call a vertex $v \in V$ *directly observed* by $u \in N(v)$ if u is in the solution, i.e., $u \in P$. The vertex $v \in V$ will be called *indirectly observed* by $u \in V$ if v is observed due to the application of OR2 onto u. An *(a,b)-branch* is a binary branch which reduces the problem measure by an amount of a in one part of the branch and by b in the other.

2 $\mathcal{NP}$-Hardness of Planar Cubic Power Dominating Set

We will reduce VERTEX COVER to PLANAR CUBIC PDS. Due to [5] VERTEX COVER remains $\mathcal{NP}$-complete on planar cubic graphs. For any planar cubic graph $G(V, E)$ and any $v \in V$ we can denominate the neighbors of v as follows: $N(v) = \{n_{v_1}, n_{v_2}, n_{v_3}\}$. The reduction works as follows: Given a planar cubic graph $G(V, E)$ introduce for every $v \in V$ the gadget T_v depicted in Figure 1, which consists of the vertices in the dotted square. For any $\{u, v\} \in E$ we can find $1 \leq b, c \leq 3$ such that $u = n_{v_b}$ and $v = n_{u_c}$. By introducing the edge $\{c_{v\,b}, c_{u\,c}\}$ we finally get $G'(V', E')$ which is planar and cubic.

Lemma 1. *G has a vertex cover of size $\leq k$ iff G' has a PDS of size $\leq k$.*

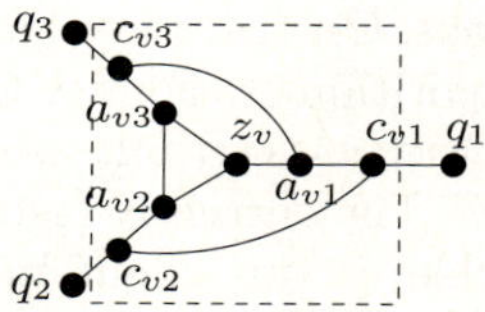

Fig. 1. The gadget T_v. The vertices q_1, q_2, q_3 correspond to vertices of the form $c_{z\,i}$ of some other gadget T_z such that $z \in V$.

According to Lemma 1 PLANAR CUBIC PDS remains $\mathcal{NP}$-hard.

3   An Exact Algorithm for Power Dominating Set

3.1   Reference Search Trees

We will introduce a new kind of search scheme for combinatorial optimization problems. These problems can usually be modeled as follows. We are given a triple $(\mathcal{U}, \mathcal{S}, c)$ such that $\mathcal{U} = \{u_1, \ldots, u_n\}$ is called the *universe*, $\mathcal{S} \subseteq \mathcal{P}(\mathcal{U})$ is the *solution space* and $c : \mathcal{P}(\mathcal{U}) \to \mathbb{N}$ is the *value function*. Generally we are looking for a $S \in \mathcal{S}$ such that $c(S)$ is minimum or maximum. We then speak of a combinatorial minimization (maximization, resp.) problem. The general search space is $\mathcal{P}(\mathcal{U})$.

The *set vector (sv_Q)* of a set $Q \in \mathcal{P}(\mathcal{U})$ is a 0/1-vector indexed by the elements of $\mathcal{U}$ such that: $sv_Q[i] = 1 \iff u_i \in Q$. We write $sv_Q \in \mathcal{S}$ when we mean $Q \in \mathcal{S}$.

A *solvec* is a $0/1/\star$-vector. We define the following partial order $\preceq$ on solvecs s_1, s_2 of length n:

$$s_1 \preceq s_2 \iff \forall 1 \leq i \leq n : (s_1[i] = \star \Rightarrow s_2[i] = \star)$$
$$\wedge (s_1[i] = d \; (d \in \{0,1\}) \Rightarrow s_2[i] \in \{d, \star\}).$$

A *branching* is a directed tree $D(V,T)$ with root $r \in V$ such that all arcs are directed from the father-vertex to the child-vertex. For a vertex $u \in V$ the term ST_u refers to the sub-tree rooted at u.

Definition 1. *A* reference search tree (rst) *for a combinatorial minimization (maximization, resp.) problem* $(\mathcal{U}, \mathcal{S}, c)$ *is a directed graph* $D(V, T \cup R)$ *together with a injective function label* $: V \rightarrow \{(z_1, \ldots, z_n) \mid z_i \in \{0,1,\star\}\}$ *with the following properties:*

1. $D(V,T)$ *is a branching.*
2. $D(V, T \cup R)$ *is acyclic.*
3. *Let* $u, v \in V(D)$ *then* u *is a descendant of* v *in* $D(V,T)$ *iff label$(u) \preceq$ label(v).*
4. *For any set vector* sv_Q *of a set* $Q \in \mathcal{P}(\mathcal{U})$ *with* $Q \in \mathcal{S}$ *and a vertex* $v \in V(D)$ *such that* $sv_Q \preceq$ *label(v) we have either one of the following properties:*
   (a) *There exists a leaf* $z \in V(ST_v)$ *such that* $c(label(z)) \leq c(sv_Q)$ *($c(label(z)) \geq c(sv_Q)$, resp.) and label$(z) \in \mathcal{S}$.*
   (b) *There exists a vertex* $x \in V(ST_v)$ *such that there is exactly one arc* $(x, y) \in R$ *and we have that there is a $0/1$-vector h with $h \preceq$ label(y), $c(h) \leq c(sv_Q)$ ($c(h) \geq c(sv_Q)$, resp.) and $h \in \mathcal{S}$.*

How can a rst be exploited algorithmically? It is important to see that in a rst all the information for finding an optimal solution is included. Ordinary search trees can be defined by omitting item (b) of Definition 1. In a search tree we skip a solution s with $s \preceq u$ for a sub-tree ST_u if we can find a solution in ST_u which is no worse. In a rst we also have the possibility to make a reference to another subtree ST_f where such a solution could be found. In ST_f it might also be the case that we have to follow a reference once more. So, the only obstacle seems to be that, if we follow reference after reference, we end up in a cycle. But this is prevented by item 2. of Definition 1. An algorithm building up an rst can eventually benefit by cutting of branches and introducing references instead.

3.2   Annotated Power Dominating Set

In what follows we assume that the vertices of the given graph $G(V, E)$ are annotated. To be precise we have a function s which assigns a label to every vertex: $s : V(G) \rightarrow \{active, inactive, blank\}$. An *active* (*inactive*, resp.) vertex has already been determined to be (not to be, resp.) part of P. For a *blank* vertex this decision has been not made yet. We will abbreviate the three attributes by $(a), (i)$ and (b). We also define $A := \{v \in V(G) \mid s(v) = (a)\}$, $I := \{v \in V(G) \mid s(v) = (i)\}$ and $B := \{v \in V(G) \mid s(v) = (b)\}$. For any given set $A \subseteq V(G)$ we can determine which vertices are already observed by applying exhaustively OR1 and OR2. Due to this we introduce $s' : V(G) \rightarrow \{(o)bserved, (u)nobserved\}$ and set

$O := \{v \in V(G) \mid s'(v) = (o)\}$ and $U := V(G) \setminus O$. The *state* of a vertex v is the tuple $(s(v), s'(v))$. During the course of the algorithm the states of the vertices (i.e., the labels s, s') will be modified in a way that they represent choices already made. We set $N^\star(v) := \{\{w, v\} \in E \mid s'(w) = (u)\}$ and $d^\star(v) := |N^\star(v)|$. $N^\star(v)$ represents the unobserved neighbors of v. Let $\Delta^\star(G) := \max_{v \in V \wedge s(v) = (b)} d^\star(v)$ and $M(G) = \{v \in B \mid d^\star(v) = \Delta^\star(G)\}$. We define $N^{(i)}(v) = N^\star(v) \cap I$ and $d^{(i)}(v) = |N^{(i)}(v)|$. We will write $d^\star_G(v)$, $N^\star_G(v)$, $d^{(i)}_G$, $N^{(i)}_G(v)$, $s_G(v)$ and $s'_G(v)$ when we are referring to a particular annotated graph G by which the functions are induced. We omit the subscript when it is clear from the context. A vertex $v \in V(G)$ such that $s'(v) = (o)$ and $d^\star(v) = 2$ will be called a *trigger*. A *triggered path* between $v_1, v_k \in V(G)$ with $s(v_1) = (b)$, $s(v_k) = (b)$ is a path $v_1, \ldots, v_k$ such that $s(v_i) = (b)$ and $d^\star(v_i) \le 2$, or v_i is a trigger for $1 < i < k$. A *triggered cycle* is a triggered path with $v_1 = v_k$. Observe that for all $u \in O$ we have $d^\star(u) \ne 1$ due to OR2.

Algorithm. In this section we present reduction rules and the algorithm. Their correctness and run time will be analyzed in the next section. We state the following reduction rules:

Isolated: Let $v \in O \cap B$ such that $d^\star(v) = 0$ then set $s(v) \leftarrow (i)$.

TrigR: Let $v \in V$ be a trigger and $s(v) = (b)$. Then set $s(v) \leftarrow (i)$.

Blank.2: Let $v \in V(G)$ with $d^\star(v) \le 2$, $v \in B \cap U$, $y \in N^\star(v)$ and $s(y) = (i)$. Then set $s(v) \leftarrow (i)$.

Obs.3: Let $v \in V(G)$ such that $v \in B \cap U$, $d^\star(v) \le 1$ and $y \in N(v)$ with $y \in I \cap O$ and $d^\star(y) \ge 3$. Then set $s(v) \leftarrow (i)$.

Trig.2: Let $v \in V(G)$ such that $v \in B \cap U$ and $d^\star(v) \le 1$. If there is a trigger u with $N^\star(u) = \{v, y\}$ and $y \in I \cap U$ then set $s(v) = (i)$.

Observe that for degree-2 vertices there is no valid contraction rule, see Figure 2(a). If we deleted u and connected x and y observation would propagate to z due to OR2. We are now ready to state Alg. 1:

Algorithm 1. An exact algorithm for POWER DOMINATING SET

1: Apply OR2 exhaustively.
2: Apply **Isolated**, **TrigR**, **Blank2**, **Obs3** and **Trig2** exhaustively.
3: Select form $M(G)$ a vertex v according to the priorities:
4:   *a)* $s(v) = (u)$.   {*We prefer unobserved vertices*}
    b) $d^{(i)}(v) < d^\star(v)$ {*We prefer vertices such that not all neighbors are inactive*}
5: **if** $d^\star(v) \ge 4$ **then**
6:    Branch on v by setting 1) $s(v) \leftarrow (i)$ and 2) $s(v) \leftarrow (a)$ in either of the branches.
7: **else if** $d^\star(v) = 3$ **then**
8:    Branch on v: 1) $s(v) \leftarrow (i)$ and 2) $s(v) \leftarrow (a)$ and for all $u \in N^\star(v)$ with $s(u) = (b)$ set $s(u) \leftarrow (i)$.
9: **else if** $d^\star(v) = 2$ **then**
10:    Branch on v by setting 1) $s(v) \leftarrow (i)$ and 2) $s(v) \leftarrow (a)$ in either of the branches.

Correctness. We will prove correctness of Alg. 1 and the reduction rules using the concept of a reference search tree (see Definition 1). We have to define $\mathcal{U} := V(G)$, $\mathcal{S} := \{S \subseteq V(G) \mid S$ is a PDS for $G\}$ and $c(Y) = |Y|$ for every $Y \in \mathcal{P}(\mathcal{U})$. The function $label : V(D) \to \{(e_1, \dots, e_n) \mid e_i \in \{(a), (i), (b)\}\}$ then expresses which vertices are no more blank, i.e., are active or inactive. Here (a) refers to 1, (i) to 0 and (b) to the $\star$-symbol defined in the function $label$ of Definition 1. According to this we set $\overline{(a)} = (i)$ and $\overline{(i)} = (a)$.

The nodes of the rst $V(D)$ represent choices made concerning the blank vertices of $V(G)$. These choices can be due to branching or to applying reduction rules. Hence there is a 1-to-1 correspondence between $V(D)$ and the application of reduction rules and branchings. According to this we will speak of *full nodes* and *flat nodes*, i.e. full nodes have two children in $D(V, T)$, flat nodes only one. In particular, nodes where reference pointers start are flat.

If we encounter a vertex $v \in V(G)$ with $s(v) = (i)$ in the current node q of the search tree we can find a second node $d_v \in V(D)$ which represents the choice made on v. That is we must have that $label(q) \preceq label(d_v)$. We can find d_v by simply going up the search tree starting from q. We sometimes indicate this relation by writing d_v^q, whereas we omit the superscript where it is clear from the context.

Prerequisites for Correctness Proofs. The correctness proofs proceed to some extent in a graphical way. For this we draw the branching (the search tree without references) $D(V, T)$ in the plane with x- and y-coordinates. If u is a point in the plane then $pos_x(u)$ denotes its x- and $pos_y(u)$ its y-coordinate. It is possible to draw $D(V, T)$ satisfying three properties.

- Firstly, if $v \in V(D)$ is a father of $u \in V(D)$ then $pos_y(v) > pos_y(u)$.
- Secondly, let $v \in V(D)$ have two children $u_v, u_{\bar{v}}$, i.e., it is a full node. $u_{\bar{v}}$ corresponds to the branch where we set $s(y) = (i)$, in u_v we decided $s(y) = (a)$ for some $y \in V(G)$. We want D to be drawn such that for all $z \in ST_{u_{\bar{v}}}$ we have $pos_x(v) > pos_x(z)$ and for all $z' \in ST_{u_v}$ we have $pos_x(z') > pos_x(v)$. Hence we may speak of $u_{\bar{v}}$ as the left and u_v as the right child of v. According to this we will refer to them as $l(v)$ and $r(v)$, respectively.
- Thirdly, let $v \in V(D)$ be a flat node with its only child v_c. Then we require that $pos_x(v) = pos_x(v_c)$.

The subsequent correctness proofs proceed as follows: Every time we skip a possible solution we show that we can insert a reference to some node $u \in V(D)$ of the search tree such that we can find a solution z with $label(z) \preceq label(u)$ which is no worse. Additionally, we show that these references always point from the left to the right (with respect to the x-coordinate) in the drawing of $D(V, T)$. This way we assure acyclicity of the final rst $D(V, T \cup R)$ which is implicitly built up by the algorithm.

Lemma 2. *Let us fix an annotated PDS instance $G(V, E)$ that corresponds to some node $q \in V(D)$ in the search tree. Let $u \in V(G)$ with $u \in I \cap O$ and $d_G^\star(u) \geq 3$. Then $d_u^q \in D(V, T)$ is a full node.*

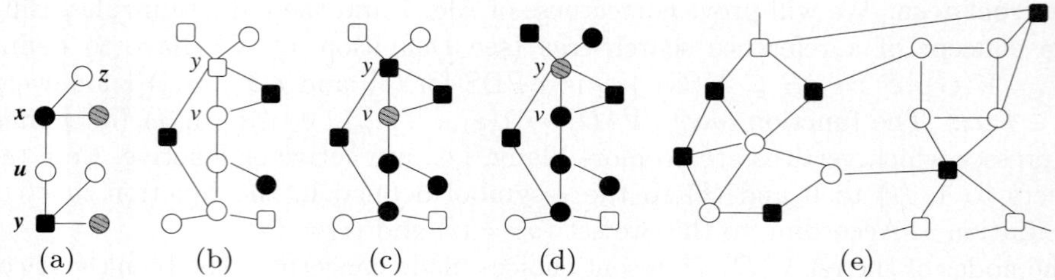

Fig. 2. Filled vertices are observed, white vertices are unobserved. Round vertices are blank, square vertices are inactive. Shaded vertices are active.

Proof. Suppose the contrary. Due to $d_G^\star(u) \geq 3$ none of the reduction rules in Alg. 1 have set $s(u) \leftarrow (i)$ and thus there is no reference starting in u. The only remaining possibility is the second part of the branch in step 7 of Alg. 1. Now suppose by setting $s(v) \leftarrow (a)$ for some $v \in V(G)$ the algorithm has set also $s(u) \leftarrow (i)$ and $s'(u) = (o)$ implicitly. We now examine the situation right before this happened. This situation is reflected by some annotated graph $G'(V, E)$. We must have $d_{G'}^\star(v) = 3$ and $s'_{G'}(u) = (u)$. Suppose at this point $s'_{G'}(v) = (u)$. From this it follows that $d_{G'}^\star(u) \geq 4$ due to our premise. But this contradicts the choice of v as branch vertex. Therefore we must have $s'_{G'}(v) = (o)$. But once more this contradicts the choice of v since we have $s'_{G'}(u) = (u)$ at that point (as step 7 applied u should have been observed by v directly). $\qquad\square$

Lemma 3. *Applying* **Blank2**, **Obs3**, **Trig2**, **TrigR**, **Isolated** *and step 7 of Alg. 1 is correct.*

Proof. We will prove the following: 1) for every vertex $v \in V(G)$ with $s(v) = (i)$ either d_v is a full node or there is a vertex h such that d_h is a full node and we inserted a reference $(v, r(d_h))$ such that if there is a solution with $s(v) = (a)$ at that point in the search tree we can find a no worse one z with $label(z) \preceq label(r(d_h))$. 2) Every reference is pointing from the left to the right in the drawing of $D(V, T)$. As step 7 makes use of this fact, it will be proven in parallel. The proof is by induction on the height s of the search tree. In case $s = 0$ nothing is to show. If $s > 0$ we will distinguish between the different operations:

Blank2. Let q be the current search tree node and, w.l.o.g., $label(q) = (e_1,\ldots, e_{l-1},(b),\ldots,(b))$ such that e_l corresponds to v and e_1 to y (with v and y we are referring to the definition of **Blank2**). Suppose **Blank2** applies to v and d_y^q is a full node (see Figure 2(b)). Suppose there is a solution $k := (e_1,\ldots,e_{l-1},(a),e_{l+1},\ldots,e_n) \preceq label(q)$ as indicated in Figure 2(c). Then also $k' := (\bar{e}_1,\ldots, e_{l-1}, (i), e_{l+1},\ldots, e_n)$ is a solution (due to $d^\star(v) \leq 2$ and OR2, see Figure 2(d)). Hence we insert a reference $(q, r(d_y^q))$ as $k' \preceq label(r(d_y^q))$ which is pointing from the left to the right. This reference means that we can find a no worse solution compared to k. We find this solution in the sub search tree $ST_{r(d_y^q)}$ as $label(k') \preceq label(r(d_y^q))$ or else we have to follow another reference to the right. Therefore we can skip the possibility of setting $s(v) \leftarrow (a)$.

If d_y^q is a flat node than due to the induction hypothesis there must be a reference (d_y^q, h') such that $h' \in V(D)$ is the right child of some $h \in V(D)$ which is a full node and (d_y^q, h') points from the left to the right. We can rule out the solution k again as k' is no worse. Due to (d_y^q, h') also k' is skipped as there must be a alternative solution z with $label(z) \preceq h'$ such that z is no worse than k'. Thus we can insert the reference (q, h') pointing also from the left to the right.

Trig2. The proof is completely analogous to the first item.

Obs3. From Lemma 2 we have that y is a full node. The correctness follows now analogously to the first part of the first item.

Step 7. We only have to consider the second part of the branch: $s(v) \leftarrow (a)$ and for all $u \in N^\star(v)$ with $s(u) = (b)$ set $s(u) \leftarrow (i)$. Let $N^\star(v) = \{a, b, c\}$. We make a case distinction concerning $d^{(i)}(v)$. Let $q \in D(V)$ be the current search tree node before branching and let $label(q) = (e_1, \ldots, e_l, (b), \ldots, (b))$, where the entries e_1 corresponds to v, e_2 to a and e_3 to b. Therefore we have $e_1 = (b)$. Assume there is a PDS $P \ni v$ with $sv_P \preceq label(q)$.

$d^{(i)}(v) = 0$ If $|P \cap N^\star(v)| \in \{2, 3\}$ then also $P \setminus \{v\}$ is a PDS due to OR2. If w.l.o.g $P \cap N^\star(v) = \{a\}$ then also $P' := P \setminus \{v\} \cup \{b\}$ is a PDS where P' is covered by the first part of the branch.

$d^{(i)}(v) = 1$ The only cases $|P \cap N^\star(v)| = 1$ and $|P \cap N^\star(v)| = 2$ can be handled analogously to the case $d^{(i)} = 0$.

$d^{(i)}(v) = 2$ W.l.o.g. $P \cap N^\star(v) = \{a\}$. Assume there is a PDS corresponding to $k := ((a), (a), (i), e_4, \ldots, e_n) \preceq label(q)$; $k' := ((i), (a), (a), e_4, \ldots, e_n)$ is then a solution, too. Suppose d_b^q is a full node. Then $k' \preceq label(r(d_b^q))$ and hence we insert a reference $(q, r(d_b^q))$. If d_b^q is a flat node there must be a reference (d_b^q, h'). Then insert (q, h').

The inserted references are pointing all from the left to the right in the drawing of $D(V, T)$. This ensures acyclicity of $D(V, T \cup R)$.   □

Note that the reduction rules treated in Lemma 3 are not valid on their own. They are only correct because they are referring to solutions which Alg. 1 definitely will consider. In other words, if we are given an annotated graph G, where the annotation is not due to Alg. 1 we cannot apply these reduction rules.

Run Time Analysis. We define the following sets: $\hat{I} = \{v \in I \mid s'(v) \neq (o), \exists u \in N^\star(v) : s(u) = (b)\}$, $\hat{O} = (O \cap B)$, $\hat{B} = B \cap U$. Here $\hat{I}$ comprises the inactive vertices, which are not observed such that they have at least one neighbor which is blank. In $\hat{O}$ we find the observed vertices for which we have not yet decided if there are active or not. Also for any $v \in \hat{O}$ we have $d^\star(v) \geq 2$ (**Isolated**). $\hat{B}$ contains the unobserved blank vertices. The measure we use in our run time is the following one:

$$\varphi(G) = |\hat{B}| + \beta \cdot |\hat{O}| + \gamma \cdot |\hat{I}| \text{ with } \beta = 0.51159, \gamma = 0.48842$$

We will now analyze the different branchings in Alg. 1. In general we can find integers ℓ, k with $\ell + k = d^\star(v)$ such that $\ell = |N^\star(v) \cap \hat{I}|$ and $k = |N^\star(v) \cap \hat{B}|$.

$d^\star(v) \geq 4$: The first case is when we have chosen a vertex v with $d := d^\star(v)$. We will explicitly analyze the case when $d = 4$. We show that any case occurring for $d > 4$ is run time upper bounded by some case when $d = 4$. First we will distinguish between the circumstances that $s'(v) = (o)$ and $s'(v) = (u)$.

$s'(v) = (u)$ In the branch where we set $s(v) \leftarrow (a)$ we get a reduction in $\varphi(G)$ of $1 + \ell \cdot \gamma + k \cdot (1 - \beta)$. This is due to v becoming observed and active, the vertices in $N^\star(v) \cap \hat{I}$ becoming inactive and observed and $N^\star(v) \cap \hat{B}$ becoming observed and blank. In the branch setting $s(v) \leftarrow (i)$ we reduce $\varphi(G)$ by at least $(1 - \gamma)$ (we obtain a greater reduction if v drops out of $\hat{I}$). As $(1 - \beta) < \gamma$ the worst case is the branch $(1 + 4(1 - \beta), 1 - \gamma)$ which is upper bounded by $\mathcal{O}^*(1.6532^n)$.

$s'(v) = (o)$ In the branch where we set $s(v) \leftarrow (a)$ we get a reduction in $\varphi(G)$ of $\beta + \ell \cdot \gamma + k \cdot (1 - \beta)$. Here we get only a reduction of β from v as it is already observed. In case $s(v) \leftarrow (i)$ the reduction is again β as v drops out of $\hat{O}$. As $(\beta + 4(1 - \beta), \beta)$ is the worst branch we have a upper bound of $\mathcal{O}^*(1.7548^n)$.

We examine now cases with $d > 4$. Here the worst case is analogously when $k = d$. But it is also no better as the case when $k = 4$ and $d = 4$, which was already considered.

$d^\star(v) = 3$: We first focus on the case where $\ell \leq 2$. As we get a reduction of one for every vertex in $N^\star(v) \cap \hat{B}$ the worst case is when $\ell = 2$. Now if $s'(v) = (u)$ then this results in a $(2 + 2 \cdot \gamma, 1 - \gamma)$ branching. If $s'(v) = (o)$ we have a $(\beta + 2 \cdot \gamma + 1, \beta)$-branching. $\mathcal{O}^*(1.7489^n)$ is an upper bound for both.

Now due to the priorities in step 4 of Alg. 1 we select a vertex v such that $\ell = 3$ with least priority. We first examine the case where $s'(v) = (u)$ and $\ell = 3$. Now suppose for all $u \in N^\star(v)$ we have $N^\star(u) \cap B = \{v\}$ (❂). Then in the branch $s(v) \leftarrow (i)$ we get an additional amount of $3 \cdot \gamma$. This is due to the fact that the vertices in $N^\star(v)$ will drop out of $\hat{I}$. Hence, we have a $(1 + 3 \cdot \gamma, (1 - \gamma) + 3 \cdot \gamma)$ branch.

Conversely, there is a $u \in N^\star(v)$ with $N^\star(u) \cap B = \{v, u_1, \ldots, u_s\}$ and $s \geq 1$ (❂). If $s'(u_1) = (o)$ then due to **TrigR** and the choice of v we have $d^\star(u_1) = 3$. In $s(v) \leftarrow (a)$ u will become a trigger and is reduced away from $\varphi(G)$ due to **TrigR**. This means we have a $(1 + 3 \cdot \gamma + \beta, 1 - \gamma)$ branch.

If $s'(u_1) = (u)$ then we have $d^\star(u_1) = 3$ due to **Blank2** and the choice of v. Also it holds that $d^{(i)}(u_1) = 3$ again by the choice of v. Hence in $s(v) \leftarrow (a)$ the $\star$-degree of u_1 drops by one. Therefore **Blank2** applies on u_1 and it will not appear in $\varphi(G)$ anymore which leads to a $(2 + 3 \cdot \gamma, 1 - \gamma)$ branch. $\mathcal{O}^*(1.6489^n)$ upper bounds both possibilities.

The second possibility for v is $s'(v) = (o)$, yielding a $(\beta + 3 \cdot \gamma, \beta + 3 \cdot \gamma)$-branch for case ❂. In case of ❂, $s'(u_1) = (o)$ is necessary or otherwise, we contradict the choice of v $(d^\star(u_1) \geq 3)$, or **Blank2** applies to u_1 $(d^\star(u_1) \leq 2)$. Again we have $d^{(i)}(u_1) = 3$. Hence, this gives a $(2\beta + 3\gamma, \beta)$-branch, as u_1 becomes a trigger. An upper bound for both is $\mathcal{O}^*(1.7488^n)$.

$d^\star(v) \le 2$:

Lemma 4. *In step 10 of Alg.1 we have:*

1. *For all $u \in V$ with $d^\star(u) \ge 3$ it follows that $s(u) = (i)$.*
2. *Let $v \in V(G)$ with $s(v) = (b)$ and $s'(v) = (u)$ chosen for branching then:*
   (a) For all $u \in N^\star(v) : s(u) = (b)$.
   (b) If $d^\star(v) \le 1$ then for all $u \in N(v) \setminus N^\star(v) : s(u) = (i)$ and $d^\star(u) = 2$.
3. *$O = O \cap I$. ($O \cap B = \emptyset$, alternatively).*

Proof. *1*. Otherwise, we have a contradiction to the choice of v. *2(a)* Otherwise, **Blank2** applies. *2(b)* Note that $s'(u) = (o)$. Suppose $s(u) = (b)$ then either **TrigR** applies ($d^\star(u) = 2$) or we have a contradiction to the choice of v ($d^\star(u) \ge 3$). From $s(u) = (a)$ it follows that $s'(v) = (o)$, a contradiction. If we had $d^\star(u) \ge 3$ and $s(u) = (i)$ then **Obs3** applies. This contradicts $s(v) = (b)$. *3*. Let $u \in O \setminus I$ then $d^\star(u) \in \{0,1,2\}$ is ruled out by **Isolated**, **OR2** and **TrigR**. If $d^\star(u) \ge 3$ then from item 1. it follows that $u \in I$, a contradiction. $\square$

Let v be the vertex chosen in step 9 of Alg.1. Let $\tilde{G} := G[B]$ and note that $B = \hat{B}$ due to Lemma 4.3. $\tilde{G}$ consists of paths and cycles formed by vertices in $\hat{B}$ due to Lemma 4.2a and the fact that for all $z \in B$ we have $d^\star(z) \le 2$ (see Figure 2(e)). v belongs to one of those components. Explore G the following way:

1. For all $u \in B$ set $visited(u) \leftarrow f$.
2. If there is $u \in N^\star(v)$ with $visited(u) = f$ then set $visited(v) \leftarrow t$ and $v \leftarrow u$.
3. If there is $t \in N(v)$ with $t \in O$ (due to Lemma 4.2b u is a trigger) such that $d^\star(t) = \{v, u\}$ and $visited(u) = f$ then set $visited(v) \leftarrow t$ and $v \leftarrow u$.
4. If one of the steps 3 or 4 applied goto 2.. Else set $visited(v) \leftarrow t$ and stop.

Let $W := \{u \in B \mid visited(u) = t\}$. W comprises the visited vertices in $\hat{B}$. Either W is path or a cycle containing at least two vertices from $\hat{B}$ (as long $|V(G)| > 1$ and due to Lemma 4.2b). Either v has a blank neighbor or it has a trigger as neighbor. Now observe that any vertex in W is equally likely to be set active: Once there is an active vertex from W the whole vertex-set W will be observed due to OR2. Also any additional trigger $t' \notin W$ which is a neighbor of some $v' \in W$ only depends on v' being observed.

Considering the branch $s(v) \leftarrow (a)$ due to exhaustively applying OR2 for any $v' \in W$ we have $N(v') \subseteq O$ afterwards. Hence W will drop out of $\hat{B}$ but will also not be included in $\hat{O}$ due to **Isolated**. Thus there is a reduction of $|W| \ge 2$.

In case $s(v) \leftarrow (i)$ due to applying **Blank2** and **Trig2** we have that $W \subset I \setminus \hat{I}$ (Lemma 4.2a/2b) and thus a reduction of $|W|$. Summing up we have a $(2,2)$ branch which we upper bound by $\mathcal{O}^*(1.415^n)$.

Note that the instances occurring at this point of Alg.1 are still $\mathcal{NP}$-hard to solve. There is a simple reduction from CUBIC PDS. For any vertex v, create a cycle C_v of length three. If $\{u, v\} \in E$, connect free vertices $x \in C_v$ and $y \in C_u$ with an inactive trigger. So, we have no alternative to continuing the branching.

Theorem 1. POWER DOMINATING SET *can be solved in time* $\mathcal{O}^*(1.7548^n)$.

We like to comment that Alg. 1 achieves a run time of $\mathcal{O}^*(1.6212^n)$ on cubic graphs. This can be seen by modifying the general analysis. Simply choose $\beta = 0.8126$ and $\gamma = 0.3286$ and skip the part where $d^\star(v) \geq 4$.

4   Conclusion and Further Perspectives

Speed-Up With Exponential Space. We precompute optimal solutions for all vertex induced subgraphs with no more than $\omega = 0.1103n$ vertices. For any subgraph G_S we create $2^{\omega n}$ instances by deciding for all $v \in V(G_S)$ if they are observed yet or not. By solving each of these instances by brute force, we spend $4^{\omega n}$ steps for any induced subgraph with predetermined observation pattern. Thus we need $\mathcal{O}^*(\binom{n}{\omega n}4^{\omega n}) \in \mathcal{O}^*(1.6493^n)$ steps for building up a table of size $\mathcal{O}(1.5275^n)$. Let $R = V(G) \setminus \{v \in O \mid N[v] \subseteq O\}$. Once we arrived at a graph G with $|R| \leq \omega n$ in Alg. 1, we can look up the rest of the solution by inspecting the table entry which is determined by $G[R]$ and $R \cap O$. Thus Alg. 1 will run in $\mathcal{O}^*(1.7548^{(1-\omega)n}) \subseteq \mathcal{O}^*(1.6493^n)$. It is important to notice that we ignored the fact that there might be active and inactive vertices in $G[R]$. The correctness follows from the fact that the state of observation of $G[V \setminus R]$ is independent of how $G[R]$ is observed. Also any solution for $G[R]$ which ignores the labels active and inactive cannot be worse than one that does not. By choosing $\omega = 0.092972$ the same algorithm solves CUBIC PDS in $\mathcal{O}^*(1.55^n)$ steps using $\mathcal{O}(1.4533^n)$ space.

Notice that this type of speed-up relies on the fact that no branching or reduction rule ever changes the (underlying) graph itself, but rather, the existing graph is annotated. This property is also important when designing improved algorithms with the help of reference search trees, since it might be tricky to argue to find a solution not worse than the ones to be expected in a particular branch of a search tree somewhere else in the tree, when the instance is (seemingly) completely changed.

Résumée. In this paper, we designed an exact algorithm for POWER DOMINATING SET consuming $\mathcal{O}^*(1.7548^n)$ time. To achieve this we made intensive use of the concept of a reference search tree. This means that we where able to cut off branches by referring to arbitrary locations in the search tree where one can find equivalent solutions. Maybe the term search-DAG expresses this property also quite well. In the long version of [2] we already applied the concept of an rst successfully. We proved the correctness of a reduction rule whose application was critical for the run time. We expect that we can exploit reference search trees further by designing exact algorithms for non-local problems or improving existent ones (e.g., CONNECTED DOMINATING SET/VERTEX COVER or MAX INTERNAL SPANNING TREE). For this kind of problems we are not allowed to delete vertices due to selecting vertices into the solution or not. We rather have to label them. Many algorithms come to decisions by respecting them. We rather try to make use of them. Let x be a labeled vertex not selected into the solution and y an unlabeled vertex. Suppose by re-labeling x (taking x into the solution) and excluding y from the solution we have a solution which is no worse to the possibility of taking y into the solution. Then we can skip this last possibility by inserting a reference. We imagine that this arguing is also possible when several vertices are involved.

References

1. Aazami, A., Stilp, M.D.: Approximation algorithms and hardness for domination with propagation. In: Charikar, M., Jansen, K., Reingold, O., Rolim, J.D.P. (eds.) RANDOM 2007 and APPROX 2007. LNCS, vol. 4627, pp. 1–15. Springer, Heidelberg (2007)
2. Fernau, H., Raible, D.: Exact algorithms for maximum acyclic subgraph on a superclass of cubic graphs. In: Nakano, S.-i., Rahman, M. S. (eds.) WALCOM 2008. LNCS, vol. 4921, pp. 144–156. Springer, Heidelberg (2008); long version available as Technical Report 08-5, Technical Reports Mathematics / Computer Science, University of Trier, Germany (2008)
3. Fomin, F.V., Grandoni, F., Kratsch, D.: Measure and conquer: domination – a case study. In: Caires, L., Italiano, G.F., Monteiro, L., Palamidessi, C., Yung, M. (eds.) ICALP 2005. LNCS, vol. 3580, pp. 191–203. Springer, Heidelberg (2005)
4. Fomin, F.V., Grandoni, F., Kratsch, D.: Solving connected dominating set faster than 2^n. In: Arun-Kumar, S., Garg, N. (eds.) FSTTCS 2006. LNCS, vol. 4337, pp. 152–163. Springer, Heidelberg (2006)
5. Garey, M.R., Johnson, D.S.: The rectilinear Steiner tree problem is NP-complete. SIAM J. Appl. Math. 32(4), 826–834 (1976)
6. Guo, J., Niedermeier, R., Raible, D.: Improved algorithms and complexity results for power domination in graphs. In: Liśkiewicz, M., Reischuk, R. (eds.) FCT 2005. LNCS, vol. 3623, pp. 172–184. Springer, Heidelberg (2005)
7. Haynes, T.W., Mitchell Hedetniemi, S., Hedetniemi, S.T., Henning, M.A.: Domination in graphs applied to electric power networks. SIAM J. Discrete Math. 15(4), 519–529 (2002)
8. Kneis, J., Mölle, D., Richter, S., Rossmanith, P.: Parameterized power domination complexity. Inf. Process. Lett. 98(4), 145–149 (2006)
9. Liao, C.-S., Lee, D.-T.: Power domination problem in graphs. In: Wang, L. (ed.) COCOON 2005. LNCS, vol. 3595, pp. 818–828. Springer, Heidelberg (2005)
10. Speckennmeyer, E.: On feedback vertex sets and nonseparating independent sets in cubic graphs. Journal of Graph Theory 3, 405–412 (1988)
11. van Rooij, J.M.M., Bodlaender, H.L.: Design by measure and conquer, a faster exact algorithm for dominating set. In: STACS 2008, volume 08001 of Dagstuhl Seminar Proceedings. Internationales Begegnungs- und Forschungszentrum fuer Informatik (IBFI), Schloss Dagstuhl, Germany, pp. 657–668 (2008)
12. Xu, G., Kang, L., Shan, E., Zhao, M.: Power domination in block graphs. Theor. Comput. Sci. 359(1-3), 299–305 (2006)

The Isolation Game: A Game of Distances

Yingchao Zhao[1,*], Wei Chen[2], and Shang-Hua Teng[3,**]

[1] State Key Laboratory of Intelligent Technology and Systems
Tsinghua National Laboratory for Information Science and Technology
Department of Computer Science and Technology, Tsinghua University
zhaoyingchao@gmail.com
[2] Microsoft Research Asia
weic@microsoft.com
[3] Boston University
steng@cs.bu.edu

Abstract. We introduce a new multi-player geometric game, which we will refer to as the *isolation game*, and study its Nash equilibria and best or better response dynamics. The isolation game is inspired by the Voronoi game, competitive facility location, and geometric sampling. In the Voronoi game studied by Dürr and Thang, each player's objective is to maximize the area of her Voronoi region. In contrast, in the isolation game, each player's objective is to position herself as far away from other players as possible in a bounded space. Even though this game has a simple definition, we show that its game-theoretic behaviors are quite rich and complex. We consider various measures of farness from one player to a group of players and analyze their impacts to the existence of Nash equilibria and to the convergence of the best or better response dynamics: We prove that it is NP-hard to decide whether a Nash equilibrium exists, using either a very simple farness measure in an asymmetric space or a slightly more sophisticated farness measure in a symmetric space. Complementing to these hardness results, we establish existence theorems for several special families of farness measures in symmetric spaces: We prove that for isolation games where each player wants to maximize her distance to her m^{th} nearest neighbor, for any m, equilibria always exist. Moreover, there is always a better response sequence starting from any configuration that leads to a Nash equilibrium. We show that when $m = 1$ the game is a potential game — no better response sequence has a cycle, but when $m > 1$ the games are not potential. More generally, we study farness functions that give different weights to a player's distances to others based on the distance rankings, and obtain both existence and hardness results when the weights are monotonically increasing or decreasing. Finally, we present results on the hardness of computing best responses when the space has a compact representation as a hypercube.

1 Introduction

In competitive facility location [4,5,7] data clustering [8], and geometric sampling [10], a fundamental geometric problem is to place a set of objects (such as facilities and

* Supported by the National Natural Science Foundation of China under grant No. 60621062 and the National Key Foundation R&D Projects under the grant No. 2004CB318108 and 2007CB311003.
** Supported by NSF grants CCR-0635102 and ITR CCR-0325630. Part of this work was done while visiting ITCS at Tsinghua University and Microsoft Research Asia Lab.

S.-H. Hong, H. Nagamochi, and T. Fukunaga (Eds.): ISAAC 2008, LNCS 5369, pp. 148–158, 2008.
© Springer-Verlag Berlin Heidelberg 2008

cluster centers) in a space so that they are mutually far away from one another. Inspired by the study of Dürr and Thang [3] on the Voronoi game, we introduce a new multi-player geometric game called *isolation game*.

In an isolation game, there are k players that will locate themselves in a space (Ω, Δ) where $\Delta(x, y)$ defines the pairwise distance among points in Ω. If $\Delta(x, y) = \Delta(y, x)$, for all $x, y \in \Omega$, we say (Ω, Δ) is symmetric. The ith player has a $(k-1)$-place function $f_i(\ldots, \Delta(p_i, p_{i-1}), \Delta(p_i, p_{i+1}), \ldots)$ from the $k - 1$ distances to all other players to a real value, measuring the farness from her location p_i to the locations of other players. The objective of player i is to maximize $f_i(\ldots, \Delta(p_i, p_{i-1}), \Delta(p_i, p_{i+1}), \ldots)$, once the positions of other players $(\ldots, p_{i-1}, p_{i+1}, \ldots)$ are given.

Depending on applications, there could be different ways to measure the farness from a point to a set of points. The simplest farness function $f_i()$ could be the one that measures the distance from p_i to its nearest player. Games based on this measure are called *nearest-neighbor games*. Another simple measure is the total distance from p_i to other players. Games based on this measure are called *total distance games*. Other farness measures include the distance of p_i to its mth nearest player, or a weighted combination of the distances from player i to other players.

Isolation games with simple farness measures can be viewed as an approximation of the Voronoi game [1,2,6]. Recall that in the Voronoi game, the objective of each player is to maximize the area of her Voronoi cell in Ω induced by $\{p_1, \ldots, p_k\}$ — the set of points in Ω that are closer to p_i than to any other player. The Voronoi game has applications in competitive facility location, where merchants try to place their facilities to maximize their customer bases, and customers are assumed to go to the facility closest to them. Each player needs to calculate the area of her Voronoi cell to play the game, which could be expensive. In practice, as an approximation, each player may choose to simply maximize her nearest-neighbor distance or total-distance to other players. This gives rise to the isolation game with these special farness measures.

The generalized isolation games may have applications in product design in a competitive market, where companies' profit may depend on the dissimilarity of their products to those of their competitors, which could be measured by the multi-dimensional features of products. Companies differentiate their products from those of their competitors by playing some kind of isolation games in the multi-dimensional feature space. The isolation game may also have some connection with political campaigns such as in a multi-candidate election, in which candidates, constrained by their histories of public service records, try to position themselves in the multi-dimensional space of policies and political views in order to differentiate themselves from other candidates.

We study the Nash equilibria [9] and best or better response dynamics of the isolation games. We consider various measures of farness from one player to a group of players and analyze their impact to the existence of Nash equilibria and to the convergence of best or better response dynamics in an isolation game. For simple measures such as the nearest-neighbor and the total-distance, it is quite straightforward to show that these isolation games are potential games when the underlying space is symmetric. Hence, the game has at least one Nash equilibrium and all better response dynamics converge. Surprisingly, we show that when the underlying space is asymmetric, Nash equilibria may not exist, and it is NP-hard to determine whether Nash equilibria exist in an isolation game. The general isolation game is far more complex even for symmetric spaces, even if we restrict our attention only to uniform anonymous isolation games. We say an isolation game is *anonymous* if for all i, $f_i()$ is invariant under the permutation of its parameters. We say an anonymous isolation game is *uniform* if $f_i() = f_j()$ for all i, j. For instance, the two potential isolation games with the nearest-neighbor or total-distance measure mentioned above are uniform anonymous games. Even these classes of games exhibit different behaviors: some subclass of games always have Nash equilibrium, some can always find better response sequences that converge to a Nash equilibrium, but some may not have Nash equilibrium and determining the existence of Nash equilibrium is NP-complete. We summarize our findings below.

First, We prove that for isolation games where each player wants to maximize her distance to her m^{th} nearest neighbor, equilibria always exist. In addition, there is always a better response sequence starting from any configuration that leads to a Nash equilibrium. We show, however, this isolation game is not a potential game — there are better response sequences that lead to cycles. Second, as a general framework, we model the farness function of a uniform anonymous game by a vector $\boldsymbol{w} = (w_1, w_2, \ldots, w_{k-1})$. Let $\boldsymbol{d_j} = (d_{j,1}, d_{j,2}, \ldots, d_{j,k-1})$ be the *distance vector* of player j in a configuration, which are distances from player j to other $k-1$ players sorted in nondecreasing order, i.e., $d_{j,1} \leq d_{j,2} \leq \ldots \leq d_{j,k-1}$. Then the utility of player j in the configuration is $\boldsymbol{w} \cdot \boldsymbol{d} = \sum_{i=1}^{k-1} (w_i \cdot d_{j,i})$. We show that Nash equilibrium exists for increasing or decreasing weight vectors $\boldsymbol{w}$, when the underlying space (Ω, Δ) satisfies certain conditions, which are different for increasing and decreasing weight vectors. For a particular version of the decreasing weight vectors, namely $(1, 1, 0, \ldots, 0)$, we show that: (a) it is not potential even on a continuous one dimensional circular space; (b) in general symmetric spaces Nash equilibrium may not exist, and (c) it is NP-complete to decide if a Nash equilibrium exists in general symmetric spaces. Combining with the previous NP-completeness result, we see that either a complicated space (asymmetric space) or a slightly complicated farness measure $((1, 1, 0, \ldots, 0)$ instead of $(1, 0 \ldots, 0)$ or $(0, 1, 0, \ldots, 0))$ would make the determination of Nash equilibrium difficult.

We also examine the hardness of computing best responses in spaces with compact representations such as a hypercube. We show that for one class of isolation games including the nearest-neighbor game as the special case it is NP-complete to compute best responses, while for another class of isolation games, the computation can be done in polynomial time.

The rest of the paper is organized as follows. Section 2 covers the basic definitions and notation. Section 3 presents the results for nearest-neighbor and total-distance isolation games. Section 4 presents results for other general classes of isolation games. Section 5 examines the hardness of computing best responses in isolation games. We conclude the paper in Section 6. The full version of the paper with complete proofs can be found in [11].

2   Notation

We use (Ω, Δ) to denote the underlying space, where we assume $\Delta(x, x) = 0$ for all $x \in \Omega$, $\Delta(x, y) > 0$ for all $x, y \in \Omega$ and $x \neq y$, and that (Ω, Δ) is bounded — there exists a real value B such that $\Delta(x, y) \leq B$ for every $x, y \in \Omega$. In general, (Ω, Δ) may not be symmetric or satisfy the triangle inequality. We always assume that there are k players in an isolation game and each player's strategy set is the entire Ω. A *configuration* of an isolation game is a vector $(p_1, p_2, \ldots, p_k)$, where $p_i \in \Omega$ specifies the position of player i. The utility function of player i is a $(k-1)$-place function $f_i(\ldots, \Delta(p_i, p_{i-1}), \Delta(p_i, p_{i+1}), \ldots)$. For convenience, we use $ut_i(c)$ to denote the utility of player i in configuration c.

We consider several classes of weight vectors in the uniform, anonymous isolation game. In particular, the *nearest-neighbor* and *total-distance* isolation games have the weight vectors $(1, 0, \ldots, 0)$ and $(1, 1, \ldots, 1)$, respectively; the *single-selection* game has vectors that have exactly one nonzero entry; the *monotonically-increasing* (or *decreasing*) games have vectors whose entries are monotonically increasing (or decreasing).

A *better response* of a player i in a configuration $c = (p_1, \ldots, p_k)$ is a new position $p_i' \neq p_i$ such that the utility of player i in configuration c' by replacing p_i with p_i' in c is larger than her utility in c. In this case, we say that c' is the *result of a better-response move* of player i in configuration c. A *best response* of a player i in a configuration $c = (p_1, \ldots, p_k)$ is a new position $p_i' \neq p_i$ that maximizes the utility of player i while player j remains at the position p_j for all $j \neq i$. In this case, we say that c' is the *result of a best-response move* of player i in configuration c.

A (pure) *Nash equilibrium* of an isolation game is a configuration in which no player has any better response in the configuration. An isolation game is *better-response potential* (or *best-response potential*) if there is a function F from the set of all configurations to a totally ordered set such that $F(c) < F(c')$ for any two configurations c and c' where c' is the result of a better-response move (or a best-response move) of some player at configuration c. We call F a *potential function*. Note that a better-response potential game is also a best-response potential game, but a best-response potential game may not be a better-response potential game. If Ω is finite, it is easy to see that any better-response or best-response potential game has at least one Nash equilibrium. Henceforth, we use the shorthand "potential games" to refer to better-response potential games.

3   Nearest-Neighbor and Total-Distance Isolation Games

In this section, we focus on the isolation games with weight vectors $(1, 0, \ldots, 0)$ and $(1, 1, \ldots, 1)$. We show that both are potential games when Ω is symmetric, but when Ω is asymmetric and finite, it is NP-complete to decide whether those games have Nash equilibria.

Theorem 1. *The symmetric nearest-neighbor and total-distance isolation games are potential games.*

The following lemma shows that the asymmetric isolation game may not have any Nash equilibrium for any nonzero weight vector. Thus, it also implies that asymmetric nearest-neighbor and total-distance isolation games may not have Nash equilibria.

Lemma 1. *Consider an asymmetric space $\Omega = \{v_1, v_2, \ldots, v_{\ell+1}\}$ with the distance function given by the following matrix with $t \geq \ell + 1$. Suppose that for every player i*

her weight vector $\mathbf{w}_i$ has at least one nonzero entry. Then, for any $2 \le k \le \ell$, there is no Nash equilibrium in the k-player isolation game.

$$
\begin{pmatrix}
\begin{array}{c|cccccc}
\Delta & v_1 & v_2 & v_3 & \ldots & v_\ell & v_{\ell+1} \\
\hline
v_1 & 0 & t-1 & t-2 & \ldots & t-\ell+1 & t-\ell \\
v_2 & t-\ell & 0 & t-1 & \ldots & t-\ell+2 & t-\ell+1 \\
\vdots & \vdots & & \ddots & & \vdots & \\
v_\ell & t-2 & t-3 & t-4 & \ldots & 0 & t-1 \\
v_{\ell+1} & t-1 & t-2 & t-3 & \ldots & t-\ell & 0
\end{array}
\end{pmatrix}
$$

Theorem 2. *It is NP-complete to decide whether a finite, asymmetric nearest-neighbor or total-distance isolation game has a Nash equilibrium.*

Proof. We first prove the case of nearest-neighbor isolation game.

Suppose that the size of Ω is n. Then the distance function Δ has n^2 entries. The decision problem is clearly in *NP*. The NP-hardness can be proved by a reduction from the Set Packing problem. An instance of the Set Packing problem includes a set $I = \{e_1, e_2, \ldots, e_m\}$ of m elements, a set $S = \{S_1, \ldots, S_n\}$ of n subsets of I, and a positive integer k. The decision problem is to decide whether there are k disjoint subsets in S. We now give the reduction.

The space Ω has $n + k + 1$ points, divided into a left set $L = \{v_1, v_2, \ldots, v_n\}$ and a right set $R = \{u_1, u_2, \ldots, u_{k+1}\}$. For any two different points $v_i, v_j \in L$, $\Delta(v_i, v_j) = 2n$ if $S_i \cap S_j = \emptyset$, and $\Delta(v_i, v_j) = 1/2$ otherwise. The distance function on R is given by the distance matrix in Lemma 1 with $\ell = k$ and $t = k+1$. For any $v \in L$ and $u \in R$, $\Delta(v, u) = \Delta(u, v) = 2n$. Finally, the isolation game has $k + 1$ players.

We now show that there exists a Nash equilibrium for the nearest-neighbor isolation game on Ω iff there are k disjoint subsets in the Set Packing instance.

First, suppose that there is a solution to the Set Packing instance. Without loss of generality, assume that the k disjoint subsets are $S_1, S_2, \ldots, S_k$. Then we claim that configuration $c = (v_1, v_2, \ldots, v_k, u_1)$ is a Nash equilibrium. In this configuration, it is easy to verify that every player's utility is $2n$, the largest possible pairwise distance. Therefore, c is a Nash equilibrium.

Conversely, suppose that there is a Nash equilibrium in the nearest-neighbor isolation game. Consider the set R. If there is a Nash equilibrium c, then the number of players positioned in R is either $k + 1$ or at most 1 because of Lemma 1. If there are $k + 1$ players in R, then every player has utility 1, and thus every one of them would want to move to points in L to obtain a utility of $2n$. Therefore, there cannot be $k + 1$ players positioned in R, which means that there are at least k players positioned in L.

Without loss of generality, assume that these k players occupy points $v_1, v_2, \ldots, v_k$ (which may have duplicates). We claim that subsets $S_1, S_2, \ldots, S_k$ form a solution to the Set Packing problem. Suppose, for a contradiction, that this is not true, which means there exist S_i and S_j among these k subsets that intersect with each other. By our construction, we have $\Delta(v_i, v_j) = 0$ or $1/2$. In this case, players at point v_i and v_j would want to move to some free points in R, since that will give them utilities of at least 1. This contradicts the assumption that c is a Nash equilibrium. Therefore, we found a solution for the Set Packing problem given a Nash equilibrium c of the nearest-neighbor isolation game.

The proof for the case of total-distance isolation game is essentially the same, with only changes in players' utility values. $\square$

4   Isolation Games with Other Weight Vectors

In this section, we study several general classes of isolation games. We consider symmetric space (Ω, Δ) in this section.

4.1   Single-Selection Isolation Games

Theorem 3. *A Nash equilibrium always exists in any single-selection symmetric game.*

Although Nash equilibria always exist in the single-selection isolation games, the following lemma shows that they are not potential games.

Lemma 2. *Let $\Omega = \{A, B, C, D, E, F\}$ contain six points on a one-dimensional circular space with $\Delta(A, B) = 15$, $\Delta(B, C) = 11$, $\Delta(C, D) = 14$, $\Delta(D, E) = 16$, $\Delta(E, F) = 13$, and $\Delta(F, A) = 12$. The five-player single-selection game with the weight vector $(0, 1, 0, 0)$ on Ω is not potential.*

Proof. Let the five players stand at A, B, C, D, and E respectively in the initial configuration. Their better response dynamics can iterate forever as shown in Figure 1. Hence this game is not a potential game. $\square$

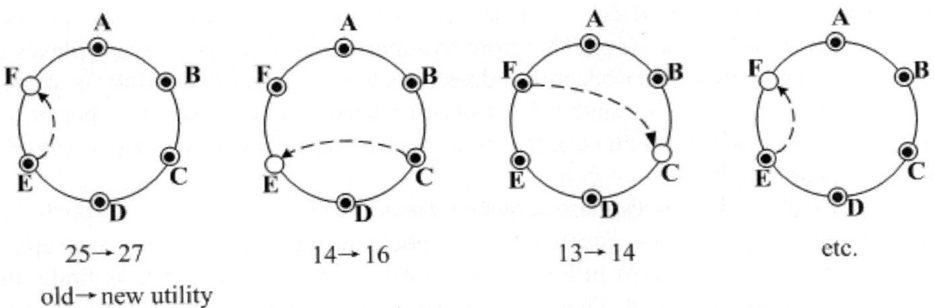

Fig. 1. An example of a better-response sequence that loops forever for a five-player isolation game with weight vector $(0, 1, 0, 0)$ in a one dimensional circular space with six points

Surprisingly, the following theorem complements the previous lemma.

Theorem 4. *If Ω is finite, then for any single-selection game on Ω and any starting configuration c, there is a better-response sequence in the game that leads to a Nash equilibrium.*

Proof. Suppose that the nonzero weight entry is the mth entry in the k-player single-selection isolation game with $m > 1$ (the case of $m = 1$ is already covered in nearest-neighbor isolation game). For any configuration $c = (p_1, \ldots, p_k)$, the utility of player i is the distance between player i and her m^{th} nearest neighbor. Let vector $\boldsymbol{u}(c) = (u_1, u_2, \ldots, u_k)$ be the vector of the utility values of all players in c sorted in nondecreasing order, i.e., $u_1 \leq u_2 \leq \ldots \leq u_k$. We claim that for any configuration c, if c is not a Nash equilibrium, there must exist a finite sequence of configurations $c = c_0, c_1, c_2, \ldots, c_t = c'$, such that c_{i+1} is the result of a better-response move of some player in c_i for $i = 0, 1, \ldots, t - 1$ and $\boldsymbol{u}(c) < \boldsymbol{u}(c')$ in lexicographic order.

We now prove this claim. Since the starting configuration $c_0 = c$ is not a Nash equilibrium, there exists a player i that can make a better response move to position p, resulting in configuration c_1. We have $ut_i(c_0) < ut_i(c_1)$. Let S_i be the set of player i's $m - 1$ nearest neighbors in configuration c_1. We now repeat the following steps to find configurations $c_2, \ldots, c_t$. When in configuration c_j, we select a player a_j such that $ut_{a_j}(c_j) < ut_i(c_1)$ and move a_j to position p, the same position where player i locates. This gives configuration c_{j+1}. This is certainly a better-response move for a_j because $ut_{a_j}(c_{j+1}) = ut_i(c_{j+1}) = ut_i(c_1) > ut_{a_j}(c_j)$, where the second equality holds because we only move the $m - 1$ nearest neighbors of player i in c_1 to the same position as i, so it does not affect the distance from i to her m^{th} nearest neighbor. The repeating step ends when there is no more such player a_j in configuration c_j, in which case $c_j = c_t = c'$.

We now show that $\boldsymbol{u}(c) < \boldsymbol{u}(c')$ in lexicographic order. We first consider any player $j \notin S_i$, either her utility does not change ($ut_j(c) = ut_j(c')$), or her utility change must be due to the changes of her distances to player i and players $a_1, a_2, \ldots, a_{t-1}$, who have moved to position p. Suppose that player j is at position q. Then $\Delta(p, q) \geq ut_i(c_1)$ because $j \notin S_i$. This means that if j's utility changes, her new utility $ut_j(c')$ must be at least $\Delta(p, q) \geq ut_i(c_1)$. For a player $j \in S_i$, if she is one of $\{a_1, \ldots, a_{t-1}\}$, then her new utility $ut_j(c') = ut_i(c') = ut_i(c_1)$; if she is not one of $\{a_1, \ldots, a_{t-1}\}$, then by definition $ut_j(c') \geq ut_i(c_1)$. Therefore, comparing the utilities of every player in c and c', we know that either her utility does not change, or her new utility is at least $ut_i(c') = ut_i(c_1) > ut_i(c)$, and at least player i herself strictly increases her utility from $ut_i(c)$ to $ut_i(c')$. With this result, it is straightforward to verify that $\boldsymbol{u}(c) < \boldsymbol{u}(c')$. Thus, our claim holds.

We may call the better-response sequence found in the above claim an epoch. We can apply the above claim to concatenate new epochs such that at the end of each epoch the vector $\boldsymbol{u}$ strictly increases in lexicographic order. Since the space Ω is finite, the vector $\boldsymbol{u}$ has an upper bound. Therefore, after a finite number of epochs, we must be able to find a Nash equilibrium, and all these epochs concatenated together form the better-response sequence that leads to the Nash equilibrium. This is clearly true when starting from any initial configuration. $\qquad\square$

4.2 Monotonically-Increasing Games

For monotonically-increasing games, we provide the following general condition that guarantees the existence of a Nash equilibrium. We say that a pair of points $u, v \in \Omega$ is a pair of *polar points* if for any point $w \in \Omega$, the inequality $\Delta(u, w) + \Delta(w, v) \leq$

$\Delta(u, v)$ holds. Spaces with polar points include one-dimensional circular space, two-dimensional sphere, n-dimensional grid with L_1 norm as its distance function, etc.

Theorem 5. *If Ω has a pair of polar points, then any monotonically-increasing isolation game on Ω has a Nash equilibrium.*

4.3   Monotonically-Decreasing Games

Monotonically-decreasing games are more difficult to analyze than the previous variants of isolation games, and general results are not yet available. In this section, we first present a positive result for monotonically-decreasing games on a continuous one-dimensional circular space. We then present some hardness result for a simple type of weight vectors in general symmetric spaces.

The following theorem is a general result with monotonically-decreasing games as its special cases.

Theorem 6. *In a continuous one-dimensional circular space Ω, the isolation game on Ω with weight vector $\boldsymbol{w} = (w_1, w_2, \ldots, w_{k-1})$ always has a Nash equilibrium if $\sum_{t=1}^{k-1}(-1)^t w_t \leq 0$.*

A monotonically-decreasing isolation game with weight vector $\boldsymbol{w} = (w_1, w_2, \ldots, w_{k-1})$ automatically satisfies the condition $\sum_{t=1}^{k-1}(-1)^t w_t \leq 0$. Hence we have the following corollary.

Corollary 1. *In a continuous one-dimensional circular space Ω, any monotonically-decreasing isolation game on Ω has a Nash equilibrium.*

We now consider a simple class of monotonically-decreasing games with weight vector $\boldsymbol{w} = (1, 1, 0, \ldots, 0)$ and characterize the Nash equilibria of the isolation game in a continuous one-dimensional circular space Ω. Although the game has a Nash equilibrium in a continuous one-dimensional circular space according to the above corollary, it is not potential, as shown by the following lemma.

Lemma 3. *Consider $\Omega = \{A, B, C, D, E, F\}$ that contains six points in a one-dimensional circular space with $\Delta(A, B) = 13$, $\Delta(B, C) = 5$, $\Delta(C, D) = 10$, $\Delta(D, E) = 10$, $\Delta(E, F) = 11$, and $\Delta(F, A) = 8$. The five-player monotonically-decreasing game on Ω with weight vector $\boldsymbol{w} = (1, 1, 0, 0)$ is not best-response potential (so not better-response potential either). This implies that the game on a continuous one-dimensional circular space is not better-response potential.*

If we extend from the one-dimensional circular space to a general symmetric space, there may be no Nash equilibrium for isolation games with weight vector $\boldsymbol{w} = (1, 1, 0, \ldots, 0)$ at all, as shown in the following lemma.

Lemma 4. *There is no Nash equilibrium for the four-player isolation game with weight vector $\boldsymbol{w} = (1, 1, 0)$ in the space with five points $\{A, B, C, D, E\}$ and the following distance matrix, where $N > 21$ (note that this distance function also satisfies triangle inequality).*

$$\begin{pmatrix} \Delta & A & B & C & D & E \\ \hline A & 0 & N-6 & N-11 & N-1 & N-6 \\ B & N-6 & 0 & N-8 & N-10 & N-1 \\ C & N-11 & N-8 & 0 & N-1 & N-6 \\ D & N-1 & N-10 & N-1 & 0 & N-10 \\ E & N-6 & N-1 & N-6 & N-10 & 0 \end{pmatrix}$$

Using the above lemma as a basis, we further show that it is NP-complete to decide whether an isolation game with weight vector $(1, 1, 0, \ldots, 0)$ on a general symmetric space has a Nash equilibrium. The proof is by a reduction from 3-Dimensional Matching problem.

Theorem 7. *In a finite symmetric space (Ω, Δ), it is NP-complete to decide the existence of Nash equilibrium for isolation game with weight vector $w = (1, 1, 0, \ldots, 0)$.*

5   Computation of Best Responses in High Dimensional Spaces

We now turn to the problem of computing the best response of a player in a configuration. A brute-force search on all points in the space can be done in $O(k \log k \sqrt{D})$, where D is the size of the distance matrix. This is fine if the distance matrix is explicitly given as input. However, it could become exponential if the space has a compact representation, such as an n-dimensional grid with the L_1 norm as the distance function. In this section, we present results on an n-dimensional hypercube $\{0, 1\}^n$ with the Hamming distance, a special case of n-dimensional grids with the L_1 norm.

Theorem 8. *In a $2n$-dimensional hypercube $\{0, 1\}^{2n}$, it is NP-complete to decide whether a player could move to a point so that her utility is at least $n - 1$ in the k-player nearest-neighbor isolation game with k bounded by $poly(n)$.*

The above theorem leads to the following hardness result in computing best responses for a general class of isolations games, with nearest-neighbor game as a special case.

Corollary 2. *It is NP-hard to compute a best response for an isolation game in the space $\{0, 1\}^{2n}$ with weight vector $w = (\underbrace{*, \ldots, *}_{c}, 1, 0, \ldots, 0)$ where c is a constant and $*$ is either 0 or 1.*

Contrasting to the above corollary, if the weight vector has only nonzero entries towards the end of the vector, it is easy to compute the best response, as shown in the following theorem.

Theorem 9. *A best response for a k-player isolation game in the space $\{0, 1\}^n$ with $w = (0, \ldots, 0, \underbrace{1, *, \ldots, *}_{c})$ can be computed in polynomial time where c is a constant, k is bounded by $poly(n)$ and $*$ is either 0 or 1.*

6 Final Remarks

The isolation game is very simple by its definition. However, as shown in this paper, the behaviors of its Nash equilibria and best response dynamics are quite rich and complex. This paper presents the first set of results on the isolation game and lays the ground work for the understanding of the impact of the farness measures and the underlying space to some basic game-theoretic questions about the isolation game. It remains an open question to fully characterize the isolation game. In particular, we would like to understand for what weight vectors, the isolation game on simple spaces, such as d-dimensional grids, hypercubes, and torus grid graphs, has potential functions, has Nash equilibria, or has converging best (better) response sequences. What is the impact of distance functions, such as L1-norm or L2-norm to these questions? We would like to know whether it is NP-hard to determine if Nash equilibria exist in these special spaces when the input is the weight vector. What can we say about other continuous spaces such as squares, cubes, balls, and spheres? For example, is there a sequence of better response dynamics that converge to a Nash equilibrium in the isolation game on the sphere with $w = (1, 1, 1, 0, \ldots, 0)$? What can we say about approximate Nash equilibria?

More concretely, In Lemma 2 we show an example in which a single-selection game with weight vector $(0, 1, 0, \ldots, 0)$ is not better-response potential in one dimensional circular space. However, we verify that the game is best-response potential. This phenomenon of being best-response potential but not better-response potential is rarely seen in other type of games. Moreover, our experiments lead us to conjecture that all games on continuous one dimensional circular space with weight vector $(0, 1, 0, \ldots, 0)$ is best-response potential. So far, we are only able to prove that in such games starting from any configuration there is always an acyclic sequence of best responses that either converge to a Nash equilibrium or is infinitely long. If the conjecture is true, we will find a large class of games that are best-response potential but not better-response potential (latter is implied by Lemma 2 for the continuous one dimensional space), an interesting phenomenon not known in other common games.

Another line of research is to understand the connection between the isolation game and the Voronoi game.

References

1. Ahn, H.-K., Cheng, S.-W., Cheong, O., Golin, M.J., Oostrum, R.: Competitive facility location: the Voronoi game. Theor. Comput. Sci. 310(1-3), 457–467 (2004)
2. Cheong, O., Har-Peled, S., Linial, N., Matousek, J.: The one-round Voronoi game. Discrete and Computational Geometry, 31, 125–138 (2004)
3. Dürr, C., Thang, N.K.: Nash equilibria in Voronoi games on graphs. In: Arge, L., Hoffmann, M., Welzl, E. (eds.) ESA 2007. LNCS, vol. 4698, pp. 17–28. Springer, Heidelberg (2007)
4. Eiselt, H.A., Laporte, G.: Competitive spatial models. European Journal of Operational Research 39, 231–242 (1989)
5. Eiselt, H.A., Laporte, G., Thisse, J.-F.: Competitive location models: A framework and bibliography. Transportation Science 27, 44–54 (1993)
6. Fekete, S.P., Meijer, H.: The one-round Voronoi game replayed. Computational Geometry: Theory and Applications 30, 81–94 (2005)

7. Francis, R.L., White, J.A.: Facility Layout and Location. Prentice-Hall, Inc., Englewood Cliffs (1974)
8. Jain, A.K., Murty, M.N., Flynn, P.J.: Data Clustering: A Review. ACM Computing Surveys 31(3) (1999)
9. Nash, J.F.: Equilibrium points in n-person games. Proceedings of the National Academy of Sciences 36(1), 48–49 (1950)
10. Teng, S.-H.: Low Energy and Mutually Distant Sampling. J. Algorithms 30(1), 52–67 (1999)
11. Zhao, Y., Chen, W., Teng, S.-H.: The isolation game: A game of distances, Microsoft Research Technical Report MSR-TR-2008-126 (September 2008)

On a Non-cooperative Model for Wavelength Assignment in Multifiber Optical Networks[*]

Evangelos Bampas, Aris Pagourtzis, George Pierrakos, and Katerina Potika

School of Elec. & Comp. Eng., National Technical University of Athens
Polytechnioupoli Zografou, 157 80 Athens, Greece
{ebamp,pagour,gpierr,epotik}@cs.ntua.gr

Abstract. We study path multicoloring games that describe situations in which selfish entities possess communication requests in a multifiber all-optical network. Each player is charged according to the maximum fiber multiplicity that her color (wavelength) choice incurs and the social cost is the maximum player cost. We investigate the price of anarchy of such games and provide two different upper bounds for general graphs—namely the number of wavelengths and the minimum length of a path of maximum disutility, over all worst-case Nash Equilibria—as well as matching lower bounds which hold even for trees; as a corollary we obtain that the price of anarchy in stars is exactly 2. We also prove constant bounds for the price of anarchy in chains and rings in which the number of wavelengths is relatively small compared to the load of the network; in the opposite case we show that the price of anarchy is unbounded.

Keywords: Selfish wavelength assignment, non-cooperative games, price of anarchy, multifiber optical networks, path multicoloring.

1 Introduction

The need for efficient access to the optical bandwidth in all-optical networks has given rise to the study of several optimization problems in the past years. The most well-studied among them is the problem of assigning a path and a color (wavelength) to each communication request in such a way that paths of the same color are edge-disjoint and the number of colors used is minimized. Nonetheless, it has become clear that the number of wavelengths in commercially available fibers is rather limited—and will probably remain such in the foreseeable future. Fortunately, the use of multiple fibers has come to the rescue. However, fibers are not unlimited either, therefore it makes sense to minimize their usage. This is particularly interesting from the customer's point of view, for example in

---

[*] This work has been funded by the project PENED 2003. The project is cofinanced 75% of public expenditure through EC–European Social Fund, 25% of public expenditure through Ministry of Development–General Secretariat of Research and Technology of Greece and through private sector, under measure 8.3 of Operational Programme "Competitiveness" in the 3rd Community Support Programme.

situations where one can hire a number of parallel fibers for a certain period and the cost depends on that number.

To this end, several optimization problems have been defined and studied, the objective being to minimize either the maximum fiber multiplicity per edge [1,2,3] or the sum of these maximum multiplicities over all edges of the graph [4,5,6]; in another scenario the allowed fiber multiplicity per edge is given and the goal is to minimize the number of wavelengths needed [7,8,5].

In this work we consider a non-cooperative model, where each request is issued by a user who tries to optimize her own fiber usage by selecting the most appropriate wavelength, taking into account other users' choice. This model is mainly motivated by the lack of centralized control in large scale networks. We assume that each user is charged according to the maximum fiber multiplicity that the user's choice incurs. More specifically, a user will be charged according to the maximum number of paths that share an edge with her and use the same wavelength. We consider as *social cost* the maximum fiber multiplicity that appears on any edge of the network. Minimizing this quantity is particularly important in cases where fibers are hired or sold as a whole, hence the maximum number of fibers needed on an edge determines the total cost; further motivation can be found in papers that address the corresponding optimization problem (see e.g. [1,2,3]). Here we focus on situations where routing is unique (acyclic topologies) or pre-determined—as happens in many practical settings, for example in cases where there are specific routing constraints such as a requirement to use lightpaths that have been set in advance, or shortest path routing.

We formulate the above model by defining the class of SELFISH PATH MULTICOLORING (S-PMC) games: the input is a graph, a set of paths, and the number of colors w. Each player controls a path in the graph and has to choose a color for that path from $\{\alpha_1, \ldots, \alpha_w\}$. A player is charged according to the maximum multiplicity of her color along her path. We consider as *social cost* the maximum color multiplicity per edge, i.e., the maximum number of paths of same color that use an edge.

Related work. Arguably, the most important notion in the theory of non-cooperative games is the *Nash Equilibrium (NE)* [9], a stable state of the game in which no player has incentive to change strategy unilaterally. A fundamental question in this theory concerns the existence of *pure* Nash Equilibria (PNE). For various games [10,11,12,13] it has been shown that a PNE exists and can usually be found with the use of potential functions. A standard measure of the worst-case quality of Nash Equilibria relative to optimal solutions is the *price of anarchy* (PoA) [14], which has been extensively studied for load balancing games [14,15] and other problems such as routing and facility location [10,16]. A second known measure related to NE is the *price of stability* (PoS), defined in [17].

S-PMC games are closely related to a variation of congestion games [18,19] where a player's cost is determined by her *maximum* latency instead of the usual cost which is the *sum* of her latencies. Next, we briefly explain the relation of those models to ours.

In [18] the authors study atomic routing games on networks, where each player chooses a path to route her traffic from an origin to a destination node, with the objective of minimizing the maximum congestion on any edge of her path. They show that these games always possess at least one optimal PNE (hence the PoS is 1) and that the PoA of the game is determined by topological properties of the network; in particular they show that the PoA is upper bounded by the length of the longest path in the player strategy sets and lower bounded by the length of the longest cycle. Some of our results extend to their model, since our model mimics traffic routing in the following sense: we may consider a multigraph, where we replace each edge with w parallel edges, one for each color. Each player's strategy set then consists of w different source-destination paths, corresponding to the w available colors in the original model. A further generalization is the model of Banner and Orda [19], where they introduce the notion of bottleneck games. In this model they allow arbitrary latency functions on the edges and consider both the case of splittable and unsplittable flows. They show existence, convergence and non-uniqueness of equilibria and they prove that the PoA for these games is unbounded. Both models are more general than ours; however our model fits better into the framework of all-optical networks for which we manage to provide, among others, smaller upper bounds on the PoA compared to the ones obtained by [18,19], as well as a better convergence rate to Nash equilibria. In [20] they study similar games and give results for restricted cases, e.g. single-commodity networks.

To the best of our knowledge selfish path multicoloring games have not been studied before. Selfish path coloring in single fiber all-optical networks have been studied in [21,22,23,24]. Bilò and Moscardelli [21] consider the convergence to Nash Equilibria of selfish routing and path coloring games. Later, Bilò et al. [22] considered different information levels of local knowledge that players may have for computing their payments in the same games and give bounds for the PoA in chains, rings and trees. The existence of Nash Equilibria and the complexity of recognizing and computing a Nash Equilibrium for selfish routing and path colorings games under several payment functions are considered by Georgakopoulos et al. [23]. In [24] upper and lower bounds of the PoA for selfish path coloring with and without routing are presented under functions that charge a player only according to her own strategy.

Our results. We first give an upper bound on the convergence rate of Nash dynamics for S-PMC games, and observe that the price of stability is always equal to 1. We also show how to efficiently compute a Nash Equilibrium of minimum social cost for S-PMC games in rooted trees, i.e. trees in which each path lies entirely on a simple path from some fixed root node to a leaf. For S-PMC games in stars, we prove that a known approximation algorithm for a related optimization problem actually gives an $\frac{1}{2}$-approximate Nash Equilibrium.

For general graphs, we obtain two upper bounds on the PoA: the first, which is not hard to show, is equal to the number of available colors. The second, which requires more involved arguments, is equal to the length of a shortest path with maximum disutility in any worst-case NE. For both bounds we provide matching lower bounds. In fact, we prove that these bounds hold even in trees.

Then, we move on to specific network topologies and show that for S-PMC games in stars PoA = 2. We also provide constant bounds on the PoA in a broad class of S-PMC games in chains and rings, namely for all games with $L = \Omega(w^2)$, where w is the number of available colors and L is the maximum load among all edges of the network. On the other hand, for any $\varepsilon > 0$ we exhibit a class of S-PMC games in chains (and rings) with $L = \Theta(w^{2-\varepsilon})$ for which the PoA is unbounded.

In order to show our upper bounds, we demonstrate path patterns that must be present in any Nash Equilibrium, while for the lower bounds we employ recursive construction techniques.

2   Definitions and Model

Given an undirected graph $G(V, E)$, a set P of simple paths defined on G, and a set $W = \{\alpha_1, ..., \alpha_w\}$ of available colors, $L(e)$ will denote the *load* of edge e, i.e. the number of paths that use edge e. The maximum of these loads will be denoted by L, i.e. $L = \max_{e \in E} L(e)$.

Given, additionally, an assignment of a color to each path we define the following:

Definition 1

1. $\mu(e, c)$ will denote the multiplicity of color c on edge e, i.e. the number of paths that use edge e and are colored with color c.
2. μ_e will denote the maximum multiplicity of any color on edge e, i.e. $\mu_e = \max_{c \in W} \mu(e, c)$.
3. $\mu_{\max}$ will denote the maximum multiplicity of any color over all edges: $\mu_{\max} = \max_{e \in E} \mu_e$.
4. $\mu(p, c)$ will denote the maximum multiplicity of color c over the edges of path p: $\mu(p, c) = \max_{e \in p} \mu(e, c)$.

It will be clear from the context which specific coloring we are referring to when we use the above notation.

The minimum $\mu_{\max}$ that can be attained by some coloring of the paths in P will be denoted by μ_{OPT}, i.e. $\mu_{\mathrm{OPT}} = \min_c \mu_{\max}$ where c ranges over all possible colorings. We note immediately the following:

Fact. No coloring can achieve a $\mu_{\max}$ smaller than $\left\lceil \frac{L}{w} \right\rceil$. Thus, $\mu_{\mathrm{OPT}} \geq \left\lceil \frac{L}{w} \right\rceil$.

We now proceed to define the class of selfish path multicoloring games and subclasses thereof.

Definition 2 (Selfish path multicoloring games). *A selfish path multicoloring game is the following strategic game defined in terms of an undirected graph G, a set P of simple paths defined on G, and an integer $w > 0$:*

- Players: *there is one player for each path in P. For simplicity, we identify a player i with the corresponding path p_i.*
- Strategies: *a strategy for player i is a color c_i chosen from the set $W = \{\alpha_1, \ldots, \alpha_w\}$ of available colors. We say that color c_i is assigned to path p_i or that path p_i is colored with color c_i. All players share the common set of strategies W.*
- Disutility: *given a strategy profile $\boldsymbol{c} = (c_1, \ldots, c_{|P|})$, the disutility $f_i : W^{|P|} \to \mathbb{N}$ of each player i is defined as follows:*

$$f_i(\boldsymbol{c}) = \mu(p_i, c_i).$$

We denote this game by $\langle G, P, w \rangle$. The class of all selfish path multicoloring games will be denoted by S-PMC.

We will use the notation S-PMC($\mathcal{G}$) to denote a subclass of S-PMC that contains only games satisfying a property $\mathcal{G}$ (for example $\mathcal{G}$ may constrain the graph on which the game is defined to belong to a specific graph class, etc.).

Following the standard definition, a strategy profile $\boldsymbol{c} = (c_1, \ldots, c_{|P|})$ is said to be a *pure Nash Equilibrium* (PNE), or simply *Nash Equilibrium* (NE), if for each player i it holds that: $f_i(c_1, \ldots, c_i', \ldots, c_{|P|}) \geq f_i(c_1, \ldots, c_i, \ldots, c_{|P|})$, for any strategy $c_i' \in W$. Moreover, following the definition of [25], we say that a strategy profile $\boldsymbol{c} = (c_1, \ldots, c_{|P|})$ is an *ε-approximate Nash Equilibrium* if for each player i it holds that: $f_i(c_1, \ldots, c_i', \ldots, c_{|P|}) \geq (1 - \varepsilon) \cdot f_i(c_1, \ldots, c_i, \ldots, c_{|P|})$, for any strategy $c_i' \in W$.

Definition 3 (Blocking edges). *If $\boldsymbol{c}$ is a strategy profile for a game $\langle G, P, w \rangle$ and $p_i \in P$, we say that edge e is an α_j-blocking edge for p_i, or that it blocks α_j for p_i, if $e \in p_i$ and $\mu(e, \alpha_j) \geq f_i(\boldsymbol{c}) - 1$. Furthermore, the $\mu(e, \alpha_j)$ paths that are colored with α_j and use edge e are called α_j-blocking paths for p_i.*

Intuitively, an α_j-blocking edge for p_i "blocks" p_i from switching to color α_j because if it did, the new disutility of path p_i would be at least $\mu(e, \alpha_j) + 1 \geq f_i(\boldsymbol{c})$, no better than its current choice. It is immediate from the definitions that the following property holds in any Nash Equilibrium of any S-PMC game:

Property 1 (Structural property of S-PMC *Nash Equilibria).* In a Nash Equilibrium, every path p must contain at least one α_j-blocking edge for p, for every color α_j.

Definition 4 (Social cost). *The* social cost *of a strategy profile $\boldsymbol{c}$ for an S-PMC game is defined as follows:* $\mathrm{sc}(\boldsymbol{c}) = \max_{e \in E} \mu_e = \mu_{\max}$.

It is straightforward to verify that the social cost of a strategy profile coincides with the maximum player disutility in that profile:

$$\mathrm{sc}(\boldsymbol{c}) = \max_{e \in E} \mu_e = \max_{p_i \in P} f_i(\boldsymbol{c})$$

We define $\hat{\mu}$ to be the maximum social cost over all strategy profiles that are Nash Equilibria: $\hat{\mu} = \max_{\boldsymbol{c} \text{ is NE}} \mathrm{sc}(\boldsymbol{c})$. Following the standard definitions, the *price of anarchy* (PoA) of a game $\langle G, P, w \rangle$ is the worst-case social cost in a Nash Equilibrium divided by μ_{OPT}, i.e.: $\mathrm{PoA}(\langle G, P, w \rangle) = \frac{\max_{c \text{ is NE}} \mathrm{sc}(c)}{\mu_{\mathrm{OPT}}} = \frac{\hat{\mu}}{\mu_{\mathrm{OPT}}}$. The *price of stability* (PoS) of a game is the best-case social cost in a NE divided by μ_{OPT}: $\mathrm{PoS}(\langle G, P, w \rangle) = \frac{\min_{c \text{ is NE}} \mathrm{sc}(c)}{\mu_{\mathrm{OPT}}}$. The price of anarchy (resp. stability) of a class of games S-PMC($\mathcal{G}$) is the maximum price of anarchy (resp. stability) among all games in S-PMC($\mathcal{G}$).

3 Existence and Computation of Nash Equilibria

We use lexicographic-order arguments similar to those in [18,19] to show that in any S-PMC game the following holds: starting from an arbitrary strategy profile any Nash dynamics converges to a Nash Equilibrium of smaller or equal social cost. The proof is omitted.

Theorem 1. *For any game* $\langle G, P, w \rangle$ *in* S-PMC:

  a. *the price of stability is* 1, *and*
  b. *any Nash dynamics converges to a Nash Equilibrium in at most* $4^{|P|}$ *steps.*

Due to Theorem 1, computing a Nash Equilibrium of minimum social cost is at least as hard as the corresponding optimization problem. As noticed in [4] this problem is **NP**-hard in general graphs, in fact even in rings and stars. Therefore, it is also **NP**-hard to compute an optimal Nash Equilibrium even in the case of rings and stars. However, we show that there exists an efficient algorithm that computes optimal Nash Equilibria for a subclass of S-PMC(TREE). Furthermore, we show that we can use a known algorithm for PATH MULTICOLORING in stars [4] to compute approximate Nash Equilibria for S-PMC(STAR) games. We will only state the theorems and omit the proofs.

Definition 5. *We define* S-PMC(ROOTED-TREE) *to be the subclass of* S-PMC *that contains games* $\langle G, P, w \rangle$ *with the following property: "G is a tree and there is a node r such that each path in P lies entirely on some simple path from r to a leaf."*

Consider the greedy algorithm that colors paths in order of non-decreasing distance from the root in such a way that the color multiplicity is the lowest possible with respect to the current partial coloring.

Theorem 2. *Given an* S-PMC(ROOTED-TREE) *game* $\langle G(V, E), P, w \rangle$ *with maximum load L as input, the greedy algorithm computes an optimal Nash Equilibrium of cost exactly* $\left\lceil \frac{L}{w} \right\rceil$.

Theorem 3. *There is a polynomial-time algorithm that computes a $\frac{1}{2}$-approximate Nash Equilibrium for any* S-PMC(STAR) *game.*

4   Tight Upper Bounds for the PoA of S-PMC Games

In this section we provide two upper bounds on the PoA of any S-PMC game and we show that both of them are tight. The first bound is determined by a property of the network, namely the number of available wavelengths. The second bound is more subtle, as it depends on the length of paths with the highest disutility in worst-case Nash Equilibria. We prove that these bounds are tight even for the class S-PMC(ROOTED-TREE), and asymptotically tight for the class S-PMC(ROOTED-TREE: $\Delta = 3$), i.e. the subclass of S-PMC(ROOTED-TREE) that contains games defined on graphs with maximum degree 3.

Lemma 1. *The price of anarchy of any* S-PMC *game* $\langle G, P, w \rangle$ *is at most* w.

Proof. Let c be a worst-case Nash Equilibrium of $\langle G, P, w \rangle$, hence $\mathrm{sc}(c) = \hat{\mu}$. Clearly, $\hat{\mu} \leq L$ and since the minimum social cost over all strategy profiles is $\mu_{\mathrm{OPT}} \geq \left\lceil \frac{L}{w} \right\rceil$, it turns out that $\mu_{\mathrm{OPT}} \geq \frac{\hat{\mu}}{w}$. This implies that $\frac{\hat{\mu}}{\mu_{\mathrm{OPT}}} \leq w$.   □

Lemma 2. *For any worst-case Nash Equilibrium* c *of an* S-PMC *game* $\langle G, P, w \rangle$ *and for any* $p_i \in P$ *with* $f_i(c) = \mathrm{sc}(c) = \hat{\mu}$, *the price of anarchy of* $\langle G, P, w \rangle$ *is at most equal to the length of* p_i.

Proof. Let $\tilde{e}$ be an edge of p_i where the color c_i chosen by p_i appears with maximum multiplicity $\hat{\mu}$: $\mu(\tilde{e}, c_i) = \hat{\mu}$. Let z denote the length of path p_i and let $e_1, \ldots, e_{z-1}$ be the edges that p uses, apart from $\tilde{e}$. For $1 \leq j \leq z - 1$, let x_j be the number of colors that are blocked for p_i on e_j and let y be the number of colors that are blocked for p_i on $\tilde{e}$ (since c is a Nash Equilibrium, it must be that $x_1 + \ldots + x_{z-1} + y \geq w - 1$).

If it is the case that $z = 1$, i.e. p_i uses only edge $\tilde{e}$, then $\tilde{e}$ must block all colors for p_i except c_i. This implies that the load of edge $\tilde{e}$ is: $L(\tilde{e}) \geq \hat{\mu} + (w-1)(\hat{\mu}-1) = w\hat{\mu} - w + 1$. Therefore, the minimum social cost over all strategy profiles satisfies: $\mu_{\mathrm{OPT}} \geq \left\lceil \frac{L(\tilde{e})}{w} \right\rceil \geq \left\lceil \hat{\mu} - \frac{w-1}{w} \right\rceil = \hat{\mu}$. We conclude that the price of anarchy in this case is equal to 1.

Now, assume that $z \geq 2$. We will prove that $L \geq 1 + \left\lceil \frac{w}{z} \right\rceil (\hat{\mu} - 1)$. First, observe that $L(\tilde{e}) \geq \hat{\mu} + y(\hat{\mu} - 1)$ and, for $1 \leq j \leq z - 1$, $L(e_j) \geq 1 + x_j(\hat{\mu} - 1)$. If $y \geq \left\lceil \frac{w}{z} \right\rceil - 1$, then $L(\tilde{e}) \geq \hat{\mu} + \left(\left\lceil \frac{w}{z} \right\rceil - 1 \right)(\hat{\mu} - 1) = 1 + \left\lceil \frac{w}{z} \right\rceil (\hat{\mu} - 1)$, therefore $L \geq 1 + \left\lceil \frac{w}{z} \right\rceil (\hat{\mu} - 1)$. If, on the other hand, $y < \left\lceil \frac{w}{z} \right\rceil - 1$, then $x_1 + \ldots + x_{z-1} \geq w - 1 - y \geq w - \left\lceil \frac{w}{z} \right\rceil + 1$. This implies that there is some x_k such that $x_k \geq \frac{w - \left\lceil \frac{w}{z} \right\rceil + 1}{z - 1} > \frac{w - \frac{w}{z} - 1 + 1}{z - 1} = \frac{w}{z}$. Since x_k is an integer, it must be that $x_k \geq \left\lceil \frac{w}{z} \right\rceil$. Therefore, $L \geq L(e_k) \geq 1 + \left\lceil \frac{w}{z} \right\rceil (\hat{\mu} - 1)$.

We conclude that in any case $L \geq 1 + \left\lceil \frac{w}{z} \right\rceil (\hat{\mu} - 1)$. So, the price of anarchy is bounded as follows:

$$\mathrm{PoA}(\langle G, P, w \rangle) = \frac{\hat{\mu}}{\mu_{\mathrm{OPT}}} \leq \frac{\hat{\mu}}{\left\lceil \frac{L}{w} \right\rceil} \leq \frac{\hat{\mu}}{\left\lceil \frac{1 + \left\lceil \frac{w}{z} \right\rceil (\hat{\mu}-1)}{w} \right\rceil} \leq z .$$

We omit the proof of the last inequality, which holds for all $\hat{\mu} \geq 2$, $w \geq 1$, and $z \geq 2$.   □

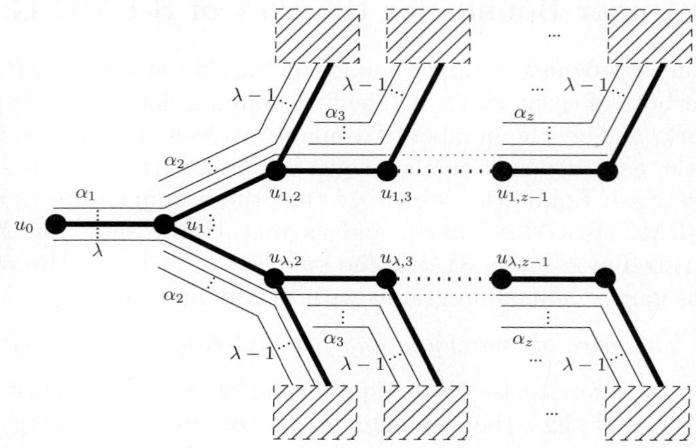

Fig. 1. The construction $A_z(\lambda)$ for the proof of Lemma 3. The thick lines represent the edges of the underlying graph, and the thin lines represent the paths defined on the graph. The color and multiplicity of each group of paths is written next to that group. Each shaded box represents a recursive copy of $A_z(\lambda - 1)$.

As an immediate corollary of Lemma 2, we derive the following upper bound on the price of anarchy:

Corollary 1. *The price of anarchy of any* S-PMC *game* $\langle G, P, w \rangle$ *is bounded as follows:*

$$\mathrm{PoA} \leq \min_{\boldsymbol{c}:\mathrm{NE} \wedge \mathrm{sc}(\boldsymbol{c})=\hat{\mu}} \ \min_{i:f_i(\boldsymbol{c})=\hat{\mu}} \ \mathrm{length}(p_i)$$

Lemma 3. *The upper bounds of Lemma 1 and Corollary 1 are tight even for the class of* S-PMC(ROOTED-TREE) *games.*

Proof. We first define a recursive construction of an S-PMC game and a Nash Equilibrium for this game. The construction is illustrated in Figure 1. For any $z \geq 1$ and $\lambda \geq 1$, let $A_z(\lambda)$ be the following S-PMC game with z available colors: there are λ paths of color α_1 and length z, starting at the "root node" u_0, which branch out into λ branches, one on each branch. Let us call these the "primary" paths for $A_z(\lambda)$. On any of the $z - 1$ edges of each such branch, one color is blocked for the primary path. The $\lambda - 1$ blocking paths of each edge branch out into an $A_z(\lambda - 1)$ game. They become primary paths for this copy of $A_z(\lambda - 1)$. The root node for the j-th recursive copy of $A_z(\lambda - 1)$ on the k-th branch is node $u_{k,j}$ (node $u_{k,1}$ is common for all branches). The base case of this recursive construction is $A_z(0)$, which is a degenerate game with no paths and no available colors, defined on a graph consisting of a single node.

Observe that for any $z \geq 1$, the construction $A_z(z)$ is an S-PMC(ROOTED-TREE) game in NE, in which all of the following are equal to z: w, L, $\mu_{\max}$, and all path lengths. By Theorem 2, the optimal strategy profile for $A_z(z)$ has social cost $\mu_{\mathrm{OPT}} = \left\lceil \frac{L}{w} \right\rceil = 1$. Therefore, the ratio $\frac{\mu_{\max}}{\mu_{\mathrm{OPT}}}$ is equal to z for this Nash Equilibrium, hence the price of anarchy is at least z. $\qquad\square$

By appropriate modification of the construction presented in Figure 1, we obtain the following:

Lemma 4. *The upper bounds of Lemma 1 and Corollary 1 are asymptotically tight even for the class of* S-PMC(ROOTED-TREE) *games with maximum degree 3.*

We summarize the results of Lemmata 1, 2, 3, and 4 in the following theorem:

Theorem 4. *The price of anarchy of any* S-PMC *game* $\langle G, P, w \rangle$ *is upper-bounded both by w and by*

$$\min_{\boldsymbol{c}:\text{NE}\wedge\text{sc}(\boldsymbol{c})=\hat{\mu}} \quad \min_{i:f_i(\boldsymbol{c})=\hat{\mu}} \text{length}(p_i).$$

These bounds are tight for the class S-PMC(ROOTED-TREE) *and asymptotically tight for the class* S-PMC(ROOTED-TREE: $\Delta = 3$).

Theorem 5. *The price of anarchy of the class* S-PMC(STAR) *is 2.*

Proof. Lemma 2 implies an upper bound of 2 on the price of anarchy since the length of any path in a star cannot be larger than 2.

For the lower bound, it can be shown that the construction of Lemma 3 can be modified to yield a family of S-PMC(STAR) games with price of anarchy 2. More specifically every game $A_2(\lambda)$ can be embedded in a star, by using additional star rays for branching. The detailed construction is omitted. $\square$

5   The Price of Anarchy on Graphs of Maximum Degree 2

In this section we study the price of anarchy of path multicoloring games on chains and rings, and we prove a constant upper bound for a broad class of S-PMC(RING) games with $L = \Omega(w^2)$. Notice that this class essentially encompasses all S-PMC(RING) games of practical importance, as the number of wavelengths is limited in practice due to technological constraints, whereas L can grow large depending on network traffic. For the sake of completeness, we show that the PoA becomes quickly unbounded if we allow the network designer to provide ample wavelengths to the users, i.e. when $L = o(w^2)$.

We begin by strengthening Property 1 to prove a more involved structural property of Nash Equilibria in S-PMC(RING) games. Let $\langle G, P, w \rangle$ be an S-PMC(RING) game. Given a coloring $\boldsymbol{c} = (c_1, \ldots, c_{|P|})$, let $P(e, \alpha_i)(\boldsymbol{c}) \subseteq P$ denote the set of paths colored with color α_i that use edge $e \in E$; by definition $|P(e, \alpha_i)(\boldsymbol{c})| = \mu(e, \alpha_i)$. For the sake of simplicity, in the rest of the section we will write $P(e, \alpha_i)$ instead of $P(e, \alpha_i)(\boldsymbol{c})$. Furthermore, let $[e_l, e_r]$ denote the clockwise arc starting at edge e_l and ending at edge e_r.

Lemma 5 (Structural property of S-PMC(Ring) NE). *Given a game in* S-PMC(RING) *and a coloring $\boldsymbol{c}$ thereof which is a Nash Equilibrium, for every edge e and color α_i there is an edge-simple arc $[e_l, e_r]$ with the following properties:*

a. *for every color $\alpha_j \neq \alpha_i$, arc $[e_l, e_r]$ contains an edge which is an α_j-blocking edge for at least half of the paths in $P(e, \alpha_i)$, and*

b. *for every edge e' of the arc $[e_l, e_r]$ it holds that $|P(e', \alpha_i) \cap P(e, \alpha_i)| \geq \left\lceil \frac{|P(e, \alpha_i)|}{2} \right\rceil$.*

Proof. Since the game is in NE, by Property 1 every path $p \in P(e, \alpha_i)$ must have at least one α_j-blocking edge, for every color $\alpha_j \neq \alpha_i$. For a fixed color $\alpha_j \neq \alpha_i$, consider the two α_j-blocking edges for some path in $P(e, \alpha_i)$ that are closest to edge e clockwise and counter-clockwise. It is not hard to see that for at least one of these two edges, call it $b(\alpha_j)$, the following property holds: the arc $[e, b(\alpha_j)]$ or the arc $[b(\alpha_j), e]$ is contained in at least $\left\lceil \frac{|P(e, \alpha_i)|}{2} \right\rceil$ of the paths in $P(e, \alpha_i)$. In case that there is only one α_j-blocking edge for all paths in $P(e, \alpha_i)$, then the property holds *a fortiori* for this edge.

For every color α_j we pick one such edge $b(\alpha_j)$. If the above property holds for arc $[e, b(\alpha_j)]$, we add $b(\alpha_j)$ to set B^+, otherwise we add it to set B^-. We now claim that a clockwise traversal of the ring starting at edge e will first encounter all edges of B^+ and then all edges of B^-. Indeed, if one edge b^- of B^- lies before one edge b^+ of B^+ on this clockwise traversal, this would imply that b^- is traversed by the $\left\lceil \frac{|P(e, \alpha_i)|}{2} \right\rceil$ paths that contain the arc $[e, b^+]$ and thus b^- should also belong to B^+.

The above discussion implies that if we define e_r to be the last edge of B^+ and e_l to be the first edge of B^- encountered in this clockwise traversal, then the edge-simple arc $[e_l, e_r]$ satisfies the conditions of the Lemma. $\square$

We now prove a constant upper bound on the price of anarchy of S-PMC(RING) games with $L = \Omega(w^2)$; denote this class by S-PMC(RING: $L = \Omega(w^2)$). This also provides an upper bound on the price of anarchy of any S-PMC(CHAIN: $L = \Omega(w^2)$) game, as every game defined on a chain can be trivially embedded in a ring topology.

We first employ the structural property of S-PMC(RING) Nash Equilibria (Lemma 5) in order to establish the existence of a heavily loaded edge in S-PMC(RING) games with $\hat{\mu} \geq w$.

Lemma 6. *In every* S-PMC(RING) *game $\langle G, P, w \rangle$ with $\hat{\mu} \geq w$ there is an edge with load at least $\frac{\hat{\mu} w}{4}$.*

Proof. Let $[e_l, e_r]_{P(e, \alpha_i)}$ be the arc that is obtained by applying Lemma 5 for path set $P(e, \alpha_i)$. We define P_1 to be the set of paths $P(\tilde{e}, \alpha_1)$ which induce the social cost $\hat{\mu}$. For $i \geq 2$ we define P_i to be the set of α_j-blocking paths for the path set P_{i-1}, for some color α_j not appearing at any of the path sets P_k, $k < i$, with the following property:

$$[e_l, e_r]_{P_i} \subseteq [e_l, e_r]_{P_{i-1}}, \tag{1}$$

if such a path set exists. If more than one path sets with the desired property exist, we arbitrarily pick one of them.

Let e_i be the α_j-blocking edge for P_{i-1}; based on the inductive definition of P_i as a set of blocking paths for path set P_{i-1} we can easily show that $\mu(e_i, \alpha_j) \geq$

$\hat{\mu} - i + 1$. Applying Lemma 5(b) for color α_j and edge e_i yields the following: for every edge $e \in [e_l, e_r]_{P_i}$ we have that $\mu(e, \alpha_j) \geq \frac{\hat{\mu}-i+1}{2}$. Furthermore, since Equation 1 holds for all $k \leq i$, the load of all edges $e \in [e_l, e_r]_{P_i}$ is at least $\sum_{\alpha_j} \mu(e, \alpha_j)$, where α_j now ranges over the colors of all path sets $P_k, k \leq i$. Hence, for every edge $e \in [e_l, e_r]_{P_i}$ we have that $L(e) \geq \sum_{\alpha_j} \mu(e, \alpha_j) \geq \sum_{k=1}^{i} \frac{\hat{\mu}-k+1}{2}$.

Let now n be the first integer for which no such path set P_n exists and consider the path set P_{n-1}. Since we are at Nash Equilibrium we know that there exist α-blocking edges for paths in P_{n-1}, for every color α. We restrict our attention to the $w - n + 1$ colors, which have not yet appeared at any P_k, for $k \leq n - 1$; let α_j be one of these colors. Consider now an α_j-blocking edge e_n such that $e_n \in [e_l, e_r]_{P_{n-1}}$ (by Lemma 5(a) such an edge must exist). We now have that, at least half of the α_j-blocking paths in $P(e_n, \alpha_j)$, i.e. at least $\frac{\hat{\mu}-n+1}{2}$ paths, extend beyond one of the edges $e_l(P_{n-1})$, $e_r(P_{n-1})$ of the arc $[e_l, e_r]_{P_{n-1}}$ (otherwise we would have picked $P(e_n, \alpha_j)$ to be P_n). This means that for at least half of these $w - n + 1$ blocking path sets, their paths leave the arc from the same edge, incurring on it an additional load of $\frac{w-n+1}{2} \cdot \frac{\hat{\mu}-n+1}{2}$.

Thus, the total load of this edge is at least $\sum_{i=1}^{n-1} \frac{\hat{\mu}-i+1}{2} + \frac{w-n+1}{2} \cdot \frac{\hat{\mu}-n+1}{2} = \frac{\hat{\mu}w}{4} + (n-1) \cdot \frac{\hat{\mu}-w+1}{4}$. Since $\hat{\mu} \geq w$ the above sum is at least $\frac{\hat{\mu}w}{4}$. $\qquad\square$

Theorem 6. *The price of anarchy of any game in the class* S-PMC(RING: $L = \Omega(w^2)$)) *is bounded by a constant.*

Proof. We distinguish between two cases:

- If $\hat{\mu} \geq w$, then by Lemma 6 we get $L \geq \frac{\hat{\mu}w}{4}$. This implies $\frac{L}{w} \geq \frac{\hat{\mu}}{4} \Rightarrow \mu_{\text{OPT}} \geq \frac{\hat{\mu}}{4} \Rightarrow \text{PoA} \leq 4$.
- If $\hat{\mu} < w$, then $\text{PoA} = \frac{\hat{\mu}}{\mu_{\text{OPT}}} \leq \frac{\hat{\mu}w}{L} < \frac{w^2}{L}$, where we used successively the facts that $\mu_{\text{OPT}} \geq \frac{L}{w}$ and $\hat{\mu} < w$. The last inequality, combined with the fact that $L = \Omega(w^2)$, implies $\text{PoA} = O(1)$. $\qquad\square$

Finally, we show that the price of anarchy can get arbitrarily large when the number of available colors increases; specifically, that it is unbounded for the classes S-PMC(CHAIN: $L = o(w^2)$) and S-PMC(RING: $L = o(w^2)$). The proof is omitted.

Theorem 7. *For any fixed* $\varepsilon > 0$ *there exists an infinite family of games in* S-PMC(CHAIN: $L = \Theta(w^{2-\varepsilon})$)) *with* $\text{PoA} = \Omega(w^{\frac{\varepsilon}{2}})$.

References

1. Andrews, M., Zhang, L.: Minimizing maximum fiber requirement in optical networks. J. Comput. Syst. Sci. 72(1), 118–131 (2006)
2. Andrews, M., Zhang, L.: Complexity of wavelength assignment in optical network optimization. In: INFOCOM 2006. IEEE, Los Alamitos (2006)
3. Andrews, M., Zhang, L.: Wavelength assignment in optical networks with fixed fiber capacity. In: Díaz, J., Karhumäki, J., Lepistö, A., Sannella, D. (eds.) ICALP 2004. LNCS, vol. 3142, pp. 134–145. Springer, Heidelberg (2004)
4. Nomikos, C., Pagourtzis, A., Zachos, S.: Routing and path multicoloring. Inf. Process. Lett. 80(5), 249–256 (2001)

5. Erlebach, T., Pagourtzis, A., Potika, K., Stefanakos, S.: Resource allocation problems in multifiber WDM tree networks. In: Bodlaender, H.L. (ed.) WG 2003. LNCS, vol. 2880, pp. 218–229. Springer, Heidelberg (2003)
6. Winkler, P., Zhang, L.: Wavelength assignment and generalized interval graph coloring. In: SODA, pp. 830–831 (2003)
7. Margara, L., Simon, J.: Wavelength assignment problem on all-optical networks with k fibres per link. In: Welzl, E., Montanari, U., Rolim, J.D.P. (eds.) ICALP 2000. LNCS, vol. 1853, pp. 768–779. Springer, Heidelberg (2000)
8. Li, G., Simha, R.: On the wavelength assignment problem in multifiber WDM star and ring networks. IEEE/ACM Trans. Netw. 9(1), 60–68 (2001)
9. Nash, J.: Non-cooperative games. The Annals of Mathematics 54(2), 286–295 (1951)
10. Fotakis, D., Kontogiannis, S.C., Koutsoupias, E., Mavronicolas, M., Spirakis, P.G.: The structure and complexity of Nash equilibria for a selfish routing game. In: Widmayer, P., Triguero, F., Morales, R., Hennessy, M., Eidenbenz, S., Conejo, R. (eds.) ICALP 2002. LNCS, vol. 2380, pp. 123–134. Springer, Heidelberg (2002)
11. Rosenthal, R.W.: A class of games possessing pure-strategy Nash equilibria. Int. J. Game Theory 2, 65–67 (1973)
12. Milchtaich, I.: Congestion games with player-specific payoff functions. Games and Economic Behavior 13, 111–124 (1996)
13. Monderer, D., Shapley, L.S.: Potential games. Games and Economic Behavior 14, 124–143 (1996)
14. Koutsoupias, E., Papadimitriou, C.H.: Worst-case equilibria. In: Meinel, C., Tison, S. (eds.) STACS 1999. LNCS, vol. 1563, pp. 404–413. Springer, Heidelberg (1999)
15. Mavronicolas, M., Spirakis, P.G.: The price of selfish routing. In: STOC, pp. 510–519 (2001)
16. Roughgarden, T., Tardos, É.: How bad is selfish routing? J. ACM 49(2), 236–259 (2002)
17. Anshelevich, E., Dasgupta, A., Kleinberg, J.M., Tardos, É., Wexler, T., Roughgarden, T.: The price of stability for network design with fair cost allocation. In: FOCS, pp. 295–304. IEEE Computer Society, Los Alamitos (2004)
18. Busch, C., Magdon-Ismail, M.: Atomic routing games on maximum congestion. In: Cheng, S.-W., Poon, C.K. (eds.) AAIM 2006. LNCS, vol. 4041, pp. 79–91. Springer, Heidelberg (2006)
19. Banner, R., Orda, A.: Bottleneck routing games in communication networks. In: INFOCOM 2006. IEEE, Los Alamitos (2006)
20. Caragiannis, I., Galdi, C., Kaklamanis, C.: Network load games. In: Deng, X., Du, D.-Z. (eds.) ISAAC 2005. LNCS, vol. 3827, pp. 809–818. Springer, Heidelberg (2005)
21. Bilò, V., Moscardelli, L.: The price of anarchy in all-optical networks. In: Kralovic, R., Sýkora, O. (eds.) SIROCCO 2004. LNCS, vol. 3104, pp. 13–22. Springer, Heidelberg (2004)
22. Bilò, V., Flammini, M., Moscardelli, L.: On Nash equilibria in non-cooperative all-optical networks. In: Diekert, V., Durand, B. (eds.) STACS 2005. LNCS, vol. 3404, pp. 448–459. Springer, Heidelberg (2005)
23. Georgakopoulos, G.F., Kavvadias, D.J., Sioutis, L.G.: Nash equilibria in all-optical networks. In: Deng, X., Ye, Y. (eds.) WINE 2005. LNCS, vol. 3828, pp. 1033–1045. Springer, Heidelberg (2005)
24. Milis, I., Pagourtzis, A., Potika, K.: Selfish routing and path coloring in all-optical networks. In: Janssen, J., Prałat, P. (eds.) CAAN 2007. LNCS, vol. 4852, pp. 71–84. Springer, Heidelberg (2007)
25. Chien, S., Sinclair, A.: Convergence to approximate Nash equilibria in congestion games. In: SODA, pp. 169–178. SIAM, Philadelphia (2007)

The Complexity of Rationalizing Matchings

Shankar Kalyanaraman* and Christopher Umans**

Computer Science Department
California Institute of Technology
Pasadena, CA 91125
{shankar,umans}@cs.caltech.edu

Abstract. Given a set of observed economic choices, can one infer preferences and/or utility functions for the players that are consistent with the data? Questions of this type are called *rationalization* or *revealed preference* problems in the economic literature, and are the subject of a rich body of work.

From the computer science perspective, it is natural to study the complexity of rationalization in various scenarios. We consider a class of rationalization problems in which the economic data is expressed by a collection of matchings, and the question is whether there exist preference orderings for the nodes under which all the matchings are *stable*.

We show that the rationalization problem for one-one matchings is NP-complete. We propose two natural notions of approximation, and show that the problem is hard to approximate to within a constant factor, under both. On the positive side, we describe a simple algorithm that achieves a $3/4$ approximation ratio for one of these approximation notions. We also prove similar results for a version of many-one matching.

1   Introduction

Given a set of consumption choices in a market, it is natural to try to infer information about the players' preferences or utility functions. This branch of consumer demand theory is known as *revealed preference theory* because consumers, by dint of the choices they make, "reveal" their preferences for various outcomes [Afr67, Die73, Sam48, Ech06] [FST04, Var82, Spr00]. It constitutes a major tool in econometric analysis used to estimate aggregate consumer demand [Afr67, Var06]. From the Computer Science perspective, this is a learning problem, and recent work initiated a study of its PAC-learnability [BV06].

Some classes of data cannot always be explained, or *rationalized* by simple (say, linear) utility functions, or even any reasonable utility function. Such settings are interesting to economists, because it becomes possible, in principle, to "test" various assumptions (e.g. that the players are maximizing a simple utility function). Several (classical and recent) results [Afr67, Var82, FST04, Ech06] in the economic literature

---

* Supported by NSF CCF-0346991, NSF CCF-0830787, BSF 2004329 and a Graduate Research Fellowship from the Social and Information Sciences Laboratory (SISL) at Caltech.
** Supported by NSF CCF-0346991, NSF CCF-0830787, BSF 2004329, a Sloan Research Fellowship, and an Okawa Foundation research grant.

S.-H. Hong, H. Nagamochi, and T. Fukunaga (Eds.): ISAAC 2008, LNCS 5369, pp. 171–182, 2008.

establish criteria for when data is *always* rationalizable, thus delineating the limits of the "testable implications" of such data.

There is an important role for Computer Science in these questions, as the feasibility of performing such tests depends on being able to answer the rationalizability question *efficiently*. In other words, given a type of economic data, and a target form for an "explanation" (preference profile, a class of utility functions, etc...), we wish to understand the *complexity* of deciding whether the data can be rationalized by an explanation of the prescribed form. To our knowledge these sort of problems have not been studied before.

Among rationalization problems, one can identify at least two broad classes of problems. Some, such as inferring utility functions from consumption data, are rather easily solved efficiently using linear programming [Afr67, Var82]. Others are more combinatorial in nature, and their complexity is not at all obvious. One recent example is the problem of inferring costs from observations of spanning trees being formed to distribute some service, say power [Özs06].

Among the combinatorial-type rationalization problems, one of the most natural is the matchings problem that we study in this paper. Here we are given a set of bipartite matchings, and we wish to determine if there are preferences for the nodes under which all of the given matchings are stable. Matchings, or more precisely "two-sided matching markets," are a central abstraction in economics, investigated in relation to the similar "marriage models" in auction and labor markets [RS90, Fle03, EO04, EY07] and from the point of view of mechanism design [Sön96] and related strategic issues [STT01]. They are also a fundamental combinatorial abstraction from the computational perspective.

1.1   Our Results

Given two sets of nodes, M ("men") and W ("women"), together with preferences for each node, the famous algorithm of Gale and Shapley [GS62] obtains a *stable matching*. We will be interested in the "reverse" question: given a set of matchings, are there preferences under which they are simultaneously stable? One may wonder why we should be given a collection of matchings instead of a single instance of a matching between the set of men and women. Indeed, we think of the men (and women) as representing instances of different *types* or populations that are matched differently in each matching and we are interested in determining the preference profiles that define these types based on the observed set of matchings. Before stating our results, we formalize the problem and introduce some terminology.

Definition 1. *Let M, W be disjoint sets of equal cardinality. A **one-one matching** μ is a bijection $\mu : M \cup W \to M \cup W$, such that for all $m \in M$, $\mu(m) \in W$, for all $w \in W$, $\mu(w) \in M$, and for all $m \in M, w \in W$, $\mu(m) = w \Leftrightarrow \mu(w) = m$.*

In the problems we consider, we will be seeking preferences for the elements of M and W, which are expressed as follows:

Definition 2. *A **preference order** for $m \in M$ (resp. $w \in W$) is a linear ordering of W (resp. M). We write $m : w > w'$ to mean that w occurs before w' in the preference order for m. A **preference profile** is a collection of preference orders for each $m \in M$ and $w \in W$.*

The "stability" of a matching with respect to a preference profile depends on the crucial notion of *blocking pair*:

Definition 3. *A* **blocking pair** *with respect to a matching μ and a preference profile $\mathcal{P}$ is a pair $(m, w) : m \in M, w \in W$ such that $\mu(m) \neq w$ and*

$$m : w > \mu(m) \text{ and } w : m > \mu(w).$$

Matching μ is **stable** *with respect to $\mathcal{P}$ if there is no blocking pair with respect to μ and $\mathcal{P}$.*

In other words, in a blocking pair (m, w) with respect to μ and $\mathcal{P}$, both people are "unhappy" with their current partner in μ and would instead prefer to be matched to each other.

Our first result is that rationalizing matchings is hard.

Theorem 1. *Given a collection of one-one matchings $\mathcal{H}$ on the sets M and W, it is NP-complete to determine if there exists a preference profile $\mathcal{P}$ such that every $\mu \in \mathcal{H}$ is stable with respect to $\mathcal{P}$.*

We call such a preference profile a *rationalization* of the matchings $\mathcal{H}$. The main gadget we use in the reduction is distilled from some fairly involved necessary and sufficient conditions for a preference profile to be a rationalization, discovered by Echenique [Ech06]. We describe the full conditions in Section 2. Our gadget is a configuration across two matchings, that looks like this:

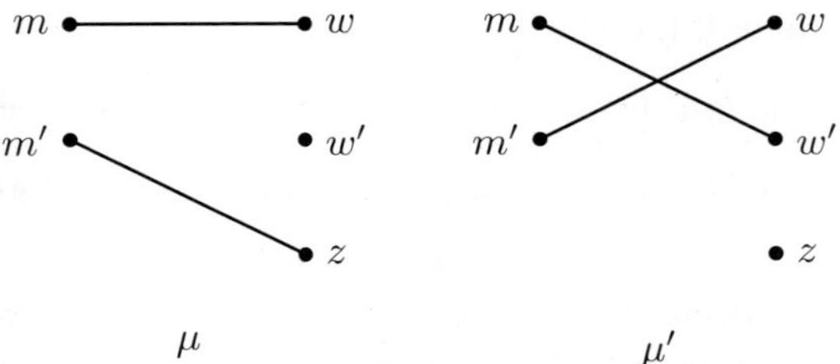

A preference profile $\mathcal{P}$ rationalizes the matchings containing this configuration only if either $m : w > w'$ and $m' : z > w$, or $m : w' > w$ and $m' : w > z$. Conversely, if these conditions hold (together with additional conditions concerning the remainder of the matchings) then $\mathcal{P}$ rationalizes the set of matchings. We use this gadget fundamentally as a Boolean choice gadget (either m prefers w over w' or w' over w), and as part of a scheme to ensure consistency (since the choice of m is tied to the choice of m').

Having ascertained that rationalizing a collection of matchings is NP-complete, we would next want to know how hard it is to solve the problem approximately. In this context, we first need to decide what exactly we mean by 'approximate' rationalization. Two notions are of particular interest: on the one hand, we can think of identifying a preference profile that rationalizes the maximum number of matchings.

Problem 1 (MAX-STABLE-MATCHINGS). Given a collection of matchings $\mathcal{H}$ on sets M, W, find a preference profile P that maximizes the number of matchings in $\mathcal{H}$ that are simultaneously rationalized by P.

This problem is hard to approximate to within some constant factor:

Theorem 2. *There is a constant $\epsilon > 0$ for which it is NP-hard to approximate* MAX-STABLE-MATCHINGS *to within a factor of* $(1 - \epsilon)$.

A second natural notion of approximation attempts to maximize "stability" among the given set of matchings at a more fine-grained level, by maximizing the number of non-blocking pairs across all matchings.

Some effort is required to make this notion of approximation meaningful. In a typical instance there will be many pairs (m, w) for which m is not matched to w in *any* of the given matchings. We say such a pair is *non-active* and pairs that are matched in some matching are *active*. It is easy to ensure that all non-active pairs are non-blocking pairs with respect to any matching, by requiring the preference profile to be *valid*:

Definition 4. *A preference profile $\mathcal{P}$ is* valid *with respect to a collection of matchings $\mathcal{H}$ if for every $m \in M$, $m : w > w'$ if (m, w) is active and (m, w') is not active, and for every $w \in W$, $w : m > m'$ if (m, w) is active and (m', w) is not active.*

In other words, each man m prefers a woman that he is matched to in some matching over women that he is never matched to, and similarly for each woman w. We argue that to have a meaningful notion of maximizing non-blocking pairs, one should consider only valid preference profiles, and therefore attempt to maximize the number of non-blocking pairs *among the active pairs* (since a valid preference profile automatically takes care of all of the non-active pairs). We are led to define the following optimization problem:

Problem 2 (MAX-STABILITY). Given a collection of matchings $\mathcal{H}$ on sets M, W, find a valid preference profile P for M, W that maximizes:

$$|\{(m, w, \mu) : (m, w) \text{ is active and is not a blocking pair with respect to } \mu, P\}|.$$

This problem is also hard to approximate to within some constant factor:

Theorem 3. *There is a constant $\epsilon > 0$ for which it is NP-hard to approximate* MAX-STABILITY *to within a factor of* $(1 - \epsilon)$.

Our proof uses the overall structure of the reduction used to prove Theorem 1 together with an explicit constant-degree expander to make aspects of the reduction robust enough to be gap-preserving.

An approximation of $3/4$ is achievable (in expectation) for this problem by a simple randomized assignment of preferences. Derandomizing via the method of conditional expectations yields:

Theorem 4. *There is a deterministic, polynomial-time approximation algorithm for* MAX-STABILITY *that achieves an approximation factor of 3/4.*

Finally, we turn to a generalization of the one-one matchings we have been considering:

Definition 5. *Let F, W be disjoint sets. A* **one-many matching** *is a pair of functions (μ, τ) with $\mu : F \to 2^W$, and $\tau : W \to F$ for which*

$$\forall w \in \mu(f), \tau(w) = f \text{ and } \forall w \in W, w \in \mu(\tau(w)).$$

Typically in economics literature, one-to-many matchings are spoken of in reference to firms and workers (or, similarly, hospitals and interns) and hence the notation of F, W is more prevalent. However, since this problem is so closely tied in with our discussion of one-to-one matchings we will continue to use the notation of "men" M and "women" W when we mention one-to-many matchings in the rest of the paper. One-many matching models have been widely studied [Rot82, Rot85].

In a one-many matching, preference order and preference profile are defined in the same way as for one-one matchings, except that each m has a linear ordering of 2^W instead of just W. Also analogous to the blocking pair for one-to-one matchings, we can define a *blocking set* and a notion of stability [EO04] for one-to-many matchings:

Definition 6. *A **blocking set** with respect to a one-many matching (μ, τ) and a preference profile $\mathcal{P}$ is a pair $(m, B) : m \in M, B \subseteq W$ such that $\mu(m) \cap B = \emptyset$ and*

$$\exists A \subseteq \mu(m) \text{ such that}$$
$$m : A \cup B > \mu(m) \text{ and } \forall w \in B \ \ w : m > \tau(w).$$

*Matching (μ, τ) is **stable*** with respect to $\mathcal{P}$ if there is no blocking set with respect to (μ, τ) and $\mathcal{P}$.*

The rationalization problem for one-many matchings is not likely to even be *in* NP, because a witness (preference profile) entails listing preference over 2^W, which is exponentially large. We are then led to consider a restricted version of the problem in which we only allow $m \in M$ to be matched to a set of cardinality at most some constant parameter ℓ. We call such matchings **one-ℓ** matchings.

The resulting rationalization problem is in NP and, we show, NP-complete:

Theorem 5. *For every fixed ℓ, given a collection of one-ℓ matchings $\mathcal{H}$ on the sets M and W, it is NP-complete to determine if there exists a preference profile $\mathcal{P}$ such that every $\mu \in \mathcal{H}$ is stable* with respect to $\mathcal{P}$.*

We can define the notion of an active pair (m, B) for one-ℓ matchings in analogy with active pairs, and also valid preference profiles as in Definition 4.

The two approximation problems arising with respect to one-ℓ matchings are hard to approximate to within some constant factor, just as in the one-one case:

Theorem 6. *There is a constant $\epsilon > 0$ for which it is NP-hard to approximate* MAX-STABLE-ONE-ℓ-MATCHINGS *to within a factor of* $(1 - \epsilon)$.

Theorem 7. *There is a constant $\epsilon > 0$ for which it is NP-hard to approximate* MAX-ONE-ℓ-STABILITY *to within a factor of* $(1 - \epsilon)$.

Please note that owing to space limitations, all our results on generalizations to the case of one-many matchings (Theorems 5-7) are proved in the full version of this paper [KU08].

2 Preliminaries

In this section, we encapsulate the working of the result for one-one matchings due to Echenique [Ech06] and provide the necessary and sufficient conditions for the existence of a preference profile that rationalizes a given collection of matchings. We start with some definitions and notations.

Definition 7. *For any two matchings $\mu, \mu' \in \mathcal{H}$, a (μ, μ')-**pivot** is a $w \in W$ such that there exist some $m_k, m_\ell \in M$ such that $\mu(m_k) = \mu'(m_\ell) = w$.*

The key to proving Theorem 1 is a result due to Echenique [Ech06] which we encapsulate in Lemma 1 which sets down necessary and sufficient conditions for the existence of a preference profile that rationalizes a given collection of matchings. We first introduce some notation that will be necessary to describe Lemma 1. Consider the directed graph G_{ij} with M as vertex set and E_{ij} as edge-set where $(m, m') \in E_{ij}$ if $\mu_i(m) = \mu_j(m')$. Let $\mathbf{C}(\mu_i, \mu_j)$ denote the set of all connected components of G_{ij}. We will denote the analogous graph obtained by considering as vertex set W as H_{ij}. The following proposition now follows from our notation and establishes a correspondence between G_{ij} and H_{ij}.

Proposition 1. *(Echenique [Ech06]) C is a connected component of G_{ij} iff $\mu_i(C)$ is a connected component of H_{ij}. Furthermore, $\mu_i(C) = \mu_j(C)$.*

Echenique [Ech06] showed the following lemma to be true.

Lemma 1. *(Echenique [Ech06]) Let $\mathcal{H} = \{\mu_1, \ldots, \mu_\ell\}$ be rationalized by preference profile $\mathcal{P}$. Consider, for all $\mu_i, \mu_j \in \mathcal{H}$ the graph G_{ij} and all $C \in \mathbf{C}_{ij}$. Then, exactly one of (1) or (2) must be true:*

$$m : \mu_i(m) > \mu_j(m) \text{ for all } m \in C \text{ and}$$
$$w : \mu_j(w) > \mu_i(w) \text{ for all } w \in \mu_i(C) \tag{1}$$
$$m : \mu_i(m) < \mu_j(m) \text{ for all } m \in C \text{ and}$$
$$w : \mu_j(w) < \mu_i(w) \text{ for all } w \in \mu_i(C) \tag{2}$$

Conversely, if $\mathcal{P}$ is a preference profile such that for all $\mu_i, \mu_j \in \mathcal{H}$ and $C \in \mathbf{C}(\mu_i, \mu_j)$, exactly one of (1) or (2) holds, and in addition:

$$m : \emptyset > w \iff w \notin \{\mu(m) | \mu \in \mathcal{H}\}$$
$$w : \emptyset > m \iff m \notin \{\mu(w) | \mu \in \mathcal{H}\}$$

where $\mu(m) = \emptyset$ would denote that m is not matched to any $w \in W$, then $\mathcal{P}$ rationalizes $\mathcal{H}$.

3 Hardness of Rationalizability of Matchings

We are given two sets M, W with $|M| = |W| = N$ and a set $\mathcal{H}$ of s matchings $\mu_1, \ldots, \mu_s : M \to W$. We show that the problem of determining whether there exists a preference profile that rationalizes $\mathcal{H}$ is NP-complete by reducing from NAE-3SAT.

3.1 Proof Outline

We give below a broad overview of the reduction used to prove Lemma 2. Our objective is to start with a set of clauses and construct matchings corresponding to them in such a way that the all-equal assignment to variables in a clause would lead to a conflicting preference relation for some element in the set of matchings. With this in mind, we build 'matching gadgets' corresponding to a given Boolean formula.

By way of example, consider a single clause $C_1 = (x_1 \vee \bar{x}_2 \vee \bar{x}_3)$. We associate with each variable x_i, the elements $m_{1i} \in M_1$, $w_{1i}, w'_{1i} \in W_1$. We will subsequently pad M_1 with dummy elements to ensure that $|M_1| = |W_1|$. For such a clause, we look up Fig. 1 to construct 10 partial matchings $\mu_1, \ldots, \mu_{10}$ involving $M_1 = \{m_{1i}|i = 1, 2, 3\} \cup \{u_1\}$ and $W_1 = \{w_{1i}, w'_{1i}|i = 1, 2, 3\} \cup \{y_1, z_1\}$. Our encoding of the truth assignment to a variable x_i in clause C_1 will then correspond to m_{1i} preferring w'_{1i} over w_{1i}, i.e. $m_{1i} : w'_{1i} > w_{1i}$ iff $x_i = 1$. The claim below gives a flavor of how the entire reduction works.

$\mu_{\ell 1}$:	$(m_i, w'_i)\ (m_j, w_i)$
$\mu_{\ell 2}$:	$(m_i, w_i)\ (m_j, y_\ell)$
$\mu_{\ell 3}$:	$(m_j, w'_j)\ (m_k, w_j)$
$\mu_{\ell 4}$:	$(m_j, w_j)\ (m_k, z_\ell)$
$\mu_{\ell 5}$:	$(m_k, w'_k)\ (u_\ell, w_k)$
$\mu_{\ell 6}$:	$(m_k, w_k)\ (u_\ell, w_j)$
$\mu_{\ell 7}$:	$(u_\ell, w_k)\ (m_i, w_j)$
$\mu_{\ell 8}$:	$(u_\ell, w_j)\ (m_i, w_i)$
$\mu_{\ell 9}$:	$(m_k, z_\ell)\ (m_i, w_j)$
$\mu_{\ell 10}$:	$(m_k, w_j)\ (m_i, w'_i)$

Fig. 1. For $C_\ell = (x_i + \bar{x}_j + \bar{x}_k)$

μ'_{p1}:	$(m_{ip}, w'_{ip})\ (v_{ip}, w_{ip})$
μ'_{p2}:	$(m_{ip}, w_{ip})\ (v_{ip}, w'_{jp})$
μ'_{p3}:	$(v_{ip}, w_{ip})\ (m_{jp}, w'_{jp})$
μ'_{p4}:	$(v_{ip}, w'_{jp})\ (m_{jp}, w_{jp})$

Fig. 2. Consistency matching for x_p occurring in clauses C_i, C_j

Claim. There exists a rationalizable preference profile for M_1, W_1 for the matchings described in Fig. 1 iff there exists a not-all-equal satisfying assignment for C_1.

Proof **(Sketch).** Suppose there exists a not-all-equal satisfiable assignment to C_1. Then, in order to show that the corresponding preference profile obtained is rationalizable, we will show that it satisfies the conditions in Lemma 1. We fix the preference for each m_{1i} between w_{1i} and w'_{1i} based on the assignment to x_i for $i = 1, 2, 3$. We set $m_{1i} : w'_{1i} > w_{1i}$ if $x_i = 1$ and $m_{1i} : w_{1i} > w'_{1i}$ otherwise. Note that since an assignment $(0, 1, 1)$ or $(1, 0, 0)$ to (x_1, x_2, x_3) is ruled out, the matchings in Table 3.1 ensure that there will be no "cycles" in the preference orders of m_{11}, m_{12}, m_{13}. Furthermore, an assignment to x_1, x_2, x_3 only fixes a preference order for all $m \in M_1$ and so we can fix a preference order for $w \in W_1$ so that there is no conflict in the preference orders for all m, w and that the conditions in Lemma 1 are satisfied.

The converse is immediate because for a rationalizable preference profile for $m \in M_1, w \in W_1$, Lemma 1 holds and hence an all-equal assignment to C_1 is not allowed. For instance, suppose (x_1, x_2, x_3) were assigned $(0, 1, 1)$ then using Lemma 1 to draw up all the preference relations we would obtain a conflict, i.e. $m_{11} : w_{12} > w'_{11}$ (applying

Lemma 1 to $\mu_{11}, \ldots, \mu_{18}$) and $m_{11} : w_{12} < w'_{11}$ (applying Lemma 1 to μ_{19}, μ_{110}). Therefore, setting each of the x_i to the values obtained depending on the preference relation for m_{1i} between w_{1i} and w'_{1i} as delineated above is a not-all-equal satisfying assignment.

In a Boolean formula with m clauses, we repeat the exercise above but use disjoint sets M_ℓ, W_ℓ for each clause C_ℓ to avoid conflicting preference orders *across* clauses. This makes it necessary for us to enforce consistency between the preference relations for $m_{\ell i}$ and $w_{\ell i}, w'_{\ell i}$ for all $\ell = 1, \ldots, m$ and the assignment to x_i. To this end, we use additional matching gadgets from Fig. 2 and an auxiliary element v_i. Again applying Lemma 1, we see that for x_1 occurring in clauses C_1, C_2 say, we must have that $m_{11} : w'_{11} > w_{11} \iff m_{21} : w'_{21} > w_{21}$.

Note that in the manner our construction of matching gadgets is set up, it is necessary for our purposes to reduce from NAE-3SAT as opposed to 3SAT because, if an all-false assignment to a clause were to lead to a conflict in preference relation for some m, w, w', then by symmetry an all-true assignment would also lead to a contradictory preference relation.

3.2   Proof of Theorem 1

The proof for Theorem 1 automatically follows from Lemma 2 which we formally state below.

Lemma 2. *Let $\mathcal{Z}$ be an instance of* NAE-3SAT *over n variables $x_1, \ldots, x_n$ and m clauses $C_0, \ldots, C_{m-1}$. Then, there exists an instance $\mathcal{Z}'$ of $O(m)$ matchings between sets M and W, $|M| = |W| = O(m + n)$ such that there exists a rationalizable preference profile for all $m \in M, w \in W$ iff there exists a not-all-equal satisfiable assignment to $x_1, \ldots, x_n$. Furthermore, these matchings can be constructed in polynomial time.*

We defer the detailed proof to the full version of this paper [KU08] but make a few remarks here summarizing the proof. First, note that the table in Fig. 1 can also be used symmetrically for the clause of type $(\bar{x}_i + x_j + x_k)$. Similar such 'matching tables' can be constructed corresponding to all the different types of clauses and are used in order to construct the partial matchings.

Finally, the remaining matchings between elements of M and W are constructed based on some simple rules to ensure that no contradictory preferences (i.e. $m : w' > w$ and $m : w > w'$) and no unintended blocking pairs occur. In the end a not-all-equal assignment for $\mathcal{Z}$ exists iff there is a rationalizing preference profile for the corresponding collection of matchings.

4   Hardness of Approximate Rationalizability of Matchings

Our next step in exploring the computational aspects of rationalizability of matchings will be to look at the complexity of 'approximate' rationalizability.

4.1   Maximizing the Number of Rationalizable Matchings

In the first setting, we wish to maximize the number of matchings that can be completely rationalized as stable by a preference profile. We argue in the theorem below that this is hard to approximate within a constant factor.

Theorem 2 states that this is hard to approximate within a constant factor. To prove Theorem 2 we show that it is NP-hard to rationalize any fixed set of matchings as captured in the lemma below.

Lemma 3. *Given a collection of matchings* $\mathcal{H} = \{\mu_1, \ldots, \mu_k\}$ *between* M *and* W *where* k *is some fixed constant, it is NP-hard to determine if there exists preferences for* $m \in M, w \in W$ *for which each of* $\mu \in \mathcal{H}$ *is a stable matching.*

From Lemma 3 (proof in full version [KU08]) it follows that it is NP-hard to approximate MAX-STABLE-MATCHINGS for $\mathcal{H}$ to within a factor of $(1-\epsilon)$ where $\epsilon = 1/(k+1)$.

Note that given a collection $\mathcal{H}$ of any two matchings, it is trivial to construct a (valid) preference profile that rationalizes $\mathcal{H}$ by arbitrarily assigning a preference for each element in M matched to W in one matching over the other and correspondingly assigning the reverse preference for elements in W.

4.2   Maximizing the Number of Non-blocking Pairs

We look at the MAX-STABILITY problem. The motivation in considering this problem as a notion of approximate rationalizability is that we are now striving to ensure that given a collection of matchings between two sets M and W, there are optimally many different pairs (m, w) for which at least one of them is happy with their current partner and has no incentive to be matched to the other.

As a preliminary exercise, we ask how well would a simple randomized assignment of preferences to $m \in M, w \in W$ perform. It turns out that this would achieve a $3/4$-approximate solution. This is the content of Theorem 4 whose proof is deferred to the full version [KU08].

It suffices to mention here that a simple randomized preference order for all $m \in M, w \in W$ achieves the $3/4$-approximation factor in expectation and can subsequently be derandomized. How much better can we do than just a random assignment of preferences? Theorem 3 as stated tells us that a constant-factor approximation is all we can hope for.

To prove the theorem, we once again construct matchings corresponding to each clause in MAX-NAE-3SAT instance Z. Recall that in proving Lemma 2 we needed to construct auxiliary matchings to ensure consistency of assignment to the variables in accordance with the preferences of the corresponding elements in the matchings. To prove hardness of approximation, we will need to establish a gap-preserving reduction by boosting the *robustness* of these consistency gadgets. We do so by augmenting the number of matchings corresponding to the consistency and argue subsequently that if there exists a preference profile that achieves at least a $(1 - \epsilon')$ fraction of stable pairs, then there exists an assignment that would satisfy at least a $(1 - \epsilon)$ fraction of the clauses. Theorem 3 then follows from the following Lemma:

Lemma 4. *Let Z be an instance of* MAX-NAE-3SAT *over n variables $x_1, \ldots, x_n$ and m clauses $C_0, \ldots, C_{m-1}$. Then, there exists a $\epsilon' < 1$ and a polynomial time reduction to an instance Z' of* MAX-STABILITY *of matchings between sets M and W, $|M| = |W| = O(m)$ such that the following is true:*

$$opt(Z) = 1 \implies opt(Z') = 1 \tag{3}$$
$$opt(Z) < 1 - \epsilon \implies opt(Z') < 1 - \epsilon' \tag{4}$$

Proof. We set up matchings corresponding to the clauses $C_0, \ldots, C_{m-1}$ as before, but now we need to work harder to boost the robustness of the consistency gadgets. Previously, we used Table 3.1 to construct additional matchings using auxiliary elements to 'link' different copies of $m_{ji}; j = 1, \ldots, m$ corresponding to a single variable x_i. It will help to conceptualize this as a graph.

For a variable x_i which occurs in some t clauses $C_{j_1}, \ldots, C_{j_t}$, we associate elements from $M, m_{j_1 i}, \ldots, m_{j_t i}$ and define the consistency graph for x_i, G_i to comprise vertex set $V_i = \{m_{j_1 i}, \ldots, m_{j_t i}\}$. An edge exists between any two vertices $(m_{j_p i}, m_{j_q i})$ if they are 'linked' together by an auxiliary element.

Then, the consistency matchings described above in Claim 3.1 correspond to a path in G_i. In order to boost the robustness, we will now replace the path in G_i by a constant-degree expander graph on t vertices. We make use of the edge expansion notion to define an expander graph: an (n, d, λ) expander graph is a d-regular graph on n vertices with the property that $|\partial(Y)|/|Y| \geq d(1-\lambda)/2$ where $Y \subseteq V_i, |Y| \leq |V_i|/2, \partial(Y)$ is the set of all edges with exactly one end-point in Y and λ is the spectral expansion parameter of the graph. In particular, the following proposition will be useful (the proof can be found in [DH05]):

Lemma 5. *For a (t, d, λ) expander graph G and all $\delta \leq (1 - \lambda)/12$, upon removing $2\delta t$ vertices from G, there exists a connected component of size at least*

$$\left(1 - \frac{4\delta}{1 - \lambda}\right) t$$

Note that the total number of occurrences of variables in all the clauses is at most $3m$, and further, that in each clause a variable corresponds to an element m matched to at most an $O(1)$ elements in W. Therefore, the total number of pairs for which a matching exists is at most $O(m)$. Since we only consider valid preference profiles, this means that the number of active pairs under consideration is also $O(m)$ say. Additionally, the total number of auxiliary elements required to construct the expander graphs in the consistency gadgets is also at most $O(m)$ and hence $|M| = O(m)$.

Since our reduction is unchanged in how a satisfying assignment will correspond to a rationalizing preference profile (and hence, all stable pairs), (3) goes through. It remains to show that (4) holds.

We shall show that if there is a valid preference profile for Z' such that there are at most an ϵ' fraction of blocking pairs, then there exists an assignment that fails to satisfy at most ϵ fraction of clauses in Z.

Suppose that there is a valid preference profile that allows at most $\epsilon' m$ blocking pairs. Note that if a pair (m, w) is a blocking pair for some matching μ, then Lemma 1 breaks down for μ. Since each matching in Z' can be identified with a clause, a blocking pair could result in the clause being unsatisfied.

For a blocking pair (m, w) for some matching μ in our reduction, we evaluate how many clauses are affected. Suppose μ corresponds to one of the matchings for clause C_ℓ. If $m \in M_\ell$ then m must be associated with some variable x_i occurring in C_ℓ, and we will label C_ℓ unsatisfiable. Otherwise, (m, w) has no effect on the satisfiability of C_ℓ.

Suppose μ corresponds to a matching constructed to ensure consistency. If $m \in M_\ell$ for some clause C_ℓ and x_i, then we delete the node $m_{\ell i}$ in G_i and as before label C_ℓ as unsatisfiable. However, now we also need to argue that (m, w) does not cause too many other clauses to be labeled unsatisfiable.

From Lemma 5 we know that deleting at most a constant fraction of vertices from G_i will result in a connected component of size at least $(1 - \frac{4\delta}{(1-\lambda)})t$. Taking the aggregate for every variable x_i and after deleting at most $\epsilon'm$ vertices from all the consistency graphs G_i together, the total sum of the largest connected components amongst all G_i will be some $(1-\epsilon)m$ where ϵ is determined by ϵ', λ and the total number of occurrences of all variables in all the clauses. Therefore, at most ϵm of these occurrences in clauses will be discarded and the corresponding ϵm clauses labeled as unsatisfiable.

MAX-NAE-3SAT is known to be APX-complete [PY91] and not approximable to within 0.917 [Zwi98].

5   Conclusions and Future Work

There are many interesting opportunities for extensions to our work on the rationalization problem for matchings. It would be interesting to tighten the constant factor in Lemma 3: is it hard even to rationalize three matchings? It would also be satisfying to tighten the hardness of approximation result in Theorem 3. We can additionally look at other (restricted) variants of the matchings problem such as many-many matchings and pose the related complexity questions.

On a more general note, the question of rationalizability per se is very tantalizing because of the mutually interesting perspectives it offers within both economics and theoretical computer science.

Acknowledgments. We are indebted to Federico Echenique for numerous invaluable discussions and for getting us started on this work.

References

[Afr67]   Afriat, S.N.: The Construction of Utility Functions from Expenditure Data. International Economic Review 8(1), 67–77 (1967)

[BV06]   Beigman, E., Vohra, R.: Learning from revealed preference. In: ACM Conference on Electronic Commerce, pp. 36–42 (2006)

[DH05]   Dwork, C., Harsha, P.: Expanders in Computer Science (CS369E) – Lecture 5, Stanford University (2005)

[Die73]   Diewert, E.: Afriat and Revealed Preference Theory. Review of Economic Studies 40(3), 419–425 (1973)

[Ech06]   Echenique, F.: What matchings can be stable? The testable implications of matching theory. Technical Report 1252, California Institute of Technology Social Science Working Papers (2006)

[EO04]   Echenique, F., Oviedo, J.: Core many-to-one matchings by fixed-point methods. Journal of Economic Theory 115(2), 358–376 (2004)

[EY07]   Echenique, F., Yenmez, M.B.: A solution to matching with preferences over colleagues. Games and Economic Behavior 59(1), 46–71 (2007)

[Fle03]   Fleiner, T.: A fixed-point approach to stable matchings and some applications. Math. Oper. Res. 28(1), 103–126 (2003)

[FST04]   Fostel, A., Scarf, H.E., Todd, M.J.: Two new proofs of Afriat's theorem. Economic Theory 24(1), 211–219 (2004)

[GS62]   Gale, D., Shapley, L.: College admissions and the stability of marriage. American Mathematical Monthly 69(1), 9–15 (1962)

[KU08]   Kalyanaraman, S., Umans, C.: The complexity of rationalizing matchings. Electronic Colloquium on Computational Complexity (ECCC) (21) (2008)

[Özs06]   Özsoy, H.: A characterization of Bird's rule, Job market paper (2006)

[PY91]   Papadimitriou, C.H., Yannakakis, M.: Optimization, approximation, and complexity classes. Journal of Computing Systems and Sciences 43(3), 425–440 (1991)

[Rot82]   Roth, A.E.: The economics of matchings: stability and incentives. Math Operations Research 7, 617–628 (1982)

[Rot85]   Roth, A.E.: The college admissions problem is not equivalent to the marriage problem. Journal of Economic Theory 35, 277–288 (1985)

[RS90]   Roth, A.E., Sotomayor, M.A.: Two-sided matching: A Study in Game-Theoretic Modeling and Analysis, 2nd edn. Econometric Society Monographs, vol. 18. Cambridge University Press, Cambridge (1990)

[Sam48]   Samuelson, P.A.: Consumption Theory in terms of Revealed Preference. Economica 15(60), 243–253 (1948)

[Sön96]   Sönmez, T.: Strategy-proofness in many-to-one matching problems. Economic Design 3, 365–380 (1996)

[Spr00]   Sprumont, Y.: On the Testable Implications of Collective Choice Theories. Journal of Economic Theory 93, 205–232 (2000)

[STT01]   Sethuraman, J., Teo, C.-P., Tan, W.-P.: Gale-Shapley stable marriage revisited: strategic issues and applications. Management Science 47(9), 1252–1267 (2001)

[Var82]   Varian, H.R.: The Nonparametric Approach to Demand Analysis. Econometrica 50(4), 945–973 (1982)

[Var06]   Varian, H.R.: Revealed Preference. In: Szenberg, M., Ramrattan, L., Gottesman, A.A. (eds.) Samuelson Economics and the Twenty-First Century, ch. 6, pp. 99–115. Oxford University Press, Oxford (2006)

[Zwi98]   Zwick, U.: Approximation algorithms for constraint satisfaction problems involving at most three variables per constraint. In: Proceedings of the 9th Annual ACM-SIAM Symposium on Discrete Algorithms, pp. 201–210. ACM-SIAM, New York (1998)

A Game Theoretic Approach for Efficient Graph Coloring[*]

Panagiota N. Panagopoulou[1,2] and Paul G. Spirakis[1,2]

[1] Computer Engineering and Informatics Department, University of Patras
[2] Research Academic Computer Technology Institute
N. Kazantzaki Str., University of Patras, GR 26500 Rion, Patras, Greece
panagopp@cti.gr, spirakis@cti.gr

Abstract. We give an efficient local search algorithm that computes a good vertex coloring of a graph G. In order to better illustrate this local search method, we view local moves as selfish moves in a suitably defined game. In particular, given a graph $G = (V, E)$ of n vertices and m edges, we define the *graph coloring game* $\Gamma(G)$ as a strategic game where the set of players is the set of vertices and the players share the same action set, which is a set of n colors. The payoff that a vertex v receives, given the actions chosen by all vertices, equals the total number of vertices that have chosen the same color as v, unless a neighbor of v has also chosen the same color, in which case the payoff of v is 0. We show:

- The game $\Gamma(G)$ has always pure Nash equilibria. Each pure equilibrium is a proper coloring of G. Furthermore, there exists a pure equilibrium that corresponds to an optimum coloring.
- We give a polynomial time algorithm $\mathcal{A}$ which computes a pure Nash equilibrium of $\Gamma(G)$.
- The total number, k, of colors used in *any* pure Nash equilibrium (and thus achieved by $\mathcal{A}$) is $k \leq \min\{\Delta_2 + 1, \frac{n+\omega}{2}, \frac{1+\sqrt{1+8m}}{2}, n - \alpha + 1\}$, where ω, α are the clique number and the independence number of G and Δ_2 is the maximum degree that a vertex can have subject to the condition that it is adjacent to at least one vertex of equal or greater degree. (Δ_2 is no more than the maximum degree Δ of G.)
- Thus, in fact, we propose here a new, efficient coloring method that achieves a number of colors satisfying (together) the known general upper bounds on the chromatic number χ. Our method is also an alternative general way of proving, constructively, all these bounds.
- Finally, we show how to strengthen our method (staying in polynomial time) so that it avoids "bad" pure Nash equilibria (i.e. those admitting a number of colors k far away from χ). In particular, we show that our enhanced method colors *optimally* dense random q-partite graphs (of fixed q) with high probability.

[*] Partially supported by the EU within the 6th Framework Programme under contract 015964 "Algorithmic Principles for Building Efficient Overlay Computers" (AEOLUS) and the ICT Programme under contract IST-2008-215270 (FRONTS), and by the General Secretariat for Research and Technology of the Greek Ministry of Development within the Programme PENED 2003.

S.-H. Hong, H. Nagamochi, and T. Fukunaga (Eds.): ISAAC 2008, LNCS 5369, pp. 183–195, 2008.

1   Introduction

Overview. One of the central optimization problems in Computer Science is the problem of *vertex coloring* of graphs: given a graph $G = (V, E)$ of n vertices, assign a color to each vertex of G so that no pair of adjacent vertices gets the same color and so that the total number of distinct colors used is minimized. The global optimum of vertex coloring (the *chromatic number*) is, in general, inapproximable in polynomial time unless a collapse of some complexity classes happens [7]. In this paper, we propose an efficient vertex coloring algorithm that is based on *local search*: Starting with an arbitrary proper vertex coloring (e.g. the trivial proper coloring where each vertex is assigned a unique color), we do local changes, by allowing each vertex (one at a time) to move to another color class of higher cardinality, until no further local moves are possible.

We choose to illustrate this local search method via a game-theoretic analysis; we do so because of the natural correspondence of the local optima of our proposed method to the pure Nash equilibria of a suitably defined *strategic game*. In particular, we view vertices of a graph $G = (V, E)$ as players in a strategic game. Each player has the same set of actions, which is a set of $|V|$ colors. In a certain profile $\mathbf{c}$ (where each vertex v has chosen a color), v gets a payoff of zero if its color is the same with the color of a neighbor of v. Else, v gets as a payoff the number of vertices having selected the same color as the color that v has chosen. In a pure Nash equilibrium of such a game (if such an equilibrium exists), no vertex can improve its payoff by unilaterally deviating. Note that, given a profile, one can compute payoffs in small polynomial time. Furthermore, a "better response" (i.e., a selfish improvement) of a vertex v, given a choice of colors by all the other vertices, can also be computed quickly by v and the only global information needed is the number of vertices per color in the graph.

In such a setting, if we start by the trivial proper coloring of G (where each v chooses its unique name as a color), then any selfish improvement sequence always produces proper colorings of G. This would give an efficient and general proper coloring heuristic provided that: (i) Pure equilibria exist; (ii) Such selfish improvement sequences reach an equilibrium in small time; and (iii) The number of colors at equilibrium is a good approximation of the chromatic number of G.

Our Results. Quite surprisingly, we show for our game that:

(1) Any selfish improvement sequence, when started with a *proper* (e.g., the trivial) coloring, always reaches an equilibrium in $O(n \cdot \alpha(G))$ selfish moves, where $\alpha(G)$ is the independence number of G. We prove this by a potential-based method [14].

(2) Any pure Nash equilibrium of the game is a *proper* coloring of G that uses a number of colors, k, bounded above by all the general known to us upper bounds on the chromatic number of G. Specifically, let $n, m, \chi(G)$ and $\omega(G)$ and $\Delta(G)$, denote the number of vertices, number of edges, chromatic number, clique number and maximum degree of G, respectively. Let $\Delta_2(G)$ be the maximum degree that a vertex v can have subject to the condition that v is adjacent to at least one vertex of degree no less than the degree of v (note that $\Delta_2(G) \leq \Delta(G)$). We show that k (in any pure Nash equilibrium) satisfies

$$k \leq \min \left\{ \Delta_2(G) + 1, \ \frac{n + \omega(G)}{2}, \ n - \alpha(G) + 1, \ \frac{1 + \sqrt{1 + 8m}}{2} \right\}.$$

Note that $\Delta_2(G) + 1$ is the bound of Stacho [16] and implies Brooks' bound [4] on $\chi(G)$. In fact, we get *constructively* all these bounds via a single polynomial time algorithm. For some of these bounds their proofs till now (in popular graph theory books, e.g. [9]) are not constructive and not based on a single unifying method.

(3) Since $\chi(G)$ is inapproximable in polynomial time (unless a collapse of complexity classes happens) it is natural to expect the existence of some pure equilibria in our game that use a number of colors k far away from $\chi(G)$. Indeed we were able to construct a class of (almost complete bipartite) graphs G which have equilibrium colorings of $k = \frac{n}{2} + 1$, while $\chi(G) = 2$. However, our selfish improvement method does not have to go to such bad equilibria. For the same class of graphs we show that a *randomized* sequence of selfish improvements achieves $k = 2$ with high probability. In fact, our class of algorithms can be started by the proper colorings achieved by the *best till now* approximation methods. Then, it may improve on them, if their output is not an equilibrium of our game.

(4) Motivated by such thoughts, we investigated the following question: What kind of polynomial time "mechanisms" (e.g., some preprocessing, a particular order of selfish moves, e.t.c.) can help our coloring method to get closer to $\chi(G)$ in certain graph classes? We managed to provide such enhanced methods that e.g. are optimal with high probability for dense random q-partite graphs.

We believe that our game and its properties can serve also as an educational tool in introducing and proving general bounds on the chromatic number.

Previous work. The problem of coloring a graph using the minimum number of colors is NP-hard [13], and the best polynomial time approximation algorithm achieves an approximation ratio of $O(n(\log \log n)^2/(\log n)^3)$ [8]. It is known [7] that the chromatic number cannot be approximated to within $\Omega(n^{1-\epsilon})$ for any constant $\epsilon > 0$, unless $\mathsf{NP} \subseteq \mathsf{co\text{-}RP}$. Several vertex coloring heuristics have been proposed in the literature, such as Brelaz's heuristic [3]. To the best of our knowledge, none of these heuristics achieves all these bounds on the total number of colors that our algorithm guarantees. Graph coloring games have been studied before, but in a very different context than here. In these games there are 2 players, who are introduced with the graph to be colored and a color bound k. A legal move of either player consists of choosing an uncolored vertex v, and assign to it any of the k colors that has not been assigned to any neighbor of v. In one variant of such a game [1] the first player which is unable to move loses the game. In another variant [1] the first player wins if and only if the game ends with all vertices colored. Further variants have also been studied by e.g. [10,6].

2   The Model

Notation. For a finite set A we denote by $|A|$ the cardinality of A. For an event E in a sample space, denote $\Pr\{E\}$ the probability of E occurring. Denote $G = (V, E)$ a simple, undirected graph with vertex set V and set of edges E. For a vertex $v \in V$ denote $N(v) = \{u \in V : \{u, v\} \in E\}$ the set of its neighbors, and let $\deg(v) = |N(v)|$ denote its degree. Let $\Delta(G) = \max_{v \in V} \deg(v)$ be the maximum degree of G. Let $\Delta_2(G) = \max_{u \in V} \max_{v \in N(u):d(v) \leq d(u)} \deg(v)$ be the maximum degree that a vertex v can have, subject to the condition that v is adjacent to at least one vertex of degree no less than $\deg(v)$. Clearly, $\Delta_2(G) \leq \Delta(G)$. Let $\chi(G)$ denote the chromatic number of G, i.e. the minimum number of colors needed to color the vertices of G such that no adjacent vertices get the same color (i.e., the minimum number of colors used by a *proper coloring* of G). Let $\omega(G)$ and $\alpha(G)$ denote the clique number and independence number of G, i.e. the number of vertices in a maximum clique and a maximum independent set of G.

The Graph Coloring Game. Given a finite, simple, undirected graph $G = (V, E)$ with $|V| = n$ vertices, we define the *graph coloring game* $\Gamma(G)$ as the game in strategic form where the set of players is the set of vertices V, and the action set of each vertex is a set of n colors $X = \{x_1, \ldots, x_n\}$. A *configuration* or *pure strategy profile* $\mathbf{c} = (c_v)_{v \in V} \in X^n$ is a combination of actions, one for each vertex. That is, c_v is the color chosen by vertex v. For a configuration $\mathbf{c} \in X^n$ and a color $x \in X$, we denote by $n_x(\mathbf{c})$ the number of vertices that are colored x in $\mathbf{c}$, i.e. $n_x(\mathbf{c}) = |\{v \in V : c_v = x\}|$. The *payoff* that vertex $v \in V$ receives in the configuration $\mathbf{c} \in X^n$ is

$$\lambda_v(\mathbf{c}) = \begin{cases} 0 & \text{if } \exists u \in N(v) : c_u = c_v \\ n_{c_v}(\mathbf{c}) & \text{else} \end{cases}.$$

A *pure Nash equilibrium* [15] (PNE in short) is a configuration $\mathbf{c} \in X^n$ such that no vertex can increase its payoff by unilaterally deviating. Let $(x, \mathbf{c}_{-v})$ denote the configuration resulting from $\mathbf{c}$ if vertex v chooses color x while all the remaining vertices preserve their colors. Then

Definition 1. *A configuration $\mathbf{c} \in X^n$ of the graph coloring game $\Gamma(G)$ is a pure Nash equilibrium if, for all vertices $v \in V$, $\lambda_v(x, \mathbf{c}_{-v}) \leq \lambda_v(\mathbf{c}) \quad \forall x \in X$.*

A vertex $v \in V$ is *unsatisfied* in the configuration $\mathbf{c} \in X^n$ if there exists a color $x \neq c_v$ such that $\lambda_v(x, \mathbf{c}_{-v}) > \lambda_v(\mathbf{c})$; else we say that v is *satisfied*. For an unsatisfied vertex $v \in V$ in the configuration $\mathbf{c}$, we say that v performs a *selfish step* if v unilaterally deviates to some color $x \neq c_v$ such that $\lambda_v(x, \mathbf{c}_{-v}) > \lambda_v(\mathbf{c})$.

The *Social Cost* $\mathsf{SC}(G, \mathbf{c})$ of a configuration $\mathbf{c} \in X^n$ of $\Gamma(G)$ is the number of distinct colors in $\mathbf{c}$, i.e., $\mathsf{SC}(G, \mathbf{c}) = |\{x \in X \mid n_x(\mathbf{c}) > 0\}|$. Given a graph G, the *Approximation Ratio* $\mathsf{R}(G)$ is the ratio of the worst, over all pure Nash equilibria of $\Gamma(G)$, Social Cost to the chromatic number: $\mathsf{R}(G) = \max_{\mathbf{c}:\mathbf{c} \text{ is a PNE}} \frac{\mathsf{SC}(G,\mathbf{c})}{\chi(G)}$.

3   Existence and Tractability of Pure Nash Equilibria

Theorem 1. *Every graph coloring game $\Gamma(G)$ possesses at least one pure Nash equilibrium, and there exists a pure Nash equilibrium $\mathbf{c}$ with $\mathsf{SC}(G, \mathbf{c}) = \chi(G)$.*

Proof. Consider any optimum coloring $\mathbf{o} = (o_v)_{v \in V} \in X^n$ of G. Then $\mathbf{o}$ uses $k = \chi(G)$ colors. For each optimum coloring $\mathbf{o}$ consider the vector $\mathbf{L_o} = (\ell_\mathbf{o}(1), \ldots, \ell_\mathbf{o}(k))$, where $\ell_\mathbf{o}(j)$ is the number of vertices that are assigned the color that is jth in the decreasing ordering of colors according to the number of vertices that use them. Let $\hat{\mathbf{o}}$ correspond to the lexicographically greatest vector $\mathbf{L_{\hat{o}}}$. We will show that $\hat{\mathbf{o}}$ is a pure Nash equilibrium. First, since $\hat{\mathbf{o}}$ is a proper coloring, all vertices receive payoff no less than 1, so no vertex has any incentive to choose a new color other than those already used. Now consider a vertex v which is assigned color $\hat{o}_v$ and let i be the coordinate that corresponds to $\hat{o}_v$ in $\mathbf{L_{\hat{o}}}$. If v had an incentive to choose a color that corresponds to the jth coordinate of $\mathbf{L_{\hat{o}}}$ for some $j < i$, then this would yield an optimum coloring that would be lexicographically greater than $\hat{\mathbf{o}}$, a contradiction. If v had an incentive to choose a color that corresponds to the jth coordinate of $\mathbf{L_{\hat{o}}}$ for some $j > i$, then it must essentially hold that $\ell_{\hat{\mathbf{o}}}(i) = \ell_{\hat{\mathbf{o}}}(j)$. So, if v deviates, this would again yield an optimum coloring that would be lexicographically greater that $\hat{\mathbf{o}}$, a contradiction. Therefore $\hat{\mathbf{o}}$ is a pure Nash equilibrium and $\mathsf{SC}(G, \hat{\mathbf{o}}) = \chi(G)$. $\square$

Lemma 1. *Every pure Nash equilibrium $\mathbf{c}$ of $\Gamma(G)$ is a proper coloring of G.*

Proof. Assume, by contradiction, that $\mathbf{c}$ is not a proper coloring. Then there exists some vertex $v \in V$ such that $\lambda_v(\mathbf{c}) = 0$. Clearly, there exists some color $x \in X$ such that $c_u \neq x$ for all $u \in V$. Therefore $\lambda_v(x, \mathbf{c}_{-v}) = 1 > 0 = \lambda_v(\mathbf{c})$, which contradicts the fact that $\mathbf{c}$ is an equilibrium. $\square$

Corollary 1. *It is NP-complete to decide whether there exists a pure Nash equilibrium of $\Gamma(G)$ that uses at most k colors.*

Proof (Sketch). Follows by reduction to the NP-complete problem of deciding whether there exists a proper coloring of a graph that uses at most k colors. $\square$

Theorem 2. *For any graph coloring game $\Gamma(G)$, a pure Nash equilibrium can be computed in $O(n \cdot \alpha(G))$ selfish steps, where n is the number of vertices of G and $\alpha(G)$ is the independence number of G.*

Proof. We define the function $\Phi : P \to \mathbb{R}$, where $P \subseteq X^n$ is the set of all configurations that correspond to proper colorings of the vertices of G, as $\Phi(\mathbf{c}) = \frac{1}{2} \sum_{x \in X} n_x^2(\mathbf{c})$, for all proper colorings $\mathbf{c}$. Fix a *proper* coloring $\mathbf{c}$. Assume that vertex $v \in V$ can improve its payoff by deviating and selecting color $x \neq c_v$. This implies that the number of vertices colored c_v in $\mathbf{c}$ is at most the number of vertices colored x in $\mathbf{c}$, i.e. $n_{c_v}(\mathbf{c}) \leq n_x(\mathbf{c})$. If v indeed deviates to x, then the resulting configuration $\mathbf{c}' = (x, \mathbf{c}_{-v})$ is again a proper coloring (vertex v can only decrease its payoff by choosing a color that is already used by one of its neighbors, and v is the only vertex that changes its color). The improvement on v's payoff will be $\lambda_v(\mathbf{c}') - \lambda_v(\mathbf{c}) = n_x(\mathbf{c}') - n_{c_v}(\mathbf{c}) = n_x(\mathbf{c}) + 1 - n_{c_v}(\mathbf{c})$. Moreover,

$$\Phi(\mathbf{c}') - \Phi(\mathbf{c}) = \frac{1}{2}\left(n_x^2(\mathbf{c}') + n_{c_v}^2(\mathbf{c}') - n_x^2(\mathbf{c}) - n_{c_v}^2(\mathbf{c})\right)$$

$$= \frac{1}{2}\left((n_x(\mathbf{c}) + 1)^2 + (n_{c_v}(\mathbf{c}) - 1)^2 - n_x^2(\mathbf{c}) - n_{c_v}^2(\mathbf{c})\right)$$

$$= n_x(\mathbf{c}) + 1 - n_{c_v}(\mathbf{c}) = \lambda_v(\mathbf{c}') - \lambda_v(\mathbf{c}).$$

Therefore, if any vertex v performs a selfish step (i.e. changes its color so that its payoff is increased) then the value of Φ is increased as much as the payoff of v is increased. Now, the payoff of v is increased by at least 1. So after any selfish step the value of Φ increases by at least 1. Now observe that, for all proper colorings $\mathbf{c} \in P$ and for all colors $x \in X$, $n_x(\mathbf{c}) \leq \alpha(G)$. Therefore $\Phi(\mathbf{c}) = \frac{1}{2}\sum_{x \in X} n_x^2(\mathbf{c}) \leq \frac{1}{2}\sum_{x \in X}(n_x(\mathbf{c}) \cdot \alpha(G)) = \frac{1}{2}\alpha(G)\sum_{x \in X} n_x(\mathbf{c}) = \frac{n \cdot \alpha(G)}{2}$. Moreover, the minimum value of Φ is $\frac{1}{2}n$. Therefore, if we allow any unsatisfied vertex (but only one each time) to perform a selfish step, then after at most $\frac{n \cdot \alpha(G) - n}{2}$ steps there will be no vertex that can improve its payoff (because Φ will have reached a local maximum, which is no more than the global maximum, which is no more than $(n \cdot \alpha(G))/2$, so a pure Nash equilibrium will have been reached. Of course, we have to start from an initial configuration that is a proper coloring so as to ensure that $\mathcal{A}$ will terminate in $O(n \cdot \alpha(G))$ selfish steps; this can be found easily since there is always the trivial proper coloring that assigns a different color to each vertex of G. $\qquad\square$

The above proof implies the following simple algorithm $\mathcal{A}$ that computes a pure Nash equilibrium of $\Gamma(G)$ (and thus a proper coloring of G):

> *Input:* Graph G with vertex set $V = \{v_1, \ldots, v_n\}$; a set of colors $X = \{x_1, \ldots, x_n\}$
> *Output:* A pure Nash equilibrium $\mathbf{c} = (c_{v_1}, \ldots, c_{v_n}) \in X^n$ of $\Gamma(G)$
>   Initialization: **for** $i = 1$ **to** n **do** $c_{v_i} = x_i$
> **repeat**
>   find an unsatisfied vertex $v \in V$ and a color $x \in X$ such that $\lambda_v(x, \mathbf{c}_{-v}) > \lambda_v(\mathbf{c})$
>   set $c_v = x$
> **until** all vertices are satisfied

I.e., at each step, $\mathcal{A}$ allows one unsatisfied vertex to perform a selfish step, until all vertices are satisfied. Note that, at each step, there may be more than one unsatisfied vertices, and more than one colors that a vertex could choose in order to increase its payoff. So actually $\mathcal{A}$ is a whole class of algorithms, since one could define a specific ordering (e.g., some fixed or some random order) of vertices and colors, and examine vertices and colors according to this order. In any case however, the algorithm is guaranteed to terminate in $O(n \cdot \alpha(G))$ selfish steps. Furthermore, each selfish step can be implemented straightforwardly in $O(n^2)$ time, since there are n vertices and n colors that each vertex can be assigned. It might be possible to improve the $O(n^2)$ complexity of a selfish step, e.g. by using appropriate data structures; this is a matter of future research and we leave it as an open question.

Let us now give a direct application of Theorem 2 to dense random graphs, and in particular consider the $G_{n,p}$ model, i.e. the class of random graphs with n vertices where each of the possible $\frac{n(n-1)}{2}$ edges occurs with probability p (for some constant $0 < p < 1$). The independence number of these graphs is known to be $(1 - o(1))\frac{\log_2 n}{\log_2(1/(1-p))}$ with high probability [2], and therefore a pure Nash equilibrium can be computed in $O(n \cdot \log_2(n))$ selfish steps, with high probability.

4   Bounds on the Total Number of Colors

Lemma 2. *In any pure Nash equilibrium of $\Gamma(G)$, the number k of total colors used satisfies $k \leq \Delta_2(G) + 1$ and hence $k \leq \Delta(G) + 1$.*

Proof. Consider a pure Nash equilibrium $\mathbf{c}$ of $\Gamma(G)$, and let k be the total number of distinct colors used in $\mathbf{c}$. If $k = 1$ then it easy to observe that G must be totally disconnected, i.e. $\Delta(G) = \Delta_2(G) = 0$ and therefore $k = \Delta_2(G) + 1$. Now assume $k \geq 2$. Let $x_i, x_j \in X$ be the two colors used in $\mathbf{c}$ that are assigned to the minimum number of vertices. W.l.o.g.[1], assume that $n_{x_i}(\mathbf{c}) \leq n_{x_j}(\mathbf{c}) \leq n_x(\mathbf{c})$ for all colors $x \notin \{x_i, x_j\}$ used in $\mathbf{c}$. Let v be a vertex such that $c_v = x_i$. The payoff of vertex v is $\lambda_v(\mathbf{c}) = n_{x_i}(\mathbf{c})$. Now consider any other color $x \neq x_i$ that is used in $\mathbf{c}$. Assume that there is no edge between vertex v and any vertex u with $c_u = x$. Then, since $\mathbf{c}$ is a pure Nash equilibrium, it must hold that $n_{x_i}(\mathbf{c}) \geq n_x(\mathbf{c}) + 1$, a contradiction. Therefore there is an edge between vertex v and at least one vertex of every other color. Hence the degree of vertex v is at least the total number of colors used minus 1, i.e. $\deg(v) \geq k - 1$. Furthermore, let u be the vertex of color $c_u = x_j$ that v is connected to. Similar arguments as above yield that u must be connected to at least one vertex of color x, for all $x \notin \{x_i, x_j\}$ used in $\mathbf{c}$. Moreover, u is also connected to v. Therefore $\deg(u) \geq k - 1$. Now:

$$\Delta_2(G) = \max_{s \in V} \max_{\substack{t \in N(s) \\ \deg(t) \leq \deg(s)}} \deg(t)$$

$$\geq \max \left\{ \max_{\substack{t \in N(v) \\ \deg(t) \leq \deg(v)}} \deg(t), \max_{\substack{t \in N(u) \\ \deg(t) \leq \deg(u)}} \deg(t) \right\}$$

$$\geq \min \{\deg(u), \deg(v)\} \geq k - 1$$

and therefore $k \leq \Delta_2(G) + 1$ as needed. $\square$

Lemma 3. *In a pure Nash equilibrium, all vertices that are assigned unique colors form a clique.*

Proof. Consider a pure Nash equilibrium $\mathbf{c}$. Assume that the colors c_v and c_u chosen by vertices v and u are unique, i.e. $n_{c_v}(\mathbf{c}) = n_{c_u}(\mathbf{c}) = 1$. Then the payoff for both vertices is 1. If there is no edge between u and v then, since $\mathbf{c}$ is an equilibrium, it must hold that $1 = \lambda_v(\mathbf{c}) \geq \lambda_v(c_u, \mathbf{c}_{-v}) = 2$, a contradiction. $\square$

---

[1] Without loss of generality.

Lemma 4. *In any pure Nash equilibrium of $\Gamma(G)$, the number k of total colors used satisfies $k \leq \frac{n+\omega(G)}{2}$.*

Proof. Consider a pure Nash equilibrium $\mathbf{c}$ of $\Gamma(G)$. Assume there are $t \geq 0$ vertices that are each assigned a unique color. These t vertices form a clique (Lemma 3), hence $t \leq \omega(G)$. The remaining $n - t$ vertices are assigned non-unique colors, so the number of colors in $\mathbf{c}$ is $k \leq t + \frac{n-t}{2} = \frac{n+t}{2} \leq \frac{n+\omega(G)}{2}$. $\square$

Lemma 5. *In any pure Nash equilibrium of $\Gamma(G)$, the number k of total colors used satisfies $k \leq \frac{1+\sqrt{1+8m}}{2}$.*

Proof. Consider a pure Nash equilibrium $\mathbf{c}$ of $\Gamma(G)$. W.l.o.g., assume that the k colors used in $\mathbf{c}$ are $x_1, \ldots, x_k$. Let V_i, $1 \leq i \leq k$, denote the subset of all vertices $v \in V$ such that $c_v = x_i$. W.l.o.g., assume that $|V_1| \leq |V_2| \leq \cdots \leq |V_k|$. Observe that, for each vertex $v_i \in V_i$, there is an edge between v_i and some $v_j \in V_j$, for all $j > i$. If not, then v_i could improve its payoff by choosing color x_j, since $|V_j| + 1 \geq |V_i| + 1 > |V_i|$. This implies that $m \geq \sum_{i=1}^{k-1} |V_i|(k - i)$ and, since $|V_i| \geq 1$ for all $i \in \{1, \ldots, k\}$, $m \geq \sum_{i=1}^{k-1}(k - i)$ or equivalently $m \geq \frac{k(k-1)}{2}$ or equivalently $k^2 - k - 2m \leq 0$, which implies $k \leq \frac{1+\sqrt{1+8m}}{2}$. $\square$

Theorem 3. *In any pure Nash equilibrium of $\Gamma(G)$, the number k of total colors used satisfies $k \leq n - \alpha(G) + 1$.*

Proof. Consider any pure Nash equilibrium $\mathbf{c}$ of $\Gamma(G)$. Let t be the maximum, over all vertices, payoff in $\mathbf{c}$, i.e. $t = \max_{x \in X} n_x(\mathbf{c})$. Partition the set of vertices into t sets $V_1, \ldots, V_t$ so that $v \in V_i$ if and only if $\lambda_v(\mathbf{c}) = i$ (note that each vertex appears in exactly one such set, however not all sets have to be nonempty). Let k_i denote the total number of colors that appear in V_i. Clearly, $|V_i| = i \cdot k_i$ and the total number of colors used in $\mathbf{c}$ is $k = \sum_{i=1}^{t} k_i$. Now consider a maximum independent set I of G. The vertices in V_1 have payoff equal to 1, therefore they are assigned unique colors, so, by Lemma 3, the vertices in V_1 form a clique. Therefore I can only contain at most one vertex among the vertices in V_1. Our goal is to upper bound the size of I. First we prove the following:

Claim 1. If there exists some $i > 1$ such that $k_i = 1$ and I contains all the vertices in V_i, then $k \leq n - \alpha(G) + 1$.

Proof of Claim 1. Let x denote the unique color that appears in V_i. Since I contains all the vertices in V_i, then it cannot contain any vertex in $V_1 \cup \cdots \cup V_{i-1}$. This is so because each vertex $v \in V_j$, $j < i$, is connected by an edge with at least one vertex of color x (otherwise v could increase its payoff by selecting x, which contradicts the equilibrium). Furthermore, each vertex in V_i has at least one neighbor of each color that appears in $V_{i+1} \cup \cdots \cup V_t$. Therefore

$$|I| = \alpha(G) \leq |V_i| + \sum_{j=i+1}^{t} |V_j| - \sum_{j=i+1}^{t} k_j = n - \sum_{j=1}^{i-1} |V_j| - k + \sum_{j=1}^{i} k_j$$

which gives $k \leq n - \alpha(G) + \sum_{j=1}^{i-1}(k_j - |V_j|) + k_i \leq n - \alpha(G) + k_i = n - \alpha(G) + 1.$ $\qquad\square$

So now it suffices to consider the case where, for all $i > 1$ such that $k_i = 1$, I does not contain all the vertices in V_i. So I contains at most $|V_i| - 1 = |V_i| - k_i$ vertices that belong to V_i. In order to complete the proof we need the following:

Claim 2. For all $i > 1$ with $k_i \neq 1$, I cannot contain more than $|V_i| - k_i$ vertices among the vertices in V_i.

Proof of Claim 2. This is clearly true for $k_i = 0$ (and hence $|V_i| = 0$). Now assume that $k_i \geq 2$. Observe that, for all vertices $v_i \in V_i$ there must exist an edge between v_i and a vertex of each one of the remaining $k_i - 1$ colors that appear in V_i (otherwise, v_i could change its color and increase its payoff by 1, which contradicts the equilibrium). Fix a color x of the k_i colors that appear in V_i. If I contains all vertices of color x, then it cannot contain any vertex of any color other than x that appears in V_i. Therefore I can contain at most $i \leq (i-1)k_i = |V_i| - k_i$ vertices among the vertices in V_i. On the other hand, if I contains at most $i - 1$ vertices of each color x that appears in V_i, then I contains again at most $(i-1)k_i = |V_i| - k_i$ vertices among the vertices in V_i. $\qquad\square$

Therefore I cannot contain more than $|V_i| - k_i$ vertices among the vertices of V_i, for all $i > 1$, plus one vertex from V_1. Therefore:

$$|I| = \alpha(G) \leq 1 + \sum_{i=2}^{t}(|V_i| - k_i) = 1 + n - |V_1| - (k - |V_1|) = n - k + 1.$$

So, in any case, $k \leq n - \alpha(G) + 1$ as needed. $\qquad\square$

The bounds given by Lemmata 2, 4, 5 and Theorem 3 imply the following:

Theorem 4. *For any graph coloring game $\Gamma(G)$ and any pure Nash equilibrium* $\mathbf{c}$ *of* $\Gamma(G)$, $\mathsf{SC}(G, \mathbf{c}) \leq \min\left\{\Delta_2(G) + 1,\ \frac{n + \omega(G)}{2},\ \frac{1 + \sqrt{1+8m}}{2},\ n - \alpha(G) + 1\right\}$.

Furthermore, since any Nash equilibrium is a proper coloring (Lemma 1) and a Nash equilibrium can be computed in polynomial time (Theorem 2):

Corollary 2. *For any graph G, a proper coloring that uses at most $k \leq \min\left\{\Delta_2(G) + 1,\ \frac{n + \omega(G)}{2},\ \frac{1 + \sqrt{1+8m}}{2},\ n - \alpha(G) + 1\right\}$ colors can be computed in $O(n^4)$ time.*

5   The Approximation Ratio

Lemma 6. *For any graph G with n vertices and m edges,*

$$\mathsf{R}(G) \leq \frac{\min\left\{\Delta_2(G) + 1,\ \frac{n + \omega(G)}{2},\ \frac{1 + \sqrt{1+8m}}{2},\ n - \alpha(G) + 1\right\}}{\max\left\{\omega(G),\ \frac{n}{\alpha(G)}\right\}}.$$

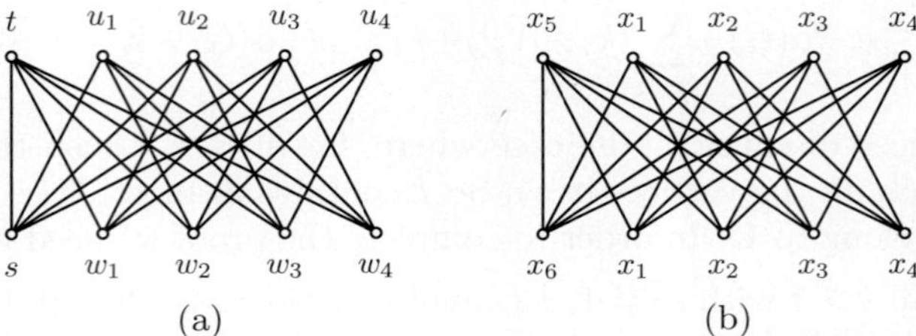

Fig. 1. (a) A graph with chromatic number 2 and (b) a Nash equilibrium using 6 colors

Proof. Follows from Theorem 4 and the fact that $\chi(G) \geq \max\{\omega(G), \frac{n}{\alpha(G)}\}$. $\square$

Lemma 7. *For any constant $\epsilon > 0$, there exists a graph $G(\epsilon)$ such that* $\mathsf{R}(G(\epsilon)) \geq n^{1-\epsilon}$ *unless* $\mathsf{NP} \subseteq \mathsf{co\text{-}RP}$.

Proof. Assume the contrary. Then there exists some constant $\epsilon > 0$ such that, for all graphs G, $\mathsf{R}(G) < n^{1-\epsilon}$. But then our selfish improvements algorithm $\mathcal{A}$ of Theorem 2 achieves, in $O(n^4)$ time, a proper coloring of G with a number of colors $k \leq \mathsf{R}(G) \cdot \chi(G)$, i.e., $k \leq n^{1-\epsilon}\chi(G)$. Thus, for all G, algorithm $\mathcal{A}$ approximates $\chi(G)$ in polynomial time with an approximation ratio $R \leq n^{1-\epsilon}$ for some constant $\epsilon > 0$. This cannot happen unless $\mathsf{NP} \subseteq \mathsf{co\text{-}RP}$ [7]. $\square$

However, can we construct a graph certificate G with unconditionally high $\mathsf{R}(G)$? The answer is yes:

Lemma 8. *We can construct a graph certificate G such that* $\mathsf{R}(G) = \frac{n}{4} + \frac{1}{2}$.

Proof. Consider a bipartite graph $G = (V, E)$ with $n = 2\kappa + 2$ vertices, $\kappa \geq 1$. Let $V = U \cup W \cup \{s, t\}$ where $U = \{u_1, \ldots, u_\kappa\}$ and $W = \{w_1, \ldots, w_\kappa\}$. The set of edges E is defined as

$$E = \{\{u_i, w_j\} \in U \times W \mid i \neq j\} \cup \bigcup_{i=1}^{\kappa}\{s, u_i\} \cup \bigcup_{i=1}^{\kappa}\{t, w_i\} \cup \{s, t\}.$$

(Figure 1(a) shows such a graph with $n = 10$ vertices.) There exists a pure Nash equilibrium that uses $\kappa + 2$ colors: vertices u_1, w_1 are colored x_1, vertices u_2, w_2 are colored x_2 e.t.c., while vertex t is colored $x_{\kappa+1}$ and vertex s is colored $x_{\kappa+2}$ (see Fig. 1(b)). This coloring is a pure Nash equilibrium since each vertex $v \in U \cup W$ receives payoff equal to 2 and the set of vertices $N(v) \cup \{v\}$ uses all colors $x_1, \ldots, x_\kappa$. Vertices s and t get payoff 1, but each of them is connected to a vertex of each of the remaining colors. The optimum coloring would use 2 colors, one to color the vertices in $U \cup \{t\}$ and another to color the vertices in $W \cup \{s\}$. Therefore $\mathsf{R}(G) \geq \frac{\kappa+2}{2} = \frac{n}{4} + \frac{1}{2}$. But $\omega(G) = 2$, so from Lemma 6 we can easily get $\mathsf{R}(G) \leq \frac{n}{4} + \frac{1}{2}$, which completes the proof. $\square$

6   On Mechanisms to Improve the Approximation Ratio

6.1   Refinements of the Selfish Steps Sequence: Randomness

The existence of the potential function $\Phi(\mathbf{c})$ assures that if we start with a proper coloring and allow at each step any single unsatisfied vertex to perform a selfish step, then a pure Nash equilibrium will be reached in polynomial time, no matter in which order the vertices are examined or which is the initial configuration. In this section we study whether there exists a sequence of selfish steps, i.e. a specific ordering of the vertices according to which the vertices are allowed to perform a selfish step, such that the Social Cost of the equilibrium reached is even less than the general bounds presented before.

Assume that, at each step, the vertex that is allowed to perform a selfish step is chosen independently and uniformly at random, among all vertices that are unsatisfied. Moreover, assume that the vertex chosen to perform a selfish step chooses a color independently and uniformly at random among the colors that can increase its payoff. Then, we can prove the following (the proof is omitted):

Proposition 1. *The random selfish steps sequence applied to the graph of Lemma 8 terminates in polynomial time at a pure Nash equilibrium that, with high probability, corresponds to an optimum coloring.*

Although Proposition 1 is rather restrictive, since it only applies to the graph of Lemma 8, we believe that the random selfish steps sequence can color other classes of graphs with a number of colors much smaller than the bounds presented previously. We expect that randomization can help in avoiding equilibria that are too far from an optimum coloring. However, we have not yet been able to prove this; this is a matter of future research and we leave it as an open problem.

6.2   Stackelberg Strategies

Consider a graph coloring game $\Gamma(G)$. Assume that there is a central authority (a *Leader*) that controls a portion $V^L \subset V$ of the vertices of $G = (V, E)$, i.e. the Leader colors the vertices in V^L and, after that, the rest of the vertices in $V \setminus V^L$ (the *followers*) are colored selfishly. The goal of the Leader is to find an assignment of colors to V^L (a *Leader's strategy*) so as to induce the followers to a pure Nash equilibrium where the total number of colors used in V is as close to the chromatic number of G as possible.

Definition 2. *For a constant $k \in \mathbb{N}$, a random balanced k-partite graph, denoted $G_{n,k,p}$, is a k-partite graph with n vertices, where the size of each vertex class is either $\lceil \frac{n}{k} \rceil$ or $\lfloor \frac{n}{k} \rfloor$, and each edge $\{u, v\}$ (such that u and v belong to different vertex classes) exists in G independently at random with probability p.*

Lemma 9. *The chromatic number of $G_{n,k,\frac{1}{2}}$ is k, with high probability.*

Proof (Sketch). Clearly, $\omega(G_{n,k,\frac{1}{2}}) \leq \chi(G_{n,k,\frac{1}{2}}) \leq k$. The proof follows by showing that, with high probability, there exists a clique of size k in $G_{n,k,\frac{1}{2}}$.   □

Theorem 5. *Consider the graph coloring game $\Gamma(G_{n,k,\frac{1}{2}})$. There exists a polynomial time computable Leader's strategy, such that with high probability the total number of colors used in the resulting pure Nash equilibrium is k.*

Proof. Let $P_1, \ldots, P_k$ denote the k vertex classes of $G_{n,k,\frac{1}{2}}$. Assume that the Leader chooses uniformly at random a subset $S \subset V$ of $|S| = c \log n$ vertices, for some constant $c > 10k$. The Leader can exhaustively search among all possible k-colorings of S in time polynomial in n, since $|S| = c \log n$. Among these possible colorings, there exists one proper coloring $\mathbf{c}^L$ that colors each vertex $s \in S \cap P_1$ with the same color x_1, each vertex $s \in S \cap P_2$ with the same color $x_2 \neq x_1$ e.t.c. In the following, assume that the Leader's strategy is $\mathbf{c}^L$.

Our next step is to show that, with high probability, each follower $v_i \in P_i \setminus S$ is connected to at least one vertex in S of color x_j, for all $j \neq i$. To do so, we use Hoeffding bounds [11] and obtain

$$\Pr\left\{\exists i, \ \exists v_i \in P_i, \ \exists j \ : \ \{v_i, v_j\} \notin E \quad \forall v_j \in S \cap P_j\right\} \leq \frac{2k}{n}.$$

So with probability at least $1 - \frac{2k}{n}$, each follower $v_i \in P_i \setminus S$ (for all $i \in \{1, \ldots, k\}$) has all the colors x_j $(j \neq i)$ in its neighborhood. But if this is the case, then the pure Nash equilibrium that will be reached by any selfish steps sequence will use the same color x_i for all $v_i \in P_i \setminus S$, for each $i = \{1, \ldots, k\}$. Therefore, with probability at least $1 - \frac{2k}{n}$, there will be k colors in the resulting pure Nash equilibrium. However, we assumed that the Leader's strategy is $\mathbf{c}^L$. This is not restrictive, since the Leader can repeatedly choose one of the possible k-colorings of S (their number is $k^{c \log n}$, i.e. polynomial in n) and then leave the followers converge to a pure Nash equilibrium. The precedent analysis shows that there exists a proper coloring $\mathbf{c}^L$ of S such that there will be k colors in the equilibrium reached by the followers, with high probability. $\qquad\square$

References

1. Bodlaender, H.L.: On the complexity of some coloring games. International Journal of Foundations of Computer Science 2(2), 133–147 (1991)
2. Bollbás, B.: The chromatic number of random graphs. Combinatorica 8, 49–55 (1988)
3. Brelaz, D.: New methods to color the vertices of a graph. Comm. ACM 22, 251–256 (1979)
4. Brooks, R.L.: On colouring the nodes of a network. Proc. Cambridge Phil. Soc. 37, 194–197 (1941)
5. Diestel, R.: Graph Theory. Springer, Heidelberg (2005)
6. Dinski, T., Zhu, X.: A bound for the game chromatic number of graphs. Discrete Mathematics 196, 109–115 (1999)
7. Feige, U., Kilian, J.: Zero knowledge and the chromatic number. Journal of Computer and System Sciences 57(2), 187–199 (1998)
8. Halldórsson, M.: A still better performance guarantee for approximate graph coloring. Information Processing Letters 45, 19–23 (1993)
9. Harary, F.: Graph theory. Addison-Wesley, Reading (1969)
10. Harary, F., Tuza, Z.: Two graph-colouring games. Bulletin of the Australian Mathematical Society 48, 141–149 (1993)
11. Hoeffding, W.: Probability inequalities for sums of bounded random variables. Journal of the American Statistical Association 58, 13–30 (1963)
12. Jensen, T.R., Toft, B.: Graph Coloring Problems. Wiley Interscience, Hoboken (1995)

13. Karp, R.M.: Reducibility among combinatorial problems. In: Complexity of Computer Computations, pp. 85–103. Plenum Press (1972)
14. Monderer, D., Shapley, L.S.: Potential games. Games and Economic Behavior 14, 124–143 (1996)
15. Nash, J.F.: Non-cooperative games. Annals of Mathematics 54(2), 286–295 (1951)
16. Stacho, L.: New upper bounds for the chromatic number of a graph. Journal of Graph Theory 36(2), 117–120 (2001)

Partitioning a Weighted Tree
to Subtrees of Almost Uniform Size

Takehiro Ito[1], Takeaki Uno[2], Xiao Zhou[1], and Takao Nishizeki[1]

[1] Graduate School of Information Sciences, Tohoku University,
Aoba-yama 6-6-05, Sendai, 980-8579, Japan
[2] National Institute of Informatics,
2-1-2 Hitotsubashi, Chiyoda-ku, Tokyo, 101-8430, Japan
takehiro@ecei.tohoku.ac.jp, uno@nii.ac.jp,
{zhou,nishi}@ecei.tohoku.ac.jp

Abstract. Assume that each vertex of a graph G is assigned a nonnegative integer weight and that l and u are integers such that $0 \le l \le u$. One wishes to partition G into connected components by deleting edges from G so that the total weight of each component is at least l and at most u. Such an "almost uniform" partition is called an (l, u)-partition. We deal with three problems to find an (l, u)-partition of a given graph: the minimum partition problem is to find an (l, u)-partition with the minimum number of components; the maximum partition problem is defined analogously; and the p-partition problem is to find an (l, u)-partition with a given number p of components. All these problems are NP-hard even for series-parallel graphs, but are solvable for paths in linear time and for trees in polynomial time. In this paper, we give polynomial-time algorithms to solve the three problems for trees, which are much simpler and faster than the known algorithms.

1 Introduction

Let G be an undirected graph, and let each vertex v of G be assigned a nonnegative integer $\omega(v)$, called the *weight* of v. Let l and u be given nonnegative integers, called the *lower bound* and *upper bound* on component sizes, respectively. We wish to partition G into connected components by deleting edges from G so that the total weights of all components are almost uniform, that is, the sum of weights of all vertices in each component is at least l and at most u for appropriately chosen bounds l and u. We call such an almost uniform partition an (l, u)-*partition* of G. In this paper, we deal with the following three partition problems to find an (l, u)-partition of a given graph G: the *minimum partition problem* is to find an (l, u)-partition of G with the minimum number of components; the *maximum partition problem* is defined analogously; and the p-*partition problem* is to find an (l, u)-partition of G with a given number p of components.

Figures 1(a) and (b) illustrate two $(5, 15)$-partitions of the same tree, where each vertex is drawn as a circle, the weight of each vertex is written inside the

S.-H. Hong, H. Nagamochi, and T. Fukunaga (Eds.): ISAAC 2008, LNCS 5369, pp. 196–207, 2008.

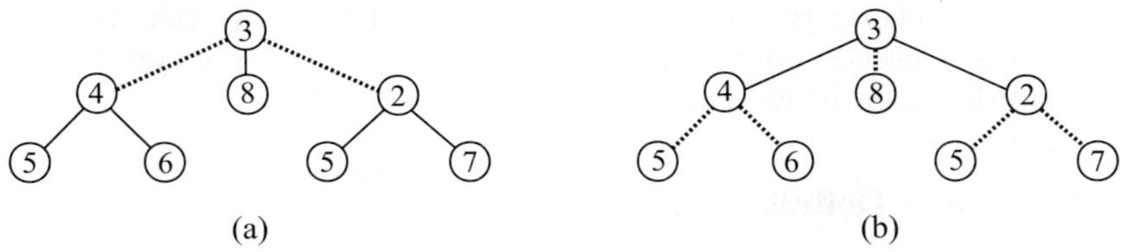

Fig. 1. (a) Solution for the minimum partition problem, and (b) solution for the maximum partition problem, where $l = 5$ and $u = 15$

circle, and the deleted edges are drawn as dotted lines. The $(5, 15)$-partition with three components in Fig. 1(a) is a solution for the 3-partition problem and for the minimum partition problem, while the $(5, 15)$-partition with six components in Fig. 1(b) is a solution for the 6-partition problem and for the maximum partition problem. The three partition problems often appear in many practical situations such as load balancing [1], image processing [4,6], paging systems of operation systems [9], and political districting [2,10].

An NP-complete problem, called the set partition problem [3], can be easily reduced in linear time to any of the three partition problems for a complete bipartite graph $K_{2,n-2}$, where n is the number of vertices in a graph. Since $K_{2,n-2}$ is a series-parallel graph, the three partition problems are NP-hard even for series-parallel graphs [5]. Therefore, it is very unlikely that the three partition problems can be solved in polynomial time even for series-parallel graphs, although the three problems can be solved in pseudo-polynomial time for graphs of bounded tree-width, including series-parallel graphs [5]. On the other hand, all the three partition problems can be solved for paths in linear time [6] and for special classes of trees, i.e., stars, worms and caterpillars, in polynomial time [8], while the p-partition problem can be solved for arbitrary trees of n vertices in time $O(p^3 n^4)$ [7].

In this paper, we show that all the three partition problems can be solved for arbitrary trees more efficiently. To be precise, we show that the p-partition problem can be solved in time $O(p^4 n)$. Since $p \leq n$, our algorithm is faster and much simpler than the known algorithm in [7]. Furthermore, our algorithm runs in linear time if p is a fixed constant. One can solve the minimum partition problem and the maximum partition problem by solving the p-partition problem for every p, $1 \leq p \leq n$. Therefore, both the minimum partition problem and the maximum partition problem can be solved in time $O(n^6)$.

2   Simple Algorithm

In this section we give a simple algorithm to solve the p-partition problem for trees of n vertices in time $O(p^4 u^2 n)$. It examines whether a given tree T has an (l, u)-partition with p subtrees or not. However, one can easily modify the algorithm so that it actually finds an (l, u)-partition with p subtrees whenever T has it. One may assume that $p \leq n$, but the upper bound u is not necessarily bounded by n. Therefore, the algorithm does not necessarily take polynomial time, but takes pseudo-polynomial time. Using this algorithm, we will give a polynomial-time algorithm in Section 3.

2.1   Terms and Definitions

We first present our idea. One may assume without loss of generality that a given tree T is a rooted tree with root r. In Fig.2(a) an (l, u)-partition of T is indicated by dotted lines. For each vertex v of T, we denote by T_v the subtree of T which is rooted at v and is induced by all descendants of v in T. Thus $T = T_r$. Every (l, u)-partition of a tree T naturally induces a partition of its subtree T_v, as illustrated in Fig.2(b). The induced partition is not necessarily an (l, u)-partition of T_v, because the component P_v containing v may not satisfy the lower bound l. However, the induced partition is an "extendable" partition of T_v; in an extendable partition of T_v, every component, except for P_v, must satisfy both the lower bound l and the upper bound u, but P_v may not satisfy either the lower bound l or the upper bound u. Thus, for a subtree T_v of T and an integer k, $0 \le k \le p - 1$, we define $S(T_v, k)$ as the set of all integers z such that z is the total weight of the component P_v in an extendable partition of T_v with $k + 1$ components. Our idea is to compute $S(T_v, k)$ from the leaves to the root r of T by means of dynamic programming. Clearly, T has an (l, u)-partition with p components if and only if $S(T_r, p - 1)$ contains an integer z such that $l \le z \le u$.

We now formally define the notion of an extendable partition of a subtree T_v of T. Let $\mathcal{P}$ be a partition of the vertex set $V(T_v)$ of T_v into nonempty subsets. $\mathcal{P}$ is called a *partition of* T_v if each subset in $\mathcal{P}$ induces a connected component (subtree) of T_v. For a set $P \subseteq V(T_v)$, we denote by $\omega(P)$ the total weight of vertices in P, that is, $\omega(P) = \sum_{x \in P} \omega(x)$. For a partition $\mathcal{P}$ of T_v, we always denote by P_v the set in $\mathcal{P}$ containing the root v of T_v. A partition $\mathcal{P}$ of T_v is *extendable* if each set $P \in \mathcal{P} \setminus \{P_v\}$ satisfies $l \le \omega(P) \le u$. Note that P_v may not satisfy $l \le \omega(P_v) \le u$.

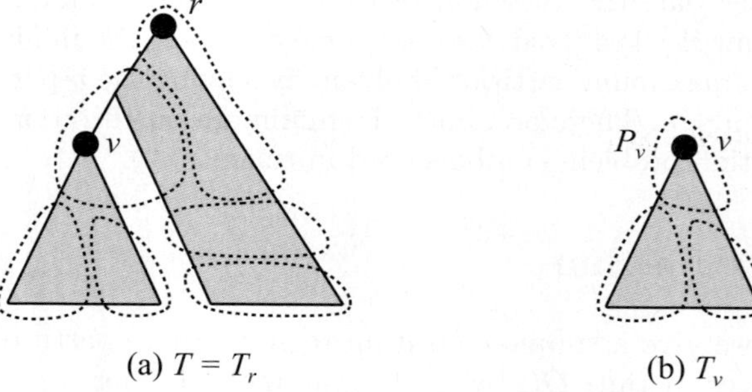

(a) $T = T_r$ (b) T_v

Fig. 2. (a) An (l, u)-partition of a tree T and (b) an extendable partition of a subtree T_v of T

We then formally define a set $S(T_v, k)$ of integers z for a subtree T_v of T and an integer k, $0 \leq k \leq p - 1$, as follows:

$$S(T_v, k) = \{z \mid T_v \text{ has an extendable partition } \mathcal{P}$$
$$\text{such that } z = \omega(P_v) \text{ and } |\mathcal{P}| = k + 1\}.$$

For a set Z of integers, let $\min(Z) = \min\{z \mid z \in Z\}$ and $\max(Z) = \max\{z \mid z \in Z\}$. We now have the following lemma.

Lemma 1. $\max(S(T_v, k)) - \min(S(T_v, k)) \leq k(u - l)$.

Proof. For each integer $z \in S(T_v, k)$, T_v has an extendable partition $\mathcal{P}$ such that $z = \omega(P_v)$ and $|\mathcal{P}| = k + 1$. Since each component $P \in \mathcal{P} \setminus \{P_v\}$ satisfies the lower bound l, we have

$$z = \omega(P_v) = \omega(V(T_v)) - \sum_{P \in \mathcal{P} \setminus \{P_v\}} \omega(P) \leq \omega(V(T_v)) - kl.$$

Thus we have

$$\max(S(T_v, k)) \leq \omega(V(T_v)) - kl.$$

Similarly, since each component $P \in \mathcal{P} \setminus \{P_v\}$ satisfies the upper bound u, we have

$$\min(S(T_v, k)) \geq \omega(V(T_v)) - ku.$$

Therefore, we have $\max(S(T_v, k)) - \min(S(T_v, k)) \leq k(u - l)$. $\qquad\square$

Our algorithm computes $S(T_v, k)$ for each vertex v of T from the leaves to the root r of T by means of dynamic programming. Since $T = T_r$, the following lemma clearly holds.

Lemma 2. *A tree T has an (l, u)-partition with p subtrees if and only if $S(T, p-1)$ contains an integer z such that $l \leq z \leq u$.*

Let v be a vertex of T, let $v_1, v_2, \cdots, v_s$ be the children of v ordered arbitrarily, and let e_i, $1 \leq i \leq s$, be the edge joining v and v_i, as illustrated in Fig.3. We denote by T_v^i the subtree of T which consists of the vertex v, the edges $e_1, e_2, \cdots, e_i$ and the subtrees $T_{v_1}, T_{v_2}, \cdots, T_{v_i}$. In Fig.3, the subtree T_v^i is indicated by a dotted line. Clearly $T_v = T_v^s$. For the sake of notational convenience, we denote by T_v^0 the subtree of a single vertex v. Therefore, $T_v = T_v^0$ if v is a leaf of T.

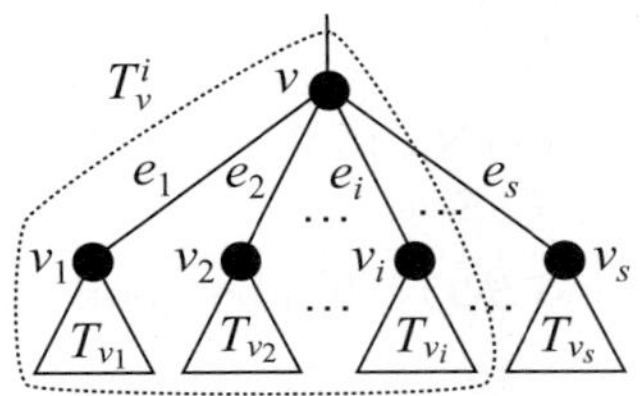

Fig. 3. Tree T_v

2.2   Algorithm

We now describe the simple algorithm.

We first compute set $S(T_v^0, k)$ for each vertex v of T and each integer k, $0 \leq k \leq p - 1$. For $k = 0$,

$$S(T_v^0, 0) = \{\omega(v)\}, \tag{1}$$

and for each integer k, $1 \leq k \leq p - 1$,

$$S(T_v^0, k) = \emptyset. \tag{2}$$

We next compute sets $S(T_v^i, k)$, $1 \leq i \leq s$, for each internal vertex v of T from the counterparts of T_v^{i-1} and T_{v_i} from the bottom to the top of T, where s is the number of the children of v. (See Fig.3.) By Lemma 2 one can know immediately from $S(T_r, p - 1)$ whether T has an (l, u)-partition with p subtrees or not.

T_v^i is obtained from T_v^{i-1} and T_{v_i} by joining v and v_i as illustrated in Fig.4. Every extendable partition $\mathcal{P}$ of T_v^i can be obtained by merging an extendable partition $\mathcal{P}'$ of T_v^{i-1} with an extendable partition $\mathcal{P}''$ of T_{v_i}. In Fig.4, extendable partitions are indicated by dotted lines. There are the following two Cases (a) and (b), and we define two sets $S^a(T_v^i, k)$ and $S^b(T_v^i, k)$ for the two cases, respectively.

Case (a): $v_i \in P_v$.

In this case, an extendable partition $\mathcal{P}$ of T_v^i can be obtained by merging an extendable partition $\mathcal{P}'$ of T_v^{i-1} with an extendable partition $\mathcal{P}''$ of T_{v_i}, as illustrated in Fig.4(a). Then, the component $P_v \in \mathcal{P}$ containing v consists of the vertices in $P_v' \cup P_{v_i}''$, where P_v' is the component in $\mathcal{P}'$ such that $v \in P_v'$ and P_{v_i}'' is the component in $\mathcal{P}''$ such that $v_i \in P_{v_i}''$. We thus define a set $S^a(T_v^i, k)$ as follows:

$$S^a(T_v^i, k) = \bigcup_{k'=0}^{k} \{z' + z'' \mid z' \in S(T_v^{i-1}, k') \text{ and } z'' \in S(T_{v_i}, k - k')\}. \tag{3}$$

Case (b): $v_i \notin P_v$.

In this case, $P_v = P_v'$, and an extendable partition $\mathcal{P}$ of T_v^i can be obtained by merging an extendable partition $\mathcal{P}'$ of T_v^{i-1} with an (l, u)-partition $\mathcal{P}''$ of T_{v_i}, as illustrated in Fig.4(b). We thus define a set $S^b(T_v^i, k)$ as follows:

$$S^b(T_v^i, k) = \bigcup S(T_v^{i-1}, k'), \tag{4}$$

where $S(T_v^{i-1}, k')$ is taken over all k', $0 \leq k' \leq k - 1$, such that $S(T_{v_i}, k - k' - 1)$ contains an integer z'', $l \leq z'' \leq u$.

From two sets $S^a(T_v^i, k)$ and $S^b(T_v^i, k)$ above, one can compute a set $S(T_v^i, k)$ as follows:

$$S(T_v^i, k) = S^a(T_v^i, k) \cup S^b(T_v^i, k). \tag{5}$$

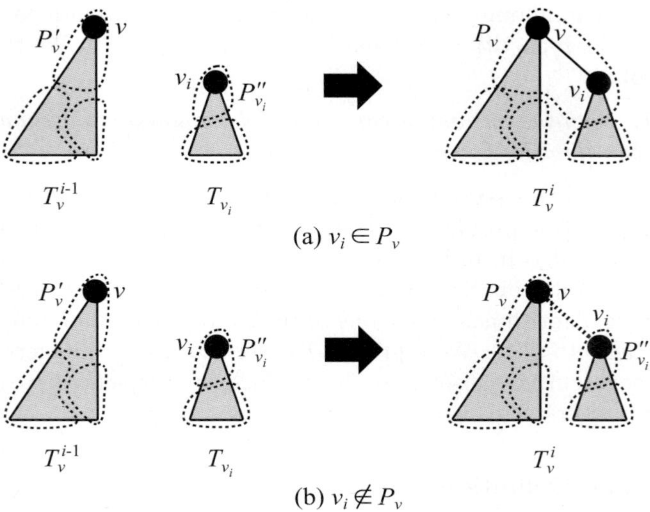

(a) $v_i \in P_v$

(b) $v_i \notin P_v$

Fig. 4. Merging a partition $\mathcal{P}'$ of T_v^{i-1} with a partition $\mathcal{P}''$ of T_{v_i} to a partition $\mathcal{P}$ of T_v^i

One may assume that $\sum_{x \in V(T)} \omega(x) \le pu$; otherwise, T has no (l, u)-partition with p subtrees. Thus $\omega(P_v) \le pu$ for any partition $\mathcal{P}$ of T_v, and hence $\max(S(T_v, k)) \le pu$. Therefore, we have $|S(T_v, k)| \le k(u - l) + 1 \le pu + 1 = O(pu)$. We represent set $S(T_v, k)$ simply by a list, whose length is $O(pu)$. Thus, one can compute a set $\{z' + z'' \mid z' \in S(T_v^{i-1}, k')$ and $z'' \in S(T_{v_i}, k - k')\}$ from two sets $S(T_v^{i-1}, k')$ and $S(T_{v_i}, k - k')$ in time $O(p^2 u^2)$, using an array of length $pu + 1$. Since k' is taken over all integers, $0 \le k' \le k \le p - 1$, in Eq. (3), one can compute $S^a(T_v^i, k)$ in time $O(p^3 u^2)$. Similarly, one can compute $S^b(T_v^i, k)$ in Eq. (4) in time $O(p^2 u)$. Thus one can compute $S(T_v^i, k)$ in Eq. (5) in time $O(p^3 u^2)$. The integer k in Eq. (5) is taken over all k, $0 \le k \le p - 1$, and there are at most $n + (n - 1)$ pairs of a vertex v and an integer i. Thus, one can recursively compute the sets $S(T_v^i, k)$ for all vertices v of T, all integers i and all integers k, $0 \le k \le p - 1$, in time $O(p^4 u^2 n)$. By Lemma 2 one can know from $S(T, p - 1)$ in time $O(pu)$ whether T has an (l, u)-partition with p components. Thus the simple algorithm takes time $O(p^4 u^2 n)$ to solve the p-partition problem for trees.

We remark that the p-partition problem can be solved in linear time for the case $l = u$. In this case, by Lemma 1 $|S(T_v^i, k)| \le 1$, and the algorithm above can be easily implemented so that it runs in linear time.

3   Polynomial-Time Algorithm

The main result of this paper is the following theorem.

Theorem 1. *The p-partition problem can be solved for a tree T in time $O(p^4 n)$, where n is the number of vertices in T and p is any positive integer.*

One can solve the minimum partition problem and the maximum partition problem by solving the p-partition problem for every p, $1 \le p \le n$. We thus have the following corollary.

Corollary 1. *Both the minimum partition problem and the maximum partition problem can be solved for a tree T in time $O(n^6)$.*

In the remainder of this section, as a proof of Theorem 1, we give an algorithm to solve the p-partition problem for trees in time $O(p^4 n)$. One may assume that $l < u$, because the algorithm in Section 2 runs in linear time if $l = u$.

The simple algorithm in Section 2 uses a DP table $S(T_v, k)$ of size $O(pu)$. Our polynomial-time algorithm is analogous to the simple algorithm, but reduces the size of a DP table to $O(p)$. We represent the set $S(T_v, k)$ of integers by at most p "maximal consecutive subsets," each of which is represented by an interval, i.e., a pair of integers. This is our main idea.

3.1   Terms and Definitions

A set Z of integers is *consecutive* (with respect to $u - l$) if, for each integer $z \in Z \setminus \{\max(Z)\}$, Z contains an integer z' such that $0 < z' - z \le u - l$. The pair $[\min(Z), \max(Z)]$ is called the *interval* of Z. Obviously Z is consecutive if $|Z| = 1$.

Lemma 3. *If both X and Y are consecutive with respect to $u - l$, then the set $Z = \{x + y \mid x \in X \text{ and } y \in Y\}$ is also consecutive with respect to $u - l$.*

Proof. If $|Z| = 1$, then Z is consecutive. One may thus assume that $|Z| \ge 2$. We shall prove that, for each integer $z \in Z \setminus \{\max(Z)\}$, Z contains an integer z' such that $0 < z' - z \le u - l$.

Clearly, $\max(Z) = \max(X) + \max(Y)$. For each integer $z \in Z \setminus \{\max(Z)\}$, there exist $x \in X$ and $y \in Y$ such that $x + y = z$. Since $z \ne \max(Z)$, either $x \ne \max(X)$ or $y \ne \max(Y)$. One may assume without loss of generality that $x \ne \max(X)$. Then, since X is consecutive, there exists an integer $x' \in X$ such that $0 < x' - x \le u - l$. Since $x' \in X$ and $y \in Y$, the integer $z' = x' + y$ is contained in Z. We now have

$$z' - z = (x' + y) - (x + y) = x' - x > 0$$

and

$$z' - z = x' - x \le u - l,$$

and hence $0 < z' - z \le u - l$.   □

Note that the interval of Z is $[\min(X) + \min(Y), \max(X) + \max(Y)]$.

For a set Z of integers, a set $Z' \subseteq Z$ is called a *maximal consecutive subset of Z* (with respect to $u - l$) if Z' is consecutive and there is no consecutive subset $Z'' \subseteq Z$ such that $Z' \subset Z''$. Set Z can be partitioned into maximal consecutive subsets of Z. We define a set $I(Z)$ of intervals, as follows:

$$I(Z) = \{[x, y] \mid [x, y] \text{ is the interval of a maximal consecutive subset of } Z\}. \quad (6)$$

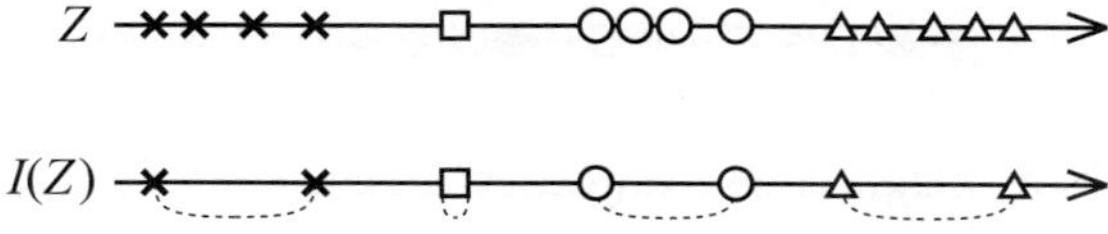

Fig. 5. Set $I(Z)$ of intervals for a set Z of integers

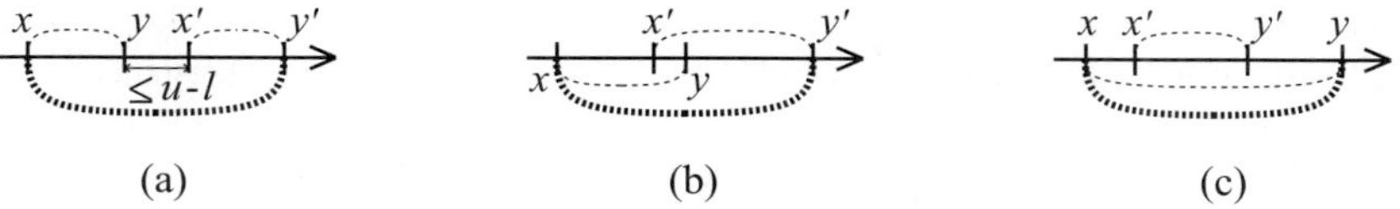

(a)   (b)   (c)

Fig. 6. Merge operation for intervals $[x, y]$ and $[x', y']$

In Fig.5, every integer in Z is represented by a cross, square, circle or triangle, all integers represented by the same symbol form a maximal consecutive subset of Z, and each interval in $I(Z)$ is indicated by a dotted curve joining two integers.

Our algorithm computes the set $I(S(T_v, k))$ of intervals in place of the set $S(T_v, k)$ of integers.

Two intervals $[x, y]$ and $[x', y']$ are *intersecting* (with respect to $u - l$) if $x \leq x'$ and $x' - y \leq u - l$. (See Fig.6.) The *merge operation for two intersecting intervals* $[x, y]$ *and* $[x', y']$ returns an interval $[x, y']$ if $y \leq y'$; otherwise, it returns the interval $[x, y]$. Figures 6(a) and (b) illustrate the merge operation for the case $y \leq y'$, and Fig.6(c) illustrates the merge operation for the case $y > y'$, where the interval obtained by the merge operation is indicated by a thick dotted curve. The *merge operation for a set Z of intervals* is to obtain a set $M(Z)$ of intervals such that no two intervals are intersecting by repeatedly applying the merge operation for two intersecting intervals in Z. (See Fig.7.) We always denote by $M(Z)$ the set obtained by the merge operation for Z.

We are now ready to define the DP table of our polynomial-time algorithm. We denote the set $I(S(T_v, k))$ of intervals simply by $I(T_v, k)$ for a subtree T_v of a tree T and an integer k, $0 \leq k \leq p - 1$. Let z be an integer, then $z \notin S(T_v, k)$ if $I(T_v, k)$ does not contain an interval $[x, y]$, $x \leq z \leq y$. However, $z \in S(T_v, k)$ does not necessarily hold even if $I(T_v, k)$ contains an interval $[x, y]$, $x \leq z \leq y$.

We estimate the size of the DP table, as follows.

Lemma 4. $|I(T_v, k)| \leq k + 1$.

Proof. If $k = 0$, then an extendable partition $\mathcal{P}$ of T_v consists of exactly one set $P_v = V(T_v)$ and hence $I(T_v, 0) = \{[\omega(V(T_v)), \omega(V(T_v))]\}$ and $|I(T_v, 0)| = 1$.

We may thus assume that $k \geq 1$. Let $I(T_v, k) = \{[x_0, y_0], [x_1, y_1], \cdots, [x_m, y_m]\}$. We shall show that $m \leq k$. One may assume without loss of generality that $x_0 \leq y_0 < x_1 \leq y_1 < \cdots < x_m \leq y_m$. Then $\min(S(T_v, k)) = x_0$, $\max(S(T_v, k)) = y_m$, and hence by Lemma 1 we have

$$y_m - x_0 \leq k(u - l). \tag{7}$$

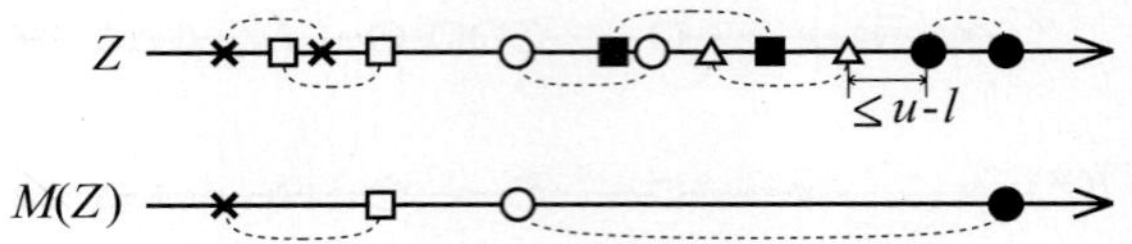

Fig. 7. Merge operation for a set Z of intervals

For each i, $0 \leq i \leq m - 1$, the two intervals $[x_i, y_i] \in I(T_v, k)$ and $[x_{i+1}, y_{i+1}] \in I(T_v, k)$ are not intersecting, and hence $u - l < x_{i+1} - y_i$. We thus have

$$m(u - l) < \sum_{i=0}^{m-1} (x_{i+1} - y_i). \tag{8}$$

Since $x_i \leq y_i$ for each i, $0 \leq i \leq m$, by Eq. (8) we have

$$m(u - l) < \sum_{i=0}^{m-1} (x_{i+1} - y_i) \leq \sum_{i=0}^{m-1} (y_{i+1} - y_i) = y_m - y_0 \leq y_m - x_0. \tag{9}$$

By Eqs. (7) and (9) we have $m < k$, and hence $|I(T_v, k)| = m + 1 \leq k$. $\square$

Our algorithm computes the set $I(T_v, k)$ for each vertex v of T from leaves to the root r of T by means of dynamic programming. We have the following lemma.

Lemma 5. *A tree T has an (l, u)-partition with p subtrees if and only if $I(T, p-1)$ contains an interval $[x, y]$ such that either* (i) $l \leq y \leq u$ *or* (ii) $x \leq u \leq y$.

Proof.
Necessity: Suppose that T has an (l, u)-partition with p subtrees. Then by Lemma 2 $S(T, p - 1)$ contains an integer z such that $l \leq z \leq u$. Therefore, $I(T, p - 1)$ contains an interval $[x, y]$ such that $x \leq z \leq y$. Then, clearly (i) or (ii) holds.

Sufficiency: Suppose that $I(T, p - 1)$ contains an interval $[x, y]$ satisfying either (i) or (ii). By Lemma 2 it suffices to show that $S(T, p - 1)$ contains an integer z such that $l \leq z \leq u$.

If the interval $[x, y] \in I(T, p - 1)$ satisfies the condition (i), that is, $l \leq y \leq u$, then the integer $y \in S(T, p - 1)$ satisfies $l \leq y \leq u$, of course.

One may thus assume that the interval $[x, y] \in I(T, p - 1)$ satisfies the condition (ii), that is, $x \leq u \leq y$. Let $W = \{w_0, w_1, \cdots, w_m\}$ be the maximal consecutive subset of $S(T, p - 1)$ such that $x = \min(W)$ and $y = \max(W)$. One may assume that $w_i < w_{i+1}$ for each i, $0 \leq i \leq m - 1$. Let $w_j \in W$ be the minimum integer such that $u \leq w_j$. If $w_j = u$, then the integer $w_j \in S(T, p - 1)$ clearly satisfies $l \leq w_j \leq u$. One may thus assume that

$$u < w_j. \tag{10}$$

Then, since $w_0 = x \leq u$, we have $w_j \neq w_0$ and hence $1 \leq j$ and $u > w_{j-1} \in S(T, p - 1)$. Since W is consecutive, we have $w_j - w_{j-1} \leq u - l$. Therefore, by Eq. (10) we have $l < w_{j-1}$. Thus we have $l < w_{j-1} < u$. $\square$

3.2 Algorithm

We now describe our polynomial-time algorithm. The algorithm recursively computes the sets $\bar{I}(T_v^i, k)$ of intervals. We will later show that $\bar{I}(T_v^i, k) = I(T_v^i, k)$.

We first compute $\bar{I}(T_v^0, k)$ for each vertex v of T and each integer k, $0 \le k \le p - 1$, similarly as in Eqs. (1) and (2). For $k = 0$, let

$$\bar{I}(T_v^0, 0) = \{[\omega(v), \omega(v)]\}, \tag{11}$$

and for each integer k, $1 \le k \le p - 1$, let

$$\bar{I}(T_v^0, k) = \emptyset. \tag{12}$$

We next compute $\bar{I}(T_v^i, k)$, $1 \le i \le s$, for each internal vertex v of T from the counterparts of T_v^{i-1} and T_{v_i}, where s is the number of the children of v. (See Figs. 3 and 4 together with Eqs. (3)–(5).) We first compute two sets $\bar{I}^a(T_v^i, k)$ and $\bar{I}^b(T_v^i, k)$, as follows:

$$\bar{I}^a(T_v^i, k) = \bigcup_{k'=0}^{k} \left\{ [x' + x'', y' + y''] : [x', y'] \in \bar{I}(T_v^{i-1}, k') \text{ and} \right.$$
$$\left. [x'', y''] \in \bar{I}(T_{v_i}, k - k') \right\}, \tag{13}$$

and

$$\bar{I}^b(T_v^i, k) = \bigcup \bar{I}(T_v^{i-1}, k') \tag{14}$$

where $\bar{I}(T_v^{i-1}, k')$ is taken over all k', $0 \le k' \le k - 1$, such that $\bar{I}(T_{v_i}, k - k' - 1)$ contains an interval $[x'', y'']$ with either $l \le y'' \le u$ or $x'' \le u \le y''$.

From $\bar{I}^a(T_v^i, k)$ and $\bar{I}^b(T_v^i, k)$ above, we compute a set $\bar{I}(T_v^i, k)$, as follows:

$$\bar{I}'(T_v^i, k) = \bar{I}^a(T_v^i, k) \cup \bar{I}^b(T_v^i, k) \tag{15}$$

and

$$\bar{I}(T_v^i, k) = M(\bar{I}'(T_v^i, k)). \tag{16}$$

3.3 Proof of Theorem 1

We first show that $\bar{I}(T_v^i, k) = I(T_v^i, k)$.

We first have the following two Lemmas 6 and 7, whose proofs are omitted due to the page limitation.

Lemma 6. *For each integer* $z \in S(T_v^i, k)$, $\bar{I}(T_v^i, k)$ *contains an interval* $[x, y]$ *such that* $x \le z \le y$.

Lemma 7. *For each interval* $[x, y] \in \bar{I}(T_v^i, k)$, *both* x *and* y *are contained in* $S(T_v^i, k)$ *and the set* $\{z \in S(T_v^i, k) \mid x \le z \le y\}$ *is consecutive with respect to* $u - l$.

Using Lemmas 6 and 7, we have the following Lemma 8, whose proof is omitted due to the page limitation.

Lemma 8. $\bar{I}(T_v^i, k) = I(T_v^i, k)$.

We then show that the algorithm takes time $O(p^4 n)$.

By Eqs. (11) and (12) one can compute the set $I(T_v^0, k) = \bar{I}(T_v^0, k)$ in time $O(p)$ for a vertex v of T and all integers k, $0 \le k \le p-1$. Therefore, $I(T_v^0, k)$ can be computed in time $O(pn)$ for all vertices v in T and all integers k, $0 \le k \le p-1$.

For $i \ge 1$, by Lemma 4 we have $|I(T_v^{i-1}, k)| \le k + 1 \le p$ and $|I(T_{v_i}, k)| \le p$. Therefore, by Eqs. (13)–(15) one can compute a (multi)set $\bar{I}'(T_v^i, k)$ from $\bar{I}(T_v^{i-1}, k')$ and $\bar{I}(T_{v_i}, k - k')$ for an internal vertex v of T, an integer i and an integer k in time $O(p^3)$.

We now explain how to compute $M(\bar{I}'(T_v^i, k))$, that is, how to execute the merge operations in Eq. (16). For each interval $[x, y] \in \bar{I}'(T_v^i, k)$, clearly

$$\omega(V(T_v^i)) - ku \le x \le y \le \omega(V(T_v^i)) - kl.$$

Thus, x and y are integers between $\omega(V(T_v^i)) - ku$ and $\omega(V(T_v^i)) - kl$, and

$$(\omega(V(T_v^i)) - kl) - (\omega(V(T_v^i)) - ku) = k(u - l).$$

We first partition the set $\bar{I}'(T_v^i, k)$ into the following k subsets J_q, $1 \le q \le k$:

$$J_q = \{[x, y] \in \bar{I}'(T_v^i, k) \mid (q - 1)(u - l) \le x - (\omega(V(T_v^i)) - ku) < q(u - l)\}$$

for each q, $1 \le q \le k - 1$, and

$$J_k = \{[x, y] \in \bar{I}'(T_v^i, k) \mid (k - 1)(u - l) \le x - (\omega(V(T_v^i)) - ku) \le k(u - l)\}.$$

The partition above can be found in time $O(p^3)$ since $|\bar{I}'(T_v^i, k)| = O(p^3)$. We then compute sets $M(J_q)$, $1 \le q \le k$. Clearly, any two intervals in $M(J_q)$ are intersecting. Therefore, if $J_q \ne \emptyset$, then $M(J_q) = \{[x_q', y_q']\}$ where $x_q' = \min\{x \mid [x, y] \in J_q\}$ and $y_q' = \max\{y \mid [x, y] \in J_q\}$. If $J_q = \emptyset$, then $M(J_q) = \emptyset$. Since $|\bar{I}'(T_v^i, k)| = O(p^3)$, one can compute all the sets $M(J_q)$, $1 \le q \le k$, in time $O(p^3)$. We then compute a set J of intervals as follows:

$$J = \bigcup_{q=1}^{k} M(J_q).$$

Then, obviously $|J| \le k \le p$ and $M(J) = M(\bar{I}'(T_v^i, k)) = \bar{I}(T_v^i, k)$. Clearly, $x_1' < x_2' < \cdots < x_k'$ although some of $x_1', x_2', \cdots, x_k'$ may be missing. Therefore, one can compute $M(J)$ from J in time $O(p)$ by merging intervals $[x_1', y_1'], [x_2', y_2'], \cdots, [x_k', y_k']$ in this order. Hence, one can compute $\bar{I}(T_v^i, k)$ in Eq. (16) in time $O(p^3)$ for an internal vertex v of T, an integer i and an integer k.

Thus, $I(T_v^i, k) = \bar{I}(T_v^i, k)$ can be computed in time $O(p^4 n)$ for all internal vertices v, all integers i and all integers k, $0 \le k \le p - 1$. Note that there are $O(n)$ pairs of v and i. Hence, one can compute the set $I(T, p - 1)$ in time $O(p^4 n)$. By Lemma 5 one can know from $I(T, p-1)$ in time $O(p)$ whether T has an (l, u)-partition with p components.

We have thus shown that the p-partition problem can be solved for trees in time $O(p^4 n)$. This completes a proof of Theorem 1.

4   Conclusions

In this paper we obtained a polynomial-time algorithm to solve the p-partition problem for trees, which is much simpler and faster than the known algorithm in [7]. Our algorithm takes time $O(p^4 n)$, and hence runs in linear time if $p = O(1)$. On the other hand, both the minimum partition problem and the maximum partition problem can be solved in time $O(n^6)$. We finally remark that our algorithm correctly solves the three partition problems even if the weights of vertices and the bounds l and u on component sizes are real numbers.

References

1. Becker, R., Simeone, B., Chiang, Y.-I.: A shifting algorithm for continuous tree partitioning. Theoretical Computer Science 282, 353–380 (2002)
2. Bozkaya, B., Erkut, E., Laporte, G.: A tabu search heuristic and adaptive memory procedure for political districting. European J. Operational Research 144, 12–26 (2003)
3. Garey, M.R., Johnson, D.S.: Computers and Intractability: A Guide to the Theory of NP-Completeness. Freeman, San Francisco (1979)
4. Gonzalez, R.C., Wintz, P.: Digital Image Processing. Addison-Wesley, Reading (1977)
5. Ito, T., Zhou, X., Nishizeki, T.: Partitioning a graph of bounded tree-width to connected subgraphs of almost uniform size. J. Discrete Algorithms 4, 142–154 (2006)
6. Lucertini, M., Perl, Y., Simeone, B.: Most uniform path partitioning and its use in image processing. Discrete Applied Math. 42, 227–256 (1993)
7. Schröder, M.: Balanced Tree Partitioning, Ph. D. Thesis, University of Karlsruhe, Germany (2001)
8. Simone, C.D., Lucertini, M., Pallottino, S., Simeone, B.: Fair dissections of spiders, worms, and caterpillars. Networks 20(3), 323–344 (1990)
9. Tsichritzis, D.C., Bernstein, P.A.: Operating Systems. Academic Press, New York (1974)
10. Williams Jr., J.C.: Political redistricting: a review. Regional Science 74, 13–40 (1995)

An Improved Divide-and-Conquer Algorithm for Finding All Minimum k-Way Cuts*

Mingyu Xiao

School of Computer Science and Engineering
University of Electronic Science and Technology of China
Chengdu 610054, China
myxiao@gmail.com

Abstract. Given a positive integer k and an edge-weighted undirected graph $G = (V, E; w)$, the *minimum k-way cut* problem is to find a subset of edges of minimum total weight whose removal separates the graph into k connected components. This problem is a natural generalization of the classical *minimum cut* problem and has been well-studied in the literature.

A simple and natural method to solve the minimum k-way cut problem is the divide-and-conquer method: getting a minimum k-way cut by properly separating the graph into two small graphs and then finding minimum k'-way cut and k''-way cut respectively in the two small graphs, where $k' + k'' = k$. In this paper, we present the first algorithm for the tight case of $k' = \lfloor k/2 \rfloor$. Our algorithm runs in $O(n^{4k-\lg k})$ time and can enumerate all minimum k-way cuts, which improves all the previously known divide-and-conquer algorithms for this problem.

Keywords: k-Way Cut, Divide-and-Conquer, Graph Algorithm.

1 Introduction

Let k be a positive integer and $G = (V, E; w)$ a connected undirected graph where each edge e has a positive weight $w(e)$. A *k-way cut* of G is a subset of edges whose removal separates the graph into k connected components, and the *minimum k-way cut* problem is to find a k-way cut of minimum total weight. The minimum k-way cut problem is a natural generalization of the classical *minimum cut* problem and has great applications in the area of VLSI system design, parallel computing systems, clustering, network reliability and finding cutting planes for the travelling salesman problems.

The minimum 2-way cut problem is commonly known as the *minimum cut* problem and can be solved in $O(mn + n^2 \log n)$ time by Nagamochi and Ibaraki's algorithm [12] or Stoer and Wagner's algorithm [19]. Another version of the minimum 2-way cut problem is the *minimum (s, t) cut* problem, which asks us to

---

* The work was done when the author was a PhD student in Department of Computer Science and Engineering, the Chinese University of Hong Kong.

S.-H. Hong, H. Nagamochi, and T. Fukunaga (Eds.): ISAAC 2008, LNCS 5369, pp. 208–219, 2008.

find a minimum cut that separates two given vertices s and t. The minimum (s, t) cut problem can be solved in $O(mn \log n^2/m)$ time by Goldberg and Tarjan's algorithm [4] and $O(\min(n^{2/3}, m^{1/2})m \log(n^2/m) \log U)$ time by Goldberg and Rao's algorithm [3], where U is the maximum capacity of the edge. Finding a minimum cut or minimum (s, t) cut is a subroutine in our algorithms. In the remainder of the paper, we use $T(n, m) = \widetilde{O}(mn)$ to denote the running time of computing a minimum cut or a minimum (s, t) cut in an edge-weighted graph.

For $k = 3$, Kamidoi et al. [8] and Kapoor [10] showed that the minimum 3-way cut problem can be solved by computing $O(n^3)$ minimum (s, t) cuts. Later, Burlet and Goldschmidt [1] improved this result to $O(n^2)$ minimum cut computations. He [6] showed that in unweighted planar graphs the minimum 3-way cut problem can be solved in $O(n \log n)$ time. Xiao [22] designed the first polynomial algorithm for finding minimum 3-way cuts in hypergraphs.

Furthermore, Kamidoi et al. [8] and Nagamochi and Ibaraki [13] proved that the minimum 4-way cut problem can be solved by computing $O(n)$ minimum 3-way cuts. Nagamochi et al. [14] extended this result for minimum $\{5, 6\}$-way cuts by showing that $T_k(n, m) = O(nT_{k-1}(n, m))$, where $k = 5, 6$, and $T_k(n, m)$ is the running time of computing a minimum k-way cut. Those results lead to $\widetilde{O}(mn^k)$ time algorithms for the minimum k-way cut problem for $k \leq 6$.

For general k, Goldschmidt and Hochbaum [5] proved that the minimum k-way cut problem is NP-hard when k is part of the input and gave the first polynomial algorithm for fixed k. The running time of their algorithm is $O(n^{k^2} T(n, m))$. Later, Kamidoi et al. [9] improved the running time to $O(n^{4k/(1-1.71/\sqrt{k})-34}T(n, m))$. Karger and Stein [11] proposed a Monte Carlo algorithm with $O(n^{2(k-1)} \log^3 n)$ running time. Recently, Thorup [20] designed an deterministic algorithm with running time $O(n^{2k})$, which is based on tree packing. Since this problem is NP-hard for arbitrary k, it is also interesting to design approximation algorithms for it. Saran and Vazirani[18] gave two simple approximation algorithms with ratio of $(2 - 2/k)$ and running time of $O(nT(n, m))$. Naor and Rabani [16] obtained an integer program formulation of this problem with integrality gap 2, and Ravi and Sinha [17] also derived a 2-approximation algorithm via the network strength method. Zhao et al. [24] proved that the approximation ratio is $2 - 3/k$ for an odd k and $2 - (3k - 4)/(k^2 - k)$ for an even k, if we compute a k-way cut of the graph by iteratively finding and deleting minimum 3-way cuts in the graph. Xiao et al. [23] determined the tight approximation ratio of a general greedy splitting algorithm, in which we iteratively increase a constant number of components of the graph with minimum cost. That result implies that the approximation ratio is $2 - h/k + O(h^2/k^2)$ for the algorithm that iteratively increases $h - 1$ components.

Most deterministic algorithms for finding minimum k-way cuts, including the two algorithms presented by Goldschmidt and Hochbaum [5] and Kamidoi et al. [9], are based on a divide-and-conquer method. The main idea is to get a minimum k-way cut by properly separating the graph into two small graphs and then finding minimum k'-way cut and k''-way cut respectively in the two small graphs, where $k' + k'' = k$. We say that cut $C = [X, \overline{X}]$ is an $(h, k - h)$-cut of

G, if there is a minimum k-way cut $C_k = [Y_1, \cdots, Y_h, Y_{h+1} \cdots, Y_k]$ of G such that $\bigcup_{i=1}^{h} Y_i = X$ and $\bigcup_{i=h+1}^{k} Y_i = \overline{X}$. Once an $(h, k-h)$-cut $C = [X, \overline{X}]$ is given, we only need to find a minimum h-way cut in induced subgraph $G[X]$ and a minimum $(k-h)$-way cut in induced subgraph $G[\overline{X}]$. Goldschmidt and Hochbaum [5] proved that there are a set S of at most $k-2$ vertices and a set T of at most $k-1$ vertices such that a minimum (S, T) cut is a $(1, k-1)$-cut. By enumerating all the possibilities of S and T, we have at most $O(n^{2k-3})$ candidates for $(1, k-1)$-cuts. Goldschmidt and Hochbaum obtained an $O(n^{k^2})$ algorithm for the minimum k-way cut problem by recursively applying this method. There are two ways to improve this method. First, we can reduce the sizes of S and T. Second, we can try to make minimum (S, T) cut a more 'balanced' cut, in other words, we want minimum (S, T) cut an $(h, k-h)$-cut such that h is close to $\frac{k}{2}$. Kamidoi et al. [9] proved that there are a set S of at most $k-2$ vertices and a set T of at most $k-2$ vertices such that a minimum (S, T) cut is a $(p, k-p)$-cut with $p = \lceil (k - \sqrt{k})/2 \rceil - 1$, and then they got an $O(n^{4k/(1-1.71/\sqrt{k})-34}T(n, m))$ algorithm for the minimum k-way cut problem. In this paper, we show that there are a set S of at most $2\lfloor \frac{k}{2} \rfloor$ vertices and a set T of at most $k-1$ vertices such that a minimum (S, T) cut is a $(\lfloor \frac{k}{2} \rfloor, \lceil \frac{k}{2} \rceil)$-cut. Based on this property, we obtain an $O(n^{4k-\lg k})$ algorithm for finding all minimum k-way cuts. Previous results as well as our result are summarized in the following table. Recently Thorup [20] designed an even faster algorithm for the minimum k-way cut problem, which is based on tree packing, but not the divide-and-conquer method.

Table 1. History of divide-and-conquer algorithms for the minimum k-way cut problem

	Goldschmidt et al. [5]	Kamidoi et al. [9]	This paper
Bounds on $|S|$ and $|T|$	$k-2$ and $k-1$	$k-2$ and $k-2$	$2\lfloor k/2 \rfloor$ and $k-1$
The min (S, T) cut	$(1, k-1)$-cut	$(p, k-p)$-cut, $p = \lceil (k - \sqrt{k})/2 \rceil - 1$	$(\lfloor k/2 \rfloor, \lceil k/2 \rceil)$-cut
Running time for the min k-way cut problem	$O(n^{k^2})$	$O(n^{4k/(1-1.71/\sqrt{k})-16})$	$O(n^{4k-\lg k})$

In this paper, we assume the original graph $G = (V, E; w)$ is a connected graph with more than k vertices. For an edge subset $E' \subseteq E$, $w(E')$ denotes the total weight of the edges in E'. Let $X_1, X_2, \cdots, X_l \subset V$ be l ($2 \leq l \leq n$) disjoint nonempty subsets of vertices, then $[X_1, X_2, \cdots, X_l]$ denotes the set of edges crossing any two different vertex sets of $\{X_1, X_2, \cdots, X_l\}$. A 2-way cut is also simply called a *cut* of the graph. Cut $[X, \overline{X}]$ is called an (S, T) *cut*, if $S \subseteq X$ and $T \subseteq \overline{X}$. Sometimes a singleton set $\{s\}$ is simply written as s and $w([X_1, X_2, \cdots, X_l])$ as $w(X_1, X_2, \cdots, X_l)$. The rest of the paper is organized as follows: We first present the simple divide-and-conquer algorithm in Section 2. Then we give the proofs of our structural results in Section 3. In the last section, we conclude with some remarks.

2 The Divide-and-Conquer Algorithm

Let $C = [X, \overline{X}]$ be a cut. Recall that cut C is an $(h, k - h)$-cut of G if there is a minimum k-way cut $C_k = [Y_1, \cdots, Y_h, Y_{h+1} \cdots, Y_k]$ of G such that $X = \sum_{i=1}^{h} Y_i$ and $\overline{X} = \sum_{i=h+1}^{k} Y_i$. Let $C_k = [Y_1, \cdots, Y_k]$ be a minimum k-way cut and $1 \leq h \leq k - 1$ an integer. By arbitrarily choosing h components $\{Y_{j_1}, Y_{j_2}, \cdots, Y_{j_h}\}$ of C_k, we get an $(h, k - h)$-cut $[\bigcup_{i=1}^{h} Y_{j_i}, \overline{\bigcup_{i=1}^{h} Y_{j_i}}]$, which is called an $(h, k - h)$-partition of C_k. Among all $(h, k - h)$-partitions, those with minimum weight are called $minimum\ (h, k - h)$-partitions of C_k and the weight of them is denoted by $w_{h,k-h}(C_k)$.

For an $(h, k - h)$-cut $[X, \overline{X}]$ of graph G, a minimum h-way cut $[Y_1, \cdots, Y_h]$ in induced graph $G[X]$ and a minimum $(k - h)$-way cut $[Z_1, \cdots, Z_{k-h}]$ in induced graph $G[\overline{X}]$ together yields a minimum k-way cut $[Y_1, \cdots, Y_h, Z_1, \cdots, Z_{k-h}]$ in the original graph G. This suggests a recursive way to solve the minimum k-way cut problem: find an $(h, k - h)$-cut $[X, \overline{X}]$ and recursively find minimum h-way and $(k - h)$-way cuts respectively in $G[X]$ and $G[\overline{X}]$.

However it is not easy to find an $(h, k - h)$-cut, even for $h = 1$. In Section 3, we will prove that for each minimum $(\lfloor \frac{k}{2} \rfloor, \lceil \frac{k}{2} \rceil)$-partition $[X, \overline{X}]$ of each minimum k-way cut, there are a set S of at most $2 \lfloor \frac{k}{2} \rfloor$ vertices and a set T of at most $k - 1$ vertices such that a minimum (S, T) cut is $[X, \overline{X}]$ (See Theorem 2). This minimum (S, T) cut is called the $nearest\ minimum\ (S, T)\ cut\ of\ S$ and can be found by using the same time of computing a maximum flow from S to T. Theorem 2 enables us to obtain the following divide-and-conquer algorithm to find minimum k-way cuts. We enumerate all possibilities of S and T and find the nearest minimum (S, T) cuts in the graph. Then we get a family Γ of at most $\binom{n}{2\lfloor \frac{k}{2} \rfloor} \times \binom{n}{k-1} < n^{2\lfloor \frac{k}{2} \rfloor + (k-1)}$ (S, T) cuts by using the same number of maximum flow computations. By Theorem 2, Γ contains all minimum $(\lfloor \frac{k}{2} \rfloor, \lceil \frac{k}{2} \rceil)$-partitions of all minimum k-way cuts. We then recursively find, for each member of Γ, minimum $\lfloor \frac{k}{2} \rfloor$-way cut in $G[X]$ and minimum $\lceil \frac{k}{2} \rceil$-way cut in $G[\overline{X}]$. The minimum ones among all k-way cuts we find will be returned as our solution. The algorithm is described in Figure 1.

The correctness of algorithm **Multiwaycut** follows from Theorem 2. Now we analyze the running time. When $k = 2$, we use Nagamochi et al.'s algorithm [15] to find all minimum cuts directly, which runs in $O(m^2 n + mn^2 \log n) = O(mT(n, m))$ time. When $k > 2$, we get the recurrence relation

$$C(n, k) \leq n^{2\lfloor \frac{k}{2} \rfloor + k - 1}(C(n, \lfloor \tfrac{k}{2} \rfloor) + C(n, \lceil \tfrac{k}{2} \rceil)) + n^{2\lfloor \frac{k}{2} \rfloor + k - 1}, \qquad (1)$$

where $C(n, k)$ is the upper bound on the number of maximum flow computations to be computed when algorithm **Multiwaycut** runs on an n-vertex graph and an integer k. It is easy to verify that $C(n, k) = O(n^{4k - \lg k - 3})$ satisfies (1) by using the substitution method.

Theorem 1. *All minimum k-way cuts can be found in $O(n^{4k - \lg k})$ time.*

Multiwaycut(G, k)

Input: A graph $G = (V, E; w)$ with nonnegative edge weights and a positive integer $k \leq |V|$.
Output: The set R of all minimum k-way cuts and the weight W of the minimum k-way cut.

1. **If** $\{k = 2\}$, **then** return the set of all the minimum cuts and the weight directly.
2. **Else** $\{k \geq 3\}$,
    Let W be $+\infty$.
    For each pair of disjoint nonempty vertex subsets S and T with $|S| \leq 2 \lfloor \frac{k}{2} \rfloor$ and $|T| \leq k - 1$, **do**
        Compute the nearest minimum (S, T) cut $C = [X, \overline{X}]$ of S.
        If $\{|X| \geq \lfloor \frac{k}{2} \rfloor$ and $|\overline{X}| \geq \lceil \frac{k}{2} \rceil\}$, **then**
            $(R_1, W_1) \longleftarrow$ **Multiwaycut**$(G[X], \lfloor \frac{k}{2} \rfloor)$.
            $(R_2, W_2) \longleftarrow$ **Multiwaycut**$(G[\overline{X}], \lceil \frac{k}{2} \rceil)$.
            If $\{W > w(C) + W_1 + W_2\}$, **then**
                $W \longleftarrow w(C) + W_1 + W_2$,
                $R \longleftarrow \{C \bigcup F_1 \bigcup F_2 \mid F_1 \in R_1, F_2 \in R_2\}$.
            Else if $\{W = w(C) + W_1 + W_2\}$, **then**
                $R \longleftarrow R \bigcup \{C \bigcup F_1 \bigcup F_2 \mid F_1 \in R_1, F_2 \in R_2\}$.
    Return (R, W).

Fig. 1. The Algorithm **Multiwaycut**(G, k)

3   Structural Properties

In this section, we prove the following key theorem, which is the foundation of our divide-and-conquer algorithm.

Theorem 2. *Let C_k be a minimum k-way $(k \geq 3)$ cut of a graph G and $[A, B]$ a minimum $(\lfloor \frac{k}{2} \rfloor, \lceil \frac{k}{2} \rceil)$-partition of C_k. Then there exits a set $S \subseteq A$ of at most $2 \lfloor \frac{k}{2} \rfloor$ vertices and a set $T \subseteq B$ of at most $k - 1$ vertices such that the nearest minimum (S, T) cut of S is $[A, B]$.*

To prove this theorem, we will derive some useful structural properties. Given two disjoint vertex sets S and T, a minimum (S, T) cut separates the graph into two components. One that contains S is called the *source part* and the other one is called the *sink part*, which contains T. For most cases, there are more than one minimum (S, T) cut. Among all minimum (S, T) cuts, the unique one that makes the source part of the maximum cardinality is called the *farthest minimum (S, T) cut* of S, and the unique one that makes the sink part of the maximum cardinality is called the *nearest minimum (S, T) cut* of S. The farthest minimum (S, T) cut of S is the same as the nearest minimum (T, S) cut of T. Ford and Fullkerson [7] proved the uniqueness of the farthest and nearest minimum (S, T) cuts by using the Max flow/Min cut theorem. We can easily get the farthest and nearest minimum (S, T) cuts in linear time based on a maximum flow from S to T. (Note that given a maximum flow, in the residual graph, let X be the set of vertices who are connected with t. Then $[V - X, X]$ is the farthest minimum isolating cut for s). These two special minimum (S, T) cuts have been discussed and used in the literature [7], [21], [5], [2]. Next, we give more structural properties of them.

Lemma 1. *Let $[X_1, \overline{X_1}]$ be the nearest minimum (S_1, T) cut of S_1 and $[X_2, \overline{X_2}]$ the nearest minimum (S_2, T) cut of S_2, if $S_1 \supseteq S_2$, then $X_1 \supseteq X_2$.*

Lemma 2. *Let $[X_1, \overline{X_1}]$ be the farthest minimum (S, T_1) cut of S and $[X_2, \overline{X_2}]$ the farthest minimum cut (S, T_2) of S, if $T_1 \supseteq T_2$, then $X_1 \subseteq X_2$.*

Lemma 1 and Lemma 2 can be proved easily by using the uniqueness of the nearest and farthest minimum (s, t) cuts.

Lemma 3. *Let $C_1 = [X_1, \overline{X_1}]$ be the nearest minimum (S_1, T) cut of S_1 and $C_2 = [X_2, \overline{X_2}]$ a minimum (S_2, T) cut. If $S_1 \subseteq X_2$, then $X_1 \subseteq X_2$.*

Proof. Let $Z = X_1 - X_2$, $Y = X_2 - X_1$, $U = X_1 \cap X_2$, and $W = \overline{X_1 \cup X_2}$ (See Figure 2). To prove $X_1 \subseteq X_2$, we only need to prove that $Z = \emptyset$. Assume to the contrary that $Z \neq \emptyset$. We show the contradiction that $[X_1 - Z, \overline{X_1} + Z]$ is a 'nearer' minimum (S_1, T) cut of S_1 than $C_1 = [X_1, \overline{X_1}]$. Obviously, we only need to prove that $w(X_1 - Z, \overline{X_1} + Z) \leq w(S_1, T)$.

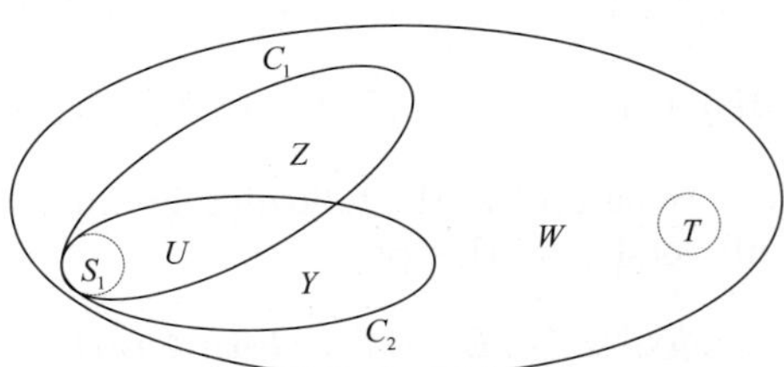

Fig. 2. Illustration for the proof of Lemma 3

Since $[U + Y + Z, W]$ is an (S_2, T) cut and $C_2 = [X_2, \overline{X_2}]$ a minimum (S_2, T) cut, we have

$$w(U + Y + Z, W) \geq w(X_2, \overline{X_2}).$$

It is clear that

$$[U + Y + Z, W] = [U + Y, W] + [Z, W]$$

and

$$[X_2, \overline{X_2}] = [U + Y, W + Z] = [U + Y, W] + [U, Z] + [Y, Z].$$

We get

$$w(U, Z) + w(Y, Z) \leq w(Z, W).$$

Therefore, $w(U, Y+W+Z) = w(U, Y+W)+w(U, Z) \leq w(U, Y+W)+w(Z, W) \leq w(U, Y+W) + w(Z, Y+W) = w(U+Z, Y+W) = w(C_1)$.

We will use the following relation between two multi-way cuts, which was proved by Xiao et al. in [23].

Proposition 1. *Given an edge-weighted graph G, and integers h and k ($2 \leq h \leq k$), then for any minimum h-way cut C_h and any k-way cut C_k of G, the following relation holds*

$$w(C_h) \leq \frac{(2k - h)(h - 1)}{k(k - 1)} w(C_k). \tag{2}$$

Kamidoi et al. [9] proved the following two important results

Proposition 2. *Given an edge-weighted graph G and two integers h and k ($1 \leq h < k$), let C_k be a minimum k-way cut in G and $w_{h,k-h}(C_k)$ the weight of the minimum $(h, k - h)$-partitions of C_k, then*

$$w_{h,k-h}(C_k) \leq \frac{2h(k - h)}{k(k - 1)} w(C_k). \tag{3}$$

Proposition 3. *Given a graph $G = (V, E)$ with at least 4 vertices, two disjoint nonempty subsets T and R of V, and an integer $p \geq 2$, let $\{s_1, s_2, \cdots, s_p\} = S \subseteq V - T \cup R$ be a set of p vertices such that, for each $i \in \{1, 2, \cdots, p\}$, there is a minimum $(S \cup R - \{s_i\}, T)$ cut $[X_i, \overline{X_i}]$ which satisfies $(T \cup \{s_i\}) \subseteq \overline{X_i}$. Let $Z = \bigcap_{1 \leq i \leq p} X_i$, $W = \bigcup_{1 \leq i < j \leq p}(\overline{X_i} \cap \overline{X_j})$, and $Y_i = \overline{X_i} - W$ ($i = 1, 2, \cdots, p$), then $C^\star = [Z, Y_1, Y_2, \cdots, Y_p, W]$ is a $(p + 2)$-way cut such that*

$$w(C^\star) + w(Z, W) + w(Y_1, Y_2, \cdots, Y_p) \leq w(X_1, \overline{X_1}) + w(X_2, \overline{X_2}). \tag{4}$$

Based on Proposition 3, we can prove the following Lemma 4. The detailed proof can be found in the full version of this paper.

Lemma 4. *Given a graph $G = (V, E)$ with at least 4 vertices, a nonempty subset of vertices $T \subset V$, and an integer $p \geq 2$, let $\{s_1, s_2, \cdots, s_p\} = S \subseteq V - T$ be a set of p vertices such that, for each $i \in \{1, 2, \cdots, p\}$, there is a minimum $(S - \{s_i\}, T)$ cut $[X_i, \overline{X_i}]$ which satisfies $(T \cup \{s_i\}) \subseteq \overline{X_i}$. Let $Z = \bigcap_{1 \leq i \leq p} X_i$, $W = \bigcup_{1 \leq i < j \leq p}(\overline{X_i} \cap \overline{X_j})$, and $Y_i = \overline{X_i} - W$ ($i = 1, 2, \cdots, p$),*

(a) : When $Z \neq \emptyset$, then $C^\star = [Z, Y_1, Y_2, \cdots, Y_p, W]$ is a $(p + 2)$-way cut such that

$$w(C^\star) + w(Z, W) + w(Y_1, Y_2, \cdots, Y_p) \leq 2w(V - T, T). \tag{5}$$

(b) : When $Z = \emptyset$ and $p \geq 3$, then $C^\star = [Y_1, Y_2, \cdots, Y_p, W]$ is a $(p + 1)$-way cut such that

$$w(C^\star) + \frac{p - 3}{p + 1} \cdot w(Y_1, Y_2, \cdots, Y_p) \leq \frac{p}{p + 1} \cdot 2w(V - T, T). \tag{6}$$

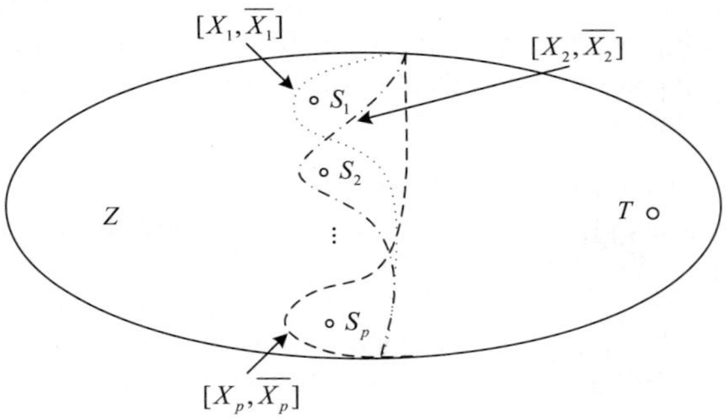

Fig. 3. Illustration for Proposition 3 and Lemma 4

Lemma 5. *Given a graph G and an integer $k \geq 3$, let w_k be the weight of a minimum k-way cut of G. For any cut $[A, B]$ in G with weight $w(A, B) < \frac{k}{2(k-1)}w_k$ (respectively, $w(A, B) < \frac{k+1}{2k-1}w_k$), there exists a set $S \subseteq A$ of at most $k - 1$ (respectively, k) vertices such that the nearest minimum (S, B) cut of S is $[A, B]$.*

Proof. Let $k' = k - 1$ or k (for the two cases respectively), we only need to consider the case that $|A| \geq k' + 1$ (when $|A| < k' + 1$, we can just let $S = A$). Our proof includes two phases. In the first phase, we prove that if the lemma does not hold, then we can find a set $S_0 \subseteq A$ of $k' + 1$ vertices such that, for each nonempty subset S_0' of S_0, the nearest minimum (S_0', B) cut $C' = [Z, \overline{Z}]$ of S_0' satisfying $S_0' \subseteq Z$ and $(S_0 - S_0' + B) \subseteq \overline{Z}$. Then S_0 is a vertex set that satisfies the conditions in Lemma 4. In the second phase, based on S_0, we will show that there is a k-way cut with weight less than w_k, which is a contradiction.

Phase 1: finding S_0. We will give a procedure to select some vertices from A into S. Initially, all the vertices in A are unmarked and S is an empty set. Once a vertex is selected into S, we mark it. Sometimes a vertex in S_0 will also be removed from S, but this vertex is still remained as marked. First, we arbitrarily select $k' + 1$ vertices in A into S. For each nonempty subset $S' \subset S$, we check the nearest minimum (S', B) cut $C' = [Z, \overline{Z}]$ of S'. If $(S - S' + B) \not\subseteq \overline{Z}$, say $a \in (S - S') \bigcap Z$, we update S by removing a from S and adding an unmarked vertex into S (When there are no more unmarked vertex in $V - S$, we just remove a from S and stop the procedure). Once S is updated, we check all nonempty subsets S''s of S again. Since A is a finite set and in each iteration, one more vertex is marked, we will find a set S of $k' + 1$ vertices that satisfies the conditions in Lemma 4 or no more unmarked vertex can be added into S. For the former case, we just let $S_0 = S$. For the later case, we show that S is a set of k' vertices such that the nearest minimum (S, B) cut of S is $[A, B]$, and thus the lemma holds. Let $S^{(1)}, S^{(2)}, \cdots, S^{(l_0)}$ be the updated sequence of S. Let $C^{(l)} = [Z_l, \overline{Z_l}]$ and $C^{(l+1)} = [Z_{l+1}, \overline{Z_{l+1}}]$ be the nearest minimum $(S^{(l)}, B)$ and $(S^{(l+1)}, B)$ cuts of $S^{(l)}$ and $S^{(l+1)}$ respectively. Since $S^{(l)} \subset Z_{l+1}$, we know that $Z_l \subseteq Z_{l+1}$ by

Lemma 3. For each $1 \leq l \leq l_0 - 1$, we have $Z_l \subseteq Z_{l+1}$. Then all the marked vertices will be in Z_{l_0}, where $[Z_{l_0}, \overline{Z_{l_0}}] = [A, B]$ is the nearest minimum $(S^{(l_0)}, B)$ cut of $S^{(l_0)}$. Furthermore, since $S^{(l_0)}$ is obtained by removing a vertex a from $S^{(l_0-1)}$, we know that the size of $S^{(l_0)}$ is k'.

Phase 2: finding a k-way cut with weight less than w_k based on S_0. Suppose $S_0 = \{s_1, \cdots, s_{k'}, s_{k'+1}\}$. Let $[X_i, \overline{X_i}]$ be the nearest minimum $(S - \{s_i\}, B)$ cut $(i = 1, \cdots, k' + 1)$, then $[X_i, \overline{X_i}]$ satisfies that $(B \cup \{s_i\}) \subseteq \overline{X_i}$. Let $Z = \bigcap_{1 \leq i \leq k'+1} X_i$, $W = \bigcup_{1 \leq i < j \leq k'+1} (\overline{X_i} \cap \overline{X_j})$, and $Y_i = \overline{X_i} - W$ $(i = 1, \cdots, k'+1)$. Next, we consider $Z = \emptyset$ and $Z \neq \emptyset$ such two cases.

When $Z = \emptyset$, it follows from Lemma 4 that $C^\star = [Y_1, Y_2, \cdots, Y_{k'+1}, W]$ is a $(k' + 2)$-way cut such that

$$w(C^\star) + \frac{k' - 2}{k' + 2} \cdot w(Y_1 + \cdots + Y_{k'+1}) \leq \frac{k' + 1}{k' + 2} \cdot 2w(A, B). \qquad (7)$$

Since $C^\star$ is a $(k' + 2)$-way cut and $k' + 2 > k$, by Proposition 1, we know that there is a k-way cut C_k with weight

$$w(C_k) \leq \frac{(2(k' + 2) - k)(k - 1)}{(k' + 2)(k' + 1)} w(C^\star). \qquad (8)$$

It follows from (7) and (8) that

$$w(C_k) \leq \frac{(2(k' + 2) - k)(k - 1)}{(k' + 2)(k' + 1)} \cdot \frac{k' + 1}{k' + 2} 2w(A, B).$$

In the case of $w(A, B) < \frac{k}{2(k-1)} w_k$, we have $k' = k - 1$. Then

$$w(C_k) \leq \frac{(k + 2)(k - 1)}{(k + 1)k} \cdot \frac{k}{k + 1} \cdot 2w(A, B) < \frac{(k + 2)k}{(k + 1)^2} w_k < w_k.$$

In the case of $w(A, B) < \frac{k+1}{2k-1} w_k$, we have $k' = k$. Then

$$w(C_k) \leq \frac{(k + 4)(k - 1)}{(k + 2)(k + 1)} \cdot \frac{k + 1}{k + 2} \cdot 2w(A, B) < \frac{2(k + 4)(k^2 - 1)}{(k + 2)^2(2k - 1)} w_k < w_k.$$

We get a contraction that C_k is k-way cut with weight less than w_k.

When $Z \neq \emptyset$, it follows from Lemma 4 that $C^\star = [Z, Y_1, Y_2, \cdots, Y_{k'+1}, W]$ is a $(k' + 3)$-way cut such that

$$w(C^\star) + w(Z, W) + w(Y_1, \cdots, Y_{k'+1}) \leq 2w(A, B). \qquad (9)$$

Suppose $w(Y_{i_0}, W) \geq w(Y_{i_1}, W) \geq w(Y_{i_2}, W) \geq \max_{i \neq i_0, i_1, i_2} \{w(Y_i)\}$. For the case of $w(A, B) < \frac{k}{2(k-1)} w_k$, we prove that $C_k = [Z, \mathcal{Y}_{-i_0-i_1}, W + Y_{i_0} + Y_{i_1}]$ is a k-way cut with weight less than w_k, where $\mathcal{Y}_{-i_0-i_1} = \{Y_1, \cdots, Y_{i_0-1}, Y_{i_0+1}, \cdots, Y_{i_1-1}, Y_{i_1+1}, \cdots, Y_{k'+1}\}$. For the case of $w(A, B) < \frac{k+1}{2k-1} w_k$, we prove that $C_k = [Z, \mathcal{Y}_{-i_0-i_1-i_2}, W + Y_{i_0} + Y_{i_1} + Y_{i_2}]$ is a k-way cut with weight less than w_k, where $\mathcal{Y}_{-i_0-i_1-i_2}$ is defined by the same way as $\mathcal{Y}_{-i_0-i_1}$.

In the case of $w(A, B) < \frac{k}{2(k-1)} w_k$, we have $k' = k-1$. Then $C_k = [Z, \mathcal{Y}_{-i_0-i_1}, W + Y_{i_0} + Y_{i_1}]$ is a k-way cut. Since $[X_i, \overline{X_i}]$ is a minimum $(S - \{s_i\}, B)$ cut, we have $w(Z, Y_i) \le w(Y_i, W)$ for each $i \in \{1, 2, \cdots, k\}$. Therefore,

$$w(Y_{i_0} + Y_{i_1}, W) \ge \frac{2}{2k}(\sum_{i=1}^{k} w(Y_i, W) + \sum_{i=1}^{k} w(Z, Y_i))$$

$$= \frac{1}{k}(w(C^\star) - w(Z, W) - w(Y_1, \cdots, Y_k)).$$

By using this inequality and (9), we get

$$w(C_k) \le w(C^\star) - w(Y_{i_0} + Y_{i_1}, W)$$

$$\le \frac{k-1}{k}w(C^\star) + \frac{1}{k}w(Z, W) + \frac{1}{k}w(Y_1, \cdots, Y_k)$$

$$\le \frac{k-1}{k}2w(A, B) < \frac{k-1}{k}\frac{k}{k-1}w_k = w_k.$$

In the case of $w(A, B) < \frac{k+1}{2k-1} w_k$, we have $k' = k$. Clearly, $C_k = [Z, \mathcal{Y}_{-i_0-i_1-i_2}, W + Y_{i_0} + Y_{i_1} + Y_{i_2}]$ is still a k-way cut. We get

$$w(Y_{i_0} + Y_{i_1} + Y_{i_2}, W) \ge \frac{3}{2(k+1)}(\sum_{i=1}^{k+1} w(Y_i, W) + \sum_{i=1}^{k+1} w(Z, Y_i))$$

$$= \frac{3}{2(k+1)}(w(C^\star) - w(Z, W) - w(Y_1, \cdots, Y_{k+1})).$$

By using this inequality and (9), we get

$$w(C_k) \le w(C^\star) - w(Y_{i_0} + Y_{i_1} + Y_{i_2}, W)$$

$$\le \frac{2k-1}{2(k+1)}w(C^\star) + \frac{3}{2(k+1)}w(Z, W) + \frac{3}{2(k+1)}w(Y_1, \cdots, Y_{k+1})$$

$$\le \frac{2k-1}{2(k+1)}2w(A, B) < \frac{2k-1}{2(k+1)}\frac{2(k+1)}{2k-1}w_k = w_k.$$

We have proved that, in both cases, there is a k-way cut with weight less than w_k. Thus, we have finished the proof.

Lemma 6. *Given a graph G and an integer $k \ge 3$, let w_k be the weight of the minimum k-way cut of G, then for any cut $[A, B]$ in G with weight $w(A, B) \le \frac{k}{2(k-1)} w_k$ (respectively, $w(A, B) \le \frac{k+1}{2k-1} w_k$), there exists a set $S \subseteq A$ of at most $k - 1$ (respectively, k) vertices such that the farthest minimum (S, B) cut of S is $[A, B]$.*

In Lemma 6, the weight of the cuts can equal $\frac{k}{2(k-1)} w_k$ or $\frac{k+1}{2k-1} w_k$ and there are the farthest minimum (S, B) cuts, instead of the nearest minimum (S, B) cuts in Lemma 5. The proof of Lemma 6 just follows the proof of Lemma 5. Note that since in Lemma 6 there are farthest minimum (S, B) cuts, we have $w(X_i, \overline{X_i}) < w(V - T, T)$ for each $i \in \{1, 2, \cdots, p\}$. Therefore, The equal signs in (5) and (6) will not hold, which guarantees that the remaining part of the proof of Lemma 5 is suitable for Lemma 6.

Now we are ready to prove Theorem 2:

Proof. By Proposition 2, we have $w(A, B) \leq \frac{2h(k-h)}{k(k-1)} w_k$, where $h = \lfloor \frac{k}{2} \rfloor$. Since $\lfloor \frac{k}{2} \rfloor \lceil \frac{k}{2} \rceil \leq (\frac{k}{2})^2$, we have $w(A, B) \leq \frac{k}{2(k-1)} w_k$ and the equal sign does not hold for odd k. By Lemma 5 and Lemma 6, we know that there is a set $S \subseteq A$ of at most $2 \lfloor \frac{k}{2} \rfloor$ vertices such that the nearest minimum (S, B) cut of S is $[A, B]$ and there is a set $T \subseteq B$ of at most $k - 1$ vertices such that the farthest minimum (T, A) cut of T is $[B, A]$. We look at the nearest minimum (S, T) cut $[X, \overline{X}]$ of S. Since $[A, B]$ is the nearest minimum (S, B) cut of S and $T \subseteq B$, we have $A \subseteq X$ by Lemma 1. Since $[B, A]$ is the farthest minimum (T, A) cut of T and $S \subseteq A$, we have $B \subseteq \overline{X}$ by Lemma 2. Therefore, $[X, \overline{X}] = [A, B]$. Thus, Theorem 2 holds.

Corollary 1. *Given a graph G and an integer $k \geq 3$, let w_k be the weight of the minimum k-way cut of G, then the number of cuts with weight less than $\frac{k}{2(k-1)} w_k$ in G is bounded by n^{2k-2} and the number of cuts with weight less than $\frac{k+1}{2k-1} w_k$ is bounded by n^{2k}.*

4   Discussion

In this paper, we presented a simple divide-and-conquer algorithm for finding minimum k-way cuts. As we mentioned in Section 1, there are two possible ways to improve the algorithm. One is to reduce the sizes of S and T, and the other one is to make the minimum (S, T) cut be a more 'balanced' $(h, k-h)$-cut. In our algorithm, $h = \lfloor \frac{k}{2} \rfloor$ and this means our $(h, k - h)$-cuts are the most 'balanced'. Our questions are: For the most 'balanced' case, can we reduce the sizes of S and T? What are the lower bounds on the two sizes? Note that if we can reduce the two sizes to $\frac{k}{2}$, then the divide-and-conquer algorithm will run in $O(n^{2k})$ time.

Nagamochi et al. [13], [14] proved that the minimum k-way cut problem can be solved by computing $O(n)$ minimum $(k - 1)$-way cuts for $k \leq 6$. Does this hold for general k? If so, then the minimum k-way cut problem can be solved in $\widetilde{O}(mn^k)$ time.

Karger and Stein [11] and Nagamochi et al. [15] have studied the bounds on the number of small cuts, which motivates the following question: Can we give nontrivial lower and upper bounds on the number of minimum k-way cuts? It is easy to get a lower bound of $O(n^k)$. Note that in a cycle consisting of n edges with equal weight, the number of minimum k-way cuts is $\binom{n}{k}$. Can better bounds be achieved?

References

1. Burlet, M., Goldschmidt, O.: A new and improved algorithm for the 3-cut problem. Operations Research Letters 21(5), 225–227 (1997)
2. Dahlhaus, E., Johnson, D., Papadimitriou, C., Seymour, P., Yannakakis, M.: The complexity of multiterminal cuts. SIAM J. Comput. 23(4), 864–894 (1994)
3. Goldberg, A.V., Rao, S.: Beyond the flow decomposition barrier. J. ACM 45(5), 783–797 (1998)

4. Goldberg, A.V., Tarjan, R.E.: A new approach to the maximum-flow problem. J. ACM 35(4), 921–940 (1988)
5. Goldschmidt, O., Hochbaum, D.: A polynomial algorithm for the k-cut problem for fixed k. Mathematics of Operations Research 19(1), 24–37 (1994)
6. He, X.: An improved algorithm for the planar 3-cut problem. J. Algorithms 12(1), 23–37 (1991)
7. Ford, J.R., Fullkerson, D.R.: Flows in networks. Princeton University Press, Princeton (1962)
8. Kamidoi, Y., Wakabayashi, S., Yoshida, N.: A divide-and-conquer approach to the minimum k-way cut problem. Algorithmica 32(2), 262–276 (2002)
9. Kamidoi, Y., Yoshida, N., Nagamochi, H.: A deterministic algorithm for finding all minimum k-way cuts. SIAM Journal on Computing 36(5), 1329–1341 (2006)
10. Kapoor, S.: On minimum 3-cuts and approximating k-cuts using cut trees. In: Cunningham, W.H., Queyranne, M., McCormick, S.T. (eds.) IPCO 1996. LNCS, vol. 1084. Springer, Heidelberg (1996)
11. Karger, D.R., Stein, C.: A new approach to the minimum cut problem. Journal of the ACM 43(4), 601–640 (1996)
12. Nagamochi, H., Ibaraki, T.: Computing edge connectivity in multigraphs and capacitated graphs. SIAM Journal on Discrete Mathematics 5(1), 54–66 (1992)
13. Nagamochi, H., Ibaraki, T.: A fast algorithm for computing minimum 3-way and 4-way cuts. Mathematical Programming 88(3), 507–520 (2000)
14. Nagamochi, H., Katayama, S., Ibaraki, T.: A faster algorithm for computing minimum 5-way and 6-way cuts in graphs. In: Asano, T., Imai, H., Lee, D.T., Nakano, S.-i., Tokuyama, T. (eds.) COCOON 1999. LNCS, vol. 1627. Springer, Heidelberg (1999)
15. Nagamochi, H., Nishimura, K., Ibaraki, T.: Computing all small cuts in an undirected network. SIAM Journal on Discrete Mathematics 10(3), 469–481 (1997)
16. Naor, J., Rabani, Y.: Tree packing and approximating k-cuts. In: Proceedings of the twelfth annual ACM-SIAM symposium on discrete algorithms (SODA 2001). Society for Industrial and Applied Mathematics, Philadelphia (2001)
17. Ravi, R., Sinha, A.: Approximating k-cuts via network strength. In: Proceedings of the thirteenth annual ACM-SIAM symposium on Discrete algorithms (SODA 2002). Society for Industrial and Applied Mathematics, Philadelphia (2002)
18. Saran, H., Vazirani, V.V.: Finding k-cuts within twice the optimal. SIAM J. Comput. 24(1), 101–108 (1995)
19. Stoer, M., Wagner, F.: A simple min-cut algorithm. J. ACM 44(4), 585–591 (1997)
20. Thorup, M.: Minimum k-way cuts via deterministic greedy tree packing. In: Proceedings of the 40th Annual ACM Symposium on Theory of Computing (STOC 2008) (2008)
21. Xiao, M.: Algorithms for multiterminal cuts. In: Hirsch, E.A., Razborov, A.A., Semenov, A., Slissenko, A. (eds.) Computer Science – Theory and Applications. LNCS, vol. 5010. Springer, Heidelberg (2008)
22. Xiao, M.: Finding minimum 3-way cuts in hypergraphs. In: Agrawal, M., Du, D.-Z., Duan, Z., Li, A. (eds.) TAMC 2008. LNCS, vol. 4978. Springer, Heidelberg (2008)
23. Xiao, M., Cai, L., Yao, A.C.: Tight approximation ratio of a general greedy splitting algorithm for the minimum k-way cut problem (manuscript, 2007)
24. Zhao, L., Nagamochi, H., Ibaraki, T.: Approximating the minimum k-way cut in a graph via minimum 3-way cuts. J. Comb. Optim. 5(4), 397–410 (2001)

On the Algorithmic Effectiveness of Digraph Decompositions and Complexity Measures

Michael Lampis[1], Georgia Kaouri[2], and Valia Mitsou[1]

[1] City University of New York
[2] National Technical University of Athens
{mlampis,vmitsou}@gc.cuny.edu, gkaouri@corelab.ntua.gr

Abstract. We place our focus on the gap between treewidth's success in producing fixed-parameter polynomial algorithms for hard graph problems, and specifically HAMILTONIAN CIRCUIT and MAX CUT, and the failure of its directed variants (directed tree-width [9], DAG-width [11] and kelly-width [8]) to replicate it in the realm of digraphs. We answer the question of why this gap exists by giving two hardness results: we show that DIRECTED HAMILTONIAN CIRCUIT is $W[2]$-hard when the parameter is the width of the input graph, for any of these widths, and that MAX DI CUT remains NP-hard even when restricted to DAGs, which have the minimum possible width under all these definitions. Our results also apply to directed pathwidth.

Keywords: Treewidth, Digraph decompositions, Parameterized Complexity.

1 Introduction

Treewidth, first introduced by Robertson and Seymour in [13], has been one of the most successful tools in the research for efficient algorithms for hard graph problems in the last 15 years. Intuitively, treewidth allows us to distinguish graphs that have a relatively simple (tree-like) structure, and exploit that structure to solve a plethora of otherwise intractable problems, usually by employing a dynamic programming technique. For an introduction to the notion of treewidth see Bodlaender's excellent survey papers [4,3,2].

One of the most celebrated theorems in the area of treewidth is Courcelle's theorem which states that every graph property that can be expressed in monadic second order logic can be decided in linear time on graphs of bounded treewidth [5]. Beginning from this starting point, algorithms for many hard graph problems have been devised using treewidth. They almost invariably have running times of the form $O(f(k) \cdot n)$, where k is the treewidth of the input graph and f some exponential or super-exponential function which represents the complexity of solving the problem exhaustively on k vertices. Thus, not only is the running time polynomial for fixed k, but also that the combinatorial explosion is confined to k. This has led treewidth to become one of the cornerstones of parameterized complexity theory, a theory which describes the distinction between algorithms with running times

S.-H. Hong, H. Nagamochi, and T. Fukunaga (Eds.): ISAAC 2008, LNCS 5369, pp. 220–231, 2008.

of the form $O(f(k) \cdot n^c)$, where c is a constant (called fixed-parameter tractable or FPT) and algorithms of the form $O(n^{g(k)})$. For an introduction to parameterized complexity see the monograph by Downey and Fellows [6] or the introductory books by Niedermeier [10] and by Flum and Grohe [7].

Several attempts have been made recently to generalize the notion of treewidth to directed graphs. The motivation behind this line of research is that, although it is possible to solve many hard problems on digraphs when the underlying undirected graph has low treewidth by using traditional tree decompositions, this approach sacrifices a great deal of generality. A problem which demonstrates this to a great degree is DIRECTED HAMILTONIAN CIRCUIT. This problem is trivial when the input graph is a DAG, but there exist DAGs of unbounded treewidth if the direction of the edges is ignored. Thus, it is desirable to come up with an alternative measure of digraph complexity which better characterizes the class of digraphs where hard problems become tractable. It should be noted at this point that HAMILTONIAN CIRCUIT admits an FPT solution with a treewidth based algorithm, therefore a logical target when defining a digraph complexity measure would be to achieve fixed-parameter tractability for DIRECTED HAMILTONIAN CIRCUIT as well.

Previous Work. The most notable variations of treewidth for digraphs that have been proposed in the past are probably directed treewidth [9], DAG-width [11] and kelly-width [8]. All these three measures can be viewed as good generalizations of treewidth in the sense that, if we take an undirected graph and replace each edge with two opposite directed edges the width of the new digraph will be the same for all three definitions and equal to the treewidth of the original graph. Directed treewidth is the most general of the three, in the sense that a graph of bounded kelly-width or DAG-width will also have bounded directed treewidth, while the converse may not be true. Also DAG-width and kelly-width are conjectured to be only a constant factor apart on any graph ([8]).

The most important positive result of directed treewidth (which can be extended to all the three measures) is an algorithm that solves DIRECTED HAMILTONIAN CIRCUIT in $O(n^k)$ time, k being the width of the input graph. Nevertheless, this algorithm is still far from the performance of the best treewidth-based algorithm for HAMILTONIAN CIRCUIT, which runs in fixed-parameter linear time. Unfortunately, the reason for this distance is not addressed in [9] or in [8] where another algorithm (of the same complexity) for this problem is given. In addition, the few already known algorithmic results on these measures don't seem to indicate that they are likely to achieve a level of success comparable to treewidth, as no FPT algorithms are known for any hard digraph problems. Of course, it could be conceivable that this is due to a lack of effort so far, since digraph decompositions have been introduced much more recently than treewidth.

A related measure is directed pathwidth. Just as pathwidth is a restriction of treewidth in the undirected case directed pathwidth is a restriction of all the previously mentioned directed measures, thus having even greater algorithmic potential. However, to the best of our knowledge no such results have been shown for directed pathwidth. In [1] it is shown that a cops-and-robber game is equivalent to directed path-width and that there always exists an (almost) optimal monotone strategy. It is worthy of note that, unlike the undirected case where treewidth and pathwidth are generalizations of different graph topologies (trees and paths respectively) in the directed case all the measures we have mentioned are based on the concept of DAGs as the simplest case.

Our Contribution. In this paper we try to address the question of whether the already proposed digraph complexity measures will be able to match the success of treewidth. Our answer is given in the form of two negative results, which show that the lack of FPT algorithms for DIRECTED HAMILTONIAN CIRCUIT and MAX DI CUT is not due to a lack of effort, but because such algorithms can not exist (under some widely believed complexity assumptions).

Our first result concerns DIRECTED HAMILTONIAN CIRCUIT which we show to be $W[2]$-hard when the parameter is the width of the input graph for any of the mentioned widths. Under the assumption that $W[2] \neq$ FPT this implies that no FPT algorithm is possible. Therefore, under this standard complexity assumption, our result implies that no significant improvement is possible for the $O(n^k)$ algorithms of [9] and [8].

Our second result concerns MAX DI CUT, for which we show APX-hardness even when we restrict the problem to DAGs and all edges have uniform weights. This is a result that is interesting in its own right, and it is rather surprising that it was not known until now, as MAX DI CUT is a widely studied problem. It is also very relevant in our case for two reasons: First, DAGs have the lowest possible width for all the widths we have mentioned, therefore our proof implies that none of them can help with MAX DI CUT. Second, using (undirected) treewidth leads to efficient FPT algorithms for both MAX CUT and MAX DI CUT. Thus, this result helps draw further contrast between the performance of treewidth and its directed variants.

Although our results are negative, they succeed in illuminating some fundamental weaknesses in the already proposed digraph measures, and thus they show the way to a possible future digraph measure that might be able to overcome them. Therefore, we believe that they serve as a starting point in a renewed search for a successful digraph complexity measure that might yet manage to at least partially match treewidth's success.

The rest of this paper is structured as follows: In Section 2 we give some necessary definitions and preliminary notions. In Section 3 we demonstrate the hardness result for DIRECTED HAMILTONIAN CIRCUIT. In Section 4 we prove the hardness of MAX DI CUT. Finally, in Section 5 we conclude with some discussion and directions to further research.

2   Definitions and Preliminaries

First, let us give the definitions of the two problems that will be our focus.

Definition 1. *The* DIRECTED HAMILTONIAN CIRCUIT *problem is that of deciding whether there exists a permutation* $(v_1, v_2, \ldots, v_n)$ *of the vertices of an input digraph* $G(V, E)$ *s.t.* $\forall i \in \{1, \ldots, n-1\}$ $(v_i, v_{i+1}) \in E$ *and* $(v_n, v_1) \in E$.

Definition 2. *The* MAX DI CUT *problem is the following: given a digraph* $G(V, E)$ *and a weight function on the edges* $w : E \to \mathbb{N}$*, find a partition of* V *into two sets* V_0 *and* V_1 *so that the weight of the edge set* $C = \{(u, v) \mid u \in V_0, v \in V_1\}$ *is maximized. That is, the objective is to maximize* $\sum_{e \in C} w(e)$*.*

MAX DI CUT was shown APX-hard in [12]. In Section 4 we show APX-hardness for the problem's restriction to DAGs. Then we show that APX-hardness also holds for the cardinality version of the problem restricted to DAGs.

We should also give the definitions of the two problems that will be the starting points of our reductions.

Definition 3. DOMINATING SET *is the problem of finding a minimum cardinality subset of vertices* D *of an undirected graph* $G(V, E)$ *s.t. any vertex in* $V \setminus D$ *has a neighbor in* D*.*

When a vertex $u \in D$ is a neighbor of a vertex v, we will say that u dominates v. We will also follow the convention of saying that any vertex in D dominates itself. We will make use of the well-known result that DOMINATING SET is $W[2]$-complete when the parameter k is the size of the dominating set we are looking for ([6]).

Definition 4. NAE3SAT *is the problem of finding a truth assignment which, for every clause of an input 3CNF formula, assigns the value true to at least one literal, and the value false to at least one literal.*

We follow the convention of saying that a clause is satisfied in the NAESAT sense, or simply satisfied, when a truth assignment assigns different truth values to two of its literals. We will mainly be concerned with the maximization version of NAE3SAT where the objective is to find a truth assignment that satisfies as many clauses as possible. This variant was shown to be APX-hard in [12].

We have already mentioned that directed pathwidth can be defined in terms of a cops-and-robber game. The game's definition is the following:

Definition 5. *The* k-cop *invisible-eager robber game is the game where* k *cops attempt to catch an invisible robber hiding in a vertex of a digraph* G*. The cops are stationed on vertices of* G *and a cop can move by removing himself from the graph and then "landing" on any other vertex. The robber can move at any time and he is allowed to follow any directed path of* G*, under the condition that he does not enter vertices occupied by stationary cops.*

We say that k cops have a monotone strategy to win this game when they have a strategy such that the robber can never visit a vertex previously occupied by a cop. In [1] it was shown that k cops have a monotone strategy on a graph G iff the graph has directed pathwidth k.

Kelly-width, DAG-width and directed treewidth have also been shown to be connected to similar games, restricted to monotone strategies. In fact, DAG-width is equivalent to the above game but with the robber being visible, while kelly-width is equivalent to the above game but with the robber only being allowed to move when a cop enters his vertex. Using the approximate connection between directed treewidth and a similar game it was shown in [8] that the directed treewidth of a graph is upper-bounded by its kelly-width multiplied by a constant.

It is not hard to infer from these results that, since the robber is stronger in the game related to directed pathwidth, a graph G will have higher pathwidth than any of the other widths. Since we are interested in proving hardness results, it will therefore suffice to show that a problem is hard for graphs of small directed pathwidth and hardness for the other widths will directly follow.

3   Directed Hamiltonian Circuit

In this section we focus on the DIRECTED HAMILTONIAN CIRCUIT problem, a problem which can be solved using directed treewidth in $O(n^k)$ time ([9]). Of course this algorithm also applies to DAG-width, kelly-width and directed pathwidth, as they are restrictions of directed treewidth. In addition, another $O(n^k)$ algorithm for this problem tailored for kelly-width is given in [8]. Thus, a significant gap exists between the performance of treewidth, which is fixed-parameter polynomial on the corresponding undirected problem and the performance of its directed variants. We show that this is a gap that can not be bridged unless $W[2] = FPT$, by demonstrating that DIRECTED HAMILTONIAN CIRCUIT is $W[2]$-hard when the parameter is any of these widths.

The hardness proof for DIRECTED HAMILTONIAN CIRCUIT will be a parameterized reduction from the naturally parameterized version of DOMINATING SET.

Theorem 1. *The parameterized versions of* DIRECTED HAMILTONIAN CIRCUIT, *where the parameter is the directed treewidth, kelly-width, DAG-width or directed pathwidth of the input graph, are* $W[2] - hard$.

Proof. We will show a parameterized reduction from the naturally parameterized version of DOMINATING SET, where the parameter k is the size of the set by constructing a digraph whose directed pathwidth is bounded by a function of k s.t. the digraph will be Hamiltonian iff the original graph had a dominating set of size k.

Suppose we are given a graph $G(V, E)$ with $V = \{1, 2, \ldots, n\}$.

Our digraph G' has vertex set $V' = V_1 \cup V_2 \cup V_3$ where

1. $V_1 = \{u_1, u_2, \ldots, u_k\}$.
2. $V_2 = \{v_1, v_2, \ldots, v_n\}$.
3. $V_3 = \{g_{(i,j,l)} \mid i \in \{1, \ldots, n\}, j \in \{0, \ldots, d(i)\}, l \in \{In, Out\}\}$. Here $d(i)$ denotes the degree of vertex i in G.

$E(G')$ consists of the following sets of directed edges

1. $E_1 = \{(u_i, v_j) \mid i \in \{1, \ldots, k\}, j \in \{1, \ldots, n\}\}$.
2. $E_2 = \{(v_i, v_{i+1}) \mid i \in \{1, \ldots, n\}\}$ (where we consider $n+1$ to be the same as 1).

3. $E_3 = \{(g_{(i,j,In)}, g_{(i,j+1,Out)}) \mid i \in \{1,\ldots,n\}, j \in \{0,\ldots,d(i)\}\}$, (where we consider $d(i) + 1$ to be the same as 0).

4. $E_4 = \{(g_{(i,j,Out)}, g_{(i,j,In)}) \mid i \in \{1,\ldots,n\}, j \in \{0,\ldots,d(i)\}\}$.

5. Finally, E_5 contains the following edges: For any vertex a of the original graph, let $j_0, j_1, \ldots, j_{d(a)}$ be the vertices of G a dominates in lexicographic order. Also, let $p(a, i)$ denote the number of vertices that dominate a vertex i and come before a in lexicographic order. Then the edge $(v_a, g_{(j_0, p(a,j_0), In)})$ and the edges $(g_{(j_i, p(a,j_i), Out)}, g_{(j_{i+1}, p(a,j_{i+1}), In)})$ for all $i < d(a)$ are included in E_5. Finally, the edges $(g_{(j_{d(a)}, p(a,j_{d(a)}), Out)}, u_i)$ for all $i \in \{1,\ldots,k\}$ are also included in E_5.

Let us now discuss the basic idea behind this construction, before we get into more details. Our digraph G' consists of three parts: a constraint part V_1, a choice part V_2 and a satisfaction part V_3. V_1 functions as a constraint part because it only has k vertices and the only edges going into V_2 originate here, thus forcing us to enter the choice part exactly k times. A Hamiltonian tour will leave V_2 k times. The vertices from which it leaves V_2 must be (as we will prove) a dominating set of G, and that is why V_2 is the choice part. Finally, V_3 is arranged in such a way that it can only be traversed in a Hamiltonian way if the choice made in V_2 is indeed a dominating set.

We will call the group of vertices $g_{(i,j,l)}$ for a specific i, the gadget C_i. The crucial part of this reduction is the way the gadget C_i works. Notice that the gadget's vertices induce a directed cycle. Also, the only way to enter this cycle is through an In vertex, and the only way to leave is through an Out vertex. Suppose that a Hamiltonian tour enters a gadget C_i m times and that $X \subseteq \{0,\ldots,d(i)\}$ is the index set of the In vertices that were used. Then it must also be the index set of the Out vertices used. To see that, suppose that $X = \{j_1, j_2, \ldots, j_m\}$ in lexicographic order. When entering from $g_{(i,j_1,In)}$ the tour has no choice but to proceed to $g_{(i,j_1+1,Out)}$. Then if $j_2 \neq j_1 + 1$ the tour must move to $g_{(i,j_1+1,In)}$, because if it were to exit this vertex would be impossible to visit in the future. Using this argument again can exclude the possibility of this part of the tour exiting through any vertex other than $g_{(i,j_2,Out)}$. Similarly, the path that starts at $g_{(i,j_2,In)}$ will exit at $g_{(i,j_3,Out)}$ and so on, with $g_{(i,j_m,In)}$ exiting through $g_{(i,j_1,Out)}$. This procedure covers all the vertices of the gadget, therefore we proved that for any set of entry vertices X the gadget can be traversed in a way that does not exclude the existence of a Hamiltonian tour of the whole graph iff X corresponds also to the exit vertices used.

Let us now make use of the property we just established. Suppose that G does not have a dominating set of size k, but that a Hamiltonian tour of G' exists. Let D be the set of choices made by the tour in V_2, i.e. the set of vertices through which the tour exits V_2. The selection of the corresponding set in G leaves some vertex not dominated, say vertex i. Consider the gadget C_i. It can not be entered directly from a vertex in V_2, since none of the vertices from which we exited V_2 corresponds to one that dominates i. Also, if all the other gadgets are traversed in a way that does not exclude a Hamiltonian cycle, we established above that the set of entry indices in each is the same as the set of exit indices. Thus, if the set of input indices into a gadget C_j corresponds to its domination by some vertices in D, the tour when exiting C_j will be led to other gadgets also corresponding to dominated vertices, and therefore it will not be lead to C_i. Thus, we have a contradiction and no Hamiltonian tour is possible.

It remains to establish the converse, namely that a dominating set of size k implies a Hamiltonian tour. Let $D = \{d_1, d_2, \ldots, d_k\}$ be a dominating set. Let us first describe the tour outside the gadgets. Starting at u_1, move to v_{d_k+1} (once again, v_{n+1} is the same as v_1) and then follow the edges in V_2 until v_{d_1} is reached. Then we exit V_2 towards the gadgets. When we reach an Out gadget vertex that points to V_1 we move to u_2. From there we move to v_{d_1+1}, then to v_{d_2} and so on. This procedure makes sure that, even though we enter V_2 only k times, all of its n vertices are covered.

Now, suppose that our path has reached a gadget vertex $g_{(i,j,In)}$. This means that we are following an edge that "belongs" to the j-th vertex of G that could dominate vertex i. Call this vertex x. Take the first vertex in D that dominates i and comes lexicographically after x. Suppose that this is the j'-th vertex that can dominate i overall. (If there is no such vertex in D that dominates i and comes after x take the first vertex in D that dominates i). Now, the traversal of gadget C_i will be $g_{(i,j,In)} \rightarrow g_{(i,j+1,Out)} \rightarrow g_{(i,j+1,In)} \rightarrow \cdots \rightarrow g_{(i,j',Out)}$, from which point we exit the gadget. (Note that j and j' need not necessarily be distinct for the above argument to work).

Let us now prove this is indeed a Hamiltonian tour. It is not hard to see that all vertices of V_1 and V_2 are visited exactly once, leaving the gadgets as the hard part of this proof. The proof will be by induction. For gadget C_1 we know that it is entered directly from V_2 only (1 is always the first vertex lexicographically that any other vertex dominates). Observe that the tour we suggested is just a restatement of the reasoning we made on how gadgets work. Then it is not hard to see that, no matter which vertices of D dominate 1, C_1 will be traversed correctly. The problem is that paths have been now "shifted", i.e. when entering from the entry point of the i-th vertex that dominates 1 we exit from the exit point of the $(i+1)$-th vertex that dominates 1. Therefore, we will make a proof by induction. Suppose that we know that the tour we suggested visits the vertices of gadgets $C_1, \ldots, C_i$ exactly once and that the entry and exit points for each gadget correspond to the vertices of D that dominate the vertex this gadget represents. Now consider the gadget C_{i+1}. It is only entered using edges that "belong" to vertices of D, because of the inductive hypothesis. Suppose that one of its vertices is visited twice. Then, one of its entry points must be used twice, meaning that an exit of a previous gadget or a vertex of V_2 is visited twice, a contradiction. Now, suppose that the vertices of D that dominate vertex $i+1$ are the j_1-th, the j_2-th, and so on of the vertices that could dominate $i+1$ overall. Our tour must visit vertex $g_{(i+1,j_1,In)}$, because the vertex of some previous gadget that point to it is used as an output point (by the inductive hypothesis). Similar arguments hold for j_2 and so on, leading us to conclude that all appropriate entry points are used and gadget C_{i+1} is traversed correctly.

Finally, what is left is to argue that G' has low directed pathwidth. Consider the following cop strategy for the robber game: Keep k cops on the vertices of V_1 at all times. Now 2 cops are enough to clean V_2, since it is just a directed cycle with edges going out to V_3 but no other incoming edges. Then, these two cops can clean gadget C_1 for the same reasons. After that, they can clean C_2 and so on, until the whole graph is clean. $\qquad\square$

4 Max Di Cut

Let us now focus on a problem of much different nature: Max Di Cut. Even though, as we saw in Section 3, no digraph complexity measure manages to provide an FPT algorithm for Directed Hamiltonian Circuit, they do succeed in providing algorithms with polynomial running times, when the width k is fixed. For Max Di Cut the situation is much worse, as we will show that the problem is NP-hard even for $k = 1$. This creates an even larger gap with the FPT performance of treewidth than we had in the case of Directed Hamiltonian Circuit.

We will prove that Max Di Cut is both NP and APX-hard, even when restricted to DAGs by showing a reduction from the maximization version of NAE3SAT.

Theorem 2. Max Di Cut *is NP-hard and APX-hard, even when restricted to DAGs.*

Proof. We give a gap-preserving reduction from NAE3SAT to Max Di Cut.

Given a NAE3SAT formula ϕ with m clauses and n variables we construct a new NAE3SAT formula ϕ' with $2m$ clauses and n variables and show that ϕ is satisfiable iff ϕ' is satisfiable (satisfaction is in the NAESAT sense). Then from ϕ' we construct a (weighted) DAG G and show that ϕ' is satisfiable iff G has a directed cut of size $46m$. Without loss of generality, we may assume that every clause of ϕ has exactly three literals (otherwise we may repeat one).

The new formula ϕ' is constructed by taking ϕ and adding to it, for every clause $(l_1 \vee l_2 \vee l_3)$, the same clause with all literals complemented. If an assignment satisfies t clauses of the original formula, it must satisfy exactly $2t$ of the $2m$ clauses of ϕ'. Note that, if we denote by f_i the number of appearances of the variable x_i in ϕ, then the same variable will appear $2f_i$ times in ϕ': f_i times as x_i and f_i times as $\neg x_i$. In other words, the positive and negative appearances of each variable in ϕ' are balanced. We will make use of this fact several times.

Let us now construct the DAG $G(V, E)$. V consists of four disjoint sets of vertices A, X, C, B. $A = \{a_1, \ldots, a_n\}$ will be a set of source vertices. $B = \{b_1, \ldots, b_{2m}\}$ will be a set of sink vertices. $X = \{x_1, x_1', x_2, x_2', \ldots, x_n, x_n'\}$ will be the set of vertices corresponding to literals of ϕ' while $C = \{c_{i,j,k} \mid i \in \{1, 2, \ldots, 2m\},\ j, k \in \{1, 2, 3\}\}$ will correspond to the clauses of ϕ'.

E consists of the following sets of weighted edges:

1. The set $E_1 = \{(a_i, x_i) \mid i \in \{1, \ldots, n\}\}$. Each of these edges has weight $6f_i$, where f_i is the total number of appearances of the variable x_i in ϕ.
2. The set $E_2 = \{(x_i, x_i') \mid i \in \{1, \ldots, n\}\}$. Each of these edges also has weight $6f_i$.
3. The set $E_3 = \{(c_{i,j,k}, b_i) \mid i \in \{1, \ldots, 2m\}, \ j, k \in \{1, 2, 3\}, j \neq k\}$. These have weight 1.
4. The set $E_4 = \{(c_{i,k,k}, c_{i,j,k} \mid i \in \{1, \ldots, 2m\}, \ j, k \in \{1, 2, 3\}, j \neq k\}$. These also have weight 1.
5. Finally, we add edges that connect vertices of the set X to the corresponding vertices of C. That is, we add the edges $\{(x_l, c_{i,j,k}), \ k \in \{1, 2, 3\}\}$ when the literal x_l appears in the j-th position of the i-th clause of ϕ', and the edges $(x_l', c_{i,j,k})$ when the literal $\neg x_l$ appears in that position. These edges have weight 2.

Suppose we are given a truth assignment that satisfies (in the NAESAT sense) t of the m clauses of ϕ. It must satisfy $2t$ of the $2m$ clauses of ϕ'. Let us partition V into V_0 and V_1. Place all vertices of A into V_0 and all vertices of B into V_1. Place the vertices of X that correspond to true literals in V_1 and the rest in V_0. Place the vertices of C that correspond to true literals in V_0 and the rest in V_1.

Let us calculate the weight of this cut. If a variable x_i is assigned the value 1 in the assignment, the edge (a_i, x_i) contributes $6f_i$ to the cut. If it is assigned 0, then x_i' is in V_1, therefore the edge (x_i, x_i') contributes $6f_i$ to the cut. Thus, the total contribution of all edges in $E_1 \cup E_2$ is $6\sum_i f_i$. Because the appearances of each variable in ϕ' are balanced, there are as many literals that took the value true as there are literals that took the value false, in any assignment. Therefore, exactly half the edges of E_3 contribute to the cut. The number of edges in E_3 is $12m$ so, a weight of $6m$ is contributed to the cut. It is not hard to see that, for a satisfied clause C_i, the edges of E_4 incident on vertices that correspond to this clause contribute exactly 2 to the cut. On the other hand, the edges of E_4 incident on vertices of C that correspond to a clause that is not satisfied will contribute 0 to the cut, since all these vertices correspond to literals with the same truth value and are therefore on the same side of the partition. Thus, we get a total of $4t$ contributed to the cut, since $2t$ clauses are satisfied. Finally, once again because of the balancing of ϕ', exactly half of the edges of E_5 contribute to the cut: those incident on vertices of X that we placed in V_0, i.e. vertices that correspond to false literals. Since the weight of each such edge is 2, this adds up to a total contribution of $2\sum_i f_i$.

Thus, the total size of the cut is $6\sum_i f_i + 6\sum_i f_i + 2\sum_i f_i + 4t$. But, since every clause of ϕ had exactly three literals, $\sum_i f_i = 3m$. Therefore, the weight of the cut is $42m + 4t$, which is equal to $46m$ when the truth assignment satisfies every clause of ϕ.

Now for the other direction, suppose we are given a partition of V into V_0 and V_1. We will show that we can transform such a cut into a cut of the previous form, thus obtaining a truth assignment. First, observe that for any optimal cut

$A \subseteq V_0$ and $B \subseteq V_1$, because it is always optimal to place a source in V_0 and a sink in V_1. Now, suppose that in the cut we are given, for some i, $x_i, x_i' \in V_0$. Then place x_i' in V_1 and this will not make the cut smaller because now the edge (x_i, x_i') contributes to the cut and its weight is exactly as much as the weight of all other edges incident on x_i'. Also, if $x_i, x_i' \in V_1$ place x_i in V_0. This can not make the cut smaller, since the only edge lost is (a_i, x_i) and its weight is the same as that of (x_i, x_i') which now enters the cut. Therefore, we have now made sure that for all i, x_i and x_i' are on different sides of the partition, without decreasing the size of our cut.

Consider now a vertex $c_{i,j,j}$. We know that there exists an edge $(x_i, c_{i,j,j})$ (or an edge $(x_i', c_{i,j,j})$) of weight 2, which is as much as the weight of all other edges incident on $c_{i,j,j}$. Therefore, if x_i (resp. x_i') is in V_0, then we can place $c_{i,j,k}$ in V_1 without decreasing the size of the cut. Otherwise, we can place $c_{i,j,j}$ in V_0, because the edge of weight 2 can not be included in the cut by changing the side of $c_{i,j,j}$ only, and therefore placing it in V_0 is not worse because this way we may also include some of the other edges in the cut. This establishes that every vertex $c_{i,j,j}$ is on a different side of the partition from its predecessor in X.

Finally, consider a vertex $c_{i,j,k}$, $j \neq k$. If its predecessor in X is in V_0 we can place it in V_1 without decreasing the size of the cut, because then the edge of weight 2 is included. Otherwise we can place it in V_0, and this will include the edge $(c_{i,j,k}, b_i)$ in the cut. This does not decrease the size of the cut, since the edge of weight 2 was not included anyway, therefore we might at most lose the other edge incident on this vertex, which also has weight 1. This establishes that each of the remaining vertices of C is also on different a different side of the partition from its predecessor in X.

Now, observe that starting with any given cut, we have transformed it into a cut of a special form, without decreasing its size. From this cut we can construct a truth assignment: set to true the literals corresponding to vertices in X that we placed in V_1. This is a valid assignment, since exactly one of x_i, x_i' is in V_1. Also, if we repeat the process of the first direction of this reduction starting from this assignment we will get the same cut. Therefore, we have shown that there is a truth assignment that satisfies t of the m clauses of ϕ iff there is a cut in the DAG G of size at least $42m + 4t$. Thus,

$$OPT_{NAESAT}(\phi) = m \Rightarrow OPT_{MDC}(G) = 46m$$

$$OPT_{NAESAT}(\phi) < (1 - \epsilon)m \Rightarrow OPT_{MDC}(G) \leq (1 - \frac{2\epsilon}{23})46m \qquad \square$$

It is not hard to extend the results of the previous theorem to the cardinality version of MAX DI CUT, that is, the version where all edges have the same weight.

Theorem 3. *Cardinality* MAX DI CUT *is NP and APX-hard, even when restricted to DAGs.*

Proof. First, observe that all the edge weights used in the proof of Theorem 2 are polynomially (in fact linearly) bounded by the size of the original NAE3SAT formula. Thus, if we extend the problem's definition to include multigraphs, we can replace every edge of weight w by w parallel edges of weight 1. It is not hard to see that this does not affect the rest of the proof.

Now, let us show how to eliminate parallel edges. For each edge (u, v) introduce a directed path of length 3 u, w_1, w_2, v where w_1 and w_2 are new vertices. Observe that, if u is assigned 0 and v is assigned 1, then it is possible to include 2 of the 3 edges of the path in the cut, by assigning 0 to w_2 and 1 to w_1. However, any other assignment to u and v ensures that at most 1 of the three edges can be included in the cut, and in fact this is always possible by assigning 0 to w_1 and 1 to w_2. Thus, it is not hard to see that the reduction's arguments can now be applied with little modification.

Corollary 1. MAX DI CUT *is NP-hard and APX-hard even when restricted to graphs of bounded directed treewidth, DAG-width, kelly-width or directed pathwidth.*

Proof. The proof is immediate, because DAGs have width 1 under the definitions of all these widths.

5   Conclusions and Further Work

Discussion of Results. In this paper we have presented two hardness results affecting all known generalizations of treewidth to digraphs as well as directed pathwidth. It may be worthwhile at this point to discuss why such results hold for the directed cousins of treewidth, when in the undirected case there has been such a huge success.

First, the hardness result for MAX DI CUT, gives us one indication why such hardness results hold. The reason is simply that for some problems DAGs are not really an "easy" topology, as trees are in the undirected case. Therefore, it would probably make sense in the future to focus research on directed treewidth variants on generalizations of a graph topology that is even simpler than a DAG. A further clue is given in this direction by the fact that DAGs (surprisingly) are the base case for both directed pathwidth and the three treewidth variants we considered. One would probably expect pathwidth and treewidth to be based on different graph topologies.

On the other hand, directed treewidth variants have had some success with path-based problems, such as DIRECTED HAMILTONIAN CIRCUIT. For such problems, DAGs usually are indeed the trivial case and it makes sense to design a width as a generalization of DAGs. However, we showed that none of the currently known widths (including directed pathwidth) is restrictive enough to provide for an FPT algorithm for DIRECTED HAMILTONIAN CIRCUIT.

Therefore, we believe that our results may suggest that in the directed case things may be more complicated and possibly no "right" complexity measure exists. On one hand, it would probably make sense to explore the possibility of a width (not based on DAGs) that can solve MAX DI CUT and similar problems, while still being more general than undirected treewidth. And on the other hand, a more realistic goal might be to attempt to refine the definition of some of the already known widths (which are based on DAGs) in order to make it restrictive enough to solve DIRECTED HAMILTONIAN CIRCUIT and related path problems in FPT time.

References

1. Barát, J.: Directed path-width and monotonicity in digraph searching. Graphs and Combinatorics 22(2), 161–172 (2006)
2. Bodlaender, H.L.: Treewidth: Algorithmic techniques and results. In: Privara, I., Ružička, P. (eds.) MFCS 1997. LNCS, vol. 1295, pp. 19–36. Springer, Heidelberg (1997)
3. Bodlaender, H.L.: Treewidth: Characterizations, applications, and computations. In: Fomin, F.V. (ed.) WG 2006. LNCS, vol. 4271, pp. 1–14. Springer, Heidelberg (2006)
4. Bodlaender, H.L.: Treewidth: Structure and algorithms. In: Prencipe, G., Zaks, S. (eds.) SIROCCO 2007. LNCS, vol. 4474, pp. 11–25. Springer, Heidelberg (2007)
5. Courcelle, B.: The Monadic Second-Order Logic of Graphs. I. Recognizable Sets of Finite Graphs. Inf. Comput. 85(1), 12–75 (1990)
6. Downey, R.G., Fellows, M.R.: Parameterized Complexity. Springer, Heidelberg (1999)
7. Flum, J., Grohe, M.: Parameterized Complexity Theory, 1st edn. Texts in Theoretical Computer Science. An EATCS Series. Springer, Heidelberg (2006)
8. Hunter, P., Kreutzer, S.: Digraph measures: Kelly decompositions, games, and orderings. In: Bansal, N., Pruhs, K., Stein, C. (eds.) SODA, pp. 637–644. SIAM, Philadelphia (2007)
9. Johnson, T., Robertson, N., Seymour, P.D., Thomas, R.: Directed tree-width. J. Comb. Theory, Ser. B 82(1), 138–154 (2001)
10. Niedermeier, R.: Invitation to Fixed Parameter Algorithms. Oxford Lecture Series in Mathematics and Its Applications. Oxford University Press, USA (2006)
11. Obdržálek, J.: Dag-width: connectivity measure for directed graphs. In: SODA, pp. 814–821. ACM Press, New York (2006)
12. Papadimitriou, C.H., Yannakakis, M.: Optimization, approximation, and complexity classes. J. Comput. Syst. Sci. 43(3), 425–440 (1991)
13. Robertson, N., Seymour, P.D.: Graph minors. II. Algorithmic Aspects of Tree-Width. J. Algorithms 7(3), 309–322 (1986)

An Efficient Scaling Algorithm for the Minimum Weight Bibranching Problem

Maxim A. Babenko[*]

Moscow State University
`max@adde.math.msu.su`

Abstract. Let $G = (VG, AG)$ be a digraph and let $S \sqcup T$ be a bipartition of VG. A *bibranching* is a subset $B \subseteq AG$ such that for each node $s \in S$ there exists a directed s–T path in B and, vice versa, for each node $t \in T$ there exists a directed S–t path in B.

Bibranchings generalize both branchings and bipartite edge covers. Keijsper and Pendavingh proposed a strongly polynomial primal-dual algorithm that finds a minimum weight bibranching in $O(n'(m+n \log n))$ time (where $n := |VG|$, $m := |AG|$, $n' := \min(|S|, |T|)$).

In this paper we develop a weight-scaling $O(m\sqrt{n} \, \log n \log(nW))$-time algorithm for the minimum weight bibranching problem (where W denotes the maximum magnitude of arc weights).

1   Introduction

In a directed graph G, the sets of nodes and arcs are denoted by VG and AG, respectively. A similar notation is used for paths, cycles, and etc.

Consider a digraph G and a fixed bipartition $S \sqcup T$ of VG. A subset $B \subseteq AG$ is called a *bibranching* if the following conditions are met:

- for each $s \in S$, set B contains a directed path from s to some node in T;
- for each $t \in T$, set B contains a directed path from some node in S to t.

Bibranchings were introduced in [Sch82]. In the present paper we study the *minimum weight bibranching problem*, which is formulated as follows:

> (BB)   *Given G, S, T, and* arc weights $w \colon AG \to \mathbb{R}_+$, *find a bibranching B whose weight $w(B)$ is minimum.*

Hereinafter we assume that every real-valued function $f \colon U \to \mathbb{R}$ is additively extended to the family of subsets of U (denoted by 2^U) by $f(A) := \sum_{a \in A} f(a)$. In particular, $w(B)$ denotes the sum of weights of arcs in B.

Minimum weight bibranchings provide a common generalization to a pair of well-known combinatorial problems. Firstly, if $S = \{s\}$ then (BB) asks for a minimum weight s-*branching* (a directed tree rooted at s that covers all nodes of G). An $O(m + n \log n)$ algorithm for the latter task is known [GGSR86]

---

[*] Supported by RFBR grants 05-01-02803 and 06-01-00122.

S.-H. Hong, H. Nagamochi, and T. Fukunaga (Eds.): ISAAC 2008, LNCS 5369, pp. 232–245, 2008.

(here $n := |VG|$ and $m := |AG|$; we are assuming throughout the paper that $n \leq m \leq n^2$). Another special case arises when graph G only contains S–T arcs (but no T–S, S–S, or T–T arcs). Then the definition of a bibranching reduces to that of a *bipartite edge cover*. The minimum weight bipartite edge cover problem (call it (EC) for brevity) seems to be harder: no strongly polynomial $o(mn)$-algorithm is known so far. Problem (EC) can be solved by finding a maximum weight bipartite matching in $O(n'(m + n \log n))$ time [FT87] (where $n' := \min(|S|, |T|)$). Keijsper and Pendavingh [KP98] generalized the latter shortest-path augmentation method to solve (BB) in the same time bound.

On the other hand, many optimization problems may be solved faster if weights are known to be integral. The corresponding algorithms are based on scaling technique and achieve time bounds that are, in a sense, better than of their strongly-polynomial counterparts. A typical example is a scaling algorithm for bipartite matching problems [GT89], which runs in $O(m\sqrt{n} \log(nW))$ time (hereinafter W denotes the maximum magnitude of weights). The latter approach can also be adopted to solve (EC) and leads to an algorithm with the same time bound.

Similar ideas are also applicable to general min-cost flow problems [GT87]. However, when the structure of dual solutions becomes non-trivial (i.e. when one needs exponentially many dual variables) the algorithm and its complexity analysis may become much more involved. A good example is the minimum weight perfect matching problem in general (non necessarily bipartite) graphs, which is solved by Gabow and Tarjan in $O(m\sqrt{n\alpha(m,n)}\log n \, \log(nW))$ time [GT91].

Since (BB) involves solving a linear program with inequalities corresponding to all possible subsets of S and T, our approach is of no exception. We present an $O(m\sqrt{n} \, \log n \log(nW))$-time weight scaling algorithm for (BB). It is based on the general notion of ε-optimality (see, e.g., [GT87]), the augmentation procedure from [KP98], and attracts some additional combinatorial ideas to deal with dual solutions during scaling steps. Also, a variant of the blocking augmentation technique [Din80] is employed.

Note that the complexity of our algorithm coincides (up to a logarithmic factor) with that of the best known scaling algorithm [GT89] for solving a special case of (BB), namely (EC).

The rest of the paper is organized as follows. Section 2 gives the needed formal background and introduces the linear programming formulation of (BB). Section 3 and Section 4 outline a high-level description of the algorithm. Section 5 bounds the number of primal and dual steps performed by the algorithm. Due to lack of space some technical proofs and implementation details are omitted and will appear in the full version of the paper.

2   Preliminaries

First, we need some additional definitions and notation. Let G be a digraph and X be a subset of nodes. Then $\delta_G^{\text{in}}(X)$ (resp. $\delta_G^{\text{out}}(X)$ and $\gamma_G(X)$) denotes the set of arcs entering X (resp. leaving X and having both endpoints in X). When it is

234   M.A. Babenko

clear from the context which graph G is meant, it is omitted from the notation. Also, when $X = \{v\}$ we drop curly braces and write just $\delta^{\mathrm{in}}(v)$ and $\delta^{\mathrm{out}}(v)$.

For a graph G and a subset $X \subseteq VG$ let $G[X]$ denote the subgraph of G induced by X (i.e. the graph obtained from G by removing all nodes in $VG - X$).

Consider a bipartition $S \sqcup T$ of VG. For a subset $X \subseteq S$ we put $\delta(X) := \delta^{\mathrm{out}}(X)$, similarly for $X \subseteq T$ we put $\delta(X) := \delta^{\mathrm{in}}(X)$. If $a \in \delta(X)$ then arc a is said to *cover* X. For a set $A \subseteq AG$, the *set of nodes covered by A* is defined as the union of the sets of nodes covered by the individual elements of A. Clearly, (BB) prompts for a minimum weight subset $B \subseteq AG$ that covers every subset in $2^S \cup 2^T$.

Let us introduce an important notion of *graph contraction*. To *contract* a set $U \subseteq VH$ in a graph H means to replace nodes in U by a single *complex node* (also denoted by U). Arcs in $\gamma(VH - U)$ are not affected, arcs in $\gamma(U)$ are dropped, and arcs in $\delta^{\mathrm{in}}(U)$ (resp. $\delta^{\mathrm{out}}(U)$) are redirected so as to enter (resp. leave) the complex node U. The resulting graph is denoted by H/U. Note that this graph may contain multiple parallel arcs. We identify arcs in H/U with their pre-images in H. If H' is obtained from H by an arbitrary sequence of contractions, then $H' = H/U_1/\ldots/U_k$ for a certain family of disjoint subsets $U_1, \ldots, U_k \subseteq VH$ (called the *maximal contracted sets*). Each node in H' corresponds to a subset of nodes in H: nodes $v \in VH - (U_1 \cup \ldots \cup U_k)$ correspond to singletons $\{v\}$, complex nodes correspond to sets U_i.

For a set $A \subseteq AG$, we shall write A_{SS}, A_{ST}, and A_{TT} to denote the sets of S–S, S–T, and T–T arcs in A, respectively. Note that any minimum weight bibranching need not contain T–S arcs (as their removal preserves the required connectivity and can only decrease the total weight). Hence, we shall assume that graph G contains no T–S arcs.

We call a bibranching B *canonical* if every its proper subset $B' \subset B$ is not a bibranching. The following observations are easy:

Fact 1. *For each bibranching B there exists and can be found in $O(m)$ time a canonical bibranching $B' \subseteq B$.*

Fact 2. *Any canonical bibranching contains at most n arcs.*

Consider the following linear program:

$$
\text{(P)} \qquad
\begin{aligned}
&\textbf{minimize} && \textstyle\sum (w(a)x(a) : a \in AG) \\
&\textbf{subject to} && x \colon AG \to \mathbb{R}_+, \\
& && x(\delta(X)) \geq 1 \quad \text{for each } X \in 2^S \cup 2^T.
\end{aligned}
$$

The program dual of (P) is:

$$
\text{(D)} \qquad
\begin{aligned}
&\textbf{maximize} && \textstyle\sum \big(\pi(X) : X \in 2^S \cup 2^T\big) \\
&\textbf{subject to} && \pi \colon 2^S \cup 2^T \to \mathbb{R}_+, \\
& && w_\pi(a) \geq 0 \quad \text{for each } a \in AG.
\end{aligned}
$$

Here $w_\pi := w - \vartheta\pi$ are the *reduced weights* of arcs w.r.t. π, and the function $\vartheta\pi \colon AG \to \mathbb{R}_+$ is defined by

$$
\vartheta\pi(a) := \sum \big(\pi(X) : a \text{ covers } X\big).
$$

It is known [Sch03] that (P) describes the upper convex hull of the incidence vectors of all bibranchings in G. Hence, finding a bibranching of the minimum weight (under the assumption $w \geq 0$) amounts to finding a 0,1-solution to (P).

Weak duality for (P) and (D) implies that $\sum_a w(a)x(a) \geq \sum_X \pi(X)$ holds for every pair of admissible solutions x and π. By strong duality, the latter turns into equality if x and π are optimal. Moreover, (P) is known [Sch03] to be *totally dual integral*, that is, if all weights w are integers then there exists an optimal dual solution π that is integer. Hence, the polyhedron determined by (P) is integral.

The complementary slackness conditions for (P) and (D) (giving an optimality criterion for solutions x and π to these programs) are viewed as:

(1)   if $x(a) > 0$ for $a \in AG$ then $w_\pi(a) = 0$;

(2)   if $\pi(X) > 0$ for $X \in 2^S \cup 2^T$ then $x(\delta(X)) = 1$.

For a set $B \subseteq AG$ and a function $\pi \colon 2^S \cup 2^T \to \mathbb{R}_+$ we say that B is π-*consistent* if $\pi(X) > 0$ implies $|B \cap \delta(X)| \leq 1$ for each $X \in 2^S \cup 2^T$. Consistency is closely related to the complementary slackness conditions, in particular, if B is a bibranching then its π-consistency is just (2).

Fact 3.   *For an arbitrary function* $\pi \colon 2^S \cup 2^T \to \mathbb{R}_+$ *and a set* $B \subseteq AG$ *one has*

$$\vartheta\pi(B) \geq \sum (\pi(X) : B \ covers \ X).$$

Additionally, if B *is* π-*consistent then the above inequality turns into equality.*

3   Algorithm

Recall that a family $\mathcal{F}$ of subsets is called *laminar* if for any $X, Y \in \mathcal{F}$ one either has $X \subseteq Y$, or $Y \subseteq X$, or $X \cap Y = \emptyset$. Also, let $\mathrm{supp}\,(f)$ denote the *support set* of a function f, i.e. $\{x \mid f(x) \neq 0\}$.

The algorithm maintains a laminar family $\mathcal{F} \subseteq 2^S \cup 2^T$ and a function $\pi \colon 2^S \cup 2^T \to \mathbb{Z}_+$. For $X \in \mathcal{F}$, the *shell* $S(X)$ of X is the graph obtained from $G[X]$ by contracting all proper maximal subsets $Y \subset X$, $Y \in \mathcal{F}$. Let $\overline{G}$ denote the graph obtained by contracting all maximal sets of $\mathcal{F}$ in G. Put $\overline{S}$ (resp. $\overline{T}$) to be the image of S (resp. T) in $\overline{G}$ under these contractions. Let $S_\pi^0(X)$ denote the subgraph of $S(X)$ consisting of arcs a with $w_\pi(a) = 0$.

For a set of nodes Y in $\overline{G}$ (or in $S(X)$ for $X \in \mathcal{F}$) we write $\widetilde{Y}$ to denote the corresponding pre-image subset in VG. When $Y = \{y\}$ we write just $\widetilde{y}$ instead of $\widetilde{\{y\}}$.

We introduce the following set of properties:

(D1) $\mathcal{F}$ is laminar and $\mathrm{supp}\,(\pi) \subseteq \mathcal{F}$;
(D2) $S_\pi^0(X)$ is strongly connected for each $X \in \mathcal{F}$;
(D3) $w_\pi(a) \geq 0$ for all $a \in AG$.

Property (D3) is just the feasibility of a dual solution π while (D1) and (D2) introduce some additional structural requirements for π.

Next, the algorithm maintains a subset $\overline{B} \subseteq A\overline{G}$, which will be referred to as a *partial bibranching*. Consider the following properties:

(P1) set $\overline{B}_{SS}$ (resp. $\overline{B}_{TT}$) forms a directed out- (resp. in-) forest in $G[\overline{S}]$ (resp. in $G[\overline{T}]$), each arc in $\overline{B}_{ST}$ covers a pair of roots of the said forests;

(P2) if $a \in \overline{B}_{SS} \cup \overline{B}_{TT}$ then $w_\pi(a) = 0$; if $a \in \overline{B}_{ST}$ then $w_\pi(a) \leq 1$;

(P3) if $v \in V\overline{G}$ is covered by more than one arc in $\overline{B}$ then $\pi(\widetilde{v}) = 0$ (c.f. (2)).

Here by an *in-* (resp. *out-*) *forest* we mean an acyclic set of arcs X such that for each node v at most one arc in X enters (resp. leaves) v. If no arc in X enters (resp. leaves) node v then v is said to be a *root* of X.

Property (P1) is required mostly by technical reasons that will become evident later. Property (P2) may be regarded as a relaxation of (1) and (P3) directly corresponds to (2).

The algorithm employs *bit scaling* and works as follows. Let $w_0 \colon AG \to \mathbb{Z}_+$ denote the input weight function. The algorithm starts with $w := 0$, $\pi := 0$, $\mathcal{F} := \emptyset$ and puts $\overline{B}$ to be an arbitrary bibranching in G obeying (P1) (the latter can be found in $O(m)$ time by an obvious routine). In case $\overline{B}$ does not exist, the algorithm halts.

Each scaling step takes a weight function w from the previous iteration, a bibranching $\overline{B} \subseteq A\overline{G}$ in $\overline{G}$, a function π, and a collection $\mathcal{F}$ altogether obeying properties (D1)–(D3), (P1)–(P3). The weights $w(a)$ are doubled and some of them are increased by 1 (namely, those having 1 at the corresponding position of the binary representation of $w_0(a)$). Changing weights w may lead to violation of the above properties so the goal of the scaling step is to restore them. The necessary details will be given in Section 4.

Lemma 1. *Suppose that the current scaling step is complete, so properties (D1)–(D3), (P1)–(P3) hold for $\mathcal{F}$, π, and $\overline{B}$. Also, let $\overline{B}$ be a bibranching in $\overline{G}$. Then there exists a canonical π-consistent bibranching B in G such that: (i) $w(B) \leq w(\overline{B})$, (ii) $w_\pi(a) = 0$ for each $a \in B_{SS} \cup B_{TT}$; and (iii) $w_\pi(a) \leq 1$ for each $a \in B_{ST}$.*

Proof. Firstly, with the help of Fact 1 set $\overline{B}$ is turned into a canonical bibranching in $\overline{G}$ by removing some arcs. Next, let $X \in \mathcal{F}$ be one of the maximal contracted sets in $V\overline{G}$. We describe procedure EXPAND(X) that extends $\overline{B}$ into the shell $S(X)$. For simplicity's sake suppose $X \subseteq S$, the other case is symmetric. Remove X from $\mathcal{F}$ thus partly uncontracting graph $\overline{G}$. Let X' denote the set of nodes in new $\overline{G}$ arising from X during this uncontraction. Let R denote the set of nodes in X' that are covered by arcs in $\overline{B}$. One has $R \neq \emptyset$ since $\overline{B}$ was a bibranching before set X got removed from $\mathcal{F}$. We grow an out-forest F in the subgraph $\overline{G}[X']$ such that: (i) R is the set of roots of F; (ii) $VF = X'$; (iii) $w_\pi(a) = 0$ holds for each $a \in AF$. Property (D2) shows that this forest always exists. Now we add the arcs of F to $\overline{B}$. Clearly, the new set $\overline{B}$ is a canonical bibranching in $\overline{G}$.

Applying EXPAND to the elements of $\mathcal{F}$ (in an appropriate order) one gets a bibranching B in G that obeys the required properties.

Weights w are iteratively scaled, as explained above, until achieving the equality $w = w_0$. Totally it takes $\lceil \log W \rceil$ scaling steps. Next, we put $t := \lfloor \log n \rfloor + 1$ and perform t additional scaling steps, doubling w each time. Finally, the algorithm applies Lemma 1 to construct the final bibranching in G.

Let us prove that this general scheme is correct. Put $\Pi := \sum_X \pi(X)$ and estimate the weight of an optimal bibranching as follows.

Lemma 2. *Property (D3) implies $w(B) \geq \Pi$ for any bibranching B in G.*

Proof. By definition B covers each subset in $2^S \cup 2^T$. Hence, by Fact 3 and (D3) one has $w(B) = w_\pi(B) + \vartheta\pi(B) \geq \sum_X \pi(X) = \Pi$, as required.

Lemma 3. *If B is a canonical bibranching in G and properties (P2) and (P3) hold (for $\overline{G} := G$, $\overline{B} := B$) then $w(B) \leq \Pi + n$.*

Proof. By Fact 2, Fact 3, and property (P2) one has $w(B) = w_\pi(B) + \vartheta\pi(B) \leq n + \sum_X \pi(X) = n + \Pi$.

Theorem 4. *The algorithm constructs a minimum weight bibranching.*

Proof. Let $\overline{B}$, w, and π denote the corresponding objects after the last scaling step. Put B to be a canonical bibranching obtained from $\overline{B}$ by Lemma 1. Let B_{min} be a minimum weight bibranching (w.r.t. w or, equivalently, w_0). One may assume by Fact 1 that B_{min} is canonical. Lemma 2 and Lemma 3 imply that $w(B_{min}) \geq \Pi$ and $w(B) \leq \Pi + n$, so $w(B_{min}) \leq w(B) \leq w(B_{min}) + n$. Recall, each of the last t scaling steps doubles arc weights. Since all initial weights w_0 are integers, $w(a)$ is divisible by 2^t for each $a \in AG$. Hence, so is $w(B)$. The choice of t implies $n < 2^t$, therefore $w(B) = w(B_{min})$, so B is optimal.

4   Scaling Step

Each scaling step consists of the following four stages: doubling stage, shell stage, ST stage, and TS stage.

First, the **doubling stage** is executed: arc weights w are multiplied by 2 and some of them are increased by 1, as described in Section 3. Also, duals π are doubled. Put $\mathcal{F} := \mathrm{supp}\,(\pi)$ and $\overline{B} := \emptyset$ (hence, any previous bibranching is discarded). As earlier, let $\overline{G}$ denote the graph obtained from G by contracting all maximal sets in $\mathcal{F}$. Obviously, properties (D1), (D3), (P1)–(P3) now hold. One needs to solve the following two tasks:

- restore property (D2) by ensuring that graphs $S_\pi^0(X)$, $X \in \mathcal{F}$, are strongly connected;
- construct a bibranching $\overline{B}$ in $\overline{G}$ obeying properties (P1)–(P3).

The **shell stage** is executed to deal with (D2). The algorithm scans the sets in $\mathcal{F}$ choosing an inclusion-wise minimal unscanned set at each iteration. Let $X \in \mathcal{F}$ be the current set to be scanned. Procedure NORMALIZE-SHELL(X) is called to adjust duals π and ensure (D2) for X or remove X from $\mathrm{supp}\,(\pi)$ (and hence also from $\mathcal{F}$).

Suppose $X \subseteq S$, the case $X \subseteq T$ is analogous. NORMALIZE-SHELL performs a series of iterations similarly to the minimum weight branching algorithm [Edm67, GGSR86].

More precisely, it maintains a directed out-forest F_S containing all nodes of X and consisting of some arcs a with $w_\pi(a) = 0$. Initially $VF_S := VS(X)$ and $AF_S := \emptyset$. If $S(X)$ is a single node graph then property (D2) is restored for X, NORMALIZE-SHELL(X) terminates. Otherwise, an arbitrary tree W in F_S is picked. Let r be the root of W. Suppose that all arcs leaving r (in $S(X)$) have positive reduced weights. Put

$$\mu_1 := \min\left(w_\pi(a) : a \in \delta^{out}_{S(X)}(r)\right), \quad \mu_2 := \pi(X), \quad \mu := \min(\mu_1, \mu_2).$$

Adjust the duals as follows:

$$\pi(\widetilde{r}) \ := \pi(\widetilde{r}) \ + \mu,$$
$$\pi(X) := \pi(X) - \mu.$$

These adjustments decrease the reduced weights of all arcs in $S(X)$ leaving r by μ. Also, $\widetilde{r}$ is added to $\mathrm{supp}\,(\pi)$. By the choice of unscanned sets for NORMALIZE-SHELL, property (D2) holds for $\widetilde{r}$ and all its subsets in $\mathcal{F}$. Set $\widetilde{r}$ is also marked as scanned, so NORMALIZE-SHELL is never called for it.

If $\pi(X) = 0$ holds after the adjustment then set X vanishes from $\mathrm{supp}\,(\pi)$, we remove X from $\mathcal{F}$ and halt NORMALIZE-SHELL(X). We also say that set X *dissolves* during the execution of the shell stage.

Now suppose that there is an arc $a \in \delta^{out}_{S(X)}(r)$ such that $w_\pi(a) = 0$. Two cases are possible. Firstly, a may connect W with another tree W' in F_S. Then, a is added to F_S thus linking W and W'. Secondly, a may connect r to a node in the very same tree W. In this case, a cycle of arcs with zero reduced weights is discovered. Let Y be the set of nodes of this cycle (in $S(X)$). Algorithm contracts Y in $S(X)$, adds $\widetilde{Y}$ to $\mathcal{F}$, and proceeds to the next iteration.

Once NORMALIZE-SHELL(X) is called for all sets $X \in \mathcal{F}$ (in an appropriate order), property (D2) gets restored. Normalization procedure for a subset $X \subseteq T$ is the same except for it considers sets δ^{out} rather than δ^{in}.

The remaining part of the scaling step builds a bibranching $\overline{B}$ in $\overline{G}$ that satisfies properties (P1)–(P3). Firstly, the **ST stage** is executed. It starts with $\overline{B} = 0$ and applies a certain augmenting path approach aiming to update $\overline{B}$ so that it covers all subsets in $2^{\overline{S}}$. Next, S and T parts are exchanged and a similar **TS stage** is executed, thus completing the scaling step. We shall only describe the ST stage since the TS stage is essentially symmetric.

Similarly to the shell stage, a directed out-forest F_S obeying $VF_S = \overline{S}$ is maintained in graph $\overline{G}$. The latter forest satisfies the following conditions:

(F1) $w_\pi(a) = 0$ holds for all $a \in AF_S$;
(F2) $w_\pi(a) > 0$ holds for each root node $r \in \overline{S}$ and an $\overline{S}$–$\overline{S}$ arc a leaving r;
(F3) $\overline{B}_{SS} \subseteq AF_S$;
(F4) if a node $v \in \overline{S}$ is not covered by $\overline{B}$ then v is a root of F_S.

Forest F_S is initially constructed by putting $VF_S := \overline{S}$ and $AF_S := \overline{B}_{SS}$ (the latter set forms a directed out-forest according to (P1)). Next, NORMALIZE-FOREST routine is applied to ensure (F2) and (F4). The latter works as follows. If (F2) fails for a root node r and an $\overline{S}$–$\overline{S}$ arc a leaving r then two cases are possible. Firstly, a may connect a tree W of F_S rooted at r with another tree W' in F_S. Then, a is added to F_S thus linking W and W'. Secondly, a may connect r to a node in the very same tree W. In this case, a cycle consisting of arcs with zero reduced weights is discovered. Let Y denote the set of nodes of this cycle (in $\overline{G}$). The algorithm puts $\overline{B} := \overline{B} \setminus \gamma(Y)$, $AF_S := AF_S \setminus \gamma(Y)$, contracts Y in $\overline{G}$, and adds $\widetilde{Y}$ to $\mathcal{F}$. Note that at this point $\pi(\widetilde{Y}) = 0$ holds.

Next, if (F4) fails for a node $v \in \overline{S}$, then v is not a root of F_S and v is not covered by $\overline{B}$. To fix this, an arc $a \in AF_S$ that covers v is fetched and added to $\overline{B}$. Property (F1) implies the validity of (P2). Also, v is covered by exactly one arc in $\overline{B}$ after the augmentation, so (P3) holds.

This completes the description of NORMALIZE-FOREST.

Once forest F_S obeying (F1)–(F4) is constructed, the algorithm builds an auxiliary digraph H. Put $VH := V\overline{G}$ and proceed as follows:

- if $a \notin \overline{B}$ is an $\overline{S}$–$\overline{T}$ arc with $w_\pi(a) = 0$ then add a to H (these arcs are called *forward*);
- if $a \in \overline{B}$ is an $\overline{S}$–$\overline{T}$ arc with $w_\pi(a) = 1$ then add a to H but change its direction to the opposite (these arcs are called *backward*).

A node $v \in \overline{S}$ is said to be *initial* if $\overline{B}$ does not cover v. By (F4) only root nodes of F_S may be initial. A node v is called *final* if any of the following cases applies:

1. $v \in \overline{T}$ and $\pi(\widetilde{v}) = 0$;
2. $v \in \overline{T}$ and v is covered by a $\overline{T}$–$\overline{T}$ arc of $\overline{B}$ (the latter is unique by (P1));
3. $v \in \overline{T}$ and v is not covered by $\overline{B}$;
4. $v \in \overline{S}$ and v is an inner node of F_S;
5. $v \in \overline{S}$ and v is covered by at least two $\overline{S}$–$\overline{T}$ arcs of $\overline{B}$.

A path in H connecting an initial node to a final node (with all intermediate nodes neither initial nor final) is called *augmenting*. Suppose for a moment that there exists an augmenting path P in H from an initial node s to a final node t. In this case, a **primal step** is possible. We construct a set $A(P) \subseteq AH$ as follows. First, we add arcs that correspond to arcs of P (both forward and backward). Second, we perform adjustments to account for the type of node t. In cases (1), (3), and (5) we do nothing. In case (2) we add to $A(P)$ the unique arc in $\overline{B}_{TT}$ that covers t. In case (4) we add to $A(P)$ the unique arc in F_S that leaves t.

The *augmentation* of $\overline{B}$ along P is performed by putting $\overline{B} := \overline{B} \triangle A(P)$ (here $\triangle$ denotes the symmetric difference).

Lemma 4. *The augmentation of $\overline{B}$ along P preserves properties (P1)–(P3), (D1)–(D3), and (F1)–(F4). The set of nodes covered in $V\overline{G}$ by $\overline{B}$ strictly increases. The set of initial nodes strictly decreases. The set of final nodes does not increase. The arc set of H decreases by AP.*

The proof is carried out by a straightforward case-splitting, so we leave details to the reader.

Note, that in order to achieve the desired time bound one cannot recompute path P from scratch each time. Taking into account the monotonicity of the sets of initial and final nodes, and the arc set of H, the *blocking augmentation* technique is applied (see, e.g., [Din80]). The latter computes, one by one, a sequence of augmenting paths and stops when there are no such paths left. The total running time of this routine is $O(m)$. Each of these paths is used for augmenting $\overline{B}$, as explained above.

A **dual step** is carried out when no more augmenting paths can be found. Let $\overline{S}_0$ (resp. $\overline{T}_0$) denote the set of nodes in $\overline{S}$ (resp. $\overline{T}$) that are reachable from initial nodes. Calculate the value of the *adjustment parameter* μ as follows:

$$
\begin{aligned}
\mu_1 &:= \min\left(\pi(\widetilde{v}) : v \in \overline{T}_0\right), \\
\mu_2 &:= \min\left(w_\pi(a) : a = (u,v) \in A\overline{G},\ u \in \overline{S}_0,\ v \in \overline{S}\right), \\
\mu_3 &:= \min\left(w_\pi(a) : a = (u,v) \in A\overline{G},\ u \in \overline{S}_0,\ v \in \overline{T} - \overline{T}_0\right), \\
\mu_4 &:= \min\left(1 - w_\pi(a) : a = (u,v) \in \overline{B},\ u \in \overline{S} - \overline{S}_0,\ v \in \overline{T}_0\right), \\
\mu &:= \min(\mu_1, \mu_2, \mu_3, \mu_4).
\end{aligned}
$$

Clearly, (P2) implies that $\mu \geq 0$. Also, $\mu_4 \in \{0, 1, \infty\}$ (moreover, case $\mu_4 = 0$ is not possible, see Lemma 5 below).

A *dual update* is performed as follows:

$$
\begin{aligned}
\pi(\widetilde{v}) &:= \pi(\widetilde{v}) + \mu \quad \text{for each } v \in \overline{S}_0, \\
\pi(\widetilde{v}) &:= \pi(\widetilde{v}) - \mu \quad \text{for each } v \in \overline{T}_0.
\end{aligned}
$$

If the dual $\pi(\widetilde{v})$ of some complex node $v \in \overline{T}_0$ drops to zero then the algorithm calls EXPAND($\widetilde{v}$) to remove $\widetilde{v}$ from $\mathcal{F}$, uncontract (partly) graph $\overline{G}$, and extend $\overline{B}$ into the shell of $\widetilde{v}$.

Also, suppose $w_\pi(a) = 0$ holds for some $a = (u,v) \in A\overline{G}$, $u \in \overline{S}_0$, $v \in \overline{S}$ after the update. Then, node u must be a root of F_S and property (F2) fails. Procedure NORMALIZE-FOREST is called to restore it.

One can see the following:

Lemma 5. *If no augmenting path exists in H then $0 < \mu < \infty$. The dual step preserves properties (P1)–(P3), (D1)–(D3), and (F1)–(F4), does not change the set of initial nodes, and does not decrease the set of reachable nodes.*

If an augmenting path arises after these changes, the dual step completes and a primal step is executed. Otherwise, the next value of μ is calculated and the process of changing π proceeds.

The algorithm changes $\overline{B}$, π, and $\mathcal{F}$ by executing primal and dual steps alternatively. It stops when $\overline{B}$ covers all nodes in $\overline{S}$. Then, parts $\overline{S}$ and $\overline{T}$ are exchanged and the TS stage runs until $\overline{B}$ covers both $\overline{S}$ and $\overline{T}$. This way, the requested bibranching $\overline{B}$ is constructed, the scaling step completes.

5   Complexity Analysis

In this section we present a sketch of the efficiency analysis. Our immediate goal is to prove an $O(\sqrt{n})$ bound for the number of primal and dual steps during each scaling step. Hereinafter w, π_0, $\mathcal{F}_0$, and $\overline{G}_0$ denote the corresponding objects after the doubling stage. Similarly, we use notation π_1, $\mathcal{F}_1$, and $\overline{G}_1$ when referring to the state immediately after the shell stage.

Lemma 6.   *There exists a canonical π_1-consistent bibranching $B_1 \subseteq AG$ obeying $w_{\pi_1}(B_1) \leq 6n$.*

Proof.   Let $\overline{B}_0$ be a bibranching in $\overline{G}_0$ that was constructed by the previous scaling step (or just an arbitrary bibranching in G for the first invocation of the scaling step). By removing an appropriate set of arcs from $\overline{B}_0$, one may assume that $\overline{B}_0$ is canonical, see Fact 1. Property (P2) and the structure of the doubling stage imply that $w_{\pi_0}(a) \leq 3$ holds for each $a \in \overline{B}_0$.

We now gradually transform graph $\overline{G}_0$ into $\overline{G}_1$ and, simultaneously, $\overline{B}_0$ into a bibranching $\overline{B}_1$ in $\overline{G}_1$ such that $w_{\pi_1}(\overline{B}_1)$ is small and (P3) holds for $\overline{B} := \overline{B}_1$, $\pi := \pi_1$. We denote the current graph by $\overline{G}$ and the current bibranching by $\overline{B}$. Initially, put $\overline{G} := \overline{G}_0$ and $\overline{B} := \overline{B}_0$. The difference between $\overline{G}_0$ and $\overline{G}_1$ is that some maximal sets in supp(π_0) could have dissolved during the shell stage. The corresponding dissolved nodes, therefore, are replaced by certain subgraphs.

We enumerate the nodes of $\overline{G}_0$, let v be the current one. If $\widetilde{v} \notin$ supp(π_0) then v is a simple node ($\widetilde{v} = \{v\}$), it remains simple in $\overline{G}_1$. No change is applied to $\overline{G}$ and $\overline{B}$. Note that the reduced weights of arcs covering v are not changed during the shell stage and, thus, do not exceed 3.

Next, suppose $\widetilde{v} \in$ supp(π_0). We assume that $\widetilde{v} \subseteq S$ (the other case is symmetric). Property (P3) implies that in $\overline{G}_0$ node v is covered by a unique arc, say $a \in \overline{B}_0$. Let the tail of arc a in G be w. Applying (D2) iteratively, we construct an out-tree W in G such that: (i) W is rooted at w; (ii) $VW = \widetilde{v}$; (iii) if $X \in$ supp(π_0) and $X \subseteq \widetilde{v}$ then X is covered by exactly one arc in $A := AW \cup \{a\}$; (iv) every arc $a \in AW$ was of zero reduced weight prior to the doubling stage. Clearly, $w_{\pi_1}(a) \leq 3$ holds for every $a \in A$.

Update $\overline{G}$ and $\overline{B}$ as follows. First, uncontract $\widetilde{v}$ completely and add AW to $\overline{B}$. Since node w is reachable from every node in $\widetilde{v}$ by arcs in AW, it follows that $\overline{B}$ remains a bibranching. Next, contract the maximal sets $X \in$ supp(π_1) such that $X \subseteq \widetilde{v}$ and update $\overline{B}$ accordingly. Let $\overline{v}$ denote the image of $\widetilde{v}$ under these contractions. (It is possible that the whole set $\widetilde{v}$ gets contracted again, this happens when $\widetilde{v} \in$ supp(π_1); in this case $\widetilde{v}$ did not dissolve during the call Normalize-Shell$(\widetilde{v})$.)

The above contractions may remove some arcs from $\overline{B}$ (more precisely, exactly those arcs whose head and tail nodes are simultaneously contained in the same maximal contracted set). However, $\overline{B}$ remains a bibranching since contraction of an arbitrary subset of S- or T-part of the graph preserves the required connectivity. Finally, we apply Fact 1 and remove all redundant arcs from $\overline{B}$ (in an arbitrary way) turning it into a canonical bibranching.

Recall that initially v was covered by the unique arc a. Now v is expanded into some set of nodes, namely $\overline{v}$, and some arcs from $\gamma(\widetilde{v})$ are added to $\overline{B}$. Canonicity implies that every node in $\overline{v}$ is covered by a unique arc from $\overline{B}$. This way, (P3) follows for $\overline{B}$.

Let us estimate the total reduced weight of all newly added arcs in $\overline{B}$. To this aim, we bound $w_{\pi_1}(A)$ (since $\overline{B}$ receives some subset of arcs from A and reduced weights of the omitted arcs are non-negative). We consider the following two subfamilies of $\operatorname{supp}(\pi_0)$ and $\operatorname{supp}(\pi_1)$:

$$\mathcal{F}_0 := \{X \in \operatorname{supp}(\pi_0) \mid X \subseteq \widetilde{v}\}, \qquad \mathcal{F}_1 := \{X \in \operatorname{supp}(\pi_1) \mid X \subseteq \widetilde{v}\}.$$

During NORMALIZE-SHELL, each time the dual variable corresponding to a set $X \in 2^S$ (resp. $X \in 2^T$) is increased by μ, the dual variable corresponding to some other set $Y \in 2^S$ (resp. $Y \in 2^T$) is decreased by the same value μ. Hence,

$$\sum \left(\pi_0(X) : X \in \mathcal{F}_0\right) = \sum \left(\pi_1(X) : X \in \mathcal{F}_1\right).$$

Also, by Fact 3 it follows that

$$w_{\pi_0}(A) = w(A) - \vartheta\pi_0(A) = w(A) - \sum\left(\pi_0(X) : X \in \mathcal{F}_0\right),$$
$$w_{\pi_1}(A) = w(A) - \vartheta\pi_1(A) \le w(A) - \sum\left(\pi_1(X) : X \in \mathcal{F}_1\right).$$

Therefore,

$$w_{\pi_1}(A) \le w_{\pi_0}(A) \le 3\,|A| = 3\,|\widetilde{v}|.$$

The above procedure is applied to each node $v \in V\overline{G}_0$ and eventually stops with $\overline{G} = \overline{G}_1$. The final set $\overline{B}$ is denoted by $\overline{B}_1$. Let us estimate its reduced weight $w_{\pi_1}(\overline{B}_1)$. First, $\overline{B}_1$ gets at most n arcs that cover simple nodes in $\overline{G}_0$; each of those arcs has a reduced weight not exceeding 3. Second, each complex node $v \in V\overline{G}_0$ generates a set of arcs with total reduced weight not exceeding $3\,|\widetilde{v}|$. Summing these bounds, one gets:

$$w_{\pi_1}(\overline{B}_1) \le 3n + 3n = 6n.$$

Finally, to get the desired bibranching B_1 in G we apply EXPAND routine and extend $\overline{B}_1$ into the maximal contracted sets of $\overline{G}_1$. This step only adds arcs of zero reduced weight w_{π_1}. Hence, $w_{\pi_1}(B_1) = w_{\pi_1}(\overline{B}_1) \le 6n$ holds.

Lemma 7. *Let* $\pi\colon 2^S \cup 2^T \to \mathbb{R}_+$ *be an arbitrary function and* $B \subseteq AG$ *be an arbitrary* π-*consistent arc set satisfying properties (D3) and (P2) (for* $\overline{G} := G$, $\overline{B} := B$*). Then*

$$\Delta(\pi, \pi_1, B) := \sum\left(\pi(X) - \pi_1(X) : X \text{ is not covered by } B\right) \le 6n + w_\pi(B).$$

Proof. Consider the duals π_1 (at the moment just before the ST stage) and the canonical bibranching B_1 constructed in Lemma 6. Put

$$Q := \sum_{a \in AG} \left(\chi^{B_1}(a) - \chi^B(a)\right)\left(w_{\pi_1}(a) - w_\pi(a)\right).$$

(Here χ^U denotes the *incidence vector* of an arc set U, i.e. the function that equals 1 on U and 0 on $AG - U$.)

Taking into account equalities $w_\pi = w - \vartheta\pi$ and $w_{\pi_1} = w - \vartheta\pi_1$ one gets

$$Q = \sum_{a \in AG} \left(\chi^{B_1}(a) - \chi^B(a)\right)\left(\vartheta\pi(a) - \vartheta\pi_1(a)\right) =$$
$$= \vartheta\pi(B_1) + \vartheta\pi_1(B) - \vartheta\pi_1(B_1) - \vartheta\pi(B).$$

Since B is π-consistent and B_1 is π_1-consistent and covers each set in $2^S \cup 2^T$, Fact 3 implies

$$Q \geq \sum\left(\pi(X) : X \text{ is covered by } B_1\right) + \sum\left(\pi_1(X) : X \text{ is covered by } B\right) -$$
$$- \sum\left(\pi_1(X) : X \text{ is covered by } B_1\right) - \sum\left(\pi(X) : X \text{ is covered by } B\right) =$$
$$= \sum\left(\pi(X) : X \text{ is not covered by } B\right) -$$
$$- \sum\left(\pi_1(X) : X \text{ is not covered by } B\right) =$$
$$= \sum\left(\pi(X) - \pi_1(X) : X \text{ is not covered by } B\right).$$

On the other hand,

$$Q = \sum_{a \in AG} \left(\chi^{B_1}(a) - \chi^B(a)\right)\left(w_{\pi_1}(a) - w_\pi(a)\right) \leq$$
$$\leq \sum_{a \in AG} \chi^{B_1}(a)w_{\pi_1}(a) + \chi^B(a)w_\pi(a) \leq$$
$$\leq w_{\pi_1}(B_1) + w_\pi(B) \leq 6n + w_\pi(B).$$

Now the claim follows by transitivity.

Lemma 8. *Each scaling step executes $O(\sqrt{n})$ primal and dual steps.*

Proof. Let $\overline{B} \subseteq A\overline{G}$ denote the current partial bibranching in the current graph $\overline{G}$ at some intermediate moment during the ST or the TS stage.

Firstly, we prove that $w_\pi(\overline{B}) \leq n$. Indeed $w_\pi(\overline{B})$ does not exceed the number of $\overline{S}$–$\overline{T}$ arcs in $\overline{B}$ (by property (P2)). The latter can only increase by 1 on each primal step. The total number of primal steps does not exceed n (since each of these steps increases the set of covered nodes).

Next, we proceed similarly to Lemma 1, apply EXPAND routine, and extend $\overline{B}$ to a π-consistent set $B \subseteq AG$. This set obeys $w_\pi(B) = w_\pi(\overline{B}) \leq n$.

Consider the sum $\Delta(\pi, \pi_1, B)$. Let X be a set such that $\pi(X) < \pi_1(X)$. It follows from the structure of the algorithm that B covers X. Indeed, when the dual $\pi(\widetilde{v})$ decreases (for $v \in V\overline{G}$), $\overline{B}$ covers v for otherwise v is a reachable final node of type (3) and the dual adjustment is not possible. Also, if $\overline{B}$ covers v then B covers $\widetilde{v}$ and all its subsets. Therefore, all terms in $\Delta(\pi, \pi_1, B)$ are non-negative.

Let us deal with the positive terms. Consider the ST stage. Initially, all nodes in $\overline{S}$ are not covered by $\overline{B}$. Then, during the course of the algorithm the number of uncovered nodes in $\overline{S}$ decreases. Let k dual steps be performed so far. For each i ($1 \leq i \leq k$) let $D_1^i, \ldots, D_{l_i}^i$ be the subsets of S that correspond to all uncovered nodes $d_1^i, \ldots, d_{l_i}^i$ in $\overline{S}$ during the i-th dual step. For each fixed i, the sets $D_1^i, \ldots, D_{l_i}^i$ are pairwise disjoint. Altogether, these sets form a laminar family

$$\mathcal{D} := \left\{ D_j^i \mid 1 \leq i \leq k, 1 \leq j \leq l_i \right\}.$$

Each uncovered node d_j^i was initial (at the corresponding moment of time) and, hence, reachable. Therefore, $\pi(D_j^i)$ was increased during the corresponding dual step.

Moreover, for each set D_j^i, $i < k$, $1 \leq j \leq l_i$, there are exactly two possibilities: (i) node d_j^i gets covered during the upcoming primal step and, hence, $D_{j'}^{i'} \cap D_j^i = \emptyset$ for each $i' > i$ and $1 \leq j' \leq l_{i'}$; or (ii) node d_j^i gets incorporated into some contracted set at the next dual step, hence, $D_j^i \subseteq D_{j'}^{i+1}$ for some $1 \leq j' \leq l_{i+1}$.

This way, sets D_j^i form a forest F and for each its tree H with root D_j^i the leafs of H (these are some sets of the form D_j^1, $1 \leq j \leq l_1$) are on the same depth i, which is equal to the height of H. Let $H_1, \ldots, H_{l_k}$ denote the set of trees in F rooted at $D_1^k, \ldots D_{l_k}^k$. All these trees are of height k. Also, for each $1 \leq i \leq k$ and $1 \leq j \leq l_k$ there exists a set D_j^i in tree H_j such that B does not cover D_j^i. Therefore, each tree H_j adds a positive term of value $\sum \mu$ to $\Delta(\pi, \pi_1, B)$ (where $\sum \mu$ denotes the sum of the dual adjustments μ performed by the algorithm). Now summing over all trees $H_1, \ldots, H_{l_k}$ one gets

$$\Delta(\pi, \pi_1, B) \geq \sum \mu \cdot l_k, \tag{3}$$

On the other hand, by Lemma 7 it follows that

$$\Delta(\pi, \pi_1, B) \leq 6n + w_\pi(B) \leq 7n \tag{4}$$

Since each dual step changes duals by at least 1, we conclude that $k \cdot l_k \leq 7n$. Hence, after $\lceil \sqrt{n} \rceil$ dual steps at most $O(\sqrt{n})$ nodes in $\overline{S}$ may remain uncovered. To cover these remaining nodes $O(\sqrt{n})$ primal steps are sufficient (as each such step decreases the number of uncovered nodes by 1).

Next, consider the state after k dual steps in the TS stage and let, as earlier, π and $\overline{B} \subseteq A\overline{G}$ denote the current duals the current partial bibranching, respectively. Put $B \subseteq AG$ to be the result of expanding $\overline{B}$. We have shown earlier that there are no negative terms in $\Delta(\pi, \pi_1, B)$. Moreover, the very same argument (with $\overline{S}$ and $\overline{T}$ exchanged) applies, so one can construct subsets $D_j^i \subseteq T$ similarly to the ST stage. Since each of these sets D_j^i corresponds to an uncovered node at some intermediate moment and the algorithm can only decrease duals of sets that are covered, it follows that none of $\pi(D_j^i)$ is changed during the ST stage. Thus, (3) and (4) hold for the TS stage as well, and the latter completes after executing $O(\sqrt{n})$ primal and dual steps.

Taking the above Lemma 8 into account and applying the ideas from [Edm67, GGSR86, FT87, KP98] one can implement the shell stage to run in $O(m \log n)$

time and both ST and TS stages to take $O(m\sqrt{n}\log n)$ time (the $\log n$ factor in the above estimates comes from the complexity of priority queue operations, maintenance of $\mathcal{F}$, and computation of reduced arc weights; we employ binary heaps and dynamic trees [ST83] for these purposes). Recall (see Section 3) that the total number of scaling steps is $\lceil \log W \rceil + \lfloor \log n \rfloor + 1 = O(\log(nW))$. Hence, we conclude as follows:

Theorem 5. *The running time of the algorithm is* $O(m\sqrt{n}\ \log n \log(nW))$.

Acknowledgements

The author is thankful to Petr Mitrichev (Faculty of Mechanics and Mathematics, Moscow State University) and also to the anonymous referees for helpful comments and suggestions.

References

[Din80]   Dinic, E.: Algorithm for solution of a problem of maximum flow in networks with power estimation. Soviet Math. Dokl. 11, 1277–1280 (1980)

[Edm67]   Edmonds, J.: Optimum branchings. J. Res. Nat. Bur. Standards 71B, 233–240 (1967)

[FT87]   Fredman, M., Tarjan, R.: Fibonacci heaps and their uses in improved network optimization algorithms. J. ACM 34(3), 596–615 (1987)

[GGSR86]   Gabow, H., Galil, Z., Spencer, T., Tarjan, R.: Efficient algorithms for finding minimum spanning trees in undirected and directed graphs. Combinatorica 6(2), 109–122 (1986)

[GT87]   Goldberg, A., Tarjan, R.: Solving minimum-cost flow problems by successive approximation. In: STOC 1987: Proceedings of the nineteenth annual ACM conference on Theory of computing, pp. 7–18 (1987)

[GT89]   Gabow, H., Tarjan, R.: Faster scaling algorithms for network problems. SIAM J. Comput. 18(5), 1013–1036 (1989)

[GT91]   Gablow, H., Tarjan, R.: Faster scaling algorithms for general graph matching problems. J. ACM 38(4), 815–853 (1991)

[KP98]   Keijsper, J., Pendavingh, R.: An efficient algorithm for minimum-weight bibranching. J. Comb. Theory, Ser. B 73(2), 130–145 (1998)

[Sch82]   Schrijver, A.: Min-max relations for directed graphs. Ann. Discrete Math. 16, 261–280 (1982)

[Sch03]   Schrijver, A.: Combinatorial Optimization. Springer, Berlin (2003)

[ST83]   Sleator, D., Tarjan, R.: A data structure for dynamic trees. Journal of Computer and System Sciences 26(3), 362–391 (1983)

The Balanced Edge Cover Problem[*]

Yuta Harada, Hirotaka Ono, Kunihiko Sadakane, and Masafumi Yamashita

Department of Computer Science and Communication Engineering,
Kyushu University, Fukuoka 812-8581, Japan
{ono,sada,mak}@csce.kyushu-u.ac.jp

Abstract. For an undirected graph $G = (V, E)$, an edge cover is defined as a set of edges that covers all vertices of V. It is known that a minimum edge cover can be found in polynomial time and forms a collection of star graphs. In this paper, we consider the problem of finding a balanced edge cover where the degrees of star center vertices are balanced, which can be applied to optimize sensor network structures, for example. To this end, we formulate the problem as a minimization of the summation of strictly monotone increasing convex costs associated with degrees for covered vertices, and show that the optimality can be characterized as the non-existence of certain alternating paths. By using this characterization, we show that the optimal covers are also minimum edge covers, have the lexicographically smallest degree sequence of the covered vertices, and minimize the maximum degree of covered vertices. Based on the characterization we also present an $O(|V||E|)$ time algorithm.

1   Introduction

1.1   Problems and Results

For an undirected graph $G = (V, E)$, an edge cover and a matching are defined as a set of edges that covers all vertices of V and a set of edges sharing no vertices each other, respectively. Both are studied well. In particular, the maximum matching problem, the problem of finding a matching whose size is maximum, has been intensively investigated and polynomial time algorithms are proposed [2,10,13], including some simple algorithms for bipartite graphs [7,8]. A minimum edge cover is an edge cover whose size is minimum, and is also known to be solvable in polynomial time, since it can be solved by simply adding uncovered vertices to a maximum matching [3,11].

Since a minimum (or minimal) edge cover is known to form a collection of star graphs (components), it is used to divide the original graph structure. In such divisions, however, the sizes of star graphs can be biased. In this paper, we consider the problem of finding a minimal edge cover in which the sizes of star components are balanced. To formulate the notion of "balanced", we introduce a cost function $f : \{1, 2, \ldots, |V|\} \to N^+$ with respect to degrees that is strictly

---

[*] This work is supported in part by the Grant-in-Aid of the Ministry of Education, Science, Sports and Culture of Japan and by the Asahi glass foundation.

S.-H. Hong, H. Nagamochi, and T. Fukunaga (Eds.): ISAAC 2008, LNCS 5369, pp. 246–257, 2008.

monotonic increasing convex, for example, $f(d) = d^2$, where d is the degree of a vertex. The problem is described as follows:

Balanced Edge Cover Problem (BEC)

Instance. A simple undirected graph $G = (V, E)$ and a strictly monotonic increasing convex function $f : N^+ \to R^+$,

Question. Find an edge cover E_c minimizing $c(E_c) = \sum_{v \in V} f(\deg_{E_c}(v))$, where $\deg_{E_c}(v) = |\{\{v, u\} \in E_c\}|$.

By the strictly monotonic increasing property and the convexity of f, the sizes of star components tend to be balanced. For example, imagine two minimal edge covers on a graph and $f(d) = d^2$. One edge cover $E_{c(1)}$ consists of star components have sizes 5, 2 and 1, and the other $E_{c(2)}$ consists of star components have sizes 3, 3 and 2. Intuitively, $E_{c(2)}$ is more balanced than $E_{c(1)}$, and indeed $E_{c(2)}$ has smaller function value $f = 3^2 + (1+1+1) + 3^2 + (1+1+1) + 2^2 + (1+1) = 30$ than $E_{c(1)}$'s $f = 5^2 + (1+1+1+1+1) + 2^2 + (1+1) + 1 + (1) = 38$, where the numbers within parentheses represent the degrees of leaf vertices. We consider the balanced edge cover problem under this setting.

In this paper, we show that (1) for any G, the set of optimal solutions for BEC is determined independently of function f as long as f is strictly monotonic increasing convex (we refer f as a *cost function*), (2) an optimal solution of BEC gives a minimum edge cover that minimizes the maximum degree with respect to covering edges, and (3) an edge cover E_c is an optimal solution of BEC if and only if the degree sequence of $G_c = (V, E_c)$ is the lexicographically smallest among all edge covers of G. Note that the notion of "balanced" comes from the equivalence to (2). These are all from a characterization of the optimality of BEC, that is, the non-existence of alternating paths of certain properties, say edge-reducing paths and cost-reducing paths. Based on the characterization, we propose two algorithms; a simple $O(\min\{|V|^{3/2}, |E|\}|E|)$-time algorithm that removes such alternating paths and an $O(|V||E|)$-time algorithm that extends the edge set so that the added edges are guaranteed not to create such alternating paths.

1.2   Related Work

The semi-matching with balanced load [4] seems to be the problem most related with BEC. For a bipartite graph $G = (U \cup V, E)$, a semi-matching is a set of edges $M \subseteq E$ such that each vertex in U is an endpoint of exactly one edge in M. The problem asks a semi-matching that is load balancing in the sense that the degrees of vertices in V with respect to M is balanced. The main difference between the problems is that in the semi-matching with balanced load, balancing the degrees of V is looked for, while BEC seeks to take balance of the degrees of all vertices (see the characterization (2) of optimal solution of our problem mentioned above), apart from the obvious difference of their instance sets; i.e., bipartite graphs and general graphs. BEC is thus a generalization of the semi-matching with balanced load. An $O(|U||E|)$-time algorithm to solve the semi-matching with balanced load is proposed in [4]. An algorithm for BEC we will propose in this paper can be transformed into an algorithm that solves the semi-matching with balanced load with the same time complexity.

Another aspect of the balanced edge cover is a special case of *jump system*, which is introduced by Bouchet and Cunningham [1], Jump systems unify and generalize matroids, integral bounded g-polymatroids and more. The balanced edge cover is considered as a constant-sum jump system. For more information, see [9] and [12].

1.3 Applications

Sensor networks consist of homogeneous sensor nodes and they communicate each other via wireless links. Some sensor network protocols organize sensor nodes into a two-level hierarchy; some are chosen as cluster heads that gather information from normal nodes [5,6]. Every normal node is assigned to exactly one cluster head. That is, all the sensor nodes are partitioned into clusters. Once a normal sensor node detects something, it transmits the information to the cluster head that it belongs to. Suppose that the sensor nodes are vertices and wireless links are edges. Then the whole topology of the network is modeled as an undirected graph and the cluster heads and their assignments form an edge cover. In this context, the balanced edge cover means that the sizes of clusters are balanced, which conserves energy of cluster heads.

The rest of the paper is organized as follows. Section 2 gives notations and basic properties of the problems. In Section 3, we give an optimality condition of balanced edge covers, which also shows the optimality of several problems. In Section 4, we propose algorithms based on the optimality condition.

2   Preliminaries

Let $G = (V, E)$ be a simple undirected graph with a vertex set V and an edge set E, where each edge is a pair $\{u, v\}$ of vertices $u, v \in V$. Throughout the paper, let $|E| = m$, $|V| = n$ for the input graph. For a subset $E' \subseteq E$ and a vertex v, we denote $\delta_{E'}(v) = \{\{u, v\} \in E'\}$ and $\deg_{E'}(v) = |\delta_{E'}(v)|$. Namely, $\delta_{E'}(v)$ represents a set of edges in E' incident with v and $\deg_{E'}(v)$ is the degree of v. The notation with the subscript deleted is used for E, i.e., $\deg(v) = \deg_E(v)$. For $G = (V, E)$, a *degree sequence* is a monotonic nonincreasing sequence of the vertex degrees of V. For $E' \subseteq E$, we analogously define *degree sequence with respect to E'* as a monotonic nonincreasing sequence of $\deg_{E'}(v)$'s of V.

An edge set $E_c \subseteq E$ is an *edge cover* of G if every vertex of V is incident with an edge of E_c. Note that an edge cover always exists if and only if G has no isolated vertices. In this paper, we assume that G is a graph without isolated vertices. Edges $e \in E_c$ and $e \notin E_c$ are called an *E_c-covering edge* and a *E_c-non-covering edge*, respectively. Norman and Rabin show that there is a polynomial time algorithm to transform a maximum matching to a minimum edge cover, and vice versa [11]. The transforming algorithm utilizes the following proposition.

Proposition 1 ([11]). *Let G be a graph without isolated vertices. Then every maximum matching is contained in a minimum edge cover, and every minimum edge cover contains a maximum matching.* □

From the proof of this result, Gallai's theorem is also shown, that is, $\nu(G) + \rho(G) = n$ for a graph G with n vertices, where $\nu(G)$ and $\rho(G)$ are the sizes of maximum matching and minimum edge cover of G, respectively [3]. By the method of Gallai's theorem, one can indeed derive a minimum edge cover from a maximum matching M in time $O(m)$, just by adding for each vertex v missed by M, an arbitrary edge incident with v. Hence a minimum edge cover can be found in an linearly equivalent running time of a maximum matching algorithm. For the maximum matching problem, $O(\sqrt{n}m)$-time algorithm is known [10,13].

We obtain the following proposition on minimal edge covers. The proof is omitted because it is not difficult.

Proposition 2. *A set of edges E_c is a minimal edge cover if and only if $G_c = (V, E_c)$ is a spanning forest of G where each connected component of E_c is a star (i.e., a $K_{1,r}$ for a natural number r).* $\qquad\square$

Let $G_s = (V, E_s)$ be a subforest of G whose connected components are all stars. By Proposition 2, E_s is a subset of some minimal edge cover E_c. For such a subforest E_s, we call a vertex v satisfying $\deg_{E_s}(v) = 1$ a *leaf vertex* (or simply *leaf*) in E_s, and call a vertex connected to a leaf vertex by an edge in E_s a *center vertex* (or simply *center*) in E_s. By definition, for a star component $K_{1,r}$ in E_s, if $r \geq 2$ then a center vertex has degree more than 1 and leaf vertices have degree 1. If $r = 1$, i.e., the star is $K_{1,1}$, only two vertices exist in the star and both of them are centers and also leaves. In addition, there may be vertices not covered by any edges in E_s, and such vertices are called *free vertices* in E_s. Let C_{E_s}, L_{E_s} and F_{E_s} be sets of centers, leaves and free vertices with respect to E_s, respectively. The union of these vertex sets contains all vertices of V, i.e. $C_{E_s} \cup L_{E_s} \cup F_{E_s} = V$. If E_s is a minimal edge cover then $F_{E_s} = \emptyset$ holds. If $L_{E_s} \cup F_{E_s} = V$ then E_s is a matching. Especially, E_s is a perfect matching if $L_{E_s} = V$ holds.

We consider to balance such an edge cover by using a strictly monotonic increasing convex function f that maps N^+ to R^+. Given an edge cover E_c and such an f (we refer f as a cost function), we define the objective function $c(E_c) = \sum_{v \in V} f(\deg_{E_c}(v))$, which is intended to avoid many covering edges concentrate a few vertices. Then the balanced edge cover problem is defined as follows: *Given a simple undirected graph $G = (V, E)$ and a strictly monotonic increasing convex function cost $: N^+ \to R^+$, find an edge cover E_c minimizing $c(E_c) = \sum_{v \in V} f(\deg_{E_c}(v))$.*

For a cost function f, we call an edge cover E_c that minimizes $c(E_c)$ a *balanced edge cover* for f. Actually, balanced edge covers do not depend on f; just the property that it is strictly monotonic increasing convex is important, as mentioned in Section 1. Hence we sometimes omit "for f", and just say "balanced" edge covers. In the following section, we give a characterization of balanced edge covers based on the non-existence of paths with certain properties, which also provides algorithms for finding them in the last section.

3   Optimality of Balanced Edge Covers

In this section, we consider the optimality of balanced edge covers. It can be characterized by paths of some properties as shown later.

Given a subforest $G_s = (V, E_s)$, a sequence of distinct edges $P = (\{v_1, v_2\}, \{v_2, v_3\}, \ldots, \{v_{k-1}, v_k\})$ is an *alternating path* with respect to E_s or an E_s-*alternating path* if edges are alternately in and not in E_s. Let E_P be a set of edges included in P. For convenience, we denote such an alternating path by a sequence of vertices $P = (v_1, \ldots, v_k)$. By definition, alternating path $P = (v_1, \ldots, v_k)$ may not be a simple path (i.e., a path where no vertex is visited more than once), and even may be a cycle ($v_1 = v_k$), but we call them paths for simplicity. Several paths that we shall introduce later are also alternating paths.

We define the notation $A \oplus B$ as the symmetric difference of sets A and B; i.e., $A \oplus B = (A \setminus B) \cup (B \setminus A)$. If P is an E_c-alternating path for an edge cover E_c, then we can obtain a set of edges $E_P \oplus E_c$ by switching E_c-covering edges and E_c-non-covering edges along P. Note that all the internal vertices $v_2, \ldots, v_{k-1}$ in P remain to be covered even after the switching operation. If both v_1 and v_k are incident with E_c-covering edges not in P or E_c-non-covering edges in P, $E_P \oplus E_c$ is also an edge cover.

Proposition 3. *For an edge cover E_c and its alternating path $P = (v_1, \ldots, v_k)$, if $\delta_{E_c \oplus E_P}(v_1) \neq \emptyset$ and $\delta_{E_c \oplus E_P}(v_k) \neq \emptyset$ hold, $E_P \oplus E_c$ is also an edge cover.* $\square$

3.1   Minimum Property

The definition of balanced edge covers themselves does not contain "minimum" explicitly, but actually they are also minimum edge covers. To show this, we introduce the notion of "edge-reducing" that characterizes the minimum edge covers.

Edge-Reducing Path. For an edge cover E_c, an E_c-alternating path $P = (v_1, v_2, \cdots, v_{2k-1}, v_{2k})$ is called an *edge-reducing path* with respect to E_c or an E_c-*edge-reducing path* if $\{v_1, v_2\}$ and $\{v_{2k-1}, v_{2k}\}$ are E_c-covering edges with $\delta_{E_c \oplus E_P}(v_1) \neq \emptyset$ and $\delta_{E_c \oplus E_P}(v_{2k}) \neq \emptyset$. For such an E_c-edge-reducing path P, $E_P \oplus E_c$ is also an edge cover by Proposition 3. We can see that $|E_P \setminus E_c| = |E_P \cap E_c| + 1$, that is, $|E_P \oplus E_c| = |E_c| - 1$, which means that the switching operation along P decreases the size of the edge cover by one. Each degree of v_1 and v_{2k} also decreases by one (or if $v_1 = v_{2k}$, it decreases by two) but other vertices are not affected. Figure 1 shows an example of a edge-reducing path, where bold lines represent the covering edges, and dotted lines represent the non-covering edges.

The following theorem shows an optimality condition of minimum edge covers and leads that a balanced edge cover is also a minimum edge cover.

Theorem 1 ([11]). *An edge cover E_c is a minimum edge cover if and only if there exists no edge-reducing path with respect to E_c.* $\square$

From this theorem, we can easily show the following corollary though we omit the proof.

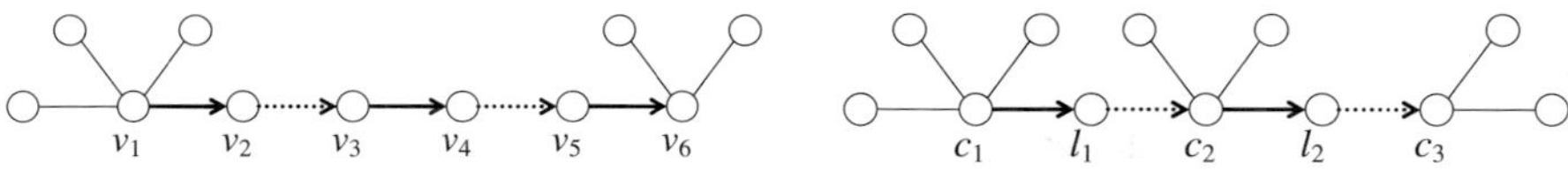

Fig. 1. Edge-Reducing Path

Fig. 2. Cost-Reducing Path

Corollary 1. *A balanced edge cover for any f is a minimum edge cover.*   □

3.2   Balance Properties

In this subsection, we focus on the optimality of balanced edge covers. We first define the cost-reducing path, in which the degree of start vertex is two larger than the one of end vertex. (A formal definition will be shown later.) Although the definition of cost-reducing paths just depends on the monotonicity and the convexity of f, we can show that it well characterizes the minimality of the total cost. Due to this, we can show that for a graph G the set of optimal solutions of BEC is determined independently of f. We then show a key lemma about the existence of the cost-reducing paths and its conditions (Lemma 1), based on which we characterize the optimality of balanced edge covers (Theorem 2).

We start to define an alternating path of a certain property.

Vertex-Alternating Path. Let $G_s = (V, E_s)$ be a subforest of G whose connected components are all stars. For such an E_s, we define a *vertex-alternating path* P with respect to E_s as an E_s-alternating path $P = (c_1, l_1, c_2, \ldots, c_{k-1}, l_{k-1}, c_k)$ such that $c_i \in C_{E_s}$, $l_i \in L_{E_s}$, $\{c_i, l_i\} \in E_s$ and $\{l_i, c_{i+1}\} \notin E_s$ hold for every i. Namely, P visits centers and leaves alternately. It should be noted that in a vertex-alternating path a center vertex can appear more than once but a leaf vertex cannot. The switching operation for a vertex-alternating path P does not change the size of edges; $|E_s \oplus E_P| = |E_s|$. For an minimum edge cover E_c, which also forms an subforest of G, if $\delta_{E_c \oplus E_P}(c_1) \neq \emptyset$ then $E_P \oplus E_c$ is also a minimum edge cover by Proposition 3.

Cost-Reducing Path. Let $G_s = (V, E_s)$ be a subforest of G. For a vertex-alternating path $P = (c_1, l_1, \ldots, l_{k-1}, c_k)$ with respect to E_s, if $\deg_{E_s}(c_1) > \deg_{E_s}(c_k) + 1$ then P is called a *cost-reducing path*. This is because by switching edges along a cost-reducing path P the total cost $c(E_s)$ is literally reduced without changing the size of edges, that is, $c(E_P \oplus E_s) < c(E_s)$. Figure 2 shows an example of a cost-reducing path. Note that this definition does not depend on cost function f. Now we show that a cost-reducing path reduces $c(E_s)$ for any strictly monotonic increasing convex function f indeed. Notice that the switching operation $E_P \oplus E_s$ changes only the degrees of c_1 and c_k. For simplicity, let d_1 and d_k denote $\deg_{E_s}(c_1)$ and $\deg_{E_s}(c_k)$ respectively. Then, $\deg_{E_P \oplus E_s}(c_1) = \deg_{E_s}(c_1) - 1 = d_1 - 1$ and $\deg_{E_P \oplus E_s}(c_k) = \deg_{E_s}(c_k) + 1 = d_k + 1$, and the condition of the cost-reducing path is $d_1 > d_k + 1$. By the convexity of f, we have

$$\frac{1}{d_1 - d_k} f(d_1) + \left(1 - \frac{1}{d_1 - d_k}\right) f(d_k) \geq f(d_k + 1),$$

$$\left(1 - \frac{1}{d_1 - d_k}\right) f(d_1) + \frac{1}{d_1 - d_k} f(d_k) \geq f(d_1 - 1).$$

The equalities do not hold because f is strictly monotonic increasing. Hence, by summing the above inequalities, we obtain $f(d_1) + f(d_k) > f(d_1 - 1) + f(d_k + 1)$, which implies that $c(E_s) > c(E_P \oplus E_s)$; $E_P \oplus E_s$ reduces the total cost.

We shall show that the non-existence of cost-reducing path for an edge cover is a necessary and sufficient condition about a balanced edge cover.

To show this, we first prove the following lemma.

Lemma 1. *Let E_c and E_c^* be two different minimum edge covers, and let $G' = (V, E_c \oplus E_c^*)$. If G' satisfies the following two conditions, then there exists a cost-reducing path with respect to E_c in G'.*

   I. There exists a center vertex $c_1 \in C_{E_c}$ satisfying $\deg_{E_c}(c_1) > \deg_{E_c^*}(c_1)$.

   II. For any c_1 of condition I, every vertex-alternating path $P = (c_1, \ldots, c_k)$ with respect to E_c in G' satisfies $\deg_{E_c^*}(c_k) \leq \deg_{E_c^*}(c_1)$.

Proof. Let E_c and E_c^* be minimum edge covers satisfying the above two conditions, and $G' = (V, E_c \oplus E_c^*)$. There is a vertex $c_1 \in C_{E_c}$ satisfying $\deg_{E_c}(c_1) > \deg_{E_c^*}(c_1)$ by the assumption. We construct a vertex-alternating path P starting from c_1 as follows: First set $P = (c_1)$. (1) We are at a center vertex $c_i \in C_{E_c}$. If both $\delta_{E_c \setminus E_c^*}(c_i) \setminus E_P \neq \emptyset$ and $\deg_{E_c}(c_1) \leq \deg_{E_c}(c_i) + 1$ are satisfied, we choose an edge $\{c_i, l_i\} \in E_c \setminus (E_c^* \cup E_P)$ and extend P by adding the edge. Note that the starting vertex c_1 satisfies both conditions. Otherwise we stop the construction of P, that is, P ends at c_i. (2) We are at a leaf vertex $l_i \in L_{E_c}$ by following an unique edge $\{c_{i-1}, l_i\} \in \delta_{E_c}(l_i)$ that does not belong to E_c^*. Then, since there exists an edge $\{l_i, c_{i+1}\} \in E_c^* \setminus E_c$ as shown below, we extend P by adding such an edge from l_i. Here, we show that such an edge always exits. First, we can see that $\delta_{E_c^* \setminus E_c}(l_i) \neq \emptyset$ holds, otherwise $\delta_{E_c^* \setminus E_c}(l_i) = \delta_{E_c^*}(l_i) \setminus \{c_{i-1}, l_i\} = \delta_{E_c^*}(l_i) = \emptyset$ holds, which contradicts that E_c^* is an edge cover. Next we show that edges in $\delta_{E_c^* \setminus E_c}(l_i)$ are connected to center vertices in C_{E_c}. Otherwise, an edge in $\delta_{E_c^* \setminus E_c}(l_i)$ forms $\{l_i, l'\}$ where $l' \in L_{E_c} \setminus C_{E_c}$. Since E_c is an edge cover, there exists an E_c-covering edge $\{l', c'\}$ where $c' \in C_{E_c}$. Note that $c' \notin L_{E_c}$ (i.e., $\deg_{E_c}(c') > 1$), because $l' \notin C_{E_c}$ (see the definition of center and leaf vertices). Then an E_c-alternating path $(c_1, l_2, \ldots, c_i, l_i, l', c')$ is an E_c-edge-reducing path, which contradicts that E_c is an minimum edge cover. By these, we can extend P by adding $\{l_i, c_{i+1}\} \in E_c^* \setminus E_c$. The above construction eventually terminates because each edge is followed by P at most once.

We then show that the constructed vertex-alternating path $P = (c_1, \ldots, c_k)$ is a cost-reducing path with respect to E_c. The extension of P stops in the following two cases: (i) $\deg_{E_c}(c_1) > \deg_{E_c}(c_k) + 1$ holds. In this case, P is obviously a cost-reducing path with respect to E_c. (ii) $\delta_{E_c \setminus E_c^*}(c_k) \setminus E_P = \emptyset$ holds. Also in this case, we can show P is a cost-reducing path. If $c_1 = c_k$ then $\deg_{E_c}(c_1) = \deg_{E_c^*}(c_1)$ holds, but this contradicts the condition I. Hence $c_1 \neq c_k$.

We can see that $|\delta_{E_c \setminus E_c^*}(c_k)| + 1 = |\delta_{E_c^* \setminus E_c}(c_k)|$, since P arrived at c_k via an edge of $E_c^* \setminus E_c$ and left at c_k via $E_c \setminus E_c^*$ (All the edges in $\delta_{E_c \setminus E_c^*}(c_k)$ are followed by P). Thus $\deg_{E_c}(c_k) < \deg_{E_c^*}(c_k)$ holds. In addition, $\deg_{E_c}(c_1) > \deg_{E_c^*}(c_1)$ and $\deg_{E_c^*}(c_k) \leq \deg_{E_c^*}(c_1)$ hold by the conditions I and II of the lemma. These three inequities yield $\deg_{E_c}(c_k) < \deg_{E_c^*}(c_k) \leq \deg_{E_c^*}(c_1) < \deg_{E_c}(c_1)$, which leads $\deg_{E_c}(c_1) > \deg_{E_c}(c_k) + 1$. This shows that P is a cost-reducing path with respect to E_c, which completes the proof. $\qquad\square$

We obtain the following theorem about the optimality of balanced edge covers by the above lemma.

Theorem 2. *For an edge cover E_c, the following (a) $\sim$ (c) are equivalent:*

(a) E_c *is a balanced edge cover for any f.*
(b) E_c *is a minimum edge cover and has no cost-reducing path.*
(c) *The degree sequence with respect to E_c is the lexicographically smallest among edge covers.*

Proof. Here we show only (a) $\Leftrightarrow$ (b) part due to the space limitation: By Corollary 1, a balanced cover is a minimum edge cover. Thus, we just show that the nonexistence of the cost-reducing path is a necessary and sufficient condition of balanced edge covers for minimum edge covers. The necessity is obvious, because otherwise the total cost $c(E_c)$ of arbitrary f can be reduced by the switching operation. Thus we concentrate the sufficiency.

For an f, let E_c be a minimum edge cover whose total cost $c(E_c)$ is not minimum, and let E_c^* be a balanced edge cover where the size of the symmetric difference $|E_c \oplus E_c^*|$ is minimum. Let $G' = (V, E_c \oplus E_c^*)$. We will show that G' satisfies the two conditions of Lemma 1; E_c always has a cost-reducing paths.

Since E_c and E_c^* are minimum edge covers, $\sum_{v \in V} \deg_{E_c}(v) = \sum_{v \in V} \deg_{E_c^*}(v) = 2\rho(G)$ holds. Also by $c(E_c) > c(E_c^*)$, there exists a vertex $c_1 \in C_{E_c}$ such that $\deg_{E_c}(c_1) > \deg_{E_c^*}(c_1)$ holds, which satisfies the condition I. Here, we show that an arbitrary vertex-alternating path from c_1 with respect to E_c in G' (say $P = (c_1, l_1, \ldots, l_{k-1}, c_k)$), satisfies the condition II, that is, $\deg_{E_c^*}(c_k) \leq \deg_{E_c^*}(c_1)$. Let $\overline{P} = (c_k, l_{k-1}, \ldots, l_1, c_1)$ denote the reverse of the path P. Then, $\overline{P}$ is an E_c^*-alternating path in G' by the definition of G'. Furthermore, we can say that when $\deg_{E_c^*}(c_k) > 1$, $\overline{P}$ is an E_c^*-vertex alternating path in G', that is, that $c_1, c_2, \ldots, c_k \in C_{E_c^*}$ and $l_1, l_2, \ldots, l_{k-1} \in L_{E_c^*}$; otherwise, there exists an $l_i \in C_{E_c^*}$, which implies $(c_k, \ldots, l_i)$ is an E_c^*-edge-reducing path in G. This contradicts E_c^* is an minimum edge cover. Thus we have $\overline{P}$ is an E_c^*-vertex alternating path if $\deg_{E_c^*}(c_k) > 1$. If $\deg_{E_c^*}(c_k) > \deg_{E_c^*}(c_1) + 1$, $\overline{P}$ is a cost-reducing path for E_c^*. This contradicts that E_c^* is an balanced edge cover. If $\deg_{E_c^*}(c_k) = \deg_{E_c^*}(c_1) + 1$ holds, $E_c^* \oplus E_{\overline{P}}$ is an edge cover and has $c(E_c^* \oplus E_{\overline{P}}) = c(E_c^*)$, also a balanced edge cover. This contradicts the definition of E_c^* because the symmetric difference between E_c and $E_c^* \oplus E_{\overline{P}}$ is smaller than $|E_c \oplus E_c^*|$. Thus $\deg_{E_c^*}(c_k) \leq \deg_{E_c^*}(c_1)$ holds for P. Therefore since G' satisfies two conditions I and II of Lemma 1, E_c has a cost-reducing path with n G'; the sufficiency is also proved. $\qquad\square$

Corollary 2. *If an edge cover E_c is a balanced edge cover, then $\max_{v \in V}\{\deg_{E_c}(v)\}$ is minimized.* $\qquad\square$

4   Algorithms

By utilizing the optimality condition shown in the previous section, we first give a simple algorithm $BEC1$, which directly uses the result of Theorem 2. Namely, whenever the algorithm finds a cost-reducing path, apply the switching operation. The algorithm is given as follows:

Algorithm $BEC1$

Step 1: Construct a minimum edge cover E_c.
Step 2: Find a cost-reducing path P with respect to E_c. If no cost-reducing path is found then output E_c.
Step 3: Let $E_c := E_c \oplus E_P$, and go to Step 2.

In Step 1, the algorithm construct a minimum edge cover E_c by the procedure introduced in Section 2. Step 3 has been already explained in Section 4. Hence we just consider how to find a cost-reducing path $P = (c_1, \ldots, c_k)$ with respect to E_c in Step 2: First, choose a vertex $c_1 \in C_{E_c}$ with maximum degree in E_c, and build an *alternating search tree* T rooted at c_1. The search tree T is a rooted tree and consists of edges in E_c directed from C_{E_c} to L_{E_c} and edges not in E_c directed from L_{E_c} to C_{E_c}, where the direction represents the relation between parents and children. Such T is constructed by applying the breadth-first search from c_1. If a vertex c_k with $\deg_{E_c}(c_1) > \deg_{E_c}(c_k) + 1$ is found in the search, the path from c_1 to c_k on T is a cost-reducing path. If such c_k is not found, we can say that there exists no cost-reducing path from c_1 by definition; we move to another vertex with maximum degree not in T and construct a new alternating search tree from the vertex. Note that the search from a vertex that once belong to an existing T is not requisite by the following Lemma 2. If all vertices of V are visited, then the algorithm terminates at this step. If no cost-reducing path exists then return E_c.

Lemma 2. *Let P be a vertex-alternating path from c_1 to c_k such that $\deg_{E_c}(c_1) \geq \deg_{E_c}(c_k)$ holds. If there is no cost-reducing path starting from c_1 then there is also no cost-reducing path from c_k.*

Proof. By contradiction. If a cost-reducing path from c_k to $c_{k'}$ exists, the vertex-alternating path P from c_1 to c_k can be extended to a vertex-alternating path from c_1 to $c_{k'}$ with $\deg_{E_c}(c_1) > \deg_{E_c}(c_{k'}) + 1$, a cost-reducing path.   □

Theorem 3. *$BEC1$ produces a balanced edge cover in $O(\min\{n^{3/2}, m\}\, m)$ time.*

Proof. We first discuss the correctness. An edge cover E_c remains minimum during the execution of $BEC1$, because the switching operation for a cost-reducing path does not change the number of edges. Then, the above explanation on Step 2 and Theorem 2 show that a balanced edge cover is obtained when it terminates.

Next, we discuss the time complexity. Step 1 is done in $O(n^{1/2}m)$ time shown in Section 2. We consider Steps 2 and 3. One iteration of Step 2 requires $O(m)$

time since each edge is checked at most once by the breadth-first search. Also a switching operation in Step 3 takes $O(m)$ time. In the total execution, each vertex v can be a starting vertex of a cost-reducing path at most $\deg(v)$ times, because once v has maximum degree for E_c, the degree of v decreases monotonically. Namely, an upper bound on the number of iterations is given as the summation of $\deg(v)$ over all the vertices, that is $\sum_{v \in V} \deg(v) = 2m$. Thus Steps 2 and 3 are iterated in $O(m)$ times. We have another upper bound $O(n^{3/2})$ on the iterations, though we omit the detail due to the space limitation. It is obtained by a similar analysis to the one in [7], which analyzes the bipartite matching can be found in $O(n^{1/2}m)$ time. Also a similar technique is shown in [4]. Therefore the total running time of $BEC1$ is $O(\min\{n^{3/2}, m\}\, m)$ time. $\square$

A Faster Algorithm. We propose another algorithm, which improves the running time. It is based on a similar technique of the matching extension algorithm [2], but in order to adapt a balanced edge cover, it keeps the non-existence of cost-reducing paths. The procedure of the algorithm is given as follows.

Algorithm $BEC2$

Step 1: Construct a maximum matching M.

Step 2: Choose a free vertex $v_1 \in F_M$ if $F_M \neq \emptyset$. Otherwise output M.

Step 3: Build an alternating search tree T rooted at v_1, where edges not in M are directed from $\{v_1\} \cup L_M$ to C_M and edges in M are directed from C_M to L_M. (Notice that the definition of this T differs from the one in $BEC1$.)

Step 4: Find a path $P = (v_1, \ldots, v_{2k})$ with $v_{2k} \in C_M$ such that $\deg_M(v_{2k})$ is as small as possible in the tree T. Such a path P is called an *augmenting path*.

Step 5: Let $M := M \oplus E_P$, and go to Step 2.

$BEC2$ constructs a maximum matching M at the beginning, and iterates extending an edge set M by finding an augmenting path P, which is a concatenation of an M-non covering edge $\{v_1, v_2\}$ of $v_1 \in F_M$ and an M-vertex-alternating path $(v_2, \ldots, v_{2k})$. In each iteration, the starting vertex v_1 of P gets covered and the size of M increases by one by the switching operation $M \oplus E_P$, because $\{v_1, v_2\} \in M \oplus E_P$ and $|E_P \setminus M| = |E_P \cap M| - 1$. The number of free vertices decreases by one and the algorithm stops if no free vertex exists. Thus, all vertices of V are covered by output M. As for its correctness, we will prove that no M-cost-reducing path is created by the extension of M in the next lemma.

Lemma 3. *No cost-reducing path exists during the execution of $BEC2$.*

Proof. Prove by contradiction. Let us call $M_0, M_1, \ldots, M_i, \ldots, M_{|F_{M_0}|}$ the sets of edges constructed in Step 5 of each iteration of $BEC2$ in this order. That is, M_0 is a maximum matching computed in Step 1, and M_i is defined by $M_{i-1} \oplus E_P$ for P found in Step 4 of i-th iteration, which satisfies $|M_i| = |M_{i-1}| + 1$. Now, we suppose that a cost-reducing path first appears after $(j-1)$-th iteration. We refer the cost-reducing path with respect to M_j as $P^* = (c_1, l_1, \ldots, l_{k-1}, c_k)$. For the previous edge set M_{j-1}, let $P' = (v', \ldots, c')$ be an augmenting path found in Step 4. Namely, $M_j = M_{j-1} \oplus E_{P'}$. Let $Q = (M_j \setminus M_{j-1}) \cap E_{P^*}$,

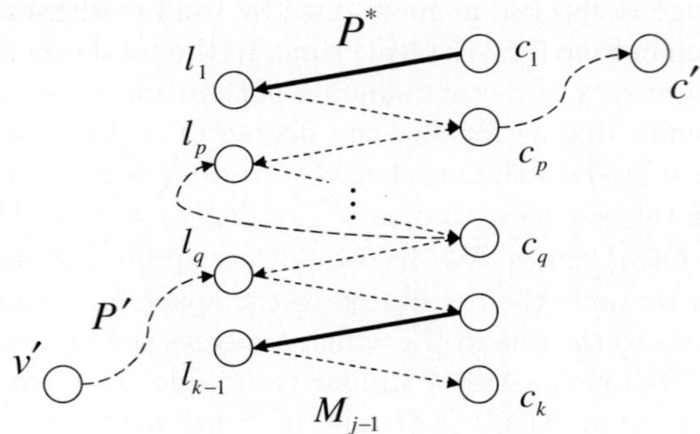

Fig. 3. P^* and P' in M_{j-1} (Lemma 3)

that is, a set of covering edges in P^* that exist in M_j but not in M_{j-1}. Here, P^* is a first cost-reducing path implies that $Q \neq \emptyset$ holds, otherwise P^* is also an M_{j-1}-cost-reducing path, a contradiction. Also, $Q \subseteq E_{P'}$ holds because P^* results from the switching operation for P'. Let $\deg_{M_j}(c_1) = d$ as the degree of starting vertex c_1 of P^*. Then $\deg_{M_{j-1}}(c_k) \leq \deg_{M_j}(c_k) \leq d - 2$ holds for the end vertex c_k because P^* is a cost-reducing path and the degrees of all vertices are nondecreasing for the switching operation. Figure 3 shows the relationship between P^* and P' in M_{j-1}. In this situation, we will show a contradiction.

Let $\{c_p, l_p\} \in Q$ be the first edge in Q that appears on P^*, that is, p is the smallest index in Q. There is a vertex-alternating path $(c_1, \ldots, l_{p-1}, c_p, \ldots, c')$ in M_{j-1} because $\{c_p, l_p\}$ is included in P' by $Q \subseteq E_{P'}$. This path is not a cost-reducing path since M_{j-1} has no cost-reducing path. Thus, $\deg_{M_{j-1}}(c') \geq \deg_{M_{j-1}}(c_1) - 1 = \deg_{M_j}(c_1) - 1 = d - 1$ holds when $c' \neq c_1$. In the case of $c' = c_1$, $\deg_{M_{j-1}}(c') = \deg_{M_j}(c') - 1 = d - 1$ holds since the degree of $c'(c_1)$ increases by one by switching of P'. Accordingly, $\deg_{M_{j-1}}(c') \geq d - 1$ holds at the end vertex c' of P'; by the definition of the augmenting path, any vertex $v \in C_{M_{j-1}}$ appearing in the alternating search tree in M_{j-1} satisfies $\deg_{M_{j-1}}(v) \geq d - 1$.

On the other hand, let $\{c_q, l_q\} \in Q$ be the last edge of Q that appear on P^*. Because it also exists in P' by $Q \subseteq E_{P'}$, an alternating path $(v', \ldots, l_q, \ldots, c_k)$ with respect to M_{j-1} exists. It is included in the alternating search tree with root v' for M_{j-1} constructed in Step 3. However, the end vertex c_k of the path satisfies $\deg_{M_{j-1}}(c_k) \leq d - 2$ as shown above, a contradiction. This completes the proof. $\square$

Theorem 4. *BEC2 produces a balanced edge cover in $O(nm)$ time.*

Proof. At the beginning of the algorithm, a maximum matching M_0 is constructed, and its size is $\nu(G)$. Steps 2 to 5 are iterated $n - 2\nu(G)$ times, which is the number of free vertices for M_0, and in each iteration one edge is added to current M. At the end of the algorithm, no free vertex is left; the obtained M is an edge cover with size $\nu(G) + (n - 2\nu(G)) = n - \nu(G)$, which is the size of a minimum edge cover, by Gallai's theorem. Considering this and Lemma 3,

*BEC*2 produces a balanced edge cover by Theorem 2, then the correctness is proved. It is not difficult to see that the running time is $O(nm)$. $\Box$

Acknowledgement

The authors are grateful to Professor Arie Tamir for his insightful comments about our problem from the view point of jump system.

References

1. Bouchet, A., Cunnigham, W.H.: Delta-matroids, jump systems, and bisubmodular polyhedra. SIAM Journal on Discrete Mathematics 8, 17–32 (1995)
2. Edmonds, J.: Paths, trees, and flowers. Canadian Journal of Mathematics 17, 449–467 (1965)
3. Gallai, T.: Über Extreme Punkt- und Kantenmengen. Annales Universitatis Scientiarum Budapestinensis de Rolando Eötvös Nominatae, Sectio Mathematica 2, 133–138 (1959)
4. Harvey, N.J.A., Ladner, R.E., Lovasz, L., Tamir, T.: Semi-matchings for bipartite graphs and load balancing. In: Dehne, F., Sack, J.-R., Smid, M. (eds.) WADS 2003. LNCS, vol. 2748, pp. 294–308. Springer, Heidelberg (2003)
5. Heinzelman, W., Chandrakasan, A., Balakrishnan, H.: Energy-Efficient Communication Protocol for Wireless Microsensor Networks. In: Proceedings of the Hawaii International Conference on System Sciences (HICSS), pp. 3005–3014 (2000)
6. Heinzelman, W., Chandrakasan, A., Balakrishnan, H.: An Application-Specific Protocol Architecture for Wireless Microsensor Networks. IEEE Transactions on Wireless Communications 1(4), 660–670 (2002)
7. Hopcroft, J.E., Karp, R.M.: An $n^{5/2}$ algorithm for maximum matchings in bipartite graphs. SIAM Journal on Computing 2, 225–231 (1973)
8. Kuhn, H.W.: The Hungarian method for the assignment problem. Naval Research Logistics Quarterly 2, 83–97 (1955)
9. Lovasz, L.: The membership problem in jump systems. Journal of Combinatorial Theory B 70, 45–66 (1997)
10. Micali, S., Vazirani, V.V.: An $O(V^{1/2}E)$ algorithm for finding maximum matching in general graphs. In: Proceedings of the 21st Annual IEEE Symposium on Foundations of Computer Science, pp. 12–27 (1980)
11. Norman, R.Z., Rabin, M.O.: An algorithm for a minimum cover of a graph. In: Proceedings of the American Mathematical Society, vol. 10, pp. 315–319 (1959)
12. Tamir, A.: Least majorized elements and generalized polymatroids. Mathematics of Operations Research 20, 583–589 (1995)
13. Vazirani, V.V.: A theory of alternating paths and blossoms for proving correctness of the $O(\sqrt{V}E)$ general graph maximum matching algorithm. Combinatorica 14, 71–109 (1994)

Firefighting on Trees: $(1 - 1/e)$–Approximation, Fixed Parameter Tractability and a Subexponential Algorithm[*]

Leizhen Cai[1], Elad Verbin[2], and Lin Yang[1]

[1] Department of Computer Science and Engineering, The Chinese University of Hong Kong, Shatin, Hong Kong SAR, China
{lcai,lyang}@cse.cuhk.edu.hk
[2] The Institute For Theoretical Computer Science, Tsinghua University, Beijing, China
elad.verbin@gmail.com

Abstract. The firefighter problem is defined as follows. Initially, a fire breaks out at a vertex r of a graph G. In each subsequent time unit, a firefighter chooses a vertex not yet on fire and protects it, and the fire spreads to all unprotected neighbors of the vertices on fire. The objective is to choose a sequence of vertices for the firefighter to protect so as to save the maximum number of vertices. The firefighter problem can be used to model the spread of fire, diseases, computer viruses and suchlike in a macro-control level.

In this paper, we study algorithmic aspects of the firefighter problem on trees, which is NP-hard even for trees of maximum degree 3. We present a $(1 - 1/e)$-approximation algorithm based on LP relaxation and randomized rounding, and give several FPT algorithms using a random separation technique of Cai, Chan and Chan. Furthermore, we obtain an $2^{O(\sqrt{n}\log n)}$-time subexponential algorithm.

1   Introduction

The Firefighter problem is a one-person's game on a graph G defined as follows. At time $t = 0$, a fire breaks out at a vertex r of G. For each time step $t \geq 1$, a firefighter protects one vertex not yet on fire (the vertex remains protected thereafter), and then the fire spreads from burning vertices (i.e., vertices on fire) to all unprotected neighbors of these vertices. The process ends when the fire can no longer spread, and then all vertices that are not burning are considered *saved*. The objective is to choose a sequence of vertices for the firefighter to protect so as to save the maximum number of vertices in the graph. The Firefighter problem was introduced by Hartnell [Har95] in 1995 and can be used to model the spread of fire, diseases, computer viruses and suchlike in a macro-control level.

The Firefighter problem is NP-hard even for trees of maximum degree 3 as shown by Finbow et al. [FKMR07]. On the other hand, Hartnell and Li [HL00]

[*] Partially supported by Earmarked Research Grant 410206 of the Research Grants Council of Hong Kong SAR, China.

S.-H. Hong, H. Nagamochi, and T. Fukunaga (Eds.): ISAAC 2008, LNCS 5369, pp. 258–269, 2008.

have proved that a simple greedy method for trees is a 0.5-approximation algorithm, and MacGillivray and Wang [MW03] have solved the problem in polynomial time for some special trees. Various aspects of the problem have been considered by Develin and Hartke [DH07], Fogarty [Fog03], Wang and Moeller [WM02], and Cai and Wang [CW07], among others. We refer the reader to a recent survey of Finbow and MacGillivray [FM07] for more information on the **Firefighter** problem.

In this paper, we study algorithmic aspects of the **Firefighter** problem on trees. Our main results are:

1. A $(1 - 1/e)$-approximation algorithm (Section 3) based on a LP-relaxation and randomized rounding. We also prove that $1 - 1/e$ is the best approximation factor one can get using any LP-respecting rounding technique with the same LP (Section 3.1).
2. Several FPT algorithms and polynomial-size kernels (Section 4 and also a summary in Table 1) when we use several different choices for the parameter k: the number of saved vertices, the number of saved leaves, and the number of protected vertices. Our FPT algorithms are based on the random separation method of Cai, Chan and Chan [CCC06].
3. A subexponential algorithm (Section 5) that runs in time $2^{O(\sqrt{n}\log n)}$. We note that an $2^{O(n^{0.33})}$-time algorithm would falsify a conjecture that there is no subexponential algorithm for SAT (see discussions in Section 5).

Table 1. Summary of FPT algorithms. The "randomized complexity" column indicates expected running time of algorithms with one-sided error, and the "deterministic complexity" column gives the worst-case running time of deterministic algorithms.

problems	randomized	deterministic	kernel size
Saving k Vertices	$O(4^k + n)$	$O(n) + 2^{O(k)}$	$O(k^2)$
Saving All But k Vertices	$O(4^k n)$	$2^{O(k)} n \log n$	open
Saving k Leaves	$O(n) + 2^{O(k)}$	$O(n) + 2^{O(k)}$	$O(k^2)$
Saving All But k Leaves	none unless $NP \subseteq RP$	NP-complete for $k = 0$	no kernel
Maximum k-Vertex Protection	$k^{O(k)} n$	$k^{O(k)} n \log n$	open

2   Definitions and Notation

We first define some terms. Let T be a rooted tree with root r which is the origin of the fire. A vertex is *protected* once it is protected by the firefighter, and *saved* if it is not burnt at the end of the game. A *strategy* for the **Firefighter** problem is a sequence $v_1, v_2, \ldots, v_t$ of protected vertices of T such that vertex $v_i, 1 \le i \le t$, is protected at time i and the fire can no longer spread to unprotected vertices at time t.

The following is the decision version of the problem we consider in the paper.

Firefighter on Trees
Instance   A rooted tree T with root r and a positive integer k.
Question   Is there a strategy for the firefighter to save at least k vertices when a fire breaks out at the root r?

Without ambiguity, we abbreviate "the Firefighter problem on trees" as "the Firefighter problem" in the rest of the paper.

We denote the subtree of T rooted at vertex v by $T(v)$, and assign the number of vertices in $T(v)$ as the weight w_v of v. For a strategy S, the *value* of S, denoted by $\|S\|$, equals the number of vertices saved by S.

Denote by L_i the set of vertices of depth i (L_0 contains only the root), and we refer to each L_i as a *level* of T. Let d_v denote the depth of v, and h denote the height of T, i.e., the depth of the deepest leaf. We write $u \preceq v$ if u is an ancestor of v.

An instance of a *parameterized problem* consists of a pair (I, k) with I being the input and k the parameter, and a parameterized problem is *fixed-parameter tractable* (FPT in short) if it admits an *FPT algorithm*, i.e., an algorithm that runs in $f(k) |I|^{O(1)}$ time for some computable function f independent of the input size $|I|$. A *kernelization* for a parameterized problem is a polynomial-time reduction that maps an instance (I, k) onto (I', k') such that (1) $|I'| \leq g(k)$ for some computable function g, (2) $k' \leq k$, and (3) (I, k) is a "Yes"-instance iff (I', k') is a "Yes"-instance. The pair (I', k') is called a *kernel* of (I, k). The kernel is *polynomial-size* if $g(k) = k^{O(1)}$. The existence of a kernel implies the existence of an FPT. The existence of an FPT implies the existence of a kernel, but not necessarily of a polynomial-size kernel.

3   A $(1 - 1/e)$–Approximation Algorithm

In this section we present a $(1 - 1/e)$-approximation algorithm for the Firefighter problem on trees, which improves the $1/2$-approximation of Hartnell and Li [HL00] (note $(1 - 1/e) \approx 0.6321$). Our algorithm, proposed by B. Alspach (see [FM07]), uses randomized rounding of an LP relaxation of a 0-1 integer program formulated by MacGillivray and Wang [MW03]. It is asked in [FM07] to investigate the performance of this algorithm, and in this section we determine the approximation ratio of the algorithm.

It is easy to see that an optimal strategy for a tree protects a vertex at level i at time i and has no need to protect descendants of a protected vertex. This observation translates into the following 0-1 integer program of MacGillivray and Wang [MW03] for a tree $T = (V, E)$, where for vertex v, x_v is a boolean decision variable such that $x_v = 1$ iff v is protected, and w_v is the number of descendants of v.

$$
\begin{aligned}
\text{maximize} \quad & \sum_{v \in V} w_v x_v \\
\text{subject to} \quad & x_r = 0 \\
& \sum_{v \in L_i} x_v \leq 1 && \text{for every level } L_i \text{ with } i \geq 1 \\
& \sum_{v \preceq u} x_v \leq 1 && \text{for every leaf } u \text{ of } T \\
& x_v \in \{0, 1\} && \text{for every vertex } v \text{ of } T
\end{aligned}
\tag{1}
$$

By relaxing the constraint $x_v \in \{0,1\}$ in the above integer program to $0 \le x_v \le 1$, we get a linear program, whose optimal solution will be denoted by OPT_{LP}. The optimal solution to the LP can be interpreted as an optimal "fractional" firefighting strategy[1].

Alspach's rounding method uses the fact that in OPT_{LP}, the values of x_v in each level of the tree sum up to at most 1, and thus they can be treated as a probability distribution. The rounding scheme is to pick the vertex to protect at each level according to this distribution. It might be the case that the fractional values in a level sum up to less than 1. In this case, with the remaining probability we choose to protect no vertex at the level (this makes no difference in the analysis). Also, it might be the case that the rounding procedure chooses to protect both a vertex v and its ancestor u. In this case, we choose to protect u rather than v, and do not protect any vertex in v's level. We call this situation an *annihilation*.

We note that the loss in rounding stems exactly from annihilations. If annihilations never occur, the expected value of the rounded strategy is at least $\|OPT_{LP}\|$ and thus the approximation ratio would be 1. On the other hand, if annihilations occur, consider a vertex v which is fully saved by the fractional strategy and consider the path of length d_v from the root r to v. In the worst case, the fractional strategy assigns a $1/d_v$-fraction of a firefighter to each vertex in this path. In this case, the probability that v is saved by the rounded strategy is equal to

$$1 - (1 - 1/d_v)^{d_v} \ge 1 - 1/e \ .$$

To turn this intuition into a full analysis, we just need to show that the above case is indeed the worst case, and that a similar behavior occurs when v is not fully saved by the fractional strategy. We do this in the following lemma.

Lemma 1. *Given any fractional strategy S_F, let S_I be the integer strategy produced by applying the randomized rounding method to S_F. Then,*

$$\mathbf{E}\left[\|S_I\|\right] \ge \left(1 - \frac{1}{e}\right) \cdot \|S_F\| \ .$$

Proof. Denote the value of the fractional strategy at v by $\tilde{x}_v$ and the value of the rounded strategy at v by x_v. Thus, x_v is an indicator random variable for whether v is protected by S_I. Similarly, for each v, define $\tilde{y}_v = \sum_{u \preceq v} \tilde{x}_v$ to be the fraction of v that is saved by S_F, and $y_v = \sum_{u \preceq v} x_v$ to indicate whether v is saved by S_I.

---

[1] A fractional firefighting strategy is a placement of fractional firefighters on the vertices so that the sum of firefighter fractions assigned to each level is at most 1, and the sum of firefighter fractions assigned to each root-to-leaf path is at most 1. For example, if a vertex v is protected by half of a firefighter, all its descendants are half-saved. If, furthermore, another vertex $u \succ v$ is protected by 0.3 of a firefighter, then all descendants of u are 0.8-saved.

Fix a vertex v, and denote by $r = v_0, v_1, v_2, \ldots, v_k = v$ the path from the root to v. By the definition of the rounding procedure, we see that

$$\mathbf{Pr}\left[y_v = 1\right] = 1 - \prod_{i=1}^{k}(1 - \tilde{x}_{v_i}).$$

We have the following bound

$$\mathbf{Pr}\left[y_v = 1\right] = 1 - \prod_{i=1}^{k}(1 - \tilde{x}_{v_i}) \geq 1 - \left(\frac{\sum_{i=1}^{k}(1 - \tilde{x}_{v_i})}{k}\right)^k = 1 - \left(\frac{k - \sum_{i=1}^{k}\tilde{x}_{v_i}}{k}\right)^k$$

$$= 1 - \left(\frac{k - \tilde{y}_v}{k}\right)^k = 1 - \left(1 - \frac{\tilde{y}_v}{k}\right)^k \geq 1 - e^{-\tilde{y}_v} \geq \left(1 - \frac{1}{e}\right)\tilde{y}_v,$$

where the first inequality follows from the inequality of the means, and the second and third inequalities follow from standard analysis, using the fact that $0 \leq \tilde{y}_v \leq 1$. Note that the sum of all $\tilde{y}_v$ is just the value of S_F. Therefore,

$$\mathbf{E}\left[\|S_I\|\right] = \sum_{v \in V} \mathbf{Pr}\left[y_v = 1\right] \geq \sum_{v \in V} \left(1 - \frac{1}{e}\right)\tilde{y}_v = \left(1 - \frac{1}{e}\right) \cdot \|S_F\|,$$

where the first equality follows from linearity of expectation. This finishes the proof of the lemma. $\qquad\square$

The above lemma implies that the expected approximation ratio of our algorithm is $(1 - 1/e)$. We can easily derandomize our algorithm by using the method of conditional expectations [AS92], which will be discussed in the full paper.

Theorem 1. *There is a deterministic polynomial-time $(1 - 1/e)$-approximation algorithm for the firefighter problem on trees.*

3.1   LP-Respecting Rounding Does Not Achieve Approximation Better Than $1 - 1/e$

We note that the best known integrality gap for MacGillivray and Wang's LP, proved by Hartke [Har06], is $\frac{16}{17}$. Thus, it is tempting to believe that the rounding method might be improvable. However, we can show that no rounding technique from a relatively rich class of rounding techniques gives an approximation ratio better than $1 - 1/e$. This means that one would have to try something very different than standard rounding methods.

A common feature of many rounding techniques in the literature is that they are *LP-respecting*. A rounding technique for the firefighter problem is called LP-respecting if it only chooses to protect vertices v with $\tilde{x}_v > 0$, and never protects any vertex v with $\tilde{x}_v = 0$. (Recall that $\tilde{x}_v$ is the value of the optimal LP solution on vertex v.) The following theorem states that *any* LP-respecting rounding technique, when used together with [MW03]'s LP, does not achieve an approximation ratio better than $1 - 1/e$. Consequently, any rounding technique that aims at getting better than $(1 - 1/e)$–approximation would have to be *LP-disrespecting*.

Theorem 2. *For any $\epsilon > 0$, there exists a tree T, and an optimal fractional solution $\tilde{S}$ for the LP on T, such that any integral strategy $S \subseteq \{v : \tilde{x}_v > 0\}$ can save no more than $(1 - 1/e + \epsilon) \cdot \|OPT_{IP}(T)\|$ vertices.*

The proof of this theorem is somewhat technical, and we defer it to the full version of the paper.

4   FPT Algorithms

In this section, we consider FPT algorithms and polynomial-size kernels for three parameterized versions of the **Firefighter** problem.

1. Saving k Vertices: The parameter k is the number of saved vertices, and we ask if there is a strategy saving at least k vertices.
2. Saving k Leaves: The parameter k is the number of saved leaves, and we wish to determine if the firefighter can save at least k leaves of the tree.
3. Maximum k-Vertex Protection: The parameter k is the number of protected vertices and we wish to find a strategy for the firefighter to protect k vertices to maximize the total number of saved vertices.

The results in this section are summarized in Table 1.

The main tool we use is the *random separation* method of Cai, Chan, and Chan [CCC06]. This technique produces randomized algorithms, which can be derandomized by using *universal sets* (see [NSS95]). A set of binary vectors of length n is (n, t)-*universal* if for every subset of size t of the indices, all 2^t configurations appear in the set. Naor et al. [NSS95] give a construction of a (n, t)-universal set of cardinality $2^t t^{O(\log t)} \log n$ in time $2^t t^{O(\log t)} n \log n$.

4.1   Saving k Vertices

First we use random separation to give an $2^{O(k)} n$-time algorithm for Saving k Vertices. Then we construct a kernel of size $O(k^2)$ for the problem. Finally we use the random separation method to solve the parametric dual problem Saving All But k Vertices in time $2^{O(k)} n$.

We start with our FPT algorithm for Saving k Vertices. Call a strategy S *satisfying* if it saves at least k vertices. The goal is then to find a satisfying strategy if one exists. First observe that if the root r has a child v with weight $w_v \geq k$ then we can protect v to solve the problem. Therefore we can assume from now on that the weight of every vertex in $V - r$ is at most $k - 1$. In particular, this means that the height of T is at most k.

The algorithm first colors each vertex of T randomly and independently by either green or red with equal probability. We call a coloring of T *good* if T has a satisfying strategy S_0 such that all vertices in S_0 are green, and all descendants of vertices in S_0 are red.

Given a good coloring of T, we can find a satisfying strategy S to T as follows:

Step 1. Find the set V_g of green vertices whose descendants are all red.
Step 2. For each level L_i ($i \geq 1$), choose from $V_g \cap L_i$ a vertex of maximum weight, and put it in S. (If $V_g \cap L_i = \emptyset$ then do nothing).

It's not hard to see that S is a satisfying strategy, as for each vertex v of S_0, v is in $V_g \cap L_{d_v}$ and can be placed in S, thus $\|S\| \geq \|S_0\|$. Therefore S is indeed a satisfying strategy, and we can find it in $O(n)$ time, given a good coloring.

However, the probability of obtaining a good coloring depends on the sum of the number of vertices in S_0 and the number of the descendants of S_0, which might be as large as $\Theta(k^2)$. We can reduce this sum to $2k$ by using the following simple existence lemma:

Lemma 2. *Suppose that every non-trivial subtree of T is of size less than k, and that there is a satisfying strategy. Then there is a satisfying strategy S_0 that saves at most $2k$ vertices.*

Proof. Let S_1 be some satisfying strategy. If S_1 saves at most $2k$ vertices, then we are done. Otherwise, we construct S_0 as follows. Since $w_v < k$ for each vertex v of S_1, we can add vertices from S_1 to S_0 in turn until S_0 saves at least k vertices. This S_0 saves at most $2k$ vertices. $\square$

Thus, if we choose S_0 not as an arbitrary satisfying strategy, but as the satisfying strategy guaranteed in Lemma 2, then the probability that a randomly-chosen coloring is good is at least $1/2^{2k}$. By choosing 4^k colorings and running steps 1 and 2 for each of the colorings, we succeed in finding a satisfying strategy, if one exists, with at least constant probability.

To derandomize the algorithm, we can use a $(n, 2k)$-universal set, and use the vectors of the set as the colorings. If a satisfying strategy exists, then at least one of the colorings will be good. Therefore we have a deterministic FPT algorithm that runs in time $4^k k^{O(\log k)} n \log n$.

We now present an $O(k^2)$-size kernel for Saving k Vertices on trees. First, note that if r has a child v with $w_v \geq k$, we can protect v to save at least k vertices. In such a case, the problem is solvable in time $O(n)$, and we can use a trivial kernel for it. From now on we assume that the weight of each child of r is at most $k - 1$ and in particular the height h of T is at most $k - 1$.

The idea behind the kernel is to ignore all but a limited number of vertices in each level of the tree. Its construction is as follows:

Step 1. If level 1 has at least k vertices then we put the k largest-weight vertices of level 1 into K_1 else we put all vertices of level 1 into K_1.
Step 2. For $i := 2$ to h, if level i has at least $2k - i$ vertices then we put the $2k - i$ largest-weight vertices of level i into K_i else we put all vertices of level i into K_i.
Step 3. The kernel is $K = \bigcup_{i=1}^{h} K_i$.

Note that the above construction of K can be performed in time $O(n)$. Also note that K has at most $k + \sum_{i=2}^{h}(2k - i) \leq 3k^2/2$ vertices. We prove that K is a kernel in the following lemma:

Lemma 3. *If T has a satisfying strategy S, then there exists a satisfying strategy $S' \subseteq K$.*

Proof. Let $S = \{v_1, v_2, \ldots, v_t\}$ be a satisfying strategy, where v_i is in level i. Define $I(S)$ to be the largest i such that $v_i \notin K_i$ (define $I(S) = \infty$ if $S \subseteq K$). We say that S is *minimal* iff the removal of any vertex from S would make S non-satisfying. We let S be a satisfying strategy that maximizes $I(S)$. Furthermore, among all of those we choose S to be a strategy which is minimal. Let $i = I(S)$. If $i = \infty$, we are finished. We assume $i < \infty$, and eventually reach a contradiction.

We will now show how to replace vertex v_i of S by a vertex in K_i to get another satisfying strategy. If $i = 1$ then $|K_1| = k$ as $v_1 \notin K_1$. Note that at most $k - 1$ vertices in K_1 are ancestors of vertices in S as $|S| \leq k$. Therefore K_1 has a vertex v that is not an ancestor of any vertex of S. Since $w_v \geq w_{v_1}$, we can remove v_1 from S and insert v, to get a satisfying strategy S' of T with $i(S') > i(S)$, contradicting the choice of S. Otherwise $i > 1$. Since $v_i \notin K_i$, we have $|K_i| = 2k - i$. Note that at most $k - i$ vertices in K_i are ancestors of vertices in S as $|S| \leq k$. Furthermore, by the minimality of S, at most $k - 1$ vertices in K_i are descendants of vertices in S (otherwise S does not need vertex v_i). Therefore K_i has at least one vertex v that is neither an ancestor nor a descendent of any vertex in S. By the definition of K_i, $w_v \geq w_{v_i}$ and we can replace v_i in S by v to get a satisfying strategy S' of T with $i(S') > i(S)$, contradicting the choice of S. $\square$

The $O(k^2)$ kernel can be easily combined with the FPT algorithm to establish the following result.

Theorem 3. *Saving k Vertices can be solved in $O(n) + 4^k k^{O(\log k)}$ time.*

For the parametric dual **Saving All But k Vertices** of **Saving k Vertices**, we can also use random separation to obtain an FPT algorithm that runs in $2^{O(k)} n \log n$ time. The main difference is that we use a random coloring to "guess" the burnt part instead of the saved part for **Saving k Vertices**. The details will be given in the full paper.

4.2   Saving k Leaves and Protecting k Vertices

In this section, we consider FPT algorithms for **Saving k Leaves** and **Maximum k-Vertex Protection**. The former uses the number of saved leaves as the parameter k, and the latter tries to save the maximum number of vertices by protecting k vertices.

We start with **Saving k Leaves**, which deals with the situation that leaves are much more valuable than internal vertices. Due to space limit, we will only sketch the main ideas of our FPT algorithm and leave the details to the full paper. Note that the parametric dual **Saving All But k Leaves** of **Saving k Leaves** is NP-complete even for $k = 0$, which was shown by Finbow et. al. [FKMR07]. Thus **Saving All But k Leaves** has no FPT algorithm unless $P \neq NP$.

To solve **Saving k Leaves**, it is possible to use the algorithm in the latter part of this section for **Maximum k-Vertex Protection**, which takes $k^{O(k)} n$ time. Here we describe an algorithm that takes time $2^{O(k)}poly(n)$. To get such an algorithm, it is tempting to try the same approach we used for the Saving k Vertices problem, but such an approach does not work: for Saving k Vertices we used the fact that if there is a strategy that saves at least k vertices, then there is a strategy whose subtrees are of total size $O(k)$. This does not hold for Saving k Leaves.

The main difficulty comes from the *snakes* in the tree. A *snake* is a path $(v_1, v_2, \ldots, v_\ell)$ with $\ell \geq 2$, such that for each $i = 1, \ldots, \ell - 1$, v_i has only a single child, v_{i+1}. (In other words, a snake is an induced path). If snakes do not exist, then a random separation algorithm similar to the one we gave for Saving k Vertices solves Saving k Leaves in time $2^{O(k)}n$. Clearly, the difficulty lies in dealing with snakes. To this end, we use a random separation approach together with an algorithm for finding a maximum matching, which is used to decide which vertex to protect inside each snake.

The first step of the algorithm is to contract each snake, and to get a snake-free tree T'. We then apply the random separation method on T' to find a satisfying (saving at least k leaves) strategy S'. Then we somehow transform S' into a satisfying strategy S of T. Note that every vertex v' in T' corresponds to either a snake or a single vertex in the original tree T, and to save the leaves that v' saves, we can protect any vertex in the snake that corresponds to v'. With this observation we can understand the transformation from S' to S as a scheduling problem, with the vertices in S' being the set of tasks, their corresponding snakes indicating the starting time and the deadline of the tasks, and each level in T being a free slot in the processing queue. We formulate this problem into a bipartite graph and find the satisfying strategy using the maximal matching algorithm for bipartite graphs. We will discuss details in the full paper.

We can use an approach similar to the one used for Saving k Vertices to obtain an $O(k^2)$-size kernel of Saving k Leaves, which can be combined with the above FPT algorithm to obtain the following result:

Theorem 4. *Saving k Leaves can be solved in $O(n) + 2^{O(k)}$ time.*

We now turn to Maximum k-Vertex Protection, the problem of protecting k vertices to save the maximum number of vertices. Note that for any tree of height k, an optimal strategy needs only protect at most k vertices. Therefore Maximum k-Vertex Protection can be also regarded as a parameterized version of the firefighter problem on trees when we take the height of a tree as the parameter k.

Theorem 5. *Maximum k-Vertex Protection can be solved in $k^{O(k)}n$ time by a randomized algorithm.*

Proof. We color the vertices randomly and independently, with probability $\frac{1}{k}$ to be green and probability $1 - \frac{1}{k}$ to be red. Let S_0 be an assumed optimal strategy. We call a coloring c *good* if all vertices in S_0 are green and all of their ancestors are red. If the coloring is good, then we can find an optimal strategy by the same procedure that we used in the algorithm for Saving k Vertices, in time $O(n)$. Now, since there are k vertices in S_0, and each of them has at most k ancestors, the probability for the coloring to be good is at least

$$\geq \left(\frac{1}{k}\right)^k \cdot \left(\frac{k-1}{k}\right)^{k^2} \geq \left(\frac{1}{k}\right)^k \left(\frac{1}{4}\right)^k = k^{-O(k)}$$

Thus, picking $k^{O(k)}$ colorings allows us to find the optimal strategy with probability at least a constant. $\qquad\square$

This algorithm cannot be derandomized by using normal universal sets while maintaining the running time, because we color each vertex green or red with unequal probabilities. We can use equal probability for green and red and then use a (n, k^2)-universal set to derandomize the algorithm, but the resulting algorithm runs in time of $2^{O(k^2)}n\log n$. To derandomize the above algorithm efficiently, Verbin [Ver] has recently introduced *asymmetric universal sets* which can be used to yield a deterministic algorithm that runs in time $k^{O(k)}n\log n$.

Question 1. *Is there an algorithm for* Maximum k-Vertex Protection *that runs in time* $2^{o(k\log k)}poly(n)$?

5   A Subexponential Algorithm

In this section we present an algorithm for exactly solving the firefighter problem on trees. The algorithm takes time $n^{O(\sqrt{n})} = 2^{O(\sqrt{n}\log n)}$ on any tree with n vertices.

The main idea is to use pruning, coupled with a careful analysis of the size of the space of feasible solutions. We note that the size of the space of feasible solutions can be $2^{\Omega(n)}$, and an exhaustive approach is clearly not sufficient. On the other hand, an exhaustive approach is good enough when the tree is somewhat balanced, and in particular if the tree is of height $O(\sqrt{n})$. Our algorithm deals with non-balanced trees by detecting parts of the tree that are small and "costly to save", and pruning such parts. By "costly to save", we mean that there are vertices on the same levels such that if we protect them, we will save many more vertices.

We now present the algorithm FF-SUBEXP, which operates recursively, and solves the firefighter problem in time $n^{O(\sqrt{n})}$. Recall that r denotes the root of T, w_v denotes the number of vertices in the subtree rooted at v, d_v denotes the depth of v, and *level i* refers to the set of all vertices of depth i. We set the parameter $k_0 = \sqrt{n}$, which is fixed throughout the execution of the algorithm, even when we call the algorithm recursively.

FF-SUBEXP works as follows. Its input is T, a tree with n vertices. Its output is an optimal firefighter strategy for T.

1. If $n \leq k_0$, run a brute-force search (taking time $O(n^{k_0})$) and return the result it gives. Otherwise, continue to step 2.
2. If r has some child v with $w_v \leq k_0$, then:
   (a) Construct a tree T' which is identical to T except that the subtree rooted at v is completely deleted. Run FF-SUBEXP recursively on T' to get an optimal strategy S' for T'.

   (b) Calculate the best strategy S'' out of the strategies that protect one vertex in each of the levels 1 through w_v, and no vertices below level w_v. Do this using the naive brute-force algorithm that takes time $O(n^{w_v+1})$.

   (c) Pick the strategy among S' and S'' which, when run on T, gives the best result. Return it. /* we will prove that this strategy is optimal in the full paper */

3. else: /* all children of the root are subtrees of size at least k_0. Here we'll do a sort of brute-force search */

   (a) Go over all children $v_1, \ldots, v_\ell$ of r. For each v_i do:

      – Find the best strategy, S_i, among all strategies that protect v_i. Do this by recursively running FF-SUBEXP on a tree with $n - w_{v_i} - \ell$ vertices, produced by deleting v_i's subtree, deleting all level-1 nodes, and making all level-2 nodes into direct descendants of r.

   (b) Pick the strategy among $S_1, \ldots, S_\ell$ that gives the best result when applied to T, and return it.

Due to the space constraint, we will defer the correctness proof and complexity analysis of the algorithm to the full paper.

It is interesting to note that the NP-hardness reduction of Finbow, King, MacGillivray and Rizzi in [FKMR07] in fact implies that an $2^{O(n^{0.33})}$-time algorithm for the firefighter problem on trees would imply an $2^{o(n)}$-time algorithm for solving 3-SAT on instances with n variables and $O(n)$ clauses. This would falsify a conjecture of Impagliazzo et al. [IP01, IPZ01]. Furthermore, the reduction of [FKMR07] also implies that an $n^{O(k^{0.99})}$-time algorithm for the firefighter problem on trees of height k would falsify the same conjecture of Impagliazzo et al. . Recall that the trivial implementation of step 2b in the algorithm takes time $O(n^k)$, which means that the implementation of that step, although naive, is likely to be close to optimal.

Question 2. *Is there an exact $2^{n^{1/3}polylog(n)}$-time algorithm for the firefighter problem on trees? Alternatively, would an $2^{n^{1/2}/polylog(n)}$-time algorithm for the problem imply an $2^{o(n)}$-time algorithm for 3-SAT with n variables and $O(n)$ clauses?*

References

[AS92]   Alon, N., Spencer, J.H.: The Probabilistic Method. Wiley, Chichester (1992)

[CCC06]   Cai, L., Chan, S.M., Chan, S.O.: Random separation: A new method for solving fixed-cardinality optimization problems. In: Bodlaender, H.L., Langston, M.A. (eds.) IWPEC 2006. LNCS, vol. 4169, pp. 239–250. Springer, Heidelberg (2006)

[CW07]   Cai, L., Wang, W.: The surviving rate of a graph (manuscript, 2007)

[DH07]   Develin, M., Hartke, S.G.: Fire containment in grids of dimension three and higher. Discrete Appl. Math. 155(17), 2257–2268 (2007)

[FKMR07] Finbow, S., King, A., MacGillivray, G., Rizzi, R.: The firefighter problem for graphs of maximum degree three. Discrete Mathematics 307(16), 2094–2105 (2007)

[FM07] Finbow, S., MacGillivray, G.: The firefighter problem: a survey (manuscript, 2007)

[Fog03] Fogarty, P.: Catching the Fire on Grids, M.Sc. Thesis, Department of Mathematics, University of Vermont (2003)

[Har06] Hartke, S.G.: Attempting to narrow the integrality gap for the firefighter problem on trees. In: Discrete Methods in Epidemiology. DIMACS Series in Discrete Mathematics and Theoretical Computer Science, vol. 70, pp. 179–185 (2006)

[Har95] Hartnell, B.: Firefighter! An application of domination. In: 24th Manitoba Conference on Combinatorial Mathematics and Computing, University of Minitoba, Winnipeg, Cadada (1995)

[HL00] Hartnell, B., Li, Q.: Firefighting on trees: how bad is the greedy algorithm? Congr. Numer. 145, 187–192 (2000)

[IP01] Impagliazzo, R., Paturi, R.: On the complexity of k-SAT. J. Comput. Syst. Sci. 62(2), 367–375 (2001)

[IPZ01] Impagliazzo, R., Paturi, R., Zane, F.: Which problems have strongly exponential complexity? J. Comput. Syst. Sci. 63(4), 512–530 (2001)

[MW03] MacGillivray, G., Wang, P.: On the firefighter problem. J. Combin. Math. Combin. Comput. 47, 83–96 (2003)

[NSS95] Naor, M., Schulman, L.J., Srinivasan, A.: Splitters and near-optimal derandomization. In: IEEE Symposium on Foundations of Computer Science, pp. 182–191 (1995)

[Ver] Verbin, E.: Asymmetric universal sets (in preparation)

[WM02] Wang, P., Moeller, S.: Fire control on graphs. J. Combin. Math. Combin. Comput. 41, 19–34 (2002)

A New Algorithm for Finding Trees with Many Leaves*

Joachim Kneis, Alexander Langer, and Peter Rossmanith

Dept. of Computer Science, RWTH Aachen University, Germany

Abstract. We present an algorithm that finds trees with at least k leaves in undirected and directed graphs. These problems are known as MAXIMUM LEAF SPANNING TREE for undirected graphs, and, respectively, DIRECTED MAXIMUM LEAF OUT-TREE and DIRECTED MAXIMUM LEAF SPANNING OUT-TREE in the case of directed graphs. The run time of our algorithm is $O(poly(|V|) + 4^k k^2)$ on undirected graphs, and $O(4^k |V| \cdot |E|)$ on directed graphs. This improves over the previously fastest algorithms for these problems with run times of $O(poly(|V|) + 6.75^k poly(k))$ and $2^{O(k \log k)} poly(|V|)$, respectively.

1 Introduction

In this paper we consider the graph theoretical problems of finding trees and spanning trees in graphs, so that their number of leaves is maximal. To be more precise, given a graph G and a number k, we are to find a (spanning) tree with at least k leaves. For undirected graphs, the terms *tree* and *spanning tree* are well-known. These terms translate to *out-tree* and *spanning out-tree* on directed graphs. Here, a (spanning) out-tree is a rooted tree, such that every leaf (every node of G) can be reached from the root via a directed path within this tree.

Being a problem that has many practical applications, e.g., in network design [10,18,21,24], it is already widely studied with regard to its complexity and approximability. All versions are APX-hard [15] and there is a polynomial time 2-approximation for undirected graphs [23] and a 3-approximation in almost linear time [20]. On cubic graphs, a 3/2-approximation was found recently [8].

In the area of parameterized algorithms, the MAXIMUM LEAF SPANNING TREE problem is very prominent. Parameterized complexity theory is an approach to explore whether hard problems can be solved exactly with a run time that comes close to polynomial time on well-behaved instances. Formally, a parameterized problem L is a set of pairs (I, k) where I is an *instance* and k the *parameter*. A parameterized problem L is called *fixed parameter tractable* and belongs to the complexity class FPT if there is an algorithm that decides membership of L in time $f(k)poly(|I|)$, where f is an arbitrary function. If the parameter is small, such an algorithm can be quite efficient in spite of the NP-hardness of the problem — in particular if f is a moderately exponential function.

---

* Supported by the DFG under grant RO 927/7-1.

S.-H. Hong, H. Nagamochi, and T. Fukunaga (Eds.): ISAAC 2008, LNCS 5369, pp. 270–281, 2008.

The parameterized version of the undirected problem is defined as follows:

Maximum Leaf Spanning Tree (MLST)
Input: An undirected graph $G = (V, E)$, a positive integer k
Parameter: k
Question: Does G have a spanning tree with at least k leaves?

It is long known that MLST $\in$ FPT because a graph G contains a k-leaf spanning tree iff G has a $K_{1,k}$ (a k-star) as a minor [13]. However, this uses the graph minor theorem from Robertson and Seymour [22] and only proves the existence of an algorithm with running time $f(k)|V|^3$. The first explicit algorithm is due to Bodlaender [3], who uses the fact that G does contain a $K_{1,k}$ as a minor if its treewidth is larger than w_k, a value that depends on k. The algorithm hence tests if the treewidth of G is bigger than w_k. In this case, the algorithm directly answers "yes". Otherwise, it uses dynamic programming on a small tree decomposition of G. The overall run time is roughly $O((17k^4)! \, |G|)$. In the following years, the run time of algorithms deciding MLST was improved further to $O((2k)^{4k} poly(|G|))$ by Downey and Fellows [11], and to $O(|G| + 14.23^k k)$ by Fellows, McCartin, Rosamond, and Stege [14]. The latter was the first algorithm with an exponential cost function $2^O(k) \cdot poly(|G|)$ and the first algorithm that employs a *small problem kernel*: In polynomial time an instance (G, k) of MLST is reduced to an equivalent instance (G', k') with $|G'| \leq f(k)$ and $k' \leq k$. Note that the existence of a small problem kernel for a parameterized problem implies that the respective problem is in FPT.

Bonsma, Brueggemann, and Woeginger [5] use an involved result from extremal graph theory by Linial and Sturtevant [19], and Kleitman and West [17] to bound the number of nodes that can possibly be leaves by $4k$. A brute force check for each k-subset of these $4k$ nodes yields a run time bound of $O(|V|^3 + 9.4815^k k^3)$. A new problem kernel of size $3.75k$ by Estivill-Castro, Fellows, Langston, and Rosamond [12] improves the exponential factor of this algorithm to 8.12^k [4]. The currently best known algorithm for MLST is due to Bonsma and Zickfeld [9], who reduce the instances to graphs without certain subgraphs called diamonds and blossoms that admit a better extremal result, obtaining a run time bound of $O(poly(|V|) + 6.75^k poly(k))$.

In the directed case, we have to distinguish between the following variants:

Directed Maximum Leaf Out-Tree (DMLOT)
Input: A directed graph $G = (V, E)$, a positive integer k
Parameter: k
Question: Does G contain a rooted out-tree with at least k leaves?

Directed Maximum Leaf Spanning Out-Tree (DMLST)
Input: A directed graph $G = (V, E)$, a positive integer k
Parameter: k
Question: Does G have a spanning out-tree with at least k leaves?

While it is easy to see that a k-leaf tree in an undirected graph can always be extended to a k-leaf spanning tree, this is not the case for directed graphs that are not strongly connected (see Figure 1 [6]).

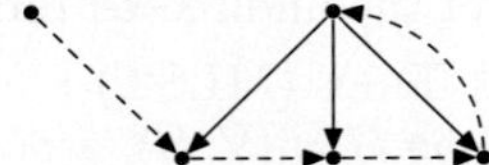

Fig. 1. A graph containing a 3-leaf out-tree, but no 3-leaf spanning tree

For both of these problems, membership in FPT was discovered only recently, since neither the graph minor theorem by Robertson and Seymour in its current shape, nor the method used by Bodlaender, nor the extremal results by Kleitman-West are applicable for directed graphs.

In the case of DMLOT, Alon, Fomin, Gutin, Krivelevich, and Saurabh [2] proved an extremal result for directed graphs, so that either a k-leaf out-tree exists, or the pathwidth of the underlying graph is bounded by $2k^2$. This allows dynamic programming, so that an overall run time bound of $2^{O(k^2 \log k)} poly(|V|)$ can be achieved, answering the long open question whether DMLOT is fixed parameter tractable. They could further improve this to $2^{O(k \log^2 k)} poly(|V|)$ and, if G is acyclic, to $2^{O(k \log k)} poly(|V|)$ [1].

The more important question, if DMLST $\in$ FPT, remained open. Only very recently, Bonsma and Dorn [6] were able to answer this question in the affirmative. Their approach is based on pathwidth and dynamic programming as well and yields a run time bound of $2^{O(k^3 \log k)} poly(|V|)$. In a subsequent paper [7], they proved that a run time of $2^{O(k \log k)} poly(|V|)$ suffices to solve both, DMLOT and DMLST.

Our Contribution

Recall that in the directed case a k-leaf out-tree cannot necessarily be extended to a k-leaf spanning out-tree even if G does contain a spanning out-tree (see Figure 1). In this paper, we use the fact that a k-leaf out-tree with root r *can* always be extended to a k-leaf spanning out-tree if G does contain a spanning out-tree rooted in r.

We develop a new algorithm that — in contrast to the prior approaches based on extremal graph theory — grows an out-tree from the root and therefore solves both DMLOT and DMLST. The algorithm recursively selects and tries two of the many possible ways to extend the tree. We prove that at least one of these recursive calls finds a k-leaf tree, if such a tree exists. The number of recursive calls can be bounded by $2^{2k} = 4^k$. The same algorithm can be used to solve MLST.

2   Preliminaries

Let $G = (V, E)$ be a graph, and let $n := |V|$ and $m := |E|$ be the number of vertices and edges, respectively. If G is undirected, we call a (spanning) tree T in G a k-*leaf (spanning) tree* iff T has at least k leaves. If G is a directed graph, a rooted out-tree T is a tree in G, such that T has a unique root $r = \mathrm{root}(T)$, and each vertex in T can be reached by a unique directed path from r in T. A k-*leaf out-tree* is an out-tree with at least k leaves, and a *k-leaf spanning out-tree* is a k-leaf out-tree that is also a spanning out-tree.

In this paper, we do not distinguish between directed and undirected graphs except when explicitly stated. The respective results and the algorithm can easily be transferred from directed graphs to undirected graphs and vice versa — in particular, if undirected graphs are seen as symmetric directed graphs, where every edge has a *reverse edge*. Such a representation is commonly used by algorithmic graph libraries like LEDA. Edges are therefore denoted by (u, v), and we universally use the terms *tree* and *spanning tree*. Without loss of generality ($k > 2$), trees in undirected graphs are assumed to be rooted.

Let T be a tree in G. $V(T)$ denotes the set of nodes of T, $E(T)$ the set of edges of T. The root, leaves, and inner nodes of T are denoted by $\mathrm{root}(T)$, $\mathrm{leaves}(T)$ and $\mathrm{inner}(T) := V(T) \setminus \mathrm{leaves}(T)$, respectively. We denote by $N(v) := \{u \in V \mid (v, u) \in E\}$ the set of all neighbors of $v \in V$, $N[v] := N(v) \cup \{v\}$, and for $U \subseteq V$ we let $N(U) := \bigcup_{u \in U} N(u)$. For a tree T and $v \in V$, we set $N_{\overline{T}}(v) := N(v) \setminus V(T)$. Similarly, $N_{\overline{T}}(U) := N(U) \setminus V(T)$ for $U \subseteq V$. For $v \in V$, let $T_v := (N[v], \bigcup_{u \in N(v)} \{(v, u)\})$ be the star rooted in v that contains all neighbors of v.

Recall that our algorithm grows a tree from the root. To do so, the algorithm further distinguishes between leaves of trees that will be leaves in the final k-leaf tree (R), and leaves that are still allowed to become inner nodes (B), when the tree is *extended* by the algorithm. This extension consists of the complete remaining neighborhood of the particular node. The resulting tree T will be such that each inner node has all of its neighbors in $V(T)$. We call such trees *inner-maximal* trees.

Definition 1. *Let $G = (V, E)$ be a graph, and let T be a tree. If $N(\mathrm{inner}(T)) \subseteq V(T)$, we call T an* inner-maximal *tree. A* leaf-labeled *tree is a 3-tuple (T, R, B), such that T is a tree, and R and B form a partition of $\mathrm{leaves}(T)$. (T, R, B) is an* inner-maximal leaf-labeled *tree, if T is inner-maximal.*

For trees $T \neq T'$, we say T' extends T, denoted by $T' \succ T$, iff $\mathrm{root}(T') = \mathrm{root}(T)$ and T is an induced subgraph of T'. If (T, R, B) is a leaf-labeled tree and T' is a tree such that $T' \succ T$ and $R \subseteq \mathrm{leaves}(T')$ (R-colored leaves of T remain leaves in T'), we say T' is an (leaf-preserving) extension of (T, R, B), denoted by $T' \succ (T, R, B)$. We say a leaf-labeled tree (T', R', B') extends a leaf-labeled tree (T, R, B), denoted by $(T', R', B') \succ (T, R, B)$, iff $T' \succ (T, R, B)$.

Lemma 1. *Let (T, R, B) be an inner-maximal leaf-labeled tree, and $T' \succ (T, R, B)$ a leaf-preserving extension of (T, R, B). Then $B \neq \emptyset$.*

Proof. Since $T \neq T'$, there is $x \in V(T')$ with $x \notin V(T)$. Let $x_1 := \mathrm{root}(T) = \mathrm{root}(T')$ and consider the path $x_1, \dots, x_l$ with $x_l = x$ from x_1 to x in T'. Since $x = x_l \notin V(T)$, there is some i such that $x_i \in V(T)$ and $x_{i+1} \notin V(T)$. It is $x_i \in \mathrm{leaves}(T) = R \cup B$, because T is inner-maximal. On the other hand, $x_i \in \mathrm{inner}(T')$, and with $R \subseteq \mathrm{leaves}(T')$, we have $x_i \in B$. $\qquad\square$

3 k-Leaf Trees Versus k-Leaf Spanning Trees

In this section, we show when and how k-leaf trees can be extended to k-leaf spanning trees. For this to work, remember that we consider trees with *at least* k leaves. In particular, we allow that the resulting spanning tree has more leaves than the originating k-leaf tree. While Lemma 2 can be considered folklore, Lemma 3 is a new contribution that significantly eases our search for k-leaf spanning trees in directed graphs.

Lemma 2. *A connected, undirected graph $G = (V, E)$ contains a k-leaf tree iff G contains a k-leaf spanning tree. Furthermore, each k-leaf tree can be extended to a k-leaf spanning tree in time $O(n + m)$.*

Proof. Let T be a tree in G with at least k leaves, and let $l := |V - V(T)|$ be the number of nodes that are not part of T. If $l = 0$, then T is a spanning tree with at least k leaves. If otherwise $l > 0$, choose $u \in V(T)$ and $v \in N_{\overline{T}}(V(T))$, such that u and v are adjacent. Let $T' := T + \{u, v\}$. It is easy to see that T' has at least as many leaves as T. Furthermore, this operation can efficiently be done with a breadth-first-search on G starting in $V(T)$, and hence after at most $O(n + m)$ steps a spanning tree with at least k leaves can be constructed from T.   □

In the undirected case, it is therefore sufficient to search for an arbitrary tree with at least k leaves. If an explicit k-leaf spanning tree is asked for, the k-leaf tree can then be extended to a spanning tree using an efficient postprocessing operation.

Lemma 2 is, however, not applicable for directed graphs (cf., Figure 1): It is easy to see that this graph contains an out-tree with three leaves, but the unique spanning out-tree contains only one leaf. If we fix the root of the trees, we obtain the following weaker result for directed graphs.

Lemma 3. *Let $G = (V, E)$ be a directed graph. If G contains a k-leaf spanning out-tree rooted in x_1, then any k-leaf out-tree rooted in x_1 can be extended to a k-leaf spanning out-tree of G in time $O(n + m)$.*

Proof. Let T be a k-leaf out-tree with $\mathrm{root}(T) = x_1$ and let $l := |V - V(T)|$ be the number of nodes that are not in T. If $l = 0$, then T is a spanning out-tree for G with at least k leaves. If $l > 0$, choose $x \in V - V(T)$ and consider a path $x_1, x_2, \ldots, x_s$ with $x_s = x$ from x_1 to x. Since G has a spanning out-tree rooted in x_1, such a path must exist in G. Furthermore, $x \notin V(T)$ and hence there is $1 \le i \le s$ such that $x_i \in V(T)$ and $x_j \notin V(T)$ for each $j = i + 1, \ldots, s$. It is easy to see that by adding the path $x_i, \ldots, x_s$ to T, the number of leaves does not decrease. Repeating this procedure yields a spanning out-tree for G that has at least k leaves. Again, this can be efficiently done with a breadth-first-search on G, which starts in T and takes time at most $O(n + m)$. See Figure 2 for an illustration.   □

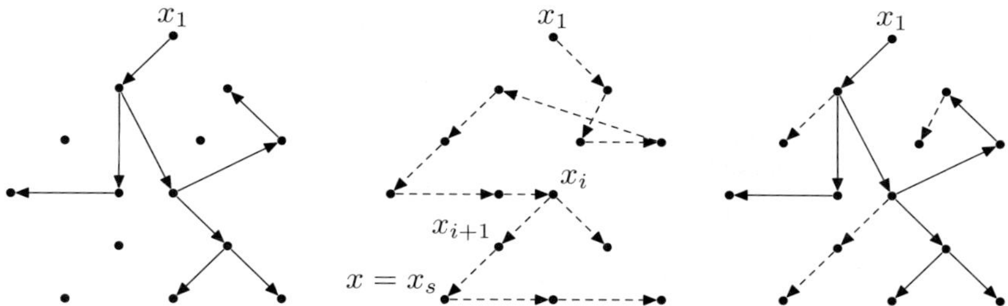

Fig. 2. How to extend a k-leaf out-tree into a k-leaf spanning out-tree: For the ease of illustration, we do not show all the edges in G. A 4-leaf out-tree with root x_1 is depicted in the first figure. The second figure shows an arbitrary spanning out-tree rooted in x_1, we chose one with two leaves. We can extend the first out-tree with edges from the spanning out-tree so that all nodes are covered.

4   The Algorithm

In this section, we introduce Algorithm 1, which given an inner-maximal leaf-labeled tree (T, R, B) recursively decides whether there is a k-leaf tree $T' \succeq (T, R, B)$. Informally, the algorithm works as follows: Choose a node $u \in B$ and recursively test whether there is a solution where u is a leaf, or whether there is a solution where u is an inner node. In the first case, u is moved from B to the set of fixed leaves R, so that u is preserved as a leaf in solutions T'. In the second case, u is considered an inner node and all of its outgoing edges to nodes in $N_{\overline{T}}(u)$ are added to T. The upcoming Lemma 4 guarantees that at least one of these two branches is successful, if a solution exists at all. In the special case that $|N_{\overline{T}}(u)| \leq 1$, we can skip the latter of the two branches by Lemma 5 and Corollary 1. Please note that the resulting algorithm is basically the same for directed and undirected graphs.

Lemma 4. *Let* $G = (V, E)$ *be a graph,* (T, R, B) *a leaf-labeled tree, and* $x \in B$.

1. *If there is no k-leaf tree T', such that $T' \succeq (T, R \cup \{x\}, B \setminus \{x\})$, then all k-leaf trees T' with $T' \succeq (T, R, B)$ have $x \in \text{inner}(T')$.*
2. *If there is a k-leaf tree T', such that $T' \succeq (T, R, B)$ and $x \in \text{inner}(T')$, then there is also a k-leaf tree $T'' \succeq (T + \{ (x, y) \mid y \in N_{\overline{T}}(x) \}, R, N_{\overline{T}}(x) \cup B \setminus \{x\})$.*

Proof. 1. This is clear, since x is either leaf or inner node in T'. 2. Let T' be a k-leaf tree, such that $T' \succeq (T, R, B)$ and $x \in \text{inner}(T')$. First note that $N_{\overline{T}}(x) \neq \emptyset$, because $x \in \text{inner}(T')$ and T is an induced subgraph of T'. Hence consider arbitrary $y \in N_{\overline{T}}(x)$. If $y \notin V(T')$, then we can construct a k-leaf tree T'' from T' by adding y and the edge (x, y). If $y \in V(T')$, but $(x, y) \notin E(T')$, consider the unique path $x_1, x_2, \ldots, x_i, y$ from $x_1 := \text{root}(T')$ to y in T'. We can now replace the edge (x_i, y) with (x, y) without decreasing the number of leaves in T': x is inner node in T' by definition, and $y \in \text{leaves}(T')$ implies $y \in \text{leaves}(T'')$. Furthermore, the connectivity of T' remains intact. Doing so iteratively for all

---

Algorithm 1. A fast algorithm for maximum leaf problems.

Algorithm MAXLEAF:
Input: Graph $G = (V, E)$, an inner-maximal leaf-labeled tree (T, R, B), $k \in \mathbf{N}$
Output: Is there a k-leaf tree $T' \succeq (T, R, B)$?
01: **if** $|R| + |B| \geq k$ **then return** "yes"
02: **if** $B = \emptyset$ **then return** "no"
03: Choose $u \in B$.
// Try branch where u is a leaf
04: **if** MAXLEAF$(G, T, R \cup \{u\}, B \setminus \{u\}, k)$ **then return** "yes"
// If u is not a leaf, it must be an inner node in all extending solutions
05: $B := B \setminus \{u\}$
06: $N := N_{\overline{T}}(u)$
07: $T := T \cup \{(u, u') \mid u' \in N\}$
// follow paths, see Lemma 5
08: **while** $|N| = 1$ **do**
09:        Let u be the unique member of N.
10:        $N := N_{\overline{T}}(u)$
11:        $T := T \cup \{(u, u') \mid u' \in N\}$
12: **done**
// Do not branch if no neighbors left, see Corollary 1
13: **if** $N = \emptyset$ **then return** "no".
14: **return** MAXLEAF$(G, T, R, B \cup N, k)$

---

neighbors y of x yields a k-leaf tree T'' with $\{(x, y) \mid y \in N_{\overline{T}}(x)\} \subseteq E(T'')$.
Therefore $T'' \succeq (T + \{(x, y) \mid y \in N_{\overline{T}}(x)\}, R, N_{\overline{T}}(x) \cup B \setminus \{x\})$. See Figure 3 for
an example. $\square$

Lemma 5. *Let $G = (V, E)$ be a graph, (T, R, B) a leaf-labeled tree and $x \in B$
with $N_{\overline{T}}(x) = \{y\}$. If there is no k-leaf tree that extends $(T, R \cup \{x\}, B \setminus \{x\})$,
then there is no k-leaf tree that extends $(T + (x, y), R \cup \{y\}, B \setminus \{x\})$.*

Proof. Let T' be a k-leaf tree that extends $(T + (x, y), R \cup \{y\}, B \setminus \{x\})$. Since
in particular $T' \succ T + (x, y) \succ T$, y is the only child of x in T', and since
$T' \succ (T + (x, y), R \cup \{y\}, B \setminus \{x\})$, y is leaf in T'. Hence y can be removed
from T', obtaining a k-leaf tree T'' with $x \in$ leaves(T''), i.e., $T'' \succ (T, R \cup
\{x\}, B \setminus \{x\})$. $\square$

Corollary 1. *Let $G = (V, E)$ be a graph, (T, R, B) a leaf-labeled tree and $x \in B$
with $N_{\overline{T}}(x) = \emptyset$. If there is a k-leaf tree that extends (T, R, B), there is a k-leaf
tree that extends $(T, R \cup \{x\}, B \setminus \{x\})$.*

Proof. Let T' be a k-leaf tree that extends (T, R, B). It is $x \in B \subseteq$ leaves(T).
Since $N_{\overline{T}}(x) = \emptyset$, we have $N(x) \subseteq V(T) \subseteq V(T')$. For each $y \in N(x)$, there is
$z \in V(T)$ with $(z, y) \in E(T)$. In particular, if $(x, y) \notin E(T)$, then $(x, y) \notin E(T')$,
since T' is a tree and $E(T) \subseteq E(T')$. Hence $x \in$ leaves(T') and $T' \succ (T, R \cup
\{x\}, B \setminus \{x\})$. $\square$

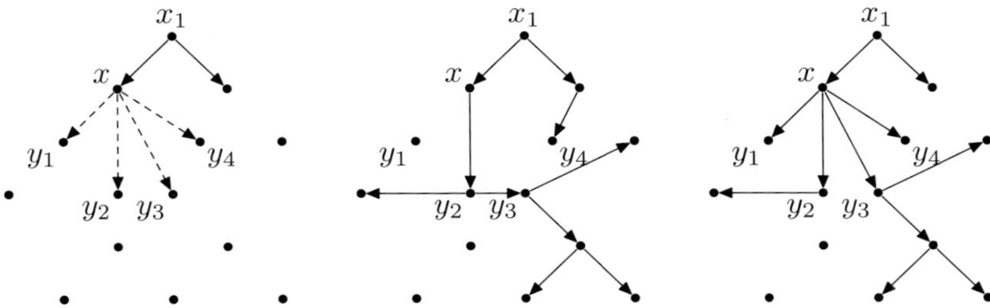

Fig. 3. The exchange argument (Lemma 4): The first figure shows a leaf-labeled tree (T, R, B) with $x \in B$. The neighborhood of x, $N_{\overline{T}}(x)$, is shown with dashed edges. The second figure shows a 5-leaf tree $T' \succ (T, R, B)$, but different choices for edges originating in x have been made: y_1 is not in T' at all, and different paths to y_3 and y_4, respectively, have been chosen. The third figure shows how the T' can be modified so that *all* $y \in N_{\overline{T}}(x)$ are children of x. This modification does not decrease the number of leaves in T': y_1 becomes a new leaf; no changes are made to the edge (x, y_2), y_3 remains inner node, and y_4 remains leaf, although it is now connected through x.

Lemma 6. *Let* $G = (V, E)$ *be a graph and let* $k > 2$. *If* G *does not contain a* k-*leaf tree,* $\text{MAXLEAF}(G, T_v, \emptyset, N(v), k)$ *returns "no" for each* $v \in V$. *If* G *contains a* k-*leaf tree rooted in* r, *Algorithm 1 returns "yes" if called as* $\text{MAXLEAF}(G, T_r, \emptyset, N(r), k)$.

Proof. We first show that all subsequent calls to MAXLEAF are always given an inner-maximal leaf-labeled tree: The star T_v is inner-maximal, and hence $(T_v, \emptyset, N(v))$ is an inner-maximal leaf-labeled tree. Let (T, R, B) be the inner-maximal tree given as argument to MAXLEAF. The algorithm chooses $x \in B$ and either fixes it as a leaf or as an inner node. If x becomes a leaf, then $(T, R \cup \{x\}, B \setminus \{x\}) \succ (T, R, B)$ is inner-maximal. If otherwise x becomes inner node, a tree T' is obtained from T by adding the nodes in $N_{\overline{T}}(x)$ as children of x, so that they are leaves. Since $N(x) \subseteq V(T')$ and $N(\text{inner}(T')) = N(\text{inner}(T)) \cup N(x) \subseteq V(T) \cup N(x) = V(T')$, the new tree T' is inner-maximal, and so is $(T', R, N_{\overline{T}}(x) \cup B \setminus \{x\})$. This step might be repeated l times while $|N_{\overline{T}}(x)| = 1$, so that we obtain a sequence of leaf-labeled trees $(T, R, B) \prec (T', R', B') \prec \cdots \prec (T^{(l+1)}, R^{(l+1)}, B^{(l+1)})$, each of them being inner-maximal for the same reason. Therefore, MAXLEAF is called with an inner-maximal leaf-labeled tree $(T^{(l+1)}, R^{(l+1)}, B^{(l+1)})$.

Whenever $\text{MAXLEAF}(G, T, R, B, k)$ returns "yes", T is a tree in G with $|\text{leaves}(T)| = |R \cup B| = |R| + |B| \geq k$. Therefore, G does contain a k-leaf tree and the algorithm never answers "yes" on no-instances. If otherwise G contains a k-leaf tree rooted in r, we use induction over $\succ$ as follows: Under the hypothesis that (T, R, B) is an inner-maximal leaf-labeled tree, such that there is a k-leaf tree $T' \succeq (T, R, B)$, we prove: Either $T = T'$, or there are (T''', R''', B''') and (T'', R'', B''), such that T''' is a k-leaf tree, $(T''', R''', B''') \succeq (T'', R'', B'') \succ (T, R, B)$ and MAXLEAF is called with (T'', R'', B''). Since G is finite, eventually MAXLEAF is called with a k-leaf leaf-labeled tree and returns "yes".

Let r be the root of some k-leaf tree T in G. Since $k > 2$, we may assume $r \in \mathrm{inner}(T)$ even when G is undirected. Consider $T' = (\{r\}, \emptyset)$. Then $(T', \emptyset, \{r\})$ is a leaf-labeled tree, and trivially $T \succ (T', \emptyset, \{r\})$. By Lemma 4, then there is also a k-leaf tree $T'' \succ (T_r, \emptyset, N(r))$.

We hence may now consider an arbitrary inner-maximal leaf-labeled tree (T, R, B) that is given a argument to MAXLEAF, such that there is a k-leaf tree $T' \succeq (T, R, B)$. If $|\mathrm{leaves}(T)| = |R \cup B| \geq k$, then (T, R, B) already is a k-leaf tree in G and the algorithm correctly returns "yes".

Otherwise, $B \neq \emptyset$ by Lemma 1, since (T, R, B) is inner-maximal. Fix an arbitrary $u \in B$. By Lemma 4, there is a k-leaf tree $T''' \succeq (T, R \cup \{u\}, B \setminus \{u\})$, or there is a k-leaf tree $T''' \succeq (T + \{\,(u, y) \mid y \in N_{\overline{T}}(u)\,\}, R, N_{\overline{T}}(u) \cup B \setminus \{u\})$.

We first assume the first case is true. Then $T''' \succeq (T, R \cup \{u\}, B \setminus \{u\}) \succ (T, R, B)$ and the call to $\mathrm{MAXLEAF}(G, T, R \cup \{u\}, B \setminus \{u\}, k)$ does satisfy the induction hypothesis for the next induction step. If however the first case is false, we know by Lemma 4, that since there is at least one k-leaf tree that extends (T, R, B) (namely $T' \succeq (T, R, B)$), there is also a k-leaf tree $T''' \succeq (T + \{\,(u, y) \mid y \in N_{\overline{T}}(u)\,\}, R, N_{\overline{T}}(u) \cup B \setminus \{u\})$. Furthermore, by Lemma 5 there is a unique sequence of vertices $v_0, v_1, \ldots, v_l$ and leaf-labeled trees $(T_0, R_0, B_0), \ldots, (T_l, R_l, B_l)$, such that $v_0 = u$, $(T_0, R_0, B_0) = (T, R, B)$, and

1. $(T_{i+1}, R_{i+1}, B_{i+1}) = (T_i + (v_i, v_{i+1}), R_i, B_i \cup N_{\overline{T_i}}(v_i) \setminus \{v_i\})$,
2. $N_{\overline{T_i}}(v_i) = \{v_{i+1}\}$ for $0 \leq i < l$,
3. $|N_{\overline{T_l}}(v_l)| \neq 1$, and
4. for each $0 \leq i \leq l$ there is a k-leaf tree $T_i' \succeq (T_i, R_i, B_i)$.

By Corollary 1, we have $N_{\overline{T_l}}(v_l) \neq \emptyset$, i.e., the algorithm does not return "no". Hence the algorithm recursively calls itself as $\mathrm{MAXLEAF}(G, T_l, R_l, B_l, k)$, where (T_l, R_l, B_l) satisfies the induction hypothesis. $\qquad\square$

Lemma 7. *Let $G = (V, E)$ be a graph and $v \in V$. The number of recursive calls of Algorithm 1 when called as $\mathrm{MAXLEAF}(G, T_v, \emptyset, N(v), k)$ for $v \in V$ is bounded by $O(2^{2k - |N(v)|}) = O(4^k)$.*

Proof. Consider a potential function $\Phi(k, R, B) := 2k - 2|R| - |B|$. When called with a leaf-labeled tree (T, R, B), the algorithm recursively calls itself at most two times: In line 4, some vertex $u \in B$ is fixed as a leaf and the algorithm calls itself as $\mathrm{MAXLEAF}(G, T, R \cup \{u\}, B \setminus \{u\}, k)$. The potential decreases by $\Phi(k, R, B) - \Phi(k, R \cup \{u\}, B \setminus \{u\}) = 1$. The while loop in lines 8–12 does not change the size of B. If, however, line 14 of the algorithm is reached, we have $|N| \geq 2$. Here, the recursive call is $\mathrm{MAXLEAF}(G, T', R, B \setminus \{u\} \cup N, k)$ for some tree T', and hence the potential decreases by $\Phi(k, R, B) - \Phi(k, R, B \setminus \{u\} \cup N) \geq 1$.

Note that $\Phi(k, R, B) \leq 0$ implies $|R + B| \geq k$. Since the potential decreases by at least 1 in each recursive call, the height of the search tree is therefore at most $\Phi(k, R, B) \leq 2k$. For arbitrary inner-maximal leaf-labeled trees (T, R, B), the number of recursive calls is hence bounded by $2^{\Phi(k, R, B)}$.

In the very first call, we already have $|B| = |N(v)|$. Hence we obtain a bound of $2^{\Phi(\emptyset, N(v))} = O(2^{2k - |N(v)|}) = O(4^k)$. $\qquad\square$

Theorem 1. MLST *can be solved in time* $O(poly(n) + 4^k \cdot k^2)$.

Proof. Let $G = (V, E)$ be an undirected graph. As Estivill-Castro et al. have shown [12], there is a problem kernel of size $3.75k = O(k)$ for MLST, which can be computed in a preprocessing that requires time $poly(n)$. Hence, $n = O(k)$.

Without loss of generality, we assume G is connected and $k > 2$. We do not know, which node $v \in V$ suffices as a root. It is however easy to see that either some $v \in V$ or one of its neighbors is root of some k-leaf spanning tree, if any k-leaf spanning tree T exists at all: If $v \in \text{leaves}(T)$, the unique predecessor u of v in T is an inner node $u \in \text{inner}(T)$. If furthermore v has minimum degree, the cost to test all $u \in N[v]$ disappears in the run time estimation.

Therefore, let $v \in V$ be a node of minimum degree. We need to call MAXLEAF with arguments $(G, T_u, R, N(u), k)$ for all $u \in N[v]$: If G contains a k-leaf tree, at least one of those u is a root of some k-leaf tree and the respective call to MAXLEAF returns "yes" by Lemma 6. Otherwise each call returns "no". By Lemma 7, the total number of recursive calls is bounded by

$$O(2^{\Phi(k,\emptyset,N(v))}) + \sum_{u \in N(v)} O(2^{\Phi(k,\emptyset,N(u))}) = O\big((d+1)2^{2k-d}\big) = O\Big(4^k \frac{d+1}{2^d}\Big).$$

It remains to show that the number of operations in each recursive call is bounded by $O(n^2) = O(k^2)$. We can assume the sets V, E, $V(T)$, $E(T)$, R, and B are realized as doubly-linked lists and an additional per-vertex membership flag is used, so that a membership test and insert and delete set operations only require constant time each.

Hence lines 1–3 and computing the new sets in lines 4 and 5 takes constant time. Computing $N_{\overline{T}}(u)$ and the new tree T takes time $O(k)$, since u has only up to k neighbors, which are tested for membership in $V(T)$ in constant time. The while loop is executed at most once per vertex $u \in V$. Each execution of the while loop can be done in constant time as well, since $|N_{\overline{T}}(u)| = 1$. Concatenating N to B in line 14 takes constant time, but updating the B-membership flag for each $v \in N$ takes up to k steps.

At this point we have shown that the overall number of operations required to decide whether G contains a k-leaf tree is bounded by $O(poly(n) + 4^k \cdot k^2)$. By Lemma 2, each k-leaf tree can be extended to a spanning tree with at least k leaves, so the decision problem MLST can be solved in the same amount of time. $\qquad\square$

Note that Algorithm 1 can easily be modified to return a k-leaf spanning tree in G within the same run time bound. In this case, an additional $O(n + m)$ postprocessing is required to extend the k-leaf tree to a k-leaf spanning tree.

Theorem 2. DMLOT *and* DMLST *can be solved in time* $O(4^k nm)$.

Proof. Let $G = (V, E)$ be a directed graph. We first consider DMLOT: If G contains a k-leaf out-tree rooted in r, MAXLEAF$(G, T_r, \emptyset, N(r), k)$ returns "yes" by Lemma 6. Otherwise, MAXLEAF$(G, T_v, \emptyset, N(v), k)$ returns "no" for all $v \in V$.

We do not know r, so we need to iterate over all $v \in V$. By Lemma 7, the total number of recursive calls is therefore bounded by

$$\sum_{v \in V} O(2^{\Phi(k,\emptyset,N(v))}) = O(n \cdot 2^{2k}) = O(4^k n).$$

What remains to show is that only $O(n+m) = O(m)$ operations are performed *on average* on each call of MAXLEAF. Consider one complete path in the recursion tree: It is easy to see, that each vertex $v \in V$ occurs at most once as the respective u in either lines 6 or 10. In particular each edge (v, w) is visited at most once per path when computing $N_{\overline{T}}(u)$. Therefore, the overall run time to solve DMLOT is bounded by $O(4^k \cdot nm)$.

To prove the run time bound for DMLST, the algorithm must be slightly modified in line 1. Here, it may only return "yes" if the leaf-labeled out-tree (T, R, B) can be extended to a k-leaf spanning out-tree. By Lemma 3, each k-leaf out-tree that shares the same root with some k-leaf spanning out-tree *can* be extended to a k-leaf spanning out-tree in time $O(n+m) = O(m)$. Thus the run time remains bounded by $O(4^k \cdot nm)$. $\qquad\square$

5 Conclusion

We solve open problems [7,16] on whether there exist $c^k poly(n)$-time algorithms for the k-leaf out-tree and k-leaf spanning out-tree problems on directed graphs. Our algorithms for DMLOT and DMLST have a run time of $O(4^k|V||E|)$, which is a significant improvement over the currently best bound of $2^{O(k \log k)} poly(|V|)$.

Since the undirected case is easier, has a linear size problem kernel, and the root of some k-leaf tree can be found faster, we can solve MLST in time $O(poly(|V|) + 4^k \cdot k^2)$, where $poly(|V|)$ is the time to compute the problem kernel of size $3.75k$. This improves over the currently best algorithm with a run time of $O(poly(|V|) + 6.75^k poly(k))$.

The question by Michael Fellows et al. from the year 2000 [14] whether there will ever be a parameterized algorithm for MLST with running time $f(k)poly(n)$, where $f(50) < 10^{20}$ unfortunately remains open, but the gap is not so big anymore.

References

1. Alon, N., Fomin, F.V., Gutin, G., Krivelevich, M., Saurabh, S.: Better algorithms and bounds for directed maximum leaf problems. In: Arvind, V., Prasad, S. (eds.) FSTTCS 2007. LNCS, vol. 4855, pp. 316–327. Springer, Heidelberg (2007)
2. Alon, N., Fomin, F.V., Gutin, G., Krivelevich, M., Saurabh, S.: Parameterized algorithms for directed maximum leaf problems. In: Arge, L., Cachin, C., Jurdziński, T., Tarlecki, A. (eds.) ICALP 2007. LNCS, vol. 4596, pp. 352–362. Springer, Heidelberg (2007)
3. Bodlaender, H.L.: On linear time minor tests with depth-first search. J. Algorithms 14(1), 1–23 (1993)
4. Bonsma, P.: Sparse cuts, matching-cuts and leafy trees in graphs. PhD thesis, University of Twente, the Netherlands (2006)

5. Bonsma, P.S., Brüggemann, T., Woeginger, G.J.: A faster FPT algorithm for finding spanning trees with many leaves. In: Rovan, B., Vojtáš, P. (eds.) MFCS 2003. LNCS, vol. 2747, pp. 259–268. Springer, Heidelberg (2003)
6. Bonsma, P.S., Dorn, F.: An FPT algorithm for directed spanning k-leaf (2007), http://arxiv.org/abs/0711.4052
7. Bonsma, P.S., Dorn, F.: Tight Bounds and Faster Algorithms for Directed Max-Leaf Problems. In: Halperin, D., Mehlhorn, K. (eds.) ESA 2008. LNCS, vol. 5193, pp. 222–233. Springer, Heidelberg (2008)
8. Bonsma, P.S., Zickfeld, F.: A 3/2-Approximation Algorithm for Finding Spanning Trees with Many Leaves in Cubic Graphs. In: Proc. of the 34th WG. LNCS. Springer, Heidelberg (to appear, 2008)
9. Bonsma, P.S., Zickfeld, F.: Spanning trees with many leaves in graphs without diamonds and blossoms. In: Laber, E.S., Bornstein, C., Nogueira, L.T., Faria, L. (eds.) LATIN 2008. LNCS, vol. 4957, pp. 531–543. Springer, Heidelberg (2008)
10. Dai, F., Wu, J.: An extended localized algorithm for connected dominating set formation in ad hoc wireless networks. IEEE Trans. Parallel Distrib. Syst. 15(10), 908–920 (2004)
11. Downey, R.G., Fellows, M.R.: Parameterized computational feasibility. In: Clote, P., Remmel, J. (eds.) Feasible Mathematics II, pp. 219–244. Birkhäuser, Boston (1995)
12. Estivill-Castro, V., Fellows, M.R., Langston, M.A., Rosamond, F.A.: FPT is P-time extremal structure I. In: Proc. of the 1st ACiD, pp. 1–41 (2005)
13. Fellows, M.R., Langston, M.A.: On well-partial-ordering theory and its applications to combinatorial problems in VLSI design. SIAM J. Discrete Math. 5, 117–126 (1992)
14. Fellows, M.R., McCartin, C., Rosamond, F.A., Stege, U.: Coordinatized kernels and catalytic reductions: An improved FPT algorithm for max leaf spanning tree and other problems. In: Kapoor, S., Prasad, S. (eds.) FST TCS 2000. LNCS, vol. 1974, pp. 240–251. Springer, Heidelberg (2000)
15. Galbiati, G., Maffioli, F., Morzenti, A.: A short note on the approximability of the maximum leaves spanning tree problem. Inf. Process. Lett. 52(1), 45–49 (1994)
16. Gutin, G., Razgon, I., Kim, E.J.: Minimum Leaf Out-Branching Problems. In: Fleischer, R., Xu, J. (eds.) AAIM 2008. LNCS, vol. 5034, pp. 235–246. Springer, Heidelberg (2008)
17. Kleitman, D.J., West, D.B.: Spanning trees with many leaves. SIAM J. Discret. Math. 4(1), 99–106 (1991)
18. Liang, W.: Constructing minimum-energy broadcast trees in wireless ad hoc networks. In: Proc. of 3rd MOBIHOC, pp. 112–122. ACM, New York (2002)
19. Linial, N., Sturtevant, D.: Unpublished result (1987)
20. Lu, H., Ravi, R.: Approximating maximum leaf spanning trees in almost linear time. J. Algorithms 29(1), 132–141 (1998)
21. Park, M.A., Willson, J., Wang, C., Thai, M., Wu, W., Farago, A.: A dominating and absorbent set in a wireless ad-hoc network with different transmission ranges. In: Proc. of the 8th MOBIHOC, pp. 22–31. ACM, New York (2007)
22. Robertson, N., Seymour, P.D.: Graph minors—a survey. In: Anderson, I. (ed.) Surveys in Combinatorics, pp. 153–171. Cambridge University Press, Cambridge (1985)
23. Solis-Oba, R.: 2-approximation algorithm for finding a spanning tree with maximum number of leaves. In: Bilardi, G., Pietracaprina, A., Italiano, G.F., Pucci, G. (eds.) ESA 1998. LNCS, vol. 1461, pp. 441–452. Springer, Heidelberg (1998)
24. Thai, M., Wang, F., Liu, D., Zhu, S., Du, D.: Connected dominating sets in wireless networks with different transmission ranges. IEEE Trans. Mob. Comput. 6(7), 721–730 (2007)

Faster Parameterized Algorithms for
Minimum Fill-In[*]

Hans L. Bodlaender[1], Pinar Heggernes[2], and Yngve Villanger[2]

[1] Department of Information and Computing Sciences, Utrecht University, P.O. Box 80.089, 3508 TB Utrecht, The Netherlands
hansb@cs.uu.nl
[2] Department of Informatics, University of Bergen, PB 7803, 5020 Bergen, Norway
{pinar.heggernes,yngve.villanger}@ii.uib.no

Abstract. We present two parameterized algorithms for the Minimum Fill-In problem, also known as Chordal Completion: given an arbitrary graph G and integer k, can we add at most k edges to G to obtain a chordal graph? Our first algorithm has running time $\mathcal{O}(k^2 nm + 3.0793^k)$, and requires polynomial space. This improves the base of the exponential part of the best known parameterized algorithm time for this problem so far. We are able to improve this running time even further, at the cost of more space. Our second algorithm has running time $\mathcal{O}(k^2 nm + 2.35965^k)$ and requires $\mathcal{O}^*(1.7549^k)$ space.

1 Introduction

The Minimum Fill-In problem asks, given as input an arbitrary graph G and an integer k, whether a chordal graph can be obtained by adding at most k new edges to G. A chordal graph is a graph without induced cycles of length at least four. This is one of the most extensively studied problems in graph algorithms, as it has many practical applications in various areas of computer science. The problem initiated from the field of sparse matrix computations, where the result of Gaussian Elimination corresponds to a chordal graph, and minimizing the number of edges in a chordal completion is equivalent to minimizing the number of non-zero elements in Gaussian Elimination [19]. Among other application areas are data-base management systems [20] and knowledge-based systems [16]. Since the problem was proved NP-complete [23], it has been attacked using various algorithmic techniques, and there exist polynomial-time approximation algorithms [17], exponential-time exact algorithms [9,10], and parameterized algorithms [14,6]. The current best bounds are $\mathcal{O}^*(1.7549)$ time and space for an exact algorithm [10], and $\mathcal{O}((m + n)4^k/(k + 1))$ time for a parameterized algorithm [6], where the $\mathcal{O}^*$-notation suppresses factors polynomial in n.

In this paper we contribute with new parameterized algorithms for the solution of the Minimum Fill-In problem. The field of parameterized algorithms, first formalized by Downey and Fellows [7], has been growing steadily and attracting

[*] Supported by the Research Council of Norway.

S.-H. Hong, H. Nagamochi, and T. Fukunaga (Eds.): ISAAC 2008, LNCS 5369, pp. 282–293, 2008.
© Springer-Verlag Berlin Heidelberg 2008

more and more attention recently [8,18]. Informally, a *parameterized algorithm* computes an exact solution of the problem at hand, but the exponential part of the running time is limited to a (hopefully small) parameter, typically an integer. For the MINIMUM FILL-IN problem, the natural parameter is the number of added edges. The number of vertices and edges of G are denoted by n and m, respectively. The first parameterized algorithms for this problem were given by Kaplan et al. and appeared more than a decade ago [14,15], with running times $\mathcal{O}(16^k m)$ and $\mathcal{O}(k^2 nm + k^6 16^k)$. A refined analysis of these algorithms by Cai gave the current best parameterized running time of $\mathcal{O}((m+n)4^k/(k+1))$ [6].

We present two algorithms that improve on the basis of the exponential part of the running time of these parameterized algorithms. Central in our algorithms is a new result, describing edges that can always be added when computing a minimum solution. Based on this result, our first algorithm is intuitive and easy to understand, and requires $\mathcal{O}(k^2 nm + 3.0793^k)$ time and polynomial space. We are able to improve the base of the exponential part even further in a second algorithm, at the cost of more space. Our second algorithm, which is more involved, requires $\mathcal{O}(k^2 nm + 2.35965^k)$ time and $\mathcal{O}^*(1.7549^k)$ space. In this extended abstract all proofs are omitted due to limited space, however a full version with all details is available [3]. In addition, we use well-known insights on chordal graphs and triangulations [12,13] without further explanations.

2   Preliminaries

All graphs in this work are undirected and simple. A graph is denoted by $G = (V, E)$, with vertex set V and edge set $E(G) = E$. For a vertex subset $S \subseteq V$, the *subgraph of G induced by S* is $G[S] = (S, \{\{v,w\} \in E \mid v, w \in S\})$. The *neighborhood* of S in G is $N_G(S) = \{v \in (V \setminus S) \mid \exists w \in S : \{v, w\} \in E\}$. We write $N_G(v) = N_G(\{v\})$ for a single vertex v, and $N_G[S] = N_G(S) \cup S$. Subscripts are omitted when not necessary. A vertex v is *universal* if $N(v) = V \setminus \{v\}$.

A vertex subset $S \subseteq V$ is a *separator* in G if $G[V \setminus S]$ has more than one connected component. A connected component C of $G[V \setminus S]$ is called *full* for S if $N(C) = S$. Vertex set S is a *minimal u, v-separator* for G if u and v are in different connected components of $G[V \setminus S]$ and S is an inclusion minimal vertex set separating u and v. The separator S is a *minimal separator* of G if there exist u and v in V such that S is a minimal u, v-separator.

A pair of vertices $\{u, v\}$ is a *non-edge* if u and v are not adjacent. For a vertex set S, we let $F(S)$ denote the set of non-edges in $G[S]$. S is a *clique* if $F(S) = \emptyset$ or $|S| = 1$. A clique is a *maximal clique* in G if it is not a proper subset of another clique in G. A set of vertices that is a clique and a separator is called a *clique separator*. A vertex v is *simplicial* if $N(v)$ is a clique.

A graph is *chordal*, or *triangulated,* if it does not contain a cycle with four or more vertices as an induced subgraph. A graph $H = (V, F)$ is a *triangulation* or *chordal completion* of a graph $G = (V, E)$ if $E \subseteq F$ and H is chordal. The edges in $F \setminus E$ are called *fill edges*. H is a *minimal triangulation* of G if there is no triangulation $H' = (V, F')$ of G where F' is a proper subset of F. A triangulation with the minimum number of edges is called a *minimum triangulation*. Every minimum triangulation is thus minimal.

A set of vertices $S \subseteq V$ is a *potential maximal clique* in G if there is a minimal triangulation H of G where S is a maximal clique in H. Potential maximal cliques play an essential role in several algorithms computing minimum triangulations and related problems, like treewidth. We use an algorithm one of [10] in our second algorithm.

A vertex set $U \subset V$ is a *moplex* if $G[U]$ is a clique, $N[v] = N[u]$ for any pair of vertices in U, and $N(U)$ is a minimal separator in G. Moplex U is *simplicial* if $G[N(U)]$ is a clique.

Proposition 1. *(Folklore) Let $G = (V, E)$ be an induced cycle, let $v_1, v_2, ..., v_n$ be the order of the vertices on the cycle, and let $H = (V, F)$ be a minimal triangulation of G, where $\{v_1, v_3\} \notin F$. Then there exists an edge $\{v_2, v\} \in F$ such that $v \in V \setminus \{v_1, v_2, v_3\}$.*

Proposition 2. *(Folklore) A set S of vertices in a graph G is a minimal u, v-separator if and only if u and v are in different full components associated to S. In particular, S is a minimal separator if and only if there are at least two distinct full components associated to S.*

Before we start describing our algorithms, we present an important new result that describes fill edges that can be safely added when computing a minimum triangulation, independent of k. This is the first result of its kind to our knowledge, and it is crucial for our further results. A generalized version of this lemma for marked graphs will be used in our algorithms.

Lemma 1. *Given a graph $G = (V, E)$, let S be a minimal separator of G such that $|F(S)| = 1$ and $S \subseteq N(u)$ for a vertex $u \in V$. Then there exists a minimum triangulation of G that has the single element of $F(S)$ as a fill edge.*

We start now the description of our algorithms, each of which will be presented in its own section. For both of our algorithms, the input is an undirected graph $G = (V, E)$ and an integer k; and each algorithm outputs either a minimum size set of fill edges of size at most k, or NO if each triangulation of G requires at least $k + 1$ edges.

3   An $\mathcal{O}^*(3.0793^k)$-Time Algorithm for Minimum Fill-In

The first algorithm that we present uses polynomial time reductions and some branching rules. In the subproblems generated by this branching algorithm, some vertices have a *marking*. As will be clear when the subproblems are analyzed, sometimes we will have the choice of adding a set of fill edges or concluding with a set of vertices that each must be incident to a fill edge. These vertices will be marked, to give the desired restriction in the solution of resulting subproblems. More precisely, subproblems can be associated with problem instances of the form (G, k, r, M) with $G = (V, E)$ a graph, k and r integers, and $M \subseteq V$ a set of *marked* vertices. For such an instance, we ask whether there exists a triangulation $H = (V, F)$ of G with $|F \setminus E| \leq k$ and $2|F \setminus E| - |M| \leq r$, such that each vertex in M is incident to a fill edge. We say that a vertex v is *marked* if $v \in M$, and r denotes the number of marks we still can place at later steps during the algorithm. From the original problem where G and k are given, the new initial problem instance is $(G, k, 2k, \emptyset)$. Any triangulation of G which requires k fill edges is also a solution to the new instance since $r = 2k$ and $M = \emptyset$.

Lemma 2. *If $(G = (V, E), k, r, M)$ has a solution with $\{v, w\}$ as a fill edge, then $((V, E \cup \{\{v, w\}\}), k - 1, r - \gamma, M \setminus \{v, w\})$ has a solution, where $\gamma \in \{0, 1, 2\}$ is the number of unmarked endpoints of $\{v, w\}$.*

At several points during our algorithm, we write: *add an edge e to F and update accordingly*. Following Lemma 2, the update consists of decreasing k by one, decreasing r by the number of unmarked endpoints of e, and removing the marks of marked endpoints of e. If we add more edges, we do this iteratively. Whenever we mark a vertex, we decrease r by one. Note that if two edges are added with a common endpoint, this endpoint is unmarked after the first addition, and thus causes a decrease of r at the second addition.

The algorithm is based on checking the existence of the structures described in the following paragraphs, and performing the corresponding actions. When a change is made to the input, we start again by checking trivial cases.

Trivial cases. First, the algorithm tests whether G is chordal and $k \geq 0$ and $r \geq 0$. If so, it returns $\emptyset$. Next, it tests if $k \leq 0$ or $r \leq -1$. If so, it returns NO.

4-Cycles. Then, the algorithm branches on induced cycles of length four (*4-cycles*). Suppose that v, w, x, y induce a 4-cycle. Then, in any triangulation, $\{v, x\}$ is an edge or $\{w, y\}$ is an edge. The algorithm recursively solves the two subcases: one where we add $\{v, x\}$ as a fill edge and update accordingly, and one where we add $\{w, y\}$ as a fill edge and update accordingly.

An invariant of the algorithm is that each 4-cycle has at least two adjacent vertices that are not marked. Initially, this holds as all vertices are unmarked. Whenever we create a 4-cycle by adding an edge, we unmark the endpoints of the added edge. Marks are only added in graphs that do not have a 4-cycle.

Note that we create in this case two subproblems. In each, k is decreased by one, and r is decreased by at least one. We will show by induction that the search tree formed by the algorithm has at most $a^k \cdot b^r$ leaves for inputs where we can use k fill edges, and place at most r marks. Thus, this case gives as condition $a^k \cdot b^r \geq 2 \cdot a^{k-1} \cdot b^{r-1}$, i.e, $ab \geq 2$.

Moplexes with marked and unmarked vertices. We use several times the following insight, following from well-known results on moplexes and triangulations (see e.g., [1]).

Lemma 3. *Let U be a moplex. There is a minimum triangulation that has a fill edge incident to each vertex in U, or there is a minimum triangulation where $N(U)$ is a clique and no fill edge is incident to any vertex in U.*

Thus, when we have a moplex that contains marked and unmarked vertices, we mark all vertices of the moplex.

Finding moplexes with unmarked vertices. Then, the algorithm tests whether there is a moplex U that contains no marked vertices. If there is no such moplex, the algorithm returns NO. Safeness of this step comes from the following lemma, which follows from the work on moplexes by Berry et al. [1], by simply considering the first moplex in a moplex elimination ordering of H.

Lemma 4. *Let $G = (V, E)$ and let $H = (V, F)$ be a minimum triangulation of G. There is a moplex U such that $F \setminus E$ has no incident edge to U, and $N(U)$ is a clique in H.*

We take such a moplex U, and let $S = N(U)$. We compute $F(S)$, i.e., the set of non-edges in the neighborhood of U.

Simplicial vertices. If $F(S) = \emptyset$, then all vertices in U are simplicial. We recurse on the instance $(G \setminus U, k, r, M)$.

By well-known theory on chordal graphs this instance is equivalent to, and a minimum set of fill edges for the new instance is also a minimum set of fill edges for, the original instance.

Moplexes missing one edge. Next, we test if $|F(S)| = 1$. By Lemma 1 it is always safe to add the edge in $F(S)$, but in some cases we need to compensate for this by removing one mark from a vertex.

Lemma 5. *Let $G = (V, E)$ be a graph and let $M \subseteq V$ be the set of marked vertices in G. Suppose there exists a minimum triangulation of G such that each vertex in M is incident to a fill edge. Let $u \in V$ be an unmarked vertex, and let $S \subseteq N(u)$ be a minimal separator $F(S) = \{\{x, y\}\}$.*

1. *If there is a unique vertex v^*, such that v^* is the last vertex on each induced path from x to y through a full component of S not containing u, then there is a minimum triangulation of G that contains the edge $\{x, y\}$ and such that each vertex in $M \setminus \{v^*\}$ is incident to a fill edge.*
2. *If there is no unique vertex v^*, such that v^* is the last vertex on each induced path from x to y through a full component of S not containing u, then there is a minimum triangulation of G that contains the edge $\{x, y\}$ and such that each vertex in M is incident to a fill edge.*

Suppose we have a moplex U, with $F(N(U))$ consisting of the single edge $\{x, y\}$. The condition of Lemma 5 now becomes: there is a vertex v^* that is the last vertex on each induced path in $G[V \setminus U]$ from x to y. A simple modification of standard breadth first search allows us to find v^* if existing in linear time. If v^* exists, we remove its marking. Then, in both cases, we add the edge $\{x, y\}$ and update accordingly.

In this case, we possibly decrease k by one, and increase r by one. This gives us as condition on the running time constants: $a \geq b$.

The condition is not symmetric. We can apply the condition with roles of x and y switched and save a marking in some cases.

Branching on moplexes. If none of the earlier tests succeeds, we arrive at the last branching, performed on a moplex U with all vertices in U unmarked.

Recall Lemma 3. It dictates which two subproblems we consider. In the first subproblem, we mark all vertices in U. In the second subproblem, we add all edges in $F(N(U))$ and update accordingly. In the first, r is decreased by $|U|$. In the second, $|F(N(U))| \geq 2$, as otherwise there is no pair of edges with a common endpoint in $F(N(U))$, so k is decreased by two. Note that there must be a vertex that is common to two elements of $F(N(U))$: if not, then suppose $\{x, y\}$ and $\{v, w\}$ are elements in $F(N(U))$, but no other combination of x, y, v, and w is an element of $F(N(U))$. Then, these four vertices form a 4-cycle, which is a contradiction. Thus, in the second subproblem, r is also decreased by at least one.

This gives us as condition for the running time analysis: $a^k \cdot b^r \geq a^{k-2} \cdot b^{r-1} + a^k \cdot b^{r-1}$, or $a^2 b \geq 1 + a^2$.

Each of these subproblems is solved recursively, and from these solutions, we then return the best one, adding $F(N(U))$ to the set returned by the second subproblem except when it returned NO.

Analyzing the running time. By standard graph algorithmic tools, each recursive call can be performed in $\mathcal{O}(nm)$ time, except that the checking for all 4-cycles before any other operation is done costs once $\mathcal{O}(m^2)$ time. We now analyze the number of recursive calls in the search tree. We start with an instance with $r = 2k$, so the running time of the algorithm is bounded by $a^k \cdot b^{2k}$. Each of the steps gave a condition on a and b, and we get as minimum $ab^2 = 3.0793$ when we set $a = 1.73205$ and $b = 1.33334$. Thus, the total running time becomes $\mathcal{O}(m^2 + nm \cdot 3.0793^k)$.

By results of [15] and [17] it is possible to reduce a given instance $(G = (V, E), k)$ of Minimum Fill-In to an equivalent instance $(G' = (V', E'), k')$ where $k' \leq k$ and $|V'| = \mathcal{O}(k^2)$, in $\mathcal{O}(k^2 nm)$ time. By preprocessing the input by such an algorithm we get an additive time cost of $\mathcal{O}(k^2 nm)$ but the size of n and m have been reduced to respectively $\mathcal{O}(k^2)$ and $\mathcal{O}(k^4)$. Thus, the time complexity for our algorithm becomes $\mathcal{O}(k^2 nm + 3.0793^k)$.

Theorem 1. *The* Minimum Fill-In *problem can be solved in* $\mathcal{O}(k^2 nm + 3.0793^k)$ *time, using polynomial space.*

4   An $\mathcal{O}^*(2.35965^k)$-Time Algorithm for Minimum Fill-In

In this section, we give a second algorithm for the Minimum Fill-In problem. This algorithm uses less time as a function of k, at the cost of exponential space as a function of k. Like the previous algorithm, we create subinstances with some vertices marked and with an additional parameter r, which is the number of marks that still can be handed out.

An important difference from the previous algorithm is that the mark is a vertex set containing the vertices which are candidates to add a fill edge incident to. The algorithm involves a more extensive analysis of subproblems, a mixing of eliminating moplexes with partitioning the graph on clique separators, resolution of cycles with four vertices, and a resolution of certain cases with the exact algorithm, recently given by Fomin and Villanger [10]. In order to properly execute the steps where we partition on clique separators, marks are *annotated*. We also allow that the algorithm returns solutions that do not respect marks. When there is a solution respecting marks with α fill edges, the algorithm may return any solution with at most α fill edges. If the algorithm returns NO, we know there is no solution that respects marks. (This is needed for the last step, where we forget the marks.)

With our algorithm description, we will also make the first steps towards the time analysis. We derive a number of conditions on a function $T(k, r)$, such that the running time of all recursive calls that originate at a node with parameters k (number of fill edges) and r (number of marks that still can be placed) is bounded by $T(k, r)$ times a function, polynomial in n, not depending on k. As the time for non-leaf nodes of the search tree is bounded by a polynomial in n times the time for leaf nodes, we only count the time at leaf nodes. We want to show that $T(k, r) \leq a^k \cdot b^k \cdot o(k)$ and derive some conditions on a and b.

The algorithm consists of carefully handling subproblems of various types. We describe in the next paragraphs which conditions are tested, in what order, and what steps are executed if a certain condition holds. First we will present some polynomial-time reduction rules. Several cases are similar to or the same as in our previous algorithm.

Trivial cases. If G is chordal and $k \geq 0$ and $r \geq 0$, then we return the empty set. If G is not chordal, and $k \leq 0$ or $r \leq -1$, we return NO.

Universal vertex. If G contains a universal vertex then we simply remove this vertex. This is safe, since no induced cycle of length at least 4 contains such a vertex.

Simplicial vertices. If an unmarked vertex is simplicial then we remove the vertex, and obtain an equivalent instance. If a marked vertex is simplicial then we return NO, since no minimal triangulation will add fill edges incident to a simplicial vertex.

Clique separators. Then, the algorithm tests if there is a clique separator. If there is a clique separator S, then let $V_1, \ldots, V_r$ be the vertex sets of the connected components of $G[V \setminus S]$. We create now r subinstances, with graphs $G[S \cup V_1]$, $\ldots$, $G[S \cup V_r]$.

Vertices in subinstances in $V \setminus S$ keep their marks. A marked vertex in S is marked in only one subinstance, containing the annotated vertex set related to the mark. We will describe this more in detail when the annotated vertices are defined.

These subroblems are now independent. First, we test for each subproblem if there is a solution with at most two fill edges. If so, we solve this in polynomial time and use this to reduce the parameter of the remaining problems.

When we have α subproblems each of whose solutions requires at least three fill edges, each can add at most $k - 3\alpha + 3$ fill edges. Thus, we need to choose a

and b such that $T(k, r) \leq \alpha \cdot T(k - 3\alpha + 3, r)$ for all α, for all $\alpha > 1$. This gives $a^{3\alpha-3} \geq \alpha$, which holds for every integer $\alpha > 1$ and $a \geq 1.3$.

A minimal separator missing one edge. The next step is similar to the steps in our previous algorithm that uses Lemma 5, but now we apply it also to vertices that do not belong to a moplex.

We test if there is an unmarked vertex v and a minimal separator S contained in $N(v)$, such that $F(S)$ contains only one edge. If we have such a minimal separator S, we add the edges in $F(S)$, test if vertex v^* described in Lemma 5 exists, remove the mark of v^*, update accordingly, and solve recursively the remaining instance.

If this instance returns NO, we return NO, otherwise we return the union of $F(S)$ and the solution found by the instance. This again gives as condition for the running time analysis: $a \geq b$.

If none of the above reduction steps applies, we consider the following branching steps.

4-Cycles. Like in the previous algorithm, we now test if there is an induced 4-cycle, and branch on the two ways of adding an edge between non-adjacent vertices in the cycle. Again, we get as condition $ab \geq 2$.

Minimal separator $\boldsymbol{S}$ with $\boldsymbol{|F(S)| \geq 3}$. Test if there is an unmarked vertex v, and a minimal separator $S \subseteq N(v)$ with $|F(S)| \geq 3$. If so, we branch on this vertex, similarly as in the previous algorithm: we create two subinstances and recurse on these, and then output the smallest fill set of these instances, treating NO as a solution of size ∞.

In one subinstance, we add all fill edges in $F(S)$, and k is decreased by $|F(S)|$. For each unmarked vertex incident to an edge of $F(S)$, r is decreased by one. For each vertex incident to $j > 1$ edges in $F(S)$, r is decreased by $j - 1$. We also remove all marks from vertices incident to edges in $F(S)$.

In the other subinstance, we mark vertex v, but we also have to define the annotation for the mark of v. Let W be a connected component of $G[V \setminus N[v]]$ not containing v such that $N(W) = S$. The connected component W exists by definition of S. The annotated vertices for the mark of v will be W. Let us justify this.

Since S is not completed into a clique there is an edge $\{x, y\} \in F(S)$ which is not used as a fill edge in the optimal solution we are searching for. Vertex set W is a full component of S, and thus there exists an induced path $u_1, u_2, ..., u_r$ from x to y only containing vertices in W. The vertex set $\{y, v, x, u_1, u_2, ..., u_r\}$ induces a cycle in G. By Proposition 1, v has a fill edge to one of the vertices in $\{u_1, u_2, ..., u_r\}$ if $\{x, y\}$ is not a fill edge. Since we do not know which one of the edges in $F(S)$ is not added when v is marked, we use W as the annotation and in this way ensure that the correct vertex is in the set.

Lemma 6. *Given a graph G, let $S \subset V$ be a clique separator, let $v \in S$ be a marked vertex, and let X be the annotation of v. Then there exists a connected component W of $G[V \setminus S]$ such that $X \subseteq W$.*

Again, we return the smallest solution found by the two subinstances, treating NO as a solution of size ∞.

Similar arguments as before show correctness of this step. In a minimum triangulation, we must either add all edges in $F(S)$ or vertex v will be incident to a fill edge. This gives: $T(k, r) \leq T(k - 3, r - 2) + T(k, r - 1)$, leading to the condition $a^3 b^2 \geq 1 + a^3 b$. In the first subinstance r is reduced by two since there are no induced 4-cycles, and thus the three edges of $F(S)$ induces a connected component. One consequence of this is that at most 4 vertices will be incident to the three edges in $F(S)$. Notice that, for any remaining minimal separator S contained in the neighborhood of an unmarked vertex, $|F(S)| = 2$.

Split the problem into two non-chordal subproblems. Let v be an unmarked vertex, let S be a minimal separator in $N(v)$, and let G' be the resulting graph where the edges $F(S)$ are added to G. We test if there are two connected components W_1 and W_2 of $G[V \setminus S]$ where $G'[N[W_1]]$ and $G'[N[W_2]]$ are non-chordal. We will then know that at least one fill edge will be required for each of the connected components W_1 and W_2 in the case where S is completed into a clique. The algorithm proceeds as follows: Check if one of the subproblems $G'[N[W_1]]$ or $G'[N[W_2]]$ can be triangulated by adding at most three fill edges. If this is the case, we get the recursive condition: $T(k, r) \leq T(k - 3, r - 2) + T(k, r - 1)$, giving $a^3 b^2 \geq 1 + a^3 b$. If not, we have the subproblems $G'[N[W_1]]$ and $G'[V \setminus N[W_1]]$, and we get the recursive condition: $T(k, r) \leq 2T(k - 5, r - 1) + T(k, r - 1)$, giving $a^5 b \geq 2 + a^5$.

Using a list of potential maximal cliques. In this case, we use another algorithm to solve the MINIMUM FILL-IN problem. This algorithm is a variant of the exact algorithm for MINIMUM FILL-IN by Fomin and Villanger [10]. Suppose none of the above holds, then we can make several observations. Let S_v be a minimal separator contained in $N(v)$ for an unmarked vertex v, and let W_v be a connected component of $G[V \setminus N[v]]$ which is full for S_v. The graph obtained by adding the two edges in $F(S_v)$ to $G[N[W_v]]$ will not be chordal, since that would imply a vertex $w \in W_v$, where $S_v \subseteq N(w)$, and thus the endpoints of an edge of $F(S_v)$ and vertices v, w would induce a 4-cycle. Since all connected components of $G[V \setminus N[v]]$ generate non-chordal subproblems by the argument above, we can notice that $G[V \setminus N[v]]$ contains exactly one connected component, since zero components would imply that v is universal, and more than one would imply that the previous described rule could be applied. A consequence of this again is that for any unmarked vertex $N(v)$ only contains one minimal separator, which we can call S_v, and there is only one connected component W_v in $G[V \setminus N[v]]$ which is full for S_v. Notice also that $G[N[v]] = G[V \setminus W_v]$ is chordal when the two edges in $F(S_v)$ are added. Before starting to describe the rule, we need more knowledge about the problem instance.

Lemma 7. *Given a graph $G = (V, E)$ where none of the rules above can be applied, let u and v be unmarked vertices, where $S_u \subset N[v]$. Then u and v are contained in the same connected component W of $G[V \setminus S_v]$.*

Lemma 8. *Given a graph $G = (V, E)$ where none of the rules above can be applied, let u and v be unmarked vertices. Then $N[u] = N[v]$, or $F(S_u) \cap F(S_v) = \emptyset$.*

Lemma 9. *Given a graph $G = (V, E)$ where none of the rules above can be applied, let v be an unmarked vertex. Then the number of unmarked vertices in S_v is at most 3.*

Lemma 10. *Given a graph $G = (V, E)$ with no induced 4-cycle or clique separator, let $H = (V, F)$ be a triangulation of G with $E \subset F$. Then only one connected component of $(V, F \setminus E)$ contains edges.*

A consequence of this lemma is that there are at most $k + 1$ vertices incident to a fill edge, and hence we can return NO at this point if there are more than $k + 1$ marked vertices. So assume there are at most $k + 1$ marked vertices.

Next step is to control the unmarked vertices. For each unmarked vertex v we know that $|F(S_v)| = 2$, and by Lemma 8 that $F(S_v) = F(S_u)$, or $F(S_v) \cap F(S_u) = \emptyset$. Partition the unmarked vertices into groups, where $F(S_v) = F(S_u)$ for all pairs u, v of vertices in the same group. Consider the minimum triangulation H, which respects all given marks. By Lemma 10 there is at most $k + 1$ vertices incident to fill edges of H. Since $N[u] = N[v]$ for a pair of unmarked vertices in the same group, we can notice that either all or none of them will have an incident fill edge in H. Remove the at most $k + 1$ groups that have incident fill edges. The number of remaining groups is at most $k/2$, since each group have two private fill edges in its neighborhood. In total there is at most $3k/2$ groups of unmarked vertices.

By Fomin et al. [9] a minimum triangulation can be computed in $\mathcal{O}((|\Delta_G| + |\Pi_G|) \cdot n^3)$ time, where Δ_G is the set of minimal separators of G, and Π_G is the set of potential maximal cliques of G. The algorithm of [9] is more powerful that stated; among all the tree decompositions that can be constructed, using provided potential maximal cliques as bags and provided minimal separators as edges between the bags, the algorithm returns the tree decomposition that introduce the minimum number of fill edges when completing every bag in the decomposition into a clique. Thus, we only have to provide a minimal separator S if $|F(S)| \leq k$ and a potential maximal clique Ω if $|F(\Omega)| \leq k$ to the algorithm if we want to solve the MINIMUM FILL-IN problem with parameter k. Previously we partitioned the unmarked vertices into $3k/2$ groups, where all vertices in the same group have the same neighborhood. When listing minimal separators and potential maximal cliques, we can treat vertices with the same closed neighborhood as a single vertex. A vertex is for instance contained in a minimal u, v-separator S if and only if it has a neighbor in the connected components of $G[V \setminus S]$ which contains respectively u and v. For potential maximal cliques there exists a similar definition. A vertex is contained in the potential maximal clique, if and only if it can reach all other vertices in the potential maximal clique by a direct edge or a path consisting only of vertices that are not contained in the potential maximal clique. Given these arguments, no vertex set that contains only a part of a group of unmarked vertices are candidates to be minimal separators or potential maximal cliques. By Lemma 9 there exists at most three unmarked vertices that have adjacency to vertices in a group, without being contained in it. This means that a minimal separator or potential maximal clique can at most contain $4\sqrt{k}$ of the $3k/2$ groups of unmarked vertices.

Lemma 11. *Given a graph G, let M be the set of marked vertices, let q be the number of groups of unmarked vertices, and let $k = |M| - 1$. Then all minimal separators and potential maximal cliques containing at most ℓ groups of unmarked vertices can be listed in $\mathcal{O}\left(1.7549^{k+1} \cdot \sum_{i=0}^{\ell} \binom{q}{i}\right)$ time and space.*

We also observe that if $H = (V, F)$ is a triangulation of $G = (V, E)$ with at most k fill edges, and $V \setminus M$ contains the set of at most $3k/2$ group, then no clique in H contains more than $4\sqrt{k}$ of the $3k/2$ groups of vertices of $V \setminus M$. So, we can list all potential maximal cliques that can contribute to a triangulation of G with at most k fill edges using Lemma 11 with $\ell = 4\sqrt{k}$ in time $\mathcal{O}^*(1.7549^k)$. Given this list, we can employ the algorithm of Fomin et al. [9] and compute a triangulation with minimum fill of G; this algorithm uses time, polynomial in n times the size of the list of potential maximal cliques.

This gives the condition: $T(k, r) \leq 1.7549^k$, and thus $a \geq 1.7549$.

Analyzing the running time. We derived a number of conditions on a and b, such that, if these hold, then by induction it follows that $T(k, r) \leq a^k b^k \cdot o(k)$. As we start with an instance with k and $r = 2k$, the running time is a polynomial in n times $T(k, 2k)$. We get as minimum $ab^2 = 2.35965$ when we set $a = 1.7549$ and $b = 1.15956$. Rounding this up allows to ignore the $o(k)$ term, and thus the algorithm requires $\mathcal{O}^*(2.35965^k)$ time. By the same arguments as the ones used for the polynomial part of the running time of our previous algorithm, we can conclude that this algorithm has running time $\mathcal{O}(k^2 nm + 2.35965^k)$, and requires $\mathcal{O}^*(1.7549^k)$ space.

Theorem 2. *The* MINIMUM FILL-IN *problem can be solved in* $\mathcal{O}(k^2 nm + 2.35965^k)$ *time, using* $\mathcal{O}^*(1.7549^k)$ *space.*

Acknowledgments. We thank Guido Diepen, Igor Razgon, Thomas van Dijk, and Johan van Rooij for useful discussions.

References

1. Berry, A., Bordat, J.: Moplex elimination orderings. Electronic notes in Discrete Mathematics 8, 6–9 (2001)
2. Berry, A., Heggernes, P., Villanger, Y.: A vertex incremental approach for maintaining chordality. Discrete Mathematics 306, 318–336 (2006)
3. Bodlaender, H.L., Heggernes, P., Villanger, Y.: Faster parameterized algorithms for Minimum Fill-In, Technical Report Reports in Informatics 375, University of Bergen, Norway (2008)
4. Bouchitté, V., Todinca, I.: Listing all potential maximal cliques of a graph. Theoret. Comput. Sci. 276, 17–32 (2002)
5. Buneman, P.: A characterization of rigid circuit graphs. Discrete Math. 9, 205–212 (1974)
6. Cai, L.: Fixed-parameter tractability of graph modification problems for heriditary properties. Inf. Process. Lett. 58, 171–176 (1996)
7. Downey, R.G., Fellows, M.R.: Parameterized complexity. Springer, New York (1999)

8. Flum, J., Grohe, M.: Parameterized Complexity Theory. Springer, New York (2006)
9. Fomin, F.V., Kratsch, D., Todinca, I.: Exact (exponential) algorithms for treewidth and minimum fill-in. In: Díaz, J., Karhumäki, J., Lepistö, A., Sannella, D. (eds.) ICALP 2004. LNCS, vol. 3142, pp. 568–580. Springer, Heidelberg (2004)
10. Fomin, F.V., Villanger, Y.: Treewidth computation and extremal combinatorics. In: Aceto, L., Damgård, I., Goldberg, L.A., Halldórsson, M.M., Ingólfsdóttir, A., Walukiewicz, I. (eds.) ICALP 2008, Part I. LNCS, vol. 5125, pp. 210–221. Springer, Heidelberg (2008)
11. Gavril, F.: The intersection graphs of subtrees in trees are exactly the chordal graphs. J. Combin. Theory Ser. B 16, 47–56 (1974)
12. Golumbic, M.C.: Algorithmic Graph Theory and Perfect Graphs. Academic Press, New York (1980)
13. Heggernes, P.: Minimal triangulations of graphs: A survey. Discrete Mathematics 306, 297–317 (2006)
14. Kaplan, H., Shamir, R., Tarjan, R.E.: Tractability of parameterized completion problems on chordal and interval graphs: Minimum fill-in and physical mapping. In: IEEE Symposium on Foundations of Computer Science, pp. 780–791 (1994)
15. Kaplan, H., Shamir, R., Tarjan, R.E.: Tractability of parameterized completion problems on chordal, strongly chordal, and proper interval graphs. SIAM J. Computing 28, 1906–1922 (1999)
16. Lauritzen, S.L., Spiegelhalter, D.J.: Local computations with probabilities on graphical structures and their applications to expert systems. J. Royal Statist. Soc. B 50, 157–224 (1988)
17. Natanzon, A., Shamir, R., Sharan, R.: A polynomial approximation algorithm for the minimum fill-in problem. SIAM J. Computing 30, 1067–1079 (2000)
18. Niedermeier, R.: Invitation to Fixed-Parameter Algorithms. Oxford University Press, Oxford (2006)
19. Rose, D.J.: Triangulated graphs and the elimination process. J. Math. Anal. Appl. 32, 597–609 (1970)
20. Tarjan, R.E., Yannakakis, M.: Simple linear-time algorithms to test chordality of graphs, test acyclicity of hypergraphs, and selectively reduce acyclic hypergraphs. SIAM J. Comput. 13, 566–579 (1984)
21. Villanger, Y.: Improved exponential-time algorithms for treewidth and minimum fill-in. In: Correa, J.R., Hevia, A., Kiwi, M. (eds.) LATIN 2006. LNCS, vol. 3887, pp. 800–811. Springer, Heidelberg (2006)
22. Walter, J.: Representations of rigid cycle graphs, PhD thesis, Wayne State University, USA (1972)
23. Yannakakis, M.: Computing the minimum fill-in is NP-complete. SIAM J. Alg. Disc. Meth. 2, 77–79 (1981)

Graph Layout Problems Parameterized by Vertex Cover

Michael R. Fellows[1], Daniel Lokshtanov[2], Neeldhara Misra[3],
Frances A. Rosamond[1], and Saket Saurabh[2]

[1] University of Newcastle,
Newcastle, Australia
{michael.fellows,frances.rosamond}@newcastle.edu.au
[2] Department of Informatics, University of Bergen,
N-5020 Bergen, Norway
{daniello,saket.saurabh}@ii.uib.no
[3] The Institute of Mathematical Sciences,
Chennai, 600 017, India
neeldhara@imsc.res.in

Abstract. In the framework of parameterized complexity, one of the most commonly used structural parameters is the *treewidth* of the input graph. The reason for this is that most natural graph problems turn out to be fixed parameter tractable when parameterized by treewidth. However, *Graph Layout* problems are a notable exception. In particular, no fixed parameter tractable algorithms are known for the CUTWIDTH, BANDWIDTH, IMBALANCE and DISTORTION problems parameterized by treewidth. In fact, BANDWIDTH remains NP-complete even restricted to trees. A possible way to attack graph layout problems is to consider structural parameterizations that are stronger than treewidth. In this paper we study graph layout problems parameterized by the size of the minimum vertex cover of the input graph. We show that all the mentioned problems are fixed parameter tractable. Our basic ingredient is a classical algorithm for INTEGER LINEAR PROGRAMMING when parameterized by dimension, designed by Lenstra and later improved by Kannan. We hope that our results will serve to re-emphasize the importance and utility of this algorithm.

1 Introduction

Parameterized complexity can be thought of as a "multivariate" approach to complexity analysis and algorithm design. In addition to the overall input size n, a secondary measurement k, the *parameter*, is also considered. In the parameterized complexity framework the central notion is *fixed parameter tractability* (FPT), defined to be solvability in time $f(k)n^c$, where f is some arbitrary function and c is a constant. For further details and an introduction to parameterized complexity we refer to [8,11,27].

In the framework of parameterized complexity, an important aspect is the *choice of parameter* for a problem. Exploring how one parameter affects the

S.-H. Hong, H. Nagamochi, and T. Fukunaga (Eds.): ISAAC 2008, LNCS 5369, pp. 294–305, 2008.

Table 1. Problem Definitions

Problem Name	Objective Function	Problem Definition
BANDWIDTH	$f_{bw}(\pi) = \max\limits_{uv \in E} |\pi(u) - \pi(v)|$	$\mathbf{bw}(G) = \min\limits_{\pi} f_{bw}(\pi)$
CUTWIDTH	$f_{cw}(\pi) = \max\limits_{1 \le i \le n} |\partial(V_i)|$	$\mathbf{cw}(G) = \min\limits_{\pi} f_{cw}(\pi)$
IMBALANCE	$f_{im}(\pi) = \sum\limits_{i=1}^{n} |L_\pi(v_i) - R_\pi(v_i)|$	$\mathbf{im}(G) = \min\limits_{\pi} f_{im}(\pi)$
DISTORTION[1]	$f_{di}(\pi) = \max\limits_{uv \in E} \sum\limits_{i=\pi(u)}^{\pi(v)-1} D(v_i, v_{i+1})$	$\mathbf{di}(G) = \min\limits_{\pi} f_{di}(\pi)$

complexity of different parameterized or unparameterized versions of the problem, often leads to non trivial combinatorics and better understanding of the problem. In general there are two kinds of parameterizations. In the first kind the parameter reflects the value of the objective function in question. The second kind, *structural parameterizations*, measure the structural properties of the input. A well developed structural parameter is the treewidth of the input graph. A celebrated result in this direction is that every problem expressible in monadic second order logic can be solved in time $O(f(t) \cdot n)$ for graphs of treewidth at most t [7]. Even though many problems become tractable when the treewidth of the input graph is bounded, there are quite a few that do not. For an example BANDWIDTH remains NP-complete even for trees. In these cases it is interesting to consider parameterizations which enforce more structure on the input than the treewidth. In this direction Fellows and Rosamond investigated how different problems behave when parameterized by the *max leaf number* of the input graph [10].

In this paper we consider parameterizing by the *vertex cover number* $(vc(G))$ of the graph. The vertex cover number of a graph G is the size of smallest set of vertices such that every edge has at least one end-point in this set. We study the graph layout problems CUTWIDTH, BANDWIDTH, IMBALANCE and DISTORTION parameterized by $vc(G)$. In a graph layout problem, we are given a graph $G = (V, E)$ as input and asked to find a permutation $\pi : V \to \{1, 2, \ldots, n\}$ that minimizes a certain problem specific objective function of π. In order to define the problems considered we need to introduce some notation. A permutation $\pi : V \to \{1, 2, \ldots, n\}$ orders the vertex set into $v_1 <_\pi v_2 <_\pi \ldots <_\pi v_n$. For every i, the set V_i is $\{v_1, \ldots, v_i\}$ and $\partial(V_i) = \{uv \mid uv \in E, u \in V_i, v \in V \setminus V_i\}$. We define $L_\pi(v)$ to be $\{u \mid u \in N(v), u <_\pi v\}$ and $R_\pi(v)$ is $\{u \mid u \in N(v), v <_\pi u\}$ where $N(v) = \{u : uv \in E\}$ is the neighborhood of v. For a pair of vertices u and v, the shortest path distance between u and v is denoted by $D(u, v)$. The precise definitions of the problems studied in the paper are given in Table 1.

Many problems in different domains can be formulated as graph layout problems. These include optimization of networks for parallel computer architectures, VLSI design, numerical analysis, computational biology, graph theory, scheduling

---

[1] The presented definition is equivalent to the original definition of distortion for embedding into line. Details are given in the section about DISTORTION.

and archaeology. In particular an algorithm for IMBALANCE is used as a starting point for many algorithms in graph drawing [19,20,28,30]. On the other hand BANDWIDTH is equivalent to the problem of minimizing bandwidth of a sparse symmetric square matrix which is useful for the storage and manipulations of these matrices, including Gaussian elimination [5,24]. CUTWIDTH was proposed as a model to minimize the number of channels in a circuit [1,25], and recently it has found applications in protein engineering [3], network reliability [21], automatic graph drawing [26], information retrieval [4], and as a subroutine in the cutting plane algorithm for TSP [17]. The problem of DISTORTION, or rather low distortion embeddings of a graph metric into simple metric spaces has proved to be a useful tool in designing algorithms in various fields. A long list of applications given in [14] includes approximation algorithms for graph and network problems, such as sparsest cut, minimum bandwidth, low-diameter decomposition and optimal group steiner trees, and online algorithms for metrical task systems and file migration problems.

Our Contributions

- We show that CUTWIDTH, BANDWIDTH, IMBALANCE and DISTORTION parameterized by the vertex cover number of the input graph are FPT. Notice that even though a graph G with $vc(G) \leq k$ has treewidth at most k, this can not be directly applied to obtain our results. The reason for this is that graph layout problems parameterized by treewidth have proven hard to cope with. In particular, the parameterized complexity of CUTWIDTH parameterized by treewidth is a non trivial problem left open in [29]. BANDWIDTH is NP-complete for trees and the parameterized complexity of IMBALANCE and DISTORTION with treewidth as parameter is unknown.
- A classical result in parameterized algorithms is that p-VARIABLE INTEGER LINEAR PROGRAMMING FEASIBLITY (p-ILP) is FPT. This powerful result, first proved by Lenstra in [23][2] and later improved by Kannan [18], is very rarely used in parameterized complexity. The only previously known examples of applications of this result in parameterized algorithms is in an FPT algorithm for the CLOSEST STRING problem [13] and in an EPTAS for MIN-MAKESPAN-SCHEDULING problem [2]. In fact, Niedermeier has explicitly asked for more applications of the result that p-ILP is FPT. In this context we quote Niedermeier [[27], Page Number:184]

  "... it remains to investigate further examples besides CLOSEST STRING where the described ILP approach turns out to be applicable. More generally, it would be interesting to discover more connections between fixed-parameter algorithms and (integer) linear programming. ... "

  We extensively use this result in all our algorithms, thus giving more examples of its applicability.

---

[2] This paper received Fulkerson Prize in 1985 for an outstanding contribution in the area of discrete mathematics.

We would like to point out that an improved version of the Lenstra/Kannan algorithm for p-ILP designed by Frank and Tardos [12] uses space polynomial in p and input size. We apply this to give a polynomial space FPT algorithm for BANDWIDTH parameterized by $vc(G)$. This gives an interesting distinction between $vc(G)$ and treewidth parameterizations, because almost all algorithms for graphs of bounded treewidth apply dynamic programming and thus need exponential space.

In Section 2, we give a brief introduction to integer linear programming parameterized by the number of variables. Sections 3, 4, 5 and 6 contain FPT algorithms for IMBALANCE, CUTWIDTH, BANDWIDTH and DISTORTION respectively. The reader is encouraged to read the section on IMBALANCE before proceeding to the later sections because this section contains a description of general scheme used in all our algorithms. Finally we conclude with some remarks and open problems in Section 7.

2   Integer Linear Programming with Few Variables

Integer linear programming (ILP) is the framework in which we will eventually formulate all the problems studied. In this section we describe the required results in this direction.

> p-VARIABLE INTEGER LINEAR PROGRAMMING FEASIBLITY (p-ILP): Given matrices $A \in \mathbb{Z}^{m \times p}$ and $b \in \mathbb{Z}^{m \times 1}$, the question is whether there exists a vector $\bar{x} \in \mathbb{Z}^{p \times 1}$ satisfying the m inequalities, that is, $A \cdot \bar{x} \leq b$. The number of variables p is the parameter.

Lenstra [23] showed that p-ILP is FPT with running time doubly exponential in p. Later, Kannan [18] provided an algorithm for p-ILP running in time $p^{O(p)}$. The algorithm uses Minkowski's Convex Body theorem and other results from Geometry of Numbers. A bottleneck in this algorithm was that it required space exponential in p. Using the method of simultaneous Diophantine approximation, Frank and Tardos [12] describe preprocessing techniques, using which it is shown that Lenstra's and Kannan's algorithms can be made to run in polynomial space. They also slightly improve the running time of the algorithm. For our purposes, we use this algorithm.

Theorem 1 ([18],[23],[12]). *p-VARIABLE INTEGER LINEAR PROGRAMMING FEASIBLITY can be solved using $O(p^{2.5p+o(p)} \cdot L)$ arithmetic operations and space polynomial in L. Here L is the number of bits in the input.*

Later, a randomized algorithm for p-ILP was provided by Clarkson, we refer to [6] for further details. The result of Lenstra was extended by Khachiyan and Porkolab [22] to semidefinite integer programming. In their work, they show that if Y is a convex set in $\mathbb{R}^k$ defined by polynomial inequalities and equations of degree at most $d \geq 2$, with integer coefficients of binary length at most l, then for fixed k, the problem of computing an optimal integral solution y^* to the problem min $\{y_k \mid y(y_1, \ldots, y_l) \in Y \cup \mathbb{Z}^k\}$ admits an FPT algorithm. Their algorithm was further improved by Heinz [16] in the specific case of minimizing a polynomial $\hat{F}$ on the set of integer points described by an inequality system $F_i \leq 0$, $1 \leq i \leq s$ where the F_i's are quasiconvex polynomials in p variables with integer coefficients. This algorithm generalizes Lenstra's algorithm. In our algorithms we need the optimization version of p-ILP rather than the feasibility version. We proceed to define the minimization version of p-ILP.

> p-VARIABLE INTEGER LINEAR PROGRAMMING OPTIMIZATION (p-OPT-ILP): Let matrices $A \in \mathbb{Z}^{m \times p}$, $b \in \mathbb{Z}^{m \times 1}$ and $c \in \mathbb{Z}^{1 \times p}$ be given. We want to find a vector $\bar{x} \in \mathbb{Z}^{p \times 1}$ that **minimizes** the objective function $c \cdot \bar{x}$ and satisfies the m inequalities, that is, $A \cdot \bar{x} \geq b$. The number of variables p is the parameter.

Now we are ready to state the theorem we will use in the later sections.

Theorem 2. [⋆]³ p-OPT-ILP *can be solved using* $O(p^{2.5p+o(p)} \cdot L \cdot \log(MN))$ *arithmetic operations and space polynomial in* L. *Here, L is the number of bits in the input, N is the maximum of the absolute values any variable can take, and M is an upper bound on the absolute value of the minimum taken by the objective function.*

3   Imbalance: The Inner Order Is Irrelevant

The solutions to all the problems considered in this paper follow the same basic scheme. The case of IMBALANCE is the simplest exhibition of this theme, and our algorithm for IMBALANCE will act as a template for the other algorithms to follow. We now proceed to give an FPT algorithm for the IMBALANCE problem parameterized by the size of the minimum vertex cover of the input graph. Our input consists of a graph $G = (V, E)$, and a vertex cover $C = \{c_1, \ldots, c_k\}$ of size k.

Fixing the order of appearce of vertices in C: We are looking for a permutation $\pi : V \to \{1, 2, \ldots, n\}$ for which $f_{im}(\pi)$ is minimized. In order to do this, we loop over all possible permutations of the vertex cover C and for each such permutation π_c, find the best permutation π of V that agrees with π_c. We say that π and π_c *agree* if for all $c_i, c_j \in C$ we have that $c_i <_\pi c_j$ if and only of $c_i <_{\pi_c} c_j$. In other words, the relative ordering π imposes on C is precisely π_c. Thus, at a cost of a factor of $k!$ in the running time we can assume that there exists an optimal permutation π such that $c_1 <_\pi c_2 <_\pi \ldots <_\pi c_k$.

Definition 1. *Let π_c be an ordering of C such that $c_1 <_{\pi_c} c_2 <_{\pi_c} \ldots <_{\pi_c} c_k$. We define C_i to be $\{c_1, c_2, \ldots, c_i\}$ for every i such that $1 \leq i \leq k$.*

Types of Vertices: Let I be the independent set $V \setminus C$. We associate a *type* with each vertex in I. A "type" is simply a subset of C.

Definition 2. *Let I be the independent set $V \setminus C$. The type of a vertex v in I is $N(v)$. For a type $S \subseteq C$ the set $I(S)$ is the set of all vertices in I of type S.*

Notice that two vertices of the same type are indistinguishable up to automorphisms of G, and that there are 2^k different types.

---

³ Proofs of results marked with [⋆] will appear in the long version of the paper.

Inner Order: Observe that every vertex of I is either mapped between two vertices of C, to the left of c_1 or to the right of c_k by a permutation π. For a permutation π we say that a vertex v is at *location* 0 if $v <_\pi c_1$ and at location i if i is the largest integer such that $c_i <_\pi v$. The set of vertices that are at location i is denoted by L_i. We define the *inner order* of π at location i to be the permutation defined by π restricted to L_i.

The task of finding an optimal permutation can be divided into two parts. The first part is to partition the set I into $L_0, \ldots, L_k$, while the second part consists of finding an optimal inner order at all locations. One should notice that partitioning I into $L_0, \ldots, L_k$ amounts to deciding *how many* vertices of each type are at location i for each i. For most layout problems, figuring out the right partitioning turns out to be more difficult than determining the inner orders once the partitioning is known. For IMBALANCE, this turns out to be particularly true as the inner orders in fact are irrelevant. The reason for this is that permuting the inner order of π at location i does not change the imbalance of any single vertex where the imbalance of a vertex v is $|L_\pi(v) - R_\pi(v)|$. Finding the optimal ordering of the vertices thus reduces to finding the right partition of I into $L_0, \ldots, L_k$. We formalize this as an instance of p-OPT-ILP.

ILP Formulation: For a type S and location i we let x_S^i be a variable that encodes the number of vertices of type S that are at location i. Also, for every vertex c_i in C we have a variable y_i that represents the imbalance of c_i. In order to represent a feasible permutation, all the variables must be non-negative. Also the variables x_S^i must satisfy that for every type S, $\sum_{i=0}^{k} x_S^i = |I(S)|$. For every vertex c_i of the vertex cover let $e_i = \big||N(c_i) \cap C_{i-1}| - |N(c_i) \cap (C \setminus C_i)|\big|$ be a constant. Finally for every $c_i \in C$ we add the constarint $y_i = e_i + \big|\sum_{\{S \subseteq C | c_i \in S\}} \big(\sum_{j=0}^{i-1} x_S^j - \sum_{j=i}^{k} x_S^j\big)\big|$.

One should notice that the last set of constraints is not a set of linear constraints. However, we can guess the sign of $y_i' = e_i + \sum_{\{S \subseteq C | c_i \in S\}} \big(\sum_{j=0}^{i-1} x_S^j - \sum_{j=i}^{k} x_S^j\big)$ for every i in an optimal solution. This increases the running time by a factor of 2^k. For every i we let t_i take the value 1 if we have guessed that $y_i' \geq 0$ and we let t_i take the value -1 if we have guessed that $y_i' < 0$. We can now replace the non-linear constraints with the linear constraints $y_i = t_i y_i'$ for every i. Finally, for every type S and location i, let z_S^i be the constant $\big||S \cap C_i| - |S \cap (C \setminus C_i)|\big|$. We are now ready to formulate the integer linear program.

$$
\min \quad \sum_{i=1}^{k} t_i \cdot y_i + \sum_{S \subseteq C} z_S^i \cdot x_S^i
$$

$$
\text{such that} \quad \sum_{i} x_S^i = |I(S)| \qquad \text{for all } i \in \{0, \ldots, k\}, S \subseteq C
$$

$$
y_i = t_i e_i + \sum_{\{S \subseteq C | c_i \in S\}} \big(\textstyle\sum_{j=0}^{i-1} t_i x_S^j - \sum_{j=i}^{k} t_i x_S^j\big) \text{ for all } i \in \{1, \ldots, k\}
$$

$$
x_S^i, \, y_i \geq 0 \qquad \text{for all } i \in \{0, \ldots, k\}, S \subseteq C
$$

Since the value of $f_{im}(\pi)$ is bounded by n^2 and the value of any variable in the integer linear program is bounded by n, Theorem 2 implies that this integer linear program can be solved in FPT time, thus implying the following theorem.

Theorem 3. *The* IMBALANCE *problem parameterized by the vertex cover number of the input graph is fixed parameter tractable.*

4 Cutwidth: The Inner Order Is Known

In the CUTWIDTH problem, we are to find the permutation of the vertices of the input graph that minimizes $f_{cw}(\pi)$, the maximum cut in the permutation. We proceed to give an FPT algorithm for minimizing $f_{cw}(\pi)$ in graphs with small vertex covers. The input is a graph $G = (C \cup I, E)$ with C being a vertex cover of size k. We define the *rank* of a vertex v with respect to a vertex set S to be $rank(S, v) = |N(v) \setminus S| - |N(v) \cap S|$. Notice that $|\partial(S \cup v)| = |\partial(S)| + rank(S, v)$.

Just as for the IMBALANCE problem, we guess the order $c_1 <_{\pi_c} \cdots <_{\pi_c} c_k$ of the vertices in C in an optimal permutation π. We consider the inner order of L_i for some i between 0 and k. Suppose $\pi(c_i) = s$, then, for any t with $s < t \leq s + |L_i|$ we have that $|\partial(V_t)| = |\partial(V_s)| + \sum_{j=s+1}^{t} rank(V_{j-1}, v_j)$. Since the set of vertices in the locations form an independent set, $rank(V_{j-1}, v_j) = rank(C_i, v_j)$ for every j between $s+1$ and t. This gives the equation $|\partial(V_t)| = |\partial(V_s)| + \sum_{j=s+1}^{t} rank(C_i, v_j)$.

Hence if we start with an optimal permutation π and reorganize the inner order at each location i to sort the vertices by rank with respect to C_i in nondecreasing order, we get another optimal ordering with a fixed inner order for each location. In such orderings the largest values of $|\partial(V_i)|$ occur either at $i = \pi(c_j) - 1$ or at $i = \pi(c_j)$ for some j between 1 and k. Since the rank of a vertex $v \in I$ with respect to C_i only depends on i and the type of v, we can use this together with the fact that $|\partial(V_t)| = |\partial(V_s)| + \sum_{j=s+1}^{t} rank(C_i, v_j)$ in order to give an integer linear programming formulation for the CUTWIDTH problem.

For every type S and location i we introduce a variable x_S^i that tells us the number of vertices of type S that are at location i. For every i between 1 and k we add a variable y_i which encodes $rank(V_{\pi(c_i)-1}, c_i)$ and the constant $e_i = |N(c_i) \cap (C \setminus C_i)| - |N(c_i) \cap C_{i-1}|$. For every type S and location i we also compute the constant e_S^i that indicates the rank of a vertex of type S with respect to C_i. Finally we need a variable c that represents the cutwidth of G. For the constraints, as for the IMBALANCE problem, we need to make sure the variables x_S^i represent a valid partitioning of I into $L_0, \ldots, L_k$. Finally we need constraints to encode the rank of the vertex cover vertices and the connection between the partitioning of I and the cutwidth c. This yields the following integer linear program:

$$\min \quad c$$

$$\text{such that } \sum_i x_S^i = |I(S)| \qquad\qquad \text{for all } S \subseteq C$$

$$y_i = e_i + \sum_{\{S \subseteq C \mid c_i \in S\}} \left(\sum_{j=i}^{k} x_S^j - \sum_{j=0}^{i-1} x_S^j \right) \text{ for all } i \in \{0, \ldots, k\}$$

$$c \geq \sum_{j=0}^{i} y_j + \sum_{j=0}^{i-1} \sum_{S \subseteq C} e_S^j \cdot x_S^j \qquad\qquad \text{for all } i \in \{1, \ldots, k\}$$

$$c \geq \sum_{j=0}^{i-1} y_j + \sum_{j=0}^{i-1} \sum_{S \subseteq C} e_S^j \cdot x_S^j \qquad\qquad \text{for all } i \in \{1, \ldots, k\}$$

$$x_S^i \geq 0 \qquad\qquad \text{for all } i \in \{0, \ldots, k\}, S \subseteq C$$

Since the value of $f_{cw}(\pi)$ is bounded by n^2 and the value of any variable in the integer linear program is bounded by n^2, Theorem 2 implies that this integer linear program can be solved in FPT time, yielding the following theorem.

Theorem 4. *The* CUTWIDTH *problem parameterized by the minimum vertex cover of the input graph is fixed parameter tractable.*

5   Bandwidth: The Inner Order Is Structured I

In the BANDWIDTH problem the aim is to minimize the function $f_{bw}(\pi) = \max_{uv \in E} |\pi(u) - \pi(v)|$. As for the previous cases we guess the ordering $c_1 <_{\pi_c} \ldots <_{\pi_c} c_k$ of the vertices in C in an optimal permutation π. Since we now are looking for the optimal permutation π that agrees with this ordering of the vertices in C, we observe that for a vertex $v \in I$ the only relevant neighbours in C are the leftmost and rightmost neighbour. We can thus delete the edges from v to all other neighbours of v. After this reduction every vertex in I has degree at most 2, and thus the number of different types is bounded by k^2 rather than 2^k.

For BANDWIDTH, we are not able to determine the inner orders a priori, contrary to the situation we had for CUTWIDTH. Instead we will show that there is an optimal permutation where the inner orderings have a specific structure. We say that an interval $[a, b]$ on the integer line is *uniform* if all vertices π maps to $[a, b]$ have the same type. A *zone* is an inclusion maximal uniform interval, and for a layout π of the vertices of G, the *zonal dimension* of π at location i, $\zeta_i(\pi)$, is the number of zones inside $[\pi(c_i) + 1, \pi(c_{i+1}) - 1]$. The zonal dimension of π is $\zeta(\pi) = \max_{i=0}^{k} \zeta_i(\pi)$. Our approach consists of two parts. First we show that there is an ordering π minimizing bandwidth such that $\zeta(\pi) \leq k^2(2k + 1) + 2k$. We then use this to show that BANDWIDTH parameterized by the size of the minimum vertex cover of the input graph is fixed parameter tractable.

Lemma 1. [⋆] *For a graph $G = (C \cup I, E)$, there is an optimal bandwidth ordering π with $\zeta(\pi) \leq k^2(2k + 1) + 2k$.*

So, how can one use Lemma 1 to give an integer linear program for the BAND-WIDTH problem? The trick is to guess the correct values of $\zeta_i(\pi)$ for every i and guess which type of vertices appears in each zone. We can do this at a cost of a factor $(3k^3)^{k+1}(k^2)^{3k^3} = k^{O(k^3)}$ in the running time. Note that the zones are ordered from left to right. We can now set up an integer linear program where the variables encode how many vertices there are in each zone. Let x_i be a variable that encodes the number of vertices in zone number i from the left. For each type $S \subseteq C$ such that $I(S)$ is nonempty, we let $Z(S)$ be the set of integers such that for each $i \in Z(S)$ we have guessed that the vertices in the zone i have type S. Let l_S and r_S be the smallest and largest numbers in $Z(S)$ respectively. Now, for an integer $1 \leq i \leq k$ we let e_i be the number of zones guessed to be to the left of c_i. Finally, for an integer i between 1 and k and a type S we define the constant $t_1(i, S)$ to be the number of vertices from C to the right of zone number l_S and to the left of c_i. Similarly, let $t_2(i, S)$ be the number of vertices from C to the left of zone number r_S and to the right of c_i. Having made the discussed guesses, we can formulate the BANDWIDTH problem as an integer linear program as follows:

$$\min \quad b$$

$$
\begin{aligned}
\text{such that } &\sum_{i \in Z(S)} x_i = |I(S)| && \text{for all } S \subseteq C : I(S) \neq \emptyset \\
&b \geq j - i - 1 + \sum_{q=e_i+1}^{e_j} x_q && \text{for all } c_i c_j \in E \\
&b \geq t_1(i, S) + \sum_{j=l_S}^{e_i} x_j && \text{for all } i \in \{1, \ldots, k\}, S \subseteq C : I(S) \neq \emptyset, c_i \in S \\
&b \geq t_2(i, S) + \sum_{j=e_i+1}^{r_S} x_j && \text{for all } i \in \{1, \ldots, k\}, S \subseteq C : I(S) \neq \emptyset, c_i \in S \\
&x_i \geq 0 && \text{for all } i \in \{0, \ldots, k\}
\end{aligned}
\tag{1}
$$

Because the value of $f_{bw}(\pi)$ is bounded from above by n and the value of any variable in the integer linear program is bounded by n, Theorem 2 implies that this integer linear program can be solved in FPT time, yielding the following theorem.

Theorem 5. [$\star$] *The* BANDWIDTH *problem parameterized by the size k of the minimum vertex cover of the input graph can be solved in time $k^{O(k^3)}n$ and polynomial space.*

6   Distortion: The Inner Order Is Structured II

In this section we consider the parametrized complexity of embedding graph metrics into the real line, parameterized by the size of the minimum vertex cover of the input graph. Given an undirected graph $G = (V, E)$, a natural metric associated with G is $M(G) = (V, D)$ where the distance function D is the shortest path distance between u and v for each pair of vertices $u, v \in V$. Given a graph metric M and another metric space M' (like real line) with distance functions D and D', a mapping $f : M \to M'$ is called an *embedding* of M into M'. The mapping f has *contraction* c_f and *expansion* e_f if for every pair of points p, q in M, $D(p, q) \leq D'(f(p), f(q)) \cdot c_f$ and $D(p, q) \cdot e_f \geq D'(f(p), f(q))$ respectively. A mapping f has *distortion* d if $(e_f \cdot c_f)$ is at most d. We say that f is *non-contracting* if c_f is at most 1. A non-contracting mapping f has *distortion* d if e_f is at most d. As observed by several authors before [9,15], the problem of finding a minimum distortion embedding of a graph metric into the line can be expressed as a problem of finding the permutation $\pi : V \to \{1, 2, \ldots, n\}$ that minimizes $f_{di}(\pi) = \max_{uv \in E} \sum_{i=\pi(u)}^{\pi(v)-1} D(v_i, v_{i+1})$.

Lemma 2 ([9]). *A graph $G = (V, E)$ has a distortion d embedding f into the real line if and only if there is a permutaion $\pi : V \to \{1, 2, \ldots, n\}$ such that $f_{di}(\pi) \leq d$.*

For a permutation π and two vertices u and v such that $u <_\pi v$ we define $D_\pi(u, v) = \sum_{i=\pi(u)}^{\pi(v)-1} D(v_i, v_{i+1})$. If $v <_\pi u$ then $D_\pi(u, v)$ is defined to be $D_\pi(v, u)$. We give a fixed parameter tractable algorithm for the DISTORTION problem parameterized by the size of the minimum vertex cover of the input graph. Our approach is similar to, albeit more involved than, the algorithm presented for the BANDWIDTH problem. As for the previous problems, we iterate over all $k!$ ways to order the vertices of C into $c_1 <_{\pi_c} \ldots <_{\pi_c} c_k$. We proceed to show that there is an optimal permutation π such that $\zeta(\pi) \leq (4k + 1)2^{2^k}$.

Lemma 3. [$\star$] *For a graph $G = (C \cup I, E)$, there is an optimal distortion ordering π with $\zeta(\pi) \leq (4k + 1)2^{2^k}$.*

Using Lemma 3 we can give an algorithm for the DISTORTION problem similar to the algorithm for BANDWIDTH. The algorithm proceeds excactly as for BANDWIDTH with the only differences being that the zonal dimension is much larger, and that one has to be careful to introduce constants that encode the distance between two consecutive vertices in the ILP. Notice that since the zonal dimension is not polynomial in k for the DISTORTION problem, we do not obtain a polynomial space algorithm.

Theorem 6. *The DISTORTION problem parameterized by the minimum vertex cover of the input graph is fixed parameter tractable.*

7   Conclusion and Discussions

In this paper we considered parameterization by vertex cover number of the graph, a structural parameter stronger than the treewidth. This enabled us to show that graph layout problems CUTWIDTH, BANDWIDTH, IMBALANCE and DISTORTION are FPT parameterized by vertex cover number of the graph. This is in contrast to the parameterization by treewidth for which the paramterized complexity of these problems is open. The structural parameterization of vertex cover number also brought forward the technique of bounded variable integer linear programming to importance. We believe that this (underused) powerful result will become one of the basic tools in classifying whether a problem is FPT, as well as in designing practical algorithms, because p-ILP is well solved for p up to 1000.

One may wonder whether there exists a problem which is not FPT for graphs of bounded vertex cover number. This in indeed the case, as LIST COLORING remains $W[1]$-hard even for graphs of bounded vertex cover number. An important graph layout problem is OPTIMAL LINEAR ARRANGEMENT where the objective is to minimize the sum of $|\partial V_i|$. We can show that this problem is in XP by giving an algorithm of time complexity $n^{f(k)}$ when parameterized by the vertex cover number of the input graph. The main difficulty we face in encoding this problem as ILP is that the objective function is not linear, but quadratic. Hence in this direction the following questions still remain unanswered.

- Is OPTIMAL LINEAR ARRANGEMENT FPT parameterized by the vertex cover number of the input graph?
- Is CUTWIDTH FPT parameterized by the treewidth of the input graph?

References

1. Adolphson, D., Hu, T.C.: Optimal linear ordering. SIAM J. Appl. Math. 25, 403–423 (1973)
2. Alon, N., Azar, Y., Woeginger, G.J., Yadid, T.: Approximation schemes for scheduling on parallel machines. J. of Scheduling 1, 55–66 (1998)
3. Blin, G., Fertin, G., Hermelin, D., Vialette, S.: Fixed-parameter algorithms for protein similarity search under RNA structure constraints. In: Kratsch, D. (ed.) WG 2005. LNCS, vol. 3787, pp. 271–282. Springer, Heidelberg (2005)
4. Botafogo, R.A.: Cluster analysis for hypertext systems. In: Proceedings of SIGIR 1993, pp. 116–125. ACM Press, New York (1993)
5. Chinn, P., Chvatalova, J., Dewdney, A., Gibbs, N.: The bandwidth problem for graphs and matrices – a survey. Journal of Graph Theory 6, 223–254 (1982)
6. Clarkson, K.L.: Las Vegas Algorithms for Linear and Integer Programming When the Dimension is Small. Journal of the Association for Computing Machinery 42(2), 488–499 (1995)
7. Courcelle, B.: The Monadic Second-Order Logic of Graphs. I. Recognizable Sets of Finite Graphs Inf. Comput. 85(1), 12–75 (1990)
8. Downey, R.G., Fellows, M.R.: Parameterized Complexity. Springer, New York (1999)
9. Fellows, M.R., Fomin, F.V., Lokshtanov, D., Losievskaja, E., Rosamond, F.A., Saurabh, S.: Parameterized Low-distortion Embeddings - Graph metrics into lines and trees, CoRR abs/0804.3028 (2008)
10. Fellows, M.R., Rosamond, F.A.: The Complexity Ecology of Parameters: An Illustration Using Bounded Max Leaf Number. In: Cooper, S.B., Löwe, B., Sorbi, A. (eds.) CiE 2007. LNCS, vol. 4497, pp. 268–277. Springer, Heidelberg (2007)
11. Flum, J., Grohe, M.: Parameterized Complexity Theory. Springer, Berlin (2006)

12. Frank, A., Tardos, E.: An Application of Simultaneous Diophantine Approximation in Combinatorial Optimization. Combinatorica 7, 49–65 (1987)
13. Gramm, J., Niedermeier, R., Rossmanith, P.: Fixed-Parameter Algorithms for CLOSEST STRING and Related Problems. Algorithmica 37(1), 25–42 (2003)
14. Gupta, A., Newman, I., Rabinovich, Y., Sinclair, A.: Cuts, trees and l_1-embeddings of graphs. Combinatorica 24(2), 233–269 (2004)
15. Heggernes, P., Meister, D., Proskurowski, A.: Minimum distortion embeddings into a path of bipartite permutation and threshold graphs. In: Gudmundsson, J. (ed.) SWAT 2008. LNCS, vol. 5124, pp. 331–342. Springer, Heidelberg (2008)
16. Heinz, S.: Complexity of integer quasiconvex polynomial optimization. Journal of Complexity 21, 543–556 (2005)
17. Junguer, M., Reinelt, G., Rinaldi, G.: The traveling salesman problem. In: Handbook on Operations Research and Management Sciences, vol. 7, pp. 225–330. North-Holland, Amsterdam (1995)
18. Kannan, R.: Minkowski's Convex Body Theorem and Integer Programming. Mathematics of Operations Research 12, 415–440 (1987)
19. Kant, G.: Drawing planar graphs using the canonical ordering. Algorithmica 16(1), 4–32 (1996)
20. Kant, G., He, X.: Regular edge labeling of 4-connected plane graphs and its applications in graph drawing problems. Theoret. Comput. Sci. 172(1-2), 175–193 (1997)
21. Karger, D.R.: A randomized fully polynomial approximation scheme for all terminal network reliability problem. In: Proceedings of STOC, pp. 11–17. ACM Press, New York (1996)
22. Khachiyan, L., Porkolab, L.: Integer Optimization on Convex Semialgebraic Sets. Discrte Computational Geometry 23, 207–224 (2000)
23. Lenstra, H.W.: Integer Programming with a Fixed Number of Variables. Mathematics of Operations Research 8, 538–548 (1983)
24. Linial, N., London, E., Rabinovich, Y.: The Geometry of Graphs and Some of its Algorithmic Applications. Combinatorica 15(2), 215–245 (1995)
25. Makedon, F., Sudborough, I.H.: Minimizing width in linear layouts. In: ICALP 1983. LNCS, vol. 154, pp. 478–490. Springer, Heidelberg (1983)
26. Mutzel, P.: A polyhedral approach to planar augmentation and related problems. In: Spirakis, P.G. (ed.) ESA 1995. LNCS, vol. 979, pp. 497–507. Springer, Heidelberg (1995)
27. Niedermeier, R.: Invitation to fixed-parameter algorithms. Oxford Lecture Series in Mathematics and its Applications, vol. 31. Oxford University Press, Oxford (2006)
28. Papakostas, A., Tollis, I.G.: Algorithms for area-efficient orthogonal drawings. Computational Geometry 9, 83–110 (1998)
29. Thilikos, D.M., Serna, M.J., Bodlaender, H.L.: Cutwidth II: Algorithms for partial w-trees of bounded degree. J. Algorithms 56(1), 25–49 (2005)
30. Wood, D.R.: Optimal three-dimensional orthogonal graph drawing in the general position model. Theoret. Comput. Sci. 299(1-3), 151–178 (2003)

A Linear Kernel for the k-Disjoint Cycle Problem on Planar Graphs

Hans L. Bodlaender[1], Eelko Penninkx[1], and Richard B. Tan[1,2]

[1] Department of Information and Computing Sciences, Utrecht University, Padualaan
14, 3584 CH, Utrecht, The Netherlands
[2] Department of Computer Science, University of Sciences & Arts of Oklahoma,
Chickasha, OK 73018, USA
hansb@cs.uu.nl, penninkx@cs.uu.nl, rbtan@cs.uu.nl

Abstract. We consider the following problem: given a planar graph
$G = (V, E)$ and integer k, find if possible at least k vertex disjoint cycles
in G. This problem is known to be NP-complete. In this paper, we give
a number of simple data reduction rules. Each rule transforms the input
to an equivalent smaller input, and can be carried out in polynomial
time. We show that inputs on which no rule can be carried out have size
linear in k. Thereby we obtain that the k-DISJOINT CYCLES problem on
planar graphs has a kernel of linear size. We also present a parameterized
algorithm with a running time of $O(c^{\sqrt{k}} + n^2)$.

1   Introduction

A general approach that is used when one wants to solve an NP-hard problem
in a practical setting is to first use *preprocessing* or *data reduction* in an attempt
to obtain a smaller but equivalent instance. A slow exact (or fixed parameter)
algorithm can then be used on the reduced instance. In some cases, it is possible
to obtain bounds on the size of a resulting instance. Techniques from fixed parameter tractability theory help to obtain such bounds. We refer to the excellent
overview paper by Guo and Niedermeier [8] for an overview of many recent results. Data reduction with an upper bound on the size of the resulting instances
is nowadays termed *kernelization*. For a kernelization algorithm, part of our input is a parameter k, and the algorithm finds in polynomial time an equivalent
input whose size is bounded by a function of the parameter. Both from a theoretical and from a practical viewpoint, it is interesting to have a small upper
bound on the size of the resulting *kernel*.

In this paper, we consider the k-DISJOINT CYCLES problem on planar graphs:
given a planar graph $G = (V, E)$ and an integer k, find a collection of at least k
cycles that are pairwise vertex-disjoint, or return that no such collection exists.
This problem is NP-complete, also on planar graphs (it contains PARTITION
INTO TRIANGLES as subproblem). We show that this problem has a kernel of
size linear in k. We give a number of simple rules that modify the given graph.
If no rule can be applied we have found an equivalent problem instance having
$O(k)$ vertices. Each rule is *safe*, i.e., when (G, k) is transformed to (G', k'), then

S.-H. Hong, H. Nagamochi, and T. Fukunaga (Eds.): ISAAC 2008, LNCS 5369, pp. 306–317, 2008.

G has a collection of k disjoint cycles if and only if G' has a collection of k' disjoint cycles, and G' is also planar.

Our result resolve an open problem by Kloks et al. [10], who give an algorithm for k-Disjoint Cycles on planar graphs that runs in $O(c^{\sqrt{k}\log k}n)$ time, and ask for a reduction to a polynomial problem kernel for the problem. We use our kernel to give an algorithm that runs in time $O(c^{\sqrt{k}} + n^2)$.

Our linear kernel for planar graphs nicely complements a very recent result by Bodlaender et al. [5], who show that there is no polynomial size kernel for the k-Disjoint Cycles problem unless the polynomial time hierarchy collapses to the third level. However, as k-Disjoint Cycles belongs to the class FPT, a kernel of exponential size does exist.

Related results are a linear kernel for Feedback Vertex Set on planar graphs [4], a cubic kernel for the same problem on general graphs [3], a linear kernel for Dominating Set on planar graphs [2, 1], and FPT-algorithms for k-Disjoint Cycles on arbitrary graphs (see [7]). Also see [9] for a generic framework for deriving linear kernels for some problems on planar graphs.

This paper is organized as follows. In Section 2, we give some preliminary definitions and results. In Section 3, we present the kernelization algorithm, the reduction rules, prove their safeness, and show that we can apply them in polynomial time. Section 4 gives proves the bound on the kernel size from which correctness of our algorithm follows. In Section 5, we state a faster parameterized algorithm for the k-Disjoint Cycles problem on planar graphs. Some final comments are made in Section 6.

2   Preliminaries

All graphs in this paper are undirected, planar, and can have parallel edges and self-loops. Consider a graph $G = (V, E)$. A *path* is a sequence of vertices and edges $v_0 e_1 v_1 e_2 \ldots e_n v_n$ such that $e_i = \{v_{i-1}, v_i\}$ and $v_i \neq v_j$ for all $1 \leq i < j \leq n$. A *cycle* is a path where $v_0 = v_n$ and all other vertices are different. The length of a path or cycle is the number of edges used. By $G[W]$ we denote the graph induced by $W \subseteq V$, defined as $G[W] = (W, E \cap W \times W)$. The *neighborhood* $N(v) = \{u \in V \setminus v : \{u, v\} \in E\}$ of a vertex $v \in V$ is the set of vertices adjacent to v. The neighborhood of a set $U \subseteq V$ is defined as $N(U) = (\cup_{u \in U} N(u)) \setminus U$.

The degree $d(v, W) = |\{e \in E : e = \{v, u\}, u \in W\}|$ of a vertex v with respect to $W \subseteq V$ is defined as the number of edges between v and vertices from W. We use $d(v)$ as a shorthand for $d(v, V)$. We sometimes use single vertices v instead of the singleton set $\{v\}$ for readability. E.g., $d(v, w) = 2$ implies that there are exactly two parallel edges $\{v, w\}$.

A tree is a connected subgraph of G that contains no cycles. Vertices on a tree with degree at least three are called *internal* vertices, leafs have degree at most one. A *forest* is a set of trees. To simplify notation we use paths, trees and cycles as if they are sets of vertices. E.g. given a tree T and a cycle C we use $G[T \cup C]$ to denote $G[V(T) \cup V(C)]$. This also allows us to use $|P|$ for the number of vertices on a path P.

When talking about disjoint cycles we always mean pairwise *vertex* disjoint cycles.

We will need the following two lemma's throughout the paper. They are stated here rather than at the time they are needed, as they are of some interest by themselves:

Lemma 1. *Consider a bipartite planar graph G with n vertices on the left side, having no parallel edges. Let x_i be the number of vertices on the right side with degree i. Then $\sum_{i \geq 3}(\frac{1}{2}i - 1)x_i \leq n - 2$.*

Proof. We assume $x_0 = x_1 = x_2 = 0$ without loss of generality. The number of vertices of G is $n + \sum_i x_i$ and the number of edges is $\sum_i ix_i$. The number of faces is at most half the number of edges, in which case every face has exactly four incident edges. Using Euler's formula $|V(G)| - |E(G)| + |F(G)| = 2$ for planar graphs we directly obtain the lemma. $\square$

Lemma 2. *Given n cycles on the plane, we can connect at most $3n - 6$ different pairs by an edge without crossings.*

Proof. We will show that the number of edges is maximum when no cycles are nested. Let $f(n)$ denote the maximum number of edges between n vertices in a planar graph not allowing self-loops or parallel edges. Then $f(n) = 3n - 6$ for $n \geq 3$, $f(2) = 1$ and $f(1) = f(0) = 0$. We compare the following two cases: (i) We have q cycles within some cycle C, and C itself has p cycles it can be connected to on the outside, (ii) we have $p + q + 1$ cycles without layering. The difference between these cases is $f(p + q + 1) - f(p + 1) - f(q + 1) \geq 0$, so we can never gain by nesting cycles. $\square$

3   Reduction

We present an algorithm that given an instance (G, k) of PLANAR k-DISJOINT CYCLES either returns true (meaning that G has at least k disjoint cycles), or returns an *equivalent* instance (G', k') where $|V(G')| = O(k)$. Two instances (G, k) and (G', k') are equivalent if G has k disjoint cycles if and only if (G', k') has k' disjoint cycles, and G' is planar if G is planar. If there are more than two parallel edges $\{v, w\}$ in G we can remove all but two without changing the problem.

The algorithm starts by applying reduction rules until none apply, resulting in an equivalent instance (G', k'). If $|V(G')| > 1209k' - 1317$ we return true, else we return this instance. The main result of this paper is the following:

Theorem 1. *For any integer $k \geq 3$ and planar graph $G = (V, E)$ where no reduction rules can be applied, $|V| \leq 1209k - 1317$, or G contains at least k disjoint cycles.*

We present eight reduction rules for the k-Disjoint Cycles problem. Each of these rules is safe, also for general graphs, and maintains planarity. A *safe* reduction rule transforms an instance (G, k) to an equivalent instance (G', k'). If we find some cycle C that is included in some optimal solution then we can *select C*, meaning that $(G', k') = (G[V \setminus C], k - 1)$. If vertex v is not included in some optimal solution then we can *remove v*, the new instance then becomes $(G[V \setminus v], k)$. Application of Rule i is only allowed if all rules $j < i$ do not apply.

3.1   Basic Rules

We present the first five rules, safeness of rules 1–4 is trivial.

Rule 1 (Self Loop Rule). *If there is an edge $\{v, v\} \in E$ then select cycle (v).*

Rule 2 (Degree Zero Rule). *If there is a vertex $v \in V$ such that $d(v) = 0$ then remove v.*

Rule 3 (Degree One Rule). *If there is a vertex $v \in V$ such that $d(v) = 1$ then remove v.*

Rule 4 (Degree Two Rule). *Suppose there is a vertex v such that $d(v) = 2$ with incident edges $\{v, w\}$ and $\{v, x\}$. Then remove v and add an edge $\{w, x\}$ if allowed (that is, if this does not create a third parallel edge $\{w, x\}$). Note that if $w = x$ a self-loop is added.*

Rule 5 (Degree Three Rule). *Suppose there is a vertex v with $d(v) = 3$ and having only two neighbors w and x where $d(v, w) = 2$. Then select cycle (v, w).*

Proof of safeness. The subgraph $G[\{v, w, x\}]$ can be used in at most one cycle, so there is always an optimal solution using cycle (v, w). $\square$

If Rules 1–5 do not apply we have the following result:

Lemma 3. *Suppose Rules 1–5 do not apply in a planar graph $G = (V, E)$. Consider a tree $T \subseteq V$ with leafs L. Then $|N(l) \setminus T| \geq 2$ for all $l \in L$, so every leaf has at least two neighbors outside of T.*

Proof. For every $l \in L$ we have $d(l, T) = 1$ and $d(l, N(T)) \geq 2$ because the Degree Two Rule does not apply. If $|N(l)| = 2$ then the Degree Three Rule would apply, so $|N(l)| \geq 3$ proving the lemma. $\square$

3.2   Advanced Rules

We now present rules that operate on *small trees*: trees T where $|N(T)| = 2$. For all rules let $N(T) = \{v, w\}$.

Rule 6 (Small Tree Rule). *Consider a small tree T with $|T| \geq 2$. If there are two disjoint cycles in $G[T \cup N(T)]$ then select those cycles, otherwise replace T by a single vertex, connected by two edges to both v and w.*

Proof of safeness. Note that T has at least two leafs, all of which are connected to both v and w because of the Degree Three Rule. Hence we always have a cycle using (v, T) and (w, T). If there are two disjoint cycles in $G[T \cup N(T)]$ then selecting both is optimal, as this subgraph can only be involved in two cycles in any solution.

If there is only one disjoint cycle in $G[T \cup N(T)]$ then there is an optimal solution where either (v, T) or (w, T) is used. Suppose not, then remove the cycle through v and replace it by (v, T) to obtain another optimal solution. This is exactly the same situation as after the reduction. □

Rule 7 (Parallel Small Tree Rule). *Consider two small trees T_1 and T_2 with $N(T_1) = N(T_2) = \{v, w\}$. Then select both cycles (v, T_1) and (w, T_2).*

Proof of safeness. The subgraph $G[T_1 \cup T_2 \cup \{v, w\}]$ contains two disjoint cycles. This is also the maximum number of cycles this subgraph can be used in any solution, so we can safely select them. □

Using Rules 1–7 we obtain the following:

Lemma 4. *Suppose Rules 1–7 do not apply in graph G. Then for any two distinct vertices $v, w \in V$ there is at most one tree T such that $N(T) = \{v, w\}$, and T consists of one vertex.*

Proof. This follows directly from the Rules 6 and 7. □

The last rule operates on subgraphs that consist of a path and a small set of adjacent vertices. We only apply this rule for $n \leq 2$.

Rule 8 (Path Rule). *Consider vertex set $W \subset V$ and path $U = (u_1, \ldots, u_m)$ such that W and U are disjoint, $n = |W|, m > n$ and $d(u_i, V \setminus W) = 2$ for all u on U. By w_j we denote the order in which the vertices of W are connected to the path. If there are n disjoint cycles in $G[W \cup U]$ then replace U by a path U' of length n where $u'_1 = u_1$ and $u'_n = u_m$ and u_i is connected by two edges to w_i for $i = 1, \ldots, n$.*

Proof of safeness. In some optimal solution the subgraph $G[U \cup W]$ is used in exactly n or $n + 1$ cycles. In the first case we can choose all cycles to be the n disjoint ones in $G[U \cup W]$ itself, in the latter case we have to use path U, implying that no internal cycle in $G[U \cup W]$ is used. This is exactly the situation as is modeled after the reduction. □

See Figure 1 for an application of this rule. We have the following:

Lemma 5. *Consider W and U as defined by the Path Rule. If $n = 1$ then the Path Rule applies if $m \geq 2$, and if $n = 2$ then Path Rule applies if $m \geq 6$.*

Proof. If $n = 1$ and $m \geq 2$ we have a disjoint cycle in $G[W \cup U]$ allowing application of the Path Rule. If $n = 2$ we cannot have a subgraph where the $n = 1$ case applies, so no two consecutive vertices on U can be connected to

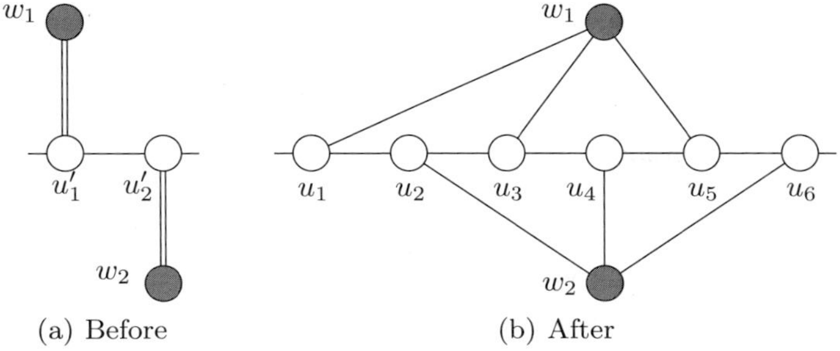

(a) Before   (b) After

Fig. 1. Application of the Path Rule where $|W| = 2$

the same vertex in W without also connecting to the other one. Without loss of generality we assume $d(u, W) = 1$ for all $u \in U$, directly showing that the longest path that does not allow the Path Rule to be applied consists of alternating the connections the vertices in W. If $m = 6$ we have two disjoint cycles as shown in Figure 1, proving the lemma. $\qquad\square$

3.3 Running Time Analysis

The total running time of the kernelization is $O(n^2)$. Due to lack of space we omit all details. We expect an actual implementation to perform very well if the Path Rule is implemented efficiently.

4 Linear Kernel

In this section we will prove Theorem 1. We use a set of disjoint cycles $\mathbb{C}$ that satisfy the following properties:

Lemma 6. *Given a planar graph $G = (V, E)$ where no reduction rules apply. Let $\mathbb{C}$ be a collection of disjoint cycles, and let $T = V \setminus \mathbb{C}$. We can find a $\mathbb{C}$ in time polynomial in $|V|$ that satisfies the following properties:*

1. *Every $C \in \mathbb{C}$ is a minimum cycle in $G[T \cup C]$,*
2. *There are no two disjoint cycles in $G[T \cup C]$ for any $C \in \mathbb{C}$,*
3. *There are no more than 5 disjoint paths between any pair of cycles in $\mathbb{C}$.*

Proof. Consider two cycles $C_1, C_2 \in \mathbb{C}$ such that the third property is invalidated. This implies that there are at least six disjoint paths between C_1 and C_2. We can then find three vertex disjoint cycles in $G[T \cup C_1 \cup C_2]$ using six of these paths. This property can be checked by using a flow algorithm thats finds a flow of size six from a source inside C_1 to a sink inside C_2 connected to the vertices of the cycle.

It is easy to see that by using iterative improvement we can find a collection of disjoint cycles satisfying all properties in polynomial time. $\qquad\square$

Let $\mathbb{C}$ be a collection of cycles satisfying Lemma 6, and let $k = |\mathbb{C}|$. Let $\mathbb{T}$ denote the forest consisting of the trees formed by vertices $V \setminus V(\mathbb{C})$. We will first show that $|V(\mathbb{T})| = O(k)$ and then $|V(\mathbb{C})| = O(k)$.

4.1   Linearity of Trees

For now assume that no cycles in $\mathbb{C}$ are nested, after the analysis we will show that this is indeed the worst-case for the size of $V(\mathbb{T})$. Let $\mathbb{T}_S$ contain all *small trees* (see Lemma 4), and let $\mathbb{T}_i \subseteq \mathbb{T} \setminus \mathbb{T}_S$ contain all trees incident to exactly i cycles in $\mathbb{C}$ that are not in $\mathbb{T}_S$.

Lemma 7. *For all $T \in \mathbb{T}$ incident to distinct $C_1, C_2 \in \mathbb{C}$ at most two leafs of T are connected to both C_1 and C_2.*

Proof. Suppose we have three or more such leafs. Each leaf connects to two cycles and some non-leaf node on T, giving a $K_{3,3}$ minor contradicting planarity.   □

Lemma 8. *There are no $C \in \mathbb{C}$ and distinct $T_1, T_2 \in \mathbb{T}$ such that $|N(T_1) \cap C| \geq 2, |N(T_2) \cap C| \geq 2$ and $|N(T_1 \cup T_2) \cap C| \geq 4$.*

Proof. Suppose for a contradiction that such T_1, T_2 and C exist. Both trees connect to some other cycle, so they are at the same side of C by our initial assumption that there is no nesting. But then we have two cycles in $G[C \cup T_1 \cup T_2]$ contradicting property 2 of Lemma 6.   □

Lemma 9. *Consider a set of one or two cycles $\mathbb{S} \subseteq \mathbb{C}$ and path $U = (u_1, \ldots, u_n)$ in $G[\mathbb{T}]$ such that every vertex on U is connected to at least one vertex in $V(\mathbb{S})$ having $N(u_i) \subseteq U \cup V(\mathbb{S})$ for $i = 2, \ldots, (n-1)$. If $|\mathbb{S}| = 1$ then $n \leq 5$, and if $|\mathbb{S}| = 2$ then $n \leq 13$.*

Proof. Let $\mathbb{S} = \{C\}$ if $|\mathbb{S}| = 1$. If we have four or more disjoint paths between U and C there are two disjoint cycles in $G[C \cup V(\mathbb{T})]$ contradicting Lemma 6. This implies that there is a subset $D \subseteq V(C)$ with $|D| \leq 3$ such that $N(D, U) = N(C, U)$. For every vertex of U we delete all edges to C with the exception of the first edge that goes to some vertex in D. Note that every vertex in U that was connected to C still is, and that all cycles in $G[U \cup D]$ are also cycles in $G[U \cup C]$.

If four consecutive vertices on U are connected to the same vertex $d \in D$ then the middle two together with d allow the Path Rule in the original graph, so at most three consecutive vertices on U connect to the same vertex $d \in D$. Because there are no two disjoint cycles in $G[U \cup D]$ we can add at most two vertices not connected to d to U, one at each side, so $|U| \leq 5$ in this case.

We only sketch the proof for $|\mathbb{S}| = 2$ due to lack of space. Let $\mathbb{S} = \{C_1, C_2\}$. Find subsets of size at most three D_1 and D_2 in $V(C_1)$ and $V(C_2)$ respectively by the same method as above. Let d_1, d_2 denote the vertices of D_1, D_2 that have most edges to U. At most four vertices on U are not incident to d_1 or d_2. Two can be incident to more than one vertex in D_1, two to more than one in D_2. Finally we have a path of at most five vertices between d_1 and d_2 because of the Path Rule. In total this are 13 vertices, proving the lemma.   □

We start our analysis of $|V(\mathbb{T})|$ by giving a bound for $|V(\mathbb{T}_S)|$. By definition for all $T \in \mathbb{T}_S$ we have $|N(T)| = 2$. Because of the Small Tree Rule also $|T| = 1$, and T is incident to exactly two vertices, both with parallel edges. We claim the following:

Lemma 10. $|V(\mathbb{T}_S)| \leq 3k - 6$.

Proof. Consider a cycle $C \in \mathbb{C}$ incident to $T \in \mathbb{T}_S$. Then $|C| = 2$, because of property 1 of Lemma 6. We also know that there are no distinct $T_1, T_2 \in \mathbb{T}_S$ such that $N(C) \subseteq N(T_1) \cup N(T_2)$, otherwise there are two disjoint cycles in $G[T_1 \cup T_2 \cup C]$ contradicting property 2 of Lemma 6.

Let x denote the number of trees in $\mathbb{T}_S$ that are incident to exactly one cycle. We claim $x \leq k$: if $x > k$ then there is some cycle $C \in \mathbb{C}$ incident to at least two such trees, implying two disjoint cycles in $G[C \cup V(\mathbb{T}_S)]$. All x cycles connected to these trees cannot be connected to any other small tree.

Now consider cycles $C_1, C_2 \in \mathbb{C}$ incident to a small tree $T \in \mathbb{T}_S$. Also we can have at most one vertex of C_1 or C_2 can be incident to small trees, else we have two disjoint cycles. Because of this and the Parallel Small Tree Rule there is at most one Small Tree that is incident to both C_1 and C_2, so the number of Small Trees incident to a pair of cycles is at most the number of edges in a planar graph having $k - x$ vertices. By setting $x = 0$ we obtain the lemma. $\square$

Lemma 11. $|V(\mathbb{T}_1)| \leq 19k$.

Proof. Consider some cycle $C \in \mathbb{C}$ and all trees $\mathbb{X} \subseteq \mathbb{T}_1$ incident to C. Let L be the vertices of $V(\mathbb{X})$ that have $d(v, V(\mathbb{X})) \leq 1$. Every node $v \in L$ is incident to at least 2 vertices in C, so $|C| \leq 4$ by property (1) of Lemma 6. If $|N(v_1, C) \cap N(v_2, C)| \geq 2$ then $|L| = 2$ because of planarity. This implies that if $|L| = 4$ then all four vertices connect to a different pair of cycle-vertices, so then $|C| = 4$. But in that case we have two disjoint cycles in $G[C \cup L]$ contradicting Lemma 6, so $|L| \leq 3$.

This means that $G[V(\mathbb{X})]$ contains at most one internal vertex v. If we remove L and v from $G[V(\mathbb{X})]$ we end up with at most three paths, all of which are incident to C, so all have size at most 5 by Lemma 9. This means that $|V(\mathbb{X})| \leq 19$, and because there are at most k cycles we obtain the lemma. $\square$

Lemma 12. $|\mathbb{T}_{\geq 2}| \leq 6k - 3$.

Proof. Consider the planar bipartite graph having a vertex for every cycle in $\mathbb{C}$, and a vertex for every tree in $\mathbb{T}_{\geq 3}$. Apply Lemma 1 to conclude that $|\mathbb{T}_{\geq 3}| \leq 2k - 2$.

Consider some maximal subset $\mathbb{T}_2' \subseteq \mathbb{T}_2$ such that $|N(T_1) \cap N(T_2) \cap C| \leq 1$ for all $C \in \mathbb{C}$, so no two trees in $\mathbb{T}_2'$ share more than one vertex on any cycle. Because $|N(T)| \geq 3$ for all $T \in \mathbb{T}_2'$, there is always some $C \in \mathbb{C}$ such that $|N(T) \cap C| = 2$. For all $C \in \mathbb{C}$ let $\mathbb{X}_C$ contain all $T \in \mathbb{T}_2'$ such that $|N(T) \cap C| \geq 2$. If $|\mathbb{X}_C| \geq 4$ we have a contradiction with Lemma 8, so $|\mathbb{X}_C| \leq 3$ for all $C \in \mathbb{C}$ and $|\mathbb{T}_2'| \leq 3k$.

Due to lack of space we only sketch the proof that $|\mathbb{T}_2 \setminus \mathbb{T}'2| \leq k - 1$: each such tree connects to at least two vertices on the same cycle that some tree from

$\mathbb{T}'_2$ also connects to. Because of planarity of G this 'separates' $\mathbb{C}$. After $k - 1$ such separations all cycles are separated, giving us the mentioned result.   □

Consider the graph G' obtained from G by removing all trees from $\mathbb{T}_S$ and $\mathbb{T}_1$, and deleting all edges between vertices on two different cycles in $\mathbb{C}$. We claim the following:

Lemma 13. *We can add $|\mathbb{T}_{\geq 2}| - 1$ edges to G' to connect all trees in $\mathbb{T}_{\geq 2}$ to a single tree without breaking planarity.*

Proof. If $p = 1$ the the lemma clearly holds. We will show that if $p > 1$ we can lower p step by step by adding an edge as follows: create the bipartite planar graph H with a vertex for every element in $\mathbb{C}$ and $\mathbb{T}$. Edges in H correspond to cycles and trees that are incident in G'. Every face of H has at least four incident vertices. We now connect two trees on a face by an edge in G' obtaining a new instance with $p - 1$ trees.   □

All that is left now is to show a bound for the size of this single big tree:

Lemma 14. $|V(\mathbb{T}_{\geq 2})| \leq 163k - 51.$

Proof. We add $|\mathbb{T}_{\geq 2}| - 1$ edges to all trees in $\mathbb{T}_{\geq 2}$ to create one tree T. Let L be the set of leafs of T, and let I be the internal nodes of T with $d(v, T) \geq 3$ for all $v \in I$. Every node in L is connected to at least two vertices outside of T because Lemma 3. Let $L_1 \subseteq L$ contain all leafs of L that connect to only one cycle, and let $L_2 = L \setminus L_1$. If $|L_1| > 3k$ then there is some $C \in \mathbb{C}$ such that at least four leafs from L_1 connect to two or more vertices on C. This implies two disjoint cycles in $G[C \cup L_1]$ contradiction property 2 of Lemma 6.

Now suppose we walk at an infinitesimal close distance around T. Let e_i denote the i'th edge we encounter. We mark a node of T if it is incident to the very first of very last edge to some cycle $C \in \mathbb{C}$. This way we mark at most $2k$ nodes. Let Q contain these marked nodes. We claim that if we encounter consecutive edges e_1 and e_2 connected to $v_1, v_2 \in T$ and $C_1, C_2 \in \mathbb{C}$ where $C_1 \neq C_2$ then at least one of v_1, v_2 is marked. Suppose not: we now can find integers $a < b < c < d$ and distinct $C_1, C_2 \in \mathbb{C}$) such that e_a and e_c connect to C_1, and e_b and e_d connect to C_2. This contradicts planarity. Note that $L_2 \subseteq Q$ because those leafs are connected to at least two different cycles.

We now decompose T in a collection of paths incident to at most two cycles as follows. First remove all incident nodes I, leaving at most $2|L| - 3$ paths (the maximum number of edges of a tree having $|L|$ leafs). Then remove all marked nodes Q splitting at most $|Q|$ paths. Finally we remove all edges added to create T from $\mathbb{T}_{\geq 2}$, splitting another $|\mathbb{T}_{\geq 2}| - 1$ paths. Adding everything together we get at most $2|L_1| + 2|L_2| - 3 + |Q| + |\mathbb{T}_{\geq 2}| - 1$ paths. Using $|L_1| \leq 3k$, $|Q| \leq 2k$ and $|L_2| \leq |Q| \leq 2k$ this resolves to at most $12k - 4$ paths. If a path is incident to more than two cycles some node on the path has to be marked. Because we removed all marked nodes we conclude that every path is incident to at most two cycles. We use Lemma 9 to bound the size of each path to 13, add the $|L| - 2 \leq 5k - 2$ internal nodes I and the $\leq 2k$ nodes in Q to the path-nodes proving the lemma.   □

Putting lemma's 10, 11 and 14 together we get $|V(\mathbb{T})| \leq 191k - 60$. By the same argument as in the proof for Lemma 2 we see that every nesting gains at most 131 vertices. As we can have at most $k - 1$ levels of nesting we get the following corollary:

Corollary 1. $|V(\mathbb{T})| \leq 322k - 191$.

4.2   Linearity of Cycles

We will now show that $|V(\mathbb{C})| = O(k)$. Because of the Degree Two Rule every vertex $v \in C \in \mathbb{C}$ is incident to at least one vertex in $V \setminus C$.

Let V_S contain all vertices on cycles $C \in \mathbb{C}$ that have $|C| \leq 4$. We group the remaining vertices $V(\mathbb{C}) \setminus V_S$ according to their minimum distance to some vertex in $V(\mathbb{T})$. By V_i we denote the subset of $V(\mathbb{C}) \setminus V_S$ having shortest-path distance i to some vertex in $V(\mathbb{T})$. Note that for every cycle not put in V_S there is no vertex in $V(\mathbb{T})$ connecting to more than one vertex on the same cycle, or we contradict property 1 of Lemma 6. We have the following:

Lemma 15. $|V_1| \leq |V(\mathbb{T})| + 19k - 48$.

Proof. Let X_i contain the vertices of $V(\mathbb{T})$ incident to exactly i cycles (and thus i vertices) in V_1. Because of Lemma 1 we have $\sum_{i \geq 3}(\frac{1}{2}i - 1)|X_i| \leq k - 2$. Under this restriction we maximize the total number of incident vertices $i|X_i|$ for $i \geq 3$. This maximum is obtained if $|X_3| = 2k - 4$ and $|X_{>3}| = 0$ giving us $6k - 12$ vertices in V_1 connected to $X_{>3}$.

Let $X_2' \subseteq X_2$ contain all vertices from X_2 such that $N(v) \cap N(w) = \emptyset$ for all different $v, w \in X_2'$. We have at most $3k - 6$ different pairs of cycles that are connected to through some vertex in X_2 by Lemma 2, and there are at most 5 disjoint paths between every connected pair. Hence X_2' contains at most $15k - 30$ vertices, connected to $30k - 60$ vertices in V_1.

All $|V(\mathbb{T})| - 17k + 34$ vertices in $V(\mathbb{T})$ that are not in X_2' or X_3 are connected to at most one vertex in V_1 that is not yet counted, so $|V_1| \leq |V(\mathbb{T})| + 19k - 48$. $\square$

Lemma 16. $|V_2| \leq 2|V_1|$.

Proof. Every vertex in V_2 is incident to at least one vertex in V_1, and a vertex in V_1 is incident to at most two vertices in V_2. $\square$

We will now count the number of vertices in $V_{\geq 3}$. Consider some cycle $C \in \mathbb{C}$ incident to c cycles and t trees. If $|N(C) \cap D| > 5$ for some incident cycle D we find some subset $W \subset V(D)$ such that $N(D) = N(W)$ and $|W| \leq 5$. If such a W does not exist we have more than 5 disjoint paths between two cycles, contradicting property 3 of Lemma 6. We delete all edges between C and $V(D) \setminus W$. This operation takes care of some nasty details when counting the number of vertices in $V_{\geq 3}$, because now $N(C) \cap D$ contains at most 5 vertices for any incident cycle D. Also note that every vertex on C that was connected to D still is.

We now define an *observer* to be either a tree or a *vertex* on some incident cycle, so we have at most $5c + 2t$ observers. We take two walks along the circle, one on the outside and one on the inside, and every time we see an edge to some observer for the very first or very last time we put the vertex of C incident to that edge in Q. The paths in $G[C \setminus Q]$ are called *segments*. Every segment is connected to at most one observer on the inside, and one on the outside. We claim the following:

Lemma 17. *The number of segments is at most* $10c + 2t$.

Proof. The number of segments is at most $|Q|$. The argument is exactly the same at in the proof of Lemma 14. $\square$

Lemma 18. $|S \cap V_{\geq 3}| \leq 5$ *for every segment* S.

Proof. If the one or two observers of S are only trees then clearly $S \cap V_{\geq 3} = \emptyset$. Suppose S has a tree and a cycle-vertex as observers, and $S \cap V_{\geq 3} \neq \emptyset$. Then at least three consecutive vertices on S are not incident to the tree, and thus incident to the cycle-vertex. But then the Path Rule applies, so we also have $S \cap V_{\geq 3} = \emptyset$ in this case.

If S is observed by only one cycle-vertex then $|S| = 1$ because of the Path Rule, and if S is observed by two cycle-vertices then $|S| \leq 5$, also because of the Path Rule. All of these vertices can be in $V_{\geq 3}$, proving the lemma. $\square$

Lemma 19. $|V_{\geq 3}| \leq 300k - 600$.

Proof. We attribute most to $V_{\geq 3}$ if we have two cycle-vertices as an observer. This can happen at most $10c$ times, leading to a total attribution of $50c$ to $V_{\geq 3}$. We have at most $3k - 6$ pairs of incident cycles by Lemma 2. Taking the sum over all cycles directly proves the lemma. $\square$

Putting Lemma 15,16 and 19 together and adding $|V_S|$ we get the following corollary:

Corollary 2. $|V(\mathbb{C})| \leq 909k - 717$.

Together with Corollary 1 this proves Theorem 1.

5   A Faster Parameterized Algorithm

We can use the $O(k)$ kernel for k-DISJOINT CYCLES on planar graphs, and a technique introduced by Dorn et al. [6] to improve the parameterized algorithm with running time $O(c^{\sqrt{k} \log k} n)$ by Kloks et al. [10]:

Theorem 2. *There is a* $O(c^{\sqrt{k}} + n^2)$ *algorithm for the* k-DISJOINT SET PROBLEM *on planar graphs.*

Proof. We first apply our kernelization algorithm in time $O(n^2)$. The resulting planar graph has $O(k)$ vertices. Next, it is possible to solve the problem on this kernel using dynamic programming on sphere cut branch decompositions in $O(c^{\sqrt{k}})$ time, employing techniques introduced by Dorn et al. [6]. Due to lack of space we omit all further details. $\square$

6 Conclusions

We presented the first kernel for the k-DISJOINT CYCLES PROBLEM on planar graphs, solving an open problem by Kloks et al. [10], and showed that this kernel is linear. This nicely complements a very recent result by Bodlaender et al. [5] that no polynomial kernel for general graphs exists for this problem, assuming the polynomial time hierarchy does not collapse.

The first question to ask is whether our analysis can be extended to broader graph classes, for example to graphs of bounded genus. It would also be of interest to obtain a much smaller kernel, say something around $10k$, or perhaps some lower bound for kernelization of this problem. Implementation would also be interesting, we expect the algorithm to obtain much smaller kernels then our crude analysis shows. This optimism is motivated by implementations done by Van Dijk [11] on the Feedback Vertex Set Problem.

References

[1] Abu-Khzam, F.N., Collins, R.L., Fellows, M.R., Langston, M.A., Suters, W.H., Symons, C.T.: Kernelization algorithms for the vertex cover problem: Theory and experiments. In: Proc. 6th ACM-SIAM ALENEX, pp. 62–69. ACM-SIAM (2004)

[2] Alber, J., Fellows, M.R., Niedermeier, R.: Polynomial-time data reduction for dominating sets. J. ACM 51, 363–384 (2004)

[3] Bodlaender, H.L.: A cubic kernel for feedback vertex set. In: Thomas, W., Weil, P. (eds.) STACS 2007. LNCS, vol. 4393, pp. 320–331. Springer, Heidelberg (2007)

[4] Bodlaender, H.L., Penninkx, E.: A linear kernel for planar feedback vertex set. In: Grohe, M., Niedermeier, R. (eds.) IWPEC 2008. LNCS, vol. 5018, pp. 160–171. Springer, Heidelberg (2008)

[5] Bodlaender, H.L., Thomassé, S., Yeo, A.: Unpublished manuscript (2008)

[6] Dorn, F., Penninkx, E., Bodlaender, H.L., Fomin, F.V.: Efficient exact algorithms on planar graphs: Exploiting sphere cut branch decompositions. In: Brodal, G.S., Leonardi, S. (eds.) ESA 2005. LNCS, vol. 3669, pp. 95–106. Springer, Heidelberg (2005)

[7] Downey, R.G., Fellows, M.R.: Parameterized Complexity. Springer, Heidelberg (1998)

[8] Guo, J., Niedermeier, R.: Invitation to data reduction and problem kernelization. ACM SIGACT News 38, 31–45 (2007)

[9] Guo, J., Niedermeier, R.: Linear problem kernels for NP-hard problems on planar graphs. In: Arge, L., Cachin, C., Jurdziński, T., Tarlecki, A. (eds.) ICALP 2007. LNCS, vol. 4596, pp. 375–386. Springer, Heidelberg (2007)

[10] Kloks, T., Lee, C.M., Liu, J.: New algorithms for k-face cover, k-feedback vertex set, and k-disjoint cycles on plane and planar graphs. In: Kučera, L. (ed.) WG 2002. LNCS, vol. 2573, pp. 282–295. Springer, Heidelberg (2002)

[11] van Dijk, T.C.: Fixed parameter complexity of feedback problems. Master's thesis, Utrecht University (2007)

How to Guard a Graph?

Fedor V. Fomin[1], Petr A. Golovach[1], Alexander Hall[2],
Matúš Mihalák[3], Elias Vicari[3,*], and Peter Widmayer[3,*]

[1] Institute of Informatics, University of Bergen, Norway
{fedor.fomin,petr.golovach}@ii.uib.no
[2] Google Switzerland
alex.hall@gmail.com
[3] Institute of Theoretical Computer Science, ETH Zurich, Switzerland
{mmihalak,vicariel,widmayer}@inf.ethz.ch

Abstract. We initiate the study of the algorithmic foundations of games
in which a set of cops has to guard a region in a graph (or digraph)
against a robber. The robber and the cops are placed on vertices of the
graph; they take turns in moving to adjacent vertices (or staying). The
goal of the robber is to enter the guarded region at a vertex with no
cop on it. The problem is to find the minimum number of cops needed
to prevent the robber from entering the guarded region. The problem is
highly non-trivial even if the robber's or the cops' regions are restricted
to very simple graphs. The computational complexity of the problem
depends heavily on the chosen restriction. In particular, if the robber's
region is only a path, then the problem can be solved in polynomial
time. When the robber moves in a tree, then the decision version of the
problem is NP-complete. Furthermore, if the robber is moving in a DAG,
the problem becomes PSPACE-complete.

1 Introduction

Chases and escapes, or pursuits and evasions, are activities which are firm parts
of our everyday or fantasy life. When young we are playing the game of tag or
the game of hide-and-seek and watch Tom & Jerry cartoons. Pursuit-evasion
problems remain fascinating for most of us even when we grow older, and it is
not surprising that in half of the Hollywood movies good guys are chasing the
bad guys, while in the remaining half bad guys are chasing the good ones.

The mathematical study of pursuit-evasion games has a long history, tracing
back to the work of Pierre Bouguer, who in 1732 studied the problem of a pirate
ship pursuing a fleeing merchant vessel. We refer to *Differential games* by Rufus
Isaac [1], the now classical book on pursuit-evasion games, published in 1965
(and reprinted by Dover in 1999), for a nice introduction to the area. One of the
general pursuit-evasion problems (mentioned in Isaac's book) is the guarding-
the-target problem. There are many variations of this problem like the deadline

* The authors are partially supported by the National Competence Center in Research
on Mobile Information and Communication Systems NCCR-MICS, a center supported by the Swiss National Science Foundation under grant number $5005 - 67322$.

S.-H. Hong, H. Nagamochi, and T. Fukunaga (Eds.): ISAAC 2008, LNCS 5369, pp. 318–329, 2008.

game, patrolling a channel, and patrolling a line. In this problem, both pursuer and evader travel with the same speed. The goal of the pursuer is to guard a target C which is an area in the plane, from attacks by the evader. In Isaacs's military conception of this problem, the evader must actually reach at least the boundary of the target to be successful in her attack.

In this paper we study the *cop-robber guarding game* (*guarding game* for short), a discrete version of the guarding-the-target problem. We use the settings from the classical pursuit-evasion game on graphs called Cops and Robbers [2,3,4].

The guarding game is played on a (directed or undirected) graph $G = (V, E)$ by two players, the *cop-player* and the *robber-player*, each having her *pawns* (*cops* or a *robber*, respectively) on the vertices of G. The focus of the game is on the *protected region* (called also the *cop-region*) $C \subsetneq V$ – the cops guard C by preventing the robber to enter the protected region without being "caught", which happens when the robber is on a vertex of C with a cop on it. The game is played in alternating turns. In the first turn the robber-player places her single robber on a vertex of $V \setminus C$, i.e., outside the protected region C. In the second turn the cop-player places her c cops on vertices of C (allowing more cops being placed at one vertex). In each subsequent turn the respective player can *move* each of her pawns to a neighboring vertex of the pawn's position (or leave it at its current position), following the rule that the cops can move only within the protected region C, and the robber can move only on vertices with no cops (thus avoiding being "caught"). At any time of the game both players know the positions of the cops and the robber in G. The goal of the cop-player is to *guard* (or *protect*) the region C, i.e., to position and move the cops such that, in any turn, the robber cannot move onto a vertex of C with no cop on it. We say that a cop on vertex v *guards* the vertex v (as the robber cannot move onto v). Naturally, the goal of the robber-player is to position and move the robber in such a way that at some point in time the robber can move onto a vertex of C with no cop on it. The region $V \setminus C$ where the robber moves before it eventually enters C is called the *robber-region* and is denoted by R.

The guarding game is a *robber-win* game if the robber-player can (at some turn) move the robber to a vertex in C with no cop on it. In this case we say that the robber-player *wins* the game. Otherwise (if the cop-player can forever prevent the robber-player to win) we say that it is a *cop-win* game, and that the cop-player *wins* the game, or that the cop-player can *guard* C. Thus, the game is fully specified by the number of cops c, and by the *board* of the game which is given by the graph $G = (V, E)$, the protected region C and the robber-region R. We denote the board by $[G; R, C]$. In this work we will be considering cases when sets R or C induce subgraphs of G with specific properties (like being a path, a tree, or DAG). By slightly abusing notations, we often will be referring to robber or cop regions as subgraphs.

Since the game is played in alternating turns starting at the first turn, the robber-player moves her robber in odd turns, and the cop-player moves her cops in even turns. Two consecutive turns $2 \cdot i - 1$ and $2 \cdot i$ are jointly referred to as a *round* i, $i \geq 1$. A *state* of the game at *time* i is given by the positions of all cops and robbers on the board after $i-1$ turns. A *strategy of a cop-player* (*strategy of a robber-player*) is a function $\mathcal{X}$ which, given the state of the game, determines the movements of the cops (the robber) in the current turn. If there are no cops (no robber) on the board, the function determines the initial positions of the cops (the robber).

The *guarding decision problem* is, given a board $[G; R, C]$, and a number c specifying the number of cops, to decide whether it is a cop-win game or a robber-win game. The *guarding problem* is, given a board $[G; R, C]$, to compute the minimum number of cops that can guard the protected region C against all possible strategies of the robber-player. We call this number the *guard number* ℓ of the game. Figure 1 depicts a board of the game. If the robber is on v_4, two cops must be on u_3 and u_4. If the robber moves to v_3 none of the two cops can reach u_1 in the next turn, hence $\ell > 2$. Later we will see that for this game $\ell = 3$.

Our Contribution. In this work, we initiate the study of algorithmic and combinatorial issues of the guarding problem. We start with the case when R is a path. We show that for such robber-regions the problem can be solved in polynomial time. The solution is based on a reduction to the minimum-cost flow problem. Furthermore, an interesting special case of the problem is when C is a path, too. In this case we establish a combinatorial result that the minimum number of cops needed to protect the path is equal to the size of a maximum independent set in a (related) comparability graph. By making use of the combinatorial result, we are able to obtain a faster solution for this case. The complexity of the problem increases even with small changes of the robber-region R. We do not know if the problem can be solved in polynomial time when R is a cycle, however, we are able to find a 2-approximation of the guard number in this case. When R is a tree, the problem becomes NP-hard and the parameterized version of the problem is $W[2]$-hard. Moreover, our reduction from MINIMUM SET COVER implies that there is a constant $\rho > 0$ such that there is no $\rho \log n$-approximation algorithm, unless $P = NP^1$. We also show that the decision version of the problem is in NP, when R is a tree. In case R is a directed acyclic graph (DAG), the problem is PSPACE-complete. However, in the general case, when R is an arbitrary graph, we do not know if the problem is in PSPACE, and we leave it as an open question. In this version, due to lack of space, some of the proofs are ommitted. We refer the reader to [12] for missing details.

Related Work. The guarding game can be seen as a member of a vast class of games called the pursuit-evasion games (see, e.g., [6] for an introduction). The discrete version of pursuit-evasion games, played on graphs, is the game called Cops and Robbers. The game was defined (for one cop) by Winkler and Nowakowski [2] and Quilliot [3] who also characterized graphs for which one cop can catch the robber. Aigner and Fromme [4] initiated the study of the problem

---

[1] If $NP \not\subseteq DTIME(n^{\text{poly} \log n})$, Lund and Yannakakis [5] show that MINIMUM SET COVER cannot be approximated within the ratio $\rho \log n$ for any $\rho < 1/4$.

with several cops. The minimum number of cops required to capture the robber is called the cop number of a graph. This problem was studied intensively and we refer to Alspach's survey [7] for references. The Cops-and-Robbers game can be seen as a special case of search games played on graphs, see the annotated bibliography [8] for further references on search and pursuit-evasion games on graphs. The computational complexity of finding the minimum number of cops in the game of Aigner and Fromme is still not settled completely. Goldstein and Reingold [9] proved that the version of the game on directed graphs is EXPTIME-complete. Also they have shown that the version of the game on undirected graphs when cops and robbers are given their initial positions is also EXPTIME-complete. They also conjectured that the problem is EXPTIME-complete on undirected graphs (without given initial positions of the players). However, until very recently even NP-hardness of the problem was open [10].

To our knowledge, cop-robber guarding games, the topic of this paper, have not been studied in mathematical and algorithmic terms before.

2   Guarding against a Path

In this section we consider the guarding problem where the protected region C induces an arbitrary graph, and the robber-region R induces a path P, i.e., the board of the game is a path P connected to C by interconnecting edges, leading to the graph G. Modeling the problem as a flow problem we can design an algorithm that computes in polynomial time the optimum number of cops that are needed to guard C when R induces a path.

Let n_1 and n_2 denote the number of vertices of R and C, respectively, and suppose that $P = (v_1, v_2, \ldots, v_{n_1})$. The restriction on the area where a robber can move before entering the protected region C allows us to consider only one strategy of the robber-player, the *direct* strategy: the robber-player places her robber on the first vertex v_1 of the path P and moves it from v_1 along P to the last vertex v_{n_1} of the path P, and enters the protected region C when possible.

Lemma 1. *If c cops can protect C against the direct strategy of the robber-player – in a guarding game where the robber-region R is a path $G[R]$ – then c cops can protect C against any strategy of the robber-player.*

Proof. Let X_i be the set of vertices of C on which the cops are placed when the robber, following the direct strategy, is on vertex v_i, $i = 1, \ldots, n_1$, of the path P. Thus, $X_1, X_2, \ldots, X_{n_1}$ describe the movements of the cops when the robber moves from v_1 to v_{n_1}. Observe now that the cop-player can move the cops from X_i to X_{i-1} if the robber moves from v_i to v_{i-1}, and the cop-player can move the cops from X_i to X_{i+1} if the robber moves from v_i to v_{i+1}, and that there is no other movement for the robber at vertex v_i. Thus, at any time the robber moves to v_i, the cops can move to X_i and the robber cannot enter C.   □

Using this lemma we prove the main result of the section.

Theorem 1. *Let $[G; R, C]$ be a board such that $G[R]$ is a path. The guarding problem on $[G; R, C]$ is solvable in time $O(n_1^2 n_2 m \cdot \log(n_1 n_2))$, where $m = |E(G)|$, $n_1 = |R|$, and $n_2 = |C|$.*

Proof. Due to the previous lemma, we can consider only the direct strategy of the robber-player. We present a (constructive) algorithm that computes the optimum number of cops and also a cop-player winning strategy. The first idea behind the algorithm is to consider the vertices of C after every round $i = 1, 2, \ldots, n_1$ when the robber is on vertex v_i. An optimal strategy with ℓ cops places the cops in round i on vertices X_i. Observe that all cops from X_i can get in one step to positions X_{i+1}, i.e., there is a matching between vertices of X_i and X_{i+1} in which the matched vertices are either identical or neighbors in C. Observe also that for any i the neighbors of v_i that lie in C have to be in X_i (otherwise the robber could move from v_i onto C and win). Let us consider one walk of the robber from vertex v_1 to vertex v_{n_1} and let us observe the movements of the cops in each round i. Consider Figure 1 for an illustration. The example shows the board consisting of C on five vertices, and a path induced by R of length four (left). The dotted lines are the interconnecting lines of R and C. For the walk of the robber from v_1 to v_4 in four rounds $1, 2, 3, 4$ we consider in each round the vertices of C (right) – four layers (copies) t_1, t_2, t_3, t_4 of C. After the moves of each round i are completed, the neighbors of v_i have to be guarded by cops (the highlighted vertices). An optimal solution is depicted by the three paths between the layers, which shows the position of each cop at every time, thus showing the number of necessary cops (three in this case) and their movements.

Fig. 1. An instance of the guarding game with the path $G[R]$ and the graph $G[C]$ to be guarded (left). The robber walks on the path $G[R]$ from vertex v_1 in round 1 (layer t_1) to vertex v_4 in round 4 (layer t_4). Three cops are necessary to guard C, and a strategy of the cops is depicted by the thick lines traversing the vertices of the auxiliary graph which consists of 4 copies of $G[C]$ (right). The highlighted vertices depict vertices on which there needs to be a cop at the respective time.

The movements of the cops can be seen as a flow between the layers. We want our algorithm to compute such a flow. We construct a new auxiliary directed graph A. If R induces a path on n_1 vertices, there are n_1 layers (copies) $t_1, \ldots, t_{n_1}$ of C in A. We connect a vertex u_j from layer t_i with a vertex u_k from layer t_{i+1} (in the direction of the increasing time steps) if $j = k$ (the cop may stay on the same vertex), or if $u_j u_k$ is an edge in $G[C]$ (the cop moves from vertex u_j

to vertex u_k). There are two additional vertices in A – source s and target t, where s is connected to every vertex of the first layer, and t is connected to every vertex of the last layer (see Figure 2). We set the capacity of each edge to be one, and the cost of each edge to be zero. We want that a minimum-cost flow from s to t reflects a guarding strategy of the cop-player. We want to make sure that the flow traverses through those vertices of layer t_i, which are neighbors of v_i. For this we duplicate every vertex of every layer, and connect the copy and the original with an edge of cost $-n_2$ if the original vertex u_j is in layer t_i and is a neighbor of v_i, otherwise we set the cost to be zero. We call an edge with cost $-n_2$ a *thick* edge. We want that a minimum-cost flow reflects an optimal cop-player strategy. For that we need that the minimum-cost flow is achieved with a minimum value of the flow (i.e., we need to minimize the amount of cops that guard C) – for this we set the cost of every edge from the source s to be one. The next lemma shows that a minimum-cost flow from s to t yields a minimum number of cops that can guard the protected region C. Finally, as computing the minimum cost flow in a graph with n vertices and m edges can be done in time $O(nm \log n)$ in our case [11], we immediately obtain the claim of the theorem. $\qquad\square$

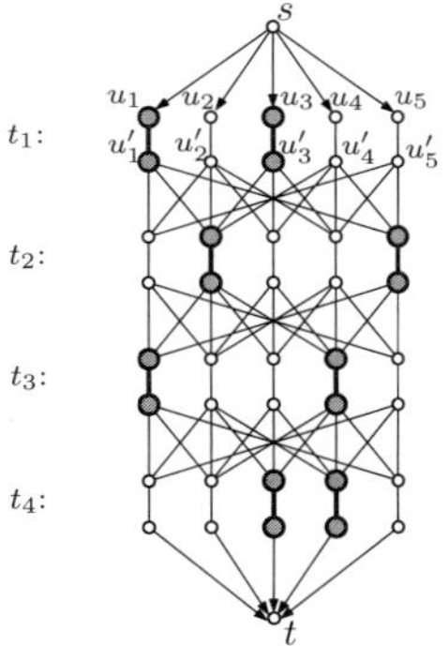

Fig. 2. The auxiliary graph A constructed from the guarding game of Figure 1. For better readability, not every edge has its direction depicted. The thick edges between the duplicated vertices denote the edges with negative cost $-n_2$.

Lemma 2. *The value of a minimum-cost flow of the auxiliary graph A, constructed from a guarding game played on $[G; R, C]$, where $G[R]$ is a path, is equal to the minimum number of cops for which the game is a cop-win game.*

Proof. Let f be a minimum-cost flow of the auxiliary graph A. As the capacities of the edges are integral, the flow f is integral, too. Moreover, every edge e has either no flow on it, or the flow f_e passing through this edge is equal to one. Also, due to the structure of the graph A (edges between every vertex u_i and its copy u'_i), every vertex u_i of every layer has at most one incoming edge with a non-zero flow and similarly every copy u'_i has at most one outgoing edge with a non-zero flow. Thus, the flow from s to t consists of k vertex-disjoint paths (with edges of non-zero flow), where k is the total flow from s to t, i.e., the value of f.

Observe first that there is no thick edge e with zero flow f_e. To see that, assume for contradiction that there is indeed such an edge e' with $f_{e'} = 0$. Then the flow f cannot be minimum, as for example the flow which uses n_2 paths where, for every i, path i goes through vertices u_i only, has smaller cost – such a flow uses all cost one edges (there are n_2 such edges), but traverses also through the edge e', which in total results into a smaller cost than the cost of f.

Thus, a strategy for the cop-player which follows the flow f guarantees that the cop-player wins. Hence, k is an upper bound on the minimum number of cops that can guard C. To see that k is also a lower bound, just observe that any optimum winning strategy for the cop-player translates directly into a flow f' in A of minimum cost: the flow f' traverses every thick edge and uses a minimum number of paths, thus it uses a minimum number of cost-one edges.   □

As we shall see in the next section, the complexity of the problem changes drastically when R is not a path. For the case R is a cycle, Theorem 1 gives possibility to receive a constant factor approximation for the guard number. The proof of the following result can be found in [12].

Corollary 1. *Let $[G; R, C]$ be a board such that $G[R]$ is a cycle. Then there is a 2-approximation polynomial time algorithm for the guarding problem.*

An interesting special case is when the robber-region and the protected regions are two paths P_1, P_2, respectively, in the underlying board G. In this case Lemma 1 applies and we consider only the direct strategy for the robber. We define an auxiliary graph $A = A(G)$ that reflects the structure of the game played on such a board. Every edge between P_1 and P_2 corresponds to a vertex of A. Two such edges (vertices in A) are connected in A by an edge, if one cop can guard the endpoints of the edges in P_2 against the direct strategy. It is possible to show that the guard number is equal to the size of a maximum independent set in A and that A is a comparability graph. Since comparability graphs are perfect [13], a maximum independent set can be computed in polynomial time [14]. We present the details in [12].

3   Hardness of Guarding

It is not difficult to guess that the problem of computing the guard number is a hard problem. In this section we provide two hardness results for quite restricted classes of boards.

We show first that the problem is NP-hard even when the robber-region is a tree. We present a reduction from SET COVER [15] (which is an NP-complete problem) to the guarding decision problem, which then shows that the guarding decision problem is NP-hard.

Theorem 2. *The guarding decision problem is NP-hard even if the robber-region is a tree $G[R]$ (or even a directed rooted tree in the case of directed graphs).*

Moreover, the parameterized version of the problem with c being a parameter is W[2]-hard.

Proof. We prove the theorem for undirected graphs. The proof for directed graphs is a simple adaptation of what follows. We describe how to reduce SET COVER to the guarding decision problem. Consider an instance of SET COVER, in which a universe $U = \{u_1, u_2, \ldots, u_n\}$ is asked to be covered by at most k sets from $S_1, S_2, \ldots, S_m \subseteq U$. In our reduction, we use U as vertices which shall be effectively guarded by the cops. Thus, U will be part of the protected region C. We add to our construction one single vertex a connected to all vertices corresponding to U by paths of length two. Denote by $b_1, b_2, \ldots, b_n$ the middle vertices of these paths. The robber-region R is then $\{a, b_1, b_2, \ldots, b_n\}$ – a tree. We finally add vertices where the cops can be placed and from which the cops can "cover" U in the same way as a solution to the SET COVER: we add one vertex S_i for every set S_i, and connect it to every vertex $u \in S_i \subseteq U$. These vertices define, together with U, the protected region C. The guarding game is defined by setting the number of cops $c = k$. Suppose that c cops can win the game. Initially, if the robber is placed on a, we can assume, w.l.o.g., that cops occupy vertices from the set $\{S_1, S_2, \ldots, S_m\}$. Obviously these vertices correspond to a set cover of size no more than $c = k$. Assume now that $X \subseteq \{S_1, S_2, \ldots, S_m\}$ is a set cover of size at most k. For every $u_i \in U$ there is $x_i \in X$ such that u_i and x_i are adjacent in our graph. Let $X_i = (X \setminus \{x_i\}) \cup \{u_i\}$ for every $i \in \{1, 2, \ldots, n\}$. We consider the cops strategy such that cops occupy vertices of X if the robber is in a, and cops occupy vertices of X_i if the robber is in b_i. Clearly, this is a winning strategy for cops.

Since our reduction is a parameterized reduction and SET COVER is a well-known W[2]-hard problem, the second claim follows immediately.[2]

The only difference for the case of directed graphs is that a is joined with $u_1, u_2, \ldots, u_n$ by directed paths, and vertices $S_1, S_2, \ldots, S_m$ are joined with $u_1, u_2, \ldots, u_n$ by arcs with the obvious orientation. $\square$

Let us note that the second claim of Theorem 2 is "tight" in some sense. It is well-known for Cops-and-Robbers games that determining whether c cops can capture the robber on a graph with n vertices can be done by a backtracking algorithm which runs in time $O(n^{O(c)})$ (thus polynomial for fixed c) [17,9,18]. A similar result holds for the guarding game. Given an integer c and a graph G on n vertices, it can be determined whether c cops can guard a given set C (and the corresponding winning strategy of c cops can be computed) by constructing the game-graph on $2\binom{|C|+c-1}{c}|R|$ nodes (every node of the game-graph corresponds to a possible position in G of c cops and one robber, taking into account two possibilities for whose turn it is), and then by making use of backtracking to find if some robber-winning position can be obtained from an initial position. The proof of the following proposition is standard and easy.

---

[2] We refer to the book of Downey and Fellows [16] for an introduction to parameterized complexity.

Proposition 1. *For a given integer $c \geq 1$ and a graph G on n vertices, the question whether the c cops can guard C can be answered in time $\left(\frac{|C|+c-1}{c}\right)^2 \cdot |R|^2 \cdot n^{O(1)} = n^{O(c)}$.*

Thus for every fixed c, one can decide in polynomial time if c cops can guard the protected region against the robber on a given graph G. But from other side the fact that our problem is $W[2]$-hard means that the existence of a $O(f(c) \cdot n^{O(1)})$-time algorithm deciding if c cops can win, where f is a function only of the parameter c, and G is a graph on n vertices, would imply that $FPT = W[2]$, which is considered to be very unlikely in parameterized complexity.

It can be easily proved that the guarding problem is difficult to approximate. We combine the (approximation preserving) reduction from the proof of Theorem 2 and the non-approximability for MINIMUM SET COVER problem [19] and obtain the following statement.

Corollary 2. *There is a constant $\rho > 0$ such that there is no polynomial time algorithm that, for every instance, approximates the minimum number of cops which are necessary to guard the protected region within a multiplicative factor $\rho \log n$, unless $P = NP$.*

For some cases we can prove that the problem is in NP (see [12] for details).

Proposition 2. *If $G[R]$ is a tree or a directed tree, then the guarding decision problem is NP-complete.*

For directed graphs the problem becomes more difficult, if we allow only a slightly more general robber-region.

Theorem 3. *The guarding decision problem is PSPACE-hard for directed graphs even if $G[R]$ is a directed acyclic graph (DAG).*

Proof. We reduce the PSPACE-complete QUANTIFIED BOOLEAN FORMULA IN CONJUNCTIVE NORMAL FORM problem [15] to the guarding decision problem. For given Boolean variables $x_1, x_2, \ldots, x_n$ and a Boolean formula $F = C_1 \wedge C_2 \wedge \cdots \wedge C_m$, where C_j is a clause, this problem asks whether the expression $\phi = Q_1 x_1 Q_2 x_2 \ldots Q_n x_n F$, where either $Q_i = \forall$ or $Q_i = \exists$, has value *true*.

Given a quantified Boolean formula ϕ, we construct an instance of a guarding game in the following way. For every quantification $Q_i x_i$ we introduce a gadget graph G_i. If $Q_i = \forall$ then we define graph $G_i(\forall)$ as the graph with the vertex set $\{u_i, v_i, x_i, \overline{x}_i, y_i, \overline{y}_i, z_i\}$ and the arc set $\{u_i y_i, y_i v_i, u_i \overline{y}_i, \overline{y}_i v_i, y_i x_i, \overline{y}_i \overline{x}_i, z_i x_i, z_i \overline{x}_i\}$. Let $W_i = \{x_i, \overline{x}_i, z_i\}$. If $Q_i = \exists$ then $G_i(\exists)$ is the graph with the vertex set $\{u_i, v_i, x_i, \overline{x}_i, y_i, a_i, b_i, z_i\}$ and the arc set $\{u_i y_i, y_i v_i, y_i a_i, a_i b_i, x_i b_i, \overline{x}_i b_i, z_i x_i, z_i \overline{x}_i\}$. In this case let $W_i = \{x_i, \overline{x}_i, b_i, z_i\}$. These graphs are joined together by gluing vertices v_{i-1} and u_i for $i \in \{2, 3, \ldots, n\}$.

In our construction we further introduce vertices $C_1, C_2, \ldots, C_m$, which correspond to clauses. A vertex x_i is joined with C_j by an arc (pointing towards C_j) if C_j contains the literal x_i, and $\overline{x}_i$ is joined with C_j if C_j contains the literal $\overline{x}_i$. The vertex v_n is connected with all vertices $C_1, C_2, \ldots, C_m$ by directed paths of length two with middle vertices $w_1, w_2, \ldots, w_m$. Denote the obtained directed graph by G, and let $C = W_1 \cup W_2 \cup \cdots \cup W_n \cup \{C_1, C_2, \ldots, C_m\}$ be the cop-region, and $R = V \setminus C$ be the robber-region. Clearly, $G[R]$ is a DAG. For the guarding decision problem we set $c = n$. Construction of G for $\phi = \forall x_1 \exists x_2 \forall x_3 (\overline{x}_1 \vee x_2 \vee \overline{x}_3) \wedge (x_1 \vee \overline{x}_2 \vee \overline{x}_3)$ is shown in Figure 3.

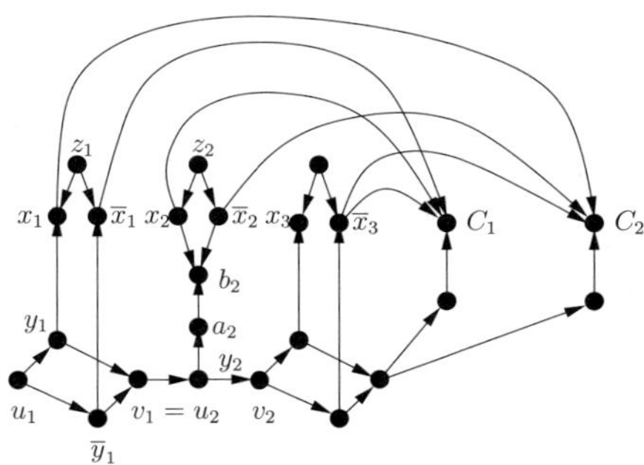

Fig. 3. Construction of G for $\phi = \forall x_1 \exists x_2 \forall x_3 (\overline{x}_1 \vee x_2 \vee \overline{x}_3) \wedge (x_1 \vee \overline{x}_2 \vee \overline{x}_3)$

Suppose that $\phi = false$. We describe a winning strategy for the robber. He chooses vertex u_1 as a starting point and then moves to vertex v_n along some directed path. If the robber comes to the vertex y_i of a graph $G_i(\forall)$ then one cop has to occupy the vertex x_i, and if the robber comes to $\overline{y}_i$ then some cop has to occupy the vertex $\overline{x}_i$. So, cops are "forced" to occupy vertices that correspond to literals. Similarly, if the robber occupies the vertex y_i in a graph $G_i(\exists)$ then at least one cop has to occupy one vertex from the set $\{x_i, \overline{x}_i, b_i\}$, and here this cop can choose between vertices x_i and $\overline{x}_i$. Since the number of cops is equal to the number of graphs $G_i(\forall)$ and $G_i(\exists)$, exactly one cop can stand on vertices of each such a gadget-graph. It follows from the condition $\phi = false$ that there is a directed path from u_1 to v_n such that if the robber moves along it to v_n then there is at least one vertex C_j which is not "covered" by cops, i.e. there are no cops on this vertex and on vertices x_i or $\overline{x}_i$ which are joined by arcs with it. Then the robber can move to this vertex and win.

Assume now that $\phi = true$ and describe a winning strategy for cops. It can be easily seen that if the robber chooses some vertex a_i or w_j as an initial position then he loses immediately. So we can assume that he occupies some vertex r on the path from u_1 to v_n. Suppose that this vertex belongs to the graph G_s ($G_s(\forall)$ or $G_s(\exists)$, it is assumed that $v_{s-1} = u_s$ belongs to G_s for $s > 1$, and v_n belongs to a virtual G_{n+1}). We place one cop on a vertex x_i or $\overline{x}_i$ for every $i < s$. The vertex is chosen arbitrarily if we consider graph $G_i(\forall)$, and if x_i is chosen then we suppose that the variable $x_i = true$ and $x_i = false$ otherwise. If $G_i(\exists)$ is considered then the choice corresponds to the value of the variable x_i: if $x_i = true$ then the vertex x_i is occupied by a cop, and $\overline{x}_i = false$ otherwise. If $r = y_s$ in $G_s(\forall)$ then one cop is placed on the vertex x_s, and if $r = \overline{y}_s$ in $G_s(\forall)$ then one cop is placed on the vertex $\overline{x}_s$. As before, $x_i = true$ in the first case and $x_i = false$ in the second. If $r = y_s$ in $G_s(\exists)$ then a cop is placed either on x_s or $\overline{x}_s$ and the choice of the vertex corresponds to the value of the variable x_s. If $r = u_s$ then we place one cop on z_s. For all $i > s$ one cop is placed on each vertex z_i. Now the robber starts to move. If he moves to the vertex y_i of $G_i(\forall)$ then the cop moves from z_i to x_i and it is assumed that the variable $x_i = true$, and if the robber moves to $\overline{y}_i$ then the cop moves to $\overline{x}_i$ and $x_i = false$. If the robber comes to y_i of $G_i(\exists)$ then the cop moves from z_i to x_i or $\overline{x}_i$, and the choice corresponds to the value of the variable x_i. Note that if the robber moves from y_i to a_i then the cop moves to b_i and cops win. So the robber is "forced" to move along some directed path from r to v_n. Since $\phi = true$, cops on graphs $G_i(\exists)$ can move in such a way that when the robber occupies the vertex v_n all vertices C_j are "covered" by cops, i.e. for every vertex C_j there is an adjacent vertex x_i or $\overline{x}_i$ which is occupied by a cop. Now the robber can move only to some vertex w_j. Cops respond by moving one cop to C_j and win. $\square$

Note that Theorem 3 (as Theorem 2) claims only "hardness", but in some cases we can prove that the problem is PSPACE-complete. The proof of PSPACE-completeness [12] is based on the fact that when the robber region is a DAG, the number of steps in the game is polynomial.

Theorem 4. *If $G[R]$ is a DAG then the guarding decision problem is* PSPACE-*complete for directed graphs.*

4   Conclusions and Future Research

In this paper we study (and define) guarding games – a discrete version of pursuit-evasion games, and a close relative of the Cops-and-Robbers games. We present several algorithmic solutions and hardness results to various families of guarding games. Our effort shall be considered as an initial attempt to under-stand the nature of guarding games, and their differences to other similar games (such as the Cops-and-Robbers games).

Concerning the complexity of the guarding game, we have shown that when the robber territory R is a DAG, the decision version of the problem is in PSPACE. We do not know if this is the case when R is a more complicated graph. An interesting related question, which remains open, concerns the number of the rounds in the game. Is it true that for a given number k and a (di)graph G, we can always decide if a subgraph of G can be guarded by k cops by playing a game with only a polynomial number of rounds? If the answer to this question is "yes", then similarly to the proof of Theorem 4 one can show that the problem is in PSPACE. Another consequence of this would be that the case when R is a cycle can be solved in polynomial time by making use of an approach similar to the path case. If the answer is "no", what is the complexity of the problem? Another open question is if on (undirected) graphs the guarding problem is PSPACE-hard.

To conclude, we feel that with this paper we barely scratched the tip of the iceberg of the universe of guarding games, but should provide insights and in-centives to motivate further research in the area.

References

1. Isaacs, R.: Differential games. A mathematical theory with applications to warfare and pursuit, control and optimization. John Wiley & Sons Inc., New York (1965)
2. Nowakowski, R., Winkler, P.: Vertex-to-vertex pursuit in a graph. Discrete Math. 43(2-3), 235–239 (1983)
3. Quilliot, A.: Some results about pursuit games on metric spaces obtained through graph theory techniques. European J. Combin. 7(1), 55–66 (1986)
4. Aigner, M., Fromme, M.: A game of cops and robbers. Discrete Appl. Math. 8(1), 1–11 (1984)
5. Lund, C., Yannakakis, M.: On the hardness of approximating minimization problems. J. of the ACM 41(5), 960–981 (1994)
6. Nahin, P.J.: Chases and escapes. The mathematics of pursuit and evasion. Princeton University Press, Princeton (2007)
7. Alspach, B.: Searching and sweeping graphs: a brief survey. Matematiche (Catania) 59(1-2), 5–37 (2006)
8. Fomin, F.V., Thilikos, D.M.: An annotated bibliography on guaranteed graph searching. Theor. Comp. Sci. 399, 236–245 (2008)
9. Goldstein, A.S., Reingold, E.M.: The complexity of pursuit on a graph. Theoret. Comput. Sci. 143(1), 93–112 (1995)
10. Fomin, F.V., Golovach, P., Kratochvíl, J.: On tractability of Cops and Robbers game. In: 5th Ifip International Conference On Theoretical Computer Science, vol. 273, pp. 171–185. Springer, Heidelberg (2008)
11. Edmonds, J., Karp, R.M.: Theoretical improvements in algorithmic efficiency for network flow problems. J. ACM 19(2), 248–264 (1972)
12. Fomin, F.V., Golovach, P.A., Hall, A., Mihalák, M., Vicari, E., Widmayer, P.: How to guard a graph? Technical Report 605, Dep. of Comp. Sc., ETH Zurich (2008), `http://www.inf.ethz.ch/research/disstechreps/techreports`
13. Brandstädt, A., Le, V.B., Spinrad, J.P.: Graph classes: a survey. SIAM Monographs on Discrete Mathematics and Applications. SIAM, Philadelphia, PA (1999)
14. Grötschel, M., Lovász, L., Schrijver, A.: The ellipsoid method and its consequences in combinatorial optimization. Combinatorica 1, 169–197 (1981)
15. Garey, M.R., Johnson, D.S.: Computers and intractability. A guide to the theory of NP-completeness, A Series of Books in the Mathematical Sciences. W. H. Freeman and Co, San Francisco (1979)
16. Downey, R.G., Fellows, M.R.: Parameterized complexity. Monographs in Computer Science. Springer, New York (1999)
17. Berarducci, A., Intrigila, B.: On the cop number of a graph. Adv. in Appl. Math. 14(4), 389–403 (1993)
18. Hahn, G., MacGillivray, G.: A note on k-cop, l-robber games on graphs. Discrete Math. 306(19-20), 2492–2497 (2006)
19. Raz, R., Safra, S.: A sub-constant error-probability low-degree test, and a sub-constant error-probability PCP characterization of NP. In: Proceedings of the 29th STOC, pp. 475–484 (1997)

Tree Decontamination with Temporary Immunity

Paola Flocchini[1], Bernard Mans[2], and Nicola Santoro[3]

[1] University of Ottawa, Ottawa, Canada
flocchin@site.uottawa.ca
[2] Macquarie University, Sydney, Australia
bernard.mans@mq.edu.au
[3] Carleton University, Ottawa, Canada
santoro@scs.carleton.ca

Abstract. Consider a tree network that has been contaminated by a persistent and active *virus*: when infected, a network site will continuously attempt to spread the virus to all its neighbours. The *decontamination* problem is that of disinfecting the entire network using a team of mobile antiviral system agents, called *cleaners*, avoiding any recontamination of decontaminated areas. A cleaner is able to decontaminate any infected node it visits; once the cleaner departs, the decontaminated node is immune for $t \geq 0$ time units to viral attacks from infected neighbours. After the *immunity time* t is elapsed, re-contamination can occur. The primary research objective is to determine the minimum *team size*, as well as the *solution strategy*, that is the protocol that would enable such a minimal team of cleaners to perform the task. The network decontamination problem has been extensively investigated in the literature, and a very large number of studies exist on the subject. However, all the existing work is limited to the special case $t = 0$. In this paper we examine the tree decontamination problem for any value $t \geq 0$. We determine the minimum team size necessary to disinfect any given tree with immunity time t. Further we show how to compute for all nodes of the tree the minimum team size and implicitly the solution strategy starting from each starting node; these computations use a total of $\Theta(n)$ time (serially) or $\Theta(n)$ messages (distributively). We then provide a complete structural characterization of the class of trees that can be decontaminated with k agents and immunity time t; we do so by identifying the *forbidden subgraphs* and analyzing their properties. Finally, we consider *generic* decontamination algorithms, i.e. protocols that work unchanged in a large class of trees, with little knowledge of their topological structure. We prove that, for each immunity time $t \geq 0$, all trees of height at most h can be decontaminated by a team of $k = \lfloor \frac{2h}{t+2} \rfloor$ agents whose only knowledge of the tree is the bound h. The proof is constructive.

Keywords: Network decontamination, tree networks, mobile agents, antiviral agents, distributed algorithm.

1 Introduction

1.1 The Problem

Among the many security threats in networked systems supporting mobile agents, one of the most predominant is the presence of extraneous mobile agents (*intruders*) that

S.-H. Hong, H. Nagamochi, and T. Fukunaga (Eds.): ISAAC 2008, LNCS 5369, pp. 330–341, 2008.

can harm the network (e.g., see [12]). Foremost examples of such harmful intruders are *viruses*: extraneous mobile agents infecting any visited site.

Consider a tree network that has been contaminated by a persistent and active *virus*: when infected, a network site will continuously attempt to spread the virus to all its neighbours. The *decontamination* problem is that of disinfecting the entire network using a team of mobile antiviral system agents, called *cleaners* injected into the system from a single site. A cleaner is able to decontaminate any infected node it visits; however, once it departs, the decontaminated node can be re-contaminated by infected neighbours. The decontamination is *t-strong* ($t \geq 0$) if once the cleaner departs, the decontaminated node is immune for t time units to viral attacks from infected neighbours. After the immunity time is elapsed, re-contamination can occur. The value t is also called *immunity time*.

The primary research objective is to determine the minimum *team size*, that is the smallest number of antiviral agents necessary to decontaminate the entire network avoiding any recontamination, as well as the *solution strategy*, that is the protocol that would enable such a minimal team of cleaners to perform the task. It is worthwhile noticing that, for a given network, both the optimal team size and the solution strategy may vary depending on the initial location of the antiviral agents; hence, both objectives are studied in both relative (i.e., with respect to a given node) and general (i.e., from the best site) terms.

The network decontamination problem, originally posed in [6], is equivalent to the *intruder capture* problem [1,5] and has been extensively studied in the literature due to the relationship between minimal team size and classical graph parameters such as tree-width and path-width (e.g., [3,7,15,25]). In particular, a large amount of studies have examined the problem when the agents are allowed to "jump" across the network (e.g., see [3,7,13,15,17,18,22,23]. Without such an extraordinary capability, the nature of the problem changes drastically [2], and it has been extensively studied for several classes of graphs under a variety of different settings (e.g., [1,5,8,9,10,11,14,15,16,20,21]).

All the existing work, in spite of their differences, is limited to the special case $t = 0$, that is assuming that a decontaminated node, in absence of an antiviral agent on site, may be *immediately* re-contaminated by infected neighbours. In this paper we examine the decontamination problem for any arbitrary value $t \geq 0$, focusing on tree networks. As mentioned before, decontamination of trees has been examined only for the case $t = 0$. It was shown that, in that case, the optimal team size and solution strategy can determined in linear time (serially) or with a linear number of messages (distributively) [1]. This must be contrasted with the fact that the problem of determining the optimal team size for general graphs is NP-hard [22] even for $t = 0$.

Prior to this work, nothing was known when $t > 0$.

1.2   Our Results

In this paper we examine the problem of decontaminating tree networks and provide a complete structural and algorithmic characterization for any arbitrary value $t \geq 0$, thus extending and generalizing the existing results for $t = 0$. In particular:

We first of all determine the minimum team size necessary to disinfect with immunity time t a tree network from a given starting node. Furthermore we show that it is possible to compute all the minimum team sizes and (implicitly) the solution strategies from all starting nodes optimally in $\Theta(n)$ time (serially) or with $\Theta(n)$ messages (distributively), regardless of the value of t.

We then provide a complete structural characterization of the class of trees that can be decontaminated with k agents and immunity time t. In particular, we identify the class $F(k,t)$ of *forbidden subgraphs* for given $k \geq 0$ and $t \geq 0$, and prove that any tree network without any such subgraph can indeed be disinfected by k agents and immunity time t. Note that, even though the case $t = 0$ have been studied before, the class $F(k, 0)$ had not been identified.

Finally, we turn to the problem of designing *generic* decontamination algorithms, i.e. protocols that are not specific to a given tree, but rather would work unchanged in a large class of trees, with little knowledge of their topological structure. Generic decontamination algorithms are useful when little information is available about the network, and thus the minimum team size cannot be computed. Trivially, a team of n cleaners can decontaminate all trees of up to n nodes for any immune time $t \geq 0$. We prove that, for each immunity time $t \geq 0$, all trees of height at most h can be decontaminated by a team of $k = \lfloor \frac{2h}{t+2} \rfloor$ agents whose only knowledge of the tree is the bound h. The proof is constructive: we provide a purely localized generic mobile agent protocols and prove that it has the claimed property.

2 Definitions and Terminology

Let $\mathcal{T} = (V, E)$ be a tree where V is the set of nodes and E is the set of edges. Let $N(u) = \{v \in V : (u,v) \in E\}$ denote the set of neighbours of node x, and let $\mathcal{T}_u$ denote $\mathcal{T}$ when rooted in $u \in V$. Let $(u,v) \in E$; we denote by $\mathcal{T}_{v \setminus u}$ the subtree of u rooted in v not containing u.

A team of mobile antiviral agents, the *cleaners*, operates in $\mathcal{T}$. Each agent has a distinct identifier, can perform local computations, can move from a node to a neighbouring one, and has limited private memory. Each agent obeys the same set of behavioural rules, and can communicate with other agents only when they are simultaneously present at the same node (face-to-face communication). The environment is synchronous; that is, it takes one unit of time for an agent to traverse a link, while local computation is considered instantaneous.

The team of cleaners is initially located at the same node (the *homebase*) and agents can move from node to neighbouring node. At any point in time each node of the network is in one of three possible states: *guarded, contaminated, clean*. A node is guarded when it contains at least one agent. Initially all nodes are contaminated except for the homebase which is guarded. A cleaner is able to decontaminate (or clean) any infected node it visits, transforming it from contaminated into guarded; once it departs, node enters state clean; the clean node is immune for t time units to viral attacks from contaminated neighbours. This immunity time is restarted at every transition of the node from a guarded state to a clean state. Once the immunity time has elapsed, a clean node becomes contaminated if one or more of its neighbours are contaminated. The solution of the problem is given by devising a strategy for the agents to move in the network in such a way that at the end all the nodes of the tree are simultaneously clean. We are interested in *monotone* decontamination strategies, i.e., where no clean node becomes contaminated.

We will denote by $\Psi_t(\mathcal{T}_r)$ the minimum number of agents necessary to perform the decontamination without recontamination of $\mathcal{T}_r$ with immunity time t starting from $r \in V$. We will denote by $\Psi_t(\mathcal{T}) = \min_{r \in V}\{\Psi_t(\mathcal{T}_r)\}$ the minimum number of agents necessary to perform the decontamination without recontamination of $\mathcal{T}$ with immunity time t. In the following, when no ambiguity arises, we will omit the subscript t to simplify the notation.

3   Minimal Team Size and Solution Strategy

3.1   Determining the Minimum Team Size and Solution Strategy

In the following we define a pair of functions: ψ that will be shown to recursively describe the minimum number of agents necessary for decontamination, and γ that will be used for the determination of ψ.

Definition 1. *Given $r \in V$, let $x \in V$, and let $\{v_1 \ldots v_d\}$ be its children in $\mathcal{T}_r$.*

$$(\psi_r(x), \gamma_r(x)) = \begin{cases} (1,1) & \text{if } x \text{ is a leaf} \\ (\psi_r(v_1) + 1, 1) \ (*reset*) & \text{if } \psi_r(v_1) = \psi_r(v_2) \text{ and } \gamma_r(v_2) > \frac{t}{2} \\ (\psi_r(v_1), \gamma_r(x) + 1) & \text{otherwise} \end{cases}$$

where, without loss of generality, v_1 and v_2 are such that $\psi_r(v_1) = \max_i\{\psi_r(v_i)\}$ and $\psi_r(v_2) = \max_{i \neq 1}\{\psi_r(v_i)\}$.

A node $x \in V$ will be called a *reset node* if $\gamma_r(x) = 1$.

The following Lemmas describe properties of ψ and γ.

Lemma 1. *Let y_1 be a child of r in $\mathcal{T}_r$. If $\gamma_r(y_1) = d$ with $d > 1$, then:*

(1) *there exists a path $y_1, \ldots, y_d$ such that $\psi_r(y_1) = \psi_r(y_i)$ for $i = 1 \ldots d$.*
(2) *if y_d is not a leaf, there exists a child u of y_d such that $\psi_r(u) = \psi_r(y) - 1$,*
(3) *for any child v of y_i with $1 \leq i < d$: either $[\psi_r(v) = \psi_r(y_1) \text{ and } \gamma_r(v) < \gamma_r(y_1)]$ or $[\psi_r(v)) < \psi_r(y_1)]$*

In the following, whenever a node $x \in V$ has more than one child, we will always indicate with x_1 and y_1 the two children such that the pair $(\psi_r(x_1), \gamma_r(x_1))$ is the lexicographically maximum among the pairs associated to the children of x and $(\psi_r(y_1), \gamma_r(y_1))$ is the lexicographically maximum among the pairs associated to the children of x excluding x_1.

Lemma 2. *Let $\mathcal{T}_r$ be a tree rooted in r. Let $\psi_r(x_1) = \psi_r(y_1)$ and $\gamma_r(x_1), \gamma_r(y_1) > \frac{t}{2}$. If a team of $\psi_r(x_1) + 1$ agents starts from r the decontamination of $\mathcal{T}_r$, one of the two subtrees $\mathcal{T}_{x_1 \backslash r}$ and $\mathcal{T}_{y_1 \backslash r}$ must be fully decontaminated before the team starts the decontamination of the other, regardless of the cleaning strategy.*

We now show that ψ recursively describes the minimum number of agents required to decontaminate a tree from a given node.

Theorem 1. *Let $x \in V$ and let $p(x)$ be its parent in $\mathcal{T}_r$. The minimum number of agents necessary to perform the decontamination of $\mathcal{T}_{x \setminus p(x)}$ from x with immunity time t is $\psi_r(x)$. In particular, the minimum number $\Psi(\mathcal{T}_r)$ of agents necessary to perform the decontamination of $\mathcal{T}_r$ from r with immunity time t is $\psi_r(r)$.*

Proof. By induction on the height of the trees $\mathcal{T}_{x \setminus p(x)}$, $x \in V$. The theorem is trivially true when the tree is composed of a single node. Assume it is true for all trees of height i and let $\mathcal{T}_{x \setminus p(x)}$ have height $i + 1$. We consider the three possible situations.

Case 1. Consider first the case when $\psi_r(x_1) = \psi_r(y_1)$ and $\gamma_r(x_1), \gamma_r(y_1) > \frac{t}{2}$. Since $\gamma_r(x_1) > \frac{t}{2}$ and $\gamma_r(y_1) > \frac{t}{2}$, by Lemma 1 there are two paths $(x, x_1, x_2, \ldots, x_a)$ and $(x, y_1, y_2, \ldots, y_b)$ in $\mathcal{T}_{x_1 \setminus x}$ and $\mathcal{T}_{y_1 \setminus x}$ of lengths respectively $a > \frac{t}{2}$ and $b > \frac{t}{2}$ from x terminating with subtrees requiring $\psi_r(y_1)$ agents each (from their roots x_a and y_b). Since, by definition of ψ, $\psi_r(r) = \psi_r(x_1) + 1$, by Lemma 2 we know that $\mathcal{T}_{x_1 \setminus x}$ and $\mathcal{T}_{y_1 \setminus x}$ must be decontaminated one after the other starting from x; w.l.g, let $\mathcal{T}_{x_1 \setminus x}$ be decontaminated before $\mathcal{T}_{y_1 \setminus x}$. However, for the $\psi_r(x_1)$ agents to reach x_a, node r stays unguarded for more than the immunity time t and is recontaminated from y_1. Thus, $\psi_r(x_1) + 1$ agents are necessary. To show that $\psi_r(x_1) + 1$ agents are indeed sufficient, we describe a cleaning algorithm with $\psi_r(x_1) + 1$ agents. One agent stays on x to guard it from recontamination, the rest of the team first cleans all the subtrees of x requiring less than $\psi_r(x_1)$ agents. When the remaining subtrees all require $\psi_r(x_1)$ agents (we know that there are at least two of those), the team moves down on one (e.g., the one containing path $\pi = (x, x_1, x_2, \ldots, x_a)$), always keeping a guard on x; whenever, along π, the agents reach a node x_i root of a subtree, $\psi_r(x_1) - 1$ agents perform the cleaning of the subtree from x_i (this number is sufficient by Lemma 1) while one agent stays on x_i. When the agents reach x_a, they clean $\mathcal{T}_{x_a \setminus x_{a-1}}$ and they all go back to x. The same strategy is followed for the remaining subtrees.

Case 2. Consider now the case when $\psi_r(x_1) > \psi_r(y_1)$. In this case $\psi_r(x_1)$ agents are clearly necessary because, by induction they are necessary for cleaning $\mathcal{T}_{x_1 \setminus x}$. They are also sufficient because they can clean $\mathcal{T}_{x \setminus p(x)}$ by first cleaning $\mathcal{T}_{y_1 \setminus x}$ (always keeping a guard on x) then moving to $\mathcal{T}_{x_1 \setminus x}$. Thus $\psi_r(x) = \psi_r(x_1)$ is the minimum number of agents to decontaminate $\mathcal{T}_{x \setminus p(x)}$ from x.

Case 3. Finally, consider the case $\psi_r(x_1) = \psi_r(y_1)$ with $\gamma_r(y_1) \leq \frac{t}{2}$. Obviously $\psi_r(x_1)$ agents are necessary also in this case because, by induction they are necessary for cleaning $\mathcal{T}_{x_1 \setminus x}$. By Lemma 1 we know that there is a path from x to a reset node u of length $\leq \frac{t}{2}$. Let $\pi = (x, \alpha, u)$ be such a path with $\alpha = (u_1, \ldots, u_c)$. Also in this case $\psi_r(x_1)$ agents are sufficient to clean $\mathcal{T}_{x \setminus p(x)}$. In fact, the team moves down on π; whenever the agents reach a node u_i root of one or more subtrees outside the path, $\psi_r(x_1) - 1$ agents perform the cleaning of each subtree from u_i while one agent keeps moving back and forth from u_i to x. Since the distance between u_i and x is smaller than $\frac{t}{2}$ each node in the path is guaranteed to be revisited within the immunity time t while the cleaning of the subtrees is performed.

3.2   Optimal Computation of Ψ

We describe how to compute the couple of values ψ and γ. The algorithm will be presented as an asynchronous distributed protocol in the classical message passing model

[24]. The serial version is easily derivable from it. The algorithm to compute $\Psi(\mathcal{T}_x)$ for each node x of the tree performs a "saturation" starting from the leaves of the tree [24]. The first step of the algorithm consists of a WAKE UP (a simple broadcast with possibly multiple initiators) with the goal of activating all the nodes. After the WAKE UP all nodes are in the state READY and the leaves start the "saturation" sending the pair of values $(1,0)$ to their parents. When a node x has received a pair of values from all its neighbours except for one (say p), it computes a pair of values (which will be shown to correspond to $\psi_p(x)$ and $\gamma_p(x)$) that it will send it to p. Node x will then change state becoming COMPUTING. Eventually all nodes become COMPUTING. When a COMPUTING node x receives the last pair of values from the remaining neighbour, it has all the necessary information to compute $\Psi(\mathcal{T}_x)$. At this point the node continues the propagation of appropriate pairs towards the other neighbours and becomes DONE. Eventually every node will receive the information from all its neighbours, compute the minimum number of agents necessary and sufficient for decontaminating the tree starting from itself, and become DONE. The rules of Algorithm are described below.

Algorithm. FIND-MIN-TEAM-SIZE **for node** x
WAKE UP (* after wake-up every node is in the state READY *)
if x is a READY leaf:
       send $(1,1)$ to parent
       become (COMPUTING)
if x is a READY non-leaf:
       receiving $(a_1, c_1), \ldots, (a_{deg(x)-1}, c_{deg(x)-1})$ from all neighbours except one (say p)
       (w.l.g., let (a_1, c_1) and (a_2, c_2) be the lexicographically max and second max resp.
       of those pairs)
          If $(a_1 = a_2)$ and $(c_1 = c_2 > \frac{t}{2})$
            send $(a_1 + 1, 1)$ to p (* Reset *)
          If $[(a_1 = a_2)$ and $(c_2 \leq \frac{t}{2})]$ or $[(a_1 \neq a_2)]$
            send $(a_1, max\{c_1, c_2\} + 1)$ to p (* Increase Counter *)
       become (COMPUTING)
if x is COMPUTING:
       receiving (a, c) from l
       (* node x has now received the information from all neighbours *)
          rename the pairs so that (a_1, c_1) and (a_2, c_2) are
          the lexicographically max and second max respectively of those pairs
          If $(a_1 = a_2)$ and $(c_1 = c_2 > \frac{t}{2})$
            $\Psi(\mathcal{T}_x) = a_1 + 1$
            send $(a_1 + 1, 1)$ to $N(x) \setminus l$
          else (* $[(a_1 = a_2)$ and $(c_2 \leq \frac{t}{2})]$ or $[(a_1 \neq a_2)]$ *)
            $\Psi(\mathcal{T}_x) = a_1$
            send $(a_1, max\{c_1, c_2\} + 1)$ to $N(x) \setminus l$
            (* note that if x is a leaf the operations above will not be performed *)
       become (DONE)

By definition of ψ and γ, it is easy to see that:

Lemma 3. *Let a node x receive a pair (a, c) from y. We have that:* $a = \psi_x(y)$ *and* $c = \gamma_x(y))$.

As a consequence of Lemma 3, a COMPUTING node x receiving the pair from the last neighbour has enough information to determine the minimum number of agents needed to decontaminate the tree starting from x and the correctness follows:

Theorem 2. *After executing Algorithm* FIND-MIN-TEAM-SIZE, *each node x is in the state DONE and knows the minimum number of agents required to decontaminate the tree starting from x.*

The total number of messages exchanged is $\Theta(n)$, where n is the number of nodes of the tree.

Theorem 3. *Algorithm* FIND-MIN-TEAM-SIZE *can be implemented serially so to run in $\Theta(n)$ time.*

4   Structural Characterization and Bounds

4.1   Forbidden Subgraphs

In this Section we give a complete characterization of the class of trees that can be decontaminated with k agents and immunity time t.

Definition 2. $(F^*(k, t))$
- $F^*(1, t)$ *is the family of rooted trees composed of two paths of length at least $\lfloor \frac{t}{2} \rfloor + 1$ joined in the root.*
- $F^*(i, t)$ *is the family of rooted trees composed of two paths of length at least $\lfloor \frac{t}{2} \rfloor + 1$ departing from the root, each terminating with a subtree of type $F^*(i - 1, t)$.*

Definition 3. $(F(k, t))$
- $F(1, t)$ *is the family of rooted trees composed of three paths of length at least $\lfloor \frac{t}{2} \rfloor + 1$ departing from the root.*
- $F(i, t)$ *is the family of rooted trees composed of three paths of length at least $\lfloor \frac{t}{2} \rfloor + 1$ departing from the root, each terminating with a subtree of type $F^*(i - 1, t)$.*

In other words, in $F^*(k, t)$ and in $F(k, t)$, between two successive branches there is always a path of length at least $\lfloor \frac{t}{2} \rfloor + 1$ (for an example see Figure 1).

In the following, we will say that a tree contains a subtree of type $F^*(i, t)$ (resp. $F(i, t)$) if it contains as a subtree the undirected version of a tree in the family $F^*(i, t)$ (resp. $F(i, t)$).

The following Lemma state forbidden conditions for decontamination when starting from a specific node of the tree.

Lemma 4. *If a tree $\mathcal{T}$ contains a subtree S of type $F^*(k, t)$, then $\mathcal{T}$ requires at least $k + 1$ agents to be decontaminated with immunity time t starting from the root of S.*

The following Theorem states forbidden conditions for decontamination regardless of the starting point.

Theorem 4. *If a tree $\mathcal{T}$ contains a subtree of type $F(k, t)$, then $\mathcal{T}$ requires at least $k+1$ agents to be decontaminated with immunity time t, regardless of the starting node.*

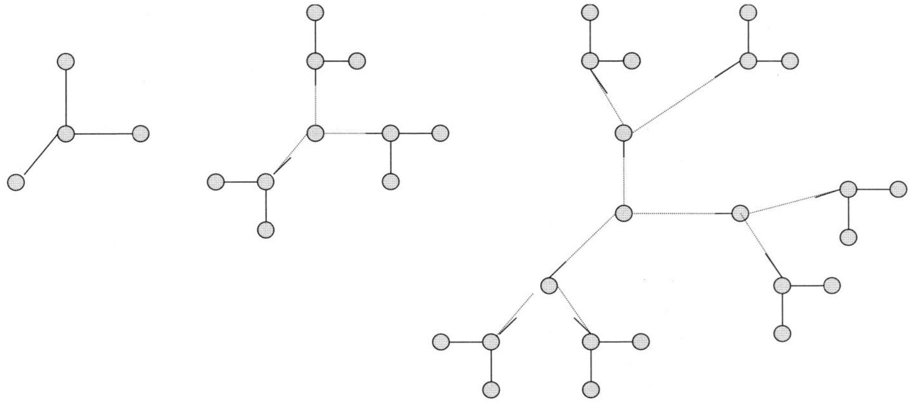

Fig. 1. Trees of type $F(1,0)$, $F(2,0)$, and $F(3,0)$

Proof. Let r be the root of a subtree of type $F(k,t)$ of $\mathcal{T}$. By definition, there exist three subtrees of type $F^*(k-1,t)$ each at distance at least $\lfloor \frac{t}{2} \rfloor + 1$ from r (see the example of $F(4,4)$ in Figure 2). Let $(r, x_1, x_2, \ldots, x_a)$, $(r, y_1, y_2, \ldots, y_b)$, $(r, z_1, z_2, \ldots, z_c)$ be the three paths and A, B, C the three subtrees respectively from x_a, y_b and z_c. By definition of ψ and by Lemma 4 we have that $\psi_r(x_1) = \psi_r(y_1) = \psi_r(z_1) = k$, and since all three paths have length at least $\lfloor \frac{t}{2} \rfloor + 1$, we have that $\gamma_r(x_1), \gamma_r(y_1), \gamma_r(z_1) > \frac{t}{2}$, which means that, starting from r, the number of agents $\Psi(\mathcal{T}_r)$ necessary to clean the tree is $k+1$. By definition of agent function, $\psi_r(x_i) = \psi_r(y_i) = \psi_r(z_i) = k+1$, thus $k+1$ agents are necessary to decontaminate the whole tree starting from all the nodes in the three paths $x_i.y_j, z_l$ ($1 < i < a, 1 < j < b, 1 < l < c$). Let the starting point be inside one of the subtrees A, B, C. Since $\psi_r(x_a) = k+1$, the team of agents coming from inside T_1 will need $k+1$ agents at node x_s to continue cleaning $T_{x_a \backslash x_{a-1}}$.

4.2   Structural Characterization

Before showing that the forbidden subgraphs described above are the only possible ones, we introduce some Lemmas. Let $\Psi(\mathcal{T}) = \min_{x \in V} \{ \Psi(\mathcal{T}_x) \}$.

Lemma 5. *Let the immunity time be t and let $\mathcal{T}$ be a tree such that $\Psi(\mathcal{T}) = k+1$. Then there exists a node r with $\Psi(\mathcal{T}_r) = k+1$ that has at least children x_1, y_1, and z_1 such that $\psi_r(x_1) = \psi_r(y_1) = \psi_r(z_1) = k$, and $\gamma_r(x_1), \gamma_r(y_1), \gamma_r(z_1) > \frac{t}{2}$.*

Lemma 6. *If $\mathcal{T}$ requires $k+1$ agents starting from node x (i.e. $\Psi(\mathcal{T}_x) = k+1$), then $\mathcal{T}$ contains a subtree of type $F^*(k,t)$ rooted in x.*

Theorem 5. *If a tree $\mathcal{T}$ does not contain a subtree of type $F(k,t)$, then it can be decontaminated with k agents from any starting node.*

Proof. We prove the theorem by showing that if $\mathcal{T}$ requires at least $k+1$ agents to be decontaminated starting from any node, then it necessarily contains a subtree of type $F(k,t)$. Let $\mathcal{T}$ be such that $\Psi(\mathcal{T}) = k+1$ and let $\Psi(\mathcal{T}_r) = k+1$. By Lemma 5

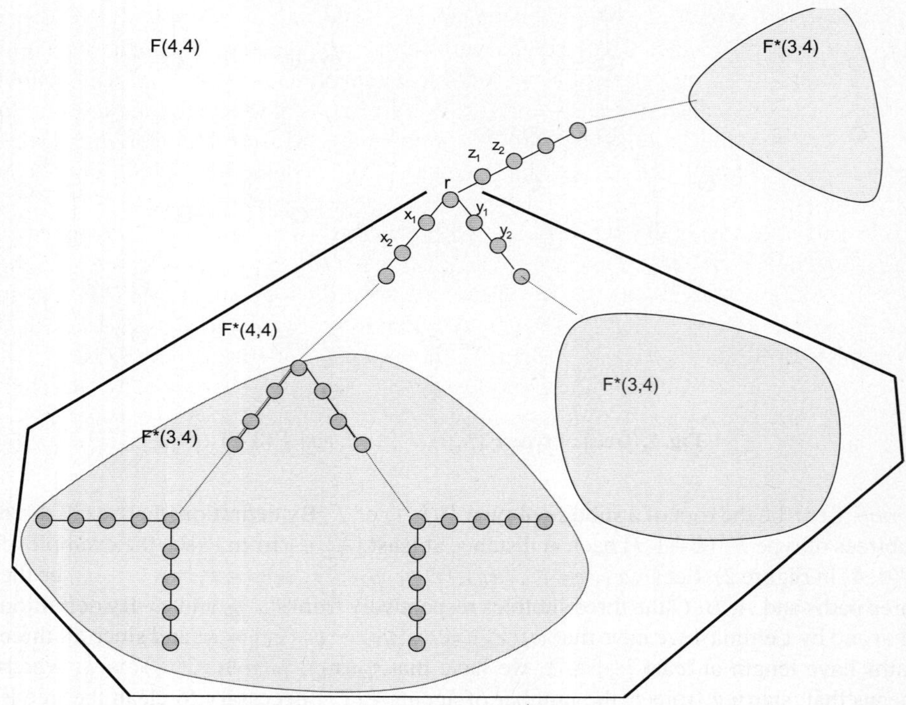

Fig. 2. A tree of type $F(4, 4)$

there exists such a node r with three paths $(r, x_1, x_2, \ldots, x_a)$, $(r, y_1, y_2, \ldots, y_b)$, and $(r, z_1, z_2, \ldots, z_c)$ terminating with three subtrees A, B, and C rooted respectively in x_a, y_b and z_c. such that $\psi_r(x_1) = \psi_r(y_1) = \psi_r(z_1) = k$, and $\gamma_r(x_1), \gamma_r(y_1), \gamma_r(z_1) > \frac{t}{2}$. This implies that $a, b, c > \frac{t}{2}$. By Lemma 6 it follows that $\mathcal{T}_{x_1 \backslash r}, \mathcal{T}_{y_1 \backslash r}$, and $\mathcal{T}_{z_1 \backslash r}$ all contain subtrees of type $F^*(k, t)$ rooted in x_1, y_1, and z_1 respectively, and the theorem is proven.

From Theorems 4 and 5 we can conclude that:

Theorem 6. *Let t be the immunity time. A tree can be decontaminated with k agents if and only if it does not contain a subtree of type $F(k, t)$.*

As a consequence of Theorem 6 we can derive the following:

Theorem 7. *There are trees of n nodes that to be decontaminated with immunity time t need a team of $\Omega(\log_3(\frac{n}{t+1}))$ cleaners.*

5   Generic Decontamination Protocol

In the classical model, to avoid recontamination of a node with contaminated neighbours, the only possibility is to guard the node (i.e., to have an agent reside on the node). When the immunity time is greater than zero, however, the situation is different. Since a decontaminated node x with contaminated neighbours and time immunity $t > 0$ can stay for t time units unguarded in contact with contaminated nodes, it is not necessary to have a stationary agent guarding it to avoid recontamination: a "guarding" agent could move away from x and come back to it within t time units to obtain the same result. In this way, a "guarding" agent can "protect" a group of agents from decontamination: for example, it could protect a path of length up to $\lfloor \frac{t}{2} \rfloor$ (i.e., containing $\lfloor \frac{t}{2} \rfloor + 1$ nodes) by oscillating (i.e., moving back and forth) from one extremity to the other.

We describe a decontamination strategy that uses this "oscillating idea" and allows to decontaminate a tree from a given starting location using $\lfloor \frac{2h}{t+2} \rfloor$ agents, where h is the height of the tree rooted in the starting location and t is the immunity time. The strategy is not optimal in terms of number of agents (achievable by using our algorithms defined in previous sections), however the algorithm presented is generic and localized.

Informally our algorithm corresponds to a simple Depth-First Traversal by one agent while all other agents protects the immunity of the nodes that have been cleaned by the leading agent traversing the tree. The algorithm follows a localized strategy as agents only require face-to-face communications.

Let us define the *node-level* as the distance of a node from the root (the root is at level 0). Let us define the *edge-level* as the node-level of the node on the edge-extremity farthest from the root.

Let $a_1, a_2, \ldots a_k$ be the agents (where $k = \lfloor \frac{2h}{t+2} \rfloor$). Each agent is responsible for $\lfloor \frac{t}{2} \rfloor$ edge-levels of the tree: more precisely, agent a_1 is responsible for levels $1, \ldots, \lfloor \frac{t}{2} \rfloor$, agent a_i is responsible for levels $i \cdot \lfloor \frac{t}{2} \rfloor + 1, \ldots, (i+1) \cdot \lfloor \frac{t}{2} \rfloor$. Let $\mathcal{A}_i$ denote the area of responsibility (or domain) of agent i.

Let agent a_k be the *decontaminating leader* of the team. All other agents travel with the decontaminating leader when it traverses nodes in their area of responsibility or above. Otherwise (i.e., when the decontaminating agent traverses the tree in lower levels), they oscillate in their respective domain.

Algorithm CLEAN   for agent a_k

Perform a Depth-First Traversal
When moving down in the domain $\mathcal{A}_i$ of agent a_i, $i < k$.
     tell agents i to $(k-1)$ to move down with you.
When moving up in the domain $\mathcal{A}_i$ of agent a_i, $i < k$.
     wait for agent a_i if necessary.
     tell agents i to $(k-1)$ to move up with you.

Algorithm CLEAN   for agents a_i $(0 < i < k)$

If a_k is moving down among nodes in $\bigcup_{j \leq i} \mathcal{A}_j$
    move down with agent a_k
    define the nodes traversed in the domain $\mathcal{A}_i$ as the active path π_i
If a_k is traversing nodes in $\mathcal{A}_{i+1}$
    OSCILLATE   on (π_i) (* where π_i is the active path in the domain of a_i *)
If a_k is moving up (backtracking) to a node in $\bigcup_{j \leq i} \mathcal{A}_j$
    move up with agent a_k.

Algorithm OSCILLATE   on (π_i) for agents a_i $(i < k)$

Continually move up and down along active path π_i.
Until find a_k ready to move up
    // returning to Algorithm CLEAN to move up with a_k.

Let us emphasize that the waiting steps of the backtrack of the leading agent a_k are required by the limited face-to-face communications. This also guarantees that the agents only oscillate on the path between the root and the leading agent a_k.

We can now prove the monoticity of the decontamination.

Lemma 7. *All nodes traversed by the agent a_k remain decontaminated.*

Proof. By construction of the Depth-First Traversal of agent a_k, let us define the *immuned path* as the path between the root and the current position of a_k. Let us first prove that the nodes on the immuned path are always *clean*. By definition of the active paths π_i, $i < k$, one can observe that an immuned path of depth d, $d \leq h$, is composed of the $\lceil \frac{d}{t/2} \rceil$ first active paths π_j defined by the respective agents a_j, $j \leq \lceil \frac{d}{t/2} \rceil$. Indeed each agent a_i, $i < k$, moves down the immuned path with agent a_k when new nodes are cleaned for the first time in their domain $\mathcal{A}_i$ (by algorithm CLEAN) and then guarantee that the path remains *clean* (by algorithm OSCILLATE) until agent a_k backtracks up in their respective domain.

It is now easy to prove that all previously traversed nodes not included in the immuned path will remain *clean*, even if they have not been visited in the last t steps. Indeed, with the Depth-First order of the traversal, clean nodes that may have infected neighbours can only be on the immuned path, and thus guarantee the on-going "protection" of the other nodes.

At the termination of the Depth-First Traversal, all the nodes of the tree have been traversed by agent a_k, and the previous lemma leads immediately to the following theorem.

Theorem 8. *Algorithm CLEAN decontaminate a tree from a given starting location r and immunity time t with $\lfloor \frac{2h}{t+2} \rfloor$ agents, where h is the height of the tree rooted in r.*

References

1. Barrière, L., Flocchini, P., Fraigniaud, P., Santoro, N.: Capture of an intruder by mobile agents. In: Proc. 14th Symp. Parallel Algorithms and Architectures (SPAA 2002), pp. 200–209 (2002)
2. Barrière, L., Fraignaud, P., Santoro, N., Thilikos, D.: Searching is not jumping. In: Bodlaender, H.L. (ed.) WG 2003. LNCS, vol. 2880, pp. 34–45. Springer, Heidelberg (2003)
3. Bienstock, D.: Graph searching, path-width, tree-width and related problems. DIMACS Series in Disc. Maths. and Theo. Comp. Sci. 5, 33–49 (1991)
4. Bienstock, D., Seymour, P.: Monotonicity in graph searching. J. Algorithms 12, 239–245 (1991)
5. Blin, L., Fraignaud, P., Nisse, N., Vial, S.: Distributed chasing of network intruders. Theoretical Computer Science 399(1-2), 12–37 (2008)

6. Breisch, R.: An intuitive approach to speleotopology. S.W. Cavers VI(5), 72–78 (1967)
7. Ellis, J., Sudborough, H., Turner, J.: The vertex separation and search number of a graph. Information and Computation 113(1), 50–79 (1994)
8. Flocchini, P., Huang, M.J., Luccio, F.L.: Decontamination of hypercubes by mobile agents. Networks (to appear, 2008)
9. Flocchini, P., Huang, M.J., Luccio, F.L.: Decontamination of chordal rings and tori using mobile agents. Int. J. of Foundation of Computer Science 18(3), 547–564 (2007)
10. Flocchini, P., Luccio, F.L., Song, L.X.: Size optimal strategies for capturing an intruder in mesh networks. In: Proc. Int. Conf. on Comm. in Computing (CIC 2005), pp. 200–206 (2005)
11. Flocchini, P., Nayak, A., Shulz, A.: Cleaning an arbitrary regular network with mobile agents. In: Chakraborty, G. (ed.) ICDCIT 2005. LNCS, vol. 3816, pp. 132–142. Springer, Heidelberg (2005)
12. Flocchini, P., Santoro, N.: Distributed Security Algorithms For Mobile Agents. In: Cao, J., Das, S. (eds.) Mobile Agents in Networking and Distributed Computing. Wiley, Chichester (2008)
13. Fomin, F., Golovach, P.: Graph searching and interval completion. SIAM J. on Discrete Mathematics 13(4), 454–464 (2000)
14. Fomin, F., Thilikos, D., Todineau, I.: Connected graph searching in outerplanar graphs. In: Proc. 7th Int. Conf. on Graph Theory (ICGT 2005) (2005)
15. Fraigniaud, P., Nisse, N.: Connected treewidth and connected graph searching. In: Correa, J.R., Hevia, A., Kiwi, M. (eds.) LATIN 2006. LNCS, vol. 3887, pp. 479–490. Springer, Heidelberg (2006)
16. Fraigniaud, P., Nisse, N.: Monotony properties of connected visible graph searching. In: Fomin, F.V. (ed.) WG 2006. LNCS, vol. 4271, pp. 229–240. Springer, Heidelberg (2006)
17. Kirousis, L., Papadimitriou, C.: Interval graphs and searching. Discrete Mathematics 55, 181–184 (1985)
18. Kirousis, L., Papadimitriou, C.: Searching and pebbling. Theoretical Computer Science 47(2), 205–218 (1986)
19. Lapaugh, A.: Recontamination does not help to search a graph. J. of the ACM 40(2), 224–245 (1993)
20. Luccio, F.L.: Intruder capture in Sierpiński graphs. In: Proc. 4th Int. Conf. on Fun with Algorithms, pp. 249–261 (2007)
21. Luccio, F., Pagli, L., Santoro, N.: Network decontamination with local immunization. Int. J. of Foundation of Computer Science 18(3), 457–474 (2007)
22. Megiddo, N., Hakimi, S., Garey, M., Johnson, D., Papadimitriou, C.: The complexity of searching a graph. J. of the ACM 35(1), 18–44 (1988)
23. Parson, T.: Pursuit-evasion in a graph. In: Theory and Applications of Graphs. Lecture Notes in Mathematics, pp. 426–441. Springer, Heidelberg (1976)
24. Santoro, N.: Design and Analysis of Distributed Algorithms. Wiley, Chichester (2007)
25. Takahashi, A., Ueno, S., Kajitani, Y.: Mixed searching and proper-path-width. Theoretical Computer Science 137(2), 253–268 (1995)
26. Yamamoto, M., Takahashi, K., Hagiya, M., Nishizaki, S.-Y.: Formalization of graph search algorithms and its applications. In: Grundy, J., Newey, M. (eds.) TPHOLs 1998. LNCS, vol. 1479, pp. 479–496. Springer, Heidelberg (1998)

Reconfiguration of Cube-Style Modular Robots Using $O(\log n)$ Parallel Moves

Greg Aloupis[1], Sébastien Collette[1,*], Erik D. Demaine[2,**],
Stefan Langerman[1,***], Vera Sacristán[3,†], and Stefanie Wuhrer[4]

[1] Université Libre de Bruxelles
{greg.aloupis,secollet,slanger}@ulb.ac.be
[2] Massachusetts Institute of Technology
edemaine@mit.edu
[3] Universitat Politècnica de Catalunya
vera.sacristan@upc.edu
[4] Carleton University
stefanie.wuhrer@gmail.com

Abstract. We consider a model of reconfigurable robot, introduced and prototyped by the robotics community. The robot consists of independently manipulable unit-square atoms that can extend/contract arms on each side and attach/detach from neighbors. The optimal worst-case number of sequential moves required to transform one connected configuration to another was shown to be $\Theta(n)$ at ISAAC 2007. However, in principle, atoms can all move simultaneously. We develop a parallel algorithm for reconfiguration that runs in only $O(\log n)$ parallel steps, although the total number of operations increases slightly to $\Theta(n \log n)$. The result is the first (theoretically) almost-instantaneous universally reconfigurable robot built from simple units.

1 Introduction

In this paper, we consider homogeneous self-reconfiguring modular robots composed of unit-cube *atoms* arranged in a grid configuration. Each atom is equipped with mechanisms allowing it to extend each face out one unit and later retract it back. Furthermore, the faces can attach/detach to faces of adjacent atoms; at all times, the atoms should form a connected mass. When groups of atoms perform the four basic atom operations (expand, contract, attach, detach) in a coordinated way, the atoms move relative to one another, resulting in a reconfiguration of the robot. Fig. 1 shows an example of such a reconfiguration. Each atom is depicted as a square, with a T-shaped arm on each side.

* Chargé de Recherches du FRS-FNRS.
** Partially supported by NSF CAREER award CCF-0347776, DOE grant DE-FG02-04ER25647, and AFOSR grant FA9550-07-1-0538.
*** Chercheur Qualifié du FRS-FNRS.
† Partially supported by projects MEC MTM2006-01267 and DURSI 2005SGR00692.

S.-H. Hong, H. Nagamochi, and T. Fukunaga (Eds.): ISAAC 2008, LNCS 5369, pp. 342–353, 2008.

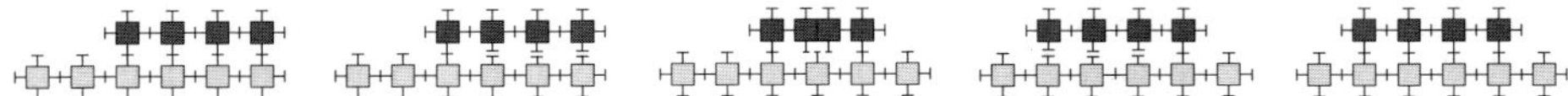

Fig. 1. Example of reconfiguring crystalline atoms: the top row of atoms is able to shift to the left, using the bottom row of atoms as a fixed base

The robotics community has implemented this model in two prototype systems: *crystalline atoms* [3,4,5] and *telecube atoms* [6,7]. In the crystalline model, the default state for atoms is expanded, while in the telecube model, the default state is contracted. Thus Fig. 1 reconfigures a crystalline robot, or an expanded telecube robot. The crystalline robots work in a single plane, forbidding expand/contract/attach/detach operations parallel to the z axis, which is the case we consider in this paper.

To ensure connectedness of the configuration space, the atoms must be arranged in *meta-modules* (or simply *modules*), which are groups of $k \times k$ atoms. Any value $k \geq 2$ suffices for universal reconfigurability [2,7]. Here the collection of atoms composing a robot must remain *connected* in the sense that its module graph (where vertices correspond to modules and edges correspond to attached modules) is connected.

The complexity of a reconfiguration algorithm can be measured by the number of *parallel steps* performed ("makespan"), as well as the total number of atom operations ("work"). In a parallel step, many atoms may perform moves simultaneously. The number of parallel steps is typically the most significant factor in overall reconfiguration time, because the mechanical actions (expansion, contraction, attachment, detachment) are the slowest part of the system.

Our main contribution in this paper is a reconfiguration algorithm that, given a source robot S and a target robot T, each composed of n atoms arranged in $k \times k$ modules for some constant k, reconfigures S into T in $O(\log n)$ parallel steps and a total of $O(n \log n)$ atom operations. This result improves upon the reconfiguration time of the algorithm presented at ISAAC 2007 [2], which takes $O(n)$ parallel steps (although only $O(n)$ total operations, and also for three-dimensional robots), as well as previous $O(n^2)$ algorithms [5,7,4].

A central assumption in our algorithm is that one atom, by contracting or expanding, can pull or push all n atoms (*linear strength*). Thus our algorithm certainly tests the structural limits of a modular robot, but on the other hand this assumption enables us to achieve reconfiguration times that are likely asymptotically optimal. The quadratic reconfiguration algorithms of [5,7,4] may be given credit for being the least physically demanding on the structure of the robot. Even the algorithm in [2] is less demanding than what we propose here, because it does not produce arbitrarily high velocities (although it still uses linear strength). Another recent algorithm [1] considers the case where atoms have only constant strength, and attains $O(n)$ parallel steps and $O(n^2)$ total operations, which is optimal in this setting. Thus the improvement in reconfiguration time obtained here *requires* a more relaxed physical model.

The main idea of our parallel algorithm is to reconfigure the given robot into a canonical form, by recursively dividing the plane into a hierarchy of square *cells* and employing a divide-and-conquer technique to merge quadruples of cells. Each merge creates a cell containing a simple structure using a constant number of moves. This structure, which fills the perimeter of a cell as much as possible, can be decomposed into a constant number of rectangular components. Because the steps to merge cells of the same level can be executed in parallel, the total number of parallel steps used to reconfigure any configuration to a simple structure is $O(\log n)$. The entire reconfiguration takes place in the smallest $2^h \times 2^h$ square containing the initial configuration, where h is an integer.

We choose to describe our algorithm in terms of the naturally expanded modules of crystalline robots. Of course, this immediately implies reconfigurability in the naturally contracted telecube model, by adding one step at the beginning and end in which all atoms expand and contract in parallel. We also expect that the individual constructions in our algorithm can be modified to directly work in the (2D) telecube model as well.

Our algorithm effectively uses modules of 4×4 atoms, but for clarity and brevity assumes that atoms initially appear in *blocks* of 32×32. Reducing the module size leads to more complicated basic operations that we have designed for use on large rectangular components. On the other hand, reducing the initial block size leads to a larger number of possible shapes that we must consider during the merge of cells. We have designed (though not rigorously analyzed) a range of algorithms for 2×2 modules with decreasing restrictions on block size. This is discussed in Section 5. However, the bulk of this paper focuses on the version that is easiest to describe.

2 Definitions

We will mainly deal with modules, not atoms, which can be viewed as lying on their own square lattice somewhat coarser than the atom lattice. Refer to Fig. 2 for examples of the following notions. In all figures, modules are depicted as squares unless mentioned otherwise. A module is a *node* if it has exactly one neighbor (a leaf node), more than two neighbors (a branching node), or exactly two neighbors not collinear with the node (a bending node). A *branch* is a straight path of (non-node) modules between two nodes (including the nodes themselves). A *cell* is a square of module positions (aligned with the module lattice), some of which may be occupied by modules. The *boundary* of a cell

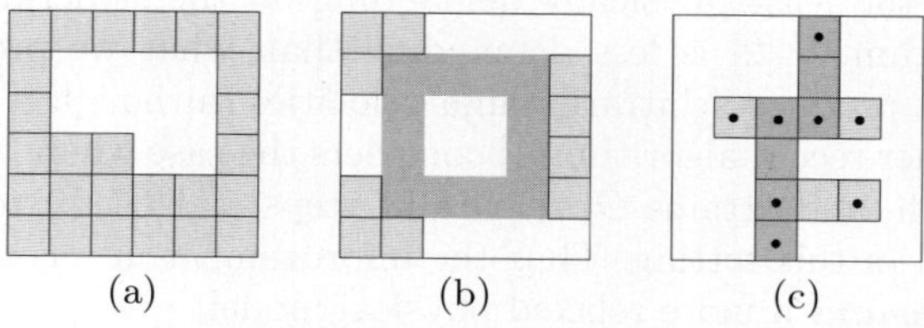

(a)   (b)   (c)

Fig. 2. Definitions; modules are depicted by squares. (a) A ring. (b) A sparse cell with five side-branches and shaded near-boundary. (c) A shaded backbone and eight nodes.

consists of all module positions touching the cell's border. For cells of sufficient size the *near-boundary* consists of all module positions adjacent to the cell's boundary. If a branch lies entirely in the boundary of a cell, we call it a *side-branch*. The configuration within a cell is a *ring* if the entire cell's boundary is occupied by modules, and all remaining modules within the cell are arranged at the bottom of the cell, filling row by row from left to right. The configuration within a cell is *sparse* if it contains only side-branches. A *backbone* is a set of branches forming a path that connects two opposite edges of a cell.

3   Elementary Moves That Use $O(1)$ Parallel Steps

Throughout this paper, whenever we describe a move, it is implied that we do not disconnect the robot and that no collisions occur. We first describe three basic module moves (*slide, compress, k-tunnel*) that are used in [2]. We omit a detailed description of how to implement these moves in terms of individual atom operations. A *compression* pushes one module m_1 into the space of an adjacent module m_2. The atoms of m_1 literally fill the spaces between those of m_2 (see Fig 3). Any part of the robot attached to m_1 will be displaced by one unit along the same direction. Two modules can occupy the same position in the module lattice. A *decompression* can be applied to such a position, as long as an adjacent position contains enough space.

A *slide* moves a module to an adjacent position , using two substrate modules. See Fig. 4a. The *k-tunnel move* compresses a leaf module into the robot, and *decompresses* another module out into a leaf position. An entire path of modules between the two leaves is involved in this move. Within each branch in this path, modules shift in the direction of the compression, and essentially transfer the compression to the next bend. Any modules attached to the branches will also shift. This issue is addressed later on. See Fig. 4b; The parameter k denotes the

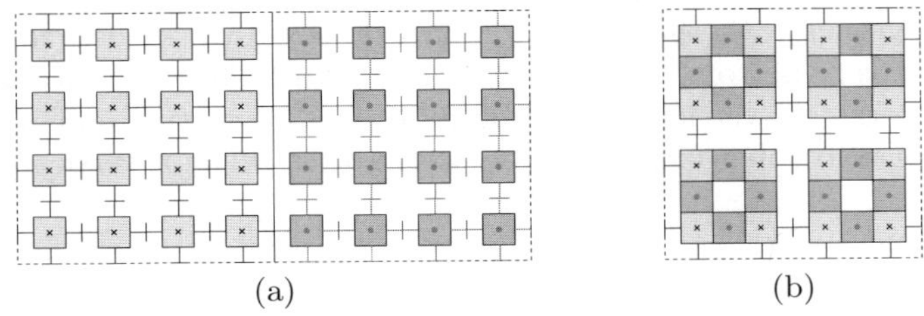

(a)   (b)

Fig. 3. Compression of two adjacent 4×4 modules into one lattice position

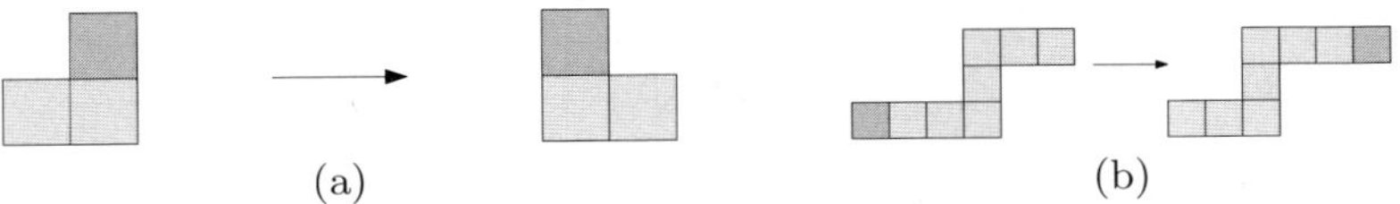

(a)   (b)

Fig. 4. (a) Slide move; (b) Tunnel move

number of branches (or bends) in the path between the two modules. The move takes $O(k)$ parallel steps, but in our uses k will always be a small constant.

We now proceed to describe new basic moves that form the basis of our reconfiguration algorithm.

3.1 Staircase Move

The *staircase move* transforms a rectangle of $k_1 \times k_2$ modules to one of dimensions $k_2 \times k_1$, both sharing the same lower-left corner C. Connectivity to the rest of the robot is maintained through the module at C, and thus that module cannot move. Without loss of generality, we can assume that $k_1 \geq k_2$; otherwise, we invert the sequence of operations described.

First, we move every row of modules to the right using a slide move with respect to the row immediately below, as in Fig. 5(b). Second, we move every column that does not touch the top or bottom border of the bounding box down using a slide move, as in Fig. 5(c). Finally, we move every row to the left using a slide move, as in Fig. 5(d). Note that the sliding motions of each step are executed in parallel. Also, each operation can be done at the atom-level, as was shown in Fig. 1. Thus the move works even if $k_2 = 1$.

If we require that the transformation between rectangles takes place within the bounding box of the source and target configurations, we can modify the above procedure without much difficulty. This modification is omitted in the present version of this paper.

3.2 Elevator Move

The *elevator move* transports a rectangle of modules by k units between two vertical strips. Fig. 6(a) shows the initial configuration in which a rectangle is to be transported vertically downward. First we detach the top half T of the rectangle from the bottom half B. Furthermore, B detaches from the vertical strip on the right. Let R be the rightmost vertical column of k atoms along the left strip, together with the atoms to the left of B. We detach R to its left, except at the very bottom, and detach R above, thus creating a corner with B. Then we contract R vertically, thereby pulling B downward half way. This is shown in Fig. 6(b), in which, however, we have let R be a vertical column of

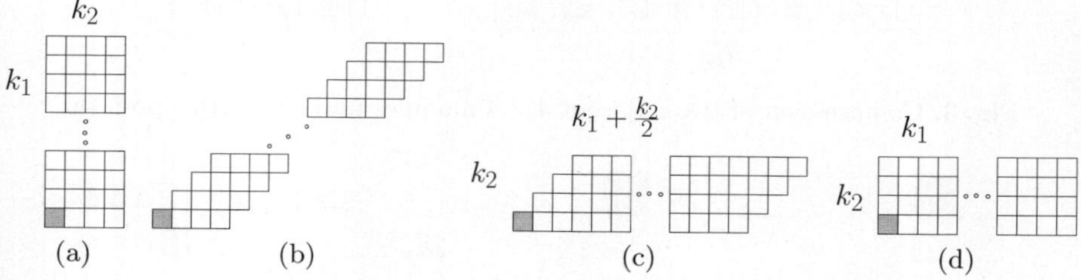

Fig. 5. Staircase move in three parallel steps. The shaded module maintains connectivity to the rest of the robot.

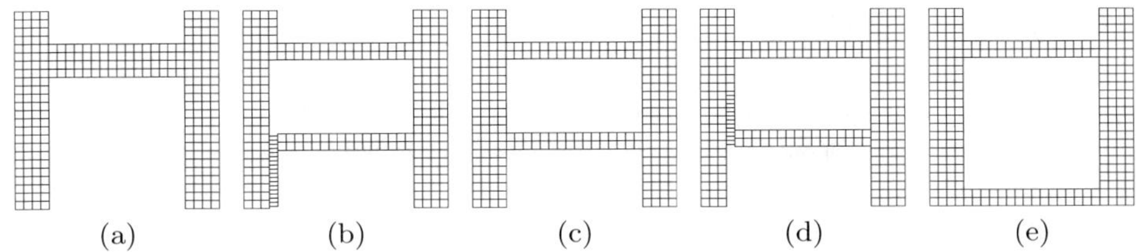

Fig. 6. Elevator move in $O(1)$ parallel steps

modules instead of atoms, due to the large width of the shape. Thus far, T has maintained the connectivity of the robot. Afterward, B attaches to the right vertical strip and detaches from R, which is now free to expand and re-attach to the top, as in Fig. 6(c). Now R detaches from the bottom and contracts upwards. It re-connects to B at the bottom, as in Fig. 6(d). In the last step, shown in Fig. 6(e), B detaches from the right side, and R expands, thereby moving B all the way to the bottom. At this point, B has reached its target position. It now assumes the role of maintaining connectivity, and the process can be repeated for T.

3.3   Corner Pop

Consider a rectangle R of $k_1 \times k_2$ module units, where without loss of generality $k_1 \leq k_2$. Let R be empty except for a strip V of modules on its left border and a strip H along the bottom. The strips form a corner, as shown in Fig. 7(a).

The *corner pop* moves the modules in R to the upper and right borders of R. During this corner pop, the modules at the top-left and bottom-right corners of R do not move. It is assumed that only these positions connect to modules outside R. Thus, this operation preserves the connectivity of the robot.

We first create two staircases of height $k_1/2$ at the two ends of H, as in Fig. 7(b). This shifts the middle of H upward. Next, we use the lower half of V to create a staircase of width $k_1/2$. Simultaneously, the rightmost staircase of H also moves so that it ends up on the right border of B, as in Fig. 7(c). We move the two remaining staircases upward, as in Fig. 7(d). Some simple cleaning up transforms this configuration into a symmetric canonical shape; see Fig. 7(e).

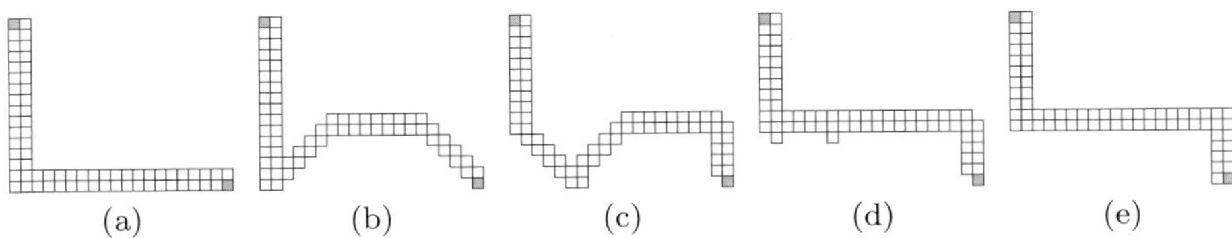

Fig. 7. Popping a corner in $O(1)$ parallel steps. The shaded modules maintain connectivity to the rest of the robot.

3.4   Parallel Tunnel Move

The *parallel tunnel move* takes as input a horizontal row H of modules together with, on the row immediately above, several smaller horizontal components that have no other connections. The top components are absorbed into H, after which H extends horizontally. Alternatively, the absorbed mass can be pushed out anywhere else on top of H, provided the target space is free. This move allows us to merge an arbitrary number of strips in the top row in $O(1)$ time.

The idea is to take all odd lattice positions along H and perform 1-tunnel moves, i.e., absorb modules from above and compress them under even positions. Then decompressing them all in parallel just expands H horizontally. Any modules remaining on top will shift over during the expansion, since they are attached to H. A gap will remain to the right of each such module, so we can repeat one more time to complete the move.

Fig. 8 illustrates half of the absorption of one module into H. Note that groups of 4 atoms move separately (they can be considered to be temporary smaller modules). As described, this procedure assumes that the bottom row is critically connected to other parts of the robot at one position, and absorbed modules are redirected away from that position. For 4×4 modules, this assumption is not required, but the minor implementation differences are omitted.

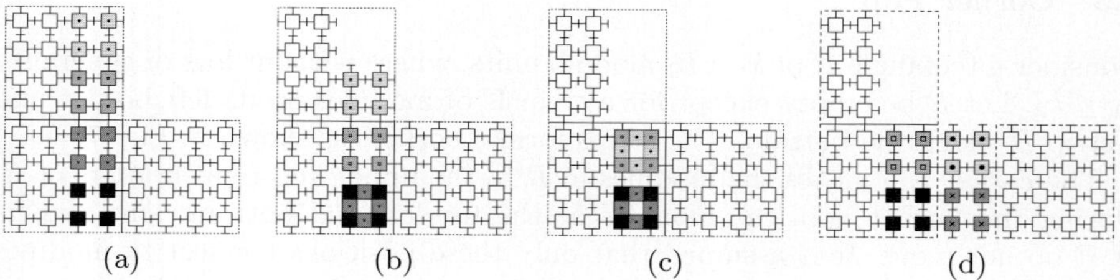

(a)   (b)   (c)   (d)

Fig. 8. Parallel tunnel move. Three 4×4 modules are involved.

4   Reconfiguration

In this section we show how to reconfigure a given robot to a canonical form with $O(\log n)$ parallel steps. Here we assume that the initial and final configurations of the robot consist of blocks of 32×32 atoms. However we will split blocks to use modules of 4×4 atoms in the intermediate configurations. Recall that the boundary has a width of four atoms.

Our divide-and-conquer algorithm proceeds as follows. Let the initial robot be placed on a grid of unit blocks (of 32×32 atoms). On this grid we construct a minimal square cell of side length 2^h that contains the initial robot (length is measured in block units). We recursively divide the cell into four subcells of length 2^{h-1}. As a base case, we take subcells of 2×2 blocks (i.e., containing 16×16 module lattice positions).

In parallel, we reconfigure each subcell within the same recursive depth, so that the resulting shape is easy to handle. Thus, by merging subcells, in $O(\log n)$

iterations we will have created a simple shape in our original square. Consider a cell M. We will use the inductive hypothesis that after merging its subcells, M will become a ring if there are enough modules, or sparse otherwise. Furthermore, if two points on the boundary of M were initially connected, the new configuration will ensure connectivity via the shortest path through its boundary.

In the base case of our induction, M has length 2. Thus we have to merge four subcells, each of which is empty or full. We will obtain a ring if there is at least one full subcell. One such subcell contains 64 modules, which suffice to cover the boundary of M. Reconfiguration can be done by tunneling each interior module iteratively (or by the lemmas that will follow). Thus our hypothesis is preserved.

Lemma 1. *Consider a cell M. If any subcell of M contained a backbone in the original configuration, then there are enough modules to create a ring in M. There are also enough modules if a path originally connected two subcell sides that belong to the boundary of M but are not adjacent.*

Proof. Consider the eight exterior sides of subcells of M as shown in Fig. 9(a). Let each of the sides M_i have length c (i.e., c modules fill the side of a subcell). The total number of modules in the boundary of M is $8c - 4$. A subcell backbone contains at least $8c$ modules and therefore suffices to cover the boundary.

Without loss of generality, suppose that a path begins on M_1 and ends at any side other than M_1, M_8, M_2. Then we have enough modules to make a ring in M, by similar counting as above. In fact to avoid having enough modules, such a path would have to remain within the lower two subcells. $\square$

Lemma 2. *Let S_1 and S_2 be adjacent sparse subcells at the top of cell M. In the original robot, there can be no path from the top border of M to the other subcells (see Fig. 9(b)).*

Proof. A path from the top to the middle of M in the initial robot would contain enough modules to make both S_1 and S_2 rings. By the pigeon-hole principle, one of the two subcells cannot be sparse. $\square$

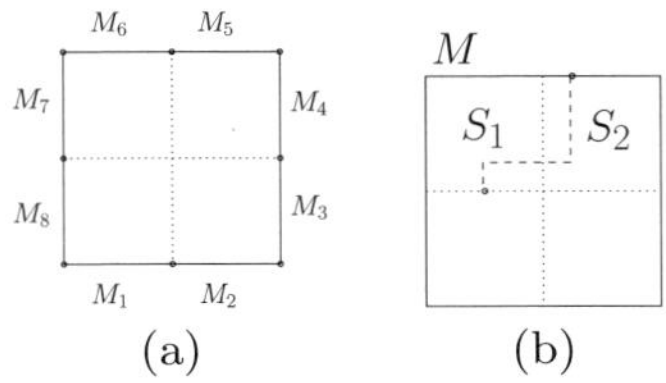

Fig. 9. Connectivity issues, in Lemmas 1 and 2

Lemma 3. *All side-branches along the common border of two cells that are rings or sparse can be merged into at most two pieces per side, with $O(1)$ moves. Furthermore each side-branch touches one end of the border.*

350   G. Aloupis et al.

Proof. If one cell is a ring then the other side can use it as a platform for a parallel tunnel move that will merge its side-branches into one piece. Otherwise, for each connected component of side-branches (of which there are at most two; one per corner) do the following.

Denote the two sides of the border by A and B. Absorb as much as possible from A to B by sliding modules from A across the border into vacant module lattice positions. Thus the component has one side-branch in B. Shift (parallel tunnel) the remainder of A towards the corner that the connected component attaches to, using B as a platform. Thus A becomes one side-branch. Now (either by a pop or by parallel-tunnel) bring back material from B to A to restore the original numbers in each cell. Thus each connected component consists of at most one side-branch from A and one from B. $\qquad\square$

Lemma 4. *Suppose B is a boundary side of a cell that has been processed according to Lemma 3. Let A be a branch that is in the near-boundary adjacent to B, and has no connectivity purpose. We can absorb A into B, or B can be filled, with $O(1)$ moves.*

Proof. By Lemma 3, B contains at most two side-branches, each attached to a corner. If no modules in B are adjacent to A, we can use a 1-tunnel to move one node (endpoint) of A into the position in B that is adjacent to the other node of A. Then the rest of A can slide into B. Otherwise, if A is adjacent to a side-branch in B, as in Fig. 10(a), we do the following. Absorb parts of A into empty positions of B, as in Fig. 10(b). Thus we create a side-branch B_1 which can be used as a platform to be extended by performing a parallel tunnel move on what remains of A. If the extension causes B_1 to reach a corner or join to another side-branch in B, then B is full; see Fig. 10(c). $\qquad\square$

For sparse cells, by repeatedly applying Lemma 4 and staircaising the remainder of A to the near-boundary side adjacent to B, we obtain the following:

Corollary 5. *If a branch A is positioned in the near-boundary of a sparse cell, either A can be fully absorbed into the boundary, or the cell will become a ring.*

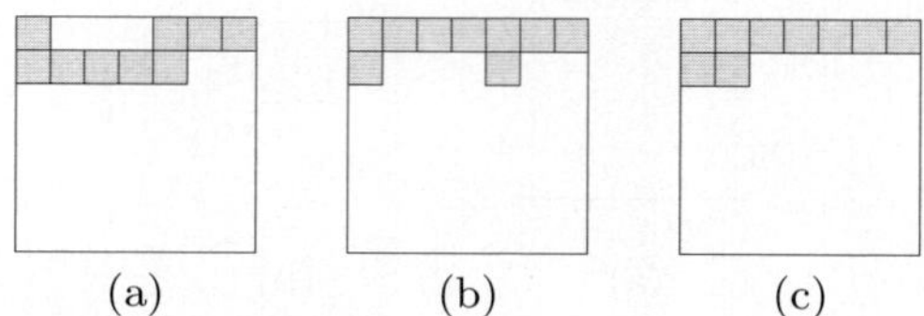

(a)   (b)   (c)

Fig. 10. Absorbing a near-boundary branch into the boundary of a cell

Let a *merged cell* contain four subcells that satisfy our induction hypothesis. That is, they are either rings or sparse, and connectivity is ensured via shortest paths along their boundaries. A merged cell becomes *well-merged* if it is reconfigured to satisfy the induction hypothesis.

Lemma 6. *Let M be a merged cell containing three or four subcell rings. Then M can become a ring using $O(1)$ moves. Thus M becomes well-merged.*

Proof. Omitted due to space restrictions.
Sketch: The outer structure of the desired ring is either in place or can be completed easily. Following this, all that remains is to organize/merge the interior modules of the subcells. □

Lemma 7. *If exactly two subcells of a merged cell M are rings, then M can become well-merged using $O(1)$ moves.*

Proof. If the two sparse subcells are adjacent, then there is no critical connectivity maintained through their common border, by Lemma 2.

Apply Corollary 5 to move side-branches in the sparse subcells to the boundary of M. There is only one module that possibly cannot be moved, in the case of two rings that exist in a diagonal configuration and must be connected. If a new ring is created, we apply Lemma 6. Now the only branches along interior borders of subcells belong to the two rings, with the possible exception of one module at the middle of M. We can use corner pops and/or staircase moves and Corollary 5 to move the interior ring sides to the boundary of M while maintaining connectivity. This happens regardless of the relative position of the rings or the presence of the extra module.

What remains is to maintain our shortest path requirement, if we still do not have a ring in M. In this case, by Lemma 1 we know that there was no initial backbone in M. Thus each connected component of robot within M "covers" at most one corner (in other words there is at least one module gap per side).

Note that the modules in the two subrings alone nearly suffice to create a ring in M. Four modules are missing. We can remove a strip of width 2 from positions where we wish to have a gap in the boundary of M, and use parallel-tunneling to position this material in the current gaps. Essentially we create a temporary ring of width 2. Then the remaining material can be moved. □

Lemma 8. *If exactly one subcell S of a merged cell M is a ring, then M can become well-merged using $O(1)$ moves.*

Proof. Without loss of generality let S be at the bottom-left of M. By Lemma 2, in the original robot there was no path from the top border of M leading to either of the bottom subcells. The same holds for the right border of M and the two left subcells. Therefore the two interior borders between the three sparse subcells do not preserve any connectivity. We may use Corollary 5 to move branches from those interior borders to the boundary of M. Finally we can do the same for the interior sides of S.

We may have to redistribute excess internal material from within S. If M has become a ring, this is easy and has been discussed previously. Otherwise, we can apply Corollary 5 to each full row of the internal ring structure. This can be required at most eight times before a ring is created.

Our shortest path connectivity requirement is preserved directly, by the fact that the internal borders where not necessary for connectivity. □

Lemma 9. *If no subcell of a merged cell M is a ring, then M can become well-merged using $O(1)$ moves.*

Proof. By Lemma 2, we know that in the original robot configuration no path existed from a side of M to either of the two subcells furthest from it. Therefore all disjoint subgraphs maintained connectivity between at most two adjacent external sides of subcells. More specifically, the first type of allowed path connects points that are separated by a corner of M but are also inside the same subcell. By induction we assume that these points are already connected along the external boundary of their subcell. The second type connects points that are on the same border side of M (possibly adjacent subcells). Again by induction we know that they are already connected along the boundary of M. Therefore our shortest path requirement is preserved.

All that remains is to remove excess material from inner borders of subcells. This material consists of one or two branches per border, each of which is connected to the boundary of M. These can be staircased and redistributed with our standard procedures.   □

Theorem 10. *Any source robot S can be reconfigured into any target robot T with $O(n \log n)$ atom operations in $O(\log n)$ parallel steps, if S and T are constructed with blocks of 32×32 atoms.*

Proof. Every cell retains the modules that it initially contained and does not interfere with the configuration of the robot outside the cell, until it is time to merge with its neighbors. A temporary exception to this occurs during Lemma 3. Therefore that step should be performed in a way so that no interference occurs (i.e., perform only this operation during one time step). At every time step, we merge groups of four cells, which by induction are either rings or sparse. By Lemmas 6–9, these four cells merge into a ring or sparse cell. Thus we construct a ring or sparse cell in $O(\log n)$ parallel time steps.

We show that the total number of operations is $O(n \log n)$. Each subcell containing m atoms can involve $O(m)$ parallel operations per time step. Because there are $O(1)$ time steps per level in the recursion, and all m_i sum to n, the total number of operations per recursion level is $O(n)$.

Now consider the bounding box B of S. We construct the smallest square B_2 of side length 2^h that contains S and has the same lower-left corner as B. Our recursive algorithm takes place within B_2. Now consider the last merge of subcells in our algorithm. The lower-left subcell L could not have contained S, because this would imply that $B_2 = L$. Therefore there must have been a path in S from the left side of B_2 leading to the two rightmost subcells (or from bottom to two topmost). This implies that S will become a ring (not sparse).

Because a ring of specific side length has a unique shape as a function of the number of modules it contains, the resulting ring in B_2 serves as a canonical form between S and T.   □

5   Discussion

The number of atoms in our modules and initial blocks can be reduced. By using 2×2 modules instead of 4×4, some of our basic operations become relatively complicated. For example, the staircase move cannot be implemented via sliding, but instead involves a form of parallel tunneling to break off strips that are one module wide, and then using those as carrying tools, etc. Corner pops also become particularly unattractive. Reducing the block size has the result that we can no longer rely only on rings and sparse cells to maintain the connectivity of any robot. We obtain a small set of orthogonal shortcut trees that must be taken into consideration when merging cells. We conjecture that reconfiguration can take place with 2×2 modules and no block restriction.

Our algorithm seems to be implementable in $O(n \log n)$ time. Each subcell contains a constant number of rectangular components, so determining their relative configuration and series of motions should require constant time. We also claim that our results extend to the case of labeled robots. This would involve a type of merge-sort using staircase moves, once a straight path of modules is constructed using our algorithm.

We have not determined if a similar result will hold in 3D, or if $O(\log n)$ steps are optimal. Such a lower bound can be given for labeled robots, by a simple Kolmogorov argument: there exist permutations that contain $\Theta(n \log n)$ bits of information. Each parallel move can be encoded in $O(n)$ bits (for each robot in order, which sides perform which operations), so we need $\Omega(\log n)$ steps.

References

1. Aloupis, G., Collette, S., Damian, M., Demaine, E.D., El-Khechen, D., Flatland, R., Langerman, S., O'Rourke, J., Pinciu, V., Ramaswami, S., Sacristán, V., Wuhrer, S.: Realistic reconfiguration of telecube and crystalline robots. In: The Eighth International Workshop on the Algorithmic Foundations of Robotics (to appear, 2008)
2. Aloupis, G., Collette, S., Damian, M., Demaine, E.D., Flatland, R., Langerman, S., O'Rourke, J., Ramaswami, S., Sacristán, V., Wuhrer, S.: Linear reconfiguration of cube-style modular robots. Computational Geometry - Theory and Applications. In: Tokuyama, T. (ed.) ISAAC 2007. LNCS, vol. 4835, pp. 208–219. Springer, Heidelberg (2007)
3. Butler, Z., Fitch, R., Rus, D.: Distributed control for unit-compressible robots: Goal-recognition, locomotion and splitting. IEEE/ASME Trans. on Mechatronics 7(4), 418–430 (2002)
4. Butler, Z., Rus, D.: Distributed planning and control for modular robots with unit-compressible modules. Intl. Journal of Robotics Research 22(9), 699–715 (2003)
5. Rus, D., Vona, M.: Crystalline robots: Self-reconfiguration with compressible unit modules. Autonomous Robots 10(1), 107–124 (2001)
6. Suh, J.W., Homans, S.B., Yim, M.: Telecubes: Mechanical design of a module for self-reconfigurable robotics. In: Proc. of the IEEE Intl. Conf. on Robotics and Automation, pp. 4095–4101 (2002)
7. Vassilvitskii, S., Yim, M., Suh, J.: A complete, local and parallel reconfiguration algorithm for cube style modular robots. In: Proc. of the IEEE Intl. Conf. on Robotics and Automation, pp. 117–122 (2002)

Squaring the Circle with Weak Mobile Robots

Yoann Dieudonné and Franck Petit

MIS Lab./Université de Picardie Jules Verne
Amiens, France

Abstract. The Circle Formation Problem (CFP) consists in the design of a protocol insuring that starting from an initial arbitrary configuration (where no two robots are at the same position), n robots eventually form a *regular n-gon* in finite time. None of the deterministic solutions proposed so far works for a system made of 4 or 3 robots. This comes from the fact that it is very difficult to maintain a geometric invariant with such a few number of robots, *e.g.*, the smallest enclosing circle, concentric cycles, properties of the convex hull, or a leader. As a matter of fact, due to the high rate of symmetric configurations, the problem was suspected to be unsolvable with 4 robots. In this paper, we disprove this conjecture. We present two non-trivial deterministic protocols that solves CFP with 4 and 3 robots, respectively. The proposed solutions do not require that each robot reaches its destination in one atomic step. Our result closes CFP for any number n (> 0) of robots in the semi-synchronous model.

Keywords: Distributed Coordination, (Uniform) Circle Formation, Mobile Robot Networks, Self-Deployment.

1   Introduction

Consider a distributed system where the computing units are *mobile weak robots* (*sensors* or *agents*), *i.e.*, devices equipped with sensors and designed to move in a two-dimensional plane. By weak, we mean that the robots are *anonymous*, *autonomous*, *disoriented*, and *oblivious*, *i.e.*, devoid of (1) any local parameter (such that an identity) allowing to differentiate any of them, (2) any central coordination mechanism or scheduler, (3) any common coordinate mechanism or common sense of direction, and (4) any way to remember any previous observation nor computation performed in any previous step. Furthermore, all the robots follow the same program (*uniform* or *homogeneous*), and there is no kind of explicit communication medium. The robots implicitly "communicate" by observing the position of the others robots in the plane, and by executing a part of their program accordingly.

In such a weak model, there has been considerable interest in the design of *deterministic* coordination protocols, *e.g.*, [14,8,12,13,7]. Among them, the *Circle Formation Problem* (CFP) consists in the design of a protocol insuring

S.-H. Hong, H. Nagamochi, and T. Fukunaga (Eds.): ISAAC 2008, LNCS 5369, pp. 354–365, 2008.

that starting from an initial arbitrary configuration (where no two robots are at the same position), n robots eventually form a circle with equal spacing between any two adjacent robots. In other words, the robots are required to form a *regular n-gon* in finite time.

First attempts to deterministically solve the CFP were presented in [2,3,1]. The CFP algorithm given in [2] is quite informal. The solutions in [3,1] work in the semi-synchronous model (SSM) [14] in which the cycles of all the robots are synchronized and their actions are atomic. They ensure only asymptotical convergence toward a configuration in which the robots are uniformly distributed on the boundary of a circle. In other words, the robots move infinitely often and never reach the desired final configuration.

The first solution leading n robots in a regular n-gon in finite time is proposed in [10]. The proposed protocol works in CORDA [11], a fully asynchronous model where the robot cycles are not required to be synchronized (as in SSM). CORDA being weaker than SSM, solutions designed in CORDA also work in SSM [12]. However, the solution in [10] works if $n \geq 5$ only. Moreover, if n is even, the robots may form a *biangular circle* in the final configuration, *i.e.*, the distance between two adjacent robots is alternatively either α or β. A deterministic CFP protocol for SSM is proposed in [5]. It works for a prime number of robots only. It assumes that every robot reaches its destination atomically — *i.e.*, no robot stops before reaching its destination. Note that this constraint is also required for the specific solution proposed in [5] for 3 robots. A general solution is given in [4] for SSM. It combines the solution in [10] and a method based on a concentric circles to eventually achieve a regular n-gon. The solution works for any number of robots, except 4, 6 and 8. Nevertheless, the solution in [4] assumes that every robot reaches its destination atomically. Finally, in [6], a deterministic CFP algorithm is given, in SSM, for any number n of robots, except 3 and 4. The robots are not assumed to reach their destination in one atomic step. The approach in [6] is based on a technique using tools from combinatorics on words and geometric properties of the *convex hull* formed by the robots. Following this work, both cases $n = 4$ and $n = 3$ remain open problems. Indeed, it is very difficult to maintain a geometric invariant with such a few number of robots, *e.g.*, the smallest enclosing circle, concentric cycles, properties of the convex hull, or a leader. As a matter of fact, due to the high rate of symmetric configurations, right now, the problem was suspected to be unsolvable with 4 robots.

In this paper, we first disprove this conjecture by presenting a non-trivial deterministic protocol that solves CFP for the case $n = 4$ (Section 3). The proposed solution does not require that each robot reaches its destination in one atomic step. Next (Section 4), we improve the solution in [5] for the case $n = 3$ by providing a protocol also making no assumption about the covered distance during a step. Since a cohort of n robots trivially always form a regular n-gon if $n \in \{1, 2\}$, this paper closes the circle formation problem for any number $n(> 0)$ of robots in SSM.

2   Preliminaries

In this section, we define the distributed system, give some definitions on geometric configurations, and specify the problem considered in this paper.

Distributed Model. We adopt the semi-synchronous model (SSM) in [14]. The *distributed system* considered in this paper consists of n robots. Each robot r_i, viewed as a point in the Euclidean plane, moves on this two-dimensional space unbounded and devoid of any landmark. Any robot can observe, compute and move with infinite decimal precision. The robots are equipped with sensors enabling to detect the instantaneous positions of the other robots in the plane. The robots agree neither on the orientation of the axes of their local coordinate system nor on the unit measure. They are *uniform* and *anonymous*, i.e, they all have the same program using no local parameter (such that an identity) allowing to differentiate any of them. They communicate only by observing the positions of the others and they are *oblivious*, i.e., none of them can remember any previous observation nor computation performed in any previous step.

Time is represented as an infinite sequence of time instants $t_0, t_1, \ldots, t_j, \ldots$ Let $P(t_j)$ be the multiset of the positions in the plane occupied by the n robots at time t_j $(j \geq 0)$. For every t_j, $P(t_j)$ is called the *configuration* of the distributed system in t_j. $P(t_j)$ expressed in the local coordinate system of any robot r_i is called a *view*. At each time instant t_j $(j \geq 0)$, each robot r_i is either *active* or *inactive*. The former means that, during the computation *step* (t_j, t_{j+1}), using a given algorithm, r_i computes in its local coordinate system a position $p_i(t_{j+1})$ depending only on the system configuration at t_j, and moves towards $p_i(t_{j+1})$—$p_i(t_{j+1})$ can be equal to $p_i(t_j)$, making the location of r_i unchanged. In the latter case, r_i does not perform any local computation and remains at the same position. In every single activation, the distance traveled by any robot r is bounded by σ_r. So, if the destination point computed by r is farther than σ_r, then r moves toward a point of distance at most σ_r. This distance may be different between two robots. The concurrent activation of robots is modeled by the interleaving model in which the robot activations are driven by a *fair scheduler*. At each instant t_j $(j \geq 0)$, the scheduler arbitrarily activates a (non empty) set of robots. Fairness means that every robot is infinitely often activated by the scheduler.

Basic Definitions and Properties. Given a set P of $n \geq 2$ points $p_1, p_2, \cdots, p_n$ in the plane, the convex hull of P, denoted $H(P)$ (H for short), is the smallest polygon such that every point in P is either on an edges of $H(P)$ or inside it. Informally, it is the shape of a rubber-band stretched around $p_1, p_2, \cdots, p_n$. The convex hull is unique and can be computed in $O(n \log n)$ time [9].

A convex hull, H, is called a (convex) quadriliteral or triangle if H forms a polygon with four or three, respectively, sides (also called edges) and vertices (or corners). In the sequel, we consider only convex quadriliterals and triangles. Let O be the *center* of H defined as follows: (1) If H forms a quadrilateral, then O is the unique point where the diagonals of H cross each other; (2) If H forms a triangle, then O is the unique point which is equidistant from each of the three sides of H. A quadrilateral is said to be *perpendicular* if and only if its diagonals are perpendicular. Otherwise, it is called a *non-perpendicular* quadrilateral.

A triangle is said to be *equilateral* if all its sides are of equal length. An *isosceles* triangle has two sides of equal length. A triangle having all sides of different lengths is said to be *scalene*. A *trapezoid* is a quadrilateral with at least one pair of opposite sides parallel. An *isosceles trapezoid* is a trapezoid whose the diagonals are of equal length. A *parallelogram* is a quadrilateral with both pairs of opposite sides parallel. A *rectangle* is defined as a parallelogram where all four of its angles are right angles. A *square* is a rectangle perpendicular quadrilateral.

The Circle Formation Problem. The problem considered in this paper, called CFP (*Circle Formation Problem*), consists in the design of a deterministic distributed protocol which arranges a group of 4 (respectively, 3) mobile robots with initial distinct positions into a *square* (resp., *equilateral triangle*) in finite time. The side length of the expected square (equilateral triangle) must be strictly greater than zero.

3   Four Robots

In this section, we present our algorithm that leads 4 mobile robots to eventually form a square. We refer to Figure 1 to explain our scheme.

Consider the convex hull H formed by the robots on the plane. If the 4 robots belong to the same line L, then H is reduced to the segment of line linking the 4 points (Figure 1, Case sL). Otherwise (the 4 robots are not aligned), there are only two possible forms for H: H forms either a quadrilateral or a triangle. If H forms a triangle, then there is a robot r which is located either inside H (Case nD-T) or between two of the three corners of the triangle (Cases nP-D and P-D). In the latter case, three out of the four robots q, r, and s are aligned on a line L (r belonging to the segment $[q, s]$), whereas the fourth robot t does not. Such a configuration is called a (*arbitrary*) *delta*. If the line L' passing through r and t is perpendicular to L, then the delta is said to be *perpendicular* (Case P-D).

3.1   Overview of the Solution

Let us first describe the main idea of our scheme. It is made of two steps: (1) Starting from an arbitrary configuration, move the robots to eventually form an arbitrary perpendicular quadrilaterals; (2) Starting from an arbitrary perpendicular quadrilateral, the robots eventually form a square.

Indeed, when the convex hull H forms an arbitrary perpendicular quadrilateral (Figure 1, Case PQ), the diagonals of H are perpendicular in O. The system eventually forms a square by sliding the closest robots away from O along their diagonal until they reach the positions which are at the same distance from O as

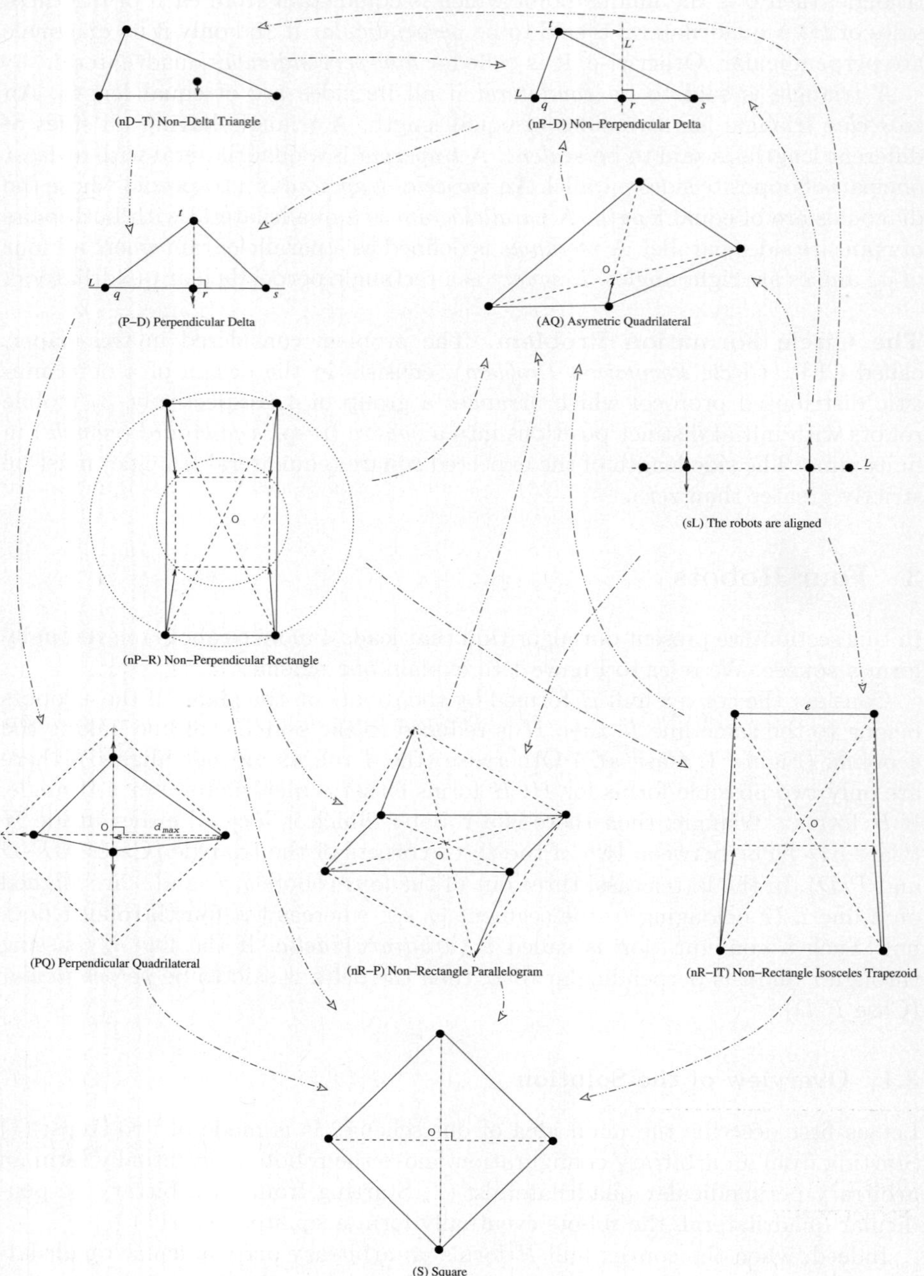

Fig. 1. General Scheme with 4 robots

the farthest ones (d_{max} in Figure 1). To reach such an arbitrary perpendicular quadrilateral, we aim to bring the system into an arbitrary delta. Starting from a perpendicular delta, Case P-D in Figure 1 (q, r, and s are aligned one a line L, the line L' passing through r and t is perpendicular to L), in one step, the system becomes an arbitrary perpendicular quadrilateral by sliding r on L' in the opposite direction of t. Starting from a non-perpendicular delta (Case nP-D), the system eventually becomes a perpendicular delta by moving t along L'', the line passing through t which is parallel to L, until L' and L become perpendicular.

Clearly, the above scheme does not cover all the possible cases. In particular, it gives no details about the "arbitrary" configurations considered in the above first item. In fact, we can detail the different classes of such "arbitrary" configurations and the corresponding moves as follows:

1. *The convex hull H forms an arbitrary quadrilateral which is not perpendicular, a rectangle, an isosceles trapezoid, nor a parallelogram.* In the sequel, such a configuration is called an *asymmetric* quadrilateral (Case AQ, in Figure 1). In that case, we will show that there always exists a robot r which is either the unique closest or the unique farthest robot from the center O of the quadrilateral. By moving either r or the opposite robot (w.r.t. to O) along its diagonal toward O, the moving robot eventually reaches O. By the way, it crosses one side of the triangle formed by the 3 other robots. The system then becomes a non-perpendicular delta, and from this point on, adopt the above behavior.

2. *The convex hull H of the 4 robots forms a symmetric non-perpendicular quadrilateral which is not reduced to a line segment.* In that case, H forms either an isosceles trapezoid (Case nR-IT) or a parallelogram (Case nR-P) — note that H can be a non-perpendicular rectangle (Case nP-R) if it is both an isosceles trapezoid and a parallelogram. In these cases, the robots move trying to form a square in one step. If they move synchronously and reach their respective positions to form a square, then they are done in one step. Otherwise, the symmetry is broken in one or two steps. Details about this cases are given below.

3. *The convex hull H forms a triangle which is not an arbitrary delta.* So, one of the four robots is located inside the triangle (Case nD-T in Figure 1). In that case, the robot r inside the triangle moves toward the closest side of the triangle — if r is at the center of the triangle, then its arbitrarily chooses one side to move on. Again, the system reaches a configuration where the cohort of robots form a delta.

4. *The 4 robots are aligned on the same line L (Case sL).* In that case, both robots r_1 and r_2 located between the two extremities of the segment formed by the 4 robots are able to move perpendicularly to L. With respect to the asynchrony, there are 5 possible resulting configurations: either a non-perpendicular quadrilateral (possibly, an isosceles trapezoid or a parallelogram) or a triangle (possibly, a non-perpendicular delta).

3.2 Details

In this subsection, we describe the basic rules of the algorithm. The main program is given in Algorithm 1. It corresponds to the global scheme shown in Figure 1. The procedures are given in Algorithms 2 to 8.

Algorithm 1. Main Program.

```
1.01    Let H be the convex hull of the four robots;
1.02    if H is not a square
1.03    then  if the four robots are located on the same line L
1.04          then  Execute Procedure φSL;
1.05          else  if H forms a triangle
1.06                then  Execute Procedure φT;
1.07                else  // The robots form a non-square quadrilateral
1.08                      if H is a perpendicular quadrilateral
1.09                      then  Execute Procedure φPQ;
1.10                      else  // The robots form a non-perpendicular quadrilateral
1.11                            if H is a non-rectangle isosceles trapezoid
1.12                            then  Execute Procedure φIT;
1.13                            elseif H is a non-rectangle parallelogram
1.14                            then  Execute Procedure φP;
1.15                            elseif H is a rectangle
1.16                            then  Execute Procedure φR;
1.17                            else  // H is an asymmetric quadrilateral
1.18                                  Execute Procedure φAQ;
```

Theorem 1. *Algorithm 1 deterministically solves the circle formation problem for 4 weak mobile robots.*

The correctness of Theorem 1 is shown in the next subsections (Lemmas 1 to 7). Due to the lack of space, the formal proofs are omitted.

Starting from a Perpendicular Quadrilateral. Each robot executes Procedure ϕ_{PQ} described in Algorithm 2 — also refer to Figure 1, Case PQ.

Algorithm 2. Procedure ϕ_{PQ}.

```
2.01    Let d be the greatest distance from O to a robot;
2.02    if my distance from myself to O is less than d
2.03    then Move away from O along my diagonal toward a position p located at distance d from O;
```

Clearly, by executing ϕ_{PQ}, H remains a perpendicular quadrilateral forever. If the perpendicular quadrilateral formed by H is a square, then for every robot r, the distance from r to O is equal to d. So, in that case, all the robots remain idle (and form a square) forever. Furthermore, by executing Procedure ϕ_{PQ}, every robot that is located at a distance $d' < d$ moves away from O on its diagonal of a distance $d'' \leq (d - d')$. So, by fairness, for each robot r, the distance from r to O is eventually equal to d. This leads to the following lemma:

Lemma 1. *In every execution starting from a configuration where H forms a perpendicular quadrilateral, both following properties hold:* (1) *The four robots eventually form a square;* (2) *whenever the four robots form a square, then they form a square forever.*

Starting from a Triangle. The case of the triangle is considered by Procedure ϕ_T, shown in Algorithm 3 — refer to Figure 1, Cases nD-T, nP-D, and P-D.

Algorithm 3. Procedure ϕ_T.

3.01	**if** H forms a delta // One out of the 4 robots is located on one side of H
3.02	**then** **if** H forms a perpendicular delta
3.03	**then** **if** I am the robot located between two corners of H
3.04	**then** Move outside H, perpendicularly to my side;
3.05	**endif**
3.06	**elseif** I am the robot which is not on the same line L than the three other robots
3.07	**then** Let L' be the line passing through me which is parallel to L;
3.08	Let L'' be the perpendicular line to L passing through the median robot on L;
3.09	Let p be the intersection of lines L' and L'';
3.10	Move toward p;
3.11	**endif**
3.12	**elseif** I am the robot which is located inside the triangle
3.13	**then** Let s be an arbitrary side of H which is the closest to myself;
3.14	Let p be the intersection of s and the perpendicular line to s passing through myself;
3.15	Move toward p;

Lemma 2. *In every execution starting in a configuration where the four robots form a triangle, both following properties hold:* (1) *The four robots eventually form a square;* (2) *whenever the four robots form a square, then they form a square forever.*

Starting from a non-perpendicular quadrilateral. In this subsection, we consider quadrilaterals which are not reduced to the segment, *i.e.*, we assume that the four corners of the quadrilateral are not aligned on a unique line L. (We consider this latter case later.) Let us first state the two following general properties — whether the quadrilateral is perpendicular or not:

Property 1. Let H be a (convex) quadrilateral. The two following properties are equivalent:

1. No corner c of H exists such that the distance from c to the center O of H is different than any distance from O to one of the three other corners;
2. H forms an isosceles trapezoid or a parallelogram.

Property 2. Given a quadrilateral H, if there exists a corner c of H such that the distance from c to the center O of H is different than the distance from O to any of the three other corners, then there exists a corner c' of H such that c' is either the unique farthest corner from O or the unique closest corner from O.

The case where the quadrilateral is perpendicular is discussed in Subsection 3.2. So, in the sequel of this subsection, we will consider *non-perpendicular* quadrilaterals only — we omit to mention the term "non-perpendicular". We first consider the case where the convex hull forms an asymmetric quadrilateral. Next, we discuss three symmetric cases: The non-rectangle isosceles trapezoid, the non-rectangle parallelogram, and the non-square rectangle. Property 1 ensures that we consider all the non-perpendicular quadrilaterals.

Starting from an asymmetric quadrilateral. The aim of Procedure ϕ_{AQ}, shown in Algorithm 4, is to reach a configuration in which the four robots form a delta — refer to Figure 1, Case AQ. Both Properties 1 and 2 ensure that if the four robots form an asymmetric quadrilateral, then there exists a robot r which is either the unique closest or the unique farthest robot from O. In the former case, r moves along its diagonal until it reaches O. While r does not reach O, it remains the closest robot from O since the other robots are not allowed to move. So, r eventually reaches O and, in this way, the four robots form a delta. In the latter case, r (the farthest robot) is not allowed to move. So, while the system remains in an irregular quadrilateral, r remains the farthest robot from O. Let r' be the the opposite robot w.r.t. O and r. Since r' is the only robot allowed to move toward O, r' eventually reaches O and, thus, the four robots form a delta. So, by the fairly repeated execution of Procedure ϕ_{AQ}:

Lemma 3. *In every execution starting from an asymmetric quadrilateral, the four robots eventually form a delta.*

Algorithm 4. Procedure ϕ_{AQ}.

4.01	**if** there exists a unique robot r which is closer from O than any other robot
4.02	**then** Let r be this robot
4.03	**else** Let r be the robot which is on the same diagonal than the farthest robot from O;
4.04	**endif**
4.05	**if** I am r **then** Move along my diagonal toward O;

Starting from a non-rectangle isosceles trapezoid. In this case, we first consider the smallest edge e among the two opposite parallel edges. Then, as illustrated in Figure 1, Case $nR\text{-}IT$, we build the unique square such that: (1) e is one of square's edge; (2) all the vertices of the square are located inside or over the isosceles trapezoid. Each of both robots which are located on the opposite side of e can move toward the closest vertex from it. The corresponding procedure, Procedure ϕ_{IT}, is is shown in Algorithm 5.

Algorithm 5. Procedure ϕ_{IT}.

5.01	Let e be the smallest edge e among the two opposite parallel edges;
5.02	Let p_1 and p_2 be the extremities of e;
5.03	Let p_3 and p_4 be the points such that p_1, p_2, p_3, and p_4 form a square;
5.04	**if** I am not located on either p_1 or p_2
5.05	**then** Move toward the closest position of me in $\{p_3, p_4\}$;

Lemma 4. *In every execution starting from a non-rectangle isosceles trapezoid, the four robots eventually form either an asymmetric quadrilateral, or a square.*

Starting from a non-rectangle parallelogram. In that case, the robots consider the unique square whose one of the two diagonals corresponds to the longest diagonal of the parallelogram. Then, each of both robots which are not located on the vertices of the square can move toward the closest vertex from it — refer to Figure 1, Case $nR\text{-}P$. Procedure ϕ_P is is shown in Algorithm 6.

Algorithm 6. Procedure ϕ_P.

6.01	Let d be the longest diagonal of H;
6.02	Let p_1 and p_2 be the extremities of d;
6.03	Let p_3 and p_4 be the points such that p_1, p_2, p_3, and p_4 from a square;
6.04	**if** I am not located on either p_1 or p_2
6.05	**then** Move toward the closest position of me among p_3 and p_4;

Lemma 5. *In every execution starting from a non-rectangle parallelogram, the four robots eventually form either an asymmetric quadrilateral, or a square.*

Starting from a non-square rectangle. Since the four robots form a rectangle, all of them are located on the boundary of a same circle C centered in 0. So the robots can compute the unique square having the following characteristics:
1. The vertices are located on the circumference of a circle centered in 0 and whose the radius is half the radius of C;
2. The square's edges are parallel to the non-square rectangle edges.

Once the unique square is computed, each robot can move toward their closest vertex from it, as depicted in Figure 1, Case nS-R. The corresponding procedure, Procedure ϕ_R, is is shown in Algorithm 7.

Algorithm 7. Procedure ϕ_R.

7.01	Let d be the diagonal size of H;
7.02	Let p_1, p_2, p_3, and p_4 be the corners of the unique square S and located on the circle
7.03	centered in 0 and of radius $\frac{d}{2}$ s.t. the edges of S are parallel to the edges of H;
7.04	Move toward the closest position of me in $\{p_1, p_2, p_3, p_4\}$;

Lemma 6. *In every execution starting from a non-square rectangle, the four robots eventually form either an asymmetric quadrilateral, a non-rectangle isosceles trapezoid, a non-rectangle parallelogram, or a square.*

Starting from a configuration where all the robots are located on the same line L. Let r_m be one of the two middle robots w.r.t. the segment formed by the 4 robots. Then, r_m considers the line L'_m which is perpendicular to L and passing through it — refer to Figure 1, Case sL. Then, r_m moves toward any point located on L' which does not belongs to L. The corresponding procedure, Procedure ϕ_{SL}, is is shown in Algorithm 8.

Algorithm 8. Procedure ϕ_{SL}.

8.01	**if** I am not an extremity of the segment formed by the four robots
8.02	**then** Let L' be the line L' passing through myself which is perpendicular to L;
8.03	Move toward an arbitrary position on L' which is not on L;

Lemma 7. *In every execution starting from a configuration where the robots are on the same line L, after one step, the convex hull H of the four robots is either a non-square quadrilateral or a triangle which is not a perpendicular delta.*

4   Three Robots

As for the case $n = 4$, consider the convex hull H formed by the robots on the plane. If the 3 robots belong to the same line L, then H is reduced to the segment of line linking the 3 points. Otherwise, H forms a non-aligned triangle. (In the following, we will omit the term "non-aligned".)

Let us consider the three following cases:

1. *The three robots form a non-equilateral isosceles triangle.* Let r be the unique robot which is placed at the unique angle different from the two others robots s and t. Let p be the position of L, the perpendicular bisector of $[s, t]$, such that p, s, and t form an equilateral triangle. Since H form an isosceles triangle, r belongs to L. So, it can move along L toward p. Clearly, while r does not reach p, it remains the single robot allowed to move. By fairness, the equilateral triangle is formed in finite time.

2. *The three robots are on the same line L.* Let s and t be the two robots located at the extremities of the segment formed by the three robots. Let r be the median robot and $d(s, r)$ $(d(t, r))$ denotes the distance between s and r (t and r). If $d(s, r) = d(t, r)$, then r can move on any position on the perpendicular bisector of $[s, t]$. After one step, the robots form an isosceles triangle, and the system behaves as in Case 1. If $d(s, r) \neq d(t, r)$, then r move toward the position p such that $d(s, p) = d(t, p)$. By fairness, r reaches p in finite time (Case 2).

3. *The three robots form a scalene triangle.* Since the three robots form a scalene triangle, the three internal angles are all different. Let r be the robots corresponding to the greatest internal angle. Then, r can move toward the intersection between the opposite side formed by the two others robots and the line passing through r which is perpendicular to the opposite side of the triangle. While the robots are not on the same line, r remains the only robots allowed to move because its internal angle increases whereas the two others internal angles decrease. By fairness, the three robot are eventually on the same line (Case 2).

The following theorem follows from the above discussion:

Theorem 2. *Three robots can deterministically solve the circle formation problem in finite time.*

5   Conclusion

We closed the circle formation problem for any number $n(n > 0)$ of robots in SSM. We proposed two non-trivial deterministic protocols solving CFP for 4 and 3 robots, respectively. The proposed solutions do not require that each robot reaches its destination in one atomic step. In a future work, we would like to address and solve the problem for any number of robots in CORDA.

References

1. Chatzigiannakis, I., Markou, M., Nikoletseas, S.: Distributed circle formation for anonymous oblivious robots. In: 3rd Workshop on Efficient and Experimental Algorithms, pp. 159–174 (2004)
2. Debest, X.A.: Remark about self-stabilizing systems. Communications of the ACM 38(2), 115–117 (1995)
3. Defago, X., Konagaya, A.: Circle formation for oblivious anonymous mobile robots with no common sense of orientation. In: 2nd ACM International Annual Workshop on Principles of Mobile Computing (POMC 2002), pp. 97–104 (2002)
4. Dieudonné, Y., Labbani, O., Petit, F.: Circle formation of weak mobile robots. In: Datta, A.K., Gradinariu, M. (eds.) SSS 2006. LNCS, vol. 4280, pp. 262–275. Springer, Heidelberg (2006)
5. Dieudonné, Y., Petit, F.: Circle formation of weak robots and Lyndon words. Information Processing Letters 104(4), 156–162 (2007)
6. Dieudonné, Y., Petit, F.: Swing words to make circle formation quiescent. In: Prencipe, G., Zaks, S. (eds.) SIROCCO 2007. LNCS, vol. 4474, pp. 166–179. Springer, Heidelberg (2007)
7. Dieudonné, Y., Petit, F.: Deterministic leader election in anonymous sensor networks without common coordinated system. In: Tovar, E., Tsigas, P., Fouchal, H. (eds.) OPODIS 2007. LNCS, vol. 4878, pp. 132–142. Springer, Heidelberg (2007)
8. Flocchini, P., Prencipe, G., Santoro, N., Widmayer, P.: Distributed coordination of a set of autonomous mobile robots. In: IEEE Intelligent Vehicule Symposium (IV 2000), pp. 480–485 (2000)
9. Graham, R.L.: An efficient algorithm for determining the convex hull of a finite planar set. Information Processing Letters 1, 132–133 (1972)
10. Katreniak, B.: Biangular circle formation by asynchronous mobile robots. In: Pelc, A., Raynal, M. (eds.) SIROCCO 2005. LNCS, vol. 3499, pp. 185–199. Springer, Heidelberg (2005)
11. Prencipe, G.: Corda: Distributed coordination of a set of autonomous mobile robots. In: Proceedings of the Fourth European Research Seminar on Advances in Distributed Systems (ERSADS 2001), pp. 185–190 (2001)
12. Prencipe, G.: Distributed Coordination of a Set of Autonomous Mobile Robots. PhD thesis, Dipartimento di Informatica, University of Pisa (2002)
13. Prencipe, G., Santoro, N.: Distributed algorithms for autonomous mobile robots. In: Sha, E., Han, S.-K., Xu, C.-Z., Kim, M.-H., Yang, L.T., Xiao, B. (eds.) EUC 2006. LNCS, vol. 4096. Springer, Heidelberg (2006)
14. Suzuki, I., Yamashita, M.: Distributed anonymous mobile robots - formation of geometric patterns. SIAM Journal of Computing 28(4), 1347–1363 (1999)

Evaluation of General Set Expressions

Ehsan Chiniforooshan[1,*], Arash Farzan[2], and Mehdi Mirzazadeh[2]

[1] Department of Computer Science and Software Engineering, Concordia University
[2] School of Computer Science, University of Waterloo

Abstract. We consider the problem of evaluating an expression over sets. The sets are preprocessed and are therefore sorted, and the operators can be any of union, intersection, difference, complement, and symmetric difference (exclusive union). Given the expression as a formula and the sizes of the input sets, we are interested in the worst-case complexity of evaluation (in terms of the size of the sets). The problem is motivated by document retrieval in search engines where a user query translates directly to an expression over the sets containing the user-entered words. Special cases of of this problem have been studied [7,6] where the expression has a restricted form. In this paper, we present an efficient algorithm to evaluate the most general form of a set expression. We show a lower bound on this problem for expressions of the form E_1, or $E_1 - E_2$ where E_1 and E_2 are expressions with union, intersection, and symmetric difference operators. We demonstrate that the algorithm's complexity matches the lower bound in these instances. We, moreover, conjecture that the algorithm works optimally, even when we allow difference and complement operations in E_1 and E_2.

1 Introduction

We consider the problem of evaluating an expression over sets with the most general form of binary operations consisting of union, intersection, difference, symmetric difference and complements. Motivated by its application in databases and search engines, we assume the sets are preprocessed and are therefore stored in the sorted order. We are interested in the worst-case complexity of evaluating the expression sensitive to the sizes of the input sets in the comparison model. We assume there are no repeated sets in the expression as otherwise, there is an easy reduction from the satisfiability problem which implies the NP-completeness of the evaluation problem.

The problem arises in evaluating search queries in text database systems; search engines maintain for each word w a set $S(w)$ consisting of all document IDs that contain w [3,11,14]. A query such as "dinner AND recipe − beef", translates directly into evaluating the expression $(S(\mathsf{dinner}) \cap S(\mathsf{recipe})) - S(\mathsf{beef})$. We note that although complicated long queries are not common from a live user, the queries and their corresponding expressions can be extremely complicated if they are automatically generated [10].

---

* Partly supported by Canada Research Chairs Program.

S.-H. Hong, H. Nagamochi, and T. Fukunaga (Eds.): ISAAC 2008, LNCS 5369, pp. 366–377, 2008.

Different variations of the problem have been studied before. The simplest case which is finding intersection or union of two sets is equivalent to the problem of merging two ordered sets of sizes m, n was studied by Hwang and Lin [9]. It gives an algorithm with a running time that matches the information theoretic lower bound $\lceil \log \binom{m+n}{n} \rceil$. They assumed sorted arrays as the input format of the sets, and as the running time can be sublinear, the output format is given as a list of cross pointers in the arrays. Brown and Tarjan [4,5] and Pugh [13] showed how more sophisticated data structures such as AVL-trees, B-trees, or skip-lists can be used as the input/output format.

Demaine, López-Ortiz, and Munro [7] studied the more general case where there are more than two sets. The expressions they considered is either the union of a number of sets or the intersection of them. They give an adaptive algorithm for this problem by defining a measure of difficulty for any instance and presenting an algorithm with a running time according to the difficulty of the given instance. Their algorithm was shown to be optimal by Barbay and Kenyon [1]. The adaptive approach was extended by Mirzazadeh [12] to expressions consisting of unions and intersections. The authors studied the worst-case complexity to evaluate an expression consisting of union and intersection operations [6] and proved matching upper and lower bounds for the worst-case complexity in the comparison model. Recently, evaluation of expression with the same form was considered in a non-comparison based model [2]. In this paper, we study the evaluation of the most general form of set expressions containing all possible binary operators of union, intersection, difference, and symmetric difference, as well as the complement operation in the comparison model.

We first present an efficient algorithm which evaluates such expressions. We prove that the algorithm is optimal by proving a lower bound in the preceding section for expressions of the form E_1, or $E_1 - E_2$ where E_1 and E_2 are expressions consisting of union, intersection, and symmetric difference operators. We, moreover, conjecture that the algorithm works optimally, even when there are difference and complement operations in E_1 and E_2.

2   Definitions and Preliminaries

In this work, we use a slightly modified version of B$^+$-trees, which we define as follows.

Definition 1. *A* partly-expanded B$^+$-tree T *is a B^+-tree in which for some internal nodes u, the subtree rooted at u is replaced with the sorted list of elements that are in that subtree. The* size *of T is the number of elements in T.*

Consequently, a B$^+$-tree or a single sorted array are special forms of a partly-expanded B$^+$-tree. Hence, we choose partly-expanded B$^+$-trees as our algorithms input/output format. This choice enables us to support the cases where the input sets are either sorted arrays, or B$^+$-trees.

An *input I* is formally defined as a pair (T, Γ), where T and Γ are as follows. T is a *set expression tree* representing the expression: every internal node v is assigned one of the operators we support (union, intersection, delta, and minus operators), namely $\pi(v)$, and each leaf v of T corresponds to an input set and is assigned an integer $\mathsf{size}(v)$. Every minus or delta internal node has exactly two children; other internal nodes have at least two children, each. Also, each node is either a *normal* or a *complement* node. We call T the *signature* of the input I. Γ is an *assignment function* that assigns a partly-expanded B$^+$-tree of size $\mathsf{size}(v)$ to each leaf v.

The *result of a node v in I*, denoted by $I(v)$, is a set defined as follows. For a leaf v, depending on v being normal or complement, $I(v)$ is the set of elements in $\Gamma(v)$ or its complement. For a normal (a complement) internal node v, $I(v)$ is the result of (the complement of the result of) the application of the operator $\pi(v)$ on results of the children of v.

It is easy to see that one can change the tree by propagating the complements up to the root such that we end with an *equivalent* expression tree: i.e. the result of the entire tree remains unaffected (e.g. trees corresponding to expressions $\overline{A} \cap B$, $\overline{A} - \overline{B}$, and $\overline{A \cup \overline{B}}$). A tree is in *least-complement form* if aside form its root, every node is a normal node. If there is no complement node in the tree, it is *complement-free*. It is easy to see that every tree has an equivalent in least-complement form and so, for every tree T we can obtain a complement-free tree which is either equivalent to T or the complement of T. Thus, in this paper, we will focus on complement-free trees.

In this paper we focus on the *comparison-based algorithms* which have oracle access to Γ: the algorithm reads the signature of the input and can later submit queries of the form (x, y) to the oracle, where x and y are elements of the input sets. Then the algorithm is told whether x is less than, equal to, or greater than y according to Γ. In such situations we say x and y are *touched* by the algorithm.

For a complement-free expression tree T, we define functions ψ^* and ψ over the set of nodes of T as follows. $\psi(v)$ is the maximum potential size of $\Gamma(v)$: For a leaf v is defined as $\psi(v) = \mathsf{size}(v)$. If v is an internal node with minus operator, $\psi(v) = \psi(u)$, for u the left child of v. For other nodes v with children $u_1, \ldots, u_k$, we define $\psi(v) = \min_{i=1}^{k} \psi(u_i)$ when v is an intersection node, and $\psi(v) = \sum_{i=1}^{k} \psi(u_i)$, when v is a union or delta node. For every node v we define $\psi^*(v) = \min \psi(u)$, where the minimum is taken over all ancestors u of v, including v itself. Intuitively, $\psi^*(v)$ is the maximum "contribution" of $\Gamma(v)$ to the result of the whole expression. Note that the values of ψ and ψ^* for all nodes of an expression tree T can be evaluated in time $O(|V_T|)$, where V_T denotes the set of nodes of T. Moreover, we assume that all the internal nodes of T have exactly two children and their children are ordered in the following way: for every internal node v,

1. If $\pi(v) \in \{\cup, \Delta\}$, the children of v are sorted in a non-increasing order from left to right based on their ψ^*, i.e. the left child has the maximum ψ^*.
2. If $\pi(v) = \cap$, the children of v are sorted in a non-decreasing order from left to right based on their ψ^*, i.e. the left child has the minimum ψ^*.

3. if $\pi(v) = -$ and the corresponding expression is $A - B$, the subtree corresponding to A is the left child of v and the subtree corresponding to B is the right child of v.

We will need the following definitions in the paper. For any node v of a tree T, we use $p(v)$ to denote the direct parent of v. Also, we use $T[v]$ to denote the subtree of T rooted at v. For any three numbers $x, y, z \geq 0$, we use $\binom{z}{x,y}$ to denote the number of ways that two sets $X \subseteq \{1, 2, \ldots, z\}$ and $Y \subseteq \{1, 2, \ldots, z\}$ can be chosen such that $|X| = x$, $|Y| = y$, and $X \cup Y = \{1, 2, \ldots, z\}$.

3   Algorithm

3.1   Basic Operations on Partly Expanded B$^+$-Trees

We first explain how we can apply the basic operations union, intersection, delta, and minus to two given partly expanded B$^+$-trees in optimal time. Note that as long as the algorithm uses edges of a partly expanded B$^+$-treeto reach other nodes of the tree for the first time rather than trying to "jump" to new nodes, a partly expanded B$^+$-tree essentially behaves analogously to a B$^+$-tree and operations can be performed in the same asymptotic cost. The reason is that once the algorithm reaches any not-expanded node for the first time, it can expand the node by creating its children (but not expanding them) in constant time. So, we explain the base algorithms for B$^+$-trees.

Evaluating union and intersection of two B$^+$-trees with n and m leaves can be performed in time $O(m \log \frac{n+m}{m})$ [7]. To apply the minus operation on B$^+$-trees, one can find the common members between two subtrees using the intersection operation and then use the following lemma (proof is omitted):

Lemma 1. *Given a B$^+$-tree with n nodes and a set of m leaves on it, the cost of deleting those nodes is $O\left(m \log \frac{n+m}{m}\right)$.*

3.2   Overview of the Algorithm

The algorithm first specifies some nodes, called "independent nodes" in the tree. A node v is defined as *independent* if v is the root or $\psi(v) \leq \psi^*(p(v))$; otherwise, v is *dependent*. The result of an independent node is evaluated entirely confined to the subtree rooted at it, and without using anything outside this subtree.

For an independent node v, the maximal subtree of $T[v]$ that includes v but has no other independent node as an internal node is called the *dependant tree of v*, and is denoted by T_v. Thus, each leaf of T_v is either a leaf of T or an independent node.

We consider independent nodes and compute their results in a bottom-up fashion. Thus, when computing the result of an independent node v, result of all leaves of the dependant tree of v are previously computed and the algorithm has access to the corresponding partly-expanded B$^+$-trees. In the following sections we discuss how the result of such a node v can be computed.

3.3 Union and Delta Independent Nodes

For an independent union or delta node v, it is easy to see that both children u_1 and u_2 of v are independent. So, the problem for such a node v reduces to computing the union or delta of two partly-expanded B^+-trees, which is discussed in Section 3.1

3.4 Intersection and Minus Independent Nodes

Suppose v is an independent node. We call the left most leaf of T_v the *key leaf of v*, and we denote it by k_v. Let K_v be the result of k_v. The result of v will be a subset of K_v. In particular, if we define K'_v to be the set of elements in K_v that also appear in the result of at least another leaf of T_v, the result of v is a subset of K'_v if $\pi(v) = \cap$ and is K_v minus a subset of K'_v if $\pi(v) = -$.

For every element e in K'_v, the algorithm makes a list of all leaves of T_v having e in their results in the following way: considering leaves of T_v, except k_v, from left to right, it computes the intersection of the result of each leaf ℓ of T_v with K_v and for every element e in the intersection, the algorithm adds ℓ to the corresponding list of e. Then, for any element e of K_v, the algorithm decides whether e is in the result of v or not, using the method described in the next theorem.

Theorem 1. *After a linear time one-time preprocessing on a set expression tree T, there is an algorithm that given, as the input, any element e and node v and the list $\mathcal{L}$ of all leaves of T_v containing e in their results in order from left to right, it can determine if e is in the result of T_v in time linear to the size of $\mathcal{L}$.*

Proof. As a one-time preprocessing the algorithm makes the following two data-structures: the first data-structure is the one by Harel and Tarjan [8] using which it is possible to find the lowest common ancestor of any two nodes n_1 and n_2 of T_v, denoted by $\mathrm{lca}(n_1, n_2)$, in constant time. The second data-structure, can determine, for a node n and an ancestor a of n, if among nodes from the path from a to n, not including a and n, there is a minus node such that n is in its right subtree or there is an intersection node. If there exists such a node in the path, we would say a *eliminates* n.

We consider an empty stack $\mathcal{S}$ and push leaves of $\mathcal{L}$ in $\mathcal{S}$, in order from left to right. After pushing each leaf, we repeat the following procedure until less than three nodes remain in the stack or no more change is made. Suppose n_1, n_2, and n_3 are the top three nodes in $\mathcal{S}$ in order (n_1 is the top one). If $\mathrm{lca}(n_1, n_3) \neq \mathrm{lca}(n_2, n_3) = n_{2,3}$, we know there will be no more node of $T[n_{2,3}]$ to be processed any more; so we pop n_1, n_2, and n_3 from $\mathcal{S}$, and push $n_{2,3}$ into $\mathcal{S}$ if and only if e will be in the result of $n_{2,3}$. This can be determined based on the operator of $n_{2,3}$. If it is a union, delta, or intersection, e is in the result of $n_{2,3}$ if $n_{2,3}$ eliminates at most one, exactly one, and none of n_2 and n_3, respectively. If $n_{2,3}$ is a minus node, then e is in the result of $n_{2,3}$ iff $n_{2,3}$ eliminates n_3 or does not eliminate n_2. We also push n_1 into $\mathcal{S}$. The running-time is linear in $\mathcal{L}$. $\square$

Finally, if $\pi(v) = \cap$, the algorithm builds a fully-expanded B^+-tree on the elements of K'_v that are in the result of v and returns it. If $\pi(v) = -$, the algorithm removes the elements of K'_v that are not in the result of v from K and returns the resulted partly-expanded B^+-tree.

3.5   Running Time

The following lemma shows the running time of the algorithm.

Lemma 2. *The total processing time to evaluate an expression tree is*

$$\sum_{v \in leaves - keys} \psi^*(v) \lg \left(\frac{\psi(v)}{\psi^*(v)} + 1 \right) + \sum_{v \in independent - keys} \psi^*(v) \lg \left(\frac{\psi^*(p(v))}{\psi^*(v)} + 1 \right), (1)$$

*where **leaves** is the set of all leaves but the left-most leaf of T, **independents** is the set of independent nodes and **keys** is the set all key leaves.*

Proof. Considering any independent node v, we prove that the time consumed to evaluate the result of T_v can be written as the sum of contribution of leaves of T_v to Equation 1. Since every leaf and independent node of the main tree is the leaf dependant tree of exactly one independent node, this will conclude the proof.

First suppose v is a union or delta node v with children u_1 and u_2, which are clearly leaves in T_v. As v is independent, $\psi^*(v) = \psi(v) = \psi^*(u_1) + \psi^*(u_2)$ and for each $i = 1, 2$, $\psi(u_i) = \psi^*(u_i)$. One of u_1 and u_2 is a non-key independent node. Thus the cost of computation for all such nodes v is bounded by $\sum_{v \in independent - keys} \psi^*(v) \lg \left\lceil \frac{\psi^*(p(v))}{\psi^*(v)} + 1 \right\rceil$.

Now consider the case when v is a minus or an intersection node. It can be seen that $\psi^*(v) = \psi^*(u_k)$ for u_k the key leaf of T_v. Thus, since every internal node of T_v other than v is dependant, for every leaf u of T_v we have $\psi^*(p(u)) = \psi^*(u_k)$. Therefore, the cost for every non-key independent leaf of T_v is the cost of taking an intersection between the result u, which is of size $\psi(u) = \psi^*(u)$ and a B^+-tree of size at most $\psi^*(u_k) = \psi^*(p(u))$. For non-key non-independent leaves of T_v also, the cost will be the cost of an intersection between a B^+-tree of size $\psi(u)$ and another one of size $\psi^*(u_k) = \psi^*(p(u)) = \psi^*(u)$. $\square$

4   Lower-Bounds

In this section, we argue on the optimality of the proposed algorithm. Similar to our previous work [6], we obtain two lower-bounds which together show the algorithm is optimal. We prove the first lower-bound in the most general case of types of trees. Although proving the second lower-bound seems difficult in the general case, we show that the algorithm is optimal on various types of expression trees. These trees consist of union, intersection and symmetric difference operator nodes with a possible difference operator as the root node. We conjecture that the second lower-bound applies even in the most general types of trees.

We fix a root result set O_T of size $\psi^*(\text{Root}(T))$. These elements stem from the leaves of the tree and come up the tree level by level according to the operation nodes. We define the notion of *proof labelling* which captures the trace-back of nodes that had an impact on the presence of an element in the final result. A proof labelling is formally defined as follows:

Definition 2. *Given a signature T, $\Lambda : V_T \mapsto 2^{O_T}$ is a proof labelling for T if it has the following properties:*

1. $\Lambda(Root(T)) = O_T$.
2. *For every vertex $v \in V_T$, $|\Lambda(v)| = \psi^*(v)$.*
3. *if $v \in V_T - \textbf{leaves}(T)$ and u_1 and u_2 are children of v, $\Lambda(u_1) \cup \Lambda(u_2) = \Lambda(v)$ if v is a union or delta node, $\Lambda(u_1) = \Lambda(u_2) = \Lambda(v)$, if v is an intersection node, and $\Lambda(u_1) = \Lambda(v)$ and $\Lambda(u_2) \subseteq \Lambda(v)$ if v is a difference node.*

Of course, it is possible that there exist more than one proof labelling for a signature T and a set O_T.

We introduce an adversary $\mathcal{B}$ which answers queries such that O_T becomes the result of the root node. The adversary $\mathcal{B}$ chooses a proof labelling Λ among all possible proof labellings and answers queries according to it. The adversary fixes the input gradually such that it is always consistent with the history of the queries it has answered.

Members of the sets associated with leaves in the input being fixed will be triples of integers which are compared in a lexicographical manner. In particular, members of O_T are of the form $(i, 0, 0)$, where $1 \leq i \leq |O_T|$ is an integer. $\mathcal{B}$ divides the sequence of members of every leaf v of T into $\psi^*(v) = |\Lambda(v)|$ consecutive *regions* of size $\left\lfloor \frac{\text{size}(v)}{\psi^*(v)} \right\rfloor$ or $\left\lceil \frac{\text{size}(v)}{\psi^*(v)} \right\rceil$. Each of $\psi^*(v)$ regions in a leaf is named after the corresponding element in the proof labelling $\Lambda(v)$: if the ith biggest member of $\Lambda(v)$ is $(a, 0, 0)$, then the ith region of v is called an *a-region*. $\mathcal{B}$ initializes the first coordinates of all members of an a-region $\mathcal{R}$ to a at the beginning. Thus, given a member x of an a-region and a member y of a b-region such that $a \neq b$, whenever a query (x, y) is submitted, $\mathcal{B}$ can answer the query without any knowledge of the second and the third coordinates.

For any region $\mathcal{R}$, the second coordinate of exactly one element of $\mathcal{R}$, which is called the *crucial member of $\mathcal{R}$*, will be zero. The strategy is to determine the second coordinates of triples of a region $\mathcal{R}$ in such a way that $\mathcal{A}$ does not touch the crucial member of $\mathcal{R}$ before touching $\log |\mathcal{R}|$ members of $\mathcal{R}$ where $|\mathcal{R}|$ denotes the length of $\mathcal{R}$. The third coordinates of non-crucial members are zero.

The first lower-bound comes directly from the following theorem.

Theorem 2. *There is an adversary that can force $\mathcal{A}$ to touch all crucial members of all regions of **deep**, where **deep** is the set of all elaves of T that have an ancestor v such that $\pi(v) = \cap$ or v is a right child of a minus node.*

We provide proof for Theorem 2 in section 4.1. An argument analogous to binary search as given in [6] yields the first lower-bound from the premise:

Theorem 3 (First Lower-Bound). *There is an adversary that forces $\mathcal{A}$ to submit at least $\mathcal{L}_1 = \sum_{v \in deep} \psi^*(v) \cdot \lg \left\lceil \frac{size(v)}{\psi^*(v)} + 1 \right\rceil$ queries, if $\mathcal{A}$ wants to touch all crucial members of all regions of* **deep.** $\square$

The third coordinates of the crucial members are determined in a manner that the algorithm $\mathcal{A}$ would have to determine which crucial members belong to the same region; we define two crucial members as *similar* if they both belong to a-regions for some $1 \leq a \leq |O_T|$.

The second lower-bound is obtained from the fact that the algorithm must eventually acquire enough knowledge to be able to determine the proof labelling the adversary has fixed. The fact is implied by the claim that the algorithm must have the knowledge to determine all similar members to any particular crucial member which we prove in section 4.2. Given this premise, one can prove as in [6] that the algorithm can determine the proof labelling. Therefore, the number of queries the algorithm submits is at least the logarithm of the number of possible proof labellings. The number of proof labellings is the product over all nodes v of the number of ways $\Lambda(v)$ can be distributed into $\Lambda(u_1)$ and $\Lambda(u_2)$ of its children which abide the definition of proof labelling. One can verify that this number is $\binom{\psi^*(v)}{\psi^*(u_1),\psi^*(u_2)}$ whether v is union, intersection, symmetric difference, or subtraction:

Theorem 4 (Second Lower-Bound). *If algorithm $\mathcal{A}$ knows the set of all similar members to any crucial member, then it has submitted at least $\mathcal{L}_2 = \log_6 \left(\prod_v \binom{\psi^*(v)}{\psi^*(u_1),\psi^*(u_2)} \right)$, where the product is taken over all internal nodes.* $\square$

Theorems 3, 4 together yield our desired lower-bound (proof is omitted):

Theorem 5 (Main Lower-Bound). *For any signature T and any deterministic comparison-based algorithm $\mathcal{A}$, if $\mathcal{A}$ knows the set of all similar members to any crucial member after an interaction with the adversary, $\mathcal{A}$ has submitted at least*

$$\sum_{v \in leaves\text{-}keys} \psi^*(v) \lg \left(\frac{\psi(v)}{\psi^*(v)} + 1 \right) + \sum_{v \in independent-keys} \psi^*(v) \lg \left(\frac{\psi^*(p(v))}{\psi^*(v)} + 1 \right) \in$$

queries. $\square$

4.1   The First Lower-Bound

We sketch the proof of theorem 2 (the first lower-bound) in this section. We prove that all crucial-members of leaves must be touched by a query. The proof we present in this section is a similar but a less sophisticated than that of the second lower-bound section 4.2.

We can transform the operation tree into an equivalent tree with only union, delta, and intersection operations and complement operation on the leaves. We fix an element $(a, 0, 0) \in O_T$ and create a subtree of the original tree, denoted by

$T^{(a)}$, that consists of nodes v whose $\Lambda(v)$ contain $(a, 0, 0)$. Moreover, we assume that all the internal nodes of $T^{(a)}$ have exactly two children (the nodes with only one children will be contracted). By the definition of proof labelling, a is in the result of T if and only if o is in the result of $T^{(a)}$.

As queries (x, y) arrive, the adversary can answer $x < y$, $x = y$, or $x > y$ only by looking at their first two coordinates if x and y do not both belong to an a-region for some $(a, 0, 0) \in O_T$ or at least one of x or y is not a crucial member. Otherwise, the first two coordinates of x and y are the same. The third coordinate of elements are set by the adversary at the first time they are touched. So, if x or y are touched before, the adversary can answer the query without any problem; otherwise, if one of them, say x, is being touched for the first time, the adversary should set the third coordinate of x first, and then answer the query.

Let x be a crucial member of an a-region of ℓ, where ℓ is a leaf of $T^{(a)}$. We say that ℓ is a-touched if the crucial member of the a-region of ℓ is touched. The adversary sets the third coordinates in a fashion that makes sure the result of $T^{(a)}$ cannot be computed unless all the leaves of $T^{(a)}$ are a-touched. It is possible, because the adversary can use the following method. For simplicity, we assume that if the expression has minus operations, we rewrite the expression so that it is minus-free but some of the sets are complement. In other words, there are no minus nodes in the tree, but some of the leaves of the tree are complement.

1. Let v be the first ancestor of ℓ that has at least another leaf, other than ℓ, that is not a-touched. If there is no such an ancestor, it means that ℓ is the only leaf of $T^{(a)}$ that is not a-touched; then,
   (a) If ℓ is a normal node, the adversary sets the third coordinate of x to zero.
   (b) Otherwise, if ℓ is a complement node, the third coordinate is set to a unique positive integer.
2. If $(\pi(v), \ell) \in \{(\cap, \text{normal}), (\cup, \text{complement}), (\triangle, \text{complement})\}$, then the adversary sets the third coordinate of x to zero.
3. Otherwise, if $(\pi(v), \ell) \in \{(\cap, \text{complement}), (\cup, \text{normal}), (\triangle, \text{normal})\}$, the adversary sets the third coordinate of x to a unique positive integer.

By answering queries in this manner, and assuming that $T^{(a)}$ has at least two leaves, the adversary guarantees at any point that the algorithm never has enough knowledge to ensure $(a, 0, 0)$ is in the root result, unless all the leaves of $T^{(a)}$ are a-touched. The fact that $T^{(a)}$ has at least two leaves if and only if $T^{(a)}$ has a minus or intersection node completes the proof of Theorem 2.

4.2   The Second Lower-Bound

In this part, we prove that, if T is of a special shape, there is an adversary $\mathcal{B}$ that can force any algorithm $\mathcal{A}$ to submit enough queries to determine the set of all similar members to any crucial member. This enables us to use Theorem 4. More precisely, we prove the following:

Theorem 6. *Given a complement-free operation signature T where difference operation is only allowed at the root, for any deterministic comparison-based algorithm $\mathcal{A}$, there exists a generous adversary $\mathcal{B}$ that can force $\mathcal{A}$ to gain enough information to determine similar members to any crucial member. A generous adversary is an adversary that gives one bit of additional information b in answering to each query (x, y) of $\mathcal{A}$: b is one if and only if both x and y are crucial members of a-regions for some $(a, 0, 0) \in O_T$.*

Note that by requiring the adversary to be generous, the number of possible answers that an algorithm can get from the adversary for each query will be six, which is constant. In order to prove Theorem 6, we define a game between $\mathcal{A}$ and the generous adversary $\mathcal{B}$ and prove that $\mathcal{B}$ has a winning strategy for the cases in which at most one difference operation is allowed and that operation must be at the root of T. We leave the question of whether $\mathcal{B}$ has a winning strategy for all expression trees open. Thus, for proving that the set expression evaluation algorithm we proposed works optimum for any expression, one has only to prove that $\mathcal{B}$ has a winning strategy in the game explained in the following proof of Theorem 6.

Proof Sketch. Similar to Section 4.1, $\mathcal{B}$ knows how to answer queries (x, y) if either of x or y is not a crucial member or if x and y are not similar, based on the first two coordinates of x and y. Our adversary in this section differs from the adversary in Section 4.1 in the way they set the third coordinate of crucial members. So, we can assume that x and y are crucial members and they both belong to a-regions for some $(a, 0, 0) \in O_T$. Similar to the proof of the first lower-bound in Section 4.1, a subtree $T^{(a)}$ of the original tree consisting of leaves with a-regions and their ancestors is created. Then, we defined a game between $\mathcal{A}$ and $\mathcal{B}$ on $T^{(a)}$: the algorithm submits queries comparing two crucial members and $\mathcal{B}$ should response in a non-contradictory manner. For simplicity, with slightly abusing the notation, when we talk about a query (ℓ_1, ℓ_2) on $T^{(a)}$, where ℓ_1 and ℓ_2 are leaves of $T^{(a)}$, we mean a query (x, y), where x is the crucial member of the a-region of ℓ_1 and y is the crucial member of the a-region of ℓ_2. Also, we may say a leaf ℓ is *set* if the third coordinate of the crucial member of its a-region is set to a positive integer; otherwise, we say ℓ is *unset*. The game finishes when $\mathcal{A}$ has a proof that the result of $T^{(a)}$ is empty or not empty. $\mathcal{B}$ aims to avoid premature ending of the game before $\mathcal{A}$ has enough information for determining the set of members similar to a crucial member of an a-region.

We keep the history of queries on crucial members of an a-region submitted by $\mathcal{A}$ in a graph $G^{(a)}$. More precisely, the vertices of $G^{(a)}$ are the leaves of $T^{(a)}$ and there is an edge between ℓ_1 and ℓ_2 in $G^{(a)}$ if and only if $\mathcal{A}$ has submitted a query (ℓ_1, ℓ_2). Moreover, the label of the edge between ℓ_1 and ℓ_2 is '$<$', '$=$', or '$>$', if the $\mathcal{B}$'s answer to the query (ℓ_1, ℓ_2) has been $<$, $=$, or $>$, respectively. Note that, because of the generousity of $\mathcal{B}$, $\mathcal{A}$ has enough information to determine similar members to a crucial member of an a-region if and only if $G^{(a)}$ is connected. Therefore, the goal of $\mathcal{B}$ is to avoid ending of the game before $G^{(a)}$ gets connected.

We can make a number of assumptions about the game. First, since repeating an already submitted query is useless, we can assume that at most $\binom{n}{2}$ queries will be submitted on $T^{(a)}$, where n is the number of leaves of $T^{(a)}$. Second, since $G^{(a)}$ is connected for $n = 1$, we can assume that $n > 1$, and hence, for every leaf ℓ of $T^{(a)}$, $\ell \in \mathsf{deep}$.

Let $(\ell_1^{(i)}, \ell_2^{(i)})$ be the ith query of $\mathcal{A}$. In order to respond to this query, $\mathcal{B}$ looks at the third coordinates of $\ell_1^{(i)}$ and $\ell_2^{(i)}$. If both $\ell_1^{(i)}$ and $\ell_2^{(i)}$ are set, then $\mathcal{B}$ can answer the query without doing anything further. If only one of them, say $\ell_1^{(i)}$, is set, then $\mathcal{B}$ answers $\ell_1^{(i)} > \ell_2^{(i)}$, without setting the third coordinate of $\ell_2^{(i)}$. This will be consistent with future queries, because of the way that $\mathcal{B}$ sets the third coordinates in the following case. The final case, which is more involved, is the case that both $\ell_1^{(i)}$ and $\ell_2^{(i)}$ are unset.

We need the following definitions: a node v of $T^{(a)}$ is called *unimportant* if

1. v is a set leaf.
2. v is a child of an unimportant node.
3. $\pi(v) \in \{\cup, \Delta\}$ and all the children of v are unimportant.
4. $\pi(v) = \cap$ and at least one of its children is unimportant.

All other nodes are called *important* nodes. Also, for two leaves ℓ and ℓ' of $T^{(a)}$ we say that ℓ is $G^{(a)}$-*equal* to ℓ' if $\ell = \ell'$ or there is a path with label '=' from ℓ to ℓ' in $G^{(a)}$. We define $f(\ell)$ to be the first grand parent v of ℓ that has a v-unlinked leaf that is not $G^{(a)}$-equal to ℓ; a v-*unlinked leaf* is an important leaf of $T^{(a)}[v]$ that is not $G^{(a)}$-equal to any node outside $T^{(a)}[v]$. Finally, we use $g(\ell)$ to denote any important leaf that is $G^{(a)}$-equal to ℓ. The answering strategy makes sure that $g(\ell)$ always exists if ℓ is unset.

Now, if $\ell_1^{(i)}$ is $G^{(a)}$-equal to $\ell_2^{(i)}$, $\mathcal{B}$ answers $\ell_1^{(i)} = \ell_2^{(i)}$ and does not set the third coordinates of them. If there is a value of $x \in \{1, 2\}$ such that $\pi(f(g(\ell_x^{(i)}))) \in \{\cup, \Delta\}$, or $\pi(f(g(\ell_x^{(i)}))) = -$ and $g(\ell_x^{(i)})$ is in the right subtree of $f(g(\ell_x^{(i)}))$, $\mathcal{B}$ answers $\ell_x^{(i)} > \ell_{3-x}^{(i)}$ and sets the third coordinate of $\ell_x^{(i)}$, and all other leaves that are $G^{(a)}$-equal to $\ell_x^{(i)}$, to $\binom{n}{2} + 1 - i$. Otherwise, $\mathcal{B}$ answers $\ell_1^{(i)} = \ell_2^{(i)}$ and does not set the third coordinates of them.

5   Conclusion

We considered the complexity of evaluating a general set expression containing complement, union, intersection, difference, and symmetric difference. These expressions are the most general types of expressions with binary and unary operations. We gave an algorithm to evaluate such expressions in the comparison model.

We argued that the algorithm performs the optimal number of comparisons by giving a matching lower bound in the cases where the expression does not contain complement and the only difference operation is at the root of the corresponding expression tree. We conjecture that the algorithm is optimal over all set expressions. We showed that first lower bound in section 4.1 applies to all types of set expressions. However, the second bound only applies to the special types of expressions as stated.

Proving the second lower bound for all types of expressions remains open as a future work.

References

1. Barbay, J., Kenyon, C.: Adaptive intersection and t-threshold problems. In: SODA, pp. 390–399 (2002)
2. Bille, P., Pagh, A., Pagh, R.: Fast evaluation of union-intersection expressions. In: Tokuyama, T. (ed.) ISAAC 2007. LNCS, vol. 4835, pp. 739–750. Springer, Heidelberg (2007)
3. Brin, S., Page, L.: The anatomy of a large-scale hypertextual Web search engine. Computer Networks and ISDN Systems 30(1–7), 107–117 (1998)
4. Brown, M.R., Tarjan, R.E.: A fast merging algorithm. J. ACM 26(2), 211–226 (1979)
5. Brown, M.R., Tarjan, R.E.: Design and analysis of a data structure for representing sorted lists. SIAM Journal on Computing 9(3), 594–614 (1980)
6. Chiniforooshan, E., Farzan, A., Mirzazadeh, M.: Worst case optimal union-intersection expression evaluation. In: Caires, L., Italiano, G.F., Monteiro, L., Palamidessi, C., Yung, M. (eds.) ICALP 2005. LNCS, vol. 3580, pp. 179–190. Springer, Heidelberg (2005)
7. Demaine, E.D., López-Ortiz, A., Munro, J.I.: Adaptive set intersections, unions, and differences. In: SODA 2000: Proceedings of the eleventh annual ACM-SIAM symposium on Discrete algorithms, pp. 743–752. Society for Industrial and Applied Mathematics, Philadelphia (2000)
8. Harel, D., Tarjan, R.E.: Fast algorithms for finding nearest common ancestors. SIAM J. Comput. 13(2), 338–355 (1984)
9. Hwang, F.K., Lin, S.: A simple algorithm for merging two disjoint linearly ordered sets. SIAM Journal on Computing 1(1), 31–39 (1972)
10. Lee, G., Park, M., Won, H.: Using syntactic information in handling natural language quries for extended boolean retrieval model (1999)
11. Mauldin, M.: Lycos: design choices in an internet search service. IEEE Expert 12(1), 8–11 (1997)
12. Mirzazadeh, M.: Adaptive comparison-based algorithms for evaluating set queries (2004)
13. Pugh, W.: A skip list cookbook. Tech. rep., University of Maryland at College Park, College Park, MD, USA (1990)
14. Witten, I.H., Moffat, A., Bell, T.C.: Managing Gigabytes: Compressing and Indexing Documents and Images. Morgan Kaufmann Publishers, San Francisco (1999)

Computing with Priced Information: When the Value Makes the Price[*]

Ferdinando Cicalese and Martin Milanič

AG Genominformatik, Faculty of Technology, Bielefeld University, Germany
{nando,mmilanic}@cebitec.uni-bielefeld.de

Abstract. We study the function evaluation problem in the priced information framework introduced in [Charikar *et al.* 2002]. We characterize the best possible extremal competitive ratio for the class of game tree functions. Moreover, we extend the above result to the case when the cost of reading a variable depends on the value of the variable. In this new value dependent cost variant of the problem, we also exactly evaluate the extremal competitive ratio for the whole class of monotone Boolean functions.

1 Introduction

Problem Statement. A function f over a set of variables $V = \{x_1, x_2, \ldots, x_n\}$ is given and we want to determine the value of f for a fixed but unknown *assignment* σ, i.e., a choice of the values for the variables of V. We are allowed to adaptively read the values of the variables. Each variable x_i has an associated non-negative cost c_{x_i} which is the cost incurred to read its value $x_i(\sigma)$. For each $i = 1, \ldots, n$, the cost c_{x_i} is fixed and known beforehand. The goal is to identify and read a minimum cost set of variables $U \subseteq V$ whose values uniquely determine the value $f(\sigma) = f(x_1(\sigma), \ldots, x_n(\sigma))$ of f w.r.t. the given assignment σ, regardless of the values of the variables not probed. We say that such a set $U \subseteq V$ is a *proof* for f with respect to the assignment σ.

An evaluation algorithm $\mathcal{A}$ for f adaptively reads the variables in V until the set of variables read so far is a proof for the value of f. Given a cost assignment $c = (c_{x_1}, \ldots, c_{x_n})$, we let $c_{\mathcal{A}}^f(\sigma)$ denote the total cost incurred by the algorithm $\mathcal{A}$ to evaluate f under the assignment σ and $c^f(\sigma)$ the cost of the cheapest proof for f under the assignment σ. We say that $\mathcal{A}$ is ρ-competitive if $c_{\mathcal{A}}^f(\sigma) \leq \rho c^f(\sigma)$, for every possible assignment σ. We use $\gamma_c^{\mathcal{A}}(f)$ to denote the competitive ratio of $\mathcal{A}$, that is, the minimum ρ for which $\mathcal{A}$ is ρ-competitive. The best possible competitive ratio for any deterministic algorithm, then, is $\gamma_c^f = \min_{\mathcal{A}} \gamma_c^{\mathcal{A}}(f)$, where the minimum is computed over all possible deterministic algorithms $\mathcal{A}$.

The extremal competitive ratio $\gamma^{\mathcal{A}}(f)$ of an algorithm $\mathcal{A}$ is defined by $\gamma^{\mathcal{A}}(f) = \max_c \gamma_c^{\mathcal{A}}(f)$. The best possible extremal competitive ratio for any deterministic algorithm is $\gamma(f) = \min_{\mathcal{A}} \gamma^{\mathcal{A}}(f)$. This is a measure of the structural complexity of f independent of a particular cost assignment and algorithm.

* This work was supported by the Sofja Kovalevskaja Award 2004 of the Alexander von Humboldt Foundation and the Bundesministerium für Bildung und Forschung.

S.-H. Hong, H. Nagamochi, and T. Fukunaga (Eds.): ISAAC 2008, LNCS 5369, pp. 378–389, 2008.

Function evaluation problems have been extensively studied in AI, particularly in the theory of computer aided games. The strategic evolution of the game is usually modeled by a so called game tree [8, 18, 16]. In a game tree (formalized in Sect. 4), each node represents a state of the game. The root is the current state. For each node/state ν, its children are the set of possible states reachable from ν given the moves available to the player moving in state ν. The possible moves for the two players are represented by alternating levels of edges. A game tree of a certain depth is built by a computer in order to explore the possible developments of the game from the current position. By assigning to each leaf-state an estimate of the "goodness" of that state for the computer-player, it is possible to evaluate all the inner states. The most fruitful move for the computer is the one corresponding to the edge from the root to its children of maximum value. In general, the evaluation of some leaf state might involve expensive computations. Since, on the other hand, not all the leaf-state evaluations are needed to compute the node of maximum value in the first level, we have here an instance of the problem of evaluating a function by only looking at a *cheap* set of its variables.

In this paper we characterize the extremal competitiveness for the class of game tree functions. Moreover, we also study the function evaluation problem when the cost of reading a variable depends on the value of the variable. The evaluation algorithm knows the cost $c_x(y)$ of reading x when $x(\sigma) = y$, for each variable x and for each value y that the variable x can take.

This above model has applications in several situations. Consider, e.g., the decision making process of a physician—or of a computer aided decision making system—who has to decide the cheapest sequence of tests to perform on a patient in order to reliably diagnose a given disease. Different tests typically involve different costs. In this framework, costs are usually understood in an extended meaning encompassing the actual monetary costs, the distress of the patient, and the possible side-effects of the tests. Also, tests' costs might be dependent on the outcome: a single lab analysis might undergo several phases, some of which are only performed depending on the result of the previous ones. Analogously, there are tests that if positive, are necessarily followed by a sequence of other tests—on which the decision maker has no alternative. In this case, the cost of the "triggering" test can be considered as the sum of the whole set, in case its outcome is positive, and only its own cost if the outcome is negative. It is then natural to consider models in which tests' costs are dependent on the outcome of the test itself. We refer the interested reader to [21] and references quoted therein for a remarkable account of several types of costs to be considered in inference procedures.

Besides the two examples above, function evaluation problems are found in a plethora of different areas both in theoretical and applied computer science like telecommunications [14], manufacturing [9], computer networks [10], satisficing search problems [11]. For more on automatic diagnosis problems and computer aided medical systems see also [1, 15] and references therein. Finally, the function evaluation problem arises in query optimization, a major issue in databases [13].

Our Results. We obtain the tight extremal competitive ratio of monotone Boolean functions in the new value dependent cost model extending the previous result of [6]. This is achieved via an adaptation of the Linear programming based approach of [6].

Outside the Boolean realm, we focus on the class of game tree functions. We obtain the tight extremal competitive ratio for game trees. In particular we show that for any game tree function f, but for special cases that we also characterize, the extremal competitiveness $\gamma(f)$ is equal to the maximum size of a *certificate* for f, i.e., of a minimal set of variables which allow to prove an upper or lower bound on the value of f, for some assignment σ. In fact, we provide a polynomial algorithm with competitiveness $\gamma(f)$ for any game tree function f. We also extend this result to the value dependent cost model. Our result significantly improves the previous best known result in [4], where a polynomial time algorithm was provided which achieves $\gamma(f)$ competitiveness over a restricted set of assignments, namely only those σ's for which exactly one variable has value $f(\sigma)$.

Related Work. Most of the earlier work on function evaluation problems was done in the classical unitary cost model for both deterministic and randomized algorithms or assuming some statistical knowledge on the values of the variables (see, e.g., [20, 19, 17, 12]). The competitive analysis scenario was proposed by Charikar *et al.* in [2] where several classes of functions were studied in this novel framework, including the class of game trees. For game trees, Charikar *et al.* [2] presented a pseudo-polynomial time algorithm with competitiveness $2\gamma_c^f$. The extremal competitiveness for game trees was also studied in [4] where a polynomial time algorithm was provided achieving competitiveness $\gamma(f)$ for any assignment σ such that there exists exactly one variable with value $f(\sigma)$. In [5] the authors showed a polynomial time algorithm with competitiveness $4\gamma_c^f$. However, to date, there was no complete and exact characterization of the optimal competitiveness for the evaluation of game trees.

All the above results are for the case when the cost is independent of the value of the variable. In fact, this is the first paper taking into account the dependency of costs on the values in the competitive analysis scenario. In [1] function evaluation with value dependent costs was also discussed, even though in the probabilistic model considered in [1] the dependency on the values can be absorbed in the distribution assigned to the values of the variables. In [5], the case of unknown costs was also considered. This is an attempt to address cases in which the algorithm has a reduced knowledge on the cost assignment. It is important to notice that the model of [5] cannot be used to solve the type of problems addressed here, and vice versa.

Due to the space limitation, some proofs are omitted. For more details, the interested reader can refer to [7].

2   Preliminaries: The Linear Programming Approach

Let $f : D \to \mathbb{R}$ be a real-valued function defined over a set of variables $V = \{x_1, \ldots, x_n\}$, where $D \subseteq \mathbb{R}^n$. For $x \in V$, let $D(x)$ denote the set of possible values that the variable x can take in the elements of the domain of f, that is, $D(x)$ is the projection of the set D on the x coordinate. For $x \in V$ and $y \in D(x)$, let $c_x(y) \geq 0$ denote the cost for querying the variable x, given that the value of x in the (unknown) assignment σ is $x(\sigma) = y$. Furthermore, let $c_x^{min} = \min\{c_x(y) : y \in D(x)\}$ and $c_x^{max} = \max\{c_x(y) : y \in D(x)\}$, for all $x \in V$. We allow that the costs of querying a certain variable are *value dependent*. In other words, the functions $c_x(y)$ are not necessarily constant as functions of y, i.e., it is possible that $c_x^{min} \neq c_x^{max}$.

We assume that a bound $r \geq 1$ is fixed (and known to the algorithm) on the maximum possible ratio between two costs of queries for a single variable. More precisely, we assume that the cost function satisfies, for all $x \in V$, $c_x^{max} \leq rc_x^{min}$. Equivalently, for every $x \in V$, and for every $y_1, y_2 \in D(x)$, we have $0 \leq c_x(y_1) \leq rc_x(y_2)$. The set of all such assignments of cost functions will be denoted by $\mathcal{C}_r(f)$. Note that without such bound, no algorithm could guarantee any competitiveness.

In order to make explicit the dependency of our results on the bound r, we shall now rephrase the definition of the competitive measures.

Definition 1. *Let* $r \geq 1$. *The r-extremal competitive ratio of a function* $f : D \rightarrow$ $\mathbb{R}$, *where* $D \subseteq \mathbb{R}^n$, *is defined as* $\gamma_r(f) = \min_{\mathbb{A}} \gamma_r^{\mathbb{A}}(f)$ *where the minimum is taken over all deterministic algorithms that evaluate* f, *and where* $\gamma_r^{\mathbb{A}}(f) =$ $\max_{c \in \mathcal{C}_r(f)} \max_{\sigma \in D} \frac{c_{\mathbb{A}}^f(\sigma)}{c^f(\sigma)}$.

It is not hard to see that every ρ-competitive algorithm for the value independent cost model is an $(r \times \rho)$-competitive algorithm in the value dependent cost model. Therefore, $\gamma_r(f) \leq r\gamma(f)$. However, we shall see that this estimate of $\gamma_r(f)$ loses an additive term of $r - 1$. For this we devise a variant of the Linear Programming Approach introduced in [6] which is adapted to the value dependent cost model. We denote this new scheme by $\mathcal{LPA}^*$.

In order to describe the $\mathcal{LPA}^*$ we shall need some new notation. Let $\mathcal{P}(f)$ denote the set of inclusion-wise minimal proofs of f, i.e., the family of sets X such that there exists at least one assignment σ with respect to which X is a proof for f, while no subset of X is. Consider the following linear program $\mathbf{LP_f}$:

$$\mathbf{LP_f} : \left\{ \text{Minimize} \sum_{x \in V} s(x) : \sum_{x \in P} s(x) \geq 1 \; \forall P \in \mathcal{P}(f) \text{ and } s(x) \geq 0 \; \forall x \in V \right\}$$

Suppose that the set of variables already read is Y. We shall denote with f_Y the *restriction* of f with respect to Y, that is, the function over $V \backslash Y$ obtained from f by fixing the values of the variables in Y as given by the valued read so far, according to the underlying fixed and unknown assignment σ. Let s_Y be a feasible solution to the linear program $\mathbf{LP_{f_Y}}$. The $\mathcal{LPA}^*$ chooses a variable u that minimizes the value of $\frac{c_x^{min}}{s_Y(x)}$. (For definiteness, we let $\frac{0}{0} := 0$. This assures that the variables of zero cost are always queried before the others.) Then, the cost assignment c is updated to a new cost assignment $\tilde{c}$ defined as follows: For $x \in V \backslash (Y \cup \{u\})$ and $y \in D(x)$, we let

$$\tilde{c}_x(y) = c_x(y) - \delta c_x(y) \quad \text{where} \quad \delta c_x(y) = c_x(y) \cdot \frac{c_u^{min}}{c_x^{min}} \cdot \frac{s_Y(x)}{s_Y(u)}. \tag{1}$$

Note that the quantities $\delta c_x(y)$ are well-defined. More importantly, the values of $\tilde{c}_x(y)$ are chosen so that $\tilde{c} \in \mathcal{C}_r(f_{\{u\}})$. (To see this, observe that equality $\tilde{c}_x(y_1)/\tilde{c}_x(y_2) = c_x(y_1)/c_x(y_2)$ holds for every $x \in V \backslash \{u\}$ and every $y_1, y_2 \in D(x)$.) The above procedure is repeated over $f_{Y \cup \{u\}}$ using the new costs $\tilde{c}$, until the value of f is determined.

The linear programming approach for the value dependent cost model is formally described in Fig. 1, where for the sake of efficiency, for each $x \in V \backslash Y$ only c_x^{min} is actually updated. An *implementation* of this meta-algorithm is then obtained by fixing

$$
\boxed{
\begin{array}{l}
\mathcal{LPA}^*(f, V, c) \\
Y \leftarrow \emptyset; \\
\textbf{While the value of } f \text{ is unknown} \\
\qquad \text{Let } s_Y \text{ be a feasible solution for } \mathbf{LP_{f_Y}}. \\
\qquad \text{Let } u \text{ be the unread variable } x \text{ that minimizes } \frac{c_x^{min}}{s_Y(x)}. \\
\qquad \text{Read}(u) \\
\qquad \textbf{For each } v \in V \setminus Y \textbf{ do } c_v^{min} \leftarrow c_v^{min} - s_Y(v) \times \frac{c_u^{min}}{s_Y(u)} \\
\qquad Y = Y \cup \{u\} \\
\textbf{End While} \\
\textbf{Return the value of } f
\end{array}
}
$$

Fig. 1. The "value dependent cost" Linear Programming Approach

the rule used to choose at each iteration the feasible solution of $\mathbf{LP_{f_Y}}$, where Y is the set of variables already probed.

Lemma 1. *Let* $\mathbb{LP}$ *be an implementation of the* $\mathcal{LPA}^*$. *For each* $Y \subset V$, *let* $s_Y(\cdot)$ *be the feasible solution used by* $\mathbb{LP}$ *when the set of variables already read is* Y. *Then, for every* $r \geq 1$,

$$
\gamma_r^{\mathbb{LP}}(f) \leq r \cdot \max_{Y \subset V} \left\{ \sum_{v \in V \setminus Y} s_Y(v) \right\} - r + 1.
$$

Proof. If f has only one variable the result holds. We assume as induction hypothesis that the result holds for every function that depends on less than n variables. Let f be a function that depends on n variables. Let $c \in \mathcal{C}_r(f)$ be a cost function such that $\gamma_c^{\mathbb{LP}}(f) = \gamma_r^{\mathbb{LP}}(f)$, and let σ be an assignment for f that maximizes the ratio $c_{\mathbb{LP}}^f(\sigma)/c^f(\sigma)$. For $U \subseteq V$, we denote $c(U) = \sum_{x \in U} c_x(x(\sigma))$. Furthermore, let X be a cheapest proof for f w.r.t. cost function c and assignment σ. Let us denote $s_\emptyset(\cdot)$ with $s(\cdot)$. It is not hard to see that the 0-cost variables do not affect the competitiveness of $\mathbb{LP}$. Then, let u be the first variable selected by $\mathbb{LP}$ with $c_u^{min} > 0$. Therefore, in particular, $c_u(u(\sigma)) > 0$ and $c_x^{min} > 0$ for all variables $x \in V$. (Here and throughout the proof, c_x^{min} denotes the value before the update.) For $x \in V \setminus \{u\}$ and $y \in D(x)$, we define the new cost function $\tilde{c}(\cdot)$ as in (1).

The total amount that the algorithm spends on f to prove the value of σ is at most the total amount of change in the costs, summed over all the variables, plus the amount that the algorithm spends on the remaining iterations, that is, the cost spent on $f_{\{u\}}$ to prove the value of $\sigma_{V \setminus \{u\}}$ with respect to the new costs $\tilde{c}(\cdot)$. In formulae:

$$
c_{\mathbb{LP}}^f(\sigma) \leq \sum_{v \in V} \delta c_v(v(\sigma)) + \tilde{c}_{\mathbb{LP}}^{f_{\{u\}}}(\sigma_{V \setminus \{u\}}) = \frac{c_u^{min}}{s(u)} \cdot \sum_{v \in V} \frac{c_v(v(\sigma))}{c_v^{min}} \cdot s(v) + \tilde{c}_{\mathbb{LP}}^{f_{\{u\}}}(\sigma_{V \setminus \{u\}}),
$$

$$
(2)
$$

where $\tilde{c}$ is the cost function defined above.

Let X' be a cheapest proof for $f_{\{u\}}$ w.r.t. cost function $\tilde{c}$ and assignment $\sigma_{V \setminus \{u\}}$. Recall that X is a cheapest proof for f w.r.t. cost function c and assignment σ. Note that $X \setminus \{u\}$ is also a proof for $f_{\{u\}}$ w.r.t. assignment $\sigma_{V \setminus \{u\}}$. Then,

$$c(X) = \sum_{v \in X} \delta c_v(v(\sigma)) + \tilde{c}(X \setminus \{u\}) \geq \frac{c_u^{min}}{s(u)} \cdot \sum_{v \in X} \frac{c_v(v(\sigma))}{c_v^{min}} \cdot s(v) + \tilde{c}(X'). \quad (3)$$

Putting together the inequalities (2) and (3) and noting that $\tilde{c}_{\mathbb{LP}}^{f_{\{u\}}}(\sigma_{V \setminus \{u\}})/\tilde{c}(X') \leq \gamma_r^{\mathbb{LP}}(f_{\{u\}})$, we have

$$\gamma_r^{\mathbb{LP}}(f) = \gamma_c^{\mathbb{LP}}(f) = \frac{c_{\mathbb{LP}}^f(\sigma)}{c(X)} \leq \max \left\{ \frac{\sum_{v \in V} \frac{c_v(v(\sigma))}{c_v^{min}} \cdot s(v)}{\sum_{v \in X} \frac{c_v(v(\sigma))}{c_v^{min}} \cdot s(v)}, \gamma_r^{\mathbb{LP}}(f_{\{u\}}) \right\}.$$

We shall now bound the first term in the maximum. In order to simplify formulas, let us write τ_v for $c_v(v(\sigma))/c_v^{min}$. We have that the first term in the maximum becomes

$$\frac{\sum_{v \notin X} \tau_v s(v) + \sum_{v \in X} \tau_v s(v)}{\sum_{v \in X} \tau_v s(v)} \leq \frac{\sum_{v \notin X} r s(v)}{\sum_{v \in X} s(v)} + 1 = \frac{r \sum_{v \in V} s(v) - r \sum_{v \in X} s(v)}{\sum_{v \in X} s(v)} + 1$$

$$\leq r \sum_{v \in V} s(v) - r + 1,$$

where the first inequality follows by $1 \leq \tau_v \leq r$; the equality by writing the summation over $V \setminus X$ as the difference between the summation over V and the one over X; the second inequality follows because $\sum_{v \in X} s(v) \geq 1$ by definition of the linear program $\mathbf{LP_f}$ and the fact that X is a minimal proof for f. Therefore we have

$$\gamma_r^{\mathbb{LP}}(f) \leq \max \left\{ r \sum_{v \in V} s(v) - r + 1, \gamma_r^{\mathbb{LP}}(f_{\{u\}}) \right\}.$$

and since $f_{\{u\}}$ depends on less than n variables, the induction hypothesis yields the desired result. $\qquad \square$

3   Monotone Boolean Functions

By virtue of the above result, it is not hard to provide an upper bound on the extremal competitiveness for monotone Boolean functions in the value dependent cost model.

Let $\Delta(f) = \max_{Y,\sigma} \left\{ \sum_{v \in V \setminus Y} s_{Y,\sigma}^*(v) \right\}$, where the maximum is taken over all possible restrictions $f_{Y,\sigma}$ of f (i.e., $f_{Y,\sigma}$ is defined by an assignment σ of the values to the variables in $Y \subset V$), and where $s_{Y,\sigma}^*(\cdot)$ denotes an optimal solution of $\mathbf{LP_{f_{Y,\sigma}}}$. Recently, Cicalese and Laber have proved in [6] that for a large class of functions, which includes all Boolean functions, $\Delta(f)$ is bounded above by $PROOF(f)$, the size of a largest minimal proof of f. In particular, in conjunction with Lemma 1 this implies that for every $r \geq 1$ and for every Boolean function f, it holds that $\gamma_r(f) \leq r \cdot PROOF(f) - r + 1$.

We shall now provide a lower bound that matches the above upper bound. For monotone Boolean functions, minimal proofs are usually referred to as *maxterms* and *minterms*. A maxterm (minterm) can be defined as a minimal set of variables such that for any σ that sets their value to 0 (1) we have $f(\sigma) = 0$ ($f(\sigma) = 1$). This is used in the following lemma which provides the matching lower bound by generalizing a construction of [2] and [3]. We use $k(f)$ and $l(f)$ to denote the size of the largest minterm and the largest maxterm of f respectively. Thus, $PROOF(f) = \max\{k(f), l(f)\}$.

Theorem 1. *Let f be a monotone Boolean function. Then $\gamma_r(f) \geq r \cdot PROOF(f) - r + 1$.*

Proof. Consider an algorithm $\mathbb{A}$ for evaluating f. We construct an assignment $\sigma^{\mathbb{A}}$ which is 'bad' for $\mathbb{A}$. Let C be a largest minterm of f, i.e., $|C| = k(f)$. For $x \in C$, we set $c_x(1) = r$, and $c_x(0) = 1$. For $x \notin C$, we set $c_x(1) = c_x(0) = 0$. For all variables in C but the last one read by $\mathbb{A}$ we let $x(\sigma^{\mathbb{A}}) = 1$. All the other variables are set to 0.

The algorithm spends $r(|C| - 1) + 1$ to prove that $f(\sigma^{\mathbb{A}}) = 0$. In fact, since C is a minterm, $\mathbb{A}$ cannot conclude that f evaluates to 0 before reading all variables in C. On the other hand, the cheapest proof costs exactly 1 since there is a maxterm of f whose intersection with C is exactly the last variable read by $\mathbb{A}$. Thus, $\gamma_r(f) \geq r(k(f)-1)+1$. By an analogous argument, one can prove that $\gamma_r(f) \geq r(l(f) - 1) + 1$ yielding the desired result. $\qquad\square$

Combining this result with the above upper bound gives the exact value of $\gamma_r(f)$ for monotone Boolean functions.

Theorem 2. *For every $r \geq 1$ and for every monotone Boolean function f, we have*

$$\gamma_r(f) = r \cdot \max\{k(f), l(f)\} - r + 1.$$

4   Game Trees

A game tree T is a tree, rooted at a node r, where every internal node has either a MIN or a MAX label and the parent of every MIN (MAX) node is a MAX (MIN) node. Let V be the set of leaves of T. Every leaf of V is associated with a real number, its value. The value of a MIN (MAX) node is the minimum (maximum) of the values of its children. The function computed by T maps the values of the leaves to the value of the root. We shall identify T with the function it computes. Thus, if f is the function computed by the game tree T, we shall also write T for f and T_Y for f_Y.

By a *minterm* (*maxterm*) of a game tree we shall understand a minimal set of leaves whose values allow to state a lower (upper) bound on the value of the game tree. More precisely, a minterm (maxterm) for a game tree T rooted at r is a minimal set C of leaves of T such that if $x(\sigma) \geq \ell\,(x(\sigma) \leq \ell)$, for each $x \in C$ then $T(\sigma) \geq \ell\,(T(\sigma) \leq \ell)$ regardless of the values of the leaves $y \notin C$. We shall use the more general term *certificate* to either refer to a minterm or to a maxterm. We shall use $\mathcal{F}_T^L$ and $\mathcal{F}_T^U$ to denote the family of all minterms and the family of all maxterms of T, respectively.

As an example, for the game tree function

$$T = \max\{\min\{x_1, x_2, x_3\}, \min\{\max\{x_4, x_5\}, x_6\}\},$$

we have $\mathcal{F}_T^U = \{\{x_1, x_6\}, \{x_2, x_6\}, \{x_3, x_6\}, \{x_1, x_4, x_5\}, \{x_2, x_4, x_5\}, \{x_3, x_4, x_5\}\}$ and $\mathcal{F}_T^L = \{\{x_1, x_2, x_3\}, \{x_4, x_6\}, \{x_5, x_6\}\}$.

These families can be obtained by the following recursive procedure:

- if r is a leaf then $\mathcal{F}_T^L = \mathcal{F}_T^U = \{\{r\}\}$,
- otherwise, let $T_1, \ldots, T_p$ be the subtrees rooted at the children of r. If r is a MIN node then $\mathcal{F}_T^U = \bigcup_{i=1}^p \mathcal{F}_{T_i}^U$ and[1] $\mathcal{F}_T^L = \prod_{i=1}^p \mathcal{F}_{T_i}^L$. If r is a MAX node, $\mathcal{F}_T^L = \bigcup_{i=1}^p \mathcal{F}_{T_i}^L$

---

[1] For all families of sets $\mathcal{F}_1, \mathcal{F}_2, \ldots, \mathcal{F}_k$ we define $\prod_i \mathcal{F}_i$ as follows: $\prod_{i=1}^k \mathcal{F}_i = \{X | X = \bigcup_{i=1}^k X_i, X_i \in \mathcal{F}_i, X_i \neq \emptyset\}$.

and $\mathcal{F}_T^U = \prod_{i=1}^p \mathcal{F}_{T_i}^U$. For the ease of notation, when the function/tree T is clear from the context we shall simply write $\mathcal{F}^U$ and $\mathcal{F}^L$ for $\mathcal{F}_T^U$ and $\mathcal{F}_T^L$.

By the above recursion, it can be verified that every maxterm and every minterm have a unique variable in common. Notice that the number of certificates of a game tree can in general be exponential in the number of leaves. Therefore, an efficient algorithm will never explicitly construct the whole families of certificates.

We shall use $k(T)$ and $l(T)$ to denote the largest minterm and maxterm of T, respectively. These quantities play a critical role in the following lower bound on the extremal competitiveness of every algorithm that evaluates a game tree (in the value dependent cost model).

Theorem 3. *Let T be a game tree. If each certificate of T has size at least 2 then*
$$\gamma_r(T) \geq r \cdot \max\{k(T), l(T)\} - r + 1.$$

Proof. The proof is exactly the same as the proof of Theorem 1. The fact that the cheapest proof is of cost 1 follows from the assumption that each certificate of T has size at least 2. $\square$

Upper Bound. We shall now employ the Linear Programming Approach for obtaining an upper bound on the $(r\text{-})$extremal competitive ratio for game trees that matches the above lower bound.

We need to introduce some more notation. Let T be a game tree on V. Consider a run of an algorithm $\mathbb{A}$ for evaluating T. Let $Y \subseteq V$ denote the set of variables read by $\mathbb{A}$ at some point during its run and let σ_Y be the assignment of real numbers to the leaves in Y corresponding to the variables read. Suppose that the restriction T_Y of T according to the assignment given by σ_Y is non-constant. Let C be a minterm (maxterm) of T. We define the (current) *value* of C as the minimum (maximum) value in σ_Y of the leaves in $Y \cap C$. We say that a minterm (maxterm) is *completely evaluated* if it is entirely contained in Y. Let LB denote the maximum value of a completely evaluated minterm (or $-\infty$, if no minterm has been completely evaluated), and let UB denote the minimum value of a completely evaluated maxterm (or ∞, if no maxterm has been completely evaluated). Note that if UB (LB) is finite, then every minterm (maxterm) has a well-defined value.

In order to study the structure of a proof for T, it is useful to express the function computed by T in terms of its certificates as follows. For every $\sigma \in \mathbb{R}^V$, we have:

$$T(\sigma) = \max_{C^L \in \mathcal{F}^L} \min\{x(\sigma) : x \in C^L\} = \min_{C^U \in \mathcal{F}^U} \max\{x(\sigma) : x \in C^U\}. \tag{4}$$

It follows that LB (UB) is the lower (upper) bound on the value of $T(\sigma)$ for any assignment that extends σ_Y. Moreover, since T_Y is assumed to be non-constant, we have $LB < UB$.

The following lemma (whose proof is omitted due to space limitations) shows that T can evaluate to any value between the two bounds.

Lemma 2. *Let $y \in \mathbb{R}$ such that $LB \leq y \leq UB$. Then, there is an assignment $\sigma \in \mathbb{R}^V$ that extends the current partial assignment σ_Y and such that $T(\sigma) = y$.*

We say that a minterm (maxterm) C is *active* if for each leaf $x \in C \cap Y$ we have $x(\sigma_Y) > LB$ $(x(\sigma_Y) < UB)$. In words, a minterm (maxterm) C is active if the evaluation of its unevaluated leaves can still lead to an improvement in the lower bound LB (upper bound UB), i.e., can provide information on the value of the game tree. Note that if all leaves of a certificate C have already been read, then C is non-active.

The following lemma characterizes the proofs of the restricted game tree T_Y. By saying that a set of variables P is a proof of (a value) y for (a function) f we mean here that P is a proof for f w.r.t. an assignment σ s.t. $f(\sigma) = y$.

Lemma 3 (Proofs of a restricted game tree). *Let $P \subseteq V \backslash Y$. Then:*

(1) **[minterm proofs]** *Suppose that UB is finite. P is a proof of UB for T_Y if and only if there is an active minterm C^L of value at least UB such that $C^L \backslash Y \subseteq P$.*

(2) **[maxterm proofs]** *Suppose that LB is finite. P is a proof of LB for T_Y if and only if there is an active maxterm C^U of value at most LB such that $C^U \backslash Y \subseteq P$.*

(3) **[combined proofs]** *Let $y \in (LB, UB)$. P is a proof of y for T_Y if and only if there is an active minterm C^L of value y^L and an active maxterm C^U of value y^U such that $LB < y^U \leq y \leq y^L < UB$ and such that $(C^L \backslash Y) \cup (C^U \backslash Y) \subseteq P$.*
If $UB = \infty$ then $y^L = \infty$ is allowed. Similarly, if $LB = -\infty$ then $y^U = -\infty$ is allowed.

Proof. We shall prove (1) and (3). Item (2) can be proved similarly as (1).

(1): First, suppose that P is a proof of UB for T_Y w.r.t. an assignment σ_P of values to the variables in P. Let $LB \leq y' < UB$, and consider the assignment σ' that agrees with σ_P on the variables in P and assigns y' to the variables in $V \backslash (Y \cup P)$. By the assumption on P, the restricted game tree T_Y evaluates to UB on σ', or, equivalently, T evaluates to UB on the assignment σ composed of σ_Y and σ'. By equation (4), there is a minterm C^L such that $\min\{x(\sigma) : x \in C^L\} = UB$. In particular, C^L is an active minterm of value UB, with $C^L \backslash Y \subseteq P$ (by the choice of y).

The other direction is considerably simpler. If there is an active minterm C^L of value at least UB such that $C^L \backslash Y \subseteq P$ then assigning UB to each variable in $P \supseteq C^L \backslash Y$ makes C^L evaluate to UB, which in turn raises the lower bound to UB, thus forcing the game tree to evaluate to UB.

(3): Let $LB < y < UB$ and suppose that P is a proof of y for T_Y w.r.t. an assignment σ_P of values to the variables in P. Similarly as above, let $LB \leq y' < y$, and consider the assignment σ' that agrees with σ_P on the variables in P and assigns y' to the variables in $V \backslash (Y \cup P)$. By the assumption on P, T_Y evaluates to y on σ', or, equivalently, T evaluates to y on the assignment σ composed of σ_Y and σ'. By equation (4), there is a minterm C^L such that $\min\{x(\sigma) : x \in C^L\} = y$. Then, C^L is an active minterm of value $y^L \geq y$, with $C^L \backslash Y \subseteq P$. Similarly, there is an active maxterm C^U of value $y^U \leq y$, with $C^U \backslash Y \subseteq P$.

For the converse direction, suppose that there is an active minterm C^L of value y^L and an active maxterm C^U of value y^U such that $LB < y^U \leq y \leq y^L < UB$ and such that $(C^L \cup C^U) \backslash Y \subseteq P$. Since C^L and C^U are active, the sets $C^L \backslash Y$ and $C^U \backslash Y$ are nonempty. Consider the assignment σ_P that assigns y to the variables in $P \supseteq (C^L \backslash Y) \cup (C^U \backslash Y)$. This makes C^L evaluate to y, which implies $T(\sigma) \geq y$ for every assignment

σ that simultaneously extends σ_Y and σ_P. At the same time, C^U gets evaluated to y, which implies $T(\sigma) \leq y$. Therefore, the value of the restricted game tree $T_{Y \cup P}$ is constantly equal to y, proving that P is a proof of y for T_Y. $\square$

For $z \in \mathbb{R}^V$, we denote $\|z\|_1 = \sum_{x \in V} |z(x)|$.

Lemma 4. *There is a solution s_Y to the $\mathbf{LP_{T_Y}}$ such that $\|s_Y\|_1 \leq \max\{k(T), l(T)\}$. Moreover, such a solution can be found in polynomial time.*

Proof. We split the proof into two cases, according to the value of UB.

Case 1. *No maxterm has been completely evaluated yet ($UB = \infty$).* In particular, all the maxterms are active, and there are no minterm proofs. Let H^U be a minimal hitting set of the family $\{C^U \backslash Y : C^U$ is an active maxterm$\}$, and let s_Y be the characteristic vector of H^U. We claim that this s_Y is a solution with the desired properties. Indeed, since there are no minterm proofs, all the minimal proofs contain a member of the family $\{C^U \backslash Y : C^U$ is an active maxterm$\}$, which implies that s_Y is a feasible solution to the linear program $\mathbf{LP_{T_Y}}$. Furthermore, it was shown in [3] that every minimal hitting set of the family $\{C^U \backslash Y : C^U$ is an active maxterm$\}$ is contained in a minterm of T. Hence $\|s_Y\|_1 = |H^U| \leq k(T)$.

Case 2. *There is a completely evaluated maxterm ($UB < \infty$).* In this case, let $\mathcal{P}_1$ denote the family of all minimal minterm proofs, and let $\mathcal{P}_2$ denote the family of all $(C^U \backslash Y)$-parts of the other (i.e., maxterm and combined) minimal proofs. By Lemma 2, the families $\mathcal{P}_1$ and $\mathcal{P}_2$ are nonempty.

Claim

> (i) $\max\{|P| : P \in \mathcal{P}_1 \cup \mathcal{P}_2\} \leq \max\{k(T), l(T)\}$.
> (ii) Every member of $\mathcal{P}_1$ intersects every member of $\mathcal{P}_2$.

Proof of Claim. Part (i) follows from the observations that every element of $\mathcal{P}_1$ is contained in a minterm, and every element of $\mathcal{P}_2$ is contained in a maxterm.

We prove (ii) by contradiction. Suppose that there is a minimal minterm proof $C_0^L \backslash Y$ and a minimal non-minterm proof $(C_1^L \backslash Y) \cup (C_1^U \backslash Y)$, with $(C_1^U \backslash Y)$ nonempty and $(C_1^L \backslash Y)$ possibly empty, such that $(C_0^L \backslash Y) \cap (C_1^U \backslash Y) = \emptyset$. Let y be the value of C_1^U. Then, by the above characterization of minimal proofs, $y < UB$.

Consider the partial assignment σ that extends the current assignment σ_Y by setting all the leaves of $C_0^L \backslash Y$ to UB, and all the leaves of $C_1^U \backslash Y$ to y. Then, the minterm C_0^L proves that the value of T at σ is at least UB, while the maxterm C_1^U proves that the value of T at σ is at most $y < UB$. This is a contradiction, and the proof of the claim is complete.

We recall the following result implicitly contained in [6].

Theorem 4 ([6]). *Let $\mathcal{A}_1, \mathcal{A}_2$ be two nonempty set families over V such that $X \cap Y \neq \emptyset$, for each $X \in \mathcal{A}_1$ and each $Y \in \mathcal{A}_2$. Then, there is a feasible solution s to the linear program $\left\{ \mathbf{Minimize}\ \|s\|_1\ \text{s.t.}\ \sum_{x \in A} s(x) \geq 1\ \forall A \in \mathcal{A}_1 \cup \mathcal{A}_2,\ \text{and}\ s(x) \geq 0\ \forall x \in V \right\}$ such that $\|s\|_1 \leq \max\{|A| : A \in \mathcal{A}_1 \cup \mathcal{A}_2\}$.*

In conjunction with the above claim, this theorem implies that there is a feasible solution s_Y to the linear program

$$\left\{ \textbf{Minimize } \|s_Y\|_1 \text{ s.t. } \sum_{x \in P} s_Y(x) \geq 1 \ \forall P \in \mathcal{P}_1 \cup \mathcal{P}_2, \text{ and } s_Y(x) \geq 0 \ \forall x \in V \backslash Y \right\}$$

such that $\|s_Y\|_1 \leq \max\{k(T), l(T)\}$.

It remains to show that s_Y is a feasible solution to $\textbf{LP}_{\textbf{T}_\textbf{Y}}$. But this follows from the fact that every minimal proof of T_Y contains a member of $\mathcal{P}_1 \cup \mathcal{P}_2$.

This concludes Case 2 and completes the proof of the existence of the desired solution s_Y. We remark that the above solution can be constructed in polynomial time [7]. □

The following result follows from Lemmas 1 and 4 and Theorem 3.

Corollary 1. *Let T be a game tree, and let $r \geq 1$. If each certificate of T has size at least 2 then $\gamma_r(T) = r \cdot \max\{k(T), l(T)\} - r + 1$. Moreover, there is a polynomial time algorithm for evaluating game trees each certificate of which has size at least 2 with optimal r-extremal competitiveness, for each $r \geq 1$.*

In the case when not all the certificates of T are of size at least 2, it is possible to improve the upper bound. We let $p(T)$ $(q(T))$ denote the number of minterms (maxterms) of T of size 1. The following theorem summarizes our findings on the $(r$-)extremal competitiveness for game trees.

Theorem 5. *Let T be a game tree. Then*

$$\gamma(T) = \begin{cases} \max\{k(T), l(T)\}, & \text{if } p(T) = q(T) = 1; \\ \max\{k(T) - q(T), l(T) - p(T)\}, & \text{otherwise.} \end{cases}$$

Furthermore, for each $r \geq 1$, we have $\gamma_r(T) = r \cdot \gamma(T) - r + 1$, and there is a polynomial time algorithm for evaluating game trees with optimal r-extremal competitiveness, for each $r \geq 1$.

5   Concluding Remarks

We believe that the value dependent cost model deserves further investigation, as called by its applications in several situations, particularly in the medical setting. The study of this model with respect to the γ_c competitiveness is a main direction for continued research. Remarkably, already the situation of AND/OR tree functions, whose certificates have a simpler structure than those of the game trees, seems to be challenging. We also remark that the existence of an optimal γ_c-competitive algorithm for game tree functions is still an open problem even in the more classical value independent cost model.

Acknowledgment. We are grateful to Mike Paterson for suggesting to us the idea of studying the value dependent cost model.

References

1. Boros, E., Ünlüyurt, T.: Diagnosing double regular systems. Annals of Mathematics and Artificial Intelligence 26(1-4), 171–191 (1999)
2. Charikar, M., Fagin, R., Guruswami, V., Kleinberg, J.M., Raghavan, P., Sahai, A.: Query strategies for priced information. Journal of Comp. and Syst. Sci. 64, 785–819 (2002)
3. Cicalese, F., Laber, E.S.: A new strategy for querying priced information. In: Proc. of the 37th Annual ACM Symposium on Theory of Computing, pp. 674–683 (2005)
4. Cicalese, F., Laber, E.S.: An optimal algorithm for querying priced information: Monotone Boolean functions and game trees. In: Brodal, G.S., Leonardi, S. (eds.) ESA 2005. LNCS, vol. 3669, pp. 664–676. Springer, Heidelberg (2005)
5. Cicalese, F., Laber, E.S.: On the competitive ratio of evaluating priced functions. In: Proc. of SODA 2006, pp. 944–953 (2006)
6. Cicalese, F., Laber, E.S.: Function evaluation via linear programming in the priced information model. In: Aceto, L., Damgård, I., Goldberg, L.A., Halldórsson, M.M., Ingólfsdóttir, A., Walukiewicz, I. (eds.) ICALP 2008, Part I. LNCS, vol. 5125, pp. 173–185. Springer, Heidelberg (2008)
7. Cicalese, F., Milanič, M.: Computing with priced information: game trees and the value dependent cost model. Technical Report 2008-03, Technical Faculty, Bielefeld University (2008); Also available from the authors' web sites
8. Diderich, C.G.: A bibliography on minimax trees. SIGACT News 24(4), 82–89 (1993)
9. Duffuaa, S.O., Raouf, A.: An optimal sequence in multicharacteristics inspection. J. Optim. Theory Appl. 67(1), 79–86 (1990)
10. Gillies, D.W.: Algorithms to schedule tasks with and/or precedence constraints. PhD thesis, Champaign, IL, USA (1993)
11. Greiner, R., Hayward, R., Jankowska, M., Molloy, M.: Finding optimal satisficing strategies for and-or trees. Artificial Intelligence 170(1), 19–58 (2006)
12. Heiman, R., Wigderson, A.: Randomized vs. deterministic decision tree complexity for read-once boolean functions. Computational Complexity 1, 311–329 (1991)
13. Hellerstein, J.M.: Optimization techniques for queries with expensive methods. ACM Transactions on Database Systems 23(2), 113–157 (1998)
14. Cox, J.L.A., Qiu, Y., Kuehner, W.: Heuristic least-cost computation of discrete classification functions with uncertain argument values. Ann. Oper. Res. 21(1-4), 1–30 (1989)
15. Qiu, Y., Cox Jr., L.A., Davis, L.: Guess-and-verify heuristics for reducing uncertainties in expert classification systems. In: Dubois, D., Wellman, M.P. (eds.) UAI, pp. 252–258. Morgan Kaufmann, San Francisco (1992)
16. Russell, S., Norvig, P.: Artificial Intelligence: A Modern Approach. Prentice Hall, Englewood Cliffs (1995)
17. Saks, M., Wigderson, A.: Probabilistic Boolean decision trees and the complexity of evaluating game trees. In: Proc. of FOCS 1986, pp. 29–38 (1986)
18. Shenoy, P.P.: Game trees for decision analysis. Theory and Decision 44, 149–171 (1998)
19. Snir, M.: Lower bounds on probabilistic linear decision trees. TCS 38(1), 69–82 (1985)
20. Tarsi, M.: Optimal search on some game trees. Journal of the ACM 30(3), 389–396 (1983)
21. Turney, P.: Types of cost in inductive concept learning. In: Proceedings of Workshop Cost-Sensitive Learning at the 17th International Conference on Machine Learning (2000)

Deductive Inference for the Interiors and Exteriors of Horn Theories[*]

Kazuhisa Makino[1] and Hirotaka Ono[2]

[1] Department of Mathematical Informatics, University of Tokyo, Tokyo, 113-8656, Japan
makino@mist.i.u-tokyo.ac.jp
[2] Department of Computer Science and Communication Engineering,
Kyushu University, Fukuoka, 819-0395, Japan
ono@csce.kyushu-u.ac.jp

Abstract. In this paper, we investigate the deductive inference for the interiors and exteriors of Horn knowledge bases, where the interiors and exteriors were introduced by Makino and Ibaraki [11] to study stability properties of knowledge bases. We present a linear time algorithm for the deduction for the interiors and show that it is co-NP-complete for the deduction for the exteriors. Under model-based representation, we show that the deduction problem for interiors is NP-complete while the one for exteriors is co-NP-complete. As for Horn envelopes of the exteriors, we show that it is linearly solvable under model-based representation, while it is co-NP-complete under formula-based representation. We also discuss the polynomially solvable cases for all the intractable problems.

1   Introduction

Knowledge-based systems are commonly used to store the sentences as our knowledge for the purpose of having automated reasoning such as deduction for them (see e.g., [1]). Deductive inference is a fundamental mode of reasoning, and usually abstracted as follows: Given the knowledge base KB, assumed to capture our knowledge about the domain in question, and a query χ that is assumed to capture the situation at hand, decide whether KB implies χ, denoted by $KB \models \chi$, which can be understood as the question: "Is χ consistent with the current state of knowledge ?"

In this paper, we consider the interiors and exteriors of knowledge base. Formally, for a given positive integer α, the α-interior of KB, denoted by $\sigma_{-\alpha}(KB)$, is a knowledge that consists of the models (or assignments) v satisfying that the α-neighbors of v are all models of KB, and the α-exterior of KB, denoted by $\sigma_{\alpha}(KB)$, is a knowledge that consists of the models v satisfying that at least one of the α-neighbors of v is a model of KB [11]. Intuitively, the interior consists of the models v that *strongly* satisfy KB, since all neighbors of v are models of KB, while the exterior consists of the models v that *weakly* satisfy KB, since at least one of the α-neighbors of v is a model of KB. Here we note that v might not satisfy KB, even if we say that it weakly satisfies KB. As mentioned in [11], the interiors and exteriors of knowledge base merit study in their

---

[*] This work is supported in part by the Grant-in-Aid of the Ministry of Education, Science, Sports and Culture of Japan and by the Asahi glass foundation.

S.-H. Hong, H. Nagamochi, and T. Fukunaga (Eds.): ISAAC 2008, LNCS 5369, pp. 390–401, 2008.

own right, since they shed light on the structure of knowledge base. Moreover, let us consider the situation in which knowledge base KB is *not perfect* in the sense that some sentences in KB are wrong and/or some are missing in KB (see also [11]).

Suppose that we use KB as a knowledge base for automated reasoning, say, duductive inference $KB \models \chi$. Since KB does not represent *real* knowledge KB^*, the reasoning result is no longer true. However, if we use the interior $\sigma_{-\alpha}(KB)$ of KB as a knowledge base and have $\sigma_{-\alpha}(KB) \not\models \chi$, then we can expect that the result is ture for real knowledge KB^*, since $\sigma_{-\alpha}(KB)$ consists of models which strongly satisfy KB. On the other hand, if we use the exterior $\sigma_{\alpha}(KB)$ of KB as a knowledge base and have $\sigma_{\alpha}(KB) \models \chi$, then we can expect that the result is ture for real knowledge KB^*, since $\sigma_{\alpha}(KB)$ consists of models which weakly satisfy KB. In this sense, the interiors and exteriors help to have *safe* reasoning.

Main problems considered. In this paper, we study the deductive inference for the interiors and exteriors of propositional Horn theories, where Horn theories are ubiquitous in Computer Science, cf. [14], and are of particular relevance in Artificial Intelligence and Databases. It is known that important reasoning problems like deductive inference and satisfiability checking, which are intractable in general, are solvable in linear time for Horn theories (cf. [3]).

More precisely, we address the following problems:

• Given a Horn theory Σ, a clause c, and an integer $\alpha > 0$, we consider the problems of deciding if deductive queries hold for the α-interior and exterior of Σ, i.e., $\sigma_{-\alpha}(\Sigma) \models c$ and $\sigma_{\alpha}(\Sigma) \models c$. It is well-known [3] that a deductive query for a Horn theory can be answered in linear time. Note that it is intractable to construct the interior and exterior for a Horn theory [11, 13], and hence a direct method (i.e., first construct the interior (or exterior) and then check a deductive query) is not possible efficiently.

• We contrast traditional formula-based (syntactic) with model-based (semantic) representation of Horn theories. The latter form of representation has been proposed as an alternative form of representing and accessing a logical knowledge base, cf. [2,4,5,7,8, 6,9,10]. In model-based reasoning, Σ is represented by a subset of its models $\mathcal{M}$, which are commonly called *characteristic models*. As shown in [7], the deductive inference can be solved in polynomial time, given its characteristic models.

• Finally, we consider Horn approximations for the exteriors of Horn theories. Note that the interiors of Horn theories are Horn, while the exteriors might not be Horn. We deal with the least upper bounds, called the *Horn envelopes* [16], for the exteriors of Horn theories.

Main results. We investigate the problems mentioned above from an algorithmical viewpoint. For all the problems, we provide either polynomial time algorithms or proofs of the intractability; thus, our work gives a complete picture of the tractability/ intractability frontier of deduction for interiors and exteriors of Horn theories. Our main results can be summarized as follows (see Figure 1).

• We present a linear time algorithm for the deduction for the interiors of a given Horn theory, and show that it is co-NP-complete for the deduction for the exteriors. Thus, the positive result for ordinary deduction for Horn theories extends to the interiors, but does not to the exteriors. We also show that the deduction for the exteriors is possible in polynomial time, if α is bounded by a constant or if $|N(c)|$ is bounded by a logarithm of the input size, where $N(c)$ corresponds to the set of negative literals in c.

- Under model-based representation, we show that the consistency problem and the deduction for the interiors of Horn theories are both co-NP-complete. As for the exteriors, we show that the deduction is co-NP-complete. We also show that the deduction for the interiors is possible in polynomial time if α is bounded by a constant, and so is for the exteriors, if α or $|P(c)|$ is bounded by a constant, or if $|N(c)|$ is bounded by a logarithm of the input size, where $P(c)$ corresponds to the set of positive literals in c.

- As for Horn envelopes of the exteriors of Horn theories, we show that it is linearly solvable under model-based representation, while it is co-NP-complete under formula-based representation. The former contrasts to the negative result for the exteriors. We also present a polynomial algorithm for formula-based representation, if α is bounded by a constant or if $|N(c)|$ is bounded by a logarithm of the input size.

	Interiors	Exteriors	Envelopes of Exteriors
Formula-Based	P	co-NP-complete*	co-NP-complete*
Model-Based	NP-complete†	co-NP-complete‡	P

*: It is polynomially solvable, if $\alpha = O(1)$ or $|N(c)| = O(\log \|\Sigma\|)$.

†: It is polynomially solvable, if $\alpha = O(1)$.

‡: It is polynomially solvable, if $\alpha = O(1)$, $|P(c)| = O(1)$, or $|N(c)| = O(\log n |chr(\Sigma)|)$.

Fig. 1. Complexity of the deduction for interiors and exteriors of Horn theories

The rest of the paper is organized as follows. In the next section, we review the basic concepts and fix notations. Sections 3 and 4 investigate the deductive inference for the interiors and exteriors of Horn theories. Section 5 considers the deductive inference for the envelopes of the exteriors of Horn theories. Most of the proofs are omitted due to space limitation. Interested readers can find the omitted parts in [12], which is a technical report version of the paper.

2 Preliminaries

Horn Theories. We assume a standard propositional language with atoms $At = \{x_1, x_2, \ldots, x_n\}$, where each x_i takes either value 1 (true) or 0 (false). A *literal* is either an atom x_i or its negation, which we denote by $\bar{x}_i$. The opposite of a literal ℓ is denoted by $\bar{\ell}$, and the opposite of a set of literals L by $\bar{L} = \{\bar{\ell} \mid \ell \in L\}$. Furthermore, $Lit = At \cup \bar{At}$ denotes the set of all literals.

A *clause* is a disjunction $c = \bigvee_{i \in P(c)} x_i \vee \bigvee_{i \in N(c)} \bar{x}_i$ of literals, where $P(c)$ and $N(c)$ are the sets of indices whose corresponding variables occur positively and negatively in c and $P(c) \cap N(c) = \emptyset$. Dually, a *term* is conjunction $t = \bigwedge_{i \in P(t)} x_i \wedge \bigwedge_{i \in N(t)} \bar{x}_i$ of literals, where $P(t)$ and $N(t)$ are similarly defined. We also view clauses and terms as sets of literals. A *conjunctive normal form* (*CNF*) is a conjunction of clauses. A clause c is *Horn*, if $|P(c)| \leq 1$. A *theory* Σ is any set of formulas; it is *Horn*, if it is a set of Horn clauses. As usual, we identify Σ with $\varphi = \bigwedge_{c \in \Sigma} c$, and write $c \in \varphi$ etc. It is known [3] that the deductive query for a Horn theory, i.e., deciding if $\Sigma \models c$ for a clause c is possible in linear time.

We recall that Horn theories have a well-known semantic characterization. A *model* is a vector $v \in \{0, 1\}^n$, whose i-th component is denoted by v_i. For a model v, let $ON(v) = \{i \mid v_i = 1\}$ and $OFF(v) = \{i \mid v_i = 0\}$. The value of a formula φ on a model v, denoted $\varphi(v)$, is inductively defined as usual; satisfaction of φ in v, i.e., $\varphi(v) = 1$, will be denoted by $v \models \varphi$. The set of models of a formula φ (resp., theory Σ), denoted by $mod(\varphi)$ (resp., $mod(\Sigma)$), and logical consequence $\varphi \models \psi$ (resp., $\Sigma \models \psi$) are defined as usual. For two models v and w, we denote by $v \leq w$ the usual componentwise ordering, i.e., $v_i \leq w_i$ for all $i = 1, 2, \ldots, n$, where $0 \leq 1$; $v < w$ means $v \neq w$ and $v \leq w$. Denote by $v \wedge w$ componentwise AND of models $v, w \in \{0, 1\}^n$, and by $Cl_\wedge(M)$ the closure of $M \subseteq \{0, 1\}^n$ under $\wedge$. Then, a theory Σ is Horn representable if and only if $mod(\Sigma) = Cl_\wedge(mod(\Sigma))$ (see [2,9]) for proofs).

Example 1. Consider $M_1 = \{(0101), (1001), (1000)\}$ and $M_2 = \{(0101), (1001), (1000), (0001), (0000)\}$. Then, for $v = (0101)$, $w = (1000)$, we have $w, v \in M_1$, while $v \wedge w = (0000) \notin M_1$; hence M_1 is not the set of models of a Horn theory. On the other hand, $Cl_\wedge(M_2) = M_2$, thus $M_2 = mod(\Sigma)$ for some Horn theory Σ.

As discussed by Kautz *et al.* [7], a Horn theory Σ is semantically represented by its characteristic models, where $v \in mod(\Sigma)$ is called *characteristic* (or *extreme* [2]), if $v \notin Cl_\wedge(mod(\Sigma) \setminus \{v\})$. The set of all such models, the *characteristic set of* Σ, is denoted by $chr(\Sigma)$. Note that $chr(\Sigma)$ is unique. E.g., $(0101) \in chr(\Sigma_2)$, while $(0000) \notin chr(\Sigma_2)$; we have $chr(\Sigma_2) = M_1$. It is known [7] that the deductive query for a Horn theory Σ from the characteristic set $chr(\Sigma)$ can be done in linear time, i.e., $O(n|chr(\Sigma)|)$ time.

Interior and Exterior of Theories. For a model $v \in \{0, 1\}^n$ and an integer $\alpha > 0$, its α-*neighborhood* is defined by

$$N_\alpha(v) = \{w \in \{0, 1\}^n \mid \|\, w - v\, \| \leq \alpha\},$$

where $\| v \|$ denotes $\sum_{i=1}^n |v_i|$. For a theory Σ and an integer $\alpha > 0$, the α-interior and α-exterior of Σ, denoted by $\sigma_{-\alpha}(\Sigma)$ and $\sigma_\alpha(\Sigma)$ respectively, are theories defined by

$$mod(\sigma_{-\alpha}(\Sigma)) = \{v \in \{0, 1\}^n \mid N_\alpha(v) \subseteq mod(\Sigma)\} \tag{1}$$

$$mod(\sigma_\alpha(\Sigma)) = \{v \in \{0, 1\}^n \mid N_\alpha(v) \cap mod(\Sigma) \neq \emptyset\}. \tag{2}$$

By definition, $\sigma_0(\Sigma) = \sigma$, $\sigma_\alpha(\Sigma) \models \sigma_\beta(\Sigma)$ for integers α and β with $\alpha < \beta$, and $\sigma_\alpha(\Sigma_1) \models \sigma_\alpha(\Sigma_2)$ holds for any integer α, if two theories Σ_1 and Σ_2 satisfy $\Sigma_1 \models \Sigma_2$.

Example 2. Let us consider a Horn theory $\Sigma = \{\bar{x}_1 \vee x_3, \bar{x}_2 \vee x_3, \bar{x}_2 \vee x_4\}$ of 4 variables, where $mod(\Sigma)$ is given by

$$mod(\Sigma) = \{(1111), (1011), (1010), (0111), (0011), (0010), (0001), (0000)\}$$

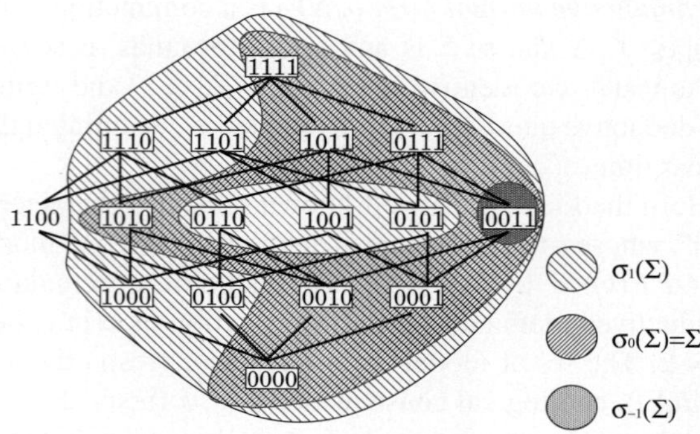

Fig. 2. A Horn theory and its interiors and exteriors

(See Figure 2). Then we have $\sigma_\alpha(\Sigma) = \{\emptyset\}$ for $\alpha \leq -2$, $\{\overline{x}_1, \overline{x}_2, x_3, x_4\}$ for $\alpha = -1$, Σ for $\alpha = 0$, $\{\overline{x}_1 \vee \overline{x}_2 \vee x_3 \vee x_4\}$ for $\alpha = 1$, and $\emptyset$ for $\alpha \geq 2$. For example, (0011) is the unique model of $mod(\sigma_{-1}(\Sigma))$, since $\mathcal{N}_1(0011) \subseteq mod(\Sigma)$ and $\mathcal{N}_1(v) \not\subseteq mod(\Sigma)$ holds for all the other models v. For the 1-exterior, we can see that all models v with $(\overline{x}_1 \vee \overline{x}_2 \vee x_3 \vee x_4)(v) = 1$ satisfy $\mathcal{N}_1(v) \cap mod(\Sigma) \neq \emptyset$, and no other such model exists. For example, (0101) is a model of $\sigma_1(\Sigma)$, since $(0111) \in \mathcal{N}_1(0101) \cap mod(\Sigma)$. On the other hand, (1100) is not a model of $\sigma_1(\Sigma)$, since $\mathcal{N}_1(1100) \cap mod(\Sigma) = \emptyset$. Notice that $\sigma_{-1}(\Sigma)$ is Horn, while $\sigma_1(\Sigma)$ is not.

Makino and Ibaraki [11] introduced the interiors and exteriors to analyze stability of Boolean functions, and studied their basic properties and complexity issues on them (see also [13]). For example, it is known [11] that, for a theory Σ and nonnegative integers α and β, $\sigma_{-\alpha}(\sigma_{-\beta}(\Sigma)) = \sigma_{-\alpha-\beta}(\Sigma)$, $\sigma_\alpha(\sigma_\beta(\Sigma)) = \sigma_{\alpha+\beta}(\Sigma)$, and

$$\sigma_\alpha(\sigma_{-\beta}(\Sigma)) \models \sigma_{\alpha-\beta}(\Sigma) \models \sigma_{-\beta}(\sigma_\alpha(\Sigma)). \tag{3}$$

For an integer $\alpha > 0$ and two theories Σ_1 and Σ_2, we have

$$\sigma_{-\alpha}(\Sigma_1 \cup \Sigma_2) = \sigma_{-\alpha}(\Sigma_1) \cup \sigma_{-\alpha}(\Sigma_2) \tag{4}$$

$$\sigma_\alpha(\Sigma_1 \cup \Sigma_2) \models \sigma_\alpha(\Sigma_1) \cup \sigma_\alpha(\Sigma_2), \tag{5}$$

where $\sigma_\alpha(\Sigma_1 \cup \Sigma_2) \neq \sigma_\alpha(\Sigma_1) \cup \sigma_\alpha(\Sigma_2)$ holds in general.

As demonstrated in Example 2, it is not difficult to see that the interiors of any Horn theory are Horn, which is, for example, proved by (4) and Lemma 1, while the exteriors might be not Horn.

3   Deductive Inference from Horn Theories

In this section, we investigate the deductive inference for the interiors and exteriors of a given Horn theory.

3.1 Interiors

Let us first consider the deduction for the α-interiors of a Horn theory: Given a Horn theory Σ, a clause c, and a positive integer α, decide if $\sigma_{-\alpha}(\Sigma) \models c$ holds. We show that the problem is solvable in linear time after showing a series of lemmas.

The following lemma is a basic property of the interiors.

Lemma 1. *Let c be a clause. Then for an integer $\alpha > 0$, we have $\sigma_{-\alpha}(c) = \bigvee_{\substack{S \subseteq c: \\ |S|=\alpha+1}} (\bigwedge_{\ell \in S} \ell) = \bigwedge_{\substack{S \subseteq c: \\ |S|=|c|-\alpha}} (\bigvee_{\ell \in S} \ell).$*

This, together with (4), implies that for a CNF φ and an integer $\alpha > 0$, we have

$$\sigma_{-\alpha}(\varphi) = \bigwedge_{c \in \varphi} \Big(\bigvee_{\substack{S \subseteq c: \\ |S|=\alpha+1}} (\bigwedge_{\ell \in S} \ell) \Big) = \bigwedge_{c \in \varphi} \Big(\bigwedge_{\substack{S \subseteq c: \\ |S|=|c|-\alpha}} (\bigvee_{\ell \in S} \ell) \Big),$$

where we regard c as a set of literals.

Lemma 2. *Let Σ be a Horn theory, and let c be a clause. For an integer $\alpha > 0$, if there exists a clause $d \in \Sigma$ such that $|N(d) \setminus N(c)| \leq \alpha - 1$ or $(|N(d) \setminus N(c)| = \alpha$ and $P(d) \subseteq P(c))$, then we have $\sigma_{-\alpha}(\Sigma) \models c$.*

Lemma 3. *Let Σ be a Horn theory, and let c be a clause. For an integer $\alpha > 0$, if* (i) *$|N(d) \setminus N(c)| \geq \alpha$ holds for all $d \in \Sigma$ and* (ii) *$\emptyset \neq P(d) \subseteq N(c)$ holds for all $d \in \Sigma$ with $|N(d) \setminus N(c)| = \alpha$, then we have $\sigma_{-\alpha}(\Sigma) \not\models c$.* $\square$

By Lemmas 2 and 3, we can easily answer the deductive queries, if Σ satisfies certain conditions mentioned in them. In the remaining case, we have the following lemma.

Lemma 4. *For a Horn theory Σ that satisfies none of the conditions in Lemmas 2 and 3, let d be a clause in Σ such that $|N(d) \setminus N(c)| = \alpha$, and $P(d) = P(d) \setminus (P(c) \cup N(c)) = \{j\}$. Then $\sigma_{-\alpha}(\Sigma) \models c \vee x_j$ holds.*

Proof. By Lemma 1, we have $\sigma_{-\alpha}(d) \models \bigvee_{i \in N(c) \cap N(d)} \bar{x}_i \vee x_j \models c \vee x_j$. This implies $\sigma_{-\alpha}(\Sigma) \models c \vee x_j$ by (4). $\square$

From this lemma, we have only to check $\sigma_{-\alpha}(\Sigma) \models c \vee \bar{x}_j$, instead of $\sigma_{-\alpha}(\Sigma) \models c$. Since $|c| < |c \vee \bar{x}_j| \leq n$, we can answer the deduction by checking the conditions in Lemmas 2 and 3 at most n times.

We can see that a straightforward implementation of the algorithm requires $O(n(\|\Sigma\| + |c|))$ time, where $\|\Sigma\|$ denotes the length of Σ, i.e., $\|\Sigma\| = \sum_{d \in \Sigma} |d|$, though we can implement a linear time algorithm by adopting a proper data structure.

Theorem 1. *Given a Horn theory Σ, a clause c and an integer $\alpha > 0$, a deductive query $\sigma_{-\alpha}(\Sigma) \models c$ can be answered in linear time, i.e., $O(\|\Sigma\| + |c|)$ time.* $\square$

3.2 Exteriors

Let us next consider the deduction for the α-exteriors of a Horn theory. In contrast to the interior case, we have the following negative result.

Algorithm 1. Deduction-Interior-from-Horn-Theory

Input: A Horn theory Σ, a clause c and an integer $\alpha > 0$.

Output: Yes, if $\sigma_{-\alpha}(\Sigma) \models c$; Otherwise, No.

Step 0. Let $N := N(c)$ and $P := P(c)$.

Step 1. /* Check the condition in Lemma 2. */
   If there exists a clause $d \in \Sigma$ such that $|N(d) \setminus N| \leq \alpha - 1$ or $(|N(d) \setminus N| = \alpha$ and $P(d) \subseteq P)$,
   then output Yes and halt.

Step 2. /* Check the condition in Lemma 3. */
   If $P(d) \subseteq N$ holds for all $d \in \Sigma$ with $|N(d) \setminus N| = \alpha$, **then** output No and halt.

Step 3. /* Update N by Lemma 4. */
   For a clause d in Σ such that $|N(d) \setminus N| = \alpha$ and $P(d) = P(d) \setminus (P \cup N) = \{j\}$, update
   $N := N \cup \{j\}$ and return to Step 1. $\square$

Theorem 2. *Given a Horn theory Σ, a clause c and a positive integer α, it is* co-NP-*complete to decide whether a deductive query $\sigma_\alpha(\Sigma) \models c$ holds, even if $P(c) = \emptyset$.* $\square$

We remark that this result can also be derived from the ones in [11].

However, by using the next lemma, a deductive query can be answered in polynomial time, if α or $N(c)$ is small.

Lemma 5. *Let Σ_1 and Σ_2 be theories. For an integer $\alpha > 0$, Then $\sigma_\alpha(\Sigma_1) \models \Sigma_2$ if and only if $\Sigma_1 \models \sigma_{-\alpha}(\Sigma_2)$.* $\square$

From Lemma 5, the deductive query for the α-interior of a theory Σ, i.e., $\sigma_\alpha(\Sigma) \models c$ for a given clause c is equivalent to the condition that $\Sigma \models \sigma_{-\alpha}(c)$. Since we have $\sigma_{-\alpha}(c) = \bigwedge_{\substack{S \subseteq c: \\ |S| = |c| - \alpha}} \left(\bigvee_{\ell \in S} \ell \right)$ by Lemma 1, the deductive query for the α-interior can be done by checking $\binom{|c|}{\alpha}$ deductions for Σ. More precisely, we have the following lemma.

Lemma 6. *Let Σ be a Horn theory, let c be a clause, and $\alpha > 0$ be an integer. Then $\sigma_\alpha(\Sigma) \models c$ holds if and only if, for each subset S of $N(c)$ such that $|S| \geq |N(c)| - \alpha$, at least $(\alpha - |N(c)| + |S| + 1)$ j's in $P(c)$ satisfy $\Sigma \models \bigvee_{i \in S} \overline{x}_i \vee x_j$.* $\square$

This lemma implies that the deductive query can be answered by checking the number of j's in $P(c)$ that satisfy $\Sigma \models \bigvee_{i \in S} \overline{x}_i \vee x_j$ for each S. Since we can check this condition in linear time and there are $\sum_{p=0}^{\alpha} \binom{|N(c)|}{p}$ such S's, we have the following result, which complements Theorem 2 that the problem is intractable, even if $P(c) = \emptyset$.

Theorem 3. *Let Σ be a Horn theory, let c be a clause, and let $\alpha > 0$ be an integer. Then a deductive query $\sigma_\alpha(\Sigma) \models c$ can be answered in $O\left(\sum_{p=0}^{\alpha} \binom{|N(c)|}{p}\right) \|\Sigma\| + |P(c)|\right)$ time. In particular, it is polynomially solvable, if $\alpha = O(1)$ or $|N(c)| = O(\log \|\Sigma\|)$.*

4   Deductive Inference from Characteristic Sets

In this section, we consider the case when Horn knowledge bases are represented by characteristic sets. Different from formula-based representation, the deductions for interiors and exteriors are both intractable, unless P=NP.

4.1   Interiors

We first present an algorithm to solve the deduction problem for the interiors of Horn theories. The algorithm requires exponential time in general, but it is polynomial when α is small.

Let Σ be a Horn theory given by its characteristic set $chr(\Sigma)$, and let c be a clause. Then for a nonnegative integer α, we have

$$\sigma_{-\alpha}(\Sigma) \models c \text{ if and only if } \sigma_{-\alpha}(\Sigma) \wedge \bar{c} \equiv 0. \tag{6}$$

Let v^* be a unique minimal model such that $c(v^*) = 0$ (i.e., $\bar{c}(v^*) = 1$). By the definition of interiors, v^* is a model of $\sigma_{-\alpha}(\Sigma)$ if and only if all v's in $\mathcal{N}_\alpha(v^*)$ are models of Σ. Therefore, for each model v in $\mathcal{N}_\alpha(v^*)$, we check if $v \in mod(\Sigma)$, which is equivalent to

$$\bigwedge_{\substack{w \in chr(\Sigma) \\ w \geq v}} w = v. \tag{7}$$

If (7) holds for all models v in $\mathcal{N}_\alpha(v^*)$, then we can immediately conclude by (6) that $\sigma_{-\alpha}(\Sigma) \not\models c$. On the other hand, if there exists a model v in $\mathcal{N}_\alpha(v^*)$ such that (7) does not hold, let $J = ON(\bigwedge_{\substack{w \in chr(\Sigma) \\ w \geq v}} w) \setminus ON(v)$. By definition, we have $J \neq \emptyset$, and we can see that

$$\sigma_{-\alpha}(\Sigma) \models \bigvee_{i \in ON(v)} \bar{x}_i \vee x_j \quad \text{for all } j \in J. \tag{8}$$

If $J \cap N(c) \neq \emptyset$, by Lemma 1 and (8), we have $\sigma_{-\alpha}(\Sigma) \models \bigvee_{i \in ON(v) \cap N(c)} \bar{x}_i$, since $|ON(v) \setminus N(c)| \leq \alpha - 1$. This implies $\sigma_{-\alpha}(\Sigma) \models c$. On ther other hand, if $J \cap N(c) = \emptyset$, then by Lemma 1 and (8), we have $\sigma_{-\alpha}(\Sigma) \models \bigvee_{i \in N(c)} \bar{x}_i \vee x_j$ for all $j \in J$. Thus, if J contains an index in $P(c)$, we can conclude that $\sigma_{-\alpha}(\Sigma) \models c$; Otherwise, we check the condition $\sigma_{-\alpha}(\Sigma) \models c \vee \bigvee_{j \in J} \bar{x}_j$, instead of $\sigma_{-\alpha}(\Sigma) \models c$. Since a new clause $d = c \vee \bigvee_{j \in J} \bar{x}_j$ is longer than c, after at most n iterations, we can answer the deductive query. Formally, our algorithm can be described as Algorithm 2.

Theorem 4. *Given the characteristic model $chr(\Sigma)$ of a Horn theory Σ, a clause c and a nonnegative integer α, a deductive query $\sigma_{-\alpha}(\Sigma) \models c$ can be answered in $O(n^{\alpha+2}|chr(\Sigma)|)$ time. In particular, it is polynomially solvable, if $\alpha = O(1)$.* $\square$

However, in general, the problem is intractable, which contrasts to the formula-model representation.

Theorem 5. *Given the characteristic set $chr(\Sigma)$ of a Horn theory Σ and a positive integer α, it is co-NP-complete to decide whether $\sigma_{-\alpha}(\Sigma)$ is consistent, i.e., $mod(\sigma_{-\alpha}(\Sigma)) \neq \emptyset$.* $\square$

This result immediately implies the following corollary.

Corollary 1. *Given the characteristic set $chr(\Sigma)$ of a Horn theory Σ, a clause c and a positive integer α, it is NP-complete to decide whether a deductive query $\sigma_{-\alpha}(\Sigma) \models c$ holds, even if $c = \emptyset$.* $\square$

Note that, different from the other hardness results, the hardness is not sensitive to the size of c.

Algorithm 2. Deduction-Interior-from-Charset

Input: The characteristic set $chr(\Sigma)$ of a Horn theory Σ, a clause c and a nonnegative integer α.

Output: Yes, if $\sigma_{-\alpha}(\Sigma) \models c$; Otherwise, No.

Step 0. Let $N := N(c)$, $d := c$ and $q := 1$.

Step 1. Let v^* be the unique minimal model such that $d(v^*) = 0$.

Step 2. **For** each v in $\mathcal{N}_\alpha(v^*)$ **do**

        If (7) does not hold,

           then let $v^{(q)} := v$, $J := ON(\bigwedge_{\substack{w \in chr(\Sigma) \\ w \geq v}} w) \setminus ON(v)$ and

               $q := q + 1$

           If $J \cap (N \cup P(c)) \neq \emptyset$, **then** output yes and halt.

           Let $N := N \cup J$ and $d := \bigvee_{i \in N} \overline{x}_i \vee \bigvee_{i \in P(c)} x_i$.

           Go to Step 1.

      end{for}

Step 3. Output No and halt. $\square$

4.2 Exteriors

Let us consider the exteriors. Similarly to the formula-based representation, we have the following negative result.

Theorem 6. *Given the characteristic set $chr(\Sigma)$ of a Horn theory Σ, a clause c and a positive integer α, it is* co-NP-*complete to decide if a deductive query $\sigma_\alpha(\Sigma) \models c$ holds.* $\square$

By using Lemma 6, we can see that the problem can be solved in polynomial time, if α or $|N(c)|$ is small. Namely, for each subset S of $N(c)$ such that $|S| \geq |N(c)| - \alpha$, let v^S denotes the model such that $ON(v^S) = S$. Then $w^S = \bigwedge_{\substack{w \in chr(\Sigma): \\ w \geq v^S}} w$ is the unique minimal model of Σ such that $ON(w^S) \supseteq S$, and hence it follows from Lemma 6 that it is enough to check if $|ON(w^s) \cap P(c)| \geq \alpha - |N(c)| + |S| + 1$. Clearly, this can be done in in $O\left(\sum_{p=0}^{\alpha} \binom{|N(c)|}{p} n |chr(\Sigma)|\right)$ time.

Moreover, if $|P(c)|$ is small, then the problem also becomes tractable, which contrasts with Theorem 2.

Lemma 7. *Let Σ be a Horn theory, let c be a clause, and α be a nonnegative integer. Then $\sigma_\alpha(\Sigma) \models c$ holds if and only if each $S \subseteq P(c)$ such that $|S| \geq |P(c)| - \alpha$ satisfies*

$$|OFF(w) \cap N(c)| \geq \alpha - |P(c)| + |S| + 1 \tag{9}$$

for all models w of Σ such that $OFF(w) \cap P(c) = S$.

Note that (9) is monotone in the sense that, if a model w satisfies (9), then all models v with $v < w$ also satisfy it. Thus it is sufficient to check if (9) holds for all *maximal* models w of Σ such that $OFF(w) \cap P(c) = S$. Since such maximal models w can be obtained from $w^{(i)}$ $(i \in S)$ with $i \in OFF(w^{(i)}) \cap P(c) \subseteq S$ by their intersection $w = \bigwedge_{i \in S} w^{(i)}$, we can answer the deduction problem in $O\left(n \sum_{p=|P(c)|-\alpha}^{|P(c)|} \binom{|P(c)|}{p} |chr(\Sigma)|^p\right)$ time.

Theorem 7. *Given the characteristic set $chr(\Sigma)$ of a Horn theory, a clause c, and an integer $\alpha \geq 0$, a deductive query $\sigma_\alpha(\Sigma) \models c$ can be answered in $O\left(n \min\{\sum_{p=0}^{\alpha} \binom{|N(c)|}{p} |chr(\Sigma)|, \sum_{p=|P(c)|-\alpha}^{|P(c)|} \binom{|P(c)|}{p} |chr(\Sigma)|^p\}\right)$ time. In particular, it is polynomially solvable, if $\alpha = O(1)$, $|P(c)| = O(1)$, or $|N(c)| = O(\log n \cdot |chr(\Sigma)|)$.*

5 Deductive Inference for Envelopes of the Exteriors of Horn Theories

We have considered the deduction for the interiors and exteriors of Horn theories. As mentioned before, the interiors of Horn theories are also Horn, while this does not hold for the exteriors. This means that the exteriors of Horn theories might lose beneficial properties of Horn theories. One of the ways to overcome such a hurdle is *Horn Approximation*, that is, approximating a theory by a Horn theory [16]. There are several methods for approximation, but one of the most natural ones is to approximate a theory by its *Horn envelope*. For a theory Σ, its *Horn envelope* is the Horn theory Σ_e such that $mod(\Sigma_e) = Cl_\wedge(mod(\Sigma))$. Since Horn theories are closed under intersection, Horn envelope is the least Horn upper bound for Σ, i.e., $chr(\Sigma_e) \supseteq chr(\Sigma)$ and there exists no Horn theory Σ^* such that $chr(\Sigma_e) \supsetneq chr(\Sigma^*) \supseteq chr(\Sigma)$. In this section, we consider the deduction for Horn envelopes of interiors of Horn theories; $\sigma_\alpha(\Sigma)_e \models c$.

5.1 Model-Based Representations

Let us first consider the case in which knowledge bases are represented by characteristic sets.

Proposition 1. *Let Σ be a Horn theory, and let α be a nonnegative integer. Then we have*

$$mod(\sigma_\alpha(\Sigma)_e) = Cl_\wedge(\bigcup_{v \in chr(\Sigma)} N_\alpha(v)). \tag{10}$$

$\square$

For a clause c, let v^* be the unique minimal model such that $c(v^*) = 0$. We recall that, for a Horn theory Φ,

$$\Phi \models c \text{ if and only if } c(\bigwedge_{\substack{v \in chr(\Phi) \\ v \geq v^*}} v) = 1. \tag{11}$$

Therefore, Proposition 1 immediately implies an algorithm for the deduction for $\sigma_\alpha(\Sigma)_e$ from $chr(\Sigma)$, since we have $chr(\sigma_\alpha(\Sigma)_e) \subseteq \bigcup_{v \in chr(\Sigma)} N_\alpha(v)$. However, for a general α, $\bigcup_{v \in chr(\Sigma)} N_\alpha(v)$ is exponentially larger than $chr(\Sigma)$, and hence this direct method is not efficient. The following lemma helps developing a polynomial time algorithm.

Lemma 8. *Let Σ be a Horn theory, let c be a clause, and let α be a nonnegative integer. Then $\sigma_\alpha(\Sigma)_e \models c$ holds if and only if the following two conditions are satisfied.*

(i) *$|OFF(v) \cap N(c)| \geq \alpha$ holds for all $v \in chr(\Sigma)$.*

(ii) *If $S = \{v \in chr(\Sigma) \mid |OFF(v) \cap N(c)| = \alpha\} \neq \emptyset$, $P(c)$ is not covered with $OFF(v)$ for models v in S, i.e., $P(c) \not\subseteq \bigcup_{\substack{v \in chr(\Sigma) \\ |OFF(v) \cap N(c)| = \alpha}} OFF(v)$.* $\square$

The lemma immediately implies the following theorem.

Theorem 8. *Given the characteristic set $chr(\Sigma)$ of a Horn theory Σ, a clause c, and an integer $\alpha \geq 0$, a deductive query $\sigma_\alpha(\Sigma)_e \models c$ can be answered in linear time.*

We remark that this contrasts with Corollary 1. Namely, if we are given the characteristic set $chr(\Sigma)$ of a Horn theory Σ, $\sigma_\alpha(\Sigma)_e \models c$ is polynomially solvable, while it is co-NP-complete to decide if $\sigma_\alpha(\Sigma) \models c$.

5.2   Formula-Based Representation

Recall that a *negative* theory (i.e., a theory consisting of clauses with no positive literal) is Horn and the exteriors of negative theory are also negative, and hence Horn. This means that, for a negative theory Σ, we have $\sigma_\alpha(\Sigma)_e = \sigma_\alpha(\Sigma)$. Therefore, we can again make use of the reduction in the proof of Theorem 2, since the reduction uses negative theories.

Theorem 9. *Given a Horn theory Σ, a clause c, and an integer $\alpha \geq 0$, it is* co-NP-*complete to decide whether $\sigma_\alpha(\Sigma)_e \models c$ holds, even if $P(c) = \emptyset$.* $\square$

However, if α or $N(c)$ is small, the problem becomes tractable by Algorithm 3.

Algorithm 3. Deduction-Envelope-Exterior-from-Horn-Theory

Input: A Horn theory Σ, a clause c and an integer $\alpha \geq 0$.

Output: Yes, if $\sigma_\alpha(\Sigma)_e \models c$; Otherwise, No.

Step 1.   /* Check if there exists a model v of Σ such that $|OFF(v) \cap N(c)| < \alpha$. */

> **For** each $N \subseteq N(c)$ with $|N| = |N(c)| - \alpha + 1$ **do**
>
>> Check if the theory obtained from Σ by assigning $x_i = 1$ for $i \in N$ is satisfiable.
>>
>> **If** so, **then** output No and halt.
>
> **end**{for}

Step 2.   /* Check if there exists a set $S = \{v \in mod(\Sigma) \mid |OFF(v) \cap N(c)| = \alpha\}$ such that $\bigcup_{v \in S} OFF(v) \supseteq P(c)$. */

> Let $J := \emptyset$.
>
> **For** each $N \subseteq N(c)$ with $|N| = |N(c)| - \alpha$ **do**
>
>> Compute a unique minimal satisfiable model v for the theory obtained from Σ by assigning $x_i = 1$ for $i \in N$ is satisfiable.
>>
>> Update $J := J \cup \{j \in P(c) \mid v_j = 0\}$.
>
> **end**{for}
>
> **If** $J = P(c)$, **then** output NO and halt.

Step 3.   Output Yes and halt. $\square$

The algorithm is based on a necessary and sufficient condition for $\sigma_\alpha(\Sigma)_e \models c$, which is obtained from Lemma 8 by replacing all $chr(\Sigma)$'s with $mod(\Sigma)$'s. It is not difficult to see that such a condition holds from the proof of Lemma 8.

Theorem 10. *Given a Horn theory Σ, a clause c, and an integer $\alpha \geq 0$, a deductive query $\sigma_\alpha(\Sigma)_e \models c$ can be answered in $O\left(\left(\binom{|N(c)|}{\alpha-1} + \binom{|N(c)|}{\alpha}\right) \|\Sigma\| + |P(c)|\right)$ time. In particular, it is polynomially solvable, if $\alpha = O(1)$ or $|N(c)| = O(\log \|\Sigma\|)$.*   □

References

1. Brachman, R.J., Levesque, H.J.: Knowledge Representation and Reasoning. Elsevier, Amsterdam (2004)
2. Dechter, R., Pearl, J.: Structure identification in relational data. Artificial Intelligence 58, 237–270 (1992)
3. Dowling, W., Galliear, J.H.: Linear-time algorithms for testing the satisfiability of propositional Horn theories. Journal of Logic Programming 3, 267–284 (1983)
4. Eiter, T., Ibaraki, T., Makino, K.: Computing intersections of Horn theories for reasoning with models. Artificial Intelligence 110, 57–101 (1999)
5. Eiter, T., Makino, K.: On computing all abductive explanations. In: Proc. AAAI 2002, pp. 62–67 (2002)
6. Kavvadias, D., Papadimitriou, C., Sideri, M.: On Horn Envelopes and Hypergraph Transversals. In: Ng, K.W., Balasubramanian, N.V., Raghavan, P., Chin, F.Y.L. (eds.) ISAAC 1993. LNCS, vol. 762, pp. 399–405. Springer, Heidelberg (1993)
7. Kautz, H., Kearns, M., Selman, B.: Reasoning with characteristic models. In: Proc. AAAI 1993, pp. 34–39 (1993)
8. Kautz, H., Kearns, M., Selman, B.: Horn approximations of empirical data. Artificial Intelligence 74, 129–245 (1995)
9. Khardon, R., Roth, D.: Reasoning with models. Artificial Intelligence 87, 187–213 (1996)
10. Khardon, R., Roth, D.: Defaults and relevance in model-based reasoning. Artificial Intelligence 97, 169–193 (1997)
11. Makino, K., Ibaraki, T.: Interior and exterior functions of Boolean functions. Discrete Applied Mathematics 69, 209–231 (1996)
12. Makino, K., Ono, H.: Deductive Inference for the Interiors and Exteriors of Horn Theories, Mathematical Engineering Technical Report, METR 2007-06, The University of Tokyo (February 2007)
13. Makino, K., Ono, H., Ibaraki, T.: Interior and exterior functions of positive Boolean functions. Discrete Applied Mathematics 130, 417–436 (2003)
14. Makowsky, J.: Why Horn formulas matter for computer science: Initial structures and generic examples. Journal of Computer and System Sciences 34, 266–292 (1987)
15. McKinsey, J.: The decision problem for some classes of sentences without quantifiers. Journal of Symbolic Logic 8, 61–76 (1943)
16. Selman, B., Kautz, H.: Knowledge compilation using Horn approximations. In: Proc. AAAI 1991, pp. 904–909 (1991)

Leaf Powers and Their Properties: Using the Trees[*]

Michael R. Fellows[1], Daniel Meister[2], Frances A. Rosamond[1], R. Sritharan[3],
and Jan Arne Telle[2]

[1] University of Newcastle, Newcastle, Australia
`michael.fellows@newcastle.edu.au`, `frances.rosamond@newcastle.edu.au`
[2] Department of Informatics, University of Bergen, Norway
`daniel.meister@ii.uib.no`, `jan.arne.telle@ii.uib.no`
[3] Department of Computer Science, The University of Dayton, Dayton, USA
`srithara@notes.udayton.edu`

Abstract. A graph G on n vertices is a *k-leaf power* ($G \in \mathcal{G}_k$) if it is
isomorphic to a graph that can be "generated" from a tree T that has n
leaves, by taking the leaves to represent vertices of G, and making two
vertices adjacent if and only if they are at distance at most k in T. We
address two questions in this paper:

(1) As k increases, do we always have $\mathcal{G}_k \subseteq \mathcal{G}_{k+1}$?
Answering an open question of Andreas Brandstädt and Van Bang Le
[2,3,1], we show that the answer, perhaps surprisingly, is "no."
(2) How should one design algorithms to determine, for k-leaf powers, if
they have some property?

One way this can be done is to use the fact that k-leaf powers have
bounded cliquewidth. This fact, plus the FPT cliquewidth approximation algorithm of Oum and Seymour [14], combined with the results of
Courcelle, Makowsky and Rotics [7], allows us to conclude that properties expressible in a general logic formalism, can be decided in FPT time
for k-leaf powers, parameterizing by k. This is wildly inefficient. We explore a different approach, under the assumption that a generating tree
is given with the graph. We show that one can use the tree *directly* to
decide the property, by means of a finite-state tree automaton. (A more
general theorem has been independently obtained by Blumensath and
Courcelle [5].)

We place our results in a general context of "tree-definable" graph
classes, of which k-leaf powers are one particular example.

1   Introduction

A graph G is a tree power if there is an integer k and a tree T on the same
vertex set as G such that two vertices are adjacent in G if and only if they are at
distance at most k in T [11]. A leaf-power graph is defined similarly except that

---

[*] This work is supported by the Research Council of Norway.

S.-H. Hong, H. Nagamochi, and T. Fukunaga (Eds.): ISAAC 2008, LNCS 5369, pp. 402–413, 2008.
© Springer-Verlag Berlin Heidelberg 2008

the vertex set of G now corresponds to the leaves of T. Leaf powers, and k-leaf powers where the power k is specified, were introduced by Nishimura, Ragde, Thilikos [13] in 2002 and have attracted a considerable amount of attention. These graphs have applications in the reconstruction of the evolutionary history of species and admit efficient algorithms for generaly difficult problems, that mainly exploit the tree structure of the underlying tree. Note that k-leaf powers can have arbitrarily large cliques, which means that the class of k-leaf powers is of unbounded treewidth. It was recently shown that a k-leaf power has cliquewidth at most $3k/2$ [10]. It was known that $(k-1)$-leaf powers are k-leaf powers for $k \in \{1,2,3,4\}$ [2], but it was an open question whether this containment relationship continued. In this paper we show that this is not the case.

Leaf-power graphs are only one example of graphs with nice algorithmic properties that rely on an underlying tree structure. In this paper, we generalise the main concepts. The definition of leaf-power graphs consists of three elements: a tree, a selection of tree vertices and an adjacency condition. For a fixed k, the selected vertices are the leaves of the tree and the adjacency condition is a distance condition. "Being a leaf, i.e., a vertex of certain degree" and "being a pair of vertices at bounded distance" are predicates that can be expressed already in limited logic. Based on these observations, we introduce the notion of tree-definable graph classes. A tree will then define a graph where the vertex set is selected by a unary graph predicate and adjacency is defined by a binary graph predicate. We show that every problem that is definable in monadic second-order logic allowing quantification over sets of vertices can be solved in linear time on a tree-definable graph class, when the representing tree is given. When allowing quantification over sets of edges the results are not so positive, and we apply a Myhill-Nerode argument to show that the set of trees generating Hamiltonian 3-leaf powers does not have finite index. The logical expressions used to generate 4-leaf powers and 5-leaf powers are very similar and yet we show that there exists a 4-leaf power which is not a 5-leaf power, thereby answering an open question of Andreas Brandstädt and Van Bang Le [2,3,1].

The paper is organised as follows. In the next section, we define the notion of *tree-definable graph class*. In Section 3, we show that well-known graph classes such as cographs and fixed leaf-power classes are tree-definable. We also show that not every 4-leaf power is a 5-leaf power. Algorithmic consequences for tree-definable graph classes are discussed in Section 4. In the final Section 5, we present open problems.

2 Tree-Definable Graph Classes

We give a very general method for defining graphs. It has three ingredients: a tree, a logical expression extracting the vertex set of the graph from the tree, and another logical expression extracting the adjacencies in the graph from the tree. In the next section, we will show for some well-known graph classes that they can be defined in this uniform way. (We have recently learned that some of our results were obtained earlier, in a very general setting, by Blumensath and Courcelle [5], where they study *rational transductions* of relational systems.)

To begin, we specify the logic that we use in this paper. An M2O *graph expression* has four types of variables, for vertices, edges, sets of vertices, sets of edges, and is formed from the following elements:

- equality for each of the four variable types
- the two predicates $x \in X$ and $e \in Y$ for x a vertex variable, X a vertex set variable, e an edge variable and Y an edge set variable and the two predicates $\mathrm{adj}(u, v)$ and $\mathrm{inc}(u, e)$ for u and v vertex variables and e an edge variable
- the connectives $\wedge, \vee, \neg$ and the quantifiers $\forall, \exists$.

Every variable type allows quantification. Clearly, $x \in X$ is true if x is interpreted by a vertex contained in X; similarly for $e \in Y$. And $\mathrm{adj}(u, v)$ is true if u and v are interpreted by adjacent, in particular different, vertices, and $\mathrm{inc}(u, e)$ is true if u is an endpoint of e, i.e., e is incident to u. A graph expression without free variables is called *graph sentence*. For an M2O graph sentence Φ we say that a graph G is a *model* for Φ, denoted as $G \models \Phi$, if Φ is true on G.

Definition 1. *Let $\Phi_1 = \Phi_1(x)$ and $\Phi_2 = \Phi_2(x_1, x_2)$ be M2O graph expressions with x, x_1 and x_2 vertex variables. For a tree T, the graph G represented by T with respect to (Φ_1, Φ_2), denoted as $G_{\Phi_1, \Phi_2}(T)$, is defined as follows:*

- $V(G) =_{\mathrm{def}} \{u \in V(T) : T \models \Phi_1(u)\}$
- $E(G) =_{\mathrm{def}} \{uv : u, v \in V(G) \text{ and } u \neq v \text{ and } T \models \Phi_2(u, v)\}$.

*We denote by $\mathcal{G}(\Phi_1, \Phi_2)$ the class of graphs defined by Φ_1, Φ_2, i.e., $\mathcal{G}(\Phi_1, \Phi_2) = \{G_{\Phi_1, \Phi_2}(T) : T \text{ a tree}\}$. We say that $\mathcal{G}(\Phi_1, \Phi_2)$ is a **tree-definable** graph class.*

Note that every tree represents a graph with respect to every pair (Φ_1, Φ_2) of graph expressions, but the represented graph may be the empty graph or may have no edges. The definition of tree-definable graph classes raises immediate questions of three types, that we will address in this paper. Let (Φ_1, Φ_2) and (Φ_1', Φ_2') be pairs of graph expressions.

1) Given a tree T, how difficult is it to compute the graph that is represented by T with respect to (Φ_1, Φ_2), i.e., the graph $G_{\Phi_1, \Phi_2}(T)$?
2) Given a graph G, how difficult is it to compute a tree T such that T represents G with respect to (Φ_1, Φ_2), i.e., such that $G = G_{\Phi_1, \Phi_2}(T)$?
3) How difficult is it to decide whether $\mathcal{G}(\Phi_1, \Phi_2)$ is contained in $\mathcal{G}(\Phi_1', \Phi_2')$, i.e., whether $\mathcal{G}(\Phi_1, \Phi_2) \subseteq \mathcal{G}(\Phi_1', \Phi_2')$?

In view of question 2 we remark that we restrict to named graphs. If graphs have unnamed vertices, question 2 should be rephrased as an isomorphism problem. Questions 1 and 2 are very natural, as they state the verification problem and a restricted version of the graph class recognition problem. Question 3 is partially motivated by a later result about the existence of efficient algorithms for certain problems. We address the three questions, among others, in different parts of this paper.

Another, in some sense more concrete question asks about graph classes that are captured by our model. Are well-known graph classes tree-definable? In the next section, we answer this question in the affirmative and give two examples. This also supports our opinion that the definition of tree-definable graph classes is indeed a natural approach to unify graph classes.

3   Examples of Tree-Definable Graph Classes

As our first example we show that all (fixed) leaf-power classes are tree-definable. For a number $k \geq 1$, a graph G is a k-*leaf power* if there is a tree T such that the vertices of G are in 1-to-1 correspondence with the leaves of T and two vertices of G are adjacent if and only if the distance of the corresponding leaves is at most k in T. Hence, for the definition of the two predicates, the vertex selection predicate should be true only for leaves, and adjacency is a simple distance condition. Let

- $L(x) =_{\mathrm{def}} \exists u \forall v (\mathrm{adj}(x, v) \rightarrow u = v)$
- $P_1(x_1, x_2) =_{\mathrm{def}} (x_1 = x_2 \vee \mathrm{adj}(x_1, x_2))$
  $P_{i+1}(x_1, x_2) =_{\mathrm{def}} \exists u (P_i(x_1, u) \wedge P_1(u, x_2))$ for all $i \geq 1$.

For a graph $G = (V, E)$, two (not necessarily different) vertices u, v of G and a number $k \geq 1$, the following holds:

- $G \models L(u)$ if and only if $d_G(u) \leq 1$
- $G \models P_k(u, v)$ if and only if there is a u, v-path of length at most k in G.

Hence, $\mathcal{G}(L, P_k)$ is the class of k-leaf powers. Using the example of k-leaf powers, we show that the inclusion relation of tree-definable graph classes is not easy to determine from the logical expressions directly, thereby settling an open problem. It is known that k-leaf powers are $(k + 1)$-leaf powers for $k \in \{1, 2, 3\}$, i.e., $\mathcal{G}(L, P_1) \subseteq \mathcal{G}(L, P_2) \subseteq \mathcal{G}(L, P_3) \subseteq \mathcal{G}(L, P_4)$ [2]. It was an open question whether this containment relationship continues. The following result shows that this is not the case.

Lemma 1. *There is a 4-leaf power that is not a 5-leaf power.*

Proof. Consider the tree T in Figure 1. Let G be the square of T, i.e., the graph on the vertices of T where two vertices are adjacent if and only if they are at distance at most 2 in T. It is clear that G is a 4-leaf power. We show that G is not a 5-leaf power. For a contradiction, assume that G is a 5-leaf power. Let G be the 5-leaf power of tree S'. Since G has no true twins, no vertex of S' is adjacent to two leaves. Let S be the subgraph of S' induced by all non-leaf vertices. It holds that G is isomorphic to an induced subgraph of the third power of S. For ease of notation, let vertices of S have the name of their corresponding vertex as in Figure 1.

Since vertices 2, 3 and 4 are pairwise adjacent in G, it holds for the distances between the three vertices: $d_S(2,3), d_S(2,4), d_S(3,4) \leq 3$. If $d_S(2,3) = 3$ then

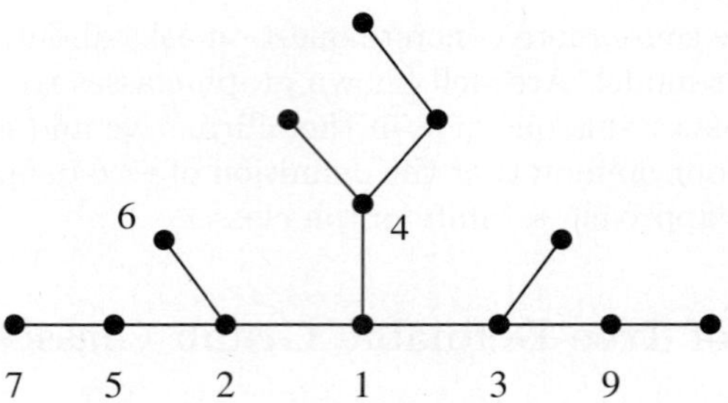

Fig. 1. The square of the depicted tree, i.e., the graph on the vertex set of the tree with vertices adjacent if they are adjacent in the tree or have a common neighbour, is a 4-leaf power but not a 5-leaf power

$d_S(2,4) \neq d_S(3,4)$. Hence, one of the three distances is at most 2. Without loss of generality, $d_S(2,3) \leq 2$. Consider the path P between 5 and 9 in S: since 5 and 9 are adjacent to 1 in G, it holds $d_S(5,9) \leq d_S(5,1) + d_S(9,1) \leq 6$. We consider different cases. Let $5 \leq d_S(5,9) \leq 6$. Then, 1 lies on P, at distance 3 to 5 or 9; by symmetry we can assume $d_S(1,5) = 3$. If 3 lies on the 1,5-path in S then 5 is adjacent also to 3. Hence, 3 does not lie on the 1,5-path in S. If the 3,9-path in S contains two vertices of the 1,5-path then $d_S(3,9) \geq 4$. Hence, 2 and 3 are contained in different connected components of $S-1$. The distance condition $d_S(2,3) \leq 2$ then shows that 2 and 3 are neighbours of 1 in S. In particular, 2 is vertex on the 1,5-path in S. This also implies that $d_S(1,9) = 3$, since $4 \leq d_S(2,9) \leq 1 + d_S(1,9)$. Then, 3 lies on the 1,9-path in S. Summarising, we obtain that the 5,9-path in S is the following: $(5,x,2,1,3,y,9)$ for some vertices x and y. Now, consider vertex 6. It is adjacent to 1, 2, 5 and non-adjacent to 3. Hence, 6 is a neighbour of x in S. Now, consider vertex 7. The neighbours of 7 are 2 and 5, but there is no tree that allows assignment of 7. Hence, $d_S(5,9) = 4$. Note that every vertex of G that corresponds to a vertex on P besides 5 and 9 is a common neighbour of 5 and 9. Since 1 is the only common neighbour of 5 and 9 in G, 2 and 3 are not contained in P. This, however, is only possible for $d_S(2,3) \geq 4$. We conclude that G cannot be a 5-leaf power.

Hence, Lemma 1 shows $\mathcal{G}(L, P_4) \not\subseteq \mathcal{G}(L, P_5)$. This example also shows that the inclusion question, given M2O predicates $\Phi_1, \Phi_2, \Phi_1', \Phi_2'$ to decide whether the inclusion $\mathcal{G}(\Phi_1, \Phi_2) \subseteq \mathcal{G}(\Phi_1', \Phi_2')$ holds, cannot be determined easily from just looking at the expressions.

As a second example, we show that cographs are tree-definable. Cographs are the P_4-free graphs, i.e., the graphs that do not contain the induced path on four vertices as induced subgraph. Our result is based on the cotree characterisation of cographs. A rooted tree whose non-leaf vertices are labelled 0 or 1 is called a *cotree*. It holds that a graph G is a cograph if and only if there is a cotree T such that the vertices of G correspond to the leaves of T and two vertices u, v of G are adjacent if and only if the least common ancestor of u and v is labelled 1.

Proposition 1. *The class of cographs is a tree-definable graph class.*

Proof. We prove the proposition in two steps. We first extend the definition of tree-definable graph classes to coloured trees, and then we show that coloured trees are captured by our model.

We need five vertex colours with the following meanings: leaf, root vertex with label 0, root vertex with label 1, non-root vertex with label 0, non-root vertex with label 1. First, we give a predicate that checks whether a given coloured tree represents a cotree correctly. This means that every vertex has a colour, there is exactly one root vertex and the leaves are exactly the vertices with the leaf colour. The colour class predicates are denoted as C_1, C_2, C_3, C_4, C_5.

- $L(x) =_{\text{def}} \exists u \forall v (\mathrm{adj}(x,v) \to u = v)$
- $Q =_{\text{def}} \forall u((L(u) \to C_1(u)) \wedge (C_1(u) \to L(u)))$
- $R =_{\text{def}} \exists u((C_2(u) \vee C_3(u)) \wedge \forall v((C_2(v) \vee C_3(v)) \to u = v))$
- $P =_{\text{def}} \forall u(C_1(u) \vee C_2(u) \vee C_3(u) \vee C_4(u) \vee C_5(u)) \wedge Q \wedge R$

It holds for a 5-coloured tree T that $T \models P$ if and only if T correctly represents a cotree, in particular, has a unique root vertex and all leaves are coloured with the unique leaf colour. It remains to define adjacency. To find the first common vertex of two leaf-root paths, we need predicates for 'path' and 'first common vertex'. The following expressions describe 'path' in the following way, precisely, predicate X is true if the vertices in Z describe a path between a and b.

- $N_{\geq 2}(x,a,b,Z) =_{\text{def}} \exists v_1 \exists v_2 (v_1, v_2 \in Z \cup \{a,b\} \wedge v_1 \neq v_2 \wedge \mathrm{adj}(x,v_1) \wedge \mathrm{adj}(x,v_2))$
- $N_{\leq 2}(x,a,b,Z) =_{\text{def}} \exists v_1 \exists v_2 \forall u((u \in Z \cup \{a,b\} \wedge \mathrm{adj}(x,u)) \to (u = v_1 \vee u = v_2))$
- $X(a,b,Z) =_{\text{def}} \forall u(u \in Z \to (N_{\geq 2}(u,a,b,Z) \wedge N_{\leq 2}(u,a,b,Z))) \wedge \exists u(u \in Z \wedge \mathrm{adj}(a,u)) \wedge \exists u(u \in Z \wedge \mathrm{adj}(b,u)).$

To determine the first common vertex of two root paths, we apply the same ideas but omit this here.

For the second step, we encode vertex colours into an uncoloured tree. The idea is to represent colours by the number of leaves adjacent to a non-leaf vertex. Predicate L verifies whether a vertex is a leaf or a non-leaf, and similar to predicates $N_{\leq 2}$ and $N_{\geq 2}$ we can formulate predicates that check whether a non-leaf vertex has exactly 1, 2, 3 or 4 or at least five adjacent leaves. The final expressions then select the non-leaf vertices with exactly one adjacent leaf to represent the vertices of the graph and two vertices are adjacent if and only if the tree is syntactically correct and the two paths to the root meet first on a 1 labelled vertex, i.e., a vertex with exactly three adjacent leaves. If the tree is not syntactically correct, the represented graph has no edges. This is also a cograph. It follows with the cograph characterisation via cotrees that cographs are tree-representable.

It is a well-known fact that the cographs are exactly the graphs of clique-width at most 2.

4   Efficiently Solvable Problems

In this section, we mainly show that the computational bottleneck for many problems on tree-definable graphs is the computation of a representing tree. We show that all problems that can be formulated in a restricted version of M2O have efficient algorithms on input a representing tree. Our result is based on Courcelle's celebrated theorem about solvability of problems on graphs of bounded treewidth [6].

We show that problems that are expressible in a restricted version of M2O logic can be solved efficiently on the representing trees as input. This restricted logic does not allow edge set variables in expressions, and we call such expressions M1O *graph expressions*.

Lemma 2. *Let Φ_1 be a unary M2O graph expression and let Φ_2 be a binary M2O graph expression. Let Ψ be an M1O graph sentence. There is an M2O graph sentence Ψ' such that for all trees T: $G_{\Phi_1,\Phi_2}(T) \models \Psi$ if and only if $T \models \Psi'$.*

Proof. We modify Ψ to obtain Ψ'. Consider the following predicate P:

$$P(X) =_{\text{def}} \exists a \exists b \forall c (a \in X \wedge b \in X \wedge \neg(a = b) \wedge (c \in X \rightarrow (c = a \vee c = b)) \wedge$$
$$\Phi_1(a) \wedge \Phi_1(b) \wedge \Phi_2(a, b))$$

where a, b, c are vertex variables and X is a vertex set variable. Let G be a graph and let U be a set of vertices of G. Then, the following holds:

- $G \models P(U)$ if and only if $U = \{u, v\}$ for some vertices u, v of G, $u \neq v$, and $G \models \Phi_1(u)$ and $G \models \Phi_1(v)$ and $G \models \Phi_2(u, v)$.

We show the claim of the lemma by induction over the definition of our graph expressions. Note that, by assumption, Ψ does not contain $e \in Y$ for e an edge variable and Y an edge set variable as subexpression. For the proof, we have to consider arbitrary graph expressions without edge set variable occurrences. For an M2O graph expression F without edge set variable occurrences we write $F = F(\alpha, \beta, \gamma)$ where α is a list of the vertex variables that occur in F, β is a list of the vertex set variables that occur in F and γ is a list of the edge variables that occur in F. Expression H denotes the equivalent expression after modification. We begin with atomic expressions.

- $F(\langle x_1, x_2 \rangle, \langle \rangle, \langle \rangle) = (x_1 = x_2)$ for x_1, x_2 vertex variables
  $H(\langle x_1, x_2 \rangle, \langle \rangle) =_{\text{def}} (x_1 = x_2)$
- $F(\langle \rangle, \langle X_1, X_2 \rangle, \langle \rangle) = (X_1 = X_2)$ for X_1, X_2 vertex set variables
  $H(\langle \rangle, \langle X_1, X_2 \rangle) =_{\text{def}} (X_1 = X_2)$
- $F(\langle \rangle, \langle \rangle, \langle y_1, y_2 \rangle) = (y_1 = y_2)$ for y_1, y_2 edge variables
  $H(\langle \rangle, \langle Y_1, Y_2 \rangle) =_{\text{def}} (Y_1 = Y_2)$
- $F(\langle x \rangle, \langle X \rangle, \langle \rangle) = (x \in X)$ for x a vertex variable and X a vertex set variable
  $H(\langle x \rangle, \langle X \rangle) =_{\text{def}} (x \in X)$
- $F(\langle x_1, x_2 \rangle, \langle \rangle, \langle \rangle) = (\text{adj}(x_1, x_2))$ for x_1, x_2 vertex variables
  $H(\langle x_1, x_2 \rangle, \langle \rangle) =_{\text{def}} \Phi_2(x_1, x_2)$

- $F(\langle x\rangle, \langle\rangle, \langle y\rangle) = (\mathrm{inc}(x,y))$ for x a vertex variable and y an edge variable
  $H(\langle x\rangle, \langle Y\rangle) =_{\mathrm{def}} (x \in Y)$

We continue with the non-atomic expressions. Clearly, it suffices to consider only $\exists$-quantified variables. By H', we denote the expression inductively obtained for F'.

- $F(\alpha, \beta, \gamma) = \exists x F'(\langle x\rangle \circ \alpha, \beta, \gamma)$ for x a vertex variable
  $H(\alpha, \beta \circ \gamma') =_{\mathrm{def}} \exists x(\Phi_1(x) \wedge H'(\langle x\rangle \circ \alpha, \beta \circ \gamma'))$
- $F(\alpha, \beta, \gamma) = \exists X F'(\alpha, \langle X\rangle \circ \beta, \gamma)$ for X a vertex set variable
  $H(\alpha, \beta \circ \gamma') =_{\mathrm{def}} \exists X(\forall a(a \in X \to \Phi_1(a)) \wedge H'(\alpha, \langle X\rangle \circ \beta \circ \gamma'))$
- $F(\alpha, \beta, \gamma) = \exists y F'(\alpha, \beta, \langle y\rangle \circ \gamma)$ for y an edge variable
  $H(\alpha, \beta \circ \gamma') =_{\mathrm{def}} \exists Y(P(Y) \wedge H'(\alpha, \beta \circ \langle Y\rangle \circ \gamma'))$
- the cases for $\wedge$, $\vee$, $\neg$ are obvious.

This completes the proof.

Theorem 1 ([6]). *Let Ψ be an M2O graph sentence. There is a linear-time algorithm that decides on input a tree T whether $T \models \Psi$.*

Corollary 1. *Let $\mathcal{G}$ be a tree-definable graph class, defined by the two M2O graph expressions Φ_1 and Φ_2. Let there be an algorithm that on input a graph G from $\mathcal{G}$ computes a tree T such that $G_{\Phi_1, \Phi_2}(T) = G$ in time $\mathcal{O}(f(n,m))$. Then, for every problem Ψ definable as an M1O graph sentence, there is an algorithm that decides Ψ on input graphs from $\mathcal{G}$ in time $\mathcal{O}(f(n,m))$.*

Proof. The algorithm is simple: on input a graph G from $\mathcal{G}$, compute a representing tree for G in time $\mathcal{O}(f(n,m))$. Note that T has at most $\mathcal{O}(f(n,m))$ vertices. Lemma 2 shows that there is an M2O graph sentence Ψ' such that $T \models \Psi'$ if and only if $G \models \Psi$. It is important to note that Ψ' does not depend on T. By Theorem 1 there exists an algorithm for deciding $T \models \Psi'$ in time linear in the size of T. Thus, there is an algorithm for deciding whether $G \models \Psi$ in time $\mathcal{O}(f(n,m))$.

With Corollary 1, the main algorithmic challenge for graph problems expressible in M1O logic is to design efficient algorithms for computing a representing tree. An important parameter determining the running time of such an algorithm is the size of the output. In general, a graph can have a large representing tree. As an example, our formulas for cographs in Proposition 1 allow trees that can be obtained from cotrees by subdividing arbitrary edges with vertices of label 0. Therefore, it is an interesting question to determine bounds on the size of representing trees. Lower bounds additionally provide lower bounds on the running times of algorithms.

Besides the questions about running time, the main question is which problems can be solved by our approach. We consider a famous example. A *Hamiltonian cycle* in a graph G is a cycle that visits every vertex of G exactly once. It is known that Hamiltonicity, the problem asking whether a graph contains a Hamiltonian cycle, is not expressible in M1O logic [12]. The essence of our approach is to *use the trees* that generate the graphs in the family (e.g., k-leaf powers). We may be interested in a property (such as Hamiltonicity) where Corollary 1 does not apply, but it still might be the case that the property (of the graphs generated by the trees) might be recognizable by a finite-state tree automaton acting on the generating trees. However, for *Hamiltonicity*, we next show that this cannot be done for k-leaf powers.

Our argument uses the "Myhill-Nerode" point of view, such as exposited in [8].

Our non-expressibility result is obtained in two steps. We first show that the problem is not finite-state for a special congruence relation. For rooted trees T_1 and T_2 with roots R_1 and R_2, respectively, the result of the operation $\oplus$ on T_1 and T_2, denoted as $T_1 \oplus T_2$, is the composition of T_1 and T_2 by gluing the two trees together on R_1 and R_2. The compound graph, that is a tree, has root the glue vertex (R_1, R_2). The relation $\sim_{\mathcal{F}}$ over a set $\mathcal{F}$ of rooted trees is defined as follows: for two rooted trees T_1 and T_2, $T_1 \sim_{\mathcal{F}} T_2$ if and only if $T_1 \oplus T \in \mathcal{F} \Leftrightarrow T_2 \oplus T \in \mathcal{F}$ for all rooted trees T. Informally, two rooted trees are considered equivalent with respect to $\mathcal{F}$ if they behave equally after compounding with the same tree.

Lemma 3. *Let $\mathcal{F}$ be the set of rooted trees that represent a 3-leaf power with a Hamiltonian cycle. Then, $\sim_{\mathcal{F}}$ does not have finite index.*

Proof. Let G be a graph whose vertex set can be partitioned into a clique A and an independent set B. Furthermore, let every vertex in A be adjacent to every vertex in B. Such graphs are called *complete split graphs*. It obviously holds that G has a Hamiltonian cycle if and only if $|A| \geq |B|$, since two vertices of B cannot appear consecutively on the cycle.

First we show that complete split graphs are 3-leaf powers. Let G be a complete split graph with vertex partition (A, B) such that A is a clique and B is an independent set. A *star* is a tree with a universal vertex. We obtain a tree for G from the star on $|B| + 1$ vertices by attaching a new leaf to every leaf of the star and $|A|$ new leaves to the universal vertex of the star. Let the new tree be rooted at the universal vertex of the star. For $r \geq 1$, denote by G_r the star on $r + 1$ vertices, which is a complete split graph, and denote by T_r the corresponding rooted tree due to the above construction. Note that, for arbitrary i and l, $T_i \oplus G_l$ is a tree for a complete split graph with independent set of size i and clique of size $l + 1$. We assume that G_l has root a universal vertex.

Second we show that all T_r are in different equivalence classes of $\sim_{\mathcal{F}}$, thus showing that $\sim_{\mathcal{F}}$ does not have finite index. Let $1 \leq i < j$. By the discussions about, $T_i \oplus G_{j-i}$ has no Hamiltonian cycle and $T_j \oplus G_{j-i}$ has a Hamiltonian cycle. Hence, $T_i \not\sim_{\mathcal{F}} T_j$.

In view of Corollary 1, the above provides an alternate proof that Hamiltonicity cannot be expressed in M1O logic. It also shows that an attempt to mimic the form of Lemma 2 by having the "extractor" graph expressions in M1O logic, does not support a conclusion that properties expressible in M2O logic are finite-state recognizable by automata acting on the generating trees. The proof of Lemma 3 is easily modified to show a similar result for any fixed $k \geq 3$. We contrast the above with a limited positive result for a problem that cannot be formulated in M1O.

Lemma 4. *Let $\mathcal{F}$ be the set of rooted trees that represent a 2-leaf power with a Hamiltonian cycle. Then, $\sim_{\mathcal{F}}$ has finite index.*

Proof. Note that 2-leaf powers are exactly the disjoint unions of complete graphs. So, it holds that a 2-leaf power contains a Hamiltonian cycle if and only if it is connected and has at least three vertices. Hence, a 2-leaf power has a Hamiltonian cycle if and only if it is represented by a star with at least three leaves. We show that $\sim_{\mathcal{F}}$ defines three equivalence classes on the set of rooted trees on at least four vertices: stars with root the centre vertex, stars with root a leaf, rooted trees that are not stars.

Let T and T' be rooted trees. First, let T not be a star. This means that there is a path of length at least 3 between two leaves in T. Then, there is a path of length at least 3 between two leaves also in $T \oplus T'$, which means that $T \oplus T'$ represents a disconnected 2-leaf power, and $T \oplus T'$ is not in $\mathcal{F}$. Second, let T be a star with root vertex a leaf. We distinguish two cases for T'. If T' has exactly one vertex then $T \oplus T'$ represents the same 2-leaf power as T, if T' has at least two vertices then $T \oplus T'$ contains two leaves at distance at least 3. Hence, all trees in the second class behave equally. Third, let T be a star with root vertex the centre. Again, we distinguish two cases. If T' is a star with root a universal vertex then $T \oplus T'$ is a star with root a universal vertex, if T' is a star with root not a universal vertex or T' is not a star then $T \oplus T'$ contains a pair of leaves at distance at least 3. Hence, the trees in this class behave equally.

We have seen that $\sim_{\mathcal{F}}$ partitions the set of rooted trees on at least four vertices into three classes. Since there is only a finite number of rooted trees on at most three vertices, $\sim_{\mathcal{F}}$ partitions the set of rooted trees into a finite number of classes, which shows the claim of the lemma.

5   Conclusions and Open Problems

We have shown that *graphs of bounded leaf-power* is not a well-behaved monotonic parameter, answering a noted open problem: there are graphs that are 4-leaf powers but not 5-leaf powers. Recently, we learned that Brandstädt and Wagner have generalized our construction of a 4-leaf power that is not a 5-leaf power, based on an early draft of the present paper [9], to construct for any $k \geq 4$ a k-leaf power that is not a k+1-leaf power [4]. But maybe it can be shown that in some sense the notion is "almost monotonic" for some suitable interpretation of *almost*.

We have also explored how the generating trees for k-leaf powers can be used directly to decide properties of graphs in these families, and have shown if a property is expressible in M1O logic, then this approach can be carried out. Such expressibility is sufficient, but not necessary for the approach to succeed. We have shown that Hamiltonicity, a property that is not expressible in M1O, is finite-state for 2-leaf powers, but not for k-leaf powers for $k \geq 3$.

The issue has been explored in the general setup that the graphs are *extracted* from the trees in some manner, and then we want to know if the trees might be exploited for finite-state recognition of properties. Our general result, Lemma 2, shows that if the extraction is by means of M2O expressions, and the properties are expressible in M1O, then the program succeeds. (We have recently learned that our general lemma seems to be a special case of results in a very general setting independently and recently published by Blumensath and Courcelle [5].)

However, this area merits much more investigation. Our main motivation is to try to find general FPT results for classes of problems, for graphs in graph families extracted from trees, much as Courcelle's Theorem does for graphs of bounded treewidth. But not just general FPT results, we also want to explore how such results can be obtained *in an efficient manner*.

Another possibility for handling the extraction of the graphs from the trees would be by means of finite-state automata. For example, consider the (somewhat artificial) parameterized family of graphs $\mathcal{F}_k$ defined where two leaves l and l' of the generating tree T are adjacent in the extracted graph $G(T)$ if there exists a cutwidth at most k layout of T with u and v consecutive. Using the approach of [8] (Theorem 6.82) one can construct a fairly efficient tree-automaton for this extraction. Is there an efficient and general way to decide properties of such graphs? Can we get good quantitative bounds on the sizes of the automata involved in these kinds of approaches?

References

1. Brandstädt, A.: Talk at GROW workshop in Eugene, USA (October 2007)
2. Brandstädt, A., Le, V.B.: Structure and linear time recognition of 3-leaf powers. Information Processing Letters 98, 133–138 (2006)
3. Brandstädt, A., Wagner, P.: On (k,ℓ)-leaf powers. In: Kučera, L., Kučera, A. (eds.) MFCS 2007. LNCS, vol. 4708, pp. 525–535. Springer, Heidelberg (2007)
4. Brandstädt, A., Wagner, P.: On k- versus (k + 1)-Leaf Powers (manuscript, 2008)
5. Blumensath, A., Courcelle, B.: Recognizability, hypergraph operations, and logical types. Information and Computation 204(6), 853–919 (2006)
6. Courcelle, B.: The monadic second order theory of graphs I: recognizable sets of finite graphs. Information and Computation 85, 12–75 (1990)
7. Courcelle, B., Makowsky, J., Rotics, U.: Linear time solvable optimization problems on graphs of bounded clique-width. Theory of Computing Systems 33(2), 125–150 (2000)
8. Downey, R., Fellows, M.: Parameterized Complexity. Springer, Heidelberg (1999)
9. Fellows, M., Meister, D., Rosamond, F., Sritharan, R., Telle, J.A.: On graphs that are k-leaf powers (manuscript, 2007)
10. Gurski, F., Wanke, E.: The Clique-Width of Tree-Power and Leaf-Power Graphs. In: Brandstädt, A., Kratsch, D., Müller, H. (eds.) WG 2007. LNCS, vol. 4769, pp. 76–85. Springer, Heidelberg (2007)
11. Kearney, P., Corneil, D.: Tree Powers. Journal of Algorithms 29(1), 111–131 (1998)

12. Makowsky, J.A.: Model theory and computer science: an appetizer. In: Handbook of Logic in Computer Science, vol. 1, pp. 763–814. Oxford University Press, New York (1992)
13. Nishimura, N., Ragde, P., Thilikos, D.M.: On graph powers of leaf-labeled trees. Journal of Algorithms 42, 69–108 (2002)
14. Oum, S., Seymour, P.: Approximating clique-width and branch-width. Journal of Combinatorial Theory, Series B 96(4), 514–528 (2006)

Deterministic Sparse Column Based Matrix Reconstruction via Greedy Approximation of SVD

Ali Çivril and Malik Magdon-Ismail

Computer Science Department, RPI, 110 8th Street, Troy, NY 12180
{civria,magdon}@cs.rpi.edu

Abstract. Given a matrix $A \in \mathbb{R}^{m \times n}$ of rank r, and an integer $k < r$, the top k singular vectors provide the best rank-k approximation to A. When the columns of A have specific meaning, it is desirable to find (provably) "good" approximations to A_k which use only a small number of columns in A. Proposed solutions to this problem have thus far focused on randomized algorithms. Our main result is a simple greedy deterministic algorithm with guarantees on the performance and the number of columns chosen. Specifically, our greedy algorithm chooses c columns from A with $c = O\left(\frac{k^2 \log k}{\epsilon^2} \mu^2(A) \ln \left(\frac{\sqrt{k} \|A_k\|_F}{\epsilon \|A - A_k\|_F} \right) \right)$ such that

$$\|A - C_{gr} C_{gr}^+ A\|_F \leq (1 + \epsilon) \|A - A_k\|_F,$$

where C_{gr} is the matrix composed of the c columns, C_{gr}^+ is the pseudo-inverse of C_{gr} ($C_{gr} C_{gr}^+ A$ is the best reconstruction of A from C_{gr}), and $\mu(A)$ is a measure of the *coherence* in the normalized columns of A. The running time of the algorithm is $O(SVD(A_k) + mnc)$ where $SVD(A_k)$ is the running time complexity of computing the first k singular vectors of A. To the best of our knowledge, this is the first deterministic algorithm with performance guarantees on the number of columns and a $(1 + \epsilon)$ approximation ratio in Frobenius norm. The algorithm is quite simple and intuitive and is obtained by combining a generalization of the well known *sparse approximation problem* from information theory with an *existence* result on the possibility of sparse approximation. Tightening the analysis along either of these two dimensions would yield improved results.

1   Introduction

Most data can be represented as an $m \times n$ matrix where the columns are objects and the rows are the features associated with them. Hence, given a matrix $A \in \mathbb{R}^{m \times n}$, one might be interested in obtaining the "important" spectral information of A by using some compressed representation. The usual approach to this problem is to take the best rank k ($k \ll \min\{m, n\}$) approximation A_k, which minimizes the error with respect to any unitarily invariant norm. A_k can be constructed from the top k singular vectors in $O(\min\{mn^2, m^2 n\})$ time. The

S.-H. Hong, H. Nagamochi, and T. Fukunaga (Eds.): ISAAC 2008, LNCS 5369, pp. 414–423, 2008.

first k singular vectors required to construct A_k can be computed efficiently using Lanczos methods. The problem with this general approach, which was also pointed out by [10] is that the singular vector representation might not be suitable to make inferences about the actual underlying data, because they are generally combinations of all the columns of the raw information in A. An example of this is the microarray data where the combinations of the column vectors have no sensible interpretation [16]. Hence, it is of practical importance to represent the approximation to A by a small number of columns of A.

1.1   Our Contributions

We give a deterministic greedy algorithm for low rank matrix reconstruction which is based on the sparse approximation of the SVD of A. We first generalize the sparse approximation problem of approximating vector [18] to one of approximating a subspace, using a small number of columns from A. We analyse a greedy algorithm which generalizes the analysis in [18]; in order to correct a minor technical error in the proof therein, we introduce a *coherence* parameter for a matrix, the *rank coherence parameter* which can be thought of as a more general and robust version of the coherence parameters defined in [23].

Our algorithm first computes the top k left singular vectors of A, and then selects columns of A in a greedy fashion so as to "fit" the space spanned by the singular vectors, appropriately scaled according to the singular values. The performance characteristics of the algorithm depend on how well the greedy algorithm approximates the optimal choice of such columns from A, and on how good the optimal columns themselves are. We give an existence result on the quality of the optimal columns, and the necessary analysis of the greedy algorithm to arrive at the following result:

Theorem 1. *The greedy algorithm chooses a column submatrix $C_{gr} \subseteq A$ with*
$$c = O\left(\frac{k^2 \log k}{\epsilon^2} \mu^2(A) \ln\left(\frac{\sqrt{k}\|A_k\|_F}{\epsilon\|A-A_k\|_F} \right) \right) \text{ columns such that}$$
$$\|A - C_{gr}C_{gr}^+ A\|_F \leq (1+\epsilon)\|A - A_k\|_F.$$

The term $\frac{k^2 \log k}{\epsilon^2}$ arises from an upper bound on the number of columns the optimal solution would choose (the existence result), and the remaining terms are contributed by the analysis of the greedy algorithm. The coherence parameter, $\mu(A)$ restricts the class of matrices for which the algorithm is useful. To the best of our knowledge, this is the first deterministic algorithm with $(1 + \epsilon)$ approximation. Note that, in order to achieve this approximation ratio, we choose more than k columns. When $\mu = O(1)$, setting $\epsilon = \sqrt{k \log k}$ and ignoring logarithmic factors, we have a $1 + \sqrt{k \log k}$ approximation ratio with $O(k)$ columns.

We believe that a result without the coherence parameter should be possible, however have not been able to construct one. In any case, improving either the upper bound on the optimal reconstruction of the singular vectors, or improving the analysis of the greedy algorithm would yield a tighter result. The running time of the algorithm is governed by the computation of the top k singular vectors, which is $O(SVD(A_k))$ and the greedy selection phase, which is $O(mnc)$.

1.2 Comparison to Related Work

With the advent of massive data sets, much work in theoretical computer science has been spent on finding algorithms for matrix reconstruction by considering a careful choice of a subset of the columns of the data matrix. The seminal paper by Frieze, Kannan and Vempala [12] gives a randomized algorithm that chooses a subset of columns $C \in \mathbb{R}^{m \times c}$ of A such that $\|A - \Pi_C A\|_F \leq \|A - A_k\|_F + \epsilon \|A\|_F$, where Π_C is a projection matrix obtained by the SVD of C and $c = poly(k, 1/\epsilon, 1/\delta)$, where δ is the failure probability of the algorithm. Subsequent work [9,8,20] introduced several improvements on the dependence of c on $k, 1/\epsilon$ and $1/\delta$ also extending the analysis to the spectral norm. Recently, the effort has been towards eliminating the additive term in the inequality thereby yielding a relative approximation in the form $\|A - \Pi_C A\|_F \leq (1 + \epsilon)\|A - A_k\|_F$. Along these lines, Deshpande et al. [5] first shows the existence of such approximations introducing a sampling technique related to the volume of the simplex defined by the column subsets of size k, without giving a polynomial time algorithm. Specifically, they show that there exists k columns with which one can get a $\sqrt{k+1}$ relative error approximation in Frobenius norm, which is tight. Later, Deshpande and Vempala [7] provides an algorithm with two steps which yields a relative approximation in expectation: first, approximate the "volume sampling" introduced in [5] by successively choosing one column at each step with carefully chosen probabilities; then, choose $O(k/\epsilon + k^2 \log k)$ columns in $O(k \log k)$ rounds in a similar fashion. The complexity of their algorithm is $O(M(k/\epsilon + k^2 \log k) + (m + n)poly(k, \epsilon))$, where M is the number of non-zero elements in A.

Recent result of Drineas et al. [10] provides two randomized algorithms for relative error approximation in Frobenius norm using "subspace sampling", i.e. selecting columns proportional to the row-norms of the matrix of top k right singular vectors. One of the algorithms chooses exactly $c = O(k^2 \log(1/\delta)/\epsilon^2)$ columns; the other chooses $c = O(k \log k \log(1/\delta)/\epsilon^2)$ columns in expectation and both of them runs in $O(SVD(A_k))$ time, i.e. the time required to compute A_k, where δ is the failure probability. Other recent approaches for the problem we consider includes random projections [21], and sampling which exploits geometric properties of high dimensional spaces [22]. [6] also considers the subspace approximation problem in general l_p norms. All of these algorithms exploit the power of randomization and they introduce a trade-off between the the number of columns chosen, the error parameter and the failure probability of the algorithm. The proof techniques presented in these papers break when the random sampling approach is sacrificed and a deterministic column selection procedure is used.

When it comes to deterministic reconstruction, no $(1 + \epsilon)$ approximation algorithms are known. The linear algebra community has developed deterministic algorithms in the framework of *rank revealing QR (RRQR) factorizations* [1] which yield some approximation guarantees in spectral norm. Given a matrix $A \in \mathbb{R}^{n \times n}$, consider the QR factorization of the form

$$A\Pi = Q \begin{pmatrix} R_{11} & R_{12} \\ 0 & R_{22} \end{pmatrix} \tag{1}$$

where $R_{11} \in \mathbb{R}^{k \times k}$ and $\Pi \in \mathbb{R}^{n \times n}$ is a permutation matrix. By the interlacing property of singular values (see [13]), $\sigma_k(R_{11}) \leq \sigma_k(A)$ and $\sigma_1(R_{22}) \geq \sigma_{k+1}(A)$. If the numerical rank of A is k, i.e. $\sigma_k(A) \gg \sigma_{k+1}(A)$, then one would like to find a permutation Π for which $\sigma_k(R_{11})$ is sufficiently large and $\sigma_1(R_{22})$ is sufficiently small. A QR factorization is said to be a rank revealing QR (RRQR) factorization if $\sigma_k(R_{11}) \geq \sigma_k(A)/p(k,n)$ and $\sigma_1(R_{22}) \leq \sigma_{k+1}(A)p(k,n)$, where $p(k,n)$ is a low degree polynomial in k and n.

Much research on finding RRQR factorizations has yielded improved results for $p(k,n)$ [1,2,4,14,15,19]. These algorithms make use of the *local maximum volume* concept and are generally complicated. Tight bounds for $p(k,n)$ can be used to give deterministic low rank matrix reconstruction with respect to the spectral norm, via the following simple fact.

Theorem 2. *Let Π_k be the matrix of first k columns of Π in (1). Then,*

$$\|A - (A\Pi_k)(A\Pi_k)^+ A\|_2 \leq p(k,n)\|A - A_k\|_2.$$

The best $p(k,n)$ was proposed by Gu and Eisenstat [14]. The authors show that there exists a permutation Π for which $p(k,n) = \sqrt{1 + k(n-k)}$. It is not known whether such a permutation can be computed in polynomial time. Instead, algorithms with $p(k,n) = \sqrt{1 + f^2 k(n-k)}$ were given which run in $O((m+n\log_f n)n^2)$ time for $f > 1$ [14]. Hence, for constant f, the approximation ratio depends on n and the running time is $O(mn^2 + n^3 \log n)$. Note that, these algorithms consider choosing exactly k columns and the results are not directly comparable to ours as they provide bounds on the spectral norm. It is not clear whether these algorithmic results can be extended to give non-trivial bounds in Frobenius norm or to choose more than k columns so as to yield $(1 + \epsilon)$ approximation.

Our results rely on a generalization of the sparse approximation problem which was formally proposed by Natarajan [18]: given $A \in \mathbb{R}^{m \times n}$, a vector $b \in \mathbb{R}^m$, and $\epsilon > 0$, find a vector $x \in \mathbb{R}^n$ satisfying $\|Ax - b\|_2 \leq \epsilon$ such that x has the fewest non-zero entries over all such vectors. This problem was also considered by Tropp [23]. Natarajan [18] proves that the problem is NP-hard and gives a greedy algorithm based on choosing the column vector from A with largest projection on b at each step. After correcting a minor technical error in his proof, his result gives that the greedy algorithm chooses at most $\lceil 18\, Opt(\epsilon/2)\mu^2(A) \ln(\|b\|_2/\epsilon) \rceil$ columns, $\mu(A)$ is a parameter defining the coherence between the normalized columns of A and $Opt(\epsilon/2)$ is the optimal number of vectors at error $\epsilon/2$. More recently, from an information theoretic point of view, Tropp [23] analyzed some previously known algorithms (e.g. *Matching Pursuit (MP)* [11,17], *Basis Pursuit (BP)* [3]) for the sparse approximation problem, showing that these algorithms perform well for dictionaries (matrices) which are close to orthonormal. A formalization of this notion is represented by the *coherence parameter* [17], which is the maximum absolute inner product betweeen two distinct column vectors. Tropp gives a natural generalization of this concept, the *cumulative coherence parameter*, which is the maximum coherence between a fixed column vector and a collection of other column vectors. Intuitively, these parameters measure how "close" the column vectors of a matrix are and smaller values indicate an *incoherent* (almost orthonormal) matrix.

1.3   Notation and Preliminaries

From now on $A \in \mathbb{R}^{m \times n}$ is the matrix we wish to reconstruct. $A_{(i)}$ denotes the ith row of A for $1 \leq i \leq m$, and $A^{(j)}$, the jth column of A for $1 \leq j \leq n$. A_{ij} is the element at i^{th} row and the jth column. Typically, we use C to denote a subset of columns of A, written $C \subset A$, i.e. C is a column submatrix of A. $span(C)$ denotes the subspace spanned by the column vectors in C. The Singular Value Decomposition of $A \in \mathbb{R}^{m \times n}$ of rank r is denoted by $A = U \Sigma V^T$ where $U \in \mathbb{R}^{m \times m}$ is the matrix of left singular vectors, $\Sigma \in \mathbb{R}^{m \times r}$ is the diagonal matrix containing the singular values of A in descending order, i.e. $\Sigma = diag(\sigma_1, \ldots, \sigma_r, 0, \ldots, 0)$ where $\sigma_1 \leq \sigma_2 \ldots \sigma_r > 0$ are the singular values of A. $V \in \mathbb{R}^{n \times n}$ is the matrix of right singular vectors. The "best" rank k approximation to A is $A_k = U_k \Sigma_k V_k^T$ where U_k and V_k are the first k columns of the corresponding matrices in the full SVD of A, and Σ_k is the $k \times k$ diagonal matrix of the first k singular values. The pseudo-inverse of A is denoted by $A^+ = V \Sigma^+ U^T$, where $\Sigma^+ = diag\left(\frac{1}{\sigma_1}, \ldots \frac{1}{\sigma_r}, 0, \ldots, 0\right)$. The Frobenius norm of A is $\|A\|_F = \sqrt{\sum_{i=1}^m \sum_{j=1}^n A_{ij}^2}$, and the spectral norm of A is $\|A\|_2 = \sigma_1(A)$. We also define the maximum column norm of a matrix A, $\|A\|_{col} = \max_{i=1}^n \{\|A^{(i)}\|_2\}$. $S^\perp$ is the space orthogonal to the space spanned by the vectors in S.

1.4   Organization of the Paper

The rest of the paper is organized as follows. In Section 2, we define a generalized version of the sparse approximation problem which asks for a small set of columns that approximates the subspace spanned by a given set of target vectors. We give a greedy algorithm along with its analysis. Section 3 gives our column based rank matrix reconstruction algorithm, which can be viewed as a special case of the generalized sparse approximation problem, where the target vectors are the left singular vectors of A.

2   Generalized Sparse Approximation

Instead of seeking sparse approximation to a single vector [18], we propose the following generalization: given matrices $A \in \mathbb{R}^{m \times n}$, a set of vectors $B \in \mathbb{R}^{m \times k}$, and $\epsilon > 0$, find a matrix $X \in \mathbb{R}^{n \times k}$ satisfying

$$\|AX - B\|_F \leq \epsilon \tag{2}$$

such that $\sum_{i=1}^n \nu_i(X)$ is minimum over all possible choices of X, where $\nu_i(X) = 1$ if the row $X_{(i)}$ contains non-zero entries, $\nu_i(X) = 0$ if $X_{(i)} = \vec{0}$. Intuitively, the problem asks for a minimum number of set of column vectors of A whose span is close to those of B.

2.1   The Algorithm

A greedy strategy for solving this problem is to choose the column v from A at each iteration, for which $\|B^T v\|_2$ is maximum, and project the column vectors of B and the other column vectors of A onto the space orthogonal to the chosen column. The algorithm proceeds greedily on these residual matrices until the norm of the residual B drops below the required threshold ϵ. Naturally, if the error ϵ cannot be attained, the algorithm will fail after selecting a maximal independent set of columns.

Greedy$(A,\ B,\ \epsilon)$

1: normalize each column of A to have norm 1.

2: $l \leftarrow 0,\ \Lambda \leftarrow \emptyset,\ A_0 \leftarrow A,\ B_0 \leftarrow B$.

3: **while** $\|B_l\|_F > \epsilon$ **do**

4:    choose $i \in \{1,\ldots,n\} - \Lambda$ such that $\|B_l^T A_l^{(i)}\|_2$ is maximum

5:    $B_{l+1}^{(j)} \leftarrow B_l^{(j)} - \left(B_l^{(j)^T} A_l^{(i)}\right) A_l^{(i)}$ for $i = 1,\ldots,k$, i.e. project $B_l^{(j)}$'s onto $\{A_l^{(i)}\}^{\perp}$.

6:    $\Lambda \leftarrow \Lambda \cup \{i\}$.

7:    $A_{l+1}^{(j)} \leftarrow A_l^{(j)} - \left(A_l^{(j)^T} A_l^{(i)}\right) A_l^{(i)}$ for $j \in \{1,\ldots,n\} - \Lambda$, i.e. project $A_l^{(j)}$'s onto $\{A_l^{(i)}\}^{\perp}$.

8:    normalize $A_{l+1}^{(j)}$ for $j \in \{1,\ldots,n\} - \Lambda$.

9:    $l \leftarrow l + 1$.

10: **end while**

11: return $C = \Lambda(A)$, the selected columns.

Fig. 1. A greedy algorithm for Generalized Sparse Approximation

We first define the coherence of a matrix.

Definition 3 (Coherence). *Given a matrix $A \in \mathbb{R}^{m \times n}$ of rank r, let $\mathbf{A}$ be the matrix A with normalized columns. Then, the rank coherence of A, $\mu(A)$ is the maximum of the inverses of the least singular value of $m \times r$ full-rank sub-matrices of $\mathbf{A}$. Namely,*

$$\mu(A) = \max_{\substack{\mathbf{C} \subseteq \mathbf{A} \\ \mathbf{C} \in \mathbb{R}^{m \times r} \\ rank(\mathbf{C}) = r}} \frac{1}{\sigma_r(\mathbf{C})}. \tag{3}$$

Remark 4. $1 \le \mu(A) < \infty$. *Small values of $\mu(A)$ indicate a matrix with near orthonormal columns.*

Theorem 5. *The number of columns chosen by Greedy is at most*

$$O\left(Opt(\epsilon/2)\mu^2(A)\ln\left(\frac{\|B\|_F}{\epsilon}\right)\right)$$

where $Opt(\epsilon/2)$ is the optimal number of columns at error $\epsilon/2$.

We will establish Theorem 5 through a sequence of lemmas. The proof follows similar reasoning to the proof in [18]. Let t be the total number of iterations of Greedy. At the beginning of the lth iteration of the algorithm, for $0 \leq l < t$, let U_l be an optimal solution to the generalized sparse approximation problem with error parameter $\epsilon/2$, i.e. U_l minimizes $\sum_{i=1}^{n} \nu_i(X)$ over $X \in \mathbb{R}^{n \times k}$ such that $\|A_l U_l - B_l\|_F \leq \epsilon/2$, where $\nu_i(X) = 1$ if the row $X_{(i)}$ contains non-zero entries, $\nu_i(X) = 0$ if $X_{(i)} = \vec{0}$. Let $N_l = \sum_{i=1}^{n} \nu_i(U_l)$ and $Q_l = A_l U_l$. Define

$$\lambda = 4 \max_{0 \leq l < t} \frac{N_l \|U_l\|_F^2}{\|B_l\|_F^2}. \tag{4}$$

The proofs of the following lemmas which essentially bound the number of iterations of the algorithm, are given in the appendix. Assuming that the Greedy has not terminated, the first lemma states that the next step makes significant progress.

Lemma 6. *For the lth iteration of Greedy,* $\|B_l^T A_l\|_{col} \geq \frac{\|B_l\|_F^2}{2\sqrt{N_l}\|U_l\|_F}$.

Thus, there exists a column in the residual A_l which will reduce the residual B_l significantly, because B_l has a large projection onto this column. Therefore, since every step of Greedy makes significant progress, there cannot be too many steps, which is the content of the next lemma.

Lemma 7. $t \leq \left\lceil 2\lambda \ln\left(\frac{\|B\|_F}{\epsilon}\right)\right\rceil$, *where t is the number of Greedy iterations.*

What remains is to bound λ. First, we will bound $\|U_l\|_F$ in terms of $\|B_l\|_F$ both of which appear in the expression for λ. Let $\pi_l = \{i | U_{l(i)} \neq \vec{0}\}$ be the indices of rows of U_l which are not all zero. Recall that these indices denote which columns are chosen by the optimal solution for A_l. Let $\tau_l = \{i_1, i_2, \ldots, i_l\}$ be the indices of the first l columns picked by the algorithm. Given an index set γ, let the set of column vectors $\{A^{(i)} | i \in \gamma\}$ be denoted by $\gamma(A)$. The proofs of the following lemmas are also in the appendix.

Lemma 8. $\pi_l(A) \cup \tau_l(A)$ *is a linearly independent set for all $l \geq 0$.*

Lemma 9. *For $0 \leq l < t$,* $\|U_l\|_F \leq \frac{3}{2}\mu(A)\|B_l\|_F$.

Proof of Theorem 5: First, we note that the number of non-zero rows in the optimal solution is non-increasing as the algorithm proceeds, that is $N_l \geq N_{l+1}$ for $l > 0$, which follows from an argument identical to the proof of Lemma 3 in [18]. Since $Opt(\epsilon/2) = N_0$, we have

$$\lambda \leq 4 \max_{0 \leq l < t} \frac{N_0 \|U_l\|_F^2}{\|B_l\|_F^2} \leq 9 Opt(\epsilon/2)\mu^2(A)$$

where the last inequality is due to the result of Lemma 9. Combining this with Lemma 7, we have that the number of iterations of the algorithm is bounded by

$$t \leq \left\lceil 18 Opt(\epsilon/2)\mu^2(A) \ln\left(\frac{\|B\|_F}{\epsilon}\right)\right\rceil$$

3   Deterministic Low-Rank Matrix Reconstruction

In this section, we give a deterministic algorithm for low rank matrix reconstruction based on the greedy approach that we have introduced and analyzed for the generalized sparse approximation problem:

LowRankApproximation(A, k)
 1: compute U_k and Σ_k of A
 2: return Greedy(A, $U_k\Sigma_k$, $\epsilon\|A - A_k\|_F$)

Fig. 2. The low-rank approximation algorithm

The algorithm first computes U_k, the top k left singular vectors of A and Σ_k the first k singular values of A, which can be performed by standard methods like Lanczos. The columns of A are then selected in a greedy fashion so as to "fit" them to the subspace spanned by the columns of $U_k\Sigma_k$. Intuitively, we select columns of A which are close to the columns of $U_k\Sigma_k$ and the analysis shows that the submatrix C of A we obtain is provably close to the "best" rank-k approximation to A. The error parameter which is given as an input to the greedy algorithm is $\epsilon\|A - A_k\|_F$. The following result provides an upper bound on the number of columns of the optimal solution at error $\epsilon\|A - A_k\|_F/2$.

Lemma 10. *There exists a column submatrix C of A with $c = O(k \log k/\epsilon^2)$ columns such that $\|U_k\Sigma_k - CC^+U_k\Sigma_k\|_F \leq \epsilon\|A - A_k\|_F/2$.*

Proof. The proof is given in the appendix due to space limitations.

We now, give the proof of Theorem 1.

Proof of Theorem 1: By the algorithm, we have

$$U_k\Sigma_k = C_{gr}C_{gr}^+U_k\Sigma_k + E,$$

for some generic error matrix E satisfying $\|E\|_F \leq \epsilon\|A - A_k\|_F$. Multiplying both sides by V_k^T, we get

$$U_k\Sigma_k V_k^T = C_{gr}C_{gr}^+U_k\Sigma_k V_k^T + EV_k^T,$$

which is clearly

$$A_k = C_{gr}C_{gr}^+A_k + EV_k^T.$$

Rearranging and adding A to the both sides of the equation, we obtain $A - C_{gr}C_{gr}^+A_k = A - A_k + EV_k^T$. Taking norms of both sides, and noting that C_{gr}^+A is the minimizer of $\|A - C_{gr}X\|_F$, we obtain

$$\|A - C_{gr}C_{gr}^+ A\|_F \leq \|A - C_{gr}C_{gr}^+ A_k\|_F$$
$$= \|A - A_k + EV_k^T\|_F$$
$$\leq \|A - A_k\|_F + \|E\|_F\|V_k^T\|_F$$
$$\leq \|A - A_k\|_F + \epsilon\sqrt{k}\|A - A_k\|_F$$
$$= (1 + \epsilon\sqrt{k})\|A - A_k\|_F$$

Third line follows due to the triangle inequality and submultiplicativity of the Frobenius norm. Fourth line is due to the fact that $\|E\|_F \leq \epsilon\|A - A_k\|_F$ and $\|V_k^T\|_F \leq \sqrt{k}$. Choosing an error parameter $\epsilon' = \epsilon/\sqrt{k}$ and combining Theorem 5 and Lemma 10 gives the desired result.

Note that, the number of columns chosen by the algorithm depends on $\mu(A)$, i.e. the structure of A. To get an idea of what this result implies when the number of columns chosen is of order k, we give the following corollary, which immediately follows upon a careful choice of error parameter.

Corollary 11. *The greedy algorithm chooses a submatrix C of $\tilde{O}(k)$ columns of A for which $\|A - CC^+A\|_F \leq \mu(A)\sqrt{k\log k}\|A - A_k\|_F$.*

Acknowledgments. We would like to thank Petros Drineas for helpful discussions.

References

1. Chan, T.F.: Rank revealing QR factorizations. Linear Algebra Appl. (88/89), 67–82 (1987)
2. Chandrasekaran, S., Ipsen, I.C.F.: On rank-revealing factorizations. SIAM J. Matrix Anal. Appl. 15, 592–622 (1994)
3. Chen, S.S., Donoho, D.L., Saunders, M.A.: Atomic decomposition by basis pursuit. SIAM Review 43(1), 129–159 (2001)
4. de Hoog, F.R., Mattheijb, R.M.M.: Subset selection for matrices. Linear Algebra and its Applications (422), 349–359 (2007)
5. Deshpande, A., Rademacher, L., Vempala, S., Wang, G.: Matrix approximation and projective clustering via volume sampling. In: SODA 2006, pp. 1117–1126. ACM Press, New York (2006)
6. Deshpande, A., Varadarajan, K.: Sampling-based dimension reduction for subspace approximation. In: STOC 2007: Proceedings of the thirty-ninth annual ACM symposium on Theory of computing, pp. 641–650. ACM, New York (2007)
7. Deshpande, A., Vempala, S.: Adaptive sampling and fast low-rank matrix approximation. In: Díaz, J., Jansen, K., Rolim, J.D.P., Zwick, U. (eds.) APPROX 2006 and RANDOM 2006. LNCS, vol. 4110, pp. 292–303. Springer, Heidelberg (2006)
8. Drineas, P., Frieze, A., Kannan, R., Vempala, S., Vinay, V.: Clustering in large graphs and matrices. In: SODA 1999: Proceedings of the tenth annual ACM-SIAM symposium on Discrete algorithms, pp. 291–299. SIAM, Philadelphia (1999)
9. Drineas, P., Kannan, R., Mahoney, M.W.: Fast monte carlo algorithms for matrices II: Computing a low-rank approximation to a matrix. SIAM Journal on Computing 36(1), 158–183 (2006)

10. Drineas, P., Mahoney, M.W., Muthukrishnan, S.: Subspace sampling and relative-error matrix approximation: column-row-based methods. In: Azar, Y., Erlebach, T. (eds.) ESA 2006. LNCS, vol. 4168, pp. 304–314. Springer, Heidelberg (2006)
11. Friedman, J.H., Stuetzle, W.: Projection pursuit regressions. J. Amer. Statist. Soc. 76, 817–823 (1981)
12. Frieze, A., Kannan, R., Vempala, S.: Fast monte-carlo algorithms for finding low-rank approximations. Journal of the Association for Computing Machinery 51(6), 1025–1041 (2004)
13. Golub, G.H., Loan, C.V.: Matrix Computations. Johns Hopkins U. Press (1996)
14. Gu, M., Eisenstat, S.C.: Efficient algorithms for computing a strong rank-revealing QR factorization. SIAM Journal on Scientific Computing 17(4), 848–869 (1996)
15. Hong, Y.P., Pan, C.T.: Rank-revealing QR factorizations and the singular value decomposition. Mathematics of Computation 58, 213–232 (1992)
16. Kuruvilla, F.G., Park, P.J., Schreiber, S.L.: Vector algebra in the analysis of genome-wide expression data. Genome Biology (3) (2002)
17. Mallat, S., Zhang, Z.: Matching pursuits with time-frequency dictionaries. IEEE Transactions on Signal Processing 41(12), 3397–3415 (1993)
18. Natarajan, B.K.: Sparse approximate solutions to linear systems. SIAM Journal on Computing 24(2), 227–234 (1995)
19. Pan, C.T., Tang, P.T.P.: Bounds on singular values revealed by QR factorizations. BIT Numerical Mathematics 39, 740–756 (1999)
20. Rudelson, M., Vershynin, R.: Sampling from large matrices: An approach through geometric functional analysis. J. ACM 54(4) (2007)
21. Sarlos, T.: Improved approximation algorithms for large matrices via random projections. In: FOCS 2006: Proceedings of the 47th Annual IEEE Symposium on Foundations of Computer Science, Washington, DC, USA, pp. 143–152. IEEE Computer Society Press, Los Alamitos (2006)
22. Shyamalkumar, N.D., Varadarajan, K.: Efficient subspace approximation algorithms. In: SODA 2007: Proceedings of the eighteenth annual ACM-SIAM symposium on Discrete algorithms, pp. 532–540. Society for Industrial and Applied Mathematics, Philadelphia (2007)
23. Tropp, J.A.: Greed is good: algorithmic results for sparse approximation. IEEE Transactions on Information Theory 50(10), 2231–2242 (2004)

Minimizing Total Flow-Time:
The Unrelated Case

Naveen Garg⋆, Amit Kumar⋆, and V.N. Muralidhara⋆⋆

Department of Computer Science and Engineering,
Indian Institute of Technology, Delhi, India
{naveen,amitk,murali}@cse.iitd.ac.in

Abstract. We consider the problem of minimizing flow-time in the un-related machines setting. We introduce a notion of (α, β) variability to capture settings where processing times of jobs on machines are not completely arbitrary and give an $O(\beta \log \alpha)$ approximation for this setting. As special cases, we get (1) an $O(k)$ approximation when there are only k different processing times (2) an $O(\log P)$-approximation if each job can only go on a specified subset of machines, but has the same processing requirement on each such machine. Further, the machines can have different speeds. Here P is the ratio of the largest to the smallest processing requirement, (3) an $O(\frac{1}{\varepsilon} \log \frac{1}{\varepsilon})$- approximation algorithm for unrelated machines if we assume that our algorithm has machines which are an ε-factor faster than the optimum algorithm's machines. We also improve the lower bound on the approximability for the problem of minimizing flow time on parallel machines from $\Omega(\sqrt{\log P / \log \log P})$ to $\Omega(\log^{1-\varepsilon} P)$ for any $\varepsilon > 0$.

1   Introduction

We consider the problem of scheduling jobs on multiple machines so as to minimize the total flow time. The flow time of a job is the total time it spends in the system and equals the difference between its completion time and release time. A job j has a processing time p_{ij} on machine i; this is the most general model of processing times and is known as the *unrelated machines model*. A special case is the *related machines model* where job j has length a_j, machine i has speed (or slowness) s_i and the processing time of job j on machine i is $a_j \cdot s_i$. When the machines are identical so that $s_i = 1$ for all i, then we have the *parallel machines model*.

Minimizing flow time is most interesting in the on line setting where the processing time of a job on various machines is revealed only when the job is released. For parallel machines an $O(\log P)$-competitive algorithm was first given

---

⋆ Work done as part of the "Approximation Algorithms" partner group of MPI-Informatik, Germany.
⋆⋆ Part of the work is supported by DST, Govt of India under Fast Track Young Scientist Scheme.

S.-H. Hong, H. Nagamochi, and T. Fukunaga (Eds.): ISAAC 2008, LNCS 5369, pp. 424–435, 2008.

by Leonardi and Raz [6] while for related machines an $O(\log^2 P)$ algorithm was given by Garg and Kumar [4], where P is the ratio of the maximum and minimum processing times. However, for unrelated machines no on line algorithm can have bounded competitive ratio and this is true even for the very restricted setting when each p_{ij} is either 1 or ∞[5]. It is for this reason that we consider offline algorithms for minimizing flow time on unrelated machines.

In this paper we present a new algorithm for this problem. The setting when the processing times p_{ij} are completely arbitrary seems very unnatural and difficult to approximate. In this paper we consider a model which tries to capture the variation in processing times which we motivate as follows. Suppose we have a cluster of heterogeneous computers (machines) on which we want to run some programs (jobs). These programs are solving different problems – matrix multiplication, sorting, sequence alignment, mpeg decoding etc. Similarly machines have different architectures – super scalars, multi-cores, vector machines – and some architectures might be better suited for a particular problem. This naturally gives rise to different processing times for different problems on different architectures; we assume that these processing times are from a set $B = \{b_1, b_2 \ldots, b_\beta\}$. Two machines which implement the same architecture could run at different speeds due to differences in clock frequency, memory etc. This results in processing time variations which would (to a large extent) be independent of the program being run. Similarly, we could be using different algorithms to solve the same problem and running them on different instances. The variation in the processing times of the resulting programs would therefore be independent of the machine on which the program is executed.

We say that processing times have an (α, β)-*variability* if the processing time of job j on machine i can be expressed as $p_{ij} = a_j \cdot b_{ij} \cdot s_i$ where $b_{ij} \in B, |B| = \beta$ and $1 \le a_j \le \alpha$. Our main result in this paper is an $O(\beta(1 + \log \alpha))$-approximation when processing times have an (α, β) variability. Note that when p_{ij} take k different values then we can always set $\alpha = 1, \beta = k$ and this gives an $O(k)$ approximation. This result was obtained independently by Sitters [7] using a different algorithm. However, our aim in this paper is to do even better by providing an algorithm which can exploit bounded variability in the processing times. As a case in point, consider the setting when job j has processing time p_j but it can be scheduled only on a subset of machines S_j. If all machines have the same speed, we call this the subset-parallel model. Garg and Kumar [5] gave an $O(\log P)$ approximation for minimizing flow time in this model, where $P = \max_j p_j / \min_j p_j$. If machine i has speed s_i then the processing time of job j on machine i (assuming $i \in S_j$) is $p_{ij} = p_j \cdot s_i$. Note that choosing $B = \{1, \infty\}$ implies a $(P, 2)$ variability on the processing times and hence our algorithm gives an $O(\log P)$-approximation. This is the first approximation algorithm for this *subset-related model*. We believe that for many other natural settings, the variations in processing times can be captured using (α, β) variability for small α, β and this would then imply better approximation guarantees using our algorithm.

The difficulty in approximating flow time for unrelated machines led us to consider the problem with resource augmentation. In this setting the approximation algorithm has machines which are an ε-factor faster than those of opt. This means that the approximation algorithm can complete $1 + \varepsilon$ units of work in 1 time unit while opt can only do 1 unit. Such a model was considered for parallel machines and Chekuri et. al. [1] showed a $O(\frac{1}{\varepsilon})$-competitive algorithm for minimizing flow time. A small modification to our main algorithm gives an $O(\varepsilon^{-1} \log \varepsilon^{-1})$-approximation for minimizing flow time on unrelated machines. Note that this result does not assume that processing times have bounded variability.

Our final result in this paper improves the lower bound on the approximation ratio for minimizing flow time on parallel machines from $\Omega(\sqrt{\log P / \log \log P})$ [5] to $\Omega(\log^{1-\varepsilon} P)$ for any $\varepsilon > 0$. Note that an $\Omega(\log P)$ lower bound is already known for the subset parallel (and hence also for unrelated) model [5].

Technical comparison with other work. The $O(\log P)$-approximation for the subset parallel model by Garg and Kumar [5] solved an LP formulation of the problem and then rounded the solution iteratively by modifying the unsplittable flow algorithm of Dinitz et al.[2].

Our approach does the rounding in one step and uses the Dinitz et al. algorithm as a black box. We are able to do this because of a different way of handling the "last jobs" on each machine. Overall this makes the algorithm and its analysis very simple.

2 Unrelated Machines

We consider the problem of minimizing average flow-time of n jobs on m machines. Job j has a release date r_j. If j is scheduled on machine i then it has to be processed for p_{ij} time units. We allow for preemption but disallow migration, i.e. a job cannot be interrupted on a machine and restarted on another machine. We now give an integer programming formulation for minimizing flow time on unrelated machines [4]. For each job j, machine i and time t, we have a variable $x_{i,j,t}$ which is 1 if machine i processes job j from time t to $t+1$, 0 otherwise. The integer program is as follows.

$$\min \sum_j \sum_i \sum_t \quad x_{i,j,t} \cdot \left(\frac{t - r_j}{p_{ij}} + \frac{1}{2} \right) \tag{1}$$

$$\sum_j x_{i,j,t} \leq 1 \qquad \text{for all machines } i \text{ and time } t \tag{2}$$

$$\sum_i \sum_t x_{i,j,t}/p_{ij} = 1 \qquad \text{for all jobs } j \tag{3}$$

$$x_{i,j,t} = 0 \qquad \text{if } t < r_j \tag{4}$$

$$x_{i,j,t} \in \{0,1\} \qquad \text{for all jobs } j, \text{ machines } i, \text{ time } t \tag{5}$$

Constraint (2) refers to the fact that a machine can process at most one job at any point of time. Equation (3) says that job j gets completed by the schedule. Equation (4) captures the requirement that we cannot process a job before its release date. It should be clear that an integral solution gives rise to a schedule where jobs can migrate across machines and may even be processed simultaneously on different machines.

The only non-trivial part of the integer program is the objective function. Let $\Delta_j(x) = \sum_i \sum_t x_{i,j,t} \cdot \left(\frac{t-r_j}{p_j} + \frac{1}{2} \right)$. So the objective function is to minimize $\Delta(x) = \sum_j \Delta_j(x)$. Let $\mathcal{S}$ be a non-migratory schedule. $\mathcal{S}$ also yields a solution to the integer program in a natural way — let $x^{\mathcal{S}}$ denote this solution. The following fact was established in [4].

Fact 1. *The total flow-time of $\mathcal{S}$ is at least $\sum_j \Delta_j(x^{\mathcal{S}})$.*

We say that the processing times have variability (α, β) if for all machines i and job j, $p_{ij} = a_j \cdot b_k \cdot s_i$ where $1 \le a_j \le \alpha$ and $b_k \in B$, where $B = \{b_1, \ldots, b_{|B|}\}$ is a set of size at most β. We impose no restriction on s_i but it is no loss of generality to assume that $s_i \ge 1$.

We say that job j on machine i is of *class* k if $p_{ij} = a_j \cdot b_k \cdot s_i$. Consider the integer program (1). For the purpose of subsequent discussion, we modify the integer program slightly by replacing the objective function $\sum_j \Delta_j(x)$ by $\sum_j \Delta'_j(x)$, where $\Delta'_j(x) = \sum_i \sum_t x_{i,j,t} \left(\frac{t-r_j}{p'_{ij}} + \frac{1}{2} \right)$. Here $p'_{ij} = p_{ij}/a_j$. It is easy to see that this modification does not decrease the cost of a solution x and increases it by at most a factor α. But this slight modification in the objective function turns out to be very useful. Consider the slots in x for a fixed machine i and class k – if we rearrange the jobs being processed on these slots (without violating release dates), the objective function does not change.

We now relax the integer program by replacing the constraints (5) by $x_{i,j,t} \ge 0$. Let x^* be an optimal fractional solution to this linear program (LP).

2.1   The Algorithm

We order the jobs in ascending order of their release times (breaking ties arbitrarily). We begin by constructing a directed graph G over which we would be running the unsplittable flow algorithm of Dinitz et al [2]. We first describe the vertex set V of G. For each machine i, job j and class k, we have a vertex $v(i, j, k)$. For each job j, we have a vertex $v(j)$ and a special source vertex s. Further we have a set of special vertices $U = \{u_1, \ldots\}$ – the size of U will be the processing time of the optimal solution divided by α (this will become clear in the construction below). We now describe the edges in G. Since we shall treat this as an instance of the unsplittable flow problem, we need to define flow values on each edge and the demand of each vertex. We describe this next:

1. Let $z^*_{i,j} = \sum_t \frac{x^*_{i,j,t}}{p'_{ij}}$. This is the total (fractional) processing of j on machine i scaled by the factor a_j. If this quantity is non-zero, and the class of j on machine i is k, we have an edge from $v(i, j, k)$ to $v(j)$ with flow $z^*_{i,j}$. There are no other edges incident at vertex $v(j)$. Note that for each job j, the LP has a constraint $\sum_i \sum_t x_{i,j,t}/p_{ij} = 1$. This implies that the total flow reaching vertex $v(j)$ is exactly a_j and this is the demand of vertex $v(j)$. The demand of all other vertices in G is zero.

2. For each j, $1 \leq j \leq n - 1$, there is an edge from $v(i, j, k)$ to $v(i, j + 1, k)$ for every i, k. So for fixed values of i and k, we can think of this as a directed path joining the corresponding vertices in the order of release dates. The edge from $v(i, j, k)$ to $v(i, j + 1, k)$ has flow given by conservation constraints.

3. We now build a bipartite sub graph between the vertices in U and the set $V' = \{v(i, 1, k) : \text{for all } i, k\}$. We order the vertices in V' by descending value of $b_k \cdot s_i$ and relabel vertices as $v_1, v_2, \ldots, v_{|V'|}$ to reflect this ordering. Let f_i be the total flow leaving vertices $v_1, \ldots, v_i$. With each vertex v_i we associate an interval $[f_{i-1}, f_i]$ (let $f_0 = 0$). Further for each vertex $u_l \in U$ we associate the interval $[\alpha \cdot (l - 1), \alpha \cdot l]$. There is an edge from u_l to v_i iff the corresponding intervals overlap and the flow on this edge is the length of the overlap. Hence the total flow leaving u_l is exactly α (except perhaps for the last vertex in U) and the total flow entering v_i is the total flow leaving it.

4. For every $u_l \in U$, there is an edge from s to u_l and the flow on these edges is given by flow conservation.

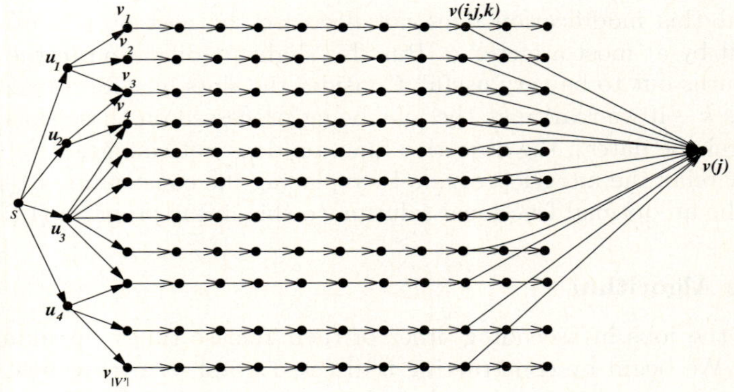

Fig. 1. The graph G constructed from the LP solution

Let f be the initial flow as described above. We will compute a flow, f^*, in which all demand from the source s to sink $v(j)$ is routed on a single path. The unsplittable flow theorem of Dinitz et al. says that this can be done such that the flow on any edge exceeds the fractional flow on that edge by at most the maximum demand, which in our case is α. The unsplittable flow f^* in graph G, allots each job to a unique machine. Once we know which jobs to schedule on which machine, we apply Shortest-Remaining-Processing-Time (SRPT) on each machines. Let us call this schedule as S. In order to analyse the flow time of this schedule, we construct another schedule S' as follows

On each machine i, we consider jobs by increasing release time and schedule them in the earliest available slots (of class k on machine i). Note that we might not be able to assign all jobs to slots since we might run out of slots. It might also happen that only a part of a job can be assigned. Let J' be the jobs that could be processed completely. After all the jobs in J' are scheduled, we schedule the jobs in $J - J'$ at the earliest possible time on the machine they were assigned to (considering these jobs in an arbitrary order). We note the flow time of S is at most the flow time of S' and hence enough to bound the flow time of S'. Let P be the total processing time of the jobs in our schedule S'.

Claim. For each machine i and class k, at most α jobs will remain unassigned in the above step.

Proof. Fix a machine i and class k. The flow through the edge $(v(i, j - 1, k), v(i, j, k))$ in f^* exceeds the original flow through this edge by at most α. Since the total available space after t equals the original flow through this edge, the amount of additional available space needed to schedule jobs released at or after t is at most α. Since at most α jobs could have used this space, the number of unassigned jobs is at most α. $\blacksquare$

The following claim was proved in [4]

Claim. Total flow time of jobs in J' is at most $\Delta'(x^*) + \beta P$.

Each of these jobs in $J - J'$ would have a flow time which is at most the total processing time on the machine. Since each machine can have at most $\alpha\beta$ of these jobs, the total increase in the flow time of schedule S' due to these jobs is at most $\alpha\beta P$. Combining the previous claim we get

Lemma 1. *Total flow time of schedule S' is at most $\Delta'(x^*) + (\alpha + 1)\beta P$.*

The following lemma bounds P.

Lemma 2. $P \leq 4 \cdot \Delta'(x^*) + 2\alpha \cdot T$ *where T is the largest processing time of a job in S'.*

Proof. Let $f(i, k)$ be the total flow leaving vertex $v(i, 1, k)$ in the flow f. It is easy to see that $\sum_{i,j,t} x^*_{i,j,t} = \sum_{i,k} s_i \cdot b_k \cdot f(i, k)$. Similarly, $P = \sum_{i,k} s_i \cdot b_k \cdot f'(i, k)$ where f' is the unsplittable flow we compute.

Claim. $\sum_{i,k} s_i \cdot b_k \cdot f'(i, k) \leq 2 \sum_{i,k} s_i \cdot b_k \cdot f(i, k) + 2\alpha T$

Proof. We shall split the sum in the left hand side as $\sum_{u_l \in U} F'(u_l)$, where $F'(u_l)$ is the portion of $\sum_{i,k} s_i \cdot b_k \cdot f'(i, k)$ where we only look at the flow $f'(i, k)$ which comes from the vertex u_l. Define $F(u_l)$ for the right hand side similarly.

First observe that the total flow leaving u_j is α in f and hence it is at most 2α in f'. Hence $F'(u_l) \leq 2F(u_{l-1})$, $l \geq 2$ – here we use the fact that if flow goes from u_l to a vertex $v(i, 1, k)$ and from u_{l-1} to $v(i', 1, k')$ then $b_{k'} s_{i'} \geq b_k s_i$. All that remains is to bound $F(u_1)$. But this is at most $2\alpha T$. $\blacksquare$

Note that $\sum_{i,j,t} x^*_{i,j,t} \leq 2\Delta'(x^*)$. This together with the above claim implies the lemma. $\blacksquare$

We guess the longest processing time of any job (this can take only mn different values) in the optimum schedule, say L. We run the above algorithm with all $p_{ij} > L$ set to infinity. This ensures that $T \leq$ opt. Hence $P \leq (4 + 2\alpha)$opt and so the flow time of schedule S' is $O(\alpha^2\beta$opt$)$.

Theorem 1. *The flow time of schedule S' is $O(\alpha^2\beta \cdot opt)$.*

Note that an (α, β) variability in processing times can also be viewed as a $(2, \beta(1 + \log\alpha)$ variability. This is because each processing time b_k in the set B can be replaced by the set $\{b_k, b_k * 2, b_k * 2^2, \ldots b_k * 2^i, \ldots b_k * 2^r\}$ where $r = \log\alpha$. Doing this would imply that $\alpha \leq 2$ and this yields the following corollary.

Corollary 1. *The algorithm in this section is an $O(\beta(1+\log\alpha))$-approximation to the problem of minimizing total flow time on unrelated machines when the processing times have an (α, β) variability.*

3   Speed Augmentation in Unrelated Machines

We now consider the setting where our algorithm has machines which are faster than the machines available to the optimum algorithm by a factor $(1 + \varepsilon)$.

We shall use the extra speed in three phases – in each phase we shall assume that we can do an extra $\varepsilon/3$ units of work in a unit time slot. We begin by rounding down all processing times to a power of $(1 + \varepsilon/3)$. Note that the optimum solution to this instance has flow time only lower than the optimum flow time for the original instance. A solution to this rounded down instance also gives us a solution to the original instance since we can use an additional speed of $\varepsilon/3$ to process each job fully – this is the first phase where we give ourselves extra speed.

We can run the algorithm from the previous section with $\alpha = 1$ and $\beta = \log_{1+\varepsilon/3} P$ where $P = \max_{i,j} p_{ij}$ is the maximum processing time. This in turn implies an approximation guarantee of $\log_{1+\varepsilon/3} P$. However, in this section we present an $O(\varepsilon^{-1} \log \varepsilon^{-1})$-approximation algorithm. We say that a job is of class k on machine i, if its processing time on this machine is $(1 + \varepsilon/3)^k$.

Our algorithm proceeds as the algorithm in the previous section (with $a_j = 1$). We solve the linear program build the graph G and compute an unsplittable flow f' – note that while solving the LP, we do not assume machines have extra speed. The unsplittable flow is used to determine the assignment of jobs to machines and apply Shortest-Remaining-Processing-Time (SRPT) on each machines. Let us call this schedule as S.

To analyse our algorithm, we compute the schedule S' as in the previous section for the jobs in J'. Now in our new schedule, consider the p_{ij} slots in which job j is scheduled on machine i. We exploit the fact that we have an additional speed $\varepsilon/3$ and move the processing of job j to the first $\frac{p_{ij}}{(1+\varepsilon/3)}$ of these p_{ij} slots – this is the second phase where we give ourselves extra speed. The remaining $\frac{p_{ij} \cdot \varepsilon/3}{(1+\varepsilon/3)}$ of p_{ij} slots is used to schedule the jobs in $J \setminus J'$. We order the jobs in ascending order of their class and schedule these jobs in the earliest possible slots. We exploit the fact that we have an additional speed $\varepsilon/3$ to bound the flow time of these jobs – this is the third phase where we give ourselves extra speed.

Let this new schedule be S''.

The following Claim uses the fact that we can get an improved result because of the extra speed.

Claim. The total flow time of jobs in J' is at most $O\left(\frac{1}{\varepsilon}\right) \cdot \sum_{j \in J'} \Delta'_j(x^{S''})$.

Now we try bound the flow time of jobs in $J \setminus J'$. For each machine i, let us denote these jobs by R_i. Let $P(R_i)$ be the total processing time of the jobs in R_i (in our schedule). Clearly $\sum_i P(R_i) \leq P \leq \mathsf{opt}$. Let T_i be the total processing time of the jobs in J' scheduled on machine i. Define ℓ as the smallest non-negative integer such that $T_i \leq \frac{(1+\varepsilon/3)^\ell}{\varepsilon^2}$. Let x be a constant such that $(1+\varepsilon/3)^x = 1/\varepsilon^2$. Note that x is $O(\frac{1}{\varepsilon} \log \frac{1}{\varepsilon})$.

We divide the jobs in R_i into three parts, depending on their class on machine i – *small*, *medium* and *large*. A job in R_i is small if it of class less than ℓ, medium if its class is between ℓ and $\ell + x$, and large if it is of class larger than $\ell + x$. We now bound the flow-time of jobs in each of these parts.

Claim. The total flow-time of small jobs in R_i is at most $O(T_i/\varepsilon)$.

Proof. Assume $\ell > 0$, otherwise we have nothing to prove. Consider a small job of class k in R_i. Note that a job of processing time $(1 + \varepsilon/3)^k$ will finish within $\frac{3}{\varepsilon} \cdot \sum_{j=0}^k (1 + \varepsilon/3)^j$ time of its release date. Hence the flow time of this job is at most $\left(\frac{1+\varepsilon}{\varepsilon^2}\right)$ times its processing time. So the total flow time of small jobs in R_i can be bounded by $O\left(\frac{1}{\varepsilon^2} \cdot \sum_{j=0} \ell(1 + \varepsilon)^\ell\right)$, which is at most a constant times $\frac{(1+\varepsilon/3)^\ell}{\varepsilon^3}$. But $T_i > \frac{(1+\varepsilon/3)^{\ell-1}}{\varepsilon^2}$. This proves the result. $\blacksquare$

Now we consider medium and large jobs. We use the following fact – the flow-time of a job of class k in R_i can be bounded by

$$T_i + \sum_{j=0}^k (1 + \varepsilon/3)^j \leq T_i + O\left(\frac{(1 + \varepsilon/3)^{k+1}}{\varepsilon}\right) \tag{6}$$

This is so because processing of such a job can be interrupted by either the jobs in J' or those in R_i of smaller class.

Claim. Total flow-time of medium jobs is $O(x \cdot T_i + P(R_i)/\varepsilon)$.

Proof. We use equation (6). Consider a job j of class k. Its processing time on machine i, $p_{ij} = (1 + \varepsilon/3)^k$. Therefore, its flow-time can be bounded by $T_i + O(p_{ij}/\varepsilon)$. Summing over all medium jobs, we see that their total flow-time is $x \cdot T_i$ plus $O(1/\varepsilon)$ times the total processing time of medium jobs. $\blacksquare$

Claim. The total flow-time of large jobs is $O(P(R_i)/\varepsilon)$.

Proof. The processing time of a large job is at least $(1+\varepsilon/3)^{\ell+x} \geq \frac{(1+\varepsilon/3)^{\ell}}{\varepsilon^2} \geq T_i$. So in equation (6), the second term dominates the first term, and so the result follows.   ∎

Combining the above claims, we get

Theorem 2. *The total flow time of schedule S is at most $O(\frac{1}{\varepsilon}\log\frac{1}{\varepsilon} \cdot opt)$.*

4   Lower Bound for the Parallel Scheduling Problem

In this section, we give lower bound for the approximation ratio of the Parallel Scheduling problem. Recall that in the Parallel Scheduling problem, all machines are identical. Each job j has the same processing time p_j on all the machines. We shall use the 3-partition problem to prove the hardness of the Parallel Scheduling problem. Recall the 3-partition problem. We are given a set X of $3m$ elements and a bound B. Each element $x \in X$ has size s_x which is an *integer*. We would like to partition the set X into m sets such that the total size of elements in any such set is exactly B. Such a partition is said to be *valid* partition. We can assume without loss of generality that the sum of the sizes of the elements in X is exactly $m \cdot B$. It is NP-complete to decide if such a valid partition exists or not [3].

In fact, we can say more about this problem. If we look at the NP-completeness proof of the 3-partition problem (see for example [3]) we can add two more constraints on the problem instance : (i) For every element x, s_x lies between $B/4$ and $B/2$. Hence if there is a valid partition, then every set in this partition must contain exactly three elements of X, and (ii) The bound B is a polynomial in m, i.e., there is a universal constant c such that B is at most m^c (for all instances on $3m$ elements). The problem remains NP-complete even in this setting. So when we talk about the 3-partition problem in this section, we shall assume that we are talking about this special case.

The following theorem was proved in [5].

Theorem 3. *Given an instance $\mathcal{I}$ of the 3-partition problem, parameters T and K, there is a poly-time computable function f, which gives an instance $f(\mathcal{I})$ of the Parallel Scheduling problem. Here T is a multiple of $K \cdot B$ (B is the bound appearing in $\mathcal{I}$). This reduction has the following properties :*

- *There are m machines in $f(\mathcal{I})$, and the job sizes lie in the range $[KB/4, KB/2]$. The total volume of jobs in $f(\mathcal{I})$ is $T \cdot m$.*
- *If $\mathcal{I}$ has a valid partition, then all jobs in $f(\mathcal{I})$ can be finished during $[0, T]$. Further the flow-time of these jobs is $O(m \cdot T)$.*
- *If $\mathcal{I}$ does not have a valid partition, then any scheduling of jobs in $f(\mathcal{I})$ must leave at least K units of unfinished volume of jobs at time T.*

Now we describe the reduction from the 3-partition problem to the Parallel Scheduling problem. Fix an instance $\mathcal{I}$ of the 3-partition problem. We construct an instance $g(\mathcal{I})$ of the Parallel Scheduling problem. This instance will have ℓ

phases, each of length T. Each phase shall invoke Theorem 3. The parameter T in the theorem will remain unchanged, but K will vary. So we shall denote by $f(\mathcal{I}, K)$ the instance returned by the theorem above when the inputs to f are $\mathcal{I}$, T and K. The parameter ℓ will be m^r, for a large enough constant r.

Let us describe phase i. In phase i we get the instance $\mathcal{I}'_i = f(\mathcal{I}, B^{\ell-i})$. All jobs in this instance releasing at time t get released at time $(t + i \cdot T)$ in $g(\mathcal{I})$, i.e., we just shift the release dates by $i \cdot T$ units of time. Let T_ℓ denote the time at which phase $\ell - 1$ (the last phase) gets over, i.e., $\ell \cdot T$. Starting from time T_ℓ, we release m jobs of length 1 at each time step t, where $t = T_\ell, \ldots, T_M$, $T_M >> T_\ell$. This completes the construction of the instance $g(\mathcal{I})$of **Parallel Scheduling**. We shall call the jobs released in phase i as jobs of *class* $\ell - i$.

Suppose $\mathcal{I}$ has a valid partitioning. Then it is easy to see that jobs in phase i of $g(\mathcal{I})$ can be processed during $[iT, (i+1)T]$. Similarly, after time T_ℓ, each new job gets done in 1 unit of time. So the total flow time is $O(m \cdot T \cdot \ell) + m \cdot (T_M - T_\ell)$, which is close to $m \cdot T_M$.

Now suppose $\mathcal{I}$ does not have a valid partitioning. We would like to argue that for any scheduling of the jobs in $g(\mathcal{I})$, the total flow time is $\Omega(T_M \cdot \ell)$. This will give a gap of ℓ/m in the approximation ratio.

Consider a fractional schedule S for the jobs in $g(\mathcal{I})$ – a fractional schedule is essentially a fractional packing which obeys the release date. We say that the fractional schedule is *valid* if it satisfies the following condition for every i, $0 \le i \le \ell - 1$ – at least $B^{\ell-i}$ volume of jobs of class $\ell - i$ is processed after phase i ends, i.e., after time $(i + 1) \cdot T$. Theorem 3 says that every integral schedule is valid. We say that a valid schedule is *minimal* if it satisfies the following stronger property : for every i, $0 \le i \le \ell - 1$, there is a job of class $\ell - i$ which has $B^{\ell-i}$ amount of processing left at time $(i + 1) \cdot T$.

Let $W(S)$ denote the jobs which are waiting at time T_ℓ in this schedule. Let $F_\ell(S)$ denote the flow-time of S if we start counting from time T_ℓ. In other words, if $N_t(S)$ is the number of jobs waiting at time t in S, then $F_\ell(S) = \sum_{t \ge T_\ell} N_t$. Since the flow-time after T_ℓ is going to dominate the total flow-time for any schedule, we should look for schedules which minimize this component of the flow-time.

The following lemma, clarifies the need for a minimal schedule.

Lemma 3. *Given any valid schedule S, there exists a (valid) minimal schedule S' such that $F_\ell(S') \le 2 \cdot F_\ell(S)$.*

Proof. Fix an i. Let $T_i = (i + 1) \cdot T$, the time at which phase i ends. Let $W_i(S)$ denote the jobs of class $\ell - i$ waiting at time T_i. Let j_i be the job in $W_i(S)$ with the latest release date. Let j' be some other job in $W_i(S)$. Suppose we process δ amount of j_i before T_i in some time slot. We now process δ amount of j' in this slot, and leave an extra δ amount of job j_i waiting at time T_i – this can always be done as long as δ is small enough. This extra volume of job j_i is processed after time T_i in the slots in which j' was being processed.

We repeat this process as long as possible. When we stop, either all of j_i is waiting at time T_i, or $W(S)$ has only one job, namely j_i. In either of these cases, we satisfy the conditions of a minimal schedule for this class. We do this for all values of i. Let S' be this schedule.

Clearly S' is minimal. Let $N_i(t)$ be the number of jobs of class $\ell - i$ waiting at time t in S, and let $N_i'(t)$ be the corresponding quantity in S'. Consider $t \geq T_i$. Then $N_i'(t) \leq N_i(t) + 1$ – indeed we are only increasing the processing of j_i beyond T_i. Further if $N_i'(t) = N_i(t) + 1$, then it must be the case that $N_i(t) \geq 1$ (because j_i is replacing the slot at time t occupied by some other job of class $\ell - i$). So we conclude that after time T_i, $N_i'(t) \leq 2N_i(t)$. Summing this over all values of i and for $t \geq T_\ell$, we get the desired result. ∎

Theorem 4. *Given any minimal schedule S for the instance $g(\mathcal{I})$, the flow-time of S is $\Omega(T_M \cdot \ell)$.*

Proof. We shall build a schedule S' from S such that $F_\ell(S') \leq F_\ell(S)$ but $W(S')$ will have jobs from each class. This will yield the desired result. However S' may not be a valid schedule and may not even obey released dates. We will construct S' in ℓ iterations. At the end of iteration i, the following invariants will hold for the schedule S' :

   (i) $F_\ell(S') \leq F_\ell(S)$.
  (ii) There is at least one job from each of the classes $1, 2, \ldots, i$ with remaining processing time at least 1 in $W(S')$.
 (iii) Jobs from classes $i + 1, \cdots, \ell$ obey release dates.
 (iv) Jobs from class $1, 2, \ldots i$ are processed after time $(\ell - i) \cdot T$.
  (v) For each $q \in \{i + 1, \ldots, \ell\}$, there is at least one job of class q which has B^q volume of processing left at time $(\ell - q + 1) \cdot T$.

We initialize S' to S – recall that jobs of class i are released during $[(\ell - i) \cdot T, (\ell - i + 1) \cdot T]$. Clearly the invariant holds at $i = 0$ because S is a minimal schedule. Suppose the invariants hold after iteration $i, i \geq 0$. We need to show that they hold at the end of iteration $i + 1$ as well. Let us consider S' at the end of iteration i. Suppose $W(S')$ has a job of class $i + 1$. Then we do not do make any changes in iteration $i + 1$. Clearly the invariants hold at the end of iteration $i + 1$ as well. So assume $W(S')$ does not have any job of class $i + 1$ at the end of iteration i. The rest of the discussion looks at S' at the end of iteration i.

We know that at time $(\ell - i) \cdot T$ at least B^{i+1} volume of jobs of class $i + 1$ are unfinished (using invariant (v) and $q = i + 1$). This volume of job must get processed during $[(\ell - i) \cdot T, \ell \cdot T]$ because $W(S')$ does not contain any job from this class. But the total volume of jobs released during $[(\ell - i) \cdot T, \ell \cdot T]$ equals $m \cdot i \cdot T$ and all these jobs are processed after time $(\ell - i) \cdot T$ (invariant (iv)). So at least B^{i+1} volume of jobs of class i or lower must be waiting at time T_ℓ.

We claim there is a class $q \in \{1, \ldots, i\}$ such that $W(S')$ has at least $\frac{B}{2} \cdot B^q + 1$ volume of class q jobs. Indeed suppose this were not the case. Then the total remaining processing time of jobs from classes $1, \ldots, i$ in $W(S')$ is at most $B/2 \cdot \left(B^i + \cdots + B^1 \right) + q$ which is at most B^{i+1}. But this contradicts the fact that at least B^{i+1} volume of jobs of class i or lower must be waiting at time T_ℓ.

So suppose such a class q exists. Let J_q be the set of jobs of class q in $W(S')$. There is a job $j \in J_q$ which has remaining processing time at least 1 (using invariant (ii)). Let us consider the jobs in $J_q - \{j\}$. These jobs have remaining processing time at least 1 because j can have size at most $\frac{B}{2} \cdot B^q$. We form a new set of jobs J' as follows. J' is initially empty. We start adding jobs from $J_q - \{j\}$ to J' – we stop as soon as the remaining processing times of jobs in J' exceeds 1. Note that the total remaining processing times of jobs in J' is at most $1 + \frac{B}{2} \cdot B^q \le B^{i+1}$. Invariant (v) implies that there is a job j of class $i+1$ which is waiting at time $(\ell - i) \cdot T$ in S'. Further this job has B^{i+1} amount of processing remaining at time $(\ell - i) \cdot T$. We process all the jobs in J' in the slots in which j is getting processed during $[(\ell - i) \cdot T, \ell \cdot T]$ – note that we will be able to finish all of J' because the processing required is at most B^{i+1} and at least B^{i+1} amount of j gets processed during this period (also observe that we may be violating the release date condition for these jobs – but our invariant conditions are not getting violated). Now we process the extra volume of job j after time T_ℓ in the slots in which jobs in J' were getting processed. Note that this cannot increase $F_\ell(S')$ (because we are replacing all the slots occupied by J' after time T_ℓ by a single job). This ends iteration $i+1$. But now $F_\ell(S') \ge \ell \cdot (T_M - T_\ell)$ because the ℓ jobs remaining at time T_ℓ will keep on waiting till time T_M .The theorem follows since $T_M >> T_\ell$. ∎

Therefore, it is NP-hard to get an $O(\ell/m)$-approximation algorithm for the **Parallel Scheduling** problem. We know that $B = m^c$ for some constant c and $\ell = m^r$ for a large enough constant r. So we get $\ell/m = m^{r-1}$. Now $P \le B/2 \cdot B^\ell \le B^{\ell+1} \le m^{cm^r+c} \le m^{rm^r}$ which implies that m^r is $\Omega(\log P / \log \log P)$. So m^{r-1} is $\Omega \left(\frac{\log P}{\log \log P} \right)^{1-\varepsilon}$, where $\varepsilon = 1/r$ is an arbitrarily small constant. But this is same as $\Omega(\log P)^{1-\varepsilon'}$ for an arbitrary small constant ε'.

References

1. Chekuri, C., Goel, A., Khanna, S., Kumar, A.: Multi-processor scheduling to minimize flow time with epsilon resource augmentation. In: STOC, pp. 363–372 (2004)
2. Dinitz, Y., Garg, N., Goemans, M.X.: On the single-source unsplittable flow problem. In: FOCS, pp. 290–299 (1998)
3. Garey, M.R., Johson, D.S.: Computers and intractability - a guide to NP-completeness. W.H. Freeman and Company, New York (1979)
4. Garg, N., Kumar, A.: Better algorithms for minimizing average flow-time on related machines. In: Bugliesi, M., Preneel, B., Sassone, V., Wegener, I. (eds.) ICALP 2006. LNCS, vol. 4051, pp. 181–190. Springer, Heidelberg (2006)
5. Garg, N., Kumar, A.: Minimizing average flow-time: Upper and lower bounds. In: FOCS, pp. 603–613 (2007)
6. Leonardi, S., Raz, D.: Approximating total flow time on parallel machines. In: STOC, pp. 110–119 (1997)
7. Sitters, R.: Minimizing flow-time for unrelated machines. In: Personal Communication (2008)

Approximating the Volume of
Unions and Intersections of
High-Dimensional Geometric Objects

Karl Bringmann[1] and Tobias Friedrich[2,3]

[1] Universität des Saarlandes, Saarbrücken, Germany
[2] Max-Planck-Institut für Informatik, Saarbrücken, Germany
[3] International Computer Science Institute, Berkeley, CA, USA

Abstract. We consider the computation of the volume of the union of high-dimensional geometric objects. While showing that this problem is #**P**-hard already for very simple bodies (i.e., axis-parallel boxes), we give a fast FPRAS for all objects where one can: (1) test whether a given point lies inside the object, (2) sample a point uniformly, (3) calculate the volume of the object in polynomial time. All three oracles can be weak, that is, just approximate. This implies that Klee's measure problem and the hypervolume indicator can be approximated efficiently even though they are #**P**-hard and hence cannot be solved exactly in time polynomial in the number of dimensions unless **P** = **NP**. Our algorithm also allows to approximate efficiently the volume of the union of convex bodies given by weak membership oracles.

For the analogous problem of the intersection of high-dimensional geometric objects we prove #**P**-hardness for boxes and show that there is no multiplicative polynomial-time $2^{d^{1-\varepsilon}}$-approximation for certain boxes unless **NP** = **BPP**, but give a simple additive polynomial-time ε-approximation.

1   Introduction

Given n bodies in the d-dimensional space, how efficiently can we compute the volume of the union and the intersection? We consider this basic geometric problem for different kinds of bodies. The tractability of this problem highly depends on the representation and the complexity of the given objects. For many classes of objects already computing the volume of one body can be hard. For example, calculating the volume of a polyhedron given either as a list of vertices or as a list of facets is #**P**-hard [6, 14]. For convex bodies given by a membership oracle one can also show that even though there can be no deterministic $(\mathcal{O}(1)d/\log d)^d$-approximation for $d \geq 2$ [2], one can still approximate the volume by an FPRAS (fully polynomial-time randomized approximation scheme). In a seminal paper Dyer, Frieze, and Kannan [7] gave an $\mathcal{O}^*(d^{23})$ algorithm, which was subsequently improved in a series of papers to $\mathcal{O}^*(d^4)$ [16] (where the asterisk hides powers of the approximation ratio and $\log d$).

S.-H. Hong, H. Nagamochi, and T. Fukunaga (Eds.): ISAAC 2008, LNCS 5369, pp. 436–447, 2008.

Volume computation of unions can be hard not only for bodies whose volume is hard to calculate. One famous example for this is *Klee's Measure Problem* (KMP). Given n axis-parallel boxes in the d-dimensional space, it asks for the measure of their union. In 1977, Victor Klee showed that it can be solved in time $\mathcal{O}(n \log n)$ for $d = 1$ [15]. This was generalized to $d > 1$ dimensions by Bentley who presented an algorithm which runs in $\mathcal{O}(n^{d-1} \log n)$, which was later improved by van Leeuwen and Wood [20] to $\mathcal{O}(n^{d-1})$. In FOCS '88, Overmars and Yap [17] obtained an $\mathcal{O}(n^{d/2} \log n)$ algorithm. This was the fastest algorithm for $d \geq 3$ until, on this years SoCG, Chan [5] presented a slightly improved version of Overmars and Yap's algorithm that runs in time $n^{d/2} 2^{\mathcal{O}(\log^* n)}$, where $\log^*$ denotes the iterated logarithm. So far, the only known lower bound is $\Omega(n \log n)$ for any d [8]. Note that the worst-case combinatorial complexity (i.e., the number of faces of all dimensions on the boundary of the union) of $\Theta(n^d)$ does not imply any bounds on the computational complexity. There are various algorithms for special cases, e.g., for hypercubes [1, 11] and unit hypercubes [4]. In this paper we explore the opposite direction and examine the problem of the union of more general geometric objects.

An important special case of KMP is the *hypervolume indicator* (HYP) [21] where all boxes are required to share a common vertex. It is also known as the "Lebesgue measure", the "S-metric" and "hyperarea metric" and is a popular measure of fitness of Pareto sets in multi-objective optimization. There, it measures the number of solutions dominated by a Pareto set. More details can be found in Section 4.

Our Results

It is not hard to see that HYP and KMP are **#P**-hard (see Theorem 1). Hence they cannot be solved in time polynomial in the number of dimensions unless **P** = **NP**. This shows that *exact* volume computation of unions is intractable for all classes of bodies that contain axis-parallel boxes.

This motivates the development of *approximation algorithms* for the volume computation of unions. This question was untackled until now – approaches exist only for discrete sets (see, e.g., Karp, Luby, and Madras [13] for an FPRAS for #DNF which is similar to our algorithm). We give an efficient FPRAS for a huge class of bodies including boxes, spheres, polytopes, schlicht domains, convex bodies determined by an oracle and all affine transformations of those objects mentioned before. The underlying bodies B just have to support the following oracle queries in polynomial time:

- POINTQUERY(x, B): Is point $x \in \mathbb{R}^d$ an element of body B (approximately)?
- VOLUMEQUERY(B): What is the volume of body B (approximately)?
- SAMPLEQUERY(B): Return a random (almost) uniformly distributed point $x \in B$.

POINTQUERY is a very natural condition which is fulfilled in almost all practical cases. The VOLUMEQUERY condition is important as it could be the case that no efficient approximation of the volume of one of the bodies itself is possible. This, of course, prevents an efficient approximation of the union of such bodies. The SAMPLEQUERY is crucial for our FPRAS. In Section 2.3 we will show that it is efficiently computable for a wide range of bodies.

Note that it suffices that all three oracles are weak. More precisely, we allow the following *relaxation* for every body B (VOL(B) denotes the volume of a body B in the standard Lebesgue measure on $\mathbb{R}^d$, more details are given in Section 2):

- POINTQUERY(x, B) answers true iff $x \in B'$ for a fixed $B' \subset \mathbb{R}^d$ with VOL$((B' \setminus B) \cup (B \setminus B')) \leq \varepsilon_P$ VOL(B).
- VOLUMEQUERY(B) returns a value V' with $(1 - \varepsilon_V)$ VOL(B) $\leq V' \leq (1 + \varepsilon_V)$ VOL(B).
- SAMPLEQUERY(B) returns only an *almost* uniformly distributed random point [10], that is, it suffices to get a random point $x \in B'$ (with B' as above) such that for the probability density f we have for every point x: $|f(x) - 1/\text{VOL}(B')| < \varepsilon_S$.

Let $P(d)$ be the worst POINTQUERY runtime of any of our bodies, analogously $V(d)$ for VOLUMEQUERY, and $S(d)$ for SAMPLEQUERY. Then our FPRAS has a runtime of $\mathcal{O}(nV(d) + \frac{n}{\varepsilon^2}(S(d) + P(d)))$ for producing an ε-approximation[1] with probability $\geq \frac{3}{4}$ if the errors of the underlying oracles are small, i.e., $\varepsilon_S, \varepsilon_P, \varepsilon_V \leq \frac{\varepsilon^2}{47n}$. For example for boxes, that is, for KMP and HYP, this reduces to $\mathcal{O}(\frac{dn}{\varepsilon^2})$ and is the first FPRAS for both problems. In Section 2.3 we also show that our algorithm is an FPRAS for the volume of the union of convex bodies.

The canonic next question is the computation of the volume of the *intersection of bodies* in $\mathbb{R}^d$. It is clear that most of the problems from above apply to this question, too. #**P**-hardness for general, i.e., not necessarily axis-parallel, boxes follows directly from the hardness of computing the volume of a polytope [6, 14]. This leaves open whether there are efficient approximation algorithms for the volume of intersection. In Section 3 we show that there cannot be a (deterministic or randomized) multiplicative $2^{d^{1-\varepsilon}}$-approximation in general, unless **NP** = **BPP** by identifying a hard subproblem. Instead we give an additive ε-approximation, which is therefore the best we can hope for. It has a runtime of $\mathcal{O}(nV(d) + \varepsilon^{-2}S(d) + \frac{n}{\varepsilon^2}P(d))$, which gives $\mathcal{O}(\frac{dn}{\varepsilon^2})$ for boxes.

2   Volume Computation of Unions

In this section we show that the volume computation of unions is #**P**-hard already for axis-parallel boxes that have one vertex at the origin, i.e., for HYP. After that we give an FPRAS for approximating the volume of the union of bodies which satisfy the three aforementioned oracles and describe several large classes of objects for which the oracles can be answered efficiently.

---

[1] We will always assume that ε is small, that is, $0 < \varepsilon < 1$.

2.1 Computational Complexity of Union Calculations

Consider the following problem: Let $\mathcal{S}$ be a set of n axis-parallel boxes in $\mathbb{R}^d$ of the form $B = [a_1, b_1] \times \cdots \times [a_d, b_d]$ with $a_i, b_i \in \mathbb{R}, a_i < b_i$. The volume of one such box is $\mathrm{VOL}(B) = \prod_{i=1}^{d}(b_i - a_i)$. To compute the volume of the union of these boxes is known as Klee's Measure Problem (KMP), while we call the problem HYP (for hypervolume) if we have $a_i = 0$ for all $i \in [d]$.

The following Theorem 1 proves $\#\mathbf{P}$-hardness of HYP and KMP. This is the first hardness result for HYP. To the best of our knowledge there is also no published result that explicitly states that KMP is $\#\mathbf{P}$-hard. However, without mentioning this implication, Suzuki and Ibaraki [19] sketch a reduction from $\#$SAT to KMP. In the following theorem we reduce $\#$MON-CNF to HYP, which counts the number of satisfying assignments of a Boolean formula in conjunctive normal form in which all variables are unnegated. While the problem of deciding satisfiability of such formula is trivial, counting the number of satisfying assignments is $\#\mathbf{P}$-hard and even approximating it in polynomial time by a factor of $2^{d^{1-\varepsilon}}$ for any $\varepsilon > 0$ is $\mathbf{NP}$-hard, where d is the number of variables (see Roth [18] for a proof).

Theorem 1. HYP *and* KMP *are* $\#\mathbf{P}$*-hard.*

Proof. To show the theorem, we reduce $\#$MON-CNF to HYP. The hardness of KMP follows immediately. Let $f = \bigwedge_{k=1}^{n} \bigvee_{i \in C_k} x_i$ be a monotone Boolean formula given in CNF with $C_k \subset [d] := \{1, \ldots, d\}$, for $k \in [n]$, d the number of variables, n the number of clauses. Since the number of satisfying assignments of f is equal to 2^d minus the number of satisfying assignments of its negation, we instead count the latter: Consider the negated formula $\bar{f} = \bigvee_{k=1}^{n} \bigwedge_{i \in C_k} \neg x_i$. First, we construct a box $A_k = [0, a_1^k] \times \cdots \times [0, a_d^k]$ in $\mathbb{R}^d$ for each clause C_k with one vertex at the origin and the opposite vertex at $(a_1^k, \ldots, a_d^k)$, where we set

$$a_i^k = \begin{cases} 1, & \text{if } i \in C_k \\ 2, & \text{otherwise} \end{cases}, \quad i \in [d].$$

Observe that the union of the boxes A_k can be written as a union of boxes of the form $B_{x_1,\ldots,x_d} = [x_1, x_1 + 1] \times \cdots \times [x_d, x_d + 1]$ with $x_i \in \{0, 1\}, i \in [d]$. Moreover, $B_{x_1,\ldots,x_d}$ is a subset of the union $\bigcup_{k=1}^{n} A_k$ iff it is a subset of some A_k iff we have $a_i^k \geq x_i + 1$ for $i \in [d]$ iff $a_i^k = 2$ for all i with $x_i = 1$ iff $i \notin C_k$ for all i with $x_i = 1$ iff $(x_1, \ldots, x_d)$ satisfies $\bigwedge_{i \in C_k} \neg x_i$ for some k iff $(x_1, \ldots, x_d)$ satisfies $\bar{f}$. Hence, since $\mathrm{VOL}(B_{x_1,\ldots,x_d}) = 1$, we have $\mathrm{VOL}(\bigcup_{k=1}^{n} A_k) = |\{(x_1, \ldots, x_d) \in \{0, 1\}^d \mid (x_1, \ldots, x_d) \text{ satisfies } \bar{f}\}|$. Thus a polynomial time algorithm for HYP would result in a polynomial time algorithm for $\#$MON-CNF, which proves the claim. $\qquad\square$

For integer coordinates it is easy to see that KMP $\in \#\mathbf{P}$ and hence that in this case KMP and HYP are even $\#\mathbf{P}$-complete.

2.2   Approximation Algorithm for the Volume of Unions

In this section we present an FPRAS for computing the volume of the union of objects for which we can answer PointQuery, VolumeQuery, and Sample-Query in polynomial time. The input of our algorithm ApproxUnion are the approximation ratio ε and the bodies $\{B_1, \ldots, B_n\}$ in $\mathbb{R}^d$ defined by the three oracles. It computes a number $\tilde{U} \in \mathbb{R}$ such that

$$\Pr\left[(1 - \varepsilon)\,\mathrm{VOL}\left(\bigcup_{i=1}^{n} B_i\right) \leq \tilde{U} \leq (1 + \varepsilon)\,\mathrm{VOL}\left(\bigcup_{i=1}^{n} B_i\right)\right] \geq \frac{3}{4}$$

in time polynomial in n, $1/\varepsilon$ and the query runtimes. The number $\frac{3}{4}$ is arbitrary and can be increased to every number $1 - \delta$, $\delta > 0$ by a factor of $\log(1/\delta)$ in the runtime by running algorithm ApproxUnion $\log(1/\delta)$ many times and returning the median of the reported values for $\tilde{U}$.

We are following the algorithm of Karp and Luby [12] which the authors used for approximating #DNF and other counting problems on discrete sets. The two main differences are that here we are handling continuous bodies in $\mathbb{R}^d$ and that we allow erroneous oracles.

Algorithm 1. ApproxUnion $(\mathcal{S}, \varepsilon, \varepsilon_P, \varepsilon_V, \varepsilon_S)$ calculates an ε-approximation of $\mathrm{VOL}(\bigcup_{i=1}^{n} B_i)$ for a set of bodies $S = \{B_1, \ldots, B_n\}$ in $\mathbb{R}^d$ determined by the oracles PointQuery, VolumeQuery and SampleQuery with error ratios $\varepsilon_P, \varepsilon_V, \varepsilon_S$.

$M := 0,\ C := 0,\ \tilde{\varepsilon} := \frac{\varepsilon - \varepsilon_V}{1 + \varepsilon_V},\ \tilde{C} := \frac{(1+\varepsilon_S)(1+\varepsilon_V)(1+\varepsilon_P)}{(1-\varepsilon_V)(1-\varepsilon_P)},\ T := \frac{24\ln(2)(1+\tilde{\varepsilon})n}{\tilde{\varepsilon}^2 - 8(\tilde{C}-1)n}$

for all $B_i \in \mathcal{S}$ **do**

   compute $V_i' := \mathrm{VolumeQuery}(B_i)$

od

$V' := \sum_{i=1}^{n} V_i'$

while $C \leq T$ **do**

   choose $i \in [n]$ with probability V_i'/V'

   $x := \mathrm{SampleQuery}(B_i)$

   repeat

      if $C > T$ **then return** $\frac{T \cdot V'}{nM}$

      choose random $j \in [n]$ uniformly

      $C := C + 1$

   until PointQuery (x, B_j)

   $M := M + 1$

od

return $\frac{T \cdot V'}{nM}$

We assume that we are given upper bounds ε_P, ε_S and ε_V for the error ratios of the oracles. In the algorithm we first compute the runtime T and then via VolumeQuery the volume V_i' for every given body B_i and their sum $V' =$

$\sum_{i=1}^{n} V_i'$. Then, in a loop, we choose a random $i \in [n]$, where we choose an i with probability $\frac{V_i'}{V'}$ and a random (almost) uniformly distributed point $x \in B_i$ via SAMPLEQUERY. Then we repeatedly choose a random $j \in [n]$ uniformly and check, whether $x \in B_j$: POINTQUERY(x, B_j) returns true iff $x \in B_j'$. If this is the case, we leave the inner loop and increase the counter M. This random variable is in the end used to calculate the result $\frac{T \cdot V'}{nM}$.

In the full version of the paper [3] we show correctness of APPROXUNION, that is, we show that it returns an ε-approximation with probability $\geq \frac{3}{4}$ and $T = \mathcal{O}(\frac{n}{\varepsilon^2})$ if $\varepsilon_S, \varepsilon_P, \varepsilon_V \leq \frac{\varepsilon^2}{47n}$. The last inequality reflects the fact that we cannot be arbitrarily accurate if the given oracles are inaccurate. If all oracles can be calculated accurately, i.e., if $\varepsilon_P = \varepsilon_S = \varepsilon_V = 0$, the algorithm runs for just $T = \frac{8 \ln(8)(1+\varepsilon)n}{\varepsilon^2}$ many steps. The runtime of APPROXUNION is clearly

$$\mathcal{O}(n \cdot V(d) + M \cdot S(d) + T \cdot P(d)) = \mathcal{O}(n \cdot V(d) + T \cdot (S(d) + P(d))),$$

where $V(d)$ is the worst VOLUMEQUERY time for any of the bodies, analogously $S(d)$ for SAMPLEQUERY and $P(d)$ for POINTQUERY. This equals $\mathcal{O}(n \cdot V(d) + \frac{n}{\varepsilon^2} \cdot (S(d) + P(d)))$ if $\varepsilon_S, \varepsilon_P, \varepsilon_V \leq \frac{\varepsilon^2}{47n}$.

For boxes all three oracles can be computed exactly in $\mathcal{O}(d)$. This implies that our algorithm APPROXUNION gives an ε-approximation of KMP and HYP with probability $\geq \frac{3}{4}$ in runtime $\mathcal{O}(\frac{nd}{\varepsilon^2})$.

2.3   Classes of Objects Supported by Our FPRAS

To find an FPRAS for the union of a certain class of geometric objects now reduces to calculating the respective POINTQUERY, VOLUMEQUERY and SAMPLEQUERY in polynomial time. We assume that we can get a random real number in constant time. Then all three oracles can be calculated in time $\mathcal{O}(d)$ for d-dimensional boxes. This already yields an FPRAS for the volume of the union of arbitrary boxes, e.g., for KMP and HYP. Note that if we have a body for which we can answer all those queries, all affine transformations of this body fulfill these three oracles, too. We will now present three further classes of geometric objects.

Generalized spheres and boxes: Let $\mathbf{B}_k$ be the class of boxes of dimension k, i.e., $\mathbf{B}_k = \{[a_1, b_1] \times \cdots \times [a_k, b_k] \mid a_i, b_i \in \mathbb{R}, a_i < b_i\}$ and $\mathbf{S}_l$ the class of spheres of dimension l. We can combine any box $B \in \mathbf{B}_k$ and sphere $S \in \mathbf{S}_{d-k}$ to get a d-dimensional object $B \times S$. Furthermore, we can permute the dimensions afterwards to get a generalized "box-sphere". In $\mathbb{R}^3$ this corresponds to boxes, spheres and cylinders. To calculate the respective VOLUMEQUERY, POINTQUERY and SAMPLEQUERY is a standard task of geometry.

Schlicht domains: Let $a_i, b_i \colon \mathbb{R}^{i-1} \to \mathbb{R}$ be continuous functions with $a_i \leq b_i$, where a_1, b_1 are constants. Let $K \subset \mathbb{R}^d$ be defined as the set of all points $(x_1, \ldots, x_d) \in \mathbb{R}^d$ such that $a_1 \leq x_1 \leq b_1, a_2(x_1) \leq x_2 \leq$

$b_2(x_1), \ldots, a_d(x_1, \ldots, x_{d-1}) \le x_d \le b_d(x_1, \ldots, x_{d-1})$. K is called a schlicht domain in functional analysis. Fubini's theorem for schlicht domains states that we can integrate a function $f \colon K \to \mathbb{R}$ by iteratively integrating first over x_d, then over x_{d-1}, $\ldots$, until we reach x_1. This way, by integrating $f \equiv 1$, we can compute the volume of a schlicht domain, as long as the integrals are computable in polynomial time, and thus answer a VOLUMEQUERY. Similarly, we can choose a random uniformly distributed point inside K: Let $K(y) = \{(x_1, \ldots, x_d) \in K \mid x_1 = y\}$. Then $K(y)$ is another schlicht domain for every $a_1 \le y \le b_1$. Assume that we can determine the volume of every such $K(y)$ and the integral $I(y) = \int_{a_1}^{y} K(x) dx$. Then the inverse function $I^{-1} \colon [0, V] \to \mathbb{R}$, where $V = \int_{a_1}^{b_1} K(x) dx$ is the volume of K, allows us to choose a y in $[a_1, b_1]$ with probability proportional to $\mathrm{VOL}(K(y))$. By this we can iteratively choose a value y for x_1 and recurse to find a uniformly random point $(y_2, \ldots, y_d)$ in $K(y)$, plugging both together to get a uniformly distributed point $(y_1, \ldots, y_d)$ in K. Hence, as long as we can compute the involved integrals and inverse functions (or at least approximate them good enough), we can answer SAMPLEQUERY. Since POINTQUERY is trivially computable – as long as we can evaluate a_i and b_i efficiently – this gives an example showing that the classes of objects that fulfill our three conditions include not only convex bodies, but also certain schlicht domains.

Convex bodies: As mentioned in the introduction, exact calculation of VOLUME-QUERY for a polyhedron given as a list of vertices or facets is #**P**-hard [6, 14]. Since there are randomized approximation algorithms (see Dyer et al. [7] for the first one) for the volume of a convex body determined by a membership oracle, we can answer VOLUMEQUERY approximately. The same holds for SAMPLEQUERY as these algorithms make use of an almost uniform sampling method on convex bodies. See Lovász and Vempala [16] for a result showing that VOLUMEQUERY can be answered with $\mathcal{O}^*(\frac{d^4}{\varepsilon_V^2})$ questions to the membership oracle and SAMPLE-QUERY with $\mathcal{O}^*(\frac{d^3}{\varepsilon_S^2})$ queries, for arbitrary errors $\varepsilon_V, \varepsilon_S > 0$ (where the asterisk hides factors of $\log(d)$ and $\log(1/\varepsilon_V)$ or $\log(1/\varepsilon_S)$). POINTQUERY can naturally be answered with a single question to the membership oracle. By choosing $\varepsilon_V = \varepsilon_S = \frac{\varepsilon^2}{47n}$, we can show [3] that APPROXUNION is an FPRAS for the volume of the union of convex bodies which uses $\mathcal{O}^*(\frac{n^3 d^3}{\varepsilon^4}(d + \frac{1}{\varepsilon^2}))$ membership queries.

Note that all above mentioned classes of geometric objects contain boxes and hence our hardness results still hold and hence an ε-approximation algorithm is the best one can hope for.

3 Volume Computation of Intersections

In this section we are considering the complement to the union problem. We show that surprisingly the volume of a intersection of a set of bodies is often much harder to calculate than its union. For many classes of geometric objects there is even no randomized approximation possible.

As the problem of computing the volume of a polytope is #**P**-hard [6, 14], so is the computation of the volume of the intersection of general (i.e., not necessarily axis-parallel) boxes in $\mathbb{R}^d$. This can be seen by describing a polytope as an intersection of halfplanes and representing these as general boxes.

Now, let us consider the convex bodies again. Trivially, the intersection of convex bodies is convex itself, and from the oracles defining the given bodies $B_1, \ldots, B_n$ one can simply construct an oracle, which answers POINTQUERY for the intersection of those objects: Given a point $x \in \mathbb{R}^d$ it asks all n oracles and returns true iff x lies in all the bodies. One could think now that we can apply the result of Dyer et al. [7] and the subsequent improvements mentioned in the introduction to approximate the volume of the intersection and get an FPRAS for the problem at hand. The problem with that is that the intersection is not "well-guaranteed": There is no point known that lies in the intersection, not to speak of a sphere inside it. However, the algorithm of Dyer et al. [7] relies vitally on the assumption that the given body is well-guaranteed and hence cannot be applied for approximating the volume of the intersection of convex bodies.

We will now present a hard subproblem which shows that the volume of the intersection cannot be approximated (deterministic or randomized) in general.

Definition 1. *Let* $p, q \in \mathbb{R}^d_{\geq 0}$. *Then* $B_p := \{x \mid 0 \leq x_i \leq p_i \; \forall i\}$ *is a* p-box, $B_{p,q} := B_p \setminus B_q$ *is a* (p, q)-box, *and* $\{B_{p,q_1}, B_{p,q_2}, \ldots, B_{p,q_n}\}$ *is a* p-set.

A (p, q)-box is basically a box where we cut out another box at one corner. The resulting object can itself be a box, too, but in general it is not even convex. It can be seen as the inverse of a box B_p relative to a larger background box B_q. Note that it is easy to calculate the union of a p-set as $\bigcup B_{p,q_i} = B_p \setminus \bigcap B_{q_i}$. On the other hand, the calculation of the intersection of a p-set is #**P**-hard as $\bigcup B_{q_i} = B_p \setminus \bigcap B_{p,q_i}$ by Theorem 1. The following theorem shows that it is not even approximable.

Theorem 2. *Let* $p, q_1, \ldots, q_n \in \mathbb{R}^d_{\geq 0}$. *Then the volume of* $\bigcap_{i=1}^n B_{p,q_i}$ *cannot be approximated (deterministic or randomized) in polynomial time by a factor of* $2^{d^{1-\varepsilon}}$ *for any* $\varepsilon > 0$ *unless* **NP** = **BPP**.

Proof. Consider again the problem #MON-CNF already defined Section 2. We use Roth's result [18] that #MON-CNF cannot be approximated by a factor of $2^{d^{1-\varepsilon}}$ unless **NP** = **BPP** and construct an approximation preserving reduction. Let $f = \bigwedge_{k=1}^n \bigvee_{i \in C_k} x_i$ be a monotone Boolean formula given in CNF with $C_k \subset [d]$, for $k \in [n]$, d the number of variables, n the number of clauses. For every clause C_k we construct a (p, q_k)-box A_k with $p = (2, \ldots, 2) \in \mathbb{R}^d$, $q_k = (q_{k,1}, \ldots, q_{k,d})$ and $q_{k,i} = 1$ if $i \in C_k$, and $q_{k,i} = 2$ otherwise.

Observe that each A_k can be written as a union of boxes of the form $B_{x_1,\ldots,x_d} = [x_1, x_1 + 1] \times \cdots \times [x_d, x_d + 1]$ with $x_i \in \{0, 1\}$. Hence, their intersection can also be written as such a union as follows:

$$\bigcap_{k=1}^n A_k = \bigcup_{(x_1,\ldots,x_d)\in S} B_{x_1,\ldots,x_d}$$

for some set $S \subset \{0,1\}^d$. Furthermore, we have $(x_1, \ldots, x_d) \in S$ iff $B_{x_1,\ldots,x_d} \subseteq A_k$ for all k iff $B_{x_1,\ldots,x_d} \cap \{x \in \mathbb{R}^d_{\geq 0} \mid x \prec q_k\} = \emptyset$ for all k iff $\exists i \in \{1, \ldots, d\} \colon x_i \geq q_{k,i}$ for all k. Since this inequality can only be satisfied if $x_i = 1$ and $q_{k,i} = 1$, which holds iff $i \in C_k$, we have that the former term holds if and only if $\exists i \in C_k \colon x_i = 1$ for all k iff $\bigvee_{j \in C_k} x_j$ is satisfied for all k iff f is satisfied. Hence, we have that the set S equals the set of satisfying assignments of f, so that

$$|\{x \in \{0,1\}^d \mid f(x)=1\}| = |S| \overset{(*)}{=} \mathrm{VOL}\Big(\bigcap_{k=1}^{n} A_k\Big) \Big/ \mathrm{VOL}\Big(B_{(0,\ldots,0)}\Big) = \mathrm{VOL}\Big(\bigcap_{k=1}^{n} A_k\Big)$$

where $(*)$ comes from the fact that $\bigcap_{k=1}^{n} A_k$ is composed of $|S|$ many boxes of equal volume and this volume is 1. Hence, we have a polynomial time reduction and inapproximability of the volume of the intersection follows. $\square$

This shows that in general there does not exist a polynomial time multiplicative ε-approximation of the volume of the intersection of bodies in $\mathbb{R}^d$. This holds for all classes of objects which include p-sets, e.g. schlicht domains (cf. Section 2.3). Though there is no multiplicative approximation, we can still give an additive approximation algorithm, that is, we can efficiently find a number $\tilde{V}$ such that

$$\Pr[V - \varepsilon \cdot V_{\min} \leq \tilde{V} \leq V + \varepsilon \cdot V_{\min}] \geq \frac{3}{4}$$

where V is the correct volume of the intersection and $V_{\min}$ is the minimal volume of any of the given bodies $B_1, \ldots, B_n$. If we could replace $V_{\min}$ by V in the equation above, we would have an FPRAS. This is not possible in general as the ratio of V and $V_{\min}$ can be arbitrarily small. Hence, such an ε-approximation is not relative to the exact result, but to the volume of some greater body. This is an additive approximation since after rescaling, so that $V_{\min} \leq 1$ we get an additive error of ε. Clearly, we get the result from above quite easily by uniform sampling in the body $B_{\min}$ corresponding to the volume $V_{\min}$. From Bernstein's inequality we know that for N proportional to $1/\varepsilon^2$ and $\tilde{V} = 1/N(Z_1 + \ldots + Z_N)$, where Z_i is a random variable valued 1, if the i-th sample point $x_i = \textsc{SampleQuery}(B_{\min})$ lies in the intersection of $B_1, \ldots, B_n$, and 0 otherwise, $\tilde{V}$ approximates V with absolute error ε. This gives an approximation algorithm with runtime $\mathcal{O}(nV(d) + \frac{1}{\varepsilon^2} S(d) + \frac{n}{\varepsilon^2} P(d))$, yielding $\mathcal{O}(\frac{dn}{\varepsilon^2})$ for boxes.

4   The Hypervolume Indicator

As an application of our results from Section 2 we now analyze the complexity of the hypervolume indicator which is widely used in evolutionary multi-objective optimization. In multi-objective optimization the aim is is to minimize (or maximize) d objective functions $f_i \colon S \to \mathbb{R}$, $1 \leq i \leq d$, over a search space $S \subseteq \mathbb{R}^d$. As these objectives are often conflicting, one does not generally search for a single optimum, but rather for a set of good compromise solutions. In order to compare different solutions, we impose an order $\preceq$ on the d-tuples

Table 1. Results for the computational complexity of the calculation of the volume of union and intersection (asymptotic in the number of dimensions d)

geometric objects	volume of the union	volume of the intersection
axis-parallel boxes	#**P**-hard + FPRAS	easy
general boxes	#**P**-hard + FPRAS	#**P**-hard
p-sets	easy	#**P**-hard + APX-hard
schlicht domains	#**P**-hard + FPRAS[2]	#**P**-hard + APX-hard
convex bodies	#**P**-hard + FPRAS	#**P**-hard

$f(x) = (f_1(x), \ldots, f_d(x))$, $x \in S$, by letting $x_1 \preceq x_2$ whenever $f_i(x_1) \leq f_i(x_2)$ for each i, $1 \leq i \leq d$. If $M \subset S$ is such that $x_1, x_2 \in M$ implies $x_1 \npreceq x_2$, then we call M a "Pareto front" and $f(M)$ a "Pareto set". Pareto fronts correspond to sets of maximal solutions to the optimization problem given by (S, f). The functions f_i are often assumed to be monotone with respect to some ordering on S whence f is bijective and the optimization problem reduces to identifying "good" Pareto sets.

How to compare Pareto sets lies at the heart of research in multi-objective optimization. One measure that has been the subject of much recent study is the so-called "hypervolume indicator" (HYP). It measures the space dominated by the Pareto set relative to a reference point $r \in \mathbb{R}^d$. For a Pareto set M, the hypervolume indicator is

$$I(M) = \mathrm{VOL}(\{x \in \mathbb{R}_{\geq 0}^d \mid \exists p \in M \text{ so that } r \preceq x \preceq p\})$$
$$= \mathrm{VOL}(\textstyle\bigcup_{p \in M} \{x \in \mathbb{R}_{\geq 0}^d \mid r \preceq x \preceq p\}).$$

To simplify the presentation we assume $r = (0, \ldots, 0)$ which can be achieved by setting $f'(x) := f(x) + r$. Using the notation of Definition 1, we can reduce the hypervolume to the well-understood union problem $I(M) = \mathrm{VOL}(\bigcup_{p \in M} B_p)$.

The hypervolume was first proposed and employed for multi-objective optimization by Zitzler and Thiele [21]. Several algorithms have been developed. It was open so far whether a polynomial algorithm for HYP is possible. Our #**P**-hardness result for HYP (Theorem 1) dashes the hope for a subexponential algorithm (unless **P** = **NP**) and motivates to examine approximation algorithms. Our algorithm APPROXUNION gives an ε-approximation of the hypervolume indicator with probability $1 - \delta$ in time $\mathcal{O}(\log(1/\delta)\, nd/\varepsilon^2)$. As its runtime is just linear in n and d, it is not only the first proven FPRAS for HYP, but also a very practical algorithm.

5   Discussion and Open Problems

We have proven #**P**-hardness for the exact computation of the volume of the union of bodies in $\mathbb{R}^d$ as long as the class of bodies includes axis-parallel boxes.

---

[2] If the integrals are computable in polynomial time (cf. Section 2.3).

The same holds for the intersection if the class of bodies contains general boxes. We have also presented an FPRAS for approximating the volume of the union of bodies that allow three very natural oracles. Very recently, there appeared a few deterministic polynomial-time approximations (FPTAS) for hard counting problems (e.g. [9]). It seems to be a very interesting open question whether there exists a deterministic approximation for the union of some non-trivial class of bodies. Since the volume of convex bodies determined by oracles cannot be approximated to within a factor that is exponential in d [2], the existence of such a deterministic approximation for the union seems implausible. It is also open whether there is a constant C so that HYP or KMP can be efficiently deterministically approximated within a factor of C?

For the intersection we proved that no multiplicative approximation (deterministic or randomized) is possible for p-sets (cf. Definition 1), but we also presented a very simple additive approximation algorithm for the intersection problem. It would be interesting to know if there is a hard class for multiplicative approximation which contains only convex bodies.

Our results are summarized in Table 1. Note the correspondence between axis-parallel boxes and p-sets. The discrete counterpart to their approximability and inapproximability is the approximability of #DNF and the inapproximability of #SAT. One implication of our results is the hardness and a practically efficient approximation algorithm for computing HYP and KMP.

Acknowledgements

We thank Joshua Cooper for suggesting the proof of Theorem 1, Ernst Albrecht for discussions on schlicht domains, and Eckart Zitzler for comments on an earlier version of this paper. This work was partially supported by a postdoctoral fellowship from the German Academic Exchange Service (DAAD).

References

[1] Agarwal, P.K., Kaplan, H., Sharir, M.: Computing the volume of the union of cubes. In: Proc. 23rd annual Symposium on Computational Geometry (SoCG 2007), pp. 294–301 (2007)

[2] Bárány, I., Füredi, Z.: Computing the volume is difficult. Discrete & Computational Geometry 2, 319–326 (1986); Announced at STOC 1986

[3] Bringmann, K., Friedrich, T.: Approximating the volume of unions and intersections of high-dimensional geometric objects (2008), http://arxiv.org/abs/0809.0835

[4] Chan, T.M.: Semi-online maintenance of geometric optima and measures. SIAM J. Comput. 32, 700–716 (2003)

[5] Chan, T.M.: A (slightly) faster algorithm for Klee's measure problem. In: Proc. 24th ACM Symposium on Computational Geometry (SoCG 2008), pp. 94–100 (2008)

[6] Dyer, M.E., Frieze, A.M.: On the complexity of computing the volume of a polyhedron. SIAM J. Comput. 17, 967–974 (1988)

[7] Dyer, M.E., Frieze, A.M., Kannan, R.: A random polynomial time algorithm for approximating the volume of convex bodies. J. ACM 38, 1–17 (1991); Announced at STOC 1989

[8] Fredman, M.L., Weide, B.W.: On the complexity of computing the measure of $\bigcup [a_i, b_i]$. Commun. ACM 21, 540–544 (1978)

[9] Halman, N., Klabjan, D., Li, C.-L., Orlin, J.B., Simchi-Levi, D.: Fully polynomial time approximation schemes for stochastic dynamic programs. In: Proc. 19th Annual ACM-SIAM Symposium on Discrete Algorithms (SODA 2008), pp. 700–709 (2008)

[10] Kannan, R., Lovász, L., Simonovits, M.: Random walks and an $O^*(n^5)$ volume algorithm for convex bodies. Random Struct. Algorithms 11, 1–50 (1997)

[11] Kaplan, H., Rubin, N., Sharir, M., Verbin, E.: Counting colors in boxes. In: Proc. 18th Annual ACM-SIAM Symposium on Discrete Algorithms (SODA 2007), pp. 785–794 (2007)

[12] Karp, R.M., Luby, M.: Monte-carlo algorithms for the planar multiterminal network reliability problem. J. Complexity 1, 45–64 (1985)

[13] Karp, R.M., Luby, M., Madras, N.: Monte-carlo approximation algorithms for enumeration problems. J. Algorithms 10, 429–448 (1989)

[14] Khachiyan, L.G.: The problem of calculating the volume of a polyhedron is enumerably hard. Russian Mathematical Surveys 44, 199–200 (1989)

[15] Klee, V.: Can the measure of $\bigcup [a_i, b_i]$ be computed in less than $O(n \log n)$ steps? American Mathematical Monthly 84, 284–285 (1977)

[16] Lovász, L., Vempala, S.: Simulated annealing in convex bodies and an $O^*(n^4)$ volume algorithm. J. Comput. Syst. Sci. 72, 392–417 (2006)

[17] Overmars, M.H., Yap, C.-K.: New upper bounds in Klee's measure problem. SIAM J. Comput. 20, 1034–1045 (1991); Announced at FOCS 1988

[18] Roth, D.: On the hardness of approximate reasoning. Artif. Intell. 82, 273–302 (1996)

[19] Suzuki, S., Ibaraki, T.: An average running time analysis of a backtracking algorithm to calculate the measure of the union of hyperrectangles in d dimensions. In: Proc. 16th Canadian Conference on Computational Geometry (CCCG 2004), pp. 196–199 (2004)

[20] van Leeuwen, J., Wood, D.: The measure problem for rectangular ranges in d-space. J. Algorithms 2, 282–300 (1981)

[21] Zitzler, E., Thiele, L.: Multiobjective evolutionary algorithms: a comparative case study and the strength Pareto approach. IEEE Trans. Evolutionary Computation 3, 257–271 (1999)

Space-Efficient Informational Redundancy

Christian Glaßer

University at Würzburg, Germany
`glasser@informatik.uni-wuerzburg.de`

Abstract. We study the relation of autoreducibility and mitoticity for polylog-space many-one reductions and log-space many-one reductions. For polylog-space these notions coincide, while proving the same for log-space is out of reach. More precisely, we show the following results with respect to nontrivial sets and many-one reductions.

1. polylog-space autoreducible $\Leftrightarrow$ polylog-space mitotic
2. log-space mitotic $\Rightarrow$ log-space autoreducible $\Rightarrow$ $(\log n \cdot \log \log n)$-space mitotic
3. relative to an oracle, log-space autoreducible $\not\Rightarrow$ log-space mitotic

The oracle is an infinite family of graphs whose construction combines arguments from Ramsey theory and Kolmogorov complexity.

1 Introduction

Our investigations are motivated by the question of whether the ability to reformulate membership questions for a set implies that the set is redundant in the sense that it can be split into two equivalent parts. This question depends on the resources available for reformulation and for splitting. In this paper we focus on notions of redundancy that have low space complexity. Suppose we are given a set A such that in log-space we can reformulate a given question $x \in A$ into an equivalent question $y \in A$. Is this enough to split A in log-space into disjoint parts A_1 and A_2 such that A, A_1, and A_2 are log-space many-one equivalent? We study this question also with respect to polylog-space.

The reformulation of questions is made precise by the notion of *autoreducibility*. The idea of splitting a set into two parts that are equivalent to the original set is formalized by the notion of *mitoticity*. Hence our main questions can be formulated as follows.

Q1. Does log-space many-one autoreducibility imply log-space many-one mitoticity?

Q2. Does polylog-space many-one autoreducibility imply polylog-space many-one mitoticity?

Trakhtenbrot [13] defined a set A to be *autoreducible* if there is an oracle Turing machine M such that $A = L(M^A)$ and M on input x never queries x. Lachlan [7] introduced the notion of mitoticity which was studied comprehensively by Ladner [9,8]. A set A is *mitotic* if there exists a recursive set S such

S.-H. Hong, H. Nagamochi, and T. Fukunaga (Eds.): ISAAC 2008, LNCS 5369, pp. 448–459, 2008.

that A, $A \cap S$, and $A \cap \overline{S}$ are Turing equivalent. Ambos-Spies [1] transferred autoreducibility and mitoticity to complexity theory and studied several versions of polynomial-time autoreducibility and polynomial-time mitoticity. A set A is *polynomial-time many-one autoreducible* if A is polynomial-time many-one reducible to A via a function f such that $f(x) \neq x$ for every x. Many-one autoreducible sets contain a local redundancy of information, since x and $f(x)$ contain the same information with respect to membership in A. A set A is *polynomial-time many-one mitotic* if there exists $S \in \mathrm{P}$ such that A, $A \cap S$, and $A \cap \overline{S}$ are polynomial-time many-one equivalent. Many-one mitoticity formalizes the aspect of informational redundancy in sets. Hence our questions $Q1$ and $Q2$ ask whether local (poly)log-space redundancy implies informational (poly)log-space redundancy. The converse implication holds in general, since mitoticity implies autoreducibility.

The question of whether local redundancy implies informational redundancy was first studied by Ladner [9,8] who showed that with respect to r.e. sets, autoreducibility and mitoticity coincide. Ambos-Spies [1] introduced the mentioned resource-bounded notions of redundancy and proved that polynomial-time Turing autoreducibility does not imply polynomial-time Turing mitoticity. Glaßer et al. [4,5] showed the same for all reducibility notions between polynomial-time 2-truth-table reducibility and polynomial-time Turing reducibility. In contrast, polynomial-time many-one autoreducibility and polynomial-time many-one mitoticity coincide [4]. The same holds for polynomial-time 1-truth-table reducibility.

In the present paper we shift the focus to notions of redundancy that have more restricted resource bounds, i.e., logarithmic or polylogarithmic space. This brings us to the study of (poly)log-space autoreducibility and (poly)log-space mitoticity. We prove that polylog-space many-one autoreducibility and polylog-space many-one mitoticity coincide. This is interesting, since it shows that even very restricted computational devices can exploit local redundancy and can transform it into informational redundancy. For log-space we show a similar, but weaker connection: Log-space many-one autoreducibility implies $(\log n \cdot \log \log n)$-space many-one mitoticity. The latter space bound can be even improved to $(\log n \cdot \log^{(c)} n)$ for any fixed constant c, where $\log^{(c)} n$ denotes the c-times composition of the log operation. On the technical side we obtain these results by developing a combination of the construction used for polynomial-time many-one reductions [4] and the repeated deterministic coin tossing by Cole and Vishkin [3].

So far we know that autoreducibility and mitoticity are equivalent with respect to unbounded Turing reductions [9,8], polynomial-time many-one reductions [4], and polylog-space many-one reductions (Corollary 3). Motivated by these equivalences, one could hope to turn the implications

$$\text{log-space many-one mitotic} \Rightarrow \text{log-space many-one autoreducible}$$

$$\text{log-space many-one autoreducible} \Rightarrow (\log n \cdot \log \log n)\text{-space many-one mitotic}$$

into a full equivalence, by replacing $(\log n \cdot \log \log n)$-space with log-space. We show that such an improvement is hard to obtain. In section 4 we discuss in detail the reason for this and we make this precise with the construction of an oracle relative to which the equivalence does not hold. This oracle construction is interesting for two reasons. First, it combines arguments from Ramsey theory and Kolmogorov complexity to make sure that log-space computable functions get lost in infinite graphs. Second, the constructed oracle separates log-space many-one autoreducibility and log-space many-one mitoticity with respect to *all* common models of log-space oracle machines. These include weak models like the ones by Ladner and Lynch [10] and Ruzzo, Simon, and Tompa [12], but also strong models like the model by Gottlob [6]. A discussion of all considered models is given in the preliminaries section.

Roughly speaking the oracle is a family of graphs whose existence follows from Ramsey theory. Each of these graphs is a ring (i.e., a connected graph with indegree and outdegree 1) where the nodes are numbers whose lengths are polynomially bounded in the size of the graph. Our witness language L is the set of all nodes that appear in some ring of the graph family. The rings are such that the successor of a given node can be computed in log-space which shows that L is log-space many-one autoreducible. In contrast, for every unbounded function t, the $t(n)$-th successor cannot be computed in log-space (where n is the size of the graph). So log-space functions can determine at most constantly many successors of a given node. Hence they see at most a constant-size part of the graph and act on the graph like a relation of constant arity. By the Ramsey theorem, we can choose our rings such that log-space machines show the same acceptance behavior on several consecutive nodes $v_1, \ldots, v_c$. So a log-space separator S puts all these nodes to the same side, either S or $\overline{S}$. For a given log-space function f we can ensure that c is large enough such that f on input v_1 can determine at most the nodes $v_1, \ldots, v_c$. The latter are all on the same side of S. So f cannot be a reduction from $L \cap S$ to $L \cap \overline{S}$. In this way we diagonalize against all f and obtain that L is not log-space many-one mitotic.

2 Preliminaries

In the paper, all variables represent natural numbers, unless they are explicitly defined in a different way. We use the following abbreviations for intervals of natural numbers: $[n, m] = \{n, n+1, \ldots, m\}$, $[n, m) = \{n, n+1, \ldots, m-1\}$, $(n, m] = \{n+1, n+2, \ldots, m\}$, $(n, m) = \{n+1, n+2, \ldots, m-1\}$. For $k \in \mathbb{Z}$ and $n \geq 1$ let $(k \bmod n)$ be the uniquely determined $m \in [0, n)$ such that $m \equiv k (\bmod n)$. Moreover, $\mathrm{sgn}(k)$ denotes the sign of k, $\mathrm{abs}(k)$ denotes the absolute value of k, and $|k|$ denotes the length of the binary representation of k. For a function f, $f^{(i)}$ denotes the i-th superposition of f, i.e., $f^{(0)}(x) = x$ and $f^{(i+1)}(x) = f(f^{(i)}(x))$. For a fixed function f, the sequence $f^{(0)}(x), f^{(1)}(x), \ldots$ is called the trajectory of x. The complement of a set A is denoted by $\overline{A}$. A set A is called *nontrivial* if $|A| \geq 2$ and $|\overline{A}| \geq 2$.

We distinguish between Turing machines and Turing transducers. Turing machines (TM for short) are used for accepting languages and so they output 0 or 1. Turing Transducers (TT for short) are machines that compute functions and hence can output arbitrary words. OTM (resp., OTT) is an abbreviation for oracle Turing machine (resp., oracle Turing transducer). If M is a TM or TT, then $M(x)$ denotes the output of M on input x. Similarly, if M is an OTM or an OTT, then $M^T(x)$ denotes the output of M on input x where T is used as oracle. If M is a TM (resp., OTM), then $L(M)$ (resp., $L(M^T)$) denotes the set of words accepted by M.

2.1   Models of Log-Space Oracle Machines

There is a canonical way to define oracle access for time-bounded machines. However, for space-bounded machines there is no such distinguished way. We briefly discuss several models of log-space oracle machines; for a detailed comparison we refer to Buss [2]. The following properties are desirable for such models: The machine should not be able to use the query tape as additional storage, but it should be able to write long strings to the query tape (e.g., its own input). Moreover, we would like that log-space computable functions are closed under composition. Not all of the presented models satisfy these conditions.

LL-Model by Ladner and Lynch [10]: The machine has one additional, one-way, write-only oracle tape which is not subject to the space bound and which is erased after asking a query.

RST-Model by Ruzzo, Simon, and Tompa [12]: Like the model by Ladner and Lynch, but additionally it is required that the machine acts deterministically while anything is written on the oracle tape. So for deterministic machines, both models are equivalent.

L-Model by Lynch [11]: The machine has an arbitrary, but fixed number of one-way, write-only oracle tapes. These tapes are not subject to the space bound and after asking a query, the corresponding tape is erased.

B-Model by Buss [2]: The machine has many one-way, write-only query tapes and a single read-write index tape which is logarithmically bounded. If i is written on the index tape, then all relativized operations (writing to a query tape, querying the oracle, erasing the query on the tape, and obtaining the answer) are with respect to tape i. If there are k active queries of maximum length m, then this considered as space $k \log m$ which must be of order $O(\log n)$.

W-Model by Wilson [14]: The machine has a one-way, write-only oracle stack which we consider to write from left to right. The machine can write several (partial) queries to the stack such that neighboring queries are separated by $\#$. If the machine enters the query state, then the characters after the rightmost $\#$ are considered to be query. After querying, the $\#$ and the query itself are erased so that the machine can continue to write the previous query at the stack. This nesting of queries may continue to any depth, but the stack contributes to the computation space as follows. If $q_1 \# q_2 \# \cdots \# q_k$ is the content of the stack, then this is considered as space $\sum_{i \in [1,k]} \max\{\log |q_i|, 1\}$ which must be of order $O(\log n)$.

G-Model by Gottlob [6]: The machine has $O(\log n)$ one-way, write-only query tapes and a single read-write index tape which is logarithmically bounded. The query tapes are not subject to the space bound. If i is written on the index tape, then all relativized operations are with respect to tape i.

For deterministic machines, the strengths of the presented models compare as follows:

$$\mathrm{LL} = \mathrm{RST} \leq \mathrm{L} \leq \mathrm{B} \leq \mathrm{G}$$

$$\mathrm{LL} = \mathrm{RST} \leq \mathrm{W} \leq \mathrm{G}$$

Throughout the paper (if not stated otherwise) we use the most powerful G-model for log-space OTMs and OTTs. Moreover, we assume the machines to have tape alphabet $\Sigma = \{0,1\}$ (which does not restrict the computational power). If we talk about the space used by such a machine, then this contains the space used by the working tape and the space used by the index tape. Recall that it does not contain the space used by query tapes. We may assume that a machine with space bound $d \log n$ has at most $d \log n$ query tapes. This latter assumption is motivated by the observation that each query tape can be used as a one-bit storage cell: For an oracle O, fix words $x_0 \notin O$ and $x_1 \in O$. A bit b can be stored in the oracle tape by writing x_b to the tape. We can read the bit b by querying the tape and writing again x_b to the tape (which was emptied by the query mechanism).

2.2   Complexity Classes, Reductions, Autoreducibility, and Mitoticity

For $s : \mathbb{N} \to \mathbb{N}$, let $\mathrm{FSPACE}(s)$ be the class of functions computable in deterministic space $O(s(n))$ and let $\mathrm{DSPACE}(s)$ be the class of languages that are decidable in deterministic space $O(s(n))$. Let FPLOG be the class of functions computable in deterministic polylog-space, i.e., $\mathrm{FPLOG} = \bigcup_{k \geq 1} \mathrm{FSPACE}(\log^k n)$. Let PLOG be the class of languages that are decidable in deterministic polylog-space, i.e., $\mathrm{PLOG} = \bigcup_{k \geq 1} \mathrm{DSPACE}(\log^k n)$. Moreover, let $\mathrm{FL} = \mathrm{FSPACE}(\log n)$ and $\mathrm{L} = \mathrm{DSPACE}(\log n)$. Observe that FPLOG and FL are closed under composition.

A set A is polylog-space many-one reducible to a set B, in notation $A \leq_{\mathrm{m}}^{\mathrm{plog}} B$, if there exists a total $f \in \mathrm{FPLOG}$ such that $x \in A \Leftrightarrow f(x) \in B$. Similarly, A is log-space many-one reducible to B, in notation $A \leq_{\mathrm{m}}^{\log} B$, if there exists a total $f \in \mathrm{FL}$ such that $x \in A \Leftrightarrow f(x) \in B$. From FL's and FPLOG's closure under composition it follows that $\leq_{\mathrm{m}}^{\log}$ and $\leq_{\mathrm{m}}^{\mathrm{plog}}$ are transitive. For integers $k \geq 1$ we write $A \leq_{\mathrm{m}}^{\log^k} B$, if there exists a total $f \in \mathrm{FSPACE}(\log^k n)$ such that $x \in A \Leftrightarrow f(x) \in B$. We write $A \leq_{\mathrm{m}}^{\log \cdot \log\log} B$, if there exists a total $f \in \mathrm{FSPACE}((\log n) \cdot \log\log n)$ such that $x \in A \Leftrightarrow f(x) \in B$. By lengths reasons, the function classes $\mathrm{FSPACE}((\log n) \cdot \log\log n)$ and $\mathrm{FSPACE}(\log^k n)$ for $k \geq 2$ are not closed under composition. Hence, the reductions $\leq_{\mathrm{m}}^{\log \cdot \log\log}$ and $\leq_{\mathrm{m}}^{\log^k}$ for $k \geq 2$ are not transitive.

Let $\leq_m^r$ be a reduction from $\{\leq_m^{\log}, \leq_m^{\text{plog}}, \leq_m^{\log \cdot \log\log}, \leq_m^{\log^k}\}$. We say A and B are $\leq_m^r$-equivalent, in notation $A \equiv_m^r B$, if $A \leq_m^r B$ and $B \leq_m^r A$. A is $\leq_m^r$-autoreducible, if $A \leq_m^r A$ via a reduction f such that $f(x) \neq x$.

A is $\leq_m^{\log}$-mitotic (resp., $\leq_m^{\text{plog}}$-mitotic), if there exists a separator $S \in \mathrm{L}$ (resp., $S \in \mathrm{PLOG}$) such that A, $A \cap S$, and $A \cap \overline{S}$ are pairwise $\leq_m^{\log}$-equivalent (resp., $\leq_m^{\text{plog}}$-equivalent).[1] A is $\leq_m^{\log^k}$-mitotic (resp., $\leq_m^{\log^k \cdot \log\log}$-mitotic), if there exists a separator $S \in \mathrm{DSPACE}(\log^k n)$ (resp., $S \in \mathrm{DSPACE}((\log^k n) \cdot \log\log n)$) such that A, $A \cap S$, and $A \cap \overline{S}$ are pairwise $\leq_m^{\log^k}$-equivalent (resp., $\leq_m^{\log^k \cdot \log\log}$-equivalent).

For an oracle O, let L^O be the class of sets decidable by a log-space OTM that has access to oracle O. Similarly, FL^O is the class of functions computable by a log-space OTT that has access to oracle O. We also need the following relativized versions of log-space reducibilities, autoreducibility, and mitoticity.

Definition 1. *Let O be an oracle and let A, B be sets of words.*

$$A \leq_m^{\log, O} B \stackrel{df}{\iff} \text{there exists } f \in \mathrm{FL}^O \text{ s.t. for all } x,$$
$$(x \in A \Leftrightarrow f(x) \in B).$$

$$A \leq_m^{\log\text{-lin}, O} B \stackrel{df}{\iff} \text{there exist } c > 0 \text{ and } f \in \mathrm{FL}^O \text{ such that for}$$
$$\text{all } x, |f(x)| \leq c|x| \text{ and } (x \in A \Leftrightarrow f(x) \in B).$$

$$A \text{ is } \leq_m^{\log, O}\text{-autoreducible} \stackrel{df}{\iff} A \leq_m^{\log, O} A \text{ via a reduction } f \text{ such that}$$
$$f(x) \neq x.$$

$$A \text{ is } \leq_m^{\log\text{-lin}, O}\text{-autoreducible} \stackrel{df}{\iff} A \leq_m^{\log\text{-lin}, O} A \text{ via a reduction } f \text{ such that}$$
$$f(x) \neq x.$$

$$A \text{ is } \leq_m^{\log, O}\text{-mitotic} \stackrel{df}{\iff} \text{there exists } S \in \mathrm{L}^O \text{ such that}$$
$$A \equiv_m^{\log, O} A \cap S \equiv_m^{\log, O} A \cap \overline{S}.$$

Proposition 1. *If L is $\leq_m^{\log}$-mitotic such that $|\overline{L}| \geq 2$, then L is $\leq_m^{\log}$-autoreducible.*

3   The Equivalence ($\leq_m^{\text{plog}}$-Autoreducible $\Leftrightarrow$ $\leq_m^{\text{plog}}$-Mitotic)

This section establishes tight connections between autoreducibility and mitoticity in the (poly)log-space setting. With respect to nontrivial set it holds that:

1. $\leq_m^{\text{plog}}$-autoreducible $\Leftrightarrow$ $\leq_m^{\text{plog}}$-mitotic
2. $\leq_m^{\log}$-autoreducible $\Rightarrow$ $\leq_m^{\log \cdot \log\log}$-mitotic

So for polylog-space many-one reductions we can prove an equivalence similar to the one that is known for polynomial-time many-one reductions [4]. However,

---

[1] This pairwise equivalence can be written as $A \equiv_m^{\log} A \cap S \equiv_m^{\log} A \cap \overline{S}$ (resp., $A \equiv_m^{\text{plog}} A \cap S \equiv_m^{\text{plog}} A \cap \overline{S}$), since $\leq_m^{\log}$ (resp., $\leq_m^{\text{plog}}$) is transitive. This is not possible in the definitions of $\leq_m^{\log^k}$-mitoticity and $\leq_m^{\log \cdot \log\log}$-mitoticity.

for log-space many-one reductions we obtain mitoticity only if we grant the reduction a little more space than $O(\log n)$. Log-space many-one autoreducibility even implies $(\log n \cdot \log^{(c)} n)$-space many-one mitoticity for every constant c.

To obtain these results, we apply a combination of the construction used in the polynomial-time many-one setting [4] and the repeated deterministic coin tossing by Cole and Vishkin [3]. In comparison with polynomial-time many-one reductions, for space-bounded many-one reductions a new difficulty comes up: Assume we are given a many-one autoreduction f that is length-preserving and let x be our input. In the polynomial-time setting, it is easy to follow the trajectory of x for $|x|$ steps. This means that if $f \in$ FP, then we can compute $f^{(|x|)}(x)$ in polynomial time. This does not work in the log-space setting. Here we can follow the trajectory only for constantly-many steps. So if $f \in$ FL and c is constant, then we can compute $f^{(c)}$ in log-space, but we cannot compute $f^{(|x|)}(x)$ (since the intermediate results $f(x), f^2(x), \ldots$ cannot be stored in log-space). Here the repeated deterministic coin tossing [3] comes into play. With this technique it is possible to construct a well-balanced separator (which is used to establish mitoticity) such that instead of $|x|$ steps we only have to follow the trajectory for $\log \log |x|$ steps. This number can actually be dropped to any fixed number of repeated log operations, i.e., $\log^{(c)} |x|$ where c is constant.

We cannot prove that log-space many-one autoreducibility is equivalent to log-space many-one mitoticity. The lack of this equivalence is not due to our particular technique. In section 4 we discuss in detail the deeper reason for the missing equivalence and we make this precise with the construction of an oracle relative to which the equivalence does not hold.

Theorem 1. *Let $k \geq 1$ be an integer and let L be a $\leq_{\mathrm{m}}^{\log^k}$-autoreducible set such that $|\overline{L}| \geq 2$. Then there exist a total $g \in$ FSPACE$((\log^{k^7} n) \cdot \log \log n)$ and a set $S \in$ DSPACE$(\log^{k^4} n)$ such that for all x,*

1. $x \in L \Leftrightarrow g(x) \in L$, and
2. $x \in S \Leftrightarrow g(x) \notin S$.

Corollary 1. *Let $k \geq 1$ be an integer and let L be a $\leq_{\mathrm{m}}^{\log^k}$-autoreducible set such that $|\overline{L}| \geq 2$. Then L is $\leq_{\mathrm{m}}^{\log^{k^7} \cdot \log\log}$-mitotic.*

Proof. From Theorem 1 we obtain $g \in$ FSPACE$((\log^{k^7} n) \cdot \log \log n)$ and $S \in$ DSPACE$(\log^{k^4} n)$ such that $(x \in L \Leftrightarrow g(x) \in L)$ and $(x \in S \Leftrightarrow g(x) \notin S)$. So $L \cap S \leq_{\mathrm{m}}^{\log^{k^7} \cdot \log\log} L \cap \overline{S}$ and $L \cap \overline{S} \leq_{\mathrm{m}}^{\log^{k^7} \cdot \log\log} L \cap S$, both via g. This shows $L \cap S \equiv_{\mathrm{m}}^{\log^{k^7} \cdot \log\log} L \cap \overline{S}$.

The following function g' witnesses $L \leq_{\mathrm{m}}^{\log^{k^7} \cdot \log\log} L \cap S$: If $x \in S$, then $g'(x) = x$ else $g'(x) = g(x)$. The following function g'' witnesses $L \cap S \leq_{\mathrm{m}}^{\log^{k^7} \cdot \log\log} L$: If $x \in S$, then $g''(x) = x$ else $g''(x) = w_1$, where w_1 is a fixed word in $\overline{L}$. This shows $L \equiv_{\mathrm{m}}^{\log^{k^7} \cdot \log\log} L \cap S$ and analogously we obtain $L \equiv_{\mathrm{m}}^{\log^{k^7} \cdot \log\log} L \cap \overline{S}$. $\square$

Corollary 2. *Let L be any set such that $|\overline{L}| \geq 2$. If L is $\leq_{\mathrm{m}}^{\log}$-autoreducible, then L is $\leq_{\mathrm{m}}^{\log \cdot \log\log}$-mitotic.*

Remark 1. Corollary 2 can be improved in the sense that for every $k \geq 1$, if L is $\leq_m^{\log}$-autoreducible and $|\overline{L}| \geq 2$, then L is mitotic with respect to many-one reductions that belong to

$$\text{FSPACE}((\log n) \cdot \underbrace{\log \log \cdots \log n}_{k \text{ times}}).$$

Corollary 3. *Let L be any set such that $|\overline{L}| \geq 2$. L is $\leq_m^{\text{plog}}$-autoreducible if and only if L is $\leq_m^{\text{plog}}$-mitotic.*

Proof. If L is $\leq_m^{\text{plog}}$-mitotic, then there exist $S \in \text{PLOG}$ and $f_1, f_2 \in \text{FPLOG}$ such that $L \cap S \leq_m^{\text{plog}} L \cap \overline{S}$ via f_1 and $L \cap \overline{S} \leq_m^{\text{plog}} L \cap S$ via f_2. By assumption, there exist different words $v, w \in \overline{L}$. The following function f' is a $\leq_m^{\text{plog}}$-autoreduction for L: If $x \in S$ and $f_1(x) \notin S$, then $f'(x) = f_1(x)$. If $x \notin S$ and $f_2(x) \in S$, then $f'(x) = f_2(x)$. Otherwise, $f'(x) = \min(\{v, w\} - \{x\})$.

The other direction follows from Corollary 1. $\square$

4   The Difficulty of ($\leq_m^{\log}$-Autoreducible $\Rightarrow$ $\leq_m^{\log}$-Mitotic)

We know that the notions of $\leq_m^P$-autoreducibility and $\leq_m^P$-mitoticity are equivalent [4]. In the preceding section we learned that with respect to log-space many-one reductions, these notions are nearly equivalent:

$$\leq_m^{\log}\text{-mitotic} \Rightarrow \leq_m^{\log}\text{-autoreducible}$$
$$\leq_m^{\log}\text{-autoreducible} \Rightarrow \leq_m^{\log \cdot \log\log}\text{-mitotic}$$

In this section we explain the reason why it is difficult to establish the full equivalence. This is done in two steps. First, in section 4.1 we describe this difficulty on an intuitive level. Second, in section 4.2 we sketch the construction of a relativized world where this difficulty becomes provable. Relative to our oracle, $\leq_m^{\log}$-autoreducibility and $\leq_m^{\log}$-mitoticity are not equivalent. This result holds with respect to all models of log-space oracle machines that were discussed in the preliminaries section.

Observe that we cannot hope to show unconditionally that $\leq_m^{\log}$-autoreducibility does not imply $\leq_m^{\log}$-mitoticity. Such a proof would separate L from P.

$$\text{L} = \text{P} \quad \Rightarrow \quad \leq_m^{\log}\text{-autoreducibility and } \leq_m^{\log}\text{-mitoticity are equivalent}$$

This is seen as follows: $\text{L} = \text{P}$ implies $\text{FL} = \text{FP}$. If A is $\leq_m^{\log}$-autoreducible, then it is $\leq_m^P$-autoreducible and hence $\leq_m^P$-mitotic [4]. So there exists a separator $S \in \text{P} = \text{L}$ such that $A \equiv_m^P A \cap S \equiv_m^P A \cap \overline{S}$. From $\text{FL} = \text{FP}$ it follows that $A \equiv_m^{\log} A \cap S \equiv_m^{\log} A \cap \overline{S}$ and hence A is $\leq_m^{\log}$-mitotic.

So in the case of log-space reductions we observe a behavior that differs from the experience we had with polynomial-time reductions. For the latter, either autoreducibility and mitoticity coincide (i.e., for $\leq_m^P$) or it is possible to separate the notions unconditionally (i.e., for all reductions between $\leq_{2-tt}^P$ and $\leq_T^P$). In contrast, with respect to log-space many-one reductions, it appears as a plausible possibility that autoreducibility and mitoticity are different, but we cannot prove this, unless we separate L from P. This is consistent with our suspicion that $\leq_m^{\log}$-autoreducibility and $\leq_m^{\log}$-mitoticity are inequivalent, but very similar notions.

4.1   Explanation on an Intuitive Level

We give an intuitive explanation of the difficulty of transforming $\leq_m^{\log}$-autoreducibility into $\leq_m^{\log}$-mitoticity. It is in the nature of such explanations that our arguments will be simplified and informal.

We say that a function $f \in$ FL *has difficult, detached cycles* if for every $g \in$ FL there exists a constant $c > 0$ such that for infinitely many x:

1. $T_x \overset{df}{=} \{f^{(0)}(x), f^{(1)}(x), \ldots, f^{(|x|-1)}(x)\}$ has cardinality $|x|$ and it holds that $f^{(|x|)}(x) = x$
2. $f^{-1}(T_x) \subseteq T_x$
3. $\forall y \in T_x, \ [g(y) \in T_x \Rightarrow g(y) \in \{f^{(0)}(y), f^{(1)}(y), \ldots, f^{(c)}(y)\}]$

Item 1 states that the trajectory of x is a cycle of length $|x|$. Item 2 says that no other arguments are mapped to T_x and therefore, the trajectory of x is not connected to other trajectories. Item 3 describes a certain hardness of f: For a given element in the trajectory, a log-space machine can only compute constantly-many successors. This is consistent with the fact that $f^{(c)}(x) \in$ FL for all $f \in$ FL and all constants c, and it is also consistent with our impression that $f^{(t(x))}(x)$ is not necessarily in FL if t is not constant.

At first glance, the property of having difficult, detached cycles might appear artificial and very strong. However, there is no reason to exclude the existence of functions $f \in$ FL that have this property and that satisfy $f(x) \neq x$. For example, with our construction below we demonstrate a relativized world in which such functions exist. So it is a reasonable hypothesis to assume the existence of such functions.

Suppose f is such a function. We use f for the construction of a set L that is $\leq_m^{\log}$-autoreducible via f, but not $\leq_m^{\log}$-mitotic. By item 3 (the hardness condition), every log-space machine can only compute a constant-size preview of f's trajectory. Therefore, every log-space computable separator S that claims to establish the $\leq_m^{\log}$-mitoticity of L can only compute such a constant-size preview of f's trajectory. This implies that with respect to the trajectory of f, every separator S of L acts like a relation of constant arity, since it depends only on constantly-many successors of the input x. From Ramsey theory (more precisely, the existence of the generalized Ramsey numbers) it follows that for every $c \geq 0$ there exists an x such that $x, f(x), \ldots, f^{(c)}(x)$ have the same membership with respect to S (i.e., either all belong to S or all belong to $\overline{S}$). Again by item 3, every log-space computable function g that claims to establish $L \cap S \leq_m^{\log} L \cap \overline{S}$ (which is needed for $\leq_m^{\log}$-mitoticity) can only compute a constant-size preview of f's trajectory. So by choosing c large enough we can enforce that either $g(x) \in \{x, f(x), \ldots, f^{(c)}(x)\}$ and hence

$$x \in S \Leftrightarrow f(x) \in S$$

or $g(x)$ does not belong to x's trajectory with respect to f in which case we can (by diagonalization) construct L such that

$$x \in L \Leftrightarrow g(x) \notin L.$$

So g is not a $\leq_{\mathrm{m}}^{\log}$-reduction from $L \cap S$ to $L \cap \overline{S}$. In this way we diagonalize against all pairs (S, g) and obtain a set L that is not $\leq_{\mathrm{m}}^{\log}$-mitotic. More precisely, our diagonalization is in such a way that we put whole trajectories inside or outside L. This implies that L is $\leq_{\mathrm{m}}^{\log}$-autoreducible via f.

4.2   Road Map for the Oracle Construction

While neglecting technical details we sketch the main arguments of the construction. In the first part, with the stagewise construction of an oracle O we create a suitable relativized environment. Then, in the second part, we use this environment and construct a language L that is $\leq_{\mathrm{m}}^{\log}$-autoreducible, but not $\leq_{\mathrm{m}}^{\log}$-mitotic.

We start with the description of stage s of the construction of O. There we diagonalize against two log-space machines M_1 (a possible log-space separator) and M_2 (a possible log-space reduction function). At the beginning we choose n large enough such that changing the oracle with respect to words of length $\geq n^2$ does not affect separations made in earlier stages. Then we choose a set $S \subseteq \Sigma^{n^2}$ such that $|S| = n$ and S has maximal Kolmogorov complexity. In particular, all subsets of S have a high Kolmogorov complexity.

Now let us observe that each $T \subseteq S$ of cardinality ≥ 2 induces a particular set $\langle T \rangle \subseteq \Sigma^{n^2}$ which can be used to define the oracle with respect to words of length n^2. This set $\langle T \rangle$ is defined as follows: If $w_0, \ldots, w_{k-1}$ are the words in T in ascending order, then the characteristic sequence of $\langle T \rangle$ (considered as a subset of Σ^{n^2}) is

$$0 \cdots 0\, 1w_1\, 0 \cdots 0\, 1w_2\, 0 \quad \cdots \quad 0\, 1w_{k-1}\, 0 \cdots 0\, 1w_0\, 0 \cdots 0$$

where the factor $1w_{i+1}$ starts after the w_i-th letter and the factor $1w_0$ starts after the w_{k-1}-th letter. So a word w of length n^2 belongs to $\langle T \rangle$ if and only if in the sequence above there is a 1 at position w.

This encoding of T has the advantage that for a given w_i, the successor w_{i+1} can be computed by a log-space machine that has access to the oracle $\langle T \rangle$. For this, the machine just has to query the words $w_i + 1, w_i + 2, \ldots, w_i + n^2$ and has to interpret the vector of answers as the word w_{i+1}. This property results in the $\leq_{\mathrm{m}}^{\log}$-autoreducibility of T and finally this will translate into the $\leq_{\mathrm{m}}^{\log}$-autoreducibility of L.

Since log-space computable functions are closed under composition, for every constant $c > 1$, there exists a log-space machine with oracle $\langle T \rangle$ that on input w_i computes the c-th next word w_{i+c}. We show that for log-space machines this c-times composition of the successor function is expensive: No log-space machine can compute successors that are farther away than a constant. Hence, there exists a constant c such that on input w_i and with access to the oracle $\langle T \rangle$ the machines M_1 and M_2 can only gain knowledge about the words $w_i, w_{i+1}, \ldots, w_{i+c}$, but not about the words $w_0, w_1, \ldots, w_{i-1}$ and $w_{i+c+1}, w_{i+c+2}, \ldots, w_{k-1}$.[2] In particular, the machines will not notice if we change the oracle with respect to the latter words. Therefore, if we consider M_1 on input w_i and with oracle $\langle T \rangle$, then this computation will not change if we replace the oracle by $\{w_i, w_{i+1}, \ldots, w_{i+c}\}$. In this sense, $M_1^{\langle T \rangle}(w_i)$ computes exactly the $(c+1)$-ary relation

$$R(w_i, w_{i+1}, \ldots, w_{i+c}) \overset{df}{=} M_1^{\langle \{w_i, w_{i+1}, \ldots, w_{i+c}\} \rangle}(w_i).$$

We now apply Ramsey theory and obtain a set $T_s = \{w_0, \ldots, w_{3c-1}\}$ such that $T_s \subseteq S$ and all words $w_0, \ldots, w_{2c-1}$ are equivalent with respect to the relation, i.e., are all inside or all outside the relation. Note that this property of T_s is very strong. It means either $w_0, \ldots, w_{2c-1} \in L(M_1^{\langle T_s \rangle})$ or $w_0, \ldots, w_{2c-1} \in \overline{L(M_1^{\langle T_s \rangle})}$. Let $O_s \overset{df}{=} \langle T_s \rangle$ which defines our oracle with respect to words of length n^2. This finishes stage s of our construction. The final oracle O is the union of all O_s.

We enter the second part, i.e., the construction of L. All remaining arguments are now relative to the oracle O. We have to construct L such that it is $\leq_{\mathrm{m}}^{\log}$-autoreducible, but not $\leq_{\mathrm{m}}^{\log}$-mitotic. On the one hand, L will be the union of sets T_s for certain s which immediately results in L's $\leq_{\mathrm{m}}^{\log}$-autoreducibility. On the other hand, we diagonalize against all possible log-space-computable separators $S = L(M_1)$ and all log-space-computable functions $f(x) = M_2(x)$ that claim to reduce $L \cap S$ to $L \cap \overline{S}$. The latter destroys $\leq_{\mathrm{m}}^{\log}$-mitoticity.

We sketch the diagonalization argument. Let s be the stage of O's construction in which we diagonalized against M_1 and M_2. In this stage we constructed the set $T_s = \{w_0, \ldots, w_{3c-1}\}$. We already observed that $M_2(w_0)$ cannot gain knowledge about the words $w_{c+1}, w_{c+2}, \ldots, w_{3c-1}$. If $M_2(w_0) \notin T_s$, then by putting T_s inside or outside L, we can enforce that f does not reduce L to L. Otherwise, $M_2(w_0) \in T_s$ and hence $M_2(w_0) \in \{w_0, \ldots, w_c\}$, since it cannot gain knowledge about the other words. However, as seen above, either $w_0, \ldots, w_{2c-1} \in L(M_1)$ or $w_0, \ldots, w_{2c-1} \in \overline{L(M_1)}$. So in this case, f does not reduce S to $\overline{S}$. Therefore, in any case, f does not reduce $L \cap S$ to $L \cap \overline{S}$. This shows that L is not $\leq_{\mathrm{m}}^{\log}$-mitotic.

Theorem 2. *There exists an oracle O relative to which $\leq_{\mathrm{m}}^{\log}$-autoreducibility does not imply $\leq_{\mathrm{m}}^{\log}$-mitoticity (i.e., there exists an L that is $\leq_{\mathrm{m}}^{\log,O}$-autoreducible, but not $\leq_{\mathrm{m}}^{\log,O}$-mitotic).*

Corollary 4. *There exists an oracle O and a language L such that all of the following holds.*

- *L is $\leq_{\mathrm{m}}^{\log\text{-lin},O}$-autoreducible by a reduction function that is computable by a log-space OTT with only one query tape.*
- *L is not $\leq_{\mathrm{m}}^{\log,O}$-mitotic.*

---

[2] This is the point where we need S and hence T to have high Kolmogorov complexity, since otherwise some information about the latter words could be contained in $w_i, w_{i+1}, \ldots, w_{i+c}$ which would allow M_1 and M_2 to obtain this information.

Corollary 5. *There exists an oracle O such that relative to O and with respect to every machine model $\mu \in \{LL, RST, L, B, W, G\}$ it holds that $\leq_m^{\log}$-autoreducibility does not imply $\leq_m^{\log}$-mitoticity.*

Proof. Let O and L be as in Corollary 4 and choose a machine model μ. From the first item of Corollary 4 it follows that L is $\leq_m^{\log,O}$-autoreducible with respect to μ. By the second item of Corollary 4, L is not $\leq_m^{\log,O}$-mitotic with respect to the G-model (which is the default machine model in this paper). Among all considered machine models, the G-model is the most powerful one. Hence, L is not $\leq_m^{\log,O}$-mitotic with respect to μ. $\qquad\qquad\square$

References

1. Ambos-Spies, K.: P-mitotic sets. In: Börger, E., Rödding, D., Hasenjaeger, G. (eds.) Rekursive Kombinatorik 1983. LNCS, vol. 171, pp. 1–23. Springer, Heidelberg (1984)
2. Buss, J.F.: Relativized alternation and space-bounded computation. Journal of Computer and System Sciences 36(3), 351–378 (1988)
3. Cole, R., Vishkin, U.: Deterministic coin tossing with applications to optimal parallel list ranking. Information and Control 70(1), 32–53 (1986)
4. Glaßer, C., Pavan, A., Selman, A.L., Zhang, L.: Redundancy in complete sets. In: Durand, B., Thomas, W. (eds.) STACS 2006. LNCS, vol. 3884, pp. 444–454. Springer, Heidelberg (2006)
5. Glaßer, C., Selman, A.L., Travers, S., Zhang, L.: Non-mitotic sets. In: Arvind, V., Prasad, S. (eds.) FSTTCS 2007. LNCS, vol. 4855, pp. 146–157. Springer, Heidelberg (2007)
6. Gottlob, G.: Collapsing oracle-tape hierarchies. In: Proceedings 11th Conference on Computational Complexity, pp. 33–42. IEEE Computer Society Press, Los Alamitos (1996)
7. Lachlan, A.H.: The priority method I. Zeitschrift für Mathematische Logik und Grundlagen der Mathematik 13, 1–10 (1967)
8. Ladner, R.E.: A completely mitotic nonrecursive r.e. degree. Trans. American Mathematical Society 184, 479–507 (1973)
9. Ladner, R.E.: Mitotic recursively enumerable sets. Journal of Symbolic Logic 38(2), 199–211 (1973)
10. Ladner, R.E., Lynch, N.A.: Relativization of questions about log space computability. Mathematical Systems Theory 10, 19–32 (1976)
11. Lynch, N.A.: Log space machines with multiple oracle tapes. Theoretical Computer Science 6, 25–39 (1978)
12. Ruzzo, W.L., Simon, J., Tompa, M.: Space-bounded hierarchies and probabilistic computations. Journal of Computer and System Sciences 28(2), 216–230 (1984)
13. Trakhtenbrot, B.: On autoreducibility. Dokl. Akad. Nauk SSSR 192(6), 1224–1227 (1970); Translation in Soviet Math. Dokl. 11(3), 814–817 (1970)
14. Wilson, C.B.: A measure of relativized space which is faithful with respect to depth. Journal of Computer and System Sciences 36(3), 303–312 (1988)

Minkowski Sum Selection and Finding

Cheng-Wei Luo[1], Hsiao-Fei Liu[1], Peng-An Chen[1], and Kun-Mao Chao[1,2,3,*]

[1] Department of Computer Science and Information Engineering
[2] Graduate Institute of Biomedical Electronics and Bioinformatics
[3] Graduate Institute of Networking and Multimedia
National Taiwan University, Taipei, Taiwan 106
kmchao@csie.ntu.edu.tw

Abstract. Let $P, Q \subseteq \mathbb{R}^2$ be two n-point multisets and $Ar \geq b$ be a set of λ inequalities on x and y, where $A \in \mathbb{R}^{\lambda \times 2}$, $r = \left[\begin{smallmatrix} x \\ y \end{smallmatrix}\right]$, and $b \in \mathbb{R}^\lambda$. Define the *constrained Minkowski sum* $(P \oplus Q)_{Ar \geq b}$ as the multiset $\{(p + q) | p \in P, q \in Q, A(p + q) \geq b\}$. Given P, Q, $Ar \geq b$, an objective function $f : \mathbb{R}^2 \rightarrow \mathbb{R}$, and a positive integer k, the MINKOWSKI SUM SELECTION problem is to find the kth largest objective value among all objective values of points in $(P \oplus Q)_{Ar \geq b}$. Given P, Q, $Ar \geq b$, an objective function $f : \mathbb{R}^2 \rightarrow \mathbb{R}$, and a real number δ, the MINKOWSKI SUM FINDING problem is to find a point (x^*, y^*) in $(P \oplus Q)_{Ar \geq b}$ such that $|f(x^*, y^*) - \delta|$ is minimized. For the MINKOWSKI SUM SELECTION problem with linear objective functions, we obtain the following results: (1) optimal $O(n \log n)$ time algorithms for $\lambda = 1$; (2) $O(n \log^2 n)$ time deterministic algorithms and expected $O(n \log n)$ time randomized algorithms for any fixed $\lambda > 1$. For the MINKOWSKI SUM FINDING problem with linear objective functions or objective functions of the form $f(x, y) = \frac{by}{ax}$, we construct optimal $O(n \log n)$ time algorithms for any fixed $\lambda \geq 1$. As a byproduct, we obtain improved algorithms for the LENGTH-CONSTRAINED SUM SELECTION problem and the DENSITY FINDING problem.

Keywords: Bioinformatics, Sequence analysis, Minkowski sum.

1   Introduction

Let $P, Q \subseteq \mathbb{R}^2$ be two n-point multisets and $Ar \geq b$ be a set of λ inequalities on x and y, where $A \in \mathbb{R}^{\lambda \times 2}$, $r = \left[\begin{smallmatrix} x \\ y \end{smallmatrix}\right]$, and $b \in \mathbb{R}^\lambda$. Define the *constrained Minkowski sum* $(P \oplus Q)_{Ar \geq b}$ as the multiset $\{(p + q) | p \in P, q \in Q, A(p + q) \geq b\}$.

In the MINKOWSKI SUM OPTIMIZATION problem, we are given P, Q, $Ar \geq b$, and an objective function $f : \mathbb{R}^2 \rightarrow \mathbb{R}$. The goal is to find the maximum objective value among all objective values of points in $(P \oplus Q)_{Ar \geq b}$. A function $f : D \subseteq \mathbb{R}^2 \rightarrow \mathbb{R}$ is said to be *quasiconvex* if and only if for all points $v_1, v_2 \in D$ and all $\gamma \in [0, 1]$, one has $f(\gamma \cdot v_1 + (1 - \gamma) \cdot v_2) \leq \max(f(v_1), f(v_2))$. Bernholt *et al.* [4] studied the MINKOWSKI SUM OPTIMIZATION problem for quasiconvex

---

* Corresponding author.

S.-H. Hong, H. Nagamochi, and T. Fukunaga (Eds.): ISAAC 2008, LNCS 5369, pp. 460–471, 2008.

objective functions and showed that their results have applications to many optimization problems arising in computational biology [1,5,8,9,14,15,16,20]. In this paper, two variations of the MINKOWSKI SUM OPTIMIZATION problem are studied: the MINKOWSKI SUM SELECTION problem and the MINKOWSKI SUM FINDING problem.

In the MINKOWSKI SUM SELECTION problem, we are given P, Q, $Ar \geq b$, an objective function $f : \mathbb{R}^2 \to \mathbb{R}$, and a positive integer k. The goal is to find the kth largest objective value among all objective values of points in $(P \oplus Q)_{Ar \geq b}$. The MINKOWSKI SUM OPTIMIZATION problem is equivalent to the MINKOWSKI SUM SELECTION problem with $k = 1$. A variety of selection problems, including the SUM SELECTION problem [3,12], the LENGTH-CONSTRAINED SUM SELECTION problem [13], and the SLOPE SELECTION problem [6,18], are linear-time reducible to the MINKOWSKI SUM SELECTION problem with a linear objective function or an objective function of the form $f(x, y) = \frac{by}{ax}$. It is desirable that relevant selection problems from diverse fields are integrated into a single one, so we don't have to consider them separately. Next, let us look at the use of the MINKOWSKI SUM SELECTION problem in practice. As mentioned above, the MINKOWSKI SUM OPTIMIZATION problem finds applications to many optimization problems arising in computational biology [1,5,8,9,14,15,16,20]. In these optimization problems, the objective functions are chosen such that feasible solutions with higher objective values are "more likely" to be biologically meaningful. However, it is not guaranteed that the best feasible solution always satisfies the needs of biologists. If the best feasible solution does not interest biologists or does not provide enough information, we still have to find the second best feasible solution, the third best feasible solution and so on until a satisfying feasible solution is finally found. As a result, it is desirable to know how to dig out extra good feasible solutions in case that the best feasible solution is not sufficient.

In the MINKOWSKI SUM FINDING problem, we are given P, Q, $Ar \geq b$, an objective function $f : \mathbb{R}^2 \to \mathbb{R}$, and a real number δ. The goal is to find a point (x^*, y^*) in $(P \oplus Q)_{Ar \geq b}$ such that $|f(x^*, y^*) - \delta|$ is minimized. This problem originates from the study of the DENSITY FINDING problem proposed by Lee et al. [11]. The DENSITY FINDING problem can be regarded as a specialization of the MINKOWSKI SUM FINDING problem with objective function $f(x, y) = \frac{y}{x}$ and find applications in recognizing promoters in DNA sequences [10,19]. In these applications, the goal is not to find the feasible solution with the highest objective value. Instead, feasible solutions with objective values close to some specific number, say δ, are thought to be more biologically meaningful and preferred.

The main results obtained in this paper are as follows.

- The MINKOWSKI SUM SELECTION problem with one constraint and a linear objective function can be solved in optimal $O(n \log n)$ time.
- The MINKOWSKI SUM SELECTION problem with two constraints and a linear objective function can be solved in $O(n \log^2 n)$ time by a deterministic algorithm and expected $O(n \log n)$ time by a randomized algorithm.
- For any fixed $\lambda > 2$, the MINKOWSKI SUM SELECTION problem with λ constraints and a linear objective function is shown to be asymptotically

equivalent to the MINKOWSKI SUM SELECTION problem with two constraints and a linear objective function.

- The MINKOWSKI SUM FINDING problem with any fixed number of constraints can be solved in optimal $O(n \log n)$ time if the objective function $f(x, y)$ is linear or of the form $\frac{by}{ax}$.

As a byproduct, we obtain improved algorithms for the LENGTH-CONSTRAINED SUM SELECTION problem [13] and the DENSITY FINDING problem [11]. Recently, Lin and Lee [13] proposed an expected $O(n \log(u - l + 1))$-time randomized algorithm for the LENGTH-CONSTRAINED SUM SELECTION problem, where n is the size of the input instance and $l, u \in \mathbb{N}$ are two given parameters with $1 \leq l < u \leq n$. In this study, we obtain a worst-case $O(n \log(u - l + 1))$-time deterministic algorithm for the LENGTH-CONSTRAINED SUM SELECTION problem. Lee, Lin, and Lu [11] showed the DENSITY FINDING problem has a lower bound of $\Omega(n \log n)$ and proposed an $O(n \log^2 m)$-time algorithm for it, where n is the size of the input instance and m is a parameter whose value may be as large as n. In this study, we give an optimal $O(n \log n)$-time algorithm for the DENSITY FINDING problem. Due to page constraint, we omit the two algorithms (see [17] for the details).

2   Preliminaries

In this section, we review some definitions and theorems. For more details, readers can refer to [4,7]. A matrix $X \in \mathbb{R}^{n \times m}$ is called *sorted* if the values of each row and each column are in nondecreasing order. Frederickson and Johnson [7] gave some results about the selection problem and the ranking problem in a collection of sorted matrices. From the results of [7], we have the following theorems.

Theorem 1. *The selection problem in a collection of sorted matrices is given a rank k and a collection of sorted matrices $\{X_1, \ldots, X_N\}$ in which X_j has dimensions $n_j \times m_j$, $n_j \geq m_j$, to find the kth largest element among all elements of sorted matrices in $\{X_1, \ldots, X_N\}$. This problem is able to be solved in $O(\sum_{j=1}^{N} m_j \log(2n_j/m_j))$ time.*

Theorem 2. *The ranking problem in a collection of sorted matrices is given an element and a collection of sorted matrices $\{X_1, \ldots, X_N\}$ in which X_j has dimensions $n_j \times m_j$, $n_j \geq m_j$, to find the rank of the given element among all elements of sorted matrices in $\{X_1, \ldots, X_N\}$. This problem can be solved in $O(\sum_{j=1}^{N} m_j \log(2n_j/m_j))$ time.*

By the recent works of Bernholt *et al.* [4], we have the following theorems.

Theorem 3. *Given a set of λ linear inequalities $Ar \geq b$ and two n-point multisets $P, Q \subseteq \mathbb{R}^2$, one can compute the vertices of the convex hull of $(P \oplus Q)_{Ar \geq b}$ in $O(\lambda \log \lambda + \lambda \cdot n \log n)$ time.*

Theorem 4. *The problem of maximizing a quasiconvex objective function f over the constrained Minkowski sum $(P \oplus Q)_{Ar \geq b}$ requires $\Omega(n \log n)$ time in the algebraic decision tree model even if f is a linear function and $Ar \geq b$ consists of only one constraint.*

3   Minkowski Sum Selection with One Constraint

In this section we study the MINKOWSKI SUM SELECTION problem and give an optimal $O(n \log n)$ time algorithm for the case where only one linear constraint is given and the objective function is also linear.

Given $P = \{(x_{1,1}, y_{1,1}), \dots, (x_{1,n}, y_{1,n})\}$, $Q = \{(x_{2,1}, y_{2,1}), \dots, (x_{2,n}, y_{2,n})\}$, a positive integer k, one constraint $L: ax + by \geq c$, and a linear objective function $f(x, y) = dx + ey$, where a, b, c, d, and e are all constants, we perform the following transformation.

1. Change the content of P and Q to $\{(ax_{1,1} + by_{1,1}, dx_{1,1} + ey_{1,1}), \dots, (ax_{1,n} + by_{1,n}, dx_{1,n} + ey_{1,n})\}$, and $\{(ax_{2,1} + by_{2,1}, dx_{2,1} + ey_{2,1}), \dots, (ax_{2,n} + by_{2,n}, dx_{2,n} + ey_{2,n})\}$, respectively.
2. Change the constraint from $ax + by \geq c$ to $x \geq c$.
3. Change the objective function from $dx + ey$ to y.

This transformation can be done in $O(n)$ time and the answer remains the same. Hence from now on, our goal becomes to find the kth largest y-coordinate on the constrained Minkowski sum of P and Q subject to the constraint $L : x \geq c$.

For ease of exposition, we assume that no two points in P and Q have the same x-coordinate and n is a power of two. The algorithm proceeds as follows. First, we sort P and Q into P_x and Q_x (P_y and Q_y, respectively) in nondecreasing order of x-coordinates (y-coordinates, respectively) in $O(n \log n)$ time. Next, we use a divide-and-conquer approach to store the y-coordinates of $(P \oplus Q)_{x \geq c}$ as a collection of sorted matrices and then apply Theorem 1 to select the kth largest element from the elements of these sorted matrices.

Now we explain how to store the y-coordinates of $(P \oplus Q)_{x \geq c}$ as a collection of sorted matrices. Let $P_x = ((x_1, y_1), \dots, (x_n, y_n))$, $Q_x = ((\bar{x}_1, \bar{y}_1), \dots, (\bar{x}_n, \bar{y}_n))$, $P_y = ((x'_1, y'_1), \dots, (x'_n, y'_n))$, and $Q_y = ((\bar{x}'_1, \bar{y}'_1), \dots, (\bar{x}'_n, \bar{y}'_n))$. We then divide P_x into two halves of equal size: $A = ((x_1, y_1), \dots, (x_{n/2}, y_{n/2}))$ and $B = ((x_{n/2+1}, y_{n/2+1}), \dots, (x_n, y_n))$. Find a point $(\bar{x}_t, \bar{y}_t)$ of Q_x such that $x_{n/2} + \bar{x}_t < c$ and t is maximized. Then divide Q_x into two halves: $C = ((\bar{x}_1, \bar{y}_1), \dots, (\bar{x}_t, \bar{y}_t))$ and $D = ((\bar{x}_{t+1}, \bar{y}_{t+1}), \dots, (\bar{x}_n, \bar{y}_n))$. The set $(P \oplus Q)_{x \geq c}$ is the union of $(A \oplus C)_{x \geq c}$, $(A \oplus D)_{x \geq c}$, $(B \oplus C)_{x \geq c}$, and $(B \oplus D)_{x \geq c}$. Because $\bar{x}_t$ is the largest x-coordinate among all x-coordinates of points in Q_x such that $x_{n/2} + \bar{x}_t < c$, we know that all points in $A \oplus C$ cannot satisfy the constraint $x \geq c$. Hence, we only need to consider points in $A \oplus D$, $B \oplus C$, and $B \oplus D$. Because P_x and Q_x are in nondecreasing order of x-coordinates, it is guaranteed that all points in $B \oplus D$ satisfy the constraint L, i.e., $B \oplus D = (B \oplus D)_{x \geq c}$. Construct in linear time $row\_vector = (r_1, r_2, \dots, r_{n/2})$ which is the y-coordinates in the subsequence of P_y resulting from removing points with x-coordinates no greater than $x_{n/2}$ from P_y. Construct in linear time $column\_vector = (c_1, c_2, \dots, c_{n-t})$ which is the y-coordinates in the subsequence of Q_y resulting from removing points with x-coordinates no greater than $\bar{x}_t$ from Q_y. Note row_vector is the same as the result of sorting B into nondecreasing order of y-coordinates, and $column\_vector$ is the same as the result of sorting D into nondecreasing order of y-coordinates. Thus, we have $\{y : (x, y) \in (B \oplus D)_{x \geq c}\} = \{y : (x, y) \in$

$B \oplus D\} = \{r_i + c_j : 1 \le i \le |row\_vector|, 1 \le j \le |column\_vector|\}$. Therefore, we can store the y-coordinates of $B \oplus D = (B \oplus D)_{x \ge c}$ as a sorted matrix X of dimensions $|row\_vector| \times |column\_vector|$ where the (i,j)-th element of X is $r_i + c_j$. Note that it is not necessary to explicitly construct the sorted matrix X, which needs $\Omega(n^2)$ time. Because the (i,j)-th element of X can be obtained by summing up r_i and c_j, we only need to keep row_vector and $column\_vector$. The rest is to construct the sorted matrices for the y-coordinates of points in $(A \oplus D)_{x \ge c}$ and $(B \oplus C)_{x \ge c}$. It is accomplished by applying the above approach recursively. The pseudocode is shown in Figure 1 and Figure 2. We now analyze the time complexity.

Algorithm ConstructMatrices$(P_x, Q_x, P_y, Q_y, L\colon x \ge c)$

Input: P_x and P_y are the results of sorting the multiset $P \subseteq \mathbb{R}^2$ in nondecreasing order of x-coordinates and y-coordinates, respectively. Q_x and Q_y are the results of sorting the multiset $Q \subseteq \mathbb{R}^2$ in nondecreasing order of x-coordinates and y-coordinates, respectively. A linear constraint $L\colon x \ge c$.

Output: The y-coordinates of points in $(P \oplus Q)_{x \ge c}$ as a collection of sorted matrices.

1    $n' \leftarrow |P_x|$; $m' \leftarrow |Q_x|$.

2   Let $P_x = ((x_1, y_1), \ldots, (x_{n'}, y_{n'}))$, $Q_x = ((\bar{x}_1, \bar{y}_1), \ldots, (\bar{x}_{m'}, \bar{y}_{m'}))$,
     $P_y = ((x'_1, y'_1), \ldots, (x'_{n'}, y'_{n'}))$, and $Q_y = ((\bar{x}'_1, \bar{y}'_1), \ldots, (\bar{x}'_{m'}, \bar{y}'_{m'}))$.

3    $\bar{x}_0 \leftarrow -\infty$.

4   **if** $n' \le 0$ or $m' \le 0$ **then**

5   **return**

6   **if** $n' = 1$ or $m' = 1$ **then**

7   Scan points in $P_x \oplus Q_x$ to find all points satisfying L and construct the sorted
      matrix for y-coordinates of these points.

8   **return** the above sorted matrix.

9   **for** $t \leftarrow m'$ down to 0 **do**

10   **if** $x_{n'/2} + \bar{x}_t < c$ **then**

11        $row\_vector \leftarrow$ subsequence of P_y being removed points with
         x-coordinates $\le x_{n'/2}$.

12        $column\_vector \leftarrow$ subsequence of Q_y being removed points with
         x-coordinates $\le \bar{x}_t$.

13        $A_x \leftarrow P_x[1, n'/2]$; $B_x \leftarrow P_x[n'/2 + 1, n']$; $C_x \leftarrow Q_x[1, t]$; $D_x \leftarrow Q_x[t+1, m']$.

14        $A_y \leftarrow$ subsequence of P_y being removed points with x-coordinates $> x_{n'/2}$.

15        $B_y \leftarrow$ subsequence of P_y being removed points with x-coordinates $\le x_{n'/2}$.

16        $C_y \leftarrow$ subsequence of Q_y being removed points with x-coordinates $> \bar{x}_t$.

17        $D_y \leftarrow$ subsequence of Q_y being removed points with x-coordinates $\le \bar{x}_t$.

18   ConstructMatrices$(A_x, D_x, A_y, D_y, L\colon x \ge c)$.

19   ConstructMatrices$(B_x, C_x, B_y, C_y, L\colon x \ge c)$.

20   **return** row_vector and $column\_vector$.

Fig. 1. The subroutine for the MINKOWSKI SUM SELECTION problem with one linear constraint and a linear objective function

---

Algorithm Selection$_1$(P, Q, L, f, k)
Input: Two multisets $P \subseteq \mathbb{R}^2$ and $Q \subseteq \mathbb{R}^2$; a linear constraint L; a linear objective
  function $f : \mathbb{R}^2 \to \mathbb{R}$; a positive integer k.
Output: The kth largest value among all objective values of points in $(P \oplus Q)_L$.
  1 Perform the input transformation described in Section 3.
  2 Sort P and Q into P_x and Q_x, respectively, in nondecreasing order of x-coordinates.
  3 Sort P and Q into P_y and Q_y, respectively, in nondecreasing order of y-coordinates.
  4 $S \leftarrow$ ConstructMatrices(P_x, Q_x, P_y, Q_y, L).
  5 **return** the kth largest element among the elements of sorted matrices in S.

Fig. 2. The main procedure for the MINKOWSKI SUM SELECTION problem with one linear constraint and a linear objective function

Lemma 1. *Given a matrix $X \in \mathbb{R}^{\mathfrak{N} \times \mathfrak{M}}$, we define the side length of X be $\mathfrak{N} + \mathfrak{M}$. Letting $T(n', m')$ be the running time of ConstructMatrices(P_x, Q_x, P_y, Q_y, L), where $n' = |P_x| = |P_y|$ and $m' = |Q_x| = |Q_y|$, we have $T(n', m') = O((n' + m') \log(n' + 1))$. Similarly, letting $M(n', m')$ be the sum of the side lengths of all sorted matrices created by running ConstructMatrices(P_x, Q_x, P_y, Q_y, L), we have $M(n', m') = O((n' + m') \log(n' + 1))$.*

Proof. It suffices to prove that $T(n', m') = O((n' + m') \log(n' + 1))$. By Algorithm ConstructMatrices in Figure 1, we have $T(n', m') \leq \max_{0 \leq i \leq m'} \{c'(n' + m') + T(n'/2, i) + T(n'/2, m' - i)\}$ for some constant c'. Then by induction on n', it is easy to prove that $T(n', m')$ is $O((n' + m') \log(n' + 1))$. $\square$

Theorem 5. *Given two n-point multisets $P \subseteq \mathbb{R}^2$ and $Q \subseteq \mathbb{R}^2$, a positive integer k, a linear constraint L, and a linear objective function $f : \mathbb{R}^2 \to \mathbb{R}$, Algorithm Selection$_1$ finds the kth largest objective value among all objective values of points in $(P \oplus Q)_L$ in $O(n \log n)$ time. Hence, by Theorem 4, Algorithm Selection$_1$ is optimal.*

Proof. Let $S = \{X_1, \ldots, X_N\}$ be the sorted matrices produced at Step 4 in Algorithm Selection$_1$. Let X_j, $1 \leq j \leq N$, be of dimensions $n_j \times m_j$ where $n_j \geq m_j$. By Lemma 1, we have $O(\sum_{i=1}^{N}(m_i + n_i)) = O(n \log n)$. By Theorem 1, the time required to find the kth largest element from the elements of matrices in S is $O(\sum_{i=1}^{N} m_i \log(2n_i/m_i))$. Since $\sum_{i=1}^{N} m_i \log(2n_i/m_i) \leq \sum_{i=1}^{N} m_i \frac{2n_i}{m_i} \leq \sum_{i=1}^{N} 2(m_i + n_i) = O(n \log n)$, the time for selecting the kth largest element from elements of matrices in S is $O(n \log n)$. Combining this with the time for the input transformation, sorting, and executing ConstructMatrices(P_x, Q_x, P_y, Q_y, L), we conclude that the total running time is $O(n \log n)$. $\square$

Using similar techniques, the following problem can also be solved in $O(n \log n)$ time. Given two n-point multisets $P \subseteq \mathbb{R}^2$ and $Q \subseteq \mathbb{R}^2$, a linear constraint L, a linear objective function $f : \mathbb{R}^2 \to \mathbb{R}$, and a real number t, the problem is to find the rank of t among all objective values of points in $(P \oplus Q)_L$, where the rank of t is equal to the number of elements in $\{y | (x, y) \in (P \oplus Q)_L, y > t\}$ plus one. The

---

Algorithm Ranking$_1$$(P, Q, L, f, t)$
Input: Two multisets $P \subseteq \mathbb{R}^2$ and $Q \subseteq \mathbb{R}^2$; a linear constraint L; a linear objective
    function $f : \mathbb{R}^2 \to \mathbb{R}$; a real number t.
Output: The rank of t among the objective values of points in $(P \oplus Q)_L$.
1 Perform the input transformation in Section 3.
2 Sort P and Q into P_x and Q_x, respectively, in nondecreasing order of x-coordinates.
3 Sort P and Q into P_y and Q_y, respectively, in nondecreasing order of y-coordinates.
4 $S \leftarrow$ ConstructMatrices(P_x, Q_x, P_y, Q_y, L).
5 **return** the rank of t among the elements of sorted matrices in S.

Fig. 3. The ranking algorithm for the Minkowski sum with one linear constraint and
a linear objective function

pseudocode is given in Figure 3. Note that in Algorithm Selection$_1$ and Algorithm
Ranking$_1$, we assume the input constraint is of the form $ax + by \geq c$. After slight
modifications, we can also cope with constraints of the form $ax + by > c$. To
avoid redundancy, we omit the details here. For ease of exposition, we assume
that Algorithm Selection$_1$ and Algorithm Ranking$_1$ are also capable of coping
with constraints of the form $ax + by > c$ in the following sections.

4   Minkowski Sum Selection with Two Constraints

In this section, we show the MINKOWSKI SUM SELECTION problem can be solved
in worst-case $O(n \log^2 n)$ time and expected $O(n \log n)$ time for the case where
two linear constraints are given and the objective function is linear.

Given $P = \{(x_{1,1}, y_{1,1}), \ldots, (x_{1,n}, y_{1,n})\}$, $Q = \{(x_{2,1}, y_{2,1}), \ldots, (x_{2,n}, y_{2,n})\}$, a
positive integer k, two constraints L_1: $a_1 x + b_1 y \geq c_1$ and L_2: $a_2 x + b_2 y \geq c_2$,
and a linear objective function $f(x, y) = dx + ey$, where $a_1, b_1, c_1, a_2, b_2, c_2, d$,
and e are all constants, we perform the following transformation.

1. Change the content of P and Q to $\{(a_1 x_{1,1} + b_1 y_{1,1}, dx_{1,1} + ey_{1,1}), \ldots,$
   $(a_1 x_{1,n} + b_1 y_{1,n}, dx_{1,n} + ey_{1,n})\}$, and $\{(a_1 x_{2,1} + b_1 y_{2,1}, dx_{2,1} + ey_{2,1}), \ldots,$
   $(a_1 x_{2,n} + b_1 y_{2,n}, dx_{2,n} + ey_{2,n})\}$, respectively.
2. Change the constraints from $a_1 x + b_1 y \geq c_1$ and $a_2 x + b_2 y \geq c_2$ to $x \geq c_1$
   and $\frac{a_2 e - b_2 d}{a_1 e - b_1 d} x + \frac{a_1 b_2 - b_1 a_2}{a_1 e - b_1 d} y \geq c_2$, respectively.
3. Change the objective function from $dx + ey$ to y.

This transformation can be done in $O(n)$ time and the answer remains the
same. Hence from now on, our goal becomes to find the kth largest y-coordinate
on the constrained Minkowski sum of P and Q subject to the constraints L_1:
$x \geq c_1$ and L_2: $ax + by \geq c_2$, where $a = \frac{a_2 e - b_2 d}{a_1 e - b_1 d}$ and $b = \frac{a_1 b_2 - b_1 a_2}{a_1 e - b_1 d}$. Note
that if the two constraints and the objective function are parallel, we cannot
use the above transformation. However, if the two constraints are parallel, this
problem can be solved in $O(n \log n)$ time. For the space limitation, we present
the algorithm for this special case in [17].

After applying the above input transformation to our problem instances, there are four possible cases: (1) $a < 0, b < 0$; (2) $a > 0, b > 0$; (3) $a < 0, b > 0$; (4) $a > 0, b < 0$. Note that the two constraints are not parallel implies $b \neq 0$. If $a = 0$, we can solve this case more easily in $O(n \log^2 n)$ time by using the same technique stated later and we omit the details here. In the following discussion we focus on Case (1), and the other three cases can be solved in a similar way.

For simplicity, we assume that n is a power of two, and each point in $(P \oplus Q)_{L_1,L_2}$ has a distinct y-coordinate. Now we are ready to describe our algorithm. First, we sort P and Q into P_y and Q_y, respectively, in nondecreasing order of y-coordinates using $O(n \log n)$ time. Let $P_y = \{(x_1', y_1'), \ldots, (x_n', y_n')\}$ and $Q_y = \{(\bar{x}_1', \bar{y}_1'), \ldots, (\bar{x}_n', \bar{y}_n')\}$. Denote by Y the sorted matrix of dimensions $n \times n$ where the (i, j)-th element is $y_i' + \bar{y}_j'$. We then run a loop where an integer interval $[l, u]$ is maintained such that the solution is within the set $\{$the hth largest element of $Y : l \leq h \leq u\}$. Initially, we set $l = 1$ and $u = n^2$. At the beginning of each iteration, we select the $\frac{u-l+1}{2}$-th largest element t of Y, which can be done in $O(n)$ time by Theorem 1. Let R be the rank of t among the objective values of points in $(P \oplus Q)_{L_1,L_2}$. Then there are three possible cases: (i) $R < k$; (ii) $R = k$; (iii) $R > k$. See Figure 4 for an illustration. If it is Case (i), then we reset l to $\frac{u-l+1}{2}$ and continue the next iteration. If it is Case (ii), then we apply the algorithm for the MINKOWSKI SUM FINDING problem (discussed in Section 5) to find the point $p = (x^*, y^*)$ in $(P \oplus Q)_{L_1,L_2,y \leq t}$ in $O(n \log n)$ time such that y^* is closest to t and return y^*. If it is Case (iii), then we reset u to $\frac{u-l+1}{2}$ and continue the next iteration.

It remains to describe the subroutine for computing R. Let $A = \{(x,y)|x < c_1 \text{ and } ax + by < c_2\}$, $B = \{(x,y)|x < c_1 \text{ and } ax + by \geq c_2 \text{ and } y > t\}$, $C = \{(x,y)|x \geq c_1 \text{ and } ax + by \geq c_2 \text{ and } y > t\}$ and $D = \{(x,y)|x \geq c_1 \text{ and } ax + by < c_2 \text{ and } y > t\}$. See Figure 5 for an illustration. First, we compute the number of points in $(P \oplus Q) \cap (A \cup B)$, say R_1, by calling Ranking$_1(P, Q, L'$: $x < c_1, f'(x,y) = y, t) - 1$. Secondly, we compute the number of points in $(P \oplus Q) \cap (A \cup D)$, say R_2, by calling Ranking$_1(P, Q, L''$: $ax + by < c_2, f'(x,y) = y, t) - 1$. Thirdly, we compute the number of points in $(P \oplus Q) \cap A$, say R_3, by calling Ranking$_1(P, Q, L''$: $ax + by < c_2, f''(x,y) = -x, c_1) - 1$. Finally, we compute the number of points in $(P \oplus Q)_{y>t}$, say R_t. It can be done by applying Theorem 2 to calculate the rank of t among the values of the elements in Y, say R_t', and set R_t to $R_t' - 1$. After getting R_1, R_2, R_3, and R_t, we can compute R by the following equation: $R = R_t - R_1 - R_2 + R_3 + 1$. Since all R_1, R_2, R_3, and R_t can be computed in $O(n \log n)$ time, the time for computing R is $O(n \log n)$. Now let us look at the total time complexity. Since the loop consists of at most $O(\log n)$ iterations and each iteration takes $O(n \log n)$ time, the total time complexity of the loop is $O(n \log^2 n)$. By combining this with the time for the input transformation and sorting, we have the following theorem.

Theorem 6. *Given two n-point multisets $P \subseteq \mathbb{R}^2$ and $Q \subseteq \mathbb{R}^2$, a positive integer k, two linear constraints L_1 and L_2, and a linear objective function $f : \mathbb{R}^2 \to \mathbb{R}$, the kth largest objective value among all objective values of points in $(P \oplus Q)_{L_1,L_2}$ can be found in $O(n \log^2 n)$ time.*

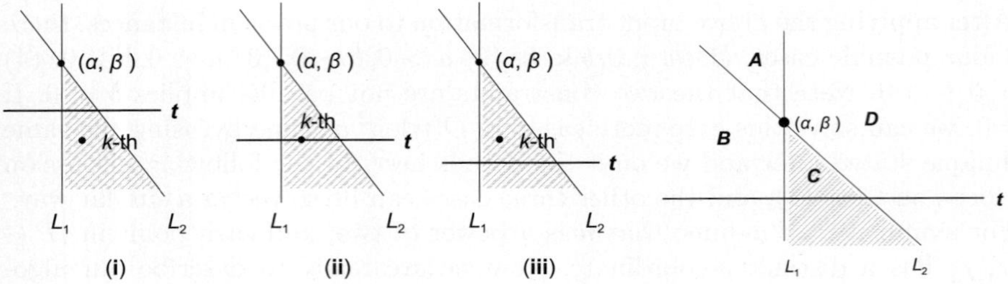

Fig. 4. The three possible cases for a given value t **Fig. 5.** The four regions above (exclusive) the line $y = t$

Theorem 7. *For linear objective functions, the* MINKOWSKI SUM SELECTION *problem with two linear constraints can be solved in expected $O(n \log n)$ time.*

Proof. Due to the space limitation, we leave the proof to [17].   □

Let λ be a fixed integer greater than two. In the following theorem, we summarize our results of the MINKOWSKI SUM SELECTION for the case where λ linear constraints are given and the objective function is linear. Due to the space limitation, we leave the proof to [17].

Theorem 8. *Let λ be any fixed integer larger than two. The* MINKOWSKI SUM SELECTION *problem with λ constraints and a linear objective function is asymptotically equivalent to the* MINKOWSKI SUM SELECTION *problem with two linear constraints and a linear objective function.*

5   Minkowski Sum Finding

In the MINKOWSKI SUM FINDING problem, given two n-point multisets P, Q, a set of λ inequalities $Ar \geq b$, an objective function $f(x, y)$ and a real number δ, we are required to find a point $v^* = (x^*, y^*)$ among all points in $(P \oplus Q)_{Ar \geq b}$ which minimizes $|f(x^*, y^*) - \delta|$. In this section, we show how to cope with an objective function of the form $f(x, y) = ax + by$ or $f(x, y) = \frac{by}{ax}$ based on the algorithms proposed by Bernholt *et al.* [4]. Instead of finding the point $v^* = (x^*, y^*)$, we would like to focus on computing the value of $|f(x^*, y^*) - \delta|$. The point $v^* = (x^*, y^*)$ can be easily constructed from the computed information. Before moving on to the algorithm, let us look at the lower bound of the problem.

Lemma 2. *The* MINKOWSKI SUM FINDING *problem with an objective function of the form $f(x, y) = ax + by$ or $f(x, y) = \frac{by}{ax}$ has a lower bound of $\Omega(n \log n)$ in the algebraic decision tree model.*

Proof. Given two real number sets $A = \{a_1, \ldots, a_n\}$ and $B = \{b_1, \ldots, b_n\}$, the SET DISJOINTNESS problem is to determine whether or not $A \cap B = \emptyset$. It is known the SET DISJOINTNESS problem has a lower bound of $\Omega(n \log n)$ in the algebraic decision tree model [2]. We first prove that the SET DISJOINT-NESS problem is linear-time reducible to the MINKOWSKI SUM FINDING problem with the objective function $f(x, y) = y$. Let $P = \{(1, a_1), \ldots, (1, a_n)\}$, $Q = \{(1, -b_1), \ldots, (1, -b_n)\}$, and (x^*, y^*) be the point in $(P \oplus Q)$ such that y^* is closest to zero. Then $(x^*, y^*) = (2, 0)$ if and only if $A \cap B \neq \emptyset$. Similarly, we can prove that the SET DISJOINTNESS problem is linear-time reducible to the MINKOWSKI SUM FINDING problem with the objective function $f(x, y) = \frac{y}{x}$. Let $P = \{(1, a_1), \ldots, (1, a_n)\}$, $Q = \{(1, -b_1), \ldots, (1, -b_n)\}$, and (x^*, y^*) be the point in $(P \oplus Q)$ such that $\frac{y^*}{x^*}$ is closest to zero. Then $(x^*, y^*) = (2, 0)$ if and only if $A \cap B \neq \emptyset$. $\qquad\square$

Now, let us look at how to cope with a linear objective function $f(x, y) = ax + by$. Without loss of generality, we assume $\delta = 0$; otherwise we may perform some input transformations first. Thus, the goal is to compute the value of $\min\{|ax + by| : (x, y) \in (P \oplus Q)_{Ar \geq b}\}$.

Lemma 3. *Divide the xy-plane into two parts: $D_1 = \{(x, y) : ax + by \geq 0\}$ and $D_2 = \{(x, y) : ax + by < 0\}$. Given two points $v_1 = (x_1, y_1)$ and $v_2 = (x_2, y_2)$ in the same part, let $v_\gamma = (x_\gamma, y_\gamma) = \gamma v_1 + (1 - \gamma) v_2$, where $\gamma \in [0, 1]$. Then we have $|ax_\gamma + by_\gamma| \geq \min(|ax_1 + by_1|, |ax_2 + by_2|)$.*

Proof. It is easy to see the lemma holds if $b = 0$. Without loss of generality, let $x_1 \geq x_2$ and $b \neq 0$. We only prove the case where both v_1 and v_2 are in D_1, and the other case can be proved in a similar way. Now consider the following two situations: (1) $|ax_1 + by_1| \leq |ax_2 + by_2|$ and (2) $|ax_1 + by_1| > |ax_2 + by_2|$. In the first situation, by $|ax_1 + by_1| \leq |ax_2 + by_2|$, $ax_1 + by_1 \geq 0$, and $ax_2 + by_2 \geq 0$, we can derive that $b(y_2 - y_1) \geq a(x_1 - x_2)$. Let $v' = (x', y')$ satisfy $ax_1 + by_1 = ax' + by'$ and $x' = x_\gamma = \gamma x_1 + (1 - \gamma) x_2$. It follows that $y' = \frac{a}{b}(1 - \gamma)(x_1 - x_2) + y_1$. By $y_\gamma = \gamma y_1 + (1 - \gamma) y_2 = (1 - \gamma)(y_2 - y_1) + y_1$, we have $by_\gamma \geq by'$. Thus, $|ax_\gamma + by_\gamma| \geq |ax' + by'| = |ax_1 + by_1|$. In the second situation, $b(y_1 - y_2) > a(x_2 - x_1)$. Let $v'' = (x'', y'')$ satisfy $ax_2 + by_2 = ax'' + by''$ and $x'' = x_\gamma = \gamma x_1 + (1 - \gamma) x_2$. It follows that $y'' = \frac{a}{b}\gamma(x_2 - x_1) + y_2$. By $y_\gamma = \gamma y_1 + (1 - \gamma) y_2 = \gamma(y_1 - y_2) + y_2$, we have $by_\gamma > by''$. Thus, $|ax_\gamma + by_\gamma| > |ax'' + by''| = |ax_2 + by_2|$. Therefore, $|ax_\gamma + by_\gamma| \geq \min(|ax_1 + by_1|, |ax_2 + by_2|)$ if $ax_1 + by_1 \geq 0$ and $ax_2 + by_2 \geq 0$. $\qquad\square$

Let $D_1 = \{(x, y) : ax + by \geq 0\}$ and $D_2 = \{(x, y) : ax + by < 0\}$. Let R_1 be the vertices of the convex hull of $(P \oplus Q)_{Ar \geq b, ax + by \geq 0}$ and R_2 be the vertices of the convex hull of $(P \oplus Q)_{Ar \geq b, ax + by < 0}$. By Theorem 3, we can compute R_1 and R_2 in $O(\lambda \log \lambda + \lambda \cdot n \log n)$ time. Let $sol_1 = \min\{|ax + by| : (x, y) \in (P \oplus Q)_{Ar \geq b} \cap D_1\}$ and $sol_2 = \min\{|ax + by| : (x, y) \in (P \oplus Q)_{Ar \geq b} \cap D_2\}$. By Lemma 3, we have $sol_1 = \min\{|ax + by| : (x, y) \in R_1\}$ and $sol_2 = \min\{|ax + by| : (x, y) \in R_2\}$. Note that both the sizes of R_1 and R_2 are bounded above by $O(\lambda \cdot n)$. Therefore, sol_1 and sol_2 are computable in $O(\lambda \cdot n)$ time by examining all points in R_1 and R_2. Finally, we have the solution is the minimum of sol_1 and sol_2. The total time complexity is $O(\lambda \log \lambda + \lambda \cdot n \log n)$, and we have the following theorem.

470   C.-W. Luo et al.

Theorem 9. *Let λ be any fixed nonnegative integer. The* MINKOWSKI SUM FINDING *problem with λ constraints and a linear objective function can be solved in optimal $O(n \log n)$ time.*

Next, we see how to cope with an objective function of the form $f(x, y) = \frac{by}{ax}$. Without loss of generality, we assume $\delta = 0$ and $a = b = 1$; otherwise we perform some input transformations first. Thus, the goal is to compute the value of $\min\{|\frac{y}{x}| : (x, y) \in (P \oplus Q)_{Ar \geq b}\}$. For technical reasons, we define $\frac{y}{x} = \infty$ if $x = 0$. A function $f : D \to (\mathbb{R} \cup \infty)$ defined on a convex subset D of $\mathbb{R}^2$ is *quasiconcave* if whenever $v_1, v_2 \in D$ and $\gamma \in [0, 1]$ then $f(\gamma \cdot v_1 + (1 - \gamma) \cdot v_2) \geq \min\{f(v_2), f(v_2)\}$.

Lemma 4. *Let $D_1 = \{(x, y) \in \mathbb{R}^2 : x \geq 0, y \geq 0\}$, $D_2 = \{(x, y) \in \mathbb{R}^2 : x \leq 0, y \geq 0\}$, $D_3 = \{(x, y) \in \mathbb{R}^2 : x \leq 0, y \leq 0\}$, and $D_4 = \{(x, y) \in \mathbb{R}^2 : x \geq 0, y \leq 0\}$. Define function $f_i : D_i \to \mathbb{R}$ by letting $f_i(x, y) = |\frac{y}{x}|$ for each $i = 1, 2, 3, 4$. Then we have function f_i is quasiconcave for each $i = 1, 2, 3, 4$.*

Proof. We only prove that f_1 is quasiconcave. The proofs for f_2, f_3, and f_4 can be derived in a similar way. Let $v_1 = (x_1, y_1) \in D_1$, $v_2 = (x_2, y_2) \in D_1$, and $x_1 \geq x_2$. Without loss of generality we may assume $x_1 > 0$ and $v_\gamma \notin \{v_1, v_2\}$. Consider the following two cases.

Case 1: $\frac{y_1}{x_1} \leq \frac{y_2}{x_2}$. Let $v' = (x', y')$ be the point which satisfies $\frac{y_1}{x_1} = \frac{y'}{x'}$ and $x' = x_\gamma = \gamma x_1 + (1 - \gamma) x_2$. By $x_1 > 0$, $x_2 \geq 0$, and $\frac{y_1}{x_1} \leq \frac{y_2}{x_2}$, we have $y_2 \geq \frac{x_2}{x_1} y_1$. It follows that $y' = (\gamma x_1 + (1 - \gamma) x_2) \frac{y_1}{x_1} = \gamma y_1 + (1 - \gamma) \frac{x_2}{x_1} y_1 \leq \gamma y_1 + (1 - \gamma) y_2 = y_\gamma$. By $0 < x' = x_\gamma$ and $0 \leq y' \leq y_\gamma$, we have $|\frac{y_\gamma}{x_\gamma}| = \frac{y_\gamma}{x_\gamma} \geq \frac{y'}{x'} = \frac{y_1}{x_1} = |\frac{y_1}{x_1}| \geq \min\{|\frac{y_1}{x_1}|, |\frac{y_2}{x_2}|\}$.

Case 2: $\frac{y_1}{x_1} > \frac{y_2}{x_2}$. The proof of this case is similar to that of Case 1. As a result, we omit this proof. $\qquad\square$

Let $D_1 = \{(x, y) \in \mathbb{R}^2 : x \geq 0, y \geq 0\}$, $D_2 = \{(x, y) \in \mathbb{R}^2 : x \leq 0, y \geq 0\}$, $D_3 = \{(x, y) \in \mathbb{R}^2 : x \leq 0, y \leq 0\}$, and $D_4 = \{(x, y) \in \mathbb{R}^2 : x \geq 0, y \leq 0\}$. Let R_i be the vertices of the convex hull of $(P \oplus Q)_{Ar \geq b} \cap D_i$ for $i = 1, 2, 3, 4$. By Theorem 3, each R_i is computable in $O(\lambda \log \lambda + \lambda \cdot n \log n)$ time. Let $sol_i = \min\{|\frac{y}{x}| : (x, y) \in (P \oplus Q)_{Ar \geq b} \cap D_i\}$ for each $i = 1, 2, 3, 4$. By Lemma 4, we have $sol_i = \min\{|\frac{y}{x}| : (x, y) \in R_i\}$ for each i. Note that the size of each R_i is bounded above by $O(\lambda + n)$. Therefore, each sol_i is computable in $O(\lambda + n)$ time by examining all points in R_i. Finally, we have the solution is $\min_{i=1}^{4} sol_i$. The total time complexity is $O(\lambda \log \lambda + \lambda \cdot n \log n)$.

Theorem 10. *Let λ be any fixed nonnegative integer. The* MINKOWSKI SUM FINDING *problem with λ constraints and an objective function of the form $f(x, y) = \frac{by}{ax}$ can be solved in optimal $O(n \log n)$ time.*

Acknowledgments

We thank the anonymous referees for helpful suggestions. Cheng-Wei Luo, Hsiao-Fei Liu, Peng-An Chen, and Kun-Mao Chao were supported in part by NSC grants 95-2221-E-002-126-MY3 and NSC 97-2221-E-002-007-MY3 from the National Science Council, Taiwan.

References

1. Allison, L.: Longest Biased Interval and Longest Non-negative Sum Interval. Bioinformatics Application Note 19(10), 1294–1295 (2003)
2. Ben-Or, M.: Lower Bounds for Algebraic Computation Trees. In: Proc. STOC, pp. 80–86 (1983)
3. Bengtsson, F., Chen, J.: Efficient Algorithms for k Maximum Sums. Algorithmica 46(1), 27–41 (2006)
4. Bernholt, T., Eisenbrand, F., Hofmeister, T.: A Geometric Framework for Solving Subsequence Problems in Computational Biology Efficiently. In: SoCG, pp. 310–318 (2007)
5. Chen, K.-Y., Chao, K.-M.: Optimal Algorithms for Locating the Longest and Shortest Segments Satisfying a Sum or an Average Constraint. Information Processing Letter 96(6), 197–201 (2005)
6. Cole, R., Salowe, J.S., Steiger, W.L., Szemeredi, E.: An Optimal-Time Algorithm for Slope Selection. SIAM Journal on Computing 18(4), 792–810 (1989)
7. Frederickson, G.N., Johnson, D.B.: Generalized Selection and Ranking: Sorted Matrices. SIAM Journal on Computing 13(1), 14–30 (1984)
8. Goldwasser, M.H., Kao, M.-Y., Lu, H.-I.: Linear-time Algorithms for Computing Maximum-density Sequence Segments with Bioinformatics Applications. Journal of Computer and System Sciences 70(2), 128–144 (2005)
9. Huang, X.: An Algorithm for Identifying Regions of a DNA Sequence that Satisfy a Content Requirement. Computer Applications in the Biosciences 10(3), 219–225 (1994)
10. Ioshikhes, I., Zhang, M.Q.: Large-Scale Human Promoter Mapping Using CpG Islands. Nature Genetics 26(1), 61–63 (2000)
11. Lee, D.T., Lin, T.-C., Lu, H.-I.: Fast Algorithms for the Density Finding Problem. Algorithmica (2007), doi:10.1007/s00453-007-9023-8
12. Lin, T.-C., Lee, D.T.: Efficient Algorithm for the Sum Selection Problem and k Maximum Sums Problem. In: Asano, T. (ed.) ISAAC 2006. LNCS, vol. 4288, pp. 460–473. Springer, Heidelberg (2006)
13. Lin, T.-C., Lee, D.T.: Randomized Algorithm for the Sum Selection Problem. Theoretical Computer Science 377(1-3), 151–156 (2007)
14. Lin, Y.-L., Jiang, T., Chao, K.-M.: Efficient Algorithms for Locating the Length-constrained Heaviest Segments with Applications to Biomolecular Sequence Analysis. Journal of Computer and System Sciences 65(3), 570–586 (2002)
15. Lin, Y.-L., Huang, X., Jiang, T., Chao, K.-M.: MAVG: Locating Non-overlapping Maximum Average Segments in a Given Sequence. Bioinformatics 19(1), 151–152 (2003)
16. Lipson, D., Aumann, Y., Ben-Dor, A., Linial, N., Yakhini, Z.: Efficient Calculation of Interval Scores for DNA Copy Number Data Analysis. Journal of Computational Biology 13(2), 215–228 (2006)
17. Luo, C.-W., Liu, H.-F., Chen, P.-A., Chao, K.-M.: Minkowski Sum Selection and Finding. CoRR abs/0809.1171 (2008)
18. Matoušek, J.: Randomized optimal algorithm for slope selection. Information Processing Letters 39(4), 183–187 (1991)
19. Ohler, U., Niemann, H., Liao, G.-C., Rubin, G.M.: Joint Modeling of DNA Sequence and Physical Properties to Improve Eukaryotic Promoter Recognition. Bioinformatics, 199–206 (2001)
20. Wang, L., Xu, Y.: SEGID: Identifying Interesting Segments in (Multiple) Sequence Alignments. Bioinformatics 19(2), 297–298 (2003)

Constructing the Simplest Possible Phylogenetic Network from Triplets[*]

Leo van Iersel[1] and Steven Kelk[2]

[1] Department of Mathematics and Computer Science, Technische Universiteit
Eindhoven, P.O. Box 513, 5600 MB Eindhoven, The Netherlands
`l.j.j.v.iersel@tue.nl`
[2] Centrum voor Wiskunde en Informatica (CWI), P.O. Box 94079, 1090 GB
Amsterdam, The Netherlands
`s.m.kelk@cwi.nl`

Abstract. A phylogenetic network is a directed acyclic graph that visualises an evolutionary history containing so-called *reticulations* such as recombinations, hybridisations or lateral gene transfers. Here we consider the construction of a simplest possible phylogenetic network consistent with an input set T, where T contains at least one phylogenetic tree on three leaves (a *triplet*) for each combination of three taxa. To quantify the complexity of a network we consider both the total number of reticulations and the number of reticulations per biconnected component, called the *level* of the network. We give polynomial-time algorithms for constructing a level-1 respectively a level-2 network that contains a minimum number of reticulations and is consistent with T (if such a network exists). In addition, we show that if T is precisely equal to the set of triplets consistent with some network, then we can construct such a network, which minimises both the level and the total number of reticulations, in time $O(|T|^{k+1})$, if k is a fixed upper bound on the level.

1 Introduction

One of the ultimate goals in computational biology is to create methods that can reconstruct evolutionary histories from biological data of currently living organisms. The immense complexity of biological evolution makes this task almost a hopeless one [17]. This has motivated researchers to focus first on the simplest possible pattern of evolution. This least complicated shape of an evolutionary history is the tree-shape. Now that treelike evolution has been extremely well studied, a logical next step is to consider slightly more complicated evolutionary scenarios, gradually extending the complexity that our models can describe. At the same time we also wish to take into account the parsimony principle, which tells us that amongst all equally good explanations of our data, one prefers the simplest one (see e.g. [8]).

For a set of taxa (e.g. species or strains), a phylogenetic tree describes (a hypothesis of) the evolution that these taxa have undergone. The taxa form the

---

[*] Part of this research has been funded by the Dutch BSIK/BRICKS project.

S.-H. Hong, H. Nagamochi, and T. Fukunaga (Eds.): ISAAC 2008, LNCS 5369, pp. 472–483, 2008.

leaves of the tree while the internal vertices represent events of genetic divergence: one incoming branch splits into two (or more) outgoing branches.

Phylogenetic networks form an extension to this model where it is also possible that two branches combine into one new branch. We call such an event a *reticulation*, which can model any kind of non-treelike (also called "reticulate") evolutionary process such as recombination, hybridisation or lateral gene transfer. In addition, reticulations can also be used to display different possible (treelike) evolutions in one figure. In recent years there has emerged enormous interest in phylogenetic networks and their application [3,9,15,17,18].

This model of a phylogenetic network allows for many different degrees of complexity, ranging from networks that are equal, or almost equal, to a tree to complex webs of frequently diverging and recombining lineages. Therefore we consider two different measures for the complexity of a network. The first of these measures is the total number of reticulations in the network. Secondly, we consider the *level* of the network, which is an upper bound on the number of reticulations per non-treelike part (i.e. biconnected component) of the network. In this paper we consider two different approaches for constructing networks that are as simple as possible. The first approach minimises the total number of reticulations for a fixed level (of at most two) and the second approach minimises both the level and the total number of reticulations, but under more heavy restrictions on the input.

Level-k phylogenetic networks were first introduced by Choy et al. [5] and further studied by different authors [11,12,14]. Gusfield et al. gave a biological justification for level-1 networks (which they call "galled trees") [6]. Minimising reticulations has been very well studied in the framework where the input consists of (binary) sequences [6,8,19]. There are also several results known already about the version of the problem where the input consists of a set of trees and the objective is to construct a network that is "consistent" with each of the input trees. Baroni et al. give bounds on the minimum number or reticulations needed to combine two trees [2] and Bordewich et al. showed that it is APX-hard to compute this minimum number exactly [4]. If restricted to level-1 networks the problem becomes polynomial-time solvable even if there are more than two input trees [10].

In this paper we also consider input sets consisting of trees, but restrict ourselves to small trees with three leaves each, called *triplets*, see Fig. 1. Triplets can for example be constructed by existing methods, such as Maximum Parsimony or Maximum Likelihood, that work accurately and fast for small numbers of taxa. Triplet-based methods have also been well-studied. Aho et al. [1] gave a polynomial-time algorithm to construct a tree from triplets if there exists a tree that is consistent with all input triplets. Jansson et al. [13] showed that the same is possible for level-1 networks if the input triplet set is *dense*, i.e. if there is a triplet for any set of three taxa. Van Iersel et al. further extended this result to level-2 networks [11]. From non-dense triplet sets it is NP-hard to construct level-k networks for any $k \geq 1$ [12,13]. From the proof of this result also follows directly that it is NP-hard to find a network consistent with a non-dense triplet

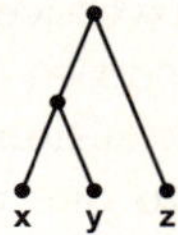

Fig. 1. One of the three possible triplets on the leaves x, y and z. Note that, as with all figures in this article, all arcs are directed downwards.

set that contains a minimum number of reticulations[1]. It is unknown whether this problem becomes easier if the input triplet set is dense.

In the first part of this paper we consider fixed-level networks and aim to minimise the total number of reticulations in these networks. In Sect. 3 we give a polynomial-time algorithm that constructs a level-1 network consistent with a dense triplet set T (if such a network exists) and minimises the total number of reticulations over all such networks. We have implemented MARLON, tested it and made it publicly available [16]. The worst case running time of the algorithm is $O(n^5)$ for n leaves (and hence $O(|T|^{\frac{5}{3}})$ with $|T|$ the input size).

In Sect. 4 we further extend this approach by giving an algorithm that even constructs a level-2 network consistent with a dense triplet set (if one exists) and again minimises the total number of reticulations over all such networks. This means that if the level is at most two, we can minimise both the level and the total number of reticulations, giving priority to the criterion that we find most important. The running time is $O(n^9)$ (and thus $O(|T|^3)$).

Constructing level-k phylogenetic networks becomes even more challenging when the level can be larger than two, even without minimising the total number of reticulations. Given a dense set of triplets, it is a major open problem whether one can construct a minimum level phylogenetic network consistent with these triplets in polynomial time. Moreover, it is not even known whether it is possible to construct a level-3 network consistent with a dense input triplet set in polynomial time. In Sect. 5 of this paper we show some significant progress in this direction. As a first step we consider the restriction to "simple" networks, i.e. networks that contain just one nontrivial biconnected component. We show how to construct, in $O(|T|^{k+1})$ time, a minimum level simple network with level at most k from a dense input triplet set (for fixed k).

Subsequently we show that this can be used to also generate general level-k networks if we put an extra restriction on the quality of the input triplets. Namely, we assume that the input set contains exactly all triplets consistent with some network. If that is the case then our algorithm can find such a network that simultaneously minimises level *and* the total number of reticulations used. The fact that in this case optimal solutions for both measures coincide, is an interesting consequence of the restriction on the input triplets. The algorithm runs in polynomial time $O(|T|^{k+1})$ if the upper bound k on the level of the network is fixed. (For $k = 1, 2$ we can use existing, optimised simple level-1

---

[1] This follows from the proof of Theorem 7 in [13], since only one reticulation is used in their reduction.

and simple level-2 algorithms as subroutines to obtain improved running times of $O(|T|)$ and $O(|T|^{\frac{8}{3}})$ respectively.) This result constitutes an important step forward in the analysis of level-k networks, since it provides the first positive result that can be used for all levels k.

2   Preliminaries

A *phylogenetic network* (*network* for short) is defined as a directed acyclic graph in which exactly one vertex has indegree 0 and outdegree 2 (the root) and all other vertices have either indegree 1 and outdegree 2 (*split vertices*), indegree 2 and outdegree 1 (*reticulation vertices*, or *reticulations* for short) or indegree 1 and outdegree 0 (*leaves*), where the leaves are distinctly labelled. A phylogenetic network without reticulations is called a *phylogenetic tree*.

A directed acyclic graph is *connected* (also called "weakly connected") if there is an undirected path between any two vertices and *biconnected* if it contains no vertex whose removal disconnects the graph. A *biconnected component* of a network is a maximal biconnected subgraph and is called *trivial* if it is equal to two vertices connected by an arc. We call an arc $a = (u, v)$ of a network N a *cut-arc* if its removal disconnects N and call it *trivial* if v is a leaf.

Definition 1. *A network is said to be a* level-k *network if each biconnected component contains at most k reticulations.*

A level-k network that contains no nontrivial cut-arcs and is not a level-$(k-1)$ network is called a *simple* level-k network[2]. Informally, a simple network thus consists of a nontrivial biconnected component with leaves "hanging" of it.

A *triplet* $xy|z$ is a phylogenetic tree on the leaves x, y and z such that the lowest common ancestor of x and y is a proper descendant of the lowest common ancestor of x and z. The triplet $xy|z$ is displayed in Fig. 1. Denote the set of leaves in a network N by L_N. For any set T of triplets define $L(T) = \bigcup_{t \in T} L_t$ and let $n = |L(T)|$. A set T of triplets is called *dense* if for each $\{x, y, z\} \subseteq L(T)$ at least one of $xy|z$, $xz|y$ and $yz|x$ belongs to T. For $L' \subseteq L(T)$, we denote by $T|L'$ the triplets $t \in T$ with $L_t \subseteq L'$. Furthermore, if $\mathcal{C} = \{S_1, \ldots, S_q\}$ is a collection of leaf-sets we use $T\nabla\mathcal{C}$ to denote the *induced* set of triplets $S_i S_j | S_k$ such that there exist $x \in S_i$, $y \in S_j$, $z \in S_k$ with $xy|z \in T$ and i, j and k all distinct.

Definition 2. *A triplet $xy|z$ is* consistent *with a network N (interchangeably: N is consistent with $xy|z$) if N contains a subdivision of $xy|z$, i.e. if N contains vertices $u \neq v$ and pairwise internally vertex-disjoint paths $u \to x$, $u \to y$, $v \to u$ and $v \to z$.*

We say that a cut-arc is a *highest cut-arc* if it is not reachable from any other cut-arc. We call a cycle containing the root a *highest cycle* and a reticulation in such a cycle a *highest reticulation*. We say that a leaf x is *below* an arc (u, v) (and

---

[2] This definition is equivalent to Definition 4 in [11] by Lemma 2 in [11].

below vertex v) if x is reachable from v. In the next section we will frequently use the set $BHR(N)$, which denotes the set of leaves in network N that is below a highest reticulation.

A subset S of the leaves is an *SN-set* (of triplet set T) if there is no triplet $xy|z$ in T with $x, z \in S$, $y \notin S$. An SN-set is called *nontrivial* if it does not contain all leaves. Furthermore, we say that an SN-set S is *maximal* (under restriction X) if there is no nontrivial SN-set (satisfying restriction X) that is a strict superset of S. Any two SN-sets of a dense triplet set T are either disjoint or one is included in the other [14, Lemma 8], which implies that there are at most $2(n-1)$ nontrivial SN-sets. All SN-sets can be found in $O(n^3)$ time [13]. If a network is consistent with T, then the set of leaves S below any cut-arc is always an SN-set, since triplets of the form $xy|z$ with $x, z \in S$, $y \notin S$, are not consistent with such a network. Furthermore, each maximal SN-set is equal to the union of leaves below one or more highest cut-arcs [11].

3   Constructing a Level-1 Network with a Minimum Number of Reticulations

Given a dense set of triplets T, the problem DMRL-k asks for a level-k network consistent with T with a minimum number of reticulations. We propose the following dynamic programming algorithm for solving DMRL-1. The algorithm considers all SN-sets from small to large and computes an optimal solution N_S for each SN-set S, based on the optimal solutions for included SN-sets. The algorithm considers both the case where the root of N_S is contained in a cycle and the case where there are two cut-arcs leaving the root. In the latter case there are two SN-sets S_1 and S_2 that are maximal under the restriction that they are a subset of S. If this is the case then the algorithm constructs a candidate for N_S by creating a root connected to the roots of N_{S_1} and N_{S_2}.

The other possibility is that the root of N_S is contained in some cycle. For this case the algorithm tries each SN-set as $BHR(N_S)$: the set of leaves below the highest reticulation. The sets of leaves below other highest cut-arcs can then be found using the property of optimal level-1 networks outlined in Lemma 1. Subsequently, an induced set of triplets is computed, where each set of leaves below a highest cut-arc is replaced by a single meta-leaf. A candidate network is constructed by computing a simple level-1 network (in $O(n^3)$ time [13]) and replacing each meta-leaf S_i by an optimal network N_{S_i} for the corresponding subset of the leaves. The optimal network N_S is then the network with a minimum number of reticulations over all candidate networks.

A structured description of the computations is in Algorithm 1. We use $f(L')$ to denote the minimum number of reticulations in any level-1 network consistent with $T|L'$. In addition, $g(L', S')$ denotes the minimum number of reticulations in any level-1 network consistent with $T|L'$ with $BHR(N) = S'$. The algorithm first computes the optimal number of reticulations. Then a network with this number of reticulations is constructed using backtracking.

Algorithm 1. Minimum Amount of Reticulation Level One Network

---

1: compute the set SN of SN-sets of T
2: **for** $i = 1 \ldots n$ **do**
3:     **for** each S in SN of cardinality i **do**
4:         **for** each $S' \in SN$ with $S' \subset S$ **do**
5:             let $\mathcal{C}$ contain S' and all SN-sets that are maximal under the restriction that they are a subset of S and do not contain S'
6:             **if** $T \nabla \mathcal{C}$ is consistent with a simple level-1 network **then**
7:                 $g(S, S') := 1 + \sum_{X \in \mathcal{C}} f(X)$
8:             **if** there are exactly two SN-sets $S_1, S_2 \in SN$ that are maximal under the restriction that they are a strict subset of S **then**
9:                 $g(S, \emptyset) := f(S_1) + f(S_2)$ $(\mathcal{C} := \{S_1, S_2\})$
10:         $f(S) := \min g(S, S')$ over all computed values of $g(S, \cdot)$
11:         store the optimal $\mathcal{C}$ and the corresponding simple level-1 network
12: construct an optimal network by backtracking.

The following property of optimal level-1 networks shows that the algorithm computes an optimal solution.

Lemma 1. *If there exists a solution to DMRL-1, then there also exists an optimal solution N, where the sets of leaves below highest cut-arcs equal either (i) $BHR(N)$ and the SN-sets that are maximal under the restriction that they do not contain $BHR(N)$, or (ii) the maximal SN-sets (if $BHR(N) = \emptyset$).*

Proof. If $BHR(N) = \emptyset$ then there are two highest cut-arcs and the sets below them are the maximal SN-sets. Otherwise, the root of N is part of a cycle.

Claim (1). Each maximal SN-set S equals either the set of leaves below a highest cut-arc or the set of leaves below a directed path P ending in the highest reticulation or in one of its parents.

Proof. If S equals the set of leaves below a single highest cut-arc then we are done. From now on assume that S equals the set of leaves below different highest cut-arcs. First observe that no two leaves in S have the root as their lowest common ancestor, since this would imply that *all* leaves are in S, because S is an SN-set. From this follows that all leaves in S are below some directed path P on the highest cycle. First assume that not all leaves reachable from vertices in P are in S. Then there are leaves x, z, y reachable respectively from vertices p_1, p_2, p_3 that are on P (in this order) with $x, y \in S$ and $z \notin S$. But this leads to a contradiction because then the triplet $xy|z$ is not consistent with N, whilst $yz|x$ and $xz|y$ cannot be in T since S is an SN-set. It remains to prove that P ends in either the highest reticulation or in one of its parents. Assume that this is not true, then there exists a vertex v on (the interior of) a path from the last vertex of P to the highest reticulation. Consider some leaf $z \notin S$ reachable from v and some leaves $x, y \in S$ below different highest cut-arcs. Then this again leads to a contradiction because $xy|z$ is not consistent with N. $\square$

First suppose that some maximal SN-set S equals the set of leaves below a directed path P ending in a parent of the highest reticulation. In this case we can modify the network by putting S below a single cut-arc, without increasing the number of reticulations. To be precise, if p and p' are the first and last vertex of P respectively and r is the highest reticulation, then we subdivide the arc entering p by a new vertex v, add a new arc (v, r), remove the arc (p', r) and suppress the resulting vertex with indegree and outdegree both equal to one. It is not too difficult to see that the resulting network is still consistent with T.

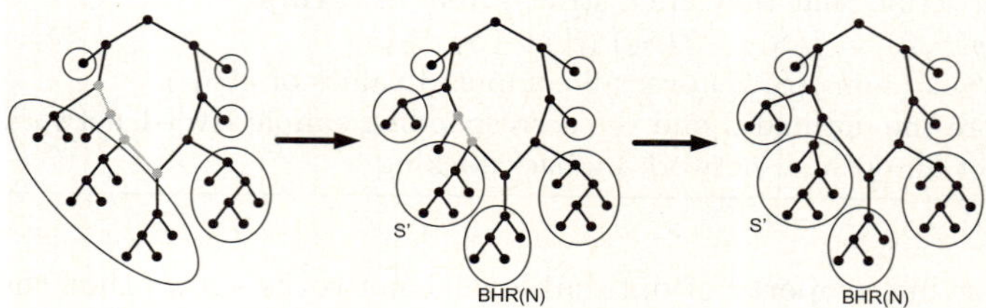

Fig. 2. Visualisation of the proof of Lemma 1. From the maximal SN-sets (encircled in the network on the left) to the sets of leaves below highest cut-arcs (encircled in the network on the right). Remember that all arcs are directed downwards.

Now suppose that some maximal SN-set S equals the set of leaves below a directed path P ending in the highest reticulation. The sets of leaves below highest cut-arcs are all SN-sets (as is always the case). One of them is equal to $BHR(N)$. If any of the others is contained in a nontrivial SN-set S' that does not contain $BHR(N)$, then the procedure from the previous paragraph can again be used to put S' below a highest cut-arc. In the resulting network the sets of leaves below highest cut-arcs are indeed equal to $BHR(N)$ and the SN-sets that are maximal under the restriction that they do not contain $BHR(N)$.

An example is given in Fig. 2. In the network on the left one maximal SN-set equals the set of leaves below the grey path. In the middle is the same network, but now we encircled $BHR(N)$ and the SN-sets that are maximal under the restriction that they do not contain $BHR(N)$. There is still an SN-set (S') below a path on the cycle (again in grey). However, in this case the network can be modified by putting S' below a single cut-arc, without increasing the number of reticulations. This gives the network to the right, where the sets of leaves below highest cut-arcs are indeed equal to $BHR(N)$ and the SN-sets that are maximal under the restriction that they do not contain $BHR(N)$.   □

Theorem 1. *Given a dense set of triplets T, algorithm MARLON constructs a level-1 network that is consistent with T (if such a network exists) and has a minimum number of reticulations in $O(n^5)$ time.*

4 Constructing a Level-2 Network with a Minimum Number of Reticulations

This section extends the approach from Sect. 3 to level-2 networks. We describe how one can find a level-2 network consistent with a dense input triplet set containing a minimum number of reticulations, or decide that such a network does not exist. Details have been omitted due to space constraints.

The general structure of the algorithm is the same as in the level-1 case. We loop though all SN-sets S from small to large and compute an optimal solution N_S for that SN-set, based on previously computed optimal solutions for included SN-sets. For each SN-set we still consider, like in the level-1 case, the possibility that there are two cut-arcs leaving the root of N_S and the possibility that this root is in a biconnected component with one reticulation. However, now we also consider a third possibility, that the root of N_S is in a biconnected component containing two reticulations.

In the construction of biconnected components with two reticulations, we use the notion of "non-cycle-reachable"-arc, or n.c.r.-arc for short, introduced in [12]. We call an arc $a = (u, v)$ an *n.c.r.-arc* if v is not reachable from any vertex in a cycle. These n.c.r.-arcs will be used to combine networks without increasing the network level. In addition, we use the notion *highest biconnected component* to denote the biconnected component containing the root.

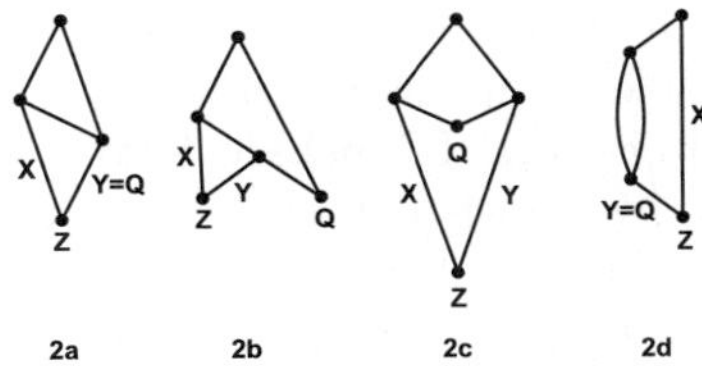

Fig. 3. The four possible structures of a biconnected component containing two reticulations

To get an intuition of the approach, consider the four possible structures of a biconnected component containing two reticulations displayed in Fig. 3. Let X, Y, Z and Q be the sets of leaves indicated in the graph that displays the form of the highest biconnected component of N_S. Observe that after removing Z in each case X, Y and Q become a set of leaves below a cut-arc and hence an SN-set (w.r.t $T|(S \setminus Z)$). In cases 2a, 2b and 2c the highest biconnected component becomes a cycle, Q the set of leaves below the highest reticulation and X and Y sets of leaves below highest cut-arcs. We will first describe the approach for these cases and show later how a similar technique is possible for case 2d.

Our algorithm loops through all SN-sets that are a subset of S and will hence at some iteration consider the SN-set Z. The algorithm removes the set Z and computes the SN-sets of $T|(S \setminus Z)$. The sets of leaves below highest cut-arcs (in some optimal solution, if one exists) are now equal to X, Y, Q

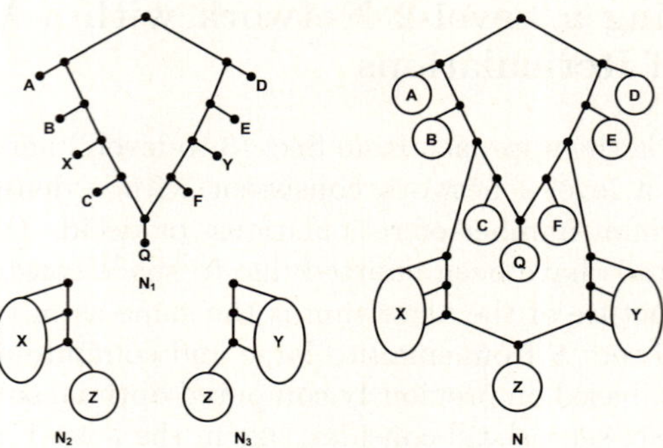

Fig. 4. Example of the construction of network N from N_1, N_2 and N_3

and the SN-sets that are maximal under the restriction that they do not contain X, Y or Q (by the same arguments as in the proof of Lemma 1). Therefore, the algorithm tries each possible SN-set for X, Y and Q and in one of these iterations it will correctly determine the sets of leaves below highest cut-arcs. Then the algorithm computes the induced set of triplets, where each set of leaves below a highest cut-arc is replaced by a single meta-leaf. All simple level-1 networks consistent with this induced set of triplets are obtained by the algorithm in [13]. Our algorithm loops through all these networks and does the following for each simple level-1 network N_1. Each meta-leaf V, not equal to X or Y, is replaced by an optimal network N_V, which has been computed in a previous iteration. To include leaves in Z, X and Y, we compute an optimal network N_2 consistent with $T|(X \cup Z)$ and an optimal network N_3 consistent with $T|(Y \cup Z)$ where in both networks Z is the set of leaves below an n.c.r.-arc. Then we combine these three networks into a single network like in Fig. 4. A new reticulation is created and Z becomes the set of leaves below this reticulation. Finally, we check for each constructed network whether it is consistent with $T|S$. The network with the minimum number of reticulations over all such networks is the optimal solution N_S for this SN-set.

Now consider case 2d. Suppose we remove Z and replace X, Y $(=Q)$ and each SN-set of $T|(S \setminus Z)$ that is maximal under the restriction that it does not contain X or Y by a single leaf. Then the resulting network consists of a path ending in a simple level-1 network, with X a child of the root and Q the child of the reticulation; and each vertex of the path has a leaf as child. Such a network can easily be constructed and subsequently one can use the same approach as in cases 2a, 2b and 2c.

Theorem 2. *Given a dense triplet set T, Algorithm MARLTN constructs a level-2 network consistent with T (if such a network exists) that has a minimum number of reticulations in $O(n^9)$ time.*

5   Constructing Networks Consistent with Precisely the Input Triplet Set

In this section we consider the problem MIN-REFLECT-k. Given a triplet set T, this problem asks for a level-k network N that is consistent with precisely those triplets in T (if such a network exists) and amongst all such solutions minimises both the level and number of reticulations used. We will show that this problem is polynomial-time solvable for each fixed k.

Given a network N let $T(N)$ denote the set of all triplets consistent with N. We say that a network N *reflects* a triplet set T if $T(N) = T$. If, for a triplet set T, there exists a network N that reflects it, we say that T is *reflective*. Note that, if N reflects T, that N is in general not uniquely defined by T. There are, for example, several distinct simple level-2 networks that reflect the triplet set $\{xy|z, xz|y, zy|x\}$.

Problem: MIN-REFLECT-k
Input: set of triplets T.
Output: level-k network N that reflects T (if such a network exists) and, ranging over all such networks, minimises both the level and the number of reticulations.

This problem might at first glance seem strangely formulated because, in general, minimising level and minimising number of reticulations are two distinct optimisation criteria. However, in the case of reflectivity it turns out that any solution that minimises the number of reticulations also minimises level.

Theorem 3. *Given a dense set of triplets T, it is possible to construct all simple level-k networks consistent with T in time $O(|T|^{k+1})$.*

We note that it is already known how to generate all simple level-1 networks consistent with T in time $O(|T|)$ [13] and all simple level-2 networks consistent with T in time $O(|T|^{\frac{8}{3}})$ [11].

Lemma 2. *Let N be any simple network. Then all the nontrivial SN-sets of $T(N)$ are singletons.*

The high-level idea behind solving MIN-REFLECT-k is that, if a network N reflects a triplet set T, the sets of leaves below highest cut-arcs are in 1:1 correlation with the maximal SN-sets of T. This is a consequence of Lemma 2. We thus use the maximal SN-sets to induce a new triplet set T', which must be reflected by some simple level-ℓ network, with $\ell \leq k$. Using Theorem 3 we can easily find such a (minimum level) simple network: if we can generate all networks *consistent* with a triplet set T, it is easy to identify which of those *reflect* T. This leads ultimately to the algorithm MINPITS (MINimum network consistent with Precisely the Input Triplet Set). The key to understanding why MINPITS produces solutions that simultaneously minimise level and the number of reticulations, lies in the fact that in any two networks N and N' that reflect T, the leaf partition induced by the highest cut-arcs of N, and the leaf partition induced by the highest cut-arcs of N', are identical. It follows that the only way in which N and N' can differ from each other, lies in the choice of simple network at each recursive step of the algorithm. But we always choose the simple network of minimum level, and it is not too difficult to see that this leads to the global minimisation of both level and the number of reticulations. Proofs and algorithms have been omitted due to space restraints.

Theorem 4. *Problem MIN-REFLECT-k can be solved in time* $O(|T|^{k+1})$, *for any fixed* k.

For $k = 1, 2$ we can actually do slightly better: running time $O(|T|)$ and $O(|T|^{\frac{8}{3}})$ respectively.

6   Conclusions and Open Questions

In this article we have shown that, for level 1 and 2, constructing a phylogenetic network consistent with a dense set of triplets that minimises the number of reticulations, is polynomial-time solvable. We feel that, given the widespread use of the principle of parsimony within phylogenetics, this is an important development, and testing on simulated data has yielded promising results. However, the complexity of finding a *feasible* solution for level-3 and higher, let alone a minimum solution, remains unknown, and this obviously requires attention. Perhaps the feasibility and minimisation variants diverge in complexity for higher k. It would be fascinating to explore this.

We have also shown, for *every* fixed k, how to generate in polynomial time all simple level-k networks consistent with a dense set of triplets. This could be an important step towards determining whether the aforementioned feasibility question is tractable for every fixed k. We have used this algorithm to show how MIN-REFLECT-k is polynomial-time solvable for fixed k. Clearly the demand that a set of triplets is exactly equal to the set of triplets in some network is an extremely strong restriction on the input. However, for small networks and high accuracy triplets such an assumption might indeed be valid, and thus of practical use. In any case, the concept of reflection is likely to have a role in future work on "support" for edges in phylogenetic networks generated via triplets. Also, the complexity of some fundamental questions like "does *any* network N reflect T?" remains unclear.

Acknowledgements

We thank Judith Keijsper, Matthias Mnich and Leen Stougie for many helpful discussions during the writing of the paper.

References

1. Aho, A.V., Sagiv, Y., Szymanski, T.G., Ullman, J.D.: Inferring a Tree from Lowest Common Ancestors with an Application to the Optimization of Relational Expressions. SIAM Journal on Computing 10(3), 405–421 (1981)

2. Baroni, M., Grünewald, S., Moulton, V., Semple, C.: Bounding the Number of Hybridisation Events for a Consistent Evolutionary History. Mathematical Biology 51, 171–182 (2005)
3. Baroni, M., Semple, C., Steel, M.: A Framework for Representing Reticulate Evolution. Annals of Combinatorics 8, 391–408 (2004)
4. Bordewich, M., Semple, C.: Computing the Minimum Number of Hybridization Events for a Consistent Evolutionary History. Discrete Applied Mathematics 155(8), 914–928 (2007)
5. Choy, C., Jansson, J., Sadakane, K., Sung, W.-K.: Computing the Maximum Agreement of Phylogenetic Networks. Theoretical Computer Science 335(1), 93–107 (2005)
6. Gusfield, D., Eddhu, S., Langley, C.: Optimal, Efficient Reconstructing of Phylogenetic Networks with Constrained Recombination. Journal of Bioinformatics and Computational Biology 2(1), 173–213 (2004)
7. He, Y.-J., Huynh, T.N.D., Jansson, J., Sung, W.-K.: Inferring Phylogenetic Relationships Avoiding Forbidden Rooted Triplets. Journal of Bioinformatics and Computational Biology 4(1), 59–74 (2006)
8. Hein, J.: Reconstructing Evolution of Sequences Subject to Recombination Using Parsimony. Mathematical Biosciences 98, 185–200 (1990)
9. Huson, D.H., Bryant, D.: Application of Phylogenetic Networks in Evolutionary Studies. Molecular Biology and Evolution 23(2), 254–267 (2006)
10. Huynh, T.N.D., Jansson, J., Nguyen, N.B., Sung, W.-K.: Constructing a Smallest Refining Galled Phylogenetic Network. In: Miyano, S., Mesirov, J., Kasif, S., Istrail, S., Pevzner, P.A., Waterman, M. (eds.) RECOMB 2005. LNCS (LNBI), vol. 3500, pp. 265–280. Springer, Heidelberg (2005)
11. van Iersel, L.J.J., Keijsper, J.C.M., Kelk, S.M., Stougie, L., Hagen, F., Boekhout, T.: Constructing Level-2 Phylogenetic Networks from Triplets. In: Vingron, M., Wong, L. (eds.) RECOMB 2008. LNCS (LNBI), vol. 4955, pp. 450–462. Springer, Heidelberg (2008)
12. van Iersel, L.J.J., Kelk, S.M., Mnich, M.: Uniqueness, Intractability and Exact Algorithms: Reflections on Level-k Phylogenetic Networks, arXiv:0712.2932v3 [q-bio.PE] (2008)
13. Jansson, J., Nguyen, N.B., Sung, W.-K.: Algorithms for Combining Rooted Triplets into a Galled Phylogenetic Network. SIAM Journal on Computing 35(5), 1098–1121 (2006)
14. Jansson, J., Sung, W.-K.: Inferring a Level-1 Phylogenetic Network from a Dense Set of Rooted Triplets. Theoretical Computer Science 363, 60–68 (2006)
15. Makarenkov, V., Kevorkov, D., Legendre, P.: Phylogenetic Network Reconstruction Approaches. In: Applied Mycology and Biotechnology. International Elsevier Series 6, Bioinformatics, pp. 61–97 (2006)
16. MARLON: Constructing Level One Phylogenetic Networks with a Minimum Amount of Reticulation, http://homepages.cwi.nl/~kelk/marlon.html
17. Morrison, D.A.: Networks in Phylogenetic Analysis: New Tools for Population Biology. International Journal for Parasitology 35(5), 567–582 (2005)
18. Moret, B.M.E., Nakhleh, L., Warnow, T., Linder, C.R., Tholse, A., Padolina, A., Sun, J., Timme, R.: Phylogenetic Networks: Modeling, Reconstructibility, and Accuracy. IEEE/ACM Transactions on Computational Biology and Bioinformatics 1(1), 13–23 (2004)
19. Song, Y.S., Hein, J.: On the Minimum Number of Recombination Events in the Evolutionary History of DNA Sequences. Journal of Mathematical Biology 48, 160–186 (2004)

New Results on Optimizing
Rooted Triplets Consistency

Jaroslaw Byrka[1], Sylvain Guillemot[2], and Jesper Jansson[3]

[1] Centrum Wiskunde & Informatica (CWI), Kruislaan 413,
NL-1098 SJ Amsterdam, Netherlands and Eindhoven University of Technology,
P.O. Box 513, 5600 MB Eindhoven, Netherlands
J.Byrka@cwi.nl
[2] INRIA Lille - Nord Europe, 40 avenue Halley, Bât.A, Park Plaza,
59650 Villeneuve d'Ascq, France
Sylvain.Guillemot@lifl.fr
[3] Ochanomizu University, 2-1-1 Otsuka, Bunkyo-ku, Tokyo 112-8610, Japan
Jesper.Jansson@ocha.ac.jp

Abstract. A set of phylogenetic trees with overlapping leaf sets is *consistent* if it can be merged without conflicts into a supertree. In this paper, we study the polynomial-time approximability of two related optimization problems called *the maximum rooted triplets consistency problem* (MAXRTC) and *the minimum rooted triplets inconsistency problem* (MINRTI) in which the input is a set $\mathcal{R}$ of rooted triplets, and where the objectives are to find a largest cardinality subset of $\mathcal{R}$ which is consistent and a smallest cardinality subset of $\mathcal{R}$ whose removal from $\mathcal{R}$ results in a consistent set, respectively. We first show that a simple modification to Wu's `Best-Pair-Merge-First` heuristic [25] results in a bottom-up-based 3-approximation for MAXRTC. We then demonstrate how any approximation algorithm for MINRTI could be used to approximate MAXRTC, and thus obtain the first polynomial-time approximation algorithm for MAXRTC with approximation ratio smaller than 3. Next, we prove that for a set of rooted triplets generated under a uniform random model, the maximum fraction of triplets which can be consistent with any tree is approximately one third, and then provide a deterministic construction of a triplet set having a similar property which is subsequently used to prove that both MAXRTC and MINRTI are NP-hard even if restricted to minimally dense instances. Finally, we prove that MINRTI cannot be approximated within a ratio of $\Omega(\log n)$ in polynomial time, unless P = NP.

1 Introduction

A *supertree method* is a method for merging an input collection of phylogenetic trees on overlapping sets of taxa into a single phylogenetic tree called a *supertree*. An input collection of trees might contain contradictory branching structure, e.g., due to errors in experimental data or because the data originates from different genes, so ideally, a supertree method should merge the input trees

S.-H. Hong, H. Nagamochi, and T. Fukunaga (Eds.): ISAAC 2008, LNCS 5369, pp. 484–495, 2008.

while keeping as much of the branching information as possible. In this paper, we investigate the computational complexity of some combinatorial problems at the core of rooted supertree methods which involve *rooted triplets*.

1.1 Problem Definitions and Notation

A *phylogenetic tree* is a rooted, unordered, distinctly leaf-labeled tree in which every internal node has at least two children, and a *rooted triplet* is a binary phylogenetic tree with exactly three leaves. From here on, each leaf in a phylogenetic tree is identified with its label. The unique rooted triplet on leaf set $\{x, y, z\}$ where the lowest common ancestor (lca) of x and y is a proper descendant of the lca of x and z (or equivalently, where the lca of x and y is a proper descendant of the lca of y and z) is denoted by $xy|z$. If $xy|z$ is an embedded subtree of a tree T, i.e., if the lca of x and y is a proper descendant of the lca of x and z in T, then $xy|z$ and T are said to be *consistent* with each other; otherwise, $xy|z$ and T are *inconsistent*. A set $\mathcal{R}$ of rooted triplets is *consistent* if there exists a phylogenetic tree T such that every $xy|z \in \mathcal{R}$ is consistent with T. The set of all rooted triplets consistent with a tree T is denoted by $rt(T)$.

Let L be a set of leaf labels. A set $\mathcal{R}$ of rooted triplets over L is called *dense* if it contains at least one rooted triplet labeled by L' for every subset L' of L of cardinality three, and *simple* if it contains at most one rooted triplet for each such subset. $\mathcal{R}$ is *minimally dense* if it is both dense and simple.

Now, we define the three problems RTC, MAXRTC, and MINRTI. For any phylogenetic tree T over a leaf set L and a set $\mathcal{R}$ of rooted triplets over L, define $C(\mathcal{R}, T) = |\mathcal{R} \cap rt(T)|$ and $I(\mathcal{R}, T) = |\mathcal{R} \setminus rt(T)|$ (the number of rooted triplets in $\mathcal{R}$ which are consistent and inconsistent with T, respectively). *The rooted triplets consistency problem* (RTC) is: Given a set $\mathcal{R}$ of rooted triplets with leaf set L, output a phylogenetic tree leaf-labeled by L which is consistent with *every* rooted triplet in $\mathcal{R}$, if one exists; otherwise, output *null*. *The maximum rooted triplets consistency problem* (MAXRTC) is: Given a set $\mathcal{R}$ of rooted triplets with leaf set L, output a phylogenetic tree T leaf-labeled by L which maximizes $C(\mathcal{R}, T)$. *The minimum rooted triplets inconsistency problem* (MINRTI) is: Given a set $\mathcal{R}$ of rooted triplets with leaf set L, output a phylogenetic tree T leaf-labeled by L which minimizes $I(\mathcal{R}, T)$. The optima for MAXRTC and MINRTI on an instance $\mathcal{R}$ are denoted by $C(\mathcal{R})$ and $I(\mathcal{R})$, respectively.

An algorithm $\mathcal{A}$ for MAXRTC is an *α-approximation algorithm* (and the *approximation ratio of $\mathcal{A}$ is at most α*) if, for every input $\mathcal{R}$, the tree output by $\mathcal{A}$ is consistent with at least $\frac{C(\mathcal{R})}{\alpha}$ of the rooted triplets in $\mathcal{R}$. Analogously, an algorithm $\mathcal{B}$ for MINRTI is a *β-approximation algorithm* (and the *approximation ratio of $\mathcal{B}$ is at most β*) if, for every input $\mathcal{R}$, the tree output by $\mathcal{B}$ is inconsistent with at most $I(\mathcal{R}) \cdot \beta$ of the rooted triplets in $\mathcal{R}$. An exact algorithm for either of MAXRTC or MINRTI automatically yields an exact algorithm for the other, but approximation ratios are not preserved, as will be demonstrated in Section 6.

Denote $n = |L|$ and $k = |\mathcal{R}|$ in the problem definitions above. (Thus, $k = O(n^3)$.) Consider a set L and a total order $>$ on L. For any non-negative integer q, let $[L]^q$ be the set of tuples $(x_1, ..., x_q) \in L^q$ with $x_1 > ... > x_q$, and let $\langle L \rangle^q$ be the set of tuples $(x_1, ..., x_q) \in L^q$ having pairwise distinct coordinates. We will alternatively view a simple triplet set $\mathcal{R}$ on L as a partial function $\mathcal{R} : \langle L \rangle^3 \to \mathbb{Z}_3$ such that for each distinct $x_0, x_1, x_2 \in L$, it holds that $\mathcal{R}(x_0, x_1, x_2) = i$ if and only if $x_{i+1}x_{i+2}|x_i \in \mathcal{R}$. Note that $\mathcal{R}$ is fully specified by its restriction to $[L]^3$.

1.2 New Results and Organization of the Paper

We first give a survey of existing results in Section 2. Then, in Section 3, we prove that a simple modification to Wu's `Best-Pair-Merge-First` heuristic [25] turns it into an approximation algorithm for MaxRTC with approximation ratio at most 3. In Section 4, we show how any approximation algorithm for MinRTI could be employed to approximate MaxRTC, and use this result to obtain the first polynomial-time approximation algorithm for MaxRTC with approximation ratio smaller than 3. In Section 5, we show that for a set of rooted triplets generated under a uniform random model, the maximum fraction of triplets which can be consistent with any tree is approximately $\frac{1}{3}$, and provide a deterministic construction of a triplet set having a similar property. Section 6 proves that MaxRTC and MinRTI are NP-hard even if restricted to minimally dense instances, which is a strengthening of the result in [14]. Section 6 also proves that (unrestricted) MinRTI cannot be approximated within a ratio of $\Omega(\log n)$ in polynomial time, unless P = NP. Finally, Section 7 discusses open problems.

2 Previous Results

This section lists known results concerning the computational complexity of RTC and MaxRTC. To our knowledge, MinRTI has not been studied before.

RTC: Aho *et al.* [1] introduced RTC and gave a recursive top-down $O(kn)$-time algorithm for the problem. It uses a so-called *auxiliary graph*, whose edges are defined by $\mathcal{R}$, to partition the current leaves into *blocks* in such a way that each block consists of all leaves which are in one subtree of the current root, and then recurses on each block[1]. Henzinger *et al.* [11] reduced the algorithm's complexity to $\min\{O(n + kn^{1/2}), O(k + n^2 \log n)\}$ time and $O(n + k \log^3 n)$ expected time by employing dynamic data structures for keeping track of the connected components in the auxiliary graph under batches of edge deletions. By replacing the dynamic graph connectivity data structures with newer ones, such as the data structure by Holm *et al.* [12], the running time of the algorithm of Aho *et al.* can immediately be further improved to $\min\{O(n + k \log^2 n), O(k + n^2 \log n)\}$ [17].

Hardness of MaxRTC: MaxRTC was proved to be NP-hard independently in [4], [15], and [25]. [5] recently observed that the reductions in [4] and [25] are in

---

[1] For any $L' \subseteq L$, the auxiliary graph $\mathcal{G}(\mathcal{R}, L')$ is the undirected graph $\mathcal{G}(\mathcal{R}, L') = (L', E)$, where E contains edge $\{x, y\}$ if and only if there is some $xy|z$ in $\mathcal{R}$ with $x, y, z \in L'$. Then, each connected component of $\mathcal{G}(\mathcal{R}, L')$ induces a block of L'. During execution, if any auxiliary graph having more than one vertex consists of just one connected component then the algorithm returns *null* and terminates.

fact L-reductions from an APX-hard problem, and hence that the general (non-dense) case of MAXRTC is APX-hard. [14] modified the reductions of [4] and [25] to prove that MAXRTC remains NP-hard even if restricted to dense inputs.

Exact algorithm for MaxRTC: Wu [25] gave an exact, dynamic-programming algorithm for MAXRTC. It runs in $O((k + n^2)3^n)$ time and $O(2^n)$ space.

Approximation algorithms for MaxRTC: The first polynomial-time approximation algorithms for MAXRTC, henceforth referred to as `One-Leaf-Split` and `Min-Cut-Split`, were proposed by Gąsieniec *et al.* in [9]. Both algorithms are greedy, top-down algorithms. `One-Leaf-Split` achieves a constant ratio approximation of MAXRTC; more precisely, it runs in $O((k + n) \log n)$ time and constructs a caterpillar tree which is guaranteed to be consistent with at least one third of the input rooted triplets. On the other hand, `Min-Cut-Split` proceeds exactly as the algorithm of Aho *et al.* [1] with two modifications: (1) the auxiliary graphs are edge-weighed; and (2) if an auxiliary graph has more than one vertex but only one connected component then instead of giving up, `Min-Cut-Split` will find a minimum weight edge cut in the auxiliary graph, delete those edges, and continue[2]. Since deleting an edge from an auxiliary graph corresponds to deleting one or more rooted triplets from $\mathcal{R}$ and since there are at most $n - 2$ recursion levels containing non-trivial auxiliary graphs in the algorithm of Aho *et al.*, it follows that if W denotes the total weight of the input rooted triplets and t the minimum total weight of triplets to remove to achieve consistency then `Min-Cut-Split` constructs a tree which is consistent with a subset of $\mathcal{R}$ whose total weight is $\geq W - (n - 2)t$. This also implies that `Min-Cut-Split` yields an $(n - 2)$-approximation algorithm for MINRTI. `Min-Cut-Split` can be implemented to run in $\min\{O(kn^2 + n^3 \log n), O(n^4)\}$ time.

Snir and Rao [24] presented a greedy, top-down, polynomial-time heuristic for MAXRTC called `MXC` which resembles `Min-Cut-Split`. The difference is that `MXC` augments the auxiliary graphs with extra edges, and whenever the algorithm of Aho *et al.* is stuck with a single connected component, instead of taking a minimum weight edge cut, `MXC` tries to find a cut that *maximizes* the ratio between the extra edges and the ordinary edges. Although the worst-case approximation ratio of `MXC` is unknown, it appears to perform very well on real data [24].

Wu [25] gave a greedy, bottom-up, polynomial-time heuristic for MAXRTC named `Best-Pair-Merge-First` which is structurally similar to the well-known UPGMA/WPGMA and Neighbor-Joining methods, described in detail in, e.g., [7]. It starts with singleton sets, each containing a single leaf label, and repeatedly merges two sets until all leaf labels are in the same set; whenever two sets A and B are merged, a new internal node is created that represents the merged set and whose two children are the (already existing) nodes representing A and B. A special scoring function determines which pair of sets to merge at each step. `Best-Pair-Merge-First` does the above six times (using six different scoring

---

[2] Semple and Steel [23] later independently developed a heuristic for merging a set of phylogenetic trees with overlapping leaf sets that uses a very similar idea, and Page [22] further modified the heuristic of Semple and Steel.

functions) and returns the best solution among those six. No theoretical analysis of the worst-case performance of `Best-Pair-Merge-First` was provided in [25], but Wu demonstrated by extensive simulations that this heuristic performs well in practice (source code in C is available from the author's webpage).

A PTAS for MaxRTC restricted to dense inputs, based on the work of [20] for the analogous unrooted (and more difficult) problem, was outlined in [16].

Miscellaneous related results: Other problems related to RTC/MaxRTC have been studied in the literature. Ng and Wormald [21] showed how to efficiently construct *all* solutions to RTC for any input set of rooted triplets. Gąsieniec *et al.* [8] considered RTC and MaxRTC for *ordered* trees. He *et al.* [10] gave algorithms for a variant of RTC/MaxRTC called *the forbidden rooted triplets consistency problem* in which the input consists of a "good" set and a "bad" set of rooted triplets, and the objective is to construct a tree which is consistent with all of the rooted triplets in the good set and none of the rooted triplets in the bad set. Recently, extensions of RTC/MaxRTC to *phylogenetic networks* (generalizations of phylogenetic trees in which certain nodes are allowed to have more than one parent) have been studied in [5,13,14,18,19].

3   A Bottom-Up 3-Approximation Algorithm for MaxRTC

Here, we modify Wu's `Best-Pair-Merge-First` heuristic [25] to achieve an approximation ratio of at most 3. Although MaxRTC already admits a polynomial-time 3-approximation by `One-Leaf-Split` (see Section 2), our new result is

---

Algorithm `Modified-BPMF`

Input: A set $\mathcal{R}$ of rooted triplets on a leaf set $L = \{\ell_1, \ell_2, \ldots, \ell_n\}$.

Output: A tree with leaf set L consistent with at least one third of the rooted triplets in $\mathcal{R}$.

1. Construct the set $\mathcal{S} = \{S_1, S_2, \ldots, S_n\}$, where each S_i is a tree consisting of a leaf labeled by ℓ_i.
2. Repeat $n - 1$ times:
   (a) For every $S_i, S_j \in \mathcal{S}$, reset $score(S_i, S_j) := 0$.
   (b) For every $xy|z \in \mathcal{R}$ such that $x \in S_i$, $y \in S_j$, and $z \in S_k$ for three different trees S_i, S_j, S_k, update $score$ as follows:
       $score(S_i, S_j) := score(S_i, S_j) + 2;$
       $score(S_i, S_k) := score(S_i, S_k) - 1;$
       $score(S_j, S_k) := score(S_j, S_k) - 1.$
   (c) Select $S_i, S_j \in \mathcal{S}$ such that $score(S_i, S_j)$ is maximum.
   (d) Create a tree S_k by connecting a new root node to the roots of S_i and S_j.
   (e) $\mathcal{S} := \mathcal{S} \cup \{S_k\} \setminus \{S_i, S_j\}$.
3. Return the tree in $\mathcal{S}$.

Fig. 1. Algorithm `Modified-BPMF`

significant because `Best-Pair-Merge-First` outperforms `One-Leaf-Split` in practice [25] and Wu left it as an open problem to derive its approximation ratio. Also, `Best-Pair-Merge-First` uses a bottom-up approach while `One-Leaf-Split` works top-down, and future work may try to incorporate both approaches.

The new algorithm is called `Modified-BPMF` and is listed in Fig. 1. Intuitively, in each iteration it looks for two currently existing trees S_i, S_j whose leaves participate in many rooted triplets of the form $xy|z$ where x belongs to S_i, y belongs to S_j, and z belongs to neither S_i nor S_j, and then merges S_i and S_j.

We now analyze the approximation ratio of `Modified-BPMF`. Let T be the final tree returned in Step 3. For any node u of T, let $\mathcal{L}[u]$ be the set of leaf labels in the subtree of T rooted at u. For each internal node u in T, denote the two children of u by u_1 and u_2, and let $\mathcal{R}(u)$ be the subset of $\mathcal{R}$ defined by $\mathcal{R}(u) = \{xy|z \in \mathcal{R} : \exists a, b, c \in \{x, y, z\}$ such that $a \in \mathcal{L}[u_1]$, $b \in \mathcal{L}[u_2]$, and $c \notin \mathcal{L}[u_1] \cup \mathcal{L}[u_2]\}$. Observe that for any two internal nodes u and v, $\mathcal{R}(u)$ and $\mathcal{R}(v)$ are disjoint. Also, each $xy|z \in \mathcal{R}$ belongs to $\mathcal{R}(u)$ for some internal node u. Thus, the internal nodes of T partition $\mathcal{R}$ into disjoint subsets. For each internal node u of T, further partition the set $\mathcal{R}(u)$ into two disjoint subsets $\mathcal{R}(u)'$ and $\mathcal{R}(u)''$ where $\mathcal{R}(u)'$ are the rooted triplets in $\mathcal{R}(u)$ which are consistent with T and $\mathcal{R}(u)'' = \mathcal{R}(u) \setminus \mathcal{R}(u)'$.

Lemma 1. $|\mathcal{R}(u)'| \geq \frac{1}{3} \cdot |\mathcal{R}(u)|$ *for each internal node u of T.*

Proof. Consider the iteration of `Modified-BPMF`$(\mathcal{R})$ in which the node u is created as a new root node for two trees S_i and S_j selected in Step 2c. Clearly, $score(S_i, S_j) \geq 0$. Moreover, by the definition of *score* in Steps 2a and 2b and the construction of T, we have $score(S_i, S_j) = 2 \cdot |\mathcal{R}(u)'| - |\mathcal{R}(u)''|$. Since $|\mathcal{R}(u)''| = |\mathcal{R}(u)| - |\mathcal{R}(u)'|$, we obtain $|\mathcal{R}(u)'| \geq \frac{1}{3} \cdot |\mathcal{R}(u)|$. $\square$

Theorem 1. *For any set $\mathcal{R}$ of rooted triplets, `Modified-BPMF`$(\mathcal{R})$ returns a tree consistent with at least one third of the rooted triplets in $\mathcal{R}$.*

Proof. Follows directly from Lemma 1 and the fact that $\mathcal{R}$ is partitioned into disjoint subsets by the internal nodes of T. $\square$

`Modified-BPMF` can be implemented to run in $O(k+n^3)$ time by using $O(k+n^2)$ time for preprocessing and then spending $O(n^2)$ time in each iteration to find the best pair of trees to merge and $O(n^2 + |\mathcal{R}(u)|)$ time in each iteration to update all relevant scores. This is faster than `One-Leaf-Split` for $k = \omega(n^3/\log n)$.

4   Approximating MaxRTC by Using MinRTI

In this section, we investigate how approximation algorithms for MinRTI can be used to approximate MaxRTC.

Theorem 2. *Suppose $\mathcal{B}$ is a β-approximation algorithm for MinRTI for some $\beta > 1$. Let $\mathcal{A}'$ be the approximation algorithm for MaxRTC which returns the best of the two approximate solutions obtained by: (1) applying `Modified-BPMF` to the input $\mathcal{R}$; and (2) applying $\mathcal{B}$ to $\mathcal{R}$ and taking the complement relative to $\mathcal{R}$. Then the approximation ratio of algorithm $\mathcal{A}'$ is at most $(3 - \frac{2}{\beta})$.*

Proof. Let $a'(\mathcal{R})$ be the number of rooted triplets in $\mathcal{R}$ consistent with the tree returned by $\mathcal{A}'$. Since $\mathcal{A}'$ returns the best of the two approximate solutions obtained by (1) and (2) above, $a'(\mathcal{R}) \geq \frac{1}{3} \cdot k$ (according to Theorem 1) and $a'(\mathcal{R}) \geq k - \beta \cdot (k - C(\mathcal{R}))$ always hold. There are two possibilities:

- $k > \frac{3\beta}{3\beta-2} \cdot C(\mathcal{R})$: Then, $a'(\mathcal{R}) \geq \frac{1}{3} \cdot k > \frac{1}{3} \cdot \frac{3\beta}{3\beta-2} \cdot C(\mathcal{R}) = \frac{1}{3-\frac{2}{\beta}} \cdot C(\mathcal{R})$.

- $k \leq \frac{3\beta}{3\beta-2} \cdot C(\mathcal{R})$: In this case, $a'(\mathcal{R}) \geq k - \beta \cdot (k - C(\mathcal{R})) = \beta \cdot C(\mathcal{R}) - (\beta - 1) \cdot k \geq \beta \cdot C(\mathcal{R}) - (\beta-1) \cdot \frac{3\beta}{3\beta-2} \cdot C(\mathcal{R}) = (\beta - \frac{(\beta-1) \cdot 3\beta}{3\beta-2}) \cdot C(\mathcal{R}) = \frac{1}{3-\frac{2}{\beta}} \cdot C(\mathcal{R})$.

In both cases, we have $a'(\mathcal{R}) \geq \frac{1}{3-\frac{2}{\beta}} \cdot C(\mathcal{R})$. $\qquad\qquad\square$

By plugging in `Min-Cut-Split` (see Section 2) into Theorem 2, one obtains:

Corollary 1. MAxRTC *admits a polynomial-time* $(3 - \frac{2}{n-2})$*-approximation.*

5   Random and Pseudorandom Triplet Sets

This section examines properties of minimally dense sets of triplets constructed in a random or pseudorandom fashion. We first show that for a triplet set, generated under a uniform random model, the maximum fraction of triplets that can be consistent is approximately one third. We then adapt a construction from [2] to obtain a deterministic construction of a triplet set having a similar property.

Let L be a set of n elements. Consider a minimally dense set $\mathcal{R}$ of rooted triplets on L generated by the following random model: for each $t \in [L]^3$, $\mathcal{R}(t)$ is a uniformly chosen random element of $\mathbb{Z}_3$. The following theorem shows that the maximum fraction of triplets which can be consistent is approximately $1/3$.

Theorem 3. *Let* $\mu = \frac{1}{3}\binom{n}{3}$. *Let* $\delta(n)$ *be any function such that* $\delta(n) = \Omega(\frac{\log n}{n})$. *With high probability:* $C(\mathcal{R}) < (1 + \delta(n))\mu$.

Proof. Fix δ. Given a binary tree T on L, we compute the probability that $C(\mathcal{R}, T)$ deviates from its expectation by a factor $1 + \delta$. Given a triplet $t \in rt(T)$, denote by $\chi(\mathcal{R}, t)$ the indicator variable which equals 1 if $t \in \mathcal{R}$ and 0 otherwise. Observe that $C(\mathcal{R}, T)$ is a sum of i.i.d. random variables: $C(\mathcal{R}, T) = \sum_{t \in rt(T)} \chi(\mathcal{R}, t)$. Since $\mathbb{E}[C(\mathcal{R}, T)] = \mu$, a straightforward application of Chernoff bounds yields: $\mathbb{P}[C(\mathcal{R}, T) > (1 + \delta)\mu] \leq exp(-c\mu\delta^2)$ for some constant c.

Now, apply union bounds to obtain: $\mathbb{P}[C(\mathcal{R}) > (1 + \delta)\mu] \leq \sum_T \mathbb{P}[C(\mathcal{R}, T) > (1 + \delta)\mu] \leq 2^{n \log n} exp(-c\mu\delta^2)$. Observe that if $\delta = \Omega(\frac{\log n}{n})$ then $c\mu\delta^2 = \Omega(n \log^2 n)$, hence the above expression tends to 0 as n tends to infinity. $\qquad\square$

In the rest of this section, we describe a deterministic construction of a minimally dense random-like triplet set. It uses the following algebraic construction which generalizes the construction of [2] by introducing an additive parameter q. The construction provides an s-coloring of the hyperedges of the complete r-uniform hypergraph, with the pseudorandom properties stated in Lemma 2 below.

Definition 1. *Consider integers $r, s > 1$, a prime p with $s \mid p-1$, and an element $q \in \mathbb{Z}_p$. Let g be a generator of $\mathbb{Z}_p^*$, let H be the subgroup of $\mathbb{Z}_p^*$ generated by g^s, and for each $i \in [s]$ let H_i be the coset Hg^i.*

For an element $j \in \mathbb{Z}_p^*$, define $[j]_p^s = i$ if $j \in H_i$, and define $[0]_p^s = 0$. Define $\phi_{p,q}^{r,s} : \mathbb{Z}_p^r \to [s]$ so that for each $j = (j_1, ..., j_r) \in \mathbb{Z}_p^r$, $\phi_{p,q}^{r,s}(j) = [j_1 + ... + j_r + q]_p^s$.

Furthermore, for any set $A \subset \mathbb{Z}_p^r$, and $j \in [s]$, write: $n_j(A) = |\{i \in A : \phi_{p,q}^{r,s}(i) = j\}|$. The following lemma from [2] states that if A arises from a cartesian product and if $|A|$ is large enough, then the fraction of hyperedges of A which have color $j \in [s]$ is approximately $1/s$.

Lemma 2 ([2]). *Let $A_1, ..., A_r$ be subsets of $\mathbb{Z}_p$, and let $A = \{i \in [\mathbb{Z}_p]^r : i_j \in A_j, j = 1...r\}$. Then for all $j \in [s]$, $|n_j(A) - |A|/s| \leq c_r |A|^{1/2} (\log |A|)^{r-1} p^{(r-1)/2}$ for some global $c_r > 0$ that depends only on r.*

We apply the construction of Definition 1 with $r = s = 3$ to obtain a minimally dense triplet set $\mathcal{R}_p$ on $\mathbb{Z}_p$ with random-like properties. More precisely, we define $\mathcal{R}_p$ so that for each distinct $x, y, z \in \mathbb{Z}_p$, $\mathcal{R}_p(x, y, z) = \phi_{p,0}^{3,3}(x, y, z)$. The next theorem shows that $\mathcal{R}_p$ is random-like: every binary tree is consistent with approximately one third of the triplets in $\mathcal{R}_p$. The proof relies on Lemma 2.

Theorem 4. *For any binary tree T on $\mathbb{Z}_p$, it holds that $|C(\mathcal{R}_p, T) - \frac{1}{3}\binom{p}{3}| \leq cp^{5/2} \log p$ for some constant c.*

Proof. Fix $z \in \mathbb{Z}_p$. Let $L_{z,1}, ..., L_{z,m}$ be the clusters hanging along the path in T from z to the root; these sets form a partition of $\mathbb{Z}_p \backslash \{z\}$. For each $i \in [m]$, let $n_{z,i}$ be the number of triplets of $\mathcal{R}_p \cap rt(T)$ of the form $xy|z$ with $x, y \in L_{z,i}$. We then have: $C(\mathcal{R}_p, T) = \sum_{z \in \mathbb{Z}_p} \sum_i n_{z,i}$.

Fix $i \in [m]$, and let $A_{z,i} = [L_{z,i}]^2$. We will show that $|n_{z,i} - |A_{z,i}|/3| \leq cp^{3/2} \log p$. Define the sets $L_{z,i}^{(1)} = \{x \in L_{z,i} : x < z\}$ and $L_{z,i}^{(2)} = \{x \in L_{z,i} : x > z\}$, and partition $A_{z,i}$ into three sets $A_{z,i}^{(1)} = [L_{z,i}^{(1)}]^2$, $A_{z,i}^{(2)} = L_{z,i}^{(2)} \times L_{z,i}^{(1)}$, $A_{z,i}^{(3)} = [L_{z,i}^{(2)}]^2$. Next, define $f : [\mathbb{Z}_p \backslash \{z\}]^2 \to \mathbb{Z}_3$ by setting $f(x, y) = \phi_{p,z}^{2,3}(x, y)$. Given $A \subset [\mathbb{Z}_p]^2, j \in \mathbb{Z}_3$, we set $n_j'(A) = |\{i \in A : f(i) = j\}|$. We then have:

$$n_{z,i} = n_1'(A_{z,i}^{(1)}) + n_2'(A_{z,i}^{(2)}) + n_3'(A_{z,i}^{(3)})$$

Since $f = \phi_{p,z}^{2,3}$, Lemma 2 applies and yields the following inequality: for each $j \in \{1, 2, 3\}$, $|n_j'(A_{z,i}^{(j)}) - |A_{z,i}^{(j)}|/3| \leq c' |A_{z,i}^{(j)}|^{1/2} (\log |A_{z,i}^{(j)}|) p^{1/2}$ for some constant c'. By using the triangle inequality and by summing over index j, we obtain: $|n_{z,i} - |A_{z,i}|/3| \leq c|A_{z,i}|^{1/2} (\log |A_{z,i}|) p^{1/2}$ for some constant c. Let $S = \sum_{z \in \mathbb{Z}_p} \sum_i |A_{z,i}|$ and $S' = \sum_{z \in \mathbb{Z}_p} \sum_i |A_{z,i}|^{1/2}$. By summing over indices z, i in the previous inequality, we obtain $|C(\mathcal{R}_p, T) - S/3| \leq cS' p^{1/2} \log p$. We conclude by observing that $S = \binom{p}{3}$ and that $S' \leq p^2$ (this last inequality following from the fact that for fixed z, $\sum_i |A_{z,i}|^{1/2} \leq \sum_i |L_{z,i}| = p - 1$). $\square$

Corollary 2. $|C(\mathcal{R}_p) - \frac{1}{3}\binom{p}{3}| \leq cp^{5/2} \log p$.

6   Hardness Results

6.1   Minimally Dense Inputs

Our first hardness result concerns the computational complexity of MaxRTC and MinRTI for *minimally dense* inputs. It is based on the deterministic construction of a minimally dense random-like triplet set given in Section 5 and is a non-trivial strengthening of the NP-hardness proof for dense inputs in [14].

Theorem 5. MaxRTC *restricted to minimally dense instances is* NP-*hard.*

Proof. We reduce the general (non-dense) case of MaxRTC (which is already known to be NP-hard [4,15,25]) to the minimally dense case, following an approach inspired by [2,3]. Starting with an arbitrary instance, the approach consists in replicating each label p times (which is called *inflating* the instance), and making the resulting instance dense by adding a pseudorandom triplet set. Formally, the reduction proceeds as follows. Consider a triplet set $\mathcal{R}$ on L given as an instance of MaxRTC. Let $n = |L|$, let p be a prime number, and let $L' = \{x_i : x \in L, i \in \mathbb{Z}_p\}$. Define the minimally dense triplet set $\mathcal{R}'$ on L' by:

1. if $\mathcal{R}(x, y, z)$ is defined then $\mathcal{R}'(x_i, y_j, z_k) = \mathcal{R}(x, y, z)$;
2. if $\mathcal{R}(x, y, z)$ is undefined and i, j, k distinct then $\mathcal{R}'(x_i, y_j, z_k) = \mathcal{R}_p(i, j, k)$, where $\mathcal{R}_p$ is the minimally dense triplet set defined in Section 5;
3. otherwise, $\mathcal{R}'(x_i, y_j, z_k)$ is an arbitrary element of $\mathbb{Z}_3$.

For $i \in \{1, 2, 3\}$, let $\mathcal{R}'_i$ be the triplet set defined by condition $i.$, so that $\mathcal{R}' = \mathcal{R}'_1 \cup \mathcal{R}'_2 \cup \mathcal{R}'_3$. Observe that $\mathcal{R}'$ is obtained by inflating $\mathcal{R}$, resulting in $\mathcal{R}'_1$, and completing the instance by a pseudorandom triplet set $\mathcal{R}'_2$ and an arbitrary triplet set $\mathcal{R}'_3$. The correctness of the reduction follows from the fact that inflating the instance multiplies the measure by a factor p^3, while completing the triplet set introduces noise which can be made small by proper choice of p, in such a way that an optimum for $\mathcal{R}$ can be recovered from an optimum for $\mathcal{R}'$. More precisely, let us introduce the following notation. Let $N_1 = n$, let $N_2 = n(n-1)$, let N_3 be the number of triples $\{x, y, z\}$ such that $\mathcal{R}(x, y, z)$ is undefined, and let $N = N_1 + 8N_2 + 27N_3$. It can be shown that:

1. $C(\mathcal{R}'_1) = p^3 C(\mathcal{R})$;
2. for each binary tree T on L', $|C(\mathcal{R}'_2, T) - N\binom{p}{3}/3| \leq cNp^{5/2} \log p$;
3. for each binary tree T on L', $C(\mathcal{R}'_3, T) \leq Np^2$.

where 2. follows from the pseudorandomness of $\mathcal{R}_p$ stated in Theorem 4.

It follows that $C(\mathcal{R}')$ is an approximation of $C(\mathcal{R}_1) + Np^3/18 = p^3 C(\mathcal{R}) + Np^3/18$ within an additive error of $cNp^{5/2} \log p$, for some constant c. Dividing by $p^3/18$, we obtain $|\frac{18C(\mathcal{R}')}{p^3} - (18C(\mathcal{R}) + N)| \leq c'Np^{-1/2} \log p$ for some constant c'. Since $N = O(n^3)$, we can choose p polynomially bounded in terms of n such that the right member is less than $\frac{1}{2}$, implying that $\lfloor \frac{18C(\mathcal{R}')}{p^3} \rfloor = 18C(\mathcal{R}) + N$.   □

Corollary 3. MaxRTC *restricted to minimally dense instances is* NP-*hard.*

Proof. Follows from Theorem 5 and the fact that MinRTI is the supplementary problem of MaxRTC.   □

6.2   Polynomial-Time Inapproximability of General MINRTI

We now establish a hardness of approximation result for MINRTI in the general case, namely a logarithmic inapproximability by reduction from HITTING SET.

Theorem 6. MINRTI *is not approximable within* $\Omega(\log n)$ *unless* $\mathsf{P} = \mathsf{NP}$.

The proof of this theorem is carried out in two steps. We first consider a *weighted version* of MINRTI, called MINRTI-W, defined as follows. Given a label set L, let $\mathcal{T}(L)$ be the set of all possible rooted triplets over L. A *weighted triplet set* on L is a function $\mathcal{R} : \mathcal{T}(L) \to \mathbb{N}$, and given a binary tree T on L, we define $I(\mathcal{R}, T) = \sum_{t \in \mathcal{T}(L) \backslash rt(T)} \mathcal{R}(t)$. The MINRTI-W problem takes a weighted triplet set $\mathcal{R}$ on L and seeks a binary tree T on L such that $I(\mathcal{R}, T)$ is minimum.

We give a measure-preserving reduction from HITTING SET to MINRTI-W (Lemma 3) and a measure-preserving reduction from MINRTI-W to MINRTI (Lemma 4). The hardness of approximation of MINRTI then follows from [6].

Due to space limitations, the proof of Lemma 3 has been omitted from this conference version of the paper. Please refer to the full version for a complete proof.

Lemma 3. *There exists a measure-preserving reduction from* HITTING SET *to* MINRTI-W.

Lemma 4. *There exists a measure-preserving reduction from* MINRTI-W *to* MINRTI.

Proof. Given a weighted triplet set $\mathcal{R}$ on L, construct an unweighted triplet set $\mathcal{R}'$ on L' where the label set L' is obtained from L by adjoining labels t_i for each $t \in \mathcal{T}(L), 1 \leq i \leq \mathcal{R}(t)$, and the triplet set $\mathcal{R}'$ consists of the triplets $xt_i|z$, $yt_i|z$ for all $t = xy|z \in \mathcal{T}(L)$, $1 \leq i \leq \mathcal{R}(t)$. The next two claims imply that the reduction is measure-preserving.

Claim 1. Given a binary tree T on L, we can construct in polynomial time a binary tree T' on L' such that $I(\mathcal{R}', T') = I(\mathcal{R}, T)$.

Proof of Claim 1. Let $<$ be an arbitrary total order on L. Starting with T, we define T' as follows: for each triplet $t = xy|z \in \mathcal{T}(L)$ with $x < y$, for each $1 \leq i \leq \mathcal{R}(t)$, insert t_i as a sibling of x. We claim that $I(\mathcal{R}', T') = I(\mathcal{R}, T)$. Indeed, consider $t = xy|z \in \mathcal{T}(L)$ with $x < y$, then: (i) if $xy|z \in rt(T)$, then for each $1 \leq i \leq \mathcal{R}(t)$, $xt_i|z, yt_i|z \in rt(T')$, hence the contribution of these triplets to $I(\mathcal{R}', T')$ is 0; (ii) if $xz|y \in rt(T)$, then for each $1 \leq i \leq \mathcal{R}(t)$, $xt_i|z \in rt(T')$ but $yt_i|z \notin rt(T')$, hence the contribution of these triplets to $I(\mathcal{R}', T')$ is equal to $\mathcal{R}(t)$; (iii) if $yz|x \in rt(T)$, the reasoning is similar.

Claim 2. Given a binary tree T' on L', we can construct in polynomial time a binary tree T on L such that $I(\mathcal{R}, T) \leq I(\mathcal{R}', T')$.

Proof of Claim 2. Consider a triplet $t = xy|z \in \mathcal{T}(L)\backslash rt(T)$. If there existed an i such that $xt_i|z \in rt(T')$ and $yt_i|z \in rt(T')$, we would obtain $xy|z \in rt(T')$, which is impossible. It follows that for each $1 \leq i \leq \mathcal{R}(t)$, one of $xt_i|z$, $yt_i|z$ is not in $\mathcal{R}(t')$, and thus the contribution of these triplets to $I(\mathcal{R}', T')$ is $\geq \mathcal{R}(t)$; in other words, setting $T = T'|L$ gives a tree such that $I(\mathcal{R}, T) \leq I(\mathcal{R}', T')$. $\square$

7 Concluding Remarks

The following table summarizes what is currently known about the polynomial-time approximability of MAXRTC and MINRTI:

	Negative results	Positive results
MAXRTC:		
general case	APX-hard ([5])	$(3 - \frac{2}{n-2})$-approx. (Section 4)
dense	NP-hard ($\downarrow$)	PTAS ([16])
minimally dense	NP-hard (Section 6.1)	PTAS ($\uparrow$)
MINRTI:		
general case	Inappr. $\Omega(\log n)$ (Section 6.2)	$(n-2)$-approx. ([9] + Section 2)
dense	NP-hard ($\downarrow$)	$(n-2)$-approx. ($\uparrow$)
minimally dense	NP-hard (Section 6.1)	$(n-2)$-approx. ($\uparrow$)

Significantly, MAXRTC can be approximated within a constant ratio of 3 in polynomial time whereas MINRTI cannot be approximated within a ratio of $\Omega(\log n)$ in polynomial time, unless P = NP.

The main open problem for MAXRTC is to determine whether it admits a constant-ratio polynomial-time approximation algorithm whose approximation ratio is asymptotically better than 3. Since MAXRTC is APX-hard [5], a PTAS is unlikely. Note that both of the 3-approximation algorithms `One-Leaf-Split` from [9] and `Modified-BPMF` in Section 3 always output a solution consistent with at least one third of the input rooted triplets and that in this sense, they are worst-case optimal [9].

We would also like to know: Is it possible to achieve a polynomial-time, polylogarithmic approximation algorithm for MINRTI? Furthermore, is there a polynomial-time, constant-ratio approximation for dense inputs? In particular, how well do the existing approximation algorithms for MAXRTC perform on MINRTI restricted to dense inputs?

Acknowledgments

We thank Leo van Iersel, Judith Keijsper, Steven Kelk, Kazuya Maemura, Hirotaka Ono, Kunihiko Sadakane, and Leen Stougie for helpful comments.

References

1. Aho, A.V., Sagiv, Y., Szymanski, T.G., Ullman, J.D.: Inferring a tree from lowest common ancestors with an application to the optimization of relational expressions. SIAM Journal on Computing 10(3), 405–421 (1981)
2. Ailon, N., Alon, N.: Hardness of fully dense problems. Information and Computation 205(8), 1117–1129 (2007)
3. Alon, N.: Ranking Tournaments. SIAM Journal of Discrete Mathematics 20(1), 137–142 (2006)
4. Bryant, D.: Building Trees, Hunting for Trees, and Comparing Trees: Theory and Methods in Phylogenetic Analysis. PhD thesis, University of Canterbury, Christchurch, New Zealand (1997)

5. Byrka, J., Gawrychowski, P., Huber, K.T., Kelk, S.: Worst-case optimal approximation algorithms for maximizing triplet consistency within phylogenetic networks (submitted, 2008)
6. Feige, U.: A Threshold of $\ln n$ for Approximating Set Cover. Journal of the ACM 45(4), 634–652 (1998)
7. Felsenstein, J.: Inferring Phylogenies. Sinauer Associates, Inc. (2004)
8. Gąsieniec, L., Jansson, J., Lingas, A., Östlin, A.: Inferring ordered trees from local constraints. In: Proc. of CATS 1998. Australian Computer Science Communications, vol. 20(3), pp. 67–76. Springer, Singapore (1998)
9. Gąsieniec, L., Jansson, J., Lingas, A., Östlin, A.: On the complexity of constructing evolutionary trees. Journal of Combinatorial Optimization 3(2–3), 183–197 (1999)
10. He, Y.J., Huynh, T.N.D., Jansson, J., Sung, W.-K.: Inferring phylogenetic relationships avoiding forbidden rooted triplets. Journal of Bioinformatics and Computational Biology 4(1), 59–74 (2006)
11. Henzinger, M.R., King, V., Warnow, T.: Constructing a tree from homeomorphic subtrees, with applications to computational evolutionary biology. Algorithmica 24(1), 1–13 (1999)
12. Holm, J., de Lichtenberg, K., Thorup, M.: Poly-logarithmic deterministic fully-dynamic algorithms for connectivity, minimum spanning tree, 2-edge, and biconnectivity. Journal of the ACM 48(4), 723–760 (2001)
13. van Iersel, L., Keijsper, J., Kelk, S., Stougie, L., Hagen, F., Boekhout, T.: Constructing level-2 phylogenetic networks from triplets. In: Vingron, M., Wong, L. (eds.) RECOMB 2008. LNCS (LNBI), vol. 4955, pp. 450–462. Springer, Heidelberg (2008)
14. van Iersel, L., Kelk, S., Mnich, M.: Uniqueness, intractability and exact algorithms: reflections on level-k phylogenetic networks (submitted, 2008)
15. Jansson, J.: On the complexity of inferring rooted evolutionary trees. In: Proc. of GRACO 2001. Electronic Notes in Discrete Mathematics, vol. 7, pp. 121–125. Elsevier, Amsterdam (2001)
16. Jansson, J., Lingas, A., Lundell, E.-M.: A triplet approach to approximations of evolutionary trees. In: Poster H15 presented at RECOMB 2004 (2004)
17. Jansson, J., Ng, J.H.-K., Sadakane, K., Sung, W.-K.: Rooted maximum agreement supertrees. Algorithmica 43(4), 293–307 (2005)
18. Jansson, J., Nguyen, N.B., Sung, W.-K.: Algorithms for combining rooted triplets into a galled phylogenetic network. SIAM Journal on Computing 35(5), 1098–1121 (2006)
19. Jansson, J., Sung, W.-K.: Inferring a level-1 phylogenetic network from a dense set of rooted triplets. Theoretical Computer Science 363(1), 60–68 (2006)
20. Jiang, T., Kearney, P., Li, M.: A polynomial time approximation scheme for inferring evolutionary trees from quartet topologies and its application. SIAM Journal on Computing 30(6), 1942–1961 (2001)
21. Ng, M.P., Wormald, N.C.: Reconstruction of rooted trees from subtrees. Discrete Applied Mathematics 69(1–2), 19–31 (1996)
22. Page, R.D.M.: Modified mincut supertrees. In: Guigó, R., Gusfield, D. (eds.) WABI 2002. LNCS, vol. 2452, pp. 537–552. Springer, Heidelberg (2002)
23. Semple, C., Steel, M.: A supertree method for rooted trees. Discrete Applied Mathematics 105(1–3), 147–158 (2000)
24. Snir, S., Rao, S.: Using Max Cut to enhance rooted trees consistency. IEEE/ACM Transactions on Computational Biology and Bioinformatics 3(4), 323–333 (2006)
25. Wu, B.Y.: Constructing the maximum consensus tree from rooted triples. Journal of Combinatorial Optimization 8(1), 29–39 (2004)

A Method to Overcome Computer Word Size Limitation in Bit-Parallel Pattern Matching

M. Oğuzhan Külekci

TÜBİTAK-UEKAE
National Research Institute of Electronics & Cryptology
41470 Gebze, Kocaeli, Turkey
kulekci@uekae.tubitak.gov.tr

Abstract. The performance of the pattern matching algorithms based on bit-parallelism degrades when the input pattern length exceeds the computer word size. Although several divide-and-conquer methods have been proposed to overcome that limitation, the resulting schemes are not that much efficient and hard to implement. This study introduces a new fast bit-parallel pattern matching algorithm that is capable of searching patterns of any length in a common bit-parallel fashion. The proposed bit-parallel length invariant matcher (BLIM) is compared with the Shift-Or and bit-parallel non-deterministic matching (BNDM) algorithms along with the standard Boyer-Moore and Sunday's quick search, which are known to be the very fast in general. Benchmarks have been conducted on natural language, DNA sequence, and binary alphabet random texts. Besides the length invariant architecture of the algorithm, experimental results indicate that on the average BLIM is 18%, 44%, and 6% faster than BNDM, which is accepted as one of the fastest algorithms of this genre, on natural language, DNA sequence and binary random texts respectively.

1   Introduction

Searching for exact or approximate occurrences of single or multiple patterns on text files is a fundamental issue that has been studied deeply during the last three decades. Besides its obvious usage in text editors, it gained great focus with the recent advances in genomics that require searching of long patterns with small alphabets. Additionally, network intrusion detection systems and anti-virus software also demand for high performance matchers.

The main idea behind the algorithms developed on off-line string searching may be grouped into two as automata theoretic approaches and sliding window techniques. In automata theoretic approaches beginning with the Aho-Corasick [1] finite state machine, most promising theoretical results have been obtained by using directed acyclic word graphs such as reverse factor and turbo reverse factor algorithms [2,3,4].

On the sliding window based approaches, the performance of searching patterns from right-to-left by Boyer-Moore (BM) [5] algorithm as oppose to the

S.-H. Hong, H. Nagamochi, and T. Fukunaga (Eds.): ISAAC 2008, LNCS 5369, pp. 496–506, 2008.

left-to-right scanning of Knuth-Morris-Pratt (KMP) [6] had gained great success and many variants of the basic BM algorithm have been proposed, where the fastest ones are assumed to be the Sunday's quick search (QS) [7] and Horspool (HOR) [8] in general.

Baeza–Yates&Gonnet [9] introduced a new way of searching aiming to benefit from the computers intrinsic bit-parallelism for better speed-ups. Their Shift-Or (SO) algorithm preprocesses the input pattern in such a way that the searching on the text is performed with bitwise operations that are very fast in computer architecture. The downside of the SO was its inability to perform shifts while passing over the text. A similar bit-parallel algorithm supporting approximate matching was proposed by Wu&Manber [10], which was implemented as the *agrep* [11] file search utility software. Navarro&Raffinot [12] combined bit-parallelism with suffix automata (BNDM). By merging the shift mechanism of the sliding window techniques with bit-parallel implementation of the non-deterministic automata, BDNM is quite fast and also flexible especially on small alphabets.

The general problem in all bit parallel algorithms is the limitation defined on the length of the input pattern, which does not permit searching of strings longer than the computer word size effectively. While searching such long patterns, bit-parallel algorithms [12,9,13] divide the input pattern into pieces that are smaller than the computer word size and perform the search process on that smaller segments accordingly. In that case the performance of the original algorithm degrades considerably.

In this study, a different way of using bit parallelism is proposed that lets searching patterns independent of their lengths, thus, given name *bit-parallel length invariant matcher* (BLIM). As oppose to the previous bit-parallel algorithms that require the pattern length not to exceed the word size, and divide the string into smaller segments if that limitation is violated, BLIM defines a unique way of handling strings of any size. Proposed architecture supports using character classes and bounded gaps on the input patterns. It is also very suitable for matching computer word size number of multiple patterns of any length.

The performance of BLIM is compared with other bit-parallel algorithms SO and BNDM, with standard BM, QS, and HOR, and also with reverse and turbo reverse factor algorithms. It is observed that BLIM is the fastest on DNA sequence searching. It shows a better performance than other bit-parallel alternatives SO and BNDM on natural language text staying very competitive with BM and QS on long strings. Thus, besides overcoming the pattern length limitation of bit-parallelism, it brings a significant improvement in speed, especially in gene searching.

2 BLIM Algorithm

Let $T = t_0 t_1 t_2 \ldots t_{n-1}$ be the text of length n, $P = p_0 p_1 p_2 \ldots p_{m-1}$ be the pattern of length m that will be scanned on T, and W denotes computer word size. The alphabet of the text will be shown by Σ, and the number of characters in the alphabet by $|\Sigma|$.

Given a pattern P, let's imagine an alignment matrix A which consist of W number of rows, where each row_i, $0 \le i < W$, contains the pattern right shifted by i characters. Thus, A has $ws = W + m - 1$ columns. Note that, throughout the study ws value will be referred as window size from now on. A sample alignment matrix corresponding to pattern $P = abaab$ is shown in Table 1 assuming computer word size $W = 8$.

Table 1. The alignment matrix generated for the sample pattern $P = abaab$ with the assumption that computer word size $W = 8$

	0	1	2	3	4	5	6	7	8	9	10	11
0	a	b	a	a	b							
1		a	b	a	a	b						
2			a	b	a	a	b					
3				a	b	a	a	b				
4					a	b	a	a	b			
5						a	b	a	a	b		
6							a	b	a	a	b	
7								a	b	a	a	b

The main idea of BLIM is to slide that ws length alignment matrix over the text, check if any possible placements of the input pattern exist in the current window via bitwise operations, and after completing the investigation of the current text portion $T[i \ldots i + ws - 1]$ right shift the window by an amount that is computed according to the immediate text character $T[i + ws]$ following the current window[1].

When the window is located on text $T[i \ldots i + ws - 1]$, BLIM visits the characters $T[i + j]$ $(0 \le j < ws)$ in an order that was previously computed at the preprocessing stage. At each character visit, the possibility of any occurrences of the pattern is checked with a bitwise *and* operation by using a mask matrix that was again precomputed. Current window is slid right by the shift amount specified by the text character $T[i + ws]$ after the current investigation is over. Thus, a shift vector of alphabet size needs to be calculated at preprocessing also. In summary, at preprocessing stage BLIM calculates the mask matrix, the shift vector and decides on the scan order according to input pattern.

2.1   Preprocessing Stage I: The Mask Matrix

Mask matrix consists of alphabet size $|\Sigma|$ rows, and ws columns. Each $Mask[ch][pos]$, where $ch \in \Sigma$, and $0 \le pos < ws$, is a bit vector of W bits as $b_{W-1}b_{W-2} \ldots b_1 b_0$. The ith bit in $Mask[ch][pos]$ is set to 0, if observing that ch at position pos is not possible for the i character right shifted placement of

---

[1] Performing shift according to the immediate text character following the current window was first proposed by Sunday in quick search algorithm [7], which is one of the fastest algorithms in BM family.

the input pattern P. Otherwise, it is equal to 1, which indicates observing that ch at that pos does not violate the placement of the i character right shifted version of the pattern. The formal definition for the actual value of the b_i in $Mask[ch][pos]$ is as follows:

$$b_i = 0 \text{ , if } (0 \le pos - i < m \text{ and } (ch \ne p_{pos-i})$$
$$b_i = 1 \text{ , otherwise}$$

An example mask matrix that is generated for the sample pattern $P = abaab$ is shown in Table 2 (in hexadecimal notation) assuming the alphabet of the text is $\Sigma = \{a, b, c, d\}$ with a computer word size $W = 8$.

Table 2. The mask matrix generated for sample pattern $P = abaab$, assuming alphabet $\Sigma = \{a, b, c, d\}$, and computer word size $W = 8$

	0	1	2	3	4	5	6	7	8	9	10	11
a	FF	FE	FD	FB	F6	ED	DB	B7	6F	DF	BF	7F
b	FE	FD	FA	F4	E9	D3	**A7**	4F	AF	3F	7F	FF
c	FE	FC	F8	F0	E0	C1	83	87	8F	1F	3F	7F
d	FE	FC	F8	F0	E0	C1	83	87	8F	1F	3F	7F

For the sample pattern $P = abaab$, Table 3 depicts the calculation and meanings of bits in $Mask[b][6]$ explicitly. It is seen that right shift of the string by 0, 1, 2, 5, and 7 characters are appropriate, as specified by the corresponding bits in $Mask[b][6]$. These possible alignments can be viewed on rows 0, 1, 2, 5, and 7 of Table 1. Note that, the observation of b at $T[i + 6]$ does not infer with the alignments beginning at $T[i]$, $T[i+1]$, and $T[i+7]$. Thus, the corresponding bits are set to 1.

Table 3. Calculation of the bit vector $Mask[b][6] = A7$ for $P = abaab$

Bit Index i	7	6	5	4	3	2	1	0
pos–i	-1	0	1	2	3	4	5	6
p_{pos-i}		a	b	a	a	b		
Bit Value	1	**0**	1	**0**	**0**	1	1	1

2.2   Preprocessing Stage II: The Shift Vector

BLIM uses the shift mechanism proposed in quick search of Sunday [7]. That is; the immediate text character following the window determines the actual shift amount. If that character is included in the pattern then $Shift[ch] = ws - k$, where $k = max\{i; p_i = ch\}$, else $Shift[ch] = ws + 1$. On the same example pattern $P = abaab$, remembering that $ws = 12$, and $\Sigma = \{a, b, c, d\}$, the shift values of the letters are given in Table 4.

If one encounters the character a at $T[i + 12]$ while the window is located at $T[i \ldots i + 11]$, then the pattern $abaab$ may begin at $T[i + 9]$, $T[i + 10]$, and

Table 4. Computing the shift values of letters for sample pattern $P = abaab$

Character	Shift Value
a	$9 = 12 - max(0, 2, 3)$
b	$8 = 12 - max(1, 4)$
c	$13 = 12 + 1$
d	$13 = 12 + 1$

$T[i+12]$. Thus, the safe shift of a should be 9. Similarly, if $b = T[i+12]$, then the shift value is 8. The characters c and d do not occur in pattern P. Observation of those bad characters at $T[i + 12]$ directly yields the next attempt to begin from $T[i + 13]$, which means a shift value of 13.

Note that such a shift mechanism does not benefit from the characters investigated in the current window. This memoryless shift function may be replaced with a more sophisticated two-dimensional shift matrix such that $Shift[ch][pos]$ would indicate the safe shift amount when one observes ch at position pos. At each character visit, the shift value should be selected as the maximum of the current value and the value of the visited letter. Although theoretically less number of characters are visited in that case, the experiments showed that the time elapsed to update shift value at each character visit downgrades the speed. Thus, it is preferred to use the simple version instead.

2.3   Preprocessing Stage III: Scan Order

The characters of the ws length window should be visited in such an order that checking all possible alignments in the current window be accomplished with minimum number of character accesses. If we traverse the window from right-to-left or left-to-right, we have to perform unnecessary checks in case of a leftmost or rightmost occurrence of the pattern. Actually, the optimum solution requires calculating the next position after each character visit, which slows down the algorithm considerably.

As a simple and efficient way of computing the scan order, BLIM algorithm checks the characters at positions $m - i, 2m - i, \ldots, km - i$, where $km - i < ws$, for $i = 1, 2, ..., m$ in order. As an example, assuming computer word size $W = 8$, the scan order of the sample pattern $P = abaab$ is $4, 9, 3, 8, 2, 7, 1, 6, 11, 0, 5, 10$. The main idea behind this ordering is to investigate the window in such a way that at each character access maximum number of alignments is checked.

2.4   Main Search Loop

BLIM algorithm is sketched in Figure 1 as a whole. It has a very simple main loop in which the current text portion is investigated very fast by using only the bitwise *and* operator. When the window is located at $T[i \ldots i + ws - 1]$, the $flag$ variable is first set to the corresponding mask value of the text character $T[i + ScanOrder[0]]$ at position $ScanOrder[0]$. Remember that $ScanOrder[0]$ is always equal to $m - 1$. The next character to be visited is defined by

$ScanOrder[1]$. The *and* operation is performed between the current flag and the mask of $T[i + ScanOrder[1]]$ defined for position $ScanOrder[1]$. The traversal of the characters continues till the *flag* becomes 0 or all the characters are visited. If *flag* becomes 0, this implies pattern does not exist on the window. Otherwise, one or more occurrences of the pattern are detected at the investigated text factor. In that case, the 1 bits of the *flag* tells the positions of occurrences exactly, e.g., say if the $3rd$ bit is 1, then pattern is detected at $T[i + 3]$. After completing the search on the current text factor, the window is slid right by the shift amount of the character at $T[i + ws]$.

A sample run of the BLIM algorithm is depicted in Table 5 assuming that pattern abaab is to be searched on a text factor $T[0\ldots11]$ = ababaabaabab.

```
Calculate Mask;
Calculate Shift;
Calculate ScanOrder;
Pad the text with m arbitrary characters;//to avoid segmentation fault
i=0;
ws=W+m-1;
while (i<n){
      flag = Mask[T[i+ScanOrder[0]]][ScanOrder[0]];
      for (j=1 ; flag & j<ws ; j++){
          flag &= MaskMatrix[T[i+ScanOrder[j]]][ScanOrder[j]];
      }
      if (flag){
          Check bits of the flag to find out detected items;
      }
      i += Shift[T[i+ws]];
}
```

Fig. 1. The BLIM algorithm

Table 5. A sample run of BLIM on text factor abababaabab for searching abaab

	j	ScanOrder[j]	Mask[ch][pos]	flag
ababaabaabab	0	4	Mask[a][4] = 11110110	11110110
ababaabaabab	1	9	Mask[b][9] = 00111111	00110110
ababaabaabab	2	3	Mask[b][3] = 11110100	00110100
ababaabaabab	3	8	Mask[a][8] = 01101111	00100100
ababaabaabab	4	2	Mask[a][2] = 11111101	00100100
ababaabaabab	5	7	Mask[a][7] = 10110111	00100100
ababaabaabab	6	1	Mask[b][1] = 11111101	00100100
ababaabaabab	7	6	Mask[b][6] = 10100111	00100100
ababaabaabab	8	11	Mask[b][11] = 11111111	00100100
ababaabaabab	9	0	Mask[a][0] = 11111111	00100100
ababaabaabab	10	5	Mask[a][5] = 11101101	00100100
ababaabaabab	11	1	Mask[b][1] = 11111101	00100100

The corresponding mask values can be reached from Table 2. At the end of that sample run, it is observed that $flag$ is 00100100, which indicates pattern is detected at positions $T[2]$ and $T[5]$ conforming to bit indices (from least significant bit to most signicicant bit) that are 1. Note that after completion of the run, the window will be slid right according to the immediate text character following the window, e.g., the shift is 9, if $T[12]$ =a, or 13 if $T[12]$ =c or d.

3   Analysis

The preprocessing of BLIM has a quadratic time complexity due to the computation of the mask matrix that is of size $|\Sigma| \times ws$. The calculation of shift vector and the scan order array are linear with the length of the input pattern.

The best shift that the algorithm can perform is $ws + 1 = W + m$, and worst is $ws - m + 1 = W$. At each attempt the algorithm must check at least $\lceil \frac{ws}{m} \rceil$ positions on the text window $T[i \ldots i + ws - 1]$ under investigation, and $ws - 1$ positions at most in case of a mismatch. Thus, the best-case complexity is $O(\lceil \frac{n}{ws+1} \rceil \times \lceil \frac{ws}{m} \rceil) \approx O(\frac{n}{m})$ assuming that best shift is performed at each attempt with minimal position checks. The worst case is when the window is slid over the text with the least shift value, and maximum character visits performed at each trial, which corresponds to a complexity of $O(\lceil \frac{n}{W} \rceil \times ws)$.

BLIM can handle character classes and bounded gaps in input patterns as well. If the pattern P contains more than one character at a position, then the corresponding cells of all those characters in the mask matrix will be filled accordingly. As an example, let's say we are searching for pattern P=ab[ab]ab, which means third character is either a or b. Then, while calculating the mask matrix, $P[2]$ will be treated not as a single letter, but a list of characters. When a gap is defined on the pattern, that means the list is composed of whole alphabet for that position.

Another advantage of BLIM is its ability to perform multiple pattern matching easily. Up to W number of patterns, whatever their sizes are, can be scanned simultaneously on the text. The main idea is to reserve each row of the alignment matrix for one of the input patterns in Table 1, and compute the mask matrix accordingly. Actually, the proposed architecture is capable of building the mask matrix according to any specified alignment matrix. Thus, it is able to handle character classes and bounded gaps on multi-patterns as well. Note that, the length invariant structure of BLIM lets to scan multi-patterns of any size again. That is more important as the total size of multiple strings is more probable to exceed the computer word size, which restricts the other bit-parallel algorithms.

4   Experimental Results

BLIM is compared with bit-parallel non-deterministic matching (BNDM), reverse factor (BDM), turbo reverse factor (TBDM), standard Boyer-Moore (BM), Sunday's quick search (QS), Horspool (HOR), and Shift-Or (SO) algorithms on natural language text, DNA sequences and binary alphabet random texts. Note that the comparisons of the algorithms shown in figures 2, 3, and 4 do not include HOR, BDM and TBDM plots. That is because QS dominates HOR in general, and the performances of BDM and TBDM in practice are poor when compared with BLIM and BNDM.

The implementations of the tested algorithms are taken from the Charras&Lecroq's handbook of exact string matching algorithms [14][2]. Slight modifications performed on QS and HOR algorithms, which let them run faster.

All text files used in experiments are 30 MB in size. During the tests, the files are totally read into the memory, and each test pattern is scanned on the data 10 times by each of the tested algorithms. The minimum timing for each algorithm is recorded to minimize the effect of the system[3]. That testing scheme is repeated at least 5 times for each data type with different 30MB files on a PC with a Intel Pentium4 2.8 Ghz CPU and 2 GB memory running Windows XP operating system[4].

Natural language test data is a collection of Turkish text from a daily newspaper. For each length of 2 to 50, 20 distinct strings are randomly selected from the text. Thus, a total of 980 test patterns are used in natural language searching experiments. The benchmark results are depicted in figure 2.

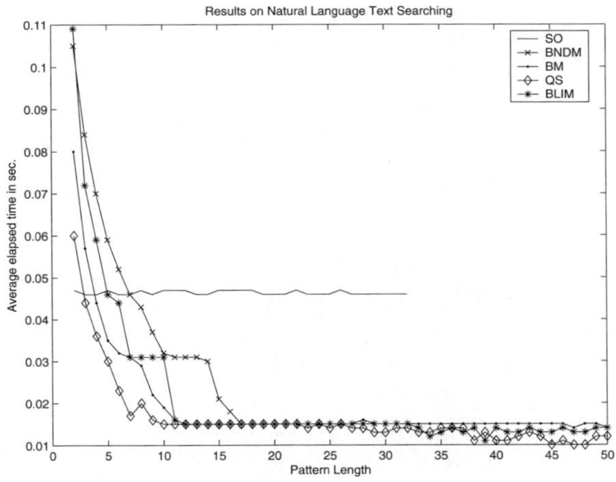

Fig. 2. Experimental results obtained on natural language text

---

[2] The codes can also be reached from the web page `http://www-igm.univ-mlv.fr/~lecroq/string`

[3] Note that time elapsed to read the data file into memory is not included in recorded timings, which means the results show the pure performance of the tested algorithms.

[4] Same experiments are also performed on Intel Xeon 3.0 Ghz machine with 3 GB of memory running Windows XP SP2, and on a Linux machine with 2 GB memory. Similar performance lines are obtained for each algorithm in each trial.

On natural language text, BLIM is on the average 18% faster than BNDM. For short patterns of length smaller than 5, SO behaves better than BLIM. Note that, although it is generally accepted that bit-parallel approaches do not give good results on natural language text, it is observed that for patterns longer than 12 characters, BLIM is very competitive with QS and BM.

The DNA sequence data used in experiments is extracted from the first human genome. Again, 20 random patterns are chosen for each length of 5 to 200 in steps of 5 making a total of 800 test patterns. Figure 3 shows the performance of the tested algorithms on DNA sequence searching.

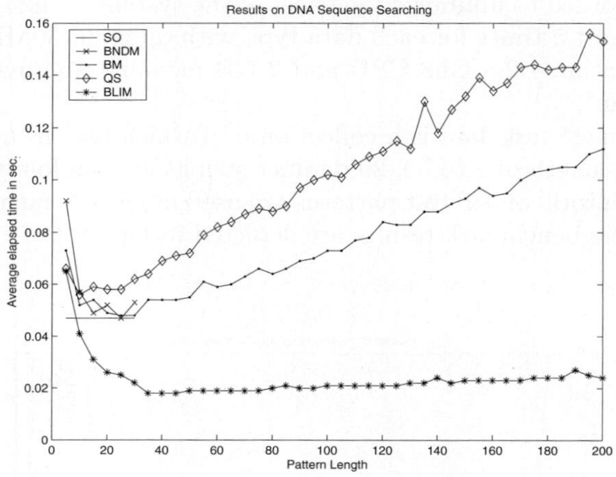

Fig. 3. Experimental results obtained on DNA sequence data

On DNA sequence searching, BM and QS are not very effective especially on longer patterns as expected. The performance of BLIM is approximately 44% better than BNDM. With that speed up, BLIM is by far the fastest technique on off-line searching of DNA sequences as BNDM is accepted as state-of-the-art off-line DNA matching in general.

Similar to DNA data, 20 random patterns with a binary alphabet $\Sigma = a, b$ are generated for each length of 5 to 200 again with a step of 5. That 800 binary alphabet strings are searched on randomly generated text files of 30 MB in size with the same alphabet.

It is observed that BLIM is only 6% better than BNDM on binary alphabet random texts. SO is very effective on short binary sequence searching. On longer strings BLIM shows a better performance than BM. Note that the speed of QS on binary random data is very bad, when compared with BM. Actually, that is because on random binary data good-suffix shift mechanism of BM gets more important than bad character shift function. As QS does not include the good suffix shift mechanism, it degrades.

In summary, BLIM outperforms the other bit-parallel alternatives BNDM and SO on both natural language text and DNA sequences. Especially on DNA

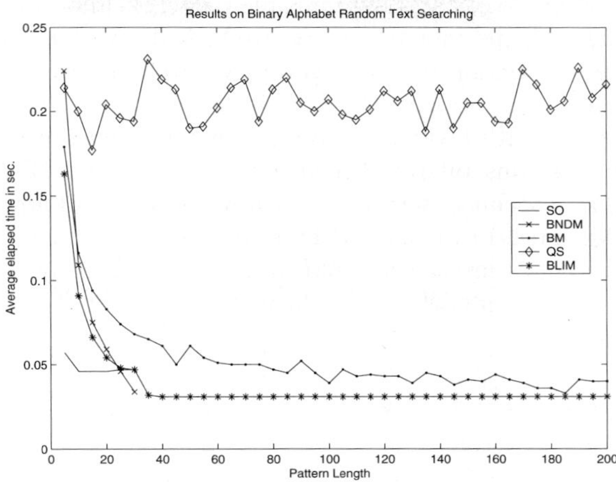

Fig. 4. Experimental results obtained on randomly generated binary alphabet text

Table 6. Average elapsed time ratio of BLIM according to other tested algorithms. Bold items represent the cases where BLIM is better.

Data Type	*Pattern Length*	$\frac{SO}{BLIM}$	$\frac{BNDM}{BLIM}$	$\frac{BM}{BLIM}$	$\frac{QS}{BLIM}$
		Average Time Ratio			
Natural Language Text	2-32	**1.83**	**1.18**	0.87	0.74
	33-50	NA	NA	**1.12**	0.90
DNA Sequence	5-30	**1.34**	**1.66**	**1.54**	**1.70**
	35-200	NA	NA	**3.75**	**5.18**
Random Text with Binary Alphabet	5-30	0.61	**1.16**	**1.30**	**2.52**
	35-200	NA	NA	**1.46**	**6.66**

pattern matching it shows a very significant improvement being approximately 1.6 times faster than BNDM. On binary random texts, the leadership of SO continues, although for long patterns BLIM is the best candidate with its length-invariant structure. The over all average speed ratios of algorithms are depicted in Table 6.

5   Conclusion

Bit-parallel pattern matching algorithms require input pattern length to be less than or equal to computer word size. If pattern length exceeds that limitation, the string is divided into pieces that are smaller than the word size, and search process is performed on that segments accordingly. In that case the performance of the algorithms degrade.

This study presented a new way of using bit-parallelism that overcomes the computer word-size limitation and permits to search patterns of any length.

The algorithm is capable of handling multiple patterns, and also using character classes and bounded gaps in patterns as well. A deep investigation of multi-pattern searching by BLIM is an on-going research, where initial results are promising.

In addition to its length invariant property, experimental results indicate that it is faster than previous bit-parallel approaches, SO and BNDM. Especially on DNA sequence searching, where up to now BNDM is known to be fastest in general, BLIM outperforms the BNDM with a 44% gain in speed. On long patterns it is 3.75 times faster than BM also. Thus, BLIM becomes a powerful tool in gene searching with both its significant improvement in speed and length-invariant structure.

References

1. Aho, A., Corasick, M.: Efficient string matching: An aid to bibliographic search. Communications of the ACM 18, 333–340 (1975)
2. Crochemore, M., Czumaj, A., Gasieniec, L., Jarominek, S., Lecroq, T., Plandowski, W., Rytter, W.: Fast practical multi-pattern matching. Information Processing Letters 71, 107–113 (1993)
3. Crochemore, M., Czumaj, A., Gasieniec, L., Jarominek, S., Lecroq, T., Plandowski, W., Rytter, W.: Speeding up two string matching algorithms. Algorithmica 12, 247–267 (1994)
4. Crochemore, M., Rytter, W.: Text Algorithms. Oxford University Press, Oxford (1994)
5. Boyer, R., Moore, J.: A fast string searching algorithm. Communications of the ACM 20, 762–772 (1977)
6. Knuth, D., Morris, J., Pratt, V.: Fast pattern matching in strings. SIAM Journal of Computing 6, 323–350 (1977)
7. Sunday, D.: A very fast substring search algorithm. Communications of the ACM 33, 132–142 (1990)
8. Horspool, N.: Practical fast searching in strings. Software – Practice and Experience 10 (1980)
9. Baeza-Yates, R., Gonnet, G.: A new approach to text searching. Communications of the ACM 35, 74–82 (1992)
10. Wu, S., Manber, U.: Fast text searching allowing errors. Communications of the ACM 35, 83–91 (1992)
11. Wu, S., Manber, U.: Agrep – a fast approximate pattern-matching tool. In: USENIX Winter, Technical Conference, pp. 153–162 (1992)
12. Navarro, G., Raffinot, M.: Fast and flexible string matching by combining bit-parallelism and suffix automata. ACM Journal of Experimental Algorithms 5, 1–36 (2000)
13. Peltola, H., Tarhio, J.: Alternative algorithms for bit-parallel string matching. In: Nascimento, M.A., de Moura, E.S., Oliveira, A.L. (eds.) SPIRE 2003. LNCS, vol. 2857, pp. 80–94. Springer, Heidelberg (2003)
14. Charras, C., Lecroq, T.: Handbook of exact string matching algorithms. King's Collage Publications (2004)

Inducing Polygons of Line Arrangements

Ludmila Scharf and Marc Scherfenberg

Institute of Computer Science, Freie Universität Berlin
{scharf,scherfen}@mi.fu-berlin.de

Abstract. We show that an arrangement $\mathcal{A}$ of n lines in general position in the plane has an inducing polygon of size $O(n)$. Additionally, we present a simple algorithm for finding an inducing n-path for $\mathcal{A}$ in $O(n \log n)$ time and an algorithm that constructs an inducing n-gon for a special class of line arrangements within the same time bound.

1 Introduction

Every simple polygon induces an arrangement of lines, simply by extending its edges. We consider the question whether every arrangement of lines in the plane has an *inducing polygon*, namely, a simple polygon P such that every line l of the arrangement $\mathcal{A}$ is collinear with an edge of P and that every edge of P is collinear with some line of $\mathcal{A}$, see Fig. 1 for an example. There are arrangements that cannot be induced by a simple polygon. Lines that all intersect in one point and lines that form a 3×2 parallel grid serve as examples of such arrangements. However, we will show that when the lines of an arrangement are *in general position*, i.e., no three lines intersect in one point, and no two lines are parallel, an inducing polygon exists and can be found in $O(n^2)$ time. From now on we consider only arrangements of lines in general position in the plane.

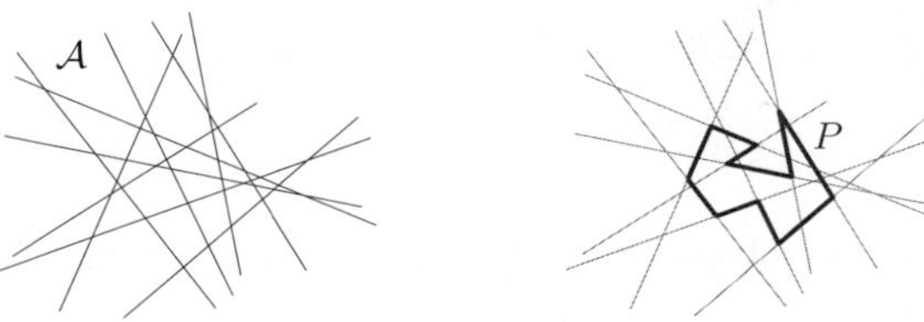

Fig. 1. Line arrangement $\mathcal{A}$ and a simple polygon P inducing $\mathcal{A}$

A stronger version of the inducing polygon problem has been addressed by Bose et al. in [1]. Namely, the authors required the inducing polygon to be an n-gon, where n is the size of the arrangement. They present an algorithm for constructing an inducing simple n-path, that is, a polygonal path that uses every line exactly once (also referred to as an *inducing polyline*) in $O(n^2)$ time. That polyline can be extended to an inducing n-gon if there exists a line such that

S.-H. Hong, H. Nagamochi, and T. Fukunaga (Eds.): ISAAC 2008, LNCS 5369, pp. 507–519, 2008.

all intersection points of the arrangement lie on one side of that line. Finally, they demonstrated that every arrangement contains a subarrangement of size $\sqrt{n-1}+1$ with an inducing $(\sqrt{n-1}+1)$-gon.

In this paper we describe an $O(n^2)$ time algorithm for finding an inducing polygon of size $O(n)$ for an arrangement of n lines. Also we show that for every arrangement that maps to a convex set of points in dual space there exists an inducing n-gon. Additionally, we present a simpler and faster algorithm for finding an inducing n-path with $O(n \log n)$ running time.

2   Inducing Polygon of Linear Size

In this section we describe an algorithm that constructs an inducing polygon P of size $O(n)$ for every arrangement $\mathcal{A}$ of n lines in general position.

Define the *envelope polygon* P_E of $\mathcal{A}$ as the polygon consisting of the finite length segments at the boundary of the unbounded faces of the arrangement [2]. A face f of $\mathcal{A}$ is a *boundary face* if f is adjacent to an edge of P_E and f is bounded, see Fig. 2(a) for an example, where the shading indicates the boundary faces of $\mathcal{A}$.

Our algorithm is based on the following observation: every line $l \in \mathcal{A}$ is either induced by P_E or crosses an edge of P_E. In the latter case l contains an edge of some boundary face f_l. We call the lines that are not induced by the polygon constructed so far *unused* lines. Intuitively, the algorithm traverses the edges of P_E, scans every edge e for unused lines that cross e and for every such line l it augments the envelope polygon by "denting in" the face f_l as depicted in Figure 2(b).

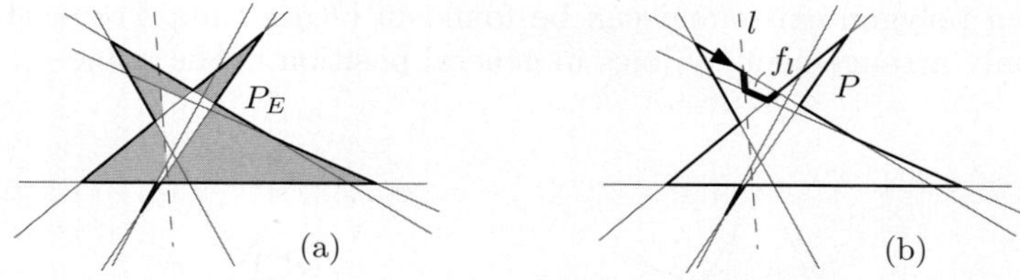

Fig. 2. (a) The envelope polygon P_E with shaded boundary faces; (b) Augmenting the the polygon P for the line l

More precisely, the algorithm constructs an inducing polygon P for $\mathcal{A}$ in the following way: First, set $P = P_E$. Then traverse the edges of P_E in clockwise order and scan every edge e for unused lines that cross it. The scanning direction for e is chosen according to the following rules:

(a)   If both end vertices v_1 and v_2 of e are reflex, the scanning direction corresponds to the traversal direction of P_E, Fig. 3(a).
(b)   If exactly one incident vertex of e is convex, the scanning direction is from the convex vertex to the reflex one, Fig. 3(b).

(c)   If both end vertices are convex, find a line l' intersecting e, such that l' contributes a convex vertex x to P_E. We split e at the intersection point with l' and set the direction for the two parts separately, each from the corresponding convex vertex towards the intersection point with l', Fig. 3(c).

Note that the line l' always exists due to the following observations: P_E has at least one convex vertex x that is not v_1 or v_2 and is not an intersection point of the lines that contain the points v_1 and v_2. From the definition of P_E and convexity of v_1 and v_2 it follows that the intersection points of the line l_1 containing e with every other line of $\mathcal{A}$ lie between or at the vertices v_1 and v_2. Especially it means that the lines, call them l' and l'', that intersect in x, cross the line l_1 in the interior of the edge e.

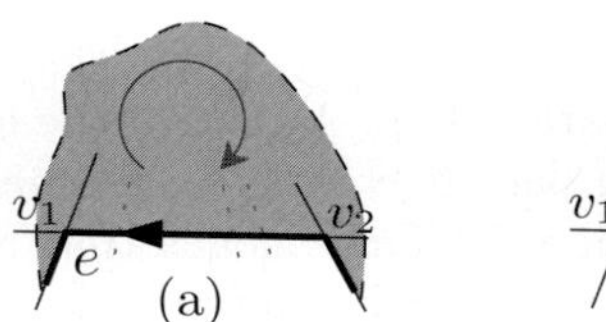
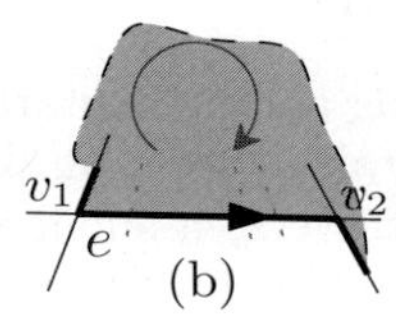
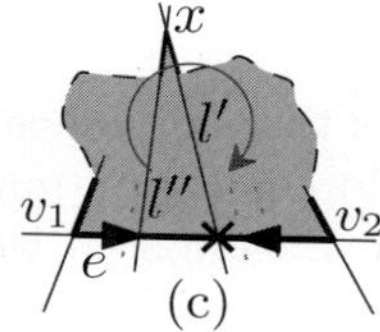

Fig. 3. Direction rules for edge traversal. The shading indicates the interior of P. The light arrow in the interior of P shows the global traversal direction of the envelope edges, and the dark big arrow on the edge e indicates the scanning direction for unused lines on e.

Next, for every unused line l that crosses e we consider the boundary face f_l that comes after l in the scanning direction of the edge. Let ∂f_l be the boundary of f_l and let e' be the edge of ∂f_l contained in e. We replace e' in P by $\partial f_l \setminus e'$ (see Fig. 2(b)) and mark all lines induced by $\partial f_l \setminus e'$, including the line l, as used.

Correctness. Due to space limitations we only sketch the correctness proof here, the details are given in [3].

We need to show that P is a polygon, P induces $\mathcal{A}$, and that P is simple.

We start with $P = P_E$. Every augmentation replaces a line segment by a polygonal chain connecting its end points. Hence P is a polygon.

Every unused line crosses an edge of P_E. Therefore, all unused lines have been handled by the time the algorithm terminates. During each augmentation step only a part of an edge e of P_E is removed from P, so we never "lose" any of the lines that P induced before the augmentation. Thus, every line in $\mathcal{A}$ is induced by P.

The algorithm starts with the envelope polygon P_E, which is a simple polygon. To show that the resulting polygon P is also simple, we first describe the possible violations of simplicity that may occur. Then, assuming that there is an augmentation step, that is, an extension of P by the boundary of one cell, violating the simplicity, we deduce a contradiction.

We say that a polygon has a *self-intersection (a)* if it has a pair of edges e_1 and e_2 intersecting in a single point v such that v is not an end point of either e_1 or e_2. A polygon has an *edge-overlap* **(b)** if there exist two edges e_1 and e_2 in the polygon such that $e_1 \subseteq e_2$. Note, that the case where two edges overlap in a line segment but neither edge contains the other is not possible by construction. A polygon has a *vertex-overlap* **(c)** if it has two coinciding vertices, see Fig. 4 for examples. Note, that the described cases are the only simplicity

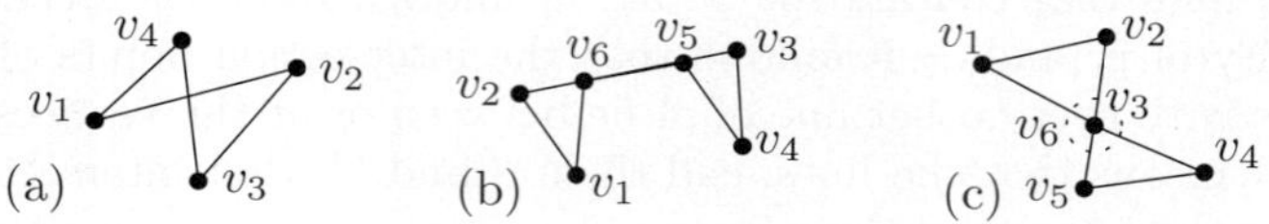

Fig. 4. (a) Self-intersection; (b) edge-overlap; (c) vertex-overlap

violations that could occur during an augmentation step. Although the edges of the envelope polygon might contain more than one edge of the arrangement, in an augmentation step we only traverse a single cell. That means that a newly added edge can either overlap completely or be completely contained in an other edge of the polygon P constructed by that time. Furthermore, since the lines of the arrangement are in general position, we cannot have a vertex of the polygon in the interior of an edge without having an edge overlap.

Case (a): It is easy to see that P has no self-intersections, since in one augmentation step P is extended by the boundary of one cell of the arrangement. Therefore, no two edges of one extension belong to the same line, and thus, the edges added during the augmentation process have no inner intersection points with any other line of the arrangement.

For the cases (b) and (c) we assume that there is an unused line l that initiates an extension step causing an edge- or a vertex-overlap for the first time during the execution of the algorithm. Let e be the edge of P_E that has been scanned when the line l was encountered, f be the corresponding face of the arrangement, and l_1 be the line containing e.

Recall, that in one extension step the boundary of f except for the edge e' contained in e is added to P, that is, we add an open polygonal chain. The scanning direction on e defines an order on the edges of $\partial f \setminus \{e'\}$, where the edge contained in l is the first one, and the other edge of f adjacent to e' is the last one.

Now we make some basic observations:

O1. All intersection points of $\mathcal{A}$ are inside or on P_E. Especially, we can say that the intersection point $l \cap l_1$ is the first one on the line l.

O2. Consider a sequence of edges $e_1, \ldots, e_k$ forming an unbounded cone and an edge e_i of that sequence. If there is a line inter- 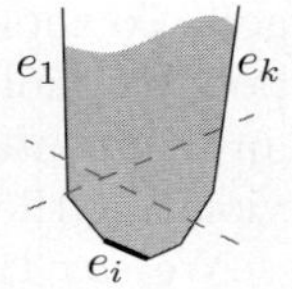
secting the chain $e_1, \ldots, e_{i-1}$ and the chain $e_{i+1}, \ldots, e_k$ then e_i is not an edge of P_E. Furthermore, if there is more than one such line, then e_i can also not be an edge of a boundary face. A similar observation holds for a vertex of the chain $e_1, \ldots, e_k$.

Case (b): Let e^* be the edge of f that is already (a part of) an edge in P. It can be shown that neither the first nor the last edge of $\partial f \setminus \{e'\}$ can be the overlapping edge. Therefore, the edge e^* must be an intermediate edge of f, as depicted in Figure 5. Let l_2 be the line containing e^*, and l_3 the line containing the edge coming after e^* in ∂f with respect to traversal direction.

Suppose e^* is an edge of the envelope polygon. Consider the possible locations of the intersection point x of l and l_3 on the line l relative to the lines l_1 and l_2. The lines l_1 and l_2 subdivide l into three sections: *(i)* Before the intersection point with l_1, i.e., outside P_E. x cannot lie in this section by Observation *O1*. *(ii)* Between the intersections with l_1 and l_2. If x is in this section, the edge e^* cannot be an edge of f. *(iii)* Behind the intersection with l_2. In this case e^* cannot be an edge of the envelope polygon by Observation *O2*.

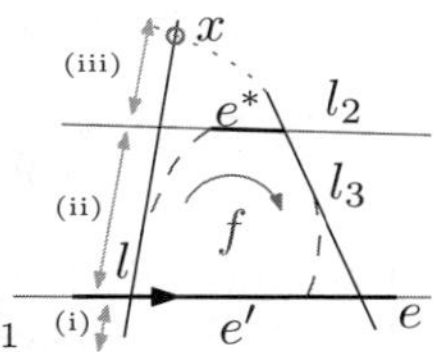

Fig. 5. Face f with an edge e^* that causes an edge-overlap. Three combinatorially different positions of the intersection point of l and l_3 on l are denoted by the double-arrows.

If e^* is an edge of a previously traversed face f', we again consider the possible positions of the intersection point of the lines l and l_3 relative to the line l_2 and to the line containing the edge of f' that is a part of an edge in P_E. In each of the possible configurations we get that either e^* is not an edge of the face f', or the face f' is not a boundary face.

From these considerations we can conclude that P has no edge-overlap.

Case (c): To exclude the possibility of a vertex-overlap, we consider possible configurations of the arrangement that might lead to a multiple usage of an intersection point.

Let v be a vertex of f, such that v was already a vertex of P by the time l was encountered. Let l_2 and l_3 denote the lines that intersect in the point v. Further since we assume a vertex-overlap, the face f contains v but does not contain the edges in P which are incident to v.

Let the scanning direction on the edge e be fixed, the opposite direction can be handled symmetrically. With respect to the chosen direction there are three possible configurations of the intersection points of lines l, l_2, l_3 with the line l_1 as

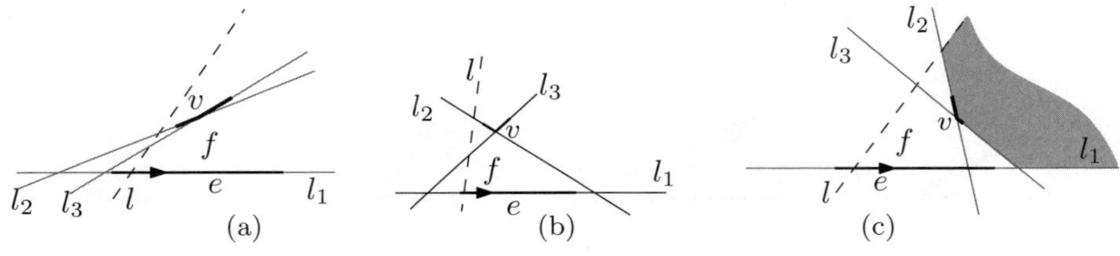

Fig. 6. Possible configurations of intersection points with lines l, l_2, l_3 on the line l_1

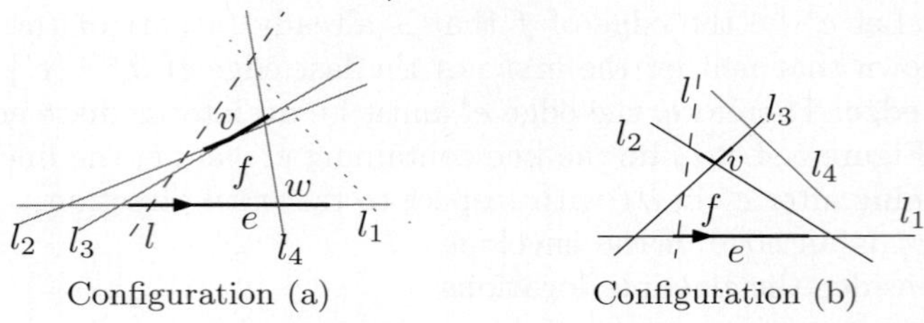

Configuration (a)   Configuration (b)

Fig. 7. Necessary continuations of configurations (a) and (b)

illustrated in Figure 6: (a) l_2 and l_3 intersect l_1 on the left of l; (b) the intersection point of l_1 with l lies between those of l_1 with l_2 and l_3; (c) the intersection points $l_2 \cap l_1$ and $l_3 \cap l_1$ are on the right of the intersection point $l \cap l_1$.

For the **configuration (a)** we observe that the line l cannot be the last one intersecting l_1. Therefore, there exists a line l_4 intersecting l_1 to the right of l as depicted in Figure 7(a). Additionally, l_4 must intersect l after l_1 by Observation *O1*, and it also may not separate the vertex v from the face f. Therefore, v cannot be a vertex of the envelope polygon, since all its incident faces are then bounded. Furthermore, if there is any other line intersecting l_1 on the right of l, then v also cannot be a vertex added in a previously performed extension step by Observation *O2*. If there were any line intersecting l_4 below l_1 and the line l_1 on the left of l, then e would not be an edge of the envelope polygon. Therefore, the vertex $w = l_1 \cap l_4$ must be a convex vertex of the envelope polygon, which contradicts the scanning direction rules.

In **configuration (b)**, if v is a vertex of the envelope polygon, there must be a line l_4 that intersects the line l_3 on the other side of v than l_1 (on the right of v in Fig. 7), because v is reflex. The line l_4 also has to cross l. The intersection point $l \cap l_4$ on the line l cannot be below line l_1 by Observation *O1*. It cannot lie between $l \cap l_3$ and $l \cap l_1$, since then v would not be a vertex in f. It also cannot lie above $l \cap l_3$, since then v is not a vertex of P_E by Observation *O2*.

Assuming that v is a vertex of a previously visited face f' we find again that there is exactly one line l_4 that intersects the line l_3 on the other side of v than l_1. Additionally we can prove that if there is a line that intersects l_2 on the same side of v as l_1, then it is the line l_4. For the line l_4 there are four possible combinatorial positions depicted in Figure 8.

In each of these subconfigurations the corresponding part of the arrangement (highlighted in Figure 8) is fixed and no other line can modify the envelope

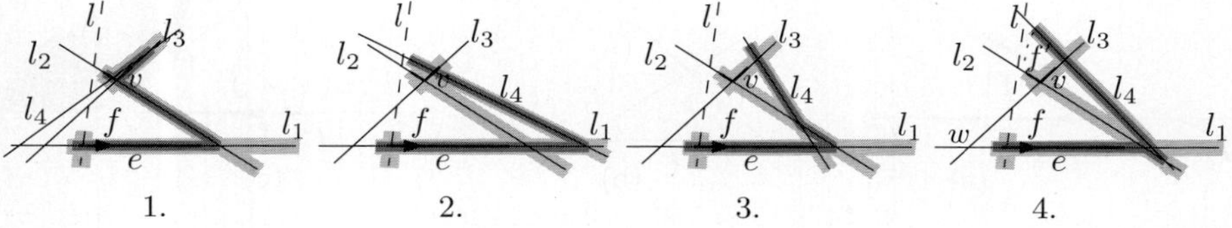

1.   2.   3.   4.

Fig. 8. Configuration (b): Possible positions of the line l_4

polygon (darker highlight). In cases 1 to 3 the right vertex of the edge e is convex in P_E, which is a contradiction to scanning direction rules. For the case 4 we can show that either we should have met the line l prior to scanning the edge e or the assumed scanning direction of e contradicts the direction rules.

Let us consider **configuration (c)**. If we assume that the vertex v is a vertex of the envelope polygon we can conclude that the point $l_1 \cap l_3$ is a convex vertex of the edge e in the envelope polygon, which contradicts the direction rules.

Now suppose that v is a vertex of a face f' traversed in one of the previous steps. Then, by Observation O2, there exists exactly one line l_4 which crosses the, so far unbounded, cone composed of lines l, l_2, l_3, l_1 (shaded region in Figure 6). Such a line l_4 intersects l_1 on the same side of the point $l_1 \cap l$ as the lines l_2 and l_3. Also there is at most one line intersecting l_2 and l_3 on the same side of v as their intersection point with l_1, that is below v in Figure 9. Consider possible locations of the intersection points $x = l_1 \cap l_4$ on l_1 and $y = l_2 \cap l_4$ on l_2: There are five combinatorial possibilities which are depicted in Figure 9:

1. x is between the intersection points $l_1 \cap l$ and $l_1 \cap l_2$, then y has to lie between v and the intersection point $l_2 \cap l_1$;
2. x is between $l_1 \cap l_2$ and $l_1 \cap l_3$, and y is on the other side of v than the point $l_2 \cap l_1$ (above v);
3. x is between $l_1 \cap l_2$ and $l_1 \cap l_3$, and y is on the other side of $l_2 \cap l_1$ than v (below l_1);
4. x is on the other side of $l_1 \cap l_3$ than $l_1 \cap l$ and $l_1 \cap l_2$ (on the right of l_3), and y is on the other side of v than $l_2 \cap l_1$ (above v);
5. x is on the other side of $l_1 \cap l_3$ than $l_1 \cap l$ and $l_1 \cap l_2$ (on the right of l_3), and y is on the other side of $l_2 \cap l_1$ than v (below l_1).

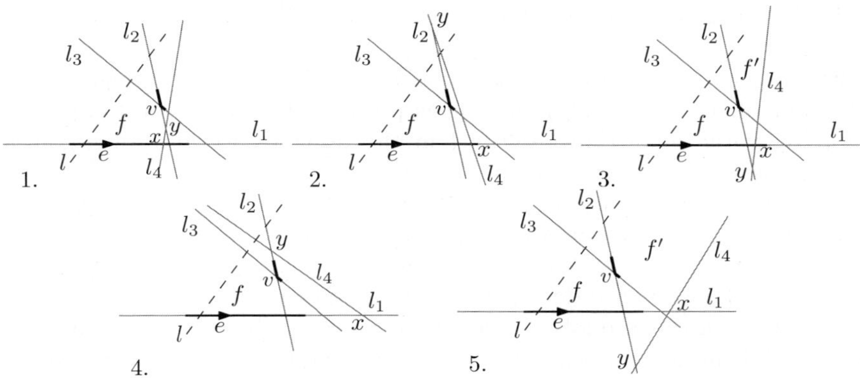

Fig. 9. Case (c): Possible positions of line l_4

In all these cases the corresponding part of the arrangement and of the envelope polygon is fixed. In cases 1, 2 and 4 the vertex incident to e, which lies in scanning direction, is convex in P_E. In these cases the traversal of the face f contradicts our direction rules.

In case 5 the envelope edge starting at the vertex y and incident to face f' has a convex vertex in P_E. According to the direction rules and the structure of f' the line causing the traversal of f' is l_1, but l_1 contributes to the envelope, which is a contradiction.

For the remaining case 3 it can be shown that if the face f' was traversed before the face f, then either we should have met l prior to scan of the edge e, or the edge e has two reflex vertices and the assumed scanning direction contradicts the rules.

We have now shown that the polygon P has no self-intersections, edge- or vertex-overlaps. Thus, the constructed polygon P is simple and comprises every line of the arrangement.

Complexity. A trivial bound for the running time of the algorithm is $O(n^2)$, since in that time we can construct and traverse the whole arrangement. The size of the constructed polygon P is $O(n)$: P consists of the edges of the envelope polygon plus the edges of some boundary faces. The size of the envelope polygon is $O(n)$ as shown by Eu et al. in [2]. This result follows from the zone theorem [4], since the envelope polygon is the zone of the infinite line. In a recent work [5] Dangelmayr et al. introduced a generalized definition of a k-zone. The k-zone of a line l in an arrangement $\mathcal{A}$ are the edges of $\mathcal{A}$ that can be connected to l by a segment intersecting at most k lines including l and the connected edge. According to this definition the zone as in [4] is a 2-zone and the boundary faces are contained in the 3-zone of the infinite line. Dangelmayr et al. showed that the complexity of a k-zone is $O(n)$ for every fixed k. It follows that the total complexity of P is linear in n.

3   Inducing n-gon: The Zigzag Algorithm

The Zigzag algorithm constructs an inducing n-gon of the arrangement $\mathcal{A}$ if the dual points of the lines in $\mathcal{A}$ lie in convex position. In contrast to the polygons generated by the algorithm described in the previous section, an n-gon is a polygon which induces every line of $\mathcal{A}$ exactly once.

Let π denote the plane containing the lines of $\mathcal{A}$, then the dual space π^* is defined as in [6]: The dual of a point $p : (a, b) \in \pi$ is the line $p^* : f(x) = ax - b$ in π^*; the dual of a line $l : f(x) = ax + b$ in π is the point $l^* : (a, -b) \in \pi^*$. We can assume w.l.o.g. that $\mathcal{A}$ does not contain a vertical line.

Define L^* as a set of duals of lines of $\mathcal{A}$. Since the lines of $\mathcal{A}$ are in general position, L^* is in general position, i.e., no three points of L^* are collinear, and all points in L^* have different x-coordinates. Let P be an inducing polygon in π with n edges. We define a *dual polygon* for P as a polygon P^* in π^* that visits the vertices of L^* in the same order as P visits the lines of $\mathcal{A}$, starting at any edge of P. P^* is uniquely defined by P and vice versa.

The Zigzag algorithm constructs a polygon P^* in the dual space. First, the convex hull of the points in L^* is computed. Then we connect the leftmost point l^*_{left} and the rightmost point l^*_{right} in L^* with an edge e. The edge e divides the

points into the *upper hull H_u* and the *lower hull H_b* of L^*. The lower hull ([6]) of
a set of points is defined as the part of the convex hull of the set that lies below
or on the line connecting the leftmost and the rightmost points of the set. The
upper hull is defined symmetrically. Starting with l_{left}^*, we connect the points of
H_b in a zig-zag manner by connecting alternately the so far unvisited points of
H_b with the largest and the smallest x-coordinate, as illustrated in Figure 10.
Thus, l_{left} is connected to the point $l_1^* \in H_b$ with the largest x-coordinate, l_1^*
is connected to the point $l_2^* \in H_b \setminus \{l_{\text{left}}, l_1^*\}$ with the smallest x-coordinate et
cetera until all points of H_b are in the path. Next, starting with l_{right}^*, the upper
hull is traversed in the "mirrored" way. Finally, we connect the end points of the
obtained path with an edge and name the polygon P^*. The corresponding path
P in primal space is an inducing n-gon for $\mathcal{A}$.

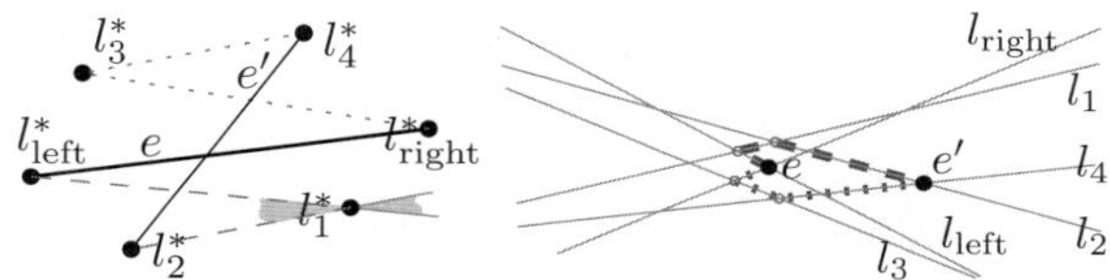

Fig. 10. The polygons P^* and P produced by the Zigzag algorithm

The time complexity of the algorithm is dominated by the construction of the
convex hull, which is $O(n \log n)$.

Correctness. By construction every point of L^* was added to P^* exactly once,
hence P is an inducing n-gon for $\mathcal{A}$. Next we show that P is simple.

The dual of a segment s in π is a left-right double wedge in π^* bounded by
the duals of the endpoints of s, see [6]. A left-right double wedge is a double
wedge not containing a vertical line. Define a *center* of a double wedge as the
intersection point of the lines that bound it. We say that two double wedges
intersect if they share a common line, see Fig. 11. In other words, two double
wedges intersect if and only if each of them contains the center of the other. It
is known ([6]) that two segments in π intersect if and only if the corresponding
double wedges intersect in π^*.

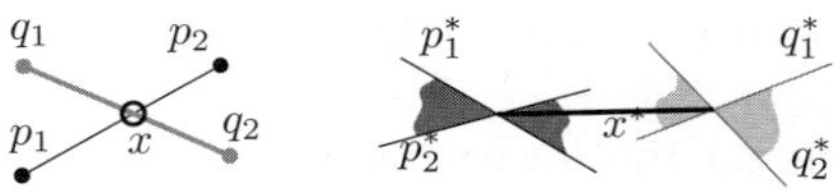

Fig. 11. Intersecting segments and their dual double wedges

Consider our polygons P and P^*. Two edges of P intersect if and only if their
dual double wedges intersect. We call two double wedges *consecutive* if they
share a bounding line. Two double wedges are consecutive if and only if they are
dual to adjacent edges of P. Hence P is simple if and only if no non-consecutive
double wedges of P^* intersect.

During the traversal of the upper and lower hull the algorithm visits alternately the so far unvisited points with the smallest and the highest x-coordinate. Hence, the x-coordinates of every three consecutively visited points p_1, p_2, p_3 form a bitonic sequence, that is the x coordinate of p_2 is either larger or smaller than the x-coordinates of both p_1 and p_2. Therefore, one part of the corresponding double wedge is bounded by the segments $\overline{p_2 p_1}$ and $\overline{p_2 p_3}$ (*inner wedge*) and the other part by the extensions of these segments in the direction of p_2 (*outer wedge*), see Fig. 12.

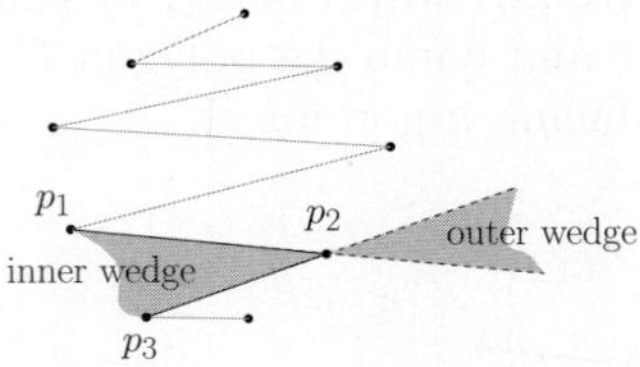

Fig. 12. Inner and outer wedge of a double wedge

Obviously, the outer wedge cannot contain any other point because it is faced outside the convex hull. The inner wedge cannot contain any other point because p_1 and p_3 are neighbors on the convex hull. Hence, two non-consecutive wedges cannot share a common line. Since the two endpoints do not lie inside any other wedge, the closing edge also does not produce double-wedges causing intersections in primal space.

4   Inducing Polyline: The Christmas Tree Algorithm

The Christmas tree algorithm constructs for an arrangement $\mathcal{A}$ of size n a simple polyline that induces every line of $\mathcal{A}$ exactly once in $O(n \log n)$ time. As the Zigzag algorithm, the Christmas tree algorithm works in the dual space π^*. In analogy to a dual polygon, the dual for a polyline P in π is defined as the polyline P^* in π^* that visits the vertices of L^* in the same order as P visits the lines of $\mathcal{A}$.

The idea of the algorithm is to traverse all points in L^* and add them to a polyline P^* in such an order that the dual polyline P in π is simple. The traversal goes as follows:

We start at any point on the lower convex hull of L^* and in direction left or right, let us say left. As long as the x-coordinate of the currently visited points is monotone decreasing (when the direction is left) or increasing (when the direction is right), the traversal goes along the lower hull (LH for short) of the so far unvisited points of L^* extended by the last visited point. If the point with the minimal or maximal x-coordinate of the convex hull is reached, the direction is reversed. Algorithm 1 formalizes the traversal in pseudo code. Figure 13 depicts an example of the polyline P^* and the corresponding inducing polyline P. Note that P^* resembles the garlands decorating a Christmas tree.

Algorithm 1. Christmas tree algorithm

---

1 set p = any point in $\mathrm{LH}[L^*]$
2 set $P^* = \{p\}$
3 set dir = left
4 **repeat**
5   **if** $\mathrm{LH}[(L^* \setminus P^*) \cup \{p\}]$ contains no point in direction dir from p **then**
6   reverse the direction in dir
7   set p = next point in $\mathrm{LH}[(L^* \setminus P^*) \cup \{p\}]$ in direction dir from p
8   set $P^* = P^* \cup \{p\}$
9 **until** $P^* = L^*$

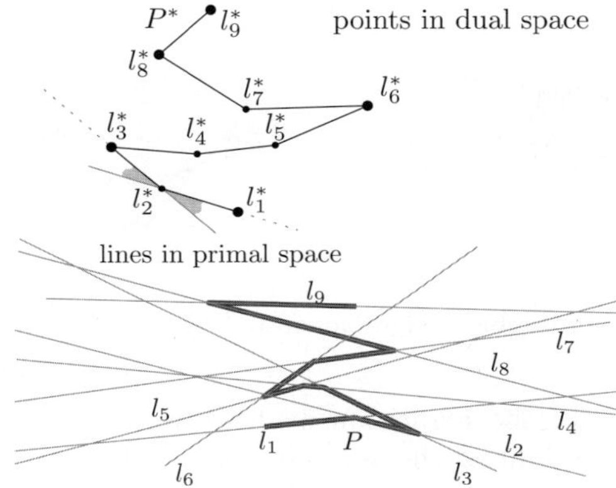

Fig. 13. The path P^* generated by the Christmas tree algorithm and the dual inducing n-path P for $\mathcal{A}$

Correctness. By construction every point of L^* was added to P^* exactly once, hence P is an inducing n-path for $\mathcal{A}$. Next we demonstrate that P is simple.

Let us consider the double wedge w centered at a point p_2, and let p_1 and p_3 denote the points visited before and after the point p_2 respectively. We analyze the position of the points in L^* that have not been visited at the time when p_2 is added to P^* and show that none of them except for p_3 is contained in w. There are two cases to distinguish: (a) the direction of the traversal stays constant at p_2 and (b) the traversal direction is reversed at p_2.

Case (a): There are two observations:
(1) The points p_1, p_2 and p_3 belong to the lower hull of the set of the so far unvisited points in L^* extended by p_2 and p_1, see Fig. 14(a).
(2) The points p_1, p_2 and p_3 form an x-monotone sequence.
From (1) and the general position of the points in L^* follows that all unvisited points except p_3 lie above the ray going from p_2 through p_1 and above the ray going from p_2 through p_3. From (1) and (2) follows that the double wedge w lies below p_1, p_2 and p_3. Thus, no double wedge can contain an unvisited center of a non-consecutive double wedge.

Case (b): The points p_2 and p_3 belong to the lower hull of the set of the so far unvisited points in L^* extended by p_2. Further, p_1, p_2 and p_3 form a bitonic sequence with respect to their x-coordinate and p_1 lies below that lower hull, see Fig. 14(b). Thus, the inner wedge of the double wedge centered at p_2 lies below the line going through p_2 and p_3. Again, it follows from the general position of the points in L^* that the inner wedge of w cannot contain an unvisited center of a non-consecutive double wedge.

Because the traversal direction changes at p_2, p_2 is the so far unvisited point with the smallest or highest x-coordinate. It follows that no other unvisited point can lie in the outer wedge of w. Since the traversal defines a total order on the points, no two non-consecutive double wedges intersect.

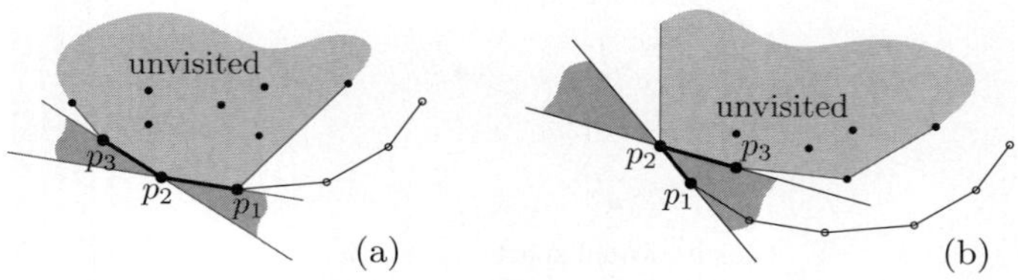

Fig. 14. Two types of double wedges generated by the Christmas tree algorithm

The running time of the algorithm is determined by a data structure for calculating the convex hull of the semi-dynamic point set starting as L^* and n delete operations on the point set in line 7 of the algorithm. The point set is semi-dynamic because the contained points are deleted only. Using the data structure of Hershberger and Suri [7] the deletions of the points can be performed in $O(\log n)$ amortized time. Thus, the overall running time for constructing an inducing polyline is $O(n \log n)$. The space required by the algorithm is linear in n.

5   Conclusions

We demonstrated that an arrangement $\mathcal{A}$ of n lines in general position in the plane has an inducing polygon of linear size that can be constructed in $O(n^2)$ time. Additionally, we presented a simpler and faster algorithm for finding a simple inducing polyline of size n for $\mathcal{A}$. Moreover, we showed that when the lines of $\mathcal{A}$ map to a set of points in convex position in dual space, an inducing n-gon for $\mathcal{A}$ can be found in $O(n \log n)$ time. Thus we widened the class of arrangements for which an inducing n-gon can be constructed. The question whether such a polygon exists for an arbitrary arrangement of lines in general position remains open.

Acknowledgements. We thank Helmut Alt, Xavier Goaoc, Hyosil Kim and Elena Mumford for fruitful discussions.

References

1. Bose, P., Everett, H., Wismath, S.: Properties of arrangement graphs. Int. J. of Computational Geometry and Applications 13(6), 447–462 (2003)
2. Eu, D., Guevremont, E., Toussaint, G.T.: On envelopes of arrangements of lines. Journal of Algorithms 21(1), 111–148 (1996)
3. Scharf, L.: An inducing simple polygon of a line arrangement. Technical Report B 08-03, Freie Universität Berlin (2008)
4. Bern, M.W., Eppstein, D., Plassman, P.E., Yao, F.F.: Horizon theorems for lines and polygons. In: Discrete and Computational Geometry: Papers from the DIMACS Special Year. DIMACS Ser. Discrete Math. and Theoretical Computer Science, vol. 6, pp. 45–66. AMS (1991)
5. Dangelmayr, C., Felsner, S., Trotter, W.T.: Intersection graphs of pseudosegments: Chordal graphs (2008) arXiv:0809.1980v1 [math.CO]
6. de Berg, M., van Kreveld, M., Overmars, M., Schwarzkopf, O.: Computational geometry: algorithms and applications. Springer, New York (1997)
7. Hershberger, J., Suri, S.: Applications of a semi-dynamic convex hull algorithm. BIT 32(2), 249–267 (1992)

Free-Form Surface Partition in 3-D

Danny Z. Chen[1,*] and Ewa Misiołek[2]

[1] Dept. of Computer Science and Engineering, Univ. of Notre Dame,
Notre Dame, IN 46556, USA
`chen@cse.nd.edu`
[2] Mathematics Department, Saint Mary's College, Notre Dame, IN 46556, USA
`misiolek@saintmarys.edu`

Abstract. We study the problem of partitioning a spherical representation S of a free-form surface F in 3-D, which is to partition a 3-D sphere S into two hemispheres such that a representative normal vector for each hemisphere optimizes a given global objective function. This problem arises in important practical applications, particularly surface machining in manufacturing. We model the spherical surface partition problem as processing multiple off-line sequences of insertions/deletions of convex polygons alternated with certain point queries on the common intersection of the polygons. Our algorithm combines nontrivial data structures, geometric observations, and algorithmic techniques. It takes $O(\min\{m^2 n \log \log m + \frac{m^3 \log^2(mn) \log^2(\log m)}{\log^3 m}, m^3 \log^2 n + mn\})$ time, where m is the number of polygons, of size $O(n)$ each, in one off-line sequence (generally, $m \leq n$). This is a significant improvement over the previous best-known $O(m^2 n^2)$ time algorithm. As a by-product, our algorithm can process $O(n)$ insertions/deletions of convex polygons (of size $O(n)$ each) and queries on their common intersections in $O(n^2 \log \log n)$ time, improving over the "standard" $O(n^2 \log n)$ time solution for off-line maintenance of $O(n^2)$ insertions/deletions of points and queries. Our techniques may be useful in solving other problems.

1 Introduction

1.1 Problem Formulation and Motivations

Constrained geometric partition is a fundamental topic in computational geometry as well as in numerous practical applications. Efficient solutions to geometric partition problems are needed in a vast array of fields, including image processing and image segmentation in computer vision [16,17], mesh decomposition in computer graphics [3,13], convex decomposition in computational geometry [4], operations research, and computer-aided design and manufacturing. Some formulations of such problems, e.g., minimizing the number of desired subsets in the partition, are NP-hard [3,15], and as a result, partitions into a fixed number of subsets have also been studied [1,11,18].

---

* The research of this author was supported in part by the National Science Foundation under Grant CCF-0515203.

S.-H. Hong, H. Nagamochi, and T. Fukunaga (Eds.): ISAAC 2008, LNCS 5369, pp. 520–531, 2008.

In this paper, we consider the problem of optimally partitioning a free-form surface F in 3-D into two sub-patches, which is an important problem in surface machining in manufacturing. A *free-form surface* is a surface that describes the shape of a manufactured product or a component (e.g., a body of a car or a part of an engine). Unlike a regular surface, a free-form surface is not described using mathematical equations; instead, it may be defined, for example, by a finite set of points on the surface and the vectors normal to the surface at those points.

In manufacturing settings, a "best" free-form surface partition is equivalent to finding a set of optimal orientations of a workpiece for surface machining using a three-, four-, or five-axis numerically controlled (NC) machine. A detailed description of the problems associated with the operations and setup of NC machining can be found in [5,18].

The problem of minimizing the number of setups has been studied [5,10,19] and has been shown to be NP-hard [6]. To deal with this computational difficulty, instead of minimizing the number of setups, one can elect to find a fixed number k of setups. In this paper, we consider partitioning a free-form surface F into two sub-patches (i.e., finding $k = 2$ setups), if possible. We also like to determine a representative direction vector for an optimal tooling direction of each sub-patch.

The following is a statement of the free-form surface partition problem. The distance between two vectors is the angle between the two vectors.

Surface Partition (SP) Problem. *Given a (discretized) free-form surface $F = (X, V)$ in 3-D, where $X = \{x_1, \ldots, x_n\}$ is a set of points on F and $V = \{v_1, \ldots, v_n\}$ is the set of vectors normal to F at the points of X, partition F into two sub-patches $F_1 = (X_1, V_1)$ and $F_2 = (X_2, V_2)$, if feasible, such that (i) the distance between the optimal representative direction vectors v_1^* and v_2^* respectively for V_1 and V_2 and all the vectors in V_1 and V_2 is as small as possible, and (ii) F is entirely accessible to the rays along the direction of v_1^* or v_2^* (i.e., F_1 is "visible" along the rays in the direction of v_1^* and F_2 is "visible" along the rays in the direction of v_2^*).*

We model the SP problem as a *spherical surface partition problem* (see [18] for the modelling details), whose key is to solve the problem of querying a dynamically changing common intersection of convex polygons. The common intersection changes as new convex polygons are added or others are deleted in an off-line sequence, which has $O(m)$ insertions/deletions of polygons of size $O(n)$ each, alternated with two types of point queries on the common intersection.

A "natural" approach for the above problem is to explicitly maintain the common intersection of convex polygons as updates occur and use the common intersection to answer queries as needed. This is what Tang and Liu did [18]. Their algorithm takes in total $O(m^2 n^2)$ time to solve the SP problem. Our approach is different in two key aspects: (1) Although the problem is of a dynamic nature, we do not handle the problem in a dynamic fashion (instead, we exploit the off-line feature of the operations); (2) we do not compute and maintain the explicit common intersection of the "active" polygons (instead, we maintain *partial* common intersections).

1.2   Our Contributions

In this paper, we present an $O(\min\{m^2 n \log\log m + \frac{m^3 \log^2(mn)\log^2(\log m)}{\log^3 m}, m^3 \log^2 n + mn\})$ time algorithm for solving the SP problem, where n is the number of sample points on the surface F and m is the number of points on the spherical representation S of F (generally, $m \leq n$). The previous best-known SP algorithm, by Tang and Liu [18], takes $O(m^2 n^2)$ time. When $m = \Theta(n)$, we achieve a nearly $O(n)$ time improvement over [18]. If $m = O(\sqrt{n})$, then the improvement is nearly $O(n^{1.5})$. Our solution combines nontrivial data structures, geometric observations, and algorithmic techniques, and is based on the following key ideas.

- We model the surface partition problem as an off-line process of convex polygon intersections with queries, instead of an on-line dynamic process as in [18]. This allows us to handle off-line sequences of polygonal intersections with queries using static data structures (e.g., time trees).
- Instead of maintaining the explicit common intersection of all "active" polygons at any moment (which is quite costly) for queries, we maintain a large (static) collection of common intersections of many subsets of polygons.
- But, performing queries on such a *collection* of common intersections (instead of on *one* common intersection) for an efficient overall algorithm then takes a much more judicious effort. We use a two-level scheme for partitioning the set of polygons into (many) subsets and come up with interesting data structures (e.g., prefix partial intersection trees) to capture all needed common intersections for any query at any time.

A key component to our algorithm is the following problem of *off-line processing of convex polygon intersections with queries.*

Off-line Process of Polygonal Intersections with Queries (OPI). *Given a set $\mathcal{P} = \{P_1, \ldots, P_m\}$ of m convex polygons of size $O(n)$ each and a set $\mathcal{I} = \{I_1, \ldots, I_m\}$ of time intervals, where $I_i = [t_i^a, t_i^d]$ represents the "life time" of the polygon P_i, i.e., t_i^a is the time when P_i is added and t_i^d is the time when P_i is deleted, process an off-line sequence of $O(m)$ operations $OP_1, OP_2, \ldots, OP_{O(m)}$, where each operation OP_i is one of the following types:*

1. *Insert a convex polygon of $\mathcal{P}$ into consideration.*
2. *Delete a convex polygon of $\mathcal{P}$ from consideration.*
3. *For a query point p_q, check if p_q is in the common intersection of the currently "alive" polygons; if not, report a point p_c on the boundary of the common intersection that is closest to p_q.*

Hershberger and Suri [12] gave an $O(n \log n)$ time algorithm for processing an off-line sequence of $O(n)$ insertions, deletions, and queries on the convex hull of points in the plane. Since maintaining the convex hull of planar points is equivalent to maintaining the common intersection of half-planes, their algorithm can be applied to OPI. A straightforward application of their algorithm to a sequence of $O(n)$ OPI operations, each update operation performing $O(n)$ point insertions or deletions (equivalent to inserting or deleting all $O(n)$ edges of a convex polygon), results in an $O(n^2 \log n)$ time solution. In comparison, our method reduces this time bound to $O(n^2 \log \log n)$ in the OPI setting. To achieve a more efficient OPI solution than simply applying [12], we utilize the following ideas.

First, we handle the off-line dynamic polygonal insertions/deletions with queries by using static data structures. Since the life times of the convex polygons (i.e., their times to be inserted in and deleted from the common intersection) are given *a priori*, we use a static *time tree* data structure so that for any time t_0, we can efficiently identify all polygons active at t_0.

Second, by using geometric observations and techniques, we query the common intersection of all polygons active at any time without explicitly computing and maintaining it. Instead, we compute and store the common intersections of various subsets of polygons, called *partial common intersections*, in a data structure called *prefix partial intersection trees*. The prefix partial intersection trees, together with the time trees, allow us to efficiently extract needed querying information for the common intersection of active polygons at any time t_0.

Third, to gain further efficiency, we partition arbitrarily the set $\mathcal{P}$ of m convex polygons into h subsets of m/h polygons each, and build the time tree and prefix partial intersection trees separately for each such subset. By carefully choosing the value of h, we can calibrate the complexity of our algorithm in terms of the parameters m and n (in this paper, $h = \frac{m}{\log^c m}$ for some constant $c \geq 3$).

Our techniques may go beyond solving the surface partition problem. For example, we efficiently solve the problem of off-line processing the common intersection of convex polygons for updates and queries, which may find other applications.

2 Our Algorithm

Our SP solution is hinged on solving $O(m)$ instances of the OPI problem. The details on how to formulate these O(m) OPI instances can be found in the full verion of the paper and in [18]. It is sufficient for us to focus on how to handle one OPI instance. Our OPI algorithm is quite involved and consists of quite a few components and steps. To ease the discussions, we first give an overview of the algorithm, and then explain its ingredients in detail.

Assume that the set of convex polygons, $\mathcal{P} = \{P_1, \ldots, P_m\}$ is stored in a set of arrays, each of which contains $O(n)$ vertices and is in clockwise order around the polygon boundary starting at the leftmost vertex. The set of time intervals, $\mathcal{I} = \{I_1, \ldots, I_m\}$, specifies the "life times" of the polygons in $\mathcal{P}$, i.e., $I_i = [t_i^a, t_i^d]$ is the time interval for P_i to be "active".

2.1 Overview of the OPI Algorithm

Our main ideas are to handle the dynamic process of polygon intersections and queries as an off-line one, and compute and use the common intersections of many subsets of polygons (instead of one common intersection for all active polygons) for queries.

We handle the off-line OPI process by using static data structures, the *time trees* and *prefix partial intersection trees* (or *PPI trees*). The *time trees* store the time intervals of $\mathcal{I}$, to allow an efficient identification of all polygons active at any given time. One could use the active polygons to compute the common intersection for a fast query, but that would be very time-consuming as in [18]. Instead, we compute the common intersections of many subsets of polygons, called *partial common intersections* (or simply *partial intersections*), and store them in the PPI trees. The set of partial intersections allows us to query the common intersection of active polygons, yielding an efficient OPI algorithm.

The partial intersections of subsets of polygons must capture needed querying information for the common intersection of all active polygons at any time, and our queries on those partial intersections must be efficient. To achieve these goals, we use a two-level scheme to partition the polygon set $\mathcal{P}$ into subsets. First, we arbitrarily partition $\mathcal{P}$ into h subsets $\mathcal{P}_1, \ldots, \mathcal{P}_h$, of $k = \frac{m}{h}$ polygons each (for a carefully chosen parameter h). Then for each subset $\mathcal{P}_i = \{P_{i_1}, \ldots, P_{i_k}\}$, we build a time tree $T_{\mathcal{I}_i}$ to store their time intervals, $\mathcal{I}_i = \{I_{i_1}, \ldots, I_{i_k}\}$. This "high-level" partition helps in efficiency of constructing our data structures.

The "low-level" partition focuses on each subset $\mathcal{P}_i$. $\mathcal{P}_i$ is further partitioned into subsets $\mathcal{P}_i^j$ based on its time tree $T_{\mathcal{I}_i}$, such that the polygons in each $\mathcal{P}_i^j$ have significant overlaps in their time intervals (and thus may be active at the same time). Each subset $\mathcal{P}_i^j$ is ordered according to the time intervals of its polygons. Every $\mathcal{P}_i^j$ is represented by a PPI tree, which stores various subsets of polygons in $\mathcal{P}_i^j$ and their common intersections. The PPI tree for $\mathcal{P}_i^j$ allows fast queries on the (implicitly represented) common intersection of any prefix subsequence of (active) polygons in $\mathcal{P}_i^j$.

The time trees and PPI trees together store the partial intersections of many subsets of polygons in $\mathcal{P}$. Each query on the common intersection of active polygons at any time is then performed on a diverse set of partial intersections, and this must be done carefully to gain efficiency. To answer a query at a time t_0, we first identify the collection of all *active* partial intersections at t_0 by searching in every time tree and its PPI trees. Then a prune-and-search process on the corresponding partial intersections is performed to find the answer to the query.

2.2   Construction and Queries of Data Structures

The static data structures, *time trees* and *prefix partial intersection trees* (PPI trees), provide means for efficiently identifying all active partial intersections for queries at any given time. As pointed out above, the partial intersections are computed and stored in PPI trees for various subsets of polygons in $\mathcal{P}$. After arbitrarily partitioning $\mathcal{P}$ into h subsets $\mathcal{P}_1, \ldots, \mathcal{P}_h$, of $k = \frac{m}{h}$ polygons each, for each subset $\mathcal{P}_i$, we construct a time tree T_i. Each time tree T_i defines (or induces) a further partition of the polygon subset $\mathcal{P}_i$ for computing partial intersections.

Time Tree. A *time tree* T_i is a standard interval tree [2,9] that stores the set $\mathcal{I}_i$ of k time intervals for the k polygons in $\mathcal{P}_i$. It is built in a top-down fashion. For $j = 1, \ldots, |T_i|$, a node v_i^j of T_i stores a subset of k_j time intervals,

$\mathcal{I}_i^j = \{I_{i_1}^j, \ldots, I_{i_{k_j}}^j\} \subseteq \mathcal{I}_i$, which are not assigned to any proper ancestor node of v_i^j in T_i and all contain a time t_i^j (see [2] for more details). This means that the polygons in $\mathcal{P}_i^j = \{P_{i_1}^j, \ldots, P_{i_{k_j}}^j\} \subseteq \mathcal{P}_i$ are all active at the time t_i^j.

The intervals in $\mathcal{I}_i^j$ at the node v_i^j are stored in two sorted arrays: L_i^j, sorted in the *increasing* order of the *left* endpoints of $\mathcal{I}_i^j$, and R_i^j, sorted in the *decreasing* order of the *right* endpoints of $\mathcal{I}_i^j$. Suppose at any time t_0, we want to use T_i to identify the set of all polygons in $\mathcal{P}_i$ active at t_0. If a node v_i^j of T_i happens to contain b^j intervals for such active polygons, then those intervals must be the first b^j elements in the "prefix" subsequence of those in L_i^j or R_i^j (depending on whether $t_0 \leq t_i^j$ or $t_0 > t_i^j$). Further, this prefix subsequence of intervals (and thus their polygons) in (say) L_i^j can be easily identified by a binary search on L_i^j in $O(\log k)$ time. Hence, we can implicitly identify the set of all active polygons in $\mathcal{P}_i$ at time t_0 by reporting the index b^j of the last interval in the prefix subsequence of L_i^j or R_i^j at each "active" node v_i^j in T_i. Since we visit at most $O(\log k)$ active nodes in T_i, the total time for identifying all active polygons in $\mathcal{P}_i$ at time t_0 is $O(\log^2 k)$. Also, since each time tree T_i (an interval tree) contains k intervals, it can be constructed in $O(k \log k)$ time and $O(k)$ space [2]. Thus, for all h time trees, we have the following result.

Lemma 1. *Given a set $\mathcal{P}$ of m polygons, the set of h time trees $T_1, \ldots, T_h$ for the subsets $\mathcal{P}_1, \ldots, \mathcal{P}_h$ of a partition of $\mathcal{P}$ can be computed in $O(m \log k)$ time and $O(m)$ space. Using the time trees, for any time t_0, we can implicitly identify the set of all polygons active at t_0 in $O(h \log^2 k)$ time.*

Since an interval tree stores each of its intervals at exactly one node, for each T_i, we have $\mathcal{I}_i^j \cap \mathcal{I}_i^{j'} = \emptyset$ for any $j \neq j'$ and $\bigcup_j \mathcal{I}_i^j = \mathcal{I}_i$. This induces a partition of the subset $\mathcal{P}_i$ of polygons for computing the partial intersections, which are stored in PPI trees.

Prefix Partial Intersection Trees. For each subset $\mathcal{P}_i^j$ in the partition of $\mathcal{P}_i$ induced by its time tree T_i, we build two *prefix partial intersection trees* (PPI trees) TL_i^j and TR_i^j. The leaves of TL_i^j, from left to right, contain the convex polygons of $\mathcal{P}_i^j$ ordered as in the array L_i^j, and the leaves of TR_i^j, from left to right, contain the convex polygons of $\mathcal{P}_i^j$ ordered as in R_i^j. Each PPI tree is built in a bottom-up fashion, so that a parent node stores a convex polygon, called *i-polygon*, that is the common intersection of the i-polygons stored at its two child nodes. The root of each PPI tree holds the common intersection of all polygons for which the tree is built. See Fig. 1(a).

Observe that the partition of polygons induced by the time trees ensures that if a polygon P_b at the bth leaf (in the left-to-right order) of a PPI tree T is active at a time t_0, then so are all polygons at the preceding $b-1$ leaves of T (i.e., all the b prefix polygons of T are active at t_0). Combining this property with the structures of the PPI trees, we are able to obtain needed information on the common intersection of any prefix subsequence of the polygons stored at the leaves of T, by looking at the partial intersections stored at multiple (in fact, $O(\log k)$) nodes of T.

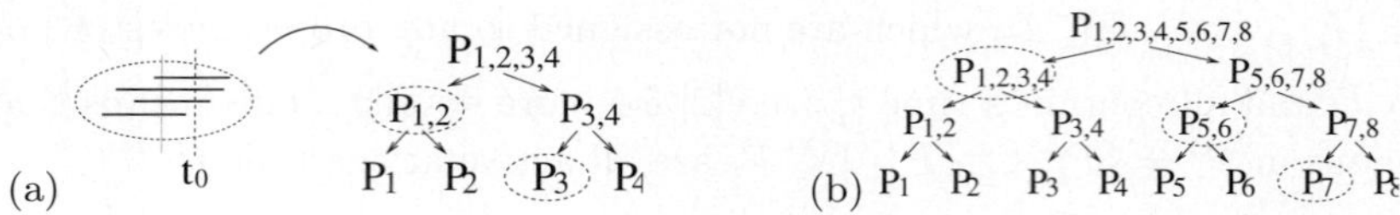

Fig. 1. Examples of PPI trees. $P_{i,j}$ denotes the intersection of polygons P_i and P_j. (a) A PPI tree TR^2 for a set of four polygons corresponding to an array R^2 of node v^2 of a time tree. Since the search for active polygons at time t_0 reports $b^2 = 3$, the two marked partial intersections are identified using TR^2. (b) A PPI tree for eight polygons. If the last prefix index is 7, the three marked polygons are identified as the partial intersections of the 7 prefix polygons.

In the example of Fig. 1(b), three partial intersections $P_{1,2,3,4}$, $P_{5,6}$, and P_7 are used for the seven prefix polygons $P_1, \ldots, P_7$. (A node $P_{i,j}$ holds the common intersection of polygons P_i and P_j.) Note that each index in $\{1, \ldots, 7\}$ appears in exactly one of the three involved partial intersections.

Suppose we want to identify the needed partial intersections for b prefix polygons $P_1, \ldots, P_b$ in a PPI tree T with k_j polygons in its leaves ($b \le k_j$). Let SP_g denote the search path in T from the root to the gth leaf (storing the polygon P_g). Let the *left fringe* of SP_g be the set of nodes of T that are the left children of the nodes on SP_g but do not belong to SP_g. (In Fig. 1(b), the left fringe of SP_7 includes the nodes $P_{1,2,3,4}$ and $P_{5,6}$.) Then for any $b < k_j$, the needed partial intersections for the b prefix polygons are those in the left fringe of SP_{b+1}. If $b = k_j$, then the (only) needed partial intersection is at the root of the PPI tree. For example, the partial intersections for the seven prefix polygons $P_1, \ldots, P_7$ in Fig. 1 are at the left fringe nodes of SP_8: $P_{1,\ldots,4}$, $P_{5,6}$, and P_7. Clearly, the left fringe of any SP_g has $O(\log k_j)$ nodes. Thus, $O(\log k_j)$ partial intersections are used for any prefix subsequence of polygons in T.

For the time bound of PPI tree construction, let $T(k_j, n)$ denote the time for building a PPI tree with k_j polygons of $O(n)$ vertices each. Then we have the following recurrence relation for $T(k_j, n)$: $T(k_j, n) = 2T(\frac{k_j}{2}, n) + ak_j n$, where $a > 0$ is a constant. The solution is $T(k_j, n) = O(k_j n \log k_j)$. The space bound for a PPI tree can be analyzed in a similar way. Hence, we have the following lemma.

Lemma 2. *Given an ordered set $\mathcal{P}_i^j$ of k_j polygons, the PPI tree TR_i^j (or TL_i^j) can be built in $O(k_j n \log k_j)$ time and space. Using TR_i^j (or TL_i^j), in $O(\log k_j)$ time, we can identify $O(\log k_j)$ partial intersections for any number of prefix polygons in $\mathcal{P}_i^j$.*

Recall that at a time tree T_i storing the time intervals of k polygons, for each node of T_i containing k_j polygons ($\sum_j k_j = k$), we build two PPI trees (TL_i^j and TR_i^j). Thus, constructing all PPI trees for a single time tree T_i takes $O(kn \log k)$ time. At any time t_0, the total number of partial intersections for one time tree T_i needed by a query at t_0 is $O(\log^2 k)$, since for each of the $O(\log k)$ active nodes of T_i at t_0, we identify $O(\log k)$ partial intersections. For h time trees, the following holds.

Lemma 3. *Given a set $\mathcal{P}$ of m polygons with $O(n)$ vertices each, the set of all PPI trees can be built in $O(mn \log k)$ time and space. Using the h time trees and their PPI trees, for any time t_0, we can identify $O(h \log^2 k)$ partial intersections for all polygons of $\mathcal{P}$ active at t_0 in $O(h \log^2 k)$ time.*

The following result is a direct consequence of Lemmas 1 and 3.

Lemma 4. *Given a set $\mathcal{P}$ of m polygons with $O(n)$ vertices each, it takes $O(mn \log k)$ time and space to construct all the time trees and PPI trees. Using these data structures, for any time t_0, we can identify all the $O(h \log^2 k)$ partial intersections active at t_0 in $O(h \log^2 k)$ time.*

2.3   Handling Queries on Common Intersections

For a query point p_q at any time t_0, we need to check (1) whether p_q is in the common intersection $\bar{P}$ of all polygons active at t_0 (the *point inclusion query*), and if not, (2) find a point on the boundary $\delta\bar{P}$ of $\bar{P}$ that is closest to p_q (the *closest point query*), provided $\bar{P} \neq \emptyset$. In the following, we call the partial intersections stored in the PPI trees as *i-polygons*.

To answer a query at time t_0, we first identify all i-polygons *active* at t_0, by searching in each time tree and its PPI trees. Let $\mathcal{P}_{t_0}$ denote the set of i-polygons active at time t_0.

Point Inclusion Query. Clearly, the common intersection $\bar{P}$ of the active polygons is the common intersection of the active i-polygons. To check whether a query point $p_q \in \bar{P}$, it is sufficient to check whether p_q is in each of the i-polygons in $\mathcal{P}_{t_0}$. This is a simple task. We perform a binary search on the vertices of each i-polygon, in $O(\log(kn))$ time. For $O(h \log^2 k)$ i-polygons, the next lemma holds.

Lemma 5. *Given any query point p_q and $O(h \log^2 k)$ i-polygons, a point inclusion query can be answered in $O(h \log^2 k \log(kn))$ time.*

If p_q is contained in every i-polygon of $\mathcal{P}_{t_0}$, then p_q is the sought point. Otherwise, some i-polygons in $\mathcal{P}_{t_0}$ do not contain p_q, and we need to proceed with the second query, i.e., finding a point p_c on the boundary $\delta\bar{P}$ of the common intersection $\bar{P}$ that is closest to p_q, or concluding that $\bar{P}$ is empty (and thus there is no feasible answer).

Closest Point Query. The problem of deciding whether the common intersection $\bar{P} = \emptyset$ and (if not) finding the closet point p_c on the boundary $\delta\bar{P}$ of $\bar{P}$ to the query point p_q is closely related to the 2-D linear programming (LP) problem [8,14]. In fact, the LP problem can be viewed as a special case of this problem, i.e., the point p_q in the LP problem is always at infinity. To solve this LP-related problem, we use several geometric observations and a prune-and-search process on a set of sorted arrays representing the boundary chains of the active i-polygons in $\mathcal{P}_{t_0}$. Due to space limit we omit the details.

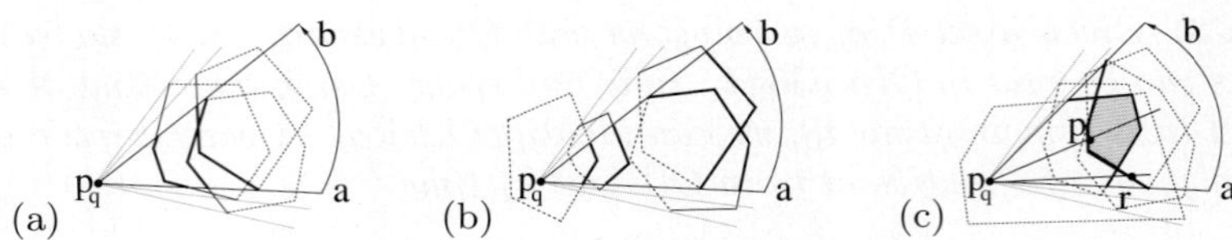

Fig. 2. (a) Three concave chains (with thick solid edges) within the angular interval $[a, b]$. (b) Four convex chains (with thick solid edges) within the angular interval $[a, b]$. (c) C^v is a subchain of U and C^n is a subchain of L. The ray $l_{p_q p}$ satisfies the statement of Lemma 6, but the ray $l_{p_q r}$ does not; hence the point r is not on the boundary $\delta \bar{P}$.

Let $\mathcal{R} = \{R_1, \ldots, R_{k_0}\}$ be the set of i-polygons in $\mathcal{P}_{t_0}$ that do not contain p_q and $\bar{R}$ be their common intersection. For any of our closest point queries, $\mathcal{R} \neq \emptyset$. Similarly, let $\mathcal{Q} = \{Q_1, \ldots, Q_{k_1}\}$ be the set of i-polygons in $\mathcal{P}_{t_0}$ that do contain p_q and $\bar{Q}$ be their common intersection. Clearly, $\bar{P} = \bar{R} \cap \bar{Q}$. For each $R_i \in \mathcal{R}$, let $C^v_{R_i}$ (resp., $C^n_{R_i}$) be the portion of the boundary chain of R_i that is visible (resp., not visible) from p_q (when the boundary of R_i is viewed as the *only* "opaque" object in the plane). and $C^n_{R_i}$ be the boundary chain of R_i that is not visible from p_q. Let the angular interval $[\alpha_i, \beta_i]$ be the set of directions in each of which a point of $C^v_{R_i}$ is visible from p_q, and $[\alpha, \beta] = \bigcap_i [\alpha_i, \beta_i]$. Note that for each $Q_i \in \mathcal{Q}$, its whole boundary is visible from p_q. Let $C^v_{Q_i}$ be the boundary chain of Q_i that is in the angular interval $[\alpha, \beta]$. Note that we only need to consider those chains in the angular interval $[\alpha, \beta]$ since $\bar{P}$ (if existing) is completely contained in this angular interval. We classify such chains of R_i's and Q_j's into two sets: The set $\mathcal{U}$ of *concave chains*, $\mathcal{U} = \{C^v_{R_i} \mid i = 1, \ldots, k_0\}$, and the set $\mathcal{L}$ of *convex chains*, $\mathcal{L} = \{C^n_{R_i} \mid i = 1, \ldots, k_0\} \cup \{C^v_{Q_j} \mid j = 1, \ldots, k_1\}$. Fig. 2(a)-(b) illustrates these two types of chains.

Since $p_q \notin \bar{P}$, we also divide the boundary $\delta \bar{P}$ of $\bar{P}$ into two chains C^v and C^n, where C^v is the chain visible from p_q and $C^n = \delta \bar{P} - C^v$. The following lemma gives a simple geometric observation.

Lemma 6. *If $\bar{P} \neq \emptyset$, then the visible chain C^v of $\delta \bar{P}$ is a subchain of the upper envelope U of the concave chains in $\mathcal{U}$ and the invisible chain C^n of $\delta \bar{P}$ is a subchain of the lower envelope L of the convex chains in $\mathcal{L}$. Moreover, any ray $l_{p_q p_d}$ originating at p_q and passing through a point p_d on $\delta \bar{P}$ intersects U before intersecting L. (See Fig. 2(c) for an illustration.)*

Finding the convex and concave chains in $\mathcal{U}$ and $\mathcal{L}$ is easy. We first do a binary search on each i-polygon $R_i \in \mathcal{R}$ to find the two tangents between p_q and R_i. The two tangents for R_i define the angular interval $[\alpha_i, \beta_i]$ for R_i. The two points at which the tangents touch R_i define the visible chain $C^v_{R_i}$ and invisible chain $C^n_{R_i}$ to p_q. Since we focus only on the angular interval $[\alpha, \beta] = \bigcap_i [\alpha_i, \beta_i]$, we find the subchains of $\{C^v_{R_i}\}_{i=1}^{k_0}$ and $\{C^n_{R_i}\}_{i=1}^{k_0}$ that are in $[\alpha, \beta]$. (To simplify the exposition, we retain the same notation for the shortened chains.) Similarly, we find the chains in $\{C^v_{Q_j}\}_{j=1}^{k_1}$ for $\mathcal{Q}$. Clearly, computing all these chains can be done in $O(\log(kn))$ time per i-polygon or in $O(h \log^2 k \log(kn))$ total time for all i-polygons in $\mathcal{P}_{t_0}$. From now on, we assume that all the concave and convex chains in $\mathcal{U}$ and $\mathcal{L}$ are available as sorted arrays of their vertices, and that $[\alpha, \beta] \neq \emptyset$.

Using $\mathcal{U}$ and $\mathcal{L}$, there are different ways to check if a given point p is on $\delta\bar{P}$ and to find a point p_c on $\delta\bar{P}$ closest to p_q (if p_c exists). For example, one could compute the envelopes U and L explicitly and then find p_c using a binary search on U and L. But, this is time-consuming, and we need a more efficient solution without computing the explicit envelopes.

We use a prune-and-search approach, in which a ray l_{α_0} originating at p_q and extending in a chosen direction α_0 in an angular interval I (initially, $I = [\alpha, \beta]$) "probes" the envelopes U and L without actually computing them. We probe U and L by finding the intersection points of l_{α_0} with all concave and convex chains. This can be done by a binary search on each chain, in totally $O(h \log^2 k \log(kn))$ time. Then, the point p_U (resp., p_L), such that the distance from p_q to p_U (resp., p_L) is the largest (resp., smallest) among all intersection points of l_{α_0} with the concave (resp., convex) chains, belongs to U (resp., L). In a similar way as for the LP algorithms [8,14], by examining some local geometry of U (resp., L) around p_U (resp., p_L) and checking the relative positions of p_U and p_L with respect to p_q (see [8,14] for more details), either we locate the sought point p_c, or conclude that $\bar{P} = \emptyset$ (and thus no feasible answer), or decide which direction to swing the ray for the next probe (i.e., we choose a subinterval $I' \subset I$, with the current α_0 as one of its endpoints, in which the next α_0 will be chosen). Once I' is established, we eliminate (prune) the portions of all chains that are in the discarded subinterval $I - I'$, and continue the search on the shortened chains.

A key to the efficiency of our above prune-and-search procedure is at selecting the ray's direction α_0 in an angular interval I so that a "significant" portion of the chains (i.e., a constant fraction of all vertices on the chains) is guaranteed to be pruned away in each probing iteration.

Our prune-and-search process on the chains is based on a weighted selection procedure [7]. Suppose the chains $C_i \in \mathcal{U} \cup \mathcal{L}$ are stored in a set of arrays A_i, each sorted in the angular order of its vertices. Clearly, there are $O(h \log^2 k)$ such sorted arrays. For each A_i, let p_i be its median element, and $w_i = |A_i|$ be the *weight* of p_i. Let M be the set of all such pairs (p_i, w_i). Then we use the *weighted median selection algorithm* [7] to find the weighted median p_{med} of M in $O(|M|) = O(h \log^2 k)$ time. Using p_{med}, we do a binary search on each A_i to prune away all its elements no bigger than p_{med} or larger than p_{med}, depending on the swinging direction of the probing ray, in $O(\log(kn))$ time (the pruning is done by simply marking the remaining subarray of A_i). Hence, every probing iteration takes $O(h \log^2 k \log(kn))$ time. It is easy to see that approximately $\frac{1}{4}$ of the total number of vertices in all arrays A_i are pruned away in each probing iteration. We continue the process recursively on the remaining subarrays of the A_i's, until only totally $O(h \log^2 k)$ vertices are left, at which point we simply compute the explicit common intersection of the $O(h \log^2 k)$ involved half-planes (in $O(h \log^2 k \log(hk))$ time), which then allows us to answer the query easily.

Let $N = O(mn)$ be the number of vertices on the $O(h \log^2 k)$ chains (with $O(kn)$ vertices per chain). The time bound $T(N)$ for the above prune-and-search process is captured by the recurrence relation $T(N) = T(\frac{3}{4}N) + ah \log^2 k \log(kn)$, where $a > 0$ is a constant. Thus, $T(N) = O(h \log^2 k \log(kn) \log(mn)) = O(h \log^2 k \log^2(mn))$.

Lemma 7. *Every closest point query can be answered in* $O(h \log^2 k \log^2(mn))$ *time.*

2.4   Putting Things Together

Based on Lemmas 1, 4, 5, and 7, our algorithm for one instance of the OPI problem, on m convex polygons of $O(n)$ vertices each, takes $O(mn \log k)$ time to build the data structures, and $O(mh \log^2 k \log^2(mn))$ time to answer the $O(m)$ queries.

Theorem 8. *Given a set $\mathcal{P}$ of m convex polygons with $O(n)$ vertices each, the OPI problem can be solved in* $O(mn \log k + mh \log^2 k \log^2(mn))$ *time, where h is the number of subsets in an arbitrary partition of $\mathcal{P}$, with each subset having $k = \frac{m}{h}$ polygons.*

We need to choose the value of h in terms of m and n to calibrate the time bound of our OPI algorithm. Note that in the surface machining settings, $m \leq n$ holds. Thus we assume $O(\log m) = O(\log n)$. In this case, let $h = \frac{m}{\log^3 m}$ (hence $k = \log^3 m$). This gives an OPI algorithm of $O(mn \log \log m + \frac{m^2 \log^2(mn) \log^2(\log m)}{\log^3 m})$ time. (In fact, we can let $h = \frac{m}{\log^c m}$ for some constant $c \geq 3$.)

However, when m is much smaller than n (even though $\log m = \Theta(\log n)$ still holds), e.g., $m = O(\sqrt{n})$, we can do even better. In this case, we avoid computing *any* partial intersections at all (since n is too large). Instead, we use only *one* time tree, in which each input convex polygon P_i is viewed as a partial intersection by itself. Then the preprocess builds only a single time tree. Every query involves $O(m)$ input polygons (which are found from the time tree). Using the techniques in Section 2.3, a query can be answered in $O(m \log^2 n)$ time. This leads to an $O(m^2 \log^2 n)$ time OPI algorithm (if the m polygons of $O(n)$ vertices each are already stored in the main memory of the computer, i.e., no time is spent on reading in the input). Since solving the SP problem involves $O(m)$ OPI instances, we have the following result.

Theorem 9. *The SP problem can be solved in* $O(\min\{m^2 n \log \log m + \frac{m^3 \log^2(mn) \log^2(\log m)}{\log^3 m}, m^3 \log^2 n + mn\})$ *time.*

Finally, we should mention that if none of its $O(m)$ OPI instances yields a feasible solution, then the SP problem instance cannot be 2-partitioned as desired.

References

1. Avis, D.: Diameter partitioning. Discrete and Computational Geometry 1(3), 265–276 (1986)
2. de Berg, M., Cheong, O., van Krefeld, M., Overmars, M.: Computational Geometry: Algorithms and Applications, 3rd edn. Springer, Heidelberg (2008)

3. Chazelle, B., Dobkin, D.P., Shouraboura, N., Tal, A.: Strategies for polyhedral surface decomposition: An experimental study. Computational Geometry: Theory and Applications 7, 327–342 (1997)
4. Chazelle, B., Palios, L.: Decomposition algorithms in geometry. In: Bajaj, C. (ed.) Algebraic Geometry and its Applications, vol. 27, pp. 419–447. Springer, Heidelberg (1994)
5. Chen, L.-L., Chou, S.-Y., Woo, T.C.: Separating and intersecting spherical polygons: Computing machinability on three-, four-, and five-axis numerically controlled machines. ACM Transactions on Graphics 12(4), 305–326 (1993)
6. Chen, L.L., Woo, T.C.: Computational geometry on the sphere for automated machining. ASME J. Mech. Des. 114(2), 288–295 (1992)
7. Cormen, T.H., Leiserson, C.E., Rivest, R.L., Stein, C.: Introduction to Algorithms, 2nd edn. MIT Press, Cambridge (2001)
8. Dyer, M.E.: Linear time algorithms for two- and three-variable linear programs. SIAM Journal on Computing 13, 31–45 (1984)
9. Edelsbrunner, H.: Dynamic data structures for orthogonal intersection queries, Report F59, Institut für Informationsverarbeitung, Technische Universität Graz, Graz, Austria (1980)
10. Gupta, P., Janardan, R., Majhi, J., Woo, T.: Efficient geometric algorithms for workpiece orientation in 4- and 5-axis NC-machining. Computer-Aided Design 28(8), 577–587 (1996)
11. Hershberger, J., Suri, S.: Finding tailored partitions. In: Proc. 5th Annual ACM Symposium on Computational Geometry, pp. 255–265 (1989)
12. Hershberger, J., Suri, S.: Off-line maintenance of planar configurations. Journal of Algorithms 21, 453–475 (1996)
13. Katz, S., Tal, A.: Hierarchical mesh decomposition using fuzzy clustering and cuts. ACM Transactions on Graphics (SIGGRAPH 2003) 22(3), 954–961 (2003)
14. Megiddo, N.: Linear programming in linear time when the dimension is fixed. Journal of ACM 31, 114–127 (1984)
15. Megiddo, N., Supowit, K.: On the complexity of some common geometric location problems. SIAM Journal on Computing 13(1), 182–196 (1984)
16. Shi, J., Malik, J.: Normalized cuts and image segmentation. IEEE Transactions on Pattern Analysis and Machine Intelligence 22(8), 888–905 (2000)
17. Sonka, M., Hlavac, V., Boyle, R.: Image Processing, Analysis, and Machine Vision, 2nd edn. PWS Pub. (1999)
18. Tang, K., Liu, Y.-J.: An optimization algorithm for free-form surface partitioning based on weighted Gaussian image. Graphical Models 67, 17–42 (2005)
19. Tang, K., Woo, T., Gan, J.: Maximum intersection of spherical polygons and workpiece orientation for 4- and 5-axis machining. Journal of Mechanical Design 114, 477–485 (1992)

Approximate Nearest Neighbor Search under Translation Invariant Hausdorff Distance[*]

Christian Knauer and Marc Scherfenberg

Institute of Computer Science, Freie Universität Berlin
{knauer,scherfen}@mi.fu-berlin.de

Abstract. The Hausdorff distance is a measure for the resemblance of two geometric objects. Given a set of n point patterns and a query point pattern Q, the nearest neighbor of Q under the Hausdorff distance is the point pattern which minimizes this distance to Q. An extension of the Hausdorff distance is the translation invariant Hausdorff distance which additionally allows the translation of the point patterns in order to minimize the distance. This paper introduces the first data structure which allows to solve the nearest neighbor problem for the *directed* Hausdorff distance under *translation* in sublinear query time in a non-heuristic manner, in the sense that the quality of the results, the performance, and the space bounds are guaranteed. The data structure answers queries for both directions of the directed Hausdorff distance with a $\sqrt{d(s-1.5)}(1+\epsilon)$-approximation factor in $O(\log \frac{n}{\epsilon})$ query time for the nearest neighbor and $O(k+\log n)$ query time for the k-th nearest neighbor for any $\epsilon > 0$. (The O-notation of the latter runtime contains terms that are quadratic in ϵ^{-1}.)

Furthermore it is shown how to find the exact nearest neighbor under the directed Hausdorff distance without transformation of the point sets within some weaker time and storage bounds.

1 Introduction

The Hausdorff distance is a well known and natural measure to quantify the similarity between two geometric objects. Therefore, it is of strong interest to have a data structure which supports the retrieval of patterns using this distance measure and its extensions. This paper presents the first data structure which solves the nearest neighbor problem for the *directed* Hausdorff distance under *translation* in sublinear query time and which is not a heuristic in the sense that the quality of the results, the performance, and the space bounds are guaranteed.

The directed Hausdorff distance $\overrightarrow{h}$ under L_∞ between two point sets Q and P is defined as

$$\overrightarrow{h}(Q,P) = \max_{q \in Q} \min_{p \in P} \|q - p\|_\infty$$

and $\overleftarrow{h}(Q,P) = \overrightarrow{h}(P,Q)$ for the opposite direction.

---

[*] Supported by the German Research Foundation (DFG), grant AL 253/5-1.

S.-H. Hong, H. Nagamochi, and T. Fukunaga (Eds.): ISAAC 2008, LNCS 5369, pp. 532–543, 2008.

When translations of the point sets are allowed, this distance measure is called the *translation invariant* Hausdorff distance $\overrightarrow{h}^{\mathrm{T}}$ defined as

$$\overrightarrow{h}^{\mathrm{T}}(Q, P) = \min_{t \in \mathbb{R}^d} \max_{q \in Q} \min_{p \in P} \|q - p + t\|_\infty \tag{1}$$

and $\overleftarrow{h}^{\mathrm{T}}(Q, P) = \overrightarrow{h}^{\mathrm{T}}(P, Q)$ respectively.

In contrast to the undirected Hausdorff distance, the directed Hausdorff distance can be used for partial matching, which also finds fractional resemblance. Because the proposed search structure allows queries for both directions of the distance measure, it can be applied in order to find point sets which contain parts of the query point set and – vice versa – point sets which *are* a part of the query point set.

Let $\mathcal{U}$ be a set of n point patterns, where each pattern $P \in \mathcal{U}$ consists of s points $p_1, ..., p_s$ in the metric space $(\mathbb{R}^d, L_\infty)$. Given a query point pattern Q, we are looking for the set $P \in \mathcal{U}$ which is nearest to Q with respect to the directed Hausdorff distance under translation. The k-th nearest neighbor problem extends this question to retrieve the point patterns $P \in \mathcal{U}$ having the k smallest Hausdorff distance to Q under translation.

The paper consists of two parts. First it is shown how an embedding can be used to solve the exact nearest neighbor problem for the directed Hausdorff distance without allowing transformations of the point patterns. This data structure employs a nearest neighbor search for single points in a high dimensional metric space under the maximum metric.

The second part of the paper extends the first structure in order to retrieve an approximate nearest point pattern under translation. It makes use of a nearest neighbor search for single points in a high dimensional Euclidean space.

Throughout this paper we will write $\mathbb{R}_2^d$ to denote the Euclidean d-space and $\mathbb{R}_\infty^d$ for the d-space with the maximum metric. Furthermore, $[n]$ denotes the set of integers $\{1, \dots, n\}$.

Table 1 shows the performance of the data structure for fixed dimension d and fixed size s of the point sets under the translation invariant Hausdorff distance and for the case when translations are not allowed. The given references describe the employed nearest neighbor search for single points in the embedding space.

The O-notation hides that the needed space is exponential in s – see Table 2 and 3 for the time and space bounds in dependance of the dimension d and the size s of the point sets. Furthermore the approximation depends on both parameters in the translation invariant case. Thus, the search structure is mainly suitable for a fast retrieval of small point sets ($s \ll n$).

2   Related Work

There are only very few results about the nearest neighbor search under the Hausdorff distance. Some of the data structures are designed for the search under the directed Hausdorff distance, some under the undirected Hausdorff distance and other under both variants of this distance function. In contrast to

Table 1. Performance of the data structure for fixed dimension and size s of the point patterns. The table shows the upper bounds for the retrieval when translation invariant case and when translations of the patterns are not allowed. † The O-notation suppresses terms that are quadratic in ϵ^{-1}. ‡Holds for $\overrightarrow{h}$ only if $\epsilon \in O(s)$.

$\overrightarrow{h}^{\mathrm{T}},$ $\overleftarrow{h}^{\mathrm{T}}$		query time	prepr. time/space	approx, factor
	approx. nns [1]	$O(\log \frac{n}{\epsilon})$	$O(n\frac{\log n}{\epsilon^{O(1)}} \log^2 \frac{n}{\epsilon})$	$\sqrt{d(s-1.5)}(1+\epsilon)$
	approx. k-nns† [2]	$O(k + \log n)$	$O(n^{O(1)})$	
$\overrightarrow{h},$ $\overleftarrow{h}$		query time	preprocessing time	space
	exact nns† [3]	$O(\log^{O(1)} n)$	$O(n \log^{O(1)} n)$	$O(n \log^{O(1)} n)$
	apprx. nns [4]	$O(\log^{O(1)} n)$	$O(n)$	$O(n^{1+\epsilon} \log^{O(1)} n)^{\ddagger}$
	app. k-nns [5]	$O((k + 1/\epsilon^{O(1)}) \log n)$	$O(n \log n)$	$O(n)$

the directed Hausdorff distance, the undirected Hausdorff distance fulfills the properties of a metric and corresponding search structures can benefit especially from the triangle inequality, e.g., by using reference points. In that aspect, the directed Hausdorff distance seems far more difficult to handle.

Braß and Knauer [6] developed a data structure for searching the nearest neighbor under the directed and undirected Hausdorff distance without allowing transformations of the point sets in $O(m \log^2 n)$ query time with size and pre-processing time in $O(n2^{cn^{2d+1}})$, where m is the size of the query point set, n is the total number of patterns and $c > 0$ is a suitable constant. With respect to our setting and notation this data structure yields a query time in $O(s \log^2(ns))$ with size and preprocessing in $O(ns2^{c(ns)^{2d+1}})$. Like the data structure proposed in this paper, Braß and Knauer's structure is capable of searching under both directions of the Hausdorff distance and needs space exponential in the size s of the point sets.

Farach-Colton and Indyk [7] gave a rather complicated approximate nearest neighbor algorithm for the undirected Hausdorff distance and stated that it is extendable to translation invariance. For $d = 2$ and $d = 3$ their data structure answers queries in $O(s^2 \log n)$ time and needs superpolynomial storage $n^{O(\log s)}$ for a constant factor approximation or $s^2 n^{1+\rho}$ space with any $\rho > 0$ for an approximation factor of $O(\log \log s)$. As the data structure proposed in this paper, their search structure is based on a nearest neighbor search in an embedding space.

Another technique which is capable of retrieving the nearest point set under the undirected Hausdorff distance makes use of the so-called vantage points [8]. The vantage point approach is a heuristic which also requires the triangle inequality.

3   Exact Nearest Neighbor Search without Transformation

Theorem 1. *Given a set $\mathcal{U}$ of n point patterns, where each pattern consists of s points in R_∞^d, and a query point pattern Q, the nearest neighbor of Q in $\mathcal{U}$ under* *the directed Hausdorff distances $\overrightarrow{h}$ and $\overleftarrow{h}$ can be determined by an L_∞-nearest neighbor search for single points in $\mathbb{R}^{sd}$.*

Proof (Theorem 1). Let us first consider the distance $\overrightarrow{h}$. The search structure makes use of the following embedding:

Each point pattern $P \in \mathcal{U}$ is represented by a set $\hat{P}$ of s^s points in the embedding space $\mathbb{R}^{sd}$. We call such a point $\hat{p}$ a representative point of P. Each $\hat{p} \in \hat{P}$ represents a different so-called variation $v \in V$ (which combines a combination and a permutation) of P's points, i.e. we *choose* and *order* s candidates $p_{v(i)}$ out of s points of P *with repetition*, yielding s^s different mappings.

$$v : [s] \rightarrow [s]. \tag{2}$$

Each variation generates a representative point $\hat{p} \in \mathbb{R}^{sd}$ by using the candidate points' coordinates as the coordinates of the representative point:

$$\begin{aligned}
\hat{p} = (\; &p_{v(1)_1}, p_{v(1)_2}, ..., p_{v(1)_d}, \\
&p_{v(2)_1}, p_{v(2)_2}, ..., p_{v(2)_d}, \\
&\quad\vdots \\
&p_{v(s)_1}, p_{v(s)_2}, ..., p_{v(s)_d} \;).
\end{aligned} \tag{3}$$

Additionally, a representative point $\hat{q} \in \mathbb{R}^{sd}$ for the query point set Q is needed. This point is derived in the same way as a representative point for a set $P \in \mathcal{U}$, but the assignment of the coordinates must correspond to any *permutation* of Q's points.

Performing a nearest neighbor search in the embedding space $\mathbb{R}^{sd}$ for the representative $\hat{q}$ of Q returns some representative point $\hat{p}$ for some point set P with the following distance:

$$\min_{\hat{p}\in\hat{P}} \|\hat{p} - \hat{q}\|_\infty$$

$$= \min_{\hat{p}\in\hat{P}} \max_{i=1}^{sd} |\hat{p}_i - \hat{q}_i|$$

Using the variation v which corresponds to $\hat{p}$ we get:

$$= \min_{v\in V} \max_{j=1}^{s} \max_{k=1}^{d} |q_{j_k} - p_{v(j)_k}|$$

$$= \min_{v\in V} \max_{j=1}^{s} \|q_j - p_{v(j)}\|_\infty$$

$$= \max_{q\in Q} \min_{p\in P} \|q - p\|_\infty$$

$$= \overrightarrow{h}(Q, P)$$

Thus, the representative $\hat{p}$ nearest to $\hat{q}$ belongs to the point set P which is the nearest neighbor of Q under the directed Hausdorff distance $\overrightarrow{h}(Q, P)$.

The data structure looks slightly different when queries in the opposite direction under $\overleftarrow{h}$ shall be answered. Instead of using s^s representative points per $P \in \mathcal{U}$ we use just one $\hat{p}$ which represents a *permutation* of P's points. On the other hand s^s queries are performed on the embedding space for a single point set Q, each of them using a different variation of Q's points in $\hat{q}$. Thus, we consider all valid point mappings from P to Q. The formal proof is equivalent. $\square$

Depending on the application, different nearest neighbor search structures for L_∞ in high dimensional spaces can be employed. Applying the exact search structure given by Gabow et al. [3, Theorem 5.1], the $\left(4 \left\lceil \log_{1+\epsilon} \log(4ds) \right\rceil + 1\right)$-approximate nearest neighbor structure of Indyk [4] and the $(1 + \epsilon)$-approximate k-nearest neighbors search structure of Arya et al. [5] yields the time and space bounds shown in Table 2.

Table 2. Time and space bounds for retrieving the exact nearest point set, the approximate nearest point set and the k nearest point sets under the directed Hausdorff distance without allowing translations

		$\overrightarrow{h}$	$\overleftarrow{h}$
exact nns [3]	query time	$O\left((ds)!\,(2\log(ns^s))^{\max(ds-1,1)}\right)$	$O\left(s^{2s}d!\,(2\log n)^{\max(ds-1,1)}\right)$
	prepr. time	$O\left(ns^{2s}d!\,(2\log(ns^s))^{\max(ds-1,1)}\right)$	$O\left(ns^{2s}d!\,(2\log n)^{\max(ds-1,1)}\right)$
	space	$O\left(ns^{2s}d!\,(2\log(ns^s))^{ds-1}\right)$	$O\left(ns^{2s}d!\,(2\log n)^{ds-1}\right)$
app. nns [4]	query time	$O\left(ds\log^{O(1)}(ns^s)\right)$	$O\left(ds^{s+1}\log^{O(1)}n\right)$
	prepr. time	$O(ns^s)$	$O(n)$
	space	$O\left(ds^{1+s(1+\epsilon)}n^{1+\epsilon}\log^{O(1)}(ns^s)\right)$	$O\left(ds^{1+s}n^{1+\epsilon}\log^{O(1)}ns\right)$
a. k-nns [5]	query time	$O\left(\left(k+\left\lceil 1+\frac{6ds}{\epsilon}\right\rceil^{ds}\right)ds\log(ns^s)\right)$	$O\left(\left(k+\left\lceil 1+\frac{6ds}{\epsilon}\right\rceil^{ds}\right)ds^{s+1}\log n\right)$
	prepr. time	$O\left(dns^{s+1}\log(ns^s)\right)$	$O(dsn\log n)$
	space	$O\left(dns^{s+1}\right)$	$O(dsn)$

4   Approximate Nearest Neighbor Search under Translation

This section extends the technique of the previous section in order to determine the nearest neighbor under the translation invariant Hausdorff distance.

Theorem 2. *Given a set $\mathcal{U}$ of n point patterns, where each pattern consists of s points in R_∞^d, and a query point pattern Q, the nearest neighbor of Q in $\mathcal{U}$ under the translation invariant directed Hausdorff distances $\overrightarrow{h}^{\mathrm{T}}$ and $\overleftarrow{h}^{\mathrm{T}}$ can be $\sqrt{d(s-1.5)}$-approximated by a nearest neighbor search for single points in $\mathbb{R}_\infty^{sd}$ for $s \geq 3$. The approximation factor is $\sqrt{d}$ in case of $s = 2$.*

Theorem 3. *Given a set $\mathcal{U}$ of n point patterns, where each pattern consists of two points in R_∞^d, and a query point pattern Q, the nearest neighbor of Q in $\mathcal{U}$ under the translation invariant directed Hausdorff distances $\overrightarrow{h}^{\mathrm{T}}$ and $\overleftarrow{h}^{\mathrm{T}}$ can be determined exactly by a nearest neighbor search for single points in $\mathbb{R}_\infty^d$.*

The proof of Theorem 2 and 3 extends the techniques shown in the previous section. It consists of three parts: First the embedding space is divided into d subspaces. With the help of these subspaces and the following Lemma 1 we realize the translation invariance. Finally all subspaces are combined into a slightly different embedding space again.

Throughout this section we denote a vector which consists only of ones by $\mathbf{1} = (1, ..., 1)$.

Lemma 1. *The distance between two parallel lines m and l in direction $\mathbf{1}$ in $\mathbb{R}_\infty^s$ can be $\sqrt{s - 1.5}$-approximated by the distance of two points in $\mathbb{R}_2^{s-1}$ for $s \geq 3$. The calculation is exact in the case of $s = 2$.*

Proof (Lemma 1). The lines m and l are represented by an arbitrary point on them. Let us use the points $a \in m$ and $b \in l$ whose s-th coordinate is 0 for this purpose. The distance

$$\|m - l\|_\infty$$

can be attained from a to an optimal translation of the point b along l

$$\|m - l\|_\infty = \min_{x \in \mathbb{R}} \|a - b + x\mathbf{1}\|_\infty .$$

With $\delta = a - b = (\delta_1, ..., \delta_s)$ we have

$$\|m - l\|_\infty = \min_{x \in \mathbb{R}} \|\delta + x\mathbf{1}\|_\infty .$$

This value is the minimum of the upper envelope of the family of functions $f_k(x) = |\delta_k + x|$ for $k = 1...s$, see Figure 1. The minimum is attained at the intersection of the two functions $|\delta_{\min} + x|$ and $|\delta_{\max} + x|$ with $\delta_{\min}$ and $\delta_{\max}$ being the minimal and maximal δ_k. Due to the slope of the functions, this happens at $-(\delta_{\min} + \delta_{\max})/2$ with the value $(\delta_{\max} - \delta_{\min})/2$. Since we do not know the maximal and minimal δ_k, we have to determine the maximal difference over all possible combinations:

$$\|m - l\|_\infty = \frac{1}{2} \max_{\substack{i,j \in [s] \\ i < j}} |\delta_i - \delta_j| .$$

Remember that we have chosen a and b such that $a_s = b_s = 0$. It follows that $\delta_s = 0$ and we can reduce the number of parameters:

$$\|m - l\|_\infty = \frac{1}{2} \max \left(\max_{\substack{i,j \in [s-1] \\ i < j}} |\delta_i - \delta_j|, \max_{i \in [s-1]} |\delta_i| \right) . \tag{4}$$

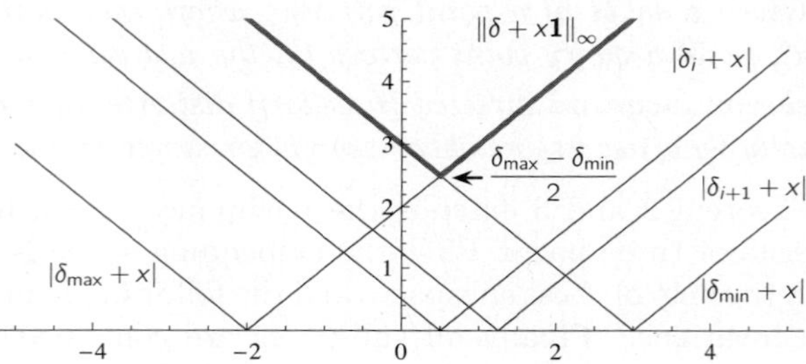

Fig. 1. The function $\min_{x\in\mathbb{R}}\|\delta + x\mathbf{1}\|_\infty$ is the minimum of the upper envelope of a family of functions

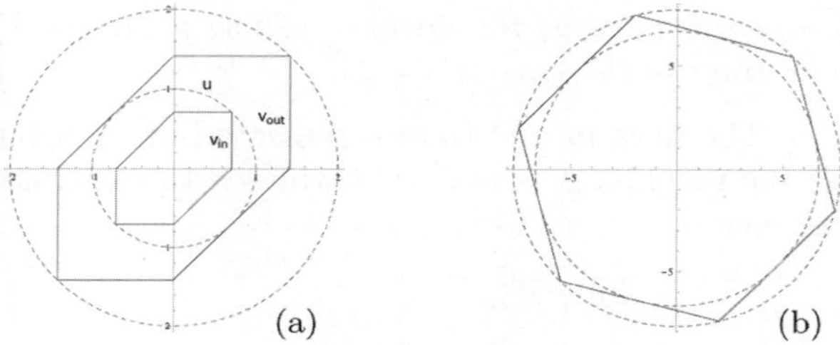

Fig. 2. (a) The points on each curve have the same distance to the origin in different metrics. The dashed curves are based on the Euclidean metric and the solid curves on the metric $\frac{1}{2}\max\left(\max_{i,j\in[s-1],i<j}|\delta_i - \delta_j|,\max_{i\in[s-1]}|\delta_i|\right)$ as defined in (4) for $s = 3$. When approximating one metric by the other the approximation factor is the quotient of the radii of the two curves in one metric. (b) The unit sphere B transformed by S for $s = 3$ (solid curve) and its circumscribed and inscribed Euclidean sphere (dashed curves).

Setting the latter equation to 1 and the point $a = 0$, this formula defines a unit sphere in $\mathbb{R}^{s-1}$ according to the L_∞-distance for lines. We denote this sphere (which is in fact a polyhedron) by B. Figure 3a shows B for $s = 4$.

We now approximate the metric given by (4) by a Minkowski-metric.

When approximating one metric by another, say M_1 by M_2, the approximation factor c can be calculated by the ratio of the radii of two unit spheres. Given a unit sphere u in the approximating metric M_2 and its circumscribed and inscribed spheres v_{in} and v_{out} in the original metric M_1, the approximation factor equals the radius of v_{out} divided by the radius of v_{in}, see Figure 4a. This holds, because a point on u with the smallest norm in M_1 lies on v_{in} and another point on u with the largest norm in M_1 lies on v_{out}. However, since both points lie on u they have the same norm in the approximation metric M_2. It is easy to see, that an alternative way of calculating c is to inscribe and circumscribe the unit sphere in the original metric by spheres in the approximation metric, as the resulting factor stays the same.

Note, that there are $2\binom{s-1}{2} + 2(s-1) = \Omega(s^2)$ facets in B, from which $\Omega(s^2)$ are point symmetric to the origin. Thus, approximating B with the unit sphere of

L_∞ – a hypercube having just $2s$ facets – is unlikely to gain good approximation factors for large s. Instead we chose L_2 for the approximation.

It remains to calculate the radii r_{in} and r_{out} of the inscribed and circumscribed Euclidean spheres of B. Before doing so, we improve the approximation factor by applying a linear transformation to the space, which transforms B into a polyhedron more similar to a Euclidean sphere. A proper transformation for that purpose is a shearing S which maps the two farthest vertices of B with different norms according to L_2 onto the same Euclidean sphere.

$$S = \begin{pmatrix} 1 & \omega & \cdots & \omega \\ \omega & 1 & \cdots & \omega \\ \vdots & \vdots & \ddots & \vdots \\ \omega & \omega & \cdots & 1 \end{pmatrix}$$

The two farthest points in B are the intersection point of the hyperplanes defined by $1 = \frac{1}{2}\delta_i$ for $i \in [s-1]$ and the intersection point of the hyperplanes defined by $1 = -\frac{1}{2}\delta_i$ for $i \in [s-1]$. These hyperplanes corresponding to the facets defined by the term $\max_{i\in[s-1]} |\delta_i|$ in equation (4). As both points have the same norm it is sufficient to consider one of them, e.g. $(2, ..., 2)$. A farthest vertex of the remaining points is $(2, ..., 2, 0)$, which is an intersection point of the hyperplanes defined by $1 = \frac{1}{2}\delta_i$ for $i \in [s-2]$ and the hyperplane $1 = \frac{1}{2}(\delta_{s-2} - \delta_{s-1})$. Adjusting the norms of the points we have

$$\left\| S \begin{pmatrix} 2 \\ \vdots \\ 2 \\ 2 \end{pmatrix} \right\|_2 = \left\| S \begin{pmatrix} 2 \\ \vdots \\ 2 \\ 0 \end{pmatrix} \right\|_2 = r_{\text{out}}.$$

One of the two valid solutions for this equation is

$$\omega = \frac{2}{(s-3)^2 - (s-2)^2} - \sqrt{\frac{4}{[(s-3)^2 - (s-2)^2]^2} + \frac{1}{(s-2)\left[(s-3)^2 - (s-2)^2\right]}}$$

yielding

$$r_{\text{out}} = \left| 2\sqrt{s-1}\,(1 + (s-2)\omega) \right|.$$

(The other solution for ω adds the square-root instead of subtracting, leading to a sharing S'. See Figure 3b/c for a comparison of S and S' for $s = 4$.)

Due to space limitations, the calculation of the radius r_{in} of the inscribed Euclidean sphere of the transformed B is omitted. It can be shown that $r_{\text{in}} = \sqrt{2}(1 - \omega)$.

It remains to calculate the approximation factor, which is the quotient of the circumscribed and the inscribed Euclidean sphere of B. Insertion and simplifying gives

$$\frac{r_{\text{out}}}{r_{\text{in}}} = \sqrt{\frac{(2s-2)(s-2)}{(2s-3)}} = \sqrt{s - \frac{3}{2} - \frac{1}{4s-6}} < \sqrt{s - \frac{3}{2}}.$$

Thus, for $s \geq 3$ the approximation factor is slightly better than $\sqrt{s - 1.5}$.

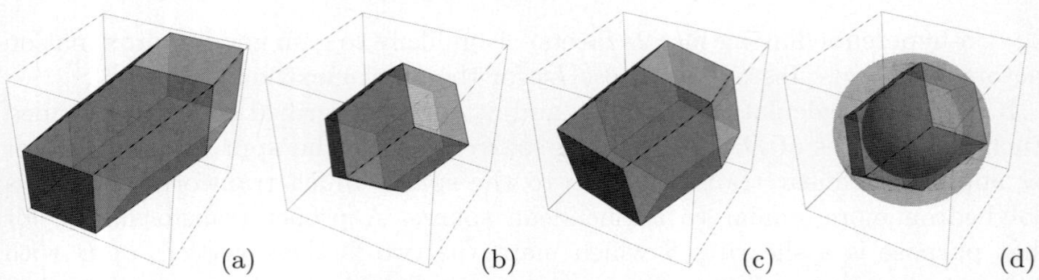

Fig. 3. All images show a variant of the unit sphere B for $s = 4$. (a) untransformed, (b) transformed by S, (c) transformed by S', which additionally mirrors B at the hyperplane with the normal vector $\mathbf{1}$, (d) transformed by S, circumscribed and inscribed by a Euclidean sphere.

In case of $s = 2$ the unit sphere B is simply defined by the L_∞-norm in $\mathbb{R}$. As this equals the one-dimensional Euclidean norm the calculation is exact. This finishes the proof of Lemma 1. $\square$

Proof (Theorem 2). As in the previous section we first show the proof for the directed Hausdorf distance $\overrightarrow{h}^{\mathrm{T}}$. Starting with the definition of the Hausdorff distance under translation $\overrightarrow{h}^{\mathrm{T}}$ by equation 1 we have

$$\overrightarrow{h}^{\mathrm{T}}(Q, P) = \min_{t \in \mathbb{R}^d} \max_{q \in Q} \min_{p \in P} \|q - p + t\|_\infty$$

Now we use the variation which minimizes the maximal point distance as defined in (2)

$$= \min_{t \in \mathbb{R}^d} \min_{v \in V} \max_{j=1}^{s} \|q_j - p_{v(j)} + t\|_\infty$$

The composition of two min-functions is commutative

$$= \min_{v \in V} \min_{t \in \mathbb{R}^d} \max_{j=1}^{s} \|q_j - p_{v(j)} + t\|_\infty$$

$$= \min_{v \in V} \min_{t \in \mathbb{R}^d} \max_{j=1}^{s} \max_{k=1}^{d} \left| q_{j_k} - p_{v(j)_k} + t_k \right|$$

The composition of two max-functions is commutative

$$\overrightarrow{h}^{\mathrm{T}}(Q, P) = \min_{v \in V} \min_{t \in \mathbb{R}^d} \max_{k=1}^{d} \max_{j=1}^{s} \left| q_{j_k} - p_{v(j)_k} + t_k \right|$$

All t_k can be chosen independently from each other. So for some function $f(t_k)$ we have

$$\min_{t \in \mathbb{R}^d} \max_{k=1}^{d} f(t_k) = \max_{k=1}^{d} \min_{t_k \in \mathbb{R}} f(t_k)$$

which allows another re-ordering of the extremum functions

$$\overrightarrow{h}^{\mathrm{T}}(Q, P) = \min_{v \in V} \max_{k=1}^{d} \min_{t_k \in \mathbb{R}} \max_{j=1}^{s} \left| q_{j_k} - p_{v(j)_k} + t_k \right| \tag{5}$$

Now, let $\mathbf{1}$ denote the s-dimensional vector $(1, ..., 1)$ and $\rho_k : \mathbb{R}^{sd} \to \mathbb{R}^s$ the projection which projects a representative point $\hat{p} \in \mathbb{R}^{sd}$ as defined in (3) into the subspace of the embedding space spanned by the coordinates of the j-th point dimension:

$$\rho_k : \hat{p} \mapsto (p_{v(1)_k}, p_{v(2)_k}, ..., p_{v(s)_k})$$

This projection lets us reformulate the Hausdorff distance in (5) in terms of representative points

$$\overrightarrow{h}^{\mathrm{T}}(Q, P) = \min_{\hat{p} \in \hat{P}} \max_{k=1}^{d} \min_{t_k \in \mathbb{R}} \|\rho_k(\hat{q}) - \rho_k(\hat{p}) + t_k \mathbf{1}\|_\infty \tag{6}$$

The inner min-function equals the L_∞-distance of $\rho_k(\hat{q})$ to the line l_k going through $\rho_k(\hat{p})$ with the slope $\mathbf{1}$

$$\min_{t_k \in \mathbb{R}} \|\rho_k(\hat{q}) - \rho_k(\hat{p}) + t_k \mathbf{1}\|_\infty$$
$$= \|\rho_k(\hat{q}) - l_k\|_\infty$$

Furthermore, this term is equal to the distance of the line l_k to the line m_k with l_k going through $\rho_k(\hat{q})$ with the slope $\mathbf{1}$

$$= \|m_k - l_k\|_\infty \tag{7}$$

According to Lemma 1 the L_∞-distance between these lines can be α-approximated by the distance between some representative points $\tilde{q}$ and $\tilde{p}$ in $\mathbb{R}_2^{s-1}$ with $\alpha = \sqrt{s - 1.5}$ for $s \geq 3$. Note, that the coordinates of the subspaces must be normalized when the smallest distance is required additionally to the nearest neighbor. Alternatively the exact Hausdorff distance of the approximate nearest neighbor can be calculated in an additional final step. Using normalized subspaces we have

$$\|\tilde{q}_k - \tilde{p}_k\|_2 \leq \|m_k - l_k\|_\infty \leq \sqrt{s - 1.5}\|\tilde{q}_k - \tilde{p}_k\|_2$$

Thus, we can α-approximate (6) with

$$\overrightarrow{h}_\alpha^{\mathrm{T}}(Q, P) = \min_{\hat{p} \in \hat{P}} \max_{k=1}^{d} \|\tilde{q}_k - \tilde{p}_k\|_2$$

where $\tilde{q}_k$ and $\tilde{p}_k$ can be calculated from $\hat{p}$ for the k-th subspace of the embedding space as described in Lemma 1.

Finally we need to approximate the composition of the maximum-function and the Euclidean metric in the last equation by a single metric. This can be achieved by using only the Euclidean metric. It holds that in $\mathbb{R}^d$ the maximum-distance can be $\sqrt{d}$-approximated by the Euclidean distance. Thus, we can $\sqrt{d}$-approximate

$$\max_{k=1}^{d} \|\tilde{q}_k - \tilde{p}_k\|_2$$

by

$$\sqrt{\sum_{k=1}^{d} \|\tilde{q}_k - \tilde{p}_k\|_2} \tag{8}$$

Table 3. Time and space bounds for retrieving the nearest point set and the k nearest point sets under the approximate and translation invariant Hausdorff distance. [†]The O-notation suppresses terms that are quadratic in ϵ^{-1} here.

	$\overrightarrow{h}^{\mathrm{T}}$	$\overleftarrow{h}^{\mathrm{T}}$
nns [1] query time	$O\left(\log \frac{n}{\epsilon} + s \log s\right)$	$O\left(s^s \log \frac{n}{\epsilon}\right)$
nns [1] prepr. time/ space	$O\left(ns^s \frac{\log ns^s}{\epsilon^{d(s-1)}} \log^2 \frac{ns^s}{\epsilon}\right)$	$O\left(n \frac{\log n}{\epsilon^{d(s-1)}} \log^2 \frac{n}{\epsilon}\right)$
k-nns [2] query time[†]	$O(k + [ds \log^2(ds)][ds + \log(ns^s)])$	$O(s^s[k + (ds \log^2(ds))(ds + \log n)])$
k-nns [2] prepr. time/ space	$O\left(ns^s \log(ds)\right)^{2d(s-1)}$	$O\left(n \log(ds)\right)^{2d(s-1)}$

Defining

$$\tilde{p} = (\ \tilde{p}_{11}, \tilde{p}_{12}, ..., \tilde{p}_{1(s-1)},$$

$$\tilde{p}_{21}, \tilde{p}_{22}, ..., \tilde{p}_{2(s-1)},$$

$$\vdots$$

$$\tilde{p}_{d1}, \tilde{p}_{d2}, ..., \tilde{p}_{d(s-1)}\ )$$

and $\tilde{q}$ in an equivalent way, we can rewrite (8) as $\|\tilde{q} - \tilde{p}\|_2$. Hence, we have a $\sqrt{d(s - 1.5)}$-approximation of the Hausdorff distance under translation:

$$\overrightarrow{h}^{\mathrm{T}}_{\mathrm{approx}}(Q, P) = \min_{\hat{p} \in \hat{P}} \|\tilde{q} - \tilde{p}\|_2$$

where $\tilde{p}$ and $\tilde{q}$ can be calculated from $\hat{p}$ and $\hat{q}$ and thus from P and Q, respectively.

Every $\tilde{p}$ maintains a reference to the point set it represents, and thus the nearest point set P can be determined.

In case of $s = 2$ the calculation of the distance between the lines can be made exactly by any Minkowski norm. This results in a $\sqrt{d}$-approximation when the Euclidean norm is used for the nearest neighbor search in the embedding space and an exact retrieval when the Maximum norm is used.

The presented embedding space is also suitable for calculating the nearest neighbor of Q under $\overleftarrow{h}^{\mathrm{T}}$ using the same modification as in the previous section. This finishes the proof of Theorem 2 and 3.   □

Having the embedding, a nearest neighbor search for single points can be applied. Using the $(1+\epsilon)$-approximate nearest neighbor structure of Har-Peled [1] and the $(1+\epsilon)$-approximate k-nearest neighbor structure of Kleinberg [2, First Algorithm] we achieve the results shown in Table 3. In all four cases the approximation factor is $c = \sqrt{d(s - 1.5)}(1 + \epsilon)$.

Future Work. It is possible to extend the presented technique for a retrieval under other geometric distance measures. It would be interesting to modify the algorithm to a constant factor approximation in terms of s or d.

Acknowledgements. Thanks to Helmut Alt for fruitful discussions.

References

1. Har-Peled, S.: A replacement for Voronoi diagrams of near linear size. Foundations of Computer Science (January 2001)
2. Kleinberg, J.M.: Two algorithms for nearest-neighbor search in high dimensions. In: STOC 1997: Proceedings of the twenty-ninth annual ACM symposium on Theory of computing, pp. 599–608. ACM, New York (1997)
3. Gabow, H.N., Bentley, J.L., Tarjan, R.E.: Scaling and related techniques for geometry problems. In: STOC 1984: Proceedings of the sixteenth annual ACM symposium on Theory of computing, pp. 135–143. ACM, New York (1984)
4. Indyk, P.: On approximate nearest neighbors under l_∞ norm. Journal of Computer and System Sciences (January 2001)
5. Arya, S., Mount, D., Netanyahu, N., Silverman, R., Wu, A.: An optimal algorithm for approximate nearest neighbor searching in fixed dimensions. Journal of the ACM (JACM) (January 1998)
6. Braß, P., Knauer, C.: Nearest neighbour search in Hausdorff distance pattern spaces. Technical report, Institut für Informatik, FU-Berlin (2001)
7. Farach-Colton, M., Indyk, P.: Approximate nearest neighbor algorithms for Hausdorff metrics via embeddings. In: Proceedings of the 40th Annual Symposium on Foundations of Computer Science (October 1999)
8. Vleugels, J., Veltkamp, R.C.: Efficient image retrieval through vantage objects. Pattern Recognition 35(1), 69–80 (2002)

Preprocessing Imprecise Points and Splitting Triangulations[*]

Marc van Kreveld[1], Maarten Löffler[1], and Joseph S.B. Mitchell[2]

[1] Department of Information and Computing Sciences
Utrecht University, The Netherlands
{marc,loffler}@cs.uu.nl
[2] Department of Applied Mathematics and Statistics
State University of New York at Stony Brook, USA
jsbm@ams.stonybrook.edu

Abstract. Given a triangulation of a set of n points in the plane, each colored red or blue, we show how to compute a triangulation of just the blue points in time $O(n)$. We apply this result to show that one can preprocess a set of disjoint regions (representing "imprecise points") in the plane having total complexity n in $O(n \log n)$ time so that if one point per region is specified with precise coordinates, a triangulation of the n points can be computed in $O(n)$ time.

1 Introduction

Computational geometry deals with computing structure for input data that is embedded in a space of two or more dimensions. The most popular input is a set of points, on which the goal is to compute a useful structure, such as a triangulation. Algorithms for such tasks have been developed many years ago and are provably fast and correct. However, a major obstacle to the practical implementation of geometric algorithms is that most theoretically sound algorithms assume that the input data is given precisely, while, in practice, the input is often taken from real-world data, and therefore has an intrinsic error. When the input is not precise, the value of the output is questionable.

In many applications, even though data is imprecise, some information about the error is known. For example, a point may be known not to be more than some ε away from a given point, to be inside a given region, or to be chosen from a known probability distribution. Several approaches have been proposed to use this additional information, varying from fuzzy methods to computing partial output that is certain to be combinatorially correct.

Here we study the situation in which each point is known to lie in a pre-described region in the plane. We are interested in preprocessing such a collection of regions, such that if the points are later given precisely, we can do certain

---

[*] This research was partially supported by the Netherlands Organisation for Scientific Research (NWO) through the project GOGO. J. Mitchell is partially supported by grants from the National Science Foundation (CCF-0431030, CCF-0528209, CCF-0729019), NASA Ames, and Metron Aviation.

S.-H. Hong, H. Nagamochi, and T. Fukunaga (Eds.): ISAAC 2008, LNCS 5369, pp. 544–555, 2008.

computations faster. This is not always possible: if all regions have a common intersection, then the precise sample could be an arbitrary point set within the intersection, so lower bounds for the classical case apply.

Specifically, the main problem we solve in this paper is that of triangulating imprecise points: given a set $S = \{P_1, P_2, \ldots, P_m\}$ of non-overlapping polygonal regions in the plane having a total of n vertices, preprocess them in such a way that when a set $\{p_1, \ldots, p_m\}$ of points is given, with $p_i \in P_i$ and each point specified by its exact coordinates and knowledge of its containing region, a triangulation of the points p_i can be computed in linear $(O(n))$ time.

In our solution, we encounter a very natural problem that we believe to be of independent interest. The problem is to *split* a triangulation: given an arbitrary triangulation of a set of n vertices in the plane, each colored red or blue, compute a triangulation of just the blue vertices. We show how to split a triangulation in linear $(O(n))$ time.

In the next section, we study the triangulation splitting problem. Then, in Section 3, we show how this result applies to preprocess a set of disjoint regions in the plane for linear-time triangulation of sample points. In Section 4, we give some concluding remarks.

Related Work. The problem of splitting an arbitrary triangulation is similar to one studied by Chazelle *et al.* [3], who show how to compute, given a *Delaunay* triangulation with red and blue vertices, the *Delaunay* triangulation of the blue vertices in linear time.

Data imprecision in computational geometry has traditionally been considered mostly in stochastic or fuzzy settings [9,17]. However, in recent years there has been a growing interest in exact models of imprecision. Guibas *et al.* [7] introduce the notion of *espilon geometry*, a framework for robust computations on imprecise points. Abellanas *et al.* [1] and Weller [21] study the *tolerance* of a geometric structure: the largest perturbation of the vertices such that the combinatorial structure remains the same. Bandyopadhyay and Snoeyink [2] compute the set of "almost-Delaunay simplices," which are the tuples of points that could define a Delaunay simplex if the entire point set is perturbed by at most $\varepsilon > 0$. Ely and Leclerc [6] and Khanban and Edalat [11] consider the epsilon geometry versions of the In-Circle predicate for Delaunay triangulation with imprecise points modeled as disks or as rectangles, respectively. Khanban and co-authors [10,12] developed a theory for returning partial Delaunay or Voronoi diagrams, consisting of the portion of the diagram that is certain. Sember and Evans [20] compute a related structure, the "guaranteed Voronoi diagram", of a set of imprecise sites. Van Kreveld and Löffler [14,15] consider the problem of determining the smallest and largest possible values for geometric extent measures—such as the diameter or convex hull area—of a set of imprecise points.

Some prior results on triangulating imprecise points are known as well. Held and Mitchell [8] consider the problem of preprocessing a set of n disjoint unit discs in $O(n \log n)$ time, such that when one point in each disc is given, the point set can be triangulated in linear time. They give a simple and practical solution. Their result can be extended to overlapping regions of different shapes, as long as the regions do not overlap more than a constant number of other regions, the regions are fat, and the sizes do not vary by more than a constant factor. In the same setting, Löffler and Snoeyink [13] show that a set of discs can be pre-processed in $O(n \log n)$ time such that the Delaunay triangulation of the points can be computed in linear time. This algorithm relies on the linear-time constrained Delaunay triangulation algorithm for polygons by Chin and Wang [4], which would not be easy to implement. The same extensions to partially overlapping fat regions are possible.

In both previous results [8,13], the extensions work only up to a class of shapes bounded by a number of constants, which end up in the running times. An interesting question is what can be done for more general regions. As mentioned earlier, we cannot hope to do any useful preprocessing if the overlap of the regions is not bounded. However, for a set of n general disjoint regions in the plane, there is hope to do more. For a set of general disjoint regions, a linear-time Delaunay triangulation algorithm is not possible, since Seidel [19] shows that even if the sorted order of a set of points is given, computing the Delaunay triangulation requires $\Omega(n \log n)$ time. If the input regions consist of a set of vertical lines, then any amount of preprocessing of the regions only yields the sorted order (in x-coordinate) of the sample points that lie on the lines.

Our results show that, although the Delaunay triangulation is out of reach, we *can* preprocess a set of disjoint regions such that *some* triangulation of a sample can be computed in linear time.

2   Splitting a Triangulation

In this section we study the following problem: given a triangulation embedded in the plane with vertices that are colored either red or blue, compute a triangulation of only the blue vertices. Figure 1 shows an example of this problem. To solve the problem, we will remove all of the red points one by one, until we have only blue points left. During this process, we will maintain a subdivision of the plane with certain properties, which allows us quickly to find new red points to remove and to remove them efficiently. We first describe this subdivision and some operations we can perform on it, and then give the algorithm and time analysis.

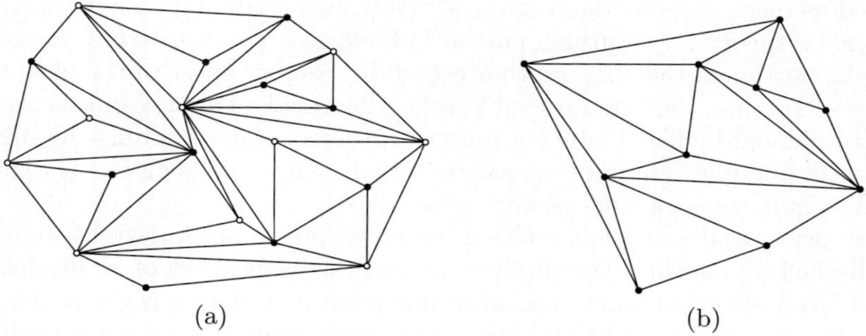

(a)   (b)

Fig. 1. (a) Example input with red (open) and blue (solid) points. (b) Example output.

2.1   Structure and Operations

During the algorithm, we will maintain a subdivision of the plane that uses the blue points and remaining red points as vertices. The subdivision is a special kind of *pseudotriangulation* (see, e.g., [16]). A pseudotriangulation is a subdivision of a convex region into *pseudotriangles*: simple polygons with exactly three convex vertices. The three convex vertices are also called the *corners* of the pseudotriangle, and the three polygonal lines connecting each pair of corners are called the *sides* of the pseudotriangle. (Note that we do not require a pseudotriangulation to be "pointy" or "minimal".)

Fig. 2. A fox is a pseudotriangle with one red vertex and one concave chain of blue vertices

In the pseudotriangulation that we maintain, we allow only two types of faces: triangles and *foxes*. A fox is a pseudotriangle that has only one side that is not a straight edge and that has all vertices blue except the one (red) vertex that is incident to the two straight sides. We call the red vertex the *chin* of the fox, and the other two convex vertices the *ears*. Figure 2 shows an example of a fox. For each fox, we store the chain of concave blue vertices in a balanced binary tree. If a pseudotriangulation has only triangles and foxes as faces, we call it *happy*.

Note that our input triangulation is happy, since it has only normal triangles. Also note that if we manage to remove all red vertices and maintain a happy subdivision, we cannot have any foxes left, since a fox has a red vertex: we are left with only normal triangles with three blue vertices, which is the required output of the algorithm.

Whenever we have a happy subdivision, we will denote the number of blue points by n and the number of remaining red points by k.

For a given red point p, let $r(p)$ be the number of red neighbors of p and $b(p)$ the number of blue neighbors of p (i.e., $r(p) + b(p)$ is the degree of p). Observe that any face to which red point p is incident is either a triangle or a fox that has p as its chin. As a consequence, the union of all faces incident to p forms a star-shaped polygon. By $c(p)$ we denote the total complexity of this polygon.

In addition to the shape restriction on the pseudotriangles, we will impose one more condition that we will maintain throughout the algorithm. For all red points p, Condition (∗) should hold.

$$b(p) \leq 2 \cdot r(p) + 3 \tag{∗}$$

This condition is not necessarily true for the input triangulation, so we will have to do an initial pass over the input triangulation to make this condition hold.

2.2 The Algorithm

First, we must make sure that all red points in the pseudotriangulation satisfy Condition $(*)$. We describe below (Lemma 1) how to simplify a red point. Applying this to each red point takes $O(n)$ time in total.

Next, we perform a sequence of reduction steps, at each step reducing the number of red points by a constant factor. At a step when we have k red points left, we want to find an independent set of $O(k)$ red points each having constant red degree. Since this step does not depend on any blue points, this can easily be done in $O(k)$ time.

Then, we remove each of the red points in the independent set by applying Lemma 3 (below). The resulting subdivision is still happy, but Condition $(*)$ may no longer hold for red points that had a red neighbor that was removed. However, we can repair this condition by applying Lemma 2 (below) to those red points, but only in the sectors where something changed. The number of red-blue edges in those sectors cannot have increased by more than the number of edges that were added in the removal step. We added no more than a constant number of red-blue edges for each removed point, so this is in total at most $O(k)$.

Simplifying a Red Point. The proof of the following lemma will appear below, after the proof of Lemma 2:

Lemma 1. *Let p be a red point in a happy pseudotriangulation. We can make Condition $(*)$ hold for p in $O(c(p))$ time.*

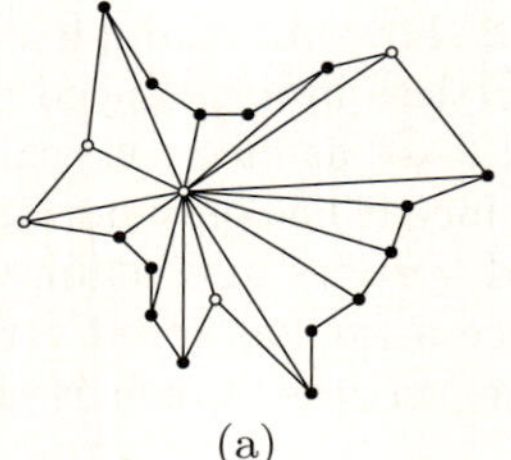 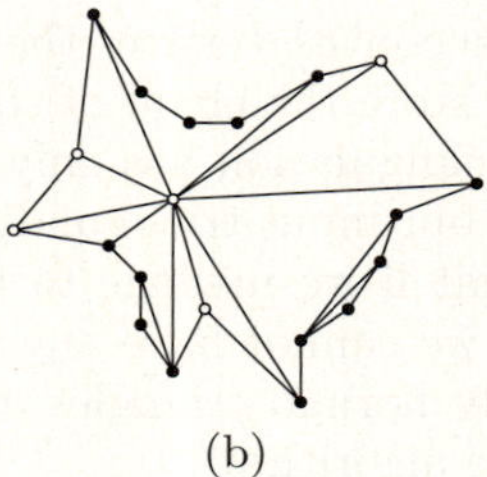

(a)   (b)

Fig. 3. (a) The pseudotriangles incident to a given red point form a star-shaped region. (b) By adding and removing the appropriate edges, we can make Condition $(*)$ hold.

Figure 3 shows an example. Since p is a red point, and the pseudotriangulation is happy, the region around p (the union of its incident cells) is a star-shaped polygon of which all red and all convex vertices are connected to p. The purpose of this step is to remove any superfluous red-blue edges incident on p. We leave all red-red edges where they are, so we do not need to consider them. Between each pair of red neighbors of p, there is a sector of the star that has only blue points (we will consider the case where p has no red neighbors separately). We will first prove the following lemma for a single sector.

Lemma 2. *Let p be a red point, and let q and r be two red neighbors of p such that there are no other red neighbors of p between them. Let b be the number* *of blue neighbors of p between q and r, and c the total number of blue points between q and r. In $O(b \log c + m)$ time or in $O(c)$ time, we can update the subdivision such that the number of red-blue edges between q and r is at most 2 if $\angle qpr \leq 180°$, and at most 3 otherwise.*

Proof: We have a sequence of blue points, some of which may be connected to p. The first and last points are always connected to p, since their neighbors are red, and any face of the subdivision with two red vertices must be a normal triangle. Now, if any other point s is also connected to p, we can almost always remove it. There are two cases.

If s is concave, we can simply remove edge ps. Both neighbors of s are blue, so the two cells ps divides must be foxes (or normal triangles with one red and two blue vertices, which are degenerate foxes). Since s is also concave, the combination of both cells is still a valid cell. In this case, we do need to concatenate the two binary trees that store the chains between the ears of the foxes. We postpone this concatenation until the end of the procedure.

If s is convex, we can remove the edge ps if the (normal) triangle formed by s and its two neighbors is empty. If this is the case, we add this completely blue normal triangle and then add edges from its other two corners to p, and recurse. If the triangle is *not* empty, then we must keep the edge ps. However, this is only possible if p itself is inside this triangle, which can happen at most once and only if the angle of the sector is at least $180°$.

After all superfluous red-blue edges have been removed, we might be left with a sequence of $O(b)$ blue chains, stored as balanced binary trees, of total complexity $O(c)$, which have to be concatenated into one big balanced binary tree. We note that this can be done in $O(b \log c)$ time by merging them one by one, or in $O(c)$ time by simpy building a new tree from scratch. $\boxtimes$

With this result, we can prove Lemma 1.

Proof: We apply Lemma 2 to all sectors, using $O(c)$ time complexity. There can be at most 2 sectors with an angle of at least $180°$, so if p has any red neighbors the number of red-blue edges after simplifying all sectors is at most $2r(p) + 2$.

If p does not have any red neighbors, we can still proceed with deleting blue edges as described above, until there are only three neighbors left. In fact, in this case we are just triangulating the star-shaped polygon that remains after removal of p.

In both cases, Condition $(*)$ follows. $\boxtimes$

Removing a Red Point

Lemma 3. *Let p be a red point in a happy subdivision with $r(p) = O(1)$, for which Condition $(*)$ holds. We can remove this point from the subdivision, and partition the gap it leaves into triangles and foxes in $O(\log c(p) + m)$ time, where m is the number of blue-blue edges formed in this step.*

Proof: Because of Condition $(*)$, we know that also $b(p) = O(1)$. We can remove p and all its incident edges, of which there are only a constant number. This results in an empty star-shaped polygon which needs to be partitioned into

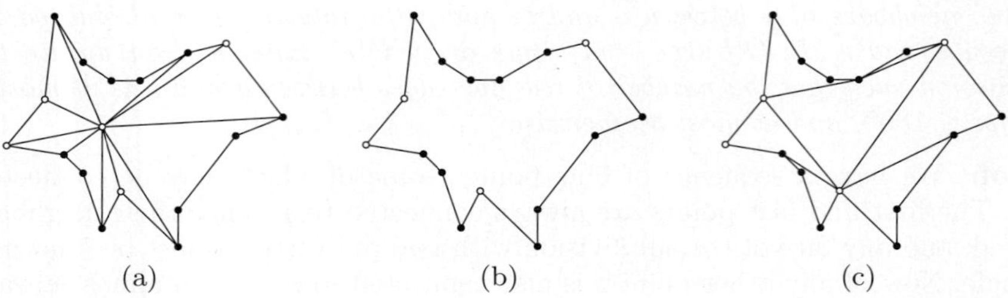

Fig. 4. (a) A red point p of constant red degree and its incident pseudotriangles. (b) The empty polygon after removing p. (c) A repseudotriangulation of the gap.

smaller cells again, see Figure 4(b). The complexity of this polygon is $c(p)$, and consists of $r(p)$ red points and at most $b(p)$ concave chains of blue points.

We will partition the gap into pseudotriangles. As described in [18], we can add geodesic shortest paths between pairs of convex vertices of a simple polygon, until a (minimal) pseudotriangulation has been found. Since we have only a constant number of convex vertices, we only need to insert a constant number of shortest paths. For a given pair of vertices, we can compute this shortest path in $O(\log c(p))$ time, because we stored the chains of concave vertices in binary trees and we can compute tangents in logarithmic time. After this procedure, the gap has been split into a constant number of pseudotriangles in $O(\log c(p))$ time, see Figure 4(c). All the pseudotriangles have only blue concave vertices.

We now apply Lemma 4 (below) to all of these pseudotriangles to obtain a partitioning of the gap into a constant number of triangles and foxes, plus any necessary number of completely blue triangles. $\boxtimes$

We now have a happy subdivision again, although Condition $(*)$ may no longer be true for some red vertices on the boundary of the gap.

Subdividing a Pseudotriangle. Since removal of a red point can lead to pseudotriangles that are not foxes or normal triangles, we describe how to subdivide a pseudotriangle:

Lemma 4. *Let T be a pseudotriangle, with the restriction that all concave vertices are blue. We can subdivide T into $O(1)$ non-blue triangles and foxes plus some number of triangles that are completely blue, in time $O(\log c + m)$ where c is the complexity of T and m is the number of blue-blue edges we produce in this step.*

Proof: If none of the sides of T have any concave vertices, then T is already a normal triangle.

If only one of the sides has concave vertices, and the corner opposite to it is blue, then we can triangulate the pseudotriangle with edges from the blue corner to all of the concave vertices, see Figure 5(a). This creates many blue triangles, and at most two triangles that involve a red point. If the corner opposite to it is red, then depending on the colors of the other two corners we either make an

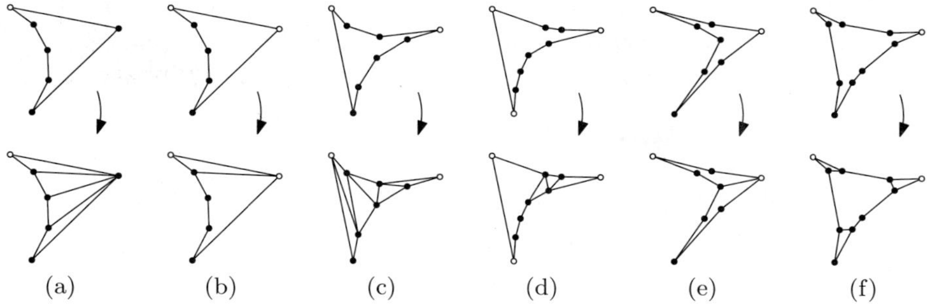

Fig. 5. We can subdivide any pseudotriangle into $O(1)$ triangles and foxes, plus a number of blue triangles

edge to the neighbors or not, see Figure 5(b). In this case, we make one fox and at most two triangles on the sides.

If two of the sides of the pseudotriangle have a concave vertex on them, consider the corner between these two sides. We can add an edge between its two neighbors, which are both blue. We can then continue adding blue-blue edges between the two chains, until this is no longer possible. The part that is left is then either a quadrilateral, see Figure 5(c), which we can simply split into two triangles, or a pseudotriangle with at most one side with concave vertices, see Figure 5(d), which we can further split in the way described above.

If all three sides have concave vertices on them, consider one of the corners. If we can make an edge between this corner and one concave vertex of the opposite side, then this splits the pseudotriangle into two pseudotriangles with both at most two sides with concave vertices, see Figure 5(e), which we can then further split as described above. We can find out whether there is such an edge in $O(\log c)$ time by extending the edges adjacent to the corner, and intersecting them with the opposite chain. If this is not possible for any corner, then we can connect the two neighbors of each corner with a blue-blue edge, see Figure 5(f). The remaining area has only blue vertices, and can be triangulated in any way. $\boxtimes$

2.3   Time Analysis

The first phase of the algorithm takes $O(n)$ time.

Then, let $1/f$ denote the fraction of the red points that remain after throwing some away in each step. Then we perform $\log_f n$ phases. At the end of the ith phase there are k/f^i red points remaining. In each phase, we spend $O(k)$ time to find an independent set. Then, we spend $O(\log c(p) + m)$ time for each element in the set. We can charge the m to the blue-blue edges that are created; since there can be at most $O(n)$ blue-blue edges and they are never removed, we spend no more than $O(n)$ time in total on them. The $c(p)$ factors are added over all elements in the independent set, and can be no more than $O(n)$ in total. In the worst case, they are divided equally, and we spend $O(k \cdot \log \frac{n}{k})$ time on removing the points. By Lemma 2, Condition $(*)$ can be repaired in a sector that was involved in a removal step in $O(\log c(p) + m)$ time, since the number of red-blue edges in such a sector is constant. There are at most $O(k)$ such sectors, so again, in the worst case all blue points are equally divided and we spend $O(k \cdot \log \frac{n}{k})$ time on repairing them.

The total time we spend is now

$$\sum_{i=1}^{\log_f n} O(\frac{n}{f^i} \log f^i) = \sum_{i=1}^{\log_f n} O(f^i \log \frac{n}{f^i}) = \sum_{k=1}^{n} O(\log \frac{n}{u(k)}),$$

where $u(k)$ is the smallest power of f larger than k. We can interprete this as the summation over k of the amount of time charged to removing one red point when there are k left. This time bound is then bounded by

$$\sum_{k=1}^{n} O(\log \frac{n}{k}) = O(\log \frac{n^n}{n!}) = O(n \log n - \log n!) = O(n).$$

3   Triangulating Imprecise Points

With the triangulation splitting result, it becomes straightforward to solve our original problem: given a set $S = \{P_1, \ldots, P_m\}$ of non-overlapping polygonal regions in the plane of total complexity n, preprocess them in such a way that when a point $p_i \in P_i$ in each region is given, a triangulation of these points can be computed in linear time.

Preprocessing

In preprocessing, we compute a triangulation of the plane of complexity $O(n)$, such that each triangle contains only (a part of) one of the input regions. Together with this, we create a list of pointers from each imprecise region P_i to the set of triangles that cover this region. If the regions S are disjoint polygons (not necessarily simple – they can have holes or multiple components), this is simple to do in $O(n \log n)$ time by computing a constrained triangulation of the vertices of the regions, as in Figure 6(b).

Reconstruction

Now, when we are given a set $\{p_1, \ldots, p_m\}$ of points such that each point p_i lies inside an input region P_i, and it is known which point lies in which region, we want to compute a triangulation of the points p_i in linear $(O(n))$ time.

To do this, we add the points to the triangulation computed in the preprocessing step. For this we need to locate each point p_i in the triangulation, which we do by simply walking through all triangles to which region P_i points (since we know that $p_i \in P_i$). Since each triangle is contained in a single region, we spend only $O(n)$ time in total. Once the triangle containing p_i is identified, we simply split it into three smaller triangles. Figure 6(c) shows an example.

Now we have a triangulation with two types of vertices: the ones from the preprocessing (red) and the ones we added (blue). We simply invoke the algorithm from Section 2 to obtain a triangulation of just the blue points.

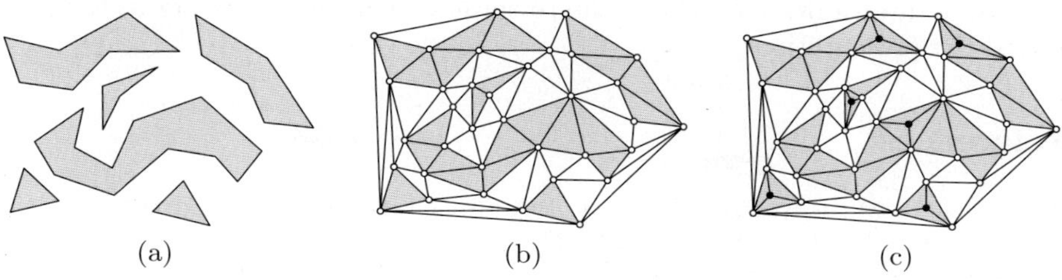

(a)   (b)   (c)

Fig. 6. (a) A set of regions in the plane. (b) A triangulation of the vertices of the regions. (c) The sample points have been added to the triangulation.

3.1   Extensions

The main improvement of our algorithm over [8] and [13] is that the input regions for the algorithm do not have to be fat or have the same size, or even be convex or connected. However, we do still assume that the regions are polygonal and do not overlap.

We can extend the approach to work also for regions that are not completely disjoint, as long as the complexity of their overlay is not too high. If we compute the overlay of the polygons and triangulate the resulting arrangement, the method will run in $O(n \log n + c)$ preprocessing and $O(ck \log k)$ reconstruction time, where c is the total complexity of the overlay, and k is the maximum number of regions that overlap in a single point.

If the input regions are not polygonal, we cannot simply triangulate their vertices, but our approach still applies if we first compute a polygonal subdivision of the plane such that each face contains one region. If the regions are convex, such a subdivision can be computed having complexity that is linear in the number of regions [5]. If the regions are not convex, the complexity may increase, depending on the exact shape of the regions.

4   Conclusions

When a set of points is unknown, but constrained by a known region for each point, it is interesting to preprocess the regions to speed up computations when the exact locations of the points become known. We give an algorithm to preprocess a set of disjoint regions in the plane in $O(n \log n)$ time, so that a sample from the regions, one point per region, can be triangulated in linear time. This time bound is optimal and improves previous results by allowing more general regions. In future work, it would be interesting to study whether similar results can be obtained in higher dimensions.

Our method is based on an algorithm we presented for splitting a triangulation in linear time. This is a very natural problem that we believe is interesting in its own right.

Acknowledgements. We thank the reviewers for helpful suggestions that improved the presentation.

References

1. Abellanas, M., Hurtado, F., Ramos, P.A.: Structural tolerance and Delaunay triangulation. Inf. Proc. Lett. 71, 221–227 (1999)
2. Bandyopadhyay, D., Snoeyink, J.: Almost-Delaunay simplices: Nearest neighbour relations for imprecise points. In: Proc. 15th ACM-SIAM Symposium on Discrete Algorithms, pp. 410–419 (2004)
3. Chazelle, B., Devillers, O., Hurtado, F., Mora, M., Sacristán, V., Teillaud, M.: Splitting a Delaunay triangulation in linear time. Algorithmica 34, 39–46 (2002)
4. Chin, F.Y.L., Wang, C.A.: Finding the constrained Delaunay triangulation and constrained Voronoi diagram of a simple polygon in linear time. SIAM J. Comput. 28(2), 471–486 (1998)
5. Edelsbrunner, H., Robison, A.D., Shen, X.: Covering convex sets with non-overlapping polygons. Discrete Math. 81, 153–164 (1990)
6. Ely, J.S., Leclerc, A.P.: Correct Delaunay triangulation in the presence of inexact inputs and arithmetic. Reliable Computing 6, 23–38 (2000)
7. Guibas, L.J., Salesin, D., Stolfi, J.: Constructing strongly convex approximate hulls with inaccurate primitives. Algorithmica 9, 534–560 (1993)
8. Held, M., Mitchell, J.S.B.: Triangulating input-constrained planar point sets. Information Processing Letters (to appear, 2008)
9. Kendall, W.S., Barndorff-Nielsen, O., van Lieshout, M.C.: Current Trends in Stochastic Geometry: Likelihood and Computation. CRC Press, Boca Raton (1998)
10. Khanban, A.A.: Basic Algorithms of Computational Geometry with Imprecise Input. PhD thesis, Imperial College, London (2005)
11. Khanban, A.A., Edalat, A.: Computing Delaunay triangulation with imprecise input data. In: Proc. 15th Canad. Conf. Comput. Geom., pp. 94–97 (2003)
12. Khanban, A.A., Edalat, A., Lieutier, A.: Computability of partial Delaunay triangulation and Voronoi diagram. In: Brattka, V., Schröder, M., Weihrauch, K. (eds.) Electronic Notes in Theoretical Computer Science, vol. 66. Elsevier, Amsterdam (2002)
13. Löffler, M., Snoeyink, J.: Delaunay triangulations of imprecise points in linear time after preprocessing. In: Proc. 24th Sympoium on Computational Geometry, pp. 298–304 (2008)
14. Löffler, M., van Kreveld, M.: Largest bounding box, smallest diameter, and related problems on imprecise points. In: Dehne, F., Sack, J.-R., Zeh, N. (eds.) WADS 2007. LNCS, vol. 4619, pp. 446–457. Springer, Heidelberg (2007)
15. Löffler, M., van Kreveld, M.: Largest and smallest convex hulls for imprecise points. Algorithmica (to appear, 2008)
16. Pocchiola, M., Vegter, G.: Pseudo-triangulations: Theory and applications. In: Proc. 12th Annu. ACM Sympos. Comput. Geom., pp. 291–300 (1996)
17. Rosenfeld, A.: Fuzzy geometry: An updated overview. Inf. Sci. 110(3-4), 127–133 (1998)
18. Rote, G., Wang, C.A., Wang, L., Xu, Y.: On constrained minimum pseudotriangulations. In: Warnow, T.J., Zhu, B. (eds.) COCOON 2003. LNCS, vol. 2697, pp. 445–454. Springer, Heidelberg (2003)

19. Seidel, R.: A method for proving lower bounds for certain geometric problems. In: Toussaint, G.T. (ed.) Computational Geometry, pp. 319–334. North-Holland, Amsterdam (1985)
20. Sember, J., Evans, W.: Guaranteed Voronoi diagrams of uncertain sites. In: Proc. 20th Canad. Conf. Comput. Geom. (2008)
21. Weller, F.: Stability of Voronoi neighborship under perturbations of the sites. In: Proc. 9th Canad. Conf. Comput. Geom., pp. 251–256 (1997)

Efficient Output-Sensitive Construction of Reeb Graphs

Harish Doraiswamy[1] and Vijay Natarajan[1,2]

[1] Department of Computer Science and Automation
[2] Supercomputer Education and Research Centre
Indian Institute of Science, Bangalore 560012, India
{harishd,vijayn}@csa.iisc.ernet.in

Abstract. The Reeb graph tracks topology changes in level sets of a scalar function and finds applications in scientific visualization and geometric modeling. This paper describes a near-optimal two-step algorithm that constructs the Reeb graph of a Morse function defined over manifolds in any dimension. The algorithm first identifies the critical points of the input manifold, and then connects these critical points in the second step to obtain the Reeb graph. A simplification mechanism based on topological persistence aids in the removal of noise and unimportant features. A radial layout scheme results in a feature-directed drawing of the Reeb graph. Experimental results demonstrate the efficiency of the Reeb graph construction in practice and its applications.

1 Introduction

The Reeb graph of a scalar function describes the connectivity of its level sets. The abstract representation of level-set topology within the Reeb graph enables development of simple and efficient methods for modeling objects and visualizing scientific data. Reeb graphs and their loop-free version, called contour trees, have a wide variety of applications including computer aided geometric design [20,25], topology-based shape matching [13], topological simplification and cleaning [11,24], surface segmentation and parametrization [12,26], and efficient computation of level sets [22]. They serve as an effective user interface for selecting meaningful level sets [2,5] and transfer functions for volume rendering [23].

1.1 Related Work

Several algorithms have been proposed for constructing Reeb graphs. However, only a few produce provably correct Reeb graphs. Shinagawa and Kunii proposed the first algorithm for constructing the Reeb graph of a scalar function defined on a triangulated 2-manifold [19]. Their algorithm explicitly tracked connected components of the level sets and has a running time of $O(n^2)$, where n is the number of triangles in the input. Cole-Mclaughlin et al. [7] improved the running time to $O(n \log n)$ by maintaining the level sets using dynamically balanced search trees. In a recent paper, Pascucci et al. [17] proposed an online algorithm that constructs the Reeb graph for streaming data. Their algorithm takes advantage of the coherency in the input to construct the Reeb graph efficiently. In the case of streaming data, where triangles are processed one after another, the algorithm essentially attaches the straight line Reeb graph corresponding to

S.-H. Hong, H. Nagamochi, and T. Fukunaga (Eds.): ISAAC 2008, LNCS 5369, pp. 556–567, 2008.

the current triangle with the Reeb graph computed so far. Even though the algorithm has a $O(n^2)$ behavior in the worst case, it performs very well in practice for 2-manifolds. However, the optimizations that result in fast incremental construction of Reeb graphs for 2-manifolds do not provide a performance benefit in higher dimensions. We adopt a simple but different approach to compute Reeb graphs that traces connected components of interval volumes (the volume between two level sets). This approach results in an algorithm that exhibits good worst-case behavior and works well in practice - while we obtain good running times for 2-manifolds, our algorithm performs better than the online algorithm for 3-manifolds.

For the special case of loop-free Reeb graphs, Carr et al. [4] described an elegant $O(v \log v)$ algorithm that works in all dimensions, where v is the number of vertices in the input. Besides the naïve $O(n^2)$ algorithm and the online algorithm, there is no known algorithm for computing Reeb graphs of manifolds in higher dimension. The presence of loops in the Reeb graph implies that its decomposition into a join and split tree, which was crucial for the efficiency of the algorithm by Carr et al., may not exist. Efficient storage and manipulation of connected components of level sets will lead to fast construction of Reeb graphs. Cole-Mclaughlin et al. [7] adopt this approach to obtain an efficient algorithm for 2-manifolds. However they exploit the unique property of one-dimensional level sets that their vertices can be ordered, and hence, their algorithm does not directly extend to higher dimension manifolds. Other algorithms for computing Reeb graphs follow a sample based approach that produces potentially inaccurate results [13,21].

1.2   Results

We present an efficient two-step algorithm for computing the Reeb graph of a piecewise-linear (PL) function in $O(n + l + t \log t)$ time, where n is the number of triangles in the input mesh, t is the number of critical points of the function, and l is the size (number of edges) of all critical level sets. The algorithm has various desirable properties. It is

- *output-sensitive*: the running time depends of the number of critical points of the function, which is equal to the number of nodes in the Reeb graph, and the size of critical level sets, which is indicative of the importance of features in the data.
- *near-optimal*: the size of critical level sets is usually $O(n)$ in practice. So, the worst-case running time is close to the optimal bound of $(n + t \log t)$ [22].
- *generic*: the algorithm works, without any modifications, for functions defined on d-manifolds and for non-manifolds.
- *simple*: the algorithm is simple to implement.

We also describe a method to simplify the Reeb graph based on an extended notion of persistence [1] that removes short leaves and cycles in the graph. Finally, we describe a feature-directed layout of the Reeb graph that serves as a useful interface for exploring and understanding three-dimensional scalar fields. We also present experimental results that demonstrate the efficiency of our algorithm.

2   Background

Let $\mathbb{M}^d$ denote a d-manifold with or without boundary. A smooth, real-valued function $f : \mathbb{M}^d \to \mathbb{R}$ is called a *Morse function* if it satisfies the following conditions [7]:

1. All critical points of f are non-degenerate and lie in the interior of $\mathbb{M}^d$.
2. All critical points of the restriction of f to the boundary of $\mathbb{M}^d$ are non-degenerate.
3. All critical values are distinct *i.e.*, $f(p) \neq f(q)$ for all critical points $p \neq q$.

The above conditions typically do not hold in practice for PL functions. However, simulated perturbation of the function [8, Section 1.4] ensures that no two critical values are equal. All multiple saddles (degenerate) can be unfolded into simple (non-degenerate) saddles by repeatedly splitting the link of the multiple saddle [9]. A total order on the vertices helps in consistently identifying the vertex with the higher function value between a pair of vertices.

2.1   Critical Points and Level Sets

Critical points of a smooth function are exactly where the gradient becomes zero. Banchoff [3] and later Edelsbrunner et al. [9] describe a combinatorial characterization for critical points of a PL function, which are always located at vertices of the mesh. The *link* of a vertex consists of all vertices adjacent to it and the induced edges, triangles, and higher-order simplices. Adjacent vertices with lower function value and their induced simplices constitute the *lower link*, whereas the adjacent vertices with higher function value and their induced simplices constitute the *upper link*. For functions defined on 2- and 3-manifolds, the critical points are classified based on the number of connected components (denoted as β_0, the zeroth Betti number) of the lower and upper link. Classification of all critical points in higher dimensions requires the computation of higher order Betti numbers.

The preimage of a real value is called a *level set*. The level set of a regular value is a $(d-1)$-manifold with or without boundary, possibly containing multiple connected components. We are interested in the evolution of level sets against increasing function value. Significant topological changes occur at critical points, whereas topology of the level set is preserved across regular points [14].

In the context of Reeb graphs, we are only interested in critical points that modify the number of level-set components. So, it is sufficient to count the number of connected components of the lower / upper link for identifying these critical points. Given a critical point c_i with function value f_i, we define a *critical level set* as the level set at a function value infinitesimally below / above f_i (i.e. $f^{-1}(f_i \pm \varepsilon)$).

For example, consider the case when $d = 3$. A level set of a 3-manifold is called an *isosurface*. Figure 1 illustrates the topology changes that occur at critical points of a 3-manifold. Specifically, the level set topology changes either by gaining/losing a component or by gaining/losing genus. The isosurface gains a component when it evolves past a minimum and loses a component when it evolves past a maximum. At 2-saddles, the local pictures in Figure 1 indicate an apparent splitting of a component into two. Global behavior of the isosurface component will determine if this is indeed a split or a reduction in genus.

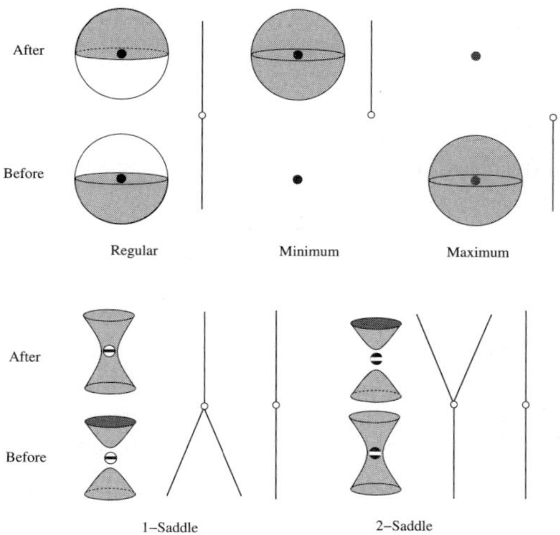

Fig. 1. Figures above show the isosurfaces before $(f^{-1}(c-\varepsilon))$ and after $(f^{-1}(c+\varepsilon))$ passing through a point with function value c and the structure of the Reeb graph at the corresponding node. Topology of the isosurface changes when it evolves past a critical point.

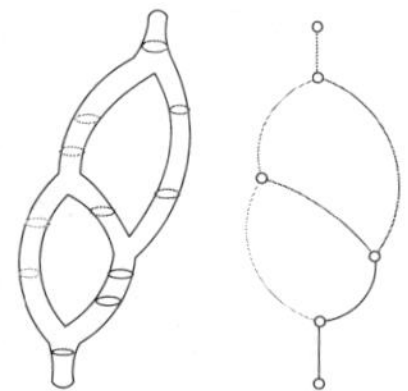

Fig. 2. Reeb graph of the height function defined on a surface with two tunnels. Reeb graph tracks the topology of level sets.

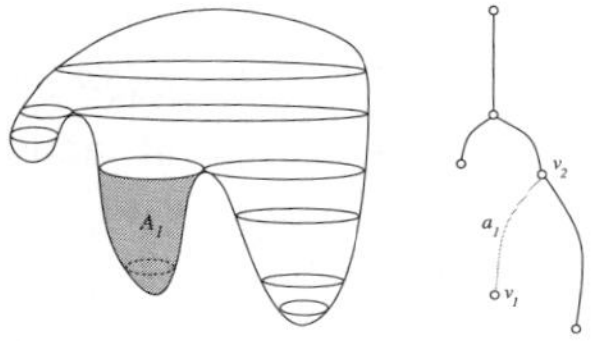

Fig. 3. Mapping between arcs in the Reeb graph and cylinders in the input

2.2 Reeb Graph

The *Reeb graph* of f is obtained by contracting each connected component of a level set to a point [18]. The Reeb graph expresses the evolution of connected components of level sets as a graph whose nodes correspond to critical points of the function, see Figure 2. Figure 1 illustrates the structure of the Reeb graph for 3-manifolds at various types of nodes. In the case of saddles, the corresponding node has degree 3 if the saddle merges/splits components, and degree 2 if it is a genus modifying saddle.

This view of the Reeb graph focuses on the mapping between components of individual level sets and nodes or points within arcs of the graph. We propose the use of an alternate mapping between nodes / arcs of the graph and components of critical level sets / equivalence classes of regular level set components. The two mappings are consistent with each other. The advantage of the alternate view is that it leads to a simple and efficient algorithm to compute the Reeb graph. For example, in Figure 3, the arc a_1 is mapped to *cylinder* A_1, a collection of regular level set components that are topologically equivalent to each other. The boundary of A_1 consists of two critical level set components. The end point v_2 of the arc originating at v_1 can be computed by tracing the cylinder from the lower boundary component to the upper component.

2.3 Input

We assume that the input manifold is represented by a triangulated mesh, the function is sampled at vertices, and linearly interpolated within each simplex. In the case of higher dimensional manifolds ($d \geq 3$), the algorithm requires only the 2-skeleton (vertices, edges, and triangles) of the mesh.

3   The Reeb Graph Algorithm

We now describe an algorithm that computes the Reeb graph of a PL function f defined on a 3-manifold. The algorithm directly extends to d-manifolds ($d \geq 2$) and non-manifolds but in order to simplify the description, we will consider the case of $d = 3$ in this section. The algorithm proceeds in two phases:

1. Locate critical points of the input and sort them based on function value.
2. Connect critical point pairs to obtain arcs of the Reeb graph.

A vertex is *regular* if it has one lower link component and one upper link component. All other vertices are *critical*. A critical point is a *maximum* if the upper link is empty and a *minimum* if the lower link is empty. Number of components of the upper and lower links are computed using a breadth first traversal of the link. We only need to locate the critical points and classify them as either a minimum, maximum or saddle.

3.1   *ls*-Graph

Tracking components of the level set requires only a 1-skeleton (vertices and edges) representation of the level set. Edges in a level set of f will pass through a set of triangles in the input mesh. We track components of level sets between two function values f_1 and f_2 ($f_1 < f_2$) by traversing triangles through which each component of the level set passes as function values are varied from f_1 to f_2. We introduce a dual graph that stores triangle adjacencies and helps track level set components. This graph $G_{ls}(V, E)$, called the *ls-graph* , is a directed graph whose nodes $V = \{t_1, t_2, \ldots, t_n\}$ corresponds to the n triangles $\{T_1, T_2, \ldots, T_n\}$ in the input mesh. Each node t_i is assigned a cost equal to the maximum over function values at the vertices of the triangle T_i. Let v_0, v_1, and v_2 ($f(v_0) < f(v_1) < f(v_2)$) be vertices of a triangle T_i. G_{ls} contains an edge between vertices t_i and t_j if triangles T_i and T_j are adjacent, unless T_i and T_j share the edge (v_0, v_1) and the cost of t_j is greater than $f(v_1)$, see Figure 4. The dotted, solid and dashed lines within the triangles indicate the edges of the level set when the function value becomes greater than $f(v_0)$, $f(v_1)$ and $f(v_2)$ respectively. An edge is directed towards the node with higher cost. Traversing an edge in G_{ls} implicitly tracks a component of a level set as function value increases. If this edge does not cross a critical value, then the traversal is equivalent to tracing a path within a cylinder. Figure 4(f) shows a configuration where the level set potentially splits and hence no edge is inserted between t_i and t_j.

3.2   Connecting the Critical Points

Our iterative algorithm uses the *ls*-graph to compute arcs in the Reeb graph. Let $\{c_1, c_2, \ldots, c_t\}$ be the ordered set of critical points with function values $\{f_1, f_2, \ldots, f_t\}$ ($f_x < f_y$ whenever $x < y$). Let L_i denote the set of triangles that contain the components of the critical level set $f^{-1}(f_i - \varepsilon)$ that are modified by c_i. The ith iteration of the algorithm connects c_i with a set of critical points c_p ($f_p > f_i$).

The *star* of a vertex consists of all simplices incident on the vertex. All simplices in the star where the function value is greater than the vertex constitute the *upper star*. Let the upper star of c_i contain k connected components ($k \leq 2$ for a Morse function). Each

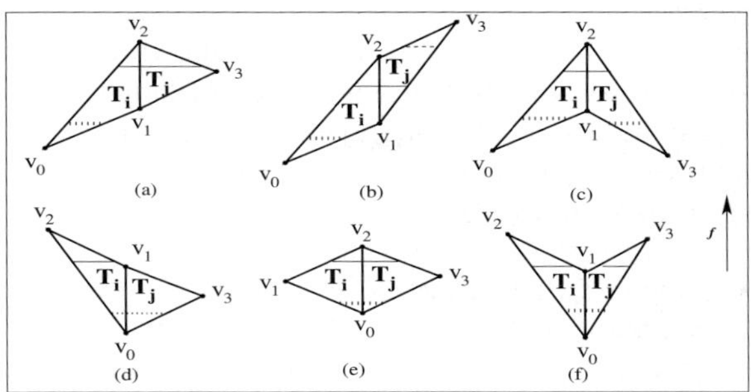
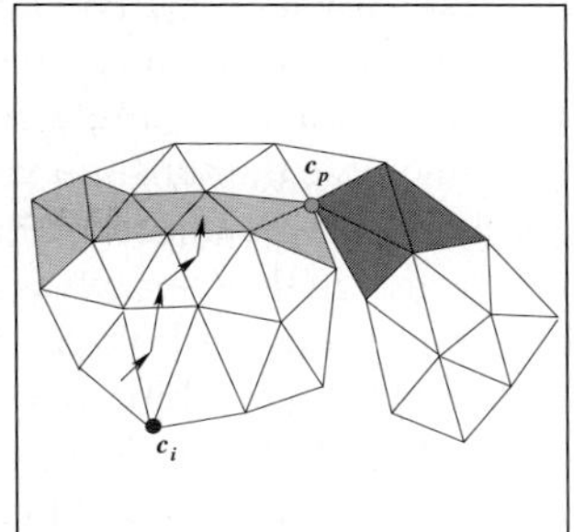

Fig. 4. The ls-graph contains an edge between t_i and t_j in all cases except the forbidden configuration in (f).

Fig. 5. Connecting critical points in the algorithm

component of the star corresponds to a possible arc in the Reeb graph starting from c_i. Initiate a search in the ls-graph beginning with a node t_{ij} corresponding to a triangle T_{ij} in the jth component ($j = 1...k$) of the upper star of c_i. We move to a higher cost node in G_{ls} at each step of the search. The search terminates when we find a node t'_{ij} with cost greater or equal to f_{i+1}. An arc connects the nodes corresponding to c_i and c_{i+1} in the Reeb graph iff the triangle T'_{ij} belongs to the set L_{i+1}. This search represents a monotone ascent through the cylinder. The search procedure for the ith iteration is illustrated in Figure 5, which shows a slice of the input mesh with the relevant triangles. Triangles in L_p are shaded, different colors indicating disjoint components of the level set.

We use a triangle-edge data structure [15] to store the input triangulation. The ls-graph is implicitly stored in this data structure because each triangle-edge pair stores a reference to neighboring triangle-edge pairs. During the search, we tag a visited node with a label $[i, j]$ if its cost is lesser or equal to f_{i+1}. If T'_{ij} does not belong to L_{i+1}, we continue the search until we reach a node with cost greater or equal to f_{i+2}, in order to determine if an arc in the Reeb graph connects c_i with c_{i+2}. We repeat the search until it is successful.

If a search initiated in the ith iteration from a node in the jth component of the upper star reaches a node with a tag $[i, j']$ ($j \neq j'$), then c_i is a genus modifying saddle and therefore the Reeb graph remains unchanged. The critical point c_i is again a genus modifying saddle if the search reaches a node, whose corresponding triangle lies in a critical isosurface component that was previously visited from a different upper star component of c_i. Note that it is impossible for the search initiated in the ith iteration to reach a vertex tagged $[i', j]$ ($i \neq i'$) for any j, because this will imply that two components of a level set merged into one at a regular vertex.

The Reeb graph is stored as an adjacency list whose nodes correspond to critical points of the function. An arc from c_i to c_p is added if the search finds a triangle in L_p. Once all critical points are processed, the adjacency list will represent the Reeb graph.

3.3 Analysis

Correctness. Let c_p ($f_i < f_p$) be a critical point such that there is an arc from c_i to c_p in the Reeb graph. So, if we track a component of the level set at function value infinitesimally above f_i and keep increasing the function value until it reaches f_p, then the topology of that component remains unchanged until the function value reaches f_p. Consider a triangle T_{ij} that contains the level set component when the tracking begins. As we increase the function value past the cost of the node t_{ij}, the level set component passes through an adjacent triangle with cost greater than that of t_{ij}. This is equivalent to the search in the ls-graph as performed by the algorithm. The algorithm continues the above procedure until we reach a node t'_{ij} with cost greater than or equal to f_p. Since the cost of the preceding node is less than f_p, an isosurface component at a function value infinitesimally below f_p will pass through the triangle T'_{ij}. Since the Reeb graph contains an arc between critical points c_i and c_p, the triangle T'_{ij} will belong to the set L_p and our algorithm will identify the arc (c_i, c_p) of the Reeb graph.

Running time. Let n be the number of triangles in the input and t be the number of critical points of the input PL function. Triangles adjacent to a given triangle can be found in $O(1)$ time using the triangle-edge data structure. The ls-graph is implicitly stored in this data structure. Critical points are located by computing the number of connected components of the lower and upper links, which can be done in $O(n)$ time using the triangle-edge data structure. Sorting the critical points takes $O(t \log t)$ time.

The sets L_i for each critical point c_i can be found by marching through the triangles that contain $f^{-1}(f_i - \varepsilon)$. This can be accomplished in $O(l)$ time, where $l = \sum_i |L_i|$, is the number of triangles in all the sets L_i. Though it is possible in theory that $l = O(n^2)$, the size of the critical level sets is usually $O(n)$ in practice.

Each node t_i in the ls-graph has at most 6 neighbors (since each triangle can be in at most two tetrahedra). Hence, the number of edges in the ls-graph is $O(n)$. During the search procedure, each node is tagged exactly once and visited at most 6 times (from each of its tagged neighbors). Thus the traversal of the graph is accomplished in $O(n)$ time. Each update of the adjacency list representation takes constant time. Total number of such updates is equal to the number of arcs in the Reeb graph. A conservative bound for the number of edges in the Reeb graph is given by the number of triangles in the input. Hence, maintaining the Reeb graph takes $O(n)$ time. Combining the above steps, we obtain an $O(n + l + t \log t)$ running time for our algorithm.

3.4 d-Manifolds and Non-manifolds

The level set of a regular value for a Morse function defined on a d-manifold is a $(d-1)$-manifold. The connectivity of a level set is represented by its 1-skeleton. Hence, similar to 3-manifolds, tracking the connected components of the level set requires only a 1-skeleton representation, which can be extracted from the 2-skeleton of the input mesh. So, the algorithm works directly on the 2-skeleton representation of d-manifolds. In the case of non-manifolds, the algorithm will again work on the 2-skeleton representation. We relax the definition of critical points to include all vertices that modify the topology of the level set. Candidate critical points are again located by counting the number of connected components of the lower and upper link.

4 Visualization of Reeb Graphs

4.1 Simplification of Reeb Graphs

Simplification is necessary for effective visualization of large and feature rich data because it aids in noise removal and creation of feature-preserving multiresolution representations. A topological feature in the input is represented by a pair of critical points, typically an arc in the Reeb graph. Unimportant features in the data can be removed by repeated cancellation of low persistence critical point pairs [10], which also leads to a multiresolution representation of the input scalar field. Features can also be ordered and removed based on geometric measures like hypervolume [5]. Existing algorithms for contour tree simplification remove critical point pairs that create / destroy a level set component. We simplify the Reeb graph using a notion of extended persistence [1] that pairs genus modifying critical points in addition to pairing component creators with destroyers.

Our approach to Reeb graph simplification is similar to the one used to simplify contour trees [5]. In addition to the *leaf pruning* and *node reduction* simplification operations, we perform an additional *loop pruning* operation on the Reeb graph. Leaf pruning removes a leaf and the incident arc from the Reeb graph. A leaf connecting to a degree-2 saddle is not pruned. Node reduction removes a degree-2 node by merging the two adjacent edges. The loop pruning operation removes a loop defined by adjacent nodes connected by two parallel arcs, from the Reeb graph. This operation is equivalent to removing one of the parallel arcs and performing node reduction on the pair of adjacent nodes.

We simplify the Reeb graph using repeated application of the three mentioned operations: (1) Perform node reduction where possible, (2) Choose the least important leaf / loop and prune it. Leaves and loops that can be pruned are stored in a priority queue ordered based on the persistence of the corresponding critical point pair. If a pruning operation results in a reducible node, then node reduction is performed immediately. All new leaves and loops created by the above operations are in turn added to the priority queue. Note that we use the simplification process as an aid for visualizing Reeb graphs and not to modify the input function. Realizing the function representing the simplified Reeb graph may require changing the topology of the input.

4.2 Reeb Graph Layout

We build upon the orrery layout proposed for contour trees [16] to obtain a layout for Reeb graphs. The extension to Reeb graphs is non-trivial because of the presence of loops. We overcome this difficulty by designing a four step layout scheme:

- First, extract a *spanning contour tree* of the Reeb graph.
- Second, compute a branch decomposition of this spanning tree.
- Third, use a radial layout scheme to embed the spanning tree in 3D.
- Finally, add the non-tree arcs to the layout.

The spanning contour tree is a spanning tree of the Reeb graph that satisfies the structural properties of a contour tree. All degree-2 nodes in this spanning tree have

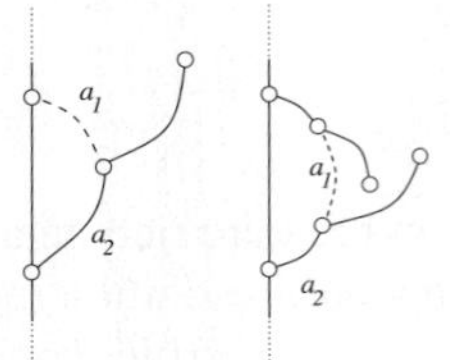

Fig. 6. The spanning contour tree of a Reeb graph is structurally similar to a contour tree. Removal of a_1 results in a spanning contour tree. Removing a_2 results in a tree with an invalid degree-2 node.

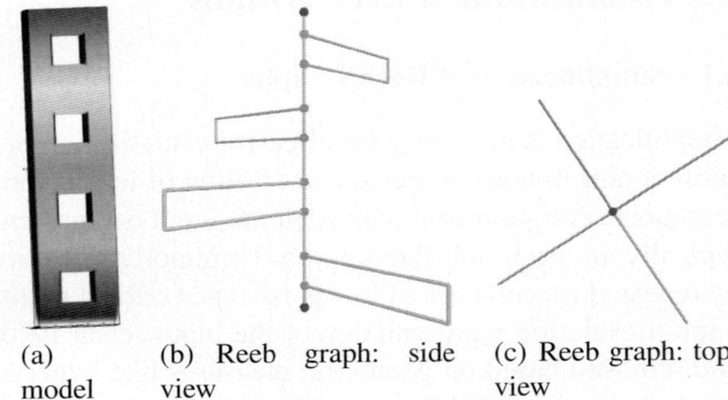

(a) model

(b) Reeb graph: side view

(c) Reeb graph: top view

Fig. 7. Radial layout of the Reeb graph of the height function on a 4-torus

Table 1. Reeb graph computation time for various 2D and 3D input. For all 2D models, solid 8-torus, and fighter, the Reeb graph was computed for the height function. In all other cases, the function is available with the data set.

Dimension	Model	#Triangles	#Critical points	Time taken (sec)	
				Our algorithm	Online algorithm
2D	bunny	40000	217	0.7	0.1
	Laurent Hand	99999	92	1.5	0.43
	Neptune	998840	1757	24.3	3.7
3D	engine	27252	160	2	0.7
	solid 8-torus	34832	18	0.56	0.14
	fighter	143881	8	2.2	28
	PMDC	237291	902	8	17
	blunt	451601	827	213	406
	nucleon	652964	2203	254	1638
	post	1243200	132	70	1671

exactly one neighbor node with higher function value and one neighbor node with lower function value. Not all spanning trees satisfy this property. For example, in Figure 6, removing arc a_1 results in a spanning contour tree. Removal of a_2 also results in a spanning tree, but one that does not correspond to a valid contour tree.

A branch decomposition is an alternate representation of a contour tree that explicitly stores the topological features and their hierarchical relationship [16]. A branch is a path between two leaves of the contour tree or a path that connects a leaf to an interior node of another branch.

All branches of the spanning contour tree are drawn as L-shaped polylines and the z-coordinate corresponds to function value. The (x, y) coordinates are computed for each branch using a radial layout scheme. The root branch is located at the origin and others are placed on concentric circles centered at the origin. All branches that connect to an interior node of the root branch are equally spaced around the origin at a fixed distance from it. Branches that connect to an interior node of a first-level branch are

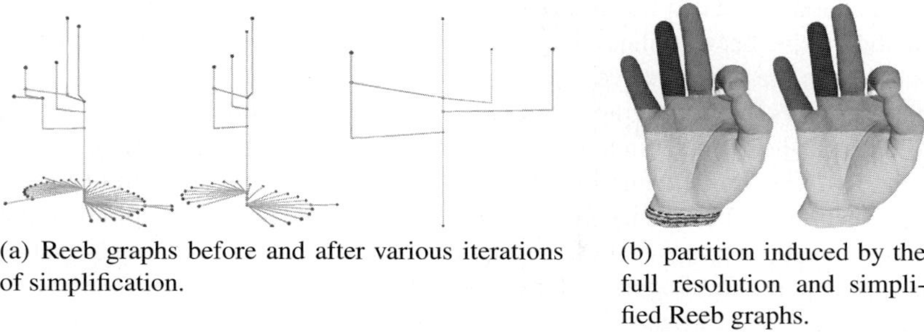

(a) Reeb graphs before and after various iterations of simplification.

(b) partition induced by the full resolution and simplified Reeb graphs.

Fig. 8. Reeb graph computed for height function on Laurent hand and induced partition on the surface

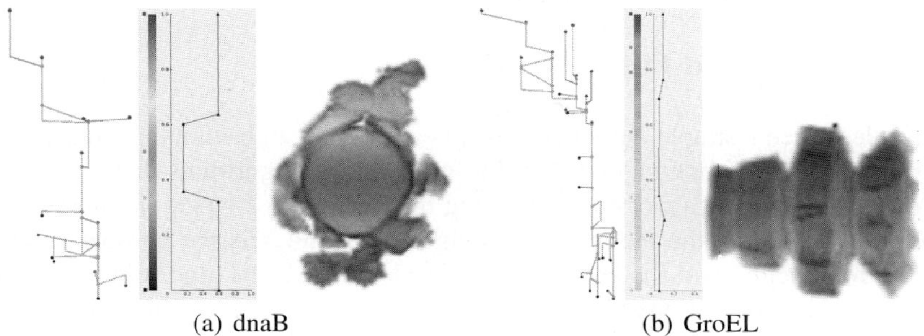

(a) dnaB

(b) GroEL

Fig. 9. Reeb graph computed for height function on volume representation of two molecules: dnaB and GroEL. The transfer function shown in the middle determines the color map.

placed in the second concentric circle within a wedge centered at a level-one branch. The angle subtended is proportional to the number of descendant branches. In order to avoid intersections when the non-tree arcs are added, we include a dummy branch for each loop arc before calculating the angular wedge subtended at each branch. Figure 7 shows the layout for the Reeb graph of a 4-torus.

5   Experimental Results

The Reeb graph construction algorithm was implemented in Java and tested on a Pentium 4, 2.4 GHz machine with 1 GB main memory. Our implementation accepts a function sampled at vertices of a simplicial mesh as input, computes the Reeb graph, and stores it as an edge list. Table 1 shows the time taken by our implementation to compute the Reeb graph for various models. We compare the performance of our algorithm with the online algorithm described in [17]. While the online algorithm performs well for 2D data, our algorithm performs substantially better for 3D data. We expect that the algorithm will also be efficient in practice for higher dimensional input. The running time depends on the number of critical points, clearly indicating the output sensitivity of our algorithm. For large datasets that do not fit in memory, our implementation can be extended similar to the contour tree algorithm described in [6]. Also, no code optimizations have been applied. An implementation in C / C++ will exhibit significant improvement in the performance.

Figure 8 shows the Reeb graph for the height function defined on the Laurent hand model. The near-horizontal branches in the Reeb graph indicate that the function is noisy near the wrist. Our implementation allows the user to interactively simplify and visualize the Reeb graph by specifying a persistence threshold. Simplification using a low persistence threshold removes these unimportant features. The remaining branches correspond to the palm and fingers and the loop corresponds to the thumb and fore-finger. Figure 9 shows the Reeb graph computed for two biological data sets: dnaB and GroEL. In both data sets, the volumetric domain represents the molecule and the scalar field is the height function. Loops in the Reeb graph indicate possible tunnels in the molecule. The Reeb graph of dnaB contains two loops and that of GroEL has five loops.

6   Conclusions and Future Work

We have described a simple output-sensitive near-optimal algorithm that constructs the Reeb graph of a PL function. Compared to prior known algorithms that run in $O(n^2)$ time, our algorithm has a worst case running time of $O(n + l + t \log t)$, where n is the number of triangles in the mesh representing the domain, t is the number of critical points of the function and l is size of all critical level sets. The algorithm works without any modification for functions defined on manifolds in any dimension, and for non-manifold domains. We have also described a method to simplify the Reeb graph based on an extended notion of persistence and provided a feature-directed layout of the Reeb graph that serves as a useful interface for exploring and understanding three-dimensional scalar fields. We have shown through our experimental results that our algorithm performs efficiently in practice. The iterations of the algorithm being independent of each other, provide an inherent scope for parallelization.

Acknowledgements. This work was supported by the Department of Science and Technology, India, under Grant SR/S3/EECE/048/2007.

References

1. Agarwal, P.K., Edelsbrunner, H., Harer, J., Wang, Y.: Extreme elevation on a 2-manifold. Disc. Comput. Geom. 36(4), 553–572 (2006)
2. Bajaj, C.L., Pascucci, V., Schikore, D.R.: The contour spectrum. In: Proc. IEEE Conf. Visualization, pp. 167–173 (1997)
3. Banchoff, T.F.: Critical points and curvature for embedded polyhedral surfaces. Am. Math. Monthly 77, 475–485 (1970)
4. Carr, H., Snoeyink, J., Axen, U.: Computing contour trees in all dimensions. Comput. Geom. Theory Appl. 24(2), 75–94 (2003)
5. Carr, H., Snoeyink, J., van de Panne, M.: Simplifying flexible isosurfaces using local geometric measures. In: Proc. IEEE Conf. Visualization, pp. 497–504 (2004)
6. Chiang, Y.-J., Lenz, T., Lu, X., Rote, G.: Simple and optimal output-sensitive construction of contour trees using monotone paths. Comput. Geom. Theory Appl. 30(2), 165–195 (2005)

7. Cole-McLaughlin, K., Edelsbrunner, H., Harer, J., Natarajan, V., Pascucci, V.: Loops in Reeb graphs of 2-manifolds. Disc. Comput. Geom. 32(2), 231–244 (2004)
8. Edelsbrunner, H.: Geometry and Topology for Mesh Generation. Cambridge Univ. Press, England (2001)
9. Edelsbrunner, H., Harer, J., Natarajan, V., Pascucci, V.: Morse-Smale complexes for piecewise linear 3-manifolds. In: Proc. Symp. Comput. Geom., pp. 361–370 (2003)
10. Edelsbrunner, H., Letscher, D., Zomorodian, A.: Topological persistence and simplification. Disc. Comput. Geom. 28(4), 511–533 (2002)
11. Guskov, I., Wood, Z.: Topological noise removal. In: Proc. Graphics Interface, pp. 19–26 (2001)
12. Hétroy, F., Attali, D.: Topological quadrangulations of closed triangulated surfaces using the Reeb graph. Graph. Models 65(1-3), 131–148 (2003)
13. Hilaga, M., Shinagawa, Y., Kohmura, T., Kunii, T.L.: Topology matching for fully automatic similarity estimation of 3d shapes. In: Proc. SIGGRAPH, pp. 203–212 (2001)
14. Matsumoto, Y.: An Introduction to Morse Theory. Amer. Math. Soc. (2002) (Translated from Japanese by Hudson K., Saito M.)
15. Mücke, E.P.: Shapes and Implementations in Three-Dimensional Geometry. PhD thesis, Dept. Computer Science, University of Illinois, Urbana-Champaign, Illinois (1993)
16. Pascucci, V., Cole-McLaughlin, K., Scorzelli, G.: Multi-resolution computation and presentation of contour trees. Tech. rep., Lawrence Livermore Natl. Lab (2005)
17. Pascucci, V., Scorzelli, G., Bremer, P.-T., Mascarenhas, A.: Robust on-line computation of reeb graphs: simplicity and speed. ACM Trans. Graph. 26(3), 58 (2007)
18. Reeb, G.: Sur les points singuliers d'une forme de pfaff complètement intégrable ou d'une fonction numérique. Comptes Rendus de L'Académie ses Séances, Paris 222, 847–849 (1946)
19. Shinagawa, Y., Kunii, T.L.: Constructing a reeb graph automatically from cross sections. IEEE Comput. Graph. Appl. 11(6), 44–51 (1991)
20. Shinagawa, Y., Kunii, T.L., Kergosien, Y.L.: Surface coding based on Morse theory. IEEE Comput. Graph. Appl. 11(5), 66–78 (1991)
21. Tung, T., Schmitt, F.: Augmented reeb graphs for content-based retrieval of 3d mesh models. In: SMI 2004: Proc. Shape Modeling Intl., pp. 157–166 (2004)
22. van Kreveld, M., van Oostrum, R., Bajaj, C., Pascucci, V., Schikore, D.R.: Contour trees and small seed sets for isosurface traversal. In: Proc. Symp. Comput. Geom., pp. 212–220 (1997)
23. Weber, G.H., Dillard, S.E., Carr, H., Pascucci, V., Hamann, B.: Topology-controlled volume rendering. IEEE Trans. Vis. Comput. Graph. 13(2), 330–341 (2007)
24. Wood, Z., Hoppe, H., Desbrun, M., Schröder, P.: Removing excess topology from isosurfaces. ACM Trans. Graph. 23(2), 190–208 (2004)
25. Shinagawa, Y., Kunii, T.L., Sato, H., Ibusuki, M.: Modeling contact of two complex objects: with an application to characterizing dental articulations. Computers and Graphics 19(1), 21–28 (1995)
26. Zhang, E., Mischaikow, K., Turk, G.: Feature-based surface parameterization and texture mapping. ACM Trans. Graph. 24(1), 1–27 (2005)

Signature Theory in Holographic Algorithms

Jin-Yi Cai[1] and Pinyan Lu[2,*]

[1] Computer Sciences Department, University of Wisconsin
Madison, WI 53706, USA
jyc@cs.wisc.edu
[2] Institute for Theoretical Computer Science, Tsinghua University
Beijing, 100084, P.R. China
lpy@mails.tsinghua.edu.cn

Abstract. Valiant initiated a theory of holographic algorithms based on perfect matchings. These algorithms express computations in terms of signatures realizable by matchgates. We substantially develop the signature theory in terms of d-realizability and d-admissibility, where d measures the dimension of the basis subvariety on which a signature is feasible. Starting with 2-admissibility, we prove a Birkhoff-type theorem for the class of 2-realizable signatures. This gives a complete structural understanding of 2-realizability and 2-admissibility. This is followed by characterization theorems for 1-realizability and 1-admissibility.

1 Introduction

It is generally conjectured that many problems in the class NP or #P are not computable in P. The prevailing view is that these problems require the processing of exponentially many potential solution fragments, by any algorithm.

However it is usually rather natural to express the answer to such a problem as a suitable exponential sum. Take for instance the canonical Boolean Satisfiability problem 3SAT. We can express the number of satisfying assignments as an exponential sum, as follows. Suppose C is an OR gate with 3 input literals in a formula Φ. We assign a *signature* $R_C = (0, 1, 1, 1, 1, 1, 1, 1)$ to C, where the entries are indexed by 3 bits $b_1 b_2 b_3 \in \{0, 1\}^3$. Here $b_1 b_2 b_3$ correspond to a truth assignment to the 3 literals. If $(R_{b_1 b_2 b_3})$ and $(R'_{b_4 b_5 b_6})$ are two signatures, then its tensor product has 64 entries, and its value indexed by $b_1 \ldots b_6 \in \{0, 1\}^6$ is the product $R_{b_1 b_2 b_3} R'_{b_4 b_5 b_6}$. If we form the tensor product of signatures for a set of m clauses $\mathbf{R} = \bigotimes_C R_C$, then $\mathbf{R}$ has $3m$ indices $(i_1, i_2, \ldots, i_{3m}) \in \{0, 1\}^{3m}$, and a particular entry has the value 1 if each OR gate is satisfied, and the value 0 otherwise. Now suppose a Boolean variable x appears in k clauses. Depending on whether it appears positively or negatively in each clause, we can assign a signature tensor of arity k, indexed by $(i_1, i_2, \ldots, i_k) \in \{0, 1\}^k$, where all but two values are 0, and the two remaining values are 1, corresponding to a consistent truth assignment. For example if x appears in C positively and in

* Supported by NSF CCF-0511679 and by the NNSF of China Grant 60553001 and the National Basic Research Program of China Grant 2007CB807900, 2007CB807901.

S.-H. Hong, H. Nagamochi, and T. Fukunaga (Eds.): ISAAC 2008, LNCS 5369, pp. 568–579, 2008.
© Springer-Verlag Berlin Heidelberg 2008

C' negatively, then we can assign $G_x = (0, 1, 1, 0)^{\mathrm{T}}$. From these we can form the tensor product for all variables $\mathbf{G} = \bigotimes_x G_x$. The indices of $\mathbf{R}$ and $\mathbf{G}$ match up one-to-one according to which x appears in which C. Then the exponential sum $\langle \mathbf{R}, \mathbf{G} \rangle = \sum_{i_1 \ldots i_j \ldots \in \{0,1\}} R_{i_1 \ldots i_j \ldots} G^{i_1 \ldots i_j \ldots}$ counts exactly the number of satisfying assignments to Φ: For each 0-1 tuple $(i_1 \ldots i_j \ldots)$, the product $R_{i_1 \ldots i_j \ldots} G^{i_1 \ldots i_j \ldots}$ is 1 when this corresponds to a consistent assignment of truth values to each variable *and* the truth assignment satisfies each clause; the product value is 0 otherwise.

This expression has exponentially many terms. The power of holographic algorithms is to evaluate such an exponential sum in *polynomial time*, for *some* combinatorial problems. This happens when suitable signatures are *realizable*.

In Valiant's theory, the notion of *realizability* has meanings at two levels. First, there is a theory of matchgate expressibility. Some signatures, such as $G_x = (0, 1, 1, 0)^{\mathrm{T}}$ are directly expressible as the *standard signature* of a planar matchgate. When this direct expressibility is possible, we can apply the Fisher-Kasteleyn-Temperley (FKT) method on planar perfect matchings [10,11,15] to evaluate the exponential sum in polynomial time. However frequently the desired signature is not directly expressible as such, e.g., $R_C = (0, 1, 1, 1, 1, 1, 1, 1)$. The main idea of holographic algorithms is to come up with suitable basis transformations in the tensor space, such that the desired signatures become expressible, as a *superposition* of standard signatures of matchgates. When realizability is achieved for all the desired signatures under some basis transformation, the FKT algorithm can still be applied, by Valiant's Holant Theorem [18], to evaluate the exponential sum in polynomial time. What happens at the heart of this "magic" is that, the superposition creates a pleasing algebraic cancelation, analogous to quantum computing.

We illustrate this cancelation process by an example. Consider the signature $R_C = (0, 1, 1, 1, 1, 1, 1, 1)$, which is not directly expressible by a matchgate. However, it so happens that there exists a matchgate with the standard signature $(\Gamma_{b_1 b_2 b_3}) = \frac{1}{4}(0, 1, 1, 0, 1, 0, 0, 1)$. Here the values 0 or 1/4 are evaluations of perfect matchings of a weighted graph called a matchgate. Now we can choose two linearly independent basis vectors $\boldsymbol{\beta} = \left[\begin{pmatrix} 1 + \omega \\ 1 - \omega \end{pmatrix}, \begin{pmatrix} 1 \\ 1 \end{pmatrix} \right]$, where $\omega = e^{2\pi i/3}$. In the tensor space spanned by $\boldsymbol{\beta}^{\otimes 3}$, we can represent the OR signature $(0, 1, 1, 1, 1, 1, 1, 1)$ by perfect matchings as follows: If we write out

$$(\boldsymbol{\beta}^{-1})^{\otimes 3} = \left(\begin{bmatrix} 1 + \omega & 1 \\ 1 - \omega & 1 \end{bmatrix}^{-1} \right)^{\otimes 3}, \text{ we can get}$$

$$(0, 1, 1, 1, 1, 1, 1, 1) \left(\begin{bmatrix} 1 + \omega & 1 \\ 1 - \omega & 1 \end{bmatrix}^{-1} \right)^{\otimes 3} = \frac{1}{4}(0, 1, 1, 0, 1, 0, 0, 1) = (\Gamma_{b_1 b_2 b_3}).$$

It follows that

$$(0, 1, 1, 1, 1, 1, 1, 1) = (\Gamma_{b_1 b_2 b_3}) \boldsymbol{\beta}^{\otimes 3}.$$

In this way each logical value 0 or 1 in $(0, 1, 1, 1, 1, 1, 1, 1)$ is expressed as a "superposition" or a "holographic mix" of perfect matching values in $(\Gamma_{b_1 b_2 b_3})$.

In order to be able to solve a particular problem in polynomial time, we would need *simultaneous* realizability (for both **R** and **G**) under the same basis transformation. Valiant [19] showed that we can count mod 7 the number of satisfying assignments to a restricted class of 3CNF formulae. The same number without mod 7 for this restricted #SAT problem is #P-complete, and to count mod 2 is NP-hard. Exactly which sum is computable in polynomial time by holographic algorithms brings us to the subject of *signature theory*.

Finding a holographic algorithm successfully for a particular problem boils down to the existence of suitable signatures in a suitable tensor space. This is the realizability problem. The requirements are specified by families of algebraic equations called Grassmann-Plücker Identities [17,1]. These families of equations are non-linear, exponential in size, and difficult to handle. But whenever we find a suitable solution, we get an exotic polynomial time algorithm. Of course the big question is whether such "freak objects" exist for any of the NP-hard problems. If not, is there a coherent explanation? To quote Valiant [18]: "Any proof of P $\neq$ NP may need to explain, and not only to imply, the unsolvability" of NP-hard problems using this approach.

In our STOC paper [4] an algebraic framework was developed which gives a satisfactory theory of *symmetric signatures*. In this framework, we defined a basis manifold $\mathcal{M}$. *A priori* the tensor space can have basis vectors of high dimensions; in [5] we have proved a general basis collapse theorem which effectively restricted the theory to the basis manifold $\mathcal{M}$ corresponding to $\mathbf{GL}_2$. Thus to Valiant's challenge above [18] what remains is the general (i.e., not necessarily symmetric) signature theory on $\mathcal{M}$. This paper is a significant step toward that.

Definition 1. *Two bases* $\beta = \left[\begin{pmatrix} n_0 \\ n_1 \end{pmatrix}, \begin{pmatrix} p_0 \\ p_1 \end{pmatrix} \right]$ *and* $\beta' = \left[\begin{pmatrix} n_0' \\ n_1' \end{pmatrix}, \begin{pmatrix} p_0' \\ p_1' \end{pmatrix} \right]$ *are equivalent, denoted by* $\beta \sim \beta'$*, iff there exist* $x, y \in \mathbf{F}^*$ *such that* $n_0' = xn_0, p_0' = xp_0, n_1' = yn_1, p_1' = yp_1$ *or* $n_0' = xn_1, p_0' = xp_1, n_1' = yn_0, p_1' = yp_0$*. The basis manifold* $\mathcal{M}$ *is* $\mathrm{GL}_2(\mathbf{F})/\sim$*.*

Definition 2. *A tensor G is* admissible *as a generator on a basis β iff $G' = \beta^{\otimes n}G$ satisfies the* parity requirements. *It is called* realizable *as a generator on a basis β iff $G' = \beta^{\otimes n}G$ satisfies both the* parity requirements *and all the* MGI. *This is equivalent to G' being the standard signature of some planar matchgate.*

Definition 3. *Let $B_{gen}(G)$ (resp. $B_{gen}^p(G)$) be the set of all possible bases in $\mathcal{M}$ for which a generator G is realizable (resp. admissible).*

Definition 4. *A generator G is called d-realizable (resp. d-admissible) for an integer $d \geq 0$ iff $B_{gen}(G) \subset \mathcal{M}$ (resp. $B_{gen}^p(G) \subset \mathcal{M}$) is an algebraic subset of dimension at least d.*

The signature theory is expressed in terms of *d*-admissibility and *d*-realizability, where $d \geq 0$ is the dimension of the algebraic variety of $\mathcal{M}$ corresponding to a signature. The higher this number d is, the more likely one can find a basis that is common to all the signatures needed for all the constraints of the problem at hand. In almost all cases where holographic algorithms are successful [18,4,6], one uses some d-realizable signature for $d > 0$. The main findings of this paper are characterizations of these highly realizable signatures.

After factoring out an equivalence relation, the basis manifold $\mathcal{M}$ has dimension 2. Thus d-realizability for $d = 2$ is the maximum possible. In fact, a 2-realizable signature is realizable on every basis in the basis manifold $\mathcal{M}$. We first prove a Birkhoff-type theorem which gives a *complete* and *explicit* characterization of the class of 2-realizable signatures (over char. 0). This turns out to be the vertices of a simplex, of which the linear span is precisely the class of 2-admissible signatures, whose dimension is the Catalan number. The 2-realizable signatures also have an explicit combinatorial interpretation in terms of planar tensor product of perfect matchings. In general the realizability of signatures is controlled by a set of algebraic equations called Matchgate Identities (MGI), a.k.a. useful Grassmann-Plücker Identities [17,1]. The proof here uses MGI *implicitly*, in the form of *explicit* Pfaffian representations. The proof of the characterization theorem for 2-realizability is rather involved. In Section 2 we give a proof outline. The main structure of the proof is a (multiply) nested induction. A key idea is the Pfaffian representation, which is purely algebraic. Another key idea is the introduction of a derivative operator ∂_j, which uses the underlying geometry of planarity. We then extend this characterization theorem to all char. $p \neq 2$.

Next we give characterization theorems concerning 1-realizability and 1-admissibility. The proof techniques are mainly algebraic. The 1-realizable signatures are much richer than the 2-realizable signatures. Many signatures used by previous holographic algorithms can be viewed as special cases of these 1-realizable signatures. The characterization theorems obtained here can give more holographic algorithms for other problems. Due to space limitation, most proofs are omitted (see [7] for more details.) The structural theory for general signatures developed here substantially move forward our understanding of the ultimate capabilities of holographic algorithms.

2   Characterization of 2-Realizability

We present our signature theory in the algebraic framework of [4]. The theory is developed in terms of d-admissibility and d-realizability. The key to the success of a holographic algorithm is to find generators and recognizers whose signatures we desire and whose realizability varieties intersect. This typically happens with at least one side having a d-realizability for $d \geq 1$. To maximize the chance of a non-empty intersection, 2-realizability is the most desirable. The central results from [4] in this regard are characterizations of 2-admissible signatures. The arity of any 2-admissible signature must be an even number $n = 2k$. The 2-admissible signatures are the solutions to a homogeneous linear equation system. The dimension of the solution space is $\frac{1}{2k}\binom{2k}{k}$, the Catalan number.

It turns out that there is a particular set of 2-admissible signatures with a clear combinatorial meaning. These are signatures of planar matchgates with

k pairs of points on the circumference of a unit disk D, constructed by planar tensor product, as shown in Figure 1.

Let P be a matchgate consisting of a path of length 2 (3 nodes and 2 edges), where we place the two end points on the circumference of D, and the two edges are weighted $+1$ and -1 respectively. This gives a planar matchgate of arity 2 with the (standard) signature $(0, +1, -1, 0)$. It is easy to verify that this signature is indeed 2-realizable. Now we can form planar tensor product [4] recursively using disjoint copies of P as the basic building block. The planar matchgate formed is also 2-realizable. Combinatorially this process is very simple: We end up with $2k$ vertices on the circumference of D, which are pair-wise matched up by k disjoint paths each with weights $+1$ and -1 on its two edges, respectively. The union of these k disjoint paths form a planar graph with a total of $3k$ vertices (planar tensor product preserves planarity, and these k paths do not cross each other). This family of matchgates with $2k$ external nodes is denoted by $\mathcal{D}_{2k}$.

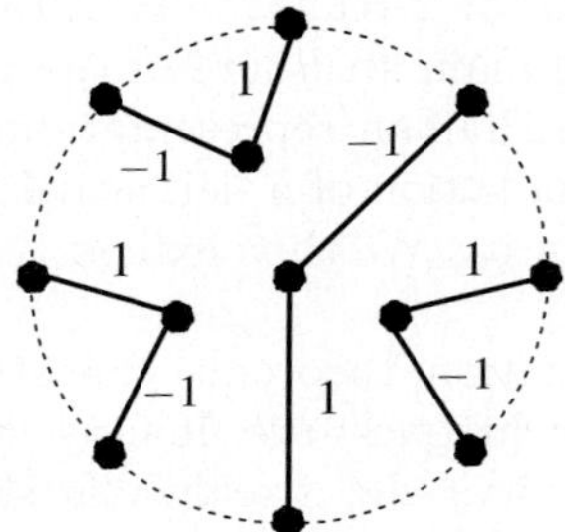

Fig. 1. A 2-realizable signature

Let $G \in \mathcal{D}_{2k}$, and let $(G^S)_{S \subset [2k]}$ be its signature. We show that (G^S) satisfies the conditions to be 2-admissible [4]. First note that each entry G^S is 0, except when S contains exactly one end point from each P. This follows from the definition of perfect matching. In particular $G^S \neq 0$ only for $|S| = k$. Now we show that $\sum_{S \subset T} G^S = 0$, for any subset $T \subset [2k]$ of cardinality $k + 1$. Since $|T| = k + 1$, there must be at least one pair $\{i, j\} \subset T$, where i and j are connected by some P of length 2 in G. Then the only *possible* non-zero terms in $\sum_{S \subset T} G^S$ come from $S = T - \{i\}$ and $S = T - \{j\}$. In order to be actually non-zero, the set $T - \{i, j\}$ must contain exactly one vertex from each pair of the other $k-1$ pairs of external nodes connected by length-2 paths. Thus either every term in $\sum_{S \subset T} G^S$ is zero, or there are exactly two non-zero terms of opposite values ± 1. Thus, $\sum_{S \subset T} G^S = 0$ for all $|T| = k + 1$.

This proof gives an explicit set of solutions to the system defining 2-admissibility [4]. this set is the Catalan number $\frac{1}{2k}\binom{2k}{k}$. We also know from the exact knowledge of the rank information derived from results of the *Kneser Graph* $\mathrm{KG}_{2k+1,k}$ [12,13,14,8,9] that this number $\frac{1}{2k}\binom{2k}{k}$ is exactly the dimension of the solution subspace. If we order the entries of the signature G^S lexicographically by its index $S \subset [2k]$, the first non-zero entry (with value

+1) occurs at the location where for each matched pair $i < j$ by a path P we assign 0 to the i-th bit and 1 to the j-th bit. This corresponds to a balanced paranthesized expression (BPE), i.e., a sequence of length $2k$ consisting of k 0's and k 1's, and any prefix has at least as many 0's as 1's (write 0 for "(" and 1 for ")"). This mapping from $\mathcal{D}_{2k}$ to BPE of length $2k$ is also reversable. By considering the submatrix whose rows are these $\frac{1}{2k}\binom{2k}{k}$ signatures from $\mathcal{D}_{2k}$ and whose columns are indexed by BPE (if we order the signatures G according to the unique corresponding BPE in lexicographic order, this submatrix is upper-triangular with 1's on the diagonal), it follows that these signatures are linearly independent. At this point the class of 2-admissible signatures is completely understood. They form the linear span of the signatures from $\mathcal{D}_{2k}$.

Theorem 1. *The set of $\frac{1}{2k}\binom{2k}{k}$ signatures from $\mathcal{D}_{2k}$ are 2-realizable, and forms a basis of the solution space of the set of all 2-admissible signatures of arity $2k$.*

Our main theorem in this section is to prove that the signatures from $\mathcal{D}_{2k}$ are precisely the class of 2-realizable signatures of arity $2k$ (over char. 0), after a scaling factor.

Theorem 2. *Up to a scalar factor, every 2-realizable signature is obtainable as a planar tensor product from $(0, 1, -1, 0)$. For arity $2k$, this is precisely the set of $\frac{1}{2k}\binom{2k}{k}$ signatures from $\mathcal{D}_{2k}$.*

Proof Outline: Since the proof of Theorem 2 is quite involved, we only present a proof outline. At the top level, the theorem is proved by an induction on the arity. Given a 2-realizable signature, we show that in a certain planar matchgate of this signature, there exist two contiguous nodes $(i, i + 1)$, which are isolated from the rest. The part on $(i, i + 1)$ makes one copy of $(0, 1, -1, 0)$. Then we apply induction on the remaining part.

However the proof for the existence of such two contiguous nodes is complicated. We first prove this under the condition that $G^{0101\cdots 01} \neq 0$. If this is true, by flipping all odd bits, we can define a new signature G_A which has the property that $G_A^{1111\cdots 11} \neq 0$. Then, from the general theory [1], we know that G_A can be realized by Pfaffians of a (weighted, but not necessarily planar) graph Γ *without* internal nodes. This transformation is a key idea of this proof and through which we bypass the difficulty of having to deal with exponentially many MGI explicitly. After that we deal with edge weights of the graph Γ rather than the entries of G. This reduces the number of variables from 2^n to $\binom{n}{2}$, and the explicit form of Pfaffian satisfies all MGI implicitly. We translate the conditions of G being 2-admissible to conditions on G_A. Then we apply these conditions on the edge weights in Γ and prove that there is one isolated edge connecting two contiguous nodes.

Then all we need to prove is that $G^{0101\cdots 01} \neq 0$. This turns out to be at least as difficult as above. We prove this by a nested induction. First we introduce *derivative operators* ∂_j which construct 2-realizable signatures of arity $n - 2$ from a 2-realizable signature of arity n. After a normalization, we use the operator and the inductive hypothesis (of the outer induction) to prove that at least one of the two values $G^{0101\cdots01}, G^{1001\cdots01}$ is non-zero. Then we prove (by the inner induction) that the case $G^{0101\cdots01} = 0, G^{1001\cdots01} \neq 0$ leads to a contradiction. This proof also uses the method of explicit Pfaffian representation.

3   1-Realizability

Section 2 gives a complete characterization of 2-realizable signatures. In this section, we study 1-realizable signatures. As discussed in [4], d-realizability for $d > 0$ is key to finding interesting holographic algorithms, since they result from a non-empty intersection of the signature varieties of both recognizers and generators. We present a structural characterization theorem for 1-realizable signatures. They are also restrictive, but they are much richer than 2-realizable signatures.

First we prove the following key lemma. This lemma plays an important role in the proof of Theorem 3. Moreover, this lemma reveals a key property of the set $B_{gen}^p(G)$, which is useful not only for the study of 1-realizable signatures.

Lemma 1. *For any G, if $T_1 = \begin{pmatrix} 1 & \alpha \\ 1 & y_1 \end{pmatrix} \in B_{gen}^p(G)$ and $T_2 = \begin{pmatrix} 1 & \alpha \\ 1 & y_2 \end{pmatrix} \in B_{gen}^p(G)$*

(for $y_1 \neq y_2$), then for all $y \in \mathbf{C} - \{\alpha\}$, $\begin{pmatrix} 1 & \alpha \\ 1 & y \end{pmatrix} \in B_{gen}^p(G)$.

Proof: If G is trivial, then the lemma is obvious. We assume G is non-trivial.

Let $B_{gen}^p(G) = V_0 \cup V_1 \subset \mathcal{M}$, and V_0 (resp. V_1) be the set defined by all the parity requirements for being an odd (resp. even) matchgate. Since G is non-trivial, we have $V_0 \cap V_1 = \emptyset$. Then there are four cases, depending on whether T_1 and T_2 are in V_0 or V_1. Here we will only present the proof for the case where both $T_1, T_2 \in V_0$. The case for $T_1 \in V_0$ and $T_2 \in V_1$ requires a different but similar proof, and is omitted here. The other two cases are similar to these two.

Let $T_1, T_2 \in V_0$. We recall the parity polynomial equation [4] for V_0 (for $|T|$ even):

$$\left\langle \bigotimes_{\sigma=1}^{n} [1, x_{[\sigma \in T]}], G \right\rangle = \sum_{\substack{0 \le i \le n-|T| \\ 0 \le j \le |T|}} x^i y^j \sum_{\substack{A \subset T^c, |A| = i \\ B \subset T, |B| = j}} G^{A \cup B} = 0. \tag{1}$$

For any $T \subset [n]$ and $|T|$ even, let

$$f_T(y) = \sum_{\substack{0 \le i \le n-|T| \\ 0 \le j \le |T|}} \alpha^i y^j \sum_{\substack{A \subset T^c, |A| = i \\ B \subset T, |B| = j}} G^{A \cup B}.$$

Then for all even T, $\deg(f_T) \le |T|$ and $f_T(y_1) = f_T(y_2) = 0$. We note that $y_1 \neq \alpha$ and $y_2 \neq \alpha$. We want to prove that f_T is identically 0 for all even T.

We prove this by induction on $|T| \ge 0$ and $|T|$ is even. The case $|T| = 0$ is obvious.

Inductively we assume this has been proved for all $|T| \le 2(k - 1)$, for some $k \ge 1$. Now $|T| = 2k$. First we prove that α is a root of $f_T(y)$ with multiplicity at least $2k - 1$. We prove this by showing that $f_T^{[r]}(\alpha) = 0$ for $0 \le r \le 2(k-1)$, where $f^{[0]} = f_T$ and $f_T^{[r]} = \frac{d}{dy}(f_T^{[r-1]})$ is the r-th derivative. We have

$$f_T^{[r]}(\alpha) = \sum_{\substack{0 \le i \le n - |T| \\ r \le j \le |T|}} r! \binom{j}{r} \alpha^i \alpha^{j-r} \sum_{\substack{A \subset T^c, |A| = i \\ B \subset T, |B| = j}} G^{A \cup B} = r! \sum_{\ell=0}^{n-r} \alpha^\ell \sum_{|S| = \ell + r} \binom{|T \cap S|}{r} G^S. \tag{2}$$

Note that the second equality follows from considering each G^S, where $|S \cap T| = j \ge r$ and $|S \cap T^c| = i$.

If r is even, we consider any T' where $|T'| = r$. Since $r \le 2(k-1)$, by induction, we have $f_{T'} \equiv 0$. Then $f_{T'}^{[r]} \equiv 0$ and $f_{T'}^{[r]}(\alpha) = 0$. On the other hand, just as in (2), we have

$$f_{T'}^{[r]}(\alpha) = r! \sum_{i=0}^{n-r} \alpha^i \sum_{|S| = i + r} \binom{|T' \cap S|}{r} G^S = r! \sum_{i=0}^{n-r} \alpha^i \sum_{|S| = i + r, S \supset T'} G^S,$$

where the second equality is due to $|T'| = r$, which implies that in the inner sum over S, the only non-zero terms are those $S \supset T'$.

Summing over all $T' \subset T$ where $|T'| = r$, we get:

$$0 = \sum_{T' \subset T, |T'| = r} f_{T'}^{[r]}(\alpha) \quad = \quad r! \sum_{i=0}^{n-r} \alpha^i \sum_{T' \subset T, |T'| = r} \sum_{|S| = i + r, S \supset T'} G^S$$

$$= r! \sum_{i=0}^{n-r} \alpha^i \sum_{|S| = i + r} \binom{|T \cap S|}{r} G^S = f_T^{[r]}(\alpha).$$

The third equality is by considering how many times each G^S appears, where $|S \cap T| \ge r$ and $|S| = i + r$.

If r is odd, we consider any T' where $|T'| = r + 1$. Since $r + 1 \le 2(k-1)$, by induction, we have $f_{T'} \equiv 0$. Then $f_{T'}^{[r]}(\alpha) = 0$. Similarly with (2), we have

$$f_{T'}^{[r]}(\alpha) = r! \sum_{i=0}^{n-r} \alpha^i \sum_{|S| = i + r} \binom{|T' \cap S|}{r} G^S$$

$$= r! \sum_{i=0}^{n-r} \alpha^i \left(\sum_{|S| = i + r, T' \subset S} (r + 1) G^S + \sum_{t \in T'} \sum_{|S| = i + r, T' \setminus S = \{t\}} G^S \right).$$

Summing over all $T' \subset T$ where $|T'| = r + 1$, we can show that this *quadruple sum* finally simplifies to $(|T| - r) f_T^{[r]}(\alpha)$. Since $|T| - r > 0$, we have $f_T^{[r]}(\alpha) = 0$. We omit the proof details.

To sum up, we proved that $f_T^{[r]}(\alpha) = 0$ for $r = 0, 1, \ldots, 2(k-1)$. So α is a root of multiplicity at least $2k - 1$. The degree of f_T is at most $2k$, and we know f_T has at least two more roots y_1 and y_2. Therefore f_T must be identically 0. This completes the proof of case 1. We omit the other cases. $\square$

This lemma says that, for any fixed $x \in \mathbf{C}$, either for all y or for at most one $y \in \mathbf{C} - \{x\}$, we have $\begin{pmatrix} 1 & x \\ 1 & y \end{pmatrix} \in B^p_{gen}(G)$.

4   Characterization Theorems of 1-Admissibility and 1-Realizability

Here we give some characterization theorems for 1-admissibility and 1-realizability of signatures. It turns out that, for a general 1-admissible signature, after omitting isolated points in $B^p_{gen}(G)$, one can show that $B^p_{gen}(G)$ is the solution for a single polynomial $F(x, y)$ on $\mathcal{M}$. Using Lemma 1, we can show that this $F(x, y)$ must be multilinear. More precisely we have the following characterization theorem of 1-admissibility. (Since we are talking about 1-admissibility or 1-realizability, in this section we will omit isolated points for both $B^p_{gen}(G)$ or $B_{gen}(G)$.)

Theorem 3. *If G is 1-admissible, then there exist three constants a, b, c such that*

$$B^p_{gen}(G) = \left\{ \left[\begin{pmatrix} n_0 \\ n_1 \end{pmatrix}, \begin{pmatrix} p_0 \\ p_1 \end{pmatrix} \right] \in \mathcal{M} \,\middle|\, an_0n_1 + b(n_0p_1 + n_1p_0) + cp_0p_1 = 0 \right\}.$$

Also for any three constants a, b, c, there exists a signature G such that the above equation holds.

Proof: We first remark that for a given a, b, c, the existence of G can be fulfilled by symmetric signatures.

If G is in fact 2-admissible, we take $a = b = c = 0$, then there is no constraint on the bases. Now we assume G is not 2-admissible. In the following proof, we use the dehomogenized coordinates $\begin{pmatrix} 1 & x \\ 1 & y \end{pmatrix} \in \mathcal{M}$. The exceptional cases are similar.

If there are two bases $\begin{pmatrix} 1 & \alpha \\ 1 & y_1 \end{pmatrix} \in B^p_{gen}(G)$ and $\begin{pmatrix} 1 & \alpha \\ 1 & y_2 \end{pmatrix} \in B^p_{gen}(G)$ $(y_1 \neq y_2)$, by Lemma 1, we have

$$\left\{ \begin{pmatrix} 1 & \alpha \\ 1 & y \end{pmatrix} \in \mathcal{M} \,\middle|\, y \in \mathbf{C} - \{\alpha\} \right\} \subset B^p_{gen}(G). \tag{3}$$

Now we prove that $B^p_{gen}(G) = \left\{ \begin{pmatrix} 1 & \alpha \\ 1 & y \end{pmatrix} \in \mathcal{M} \,\middle|\, y \in \mathbf{C} - \{\alpha\} \right\}$. If not, we assume for a contradiction that $\begin{pmatrix} 1 & u \\ 1 & v \end{pmatrix} \in B^p_{gen}(G)$ and $u, v \neq \alpha$: $\begin{pmatrix} 1 & u \\ 1 & v \end{pmatrix}$ is equivalent to $\begin{pmatrix} 1 & v \\ 1 & u \end{pmatrix}$ on $\mathcal{M}$). Under this assumption, we prove that G is 2-admissible. For any basis $T = \begin{pmatrix} 1 & x_0 \\ 1 & y_0 \end{pmatrix} \in \mathcal{M}$, if $x_0 = \alpha$ or $y_0 = \alpha$ then we know $T \in B^p_{gen}(G)$. Now we assume $x_0, y_0 \neq \alpha$. Since $\begin{pmatrix} 1 & u \\ 1 & v \end{pmatrix} \in B^p_{gen}(G)$ and $\begin{pmatrix} 1 & u \\ 1 & \alpha \end{pmatrix} \in B^p_{gen}(G)$ by (3), it follows from Lemma 1 that for any $t \neq u$, we have

$$\begin{pmatrix} 1 & u \\ 1 & t \end{pmatrix} \in B^p_{gen}(G) \tag{4}$$

So if $x_0 = u$ or $y_0 = u$, we have $T \in B^p_{gen}(G)$. Similarly if $x_0 = v$ or $y_0 = v$, we also have $T \in B^p_{gen}(G)$. Now we further assume $x_0, y_0 \notin \{u, v\}$. Then we have $\begin{pmatrix} 1 & u \\ 1 & y_0 \end{pmatrix} \in B^p_{gen}(G)$ by (4) and $\begin{pmatrix} 1 & \alpha \\ 1 & y_0 \end{pmatrix} \in B^p_{gen}(G)$ by (3). By Lemma 1, we have $\begin{pmatrix} 1 & x_0 \\ 1 & y_0 \end{pmatrix} \in B^p_{gen}(G)$. Since this is true for any $T = \begin{pmatrix} 1 & x_0 \\ 1 & y_0 \end{pmatrix} \in \mathcal{M}$, we conclude that G is 2-admissible, which we assumed not to be. Therefore if G is not 2-admissible and if $\begin{pmatrix} 1 & \alpha \\ 1 & y_1 \end{pmatrix} \in B^p_{gen}(G)$ and $\begin{pmatrix} 1 & \alpha \\ 1 & y_2 \end{pmatrix} \in B^p_{gen}(G)$ (for $y_1 \neq y_2$), then $B^p_{gen}(G) = \left\{ \begin{pmatrix} 1 & \alpha \\ 1 & y \end{pmatrix} \in \mathcal{M} \,\middle|\, y \in \mathbf{C} - \{\alpha\} \right\}$. We can let $a = \alpha^2, b = -\alpha, c = 1$ in the theorem.

Now we can assume $B^p_{gen}(G)$ does not contain two bases of the above form. More precisely, for a basis $\begin{pmatrix} 1 & x \\ 1 & y \end{pmatrix} \in \mathcal{M}$, whenever we fix a x, there exist at most one y, such that $\begin{pmatrix} 1 & x \\ 1 & y \end{pmatrix} \in B^p_{gen}(G)$. This is also true for any fixed y. On the other hand, if we disregard at most finitely many points, it can be shown using Hilbert's *Nullstellensatz* that, to be 1-admissible, there exists a single polynomial $F(x, y) \in \mathbf{C}[x, y]$ such that

$$B^p_{gen} = \left\{ \begin{pmatrix} 1 & x \\ 1 & y \end{pmatrix} \in \mathcal{M} \,\middle|\, F(x, y) = 0 \right\}.$$

We omit the proof of this claim. Furthermore we will assume $F(x, y)$ is of minimal degree. In particular, we may assume $F(x, y)$ is square-free.

W.o.l.o.g., assume $d = \deg_y F \geq \deg_x F$. Clearly $d \geq 1$. Otherwise, $F(x, y)$ is a constant, which is absurd. Write

$$F(x, y) = f_d(x)y^d + f_{d-1}(x)y^{d-1} + \cdots + f_0(x), \tag{5}$$

where $f_i(x) \in \mathbf{C}[x]$, $\deg f_i \leq d$, for all $0 \leq i \leq d$, and f_d is not identically zero. For any x_0 such that $f_d(x_0) \neq 0$, we can write

$$F(x_0, y) = f_d(x_0) \left(y^d + \frac{f_{d-1}(x_0)}{f_d(x_0)} y^{d-1} + \cdots + \frac{f_0(x_0)}{f_d(x_0)} \right). \tag{6}$$

This polynomial in y has d roots in $\mathbf{C}$ counting multiplicity, but does not have two distinct roots. Therefore, there exists $\alpha \in \mathbf{C}$ such that

$$F(x_0, y) = f_d(x_0)(y + \alpha)^d. \tag{7}$$

If we compare the expressions in (6) and (7), we get for all $1 \le k \le d$, $\binom{d}{k}\alpha^k = \frac{f_{d-k}(x_0)}{f_d(x_0)}$. It follows that $\binom{d}{k}\left(\frac{f_{d-1}(x_0)}{\binom{d}{1}f_d(x_0)}\right)^k = \frac{f_{d-k}(x_0)}{f_d(x_0)}$, for $1 \le k \le d$.

Writing in terms of polynomials, for all $1 \le k \le d$,

$$\binom{d}{k}\frac{f_{d-1}^k(x)}{d^k} = f_{d-k}(x)f_d^{k-1}(x), \tag{8}$$

holds for infinitely many $x \in \mathbf{C}$, and therefore holds identically, as polynomials in $\mathbf{C}[x]$.

It follows that, in $\mathbf{C}[x, y]$,

$$f_d^{d-1}(x) \cdot F(x, y) = \left(f_d(x)y + \frac{f_{d-1}(x)}{d}\right)^d. \tag{9}$$

Assume for a contradiction that $d \ge 2$. Take $k = 2$ in (8), we get $f_d(x)|f_{d-1}(x)$ in $\mathbf{C}[x]$. Also for all $k \ge 1$, $f_{d-k}(x) = \frac{\binom{d}{k}}{d^k}f_{d-1}(x)\left(\frac{f_{d-1}(x)}{f_d(x)}\right)^{k-1}$, and therefore $f_{d-1}(x)|f_{d-k}(x)$ in $\mathbf{C}[x]$. In particular $f_d(x)|f_{d-k}(x)$ for all $k \ge 1$, which implies that $f_d(x)|F(x, y)$ in $\mathbf{C}[x, y]$. If $\deg f_d(x) \ge 1$, then for a root x of f_d, there would have been infinitely many zero of $F(x, y)$. Since this is not the case, we must have $\deg f_d(x) = 0$, i.e., $f_d(x)$ is a non-zero constant $c \in \mathbf{C}$.

It follows that

$$F(x, y) = c\left(y + \frac{f_{d-1}(x)}{cd}\right)^d.$$

But $F(x, y)$ is square-free in $\mathbf{C}[x, y]$, it follows that $d = 1$ after all.

So back to (9) we obtain $F(x, y) = f_1(x)y + f_0(x)$, and $\deg f_1, \deg f_0 \le 1$. Therefore $F(x, y)$ is of the form $a + bx + b'y + cxy$. By symmetry on x and y in $\mathcal{M}$, we get $b = b'$.

$\square$

Now we can prove the characterization theorem for 1-realizability.

Theorem 4. *If G is 1-realizable, then there exist three constants a, b, c such that*

$$B_{gen}(G) = \left\{\left[\binom{n_0}{n_1}, \binom{p_0}{p_1}\right] \in \mathcal{M} \,\middle|\, an_0n_1 + b(n_0p_1 + n_1p_0) + cp_0p_1 = 0\right\}.$$

Also for any three constants a, b, c, there exists a signature G such that the above equation holds.

Proof: Again, we first remark that for a given a, b, c, the existence of G can be fulfilled by symmetric signatures.

Since G is 1-realizable, G is also 1-admissible. There are two cases: if G is in fact 2-admissible, then as a 1-realizable signature, G is at least realizable on some bases. G is indeed a 2-realizable signature. In this case we take $a = b = c = 0$.

If G is 1-admissible but not 2-admissible, then in Theorem 3 we must have a non-zero triple (a, b, c), defining $B_{gen}^p(G)$ as a 1-dimensional variety. We claim that, for any $T \in B_{gen}^p(G)$, all the MGI of $T^{\otimes n}G$ must vanish. Otherwise $B_{gen}(G)$ cannot have dimension 1. Since all MGI are satisfied for any $T \in B_{gen}^p(G)$, we get $B_{gen}(G) = B_{gen}^p(G)$. Theorem 4 now follows from Theorem 3. $\square$

References

1. Cai, J.-Y., Choudhary, V.: Some Results on Matchgates and Holographic Algorithms. In: Bugliesi, M., Preneel, B., Sassone, V., Wegener, I. (eds.) ICALP 2006. LNCS, vol. 4051, pp. 703–714. Springer, Heidelberg (2006)
2. Cai, J.-Y., Choudhary, V.: Valiant's Holant Theorem and Matchgate Tensors. Theoretical Computer Science 384(1), 22–32 (2007)
3. Cai, J.-Y., Lu, P.: On Symmetric Signatures in Holographic Algorithms. In: Thomas, W., Weil, P. (eds.) STACS 2007. LNCS, vol. 4393, pp. 429–440. Springer, Heidelberg (2007)
4. Cai, J.-Y., Lu, P.: Holographic Algorithms: From Art to Science. In: The proceedings of STOC, pp. 401–410 (2007)
5. Cai, J.-Y., Lu, P.: Holographic Algorithms: The Power of Dimensionality Resolved. In: Arge, L., Cachin, C., Jurdziński, T., Tarlecki, A. (eds.) ICALP 2007. LNCS, vol. 4596, pp. 631–642. Springer, Heidelberg (2007)
6. Cai, J.-Y., Lu, P.: Holographic Algorithms With Unsymmetric Signatures. In: ACM-SIAM Symposium on Discrete Algorithms (SODA), pp. 54–63 (2008)
7. Cai, J.-Y., Lu, P.: Signature Theory in Holographic Algorithms, http://pages.cs.wisc.edu/~jyc/papers/signature-thy.pdf
8. Graham, R.L., Li, S.-Y.R., Li, W.-C.W.: On the Structure of t-Designs. SIAM J. on Algebraic and Discrete Methods 1, 8 (1980)
9. Linial, N., Rothschild, B.: Incidence Matrices of Subsets–A Rank Formula. SIAM J. on Algebraic and Discrete Methods 2, 333 (1981)
10. Kasteleyn, P.W.: The statistics of dimers on a lattice. Physica 27, 1209–1225 (1961)
11. Kasteleyn, P.W.: Graph Theory and Crystal Physics. In: Harary, F. (ed.) Graph Theory and Theoretical Physics, pp. 43–110. Academic Press, London (1967)
12. Kneser, M.: "Aufgabe 360". Jahresbericht der Deutschen Mathematiker-Vereinigung, 2. Abteilung 58, 27 (1955)
13. Lovász, L.: Kneser's conjecture, chromatic number, and homotopy. Journal of Combinatorial Theory, Series A 25, 319–324 (1978)
14. Matoušek, J.: A combinatorial proof of Kneser's conjecture. Combinatorica 24(1), 163–170 (2004)
15. Temperley, H.N.V., Fisher, M.E.: Dimer problem in statistical mechanics – an exact result. Philosophical Magazine 6, 1061–1063 (1961)
16. Valiant, L.G.: Quantum circuits that can be simulated classically in polynomial time. SIAM Journal of Computing 31(4), 1229–1254 (2002)
17. Valiant, L.G.: Expressiveness of Matchgates. Theoretical Computer Science 281(1), 457–471 (2002)
18. Valiant, L.G.: Holographic Algorithms (Extended Abstract). In: Proc. 45th IEEE Symposium on Foundations of Computer Science, pp. 306–315 (2004); A more detailed version appeared in ECCC Report TR05-099
19. Valiant, L.G.: Accidental Algorithms. In: Proc. 47th Annual IEEE Symposium on Foundations of Computer Science, pp. 509–517 (2006)

The Complexity of SPP Formula Minimization

David Buchfuhrer[*]

Computer Science Department
California Institute of Technology
Pasadena, CA 91125
dave@cs.caltech.edu

Abstract. Circuit minimization is a useful procedure in the field of logic synthesis. Recently, it was proven that the minimization of $(\vee, \wedge, \neg)$ formulae is hard for the second level of the polynomial hierarchy [BU08]. The complexity of minimizing more specialized formula models was left open, however. One model used in logic synthesis is a three-level model in which the third level is composed of parity gates, called SPPs. SPPs allow for small representations of Boolean functions and have efficient heuristics for minimization. However, little was known about the complexity of SPP minimization. Here, we show that SPP minimization is complete for the second level of the Polynomial Hierarchy under Turing reductions.

1 Introduction

Circuit minimization problems are an interesting class of problems contained in the second level of the Polynomial-Time Hierarchy (PH). Given a circuit C computing a function f, the goal is to find the smallest circuit C' also computing f. The complexity of circuit minimization depends heavily upon the model of circuit being minimized as well as how the size of a circuit is calculated. Here we use boolean formulae as our model. The details of this model and the size function are defined later.

As a simple example, take the formula $a \to b$. It could be expressed either by the formula $(a \wedge b) \vee \neg a$ or by the simpler and more common $\neg a \vee b$. Since we are measuring the size by the number of occurrences of variables, the size of the first formula is 3, while the size of the second is 2. Even with this simple example, the fact that the first formula is not minimal is not immediately clear without the knowledge that it computes $a \to b$. With larger, more complicated formulae, the problem becomes even more difficult.

Circuit minimization has practical applications in fields such as logic synthesis and hardware design. Two popular variants of circuit minimization are unbounded-depth and constant depth formula minimization over $(\wedge, \vee, \neg)$ formulae. The 2-level (DNF) version was proven Σ_2^P complete in [Uma98]. More

---

[*] Supported by NSF CCF-0346991, CCF-0830787 and BSF 2004329.

S.-H. Hong, H. Nagamochi, and T. Fukunaga (Eds.): ISAAC 2008, LNCS 5369, pp. 580–591, 2008.

recently, the constant depth version for all constants greater than 2 as well as the unlimited depth version were proven to be hard for the second level of the PH under Turing reductions in [BU08]. More recently, three-level formula minimization in which the first level is an OR gate, the second level is AND gates, and the third level is composed of XOR gates, has been introduced [LP99]. Such a formula is referred to as a Sum of Pseudoproducts (SPP) in the literature.

Definition 1 (SPP Formulae). *An SPP formula is the disjunction of the conjunction of parity formulae. In other words, an SPP formula contains a single OR gate of unlimited fan-out, to which the inputs are unlimited fan-out AND gates which in turn take unlimited fan-out XOR gates as input. The size of an SPP formula is the number of times the input variables appear in it.*

Rather than allow negations of input variables, we allow the XOR gates to take constants as inputs. Adding a true input is equivalent to negating one of the input variables and does not increase the formula size. The number of true inputs thus determines whether an XOR gate computes odd or even parity on the input variables. The sub-formula computed by an AND gate is referred to as a pseudoproduct.

Definition 2 (Pseudoproduct). *A pseudoproduct is the conjunction of XORs.*

Note that when we refer to XORs above, we mean the function computed by an XOR gate rather than the gate itself. Both meanings are used throughout this paper.

SPP formula minimization has been a recent subject of study in the field of logic synthesis [BCDV08, CB02, Cir03]. Some research has found inefficient but exact minimization algorithms, while other research is directed toward finding more efficient heuristics. Little has been known about the formal complexity of SPP minimization. Characterizing the complexity of SPP minimization is important, as SPPs are well-suited for use in practical situations such as representation of arithmetic functions [LP99, JSA97].

Note that SPP formulae are a generalization of DNF formulae. Recall that a DNF formula is simply the disjunction of several terms, each of which is the conjunction of some of the input variables and their negations. One reason that SPP formulae are useful in logic synthesis is that they allow for much smaller representations of many Boolean functions than DNF formulae. The simplest such example is that of the parity function. Since SPP formulae contain XOR gates of unlimited fan-out, parity only requires a single XOR gate with a linear number of inputs. By contrast, a CNF or DNF for parity requires exponential size.

We formally define the SPP minimization problem as

Problem 1 (Minimum SPP Formula (MSF)). Given an SPP formula S and an integer k, is there an SPP formula of size at most k which is equivalent to S?

Here, we show that the minimization of SPP formulae is hard for the second level of the PH under Turing reductions. This result provides the theoretical backing for work on inefficient exact minimization algorithms [Cir03] and heuristics

[BCDV08]. It also closes a gap in our knowledge of the complexity of circuit minimization. Σ_2^P hardness demonstrates that polynomial time algorithms making use of NP or coNP queries cannot exist unless the PH collapses. This is an important result to those working in logic synthesis, as queries to problems such as the coNP-hard tautology problem are often used in the design of algorithms and heuristics.

1.1   Description of the Reduction

This section contains a high-level description of the reduction used to show that Problem 1 is Σ_2^P-complete under Turing reductions, in order to facilitate understanding of a more technical description later. We also compare the reduction in this paper to a related reduction in [BU08].

The problem we reduce from is MODIFIED SUCCINCT SET COVER, which was shown to be Σ_2^P-complete in [BU08].

*Problem 2 (*MODIFIED SUCCINCT SET COVER (MSSC)*).* Given a DNF formula D on variables $v_1, v_2, \ldots, v_m, x_1, x_2, \ldots, x_n$ and an integer k, is there a subset $I \subseteq \{1, 2, \ldots n\}$ with $|I| \leq k$ and for which $D \vee \bigvee_{i \in I} \neg x_i \equiv (\bigvee_{i=1}^m \neg v_i \vee \bigvee_{i=1}^n \neg x_i)$?

Less formally, the goal is to determine whether there exists a small subset I of the x_i variables, for which $\bigvee_{i \in I} x_i$ accepts every assignment not accepted by D, except for the all true assignment. In this way, MSSC can be seen as a succinct form of the set cover problem, in which the points accepted by D and the formulae $\neg x_i$ are the sets which form the instance.

In order to reduce MSSC to MSF, we attempt to split the SPP formula of the MSF instance into one portion which computes D and another which accepts $\bigvee_i \neg x_i$. To accomplish this, we add a variable z, the value of which determines whether the formula computes D on the rest of the variables, or $D \vee \bigvee_i \neg x_i$. With this variable added, the formula should look like $D \vee (z \wedge \bigvee_i \neg x_i)$. Because SPP formulae only have one OR gate at the top level, the particular form is more similar to $D \vee \bigvee_i (z \wedge \neg x_i)$.

The difficult direction of the reduction is showing that if the MSSC instance is negative, so is the MSF instance. In order to prevent a small equivalent SPP formula from existing if the MSSC is negative, we attach different weights to the variables. Rather than simply counting each occurrence of a variable as 1 toward the size of the formula, we wish to count each variable v as $w(v)$ to add flexibility to our reduction.

In order to increase the size contribution of each variable, we replace each occurrence of each variable v with $w(v)$ new variables, so they contribute $w(v)$ toward the size at each occurrence of v. This is achieved exactly by replacing each variable v with $v_1 \oplus v_2 \oplus \cdots \oplus v_{w(v)}$. The fact that this transformation preserves minimum formula size is shown by Lemma 3.

We weight the z variable in a less general way that is better suited for achieving the desired form. We replace z by $z_1 \wedge \cdots \wedge z_\ell$, where ℓ is an integer value particular to the instance of MSSC. In order to avoid confusion, we do not refer to this as weighting in the more technical description of the reduction, and simply state the reduction explicitly with variables $z_1, \ldots, z_\ell$.

The reduction approach we use in this paper is superficially similar to one contained in [BU08], which proves that minimization of $(\vee, \wedge, \neg)$ formulae of both fixed depth ≥ 3 and unrestricted depth are Σ_2^P-complete under Turing reductions. Consider the depth-3 version, which is most similar to SPP minimization.

Problem 3 (MINIMUM EQUIVALENT DEPTH 3 EXPRESSION (MEE_3)). Given a depth 3 Boolean $(\vee, \wedge, \neg)$ formula F and an integer k, is there an equivalent depth 3 formula of size at most k?

Here, the depth ignores NOT gates. The reduction in [BU08] showing that MEE_3 is Σ_2^P-complete under Turing reductions also reduced from MSSC, and used the same basic strategies. The original variables were given different weights to avoid the formula being too small when the initial MSSC instance is negative. Extra variables were added which split the formula into a portion computing D and a portion computing $D \vee \bigvee_i \neg x_i$, playing a similar role to the z_i variables in this paper, but with an altogether different form. In the SPP reduction shown in this paper, we create a formula with a form similar to $D \vee \bigvee_i (z \wedge \neg x_i)$. By contrast, in [BU08] the formula resulting from the reduction is more similar to $D \vee (z \wedge \bigvee_i \neg x_i)$.

The two above forms are different ways of expressing the same formula. The big difference, however, is the number of occurrences of the z variable. In [BU08], the z variable is given such a heavy weight that it must appear only once in a minimal formula. The entire proof depends on this fact. This is not possible with an SPP formula, however, as the z would then appear in only one pseudoproduct, and this pseudoproduct would thus be responsible for computing $\bigvee_i \neg x_i$. Lemma 1 demonstrates why this prevents any such reduction from working.

Lemma 1. *No pseudoproduct can compute $\neg a \vee \neg b$.*

A proof of this lemma is contained in the full version of this paper.

Lemma 1 is important because it demonstrates that a single pseudoproduct is not powerful enough to accept formulae like $\bigvee_i (z \wedge \neg x_i)$, which can be restricted to $\neg x_i \vee \neg x_j$ by restricting z to true and all other variables except x_i and x_j to false.

Thus, we require several copies of the z variable in any reduction from MSSC to MSF using the general approach developed in [BU08]. In a depth-3 Boolean formula like those under consideration in the [BU08] reduction from MSSC to MEE_3, the sub-formula containing the z variable is a DNF, and is therefore not limited in computational capability like a pseudoproduct.

Since the z variable cannot occur only once, the relative weights must be completely different from those used in [BU08], in order to allow for an unknown number of z variables. Rather than a single z variable carrying more weight than the rest of the formula combined, all the z variables combined carry less weight than any other single variable. This is a huge change, and just the existence of multiple copies of the z variables alone completely destroys many of the proof techniques used in [BU08]. We are able to get a handle on the structure of the pseudoproducts containing the z variables by developing new proof techniques. One key structural lemma used in obtaining our results is Lemma 2.

Lemma 2. *If a pseudoproduct computes the conjunction $z_1, \ldots, z_n$ of n variables, the pseudoproduct contains at least n XORs.*

The proof of Lemma 2 is contained in the full version of this paper.

2   Results

In this section, we present new classifications for the complexity of problems related to SPP formulae. We begin in Section 2.1 with a few simple results, then continue in Section 2.3 with the proof of the Σ_2^P completeness of MSF under Turing reductions.

2.1   Complexity of Related Problems

In this section, we note some complexity results that remain unchanged from the corresponding results for related DNF problems.

Problem 4 (SPP Equivalence). Given two SPP formula S, T, do both S and T compute the same function?

Proposition 1. *SPP Equivalence is coNP-complete.*

While the proof is quite simple, this result is important to note, as not all types of logic formulae are known to share this result. For example, the equivalence of two ordered binary decision diagrams (OBDDs) is decidable in P [Bry86]. The proof is contained in the full version of this paper.

Problem 5 (DNF Irredundancy). Given a DNF formula T and an integer k, does there exist an equivalent formula T' consisting of the disjunction of at most k terms from T?

The irredundancy problem is important because it is often used in heuristics for DNF minimization problems, despite having been proven Σ_2^P hard to even approximate in [Uma99]. We see here that the corresponding problem for SPP formulae shares the same complexity classification.

Problem 6 (SPP Irredundancy). Given an SPP S and an integer k, is there an equivalent SPP S' composed of the disjunction of k pseudoproducts from S?

Theorem 1. *SPP Irredundancy is Σ_2^P hard to approximate to within a factor of n^ϵ for some $\epsilon > 0$, where n is the input size.*

A proof of this theorem is contained in the full version of this paper.

Prime implicants are also important to DNF minimization, as a minimum DNF is composed only of prime implicants.

Definition 3 (Prime Implicant). *A prime implicant of a DNF D is a term T which implies D and does not imply any other term which implies D.*

An implicant $x_1 \wedge \cdots \wedge x_n$ of D is prime iff there is no k such that $x_1 \wedge \cdots \wedge x_{k-1} \wedge x_{k+1} \wedge \cdots \wedge x_n$ is also an implicant of D. Since prime implicants are necessary for DNF minimization, it is important to be able to determine whether a term is a prime implicant.

Problem 7 (IS-PRIMI). Given a DNF D and a term T, is T a prime implicant of D?

IS-PRIMI was proven DP-complete independently in [GHM08] and [UVSV06].

Prime pseudoproducts are a generalization of prime implicants to SPP formulae.

Definition 4 (Prime Pseudoproduct). *A prime pseudoproduct of an SPP S is a pseudoproduct P which implies S, but does not imply any other pseudoproduct which implies S.*

Prime pseudoproducts characterize SPP with a minimum number of XOR gates [LP99] rather than minimum SPP formulae as defined in this paper. Still, they are important to the study of SPP formulae, so we consider the complexity of determining whether a pseudoproduct is prime.

Problem 8 (SPP Prime Pseudoproduct (SPP-PP)). Given an SPP S and a pseudoproduct P, is P a prime pseudoproduct of S?

We show that SPP-PP is DP-hard using the same reduction used in [GHM08] to prove that IS-PRIMI is DP-hard. The reduction is from SAT-UNSAT.

Problem 9 (SAT-UNSAT). Given CNF formulae α, β, is α satisfiable and β unsatisfiable?

Theorem 2. *SPP-PP is DP-hard.*

A proof of this theorem is contained in the full version of this paper.

Note that we only achieve a hardness result here and not a completeness result. One key feature of IS-PRIMI that is used to prove containment in DP is that in order to check that $a_1 \wedge \cdots \wedge a_n$ is prime, we need only verify that the n terms obtained by removing one of the a_is do not imply f. With pseudoproducts, it is unclear how to check primality in NP. Thus, we only have a lower-bound complexity result.

In this section, we have seen that complexity results for equivalence and irredundancy of SPP formulae can be inferred directly from the corresponding results for DNF formulae because DNF formulae are a special case of SPP formulae. These results were simple because the problems only depend on the formulae taken as inputs, and do not depend on properties of any other formula. This preserves the DNF versions as special cases of the SPP versions. In general, formula minimization problems differ in that the problem instance is compared to all formulae of size less than k, necessitating a different reduction when the type of formulae under consideration is changed. So the result that DNF formula minimization is Σ_2^P-complete [Uma01] does not immediately carry over to SPP formula minimization.

The Σ_2^P-completeness result does not directly carry over from DNFs to SPPs, but one might hope that the proof techniques used would, as happened with the DP-hardness proofs of IS-PRIMI and SPP-PP. However, the proof of DP-hardness for prime implicants had the convenient feature that the DNF term created by the reduction is a prime implicant iff it is a prime pseudoproduct. The proof of DNF minimization being Σ_2^P-complete does not have the analogous feature that the DNF produced is a minimum DNF iff it is a minimum SPP. So a new reduction is necessary.

2.2 Weighted Variables

Normally, we measure the size of an SPP formula by the number of occurrences of variables within it. Each occurrence counts as 1 toward the total size. Another notion of size might be to assign a positive integer weight $w(v)$ to each variable v, and count each occurrence of the variable v as $w(v)$ toward the total size. We will show a transformation from positive integer weights to the normal size measure in which each variable counts as 1, which preserves minimum formula size.

Since the SPP formulae in consideration have unlimited fanout XOR gates, we can replace a variable v of weight $w(v)$ with the XOR of $w(v)$ new variables, $v_1 \oplus \cdots \oplus v_{w(v)}$ in order to eliminate the need for directly assigning weights to variables. This weighting is achieved by adding the new v_i variables to each XOR gate containing v, and removing v completely.

Given an SPP formula F with associated weight function $w(v)$, we call F the *weighted* version of the formula. Let F' be the formula obtained by replacing every variable v with the XOR of $w(v)$ new variables. We call F' the *expanded* version of the formula. Note that if F computes f and F' computes f', $f(x_1 \oplus \cdots \oplus x_{w(x)}, y_1 \oplus \cdots \oplus y_{w(y)}, \ldots)$ is equivalent to $f'(x_1, \ldots, x_{w(x)}, y_1, \ldots, y_{w(y)}, \ldots)$. We denote the weighted size of F with weights $w(v)$ by $|F|_w$, and we denote the size of F' by $|F'|$.

The expansion described above differs from the expansion strategy used in [BU08], in which a conjunction of $w(v)$ new variables was used rather than an XOR. Our strategy has the advantages that it handles negation without requiring a possible increase in the depth of the formula and can be easily applied to SPP formulae. Lemma 3 demonstrates that applying this simple transformation is equivalent to positive integer weighting. Before we get to Lemma 3, we define the weighted version of MSF.

Problem 10 (Weighted MSF (W-MSF)). Given an SPP formula S, a list of weights w for each variable in unary, and an integer k, is there a formula S' equivalent to S for which $|S'|_w \le k$?

We refer to a formula which is minimum with respect to w as a w-minimum formula.

Lemma 3. *Let f be a Boolean function and w be a positive integer weighting function for the variables of f. If F is a w-minimum formula for f and F' is a minimum formula equivalent to the expanded version of F, then $|F|_w = |F'|$.*

A proof of this lemma is contained in the full version of this paper.

Now, we apply Lemma 3 to show that W-MSF and MSF are poly-time equivalent.

Theorem 3. *There exist polynomial-time reductions between W-MSF and MSF.*

This follows pretty quickly from Lemma 3 and a detailed proof is contained in the full version of this paper.

Since these two problems are polynomial-time equivalent, we only need to prove that one of them is Σ_2^P complete under Turing reductions to see that they both are. We will show that MSF is Σ_2^P complete under Turing reductions via W-MSF.

2.3   Main Result

In this section, we show that Problem 1, the SPP minimization problem, is Σ_2^P complete under Turing reductions by reducing from MSSC to W-MSF. The following steps describe how a Turing machine with access to an oracle for W-MSF can solve MSSC.

- We are given an instance $\langle D, x_1, x_2, \ldots, x_n, k \rangle$ of MSSC, where D is a DNF formula over variables $v_1, \ldots, v_m, x_1, \ldots, x_n$.
- Set $\ell = 2k(k+1)(|D|+1)$.
- Create new variables $z_1, \ldots, z_\ell$.
- Set the weight function w such that $w(v_i) = 2k^2\ell$, $w(x_i) = k\ell$ and $w(z_i) = 1$ for all i.
- Using binary search, find the size of the w-minimum SPP formula for D using $O(\log |D|_w)$ oracle calls. Call this size τ.
- Set $S = D \vee (Z \wedge \neg x_1) \vee (Z \wedge \neg x_2) \vee \cdots \vee (Z \wedge \neg x_n)$ where $Z = z_1 \wedge z_2 \wedge \cdots \wedge z_\ell$.
- Finally, we make one last oracle call to determine whether there is an SPP formula S' equivalent to S for which $|S'|_w \leq \tau + k(k+1)\ell$. Accept iff such an S' exists.

Remark 1. Since this reduction only uses logarithmically many adaptive oracle calls, standard techniques can be used to turn this into a non-adaptive reduction with polynomially many oracle calls.

A helpful lemma was proven in [BU08]:

Lemma 4. *Let $t_1, t_2, \ldots, t_n$ be a set of variables, and S a set of assignments of true/false values to $t_1, \ldots, t_n$. A minimum formula accepting at least S and not the all-true assignment to variables $t_1, t_2, \ldots, t_n$ is of the form $\bigvee_{i \in I} \neg t_i$ for some $I \subseteq \{1, 2, \ldots, n\}$.*

Although this lemma was only proven for $(\vee, \wedge, \neg)$-formula in [BU08], it holds true for weighted SPP as well using the same proof. The proof in both cases relies on transforming a formula F accepting at least S but not the all-true assignment into the form $\bigvee_{i \in I} \neg t_i$, by setting I as the set of all variables mentioned in F. Everything in S will still be accepted, the all-true assignment will still be rejected, and the size is not increased.

One final necessary lemma will allow us to determine the placement of the z_i variables in S' when $|S'| \leq \tau + k(k+1)\ell$. Before proving this lemma, it is necessary to define the notion of a restriction.

Definition 5 (Restriction). *A restriction of a function f to a partial assignment σ is simply the remaining function on all other variables when the variables contained in σ are fixed according to σ. Similarly, a restriction of a formula F to σ results in the formula F with occurrences of variables in σ replaced with values according to σ.*

A restriction will reduce the size of the formula, as constants do not count toward formula size. Note that if we restrict an SPP formula, the resulting formula is also an SPP formula.

Lemma 5. *Let S' be an equivalent SPP formula to the SPP formula S created by the Turing reduction described at the beginning of Section 2.3. Let U be the subset of pseudoproducts in S' that accept an assignment not accepted by D. Let H_P be the set of z_i variables in pseudoproduct P which occur in some XOR in P that contains only constants and z_i variables. Either $|S'|_w > \tau + k(k+1)\ell$ or there exists some index i^* such that $z_{i^*} \in H_P \ \forall P \in U$.*

Proof. Suppose that there does not exist an index i^* such that $z_{i^*} \in H_P \ \forall P \in U$.

Consider a pseudoproduct $P \in U$. Let $H_P{}^C$ denote the complement of H_P within the set of z_i variables. So $H_P{}^C$ is the set of z_i variables that never appear in an XOR in P consisting of only z_i variables and constants. Since $P \in U$, there exists some assignment σ accepted by P but not by D.

Let P' be the restriction of P to σ on all variables which are not members of $H_P{}^C$. Thus, P' computes some function on the variables in $H_P{}^C$. Recall that $S = D \vee \bigvee_i (z_1 \wedge \cdots \wedge z_\ell \wedge \neg x_i)$. Since σ is not accepted by D, S becomes $\bigwedge_{z_i \in H_P{}^C} z_i$ under the restriction to σ on all variables outside of $H_P{}^C$.

P' cannot accept anything not accepted by S', so P' can only accept when $\bigwedge_{z_i \in H_P{}^C} z_i$ is true. Furthermore, it must accept in this case, as P accepted σ. So P' computes exactly $\bigwedge_{z_i \in H_P{}^C} z_i$. Thus, by Lemma 2, there are at least $|H_P{}^C|$ XORs in P'. Note that this does not include XORs which contain only constants. These can be safely ignored, as they must all reduce to true since P' does not compute a constant function.

Each of the $|H_P{}^C|$ non-trivial XORs in P' corresponds to an XOR in P that contains a variable other than one of the z_is. This follows from the definition of $H_P{}^C$ and the fact that each non-trivial XOR in P' contains a member of $H_P{}^C$. Since all variables other than the z_i are weighted at least ℓ, $|P|_w \geq \ell|H_P{}^C|$. Furthermore, since no z_i is contained in $H_P \ \forall P \in U$, $\bigcup_{P \in U} H_P{}^C = \{z_1, \ldots, z_\ell\}$.

So $\sum_{P \in U} |H_P{}^C|\ell \geq \ell^2$, which is important because $|S'|_w \geq \sum_{P \in U} |P|_w \geq \sum_{P \in U} |H_P{}^C|\ell$. Thus, we have $|S'|_w \geq \ell^2$, which since $\ell = 2k(k+1)(|D|+1)$, is greater than $2k(k+1)\ell|D| + k(k+1)\ell$. We know $2k(k+1)\ell|D| > \tau$ since $w(v) \leq 2k^2\ell < 2k(k+1)\ell$, so $\tau \leq |D|_w < 2k(k+1)\ell|D|$ for every variable v. Thus, $|S'|_w > \tau + k(k+1)\ell$.

Therefore, either there is some index i^* such that $z_{i^*} \in H_P \; \forall P \in U$ or $|S'| > \tau + k(k+1)\ell$. $\qquad\qquad\square$

The main result that W-MSF is Σ_2^P-complete under Turing reductions, and therefore so is MSF, follows from Theorem 4.

Theorem 4. *Let S be the SPP formula resulting from the Turing reduction at the beginning of Section 2.3. The W-MSF instance $\langle S, w, \tau + k(k+1)\ell \rangle$ is positive iff the original MSSC instance was positive as well.*

Proof. First, we will show that a positive MSSC instance implies a positive W-MSF instance. Let $I \subseteq [n]$ such that $|I| \leq k$ and $D \vee \bigvee_{i \in I} \neg x_i \equiv \bigvee_{i=1}^{m} \neg v_i \vee \bigvee_{i=1}^{n} \neg x_i$. Since the MSSC instance is positive, such an I exists. Let $\widehat{D}$ be a w-minimum formula equivalent to D. Note $|\widehat{D}|_w = \tau$. We construct a formula $S' = \widehat{D} \vee \left[\bigvee_{i \in I} \left(\neg x_i \wedge \bigwedge_{j=1}^{\ell} z_j \right) \right]$. which is equivalent to S since when we restrict such that $\bigwedge_{j=1}^{\ell}$ is false, it computes D and when we restrict such that $\bigwedge_{j=1}^{\ell}$ is true, it computes $D \vee \bigvee_{i \in I} \neg x_i \equiv \bigvee_{i=1}^{m} \neg v_i \vee \bigvee_{i=1}^{n} \neg x_i$. Thus, S' and S compute the same functions under any restriction to the z_i variables, so $S' \equiv S$. Note that $|S'|_w = |\widehat{D}|_w + \sum_{i \in I}(w(x_i) + \ell) = \tau + k(k\ell + \ell) = \tau + k(k+1)\ell$. Thus, a positive instance of MSSC implies a positive instance of W-MSF.

Now we prove that a negative instance of MSSC implies a negative instance of W-MSF. We will show that if the MSSC instance is negative, so is the W-MSF instance by first restricting to the portion of the formula computing D via the upcoming Claim 1 to achieve a τ lower bound. We then show that the rest of S' is of size greater than $k(k+1)\ell$ by the upcoming Claim 2.

Suppose that the instance of MSSC is negative. Assume by way of contradiction that the W-MSF instance is positive. By this assumption, there exists some formula S' which is equivalent to S and for which $|S'|_w \leq \tau + k(k+1)\ell$. Let U be the subset of pseudoproducts in S' that accept an assignment to the input variables not accepted by D. Let H_P be defined as in Lemma 5 to be the set of z_i variables that occur in some XOR in a pseudoproduct P containing only z_i variables and constants. By Lemma 5, there is some index i^* such that $z_{i^*} \in H_P$ $\forall P \in U$, since $|S'|_w \leq \tau + k(k+1)\ell$.

Claim 1. *Every member of U becomes false under the restriction $z_i = $ true $\forall i \neq i^*$ and $z_{i^*} = $ false.*

A proof of this claim is contained in the full version of this paper.

Let S^* be the restriction of S' to $z_i = $ true $\forall i \neq i^*$ and $z_{i^*} = $ false. By Claim 1, S^* only depends on pseudoproducts outside of U. Also, S^* must be equivalent to D, since we have restricted such that $z_1 \wedge \cdots \wedge z_\ell$ is false and $S' \equiv S = D \vee \bigvee_i (z_1 \wedge \cdots \wedge z_\ell \wedge \neg x_i)$. Thus, since $S^* \equiv D$, $|S^*|_w \geq \tau$.

Claim 2. *The total weighted size of the pseudoproducts in U is greater than $k(k+1)\ell$.*

This claim follows from Lemma 4 and the MSSC instance being false. A detailed proof is contained in the full version of this paper.

The total size of S' can now be calculated. The total size of all the pseudoproducts outside of U is at least τ. By Claim 2, the total size of the pseudoproducts in U is greater than $k(k+1)\ell$. Thus,

$$|S'|_w > \tau + k(k+1)\ell,$$

contradicting $|S'|_w \leq \tau + k(k+1)\ell$. So a negative instance of MSSC implies a negative instance of W-MSF, completing the proof.   □

Thus, W-MSF is Σ_2^P-complete under Turing reductions, and therefore so is MSF by Theorem 3.

3   Conclusions and Open Problems

In this paper, we have resolved several problems relating to the complexity of SPP formulae. We have proven that formula equivalence remains coNP-complete and that irredundancy remains Σ_2^P-hard to approximate to within n^ϵ for some $\epsilon > 0$ when we generalize DNF formulae to SPP formulae. We also saw that when we generalize IS-PRIMI to SPP-PP, the DP-hardness result remains. Since we do not show SPP-PP $\in$ DP, the complete characterization of SPP-PP remains an open problem. Most importantly, we have resolved the complexity of the MSF problem, providing critical theoretical background to an important area of research in the field of logic synthesis.

The reduction used in this paper is a Turing reduction. A Turing reduction is needed to find the minimum formula size for D, which is critical to the reduction. So finding a many-one reduction remains an open problem.

Approximability of SPP minimization has not been addressed here, and remains an interesting open problem. The nature of the reduction given in this paper precludes direct use toward an inapproximability result. This is because the way in which the variables are weighted causes the part of the formula computing D to vastly outweigh the part determining how many x_i variables are necessary. This makes it impossible to use the reduction to approximate the number of x_i variables necessary. If this could be reversed such that each x_i variable carried greater weight in the problem size than the portion computing D, it would lead to both approximability results and a many-one reduction.

Rather than attempt to find a minimum SPP formula, several researchers have further restricted the possible solution space by restricting the fan-out of the XOR gates to a constant [BCDV08, CB02, Cir03].

Definition 6 (k-SPP Formulae). *A k-SPP formula is an SPP formula in which the XOR gates have fan-out of at most k.*

While a k-SPP formula may require larger size than an SPP formula, they are sufficiently powerful to be exponentially smaller than the smallest equivalent DNF in some cases [Cir03].

Although k-SPP formulae are an active area of research [BCDV08], we have not resolved their complexity here. However, this work is an important first step toward determining the complexity of k-SPP, since so little was known about the complexity of either SPP or k-SPP minimization.

References

[BCDV08]   Bernasconi, A., Ciriani, V., Drechsler, R., Villa, T.: Logic minimization and testability of 2-SPP networks. IEEE Transactions on Computer-Aided Design of Integrated Circuits and Systems (to appear, 2008)

[Bry86]   Bryant, R.E.: Graph-based algorithms for boolean function manipulation. IEEE Trans. Computers 35(8), 677–691 (1986)

[BU08]   Buchfuhrer, D., Umans, C.: The complexity of Boolean formula minimization. In: Aceto, L., Damgård, I., Goldberg, L.A., Halldórsson, M.M., Ingólfsdóttir, A., Walukiewicz, I. (eds.) ICALP 2008, Part I. LNCS, vol. 5125, pp. 24–35. Springer, Heidelberg (2008)

[CB02]   Ciriani, V., Bernasconi, A.: 2-SPP: a practical trade-off between SP and SPP synthesis. In: 5th International Workshop on Boolean Problems (IWSBP 2002), pp. 133–140 (2002)

[Cir03]   Ciriani, V.: Synthesis of SPP three-level logic networks using affine spaces. IEEE Trans. on CAD of Integrated Circuits and Systems 22(10), 1310–1323 (2003)

[GHM08]   Goldsmith, J., Hagen, M., Mundhenk, M.: Complexity of DNF minimization and isomorphism testing for monotone formulas. Information and Computation 206(6), 760–775 (2008)

[JSA97]   Jacob, J., Sivakumar, P.S., Agrawal, V.D.: Adder and comparator synthesis with exclusive-or transform of inputs. In: VLSI Design, pp. 514–515. IEEE Computer Society, Los Alamitos (1997)

[LP99]   Luccio, F., Pagli, L.: On a new Boolean function with applications. IEEE Trans. Computers 48(3), 296–310 (1999)

[Uma98]   Umans, C.: The minimum equivalent DNF problem and shortest implicants. In: FOCS, pp. 556–563 (1998)

[Uma99]   Umans, C.: Hardness of approximating Σ_2^p minimization problems. In: FOCS, pp. 465–474 (1999)

[Uma01]   Umans, C.: The minimum equivalent DNF problem and shortest implicants. J. Comput. Syst. Sci. 63(4), 597–611 (2001)

[UVSV06]   Umans, C., Villa, T., Sangiovanni-Vincentelli, A.L.: Complexity of two-level logic minimization. IEEE Trans. on CAD of Integrated Circuits and Systems 25(7), 1230–1246 (2006)

Understanding a Non-trivial Cellular Automaton by Finding Its Simplest Underlying Communication Protocol[*]

Eric Goles[1,2], Cedric Little[1], and Ivan Rapaport[2,3]

[1] Facultad de Ciencias y Tecnologías, Universidad Adolfo Ibáñez, Chile
[2] Centro de Modelamiento Matemático (UMI 2807 CNRS), Universidad de Chile
[3] Departamento de Ingeniería Matemática, Universidad de Chile

Abstract. In the present work we find a *non-trivial* communication protocol describing the dynamics of an elementary CA, and we prove that there are *no simpler* descriptions (protocols) for such CA. This is, to our knowledge, the first time such a result is obtained in the study of CAs. More precisely, we divide the cells of Rule 218 into two groups and we describe (and therefore understand) its global dynamics by finding a protocol taking place between these two parts. We assume that $x \in \{0,1\}^n$ is given to Alice while $y \in \{0,1\}^n$ is given to Bob. Let us call $z(x,y) \in \{0,1\}$ the result of the dynamical interaction between the cells. We exhibit a protocol where Alice, instead of the n bits of x, sends $2\lceil \log(n) \rceil + 1$ bits to Bob allowing him to compute $z(x,y)$. Roughly, she sends 2 particular positions of her string x. By proving that any one-round protocol computing $z(x,y)$ must exchange at least $2\lceil \log(n) \rceil - 5$ bits, the optimality of our construction (up to a constant) is concluded.

1   Introduction

The process of understanding and classifying cellular automata (CAs) has been carried out mainly by researchers belonging to the dynamical systems community [2,9,14]. This interest can be explained on one hand by the simple fact that CAs *are* discrete dynamical systems and, on the other hand, by the impact of Wolfram's classification [21], which is an "empirical categorization of space-time pattterns into four classes loosely based on an analogy with those found in continous state dynamical systems"; nevertheless, this classification "has resisted numerous attempts at formalizations" [7].

We claim that CAs are extremely complex (highly non linear) objects and therefore the language of computer science appears to be particularly suitable for studying them. More precisely, our approach is to divide the cells into (two) groups in order to describe the dynamics by finding simple communication protocols taking place between these parts.

[*] Partially supported by Programs Conicyt "Anillo en Redes", Fondap, Basal-CMM, Fondecyt 1070022 and Instituto Milenio ICDB.

S.-H. Hong, H. Nagamochi, and T. Fukunaga (Eds.): ISAAC 2008, LNCS 5369, pp. 592–604, 2008.
© Springer-Verlag Berlin Heidelberg 2008

Obviously, this is not the first time CAs are analyzed from a (theoretical) computer science point of view. The algorithmic approach has always been present. In fact the model itself was invented in the 1950's as a tool to study self-reproduction [16]. And more recently researchers have tackled different algorithmic problems ranging from the intrinsic universality and the complexity of predicting [4,11,12,15,17,20] to the decidability/complexity of different dynamical systems properties [1,3,6,8]. But the present work is, to our knowledge, the first one where non-trivial protocols are discovered in the dynamics itself (in a previous paper the connection between CAs and communication complexity began to be explored [5]; nevertheless, in that work, instead of understanding the CAs behavior, the main interest was to give a formal classification; in fact, proofs were given just for simple cases).

1.1 Basics

An (elementary) CA is defined by a local function $f : \{0,1\}^3 \to \{0,1\}$, which maps the state of a cell and its two immediate neighbors to a new cell state. There are $2^{2^3} = 256$ CAs and each of them is identified with its Wolfram number $\omega = \sum_{a,b,c \in \{0,1\}} 2^{4a+2b+c} f(a,b,c)$ (see [21,22]). Sometimes, instead of expliciting function f, we refer to f_ω. The dynamics is defined in the one-dimensional cellspace. Following the CAs paradigm, all the cells change their states synchronously according to f. This endows the line of cells with a global dynamics whose links with the local function are still to be understood. After n time steps the value of a cell depends on its own initial state together with the initial states of the n immediate left and n immediate right neighbor cells. More precisely, we define the n-th iteration $f^n : \{0,1\}^{2n+1} \to \{0,1\}$ recursively: $f^1(z_{-1}, z_0, z_1) = f(z_{-1}, z_0, z_1)$ and, for $n \geq 2$,

$$f^n(z_{-n} \ldots z_1, z_0, z_1 \ldots z_n) = f^{n-1}(f(z_{-n}, z_{-n+1}, z_{-n+2}) \ldots f(z_{n-2}, z_{n-1}, z_n)).$$

This work is motivated by the following idea: if we were capable of giving a simple description of f^n (for arbitrary n) then we would have understood the behavior of the corresponding CA.

1.2 Representation

The first step is to represent f^n as two families of 0-1 matrices depending on whether the central cell begins in state $c = 0$ or $c = 1$. More precisely, the square matrices $M_f^{c,n}$ of size 2^n are defined as follows (see Figure 1).

$$M_f^{c,n}(x,y) := f^n(x, c, y) \text{ with } x = x_n \ldots x_1 \text{ and } y = y_1 \ldots y_n \text{ in } \{0,1\}^n.$$

Note that the first matrix of each family, standing for $n = 1$, completely defines the local function. One can think of these matrices as seeds for the families. We should emphasize also that the space-time diagram shows the evolution of only a *single configuration*, while the matrix covers *all configurations*.

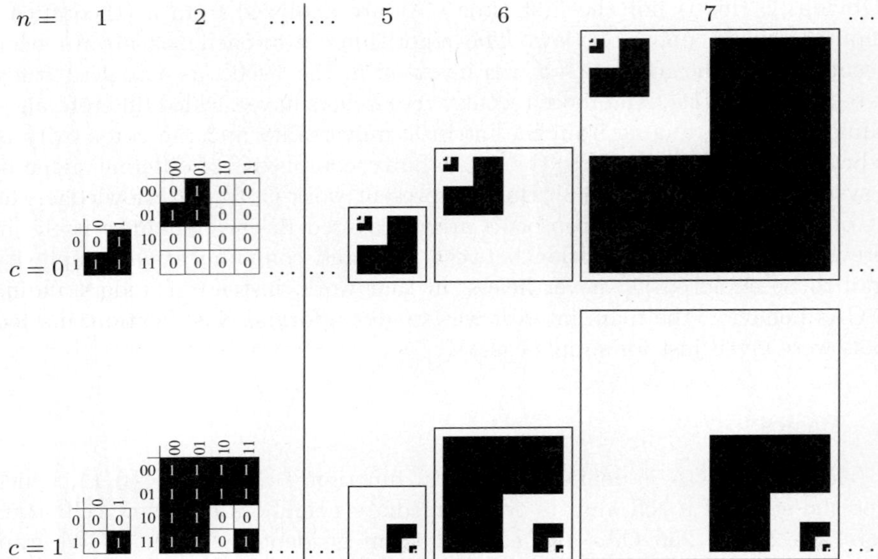

Fig. 1. The two families of binary matrices $M_{f_{178}}^{c,n}$ of Wolfram Rule 178

1.3   Interpretation

This step is obviously the most difficult. Here we try to prove and interpret the behavior of $M_f^{c,n}$ for arbitrary values of n. Fortunately, these 0-1 matrices reveal themselves to be a striking representation. For instance, let us consider Rule 105. In Figure 2 we show on the left the space-time diagram of Rule 105 for some initial configuration, and on the right the matrix $M_{f_{105}}^{0,6}$. In contrast with the space-time diagram, the matrix looks simple. In fact, as we are going to see later, the simplicity of a matrix $M_f^{c,n}$ is related to the simplicity of the communication protocol that computes f^n. Therefore, assuming that $x = x_n \ldots x_1 \in \{0,1\}^n$ is given to one party (say Alice) and that $y = y_1 \ldots y_n \in \{0,1\}^n$ is given to another party (say Bob), we are going to look for the simplest communication protocols that compute both $f^n(x,0,y)$ and $f^n(x,1,y)$.

1.4   Our Contribution

Let $d(M)$ be the number of different rows of a matrix M. In [5] the only CAs we managed to explain were those we called *bounded* (where $d(M_f^{c,n})$ was constant) and *linear* (where $d(M_f^{c,n})$ grew as $\Theta(n)$). All the other CAs were grouped together using a mainly experimental criterion. We conjectured the existence of *polynomial* and *exponential* classes. In the present work we prove the existence of a CA for which $d(M_f^{c,n})$ grows as $\Theta(n^2)$.

Linear and bounded rules are easy to explain in terms of communication protocols. This is the case of Rule 178 of Figure 1 (this particular rule has just been studied by D. Regnault [18] using percolation theory and considering

 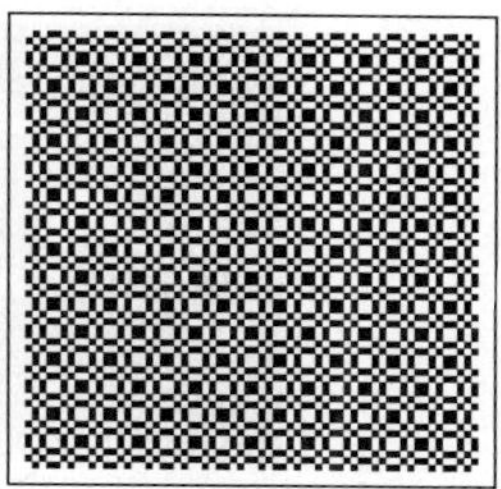

Fig. 2. A space time diagram for Rule 105 (left) and matrix $M^{0;6}_{f_{105}}$ (right). In the diagram every row is a configuration and time goes upward; grey cells represent states which are undetermined from the bottom (initial) configuration.

asynchronicity; we belive that the linearity of the rule and the fact that it is amenable to other types analysis is not a coincidence).

This paper shows that as soon as we move up in the hierarchy the underlying protocols become rather sophisticated. In fact, for the CA we treat here (Rule 218), we prove that if $c = 0$ then Alice needs to send 2 positions of her string ($2\log(n)$ bits). The quantitative relation between *the one-round communication complexity* and *the number of different rows* will be explained later. But roughly, the first is the logarithm of the second. Therefore, sending $2\log(n)$ bits is equivalent to having $\Theta(n^2)$ different rows. The difference between sending 1 position ($\Theta(n)$ behavior) and 2 positions ($\Theta(n^2)$ behavior) is huge. The reader can verify this by comparing the cases $c = 0$ and $c = 1$.

We think Rule 218 is one of the few CAs for which a non-trivial behavior can be proven. Experimentally, we do not find many candidates in a class $\Theta(n^k)$ with $k \geq 2$. This could imply that there are no other CAs with simple descriptions (shortcuts). We should also point out that, if more than 2 states were allowed, we could build CAs with arbitrary complexity. In fact, in [5] it is shown how to construct a 3 state CA exhibiting a $\Theta(n^3)$ behavior. But in the present work we are dealing with the *inverse problem*.

1.5   Rule 218

The local function of CA Rule 218 is the following:

$$\begin{array}{cccccccc}
0 & 1 & 0 & 1 & 1 & 0 & 1 & 1 \\
000 & 001 & 010 & 011 & 100 & 101 & 110 & 111
\end{array}$$

Its global dynamics is represented by the two matrices of Figure 3 and by the space-time diagrams of Figure 4 (Rule 218 and Rule 164 are the same; 0s behave as 1s and viceversa). We encountered Rule 218 when trying to find a (kind of) double-quiescent palindrome-recognizer. Despite the fact that it belongs to class 2 (according to Wolfram's classification), it mimics Rule 90 (class 3) for very particular initial configurations.

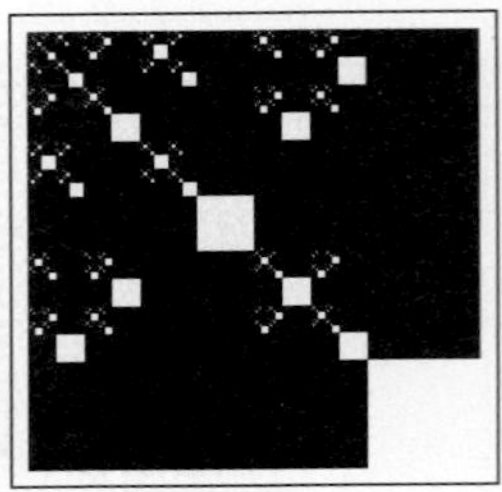 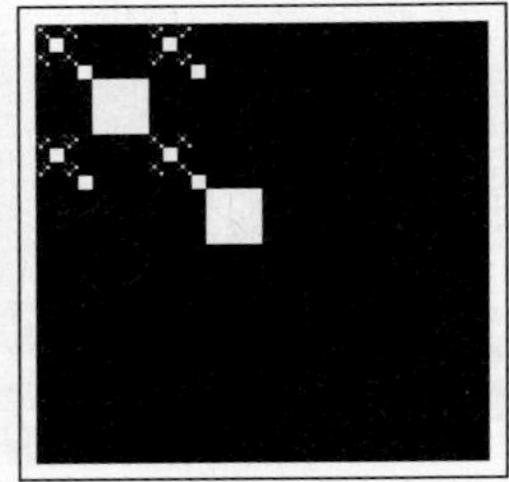

Fig. 3. $M_{f_{218}}^{0,9}$ (left) and $M_{f_{218}}^{1,9}$ (right)

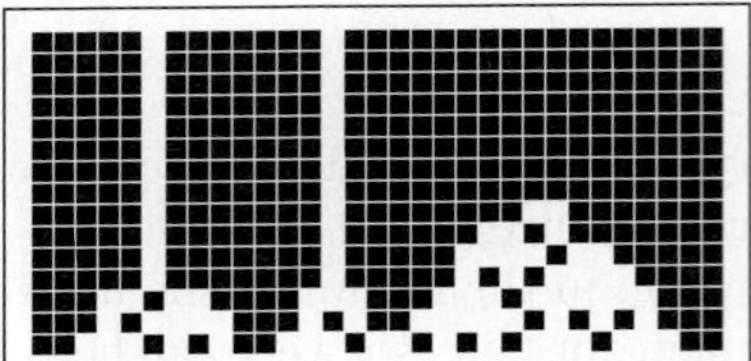 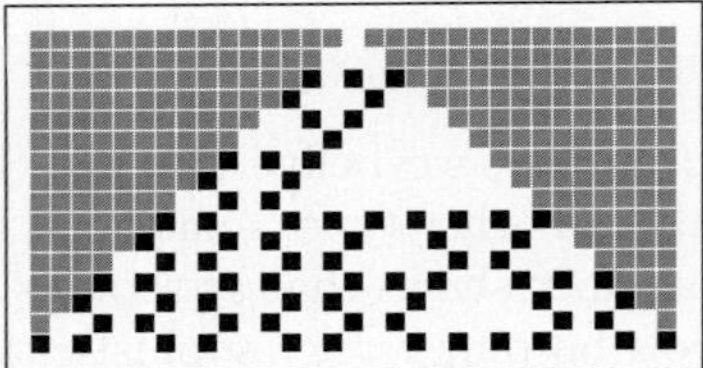

Fig. 4. Two space-time diagram for Rule 218. Every row is a configuration and time goes upward. Grey cells represent states which are undetermined from the bottom (initial) configuration.

Authors in [13] were surprised when they found, "unexpectedly", that the rule exhibited $1/f^\alpha$ spectra. Rule 218 has also been proposed as a symmetric cipher [19]. Nevertheless, it should be clear that the most relevant aspect of Rule 218 is its behavior in this communication context. More precisely, Rule 218 seems to be one of the most complex CAs for which a reasonable protocol can be found.

2   Two-Party Protocols

The communication complexity theory studies the information exchange required by different actors to accomplish a common computation when the data is initially distributed among them. To tackle that kind of questions, A.C. Yao [23] suggested the two-party model: two persons, say Alice and Bob, are asked to compute together $f(x, y)$, where Alice knows x only and Bob knows y only (x and y belonging to finite sets). Moreover, they are asked to proceed in such a way that the cost –the total number of exchanged bits– is minimal in the worst case. Different restrictions on the communication protocol lead to different communication complexity measures. Whereas most studies concern the many-round communication complexity, we focus only on the one-round.

Definition 1 (One-round communication complexity). *A protocol $\mathcal{P}$ is an AB-one-round f-protocol if only Alice is allowed to send information to Bob,* *and Bob is able to compute the function solely on its input and the received information. The cost of the protocol $c_{AB}(\mathcal{P})$ is the (worst case) number of bits Alice needs to send. Finally, the AB-one-round communication complexity of a function f is $c_{AB}(f) = c_{AB}(\mathcal{P}^*)$, where $\mathcal{P}^*$ is an AB-one-round f-protocol of minimum cost. The BA-one-round communication complexity is defined in the same way.*

The following fact throws light on the interest of the one-round communication complexity theory for our purpose: we can infer the exact cost of the optimal AB-one-round protocol by just counting the number of different rows in the matrix.

Fact 1 ([10]). *Let f be a binary function of $2n$ variables and $M_f \in \{0,1\}^{2^n \times 2^n}$ its matrix representation, defined by $M_f(x,y) = f(xy)$ for $x, y \in \{0,1\}^n$. Let $d(M_f)$ be the number of different rows in M_f. We have $c_{AB}(f) = \lceil \log(d(M_f)) \rceil$.*

Example 1. Consider Rule 90, which is defined as follows: $f(a,b,c) = a + c$ (the sum is mod 2). This is an additive rule and it satisfies the superposition principle. More precisely, for every $x_n \ldots x_1 \in \{0,1\}^n$, $\tilde{x}_n \ldots \tilde{x}_1 \in \{0,1\}^n$, $y_1 \ldots y_n \in \{0,1\}^n$, $\tilde{y}_1 \ldots \tilde{y}_n \in \{0,1\}^n$, $c, \tilde{c} \in \{0,1\}$:

$$f^n(x_n \ldots x_1, c, y_1 \ldots y_n) + f^n(\tilde{x}_n \ldots \tilde{x}_1, \tilde{c}, \tilde{y}_1 \ldots \tilde{y}_n) = f^n(x_n + \tilde{x}_n \ldots, c + \tilde{c}, \ldots y_n + \tilde{y}_n).$$

Therefore, there is a simple one-round communication protocol. Alice sends one bit b to Bob. The bit is $b = f^n(x_n \ldots x_1, c, 0 \ldots 0)$. Then Bob outputs $b + f^n(0 \ldots 0, 0, y_1 \ldots y_n)$. The same superposition principle holds for Rule 105 of Figure 2. This simple protocol (together with Fact 1) explains why the number of different rows is just 2.

3   The Protocols of Rule 218

Since Rule 218 is symmetric we are going to assume, w.l.g., that Alice is the party that sends the information. Moreover, we are going to refer simply to one-round protocols or one-round communication complexity (because the AB and BA settings are in this case equivalent). We denote f_{218} simply by f.

Notice that we can easily extend the notion of t iterations to blocks of size bigger than $2t + 1$. In fact, for every $m \geq 2t + 1$ and every finite configuration $z = z_1 \ldots z_m \in \{0,1\}^m$ we define $f^0(z) = z$,

$$f^1(z) = (f(z_1, z_2, z_3), \ldots, f(z_{m-2}, z_{m-1}, z_m)) \in \{0,1\}^{m-2}$$

and, recursively, $f^t(z) = f^{t-1}(f(z)) \in \{0,1\}^{m-2t}$.

Let $c \in \{0,1\}$. Let $x, y \in \{0,1\}^n$. From now on in this section, in order to simplify the notation, we are always assuming that these arbitrary values (i.e., n, c, x, y) have already been fixed.

Definition 2. *We say that a word in $\{0,1\}^*$ is additive if the 1s are isolated and every consecutive couple of 1s is separated by an odd number of 0s.*

Lemma 1. *If $xcy \in \{0,1\}^{2n+1}$ is additive, then $f^n(x,c,y) = f^n(x,c,0^n) + f^n(0^n,0,y)$.*

Proof. Rule 218 is "almost" the same as Rule 90 which is defined as follows: $(b_{-1}, b_0, b_1) \rightarrow b_{-1} + b_1$. The only case where the two rules differ is when $b_{-1}b_0b_1 = 111$. But for additive configurations the pattern 111 never appears and therefore its dynamics corresponds to the one of Rule 90. This rule is additive and therefore the superposition principle applies. $\qquad\square$

Notation 1. *Let α be the maximum index i for which $x_i \ldots x_1 c$ is additive. Let β be the maximum index j for which $cy_1 \ldots y_j$ is additive. Let $x' = x_\alpha \ldots x_1 \in \{0,1\}^\alpha$ and $y' = y_1 \ldots y_\beta \in \{0,1\}^\beta$.*

Notation 2. *Let l be the minimum index i for which $x_i = 1$. If such index does not exist we define $l = 0$. Let r be the minimum index j for which $y_j = 1$. If such index does not exist we define $r = 0$.*

3.1   The Lemmas

In this subsection we present all the lemmas we need in order to conclude the correctness of the protocols. These protocols are going to be presented in the next subsection. One could therefore begin by reading subsection 3.2 and check the lemmas later.

Lemma 2. $f^n(x,c,y) = f^n(1^{n-\alpha}x', c, y) = f^n(x,c,y'1^{n-\beta}) = f^n(1^{n-\alpha}x', c, y'1^{n-\beta})$.

Proof. By symmetry it is clear that it is enough to prove $f^n(x,c,y) = f^n(1^{n-\alpha}x', c, y)$. If $\alpha = n$ then it is direct. If $\alpha < n$ then there is a non-negative integer s such that $x_{\alpha+1} \ldots x_{\alpha-2s} = 10^{2s}1$ (notice that s could be 0). It follows that $f^s(10^{2s}1) = 11$. Notice that a word 11 acts as a wall through which information does not flow. In fact, for all $b \in \{0,1\}$, $f(b,1,1) = f(1,1,b) = 1$. Therefore we conclude that the result is independent of the information to the left of position $\alpha + 1$ and we can assume, w.l.g, that $x_n \ldots x_{\alpha+1} = 1^{n-\alpha}$. $\qquad\square$

Definition 3. *A string z is called left additive if it satisfies one of the two following conditions: either (i) $z = 0 \ldots 0$, or (ii) z is additive while $1z$ is not. For the right additivity definition we replace $1z$ by $z1$.*

Lemma 3. *Let $1 \leq s \leq n$. Let $z \in \{0,1\}^{2n+1-s}$. If z is left additive then $f(1^s z) = 1^s u$ with $u \in \{0,1\}^{2n-1-s}$ being left additive. If z is right additive then $f(z1^s) = u1^s$ with $u \in \{0,1\}^{2n-1-s}$ being right additive.*

Proof. We will prove the left additivity case (the right case is analogous). First we need to prove that the block of 1s moves to the right (see Figure 5 a). More precisely, that $f(1, z_1, z_2) = 1$. We know that $1z_1z_2 \neq 101$ because in that case $1z$ would have been additive. Therefore $f(1, z_1, z_2) = 1$. On the other hand, since

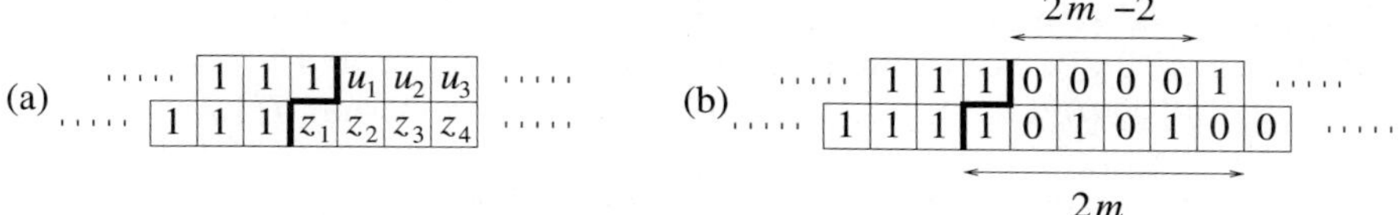

Fig. 5. a. $f(1^s z) = 1^s u$. **b.** $0^{2m-2}1$ is a prefix of u.

$f(z) = u$, we know that u is additive. Now we need to prove that $u = 0 \ldots 0$ or that $1u$ is not additive. Let us analyze two cases.

Case $z_1 = 0$. If $z = 0 \ldots 0$ then $u = 0 \ldots 0$. The other possibility is that z_1 belongs to an even length block of 0s (bounded by two 1s). If the length is 2 then $u_1 = f(z_1, z_2, z_3) = f(0, 0, 1) = 1$ and therefore $1u$ is not additive. If the length is even but bigger than two then the block shrinks in its two extremities and it remains even. Therefore, $1u$ also is not additive.

Case $z_1 = 1$. By the additivity of z we know that $z_2 = 0$. If $z_3 = 0$ then $u_1 = 1$ and therefore $1u$ is not additive. Hence let us assume $z_3 = 1$ (see Figure 5 b). Since $z_1 z_2 z_3 = 101$ we must consider three cases: $z = (10)^m$ with $m \geq 2$; $z = (10)^m 1$ with $m \geq 1$; and the case where $(10)^m 0$ is a prefix of z (with $m \geq 2$). In the first two cases $u = 0 \ldots 0$. In the third case $0^{2m-2}1$ is a prefix of u and therefore $1u$ is not additive. $\qquad\square$

Lemma 4. *Let* $1 \leq s \leq n$. *Let* $z \in \{0,1\}^{2n+1-s}$. *If* z *is left additive then* $f^n(1^s z) = 1$. *If* z *is right additive then* $f^n(z 1^s) = 1$.

Proof. Direct from Lemma 3. Let us just consider the left additivity case. It is clear that the state of the leftmost cell (which is a 1 because $s \geq 1$) propagates to the right. Therefore, $f^n(1^s z) = 1$. $\qquad\square$

Lemma 5. *If* $x'cy'$ *is additive, then*

1. *If* $|\alpha - \beta| \geq 1$ *then* $f^n(1^{n-\alpha}x', c, y'1^{n-\beta}) = 1$.
2. *If* $\alpha = \beta = k$ *then* $f^n(1^{n-\alpha}x', c, y'1^{n-\beta}) = f^k(x', c, y') = f^k(x', c, 0^k) + f^k(0^k, 0, y')$.

Proof.
1. Let us assume, w.l.g., that $\alpha < \beta$. Consider $z = f^{\alpha+1}(1^{n-\alpha}x', c, y'1^{n-\beta})$. The number of bits of z is $2n + 1 - 2(\alpha + 1) = 2(n - \alpha - 1) + 1$. If we denote $z = z_{-(n-\alpha-1)} \ldots z_0 \ldots z_{(n-\alpha-1)}$ it follows that

$$z_{-(n-\alpha-1)} \ldots z_0 = f^{\alpha+1}(1^{n-\alpha}x'_\alpha \ldots x'_1 c y'_1 \ldots y'_{\alpha+1}).$$

By Lemma 4 we conclude that $z_{-(n-\alpha-1)} \ldots z_0 = 1^{n-\alpha}$. If $\alpha = n - 1$ then $z_0 = 1 = f^n(1^{n-\alpha}x', c, y'1^{n-\beta})$ and the result is concluded. If $\alpha < n - 1$ then $z_{-1}z_0 = 11$. Since such a wall of size two never changes we conclude that the central cell will always remain in state 1.

2. If $k = n$ then it is direct. Suppose $k < n$. Consider the configuration $z = f^k(1^{n-k}x', c, y'1^{n-k})$. We conclude from Lemma 4 that $z = 1^{n-k}b1^{n-k}$, where $b = f^k(x', c, y') \in \{0, 1\}$. The results follows from the fact that $f(1, b, 1) = b$. $\quad\square$

Remark 1. The purpose of the following lemmas is to treat the case where $x'0y'$ is not additive (because $x'1y'$ is always additive). Therefore, we are interested in the case where an even length block of 0s between two consecutive 1s appear in $x'0y'$. In other words by recalling Notation 2, when $|l + r - 1|$ is even or, equivalently, when $r \neq l \pmod 2$.

Lemma 6. *If $r \neq 0$, $l \neq 0$, $|l + r - 1|$ is even and $l \geq r - 1$, then*

$$f^n(x, 0, y) = \begin{cases} f^n(1^{n-l+1}0^{l-1}, 0, y) & \text{if } l \geq r + 3, \\ 1 & \text{if } |l - r| = 1. \end{cases}$$

Proof. Notice that $x_l \ldots x_1 0 y_1 \ldots y_r = 10^{l+r-1}1$. So $f^{\frac{l+r-1}{2}}(x_l \ldots x_1 0 y_1 \ldots y_r) = 11$. If $l > r$ then this 11 wall (through which information can not flow) will be located on the left side of the center cell. It follows that the final result will not depend on $x_n \ldots x_{l+1}$ (if $l = n$ this is just the empty word). Then we can assume, w.l.g, that $x_n \ldots x_{l+1} = 1^{n-l}$.

For the particular cases $l = r + 1$ and $l = r - 1$, the 11 wall will appear precisely in the center (and the result corresponds to 1). $\quad\square$

Lemma 7. *If $r \neq 0$, $l \neq 0$, $|l + r - 1|$ is even and $l \leq r - 3$, then*

$$f^n(x, 0, y) = \begin{cases} f^\alpha(x', 0, 0^\alpha) & \text{if } r = \alpha + 1, \\ 1 & \text{if } r \neq \alpha + 1. \end{cases}$$

Proof. Since $r > l$, by the same argument used in the proof of Lemma 6, we know that the result does not depend on $y_{r+1} \ldots y_n$ and we can therefore assume that $y_{r+1} \ldots y_n = 1^{n-r}$. On the other hand, from Lemma 2, we can assume that $x_n \ldots x_{\alpha+1} = 1^{n-\alpha}$. It follows from the two previous remarks that

$$f^n(x, 0, y) = f^n(1^{n-\alpha}x_\alpha \ldots x_1, 0, 0^{r-1}1^{n-r+1}).$$

Case $r = \alpha + 1$. Let us denote

$$z_{-(n-\alpha)} \ldots z_0 \ldots z_{n-\alpha} = f^\alpha(x, 0, y) = f^\alpha(1^{n-\alpha}x_\alpha \ldots x_1, 0, 0^\alpha 1^{n-\alpha}).$$

Let us compute $z_{-2}z_{-1}z_0z_1z_2$ (if $\alpha = n - 1$ we only consider $z_{-1}z_0z_1$). It follows that $z_{-2} = f^\alpha(11x_\alpha \ldots x_3, x_2, x_10^{\alpha-1})$, $z_{-1} = f^\alpha(1x_\alpha \ldots x_2, x_1, 0^\alpha)$, $z_1 = f^\alpha(x_{\alpha-1} \ldots x_10, 0, 0^{\alpha-1}1)$, $z_2 = f^\alpha(x_{\alpha-2} \ldots x_100, 0, 0^{\alpha-2}11)$. From Lemma 4, $z_{-2} = z_{-1} = z_1 = z_2 = 1$ (for $z_1 = 1$ and $z_2 = 1$ recall that $|l - r + 1|$ is even). Therefore, the pattern $11z_011$ appears in the center. Since $z_0 = f^\alpha(x_\alpha \ldots x_1, 0, 0^\alpha)$ the results follows (for the particular case $\alpha = n - 1$ we have that $f^n(x, 0, y) = f(z_{-1}, z_0, z_1) = f(1, z_0, 1) = z_0$ and the same conclusion is obtained).

Case $r > \alpha + 1$. Let us denote

$$z_{-(n-\alpha-1)} \cdots z_0 \cdots z_{n-\alpha} = f^{\alpha+1}(x,0,y) = f^{\alpha+1}(1^{n-\alpha}x_\alpha \ldots x_1, 0, 0^{r-1}1^{n-r+1}).$$

The result follows because $z_{-1}z_0 = 11$. In fact, $z_{-1} = f^{\alpha+1}(11x_\alpha \ldots x_2, x_1, 0^\alpha)$ and $z_0 = f^{\alpha+1}(1x_\alpha \ldots x_1, 0, 0^{\alpha+1})$. In both cases we apply Lemma 4.

Case $r < \alpha + 1$. Let us denote

$$z_{-(n-r)} \cdots z_0 \cdots z_{n-r} = f^r(x,0,y) = f^r(1^{n-\alpha}x_\alpha \ldots x_1, 0, 0^{r-1}1^{n-r+1}).$$

The result follows because $z_0 z_1 = 11$. In fact, $z_0 = f^r(x_r \ldots x_1, 0, 0^{r-1}1)$ and $z_1 = f^r(x_{r-1} \ldots x_1 0, 0, 0^{r-2}11)$. In both cases we apply Lemma 4 (right additivity because $l + r - 1$ is even). In the particular case where $r = \alpha = n$ we have $f^n(x,0,y) = z_0 = 1$. $\qquad\qquad\square$

3.2   The Protocols

3.2.1   When $c = 0$

We are going to define a one-round protocol $\mathcal{P}_0$ for the case where the central cell begins in state 0. Recall the Alice knows x and Bob knows y. $\mathcal{P}_0$ goes as follows. Alice sends to Bob α, l, and $a = f^\alpha(x',0,0^\alpha)$. The number of bits is therefore $2\lceil \log(n) \rceil + 1$.

If $l = 0$ then Bob knows (by definition of l) that $x = 0^n$ and he outputs $f^n(0^n,0,y)$. If $r = 0$ his output depends on α. If $\alpha = n$ he outputs a (Lemma 1) and if $\alpha < n$ he outputs 1 (Lemma 5). We can assume now that neither l nor r are 0. The way Bob proceed depends mainly on the parity of $|l + r - 1|$.

Case $|l+r-1|$ **is odd.** In this case $x'0y'$ is additive and Bob can apply Lemma 5. In fact, if $|\alpha - \beta| \geq 1$ he outputs 1. If $\alpha = \beta = k$ he outputs $a + f^k(0^k, 0, y')$.

Case $|l + r - 1|$ **is even.** Bob compares r with l. If $l \geq r - 1$ then he applies Lemma 6. More precisely, he outputs $f^n(1^{n-l+1}0^{l-1}, 0, y)$ if $l \geq r + 3$ and 1 otherwise. If $l \leq r - 3$ then he applies Lemma 7. More precisely, he outputs $a = f^\alpha(x',0,0^\alpha)$ if $r = \alpha + 1$ and 1 otherwise.

Proposition 1. *$\mathcal{P}_0$ is a one-round f-protocol for $c = 0$ with cost $2\lceil \log(n) \rceil + 1$.*

3.2.2   When $c = 1$

We are going to define a one-round protocol $\mathcal{P}_1$ for the case where the central cell begins in state 1. Alice sends to Bob α and $a = f^\alpha(x',1,0^\alpha)$. The number of bits is therefore $\lceil \log(n) \rceil + 1$. Notice that $x'1y' \in \{0,1\}^{\alpha+\beta+1}$ is additive and therefore Bob applies Lemma 5. More precisely, if $\alpha \neq \beta$ then $f^n(x,1,y) = 1$. On the other hand, if $\alpha = \beta = k$ then $f^n(x,1,y) = f^k(x',1,y')$ and Bob outputs $a + f^k(0,1,y')$.

Proposition 2. *$\mathcal{P}_1$ is a one-round f-protocol for $c = 1$ with cost $\lceil \log(n) \rceil + 1$.*

4   Optimality

In this section we exhibit lower bounds for $d(M_f^{c,n})$, the number of different rows of $M_f^{c,n}$. If these bounds appear to be tight then, from Fact 1, they can be used for proving the optimality of our protocols.

4.1   Case $c = 0$

Consider the following subsets of $\{0,1\}^n$. First, $S_3 = \{1^{n-3}000\}$. Also, $S_5 = \{1^{n-5}00000, 1^{n-5}01000\}$. In general, for every $k \geq 2$ such that $2k+1 \leq n$, we define $S_{2k+1} = \{1^{n-2k-1}0^{2k+1}\} \cup \{1^{n-2k-1}0^a10^b \mid a$ odd, b odd, $b \geq 3$, $a + b = 2k\}$.

Lemma 8. *Let $x_n \ldots x_1 \in S_{2k+1}$ and $\tilde{x}_n \ldots \tilde{x}_1 \in S_{2\tilde{k}+1}$ with $k \neq \tilde{k}$. It follows that the rows of $M_f^{c,n}$ indexed by $x_n \ldots x_1$ and $\tilde{x}_n \ldots \tilde{x}_1$ are different.*

Proof. We can first easily prove (by induction on n) that every $z_n \ldots z_1 \in \{0,1\}^n$ satisfies $f^n(z_n \ldots z_1, 0, z_1 \ldots z_n) = 0$. Let $x_n \ldots x_1 \in S_{2k+1}$ and $\tilde{x}_n \ldots \tilde{x}_1 \in S_{2\tilde{k}+1}$ (with $k \neq \tilde{k}$). From Lemma 5, $f^n(x_n \ldots x_1, 0, \tilde{x}_1 \ldots \tilde{x}_n) = f^n(\tilde{x}_n \ldots \tilde{x}_1, 0, x_1 \ldots x_n) = 1$. $\qquad\square$

Lemma 9. *Let $x = x_n \ldots x_1$, $\tilde{x} = \tilde{x}_n \ldots \tilde{x}_1 \in S_{2k+1}$ with $x \neq \tilde{x}$. It follows that there exists $y = y_1 \ldots y_n \in \{0,1\}^n$ such that $f^n(x, 0, y) \neq f^n(\tilde{x}, 0, y)$.*

Proof. Assume, w.l.g., that for some odd number $b \geq 3$ the word 10^b is a suffix of x while 0^{b+2} is a suffix of $\tilde{x}$ (i.e., such b can be at most $n - 4$). Let $y = 0^{b-3}101^{n-b+1}$. Let $z_{-(n-b+1)} \ldots z_0 \ldots z_{n-b+1} = f^{b-1}(x, 0, y)$. Then $z_2 = f^{b-1}(0^{2b-5}1011)$ and $z_{-2} = f^{b-1}(10^{b-2}, 0, 0^{b-1})$. It is direct that $z_{-2} = 1$. On the other hand, from Lemma 4 (right additivity), we know that $z_2 = 1$. For $z_{-1}z_0z_1$ notice that $z_{-1}z_0z_1 = f^{b-1}(0^{2b-2}101)$. In this case the dynamics is such that the configuration 0^*101 reappears every 2 steps. Since $b-1$ is even we conclude that $z_{-1}z_0z_1 = 101$. So we have proven that the pattern 11011 appears in the center and therefore $f^n(x, 0, y) = 0$.

Let $\tilde{z}_{-(n-b-2)} \ldots \tilde{z}_0 \ldots \tilde{z}_{n-b-2} = f^{b+2}(\tilde{x}, 0, y)$. Then $\tilde{z}_1 = f^{b+2}(0^{2b-1}101111)$ and $\tilde{z}_0 = f^{b+2}(0^{b+2}, 0, 0^{b-3}10111)$. From Lemma 4, $\tilde{z}_0 = \tilde{z}_1 = 1$ and therefore the wall 11 appears in the center. Since this means that $f^n(\tilde{x}, 0, y) = 1$, the lemma is proven. $\qquad\square$

Proposition 3. *The cost of any one-round f-protocol for $c = 0$ is at least $2\lceil \log(n) \rceil - 5$.*

Proof. From Lemmas 8 and 9, the number of different rows in $M_f^{0,n}$ is $\sum_{3 \leq 2k+1 \leq n} |S_{2k+1}| = \sum_{i=1}^{\lceil \frac{n}{2} \rceil - 1} i$. For sufficiently large n the sum is lower bounded by $\frac{1}{16}n^2$. Therefore $d(M_f^{0,n}) \geq \lceil 2\log(n) - 4 \rceil \geq 2\lceil \log(n) \rceil - 5$. $\qquad\square$

4.2   Case $c = 1$

Proposition 4. *The cost of any one-round f-protocol for $c = 1$ is at least* $\lceil \log(n) \rceil$.

Proof. Consider the set $T = \{1^{n-k}0^k | 1 \leq k \leq n\}$. All we need to prove is that the rows indexed by any two different strings in T are different (because $|T| = n$).

Let $x = 1^{n-a}0^a$ and $\tilde{x} = 1^{n-\tilde{a}}0^{\tilde{a}}$ with $1 \leq a < \tilde{a} \leq n$. It is easy to prove (by induction on n) that $f^n(x, 1, 0^a 1^{n-a}) = f^n(\tilde{x}, 1, 0^{\tilde{a}} 1^{n-\tilde{a}}) = 0$. It remains to prove that $f^n(x, 1, 0^{\tilde{a}} 1^{n-\tilde{a}}) = f^n(\tilde{x}, 1, 0^a 1^{n-a}) = 1$.

By Lemma 5 we directly conclude that $f^n(x, 1, 0^{\tilde{a}} 1^{n-\tilde{a}}) = 1$ except for the case when $\tilde{a} = a + 1$ and a is odd. Let us therefore treat this last case now. First notice that $f^{a+1}(1^{n-a}0^a, 1, 0^{a+1}1^{n-a-1}) = 1^{n-a-1} f^{a+1}(10^a, 1, 0^{a+1})1^{n-a-1}$.

Therefore, by additivity (a is odd) and by the fact that $f(1, b, 1) = b$ for all $b \in \{0, 1\}$, the final result is

$$f^{a+1}(10^a, 1, 0^{a+1}) = f^{a+1}(00^a, 1, 0^{a+1}) + f^{a+1}(10^a, 0, 0^{a+1}) = 0 + 1 = 1. \qquad \square$$

References

1. Bernardi, V., Durand, B., Formenti, E., Kari, J.: A new dimension sensitive property for cellular automata. In: Fiala, J., Koubek, V., Kratochvíl, J. (eds.) MFCS 2004. LNCS, vol. 3153, pp. 416–426. Springer, Heidelberg (2004)
2. Blanchard, F., Maass, A.: Dynamical properties of expansive one-sided cellular automata. Israel Journal of Mathematics 99, 149–174 (1997)
3. D'amico, M., Manzini, G., Margara, L.: On computing the entropy of cellular automata. Theoretical Computer Science 290(3), 1629–1646 (2003)
4. Cook, M.: Universality in elementary cellular automata. Complex Systems 15(1), 1–40 (2004)
5. Durr, C., Rapaport, I., Theyssier, G.: Cellular automata and communication complexity. Theoretical Computer Science 322(2), 355–368 (2004)
6. Formenti, E., Kurka, P.: A search algorithm for the maximal attractor of a cellular automaton. In: Thomas, W., Weil, P. (eds.) STACS 2007. LNCS, vol. 4393, pp. 356–366. Springer, Heidelberg (2007)
7. Hanson, J.E., Cruchfield, J.P.: Computational mechanics of cellular automata: An example. Physica D 103, 169–189 (1997)
8. Kari, J.: The nilpotency problem of one-dimensional cellular automata. SIAM Journal on Computing 21(3), 571–586 (1992)
9. Kurka, P.: Languages, equicontinuity and attractors in cellular automata. Ergodic Theory and Dynamical Systems 17, 417–433 (1997)
10. Kushilevitz, E., Nisan, N.: Communication Complexity. Cambridge University Press, Cambridge (1997)
11. Lindgren, K., Nordahl, M.G.: Universal computation in simple one-dimensional cellular automata. Complex Systems 4, 299–318 (1990)
12. Moore, C.: Predicting non-linear cellular automata quickly by decomposing them into linear ones. Physica D 111, 27–41 (1998)
13. Nagler, J., Claussen, J.C.: $1/f^\alpha$ spectra in elementary cellular automata and fractal signals. Physical Review E 71, 067103 (2005)

14. Nasu, M.: Nondegenerate q-biresolving textile systems and expansive automorphisms of onesided full shifts. Transactions of the American Mathematical Society 358, 871–891 (2006)
15. Neary, T., Woods, D.: P-completeness of cellular automaton Rule 110. In: Bugliesi, M., Preneel, B., Sassone, V., Wegener, I. (eds.) ICALP 2006. LNCS, vol. 4051, pp. 132–143. Springer, Heidelberg (2006)
16. von Neumann, J.: The theory of self reproducting cellular automata. University of Illinois Press, Urbana (1967)
17. Ollinger, N.: The quest for small universal cellular automata. In: Widmayer, P., Triguero, F., Morales, R., Hennessy, M., Eidenbenz, S., Conejo, R. (eds.) ICALP 2002. LNCS, vol. 2380, pp. 318–329. Springer, Heidelberg (2002)
18. Regnault, D.: Directed percolation arising in stochastic cellular automata. In: MFCS 2008 (to appear, 2008)
19. Srisuchinwong, B., York, T.A., Tsalides, P.: A symmetric cipher using autonomous and non-autonomous cellular automata. In: Proc. of Global Telecommunication Conference, vol. 2, pp. 1172–1177. IEEE, Los Alamitos (1995)
20. Umeo, H., Morita, K., Sugata, K.: Deterministic one-way simulation of two-way real-time cellular automata and its related problems. Information Processing Letters 14(4), 158–161 (1982)
21. Wolfram, S.: Universality and Complexity in Cellular Automata. Physica D 10, 1–35 (1984)
22. Wolfram, S.: A New Kind of Science. Wolfram Media, Illinois (2002)
23. Yao, A.C.: Some complexity questions related to distributed computing. In: Proc. of 11th ACM Symposium on Theory of Computing, pp. 209–213 (1979)

Negation-Limited Inverters of Linear Size

Hiroki Morizumi[1] and Genki Suzuki[2]

[1] Graduate School of Information Sciences, Tohoku University
Sendai 980-8579, Japan
`morizumi@ecei.tohoku.ac.jp`
[2] School of Engineering, Tohoku University
Sendai 980-8579, Japan
`suzuki@nishizeki.ecei.tohoku.ac.jp`

Abstract. An inverter is a circuit which outputs $\neg x_1, \neg x_2, \ldots, \neg x_n$ for any Boolean inputs $x_1, x_2, \ldots, x_n$. Beals, Nishino and Tanaka have given a construction of an inverter which has size $O(n \log n)$ and depth $O(\log n)$ and uses $\lceil \log(n+1) \rceil$ NOT gates. In this paper we give a construction of an inverter which has size $O(n)$ and depth $\log^{1+o(1)} n$ and uses $\log^{1+o(1)} n$ NOT gates. This is the first negation-limited inverter of linear size using only $o(n)$ NOT gates.

1 Introduction

No superlinear lower bound has been known so far on the size of Boolean circuits computing an explicit Boolean function, while exponential lower bounds have been known on the size of *monotone* circuits, which consist of only AND and OR gates and do not contain NOT gates [2,5,16]. It is natural to ask: what happens if a limited number of NOT gates are allowed? This motivates us to study the *negation-limited* circuit complexity under various scenarios [3,4,6,10,18,21].

An *inverter* is a circuit which outputs $\neg x_1, \neg x_2, \ldots, \neg x_n$ for any n Boolean inputs $x_1, x_2, \ldots, x_n$. One can easily construct an inverter by placing n NOT gates in a row. We wish to construct an inverter by using fewer than n NOT gates and an arbitrary number of AND gates and OR gates. It is known that $\lceil \log(n+1) \rceil$ NOT gates are necessary and sufficient to construct an inverter [13]. (All logarithms in the paper are base 2.) Beals, Nishino and Tanaka constructed an inverter of size $O(n \log n)$ and depth $O(\log n)$ using $\lceil \log(n+1) \rceil$ NOT gates [6], while no inverter of linear size was known even for the case where the number of NOT gates is $o(n)$. In the paper, we give a construction of an inverter which has size $O(n)$ and depth $\log^{1+o(1)} n$ and uses $\log^{1+o(1)} n$ NOT gates. The results above are summarized in Table 1.

It is known that a construction of a negation-limited inverter implies a relation between negation-limited circuits and *general* circuits, which are circuits with no limitation of the number of NOT gates, for these sizes. More precisely, if there are a general circuit of size s_1 computing an arbitrary Boolean function f and an inverter of size s_2 with r NOT gates, then there is a circuit of size $2s_1 + s_2$ with at most r NOT gates computing f. In the paper, we give such a relation

S.-H. Hong, H. Nagamochi, and T. Fukunaga (Eds.): ISAAC 2008, LNCS 5369, pp. 605–614, 2008.
© Springer-Verlag Berlin Heidelberg 2008

Table 1. Negation-limited inverters

	# of NOTs	size	depth
trivial	n	n	1
this paper	$\log^{1+o(1)} n$	$O(n)$	$\log^{1+o(1)} n$
Beals, Nishino, Tanaka [6]	$\lceil \log(n+1) \rceil$	$O(n \log n)$	$O(\log n)$

which is implied by our construction of inverters, and discuss two implications of the relation.

2   Preliminaries

In the paper, a *circuit* is a combinational circuit which consists of AND gates of fan-in two, OR gates of fan-in two and NOT gates. The *size* of a circuit C is the number of gates in C, and denoted by $size(C)$. The *depth* of C is the length (number of gates) of the longest path in C, and denoted by $depth(C)$. We denote by $not(C)$ the number of NOT gates in C.

The function $\log^{(i)} x$ is defined recursively for nonnegative integers i and real numbers x as follows:

$$\log^{(i)} x$$
$$= \begin{cases} x & \text{if } i = 0; \\ \log(\log^{(i-1)} x) & \text{if } i > 0 \text{ and } \log^{(i-1)} x > 0; \\ \text{undefined} & \text{if } (i > 0) \text{ and } (\log^{(i-1)} x \leq 0 \text{ or } \log^{(i-1)} x \text{ is undefined}). \end{cases}$$

The iterated logarithm function $\log^* x$ is defined as follows:

$$\log^* x = \min \left\{ i \geq 0 : \log^{(i)} x \leq 1 \right\}.$$

In our construction, we use the following negation-limited inverter.

Proposition 1. *(Beals, Nishino and Tanaka [6]) There is an inverter of size $O(n \log n)$ and depth $O(\log n)$ using $\lceil \log(n+1) \rceil$ NOT gates.*

3   Negation-Limited Inverter

In this section, we construct an inverter of size $O(n)$ and depth $\log^{1+o(1)} n$ using $\log^{1+o(1)} n$ NOT gates.

3.1   Result

In fact we give a construction of an inverter with size $O(n)$ such that both the depth and the number of NOT gates are a little smaller than $\log^{1+o(1)} n$.

To state our result in detail, we introduce the following notation. We denote $\underbrace{\log x \log \log x \log \log \log x \cdots}_{i}$ simply by $\log^{[i]} x$. More precisely, we define $\log^{[i]} x$ for nonnegative integers i as follows:

$$\log^{[i]} x = \begin{cases} 1 & \text{if } i = 0; \\ \log x & \text{if } i = 1; \\ \log^{[i-1]} x \cdot \log^{(i)} x & \text{if } i > 1 \text{ and } \log^{(i)} x \text{ is defined}; \\ \text{undefined} & \text{if } i > 1 \text{ and } \log^{(i)} x \text{ is undefined.} \end{cases}$$

The main result of this paper is the following.

Theorem 1. *There is an inverter of size $O(n)$ and depth $O(\log^{[\log^* n - c]} n)$ using $\log^{[\log^* n - c]} n$ NOT gates for an arbitrary constant $c \geq 1$.*

Note that both the depth and the number of NOT gates in Theorem 1 are $\log^{1+o(1)} n$, which can be confirmed as follows:

$$O(\log^{[\log^* n - c]} n) \leq c' \underbrace{\log n \log \log n \log \log \log n \cdots}_{\log^* n - c}$$
$$\leq c' \log n \cdot (\log \log n)^{\log^* n - (c+1)}$$
$$\leq \log n \cdot c' 2^{\log \log \log n \cdot \log^* n}$$
$$= \log n \cdot 2^{o(\log \log n)}$$
$$= \log n \cdot \log^{o(1)} n$$
$$= \log^{1+o(1)} n,$$

where c' is some constant. In the rest of this section, we prove Theorem 1.

3.2   Proof of Theorem 1

The inverter in Proposition 1 has size $O(n \log n)$ and depth $O(\log n)$ and uses $\lceil \log(n+1) \rceil$ NOT gates. Let c_s and c_d be the hidden constants in each of the size $O(n \log n)$ and the depth $O(\log n)$. We call inverters which have n inputs $x_1, x_2, \ldots, x_n$ n-input inverters. Our inverter is recursively constructed, based on the following lemma.

Lemma 1. *Let $n \geq n_0$ be an integer for some constant n_0. Let i be an integer such that $1 \leq i \leq \log^* n - c_0$, for some constant c_0. If, for all $m \geq n/(4 \log n)$, there is an m-input inverter I such that*

$$size(I) \leq 2c_s m \log^{(i)} m,$$
$$depth(I) \leq c_d (\log^{[i-1]} m)(\log^{(i)} m + 4),$$
$$not(I) \leq (\log^{[i-1]} m)(\log^{(i)} m + 4),$$

then there is an n-input inverter I' such that

$$size(I') \leq 2c_s n \log^{(i+1)} n,$$
$$depth(I') \leq c_d (\log^{[i]} n)(\log^{(i+1)} n + 4),$$
$$not(I') \leq (\log^{[i]} n)(\log^{(i+1)} n + 4).$$

Using Lemma 1, one can easily prove Theorem 1.

Proof of Theorem 1. We assume that the constant c in Theorem 1 satisfies that $c \geq c_0 - 2$ and $c \geq 3$, where c_0 is the constant in Lemma 1. It is enough that we give a proof on the assumptions, since an increase of c strengthens the condition of the depth and the number of NOT gates in Theorem 1.

The m_1-input inverter I_1 in Proposition 1 satisfies

$$
\begin{aligned}
size(I_1) &\leq c_s m_1 \log m_1, \\
depth(I_1) &\leq c_d \log m_1, \\
not(I_1) &= \lceil \log(m_1 + 1) \rceil,
\end{aligned}
$$

for all $m_1 \geq n^{1/\log^* n}$. Therefore, by Lemma 1 for $i = 1$, there is an m_2-input inverter I_2 such that

$$
\begin{aligned}
size(I_2) &\leq 2 c_s m_2 \log \log m_2, \\
depth(I_2) &\leq c_d \log m_2 (\log \log m_2 + 4), \\
not(I_2) &\leq \log m_2 (\log \log m_2 + 4),
\end{aligned}
$$

for all $m_2 \geq n^{2/\log^* n}$. By applying Lemma 1 for each i, $1 \leq i \leq \log^* n - c - 2$, one can know that there is an m_*-input inverter I_* such that

$$
\begin{aligned}
size(I_*) &\leq 2 c_s m_* \log^{(\log^* n - c - 1)} m_*, \\
depth(I_*) &\leq c_d (\log^{[\log^* n - c - 2]} m_*)(\log^{(\log^* n - c - 1)} m_* + 4), \\
not(I_*) &\leq (\log^{[\log^* n - c - 2]} m_*)(\log^{(\log^* n - c - 1)} m_* + 4),
\end{aligned}
$$

for all $m_* \geq n^{(\log^* n - c - 2)/\log^* n}$. Since $n \geq n^{(\log^* n - c - 2)/\log^* n}$, there is an n-input inverter I such that

$$
\begin{aligned}
size(I) &\leq 2 c_s n \log^{(\log^* n - c - 1)} n \\
&\leq 2 c_s n \cdot \left. 2^{2^{2^{\cdot^{\cdot^{\cdot^2}}}}} \right\} c \\
&= O(n), \\
depth(I) &\leq c_d (\log^{[\log^* n - c - 2]} n)(\log^{(\log^* n - c - 1)} n + 4) \\
&\leq c_d \log^{[\log^* n - c]} n \\
&= O(\log^{[\log^* n - c]} n), \\
not(I) &\leq (\log^{[\log^* n - c - 2]} n)(\log^{(\log^* n - c - 1)} n + 4) \\
&\leq \log^{[\log^* n - c]} n. \qquad \square
\end{aligned}
\tag{1}
$$

As shown in Eq. (1), the hidden constant of the size $O(n)$ in Theorem 1 may be extremely large. Our construction is only of theoretical interest.

3.3   Proof of Lemma 1

Proof of Lemma 1. We give a construction of the inverter I' in Lemma 1. We can assume that n is enough large since we choose some enough large constant as n_0 in Lemma 1.

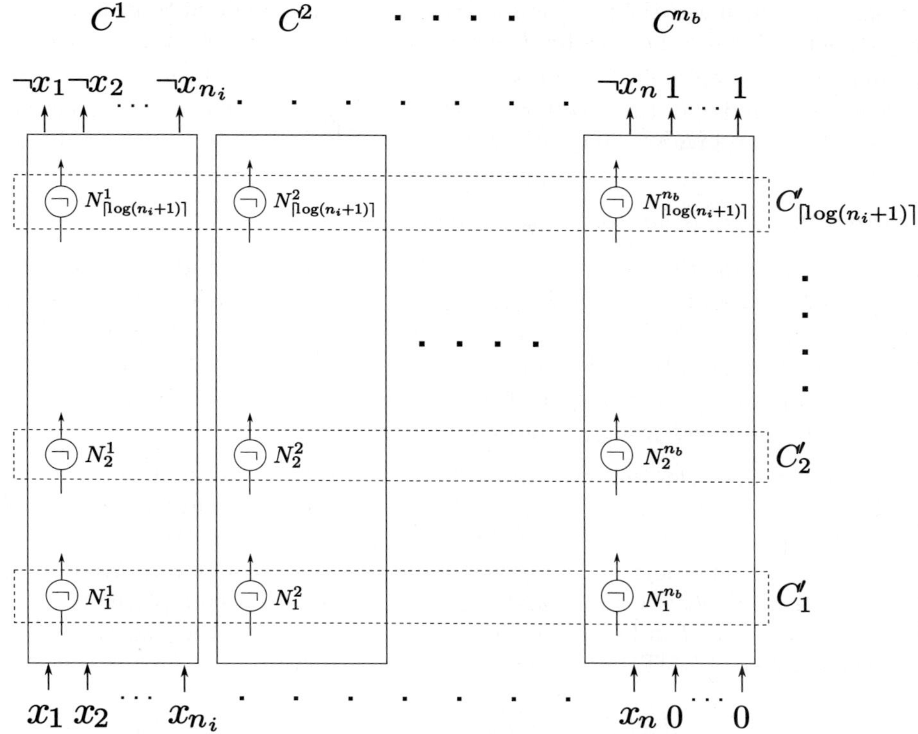

Fig. 1. Overall structure of the inverter I'

See Fig. 1. We first partition n inputs $x_1, x_2, \ldots, x_n$ of the inverter I' to $n_b = \lceil n/3\lceil \log^{(i)} n\rceil\rceil$ blocks so that each block contains $n_i = 3\lceil \log^{(i)} n\rceil$ inputs. If necessary, we pad the last block with constants 0's so that the block contains exactly n_i inputs. For the kth block of inputs, $1 \le k \le n_b$, we construct an inverter C^k for the inputs in the block. As C^k we use the inverter in Proposition 1. We choose the constant c_0 in Lemma 1 so that the n_i-input inverter of Proposition 1 exists. The input of C^k are $x_{n_i(k-1)+1}, \ldots, x_{n_i k}$ where for each $l > n$, x_l is replaced to 0. C^k is represented by a rectangle of solid lines in Fig. 1. The set of inverters $C^1, C^2, \ldots, C^{n_b}$ forms an inverter for $x_1, x_2, \ldots, x_n$. By Proposition 1, for $1 \le k \le n_b$, the size and the depth of each C^k are as follows.

$$size(C^k) \le c_s n_i \log n_i. \tag{2}$$

$$depth(C^k) \le c_d \log n_i. \tag{3}$$

Next we replace NOT gates in C^k's to several inverters as follows. By Proposition 1, each C^k contains $\lceil \log(n_i+1)\rceil$ NOT gates. Let N^k_j be the jth NOT gate from the input side in C^k for $1 \le j \le \lceil \log(n_i + 1)\rceil$ and $1 \le k \le n_b$. More precisely, we choose each number j so that there is no path from the output of N^k_j to the input of $N^k_{j'}$ if $j' \le j$. For each j, we replace n_b NOT gates $N^1_j, N^2_j, \ldots, N^{n_b}_j$

to one n_b-input inverter C_j'. C_j' is represented by a rectangle of broken lines in Fig. 1. As C_j' we use the inverter I in Lemma 1. The n_b-input inverter I exists, since $n_b = \lceil n/3\lceil \log^{(i)} n \rceil \rceil \geq n/(4\log n)$.

The construction of the inverter I' has been completed. The obtained circuit I' behaves as an inverter since at replacement no loop occurs in the circuit by the following lemma.

Lemma 2. *There is no loop in the circuit I'.*

Proof. We assume that there is a loop in I' and show a contradiction. Let l be a loop in I'. Since all C^k's, $1 \leq k \leq n_b$, before replacement have no loop, l must be through some C_j''s, $1 \leq j \leq \lceil \log(n_i + 1) \rceil$. Let $C_{j_1}', C_{j_2}', \ldots, C_{j_p}'$ be all C_j''s which l is through in the order. After C_{j_p}', l goes to C_{j_1}', since l is a loop. Since there is no path from the output of N_j^k to the input of $N_{j'}^k$ if $j' \leq j$, $j_i < j_{i+1}$ for $1 \leq i \leq p - 1$, and $j_p < j_1$. Thus a contradiction happens. $\square$

In the rest, we confirm that the circuit I' satisfies the condition of the size, the depth and the number of NOT gates. The size, the depth and the number of NOT gates of I' are obtained from the ones of C^k's and C_j''s. We have already shown $size(C^k)$ and $depth(C^k)$ in Eq. (2) and Eq. (3), respectively. We do not have to consider $not(C^k)$ since all NOT gates in C^k's have been replaced to C_j''s. By the condition of I in Lemma 1, for $1 \leq j \leq \lceil \log(n_i + 1) \rceil$, the size, the depth and the number of NOT gates of each C_j' are as follows.

$$size(C_j') \leq 2c_s n_b \log^{(i)} n_b. \tag{4}$$

$$depth(C_j') \leq c_d(\log^{[i-1]} n_b)(\log^{(i)} n_b + 4). \tag{5}$$

$$not(C_j') \leq (\log^{[i-1]} n_b)(\log^{(i)} n_b + 4). \tag{6}$$

By Eq. (2) to Eq. (6), the size, the depth and the number of NOT gates of I' are as follows. In the following, we use the assumption that n is enough large.

$$\begin{aligned}
size(I') &\leq n_b \cdot size(C^k) + \lceil \log(n_i + 1) \rceil \cdot size(C_j') \\
&\leq n_b \cdot c_s n_i \log n_i + \lceil \log(n_i + 1) \rceil \cdot 2c_s n_b \log^{(i)} n_b \\
&\leq c_s n_b(n_i + 2\log^{(i)} n_b)\lceil \log(n_i + 1) \rceil \\
&\leq c_s(\lceil n/3\lceil \log^{(i)} n \rceil \rceil)(3\lceil \log^{(i)} n \rceil + 2\log^{(i)} n)\lceil \log(3\lceil \log^{(i)} n \rceil + 1) \rceil \\
&\leq c_s((n/3\log^{(i)} n) + 1)(5\log^{(i)} n + 3)(\log^{(i+1)} n + 3) \\
&\leq c_s(11/6)n(\log^{(i+1)} n + 3) \\
&\leq 2c_s n \log^{(i+1)} n.
\end{aligned}$$

$$\begin{aligned}
depth(I') &\leq depth(C^k) + \lceil \log(n_i + 1) \rceil \cdot depth(C_j') \\
&\leq c_d \log n_i + \lceil \log(n_i + 1) \rceil \cdot c_d(\log^{[i-1]} n_b)(\log^{(i)} n_b + 4) \\
&\leq c_d \log n_i + c_d(\log^{[i-1]} n)(\log^{(i)} n + 4) \cdot \lceil \log(3\lceil \log^{(i)} n \rceil + 1) \rceil \\
&\leq c_d(\log^{[i-1]} n)(\log^{(i)} n + 4)(\log^{(i+1)} n + 3)
\end{aligned} \tag{7}$$

$$\leq c_d(\log^{[i-1]} n)(\log^{(i)} n)(\log^{(i+1)} n + 4)$$
$$= c_d(\log^{[i]} n)(\log^{(i+1)} n + 4).$$

$$not(I') \leq \lceil \log(n_i + 1)\rceil \cdot not(C'_j)$$
$$\leq \lceil \log(n_i + 1)\rceil \cdot (\log^{[i-1]} n_b)(\log^{(i)} n_b + 4)$$
$$\leq (\log^{[i-1]} n)(\log^{(i)} n + 4) \cdot \lceil \log(3\lceil \log^{(i)} n\rceil + 1)\rceil$$
$$\leq (\log^{[i-1]} n)(\log^{(i)} n + 4)(\log^{(i+1)} n + 3)$$
$$\leq (\log^{[i-1]} n)(\log^{(i)} n)(\log^{(i+1)} n + 4)$$
$$= (\log^{[i]} n)(\log^{(i+1)} n + 4).$$

At Eq. (7), the fact that $C^1, C^2, \ldots, C^{n_b}$ have the same construction is used. $\square$

As proved above by Lemma 2, the constructed circuit I' has no loop, which is guaranteed by replacing NOT gates $N_j^1, N_j^2, \ldots, N_j^{n_b}$ to one inverter C'_j for each j. If we replace NOT gates including two NOT gates $N_{j_1}^k$ and $N_{j_2}^k$ for some k to one inverter, the obtained circuit has a loop.

4   General Circuits and Negation-Limited Circuits

In this section, we give a relation between general circuits and negation-limited circuits for these sizes and discuss two implications of the relation. We denote by $size(f)$ the size of the smallest circuit computing a function f and denote by $size_r(f)$ the size of the smallest circuit with at most r NOT gates computing f.

See Fig. 2. Let C be a general circuit computing an arbitrary Boolean function f. The well-known technique based on DeMorgan's laws can move all NOT gates in the circuit C to the inputs side with at most twice increase of the size. If we use an inverter with r NOT gates to obtain $\neg x_1, \neg x_2, \ldots, \neg x_n$, the overall circuit includes only r NOT gates. Beals, Nishino and Tanaka used the inverter of Proposition 1 in the conversion of circuits and obtained the following corollary:

Corollary 1. *(Beals, Nishino and Tanaka [6]) For every function f,*

$$size_{\lceil \log(n+1)\rceil}(f) \leq 2size(f) + O(n \log n).$$

Our construction gives the following corollary:

Corollary 2. *For every function f,*

$$size_{\log^{1+o(1)} n}(f) \leq 2size(f) + O(n).$$

$O(n \log n)$ in Corollary 1 and $O(n)$ in Corollary 2 are the size of the inverter which is used in each conversion.

Corollary 2 is useful to prove upper bounds on the size for negation-limited circuits computing some functions. Consider sorters for example. A *sorter* is a circuit which sorts the inputs $x_1, x_2, \ldots, x_n$, i.e., outputs $y_1, y_2, \ldots, y_n$ such

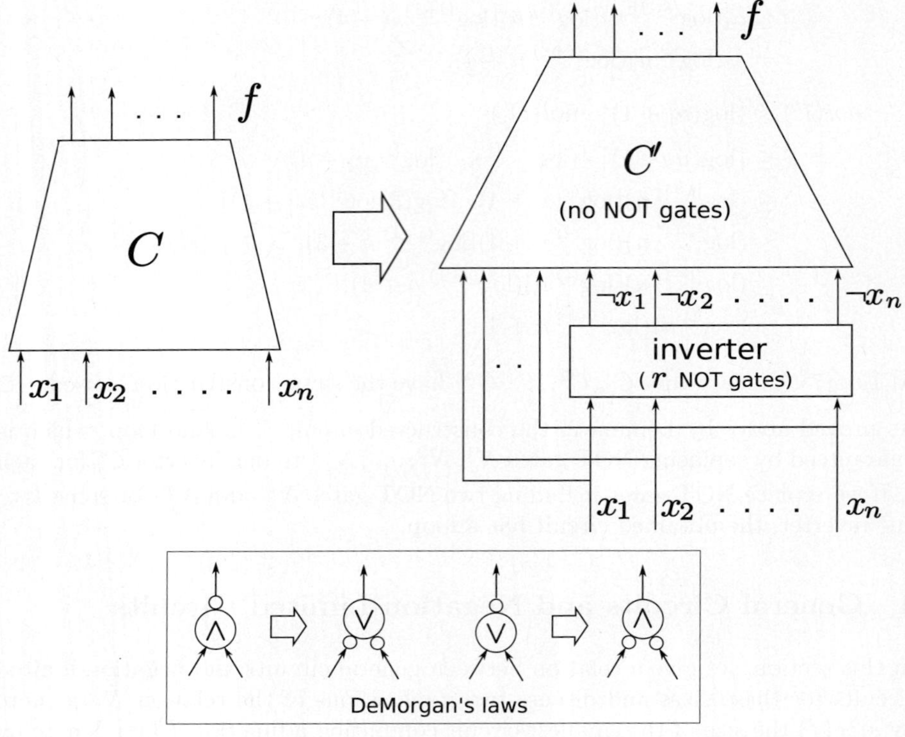

Fig. 2. The conversion from a general circuit to a negation-limited circuit

that $y_1 \geq y_2 \geq \cdots \geq y_n$ and $\Sigma x_i = \Sigma y_i$. Negation-limited sorters or circuits computing the merging function, which can be regarded as restricted sorters, have been studied in several papers [4,14,19]. Although if the number of NOT gates is not limited, then there is a sorter of size $O(n)$ (Theorem 4.1 at p. 76 of [24]), it was not known whether there is a sorter of size $O(n)$ using $o(n)$ NOT gates. Corollary 2 immediately gives the following theorem:

Theorem 2. *There is a sorter of size $O(n)$ using $\log^{1+o(1)} n$ NOT gates.*

Although by Corollary 1 we can prove that there is a sorter of size $O(n \log n)$ using $\lceil \log(n+1) \rceil$ NOT gates, it has been known that a sorter of size $O(n \log n)$ exists even if NOT gates are not allowed to use [1].

Another implication of Corollary 2 is an approach to prove a superlinear size lower bound for general circuits. To prove a superlinear lower bound on the size of general circuits computing an explicit function is one of the most challenging open problems in computational complexity theory. By Corollary 2, when one try to prove a superlinear size lower bound for general circuits, it is enough to prove a superlinear size lower bound for circuits using at most $\log^{1+o(1)} n$ NOT gates. At the case of Corollary 1, an $\omega(n \log n)$ size lower bound for circuits using at most $\lceil \log(n+1) \rceil$ NOT gates is needed. This difference between superlinear and $\omega(n \log n)$ lower bounds may be meaningful, since $\Omega(n \log n)$ lower bounds are known for some n-input n-output functions [11,12,15], although the lower bounds are only for monotone circuits. Even for n-output functions, no superlinear size lower bound for general circuits have been known.

5   Conclusion

In this paper we gave a construction of an inverter of size $O(n)$ and depth $\log^{1+o(1)} n$ using $\log^{1+o(1)} n$ NOT gates. The following question by Turán is known as an open problem [6]: is the size of any inverter with $O(\log n)$ NOT gates and $O(\log n)$ depth superlinear? Our construction implies that one can construct an inverter of linear size if one can use a little more than $O(\log n)$ NOT gates and the depth can be a little bigger than $O(\log n)$.

A natural next step towards a solution of Turán's question is reducing the number of NOT gates to $O(\log n)$ or the depth to $O(\log n)$. Our construction has a recursive structure. Although we may be able to improve the number of NOT gates or the depth to a smaller number than Theorem 1 by a similar recursive structure, it is difficult to reduce it to $O(\log n)$. To achieve $O(\log n)$, another construction is needed.

Acknowledgments

We would like to sincerely thank Professor Takao Nishizeki who read the draft of this paper and gave us valuable comments and suggestions.

References

1. Ajtai, M., Komlós, J., Szemerédi, E.: An $O(n \log n)$ sorting network. In: Proc. of 15th STOC, pp. 1–9 (1983)
2. Alon, N., Boppana, R.B.: The monotone circuit complexity of Boolean functions. Combinatorica 7(1), 1–22 (1987)
3. Amano, K., Maruoka, A.: A superpolynomial lower bound for a circuit computing the clique function with at most $(1/6) \log \log n$ negation gates. SIAM J. Comput. 35(1), 201–216 (2005)
4. Amano, K., Maruoka, A., Tarui, J.: On the negation-limited circuit complexity of merging. Discrete Applied Mathematics 126(1), 3–8 (2003)
5. Andreev, A.E.: On a method for obtaining lower bounds for the complexity of individual monotone functions. Sov. Math. Doklady 31(3), 530–534 (1985)
6. Beals, R., Nishino, T., Tanaka, K.: On the complexity of negation-limited Boolean networks. SIAM J. Comput. 27(5), 1334–1347 (1998)
7. Fischer, M.J.: The complexity of negation-limited networks - a brief survey. In: Brakhage, H. (ed.) GI-Fachtagung 1975. LNCS, vol. 33, pp. 71–82. Springer, Heidelberg (1975)
8. Fischer, M.: Lectures on network complexity, Technical Report 1104, CS Department, Yale University (1974, revised 1996)

9. Iwama, K., Morizumi, H., Tarui, J.: Negation-limited complexity of parity and inverters. In: Asano, T. (ed.) ISAAC 2006. LNCS, vol. 4288, pp. 223–232. Springer, Heidelberg (2006)
10. Jukna, S.: On the minimum number of negations leading to super-polynomial savings. Inf. Process. Lett. 89(2), 71–74 (2004)
11. Lamagna, E.A.: The complexity of monotone networks for certain bilinear forms, routing problems, sorting, and merging. IEEE Trans. Computers 28(10), 773–782 (1979)
12. Lamagna, E.A., Savage, J.E.: Combinational complexity of some monotone functions. In: Proc. 15th Ann. IEEE Symp. on Switching and Automata Theory, pp. 140–144 (1974)
13. Markov, A.A.: On the inversion complexity of a system of functions. J. ACM 5(4), 331–334 (1958)
14. Morizumi, H., Tarui, J.: Linear-size log-depth negation-limited inverter for k-tonic binary sequences. In: Cai, J.-Y., Cooper, S.B., Zhu, H. (eds.) TAMC 2007. LNCS, vol. 4484, pp. 605–615. Springer, Heidelberg (2007)
15. Pippenger, N., Valiant, L.G.: Shifting graphs and their applications. J. ACM 23(3), 423–432 (1976)
16. Razborov, A.A.: Lower bounds on the monotone complexity of some Boolean functions. Sov. Math. Doklady 31, 354–357 (1985)
17. Tanaka, K., Nishino, T.: On the complexity of negation-limited Boolean networks. In: Proc. of 26th STOC, pp. 38–47 (1994)
18. Santha, M., Wilson, C.: Limiting negations in constant depth circuits. SIAM J. Comput. 22(2), 294–302 (1993)
19. Sato, T., Amano, K., Maruoka, A.: On the negation-limited circuit complexity of sorting and inverting k-tonic sequences. In: Chen, D.Z., Lee, D.T. (eds.) COCOON 2006. LNCS, vol. 4112, pp. 104–115. Springer, Heidelberg (2006)
20. Sung, S., Tanaka, K.: Lower bounds on negation-limited inverters. In: Proc. of 2nd DMTCS: Discrete Mathematics and Theoretical Computer Science Conference, pp. 360–368 (1999)
21. Sung, S., Tanaka, K.: An exponential gap with the removal of one negation gate. Inf. Process. Lett. 82(3), 155–157 (2002)
22. Sung, S., Tanaka, K.: Limiting negations in bounded-depth circuits: an extension of Markov's theorem. In: Ibaraki, T., Katoh, N., Ono, H. (eds.) ISAAC 2003. LNCS, vol. 2906, pp. 108–116. Springer, Heidelberg (2003)
23. Tanaka, K., Nishino, T., Beals, R.: Negation-limited circuit complexity of symmetric functions. Inf. Process. Lett. 59(5), 273–279 (1996)
24. Wegener, I.: The Complexity of Boolean Functions. Wiley-Teubner Series in Computer Science (1987)

3-Message NP Arguments in the BPK Model with Optimal Soundness and Zero-Knowledge

Giovanni Di Crescenzo[1] and Helger Lipmaa[2]

[1] Telcordia Technologies, Piscataway, NJ, USA
giovanni@research.telcordia.com
[2] Cybernetica AS, Estonia
http://research.cyber.ee/~lipmaa

Abstract. Under sub-exponential time hardness assumptions, we show that any language in NP has a 3-message argument system in the bare public key (BPK) model, that satisfies resettable zero-knowledge (i.e., it reveals no information to any cheating verifier that can even reset provers) and bounded-resettable soundness (i.e., a verifier cannot be convinced of a false theorem, even if the cheating prover resets the verifier up to a fixed polynomial number of sessions). Our protocol has essentially optimal soundness among 3-message protocols (in that all stronger known soundness notions cannot be achieved with only 3 messages) and zero-knowledge (in that it achieves the strongest known zero-knowledge notion). We also show an extension of this protocol so that it achieves poly-logarithmic communication complexity, although under very strong assumptions.

Keywords: Zero-knowledge arguments, resettable zero-knowledge, resettable soundness, bare public-key model for zero-knowledge protocols.

1   Introduction

Zero knowledge proofs, introduced in the seminal paper [GMR89], have received much attention from the research literature due to their significance and applications to cryptography and computational complexity. Informally, a zero-knowledge proof is a method for a prover to convince a verifier that a theorem of the type "$x \in L$", x being a string and L being a language, is true without revealing any additional information. Zero-knowledge proofs are interactive protocols, with three requirements: *completeness*, typically easy to satisfy, saying that if prover and verifier follow the protocol, then the verifier accepts with very high probability; *soundness*, saying that a (cheating) prover not following the protocol can make the verifier accept only with very small probability; *zero-knowledge*, saying that a (cheating) verifier not following the protocol cannot obtain any information about a true statement. Since their introduction, the following increasingly stronger variants of both soundness and zero-knowledge requirements have been studied: *one-time*, meaning that the protocol remains sound/zero-knowledge if a given theorem is proved at most once within a sequence of protocol executions; *sequential*, where the protocol remains sound/zero-knowledge even if a sequence of protocol executions are performed; *concurrent*, where the protocol remains sound/zero-knowledge even if the cheating prover/verifier can force an arbitrary concurrent scheduling of multiple protocol executions; *resettable*, where the protocol remains

S.-H. Hong, H. Nagamochi, and T. Fukunaga (Eds.): ISAAC 2008, LNCS 5369, pp. 615–627, 2008.

sound/zero-knowledge even if the cheating prover/verifier can perform the previous concurrent attack and additionally arbitrarily reset the honest verifier/prover.

In the "standard" interactive protocol model (i.e., without additional setup infrastructures or network assumptions), concurrent (and thus, resettable) zero-knowledge arguments for non-trivial languages with black-box simulation require a super-constant number of messages (see, e.g. [CKPR01]). On the other hand, only a few zero-knowledge arguments with non-black-box simulation exist (see, e.g. [Bar01]) and the current best protocol requires 5 messages and is only bounded-resettable zero-knowledge. Designing 3-message zero-knowledge protocols with black-box simulation is only possible for BPP languages [GK90] and is still open in the case of non-black-box simulation. Given this state of the art in the standard model, other models are being studied to achieve constant-message resettable or concurrent zero-knowledge protocols for all NP languages. Among these, the model that seems to have the minimal set-up or network assumptions is the bare public-key (BPK) model [CGGM00], where verifiers register their public key in a public file during a set-up stage, and there is no interactive preprocessing stage, trusted third party, trusted string, or assumption on the asynchronicity of the network. In this model, a 3-message, one-time soundness, resettable zero-knowledge argument system for all NP languages was given in [MR01a]. This was improved in [ZDLZ03], whose main protocol additionally satisfies a bounded version of concurrent soundness. Recently, [APV05] proved limitations on further improving the soundess of such protocols, by showing that sequential soundness in 3 messages can only be achieved for languages in BPP. Other results in the BPK model mainly focus on $\geq$ 4-message protocols (see, e.g., [CGGM00, BGG+01, DPV04a, DV05]) or 3-message protocols in more stringent versions of the BPK model (see, e.g., [MR01b, DPV04b]).

Our results and comparison with previous work. We study argument systems for NP languages in the BPK model with resettable attacks from provers and verifiers. We define a new and natural notion of *bounded-resettable soundness*, which means, informally, soundness against provers that can reset verifiers and force them to continue using the same randomness for any *fixed* polynomial number of times (as opposed to an arbitrary polynomial, as in resettable soundness).

Our main result is a 3-message argument system for any NP language, that satisfies resettable zero-knowledge and bounded-resettable soundness, under subexponential-time hardness assumptions. Our argument is black-box zero-knowledge and therefore message-optimal unless NP is in BPP due to a lower bound from [MR01a]. Extending results from [MR01a], we obtain that bounded-resettable soundness is strictly stronger than bounded-concurrent soudness. Then we also show that bounded-resettable soundness is strictly weaker than sequential soundness. Thus, our protocol bridges the gap between the two previous best results for 3-message resettable zero-knowledge arguments in the BPK model: the 3-message bounded-concurrent soundness, resettable zero-knowledge argument system from [ZDLZ03] (of which our protocol is also conceptually much simpler) and the impossibility for 3-message sequential soundness and resettable zero-knowledge argument systems for languages not in BPP [APV05]. (See also Figure 1 for more details.) Thus, our protocol has essentially optimal soundness (in that the next stronger soundness notion, sequential soundness, cannot be achieved in 3 messages) and optimal zero-knowledge (in that resettable zero-knowledge is the currently strongest zero-knowledge notion known). We also show an application of this protocol to constructing a 3-message, polylogarithmic communication complexity argument system for any NP language with bounded-resettable soundness and resettable zero-knowledge, under additional assumptions (including very strong assumptions, called extractability assumptions, for which the question of determining whether they are false or reasonable is still unsettled.) This is obtained by combining our main protocol with a 2-message protocol in [DL08]. Many formal proofs are omitted due to the space limit.

Paper	Model	Number of messages	Type of Soundness	Type of bounded Soundness	Resettable Zero-Knowledge
[MR01a]	BPK	3	One-time	One-time	Resettable
[MR01b]	UPK	3	Concurrent	Concurrent	Resettable
[ZDLZ03]	WPK	3	Concurrent	Concurrent	Resettable
[DPV04b]	cBPK	3	Concurrent	Concurrent	Resettable
[ZDLZ03]	BPK	3	One-Time	Concurrent	Resettable
this paper	BPK	3	One-Time	Resettable	Resettable
impossible	BPK	3	Sequential	Sequential	Resettable
impossible	any	any	Resettable		

Fig. 1. Our result and comparison with previous 3-message NP arguments and impossibility results. Note the Upperbounded-BPK (UPK), Weak-BPK (WPK), Counter-BPK (cBPK), models make further and stronger assumptions over the BPK model.

2   Definitions

We recall known definitions of the BPK model, and the notions of soundness and zero-knowledge used in the paper, including a novel definition for bounded-resettable soundness in the BPK model. (In the process, we use a simpler set of notations and model both provers and verifiers' resetting attacks as appropriate oracle queries.) Finally, we recall three main cryptographic tools that we will use in our constructions. (We only assume familiarity with pseudo-random function families [GGM86].)

Model description. The BPK model can be seen as a relaxed version of two previously considered models in Cryptography: the Public-Key Infrastructure model, and the Preprocessing model. One main difference with the Preprocessing model is that in the BPK model the preprocessing phase is reduced to users non-interactively posting public keys on a public file, rather than a potentially long interaction between prover and verifier. One main difference with the Public-Key Infrastructure model is that in the BPK model only verifiers' public keys are used in protocols. Formally, the BPK model makes the following assumptions. (1) There exists a public file F that is a collection of records, each containing a public key. (2) An (honest) prover is an interactive deterministic polynomial-time Turing machine that takes as input a security parameter 1^n, F, an n-bit string x, such that $x \in L$, for some language L, an auxiliary input y, a reference to an entry of F and a random tape. (3) An (honest) verifier V is an interactive deterministic polynomial-time Turing machine that works in the following two stages: on input a security parameter 1^n and a random tape, V generates a key pair (pk, sk) and stores the public key pk in one entry of the file F; later, V takes as input the secret key sk, a statement $x \in L$ and a random string, and outputs "accept" or "reject" after performing an interactive protocol with a prover. (4) The first interaction between a prover and a verifier starts after all verifiers have completed their first stage.

Malicious provers in the BPK model. Let p be a positive polynomial and P^* be a probabilistic polynomial-time algorithm that, given as input security parameter 1^n, and public key pk of V, may perform $\leq p(n)$ interactive protocols with V.

P^* is a *p-resetting* (malicious) prover if it makes queries to oracle V, where the number of queries is at most $p(n)$, and queries and replies are defined as follows. Each query is a triple (i, x, tr), where tr may be a prefix of the transcript so far for a conversation between P^* and V on input x and random tape r_i, followed by a (unrestricted) message from P^*. An answer to this query is V's message on input r_i as a random tape, the n-bit language instance x common to P^* and V; and transcript so far tr. (Note that according to this definition P^* can perform a reset action by asking a query q' to V containing the same r_i, x as in a previous query q, and a trancript that is a prefix of the transcript in q, followed by a freshly-computed message from P^*. Also, note that we are parameterizing as p the number of queries by P^* rather than the number of resetted sessions, however note that this formalization suffices as the number of sessions is at most the number of queries by P^*.)

Given a p-resetting malicious prover P^* and an honest verifier V, a *p-resetting attack* is performed as follows: 1) the first stage of V is run on input 1^n and a random string, so that a pair (pk, sk) is obtained; 2) $p(n)$ independently distributed random tapes $r_1, \ldots, r_{p(n)}$ are generated, and P^* is run on input 1^n and pk, making at most $p(n)$ queries to oracle V, each query being of the type $(i, \cdot, \cdot)$, for some $i \in \{1, \ldots, p(n)\}$, meaning that random tape r_i should be used by V in answering the query.

Definition 1. Let L be a language. A pair (P, V) satisfies *completeness* over L in the BPK model if for all $x \in L$, the probability that the following happens is negligible: 1) the first stage of V is run on input 1^n and a random string so that a pair (pk, sk) is obtained; 2) P is run on input 1^n, pk, and an n-bit string x, and interacts with V that takes as input pk, sk, x. 3) At the end of the interaction, V outputs: "reject". This definition naturally extends to $\{(P, V)_p\}$, i.e., when an additional parameter p is shared by P and V.

Definition 2. Let p be a polynomial. A pair $(P, V)_p$ satisfies the *p-resettable soundness* in the BPK model, if for all p-resetting malicious provers P^*, for any false statement "$x \in L$" the probability that in an execution of a p-resetting attack V outputs "accept" for such a statement is negligible in n. We say that $\{(P, V)_p\}$ satisfies *bounded-resettable soundness* if for any polynomial p (called the *bounded-resettability parameter*), $(P, V)_p$ satisfies *p-resettable soundness*.

Malicious verifiers in the BPK model. We say that V^* is a *p-resetting* (malicious) verifier if it acts as follows. In the key-generation stage, it posts (at least) one public key pk in the public file. In the proof stage, it can interact by making at most $p(n)$ queries to oracle P, where queries and replies are defined as follows. First, let $r_1, \ldots, r_{p(n)}$ be independent random tapes for P. Then, each query can be written as a triple (i, x, tr), where r_i, for some $i \in \{1, \ldots, p(n)\}$, is intended to be P's randomness; x is intended to be the n-bit common input to P and V^*; and tr is intended to be the transcript so far for a conversation between P and V^* on input x. To this query, P replies with an answer, to be intended as a continuation of transcript tr. (Note that according to this definition V^* can perform a reset action by asking a query q' to P containing the same i, x as in a previous query q, and a trancript that is a prefix of the transcript in q followed by a freshly-computed message from V^*.)

Definition 3. We say that (P, V) is *resettably-zero-knowledge* over L if for all positive polynomials p, for any p-resetting verifier $V^\star$, there exists a probabilistic polynomial-time algorithm $S_{V^\star}$, called the *simulator*, such that for all distinct $x_1, \ldots, x_{p(n)} \in L$, the probability distributions $\{\text{view}_{V^\star}^P(\bar{x})\}$ and $\{S_{V^\star}(\bar{x})\}$ are computationally indistinguishable, where $\{\text{view}_{V^\star}^P(\bar{x})\}$ is the distribution of the transcript seen by $V^\star$ on its input tape (i.e., $\bar{x} = x_1, \ldots, x_{p(n)}$), random tape and communication tape during its interaction with P. This definition naturally extends to $\{(P, V)_p\}$, i.e., when an additional parameter p is shared by P and V.

Cryptographic Tools. We review some cryptographic primitives that are used in the rest of the paper: digital signature schemes secure against subexponential-time adversaries, commitment schemes secure against subexponential-time adversaries and extractable by subexponential-time algorithms, and 2-message public-coin witness-indistinguishable proof systems (or, their slightly stronger version, called ZAPs).

Definition 4. Let k be a security parameter, and let σ be a constant such that $0 < \sigma < 1$. A σ-*secure digital signature scheme* SS is a triple of (probabilistic) polynomial-time algorithms SS $= $ (G, Sig, Ver) satisfying: (*Correctness*) For all messages $m \in \{0,1\}^k$, $\Pr[\,(\text{pk}, \text{sk}) \leftarrow \text{G}(1^k); \hat{m} \leftarrow \text{Sig}(m, \text{pk}, \text{sk}) : \text{Ver}(m, \hat{m}, \text{pk}) = 1\,] = 1$; (*Unforgeability*) For all algorithms A running in time $o(2^{k^\sigma})$, it holds that $\Pr[\,(\text{pk}, \text{sk}) \leftarrow \text{G}(1^k); (m, \hat{m}) \leftarrow A^{\mathcal{O}(\text{pk},\text{sk})}(\text{pk}) : m \notin \text{Query and } \text{Ver}(m, \hat{m}, \text{pk}) = 1\,]$ is negligible in k where $\mathcal{O}(\text{pk}, \text{sk})$ is a *signature oracle* that on input a message returns as output a signature of the message and Query is the set of messages for which A has requested a signature from $\mathcal{O}$.

Signature schemes secure against polynomial-time adversaries exist under the assumption of the existence of one-way functions secure against polynomial-time adversaries [R90, NY91]. We note that the same reduction applies if one considers both primitives with respect to subexponential-time adversaries.

Definition 5. Let k be a security parameter, and let α, be a constant such that $0 < \alpha < 1$. An α-*extractable commitment scheme* is a pair of (probabilistic) polynomial-time algorithms (Com, Rec) satisfying: (*Correctness*) For all $b \in \{0, 1\}$ and for all k, $\Pr[\,(\text{com}, \text{dec}) \leftarrow \text{Com}(b, 1^k) : \text{Rec}(\text{com}, \text{dec}, b) = 1\,] = 1$; (*Perfect Binding*) For all k, and any string com, there exists at most one bit $b \in \{0, 1\}$ such that $\text{Rec}(\text{com}, \text{dec}_b, b) = 1$, for some string dec_b; (*Computational Hiding*) The distributions $\{[(\text{com}, \text{dec}) \leftarrow \text{Com}(0, 1^k) : \text{com}]\}_{k>0}$ and $\{[(\text{com}, \text{dec}) \leftarrow \text{Com}(1, 1^k) : \text{com}]\}_{k>0}$

are indistinguishable from algorithms running in polynomial time; (*Extractability*) There exists an extractor algorithm E running in time 2^{k^α} such that, for all commitments com computed by a probabilistic polynomial-time committer adversary A, if A succeeds in decommitting com as b with non-negligible probability, then the probability that $E(\text{com}) \neq b$ is negligible.

The above definitions can be easily extended to the case in which we wish to commit to a string (instead of committing to a bit). Such a commitment scheme exists, for instance under the assumption that there exist permutations that are one-way with respect to polynomial-time adversaries but such that they can be inverted in subexponential time. We note that this type of schemes were already used in [Pas03] to achieve straight-line extractability in superpolynomial time.

Finally, we review the notion of a ZAP, introduced in [DN00]. Informally, a ZAP is a 2-message public-coin witness-indistinguishable proof system where the soundness and witness-indistinguishable requirements hold even if a polynomial number of proofs are given using the same message from the verifier, and the n-bit statements to be proved are chosen after seeing the verifier's message.

Definition 6. Let L be a language in NP. A triple of polynomial-time algorithms (ZG, ZP, ZV) is a *ZAP* for L iff it satisfies: (*Completeness*) Given a witness y for "$x \in L$" and $z = ZG(1^k, L)$ then $ZV(x, z, ZP(x, y, z)) = 1$ with probability 1; (*Soundness*) With overwhelming probability over choice of $z = ZG(1^k, L)$, there exists no $x' \notin L$ no string π' such that $ZV(x', z, \pi') = 1$; (*Witness-Indistinguishability*) Let y_0, y_1 be witnesses for "$x \in L$. Then, for each z, the distribution $\{ZP(x, y_0, z)\}$ and $\{ZP(x, y_1, z)\}$ are indistinguishable by non-uniform probabilistic polynomial time algorithms.

In [DN00] the existence of a ZAP for an NP-complete language is proved to be equivalent to the existence of non-interactive zero-knowledge proofs for an NP-complete language; thus, the existence of ZAPs is implied by the existence of one-way trapdoor permutations or verifiable random functions. An important property of ZAPs, noted in [BGG+01], is that they can be easily modified so to satisfy resettable soundness and resettable witness-indistinguishability.

3   Bounded-Resettable Soundness vs. Sequential Soundness

We start our analysis by studying how the bounded-resettable soundness notion compares with other known or natural types of soundness notions in the BPK model. Recall that in [MR01a] it was proved that resettable soundness strictly implies concurrent soundness, which, in turn, strictly implies sequential soundness. It is not hard to verify that the same techniques can be used to prove that bounded-resettable soundness strictly implies bounded concurrent soundness, which, in turn, strictly bounded sequential soundness. More interestingly, we observe that bounded-resettable soundness does not imply sequential soundness. Specifically, we obtain the following

Theorem 1. Let L be a language and p be a polynomial. If there exists an argument $(P, V)_p$ for L satisfying p-resettable soundness, and there exists a digital signature scheme secure against chosen message attack, then there exists an argument $(P', V')_p$ for L satisfying p-resettable sondness that does not satisfy sequential soundness. Moreover, if $(P, V)_p$ has a constant number of messages, then the same holds for $(P', V')_p$.

We note that a much weaker fact (specifically, that bounded sequential soundness does not imply sequential soundness) was proved in [ZDLZ03] assuming the existence of families of pseudo-random functions.

Remarks. This theorem clarifies that there is a gap between possibility and impossibility results for 3-message resettable zero-knowledge arguments for NP in the BPK model. Specifically, the notion of bounded-resettable soundness is strictly stronger than the notion of bounded-concurrent soundness achieved by the main protocol in [ZDLZ03], as observed above, and, as a consequence of Theorem 1, strictly weaker than sequential soundness, which is impossible to achieve in 3 messages resettable zero-knowledge arguments for non-trivial languages in the BPK model [APV05]. Thus, a natural question (which we solve in the next section) is the existence, or not, of 3-message resettable zero-knowledge arguments for NP languages with bounded-resettable soundness.

4   Bounded-Resettable Soundness in 3 Messages

In this section we present a 3-message argument for any NP language in the public-key model that satisfies resettable zero-knowledge and bounded-resettable soundness, under subexponential-time hardness assumptions. Formally, we obtain the following

Theorem 2. Let L be a language in NP. Assuming the existence of α-extractable commitment schemes, σ-secure signature schemes, for $0 < \alpha < \sigma < 1$, and ZAPs for NP-complete languages, there exists a 3-message argument system for L that satisfies completeness, bounded-resettable soundness, and resettable zero-knowledge.

The above result improves the best previous 3-message results in this model (i.e., [MR01a, ZDLZ03]) by the (stronger) level of soundness achieved, and in the case of [DPV04b, MR01b] by the generality of the model. We also note two additional interesting properties of the above argument system: the length of the public key does not depend on the bounded-resettability parameter p, and only the last message depends on the statement being proved, so that it can be used to actually prove m on-line and independent theorems in $m + 2$ messages, for $m \geq 1$ (instead of $2m + 1$, as for protocols in [ZDLZ03, MR01a]). We now proceed with the proof of Theorem 2.

Informal description. Our argument system is based on the very often used 'OR-based paradigm' for zero knowledge, first introduced by [FLS99] in the non-interactive model for zero-knowledge proofs. According to the natural application of this paradigm in interactive models, the prover proves to the verifier that either the original statement is true or some other statement, obtained from the transcript τ of the communication so far, is true. Specifically, the prover creates a certain NP statement st_τ having the following two properties: 1) if τ is generated by honest prover and verifier, then with high probability statement st_τ is false; 2) for any probabilistic polynomial-time verifier, there exists an efficient simulator that generates a transcript τ which is computationally indistinguishable from the analogue transcript generated during a real execution of the protocol between prover and this verifier, and such that st_τ is true. Then, formally, in order to prove the NP statement input '$x \in L$', a prover will rather prove the statement '$(x \in L) \vee st_\tau$'. We now define the auxiliary language T_L that we are going to use.

Definition 7. *Let $q = p(n) + 1$. The triple (x, com, pk) belongs to the language T_L if $x \in L$ or there exist $m_1, \ldots, m_q, sig_1, \ldots, sig_q, dec, r_1, r_2, \mathrm{sk}$ such that*

1. *$(\mathrm{pk}, \mathrm{sk}) = \mathrm{G}(1^k, r_1)$;*
2. *$m_i \neq m_j$, for all distinct $i, j \in \{1, \ldots, q\}$;*
3. *$\mathrm{Ver}(m_i, sig_i, \mathrm{pk}) = 1$ for $i = 1, \ldots, q$, and*
4. *$(com, dec) = \mathrm{Com}((m_1, \ldots, m_q, sig_1, \ldots, sig_q), 1^k, r_2)$.*

Informally speaking, triple (x, com, pk) belongs to T_L if x belongs to L or if com is the commitment of $p(n) + 1$ distinct n-bit strings and $p(n) + 1$ valid signatures sig_i (with respect to pk) for these strings.

Tools and assumptions. In our construction we assume the existence of the following cryptographic tools.

1. a σ-secure digital signature scheme $\mathrm{SS} = (\mathrm{G}, \mathrm{Sig}, \mathrm{Ver})$, for $0 < \sigma < 1$;
2. an hiding, perfectly-binding and α-extractable commitment scheme $(\mathrm{Com}, \mathrm{Rec})$, where $0 < \alpha < \sigma$;
3. a ZAP (ZG, ZV, ZP) for the language T_L;
4. a pseudo-random family of functions $\mathcal{F}$ (this follows from the existence of item 1).

High-level overview. Let k be the security parameter. The public entry of a verifier contains a public key pk for the σ-secure signature scheme. The actual proof that $x \in L$ consists of a first message where the prover sends a random message m to the verifier. The verifier replies with a signature sig of m and the first message z_1 of a ZAP for language T_L. The prover continues iff sig is a valid signature of m, and it constructs a commitment com to $2p + 2$ strings $m_i = sig_i = 0^n$, for $i = 1, \ldots, p$, using the perfectly-binding and α-extractable commitment scheme. Finally the prover computes the second message x_2 of the ZAP in which she proves that the triple (x, com, pk) belongs to T_L. All three messages by prover or verifier are computed using pseudo-randomness generated as follows: first, the prover (or verifier) picks a fixed-length random seed from its random tape, and then it generates a long pseudo-random string by running the pseudo-random function on input the seed, the input common to prover and verifier, any secret inputs, and the transcript of the conversation so far. This generates sufficiently long randomness to be used in the rest of each step. Finally, the verifier accepts if the second message of the ZAP is accepting.

Let us now informally argue the properties of our construction. We start with the bounded-resettable soundness requirement. Since we assume that $x \notin L$, by the soundness of the ZAP used, the only way that a prover can make V accept with non-negligible probability is if $(x, com, \mathrm{pk}) \in T_L$ because of the fact that com is a commitment to $p+1$ distinct messages and $p + 1$ valid signatures of them, according to the signature public key pk. Since the prover uses an α-extractable commitment scheme to compute commitment key com, there exists an extractor running in time $O(2^{n^\alpha})$ that can compute the messages and signatures committed through com. We then note that any prover that resets at most $p(n)$ times the verifier, is sent from the verifier signatures for at most $p(n)$ distinct messages. Therefore, the extractor and the prover can be composed to obtain an algorithm that runs in time $O(2^{n^\alpha})$ and breaks the σ-secure signature scheme. Note that we base the proof on a subexponential-time hardness assumption. Such a telescopic use of the hardness of different cryptographic assumptions is referred to as *complexity leveraging* and is often being used in proofs of the soundness requirements for proposed resettable zero-knowledge arguments in the public-key model. The use of an *extractable* commitment along with a ZAP is also discussed and used in [Pas03].

We now consider the resettable zero knowledge requirement. We first note that the prover uses, as randomness to compute its messages, pseudo-random bits computed as function of the common and secret inputs and the messages received until then. Then any reset operations from the verifier will result in either: (1) a new session of the same protocol being replayed with identical transcript and the prover reusing the same randomness; or (2) a new session of the same protocol being replayed with identical first message from the prover but different remaining part of the transcript. While this latter case may give problems in obtaining the zero-knowledge property for protocols of the 'cut-and-choose' type[1], it does not in our case, as we use a different type of protocol: a 'non-committing' first message from the prover followed by a ZAP. A problem could still arise if the verifier happens to see correlated prover's answer on the same first message for the ZAP, but this does not happen as the prover computes its second message by using the output of the pseudo-random function on a new input (in fact, a new input and a new seed) as a random tape, which will be indistinguishable from a new random string to the verifier, and we can then use the witness-indistinguishability properties of ZAPs. Then, the strategy of the simulator is that of rewinding verifier V^* so to obtain $p(n) + 1$ signatures for distinct messages sent by the simulator, while playing as the prover P. Here, we also avoid the subtle simulation problem solved in [GK90] due to the cheating verifier V^* aborting with some non-zero probability. We can do that because the first message has the same distribution before and after the simulator rewinds verifier V^*, and thus the expected number of rewindings remains polynomial. After $p(n) + 1$ signatures are obtained, the simulator can run the prover's program as it has a witness for the statement $(x, com, \mathrm{pk}) \in T_L$. A formal description of this protocol (based on parameter p) is given on Fig. 2.

5   Application to Polylogarithmic-Communication Arguments

In this section we consider the problem of reducing the communication complexity of the 3-message protocol from Section 4. An intriguing question is whether there exists a protocol with the same soundness and zero-knowledge properties, that further has communication complexity polylogarithmic in the input length.

---

[1] By 'cut-and-choose' protocol, we denote here all protocols where the prover sends a commitment, the verifier reveals a challenge and the prover sends an answer Most zero-knowledge protocols in the literature are of this type or contain a subprotocol of this type.

KEY GENERATION PHASE.

V: On input a security parameter k, run the key generator G for the σ-secure signature scheme on input 1^k obtaining the pair (spk, ssk). Let pk $= spk$ and sk $= ssk$.

PROOF PHASE.

Common input: security parameter k, public file $F = \{(\text{pk}, \ldots)\}$, instance x.
P's private input: a witness y for $x \in L$.
V's private input: private key sk.

P(message 1):
  1. randomly choose seed s;
  2. compute $w = \mathcal{F}(s, x|y|\text{pk})$ and use w below as randomness;
  3. randomly choose $m \leftarrow \{0,1\}^n$ and send it to V;

V(message 2):
  1. randomly choose seed s';
  2. compute $w' = \mathcal{F}(s', x|y|\text{pk}|m)$ and use w' below as randomness;
  3. compute $sig = \text{Sig}(m, \text{sk})$ and $z_1 = ZG(1^k; T_L)$;
  4. send sig, z_1 to P;

P(message 3):
  1. if $\text{Ver}(m, sig, \text{pk}) \neq 1$ then halt;
  2. randomly choose seed s'';
  3. compute $w'' = \mathcal{F}(s'', x|y|\text{pk}|m|sig|z_1)$ and use w'' below as randomness;
  4. let m' be the $(p(n)+1)$-tuple $(0^k, \ldots, 0^k)$ and let $sig' = m'$;
  5. compute commitment and decommitment keys $(com, dec) = \text{Com}(m'|sig', 1^k)$ and $z_2 = ZP(x, com, \text{pk}; y; z_1)$;
  6. send com, z_2 to V;

V(decision):
  1. verify that the ZAP is valid by checking that ZV on input instance (x, com, pk) and transcript (z_1, z_2), returns 1.

Fig. 2. The 3-message bounded resettably sound and resettable zero-knowledge argument system for $\mathcal{NP}$ in the BPK model

The previous result coming closer to solving this problem is the polylogarithmic-communication zero-knowledge argument system from [K91]. We could think of two variations of this protocol that can be implemented so to require only 4-messages in the standard model and can even be compressed to 3 messages in the BPK model. However, both resulting protocols do not even satisfy 2-resettable soundness, as a cheating prover can reset the honest verifier after learning the verifier's challenge and then prepare a first message for which he can answer that challenge. We then try a different approach, based on a polylogarithmic-communication 2-message argument system. By combining such an argument system with the protocol from Section 4, we obtain.

Theorem 3. Let L be a language in NP. There exists a 3-message argument system for L that satisfies completeness, bounded-resettable soundness, resettable zero-knowledge, and has communication complexity polylogarithmic in the input length, if the following assumptions hold: the existence of α-extractable commitment schemes, of σ-secure signature schemes, for $0 < \alpha < \sigma < 1$, and of 2-message argument systems for an NP-complete language with communication complexity polylogarithmic in the input length.

We stress that currently the only way to instantiate the latter assumption in the above result is to use the argument system of [DL08]. Their argument system uses several standard assumptions (like the existence of collision-intractable hash function families and polylogarithmic-communication computationally private information retrieval schemes) but also one very strong knowledge-based assumption called an extractable-algorithm assumption in [DL08]. As already observed in [DL08] as well as previous papers using similar assumptions, determining whether extractable-algorithm assumptions are false or reasonable remains an interesting question. We now proceed with the proof of Theorem 3.

An informal description. Our argument system is obtained by modifying the protocol in Section 4 as follows: first, using the input length n as the protocol's security parameter, each used cryptographic primitive is deployed with security parameter polylogarithmic in n; second, instead of requiring the prover to send (the long) ZAP answer z_2 and verifier to check it, prover and verifier apply the polylogarithmic-communication 2-message argument system, denoted as (A, B), to prove/verify that such z_2 exists (here, any randomness needed by algorithm A is still drawn from w''). We formally define language $metaT_L$ as follows: the 4-tuple $(x, com, \mathrm{pk}, z_1)$ belongs to the language $metaT_L$ if $x \in L$ or there exists z_2 such that $ZV(x', z_1, z_2) = 1$, where $x' = (x, com, \mathrm{pk})$. Then, instead of proving that $(x, com, \mathrm{pk}) \in T_L$, the prover proves that $(x, com, \mathrm{pk}, z_1) \in metaT_L$. Note that $metaT_L \in$ NP.

Let us now informally argue the properties of our construction. The polylogarithmic communication complexity follows from the analogous property of protocol (A,B), when $z_i = \emptyset$ (i.e., using a 1-message ZAP [GOS06]), and by inspection of the rest of the protocol in Fig. 2, where each primitive is used with security parameter polylogarithmic in n. For the bounded-resettable soundness property, assume that $x \notin L$ and the verifier accepts. The messages obtained during a bounded-resettable attack do not help the attacker to violate the soundness of (A, B) and thus, by the soundness of (A, B), we can extract z_2 that makes the verifier accept in a protocol almost identical to the protocol in Fig. 2 (the only addition being that the verifier further sends B's message). As the latter also satisfies bounded-resettable soundness, we reach the desired contradiction. Finally, the proof of the resettable zero knowledge requirement is directly obtained as an extension of the same proof for the protocol in Fig. 2. Specifically, we use the same simulator's strategy of rewinding verifier V^* so to obtain $p(n)+1$ signatures for distinct messages, which gives the simulator a witness for statement $(x, com, \mathrm{pk}) \in T_L$, and, using ZP, a witness for statement $(x, com, \mathrm{pk}, z_1) \in metaT_L$. Thus the simulator can run prover A's algorithm to successfully complete the simulation of message z_2.

Acknowledgement. The second author was supported by Estonian Science Foundation, grant #6848, European Union through the European Regional Development Fund and the 6th Framework Programme project AEOLUS (FP6-IST-15964).

References

[APV05]   Alwen, J., Persiano, G., Visconti, I.: Impossibility and Feasibility Results for Zero Knowledge with Public Keys. In: Shoup, V. (ed.) CRYPTO 2005. LNCS, vol. 3621. Springer, Heidelberg (2005)

[Bar01]   Barak, B.: How to Go Beyond the Black-Box Simulation Barrier. In: Proc. of IEEE FOCS 2001 (2001)

[BGG+01]   Barak, B., Goldreich, O., Goldwasser, S., Lindell, Y.: Resettably-Sound Zero-Knowledge and its Applications. In: Proc. of IEEE FOCS 2001 (2001)

[BDMP91]   Blum, M., De Santis, A., Micali, S., Persiano, G.: Non-Interactive Zero-Knowledge. SIAM Journal on Computing 20 (1991)

[CGGM00]   Canetti, R., Goldreich, O., Goldwasser, S., Micali, S.: Resettable Zero-Knowledge. In: Proc. of ACM STOC 2000 (2000)

[CKPR01]   Canetti, R., Kilian, J., Petrank, E., Rosen, A.: Black-Box Concurrent Zero-Knowledge Requires $\omega(\log n)$ Rounds. In: Proc. of ACM STOC 2001 (2001)

[DL08]   Di Crescenzo, G., Lipmaa, H.: Succinct NP Proofs from an Extractability Assumption. In: Beckmann, A., Dimitracopoulos, C., Löwe, B. (eds.) CiE 2008. LNCS, vol. 5028. Springer, Heidelberg (2008)

[DPV04a]   Di Crescenzo, G., Persiano, G., Visconti, I.: Constant-Round Resettable Zero Knowledge with Concurrent Soundness in the Bare Public-Key Model. In: Franklin, M. (ed.) CRYPTO 2004. LNCS, vol. 3152. Springer, Heidelberg (2004)

[DPV04b]   Di Crescenzo, G., Persiano, G., Visconti, I.: Improved Setup Assumptions for 3-Round Resettable Zero Knowledge. In: Lee, P.J. (ed.) ASIACRYPT 2004. LNCS, vol. 3329, pp. 530–544. Springer, Heidelberg (2004)

[DV05]   Di Crescenzo, G., Visconti, I.: Concurrent Zero-Knowledge in the Public-Key Model. In: Caires, L., Italiano, G.F., Monteiro, L., Palamidessi, C., Yung, M. (eds.) ICALP 2005. LNCS, vol. 3580. Springer, Heidelberg (2005)

[DN00]   Dwork, C., Naor, M.: Zaps and their applications. In: Proc. of IEEE FOCS 2000 (2000)

[FLS99]   Feige, U., Lapidot, D., Shamir, A.: Multiple Non-Interactive Zero Knowledge Proofs Under General Assumptions. SIAM J. on Computing 29 (1999)

[GGM86]   Goldreich, O., Goldwasser, S., Micali, S.: How to Construct Random Functions. Journal of the ACM 33 (1986)

[GK90]   Goldreich, O., Krawczyk, H.: On the Composition of Zero-Knowledge Proof Systems. In: Proc. of ICALP 1990 (1990)

[GMR89]   Goldwasser, S., Micali, S., Rackoff, C.: The Knowledge Complexity of Interactive Proof-Systems. SIAM J. on Computing 18 (1989)

[GOS06]   Groth, J., Ostrovsky, R., Sahai, A.: Non-Interactive Zaps and New Techniques for NIZK. In: Dwork, C. (ed.) CRYPTO 2006. LNCS, vol. 4117, pp. 97–111. Springer, Heidelberg (2006)

[K91]   Kilian, J.: A Note on Efficient Zero-Knowledge Proofs and Arguments. In: Proc. of ACM STOC 1991 (1991)

[NY91]   Naor, M., Yung, M.: Universal One-Way Hash Functions and Applications. In: Proc. of ACM STOC 1991 (1991)

[RK99]   Richardson, R., Kilian, J.: On the Concurrent Composition of Zero-Knowledge Proofs. In: Stern, J. (ed.) EUROCRYPT 1999. LNCS, vol. 1592. Springer, Heidelberg (1999)

[MR01a]   Micali, S., Reyzin, L.: Soundness in the Public-Key Model. In: Kilian, J. (ed.) CRYPTO 2001. LNCS, vol. 2139. Springer, Heidelberg (2001)

[MR01b]   Micali, S., Reyzin, L.: Min-Round Resettable Zero-Knowledge in the Public-key Model. In: Pfitzmann, B. (ed.) EUROCRYPT 2001. LNCS, vol. 2045. Springer, Heidelberg (2001)

[Pas03]   Pass, R.: Simulation in Quasi-Polynomial Time and Its Applications to Protocol Composition. In: Biham, E. (ed.) EUROCRYPT 2003. LNCS, vol. 2656. Springer, Heidelberg (2003)

[R90]   Rompel, J.: One-Way Functions are Necessary and Sufficient for Digital Signatures. In: Proc. of ACM STOC 1990 (1990)

[ZDLZ03]   Zhao, Y., Deng, X., Lee, C.H., Zhu, H.: Resettable Zero-Knowledge in the Weak Public-Key Model. In: Pfitzmann, B. (ed.) EUROCRYPT 2001. LNCS, vol. 2045. Springer, Heidelberg (2001)

A Complete Approximation Algorithm for Shortest Bounded-Curvature Paths

Jonathan Backer and David Kirkpatrick

University of British Columbia
`backer@cs.ubc.ca, kirk@cs.ubc.ca`

Abstract. We address the problem of finding a polynomial-time approximation scheme for shortest bounded-curvature paths in the presence of obstacles. Given an arbitrary environment $\mathcal{E}$ consisting of polygonal obstacles, two feasible configurations, a length ℓ, and an approximation factor ϵ, our algorithm either (i) verifies that every feasible bounded-curvature path joining the two configurations is longer than ℓ or (ii) constructs such a path Π whose length is at most $(1 + \epsilon)$ times the length of the shortest such path. The run time of our algorithm is polynomial in n (the total number of obstacle vertices and edges in $\mathcal{E}$), m (the bit precision of the input), ϵ^{-1}, and ℓ.

For general polygonal environments, there is no known upper bound on the length, or description, of a shortest feasible bounded-curvature path as a function of n and m. Furthermore, even if the length and description of a shortest path are known to be linear in n and m, finding such a path is known to be NP-hard [14].

Previous results construct $(1 + \epsilon)$ approximations to the shortest ϵ-robust bounded-curvature path [11,3] in time that is polynomial in n and ϵ^{-1}. (Intuitively, a path is ϵ-robust if it remains feasible when simultaneously twisted by some small amount at each of its environment contacts.) Unfortunately, ϵ-robust solutions do not exist for all problem instances that admit bounded-curvature paths. Furthermore, even if a ϵ-robust path exists, the shortest bounded-curvature path may be arbitrarily shorter than the shortest ϵ-robust path. In effect, these earlier results confound two distinct sources of problem difficulty, measured by ϵ^{-1} and ℓ. Our result is not only more general, but it also clarifies the critical factors contributing to the complexity of bounded-curvature motion planning.

1   Introduction

We are interested in planning the collision-free motion of vehicles with a restricted turning radius r in the presence of arbitrary polygonal obstacles (see Fig. 1(a)). As in the bulk of previous work, we consider paths traced by a fixed point on the vehicle (say, the mid-point of its rear axle), we prohibit reversal, and we permit arbitrary changes in motion curvature as long as the curvature remains bounded (in effect, the steering wheel has a restriction on how far, but not how fast, it can turn). Specifically, we insist that the path be continuous and

S.-H. Hong, H. Nagamochi, and T. Fukunaga (Eds.): ISAAC 2008, LNCS 5369, pp. 628–643, 2008.

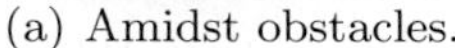

(a) Amidst obstacles.

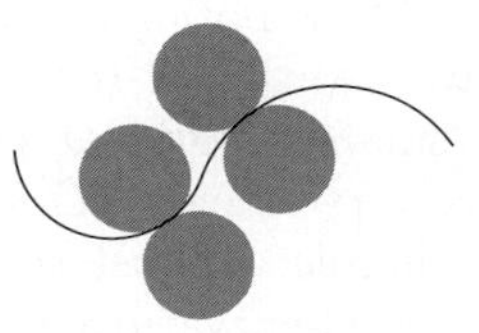

(b) Sandwiched.

Fig. 1. A bounded-curvature path amidst obstacles (left), and a bounded-curvature path is wedged between constant-radius circles (right)

that $|\Delta\phi/\Delta l| < r^{-1}$ between any two points on the path, where $\Delta\phi$ is the change in forward direction measured in radians and Δl is the distance traveled between the points. Intuitively, this restriction means that every point on the path can be locally sandwiched between two circles of radius r (see Fig. 1(b)). By suitable scaling, *we assume that* $r = 1$. This is a standard assumption throughout the literature, but it is essential to note because path length varies with scaling.

An (instantaneous) *configuration* of a point tracing a bounded-curvature path is specified by its position and forward direction of motion. A problem instance is specified with a start configuration α_s, a terminal configuration α_t, and a set of polygonal obstacles $\mathcal{E}$, that we refer to collectively as the *environment*. A path that is disjoint from the interior of all obstacles in $\mathcal{E}$ is said to be *feasible*. Hereafter we restrict our attention to feasible paths, so when we refer to a bounded-curvature path it is understood that it is feasible.

A solution to a problem instance is a feasible bounded-curvature path from α_s to α. By environment *feature*, we mean one of the *corners* (vertices) or *walls* (edges) We denote by n the total number of features. We assume that continuous data (e.g. configurations, corner co-ordinates, ℓ, and ϵ) are represented using fixed-point notation with a maximum of m digits.

Determining if a given problem instance admits a solution is decidable: Fortune and Wilfong present an algorithm that is single-exponential in time and space $(2^{(n+m)^{O(1)}})$ [10]. They use a very compact representation for paths of finite but unbounded length. This representation prevents path recovery, so their algorithm only returns a yes/no answer. More recently, Backer and Kirkpatrick describe an algorithm that will construct a solution path in time polynomial in n, m, and the descriptive complexity τ (essentially, the number of turns) of the simplest bounded-curvature path, if a solution path exists [4]. However, since there is no known upper bound on τ as a function of n and m, nor is there a relationship between τ and the length of the *shortest* bounded-curvature path, this algorithm provides neither a general decision procedure nor an approximation scheme for shortest bounded-curvature paths.

Finding a shortest bounded-curvature path in the midst of arbitrary polygonal obstacles is NP-hard [14], even if the length and description of a shortest path are known to be linear in n and m. Previous work on approximation algorithms has provided a $(1 + \epsilon)$-approximation to the shortest ϵ-robust bounded-curvature path [11,3]. (The notion of ϵ-robustness is described precisely in the next section; intuitively, a path is ϵ-robust if remains feasible when simultaneously twisted by some small amount at each of its environment contacts.) Although these results are very efficient ($O(n^2\epsilon^{-4}\log n)$ in [3]), the restriction to ϵ-robust paths is non-negligible. Robust solutions do not exist for all problem instances that admit bounded-curvature paths. Furthermore, even if a ϵ-robust path exists, the shortest bounded-curvature path may be arbitrarily shorter than the shortest ϵ-robust path. Finally, these approaches do not really address the difficulty exposed by the NP-hardness result because the only feasible paths in the associated reduction have very low robustness ($2^{-\Omega(n)}$); in other words, it is completely consistent with existing approximation algorithms that no general polynomial-time approximation scheme exists for shortest bounded-curvature paths.

In this paper, we describe an algorithm that unifies and extends previous work on the feasibility and approximation of bounded-curvature paths. Given a parameter ℓ, the our algorithm either (i) verifies that every feasible bounded-curvature path has length greater than ℓ or (ii) returns a feasible bounded-curvature path that is at most $(1 + O(\epsilon))$ times longer than the shortest such path. The run time of the algorithm is polynomially bounded[1] in n, m, ϵ^{-1}, and ℓ. (It follows, of course, that we could produce a true $(1 + \epsilon)$ approximation by scaling ϵ appropriately. In fact, we will assume that our ϵ satisfies $\epsilon^{-1} \geq \lambda n^2 \ell$, for some sufficiently large constant λ.) Our new result exchanges a polynomial dependence on τ, the minimal path description, for a polynomial dependence on the minimal path length: in practice, the desired approximation Π can be found in time dependent on its descriptive complexity but, since its length may significantly exceed its description, we cannot avoid considering paths of much higher descriptive complexity to verify that Π has length at most $(1 + \epsilon)$ times the length of the shortest path.

In the remainder of this paper, we first describe a path normalisation procedure that takes a shortest path and deforms it without increasing its length by more than a factor of $(1 + \epsilon)$. After normalisation, the path has a combinatorial description. This allows us to find a path by enumerating a finite (but potentially exponential-size) space. We next describe the redundancy used to filter the enumeration, which results in an efficient algorithm. This redundancy allows us to associate a succinct set of configurations with every environment feature. These configurations serve as an adequate set of potential checkpoints in the construction of a path. The total size of these configuration sets is a dominant factor in the complexity of our algorithm. Thus, the critical step in our analysis is the demonstration that we can maintain a polynomial bound on the size of these configuration sets while still achieving a good approximation. We conclude with several interesting open questions.

---

[1] We make no attempt here to minimise the degree of the polynomial bound. In fact, we frequently sacrifice polynomial factors for the sake of simplicity in exposition.

2   Path Normalisation

Prior algorithmic results in bounded-curvature motion planning either normalise a given feasible path [10,5,4] or characterise shortest feasible paths [2,1,6,3,11]. The resulting structure enables a systematic search for a path and narrows the space of paths considered. In this section, we combine these approaches by normalising a given shortest feasible path.

A shortest bounded-curvature path is composed of arcs of a unit-radius circle (*C-segments*) and straight-line segments (*L-segments*)[9,10]. We distinguish L-segments of length 0, denoted L_0, as *degenerate L-segments*. Similarly, for reasons that will become clear, we distinguish C-segments of lengths 0 or π, denoted C_0 and C_π respectively, as degenerate C-segments. The *(segment) structure* of a segmented path is the string in $\{C, L\}^*$ that describes the sequence of segment types of which it is composed. We say that a segmented path has *type Γ* if the structure of the path is Γ, where several of the segments may be degenerate.

Theorem 1. *[9] In the absence of obstacles, a shortest bounded-curvature path is type CLC or CCC. If its structure has type CCC, the middle C-segment has length at least π.*

We call a bounded-curvature path with the properties set out in Theorem 1 a *(Dubins) jump*. Let Π be a bounded curvature path. A *contact configuration* θ of Π is a configuration where Π just touches the environment boundary (i.e. a point on Π arbitrarily close to θ does not touch the boundary). In the midst of arbitrary polygonal obstacles, any shortest bounded-curvature path can be expressed as a sequence of jumps.

Corollary 1. *[11] Let Π be any shortest bounded-curvature path in the midst of polygonal obstacles. When we split Π at its contact configurations, each resulting contact-free subpath is a Dubins jump.*

We call any path that satisfies the above property a *Dubins path*. Each Dubins path Π has a succinct representation $\langle \theta_0, \theta_1, \ldots, \theta_{k+1} \rangle$ where θ_0 is the start configuration of Π, θ_{k+1} is the terminal configuration of Π, and $\theta_1, \ldots, \theta_k$ are the internal contact configurations of Π in order from its beginning to end. This representation reduces the problem of finding paths to that of determining which contact configurations can be reached from a given start configuration. Fig. 2 represents a continuous range of Dubins paths from the same start configuration to the same terminal configuration. Notice that there is a continuous range of contact configurations, but in each case, only one configuration represents a path of minimal length. Finding a shortest path requires identifying this single configuration, but (as we shall see) approximating it only requires finding a nearby configuration.

To this point, optimality has restricted our attention to a continuous space of Dubins paths. To further restrict our attention to a countable set of paths, we impose additional constraints: we *perturb* contact configurations while keeping each such configuration θ_i in contact with every feature that it touches. Even

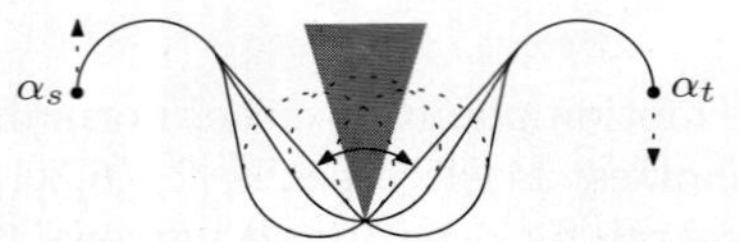

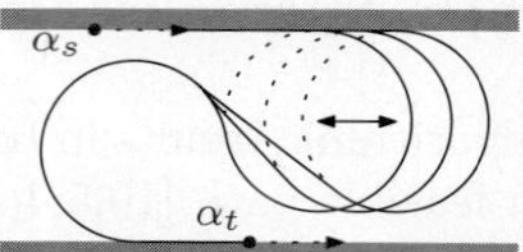

(a) Contact configuration can pivot about a corner.

(b) Contact configuration can slide along a wall.

Fig. 2. Contact configuration degree of freedom

with this restriction, θ_i may still have one degree of freedom: when θ_i contacts a single corner, the position of θ_i is fixed, but its orientation may vary (see Fig. 2(a)); when θ_i contacts the interior of a wall, the direction of θ_i is fixed, but its position may vary (see Fig. 2(b)). These contact-preserving motions (pivoting and sliding, respectively) are the only ways that we deform paths. When we pivot, the perturbation distance is measured in radians; when we slide, the the perturbation is measured in standard distance units.

Using this notion of perturbation, we can provide a more precise definition of the notion of ϵ-robustness that is central to previous approximation results. A jump between two contact configurations is *ϵ-robust* if every simultaneous and independent ϵ-perturbation of its endpoint configurations can be joined by another feasible jump of the same structure type. More generally, a Dubins path is *ϵ-robust* if each if its constituent jumps is ϵ-robust. Note that a jump is non-robust if, for some ϵ-perturbation of each of its endpoint configurations either (i) the resulting configuration pair cannot be joined by a jump without violating feasibility, or (ii) the resulting configuration pair can be joined by a feasible jump but only by undergoing a structural change. It follows from condition (ii) that the middle segment of every ϵ-robust jump must be non-degenerate.

The next lemma makes precise the intuition that small deformations of a jump J result in at most a small increase its length $\|J\|$.

Lemma 1. *[3] Let J and J' be any pair of similar-type jumps between the same two contact configurations. Then $\big|\|J'\| - \|J\|\big|$ is bounded by the sum of the perturbation distances between the corresponding endpoints of J and J'.*

If a jump J between two contact configurations is ϵ-robust, $\|J\|$ must be at least ϵ. It follows from the lemma above that if we perturb the source and destination configurations of J by at most $O(\epsilon^2)$, the length of the result is at most $\|J\| + \epsilon^2 \le (1+\epsilon)\|J\|$. Therefore, simultaneously and arbitrarily perturbing each contact on an ϵ-robust path Π by $O(\epsilon^2)$ causes the length of Π to increase by a factor of at most $(1 + \epsilon)$.

Previous algorithms that find a $(1 + \epsilon)$ approximation to the shortest ϵ-robust path can be described as follows [11,3]: (i) sample the space of possible contact configurations with $O(\epsilon^2)$ granularity, then (ii) find the shortest Dubins path Π where each contact configuration is a sampled configuration. To see that Π exists and is a good approximation, let Π' be the shortest ϵ-robust path. We can perturb each internal contact configuration of Π' by at most $O(\epsilon^2)$ to get Π'' where each contact configuration of Π'' is a sampled configuration. Then Π'' is a $(1 + \epsilon)$ approximation of Π' by Lemma 1. Therefore, Π is a $(1 + \epsilon)$ approximation of Π' because Π is no longer than Π'' by our choice of Π.

We now describe a new type of normalisation that copes with the degeneracies that prevent a Dubins path Π from being ϵ-robust. The general approach is similar to what we used when finding a feasible path [4]. However, it is complicated by that fact that, in order to consider only approximately-shortest paths, we must now deal with a family of paths whose form is less restrictive. As with the previous approximation approaches, we first uniformly sample the space of possible contact configurations, at each of the $O(n)$ obstacle features, with granularity $O(\epsilon^2)$. In addition, we identify, for each obstacle feature, the $O(n)$ contact configurations that lead, by a *doubly-degenerate* jump (a jump with at least two degenerate segments), to a contact configuration on some other feature. Together, we refer to these sampled or doubly-degenerate configurations, in addition to the start and terminal configurations of Π, as *anchor* configurations. There are $O(n^2 + n\epsilon^{-2})$ such configurations in total.

Next, we continuously deform Π while preserving all of its contacts and segment degeneracies. The goal is to exercise any residual degrees of freedom in the path to bring its internal contact configurations as close as possible to an anchor configuration. If some contact configuration cannot be perturbed directly to an anchor configuration, one of its incident jumps must either make a new environment contact (on its middle segment) or become *singly-degenerate* (i.e. at least one of its segments becomes degenerate). In the former case we split the jump into two singly-degenerate jumps at the new contact.

We refer to the degenerated jumps that arise in the normalisation process as Dubins *hops*, and to the functional relationship between the motions of contact configurations that are joined by a hop as a *link*. In general, a path consists of a set of maximal length disjoint *chains* of linked contact configurations, Any chain containing an anchor configuration is said to be *anchored*.

The preservation of segment degeneracies ensures that all hops are preserved under further deformation of the path. Thus, once a chain becomes anchored it remains anchored and its constituent contact configurations cannot be further perturbed. It follows that each normalisation step results in an unanchored chain either (i) merging with some other chain, or (ii) becoming anchored. In either case, the total number of unanchored chains is reduced. Consequently, the normalisation process eventually terminates in a path Π' each of whose contact configurations is anchored. We refer to any such path as an ϵ-*discrete path*.

In the following example, we talk about perturbing circles rather than their underlying contact configurations because underlying circles are easier to visualise. We assume that ϵ is quite large so that we can focus on the changes that can occur before a configuration becomes anchored. In Fig. 3(a), we start by pivoting an underlying circle X at corner x until X kisses the previous underlying circle W. At this point, X is *linked* to W (by a hop of type CL_0C); it is not possible to pivot X further, if W remains stationary. So we simultaneously pivot

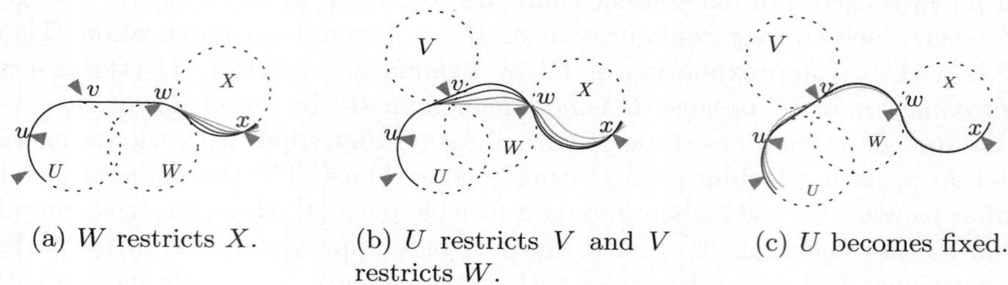

(a) W restricts X.   (b) U restricts V and V restricts W.   (c) U becomes fixed.

Fig. 3. As we twist the underlying circle X in a counterclockwise direction, underlying circles restrict each other, which forces us to simultaneously perturb the underlying circles to make further progress

W and X, maintaining the WX-link throughout. We do this until the L-segment on the jump from u to w touches the corner v. At this point, further pivoting would make the jump from u to w infeasible. So we split the jump from u to w into two jumps: one from u to v and the other from v to w. We illustrate this by adding an underlying circle V at v to Fig. 3(b). At this point, we say that V is linked to W (by a hop of type C_0LC) and U is linked to V (by a hop of type CLC_0). We maintain the UV-, VW-, and WX-links by simultaneously pivoting U, V, W, and X. In Fig. 3(c), the underlying circle U eventually touches another obstacle at which point the contact configuration at u becomes an anchor.

We summarise our new normalisation result in the following:

Theorem 2. *Let Π be an arbitrary bounded-curvature path of length $\|\Pi\|$ from a configuration α_s to a configuration α_t. Then there exists an ϵ-discrete path Π' from α_s to α_t with length at most $\|\Pi\|(1 + O(n\epsilon^2))$ and $O(n\|\Pi\|)$ contact configurations.*

Proof: We have already outlined the argument that our normalisation results in a path Π' in which all contact configurations are anchored. To bound the length of Π', we first observe that the shortest feasible bounded-curvature path Π from α_s to α_t has $O(n\|\Pi\|)$ contact configurations. The intuition behind this claim is that the shortest bounded-curvature path from a feature to the same feature is a unit radius circle. Hence, if we split Π into subpaths of length 2π, each subpath touches each of the n features at most once.

This reasoning is precise for corners, but we must refine it for walls. Notice that once Π reaches a wall w at a configuration θ, Π can follow w to reach other contact configurations at w in the direction of θ. Hence, once Π leaves w after θ, it must loop back (travel at least 2π) before reaching w in the direction of θ because otherwise (as a shortest path) Π should not have left w. This implies that each subpath of length at most 2π has $O(1)$ contact configurations at any particular wall.

The argument that the path Π' has $O(n\|\Pi\|)$ contact configurations relies on our earlier observation that during normalisation every new contact configuration that is introduced coincides with a reduction in the total number of unanchored chains. Since the path Π has at most $O(n\|\Pi\|)$ unanchored chains, it follows that the normalised path Π' also has $O(n\|\Pi\|)$ contact configurations.

Since no contact configuration on Π' is perturbed by more than ϵ^2 during the normalisation process, it follows from Lemma 1 that none of the jumps that are modified by the normalisation process is stretched by more than ϵ^2. Thus the total increase in path length is $O(\epsilon^2 n\|\Pi\|)$. $\qquad\square$

An important property of all ϵ-discrete paths is that they admit a combinatorial description: each such path consists of a sequence of contact configurations each of which is either an anchor configuration or is reachable from an anchor configuration by a sequence of Dubins hops.

3 Systematic Search for Shortest ϵ-Discrete Paths

In this section, we describe an algorithm that constructs a feasible bounded-curvature path that is at most $(1 + \epsilon)$ times longer than the shortest feasible bounded-curvature path, provided that path has length at most ℓ. By Theorem 2, it suffices to construct a shortest ϵ-discrete path with at most $O(n\ell)$ contact configurations. We can systematically search for such a path because ϵ-discrete paths have a combinatorial description.

To structure our search, we note that each ϵ-discrete path Π is the concatenation of one or more subpaths that are (internally) free of anchor configurations. This condition implies additional structure on these subpaths: they each consist of a sequence of hops constrained by the start configuration α of the subpath followed by a sequence of hops constrained by the terminal configuration of the subpath. We call any such subpath an (α, α')-*linkage*. Since every ϵ-discrete path is a sequence of linkages, it suffices to find a shortest (α, α')-linkage with at most τ contact configurations, for every pair of anchor configurations α, α', and every τ bounded by some constant times $n\ell$.

Fig. 4 outlines a systematic way to find the shortest (α, α')-linkage with at most τ contact configurations, for every pair of anchor configurations α, α'. For each feature F, we define the set $\Phi_h^{\alpha,F}$ to be the set of all configurations at F that can be reached by a sequence of h hops from the anchor configuration α. The procedure PROPAGATE builds the sets $\Phi_h^{\alpha,F}$ incrementally: PROPAGATE($\Phi_{h-1}^{\alpha,F}$) adds a configuration ϕ' to $\Phi_h^{\alpha,G}$ whenever there exists a configuration ϕ in $\Phi_{h-1}^{\alpha,F}$ such that ϕ' can be reached from ϕ by a single hop.

Since checking if a given L- or C-segment avoids all obstacles can be done in $O(n)$ time, it is straightforward to confirm that procedure ALLLINKAGES(τ) can be implemented to run in time that is polynomial in the total number of configurations generated by propagation.

Since there several features to which a hop from one contact configuration can potentially go, the number of contact configurations that can be reached from an anchor configuration by a sequence of h hops may grow exponentially with h. Hence, the explicit generation of contact configurations described in the procedure above may not be realised in polynomial time. Fortunately, it suffices to discover only a *single* (α, α')-linkage that is approximately the same

---

ALLLINKAGES(τ)

1 **for** all anchor configurations α
    do for $h = 1$ to τ
        do for all environment features F
            do PROPAGATE($\Phi_{h-1}^{\alpha,F}$)
2 **for** all pairs of anchor configurations α, α'
    do for all pairs of environment features F, F'
        do for all $h \leq \tau$
            do determine if there exist configurations ϕ in $\Phi_h^{\alpha,F}$
                and ϕ' in $\Phi_{\tau-h}^{\alpha',F'}$ such that:
                (i) the jump joining ϕ and ϕ' is collision-free, and
                (ii) the (α, α')-linkage through ϕ and ϕ' is shorter than
                the shortest (α, α')-linkage found so far.

Fig. 4. Procedure for the construction of all shortest (α, α')-linkages with τ contact configurations

length as the shortest (α, α')-linkage. This allows us to identify and propagate only a subset of reachable configurations at each feature in each propagation phase provided that the subset is guaranteed (somehow) to capture at least one such (α, α')-linkage. The intuition is that if two reachable configurations at a feature F are sufficiently similar then it should be possible to replace one by the other in all paths that use the first without sacrificing feasibility, or significantly increasing the length, of the path. While this intuition is basically correct, the details of what constitutes "sufficiently similar", as well as the analysis of the impact of this redundancy elimination, are somewhat involved.

4 Redundancy among Contact Configurations

By allowing a factor $(1 + \epsilon)$ increase in length, we were able to restrict our search to a finite space of paths (for a fixed ℓ). In this section, we use a further potential increase in length to restrict our search to a space of paths that is polynomially bounded in size.

At the core of our approach is a simple form of path subsumption. Let H and H' be two hops of the same type from feature F to feature G. We call a feasible jump from the source configuration of H to the destination configuration of H' an H/H'- *splice*. The solid curves in Fig. 5 illustrate three hops of type CL_0C from corner F to corner G. In this example, the broken lines in Fig. 5 denote an H_1/H_2-splice and an H_1/H_3-splice. In this instance, H_1 subsumes H_2 because the H_1/H_2-splice is feasible. On the other hand, H_1 does not subsume H_3 because the H_1/H_3-splice is blocked by an obstacle.

We now apply this notion of redundancy to narrow our search for approximately shortest (α, α')-linkages. Suppose that H_1 and H_2 in Fig. 5 are hops that extend chains P_1 and P_2 (the *prefixes* of H_1 and H_2, respectively) from the

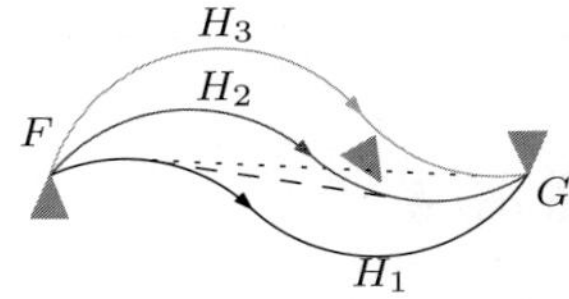

Fig. 5. The splice from H_1 to H_2 makes H_2 redundant

anchor configuration α. Suppose further that P_2 and H_2 occur on some (α, α')-linkage Π. Consider the path Π' from α to α' formed by replacing P_2 with P_1 and H_2 with the H_1/H_2-splice. Since Π' has one unanchored chain (resulting from the broken link corresponding to H_2) it is not ϵ-discrete. However, if we renormalise Π' we get an ϵ-discrete path Π'' that has at most one more contact configuration than Π'. The chain P_1 is a prefix of Π'' because P_1 is a prefix of Π' and renormalisation does not alter contact configurations that are already anchored.

In order to use this subsumption for approximation, we are careful to ensure that Π'' is no more than $2\epsilon^2$ longer than Π. To guarantee this, it suffices to ensure that P_1 is at most ϵ^2 longer than P_2 and the splice from H_1 to H_2 is at most ϵ^2 longer than H_2. The latter follows immediately from Lemma 1 if we assume that the endpoints configurations of H_1 and H_2 are separated by perturbation distance at most ϵ^2.

Using this form of subsumption, we construct subsets $\hat{\Phi}_h^{\alpha,F}$ of the sets $\Phi_h^{\alpha,F}$. Our new configuration sets $\hat{\Phi}_h^{\alpha,F}$ are built by iterative propagation as before. However, preceding every second propagation step we now filter the sets $\hat{\Phi}_h^{\alpha,F}$. We refer to the configurations in $\hat{\Phi}_h^{\alpha,F}$ that were not removed by filtering as *viable* configurations. When we filter $\hat{\Phi}_h^{\alpha,F}$ we guarantee the following property:

Completeness Property: Let θ_2 be any configuration that was removed from $\hat{\Phi}_h^{\alpha,F}$ by filtering and suppose that H_2 is a feasible hop from feature F to feature G starting from configuration θ_2. Then there exists a feasible hop H_1, of the same type as H_2, from F to G starting from a viable configuration θ_1 of $\hat{\Phi}_h^{\alpha,F}$ such that

    1. the prefix from α to θ_1 is at most ϵ^2 longer than the prefix from α to θ_2,
    2. the splice from H_1 to H_2 is at most ϵ^2 longer than H_2, and
    3. the splice from H_1 to H_2 is feasible.

By Theorem 2, we only need to consider ϵ-discrete paths with $O(n\ell)$ contact configurations. However, since we now restrict our search to paths that use only *viable* configurations, we need to ensure not only that sufficiently short such paths exist but also that their number of contact configurations remains bounded.

As before, we start with $O(n\ell)$ configuration-propagation phases. However, before every second phase of propagation, we filter our sets of contact configurations that are reached by viable chains (chains of viable configurations). We

---

$\textsc{AllViableLinkages}(\tau)$

  1 **for** all anchor configurations α
       do for $h = 1$ to τ
            do for all environment features F
                do if (h is even)
                      then $\hat{\Phi}_{h-1}^{\alpha,F} \leftarrow \textsc{Filter}(\hat{\Phi}_{h-1}^{\alpha,F})$
                    $\textsc{Propagate}(\hat{\Phi}_{h-1}^{\alpha,F})$
  2 **for** all pairs of anchor configurations α, α'
       do for all pairs of environment features F, F'
            do for all $h \leq \tau$
                do determine if there exist configurations ϕ in $\hat{\Phi}_{h}^{\alpha,F}$
                    and ϕ' in $\hat{\Phi}_{\tau-h}^{\alpha',F'}$ such that:
                    (i) there is a collision-free (ϕ, ϕ')-linkage λ
                    with at most three hops, and
                    (ii) the (α, α')-linkage through ϕ and ϕ' using λ
                    is shorter than the shortest (α, α')-linkage found so far.

Fig. 6. Procedure for the construction of all shortest viable (α, α')-linkages with τ contact configurations

construct minimum-length viable (α, α')-linkages using only these viable chains. Our modified procedure is described in Fig. 6.

Since every ϵ-discrete path is a sequence of linkages, it suffices to show that (i) the set of all viable contact configurations is bounded by some polynomial function of n, ℓ, ϵ^{-1} and m, and (ii) the viable (α, α')-linkages constructed by the procedure above are at most a factor $(1 + \epsilon)$ longer than the corresponding shortest (α, α')-linkage that is not constrained to use only viable contact configurations. These facts are established in the next section. We summarise our central result in the following:

Theorem 3. *Suppose that there exists a bounded-curvature path Π from a start configuration α_s to a terminal configuration α_t with length at most ℓ. Then there is an ϵ-discrete path from α_s to α_t with length at most $\|\Pi\|(1 + O(\epsilon))$ and $O(n\|\Pi\|)$ contact configurations that can be constructed from the viable (α, α')-linkages produced by the procedure $\textsc{AllViableLinkages}(n\ell)$.*

5   Filtering Redundant Configurations

In this section, we outline how to filter a set $\Phi_{h}^{\alpha,F}$ of contact configurations, while maintaining the Completeness Property. The idea is to partition $\Phi_{h}^{\alpha,F}$ in such a way that it suffices to choose just one element from each partition to belong to $\hat{\Phi}_{h}^{\alpha,F}$. To maintain the Completeness Property, we must satisfy three distinct conditions. Accordingly, we split $\Phi_{h}^{\alpha,F}$ into smaller subsets where it is easy to satisfy one of the conditions.

To satisfy the first condition, we partition $\Phi_h^{\alpha,F}$ into subsets $\Phi_h^{\alpha,F}[i]$ such that the lengths of the chains from α to configurations in $\Phi_h^{\alpha,F}[i]$ lie in the range $[(i-1)\epsilon^2, i\epsilon^2)$. Since we restrict our attention to paths of length at most $\ell(1 + O(\epsilon))$, the number of resulting partitions is at most $\ell(1 + O(\epsilon))/\epsilon^2$.

To satisfy the second condition, we uniformly partition the space of contact configurations at all features with granularity ϵ^2, as before, to partition $\Phi_h^{\alpha,F}[i]$ into subsets $\Phi_h^{\alpha,F}[i][j]$. The second condition then follows directly from Lemma 1.

To satisfy the third condition, we further partition $\Phi_h^{\alpha,F}[i][j]$ so that for any pair of similar-type feasible hops H_1 and H_2 that start from configurations in a given partition $\Phi_h^{\alpha,F}[i][j][k]$ there exists a splice between H_1 and H_2. In particular, the partitions $\Phi_h^{\alpha,F}[i][j][k]$ are sufficiently narrow that any pair of feasible hops, of the same type and destination feature, originating in a given partition are *homotopic* (i.e. they do not straddle any obstacles). Unfortunately, this alone is not sufficient to guarantee that a splice exists. The essential question that remains to be addressed is: how close (in terms of the perturbation distance of their corresponding endpoints) do two hops, of the same type and homotopy class, have to be in order to guarantee that there exists a splice between them?

For some hop types there is essentially no separation constraint, that is a splice exists between all relevant hop pairs. For others, specifically hops of the form CL_0C that we have used in our illustrations, there is no fixed bound that will guarantee the existence of a splice. Fortunately, in all cases we can derive a *relative* separation bound that guarantees the existence of a splice. Specifically,

Lemma 2. *Let H_1 and H_2 be any pair of similar-type Dubins hops between the same two features F and G. Then a splice exists between H_1 and H_2 provided that, at both F and G, the perturbation distance δ between the endpoint configurations of H_1 and H_2 is at most the square root of the distances from each of these configurations to their closest anchor configuration.*

The proof of Lemma 2 is somewhat involved. At its core is an argument about the quadratic convergence of a sequence of splice-coverage regions, which we bound by a geometric reduction to a simple Newton-Raphson root-finding process. A sketch of this argument appears in [4]; the full details will be presented in an expanded version.

Of course, even this relative separation bound does not preclude the existence of an arbitrarily large set of similar hops that admit no splices. However (and finally), we can use our assumption about bounded input precision to prove that any two distinct anchored contact configurations must be reasonably well separated.

Lemma 3. *Let ϕ and ϕ' be distinct contact configurations at some environment feature F that belong to anchored chains with at most k links. Then, the perturbation distance between ϕ and ϕ' is at least $2^{-O(c^k m)}$ for some positive constant c, where m is number of bits of the fixed-point input and n is the number of obstacle features.*

Proof: The constraints between linked contact configurations can be expressed as a system of equations involving $+$, $-$, $\times$, $\sqrt{\ }$, and integers, that relate the co-ordinates of underlying circles. Given such a system, Burnikel et al. describe how to construct a quotient of irreducible polynomials with integer coefficients such that the result of the arithmetic expression corresponds to a root of the quotient [7]. The magnitude of a non-zero root is bounded above and below by the degrees and coefficients of both polynomials. Rather than compute the polynomials, they recursively compute the corresponding root bounds directly from the expression.

Using a natural representation for configurations, Burnikel et al.'s technique applied to linked jumps bounds the difference between distinct contact configurations on ϵ-discrete paths, with the weak assumption that real values are represented in fixed-point with m digits. $\qquad\qquad\square$

It follows from Lemma 2 that the perturbation width of the partitions $\Phi_h^{\alpha,F}[i][j][k]$ converge quadratically to zero. By Lemma 3, we need not consider partitions of width less than $2^{-O(c^k m)}$, for some positive constant c. Thus, there is a polynomial bound on the total number of partitions $\Phi_h^{\alpha,F}[i][j][k]$. By construction, the set $\hat{\Phi}_h^{\alpha,F}$ needs to contain at most one configuration from each partition. It remains to argue that these viable configurations suffice to construct near-optimal length ϵ-discrete paths. In particular, we argue that every minimum-length ϵ-discrete path, with at most $n\ell$ jumps, is well-approximated by some sequence of viable linkages containing $O(n\ell)$ jumps in total.

To allow us to quantify the degree to which a given ϵ-discrete path differs from a viable such path, we define the *viability* of an anchored chain to be the number of jumps on its longest even-length prefix that consists entirely of viable configurations, and the viability of an ϵ-discrete path Π, denoted $v(\Pi)$, as the sum of the viabilities of all of its anchored chains. We have observed that splicing, and subsequent renormalisation, can be used to replace non-viable configurations on a path by viable ones. However, this comes at a cost: it increases both the path length and the total number of jumps. The associated trade-off between the length, number of jumps and viability of ϵ-discrete paths is captured by the following potential function $\mathcal{P}$ defined on all ϵ-discrete paths: Let L be any linkage consisting of a two anchored chains with a total of r internal contact configurations. Then the *potential* of L, denoted $\mathcal{P}(L)$, is given by $\|L\|/\epsilon^2 + \xi n\ell(r - v(L))$, for some suitably chosen constant ξ. More generally, the potential, $\mathcal{P}(\Pi)$, of an ϵ-discrete path Π is just the sum of the potentials of all if its constituent linkages.

Let Π be any minimum length ϵ-discrete path. Recall that Π consists of two chains, with $O(n\|\Pi\|)$ internal contact configurations, anchored to its end configurations α and α'. Thus, $\mathcal{P}(\Pi) < \|\Pi\|/\epsilon^2 + O(n^2\ell\|\Pi\|)$. Hence, if Π^* is any minimum potential ϵ-discrete path joining α and α', then $\|\Pi^*\| \leq \epsilon^2\mathcal{P}(\Pi)$ (otherwise $\mathcal{P}(\Pi^*) > \mathcal{P}(\Pi)$). Hence, $\|\Pi^*\| \leq \|\Pi\| + O(n^2\ell\|\Pi\|\epsilon^2) \leq \|\Pi\|(1 + O(\epsilon))$, and it remains only to argue that Π^* is well-approximated by viable linkages.

Suppose that Π^* contains an (α, α')-linkage L that starts with an anchored chain $\alpha = \theta_0, \theta_1, \ldots, \theta_r$, whose length exceeds its viability by at least three (the case where L ends with such an anchored chain is symmetric). Let F_i denote the feature associated with configuration θ_i. Let θ_{2j} be the last configuration on the longest even-length viable prefix of this chain. It follows that θ_{2j+1} belongs to $\hat{\Phi}_{2j+1}^{\alpha, F_{2j+1}}$, but θ_{2j+2} was filtered from the set $\hat{\Phi}_{2j+2}^{\alpha, F_{2j+2}}$. Hence, by the Completeness Property, there is a configuration θ'_{2j+2} in $\hat{\Phi}_{2i+2}^{\alpha, F_{2j+2}}$ and a configuration θ'_{2j+3} at F_{2j+3} such that there is a feasible splice between the hop H', joining θ'_{2i+2} and θ'_{2i+3}, and the hop H, joining θ_{2i+2} and θ_{2i+3}. If we perform this splice, and renormalise the resulting path the linkage L is replaced by a new ϵ-discrete path $\hat{P}$ from α to α' (and the remainder of Π is unaltered). By construction, this new path $\hat{P}$ has at most one more configuration than L, has length $O(n\|L\|\epsilon^2)$ more than $\|L\|$, and has viability at least two more than L. Hence $\mathcal{P}(\hat{P}) - \mathcal{P}(L) \leq O(n\|L\|) - \xi n\ell < 0$, for large enough ξ, contradicting the minimality of Π^*.

It follows that every linkage L on Π^* is composed of two almost completely viable anchored chains. But this means that L itself is discovered by procedure ALLVIABLELINKAGES. (This was the reason why the procedure looks for *three-hop bridges* between viable configurations). Hence, Π^* (or some other, even shorter, ϵ-discrete path) can be constructed from the linkages identified by procedure ALLVIABLELINKAGES.

6   Conclusion

We have presented an approximation algorithm for finding a bounded-curvature path amidst polygonal obstacles. We used the approximation factor to reduce the number of paths that we consider through path normalisation and filtering before propagation. The run time of our algorithm is polynomially bounded because filtering results in a polynomially sized set of configurations and we only need to propagate and filter a polynomially bounded number of times.

Although the run time of our algorithm is a function of input precision m, we have still exploited the Real RAM model of computation in our analysis. This is similar to the $(1+\epsilon)$ approximation algorithm by developed by Papadimitriou for finding the shortest unconstrained path amidst arbitrary polyhedra in three dimensions [12]. Subsequent work has shown that Papadimitriou's approach works in polynomial time in a bit model of computation [8]. It remains to be seen if our algorithm also has a polynomially bounded complexity in a bit model.

The (implicit) polynomial bound on the run time of our algorithm is high degree. In this paper, we have not made any attempts to optimise the asymptotic run time. However, the subtleties that lead to a high complexity bound are real issues that any complete algorithm must address. Therefore, it is not clear if any complete algorithm can be asymptotically competitive with the fast $O(n^2\epsilon^{-4}\log n)$ approximation algorithm for ϵ-robust paths.

Note that the worst case paths that a complete algorithm must find are unrealistic for real-world robotic applications because they require touching obstacles and turning with great precision. The restriction to ϵ-robust paths reduces the precision required, but paths must still brush up against the environment. A bounded-curvature path is ϵ-safe if the center of an ϵ-radius ball can trace the path without the ball touching any obstacles. This definition is distinct from the notion of robustness: an ϵ-robust path is 0-safe because it is Dubins, and there exist ϵ-safe paths in environments that only permit 0-robust solutions. Finding an ϵ-safe path reduces to problems that we have already considered by taking a Minkowski sum of the obstacles with an ϵ-radius ball. However, if we are willing to find an approximation to an ϵ-safe path that is less safe (say $\epsilon/2$-safe), the problem may become easier: there is a non-uniform discretisation approach to finding an approximation to the shortest ϵ-robust bounded-curvature path in three or more dimensions [15], but the returned path is not ϵ-safe. Finding an approximation to the shortest ϵ-safe bounded-curvature path in two dimensions remains open.

Although we express the run-time of our algorithm in terms of the length ℓ, we use ℓ only to bound the number of turns that we must consider. The number of turns is a natural measure of the descriptive complexity of a Dubins path, and we obtain tighter results when we rephrase our results in terms of this measure. However, there are no known bound on this quantity in arbitrary polygonal environments, which poses an interesting open problem. If the number of turns can grow super-polynomially in the input size, finding a minimum-turn path is another interesting open problem. This is analogous to finding a minimum-link path when the path is unconstrained.

Shortest bounded-curvature paths with reversals are also composed of C-segments and L-segments [13]. The feasibility problem is less interesting for this class of paths because with reversals, bounded-curvature paths can arbitrarily approximate any given continuous path by using a large number of turns. However, the shortest path and minimum-turn motion planning variants are interesting open problems when reversals are allowed.

References

1. Agarwal, P.K., Biedl, T., Lazard, S., Robbins, S., Suri, S., Whitesides, S.: Curvature-constrained shortest paths in a convex polygon. SIAM J. Comput. 31(6), 1814–1851 (2002)
2. Agarwal, P.K., Raghavan, P., Tamaki, H.: Motion planning for a steering-constrained robot through moderate obstacles. In: ACM symposium on Theory of computing, pp. 343–352. ACM Press, New York (1995)
3. Agarwal, P.K., Wang, H.: Approximation algorithms for curvature-constrained shortest paths. SIAM J. Comput. 30(6), 1739–1772 (2001)
4. Backer, J., Kirkpatrick, D.: Finding curvature-constrained paths that avoid polygonal obstacles. In: SCG 2007: Proceedings of the twenty-third annual symposium on Computational geometry, pp. 66–73. ACM, New York (2007)
5. Bereg, S., Kirkpatrick, D.: Curvature-bounded traversals of narrow corridors. In: SCG 2005: Proceedings of the twenty-first annual symposium on Computational geometry, pp. 278–287. ACM Press, New York (2005)

6. Boissonnat, J.-D., Lazard, S.: A polynomial-time algorithm for computing shortest paths of bounded curvature amidst moderate obstacles. Internat. J. Comput. Geom. Appl. 13(3), 189–229 (2003)
7. Burnikel, C., Funke, S., Mehlhorn, K., Schirra, S., Schmitt, S.: A separation bound for real algebraic expressions. In: Meyer auf der Heide, F. (ed.) ESA 2001. LNCS, vol. 2161, p. 254. Springer, Heidelberg (2001)
8. Choi, J., Sellen, J., Yap, C.-K.: Approximate euclidean shortest path in 3-space. In: SCG 1994: Proceedings of the tenth annual symposium on Computational geometry, pp. 41–48. ACM Press, New York (1994)
9. Dubins, L.E.: On curves of minimal length with a constraint on average curvature, and with prescribed initial and terminal positions and tangents. Amer. J. Math. 79, 497–516 (1957)
10. Fortune, S., Wilfong, G.: Planning constrained motion. Ann. Math. Artificial Intelligence 3(1), 21–82 (1991); Algorithmic motion planning in robotics
11. Jacobs, P., Canny, J.: Planning smooth paths for mobile robots. In: Nonholonomic Motion Planning, pp. 271–342. Kluwer Academic, Norwell (1992)
12. Papadimitriou, C.H.: An algorithm for shortest-path motion in three dimensions. Inform. Process. Lett. 20(5), 259–263 (1985)
13. Reeds, J.A., Shepp, L.A.: Optimal paths for a car that goes both forwards and backwards. Pacific J. Math. 145(2), 367–393 (1990)
14. Reif, J., Wang, H.: The complexity of the two dimensional curvature-constrained shortest-path problem. In: Workshop on the Algorithmic Foundations of Robotics, Natick, MA, USA, pp. 49–57. A. K. Peters, Ltd. (1998)
15. Reif, J., Wang, H.: Nonuniform discretization for kinodynamic motion planning and its applications. SIAM J. Comput. 30(1), 161–190 (2000)

Detecting Commuting Patterns by Clustering Subtrajectories*

Kevin Buchin[1], Maike Buchin[1], Joachim Gudmundsson[2], Maarten Löffler[1], and Jun Luo[1]

[1] Inst. for Information and Computing Sciences, Utrecht University, The Netherlands
{buchin,mbuchin,loffler,ljroger}@cs.uu.nl
[2] NICTA, Sydney, Australia
joachim.gudmundsson@nicta.com.au

Abstract. In this paper we consider the problem of detecting *commuting patterns* in a trajectory. For this we search for similar subtrajectories. To measure spatial similarity we choose the Fréchet distance and the discrete Fréchet distance between subtrajectories, which are invariant under differences in speed. We give several approximation algorithms, and also show that the problem of finding the 'longest' subtrajectory cluster is as hard as MaxClique to compute and approximate.

1 Introduction

Technological advances of location-aware devices, surveillance systems and electronic transaction networks produce more and more opportunities to trace moving entities. Consequently, a variety of disciplines including geography, market research, data base research, animal behavior research, surveillance, security and transport analysis shows an increasing interest in movement patterns of entities moving in various spaces over various times scales [9,14]. In the case of animals, movement patterns can be viewed as the spatio-temporal expression of behaviors, e.g. flocking sheep or the seasonal migration of birds. In a transportation context, such pattern could be a traffic jam or a commuting route.

In this paper we will focus on the problem of detecting *commuting patterns* in a trajectory. For this we search for similar subtrajectories. To measure spatial similarity we choose the Fréchet distance and the discrete Fréchet distance between subtrajectories. Both of these are invariant under differences in speed: for instance, in a transportation context, this allows to detect a commuting pattern even in the presence of different traffic conditions and varying means of transport. We will also consider how to detect a common movement pattern of

* This research was initiated during the GADGET Workshop on Geometric Algorithms and Spatial Data Mining, funded by the Netherlands Organisation for Scientific Research (NWO) under BRICKS/FOCUS grant number 642.065.503. The research was further supported by NWO through the GADGET and GOGO projects, and by the Australian Government's Backing Australia's Ability initiative, in part through the Australian Research Council.

S.-H. Hong, H. Nagamochi, and T. Fukunaga (Eds.): ISAAC 2008, LNCS 5369, pp. 644–655, 2008.
© Springer-Verlag Berlin Heidelberg 2008

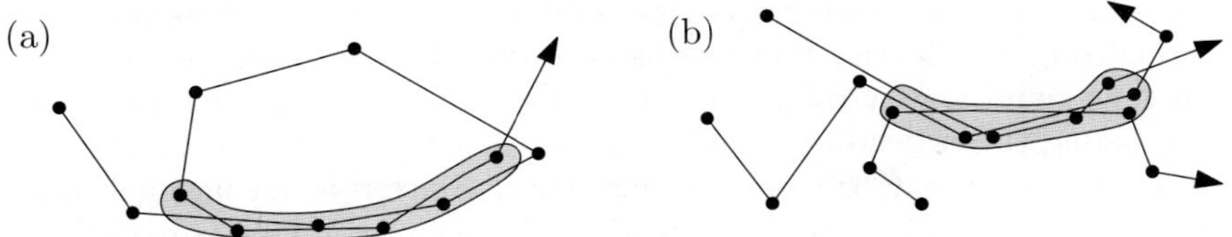

Fig. 1. (a) The two subtrajectories within the shaded region form a subtrajectory cluster for $m = 2$ if the length of the longest subtrajectory is longer than ℓ and the distance between the two subtrajectories is at most d. (b) Illustrating a subtrajectory cluster for $m = 3$ in the case when a set of trajectories is given as input.

a group of entities. That is, we want to find similar subtrajectories in a given set of trajectories. This problems can be handled using the same approach as for one entity. We will focus on the problem for one entity, but also discuss the changes for a group of entities. Figure 1 shows examples for the two different problem statements, i.e., similar subtrajectories of a single trajectory and of several trajectories.

More formally, the input is a moving point object, called *entity*, whose location is known at n consecutive time-steps. Thus, the trajectory of an entity is a polygonal curve that can self-intersect. We assume that an entity moves between two consecutive time steps on a straight line. Given the trajectory T of an entity in the plane, an integer $m > 0$ and two positive real values ℓ and d, we define a commuting pattern as a set of m subtrajectories of T, where the time intervals of two subtrajectories overlap in at most a point, the subtrajectories are within distance d from each other, and at least one subtrajectory has length ℓ between its first and last vertex. See Fig. 1 for an example. As mentioned above, we will use the Fréchet distance and the discrete Fréchet distance as distance measures, which are natural distance measures for curves and polygonal curves.

Recently there has been considerable research in the area of analyzing and modeling spatio-temporal data [2,3,13,16]. In the database community, research has mainly focused on indexing databases so that basic spatio-temporal queries concerning the data can be answered efficiently. Typical queries are spatio-temporal range queries, spatial or temporal nearest neighbors [15,18]. Vlachos *et al.* [19] state that in the area of spatio-temporal analysis it is an important task to detect commuting patterns of a single entity, or to find objects that moved in a similar way. Thus, an efficient clustering algorithm for trajectories, or subtrajectories, is essential for such data analysis tasks. Gaffney *et al.* [11,10] proposed a model-based clustering algorithm for trajectories. In this algorithm, a set of trajectories is represented using a regression mixture model. Specifically, their algorithm is used to determine the cluster memberships and it only clusters whole trajectories. Mamoulis *et al.* [17] used a different approach to detect periodic patterns. They assumed that the trajectories are given as a sequence of spatial regions, for example ABC would denote that the entity started at region A and then moved to region C via region B. Using this model data mining tools such as association rule mining can be used. Vlachos *et al.* [19] also looked at discovering similar trajectories of moving objects. They mainly focused on formalizing a similarity function based on the longest common subsequence which they claim is very robust to noise. However, their approach matches the vertices along the trajectories which requires that the vertices along the trajectories are synchronized, or almost synchronized. Since it is not unusual that the coordinates are recorded with a frequency of five per second it has recently been argued [6,12] that this is an unreasonable assumption since trajectories will have to be compressed (simplified) to allow fast computations. In this paper we will make no such assumptions on the input.

The paper is organized as follows. In the next section we introduce the Fréchet distance formally define the problem considered in this paper, and show hardness results for the problem. In Sect. 3 we present approximation algorithms. For omitted proofs we refer to the full version of this paper [4].

2   Preliminaries and Hardness

In this paper we will present algorithms that find clusters of subtrajectories of a given trajectory. The idea is to select a reference subtrajectory and then to find all subtrajectories that are close to this using the Fréchet distance. The algorithms can be extended to handle subtrajectory clustering in the case when the input is a set of trajectories and the aim is to find the longest subtrajectory cluster where each subtrajectory in the cluster is a subtrajectory of an input trajectory and no two trajectories in the cluster belong to the same input trajectory. Specifying exactly which of the patterns should be reported is often a subject for discussion. In this paper we consider the problems of finding the longest cluster of a fixed size and the largest cluster of a fixed length. We say that a cluster C of (sub)trajectories has length at least ℓ if there exists a (sub)trajectory in C whose length between the first and last vertex is at least ℓ.

Definition 1. *Given a trajectory T with n vertices, a subtrajectory cluster $C_T(m, \ell, d)$ for T of length ℓ consists of at least m non-identical subtrajectories $T_1, \ldots, T_m$ of T such that the time intervals for two subtrajectories overlap in at most a point, the distance between the subtrajectories is at most d, and at least one subtrajectory has length ℓ.*

Problem 1. (SUBTRAJECTORY CLUSTER — $\mathsf{SC}(m, \ell, d)$)
Given a trajectory T, the subtrajectory cluster problem $\mathsf{SC}(m, \ell, d)$ is the decision problem: does there exist a subtrajectory cluster with these parameters? We also define the related optimization problems $\mathsf{SC}(\max, \ell, d)$, $\mathsf{SC}(m, \max, d)$, and $\mathsf{SC}(m, \ell, \min)$, which keep two parameters fixed and aim to maximize or minimize the third.

A variant of this problem is to find a subtrajectory cluster in a given set of trajectories, as shown in Fig. 1(b). This variant has two further subvariants: for each trajectory at most one subtrajectory is allowed, or several. By concatenating the given set of trajectories to one trajectory, any instance of this problem can be made into an instance of the subtrajectory cluster problem. For the subvariant where only one subtrajectory of each trajectory is allowed, this restriction has to be enforced, as well. This subvariant can be seen as a partial matching problem of curves. Partial matching problems for two curves using the Fréchet distance that have been considered are matching a curve to a subcurve of a different trajectory [1] and matching two curves where several subcurves may contribute to the matching [5]. In Definition 1 we omitted to define the distance metric and other distance functions than the Fréchet distance could be used.

The Fréchet distance is a distance measure for continuous shapes such as curves and surfaces, and is defined using reparameterizations of the shapes. Since it takes the continuity of the shapes into account it is generally regarded as being a more appropriate distance measure than the Hausdorff distance for curves. For polygonal curves, the discrete Fréchet distance is a natural variant of the Fréchet distance. In this paper we will use the Fréchet distance [1] and discrete Fréchet distance [8]. For a set of $m > 2$ curves there is a natural extension due to Dumitrescu and Rote [7].

Definition 2. *For a set of m curves $\mathcal{F} = \{f_1, \ldots, f_m\}$, $f_i : [a_i, a_i'] \to \mathbb{R}^2$ we define the Fréchet distance as*

$$\delta_F(\mathcal{F}) = \inf_{\substack{\alpha_1 : [0,1] \to [a_1, a_1'] \\ \vdots \\ \alpha_m : [0,1] \to [a_m, a_m']}} \max_{\substack{t \in [0,1] \\ 1 \le i,j \le m}} |f_i(\alpha(t)) - f_j(\beta(t))|,$$

where $\alpha_1, \ldots, \alpha_m$ range over continuous and increasing functions with $\alpha_i(0) = a_i$ and $\alpha_i(1) = a_i'$, $i = 0, \ldots, m$.

In this paper we will also consider the discrete Fréchet distance, which is a variant of the Fréchet distance for polygonal curves. For the discrete Fréchet distance only distances between vertices are computed. It is defined using *couplings* of the vertices of the curves [8]. Consider two polygonal curves P, Q in $\mathbb{R}^c$ given by the sequences of their vertices $\langle p_1, \ldots, p_n \rangle$ and $\langle q_1, \ldots, q_m \rangle$, respectively. A coupling C of the vertices of P, Q is a sequence of pairs of vertices $C = \langle C_1, \ldots, C_k \rangle$ with $C_r = (p_i, q_j)$ for all $r = 1, \ldots, k$ and some $i \in \{1, \ldots, n\}, j \in \{1, \ldots, m\}$, fulfilling

- $C_1 = (p_1, q_1)$ and $C_k = (p_n, q_m)$
- $C_r = (p_i, q_j) \Rightarrow C_{r+1} \in \{(p_{i+1}, q_j), (p_i, q_{j+1}), (p_{i+1}, q_{j+1})\}$ for $r = 1, \ldots, k-1$.

Definition 3. *Let P, Q be two polygonal curves in $\mathbb{R}^c$. Let $| \cdot |$ denote an underlying norm on $\mathbb{R}^c$. Then the discrete Fréchet distance $\delta_{dF}(f, g)$ is defined as*

$$\delta_{dF}(f, g) = \min_{coupling\ C} \max_{(p_i, q_j) \in C} |p_i - q_j|,$$

where C ranges over all couplings of the vertices of f and g.

The study of the Fréchet distance from a computational point of view was initiated by Alt and Godau [1]. They showed that the Fréchet distance between two curves with m and n segments respectively can be computed in $O(mn \log mn)$ time, and the associated decision question can be answered in $O(mn)$ time assuming the model of computation can do square roots in constant time. In the following we will assume that the model of computation can do this for all results concerning the Fréchet distance. This is not necessary for the discrete Fréchet distance. The study of the discrete Fréchet distance was started by Eiter and Mannila [8]. They gave a simple dynamic programming algorithm for computing the discrete Fréchet distance in $O(mn)$ time. Sometimes, we will use the term continuous Fréchet distance to contrast the Fréchet distance to the discrete Fréchet distance.

In this paper we will focus on the optimization version of Problem 1 where the length ℓ is maximized, i.e., the longest subtrajectory cluster $\mathsf{SC}(m, \max, d)$ for fixed parameters m and d. The general decision problem $\mathsf{SC}(m, \ell, d)$ is NP-complete (see full version [4]), so the optimization variants are also NP-hard. Even more, a $(2 - \phi)$-distance approximation for any $0 < \phi \leq 1$ of $\mathsf{SC}(m, \max, d)$ (and for $\mathsf{SC}(\max, \ell, d)$) is NP-hard, and any distance approximation is 3-SUM hard. We speak of a *distance* approximation when we approximate d. For instance, assume an optimal solution of $\mathsf{SC}(m, \max, d)$ has length ℓ^*; a c-distance approximation algorithm for this problem would return a cluster $C_T(m, \ell^*, cd)$ of length at least ℓ^* and with m subtrajectories of T within Fréchet distance $c \cdot d$ from each other. Since a $(2 - \phi)$-distance approximation is NP-hard, we will turn our attention to 2-distance approximation algorithms.

In the following, for an instance of the SC problem we will always use n to denote the number of vertices in the given trajectory. In the case of $\mathsf{SC}(\max, \ell, d)$ we will denote by m the maximum number of trajectories in a cluster and by ℓ the length of a cluster, in the case of $\mathsf{SC}(m, \max, d)$ we will denote by ℓ the number of vertices of a (reference) curve in the cluster and by m the number of curves in a cluster.

3   A 2-Distance Approximation Using the Fréchet Distance

First we briefly outline the *free space diagram* of two polygonal curves f and g, which is a geometric data structure introduced by Alt and Godau [1] for computing the Fréchet distance. Let f be a polygonal curve with n vertices $p_1, \ldots, p_n$. We use ϕ_f to denote the following natural parameterization of f. The map $\phi_f : [1, n] \to \mathbb{R}^c$ maps $i \in \{1, \ldots, n\}$ to p_i and interpolates linearly in between vertices, where c is any dimension. The free space diagram of two polygonal curves f and g, with n and m vertices, respectively, is the set

$$F_d(f, g) = \{(s, t) \in [1, n] \times [1, m] : |\phi_f(s), \phi_g(t)| \leq d\}$$

which describes all tuples (s, t) such that the points $\phi_f(s)$ and $\phi_g(t)$ have Euclidean distance at most d. See Fig. 3 for an example of a free space diagram. The complexity of the free space diagram is $O(nm)$ and it can be constructed in time $O(nm)$. Alt and Godau [1] showed that the Fréchet distance between f and g is less than d if and only if there exists a monotone path in the free space diagram $F_d(f, g)$ from $(0, 0)$ to (n, m).

For the discrete Fréchet distance the free space diagram consists of the nm grid points. Each grid point (x, y) is labeled free or not depending on whether $|\phi_f(x) - \phi_g(y)|$ is less than or equal to the distance d or not. A monotone path is a sequence of free grid points from $(0, 0)$ to (n, m) where a grid point (x, y) is followed either by $(x + 1, y)$, $(x, y + 1)$ or $(x + 1, y + 1)$ [8]. Analogous to the continuous Fréchet distance, it holds that the discrete Fréchet distance is less than d if and only if there is a monotone path in the free space diagram.

Cluster Curves in the Free Space. In our application, we have just one polygonal curve: this input trajectory T. Now consider the free space diagram $F_d(T, T)$ of two copies of T, and distance value d. Let $s, t \in [1, n]$ be two points in time for the trajectory T, and let l_s and l_t be the two vertical lines in $F_d(T, T)$ corresponding to them. For the discrete Fréchet distance, we assume that $s, t \in \{1, \ldots, n\}$, i.e., l_s and l_t fall on vertices in $F_d(T, T)$, see Fig. 2. For simplicity we assume that the x- and y-axis of the diagram goes from 1 to n, i.e., from v_1 to v_n.

Definition 4. *We say that there are m cluster curves between l_s and l_t in $F_d(T, T)$ if and only if there are m non-identical curves starting at l_s and ending at l_t such that:*

- *each curve is monotonically increasing in both coordinates from l_s to l_t, and*
- *the y-coordinates of two curves overlap in at most a point.*

Figure 2 shows an example of $F_d(T, T)$, and the corresponding cluster of subtrajectories of T.

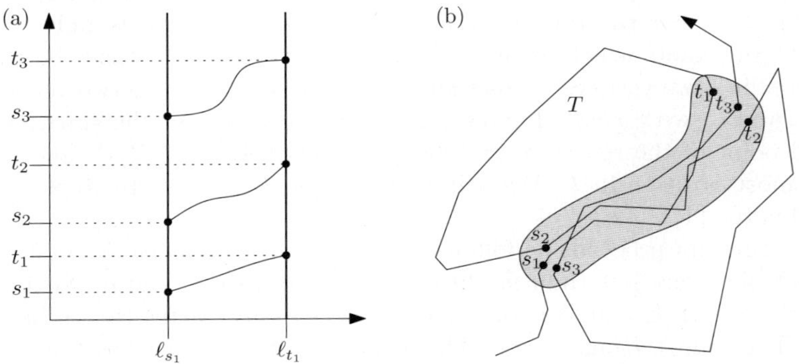

Fig. 2. (a) There exists a trajectory cluster with three subtrajectories between s and t (b) A possible trajectory corresponding to the cluster curves. The subtrajectory s_1, t_1 is one of the subtrajectories.

We will make use of the two important implications in Lemma 2, which are necessary for the algorithm in Sect. 3. For this, we use the following lemma of Alt and Godau [1].

Lemma 1 (Alt and Godau [1]). *Two polygonal curves f and g in the plane, with n and m vertices respectively, have Fréchet distance at most d, if and only if there exists a curve in the corresponding free space diagram $F_d(f, g)$ from $(0, 0)$ to (n, m) that is monotone in both coordinates.*

Lemma 2. *Let T be trajectory and T' the subtrajectory of T that starts at s and ends at t, where $t - s = \ell$. Then the following holds:*

- *If there exist m cluster curves within the free space diagram $F_d(T, T)$ between l_s and l_t, then there exists a subtrajectory cluster $C_T(m, \ell, 2d)$ containing T'.*
- *If there exists a subtrajectory cluster $C_T(m, \ell, d)$ containing T', then there exists a set of m cluster curves within the free space diagram $F_d(T, T)$ between l_s and l_t.*

Lemma 2 implies that if there is no subtrajectory cluster $C_T(m, \ell, 2d)$ containing T' then there do not exist m cluster curves between t_a and t_b. The simplest variant of the algorithm is using the discrete Fréchet distance. We first present and analyze this variant in Sect. 3.1 in full detail and then extend it to the continuous Fréchet distance in Sect. 3.2 (with the details in the full version [4]).

3.1 Discrete Fréchet Distance

In this section we describe and analyze the algorithm for the discrete Fréchet distance.

Algorithm. Recall that as input we are given a trajectory T with n vertices, an integer m and a positive real value d. Consider the free space diagram $F = F_d(T, T)$ of T. For the discrete Fréchet distance, this consists only of the at most n^2 grid points which lie in free space, i.e., where the distance between the corresponding two vertices is less than or equal to d. We sweep two vertical lines in F: l_s and l_t, with x-coordinates s and t, which represent the start point and the end point of the reference trajectory, respectively. Initially l_s and l_t are at the leftmost position in F. We will sweep l_s and l_t in discrete steps along the x-coordinates $\{1, \ldots, n\}$.

The algorithm proceeds as follows. Move l_t to the right to the first position for which there are less than m cluster curves between l_s and l_t. Next move l_s to the right until l_s equals l_t or there are at least m cluster curves between l_s and l_t. Then move l_t again, etc. This continues until l_s reaches the rightmost position in F. During the sweep we maintain the longest length of a reference trajectory for which there exist at least m cluster curves between its endpoints. It is easily seen that the two sweep lines of l_s and l_t move monotonically from left to right over n event points. This implies that only $O(n)$ possible pairs of start and endpoints are considered.

What remains to be done is to build and update a data structure that allows us to check if there are m monotone curves between l_s and l_t that overlap in at most a point along the y-axis. For this, we will store the free space between l_s and l_t as directed, labeled graph on the vertices in the free space. When updating l_t, we add a new column of the free space. When updating l_s, we delete the leftmost column. For checking whether m curves exist between l_s and l_t, we will search this graph. We will see that this data structure can be maintained in $O(n)$ time per step and using $O(\ell n)$ space where ℓ denotes the maximum number of vertices of a reference trajectory occurring in a cluster of m curves.

Data structure. We store the free space between l_s and l_t as a directed, labeled graph on the vertices in the free space (see Fig. 3). We store a directed edge from p to p' if p, p' lie in the free space of the same cell and p is to the left or above p'. We label each edge by the smallest x-coordinate reachable by a path along this edge. This graph can be efficiently computed column-by-column from left to right, and bottom to top within a column. The directed edges encode the free space. The edge labels are easy to compute: An edge pointing to a vertex without outgoing edges is labeled with the x-coordinate of that vertex. An edge pointing to a vertex p with outgoing edges is labeled by the smallest label of the outgoing edges of p. Since a vertex has outdegree at most three, an edge label can be computed in constant time.

Update. We can update this graph in $O(n)$ time per update: When l_t is moved one position to the right, we need to compute the new column of the free space which has complexity $O(n)$. When l_s is moved one position to the right, we simply delete the currently leftmost column of the free space.

Query. Using our data structure, we can determine whether there are m cluster curves between l_s and l_t. We do this by greedily searching for monotone paths from l_t to l_s. Let i be the x-coordinate of l_s. We start at the topmost vertex on l_t which has an outgoing edge labeled less or equal than i. In each vertex we follow the topmost edge which again has a label less or equal than i. This ends on a vertex (i, j) on l_s. We continue with the topmost vertex on l_t at height at most j which as before has an outgoing edge labeled less or equal than i. We stop when we have found m curves, or no more vertices on l_t with an outgoing edge labeled less or equal than i exist. When we find a horizontal path in the free space, i.e., a path from (i', j) on l_t to (i, j) on l_s, we continue with the topmost vertex on l_t at height at most $j - 1$ (else we would continue finding the same horizontal path). Note that the edge labels prevent us from going into dead ends in the graph.

Since we need to ensure that the reference subtrajectory is part of the output cluster, must take care that the vertical span $[s, t]$ is not used in any of the curves. During the query, if the y coordinate of current vertex gets smaller than t, then we do not increase the number of cluster curves and we skip down to s and continue searching for cluster curves from the vertex (t, s).

This query takes $O(n + m\ell)$ time, because the m curves will in total consist of at most n edges directed downward and $m\ell$ edges directed to the left (where we count diagonal edges as either down or left directed). There may not be more than n edges directed downward since the m curves cannot overlap in more than a point.

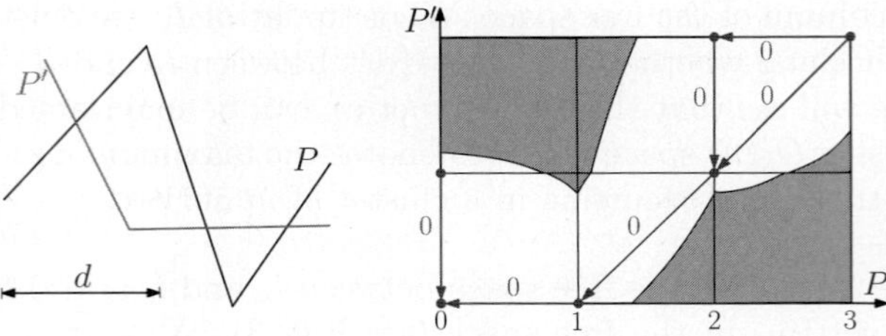

Fig. 3. Illustrating the associated data structure

Analysis. The description of the algorithm immediately gives the following lemma. Note that in practice, $m\ell$ should be of order n. Note that in practice, $m\ell$ should be of order n, i.e., the length ℓ is roughly the total length of T divided by the number of curves in the cluster, that is n/m.

Lemma 3. *The data structure of the algorithm of size $O(n\ell)$ can be maintained in $O(n)$ time per update such that number-of-cluster-curves queries can be answered in $O(n + m\ell)$ time.*

In total we have the following theorem.

Theorem 1. *A 2-distance approximation of the $\mathsf{SC}(m, \max, d)$ problem can be computed in $O(n^2 + nm\ell)$ time and $O(n\ell)$ space using the discrete Fréchet distancewhere n denotes the number of vertices of the trajectory and ℓ denotes the maximum number of vertices occurring on a reference trajectory in a cluster of m curves.*

3.2   Fréchet Distance

In this section we extend the algorithm to the continuous Fréchet distance. We first consider the case where subtrajectories in a cluster start and end at vertices and then further extend to arbitrary subtrajectories. Note that in our application it is realistic to assume that the subtrajectories in a cluster start and end at a vertex since a commuting route most often is between two places where the entity will stop and spend some time (i.e., work and home), hence generating data points.

All Subtrajectories Start and End at Vertices. We first consider the continuous Fréchet distance with the restriction that all subtrajectories start and end at vertices. In $F_d(T, T)$ each *cell* corresponds to two line segments of T and the free space in one cell is the intersection of the cell with an ellipse, possibly degenerated to the space between two parallel lines [1]. There are at most eight intersection points of the boundary of the cell with free space. We call these intersection points *critical points*. For each critical point, we put a vertex on it. Furthermore, we propagate all critical points in the free space of the corresponding row or column, respectively. That is, we propagate critical points on vertical cell boundaries horizontally, and critical points on horizontal cell boundaries vertically. In the following, we use R to denote the set of propagated critical points and vertices of the free space. Using the propagated critical points and convexity of the free space in one cell we get the following. If there is a monotone path from v to w for two vertices of v, w of R then there is a monotone polygonal path from v to w having only vertices from R. By this it suffices to compute monotone polygonal paths using vertices of R. We can extend the data structure and algorithm for the discrete Fréchet distance to work with vertices of R instead of only vertices of the free space. The size of R is $O(n^3)$ since there are a linear number of critical points in each cell. However, in our algorithm we only store and update the part of the free space between the sweep lines l_s and l_t. With this, we will get a space complexity of $O(n\ell^2)$. For the query we are only interested in monotone paths between critical points on l_s and l_t, i.e., only between critical points on vertical cell boundaries.

Reference Subtrajectory Starts and Ends at Vertices. In this section, we consider the continuous Fréchet distance with the restriction that only the reference subtrajectory has to start and end at vertices. The main modification to the algorithm we need is that we need to perform the query on all points and not only on the vertices of the free space. This does not change the time and space complexity of the algorithm.

Theorem 2. *A 2-distance approximation of the* $\mathsf{SC}(m, \max, d)$ *problem can be computed in* $O(n^2\ell)$ *time and* $O(n\ell^2)$ *space using the Fréchet distance in the case when the all or only the reference subtrajectory must start and end at vertices of the trajectory where n denotes the number of vertices of the trajectory and ℓ denotes the maximum number of vertices occurring on a reference trajectory in a cluster of m curves.*

Arbitrary Subtrajectories. We now consider the case where all subtrajectories (including the reference subtrajectory) may start and end at arbitrary points on the given trajectory. The algorithm requires two major changes: (1) we add more critical points and event points and (2) we also need to search for optimal solutions in between event points. The new critical and event points are points where free space starts or ends within a cell. For the searching in between event points we need to do computation on the quadratic curves describing the cell boundaries of a free space cell. To our knowledge, this is the first algorithm explicitly working on the boundaries of free space cells. A complete description of the algorithm can be found in the full version [4].

Theorem 3. *A 2-distance approximation of the* $\mathsf{SC}(m, \max, d)$ *problem can be computed in time* $O(n^3 m 2^{\alpha(n/m)}(\log n \log(n/m) + m))$ *using* $O(n\ell^2 + nm 2^{\alpha(n/m)})$ *space using the continuous Fréchet distance for arbitrary subtrajectories. In these bounds, n denotes the size of the given trajectory and ℓ the maximum number of vertices occurring on a reference trajectory in a cluster of m curves, and we assume that we can find the local maxima and intersections of algebraic functions of degree 4 in constant time.*

3.3   Extensions

The algorithms can be modified to handle other settings (see full version [4]), in particular the cases when

- the input is a set S of trajectories and the aim is to find the longest subtrajectory cluster of S, allowing either one or several subtrajectories of each trajectory in S,
- the number of subtrajectories in a cluster is maximized provided the length of the cluster is fixed, and
- the weak Fréchet distance is used as a distance measure.

3.4   A Note on the Experiments

To test our ideas we implemented two simplified versions of the algorithm for the continuous Fréchet distance and tested them on a set of generated trajectories whose complexity varied from 50 to 25,000 vertices. We measured the time required to run different experiments as well as the amount of memory each experiment consumed. The running time increased rapidly with the complexity of the trajectory (see full version [4]). To test the usefulness of our algorithm we tested it on a real trajectory. The trajectory was obtained by carrying a GPS for one month in a 50×50 km region and it generated 82,160 time-points. Running the memory efficient algorithm without any optimization options took approximately 7 hours and used 42 MB of memory. This is not manageable in practice but there are many simple ideas that can be used to improve the running time. A simple and effective improvement is to simplify the trajectory [12]. Using a distance threshold of 10 meters compressed the trajectory to 5,228 time-steps and is done in about 170ms. With the compressed trajectory the memory efficient algorithm computed the clusters in roughly 2 minutes using 2.65 MB of memory where the distance threshold d was set to 15 meters.

Acknowledgments

The authors would like to thank Matthew Lowry for introducing us to the problem and giving several insightful comments, Günter Rote and Thomas Wolle for helpful discussions, and Cahya Ong for the implementation.

References

1. Alt, H., Godau, M.: Computing the Fréchet distance between two polygonal curves. Internat. J. Comput. Geom. Appl. 5, 75–91 (1995)
2. Andersson, M., Gudmundsson, J., Laube, P., Wolle, T.: Reporting leadership patterns among trajectories. GeoInformatica 12(4), 497–528 (2008)
3. Buchin, K., Buchin, M., Gudmundsson, J.: Detecting single file movement. In: Proc. 16th ACM SIGSPATIAL Internat. Conf. Advances in Geographic Information Systems (to appear, 2008)

4. Buchin, K., Buchin, M., Gudmundsson, J., Löffler, M., Luo, J.: Detecting commuting patterns by clustering subtrajectories. Tech. Rep. UU-CS-2008-029, Utrecht University (2008)
5. Buchin, K., Buchin, M., Wang, Y.: Exact algorithm for partial curve matching via the Fréchet distance. In: Proc. 20th ACM-SIAM Symp. Discrete Algorithms (to appear, 2009)
6. Cao, H., Wolfson, O., Trajcevski, G.: Spatio-temporal data reduction with deterministic error bounds. The VLDB Journal 15(3), 211–228 (2006)
7. Dumitrescu, A., Rote, G.: On the Fréchet distance of a set of curves. In: Proc. 16th Canad. Conf. Comput. Geom. pp. 162–165 (2004)
8. Eiter, T., Mannila, H.: Computing discrete Fréchet distance. Tech. Rep. CD-TR 94/64, Information Systems Department, Technical University of Vienna (1994)
9. Frank, A.U.: Socio-economic units: Their life and motion. In: Frank, A.U., Raper, J., Cheylan, J.P. (eds.) Life and motion of socio-economic units, GISDATA, vol. 8, pp. 21–34. Taylor & Francis, London (2001)
10. Gaffney, S., Robertson, A., Smyth, P., Camargo, S., Ghil, M.: Probabilistic clustering of extratropical cyclones using regression mixture models. Climate Dynamics 29(4), 423–440 (2007)
11. Gaffney, S., Smyth, P.: Trajectory clustering with mixtures of regression models. In: Proc. 5th ACM SIGKDD Internat. Conf. Knowledge Discovery and Data Mining, pp. 63–72 (1999)
12. Gudmundsson, J., Katajainen, J., Merrick, D., Ong, C., Wolle, T.: Compressing spatio-temporal trajectories. In: Proc. 18th Internat. Symp. Algorithms and Computation, pp. 763–775 (2007)
13. Gudmundsson, J., Laube, P., Wolle, T.: Movement Patterns in Spatio-Temporal Data. In: Encyclopedia of GIS, pp. 726–732. Springer, Heidelberg (2008)
14. Güting, R.H., Schneider, M.: Moving Objects Databases. Morgan Kaufmann Publishers, San Francisco (2005)
15. Hadjieleftheriou, M., Kollios, G., Tsotras, V.J., Gunopulos, D.: Indexing spatio-temporal archives. The VLDB Journal 15(2), 143–164 (2006)
16. Laube, P., van Kreveld, M., Imfeld, S.: Finding REMO – detecting relative motion patterns in geospatial lifelines. In: Proc. 11th Internat Symp. Spatial Data Handling, pp. 201–214 (2004)
17. Mamoulis, N., Cao, H., Kollios, G., Hadjieleftheriou, M., Tao, Y., Cheung, D.: Mining, indexing, and querying historical spatiotemporal data. In: Proc. 10th Internat Conf. Knowledge Discovery and Data Mining, pp. 236–245 (2004)
18. Sâltenis, S., Jensen, C.S., Leutenegger, S.T., Lopez, M.A.: Indexing the positions of continuously moving objects. In: Proc. ACM SIGMOD Internat Conf. Management of Data, pp. 331–342 (2000)
19. Vlachos, M., Gunopulos, D., Kollios, G.: Discovering similar multidimensional trajectories. In: Proc. 18th Internat Conf. Data Engineering, pp. 673–684 (2002)

On the Stretch Factor of Convex Delaunay Graphs

Prosenjit Bose[1], Paz Carmi[1], Sébastien Collette[2,*], and Michiel Smid[1]

[1] School of Computer Science[**], Carleton University, Ottawa, Ontario,
K1S 5B6, Canada
[2] Computer Science Department, Université Libre de Bruxelles, CP212,
Bvd du Triomphe, 1050 Brussels, Belgium

Abstract. Let C be a compact and convex set in the plane that contains the origin in its interior, and let S be a finite set of points in the plane. The Delaunay graph $DG_C(S)$ of S is defined to be the dual of the Voronoi diagram of S with respect to the convex distance function defined by C. We prove that $DG_C(S)$ is a t-spanner for S, for some constant t that depends only on the shape of the set C. Thus, for any two points p and q in S, the graph $DG_C(S)$ contains a path between p and q whose Euclidean length is at most t times the Euclidean distance between p and q.

1 Introduction

Let S be a finite set of points in the plane and let G be a graph with vertex set S, in which each edge (p, q) has a weight equal to the Euclidean distance $|pq|$ between p and q. For a real number $t \geq 1$, we say that G is a *t-spanner* for S, if for any two points p and q of S, there exists a path in G between p and q whose Euclidean length is at most $t|pq|$. The smallest such t is called the *stretch factor* of G. The problem of constructing spanners has received much attention; see Narasimhan and Smid [12] for an extensive overview.

Spanners were introduced in computational geometry by Chew [3,4], who proved the following two results. First, the L_1-Delaunay graph, i.e., the dual of the Voronoi diagram for the Manhattan metric, is a $\sqrt{10}$-spanner. Second, the Delaunay graph based on the convex distance function defined by an equilateral triangle, is a 2-spanner. We remark that in both these results, the stretch factor is measured in the Euclidean metric. Chew also conjectured that the Delaunay graph based on the Euclidean metric, is a t-spanner, for some constant t. (If not all points of S are on a line, and if no four points of S are cocircular, then the Delaunay graph is the well-known Delaunay *triangulation*.) This conjecture was proved by Dobkin *et al.* [8], who showed that $t \leq \pi(1 + \sqrt{5})/2$. The analysis was improved by Keil and Gutwin [9], who showed that $t \leq \frac{4\pi\sqrt{3}}{9}$.

* Chargé de recherches du F.R.S.-FNRS.
** Supported by NSERC.

In this paper, we unify these results by showing that the Delaunay graph based on any convex distance function has bounded stretch factor.

Throughout this paper, we fix a compact and convex set C in the plane. We assume that the origin is in the interior of C. A *homothet* of C is obtained by scaling C with respect to the origin, followed by a translation. Thus, a homothet of C can be written as

$$x + \lambda C = \{x + \lambda z : z \in C\},$$

for some point x in the plane and some real number $\lambda \geq 0$. We call x the *center* of the homothet $x + \lambda C$.

For two points x and y in the plane, we define

$$d_C(x, y) := \min\{\lambda \geq 0 : y \in x + \lambda C\}.$$

If $x \neq y$, then this definition is equivalent to the following: Consider the translate $x + C$ and the ray emanating from x that contains y. Let y' be the (unique) intersection between this ray and the boundary of $x + C$. Then

$$d_C(x, y) = |xy|/|xy'|.$$

The function d_C is called the *convex distance function* associated with C. Clearly, we have $d_C(x, x) = 0$ and $d_C(x, y) > 0$ for all points x and y with $x \neq y$. Chew and Drysdale [5] showed that the triangle inequality $d_C(x, z) \leq d_C(x, y) + d_C(y, z)$ holds. In general, the function d_C is not symmetric, i.e., $d_C(x, y)$ is not necessarily equal to $d_C(y, x)$. If C is symmetric with respect to the origin, however, then d_C is symmetric.

Let S be a finite set of points in the plane. For each point p in S, we define

$$V'_C(p) := \{x \in \mathbb{R}^2 : \text{ for all } q \in S, d_C(x, p) \leq d_C(x, q)\}.$$

If C is not strictly convex, then the set $V'_C(p)$ may consist of a closed region of positive area with an infinite ray attached to it. For example, in figure 1, the set $V'_C(a)$ consists of the set of all points that are on or to the left of the leftmost zig-zag line, together with the infinite horizontal ray that is at the same height as the point a. Also, the intersection of two regions $V'_C(p)$ and $V'_C(q)$, where p and q are distinct points of S, may have a positive area. As a result, the collection $V'_C(p)$, where p ranges over all points of S, does not necessarily give a subdivision of the plane in which the interior of each cell is associated with a unique point of S. In order to obtain such a subdivision, we follow the approach of Klein and Wood [10] (see also Ma [11]): first, infinite rays attached to regions of positive area are not considered to be part of the region. Second, a point x in $\mathbb{R}^2$ that is in the interior of more than one region $V'_C(p)$ is assigned to the region of the lexicographically smallest point p in S for which $x \in V'_C(p)$.

To formally define Voronoi cells, let $\prec$ denote the lexicographical ordering on the set of all points in the plane. Let $p_1 \prec p_2 \prec \ldots \prec p_n$ be the points of S, sorted according to this order. Then the *Voronoi cells* $V_C(p_i)$ of the points of S are defined as $V_C(p_1) := cl(int(V'_C(p_1)))$ and, for $1 < i \leq n$,

$$V_C(p_i) := cl\left(int\left(V_C'(p_i) \setminus \left(\bigcup_{j<i} V_C(p_j)\right)\right)\right),$$

where $cl(X)$ and $int(X)$ denote the closure and the interior of the set $X \subseteq \mathbb{R}^2$, respectively.

Thus, in figure 1, the Voronoi cell $V_C(a)$ consists only of the set of all points that are on or to the left of the leftmost zig-zag line; the infinite horizontal ray that is at the same height as the point a is not part of this cell.

The *Voronoi diagram* $VD_C(S)$ of S with respect to C is defined to be the collection of Voronoi cells $V_C(p)$, where p ranges over all points of S. An example is given in figure 1.

As for the Euclidean case, the Voronoi diagram $VD_C(S)$ induces Voronoi cells, Voronoi edges, and Voronoi vertices. Each point in the plane is either in the interior of a unique Voronoi cell, in the relative interior of a unique Voronoi edge, or a unique Voronoi vertex. Each Voronoi edge e belongs only to the two Voronoi cells that contain e on their boundaries. Observe that Voronoi cells are closed.

The Delaunay graph is defined to be the dual of the Voronoi diagram:

Definition 1. *Let S be a finite set of points in the plane. The* Delaunay graph *$DG_C(S)$ of S with respect to C is defined to be the dual of the Voronoi diagram $VD_C(S)$. That is, the vertex set of $DG_C(S)$ is S and two distinct vertices p and q are connected by an edge in $DG_C(S)$ if and only if the Voronoi cells $V_C(p)$ and $V_C(q)$ share a Voronoi edge.*

For example, the Delaunay graph $DG_C(S)$ for the point set in figure 1 consists of the five edges (a, b), (a, d), (b, c), (b, d), and (d, e).

We consider the Delaunay graph $DG_C(S)$ to be a geometric graph, which means that each edge (p, q) is embedded as the closed line segment with endpoints p and q.

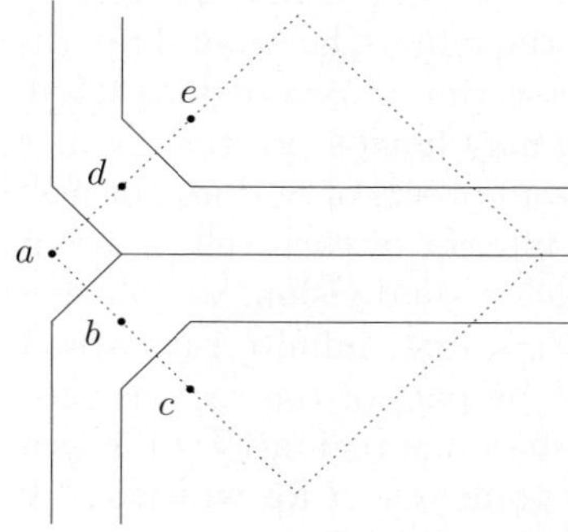

Fig. 1. The Voronoi diagram $VD_C(S)$ for the set $S = \{a, b, c, d, e\}$. The set C is the square as indicated by the dotted figure; the origin is at the center of C.

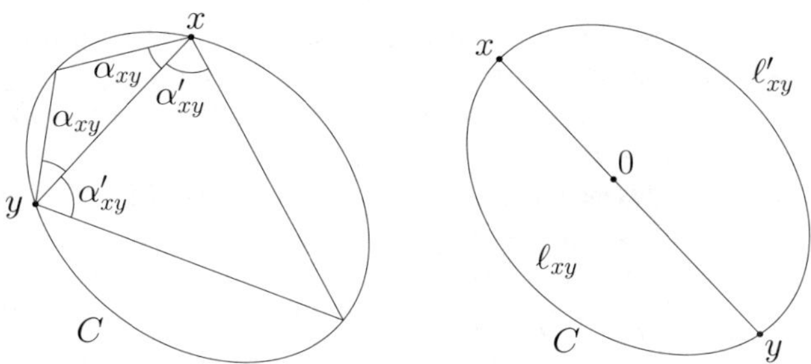

Fig. 2. The two parameters associated with C

Before we can state the main result of this paper, we introduce two parameters whose values depend on the shape of the set C. Let x and y be two distinct points on the boundary ∂C of C. These points partition ∂C into two chains. For each of these chains, there is an isosceles triangle with base xy and whose third vertex is on the chain. Denote the base angles of these two triangles by α_{xy} and α'_{xy}; see figure 2 (left). We define

$$\alpha_C := \min\{\max(\alpha_{xy}, \alpha'_{xy}) : x, y \in \partial C, x \neq y\}.$$

Consider again two distinct points x and y on ∂C, but now assume that x, y, and the origin are collinear. As before, x and y partition ∂C into two chains. Let ℓ_{xy} and ℓ'_{xy} denote the lengths of these chains; see figure 2 (right). We define

$$\kappa_{C,0} := \max\left\{\frac{\max(\ell_{xy}, \ell'_{xy})}{|xy|} : x, y \in \partial C, x \neq y, \text{ and } x, y, \text{ and } 0 \text{ are collinear}\right\}.$$

Clearly, the convex distance function d_C and, therefore, the Voronoi diagram $VD_C(S)$, depends on the location of the origin in the interior of C. Surprisingly, the Delaunay graph $DG_C(S)$ does not depend on this location; see Ma [11, Section 2.1.6]. We define

$$\kappa_C := \min\{\kappa_{C,0} : \ 0 \text{ is in the interior of } C\}.$$

In this paper, we will prove the following result:

Theorem 1. *Let C be a compact and convex set in the plane with a non-empty interior, and let S be a finite set of points in the plane. The stretch factor of the Delaunay graph $DG_C(S)$ is less than or equal to*

$$t_C := \begin{cases} 2\kappa_C \cdot \max\left(\frac{3}{\sin(\alpha_C/2)}, \kappa_C\right) & \text{if } DG_C(S) \text{ is a triangulation,} \\ 2\kappa_C^2 \cdot \max\left(\frac{3}{\sin(\alpha_C/2)}, \kappa_C\right) & \text{otherwise.} \end{cases}$$

Thus, for any two points p and q in S, the graph $DG_C(S)$ contains a path between p and q whose Euclidean length is at most t_C times the Euclidean distance between p and q.

We emphasize that we do *not* make any "general position" assumption; our proof of Theorem 1 is valid for *any* finite set of points in the plane.

Throughout the rest of this paper, we assume that the origin is chosen in the interior of C such that $\kappa_C = \kappa_{C,0}$.

The rest of this paper is organized as follows. In Section 2, we prove some basic properties of the Delaunay graph which are needed in the proof of Theorem 1. Section 3 contains a proof of Theorem 1. This proof is obtained by showing that the Delaunay graph satisfies the "diamond property" and a variant of the "good polygon property" of Das and Joseph [6]. The proof of the latter property is obtained by generalizing the analysis of Dobkin *et al.* [8] for the lengths of so-called one-sided paths.

Due to the lack of space, some proofs are omitted or are sketched.

2 Some Properties of the Delaunay Graph

Recall that in the Euclidean Delaunay graph, if two points p and q of S are connected by an edge, then there exists a disk having p and q on its boundary that does not contain any point of S in its interior. The next lemma generalizes this result to the Delaunay graph $DG_C(S)$.

Lemma 1. *Let p and q be two points of S and assume that (p, q) is an edge in the Delaunay graph $DG_C(S)$. Then, the following are true.*

1. *The line segment between p and q does not contain any point of $S \setminus \{p, q\}$.*
2. *For every point x in $V_C(p) \cap V_C(q)$, there exists a real $\lambda > 0$ such that*
    (a) the homothet $x + \lambda C$ contains p and q on its boundary, and
    (b) the interior of $x + \lambda C$ does not contain any point of S.

Due to space constraints, the proof is omitted.

As can be seen in figure 1, Voronoi cells are, in general, not convex. They are, however, star-shaped:

Lemma 2. *Let p be a point of S and let x be a point in the Voronoi cell $V_C(p)$. Then the line segment xp is completely contained in $V_C(p)$.*

Due to space constraints, the proof is omitted.

It is well known that the Euclidean Delaunay graph is a plane graph; see, for example, de Berg *et al.* [7, page 189]. The following lemma states that this is true for the Delaunay graph $DG_C(S)$ as well.

Lemma 3. *The Delaunay graph $DG_C(S)$ is a plane graph.*

Even though this fact seems to be well known, we have not been able to find a proof in the literature. The full proof is given in the full version of this paper.

3   The Stretch Factor of Delaunay Graphs

In this section, we will prove Theorem 1. First, we show that the Delaunay graph $DG_C(S)$ satisfies the diamond property and a variant of the good polygon property of Das and Joseph [6]. According to the results of Das and Joseph, this immediately implies that the stretch factor of $DG_C(S)$ is bounded. In fact, we will obtain an upper bound on the stretch factor which is better than the one that is implied by Das and Joseph's result.

3.1   The Diamond Property

Let G be a plane graph with vertex set S and let α be a real number with $0 < \alpha < \pi/2$. For any edge e of G, let Δ_1 and Δ_2 be the two isosceles triangles with base e and base angle α; see figure 3. We say that e satisfies the α-*diamond property*, if at least one of the triangles Δ_1 and Δ_2 does not contain any point of S in its interior. The graph G is said to satisfy the α-*diamond property*, if every edge e of G satisfies this property.

Lemma 4. *Consider the value* α_C *that was defined in Section 1. The Delaunay graph* $DG_C(S)$ *satisfies the* α_C-*diamond property.*

Proof. Let (p, q) be an arbitrary edge of $DG_C(S)$ and let x be any point in the relative interior of $V_C(p) \cap V_C(q)$. By Lemma 1, there exists a real number $\lambda > 0$ such that p and q are on the boundary of the homothet $x + \lambda C$ and no point of S is in the interior of $x + \lambda C$. The points p and q partition $\partial(x + \lambda C)$ into two chains. For each of these chains, there is an isosceles triangle with base pq and whose third vertex is on the chain. We denote the base angles of these two triangles by β and γ; see figure 4. We may assume without loss of generality that $\beta \geq \gamma$. Let a denote the third vertex of the triangle with base angle β. If we translate $x + \lambda C$ so that x coincides with the origin and scale the translated homothet by a factor of $1/\lambda$, then we obtain the set C. This translation and scaling does

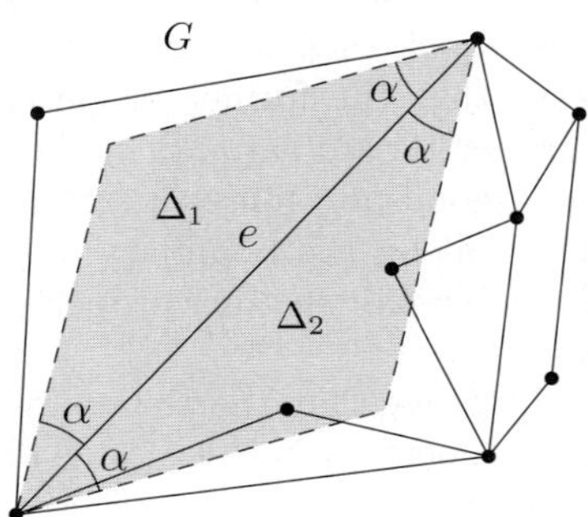

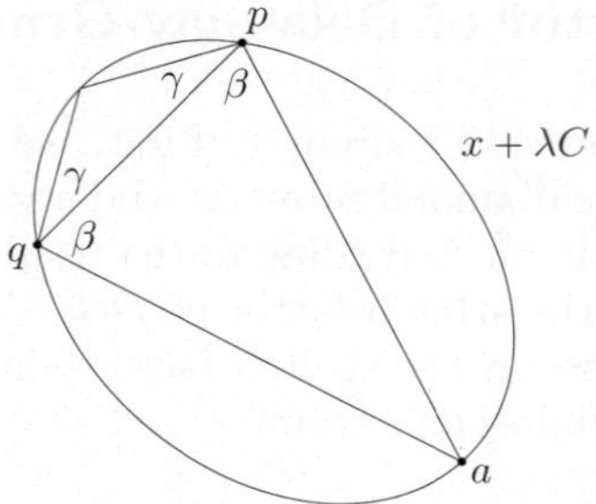

Fig. 4. Illustrating the proof of Lemma 4

not change the angles β and γ. Thus, using the notation of Section 1 (see also figure 2), we have $\{\beta, \gamma\} = \{\alpha_{pq}, \alpha'_{pq}\}$. The definition of α_C then implies that

$$\alpha_C \leq \max(\alpha_{pq}, \alpha'_{pq}) = \beta.$$

Let Δ be the isosceles triangle with base pq and base angle α_C such that a and the third vertex of Δ are on the same side of pq. Then Δ is contained in the triangle with vertices p, q, and a. Since the latter triangle is contained in $x + \lambda C$, it does not contain any point of S in its interior. Thus, Δ does not contain any point of S in its interior. This proves that the edge (p, q) satisfies the α_C-diamond property. $\qquad\square$

3.2   The Visible-Pair Spanner Property

For a real number $\kappa \geq 1$, we say that the plane graph G satisfies the *strong visible-pair κ-spanner property*, if the following is true: For every face f of G, and for every two vertices p and q on the boundary of f, such that the open line segment joining p and q is completely in the interior of f, the graph G contains a path between p and q having length at most $\kappa|pq|$. If for every face f of G and for every two vertices p and q on the boundary of f, such that the line segment pq does not intersect the exterior of f, the graph G contains a path between p and q having length at most $\kappa|pq|$, then we say that G satisfies the *visible-pair κ-spanner property*. Observe that the former property implies the latter one. Also, observe that these properties are variants of the *κ-good polygon* property of Das and Joseph [6]: The κ-good polygon property requires that G contains a path between p and q that is along the boundary of f and whose length is at most $\kappa|pq|$; in the (strong) visible-pair spanner property, the path is not required to be along the boundary of f.

In this subsection, we will prove that the Delaunay graph $DG_C(S)$ satisfies the visible-pair κ_C-spanner property, where κ_C is as defined in Section 1. This claim will be proved by generalizing results of Dobkin *et al.* [8] on so-called one-sided paths.

Let p and q be two distinct points of S and assume that (p, q) is not an edge of the Delaunay graph $DG_C(S)$. Consider the Voronoi diagram $VD_C(S)$. We consider the sequence of points in S whose Voronoi cells are visited when the line segment pq is traversed from p to q. If pq does not contain any Voronoi vertex, then this sequence forms a path in $DG_C(S)$ between p and q. Since, in general, Voronoi cells are not convex, it may happen that this path contains duplicates. In order to avoid this, we define the sequence in the following way.

In the rest of this section, we will refer to the line through p and q as the X-axis, and we will say that p is to the *left* of q. This implies a *left-to-right* order on the X-axis, the notion of a point being *above* or *below* the X-axis, as well as the notions *horizontal* and *vertical*. (Thus, conceptually, we rotate and translate all points of S , the set C, the Voronoi diagram $VD_C(S)$, and the $DG_C(S)$, such that p and q are on a horizontal line and p is to the left of q. Observe that $VD_C(S)$ is still defined based on the lexicographical order of the points of S before this rotation and translation.) In the following, we consider the (horizontal) line segment pq. If this segment contains a Voronoi vertex, then we imagine moving pq vertically upwards by an infinitesimal amount. Thus, we may assume that pq does not contain any Voronoi vertex of the (rotated and translated) Voronoi diagram $VD_C(S)$.

The first point in the sequence is $p_0 := p$. We define $x_1 \in \mathbb{R}^2$ to be the point on the line segment pq such that $x_1 \in V_C(p_0)$ and x_1 is closest to q.

Let $i \geq 1$ and assume that the points $p_0, p_1, \ldots, p_{i-1}$ of S and the points $x_1, \ldots, x_i$ in $\mathbb{R}^2$ have already been defined, where x_i is the point on the line segment pq such that $x_i \in V_C(p_{i-1})$ and x_i is closest to q. If $p_{i-1} = q$, then the construction is completed. Otherwise, observe that x_i is in the relative interior of a Voronoi edge. We define p_i to be the point of $S \setminus \{p_{i-1}\}$ whose Voronoi cell contains x_i on its boundary, and define x_{i+1} to be the point on the line segment pq such that $x_{i+1} \in V_C(p_i)$ and x_{i+1} is closest to q.

Let $p = p_0, p_1, \ldots, p_k = q$ be the sequence of points in S obtained in this way. By construction, these $k+1$ points are pairwise distinct and for each i with $1 \leq i \leq k$, the Voronoi cells $V_C(p_{i-1})$ and $V_C(p_i)$ share an edge. Therefore, by definition, (p_{i-1}, p_i) is an edge in $DG_C(S)$. Thus, $p = p_0, p_1, \ldots, p_k = q$ defines a path in $DG_C(S)$ between p and q. We call this path the *direct path* between p and q. If all points $p_1, p_2, \ldots, p_{k-1}$ are strictly on one side of the line through p and q, then we say that the direct path is *one-sided*.

We will show in Lemma 6 that the length of a one-sided path is at most $\kappa_C |pq|$. The proof of this lemma uses a geometric property which we prove first.

Let C' be a homothet of C whose center is on the X-axis, and let x and y be two points on the boundary of C' that are on or above the X-axis. The points x and y partition the boundary of C' into two chains. One of these chains is completely on or above the X-axis; we denote this chain by $arc(x, y; C')$. The length of this chain is denoted by $|arc(x, y; C')|$.

For two points x and y on the X-axis, we write $x <_X y$ if x is strictly to the left of y, and we write $x \leq_X y$ if $x = y$ or $x <_X y$.

We now state the geometric property, which is illustrated in figure 5. Recall the value κ_C that was defined in Section 1.

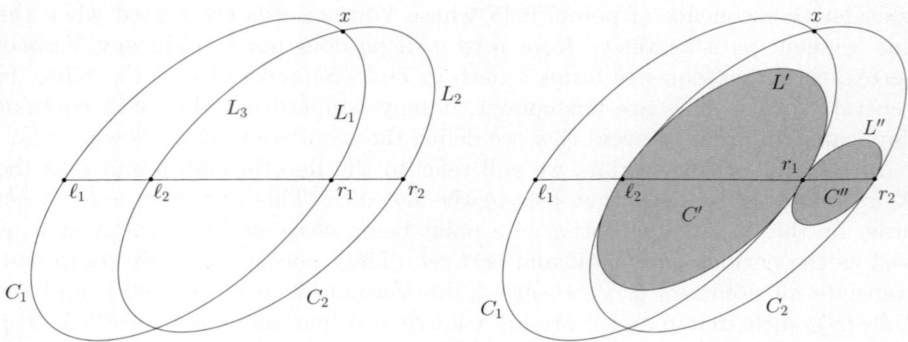

Fig. 5. Illustrating the proof of Lemma 5

Lemma 5. *Let $C_1 = y_1 + \lambda_1 C$ and $C_2 = y_2 + \lambda_2 C$ be two homothets of C whose centers y_1 and y_2 are on the X-axis. Assume that $\lambda_1 > 0$, $\lambda_2 > 0$, and $y_1 <_X y_2$. For $i = 1, 2$, let ℓ_i and r_i be the leftmost and rightmost points of C_i on the X-axis, respectively. Assume that $r_1 \leq_X r_2$ and $\ell_1 \leq_X \ell_2 <_X r_1$. Let x be a point that is on the boundaries of both C_1 and C_2 and on or above the X-axis. Let $L_1 = |arc(x, r_1; C_1)|$ and $L_2 = |arc(x, r_2; C_2)|$. Then*

$$L_2 \leq L_1 + \kappa_C |r_1 r_2|.$$

Proof. We define $L_3 = |arc(\ell_2, x; C_2)|$. Let C' be the homothet of C whose center is on the X-axis such that the intersection between C' and the X-axis is equal to the line segment $\ell_2 r_1$, and let $L' = |arc(\ell_2, r_1; C')|$; see Fig. 5. Observe that, for $\lambda := |\ell_2 r_1|/|\ell_2 r_2|$, C' is obtained from C_2 by a scaling by a factor of λ. Thus, since $|arc(\ell_2, r_2; C_2)| = L_2 + L_3$, we have $L' = \lambda(L_2 + L_3)$.

Let C'' be the homothet of C whose center is on the X-axis such that the intersection between C'' and the X-axis is equal to the line segment $r_1 r_2$, and let $L'' = |arc(r_1, r_2; C'')|$. Since C'' is obtained from C_2 by a scaling by a factor of $1 - \lambda$, we have $L'' = (1 - \lambda)(L_2 + L_3)$.

Thus, we have $L' + L'' = L_2 + L_3$. By convexity, we have $C' \subseteq C_1 \cap C_2$. Then it follows, again from convexity (see Benson [1, page 42]), that $L' \leq L_1 + L_3$. Thus, we have $L_2 + L_3 = L' + L'' \leq L_1 + L_3 + L''$, which implies that $L_2 \leq L_1 + L''$. Since, by the definition of κ_C, $L'' \leq \kappa_C |r_1 r_2|$, the proof is complete. $\square$

We are now ready to prove an upper bound on the length of a one-sided path.

Lemma 6. *If the direct path between p and q is one-sided, then its length is at most $\kappa_C |pq|$.*

Due to space constraints, we only give a proof sketch.

Proof (Sketch). As above, we assume that p and q are on the X-axis and that p is to the left of q. Consider the direct path $p = p_0, p_1, \ldots, p_k = q$ in $DG_C(S)$ and the sequence $x_1, x_2, \ldots, x_k$, as defined above. Since the direct path is one-sided, we may assume without loss of generality that the points $p_1, p_2, \ldots, p_{k-1}$ are strictly above the X-axis.

Recall that, for each i with $1 \leq i \leq k$, x_i is in the relative interior of $V_C(p_{i-1}) \cap V_C(p_i)$ and x_i is on the line segment pq. Therefore, by Lemma 1, if we define $\lambda_i := d_C(x_i, p_{i-1})$ (which is equal to $d_C(x_i, p_i)$), then the homothet $C_i := x_i + \lambda_i C$ contains p_{i-1} and p_i on its boundary and no point of S is in its interior.

Remark that the length of the upper boundary of the union of the C_i is an upper bound on the length of the one-sided path. If there are k edges, then the upper boundary is composed of portions of boundaries of k homothets. Now we proceed by induction on the number of edges (and thus of homothets):

For the base case, i.e., when there is one edge, the path is bounded by the upper boundary of a single homothet, and thus the fact that the length of the path between p and q is at most $\kappa_C |pq|$ follows from the definition of κ_C.

Assume it is true for all paths up to length k. By Lemma 5, a case analysis shows that when we add one more edge to the path, its length is not increasing by more than κ_C times the length of the projection of the added edge on the X-axis. This implies that the whole path has length less than $\kappa_C |pq|$. $\square$

We are now ready to prove that the Delaunay graph satisfies the visible-pair spanner property:

Lemma 7. *The Delaunay graph $DG_C(S)$ satisfies the visible-pair κ_C-spanner property.*

Proof. Recall from Lemma 3, that the graph $DG_C(S)$ is plane. It suffices to prove that $DG_C(S)$ satisfies the strong visible-pair κ_C-spanner property. Let f be a face of G and let p and q be two vertices on f such that the open line segment between p and q is contained in the interior of f. We have to show that there is a path in $DG_C(S)$ between p and q whose length is at most $\kappa_C |pq|$.

As before, we assume that p and q are on the X-axis and that p is to the left of q. Consider the direct path $p = p_0, p_1, \ldots, p_k = q$ in $DG_C(S)$ and the sequence $x_1, x_2, \ldots, x_k$, as defined in the beginning of this section. We will show that the direct path is one-sided. The lemma then follows from Lemma 6.

Since the open line segment between p and q is in the interior of f, none of the points $p_1, \ldots, p_{k-1}$ is on the closed line segment pq. Assume that for some i with $1 \leq i < k$, p_i is on the X-axis. Then p_i is either strictly to the left of p or strictly to the right of q. We may assume without loss of generality that p_i is strictly to the right of q. Consider the point x_i and the homothet $C_i = x_i + \lambda_i C$ as in the proof of Lemma 6. Since x_i is on pq and in the interior of C_i, and since p_i is on the boundary of C_i, it follows from convexity that q is in the interior of C_i, which is a contradiction. Thus we have shown that none of the points $p_1, \ldots, p_{k-1}$ is on the X-axis.

Assume that the direct path is not one-sided. Then there is an edge (p_{i-1}, p_i) on this path such that one of p_{i-1} and p_i is strictly below the X-axis and the other point is strictly above the X-axis. Let z be the intersection between $p_{i-1}p_i$ and the X-axis. By assumption, z is not on the open line segment joining p and q, and by Lemma 1, $z \neq p$ and $z \neq q$. Thus, z is either strictly to the left of

p or strictly to the right of q. We may assume without loss of generality that z is strictly to the right of q. Consider again the point x_i and the homothet $C_i = x_i + \lambda_i C$ as in the proof of Lemma 6. This homothet contains the points x_i, p_{i-1} and p_i. Thus, by convexity, C_i contains the triangle with vertices x_i, p_{i-1}, and p_i. Since q is in the interior of this triangle, it follows that q is in the interior of C_i, which is a contradiction. $\qquad\square$

3.3   The Proof of Theorem 1

Das and Joseph [6] have shown that any plane graph satisfying the diamond property and the good polygon property has a bounded stretch factor. The analysis of the stretch factor was slightly improved by Bose *et al.* [2]. A close inspection of the proof in [2] shows that the following holds: Let G be a geometric graph with the following four properties:

1. G is plane.
2. G satisfies the α-diamond property.
3. The stretch factor of any one-sided path in G is at most κ.
4. G satisfies the visible-pair κ'-spanner property.

Then, G is a t-spanner for

$$t = 2\kappa\kappa' \cdot \max\left(\frac{3}{\sin(\alpha/2)}, \kappa\right).$$

We have shown that the Delaunay graph $DG_C(S)$ satisfies all these properties: By Lemma 3, $DG_C(S)$ is plane. By Lemma 4, $DG_C(S)$ satisfies the α_C-diamond property. By Lemma 6, the stretch factor of any one-sided path in $DG_C(S)$ is at most κ_C. By Lemma 7, $DG_C(S)$ satisfies the visible-pair κ_C-spanner property. If $DG_C(S)$ is a triangulation, then obviously, $DG_C(S)$ satisfies the visible-pair 1-spanner property. Therefore, we have completed the proof of Theorem 1.

4   Concluding Remarks

We have considered the Delaunay graph $DG_C(S)$, where C is a compact and convex set with a non-empty interior and S is a finite set of points in the plane. We have shown that the (Euclidean) stretch factor of $DG_C(S)$ is bounded from above by a function of two parameters α_C and κ_C that are determined only by the shape of C. Roughly speaking, these two parameters give a measure of the "fatness" of the set C.

Our analysis provides the first generic bound valid for any compact and convex set C. In all previous works, only special examples of such sets C were considered. Furthermore, our approach does not make any "general position" assumption about the point set S.

Note that for the Euclidean Delaunay triangulation, we derive an upper bound on the stretch factor of ≈ 24.6, which is worse than the currently best known upper bound ($4\pi\sqrt{3}/9 \approx 2.42$ as proved by Keil and Gutwin [9]). We leave open the problem of improving our upper bound. In particular, is it possible to generalize the techniques of Dobkin *et al.* [8] and Keil and Gutwin [9]?

References

1. Benson, R.V.: Euclidean Geometry and Convexity. McGraw-Hill, New York (1966)
2. Bose, P., Lee, A., Smid, M.: On generalized diamond spanners. In: Dehne, F., Sack, J.-R., Zeh, N. (eds.) WADS 2007. LNCS, vol. 4619, pp. 325–336. Springer, Heidelberg (2007)
3. Chew, L.P.: There is a planar graph almost as good as the complete graph. In: Proceedings of the 2nd ACM Symposium on Computational Geometry, pp. 169–177 (1986)
4. Chew, L.P.: There are planar graphs almost as good as the complete graph. Journal of Computer and System Sciences 39, 205–219 (1989)
5. Chew, L.P., Drysdale, R.L.: Voronoi diagrams based on convex distance functions. In: Proceedings of the 1st ACM Symposium on Computational Geometry, pp. 235–244 (1985)
6. Das, G., Joseph, D.: Which triangulations approximate the complete graph? In: Djidjev, H.N. (ed.) Optimal Algorithms. LNCS, vol. 401, pp. 168–192. Springer, Heidelberg (1989)
7. de Berg, M., van Kreveld, M., Overmars, M., Schwarzkopf, O.: Computational Geometry: Algorithms and Applications, 2nd edn. Springer, Berlin (2000)
8. Dobkin, D.P., Friedman, S.J., Supowit, K.J.: Delaunay graphs are almost as good as complete graphs. Discrete & Computational Geometry 5, 399–407 (1990)
9. Keil, J.M., Gutwin, C.A.: Classes of graphs which approximate the complete Euclidean graph. Discrete & Computational Geometry 7, 13–28 (1992)
10. Klein, R., Wood, D.: Voronoi diagrams based on general metrics in the plane. In: Cori, R., Wirsing, M. (eds.) STACS 1988. LNCS, vol. 294, pp. 281–291. Springer, Heidelberg (1988)
11. Ma, L.: Bisectors and Voronoi Diagrams for Convex Distance Functions. Ph.D. thesis, Department of Computer Science, FernUniversität Hagen, Germany (2000)
12. Narasimhan, G., Smid, M.: Geometric Spanner Networks. Cambridge University Press, Cambridge (2007)

Covering a Simple Polygon by Monotone Directions

Hee-Kap Ahn[1], Peter Brass[2], Christian Knauer[3], Hyeon-Suk Na[4],
and Chan-Su Shin[5]

[1] Department of Computer Science and Engineering, POSTECH, Korea
`heekap@postech.ac.kr`
[2] Department of Computer Science, City College, New York, USA
`peter@cs.ccny.cuny.edu`
[3] Institute of Computer Science, Free University Berlin, Germany
`knauer@inf.fu-berlin.de`
[4] School of Computing, Soongsil University, Seoul, Korea
`hsnaa@ssu.ac.kr`
[5] School of Electrical and Information Engineering,
Hankuk University of Foreign Studies, Korea
`cssin@hufs.ac.kr`

Abstract. In this paper we study the problem of finding a set of k directions for a given simple polygon P, such that for each point $p \in P$ there is at least one direction in which the line through p intersects the polygon only once. For $k = 1$, this is the classical problem of finding directions in which the polygon is monotone, and all such directions can be found in linear time for a simple n-gon. For $k > 1$, this problem becomes much harder; we give an $O(n^5 \log^2 n)$-time algorithm for $k = 2$, and $O(n^{3k+2})$-time algorithm for $k \geq 3$. These results are the first on the generalization of the monotonicity problem.

1   Introduction

A polygon P is said to be *monotone* in a direction α if every line in direction α intersects P in at most one connected component. Determining if a given polygon is monotone is a well-studied problem. Preparata and Supowit [7] presented a linear time algorithm to find all directions in which a given polygon is monotone. Rappaport and Rosenbloom [8] gave a linear time algorithm to determine whether the boundary of a simple polygon can be decomposed into two monotone chains.

In this paper we consider a generalization of the monotonicity problem: given a simple polygon P, find a set $\mathcal{D}$ of k directions such that each point $p \in P$ is *covered* by at least one direction of $\mathcal{D}$, in the sense that the line through p in this direction intersects P in one connected component. A simple polygon having such a set $\mathcal{D}$ of k directions is called k-*monotone* for $\mathcal{D}$.

Only few generalization of the monotonicity has been known so far. Bose and Kreveld [1][2] studied rotational plane sweep on a simple polygon with the restriction that the sweep line intersects the polygon in at most one connected

S.-H. Hong, H. Nagamochi, and T. Fukunaga (Eds.): ISAAC 2008, LNCS 5369, pp. 668–679, 2008.

component. Monotonicity of three dimensional polyhedron was introduced by Toussaint [9], and studied by Ha, Yoo and Hahn [4]. Up to our best knowledge, our results are the first on the generalized k-monotonicity problem for the simple polygon.

Our results. Our results are summarized as follows:

Theorem 1. *Given a direction α and a simple polygon P with n vertices, we can test in $O(n^2 \log^2 n)$ time if there is a direction β such that P is 2-monotone for $\{\alpha, \beta\}$.*

Theorem 2. *Given a simple polygon P with n vertices, we can find in $O(n^5 \log^2 n)$ time all pairs of two directions for which P is 2-monotone. If no such pair is found, then P is not 2-monotone.*

Theorem 3. *Given a simple polygon P with n vertices and a fixed number $k \geq 3$, we can find in $O(n^{3k+2})$ time all sets of k directions for which P is k-monotone. If no such set is found, then P is not k-monotone.*

The remainder of this paper is organized as follows: We first consider the case $k = 2$ in Section 3. In Section 3.1 and Section 3.2, we give algorithms proving Theorems 1 and 2, respectively. We then consider the general case $k \geq 3$ and give an algorithm for Theorem 3 in Section 4.

2   Preliminaries

Throughout the paper, $P = (v_1, \ldots, v_n)$ is a simple polygon with n vertices $v_1, \ldots, v_n$ ordered counterclockwise. We denote by ∂P the boundary of the polygon P. Without loss of generality, we assume that the k directions lie in the range $[0, \pi)$.

There are three different notations of a line: we use $\ell(p, q)$ to denote the line through two points p and q, $\ell(e)$ to denote the line containing an edge or a segment e, and $\ell_\gamma(p)$ to denote the line through p in direction γ. We use $\overline{pq}$ to denote the line segment connecting two points p and q. Note that a line defines a direction (or an angle) in the angle range $[0, \pi)$, so we will identify the line with its direction. Similarly, an edge or a segment in P will be identified with their directions in the range $[0, \pi)$. A direction γ is said to *cover* a point p in P if $\ell_\gamma(p)$ intersects P in one connected component.

Before proving the results, we first show that the candidate directions for which a polygon is k-monotone are not necessarily directions defined by edges or pairs of vertices. Consider a polygon with three reflex vertices u, u' and v as in Figure 1(a). To cover the points in the neighborhood of u, we must set the first direction α to be between the directions of two edges incident to u. Indeed, any α in the range covers the whole polygon, except the points in the gray region lying below $\ell_\alpha(v)$, as shown in Figure 1. If there is a second direction to cover this gray region, then the polygon is 2-monotone. Take α in the middle of directions of two edges incident to u, and define β to be the direction defined by u and the leftmost

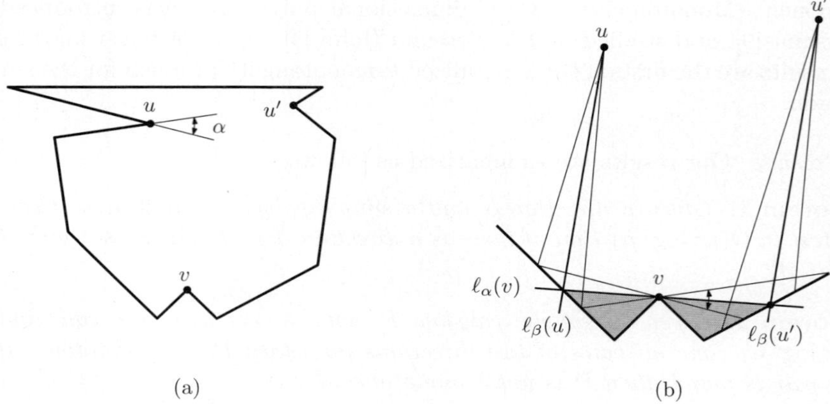

Fig. 1. There is only one pair of directions for which the polygon is 2-monotone. These directions are not defined by edges or pairs of vertices.

intersection of $\ell_\alpha(v)$ with the polygon boundary. Set the third reflex vertex u' to lie on the line in direction β that passes through the rightmost intersection of $\ell_\alpha(v)$ with the polygon boundary. Then it is not difficult to see that the polygon is 2-monotone for $\{\alpha, \beta\}$ and that it is the only pair of directions for which the polygon is 2-monotone; for any other α, any direction β cannot cover all points in the interior of the polygon. Clearly, these directions are not defined by any edges or pairs of vertices.

3   Monotonicity for Two Directions

Given a direction α, let P_α be the set of points $p \in P$ such that $\ell_\alpha(p) \cap P$ is one connected component. In other words, P_α is the set of all the points in P that are covered by α. Then $P \setminus P_\alpha$ is a collection of strips whose boundaries are parallel to α. We call the closure of such a strip an α-*bad strip*. If two α-bad strips share a vertex, we consider their union as one α-bad strip. The boundary of an α-bad strip consists of at most two sides which are parallel to α, and at most two polygonal chains from ∂P which connect endpoints of two sides. Then P is partitioned into several strips parallel to α, which are α-bad and not α-bad strips appearing in alternating order along y-axis, as illustrated in Figure 2. The polygon in Figure 2 has five parallel strips, two of which are α-bad strips S and S'; the strip S has two sides parallel to α, one of which passes through u. Its left chain is the portion of ∂P between the left endpoints of the two sides.

   If P is 2-monotone for α and some other direction β, every point p in α-bad strips must be covered by β, i.e., $\ell_\beta(p)$ must intersect P in one connected component. Thus we examine first a way of finding a direction β for given α such that every α-bad strip of P is covered by β, and then extend it to find two directions $\{\alpha, \beta\}$ for which P is 2-monotone.

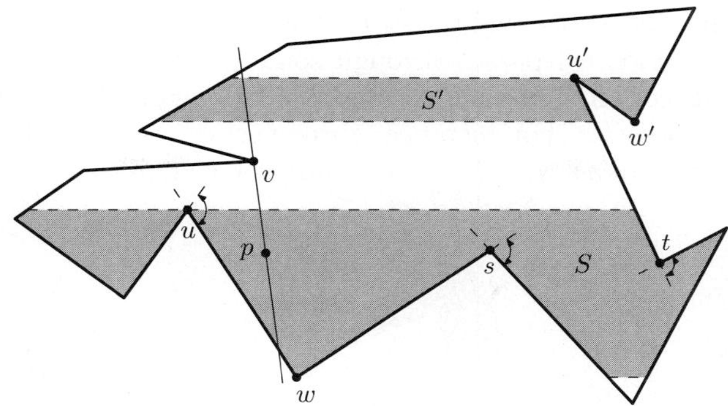

Fig. 2. For the horizontal direction α, the shaded regions in P are α-bad strips

3.1   Finding a Direction β for a Fixed Direction α

This section is devoted to the proof of Theorem 1: for given α, test in $O(n^2 \log^2 n)$ time if there is a direction β such that P is 2-monotone for $\{\alpha, \beta\}$.

We first compute the α-bad strips by plane sweep in $O(n \log n)$ time. The second direction must cover all points in the α-bad strips. We say a direction β is *forbidden* if there is a point p in an α-bad strip such that $\ell_\beta(p)$ intersects P in more than one connected component. So for every point p in an α-bad strip, we can define the set of angle intervals of the forbidden directions β such that $\ell_\beta(p)$ intersects P in more than one connected component. From this, we can solve the problem by computing the union of the forbidden intervals over all points in the α-bad strips and checking if the union is the whole angle space. If the union is the whole angle space, there is no valid second direction.

In the sequel, we will define an angle interval $f_\alpha(S, v)$ of forbidden directions for any pair of an α-bad strip S and a reflex vertex v of P. In Lemma 4 we will show that the union of $f_\alpha(S, v)$ for all pairs (S, v) equals to the union of $f_\alpha(p)$ for all points p in the α-bad strips. Without loss of generality, we assume that α is the horizontal direction.

Forbidden intervals. We define the forbidden interval $f_\alpha(S, v)$ for a pair (S, v) where S is an α-bad strip and v is a reflex vertex of P.

Let e_1 and e_2 be the edges incident to v, where e_1 appears right before e_2 in counterclockwise order. Two extensions $\ell(e_1)$ and $\ell(e_2)$ partition the region around v into four wedges as shown in Figure 3(a); the one containing the outside of P is denoted by $W_4(v)$, and by $W_2(v)$ its diagonally faced wedge. The remaining two wedges are denoted by $W_1(v)$ and $W_3(v)$, where $W_1(v)$, $W_2(v)$, $W_3(v)$, $W_4(v)$ are ordered in counterclockwise. The union of two wedges $W_1(v)$ and $W_3(v)$ forms a *double wedge* $W_{13}(v)$. Depending on the geometric relation between the wedges and S, we define the following three types of forbidden intervals.

Type A: This is the case that $v \in S$. We set $f_\alpha(S, v)$ to be the (open) angle interval of $W_{13}(v)$. Any direction $\beta \in f_\alpha(S, v)$ is forbidden because we can translate the line $\ell_\beta(v)$ slightly so that it intersects both edges e_1 and e_2 incident to v and still passes through some point in S, which means that the line in direction β passing through that point in S intersects P in two or more components. For instance, three vertices u, s, t in Figure 2 define forbidden intervals of type A for the same α-bad strip S.

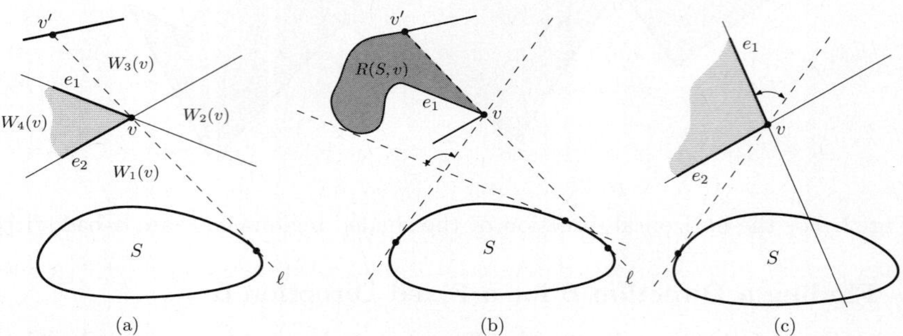

Fig. 3. Forbidden interval of Type B and Type C for a pair (S, v)

For the other two types, let $v \in P \setminus S$. By the definition of α-bad strip, any point of P having y-coordinate between the minimum and maximum y-coordinates of points in S must be contained in S. Thus the y-coordinate of v must be either higher or lower than that of any point in S. Without loss of generality, assume that v is above S.

Type B: This is the case that $S \cap W_2(v) = \emptyset$ as in Figure 3(a) and 3(b). Let ℓ be the line through v and tangent to S such that e_1 and S lie in the same side of ℓ. Let v' be the first intersection of the ray from v with ∂P in the upward direction of ℓ. If the segment $\overline{vv'}$ is inside P, then it cuts P into two pieces, and we denote the one not containing S by $R(S, v)$. No point of $R(S, v)$ is visible from any point in S, i.e., for any pair of points $p \in S$ and $q \in R(S, v)$, $\ell(p, q)$ intersects P in more than one connected components and thus the direction is forbidden. So we take the double wedge containing $R(S, v)$ and S whose boundaries are two (crossing) tangent lines between them, as in Figure 3(b), and define $f_\alpha(S, v)$ as the (open) angle interval of this double wedge.

Type C: This is the case that $S \cap W_2(v) \neq \emptyset$ as in Figure 3(c). Define $f_\alpha(S, v)$ as the (open) angle interval of the directions β such that $\ell_\beta(v)$ is in the double wedge $W_{13}(v)$ and intersects S. We note that if $\ell(e_1)$ and $\ell(e_2)$ both interest S, then the forbidden interval consists of two sub-intervals. We note that if S is completely inside $W_2(v)$, then there is no forbidden interval, so we set $f_\alpha(S, v) = \emptyset$. As in Type A, using the translation argument, we can show that every direction in $f_\alpha(S, v)$ is forbidden.

Lemma 4. *Given α, let $\mathcal{F}_\alpha$ be the union of $f_\alpha(S, v)$ for all pairs of α-bad strips S and reflex vertices v of P. Then $\mathcal{F}_\alpha$ is the union of all forbidden second directions.*

Proof. We have seen that any direction in $\mathcal{F}_\alpha$ is forbidden in the type definition, so what remains is to show that any forbidden direction for the points in α-bad strips is contained in $\mathcal{F}_\alpha$. Let β be a forbidden second direction. Then there is a point p in an α-bad strip S such that the line $\ell_\beta(p)$ intersects P in more than one segment. Among the intersection points between $\ell_\beta(p)$ and ∂P, let q be the closest point from p that are not visible from p, and let p' be the closest point of ∂P from p lying on $\overline{pq}$. (Note that p' and p may be identical.)

We now consider the geodesic shortest path from p to q in P. Since q is not visible from p, it consists of more than one segment and all its vertices, except p' and q, are reflex. Let v be a reflex vertex in the path such that $\ell_\beta(v)$ is tangent to the path. Without loss of generality, assume that p and q lie in the left side of $\ell_\beta(v)$. We have two cases depending on whether q is in S or not.

Case 1: q is in S. We claim that the reflex vertex v must be in S; otherwise, v were strictly above (resp., below) the upper (resp., lower) side of S and a horizontal line slightly above (resp., below) the side would intersect P in more than one connected component, which contradicts the maximality of the α-bad strip S. Hence this pair (S, v) defines the forbidden interval $f_\alpha(S, v)$ of Type A, being the angle interval of $W_{13}(v)$. Since the edges incident to v are both on the same side of $\ell_\beta(v)$, β is contained in $W_{13}(v)$ and thus in $f_\alpha(S, v)$.

Case 2: q is in $P \setminus S$. Let e_1 and e_2 be two edges incident to v, where they appear in counterclockwise order. If $\ell_\beta(v)$ does not intersects S, then S lies completely in the left side of $\ell_\beta(v)$ and $R(S, v)$ is defined. Since we defined q as the closest invisible point from p among the points in $\ell_\beta(p) \cap \partial P$, one portion of ∂P between p' and q is completely enclosed by the segment $\overline{pq}$ and the geodesic path from p to q. Note that v lies between p' and q along that portion, implying that q lies in $R(S, v)$. So the direction of $\overline{pq}$, β, is contained in $f_\alpha(S, v)$ of Type B. Now consider the case that $\ell_\beta(v)$ intersects S. Then the line $\ell_\beta(v)$ separates p from $W_2(v)$, and β is included in an angle interval $f_\alpha(S, v)$ either of Type C or of Type B, depending on whether $W_2(v)$ intersects S or not. This completes the proof of Lemma 4. $\square$

Computing $\mathcal{F}_\alpha$. We now describe how to compute $\mathcal{F}_\alpha$. To answer the visibility queries quickly, we preprocess P in $O(n \log n)$ time so that ray shooting queries can be answered in $O(\log n)$ time [5]. In addition, we construct a data structure on the vertices $v_1, \ldots, v_n$ of P in $O(n \log n)$ time such that for each pair (i, j) with $i < j$, the convex hull of the vertices $\{v_i, v_{i+1} \ldots, v_j\}$ can be (implicitly) constructed in $O(\log^2 n)$ time [6].

We first compute all the α-bad strips S of P and their convex hulls $\mathrm{CH}(S)$, by the plane sweep. This takes $O(n \log n)$ time since the total complexity of the strips is linear. Next, we compute $f_\alpha(S, v)$ for each pair of an α-bad strip S and a reflex vertex v of P as follows: The type of $f_\alpha(S, v)$ can be determined in $O(\log n)$ time, by checking if v is in S and then if the extension of the incident edges of v intersects S. Depending on the type of $f_\alpha(S, v)$, we use different methods as follows:

Type A: For this type, $f_\alpha(S, v)$ is the angle interval of $W_{13}(v)$ and thus can be computed in $O(1)$ time.

Type B: For this type, we need to compute $R(S, v)$ and the two tangent lines between $R(S, v)$ and S, as in Figure 3(b). To do this, we first compute two tangent lines from v to S (in fact, from v to the convex hull of S) in $O(\log n)$ time. To determine the cutting segment vv' that defines $R(S, v)$, we perform a ray shooting query from v along the tangent line (which has S and the incoming edge of v in the same side) in $O(\log n)$ time. If v' is a point on the edge (v_i, v_{i+1}) and v is equal to v_j for $i < j$, we determine the convex hull $\mathrm{CH}(R(S, v))$ of $\{v_{i+1}, \ldots, v_j\} \cup \{v'\}$ in $O(\log^2 n)$ time, by inserting v' into the pre-computed convex hull of $\{v_{i+1}, \ldots, v_j\}$. We then compute the two tangents between $\mathrm{CH}(R(S, v))$ and $\mathrm{CH}(S)$ in $O(\log n)$ time by binary searching.

Type C: For this type, we only need to compute the tangents from v to $\mathrm{CH}(S)$ in $O(\log n)$ time by binary searching.

For each pair (S, v), we compute $f_\alpha(S, v)$ in $O(\log^2 n)$ time, so computing all such intervals costs $O(n^2 \log^2 n)$ time. Finally we compute the union of the forbidden intervals in $O(n^2 \log n)$ time, by using a simple greedy algorithm after sorting the interval endpoints, and check if the union is the whole angle space in $O(1)$ time. This completes the description of the algorithm and the proof of Theorem 1.

3.2   Finding Two Directions α and β

This section is devoted to the proof of Theorem 2: for given simple polygon P, compute in $O(n^5 \log^2 n)$ time all pairs $\{\alpha, \beta\}$ of two directions for which P is 2-monotone.

We initially set α to be the horizontal direction, and compute the interval system $\mathcal{F}_\alpha$ by the procedure described in the previous section. We then perform a sweep over the angle space of α, and maintain the system $\mathcal{F}_\alpha$ during the sweep. Of course, since the exact values of the interval endpoints change continuously, what we maintain is the combinatorial description of the intervals of $\mathcal{F}_\alpha$, i.e., the ordered sequence of the interval endpoints.

Let S be an α-bad strip in P. The boundary between S and its neighboring strip is a line ℓ with direction α through a vertex w (refer to Figure 2), and there is a maximal segment s of positive length in $\ell \cap P$. We call this segment s a *lid* of S and the vertex w an *owner* of S. In general, there are two (top and bottom) lids and two (top and bottom) owners of an α-bad strip. In Figure 2 (for horizontal α), there are two α-bad strips: S with owners u and w, and S' with owners u' and w'. We denote by $T(\alpha)$ the set of endpoints of the lids of all α-bad strips in P, and by V the set of vertices of P. Then we can show the following lemma easily.

Lemma 5. *Let S be an α-bad strip and v be a reflex vertex of P. The endpoints of the forbidden interval $f_\alpha(S, v)$ are defined by a pair of points from $V \cup T(\alpha)$.*

Data structures and preprocessing. As mentioned before, we maintain $\mathcal{F}_\alpha$ during the angular sweep of α. Lemma 5 implies that what we need to maintain are: (i) the owners, lids, and convex hull of every α-bad strip S, (ii) the convex hull of $R(S, v)$ for every pair (S, v) of an α-bad strip S and a reflex vertex v that defines $R(S, v)$, and finally (iii) all *generating lines* defined by pairs of points from $V \cup T(\alpha)$. Dynamic data structures that can be used to maintain them are:

- $\mathcal{F}_\alpha$: During the sweep, we maintain all forbidden intervals $f_\alpha(S, v)$ and their union $\mathcal{F}_\alpha$, using the data structure $\mathcal{I}$, due to Cheng and Janardan [3]. This structure maintains the union of a set of m intervals, under insertions and deletions of intervals in $O(\log m)$ time. The union of the intervals can be listed in $O(k)$ time, if there are k components. In particular, it can be tested in $O(1)$ time whether the union is trivial.
- S, $R(S, v)$, $CH(S)$ and $CH(R(S, v))$: Each of bad strips S or $R(S, v)$ can be represented as a simple polygon with a set of vertices or lid-endpoints on the boundary in counterclockwise order. The data structure $\mathcal{C}$, due to Overmars and van Leeuwen [6], maintains the set of all bad strips S and $R(S, v)$ where computing their convex hulls or answering various queries (such as computing tangents) can be done in $O(\log^2 n)$ time.
- Next event α: The standard priority queue $\mathcal{Q}$ is used to store the values of α at which the interval system changes combinatorially. This queue is maintained dynamically in $O(\log n)$ time per insertion and deletion.

As a preprocessing step, we build a static data structure in $O(n \log n)$ time that supports visibility queries in $O(\log n)$ time. This can be used when we need to determine the first point at which a ray from a vertex of P hits ∂P. We build another data structure of Overmars and van Leeuwen [6] in $O(n \log n)$ time that enables us to (implicitly) determine the convex hull of $\{v_i, \ldots, v_j\}$ for any pair of vertices v_i and v_j in $O(\log^2 n)$ time.

Events. We define two types of events at which the combinatorial structure of the interval system $\mathcal{F}_\alpha$ in $\mathcal{I}$ or the associated information(e.g. owners or lids) of bad strips in $\mathcal{C}$ change. The first type of events, called *strip events*, happens when a bad strip is created, deleted, split into two, or two bad strips are merged into one, or the owner or lid-endpoints of the strip are changed. Any of these events result in creation/deletion of strips S and $R(S, v)$ in $\mathcal{C}$ and creation/deletion of intervals in $\mathcal{I}$. The second type of events, called *interval events*, happens when the combinatorial description of intervals in $\mathcal{I}$ changes.

(1) Strip events: These events are defined whenever a combinatorial change arises in the dynamic structure $\mathcal{C}$: bad strips are created, deleted, split into two, merged into one, or the owners or lid-endpoints are changed.

- Merge/Split: Two strips S_1 and S_2 in $\mathcal{C}$ can be merged into a larger strip S when the top owner of S_1 and the bottom owner of S_2 lie on a line with direction α. At this event, we remove two bad strips S_1 and S_2 from $\mathcal{C}$, and then create a new strip S in $\mathcal{C}$.
- Owner/Lid-endpoint Change: An owner of S can change if the old and new owners are on a line with direction α. At this event, we remove S from $\mathcal{C}$ and create the strip S with the new owner. An endpoint of a lid of S can move from one edge to another if the owner of S and a vertex of P are on

a line with direction α. For this event, we insert this vertex as a new vertex into S of $\mathcal{C}$ and update all information of S.

- Creation/Deletion: Each event listed above results in some creation and deletion of bad strips in $\mathcal{C}$. Creating or deleting a strip S is followed by inserting/deleting $R(S, v)$ for all reflex vertices v into/from $\mathcal{C}$, angle intervals into/from $\mathcal{I}$ and event moments α into/from $\mathcal{Q}$.

The first two events arise when two vertices of P lie on a line with direction α, and such a event calls $O(1)$ creation or deletion of bad strips in $\mathcal{C}$. Thus we execute $O(n^2)$ creation or deletion of bad strips in $\mathcal{C}$. Creation or deletion of a bad strip causes $O(n)$ insertion or deletion of $R(S, v)$ in $\mathcal{C}$ and thus $O(n)$ interval operations in $\mathcal{I}$ and $O(n)$ operations in $\mathcal{Q}$. Therefore, the total number of operations in $\mathcal{C}$, $\mathcal{I}$ and $\mathcal{Q}$ that are caused by strip events during the sweep of α is $O(n^3)$.

(2) Interval events: These events arise when the combinatorial structure of the interval system in $\mathcal{I}$ changes. Every interval $f_\alpha(S, v) = (x_\alpha, y_\alpha)$ in $\mathcal{I}$ is defined by two generating lines $\ell(x_\alpha)$ and $\ell(y_\alpha)$, where x_α and y_α represent the endpoints of this interval. According to Lemma 5, these lines $\ell(x_\alpha)$ and $\ell(y_\alpha)$ are determined by two points in $V \cup T(\alpha)$. During the sweep of α, the combinatorial change of the interval system in $\mathcal{I}$ arises in three different ways.

- Type of $f_\alpha(S, v)$ changes: Transition between Type A and Types B or C happens when α is equal to the direction of one of the edges e_1 and e_2 incident to v. There are $O(n)$ such directions. Transition from Type B to Type C occurs in the following way. Assume that v is above S and $\ell(e_1)$ intersects ∂P at edge e. Let $p_v := \ell(e_1) \cap e$ (i.e., p_v is the first intersection of the ray from v along $\ell(e_1)$ with ∂P). Let w be the top owner of S and $t_w(\alpha) := \ell_\alpha(w) \cap e$. Let α' be the direction of $\ell(w, p_v)$. As α approaches to α' in counterclockwise direction, $t_w(\alpha)$ approaches p_v. At the moment that $\alpha = \alpha'$, i.e., that $t_w(\alpha)$ arrives at p_v, the type of $f_\alpha(S, v)$ changes from Type B to Type C. The angle α' is defined by the owner w of S and the intersection p_v between $\ell(e_1)$ and e. Since there are at most n owners and at most n intersection points, the number of such transitions is $O(n^2)$.
- Combinatorial description of $\ell(x_\alpha)$ and $\ell(y_\alpha)$ changes while the type of $f_\alpha(S, v)$ remains the same: If $f_\alpha(S, v)$ is of Type A, then $\ell(x_\alpha)$ and $\ell(y_\alpha)$ are extended lines of the edges incident to v, so they do not change as long as the interval type remains the same. The case that $f_\alpha(S, v)$ is of Type C applies the similar argument as that of Type B, so we consider only when $f_\alpha(S, v)$ is of Type B. Line $\ell(x_\alpha)$ (and $\ell(y_\alpha)$) is a tangent between S and $R(S, v)$, where $\ell(x_\alpha)$ passes through two tangential points $p(x_\alpha) \in \mathrm{CH}(S)$ and $q(x_\alpha) \in \mathrm{CH}(R(S, v))$. The point $p(x_\alpha)$ can be either a vertex of P or a point of $T(\alpha)$ whereas $q(x_\alpha)$ is a vertex of P. We have three different situations as follows.
    - While α rotates from α_1 to α_3, $\mathrm{CH}(S)$ remains unchanged. For $\alpha \leq \alpha_2$, $p(x_\alpha)$ is the vertex u, but $p(x_\alpha)$ becomes a point of $T(\alpha)$ for $\alpha \geq \alpha_2$, so the combinatorial description of $\ell(x_\alpha)$ changes at $\alpha = \alpha_2$. The direction

α_2 is determined by the owner of S and the intersection $\ell(q(x_\alpha), u) \cap e$. Since e is fixed for a strip S during $\alpha_1 \leq \alpha < \alpha_3$, there are $O(n^2)$ directions for each S and thus $O(n^3)$ events in total.

- $\ell(x_\alpha)$ can change combinatorially when $\mathrm{CH}(S)$ changes combinatorially. While α rotates from α_1 to α_3, $\mathrm{CH}(S)$ has a new edge e' on its boundary. So the combinatorial description of $\ell(x_\alpha)$ at $\alpha = \alpha_2$ has changed. However, this event is also detected in the strip event caused by lid-endpoint changes. (The direction α_2 is defined by a strip owner and a vertex, so there are $O(n^2)$ such directions for a single strip and thus $O(n^3)$ in total.)

- $\ell(x_\alpha)$ can change when $\mathrm{CH}(R(S, v))$ changes combinatorially. It is easy to see that a change of $\mathrm{CH}(R(S, v))$ for a fixed pair (S, v) happens $O(n)$ times, according to the directions defined by v and the other vertices in $R(S, v)$. Since there are $O(n^2)$ pairs (S, v), we have $O(n^3)$ such events. In addition, we update the information of $R(S, v)$ and $\mathrm{CH}(R(S, v))$ in $\mathcal{C}$ whenever they have changes.

- The order of intervals changes in $\mathcal{I}$: This event happens when the endpoints of two intervals change their relative (cyclic) order. To update the interval system $\mathcal{I}$, we first delete both intervals from $\mathcal{I}$ and then insert two new intervals whose endpoints reflect the new order. Whenever an interval is inserted or deleted in $\mathcal{I}$ (e.g., a new strip causes $O(n)$ insertion of $R(S, v)$ into $\mathcal{C}$ and thus $O(n)$ insertion of new intervals into $\mathcal{I}$), we need to compute all the moments α that the order between the new interval and the others of $\mathcal{I}$ is changed. There are $O(n^2)$ intervals in $\mathcal{I}$, so insertion/deletion of an interval generates $O(n^2)$ ordering change moments and we need to push these α into $\mathcal{Q}$. We have seen that $O(n^3)$ insertion or deletion of intervals in $\mathcal{I}$ can happen by strip events and the above two interval events, so we may need $O(n^5)$ ordering change events in $\mathcal{I}$.

In total, during the whole sweep there are $O(n^3)$ operations in $\mathcal{C}$ and $O(n^5)$ operations in $\mathcal{I}$ and $\mathcal{Q}$.

Overall algorithm. After the preprocessing step, we set α to be the horizontal direction and compute all types of events for this configuration. We then initialize the data structures $\mathcal{I}$, $\mathcal{C}$, and the event queue $\mathcal{Q}$. While $\mathcal{Q}$ is not empty, we repeat this process: we extract the smallest angle α from $\mathcal{Q}$, and update the data structures according to the event type of α in $O(\log^2 n)$ time. After that, we test in $O(1)$ time if the updated intervals cover the whole angle space. The total number of operations needed in the whole process is $O(n^5)$, so the algorithm runs $O(n^5 \log^2 n)$ time. This completes the proof of Theorem 2.

4 Monotonicity for $k \geq 3$ Directions

This section is devoted to the proof of Theorem 3: for fixed $k \geq 3$ and given simple polygon P with n vertices, compute in $O(n^{3k+2})$ time all sets of k directions for which P is k-monotone.

The general strategy is that we divide the space of all *direction k-tuples* into cells, where we can show that in each cell, either all direction k-tuples cover the polygon in this way, or none of them does. Then we just have to test a sample direction k-tuple in each cell to decide whether there is some k-tuple that covers the polygon. Our cell decomposition is generated by $O(n^3)$ hyper-surfaces in that space of k-tuples; they divide this k-dimensional space into $O(n^{3k})$ cells. By this, we reduce the existence problem of a direction k-tuple covering the polygon to $O(n^{3k})$ decision problems, each deciding in $O(n^2)$ time if a specific direction k-tuple covers the polygon. So the total complexity of the algorithms will be $O(n^{3k+2})$.

We define three types of hyper-surfaces for the subdivision of the k-dimensional space of direction k-tuples:

- For $a \in \{1, \ldots, k\}$, S_{abc} is the set of $(\varphi_1, \ldots, \varphi_k)$ where the direction φ_a coincides with the direction of the line through the polygon vertices v_b and v_c.
- For $b, d \in \{1, \ldots, k\}$, T_{abcd} is the set of $(\varphi_1, \ldots, \varphi_k)$ where the line through v_a with direction φ_b intersects the line through v_c with direction φ_d in a point on the polygon boundary.
- For $b, d, f \in \{1, \ldots, k\}$, R_{abcdef} is the set of $(\varphi_1, \ldots, \varphi_k)$ where the line through v_a with direction φ_b intersects the line through v_c with direction φ_d in a point on the line through v_e with direction φ_f.

There are $O(n^2 k)$ hyper-surfaces of the first type, $O(n^2 k^2)$ of the second type, and $O(n^3 k^3)$ of the third type, so for fixed k the total number of surfaces in our arrangement is $O(n^3)$.

We now have to show that within each cell of the arrangement, the polygon is either covered for all $(\varphi_1, \ldots, \varphi_k)$, or not covered for any $(\varphi_1, \ldots, \varphi_k)$. Suppose that the polygon is covered for $\Phi = (\varphi_1, \ldots, \varphi_k)$, and not covered for $\Psi = (\psi_1, \ldots, \psi_k)$ in the same cell. Consider any path from Φ to Ψ. Along this path there is a last stage in which the entire polygon is covered, and a first point u that will be uncovered. There are several possibilities how a point can become uncovered. Each boundary of a region covered by a direction is either an edge of the polygon or a line in that direction through a vertex of the polygon. So the uncovered region around u must be bounded by lines, which are either polygonal edges or lines in one of the directions through a vertex of the polygon. We can distinguish the following cases:

- If u is a vertex, then in the moment u becomes uncovered the line through u goes through another vertex, so the path from Φ to Ψ crosses a surface of the first type.
- If u is a point on an edge of the polygon, and
  - the region becoming uncovered is bounded by a line parallel to the polygonal edge, moving away from it, then the path from Φ to Ψ again crossed a surface of the first type, or
  - the region becoming uncovered is bounded by two lines intersecting the polygonal edge, then in the moment that u becomes uncovered, these

two lines intersect each other on the polygonal boundary, so the path from Φ to Ψ crosses a surface of the second type.

- If u is an interior point of the polygon, and
  - the region becoming uncovered is bounded by two parallel lines moving away from each other, then the path from Φ to Ψ crosses a surface of the first type, or
  - the region becoming uncovered is bounded by three lines, then in the moment that u became uncovered, these three lines intersect one another in a point, so the path from Φ to Ψ crosses a surface of the third type.

Therefore it is sufficient to check one sample point from each cell of this arrangement. To test whether a given k-tuple of directions actually covers a given n-gon, we just construct the arrangement of all the lines of these directions through all polygon vertices. This arrangement has $O(n^2 k^2)$ cells and can be constructed in that time, and all potential boundaries of uncovered regions are among these lines. So we just have to check whether each cell is covered. This finishes the proof of Theorem 3.

References

1. Bose, P., van Kreveld, M.: Computing nice sweeps for polyhedra and polygons. In: 16th Canadian Conference on Computational Geometry, pp. 108–111 (2004)
2. Bose, P., van Kreveld, M.: Generalizing monotonicity: On recognizing special classes of polygons and polyhedra by computing nice sweeps. International Journal of Computational Geometry and its Applications 15(6) (2005)
3. Cheng, S.W., Janardan, R.: Efficient maintenance of the union of intervals on a line, with applications. Journal of Algorithms 12, 57–74 (1991)
4. Ha, J.S., Yoo, K.H., Hahn, J.K.: Characterization of polyhedron monotonicity. Computer-Aided Design 38(1), 48–54 (2006)
5. Hershberger, J., Suri, S.: A pedestrian approach to ray shooting: Shoot a ray, take a walk. Journal of Algorithms 18, 403–431 (1995)
6. Overmars, M., van Leeuwen, J.: Maintenance of configurations in the plane. Journal of Computer and System Sciences 23, 166–204 (1981)
7. Preparata, F., Supowit, K.: Testing a simple polygon for monotonicity. Information Processing Letters 12(4), 161–164 (1981)
8. Rappaport, D., Rosenbloom, A.: Moldable and castable polygons. Computational Geometry: Theory and Applications 4, 219–233 (1994)
9. Toussaint, G.T.: Movable separability of sets. In: Computational Geometry, pp. 335–375. North-Holland, Netherlands (1985)

On the Stability of Web Crawling and Web Search

Reid Anderson[1], Christian Borgs[1], Jennifer Chayes[1],
John Hopcroft[2], Vahab Mirrokni[3], and Shang-Hua Teng[4]

[1] Microsoft Research
[2] Cornell University
[3] Google Research
[4] Boston University

Abstract. In this paper, we analyze a graph-theoretic property motivated by web crawling. We introduce a notion of stable cores, which is
the set of web pages that are usually contained in the crawling buffer
when the buffer size is smaller than the total number of web pages. We
analyze the size of core in a random graph model based on the bounded
Pareto power law distribution. We prove that a core of significant size
exists for a large range of parameters $2 < \alpha < 3$ for the power law.[1]

1 Introduction

Since the World Wide Web is continually changing, search engines [1] must
repeatedly crawl the web, and update the web graph. In an ideal world one would
search, discover, and store all web pages on each crawl. However, in practice
constraints allow indexing and storing only a fraction of the web graph [3]. This
raises the question as to what fraction of the web one needs to crawl in order to
maintain a relatively stable set of pages that contains all sufficiently important
web pages.

When a link to a web page is encountered, the page is said to be *discovered*.
When the page is retrieved and explored for its links, it is said to be *explored*.
Thus, we can partition the web into three types of pages. (1) Pages the crawl
has explored; (2) Pages the crawl has discovered but not explored; and (3) All
other pages.

Web crawling can be viewed as a dynamic process over the entire web graph.
As time goes by, these three sets change dynamically, depending on the crawling
algorithm as well as the space/time constraints. Let S_t be the set of pages that
have been discovered and explored at time t. The search engine typically ranks
pages in S_t. When users make queries during phase t, only pages from S_t are returned. Presumably these are the pages that were deemed sufficiently important
to index and are used to answer queries. Let C_t be the set of pages that have
been discovered but not explored at time t. At this point the edges from pages
in C_t are not known since the search has not crawled these pages.

---

[1] Most of this work was done while the authors were at Microsoft Research.

S.-H. Hong, H. Nagamochi, and T. Fukunaga (Eds.): ISAAC 2008, LNCS 5369, pp. 680–691, 2008.

Consider the subgraph of the web that consists of all web pages in $S_t \cup C_t$ and all directed edges from pages in S_t ending in either S_t or C_t. At the next stage the search engine would calculate the page rank of all pages in this graph and select the set of size b of pages of highest page rank to be S_{t+1}. The pages in S_{t+1} would then be explored to produce a new set C_{t+1} of pages reachable from pages in S_{t+1} but which are not in S_{t+1}. Here b is determined by the available amount of storage.

The space constraints immediately raise several basic questions in web crawling and web search. An important question is how large b needs to be in order for the search engine to maintain a core that contains all sufficiently important pages from the web. Assuming the web is on the order of 100 billion pages, is a buffer of size of 5 billion sufficient to ensure that the most important 100 million pages are always in the buffer and hence available to respond to queries? In general, what percent of pages are stable in the sense that they are always in the buffer? What percent of pages are regularly moving in and out of the buffer? What percent of the buffer is just random pages? What is the relationship between the importance of a page (such as high page rank or high degree) with the frequency that page is in the buffer? These questions are particular interesting when the graph is changing with the time? How frequently must we do a crawl where frequency is measured by the percentage of the graph that changed between the crawls? How should we design high a quality crawl given the space and time constraints? For example, should we completely explore the highest page ranked pages or should we explore some fraction of links from a larger number of pages, where the fraction is possibly determined by the page rank or degree? How accurately do we need to calculate the page rank in order to maintain the quality? Could we substitute in-degree for page rank?

Clearly the theoretical answers to above questions depend on how the underlying graph is modeled. We investigate the behaviors of the web crawling process and web crawling dynamics. The motivation behind our investigation is to design better and robust algorithms for web crawling and web search.

We will focus on the stability of web crawling and its potential impact to web search. In particular, we analyze a graph-theoretic property that we discovered based on our initial experiments involving web crawling dynamics.

Suppose b pages are stored at each stage of the crawling. We have observed that, for various choices of **EXPLORE** and **STORE** functions and large enough b, the sets S_t do not converge. However, there exists t_0 such that $\left(\cap_{t \geq t_0}^{\infty} S_t \right)$ converges to a non-empty set. We call $K = \left(\cap_{t \geq t_0}^{\infty} S_t \right)$ the *core* of the crawling dynamics with (**EXPLORE**, **STORE**) transitions. Naturally, the size of the core depends on b as well as on (**EXPLORE**, **STORE**). When b is small, the core might be empty. Naturally, when $b = |V|$, the whole graph is the core. When $b = 1$, we do not expect a non-empty core.

In this paper, we consider a simplified crawling algorithm with limited space. Let

$$C_t = \textsf{EXPLORE}(S_{t-1}) = \{ v \, | (u \rightarrow v) \in E, \text{for some } u \in S_{t-1} \} - S_{t-1},$$

be the set of direct neighbors of S_{t-1}. For each page $v \in S_{t-1} \cup C_t$, let $\Delta_t(v)$ be the number of links from pages in S_{t-1} to v. Then, $\mathsf{STORE}(S_{t-1}, C_t)$ is the set of b pages with the largest Δ_t value, where ties are broken according to some predefined rule.

We analyze the size of the core in a random graph model based on the bounded Pareto power law distribution [2,4]. We prove that a core of significant size exists when the power law parameter α lies in the range $[2:3)$.

2   The Core

Web crawling defines a sequence B_0, ..., B_t,...,B_∞, where B_t is the content of the buffer at time t. If a page enters the buffer at some stage and stays in the buffer after that, then we say the page is in the *core* of of the sequence.

For example, suppose the web graph is fixed and the crawl process is deterministic, then since the number of the subsets of web pages of size b is finite, the above sequence eventually become periodic. In other words, there exist a t_0 and p such that $B_{t_0} = B_{t_0+p}$. In this case, the *core* of the sequence is equal to $\cap_{t=t_0}^{t_0+p} B_p$. When the graph is fixed, but the web crawling is stochastic, we define the core as those pages that stay in the buffer with high probability.

In the rest of the paper, we assume B_0 is a set of size b uniformly chosen from the vertices of the input graph. The core of this graph is then defined according to the sequence produced by the crawling process.

3   Bounded Pareto Degree Distributions and Its Graphs

One of the objectives of this paper is to estimate the core size as a function of b, for a directed graph G. Naturally, this quantity depend on G and the initial set B_0. It is well known that the web graph has power law degree distribution [5]. To present our concepts and results as clearly as possible we use the following "first-order approximation" of the power-law graphs with bounded Pareto degree distributions. We first define a degree vector, which is the expected degree of a bounded Pareto degree distribution [2]. Then, we consider random graphs with expected degrees specified by the degree vector. We will show the core size depends on both the degree distribution and on the size of the buffer.

The expected number of vertices of degree k in the bounded Pareto degree distribution with parameters $\alpha > 1$ and positive integer n is $Cnk^{-\alpha}$ where $C = 1/\left(\sum_{x=1}^{\infty} x^{-\alpha}\right)$. We can construct a typical degree sequence as follows. Let h_i be the largest integer such that $\sum_{k=h_i}^{\infty} Cnk^{-\alpha} \geq i$. The sequence starts with the highest degrees $(h_1, ..., h_i, ...)$. Note that h_i is approximately

$$\left(\frac{C}{\alpha-1}\right)^{1/(\alpha-1)} \left(\frac{n}{i}\right)^{1/(\alpha-1)}$$

To construct the degree vector $\mathbf{d}_{\alpha,n}$, we start at the right with the degree one vertices and work to the left. Let k_0 be the smallest integer such that $Cnk^{-\alpha} < 1$.

- For each $1 \leq k < k_0$, from right to left, assign the k to the next $Cnk^{-\alpha}$ entries in $\mathbf{d}_{\alpha,n}$. To be more precise, when $Cnk^{-\alpha}$ is not an integer, we first assign k to $\lfloor Cnk^{-\alpha} \rfloor$ entries, and then, with probability $Cnk^{-\alpha} - \lfloor Cnk^{-\alpha} \rfloor$, add one more entry with value k. Suppose this step assigns n' entries.
- For $j = 1 : n - n'$, assign the value h_j to $\mathbf{d}_{\alpha,n}[n - n' - j]$.

In other words, $\mathbf{d}_{\alpha,n}$ is a sorted vector of expected degrees, from the largest to the smallest. In this vector, the smallest degree k that appear s times approximately solves $Cnk^{-\alpha} = s$, implying $k \approx s^{-1/\alpha} (Cn)^{1/\alpha}$.

Note that for $\alpha > 2$, the expected number of edges is proportional to n and for $\alpha = 2$ the expected number of edges is proportional to $n \log n$. That is,

$$E\left[\|\mathbf{d}_{\alpha,n}\|_1\right] = \begin{cases} \Theta(n) & \text{if } \alpha > 2, \text{ and} \\ \Theta(n \log n) & \text{if } \alpha = 2. \end{cases}$$

The graph we analyze has n vertices, labeled 1 to n, and is generated by the following random process: Let $m = \|\mathbf{d}_{\alpha,n}\|_1$. Independently choose m directed edges, by first selecting a vertex i randomly according to $\mathbf{d}_{\alpha,n}$, and then choosing another vertex j randomly, also according to $\mathbf{d}_{\alpha,n}$. Note that this graph model allows multiple edges and self-loops.

Call a random graph from this distribution a random (α, n)-*BBPL* graph. This class of graphs has several statistical properties. The expected number of vertices with in-degree 0 is highly concentrated around $\sum_{k=1} e^{-k} Cnk^{-\alpha}$, which is a constant fraction of n.

Lemma 1. *For $h \geq 3$, the expected number of vertices with in-degree h or larger in a random (α, n)-BPPL graph is highly concentrated around*

$$C \sum_{k=1}^{\Theta(n^{1/(1-\alpha)})} \binom{n}{h} \left(\frac{k}{n}\right)^h \frac{n}{k^\alpha} \leq \Theta\left(\frac{n}{h^{(\alpha-1)(1+1/(2h))}}\right).$$

4   Estimating the Core Size for Power-Law Graphs

Consider a buffer B of b vertices. These vertices induce a graph which we will refer to as the *buffer-induced graph* of B. The buffer-induced degree of a vertex is its degree in the buffer-induced graph.

4.1   A Simple Thought Process

When the buffer size is a fraction of n, a vertex with constant degree may have buffer-induced degree of 0 and thus may drop out of the buffer. This implies that the core size might be $o(n)$, depending on the tie breaking rule. However, in this section, we show that for any ϵ, the core size is $\Omega(n^{1-\epsilon})$.

Suppose we do a crawl with a buffer large enough to hold every vertex of in-degree at least one in the web. This does not implies that the vertices with indegree at least in the original graph may stay or enter the buffer, since its buffer-induced may be 0. We show with high probability, the two highest degree vertices, to be called 1 and 2, are mutually connected and will enter the buffer in step 1. The fact that they are mutually connected means that they will always remain in the buffer. In subsequent steps, all vertices reachable from 1 and 2 will be added to the buffer and will remain in the buffer. Thus, the core contains all these vertices. We now give a lower bound on the expected number of vertices reachable from 1 and 2, which provides a lower bound on the core size.

Lemma 2. *The probability that vertices 1 and 2 are mutually connected to each other in a random (α, n)-BPPL graph is*

$$1 - e^{\Theta\left(-n^{\frac{3-\alpha}{\alpha-1}}\right)}.$$

Proof. Let m be the number of edges in a random (α, n)-BPPL graph. Thus, $m = ||\mathbf{d}_{\alpha,n}||_1$ and is linear in n. The probability that 1 and 2 are not mutually connected to each other is at most

$$\left(1 - \Theta\left(\frac{n^{1/(\alpha-1)}}{m} \frac{n^{1/(\alpha-1)}}{m}\right)\right)^m = e^{-\Theta\left(n^{\frac{3-\alpha}{\alpha-1}}\right)}.$$

$\square$

With a relatively small buffer of size $\Theta(n^{1-1/(\alpha-1)} \log n)$ containing randomly chosen vertices, the probability that vertices 1 and 2 will be in the buffer in the next step is high. Note that for $\alpha = 3$, this buffer size is only $\Theta(\sqrt{n})$.

Lemma 3. *Suppose $G = (V, E)$ is a random (α, n)-BPPL graph and S is a set of b randomly chosen vertices of V. There exists a constant c such that if $b \geq cn^{1-1/(\alpha-1)} \log n$, then with high probability, there are edges from vertices in S to both vertices 1 and 2.*

Proof. Because in our model, the expected degree of each vertex is at least 1, $\sum_{u \in S} \mathbf{d}_{\alpha,n}[u] \geq b$. The expected indegrees of vertices 1 and 2 are $\Theta(n^{1/(\alpha-1)})$. Their expected indegrees counting only edges from S are

$$\Theta\left(\frac{n^{1/(\alpha-1)}}{n}\right) \sum_{u \in S_0} \mathbf{d}_{\alpha,n}[u] = \Theta\left(n^{1/(\alpha-1)} \frac{b}{n}\right) \geq \Theta(c \log n).$$

As this bound is highly concentrated, when c is large enough, with high probability (e.g., $1 - n^{-\Theta(c)}$), the buffer-induced in-degrees of vertices 1 and 2 are larger than 1. $\square$

Lemma 4. *For $2 < \alpha < 3$, with high probability, the number of vertices reachable from $\{1, 2\}$ in a random (α, n)-BPPL graph G is $\Omega(n^{1-\epsilon})$.*

Here we will sketch an outline of the proof but skip some technical details since we will give a stronger result later with all the details. To better illustrate the analysis, instead of writing a proof for all $\alpha : 2 < \alpha < 3$, we choose a typical value in this range and provide an explicit derivation. The proof is easily adapted to handle all $\alpha : 2 < \alpha < 3$. Our choice of "typical" value is $\alpha = 11/4$. In this case, note that the expected degree of vertex 1 is $\Theta(n^{4/7})$. The expected number of vertices directly reachable from $\{1, 2\}$ is

$$\Theta\left(\sum_k \left[1 - \left(1 - \frac{k}{n}\right)^{n^{4/7}}\right]\frac{n}{k^{11/4}}\right) = \Theta(n^{4/7}).$$

The expected total degree of the nodes directly reachable from $\{1, 2\}$ is

$$\Theta\left(\sum_k k\left[1 - \left(1 - \frac{k}{n}\right)^{n^{4/7}}\right]\frac{n}{k^{11/4}}\right) = \Theta\left(\int_1^{n^{3/7}} \frac{n^{4/7}}{k^{3/4}}\right) = \Theta(n^{19/28})$$

Let $S_0 = \{1, 2\}$. Let S_t be the set defined by the set of vertices t hops away from $\{1, 2\}$. Let $\Delta(S_t)$ be their expected degree. We thus have

$$E[|S_1|] = \Theta(n^{4/7}), \text{ and } E[|\Delta(S_1)|] = \Theta(n^{19/28}).$$

The key to the analysis is that $E[|\Delta(S_1)|]$ is magnitudely larger than $E[|S_1|]$, which means that the frontiers of the Breadth-First Search starting from $\{1, 2\}$ have good expansions. There are two types of out-links from S_t: the edges to $S_0 \cup ... \cup S_t$ and the edges to S_{t+1}. We now bound the expected size of S_2 and the expected total degree $\Delta(S_2)$ of S_2. A similar analysis can be extended to any t.

Let $F_1 = \{v \mid (u \to v) \in E, \text{for some } u \in S_1\}$ and let $B_1 = F_1 \cap (S_0 \cup S_1)$. We have $S_2 = F_1 - B_1$. Note that $E[|B_1|] \leq E[|S_1|] + 2 = \Theta(n^{4/7})$. Thus,

$$E[|S_2|] = E[|F_1|] - E[|B_1|]$$

$$= \left(\sum\left[1 - \left(1 - \frac{k}{n}\right)^{n^{19/28}}\right]\frac{n}{k^{11/4}}\right) - E[|B_1|]$$

$$= \Theta\left(\int\left[1 - \left(1 - \frac{k}{n}\right)^{n^{19/28}}\right]\frac{n}{k^{11/4}}\right) - E[|B_1|]$$

$$= \Theta\left(\int_1^{n^{9/28}}\left[n^{19/28}\left(\frac{k}{n}\right)\right]\frac{n}{k^{11/4}}\right) - E[|B_1|]$$

$$= \Theta(n^{19/28}) - \Theta(n^{4/7}) = \Theta(n^{19/28}).$$

We now bound $E[\Delta(S_2)]$, which is $E[\Delta(F_1)] - E[\Delta(B_1)]$. Because $B_1 = F_1 \cap (S_0 \cup S_1)$, $E[\Delta(B_1)] \leq E[\Delta(S_0) + \Delta(S_1)] = \Theta(n^{19/28})$. Thus,

$$E[|\Delta(S_2)|] = E[|\Delta(C_1)|] - E[|\Delta(B_1)|]$$

$$= \left(\sum k \left[1 - \left(1 - \frac{k}{n}\right)^{n^{19/28}}\right] \frac{n}{k^{11/4}}\right) - E[|\Delta(B_1)|]$$

$$= \Theta\left(\int \left[1 - \left(1 - \frac{k}{n}\right)^{n^{19/28}}\right] \frac{n}{k^{7/4}}\right) - E[|B_1|]$$

$$= \Theta\left(\int_1^{n^{9/28}} \left[n^{19/28}\left(\frac{k}{n}\right)\right] \frac{n}{k^{7/4}}\right) - E[|B_1|]$$

$$= \Theta(n^{85/112}) - \Theta(n^{19/28}) = \Theta(n^{85/112}).$$

Note that S_2 still has a polynomial expansion. So S_3 will continue to grow. As these bounds a highly concentrated, by repeating this argument a constant number of times, to be formalized in the next subsection, we can show that the expected number of vertices reachable from $\{1, 2\}$ is $\Omega(n^{1-\epsilon})$ for any $\epsilon > 0$.

4.2   Crawling with Buffer of Size Constant Fraction of n

We now consider the case when the buffer is too small to contain all vertices of in-degree 1. Let h be an integer such that the buffer is large enough to contain all vertices of in-degree at least $h - 1$. We will use the following structure to establish a lower bound on the core size: Let $S_0 = [1 : h]$. Let the h-PYRAMID of S_0, denoted by PYRAMID(S_0), be the following subgraph. For each i, let

$$S_i = \text{NEIGHBORS}(S_{i-1}) - \cup_{j=1}^{i-1} S_j.$$

Then, PYRAMID(S_0) is the subgraph induced by $\cup_i S_i$.

We will use the following lemma whose proof is straightforward.

Lemma 5. *Suppose $G = (V, E)$ is a directed graph and S_0 is a subset of V of size b. If there is a t_0 such that S_{t_0} contains a subset C_0 satisfying that the indegree of every vertex in C_0 in the induced subgraph $G(C_0)$ over C_0 is at least h, then PYRAMID(S_0) is in the core if b is larger that the number of vertices in G whose indegrees are h or more.*

Below, we will show the h highest degree vertices form a clique. Furthermore, if we start with a random set of b vertices, then with high probability, these h vertices get in the buffer in the first step and will remain there. Again, we will focus on $\alpha = 11/4$. Let CLIQUE(h) be the event that the subgraph induced by $[1 : h]$ is a complete direct graph. We use $[A]$ to denote that an event A is true.

Lemma 6. *In a random (α, n)-BBPL graph G with $\alpha = 11/4$,*

$$Pr[[\text{CLIQUE}(h)]] \geq 1 - e^{-\frac{n^{1/7}}{h^{8/7}}}.$$

Proof. The expected degree of vertex i is $\Theta\left(\left(\frac{n}{i}\right)^{4/7}\right)$. By a union bound,

$$Pr[[\text{not CLIQUE}(h)]] \le \sum_{i,j \le h} \left(1 - \frac{\left(\frac{n}{i}\right)^{4/7} \left(\frac{n}{j}\right)^{4/7}}{n}\right)^n \le \sum_{i,j \le h} e^{-\frac{n^{1/7}}{(ij)^{4/7}}} \approx e^{-\frac{n^{1/7}}{h^{8/7}}}$$

$\square$

Note that if $h < n^{1/8}$, then with high probability, $[1 : h]$ induces a complete directed clique.

Lemma 7. *Let $G = (V, E)$ be a random (α, n)-BPPL graph, for $2 < \alpha < 3$. With high probability, there exists a constant c, such that for any h (not necessarily a constant), for a set of $b \ge cn^{1-1/(\alpha-1)}h^{1+1/(\alpha-1)} \log n$ randomly chosen vertices S, the buffer-induced in-degrees of vertices $1,..., h$ are larger than h.*

Proof. Note that at least 1, we have $\sum_{u \in S} \mathbf{d}_{\alpha,n}[u] \ge b$. The expected in-degrees of vertices $1,..., h$ are bounded by

$$\Theta\left(\frac{n^{1/(\alpha-1)}}{n}\right) \sum_{u \in S} \mathbf{d}_{\alpha,n}[u] = \left(\frac{n}{h}\right)^{1/(\alpha-1)} \frac{b}{n} \ge ch \log n.$$

As this bound is highly concentrated, thus if c is large enough, with high probability (e.g., $1 - n^{-\Theta(c)}$), the buffer-induced in-degrees of vertices $1,...,h$ are at least h. $\square$

Lemma 8. *In a random BPPL graph with parameters n and $\alpha = 11/4$, the total expected degrees of vertices $[1 : h]$ is $\Theta\left(n^{4/7}h^{3/7}\right)$.*

Proof. $\sum_{x=1}^{h} \left(\frac{n}{x}\right)^{4/7} = \Theta(n^{4/7}h^{3/7})$. $\square$

We now analyze the size of pyramid.

Lemma 9. *Let T_0 be a subset of vertices of $[1 : n]$ whose expected total degree is $\Theta(n^\gamma)$ and $|T_0| = o(n^\gamma)$, for a constant γ. Suppose T_0 has n^γ random outgoing edges chosen according to the BPPL distribution with parameters n and $\alpha = 11/4$. Let $T_1 = \text{NEIGHBORS}(T_0) - T_0$. Then, for any constant $h \ge 2$,*

$$E[|T_1|] \ge \Theta\left(n^{(7/4)\gamma - 3/4}\right).$$

Proof. Let $T_1' = \text{NEIGHBORS}(T_0)$. So, $T_1 = T_1' - T_0$. For $k \le n^{1-\gamma}$ and $h \ll n^\gamma$, the probability that a weight k vertex receives at least h edges from T_0 is

$$\binom{n^r}{h} \left(\frac{k}{n}\right)^h = \Theta\left(\min\left((n^\gamma)^h \left(\frac{k}{n}\right)^h, 1\right)\right) \tag{1}$$

Thus, as $h \geq 2$, the expected size of T_1 is

$$E[|T_1|] = E[|T_1'|] - E[|T' \cap T_0|]$$

$$= \Theta \left(\sum_1^{n^{1-\gamma}} \frac{n}{k^{11/4}} \cdot n^{\gamma h} \left(\frac{k}{n} \right)^h \right) - E[|T' \cap T_0|]$$

$$= \Theta \left(n^{h\gamma - h + 1} \int_1^{n^{1-\gamma}} k^{h - 11/4} \right) - o(n^\gamma) \tag{2}$$

$$= \Theta \left(n^{h\gamma - h + 1} n^{(1-\gamma)(h - 7/4)} \right) = \Theta \left(n^{(7/4)\gamma - 3/4} \right). \qquad \square$$

This bound is independent of h. If $\gamma = 4/7$, then $(7/4)\gamma - 3/4 = 1/4$. We would like to remark that if $h = 1$, then the calculation follows from what we did in the previous subsection. The integral there was a constant but is not here. The next lemma bounds $\Delta(T_1)$, the expected total degrees of vertices in T_1.

Lemma 10. *Let T_0 be a subset of vertices of $[1 : n]$ whose total expected degrees is $\Theta(n^\gamma)$ and $|T_0| = o(n^\gamma)$, for a constant γ. Suppose T_0 has n^γ random outgoing edges chosen according to the BPPL distribution with parameters n and $\alpha = 11/4$. Let $T_1 = \mathrm{NEIGHBORS}(T_0) - T_0$. Then, for any constant $h \geq 2$,*

$$E[\Delta(T_1)] \geq \Theta \left(n^{(3/4)\gamma + 1/4} \right).$$

Proof. Let $T_1' = \mathrm{NEIGHBORS}(T_0)$. We have $T_1 = T_1' - T_0$.

$$E[\Delta(T_1)] = E[\Delta(T_1')] - E[\Delta(T_1' \cap T_0)]$$

$$\geq \Theta \left(\sum_1^{n^{1-\gamma}} k \frac{n}{k^{11/4}} \cdot n^{\gamma h} \left(\frac{k}{n} \right)^h \right) - |\Delta(T_0)|$$

$$= \Theta \left(n^{h\gamma - h + 1} \int_1^{n^{1-\gamma}} k^{h - 7/4} \right) - \Theta(n^\gamma) \tag{3}$$

$$= \Theta \left(n^{h\gamma - h + 1} n^{(1-\gamma)(h - 3/4)} \right) = \left(n^{(3/4)\gamma + 1/4} \right). \qquad \square$$

We now apply Lemmas 9, 10 and 6, to prove the following theorem. Because we need to apply these lemmas iteratively, we need to know the concentration of the bounds in Lemmas 9 and 10. Recall the the original Hoeffding bounds states that if X_i are independent random variables in $[0, 1]$ (not necessarily binary) and $S = \sum X_i$, then

$$\Pr\left[S > (1 + \lambda)E \right] \leq e^{-\lambda^2 E[S]/2} \tag{4}$$

$$\Pr\left[S < (1 - \lambda)E \right] \leq e^{-\lambda^2 E[S]/3}. \tag{5}$$

In Lemma 9, the bound of $E[|T_1|]$ is the sum of random 0 and 1 variables. We use the standard Chernoff bound to show that the sum is exponentially concentrated, i.e., with probability $1 - e^{-n^{\Theta(1)}}$. The bound of $\mathrm{E}\,[\Delta(T_1)]$ in Lemma 10 is no longer the sum of random 0/1 variables or random variables whose value is in the range of $[0, 1]$. We need the following restatement of Heoffding bound: If X_i are independent random variables in $[0, A]$ and $S = \sum X_i$, then

$$\Pr\,[S > (1 + \lambda)E] \leq e^{-\lambda^2 \mathrm{E}[S/A]/2} \tag{6}$$

$$\Pr\,[S < (1 - \lambda)E] \leq e^{-\lambda^2 \mathrm{E}[S/A]/3}. \tag{7}$$

To obtain a concentration bound, we observe that k in Equation (3) is in the range of $[1 : n^{1-\gamma}]$. Thus, the bound in Equation (3) is the sum of random variables in range $[1 : n^{1-\gamma}]$. So, as long as $n^{1-\gamma} \ll n^{(3/4)\gamma+1/4}$, we can use this restatement of Hoeffding bound to an $1 - e^{-n^{\Theta(1)}}$ concentration. In our argument below that uses Lemma 10, we will have $\gamma \geq 4/7$. Thus, all our bounds are exponentially concentrated.

Theorem 1 (Size of Pyramid: h is a constant). *Let $G = (V, E)$ be a random (α, n)-BPPL graph with $\alpha = 11/4$. For any constant h, let S_0 be a random set of size b, where $b = \Theta\left(\frac{n}{h^{\alpha-1}(1+1/(2h))}\right)$. Then, for any constant $\epsilon > 0$, the expected size of* PYRAMID(S_0) *is* $\Theta(n^{1-\epsilon})$.

Proof. Because S_0 is a random set of b elements, by Lemma 7, with high probability, S_1 contains $[1 : h + 1]$. Let $\gamma = 4/7$ and $\beta = 3/7$, i.e., $\gamma = 1 - \beta$. By Lemmas 9 and 10, we have that the expected value of $|S_2|$ and $\Delta(S_2)$ are

$$\left[\Theta\left(n^{1-\frac{7}{4}\beta}\right), \Theta\left(n^{1-\frac{3}{4}\beta}\right)\right] \tag{8}$$

By iteratively applying this analysis, for any constant t, the expected values of $|S_t|$ and $\Delta(S_t)$ are

$$\left[\Theta\left(n^{1-\frac{7}{4}\left(\frac{3}{4}\right)^{t-1}\beta}\right), \Theta\left(n^{1-\left(\frac{3}{4}\right)^{t}\beta}\right)\right]$$

Moreover, these random variables are highly concentrated. Thus, for $t = \lceil \log_{4/3} \epsilon \rceil$, we have $\mathrm{E}\,[|\mathrm{PYRAMID}(S_0)|] \geq \mathrm{E}\,[|S_t|] = \Theta(n^{1-\epsilon})$. $\qquad\square$

By Lemma 1, if we set buffer size $b = \Theta\left(\frac{n}{h^{\alpha-1}(1+1/(2h))}\right)$, with a sufficiently large constant, every vertex that receives at least h votes will be in the buffer.

Theorem 2. *For any constants $2 < \alpha < 3$, $0 < c < 1$, and $\epsilon > 0$, with high probability, our crawling process with buffer size $b = cn$ starting on a randomly chosen set S_0 of vertices of a random (α, n)-BPPL graph G has a core of expect size $\Theta(n^{1-\epsilon})$.*

4.3   As Buffer Becomes Even More Smaller

When h is a function of n, e.g., $h = n^\delta$, we need to be a little more careful. But, our analysis can still be extended to establish the following theorem similar to Theorem 1.

Theorem 3 (*h*: $h = n^{\Theta(1)}$). *Let $G = (V, E)$ be a random (α, n)-BPPL graph with $\alpha = 11/4$. For any $\delta \leq 1/8$ and $\epsilon > 0$, letting $h = n^\delta$, if our crawling process starting with a random set S_0 of size $b = \Theta\left(\frac{n}{h^{\alpha-1}(1+1/(2h))}\right)$ has a core with expect size $\Theta(n^{1-\frac{7}{4}\delta-\epsilon})$.*

Proof. The main difference is that if $h = n^\delta$, then the estimation of the probability that a weight k vertex receives at least h edges from S_0 is

$$\Theta\left(\left(\frac{en^\gamma}{h}\right)^h \left(\frac{k}{n}\right)^h\right) = \Theta\left((en^{\gamma-\delta})^h \left(\frac{k}{n}\right)^h\right) \tag{9}$$

instead of Equation (1) used in the proof of Lemma 9. With the help of this bound, the bound of Equation of 2 becomes

$$E[|S_1|] = E[|S_1'|] - E[|S' \cap S_0|] = \Theta\left(n^{(7/4)\gamma-\frac{3}{4}\delta-3/4}\right), \tag{10}$$

and the bound of Equation of 3 becomes

$$E[\Delta(S_1)] = E[\Delta(S_1')] - E[\Delta(S_1' \cap S_0)] = \Theta\left(n^{(3/4)\gamma-\frac{7}{4}\delta+1/4}\right). \tag{11}$$

Applying these bounds in the analysis of Theorem 1, setting $\gamma = 1 - \beta = 4/7$, the expected value of $|S_2|$ and $\Delta(S_2)$ are

$$\left[\Theta\left(n^{1-\frac{3}{4}\delta-\frac{7}{4}\beta}\right), \Theta\left(n^{1-\frac{7}{4}\delta-\frac{3}{4}\beta}\right)\right]. \tag{12}$$

By iteratively applying this analysis, if $\delta \leq 1/8$ (which ensures that Proposition 6 holds), then for any constant t, the expected values of $|S_t|$ and $\Delta(S_t)$ are

$$\left[\Theta\left(n^{1-\frac{7}{4}\delta-\frac{7}{4}\left(\frac{3}{4}\right)^{t-1}\beta}\right), \Theta\left(n^{1-\frac{7}{4}\delta-\left(\frac{3}{4}\right)^t\beta}\right)\right].$$

Again, these random variables are highly concentrated. Thus, the core is at least $n^{1-(7/4)\delta-\epsilon}$ for all $\epsilon > 0$, i.e., for large enough t, the expected values of $|T_t|$ and $\Delta(T_t)$ are

$$\left[\Theta\left(n^{1-\frac{7}{4}\delta-\epsilon}\right), \Theta\left(n^{1-\frac{7}{4}\delta-\epsilon}\right)\right]. \qquad \square$$

4.4   Discussion

First of all, in our proof, we in fact consider the graph generated by the BBPL process and remove the multiple edges and self-loops. If we use the self-loops and multiple edges, we can further simplify the proof by starting with vertex 1 only, because its self-loop contribution is sufficient to keep it in the buffer. In other words, we do not need to start with an h-clique.

Our analysis can be easily modified to apply to the following family of random graphs: For vertex i, we add $\mathbf{d}_{\alpha,n}$ outward edges whose endpoints are chosen according to $\mathbf{d}_{\alpha,n}$. Again, in this model, we can remove self-loops and multiple edges. All our lemmas and theorems can be extended to this model. The analysis can also be extended to the following model with exact in and out degree. Let A and B be the array of length $\|\mathbf{d}_{\alpha,n}\|_1$, in which there are $\mathbf{d}_{\alpha,n}(i)$ entries with value i. Now randomly permute A and B, and add a directed edge from $A(i)$ to $B(i)$. Again, this graph may have multiple edges and self loops.

5   Final Remarks on Experiments and Future Directions

This paper is a step towards modeling web processing with limited space and time. Its objective is to provide some theoretical intuition indicating why table cores of non-trivial size exist. However, the models we consider here, both in terms of the crawling process and in terms of the graphical models, are in some respects unlike these usually encountered in practice. We have conducted limited experiments with some other models of power law graphs, for example, as discussed in [4] as well as some segments of web graphs. These experiments have shown the existence of non-trivial stable cores.

As the next step of this research, we would like to extend our result to other more realistic power-law models. The following are a few examples. (1) This is a growth model. Start with one node and at time t do the following based on a uniform three-way coins: (i) add a new node and connect it from a link from the existing nodes according the out degree distribution (plus some constant); (ii) add a new node and connect it to a link from the existing nodes according the in degree distribution (plus some constant); and (iii) choose a vertex according to the out degree and a vertex according to in degree, and insert this edge. (2)Given two vectors, IN and OUT and an integer m. Repeat m times, at each time, choose a vertex according to the out degree and a vertex according to the in degree, and insert this edge. In this model, we would like to study the graph based on the properties of IN and OUT, such as, IN and OUT follows some kind of power law. For example, in this paper, we analyze a particular (IN, OUT) pair. We would like to analyze the process for a larger family of (IN, OUT) distributions. (3) Start with one node and at time t, insert one new vertex and three edges. One out of the new vertex and one into the new vertex, and of course, according the in or out degree. (4) Other models in Chung and Lu's book.

References

1. Brin, S., Page, L.: The anatomy of a large-scale hypertextual Web search engine. Comput. Netw. ISDN Syst. 30, 107–117 (1998)
2. Bollobas, B., Riordan, O., Spencer, J., Tusnady, G.: The degree sequence of a scale-free random process. Random Structures and Algorithms 18, 279–290 (2001)
3. Castillo, C.: Effective Web Crawling, Ph.D. Thesis, University of Chile (2004)
4. Chung, F., Lu, L.: Complex Graphs and Networks. AMS (2007)
5. Faloutsos, C., Faloutsos, M., Faloutsos, P.: On power-law relationships of the internee topology. In: Proc. SIGCOMM (1999)

Average Update Times for Fully-Dynamic All-Pairs Shortest Paths

Tobias Friedrich[1,2] and Nils Hebbinghaus[1]

[1] Max-Planck-Institut für Informatik, Saarbrücken, Germany
[2] International Computer Science Institute, Berkeley, CA, USA

Abstract. We study the fully-dynamic all pairs shortest path problem for graphs with arbitrary non-negative edge weights. It is known for digraphs that an update of the distance matrix costs $\tilde{\mathcal{O}}(n^{2.75})^1$ worst-case time [Thorup, STOC '05] and $\tilde{\mathcal{O}}(n^2)$ amortized time [Demetrescu and Italiano, J.ACM '04] where n is the number of vertices. We present the first average-case analysis of the undirected problem. For a random update we show that the expected time per update is bounded by $\mathcal{O}(n^{4/3+\varepsilon})$ for all $\varepsilon > 0$.

Keywords: Dynamic graph algorithms, shortest paths, average-case analysis, random graphs.

1 Introduction

Dynamic graph algorithms maintain a certain property (e. g., connectivity information) of a graph that changes (a new edge inserted or an existing edge deleted) dynamically over time. They are used in a variety of contexts, e. g., operating systems, information systems, database systems, network management, assembly planning, VLSI design and graphical applications. An algorithm is called *fully-dynamic* if both edge weight increases and edge weight decreases are allowed. While a number of fully dynamic algorithms have been obtained for various properties on undirected graphs (see [6]), the design and analysis of fully-dynamic algorithms for directed graphs has turned out to be much harder (e. g., [9, 13, 15, 16]).

In this article, we consider the fully-dynamic *all-pairs shortest path problem* (APSP) for undirected graphs, which is one of the most fundamental problems in dynamic graph algorithms. The problem has been studied intensively since the late sixties (see [4] and references therein). We are interested in algorithms that maintain a complete distance matrix as edges are inserted or deleted. The static directed APSP problem can be solved in $\mathcal{O}(mn + n^2 \log n)$ time [8] where n is the number of vertices and m is the number of edges. This gives $\mathcal{O}(n^3)$ per update in the worst-case for a static recomputation from scratch. The first major improvement that is provably faster than this only worked on digraphs with small integer weights. King [10] presented a fully-dynamic APSP algorithm

---

[1] Throughout the paper, we use $\tilde{\mathcal{O}}(f(n))$ to denote $\mathcal{O}(f(n)\ \text{polylog}(n))$.

S.-H. Hong, H. Nagamochi, and T. Fukunaga (Eds.): ISAAC 2008, LNCS 5369, pp. 692–703, 2008.

for general directed graphs with positive integer weights less than C that supported updates in $\mathcal{O}(n^{2.5}\sqrt{C\log n})$. In the remainder of the paper, we will only consider non-negative real-valued edge weights. Demetrescu and Italiano pursued this problem in a series of papers and showed that it can be solved in $\mathcal{O}(n^2 \log^3 n)$ amortized time per update. [4]. This has been slightly improved to $\mathcal{O}(n^2(\log n + \log^2((m+n)/n)))$ amortized time per update by Thorup [17]. In [18], Thorup showed a worst-case update time of $\tilde{\mathcal{O}}(n^{2.75})$.

We are interested in *expected update times*. The only known result for this is for the undirected, unweighted, decremental, randomized, and approximate version of the APSP problem. Roditty and Zwick [14] showed for this setting an expected amortized time of $\tilde{\mathcal{O}}(n)$. For our setting of the problem on *undirected graphs with arbitrary non-negative edge weights*, there is nothing known about the average-case update times. We analyze a variant of Demetrescu and Italiano's algorithm described in Section 3. Let $\mathcal{R}(p)$ denote the expected runtime of our algorithm for a single random edge update of a random graph $G \in \mathcal{G}(n,p)$. Let $\varepsilon, \varepsilon' > 0$. For arbitrary p, we can show $\mathcal{R}(p) = \mathcal{O}(n^{4/3+\varepsilon})$. However, for most p we can prove that the runtime is actually much smaller. The above bound is best only at the phase transition around $pn = 1$, i.e., when the size of the largest component rapidly grows from $\Theta(\log n)$ to $\Theta(n)$. When the graph is sparser, our algorithm is much faster. In this case, we can show $\mathcal{R}(p) = \mathcal{O}(n^{2/3+\varepsilon})$ for $pn \leq 1 - n^{-1/3}$ and $\mathcal{R}(p) = \mathcal{O}(n^\varepsilon)$ for $pn < 1/2$. Similarly, the algorithm becomes faster when the graph has passed the critical window. We show $\mathcal{R}(p) = \mathcal{O}(n^\varepsilon/p)$ for $pn \geq 1 + \varepsilon'$. The final result is given in Theorem 11. Additionally to these asymptotic upper bounds on the expected runtime, we also examined the empirical average runtime. Interestingly, this also shows that the update costs are first increasing and later decreasing when more edges are inserted. This corresponds well with the above phase distinction for the asymptotic bounds.

The remainder of this paper is organized as follows. The next section presents all necessary graph theoretical notations. In Section 3 we present our algorithm. In Section 4 we prove a number of random graph properties which are then used in Section 5 to show the asymptotic bounds. The last section presents some empirical results.

2   Preliminaries

Demetrescu and Italiano [5] performed several experiments on directed random graphs. We want to bound the expected runtime of random updates on a random graph of a very similar algorithm. We utilize the random graph model $\mathcal{G}(n,p)$ introduced and popularized by Erdős and Rényi [7]. The $\mathcal{G}(n,p)$ model consists of a graph with n vertices in which each edge is chosen independently with probability p. In our model, a random update first chooses two vertices x and y $(x \neq y)$. Then, with (fixed) probability δ, it inserts the edge (x,y) with a random weight $w \in [0,1]$. If the edge was already in the graph, it changes its weight to w. Otherwise, with probability $1-\delta$, a random update deletes the edge (x,y) if (x,y) is in the graph (otherwise, it does nothing). Note that T random updates on a random graph $G \in \mathcal{G}(n,p)$ lead to a graph with edges present with probability $p' = \delta + (p - \delta)(1 - 1/\binom{n}{2})^T$, but not necessarily mutually independent.

Throughout the paper, we use the following notations:

- $G = (V, E)$ is an undirected graph with arbitrary non-negative edge weights.
- $\Delta := \max_{v \in V} \deg(v)$ is the maximum degree of the graph G.
- $\mathrm{dist}(x, y)$ (distance) is the length of the shortest path from x to y.
- $\mathrm{diam}(G)$ (diameter) is the greatest distance between any two vertices of one component.
- $C(x)$ denotes the component that contains the vertex x.
- $C(G)$ denotes the largest component of G.
- w_{xy} denotes the weight of an edge (x, y).
- $\pi_{xy} = \langle u_0, u_1, \ldots, u_k \rangle$ is a path from vertex $x = u_0$ to vertex $y = u_k$, i.e., a sequence of vertices such that with $(u_i, u_{i+1}) \in E$ for each $0 \leq i < k$ (no repeated edges).
- $w(\pi_{xy}) = \sum_{i=0}^{k-1} w_{u_i u_{i+1}}$ is the weight of a path.
- $\pi_{xy} \circ \pi_{yz}$ denotes the concatenation of two paths π_{xy} and π_{yz}.
- $\ell(\pi_{xy})$ denotes the subpath π_{xa} of π_{xy} such that $\pi_{xy} = \pi_{xa} \circ \langle a, y \rangle$.
- $r(\pi_{xy})$ denotes the subpath π_{by} of π_{xy} such that $\pi_{xy} = \langle x, b \rangle \circ \pi_{by}$.

We assume without loss of generality that there is only one shortest path between each pair of vertices in G. Otherwise, ties can be broken as discussed in Section 3.4 of [4].

3   Algorithm

We will now describe our algorithm. It is a slight modification of the algorithm of Demetrescu and Italiano [5] as our aim is an average-case analysis of the undirected problem while they were interested in the amortized costs for directed graphs.

The main tool Demetrescu and Italiano [4] very cleverly introduced and applied is the concept of "locally shortest paths". A path π_{xy} is *locally shortest* if every proper subpath is a shortest path or it consists of only a single vertex. The algorithm maintains the following data structures:

- w_{xy} weight of edge (x, y)
- P_{xy} priority queue of the locally shortest paths from x to y (priority $w(\pi_{xy})$)
- P^*_{xy} shortest path from x to y
- $L(\pi_{xy})$ set of left-extensions $\langle x', x \rangle \circ \pi_{xy}$ of π_{xy} that are locally shortest paths
- $L^*(\pi_{xy})$ set of left-extensions $\langle x', x \rangle \circ \pi_{xy}$ of π_{xy} that are shortest paths
- $R(\pi_{xy})$ set of right-extensions $\pi_{xy} \circ \langle y, y' \rangle$ of π_{xy} that are locally shortest paths
- $R^*(\pi_{xy})$ set of right-extensions $\pi_{xy} \circ \langle y, y' \rangle$ of π_{xy} that are shortest paths

Note that $P^*_{xy} \subseteq P_{xy}$ and that every minimum weight path in P_{xy} is also a shortest path. Each path $\pi_{xy} \in P_{xy}$ is stored implicitly with constant space by just storing two pointers to the subpaths $\ell(\pi_{xy})$ and $r(\pi_{xy})$.

$\textsc{Update}(u, v, w)$

    $\triangleright$ Phase 1: Delete edge (u, v) and all paths containing edge (u, v),
                  store pairs of vertices affected by the update in list A

1   **if** $(u, v) \in P_{uv}$ **then**

2   $Q \leftarrow \{(u, v)\}$

3   **while** $Q \neq \emptyset$ **do**

4   extract any π_{xy} from Q

5   remove π_{xy} from P_{xy}, $L(r(\pi_{xy}))$, and $R(\ell(\pi_{xy}))$

6   **if** $\pi_{xy} \in P^*_{xy}$ **then**

7   remove π_{xy} from P^*_{xy}, $L^*(r(\pi_{xy}))$, and $R^*(\ell(\pi_{xy}))$

8   add (x, y) to A

9   add all paths in $L(\pi_{xy})$ to Q

10   add all paths in $R(\pi_{xy})$ to Q

    $\triangleright$ Phase 2: Insert edge (u, v) with weight w

11 **if** $w < \infty$ **then**

12   add (u, v) to A

13   add (u, v) to P_{uv}, $L(\pi_{vv})$, and $R(\pi_{uu})$

    $\triangleright$ Phase 3: Scan pairs in A

14 **while** $A \neq \emptyset$ **do**

15   extract any pair (x, y) from A

16   add π_{xy} with minimum $w(\pi_{xy})$ to H (if any)

    $\triangleright$ Phase 4: Propagation loop

17 **while** $H \neq \emptyset$ **do**

18   extract path π_{xy} with minimum $w(\pi_{xy})$ from H

19   **if** $w(\pi_{xy})$ is larger than the smallest weight in P_{xy} **then continue**

20   add P^*_{xy} to Q

21   add π_{xy} to P^*_{xy}, $L^*(r(\pi_{xy}))$, and $R^*(\ell(\pi_{xy}))$

22   **for each** $\pi_{x'b} \in L^*(\ell(\pi_{xy}))$ **do**

23   **if** $(x', x) \circ \pi_{xy} \in L(\pi_{xy})$ **then continue**

24        $\pi_{x'y} \leftarrow (x', x) \circ \pi_{xy}$

25        $w(\pi_{x'y}) \leftarrow w_{x'x} + d_{xy}$

26        $\ell(\pi_{x'y}) \leftarrow \pi_{x'b},\ r(\pi_{x'y}) \leftarrow \pi_{xy}$

27   add $\pi_{x'y}$ to $P_{x'y}$, $L(\pi_{xy})$, $R(\pi_{x'b})$, and H

28   **for each** $\pi_{ay'} \in R^*(r(\pi_{xy}))$ **do**

29   **if** $\pi_{xy} \circ (y, y') \in R(\pi_{xy})$ **then continue**

30        $\pi_{xy'} \leftarrow \pi_{xy} \circ (y, y')$

31        $w(\pi_{xy'}) \leftarrow d_{xy} + w_{yy'}$

32        $\ell(\pi_{xy'}) \leftarrow \pi_{xy},\ r(\pi_{xy'}) \leftarrow \pi_{ay'}$

33   add $\pi_{xy'}$ to $P_{xy'}$, $L(\pi_{ay'})$, $R(\pi_{xy})$, and H

    $\triangleright$ Phase 5: Delete all LSPs π that stopped being LSP
                because $\ell(\pi)$ or $r(\pi)$ stopped being SP

34 **while** $Q \neq \emptyset$ **do**

35   extract any π_{xy} from Q

36   **for each** $\pi_{x'y} \in L(\pi_{xy})$ **do**

37   remove $\pi_{x'y}$ from $R((x', x) \circ \ell(\pi_{x,y}))$ and $L(\pi_{x,y})$

38   **for each** $\pi_{xy'} \in R(\pi_{xy})$ **do**

39   remove $\pi_{xy'}$ from $L(r(\pi_{x,y}) \circ (y, y'))$ and $R(\pi_{x,y})$

Fig. 1. The slightly modified APSP algorithm of Demetrescu and Italiano [5]

The pseudo-code of our algorithm is given in Figure 1. The first four phases are equivalent to [5]. We will just describe them briefly. A detailed description can be found in [4]. In the first phase, the algorithm deletes from the data structure all the paths that would stop being locally shortest if we deleted the edge (u, v). In doing so it stores the pairs of the endpoints of the affected paths in the temporary list A. In the following phase it adds the edge (if it is an insert or update operation) to the data structures. The third phase initializes the heap H with the minimum weight paths π_{xy} for all $(x, y) \in A$. In the fourth phase the algorithm repeatedly extracts the cheapest path π_{xy} from H. The first extracted path for each pair (x, y) must be a shortest path. If this is the case, the path is stored in the data structures. To propagate this information, also its left- and right-extensions are updated and added to H to find all further extensions.

The amortized number of new locally shortest paths can be $\Omega(n^3)$ per update. To allow a better worst-case performance, Demetrescu and Italiano [4] had to delay the update of the data structure in a very clever way. Their data structure can contain paths in P_{xy} which are not locally shortest anymore. We avoid this with the fifth phase. There, all locally shortest paths which stopped being locally shortest because one of their two subpaths stopped being shortest path are detected and deleted.

We analyze the expected time for the algorithm to insert a randomly chosen edge e in the graph $G \in \mathcal{G}(n, p)$ and maintain the sets of shortest path and locally shortest path. The weights of e and of the edges in G are chosen uniformly at random from the set $[0, 1]$.

4   Random Graph Properties

To bound the runtime of our algorithm in the next section, we first provide some properties of random graphs $G \in \mathcal{G}(n, p)$. The main result of this section will be Theorem 9. It bounds the quantity $\mu(p)$ which we define as the expected number of locally shortest paths and shortest paths passing a fixed edge of G. Let LSP denote the set of all locally shortest paths and SP the set of all shortest paths in G.

The following four lemmas are well-known.

Lemma 1 (Bollobás [2]). *Let $G \in \mathcal{G}(n, p)$ with $pn < 1/2$. Then,* $\mathbf{Pr}\left[|C(G)| \leq 20 \log n\right] = 1 - \mathcal{O}(n^{-2})$.

Lemma 2 (Bollobás [2]). *For every $\alpha > 0$ and $G \in \mathcal{G}(n, p)$ with $pn = \alpha \log n$,* $\mathbf{Pr}\left[G \text{ is connected}\right] = 1 - \mathcal{O}(n^{1-2\alpha})$.

Lemma 3 (Chung and Lu [3]). *For every $\varepsilon > 0$ and $G \in \mathcal{G}(n, p)$ with $pn = 1 + \varepsilon$,* $\mathbf{Pr}\left[\text{diam}(G) \leq 2 \log n\right] = 1 - o(n^{-1})$.

Lemma 4 (Nachmias and Peres [12]). *Let $x \in G$ and $G \in \mathcal{G}(n, p)$ with* $pn \leq 1 + n^{-1/3}$. *Then,* $\mathbf{Pr}\left[|C(x)| > 2n^{2/3}\right] = \mathcal{O}(n^{-1/3})$.

The following lemma gives a general upper bound on the expected diameter of a random graph $G \in \mathcal{G}(n,p)$ for arbitrary p. Recall that we defined the diameter of a disconnected graph as the maximum diameter of its components.

Lemma 5. *Let $G \in \mathcal{G}(n,p)$. Then,* $\mathbf{E}\left[\operatorname{diam}(G)\right] = \mathcal{O}(n^{1/3})$.

Proof. Let G be a complete graph on n vertices with edge weights uniformly distributed at random in $[0,1]$. Let $G_{\leq p} = (V, E_{\leq p})$ be the subgraph of G containing all vertices but only those edges with weight less or equal p. Then $G_{\leq p}$ is a $\mathcal{G}(n,p)$-graph. We apply Kruskal's algorithm [11] for the construction of a minimum spanning forest of G, i.e., we look at the edges in increasing weight order and integrate every edge that does not introduce a cycle in the current edge set. We stop this process if the current edge has weight greater than p and denote the obtained subset of edges $E_{\mathrm{Kruskal}, \leq p}$. Let us also denote the edge set of the spanning forest which is returned from the completed Kruskal algorithm by E_{Kruskal}. By Addario-Berry, Broutin, and Reed [1], the expected diameter of $G_{\mathrm{Kruskal}} := (V, E_{\mathrm{Kruskal}})$ is of order $\Theta(n^{1/3})$. Clearly, $E_{\mathrm{Kruskal}, \leq p} \subseteq E_{\mathrm{Kruskal}} \cap E_{\leq p}$. As $G_{\mathrm{Kruskal}, \leq p} := (V, E_{\mathrm{Kruskal}, \leq p})$ is a minimum spanning forest of $G_{\leq p}$, we get

$$\mathbf{E}\left[\operatorname{diam}(G_{\leq p})\right] \leq \mathbf{E}\left[\operatorname{diam}(G_{\mathrm{Kruskal}, \leq p})\right] \leq \mathbf{E}\left[\operatorname{diam}(G_{\mathrm{Kruskal}})\right] = \mathcal{O}(n^{1/3}). \quad \square$$

To prove the desired bound on $\mu(p)$ we also need the following three technical lemmas.

Lemma 6. *Let $G \in \mathcal{G}(n,p)$ with $pn \geq 4 \log n$. Then every shortest path in G has weight $\mathcal{O}(\frac{\log^2 n}{n})$ with probability $1 - \mathcal{O}(n^{-2})$.*

Proof. Let us consider two random graphs $G_1, G_2 \in \mathcal{G}(n, \frac{2 \log n}{n})$ on the same set of vertices and let $G_{\cup}$ be the union of G_1 and G_2 (union of the edge sets). Then we get $G_{\cup} \in \mathcal{G}(n, \frac{4 \log n}{n} - \frac{4 \log^2 n}{n^2})$. By Lemmas 2 and 3, G_i is connected and and $\operatorname{diam}(G_i) \leq 2 \log n$ with probability $1 - \mathcal{O}(n^{-1})$ for $i = 1, 2$. As the two random graphs are chosen independently, at least one of them is connected and has diameter $\mathcal{O}(\log n)$ with probability $1 - \mathcal{O}(n^{-2})$. By construction, this also holds for $G_{\cup}$. Therefore all $G \in \mathcal{G}(n,p)$ with $pn \geq 4 \log n$ are connected and have a diameter of order $\mathcal{O}(\log n)$ with probability $1 - \mathcal{O}(n^{-2})$.

We now prove that every shortest path in G has a total weight of $\mathcal{O}(\frac{\log^2 n}{pn})$ with probability $1 - \mathcal{O}(n^{-2})$. Let us consider the subgraph $G_0 = (V, E_{\leq \frac{4 \log n}{pn}})$ consisting of all vertices but only those edges of G with weight at most $\frac{4 \log n}{pn}$. This is a random graph $\mathcal{G}(n, \frac{4 \log n}{n})$ with weights chosen uniformly at random from $[0, \frac{4 \log n}{pn}]$. As we have shown above, G_0 is connected and has diameter of order $\mathcal{O}(\log n)$ with probability $1 - \mathcal{O}(n^{-2})$. This implies that every shortest path in G_0 has a total weight of $\mathcal{O}(\frac{\log^2 n}{pn})$ with probability $1 - \mathcal{O}(n^{-2})$. This upper bound also holds with the same probability for all shortest paths in G. $\quad \square$

Lemma 7. *Let $G \in \mathcal{G}(n, p)$. The subgraph $G_{\mathrm{SP}} = (V, E_{\mathrm{SP}})$ of all edges that are shortest paths in G fulfills $\Delta(G_{\mathrm{SP}}) \leq n^\varepsilon$ with probability $1 - \mathcal{O}(n^{-2})$. In particular, $|\mathrm{LSP}| \leq |\mathrm{SP}| n^\varepsilon$ with probability $1 - \mathcal{O}(n^{-2})$.*

Proof. Let us first consider the case $pn \geq 4 \log n$. We know from Lemma 6 that all elements of E_{SP} have weight $\mathcal{O}(\frac{\log^2 n}{pn})$ with probability $1 - \mathcal{O}(n^{-2})$. Thus, G_{SP} is a subgraph of a random graph in $\mathcal{G}(n, \frac{\log^2 n}{n})$. Therefore, we can prove the first claim of the lemma by bounding $\Delta(G')$ for $G' \in \mathcal{G}(n, \frac{\log^2 n}{n})$. Using Stirling's formula, the probability for a vertex in G' to have a degree greater or equal n^ε is at most

$$\binom{n}{n^\varepsilon} \left(\frac{\log^2 n}{n} \right)^{n^\varepsilon} \leq \frac{(\log^2 n)^{n^\varepsilon}}{\sqrt{2\pi n^\varepsilon} \left(\frac{n^\varepsilon}{e} \right)^{n^\varepsilon}} \leq n^{-\varepsilon n^{\varepsilon}/2}$$

for n large enough. Hence, the probability for $\Delta(G')$ to be greater or equal n^ε is at most

$$1 - \left(1 - n^{-\varepsilon n^{\varepsilon}/2} \right)^n \leq 1 - \left(\left(1 - n^{-\varepsilon n^{\varepsilon}/2} \right)^{n^{\varepsilon n^{\varepsilon}/2} - 1} \right)^{2n^{1 - \varepsilon n^{\varepsilon}/2}}$$

$$\leq 1 - e^{-2n^{1 - \varepsilon n^{\varepsilon}/2}} \leq n^{-n^{\varepsilon/3}},$$

where we used $2n^{1 - \varepsilon n^{\varepsilon}/2} \leq n^{-n^{\varepsilon/3}}$ for n large enough and $1 + x \leq e^x$ for all $x \in \mathbb{R}$. This proves $\Delta(G_{\mathrm{SP}}) \leq n^\varepsilon$ with probability $1 - \mathcal{O}(n^{-2})$. The second claim is a consequence of the fact, that each locally shortest path from vertex x to vertex y is uniquely determined by its first and also by its last edge. Moreover, every locally shortest path with at least 2 edges starts and ends with edges that are shortest paths themselves. Thus, there are at most $\mathcal{O}(n^\varepsilon)$ locally shortest paths for each shortest path with probability at least $1 - \mathcal{O}(n^{-2})$, which proves the lemma for $pn \geq 4 \log n$.

Let $pn < 4 \log n$. We consider $G' \in \mathcal{G}(n, \frac{4 \log n}{n})$ with edge weights chosen randomly in $[0, \frac{4 \log n}{pn}]$. Although the weights of this graph are scaled up by the factor $\frac{4 \log n}{pn}$, we get $\Delta(G'_{\mathrm{SP}}) \leq n^\varepsilon$ with probability $1 - \mathcal{O}(n^{-2})$, since the scaling has no effect on the subgraph G'_{SP} of all shortest path edges in G'. Now the subgraph G of all edges of G' with weight less or equal 1 is a $\mathcal{G}(n, p)$-graph with edge weights chosen uniformly at random from $[0, \frac{4 \log n}{pn}]$. Hence, every edge that is a shortest path in G is also a shortest path in G'. With this we get $\Delta(G_{\mathrm{SP}}) \leq n^\varepsilon$ with probability $1 - \mathcal{O}(n^{-2})$. The second claim follows with the same arguments as in the case $pn \geq 4 \log n$. $\qquad\square$

Lemma 8. *Let $G \in \mathcal{G}(n, p)$ with $pn \geq 1/2$. For all $\varepsilon > 0$, $\mu(p) = \mathcal{O}\big(\mathbf{E}\left[\mathrm{diam}(G)\right] |\mathrm{SP}| n^{\varepsilon - 2} / p\big)$.*

Proof. We consider the subgraph $G' = (V, E_{<1/(2pn)})$ containing all vertices of G but only those edges with weight less than $1/(2pn)$. Then by Lemma 1 $G \in \mathcal{G}(n, 1/(2n))$ and the largest component of G' is of order $\mathcal{O}(\log n)$ with probability $1 - \mathcal{O}(n^{-2})$. Thus, every path in G' contains $\mathcal{O}(\log n)$ edges with probability $1 - \mathcal{O}(n^{-2})$. The expected weight of the heaviest element of LSP is at most $\mathbf{E}\left[\text{diam}(G)\right]$. Moreover, in the case $p \geq \frac{4 \log n}{n}$ the expected weight of the heaviest element of LSP is at most $\mathcal{O}(\frac{\log^2 n}{pn})$ as shown in Lemma 6. Thus, in expectation the largest number of edges with a weight greater or equal $1/(2pn)$ in an element of LSP is of order $\mathcal{O}(\mathbf{E}\left[\text{diam}\, G\right] pn)$ and $\mathcal{O}(\log^2 n)$ if $p \geq \frac{4 \log n}{n}$. This implies an upper bound of $\mathcal{O}(\mathbf{E}\left[\text{diam}\, G\right] \log^2 n)$ for the maximal number of edges with weight greater or equal $1/(2pn)$ in locally shortest paths of G in expectation for all $p \geq 1/(2pn)$.

By Lemma 7 we know that the bound $|\text{LSP}| = \mathcal{O}(|\text{SP}|n^{\varepsilon/2})$ is violated with probability $\mathcal{O}(n^{-2})$. In this case we can estimate the number of locally shortest paths and shortest paths in G by $\mathcal{O}(n^3)$ (the first edge and the other endpoint of a locally shortest path determines the path uniquely) and the length of this paths trivially by $n - 1$. Since the probability for this event is $\mathcal{O}(n^{-2})$, the contribution to the expected number of edges in the multiset of all edges of all (locally) shortest paths is $\mathcal{O}(n^2)$. If $|\text{LSP}| = \mathcal{O}(|\text{SP}|n^{\varepsilon/2})$, the maximal number of edges with weight greater or equal $1/(2pn)$ in locally shortest paths of G is $\mathcal{O}(\mathbf{E}\left[\text{diam}\, G\right] \log^2 n)$ in expectation. Now in every (locally) shortest path in G there can only be consecutive parts of edges of G' of order $\mathcal{O}(\log n)$ and they must be followed by an edge with weight greater or equal $1/(2pn)$. Since there can only be $\mathcal{O}(\mathbf{E}\left[\text{diam}\, G\right] \log^2 n)$ of these edges in the path in expectation, the total number of edges in the longest of all (locally) shortest paths is $\mathcal{O}(\mathbf{E}\left[\text{diam}\, G\right] \log^3 n)$ in expectation. Thus, the multiset of all edges of all (locally) shortest paths contains $\mathcal{O}(|\text{SP}|\mathbf{E}\left[\text{diam}\, G\right] n^{\varepsilon/2} \log^3 n)$ edges. By Chernoff bounds G has $\Theta(pn^2)$ edges. Therefore, the average number of (locally) shortest paths through a fixed edge is $\mathcal{O}(\mathbf{E}\left[\text{diam}\, G\right] |\text{SP}|n^{\varepsilon-2}/p)$. $\square$

We are now well-prepared to prove the main theorem of this section. It bounds $\mu(p)$ which is the expected number of locally shortest paths and shortest paths passing a fixed edge.

Theorem 9. *Let $G \in \mathcal{G}(n, p)$. For all $\varepsilon, \varepsilon' > 0$,*

    (i) $\mu(p) = \mathcal{O}(1)$ *for $pn < 1/2$,*
    (ii) $\mu(p) = \mathcal{O}(n^{2/3})$ *for $1/2 \leq pn \leq 1 - n^{-1/3}$,*
    (iii) $\mu(p) = \mathcal{O}(n^{1+\varepsilon})$ *for $1 - n^{-1/3} \leq pn \leq 1 + n^{-1/3}$,*
    (iv) $\mu(p) = \mathcal{O}(n^{4/3+\varepsilon})$ *for $1 + n^{-1/3} \leq pn \leq 1 + \varepsilon'$,*
    (v) $\mu(p) = \mathcal{O}(n^{\varepsilon}/p)$ *for $1 + \varepsilon' \leq pn$.*

Proof. (i) We bound $\mu(p)$ by the total number of paths passing a fixed edge. Let us first estimate the expected number of paths of a fixed length k in G going through a fixed edge. There are k possible positions of the fixed edge in a path of length k. Furthermore, we can choose the remaining $k - 1$ vertices of such a path in $\prod_{i=1}^{k-1}(n - i)$ different ways. Since every $e \in \binom{[n]}{2}$ is an edge of G with probability p, the expected number of paths in G that go through a fixed edge is bounded above by

$$\sum_{k=1}^{n-1} kp^{k-1} \prod_{i=1}^{k-1}(n-i) \leq \sum_{k=1}^{n-1} k(pn)^{k-1} \leq (pn)^{-1}\sum_{k=1}^{n} k(pn)^{k}.$$

Thus, the expected number of paths in G going through a fixed edge is at most

$$(pn)^{-1}\sum_{k=1}^{n}\sum_{i=k}^{n}(pn)^{k} = \sum_{k=1}^{n}\frac{(pn)^{k-1}-(pn)^{n}}{1-pn} \leq \frac{1-(pn)^{n}}{(1-(pn))^{2}} \leq \frac{1}{(1-pn)^{2}}.$$

Thus, $\mu(p) = \mathcal{O}(1)$ for $pn \leq 1/2$.

(ii) Using the bound in (i), we get $\mu(p) = \mathcal{O}(1/(1 - pn)^2) = \mathcal{O}(n^{2/3})$ for $pn \leq 1 - n^{-1/3}$.

(iii) Applying Lemma 4, we get $|\text{SP}| = \mathcal{O}(n^{5/3})$ with probability $1 - O(n^{-1/3})$ and $|SP| = \mathcal{O}(n^2)$ otherwise. Combining this with Lemma 8 and Lemma 5 gives $\mu(p) = O(n^{1+\varepsilon})$.

(iv) By Lemma 5, the expected diameter of G is $\mathcal{O}(n^{1/3})$. Thus, Lemma 8 yields $\mu(p) = \mathcal{O}(n^{4/3+\varepsilon})$.

(v) By Lemma 3, we get $\mathbf{E}\left[\operatorname{diam} G\right] = O(\log n)$. Now Lemma 8 yields $\mu(p) = \mathcal{O}(n^{\varepsilon}/p)$. $\qquad\Box$

5   Runtime Analysis

In this section we describe the runtime of our algorithm in terms of the parameter $\mu(p)$. With this, the main result Theorem 11 is an immediate corollary of the bounds on $\mu(p)$ from the previous section.

Theorem 10. *Let $G \in \mathcal{G}(n,p)$. The expected runtime of our algorithm for a random edge update on G is $\mathcal{O}(\mu(p)\,n^{\varepsilon})$.*

Proof. To bound the runtime, we will use the quantity $\mu(p)$ which is the expected number of locally shortest paths and shortest paths through a fixed edge e of G. If the algorithm performs the deletion of the edge e, this is exactly the number of paths that stop being (locally) shortest. In the case of the insertion, we get (almost) the same picture by making a backwards analysis. Instead of the insertion of e to $G = (V, E)$ we can also investigate the deletion of e from the graph $G' = (V, E \cup \{e\})$. Therefore the quantity $\mu(p)$ is also the expected number of paths in G' that start being (locally) shortest. The slight modification that G' contains one edge more than G has no consequence for the order of $\mu(p)$.

We bound the runtime of all five phases separately. The algorithm is running through the first phase, only if the considered edge e is already in the graph and has to be deleted or updated. If this is the case, the algorithm goes through the while loop for every locally shortest path of G that contains the edge e at most twice, since a locally shortest path can be added to Q as a left- and as a right-extension. The only part of the while loop with more than constant runtime is the removing of the path π_{xy} from the lists P_{xy}, $L(r(\pi_{xy}))$, and $R(\ell(\pi_{xy}))$. Since every locally shortest path is uniquely determined by the first (respectively the last edge), we can bound the runtime of phase 1 by $\mathcal{O}(\mu(p)n^{\varepsilon})$ using Lemma 7.

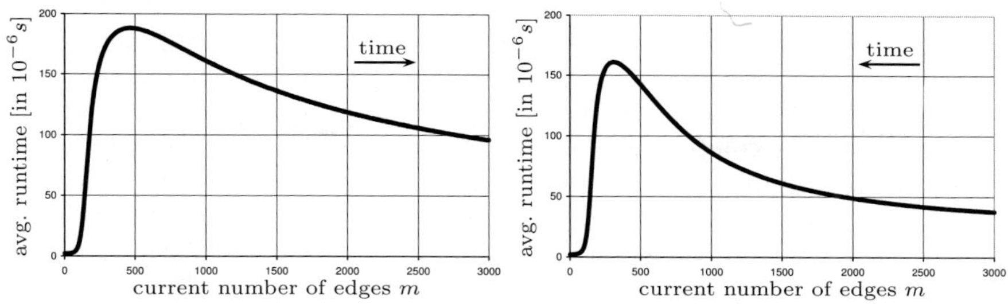

(a) Insertion of 3000 random edges.   (b) Deletion of 3000 random edges.

Fig. 2. Experimental results for the algorithm of Demetrescu and Italiano [5]. We start with an empty graph with $n = 100$ vertices and insert 3000 random edges. (a) shows the measured runtimes depending on the number of inserted edges. Analogously, (b) shows the measured runtimes for the deletion of 3000 edges in a random order till the empty graph is obtained again. The horizontal axes describe the current number of edges m. The vertical axes show the measured runtimes averaged over three million runs.

The runtime of the second phase is constant. The while loop in the third phase has an expected length of $\mathcal{O}(\mu(p))$. Since adding the path π_{xy} to the priority queue H costs $\mathcal{O}(\log n)$, the runtime of phase 3 is $\mathcal{O}(\mu(p)\log n)$.

For the analysis of the runtime of phase 4 it is crucial to observe that every line in the for loops as well as every other line is executed $\mathcal{O}(\mu(p))$ times in expectation. Moreover, the algorithm has to add the extended paths $\pi_{x'y}$ and $\pi_{xy'}$ to lists of locally shortest path and the priority queue H which is done in time $\mathcal{O}(n^\varepsilon)$ in every execution. Thus, the runtime of the algorithm in phase 4 is $\mathcal{O}(\mu(p)n^\varepsilon)$.

If the algorithm performs an insertion or an update, the set Q in phase 5 contains all shortest paths of G that stop being shortest. If the algorithm performs a deletion, Q is empty. Thus, the algorithm is running through the while loop $\mathcal{O}(\mu(p))$ times in expectation. The for loops are both performed $\mathcal{O}(n^{\varepsilon/2})$ times using Lemma 7 in the same way as in the beginning of this proof but with $\varepsilon/2$ instead of ε. In the same way, we can bound the expected runtime of the lines in the for loops by $\mathcal{O}(n^{\varepsilon/2})$. Altogether this gives an expected runtime of $\mathcal{O}(\mu(p)n^\varepsilon)$ in phase 5. ☐

With this we can now conclude our main result.

Theorem 11. *Let $\mathcal{R}(p)$ denote the expected runtime for an edge update in a graph $G \in \mathcal{G}(n, p)$. For all $\varepsilon, \varepsilon' > 0$ we have shown that*

$$
\begin{aligned}
&(i)\ \ \mathcal{R}(p) = \mathcal{O}(n^\varepsilon) &&\text{for } pn < 1/2,\\
&(ii)\ \ \mathcal{R}(p) = \mathcal{O}(n^{2/3+\varepsilon}) &&\text{for } pn \le 1 - n^{-1/3},\\
&(iii)\ \ \mathcal{R}(p) = \mathcal{O}(n^{1+\varepsilon}) &&\text{for } pn \le 1 + n^{-1/3},\\
&(iv)\ \ \mathcal{R}(p) = \mathcal{O}(n^{4/3+\varepsilon}) &&\text{for } pn \le 1 + \varepsilon',\\
&(v)\ \ \mathcal{R}(p) = \mathcal{O}(n^\varepsilon/p) &&\text{for } pn \ge 1 + \varepsilon'.
\end{aligned}
$$

Let us give an intuition how the properties of $\mathcal{G}(n,p)$ change when more and more edges are inserted and how this affects $\mathcal{R}(p)$. In the early stage (i) of the random graph process, the graph consists of many small components of size $O(\log n)$ which are trees or unicyclic. There, it is very fast to update edges. Soon after in stage (ii), the components become larger and it becomes likely for a new edge to connect two of them. Therefore, the expected number of new (locally) shortest paths increases significantly. In stage (iii) and (iv) a giant component grows and the algorithm has to update many (locally) shortest paths whenever the giant component catches other components of the graph. In (v) the last isolated vertex joins the giant component and the graph becomes connected. As the process evolves, the minimum degree and the connectivity grows and it becomes less and less likely that an inserted edge is a shortest path. Thus, also the expected insertion costs are going down.

6   Empirical Observations

To show that the theoretically observed behavior indeed occurs in practice, we also performed some experiments. For this, we used the original algorithm of Demetrescu and Italiano [5] available from www.dis.uniroma1.it/~demetres/experim/dsp/[2]. As the number of locally shortest paths between any pair of nodes has been reported to be very small [5], we assume that the experimental performance of our algorithm described in Section 3 should be similar to that of Demetrescu and Italiano.

We start with an empty graph with $n = 100$ vertices and add 3000 edges in a random order. Figure 2(a) shows the measured runtimes per insertion averaged over three million runs. Afterwards, we examine the opposite direction and remove all edges in a random order. The measured average runtimes per deletion are shown in Figure 2(b). Note that as predicted in Theorem 11, both charts identify the largest update complexity shortly after the critical window.

Acknowledgements

Thanks are due to Daniel Johannsen, Frank Neumann, and Yuval Peres for various helpful discussions. This work was partially supported by a postdoctoral fellowship from the German Academic Exchange Service (DAAD).

References

[1] Addario-Berry, L., Broutin, N., Reed, B.: The diameter of the minimum spanning tree of a complete graph. In: Proc. Fourth Colloquium on Mathematics and Computer Science Algorithms, Trees, Combinatorics and Probabilities (2006)

---

[2] Analogous to the experiments of Demetrescu and Italiano [5], we used the D-LHP code without smoothing. All experiments are conducted on 2.4 GHz Opteron machines running Debian GNU/Linux.

[2] Bollobás, B.: Random Graphs. Cambridge Univ. Press, Cambridge (2001)

[3] Chung, F., Lu, L.: The diameter of sparse random graphs. Advances in Applied Mathematics 26, 257–279 (2001)

[4] Demetrescu, C., Italiano, G.F.: A new approach to dynamic all pairs shortest paths. J. ACM 51, 968–992 (2004)

[5] Demetrescu, C., Italiano, G.F.: Experimental analysis of dynamic all pairs shortest path algorithms. ACM Trans. Algorithms 2, 578–601 (2006)

[6] Eppstein, D., Galil, Z., Italiano, G.F.: Dynamic graph algorithms. In: Algorithms and Theory of Computation Handbook, ch. 8 (1999)

[7] Erdős, P., Rényi, A.: On random graphs. Publ. Math. Debrecen. 6, 290–297 (1959)

[8] Fredman, M.L., Tarjan, R.E.: Fibonacci heaps and their uses in improved network optimization algorithms. J. ACM 34, 596–615 (1987)

[9] Frigioni, D., Marchetti-Spaccamela, A., Nanni, U.: Fully dynamic shortest paths and negative cycles detection on digraphs with arbitrary arc weights. In: Bilardi, G., Pietracaprina, A., Italiano, G.F., Pucci, G. (eds.) ESA 1998. LNCS, vol. 1461, pp. 320–331. Springer, Heidelberg (1998)

[10] King, V.: Fully dynamic algorithms for maintaining all-pairs shortest paths and transitive closure in digraphs. In: Proc. 40th Annual Symposium on Foundations of Computer Science (FOCS 1999), pp. 81–91 (1999)

[11] Kruskal, J.B.: On the shortest spanning subtree of a graph and the traveling salesman problem. Proc. of the AMS 7, 48–50 (1956)

[12] Nachmias, A., Peres, Y.: The critical random graph, with martingales. Israel Journal of Mathematics (2008) arXiv:math/0512201

[13] Ramalingam, G., Reps, T.W.: On the computational complexity of dynamic graph problems. Theor. Comput. Sci. 158, 233–277 (1996)

[14] Roditty, L., Zwick, U.: Dynamic approximate all-pairs shortest paths in undirected graphs. In: Proc. 45th Annual Symposium on Foundations of Computer Science (FOCS 2004), pp. 499–508 (2004a)

[15] Roditty, L., Zwick, U.: A fully dynamic reachability algorithm for directed graphs with an almost linear update time. In: Proc. 36th Annual ACM Symposium on Theory of Computing (STOC 2004), pp. 184–191 (2004b)

[16] Roditty, L., Zwick, U.: On dynamic shortest paths problems. In: Albers, S., Radzik, T. (eds.) ESA 2004. LNCS, vol. 3221, pp. 580–591. Springer, Heidelberg (2004)

[17] Thorup, M.: Fully-dynamic all-pairs shortest paths: Faster and allowing negative cycles. In: Hagerup, T., Katajainen, J. (eds.) SWAT 2004. LNCS, vol. 3111, pp. 384–396. Springer, Heidelberg (2004)

[18] Thorup, M.: Worst-case update times for fully-dynamic all-pairs shortest paths. In: Proc. 37th Annual ACM Symposium on Theory of Computing (STOC 2005), pp. 112–119. ACM Press, New York (2005)

Computing Frequency Dominators and Related Problems

Loukas Georgiadis[*]

Hewlett-Packard Laboratories, Palo Alto, CA, USA
`lgeorg@uowm.gr`

Abstract. We consider the problem of finding *frequency dominators* in a directed graph with a single source vertex and a single terminal vertex. A vertex x is a *frequency dominator* of a vertex y if and only if in each source to terminal path, the number of occurrences of x is at least equal to the number of occurrences of y. This problem was introduced in a paper by Lee et al. [11] in the context of dynamic program optimization, where an efficient algorithm to compute the frequency dominance relation in reducible graphs was given. In this paper we show that frequency dominators can be efficiently computed in general directed graphs. Specifically, we present an algorithm that computes a sparse ($O(n)$-space), implicit representation of the frequency dominance relation in $O(m + n)$ time, where n is the number of vertices and m is the number of arcs of the input graph. Given this representation we can report all the frequency dominators of a vertex in time proportional to the output size, and answer queries of whether a vertex x is a frequency dominator of a vertex y in constant time. Therefore, we get the same asymptotic complexity as for the regular dominators problem. We also show that, given our representation of frequency dominance, we can partition the vertex set into *regions* in $O(n)$ time, such that all vertices in the same region have equal number of appearances in any source to terminal path. The computation of regions has applications in program optimization and parallelization.

1 Introduction

Let $G = (V, A, s, t)$ be a flowgraph with a distinguished source vertex s and terminal vertex t, which is a directed graph such that every vertex is reachable from s and reaches t. The concept of *frequency dominators* was introduced in [11] with an application in dynamic program optimization. Formally, vertex x is a frequency dominator of vertex y if and only if in each s-t path, the occurrences of x are at least as many as the occurrences of y. We will denote by *fdom(y)* the set of frequency dominators of y. Frequency dominators are related to the well-studied problem of finding *dominators*, which has applications in several areas [9]. A vertex x *dominates* a vertex y if and only if every s-y path contains x. Similarly, a vertex x *postdominates* a vertex y if and only if every y-t path contains x.

[*] Current address: Informatics and Telecommunications Engineering Department, University of Western Macedonia, Kozani, Greece.

S.-H. Hong, H. Nagamochi, and T. Fukunaga (Eds.): ISAAC 2008, LNCS 5369, pp. 704–715, 2008.

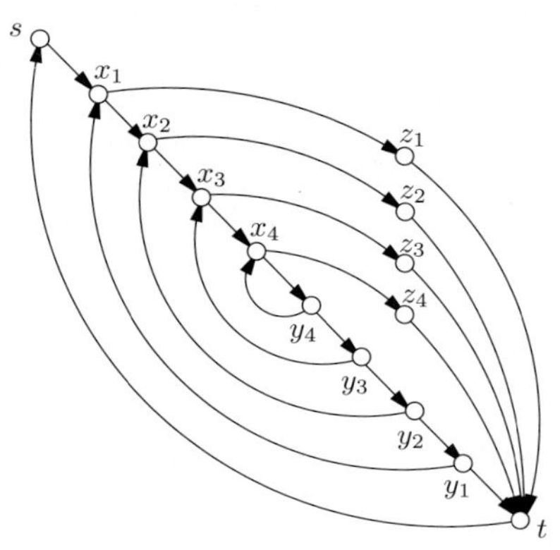

vertex y	$fdom(y)$
s	$\{s, x_1, t\}$
x_1	$\{x_1\}$
x_2	$\{x_2\}$
x_3	$\{x_3\}$
x_4	$\{x_4\}$
y_1	$\{x_1, x_2, x_3, x_4, y_1, y_2, y_3, y_4\}$
y_2	$\{x_2, x_3, x_4, y_2, y_3, y_4\}$
y_3	$\{x_3, x_4, y_3, y_4\}$
y_4	$\{x_4, y_4\}$
z_1	$\{s, x_1, z_1, t\}$
z_2	$\{s, x_1, x_2, z_2, t\}$
z_3	$\{s, x_1, x_2, x_3, z_3, t\}$
z_4	$\{s, x_1, x_2, x_3, x_4, z_4, t\}$
t	$\{s, x_1, t\}$

Fig. 1. The frequency dominators of a family of reducible graphs with $n = 3k + 2$ vertices and $m = 5k + 2$ arcs. (In this figure $k = 4$.) For each z_i, $fdom(z_i)$ contains a subset of the x_j's. However, no pair of x_j's is related under $fdom$. Hence, any transitive reduction of the $fdom$ relation without Steiner nodes requires $\Theta(k^2)$ arcs.

This is equivalent to stating that y dominates x in the reverse flowgraph $G^r = (V, A^r, s^r, t^r)$, where $A^r = \{(x, y) \mid (y, x) \in A\}$, $s^r = t$, and $t^r = s$. We will denote by $dom(y)$ the set of dominators of y, and by $dom^r(y)$ the set of postdominators of y.

Both the dominance and the frequency dominance relations are reflexive and transitive, but only the former is antisymmetric. The transitive reduction of dom is the *dominator tree*, which we denote by D. This tree is rooted at s and satisfies the following property: For any two vertices x and y, x dominates y if and only if x is an ancestor of y in D [1]. For any vertex $v \neq s$, the *immediate dominator* of v, denoted by $d(v)$, is the parent of v in D. It is the unique vertex that dominates v and is dominated by all the vertices in $dom(v) \setminus \{v\}$. On the other hand, a transitive reduction of $fdom$ can have complex structure; refer to Figure 1 for an example.

Lee et al. [11] presented an efficient algorithm for computing frequency dominators in *reducible* graphs. A flowgraph G is reducible when the repeated application of the following operations

(i) delete a loop (v, v);
(ii) if (v, w) is the only arc entering $w \neq s$ delete w and replace each arc (w, x) with (v, x),

yields a single node. Equivalently, G is reducible if every loop has a single entry vertex from s. Reducible graphs are useful in program optimization as they represent the control flow in structured programs, i.e., in programs that use a restricted set of constructs, such as IF-THEN and WHILE-DO. However, the use of GOTO constructs, or the transformations performed by some optimizing compilers can often produce irreducible graphs. Although there are techniques that can transform an irreducible graph to an equivalent reducible graph, it has been shown that there exist irreducible graphs such that any equivalent reducible graph must be exponentially larger [5]. The algorithm of Lee et al. can compute $fdom(y)$, for any vertex y, in $O(|fdom(y)|)$ time, after a preprocessing phase. This preprocessing phase consists of computing the dominators, postdominators, and the loop structure of G. Each of these computations can be performed in $O(m+n)$ time [4,8], or in $O(m\alpha(m,n))^1$ time with simpler and more practical algorithms [12,14].

Another concept related to frequency dominators, that has also been used in optimizing and parallelizing compilers, is that of *regions* [3]. Two vertices x and y are in the same region if and only if they appear equal number of times in any s-t path. Hence, x and y are in the same region if and only if $x \in fdom(y)$ and $y \in fdom(x)$. Regions define an equivalence relation on the flowgraph vertices, therefore we say that x and y are *equivalent* if and only if they belong to the same region. Johnson et al. [10] showed that the problem of finding regions can be solved by computing the *cycle equivalence* relation: Two vertices are cycle equivalent if and only if they belong to the same set of cycles. Cycle equivalence for edges can be defined similarly. In [10], Johnson et al. show that vertex cycle equivalence can be easily reduced to edge cycle equivalence in an *undirected* graph. Then, they develop an $O(m+n)$-time algorithm, based on depth-first search, for computing edge cycle equivalence.

1.1   Overview and Results

In Section 2 we present an efficient algorithm for computing frequency dominators in general flowgraphs. This algorithm is a generalization of the algorithm of Lee et al. for reducible graphs, and achieves the same, asymptotically optimal, time and space bounds. Then, in Section 3, we show that the equivalence relation can be derived from our representation of the frequency dominance relation in $O(n)$ time. Our main results are summarized by the next theorem.

Theorem 1. *Let $G = (V, A, s, t)$ be any flowgraph with $|V| = n$ vertices and $|A| = m$ arcs. We can compute an $O(n)$-space representation of the frequency dominance relation of G in $O(m+n)$ time, such that:*

  - *For any vertex y we can report all the frequency dominators of y in $O(|fdom(y)|)$ time,*
  - *for any two vertices x and y, we can test in constant time whether $x \in fdom(y)$, and*
  - *we can compute the regions of G in $O(n)$ time.*

Some of the methods we use to obtain these results may be useful in other graph problems related to program optimization.

Finally, in Section 4, we describe a general framework that can express the problems we considered in Sections 2 and 3 as a type of reachability problems

---

1 $\alpha(m,n)$ is a functional inverse of Ackermann's function and is very slow-growing [15].

in graphs representing transitive binary relations, and for which an efficient reporting algorithm is required. We believe that this formulation gives rise to questions that deserve further investigation.

1.2   Preliminaries and Notation

We assume that G is represented by adjacency lists. With this representation it is easy to construct the adjacency list representation of the reverse flowgraph G^r in linear-time. Given a depth-first search (DFS) tree T of G rooted at s, the notation "$v \overset{*}{\to} u$" means that v is an ancestor of u in T and "$v \overset{+}{\to} u$" means that v is a proper ancestor of u in T. An arc $a = (u, v)$ in G is a *DFS-tree arc* if $a \in T$, a *forward arc* if $u \overset{+}{\to} v$ and $a \notin T$, a *back arc* if $v \overset{+}{\to} u$, and a *cross arc* otherwise (when u and v are unrelated in T). For any rooted tree T, we let $p_T(v)$ denote the parent of v in T, and we let $T(v)$ denote the subtree of T rooted at v. Finally, for any function or set f defined on G, we denote by f^r the corresponding function or set operating on G^r.

2   Efficient Computation of the *fdom* Relation

First we state some useful properties of frequency dominators. As noted in [10,11], it is convenient to assume that G contains the arc (t, s); adding this arc does not affect the *fdom* relation and allows us to avoid several boundary cases. Henceforth we assume that $(t, s) \in A$, which implies that G is strongly connected. In particular, note that there is a cycle containing all vertices in $dom(y) \cup dom^r(y)$, for any vertex y. We also have the following characterization of the *fdom* relation.

Lemma 1 ([11]). *For any vertices x and y, $x \in fdom(y)$ if and only if x belongs to all cycles containing y.*

Obviously, reversing the orientation of all arcs in a directed graph does not change the set of cycles that contain a vertex. Thus, from Lemma 1 we get:

Lemma 2. *For any vertex y, $fdom(y) = fdom^r(y)$.*

Our algorithm will be able to identify easily a subset of *fdom(y)* using certain structures that are derived from G. The remaining vertices in *fdom(y)* will be found by performing the symmetric computations on G^r.

Let T be a depth-first search tree of G rooted at s. We identify the vertices by their preorder number with respect to the DFS: $v < w$ means that v was visited before w. For any vertex $v \neq s$, the *head* of v is defined as

$$h(v) = \max\{u : u \neq v \text{ and there is a path in } G \text{ from } v \text{ to } u$$
$$\text{containing only vertices in } T(u)\},$$

and we let $h(s) = null$. Informally, $h(v)$ is the deepest ancestor u of v in T such that G contains a v-u path visiting only vertices in $T(u)$. Note that the h function is well-defined since G is strongly connected. The heads define a tree H, called the *interval tree* of G, where $h(v) = p_H(v)$. It follows that the vertices in each subtree $H(v)$ of H induce a strongly connected subgraph of G, which contains only vertices in $T(v)$. Tarjan [16] gave a practical algorithm for computing the interval tree in $O(m\alpha(m, n))$ time using nearest common ancestor computations. A more complicated linear-time construction was given by Buchsbaum et al. [4].

Now we also define

$$\hat{h}(v) = \begin{cases} v, & \text{if there is a back arc entering } v, \\ h(v), & \text{otherwise.} \end{cases}$$

We say that a vertex v is *special* if $\hat{h}(v) = v$. The following lemma provides necessary conditions for frequency dominance.

Lemma 3. *If $x \in fdom(y)$ then the following statements hold:*

(a) $x \in dom(y) \cup dom^r(y)$,
(b) $x \in H(\hat{h}(y))$, *and*
(c) $x \in H^r(\hat{h}^r(y))$.

Proof. The proof follows directly from the definitions. For (a), consider a vertex x that neither dominates nor postdominates y. Then, there exist two paths P and Q, from s to y and from y to t respectively, that both avoid x. The catenation of P and Q followed by (t, s) is a cycle that passes through y but not x, hence $x \notin fdom(y)$ by Lemma 1. In order to show (b), suppose $x \in fdom(y)$ but $x \notin H(\hat{h}(y))$. Since $H(\hat{h}(y))$ induces a strongly connected subgraph, there is a cycle C passing through y that contains only vertices in $H(\hat{h}(y))$. Then, $x \notin C$ which contradicts Lemma 1. Finally, (c) is implied by (b) and Lemma 2.   □

Similarly to the algorithm given in [11], our algorithm is also based on computing dominators and postdominators. Here, however, we make explicit use of the reverse graph and of properties of depth-first search. In particular, we will need the following fact.

Lemma 4 ([13]). *Any path from x to y with $x < y$ contains a common ancestor of x and y.*

Now we provide sufficient conditions for frequency dominance.

Lemma 5. *If w is in $H(\hat{h}(v))$ and dominates v then w is a frequency dominator of v.*

Proof. Let $w \in dom(v)$ and $w \in H(\hat{h}(v))$. Notice that w satisfies $\hat{h}(v) \xrightarrow{*} w \xrightarrow{*} v$. Therefore, the lemma is trivially true when $\hat{h}(v) = v$. Now suppose $\hat{h}(v) = h(v)$. For contradiction, assume that w is not in $fdom(v)$. Then, Lemma 1 implies that there is a cycle C containing v but not w. Let z be the minimum vertex on C. Note that by Lemma 4, z is an ancestor of v. Hence, v belongs to $H(z)$ and by the definition of $h(v)$ we have that $z \xrightarrow{*} h(v)$. Also, $w \neq z$ (because $w \notin C$), thus $z \xrightarrow{+} w$. Now observe that the DFS-tree path from s to z followed by the part of C from z to v is an s-v path avoiding w. This contradicts the fact that $w \in dom(v)$. $\qquad\square$

Now, from Lemma 5 and Lemma 2 we get:

Corollary 1. *Let w be vertex such that*

(a) $w \in H(\hat{h}(v))$ and $w \in dom(v)$, or
(b) $w \in H^r(\hat{h}^r(v))$ and $w \in dom^r(v)$.

Then w is a frequency dominator of v.

It is easy to show, using Lemma 3, that the above conditions suffice to compute frequency dominators. (See Lemma 6 below.) Note that in general $dom(v) \cap dom^r(v) \supseteq \{v\}$, so several vertices may satisfy both (a) and (b) of Corollary 1. Also, we point out that a vertex w that satisfies $w \in dom^r(v)$ and $w \in H(\hat{h}(v))$ may not be a frequency dominator of v. (Consider $w = e$ and $v = c$ in Figure 2 for an example.)

Preprocessing. In the preprocessing phase we compute the interval tree H, the dominator tree D, and the $\hat{h}$ function in G. We also compute the corresponding structures in G^r. For the query algorithm we need to test in constant time if two vertices are related in one of the trees that we have computed. To that end, we assign to the vertices in each tree $T \in \{H, H^r, D, D^r\}$ a preorder and a postorder number[2], denoted by pre_T and $post_T$ respectively. Then x is an ancestor of y in T if and only if $pre_T(x) \le pre_T(y)$ and $post_T(x) \ge post_T(y)$. Calculating these numbers takes $O(n)$ time.

Queries. To find $fdom(y)$ we report y and all the proper ancestors x of y in D such that x is a descendant of $\hat{h}(y)$ in H. To that end, we visit the dominators of y starting from $d(y)$ and moving towards s. At each visited vertex x we test whether $x \in H(\hat{h}(y))$, which can be done in constant time using the pre_H and $post_H$ numberings. We show below that we can stop the reporting process as soon as we find a dominator of y not in $H(\hat{h}(y))$. (See Lemma 6.) Similarly, we report all the proper ancestors x of y in D^r such that x is a descendant of $\hat{h}^r(y)$ in H^r. Therefore, we can answer the reporting query in $O(|fdom(y)|)$ time.

To test whether $x \in fdom(y)$ in constant time, it suffices to test if condition (a) or condition (b) of Corollary 1 holds. Again this is easily accomplished with the use of the preorder and postorder numbers for D and D^r.

Figure 2 illustrates the computations performed by our algorithm. Next, we show that the algorithm finds all the frequency dominators of a query vertex.

Lemma 6. *The query algorithm is correct.*

---

[2] A preorder traversal of a tree visits a vertex before visiting its descendants; a postorder traversal visits a vertex after visiting all its descendants. The corresponding numberings are produced by assigning number i to the ith visited vertex.

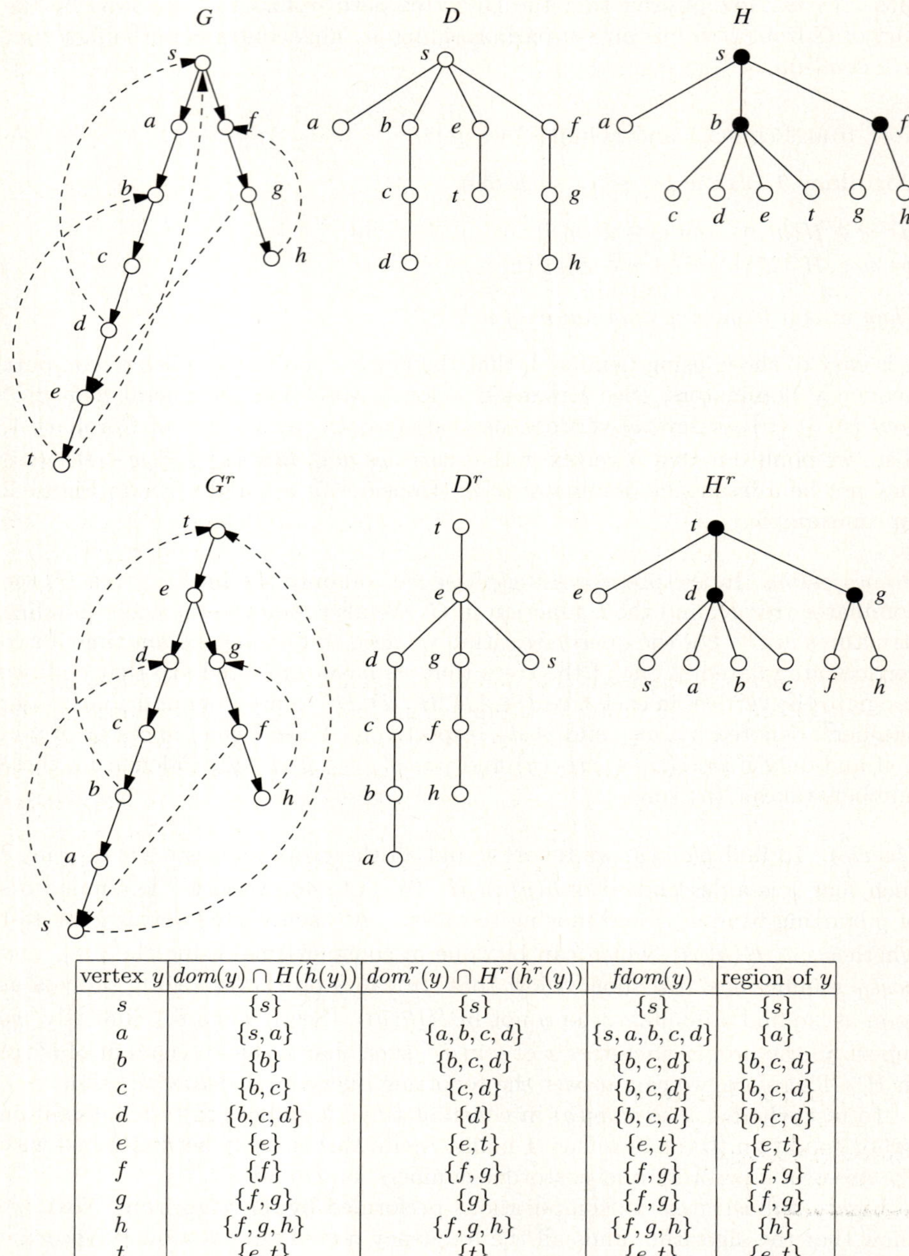

vertex y	$dom(y) \cap H(\hat{h}(y))$	$dom^r(y) \cap H^r(\hat{h}^r(y))$	$fdom(y)$	region of y
s	$\{s\}$	$\{s\}$	$\{s\}$	$\{s\}$
a	$\{s,a\}$	$\{a,b,c,d\}$	$\{s,a,b,c,d\}$	$\{a\}$
b	$\{b\}$	$\{b,c,d\}$	$\{b,c,d\}$	$\{b,c,d\}$
c	$\{b,c\}$	$\{c,d\}$	$\{b,c,d\}$	$\{b,c,d\}$
d	$\{b,c,d\}$	$\{d\}$	$\{b,c,d\}$	$\{b,c,d\}$
e	$\{e\}$	$\{e,t\}$	$\{e,t\}$	$\{e,t\}$
f	$\{f\}$	$\{f,g\}$	$\{f,g\}$	$\{f,g\}$
g	$\{f,g\}$	$\{g\}$	$\{f,g\}$	$\{f,g\}$
h	$\{f,g,h\}$	$\{f,g,h\}$	$\{f,g,h\}$	$\{h\}$
t	$\{e,t\}$	$\{t\}$	$\{e,t\}$	$\{e,t\}$

Fig. 2. Computation of the *fdom* relation using dominator trees and interval trees. The DFS-tree arcs of G and G^r are solid and remaining arcs are dashed. Special vertices are filled in the interval trees.

Proof. From Corollary 1 we have that all the vertices that are reported during a query for $fdom(y)$ are indeed frequency dominators of y. Now we argue that any frequency dominator x of y that is not found using condition (a) of Corollary 1 is found using condition (b). In this case, Lemma 3(a) implies that x is a postdominator of y. This fact combined with Lemma 3(c) means that condition (b) of Corollary 1 applies.

We also need to argue that we can stop the search procedure for the vertices in $fdom(y)$ as soon as we find the deepest dominator of y that is not in $H(\hat{h}(y))$. To that end, note that for any vertex z such that $\hat{h}(y) \xrightarrow{*} z \xrightarrow{*} y$, $z \in H(\hat{h}(y))$. Therefore, if $x \in dom(y)$ and $x \notin H(\hat{h}(y))$ then $x \xrightarrow{+} \hat{h}(y)$, which proves our claim. $\qquad\square$

Strict frequency dominance. Let us call vertex x a *strict frequency dominator* of vertex y if and only if in each s-t path containing y, the number of appearances of x is strictly greater than the number of appearances of y. E.g., in Figure 2, f and g are strict frequency dominators of h. We can show that such an x must satisfy both conditions (a) and (b) of Corollary 1. Hence, with our $fdom$ structure we can test if x is a strict frequency dominator of y in constant time, and report all the strict frequency dominators of y in $O(|fdom(y)|)$ time. Note, however, that this bound is not proportional to the size of the output.

Inverse frequency dominance. Let $fdom^{-1}(x)$ denote the set of vertices y such that $x \in fdom(y)$. The problem of reporting $fdom^{-1}(x)$ can be reduced to the following task. We are given an arbitrary rooted tree T with n vertices, where each vertex x is assigned an integer label $\ell(x)$ in $[1, n]$, and we wish to support the following type of queries: For a vertex x and a label j, report all vertices $y \in T(x)$ with $\ell(y) \geq j$. Let k be the number of such vertices. We can achieve an $O(k \log n)$ query time (using $O(n)$ space), thus getting an $O(|fdom^{-1}(x)| \log n)$ bound for reporting $fdom^{-1}(x)$.

3   Computing Regions

Recall that two vertices x and y are equivalent if and only if they appear equal number of times in any s-t path. Hence, x and y are equivalent if and only if $x \in fdom(y)$ and $y \in fdom(x)$. As in Section 2, we assume that G contains the arc (t, s). Then, by Lemma 1 it follows that x and y are equivalent if and only if they appear in the same set of cycles.

The structure of Section 2 clearly supports constant time queries that ask if two vertices are equivalent. Therefore, it is also straightforward to find the equivalence class of a given vertex y in $O(|fdom(y)|)$ time. Still, with this method we can only guarantee an $O(n^2)$ bound for computing the complete equivalence relation for all vertices (i.e., the regions of G). In this section we show how to achieve this computation in overall $O(n)$ time, with a somewhat more involved manipulation of the structure of Section 2.

We start with a technical lemma that relates the structure of the dominator tree to that of the interval tree.

Lemma 7. *Let v be any vertex other than s. Then, all vertices w in $H(v)$ such that $d(w) \xrightarrow{+} v$ have the same immediate dominator $x = d(v)$.*

Proof. Let x be the minimum vertex that is an immediate dominator of any vertex in $H(v)$. Since $v \neq s$, $d(v)$ is a proper ancestor of v and so $x \xrightarrow{+} v$. Let w be a vertex such that $d(w) = x$. Then, w is not dominated by any vertex z that satisfies $x \xrightarrow{+} z \xrightarrow{+} w$, thus for each such z there exists a path P from x to w that avoids z. Also, since the vertices in $H(v)$ induce a strongly connected subgraph, then for any $y \in H(v)$ there is a path Q from w to y that contains only vertices in $H(v)$. Therefore, the catenation of P and Q is a path from x to y avoiding z. This implies that either $d(y) = x$ or $d(y) \in H(v)$. Setting $y = v$ gives $d(v) = x$. $\square$

From the above lemma we immediately get:

Corollary 2. *The subgraph of D induced by the vertices in $H(v)$ and $d(v)$ is a tree rooted at $d(v)$.*

We begin with an initial partition of the vertices, which after a refinement process (described in Section 3.1) will produce the actual equivalence classes. Our first step is to label each vertex w in D with $\hat{h}(w)$. Then, each vertex has a unique label and two vertices are equivalent only if they have the same label. Let $L(v)$ denote the set of vertices labeled with v. Note that $L(v) = \emptyset$ if v is not special; otherwise $L(v)$ consists of v and the children of v in H that are not special. Our goal is to compute the equivalent vertices for each $L(v)$ separately, using information from the dominator tree D. To that end, let w be any vertex in $L(v)$. We define $\hat{d}(w)$ to be the nearest proper ancestor z of w in D with label v. If such a z does not exist for w, then we set $\hat{d}(w) = d(v)$. A simple and efficient way to compute the $\hat{d}$ function is by using *path compression*: Starting from the parent of w in D, we follow parent pointers until we reach a vertex z which is either the first vertex in $L(v)$ or $d(v)$. Then, we set $\hat{d}(w) = z$ and make z the parent of all the visited vertices. In order to guarantee that this process returns the correct $\hat{d}(w)$ for all w, we need to process the sets $L(v)$ in an appropriate order. We argue that it suffices to process the special vertices by decreasing depth in H.

Lemma 8. *Suppose that for each special vertex v, we process $L(v)$ after we have processed $L(u)$ for all special vertices $u \in H(v)$. Then, the path compression algorithm computes the $\hat{d}$ function correctly.*

Proof. Let v be the currently processed special vertex. We denote by $\hat{D}$ the dominator tree after the changes due to the path compressions performed so far, just before processing v. Let u be a special vertex in $H(v)$. Then Corollary 2 implies that when the path compression algorithm was applied to $L(u)$, only vertices in $H(u) \cup \{d(u)\}$ were visited. Moreover, Lemma 7 and the fact that $H(u) \subset H(v)$ imply $d(v) \xrightarrow{*} d(u)$. Also note that the parent of $d(u)$ did not change. Now, since v is an ancestor of u in H, there is no vertex in $H(u)$ with label v. Hence, for any $y \in L(v)$ and $x \in L(v) \cup \{d(v)\}$, x is an ancestor of y in $\hat{D}$ if and only if it is an ancestor of y in D. The claim follows. $\square$

The worst-case running time of this algorithm is $O(n \log n)$ [17], but in practice it is expected to be linear [9]. We can get an actual $O(n)$ bound for computing the $\hat{d}$ function via a simple reduction to the *marked ancestors problem* [2]. In the marked ancestors problem we are given a rooted tree T, the vertices of which can be either *marked* or *unmarked*. We wish to support the following operations: mark or unmark a given vertex, and find the nearest marked ancestor of a query vertex. In our case $T = D$ and initially all vertices are marked. Again we process the special vertices by decreasing depth in H. When we process a special vertex v we perform a nearest marked ancestor query for all $w \in L(v)$; we set $\hat{d}(w)$ to be the answer to such a query for w. Then we unmark all vertices in $L(v)$. It follows from similar arguments as in the proof of Lemma 8 that this process computes the correct $\hat{d}$ function. Also, since we never mark any vertices, this process corresponds to a special case of the disjoint set union (DSU) problem, for which the result of Gabow and Tarjan [7] gives constant amortized time complexity per operation. Since we perform two DSU operations per vertex, the $O(n)$ time bound follows.

Now it remains to show how to compute the equivalence classes for each $L(v)$ using the $\hat{d}$ function.

3.1   Equivalence Classes in $L(v)$

Let D_v be the graph with vertex set $L(v) \cup \{d(v)\}$ and edge set $\{(\hat{d}(w), w) \mid w \in L(v)\}$. From Corollary 2 we have that D_v is a tree rooted at $d(v)$. Let D'_v be the forest that results after removing $d(v)$ and its adjacent edges from D_v; the trees in this forest are rooted at the children of $d(v)$ in D_v. Let w be any vertex in D'_v. Note that by condition (a) of Corollary 1 we have that all ancestors of w in D'_v are in $fdom(w)$. The next lemma allows us to compute the equivalence classes in each tree of D'_v separately.

Lemma 9. *Let x be a vertex in D, and let x_1 and x_2 be two distinct children of x in D. Then no pair of vertices $y_1 \in D(x_1)$ and $y_2 \in D(x_2)$ are equivalent. Also x can be equivalent with at most one of y_1 and y_2.*

Proof. Let y and z be two equivalent vertices. Then $y \in fdom(z)$ and $z \in fdom(y)$. Since the *dom* relation is antisymmetric, Lemma 3(a) implies (with no loss of generality) that $y \in dom(z)$ and $z \in dom^r(y)$. Hence, the vertices y_1 and y_2, defined in the statement of the lemma, cannot be equivalent. So, they also cannot be both equivalent with x. $\square$

Now we can consider a single tree T in D'_v. Lemma 9 also implies that only vertices with ancestor-descendant relation in T can be equivalent. Our plan is to process T bottom-up, and at each vertex w of T test if $w \in fdom(p_T(w))$. If the outcome of the test is true then we add $p_T(w)$ to the same equivalence class as w, since we already know that $p_T(w) \in fdom(w)$. If, on the other hand, the outcome of the test is false, then the next lemma infers that no other vertex can be in the same equivalence class with w.

Lemma 10. *Let T be a tree of D'_v, and let w be a non-root vertex in T. If w is not equivalent with $p_T(w)$ then w is not equivalent with any of its proper ancestors in T.*

Proof. For contradiction, assume w is equivalent with a proper ancestor z of $u = p_T(w)$ in T. Notice that $v \xrightarrow{*} z \xrightarrow{+} u$. Then, we either have a cycle C_{-u} through w and z but not u, or a cycle C_u through u but not w and z. The first case contradicts the fact that $u \in fdom(w)$. In the second case, let x be the minimum vertex on C_u. From Lemma 4 and the fact that $u \in L(v)$, we have $x \xrightarrow{*} v \xrightarrow{*} z$. But then, the DFS-tree path from x to u, followed by the part of C_u from u to x, forms a cycle through z that does not contain w. This contradicts the fact that w and z are equivalent. $\qquad\square$

4   A General Framework

We can show that the problems we have considered in this paper can be formulated in more general terms as follows. We are given a collection $\mathcal{G}$ of k directed graphs $G_i = (V_i, A_i)$, $1 \le i \le k$, where each graph G_i represents a transitive binary relation R_i over a set of elements $U \subseteq V_i$. That is, for any $a, b \in U$, we have aR_ib if and only if b is reachable from a in G_i. Let R be the relation defined by: aRb if and only if aR_ib for all $i \in \{1, \ldots, k\}$. I.e., b is reachable from a in all graphs in $\mathcal{G}$. (Note that R is also transitive.) For instance, consider the relation that is defined by the pairs (a, b) of vertices of a directed graph, such that a is both a dominator and a postdominator of b. In this case we have $\mathcal{G} = \{D, D^r\}$ (where the edges of the dominator trees are directed from parent to child).

Our goal is to find an efficient representation of R with a data structure that, for any given $b \in U$, can report fast all elements a satisfying aRb. Another related problem is to bound the combinatorial complexity of R, i.e., the size of a directed graph $G = (V, A)$, with $U \subseteq V$, such that for any $a, b \in U$, aRb if and only if b is reachable from a in G. Some interesting results can be obtained for several special cases. For instance, in the simple case where $k = 2$ and G_1 and G_2 are (directed) paths, it can be shown that the size of G is $O(n \log n)$, and moreover that there exist paths that force any G to have $\Omega(n \log n)$ size [18]. On the other hand, we can represent R with a Cartesian tree in $O(n)$ space, and answer reporting queries in time proportional to the number of reported elements [6].

Acknowledgements. We would like to thank Bob Tarjan, Renato Werneck, and Li Zhang for some useful discussions.

References

1. Aho, A.V., Sethi, R., Ullman, J.D.: Compilers: Principles, Techniques, and Tools. Addison-Wesley, Reading (1986)
2. Alstrup, S., Husfeldt, T., Rauhe, T.: Marked ancestor problems. In: Proc. 39th IEEE Symp. on Foundations of Computer Science, p. 534
3. Ball, T.: What's in a region?: or computing control dependence regions in near-linear time for reducible control flow. ACM Lett. Program. Lang. Syst. 2(1-4), 1–16 (1993)
4. Buchsbaum, A.L., Georgiadis, L., Kaplan, H., Rogers, A., Tarjan, R.E., Westbrook, J.R.: Linear-time algorithms for dominators and other path-evaluation problems. SIAM Journal on Computing (to appear), http://arxiv.org/abs/cs.DS/0207061
5. Carter, L., Ferrante, J., Thomborson, C.: Folklore confirmed: reducible flow graphs are exponentially larger. In: Proc. 30th ACM SIGPLAN-SIGACT Symp. on Principles of Programming Languages, pp. 106–114 (2003)
6. Gabow, H.N., Bentley, J.L., Tarjan, R.E.: Scaling and related techniques for geometry problems. In: Proc. 16th ACM Symp. on Theory of Computing, pp. 135–143 (1984)
7. Gabow, H.N., Tarjan, R.E.: A linear-time algorithm for a special case of disjoint set union. Journal of Computer and System Sciences 30(2), 209–221 (1985)
8. Georgiadis, L., Tarjan, R.E.: Finding dominators revisited. In: Proc. 15th ACM-SIAM Symp. on Discrete Algorithms, pp. 862–871 (2004)
9. Georgiadis, L., Tarjan, R.E., Werneck, R.F.: Finding dominators in practice. Journal of Graph Algorithms and Applications (JGAA) 10(1), 69–94 (2006)
10. Johnson, R., Pearson, D., Pingali, K.: The program structure tree: computing control regions in linear time. SIGPLAN Not 29(6), 171–185 (1994)
11. Lee, B., Resnick, K., Bond, M.D., McKinley, K.S.: Correcting the dynamic call graph using control-flow constraints. In: Krishnamurthi, S., Odersky, M. (eds.) CC 2007. LNCS, vol. 4420, pp. 80–95. Springer, Heidelberg (2007); Full paper appears as technical report TR-06-55, The University of Texas at Austin, Department of Computer Sciences (2006)
12. Lengauer, T., Tarjan, R.E.: A fast algorithm for finding dominators in a flowgraph. ACM Transactions on Programming Languages and Systems 1(1), 121–141 (1979)
13. Tarjan, R.E.: Depth-first search and linear graph algorithms. SIAM Journal on Computing 1(2), 146–159 (1972)
14. Tarjan, R.E.: Testing flow graph reducibility. In: Proceedings of the Fifth Annual ACM Symposium on Theory of Computing, pp. 96–107 (1973)
15. Tarjan, R.E.: Efficiency of a good but not linear set union algorithm. Journal of the ACM 22(2), 215–225 (1975)
16. Tarjan, R.E.: Edge-disjoint spanning trees and depth-first search. Acta Informatica 6(2), 171–185 (1976)
17. Tarjan, R.E., van Leeuwen, J.: Worst-case analysis of set union algorithms. Journal of the ACM 31(2), 245–281 (1984)
18. Zhang, L.: Personal communication (2008)

Computing Best Swaps in
Optimal Tree Spanners[*]

Shantanu Das, Beat Gfeller, and Peter Widmayer

Institute of Theoretical Computer Science, ETH Zurich, Switzerland

Abstract. In a densely connected communication network, represented by a graph G with nonnegative edge-weights, it is often advantageous to route all communication on a sparse, spanning subnetwork, typically a spanning tree of G. With the communication overhead in mind, we consider a spanning tree T of G which guarantees that for any two nodes, their distance in T is at most k times their distance in G, where k, called the stretch, is as small as possible. Such a spanning tree which minimizes the stretch is called an optimal tree spanner, and it can be used for efficient routing. However, for a communication tree, the failure of an edge is catastrophic; it disconnects the tree. Functionality can be restored by connecting both parts of the tree with another edge, while leaving the two parts themselves untouched. In situations where the failure can be repaired rapidly, such a quick fix is preferred over the recomputation of an entirely new optimal tree spanner, because it is much closer to the previous solution and hence requires far fewer adjustments in the routing scheme. We are therefore interested in the problem of finding for any possibly failing edge in the spanner T a best swap edge to replace it. The objective here is naturally to minimize the stretch of the new tree. We show how all these best swap edges can be computed in total time $O(m^2 \log n)$ in graphs with arbitrary nonnegative edge weights. For graphs with unit weight edges (also called unweighted graphs), we present an $O(n^3)$ time algorithm. Furthermore, we present a distributed algorithm for computing the best swap for each edge in the spanner.

1   Introduction

In a typical communication network, there are often more links (i.e., communication channels) available than what is useful for performing most computations. The presence of these additional links makes it possible to deal with failures in the network. However, at any given time, only a subset of the links are actually used for communication. For efficiency reasons, it is beneficial to maintain such a sub-network and route all communication through this subnet. We represent the original network as a connected, undirected graph $G = (V, E)$, and the subnet is a spanning tree T of this graph G. Instead of using any arbitrary spanning tree of

---

[*] Work partially supported by NCCR-MICS, a center supported by the Swiss NSF under grant number 5005 – 67322, and by the Swiss SBF under contract no. C05.0047 within COST-295 (DYNAMO) of the European Union.

S.-H. Hong, H. Nagamochi, and T. Fukunaga (Eds.): ISAAC 2008, LNCS 5369, pp. 716–727, 2008.

G, we prefer one which has certain desirable properties supporting the required computations. In this paper, we measure the overhead for communication as a result of using the subnet T instead of the original network G as the largest multiplicative increase in distance that any pair of nodes experiences. Thus, we use a tree T which minimizes the maximum stretch between any two nodes in T, where the stretch between nodes $a, b \in T$ is the ratio of their distance in T over their distance in G. Such a tree is called an optimal tree spanner of G. Tree spanners are used for routing in communication networks because they achieve a good tradeoff between the lengths of communication paths and the sizes of routing tables needed [15].

A critical problem with which we are confronted in this context is what happens when one of the links, say edge $e \in T$ fails, thereby disconnecting the tree. There are at least two possible (and extreme) solutions to this: (1) recomputing an entirely new optimal tree spanner for $G - e$, or (2) replacing just the failing edge e by another edge (called a swap edge) that connects the two disconnected parts of $T - e$ in a best possible way, i.e., so that the stretch of the resulting tree is as small as possible. For temporary network failures, the second approach is much better suited than the first, because it is more efficient to use a swap edge for the duration of the failure, so that we can quickly revert back to the original spanner T, once the fault has been repaired. Furthermore, this approach needs only a very small adjustment of routing tables and has therefore attracted research attention in recent years for simpler spanning trees, under the name of "on-the-fly rerouting" [6, 8, 10]. As an aside, note also that an entirely new optimal tree spanner might not only require a total replacement of all routing table entries, but is in addition NP-hard to find.

When choosing a swap edge for a failing edge e, it is natural to use the same criterion as before, that is, minimizing the stretch. We always measure the stretch of a tree with respect to distances in the original graph G, and not with respect to distances in the (transient) fault-free subgraph $G - e^1$. Interestingly enough, by merely going for the best swap, for unweighted graphs we are guaranteed to find a tree that is quite good also in comparison with an entirely new optimal tree spanner: We show that the stretch of a new tree T' obtained by adding a best swap edge is at most twice that of an optimal tree spanner of $G - e$ (again, measured w.r.t. distances in G). In order to quickly recover from an arbitrary edge failure, we pre-compute the best swap edge for each possible failing edge. The problem we consider in this paper is that of efficiently computing for every edge in the tree T a best swap edge. This *All-Best-Swaps* (ABS) problem has been studied for the cases when the tree T is a minimum spanning tree (MST), a shortest paths tree (SPT), or a minimum diameter spanning tree (MDST), with the corresponding different optimization criteria. Our paper is the first one to study the problem of finding all best swaps for an optimal tree spanner. This problem appears to be considerably more difficult than the previously studied ones; none of the techniques used in earlier studies are applicable to our problem.

---

[1] Note that this definition for the stretch of a tree is an upper bound for the stretch measured with respect to $G - e$.

Our Contributions. We first present and analyze a brute-force algorithm for solving the problem in Section 2. This algorithm requires $O(m^2 n)$ time for a graph having n vertices and m edges. In Section 3, we describe a more efficient algorithm that reduces the time complexity to $O(m^2 \log n)$ and requires $O(m)$ space. We also present an $O(n^3)$ time and $O(n^2)$ space solution for unweighted graphs in Section 4. Finally, in Section 5, we show how to compute all the best swaps of a tree spanner in a distributed fashion. Our distributed algorithm solves the problem in $O(D)$ time with a communication cost of $O(n^* \log n)$, where D is the diameter of T and n^* is the size of the transitive closure of T. Due to lack of space, some details of our results are omitted and can be found in [3].

Related Work. The concept of graph spanners was introduced in [14] where the authors used it to construct good synchronizers for communication networks. Peleg and Upfal [15] showed that using spanners as a subnet for routing helps in optimizing both the route lengths and the space required for storing routing information. Graph spanners (and in particular sparse spanners) are useful in many applications such as designing communication networks, distributed systems, parallel computers and also in motion planning [2].

The problem of finding a tree spanner that minimizes the maximum stretch, called the MMST problem, was shown to be NP-hard [2]. It can be approximated with ratio $O(\log n)$ in unweighted graphs [5].

As mentioned before, the All-Best-Swaps problem has been studied earlier for different optimization criteria, where the original tree is either a minimum spanning tree (MST) or a shortest paths tree (SPT) or a minimum diameter spanning tree (MDST). For a MST, all the best swap edges can be found in $O(m)$ time [4]. When the original tree is a SPT and the objective is minimizing the (maximum or total) distance from a fixed root, the ABS problem has been solved in $O(n^2)$ time [12]. While these solutions are centralized, there also exist distributed solutions for computing best swaps in both the SPT [8] and the MST [7], where the efficiency is measured in terms of the communication cost in performing the computation. For minimum spanning trees, a stronger version of the problem was studied in [7] where the failures were assumed to occur at the nodes of the network, thereby disabling several edges at the same time.

In the case of minimum diameter spanning trees (MDST), there exists a centralized solution [9] that requires $O(m \log n)$ time and $O(m)$ space. The distributed version of the problem was solved in [10] using $O(\max\{n^*, m\})$ messages of constant size, where n^* is the size of the transitive closure of the tree, when the edges are directed towards the node initiating the computation.

2   Computing All Best Swaps

2.1   Some Definitions and Properties

We use the following definitions and notations throughout this paper.

1. A communication network is a 2-edge-connected, undirected graph $G = (V, E)$, with $n = |V|$ vertices and $m = |E|$ edges. Each edge $e \in E$ has a non-negative real *length* $|e|$.

2. The length $|\mathcal{P}|$ of a path $\mathcal{P} = \langle p_1, \ldots, p_r \rangle$ is the sum of the lengths of its edges, and the *distance* $d_H(x, y)$ between any two vertices x, y in a graph H is the length of a shortest path in H between x and y.

3. The *stretch* of a spanning tree T of G is

$$\max_{x, y \in V} \{d_T(x, y)/d_G(x, y)\}$$

   An *optimal tree spanner* is a spanning tree with minimum stretch.

4. We often consider a spanning tree T to be rooted at some node u. For each node $x \neq u$, we then denote the *parent* of x by $p(x)$ and the set of its *children* by $C(x)$. Furthermore, let $T_x = (V(T_x), E(T_x))$ be the subtree of T rooted at x, including x.

5. The removal of any edge $e = (x, y)$ from T partitions the spanning tree into two disjoint trees T^x and T^y, where T^x contains node x and T^y contains y. A *swap edge* f for e is any edge in $E \backslash E_T$ that (re-)connects T^x and T^y, i.e., for which $T_{e/f} := (V, E_T \backslash \{e\} \cup \{f\})$ is a spanning tree of $G - e := (V, E \backslash \{e\})$. Let $S(e)$ be the set of swap edges for e. A *best swap edge* for e is any edge $f \in S(e)$ for which the stretch of $T_{e/f}$, defined as $\max_{x, y \in V} \{d_{T_{e/f}}(x, y)/d_G(x, y)\}$, is minimum. Any edge $f \in E \backslash E_T$ is called a *candidate swap edge*, as it is a swap edge for at least one edge in T.

6. The *All-Best-Swaps* problem for a given graph G and a given optimal tree spanner T of G consists of finding for every edge $e \in E_T$ a best swap edge.

The following property simplifies the computation of the stretch for a given swap edge f with respect to a given failing edge e:

Observation 1 (follows from Lemma 16.1.1 in [13]). *Let $G = (V, E)$ be a weighted, undirected graph and let T be an optimal tree spanner of G with stretch k. Consider the spanning tree $T_{e/f}$ which results in deleting edge e from T and inserting f instead. Let $a, b \in V$ be two non-adjacent nodes in G, i.e., $(a, b) \notin E$, whose stretch in $T_{e/f}$ is $k' > k$. Then, there exists an edge $(x, y) \in E$ such that the stretch of the pair (x, y) in $T_{e/f}$ is at least k'.*

This observation motivates the following concepts, which we use for the description of our algorithms:

1. Each pair of nodes a, b with $(a, b) \in E$ is called a *stretch pair*. A stretch pair $g = (a, b)$ is *relevant* for measuring the stretch of any swap edge replacing a given failing edge e if the cycle which g forms with T contains the edge e (in other words, if g is also a swap edge for e).

2. For any edge $e \in T$ and any swap edge $f = (u, v) \in E \backslash E_T$ for e, the *stretch of a relevant stretch pair* (a, b) is the ratio of the distance between a and b in $T_{e/f}$, over the distance between them in G (i.e. $d_{T_{e/f}}(a, b) \, / \, d_G(a, b)$). We can then express the stretch of $T_{e/f}$ as the maximum over the stretches of all the stretch pairs which are relevant for f replacing e.

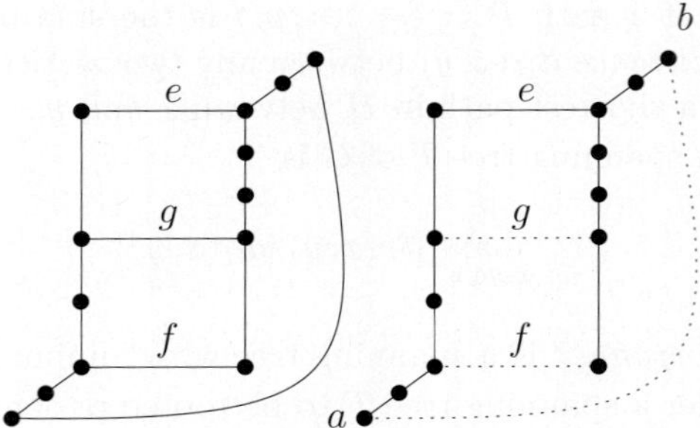

Fig. 1. An example showing that minimizing detour length does not minimize the stretch: On the left side, the 2-edge-connected graph G is shown, and on the right side the given tree spanner with stretch 8 is shown. Assuming that all edges have equal weight, the swap edge f minimizes the stretch to the value 9. However, the swap edge minimizing the detour length is g, which yields a stretch of 10 (attained by the stretch pair (a, b)), worse than choosing swap edge f.

2.2   Naive Approach

To compute the best swap edge for an edge $e \in T$, we need to compare all possible candidate swap edges that are relevant for the failing edge e. Unfortunately, there is no straightforward way of selecting the best among these candidates without evaluating each possible candidate. A simple trick such as choosing the swap edge minimizing the detour around the failure does not always give the optimal solution. For instance, see the counter-example shown in Fig. 1 (This example can be generalized to obtain an arbitrarily large difference between the stretch for the best swap f and the minimum detour edge g.)

The brute-force method for solving the *All-Best-Swaps* problem in a tree spanner is to compute the stretch of $T_{e/f}$ for every edge pair (e, f) where $e \in T$ and f is a candidate swap edge for replacing edge e. There are $O(nm)$ such pairs and for each pair (e, f), the stretch of the tree $T_{e/f}$ can be computed in $O(m)$ time due to the following lemma.

Lemma 1. *After preprocessing in time $O(mn + n^2 \log n)$, for any failing edge $e \in E_T$, swap edge $f \in E \backslash E_T$, and relevant stretch pair (u, v), the stretch of (u, v) in $T_{e/f}$ can be computed in $O(1)$ time.*

The following preprocessing is required for the above computation. First we obtain distances between all pairs of nodes in G in time $O(nm + n^2 \log n)$, using the standard "all-pairs shortest paths" algorithm. Next, we root the tree T at an arbitrary node r and compute the "to-root" distance $d_T(r, v)$ for each node $v \in V(G)$, with a single pre-order traversal of T. Finally we construct a data structure which provides the nearest common ancestor of any two given nodes in constant time. (Such a data structure can be computed in $O(n)$ time, for example using the method described in [11].)

After the preprocessing, we consider each relevant stretch pair (u, v), i.e. each $(u, v) \in E \backslash E_T$ where u and v lie on different sides of the failing edge[2]. For each such pair (u, v) we have $d_{T_{e/f}}(u, v) = d_T(u, p) + |f| + d_T(q, v)$. where $f = (p, q)$ and p lies on u's side of the cut induced by e. In general, $d_T(u, p) = d_T(v, \mathrm{nca}(u, p)) + d_T(p, \mathrm{nca}(u, p))$. Both of these two terms can be computed as the absolute difference between the "to-root" distance of the two nodes involved. To summarize, $d_{T_{e/f}}(u, v)$ and thus, the stretch of (u, v) can be computed in constant time, for each of the $O(m)$ relevant stretch pairs, for a particular e and f. This implies the following:

Theorem 1. *The* All-Best-Swaps *problem in a tree spanner can be solved in* $O(nm^2)$ *time.*

In the following, we present some techniques to reduce this time complexity.

3 An $O(m^2 \log n)$ Time Solution for Weighted Graphs

In the following, we describe an algorithm which computes all best swap edges of a tree spanner in $O(m^2 \log n)$ time and $O(m + n)$ space.

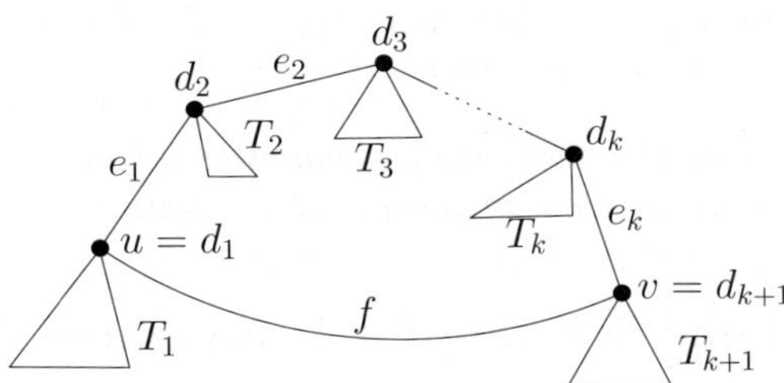

Fig. 2. The cycle that a non-tree edge f forms with the given spanning tree

The idea of the algorithm (called *BestSwaps*) is sketched in the following. We consider each potential swap edge $f \in E \backslash E_T$ separately, focusing on the cycle which $f = (u, v)$ forms with T (see Fig. 2). This cycle consists of the edges $e_1, e_2, \ldots e_k$ which form a path in T. Note that all nodes of V which are not on the path in T from $u = d_1$ to $v = d_{k+1}$ lie in some subtree T_i of T which is rooted at node d_i. For a given failing edge $e_i = (d_i, d_{i+1})$ for which f is a relevant swap edge, the set of relevant stretch pairs contain all pairs(edges) where one endpoint lies in some tree attached to any of $d_1, \ldots, d_i$, and the other endpoint lies in some tree attached to any of $d_{i+1}, \ldots, d_{k+1}$. We associate with each node the index i of the subtree T_i containing it. For any edge $(a, b) \in E \setminus E_T$, this defines an order of the endpoints: we say that a is the *lower* endpoint if its index is smaller than the index of b. In order to evaluate f as a potential swap edge,

<hr>

[2] This can be checked using a "preorder/inverted preorder" labelling. For details, see e.g. [8], Section 3.2.

we need to compute the stretch for every relevant stretch pair with respect to f and some failing edge e_i. Notice that irrespective of the failing edge, any relevant stretch pair for f, would be an edge $(a, b) \in E \setminus E_T$ where the endpoints a and b would have different indices. We maintain a data structure H described below which stores these relevant stretch pairs.

We consider the potential failing edges $e_1, e_2, \ldots, e_k$, in that order and evaluate f as potential best swap with respect to each e_i in turn. To that end, observe the following: if $S(e_{i-1})$ is the set of relevant stretch pairs when considering f as a swap for e_{i-1}, then $S(e_i)$, the set of relevant stretch pairs when considering f as a swap for e_i, is $S(e_i) = \big(S(e_{i-1}) \cup Start_i\big) \setminus End_i$, where $Start_i$ is the set of stretch pairs whose lower endpoint is d_i, and End_i is the set of stretch pairs whose upper endpoint is d_i. We store the set $S(e_i)$ in our data structure H and update it as we move from e_i to e_{i+1}. To compute $S(e_i)$ from $S(e_{i-1})$, all stretch pairs that become relevant are added to H and all stretch pairs that become irrelevant are deleted from H. The data structure H we use to store the set $S(e_i)$ can be implemented as a priority queue (or heap) where priority is defined by the stretch value. The largest element in H yields the worst stretch pair for f replacing e_i. We simply check whether this value is smaller than the stretch of the current best swap edge for e_i (we maintain these in a separate data structure) and update the current swap edge for e_i if required. Once we have repeated the above process for each edge $f \in E \setminus E_T$, we have obtained for each edge in T a best swap edge. We have:

Theorem 2. *The Algorithm* BestSwaps *computes all the best swap edges of a tree spanner in $O(m^2 \log n)$ time and using $O(m)$ space.*

4   An $O(n^3)$ Time Solution for Unweighted Graphs

In this section we consider a dynamic programming approach for computing all best swaps in an unweighted graph. We compute the best swap edge for each of the $n - 1$ edges of T in a separate computation, each requiring $O(n^2)$ time and $O(n^2)$ space. For each failing edge $e = (l, r)$, we root the two subtrees of $T - e$, denoted by T^l (for "left") and T^r (for "right"), at the nodes l and r, respectively. Recall that by Observation 1, the stretch of a swap edge $f = (a, b)$ is obtained at some stretch pair x, y, whose stretch is $d_{T_{e/f}}(x, y)/d_G(x, y)$. In unweighted graphs, $d_G(x, y) = 1$ and hence the maximum stretch is obtained by the stretch pair x, y for which $d_{T_{e/f}}(x, y)$ is maximum[3]. Furthermore, for $a, x \in T^l$ and $b, y \in T^r$ we have $d_{T_{e/f}}(x, y) = d_{T_{e/f}}(x, a) + |(a, b)| + d_{T_{e/f}}(b, y) = d_{T-e}(x, a) + |(x, y)| + d_{T-e}(b, y)$. Therefore, the stretch of a swap edge f is equal to the length of a longest simple path from a to b in G, using only edges of $T - e$ plus either exactly one candidate swap edge $(x, y) \in E \setminus E_T$, or the edge e[4]. In the following, we call paths of this nature the *stretch paths* of the node

---

[3] Recall that any stretch pair's endpoints are by definition connected by an edge in G.
[4] We have to include e here because the stretch is measured with respect to G, not with respect to $G - e$.

pair a, b. In our approach, we compute the length of a longest stretch path for each of the $O(n^2)$ node pairs a, b, even for those which are not linked by an edge in G. It turns out that by partitioning the set of all stretch paths into nine different types, and by computing the length of the longest stretch paths of a particular type for each node pair a, b in a suitable order, all these lengths can be computed in $O(n^2)$ time by dynamic programming. That is, the length of a longest stretch path of a type i for a given node pair a, b can be computed in constant time, given only information that was previously computed. In the following, we describe this approach in detail.

The *type* of a stretch path $\mathcal{P}$ depends on which of the edges incident to $a \in T^l$ and $b \in T^r$ it includes. If $\mathcal{P}$ contains the edge $(a, p(a))$, we say it goes *up* on the left side. If $\mathcal{P}$ contains an edge (a, q) for some $q \in C(a)$, we say it goes *down* on the left side. Furthermore, if $\mathcal{P}$ uses a candidate swap edge incident to a (and hence does not contain any other edge from T^l), we say it *stays* at a. The corresponding definitions hold for the right side of stretch paths. Hence, we have the following nine types of paths (where the first word corresponds to the left side of the path, and the second to the right side): Stay-Stay, Stay-Down, Down-Stay, Stay-Up, Up-Stay, Down-Down, Down-Up, Up-Down, Up-Up. For each TypeA-TypeB and each node pair a, b, we denote by TypeA-TypeB(a, b) the length of a longest stretch path from a to b of type TypeA-TypeB. If no stretch path from a to b of type TypeA-TypeB exists, then we define TypeA-TypeB$(a, b) := -\infty$.

We compute the longest path of each type with an inductive computation (dynamic programming) requiring $O(n^2)$ time. To that end, we first explain the necessary recursive equations. We start with Stay-Stay paths: for a given node pair a, b, the only possible path of that type is composed of the edge (a, b) (if present). Thus, we have

$$\text{Stay-Stay}(a, b) = \begin{cases} 1 & \text{if } (a, b) \in E \backslash E_T \\ -\infty & \text{otherwise.} \end{cases}$$

Clearly, Stay-Stay(a, b) for all $a, b \in V$ can be obtained in $O(n^2)$ time.

The types Stay-Down, Down-Stay, Stay-Up and Up-Stay satisfy simple recursive equations, and can be obtained in $O(n^2)$ time as well. Due to lack of space, we omit these equations here. Consider now a Down-Down stretch path from a to b (see Fig. 3(i)). We have:

$$\text{Down-Down}(a, b) = 1 + \max \left\{ \begin{array}{l} \max_{q \in C(a)} \{\text{Stay-Down}(q, b), \text{Down-Down}(q, b)\}, \\[2mm] \max_{q' \in C(b)} \{\text{Down-Stay}(a, q'), \text{Down-Down}(a, q')\} \end{array} \right\}.$$

In order to write a dynamic program corresponding to this recursion, the node pairs a, b must be ordered such that both the a's and the b's occur in a postorder (note that this is easily possible). Hence, Down-Down(a, b) for all $a, b \in V$ can be computed in $O(n^2)$ time.

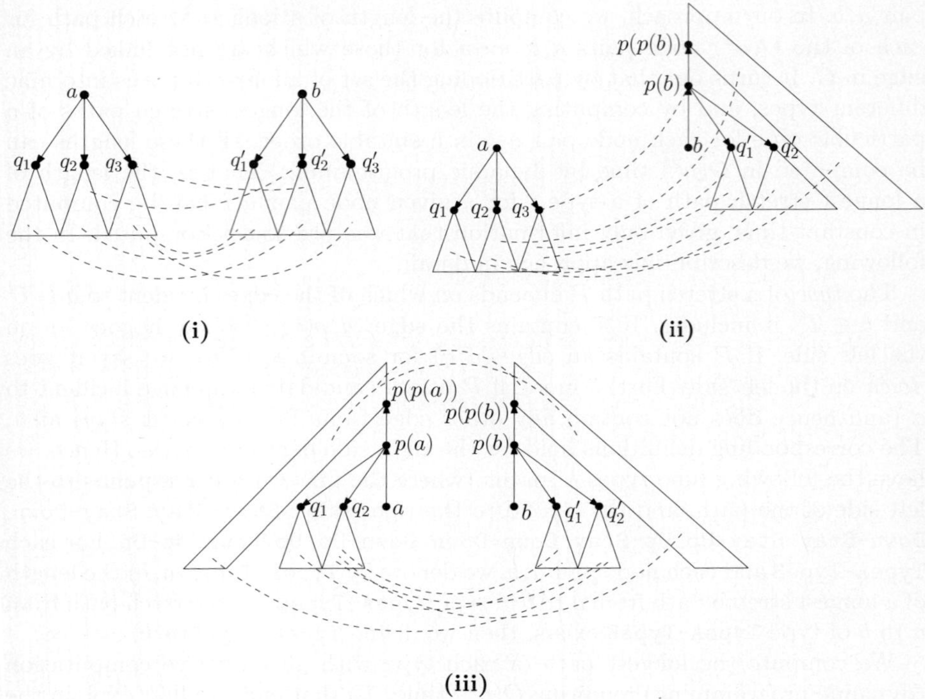

Fig. 3. Three of the possible stretch path types of the node pair a, b: (i) Down-Down, (ii) Down-Up, (iii) Up-Up

Next, let us focus on the Down-Up paths (see Fig. 3(ii)). Here, we have

$$\text{Down-Up}(a,b) = \max\left\{ \begin{array}{l} 1 + \text{Down-Stay}(a, p(b)), \quad 1 + \text{Down-Up}(a, p(b)), \\[2mm] 2 + \max_{q' \in C(p(b)), q' \neq b} \text{Down-Down}(a, q') \end{array} \right\}.$$

We omit the equation for $\text{Up-Down}(a,b)$, which is completely symmetric. By considering all pairs $a, b \in V$ such that the b's occur in preorder, $\text{Down-Up}(a,b)$ and $\text{Up-Down}(a, p(b))$ is obtained in $O(n^2)$ time.

Finally, the length of a longest Up-Up stretch path for a, b can be expressed as (see Fig. 3(iii))

$$\text{Up-Up}(a,b) = \max\left\{ \begin{array}{l} 1 + \text{Up-Stay}(a, p(b)), \quad 1 + \text{Up-Up}(a, p(b)), \\[2mm] 1 + \text{Stay-Up}(p(a), b), \quad 1 + \text{Up-Up}(p(a), b), \\[2mm] 2 + \max_{q' \in C(p(b)), q' \neq b} \text{Up-Down}(a, q'), \quad 2 + \max_{q \in C(p(a)), q \neq a} \text{Down-Up}(q, b) \end{array} \right\}.$$

To obtain $\text{Up-Up}(a,b)$ for all $a, b \in V$ in $O(n^2)$ time, the pairs are considered in an order in which both the a's and the b's occur in preorder. Each of these dynamic programs fills an $(n \times n)$-matrix, and thus needs $O(n^2)$ space. As mentioned in the beginning, we repeat these computations for each of the $O(n)$ edges $e \in E_T$. Then, the algorithm computes, for each swap edge candidate $f = (u, v)$, the stretch of $T_{e/f}$ as

$$
\max \left\{
\begin{array}{lll}
\texttt{Stay-Stay}(u,v), & \texttt{Stay-Down}(u,v), & \texttt{Down-Stay}(u,v), \\
\texttt{Stay-Up}(u,v), & \texttt{Up-Stay}(u,v), & \texttt{Down-Down}(u,v), \\
\texttt{Down-Up}(u,v), & \texttt{Up-Down}(u,v), & \texttt{Up-Up}(u,v)
\end{array}
\right\},
$$

in constant time. After each computation, we can delete the computed matrix from memory, only storing the best swap edge found for the considered failing edge e. Thus, the total space complexity of our approach is $O(n^2)$. In short, we have the following:

Theorem 3. *In unweighted graphs, all best swap edges of a tree spanner can be computed in $O(n^3)$ time and $O(n^2)$ space.*

5   A Distributed Solution to All-Best-Swaps for Spanners

In this section, we consider the scenario when each node in the network has only local information about the network. We are interested in a distributed algorithm which enables each node to compute the information required by it to modify its local routing table whenever any of the edges in T fails. More specifically, a node v having incident edges $e_1, e_2, \ldots e_t \in E_T$ must compute the best swap edges $f_1, f_2, \ldots f_t$ corresponding to these incident edges. Note that the information about the best swap edge for any particular edge $e \in T$, needs to be stored only at the endpoints of e [10]. For the following, let n^* denote the size of the transitive closure of the tree T, when the edges are directed towards the root node.

We make the following assumptions:

- The tree T is rooted at fixed node r, which is know to every node $x \in G$.
- Each node $x \in G$ knows which of its incident edges connect to its children in T and which edge connects to its parent in T.
- Each node x knows the weights of all edges in G that are incident to it.
- Each node x knows the distance $d_T(x, v)$ to any node $v \in T$. (Note that this can be easily pre-computed using $O(n^*)$ messages.)
- The nodes of the tree T are labelled in such a way that given the labels of two nodes $a, b \in T$, it is possible to find the label of the nearest common ancestor of a and b (i.e. $nca(a, b)$). This can be done using labels of size $O(\log n)$ as shown in [1].

Given a rooted spanning tree T of G and any node $v \in T$, we define T_v as the subtree of T that is rooted at v and we denote by $Start(T, v)$ the set of all non-tree edges whose one end-point is in T_v and the other end-point is outside

T_v. At each node x and for each edge $e = (x, y) \in T$, we wish to compute a list of all swap edges with one end-point in T^x and the other endpoint in T^y. If x is a child of y, then this list is same as $Start(T, x)$. Otherwise this list is same as $Start(T, y)$.

The idea of our algorithm is to compute all these lists using the converge-cast technique on the tree, once propagating information from the leaves of T up to the root and then from the root back to the leaves again. Notice that each leaf node already has the required information.

From the above computation, node x has a list of all potential swap edges connecting T_x to T_y (along with their weights). Given any two such edges $f = (u, v)$ and $g = (a, b)$, it is possible to compute the stretch $d_{T_{e/f}}(a, b)$ as follows:

$$d_{T_{e/f}}(a, b) = (d_T(a, u) + |f| + d_T(v, b))/|g|$$

$d_T(a, u)$ can be computed as $d_T(a, u) = d_T(a, x) + d_T(x, u) - 2 \cdot d_T(x, nca(a, u))$. Similarly $d_T(v, b)$ can also be computed. Thus node x can compute the stretch of every swap edge and thus determine the best swap edge for e.

Complexity of the Algorithm. The above algorithm requires only $O(D)$ time for trees of diameter D (measured as the number of "hops"), because the message from the farthest leaf has to reach the root and vice versa. Note that the only information exchanged between the nodes is the list of candidate swap edges. Information about each swap edge $f = (u, v)$ travels only along the path in T from u to v. So the overall communication complexity is $O(n^*)$ times the size of node labels (and edge weights).

6   Swapping Versus Recomputing a New Tree Spanner

In this section, we investigate how a best swap tree compares with a newly computed optimal tree spanner of $G - e$, with respect to the maximum stretch. We show that at least for unweighted graphs, the stretch is at most twice as large in the swap tree as in the tree spanner.

Lemma 2. *For any failing edge e of an optimal tree spanner of an* unweighted *graph G, the maximum stretch of the* swap tree, *measured w.r.t. distances in G, is at most 2 times larger than the stretch of an optimal tree spanner of $G - e$, also measured w.r.t. G. The bound of 2 is tight.*

Proof. Let T be the optimal tree spanner of G, let k be the maximum stretch of T, and let T' be the best swap tree when e fails. Let (a, b) be the pair of nodes for which the stretch with respect to T' is maximum, i.e.

$$(a, b) = \arg \max_{(i,j) \in E} \frac{d_{T'}(i, j)}{d_G(i, j)}.$$

Further, let (u', v') be the best swap edge for e. We have

$$\frac{d_{T'}(a, b)}{d_G(a, b)} \leq \frac{d_T(a, b) - |(u, v)| + d_T(u', v') + |(u', v')|}{d_G(a, b)} \leq k + \frac{d_T(u', v')}{d_G(a, b)} \leq 2k.$$

For any other spanning tree of G (including the optimal spanner of $G - e$), the stretch must be at least k, and hence the result follows. It is easy to construct an example that achieves the bound of 2 (see [3]). $\square$

References

[1] Alstrup, S., Gavoille, C., Kaplan, H., Rauhe, T.: Nearest common ancestors: a survey and a new algorithm for a distributed environment. Theory Comput. Syst. 37(3), 441–456 (2004)

[2] Cai, L., Corneil, D.: Tree Spanners. SIAM J. Discr. Math. 8(3), 359–387 (1995)

[3] Das, S., Gfeller, B., Widmayer, P.: Computing Best Swaps in Optimal Tree Spanners. Technical Report 607, ETH Zurich (September 2008),
http://www.inf.ethz.ch/research/disstechreps/techreports

[4] Dixon, B., Rauch, M., Tarjan, R.: Verification and Sensitivity Analysis of Minimum Spanning Trees in Linear Time. SIAM J. Comp. 21(6), 1184–1192 (1992)

[5] Emek, Y., Peleg, D.: Approximating Minimum Max-Stretch spanning Trees on unweighted graphs. In: SODA 2004: Proceedings of the fifteenth Annual ACM-SIAM Symposium on Discrete Algorithms, Philadelphia, PA, USA, 2004. Society for Industrial and Applied Mathematics, pp. 261–270 (2004)

[6] Flocchini, P., Enriques, A., Pagli, L., Prencipe, G., Santoro, N.: Point-of-failure Shortest-path Rerouting: Computing the Optimal Swap Edges Distributively. IEICE Transactions on Information and Systems E89-D(2), 700–708 (2006)

[7] Flocchini, P., Enriquez, T., Pagli, L., Prencipe, G., Santoro, N.: Distributed Computation of All Node Replacements of a Minimum Spanning Tree. In: Kermarrec, A.-M., Bougé, L., Priol, T. (eds.) Euro-Par 2007. LNCS, vol. 4641, pp. 598–607. Springer, Heidelberg (2007)

[8] Flocchini, P., Pagli, L., Prencipe, G., Santoro, N., Widmayer, P.: Computing All the Best Swap Edges Distributively. In: Higashino, T. (ed.) OPODIS 2004. LNCS, vol. 3544, pp. 154–168. Springer, Heidelberg (2005)

[9] Gfeller, B.: Faster Swap Edge Computation in Minimum Diameter Spanning Trees. In: Halperin, D., Mehlhorn, K. (eds.) ESA 2008. LNCS, vol. 5193, pp. 454–465. Springer, Heidelberg (2008)

[10] Gfeller, B., Santoro, N., Widmayer, P.: A Distributed Algorithm for Finding All Best Swap Edges of a Minimum Diameter Spanning Tree. In: Pelc, A. (ed.) DISC 2007. LNCS, vol. 4731, pp. 268–282. Springer, Heidelberg (2007)

[11] Harel, D., Tarjan, R.: Fast Algorithms for Finding Nearest Common Ancestors. SIAM Journal on Computing 13(2), 338–355 (1984)

[12] Nardelli, E., Proietti, G., Widmayer, P.: How to Swap a Failing Edge of a Single Source Shortest Paths Tree. In: Asano, T., Imai, H., Lee, D.T., Nakano, S.-i., Tokuyama, T. (eds.) COCOON 1999. LNCS, vol. 1627, pp. 144–153. Springer, Heidelberg (1999)

[13] Peleg, D.: Distributed computing: a locality-sensitive approach. Society for Industrial and Applied Mathematics, Philadelphia, PA, USA (2000)

[14] Peleg, D., Ullman, J.: An optimal synchronizer for the hypercube. In: PODC 1987: Proceedings of the sixth annual ACM Symposium on Principles of Distributed Computing, pp. 77–85. ACM, New York (1987)

[15] Peleg, D., Upfal, E.: A trade-off between space and efficiency for routing tables. J. ACM 36(3), 510–530 (1989)

Covering a Point Set by Two Disjoint Rectangles*

Hee-Kap Ahn[1] and Sang Won Bae[2]

[1] Department of Computer Science and Engineering, POSTECH, Korea
heekap@postech.ac.kr
[2] Division of Computer Science, KAIST, Korea
swbae@tclab.kaist.ac.kr

Abstract. Given a set S of n points in the plane, the disjoint two-rectangle covering problem is to find a pair of disjoint rectangles such that their union contains S and the area of the larger rectangle is minimized. In this paper we consider two variants of this optimization problem: (1) the rectangles are free to rotate but must remain parallel to each other, and (2) one rectangle is axis-parallel but the other rectangle is allowed to have an arbitrary orientation. For both of the problems, we present $O(n^2 \log n)$-time algorithms using $O(n)$ space.

1   Introduction

For a set S of n points in the plane, the *disjoint two-rectangle covering problem* is to find a pair of disjoint rectangles with arbitrary orientations such that the union of the rectangles contains all the points in S and the area of the larger rectangle is minimized. This is a fundamental optimization problem that deals with covering a point set S in the plane by two geometric objects of the same type. The surveys by Agarwal and Sharir [1] and by Segal [12] provide comprehensive reviews on a list of such problems.

More specifically, the disjoint two-rectangle covering is a generalization of the axis-parallel two-rectangle covering problem in which the two rectangles are restricted to be axis-parallel. Bespamyatnikh and Segal [3] studied the restricted version of the problem and presented a simple $O(n \log n)$ time algorithm that finds the optimal axis-parallel covering. They also extended the result into higher dimensions and presented an $O(n \log n + n^{d-1})$ time algorithm for the problem in d-dimensional space.

For arbitrary orientations, Jaromczyk and Kowaluk [7] gave an $O(n^2)$ time algorithm for the two-square covering problem with the restriction that the two squares are congruent and parallel to each other. Later, Katz, Kedem and Segal [8] considered the discrete rectilinear two-center problem: find two squares covering the point set S such that their centers are constrained to be at points in S and the area of the larger square is minimized. They presented algorithms for three variants of this problem: when two squares are axis-parallel, an $O(n \log^2 n)$-time/$O(n)$-space algorithm; when two squares are parallel to each other but not necessarily axis-parallel, an $O(n^2 \log^4 n)$-time/$O(n^2)$-space algorithm; when both can rotate independently, an $O(n^3 \log^2 n)$-time/$O(n^2)$-space algorithm. Recently Saha and Das considered the two-rectangle covering problem with

* This work was supported by the Korea Research Foundation Grant funded by the Korean Government (MOEHRD, Basic Research Promotion Fund) (KRF-2007-331-D00372) and by the Brain Korea 21 Project.

S.-H. Hong, H. Nagamochi, and T. Fukunaga (Eds.): ISAAC 2008, LNCS 5369, pp. 728–739, 2008.
© Springer-Verlag Berlin Heidelberg 2008

restriction that the two rectangles are parallel to each other, and presented an algorithm that finds an optimal two-rectangle covering in time $O(n^3)$ using $O(n^2)$ space [11].

In this paper, we present an $O(n^2 \log n)$ time algorithm that finds an optimal disjoint parallel two-rectangle covering of S in arbitrary orientation, which improves the result of Saha and Das [11]. We also consider a variant of the problem in which one rectangle is axis-parallel and the other is allowed to have an arbitrary orientation while they remain disjoint. We present an $O(n^2 \log n)$ time algorithm for this variant. Both of our algorithms presented in this paper use only linear space. The general approach to most of optimal covering problems is first to solve the corresponding decision problem, then to apply an optimization scheme, such as the sorted matrices technique [5], the expander-based technique [9], or parametric search [10]. In contrast, our algorithms are rather intuitive: based on a few geometric observations and analysis of the area functions of rectangles, they capture combinatorial changes in the configuration carefully and maintain the optimal two-rectangle covering during rotation.

We also study another variant of the problem in which both rectangles are allowed to have arbitrary orientations independently while they remain disjoint. However, it seems that the same approach does not apply to this generalized case; this is mainly because the area functions are too complicated to analyze. We discuss about this issue at the end of the paper.

2   Covering Points by Two Disjoint Parallel Rectangles

Throughout this paper, a (directed) line ℓ or a rectangle is called θ-*oriented* if it is parallel (or directed) to an orientation θ. We denote by $B_\theta(P)$ the θ-oriented bounding box of a point set P.

Let S be a given set of n points in the plane. We assume that no three points lie on a line. Consider a fixed orientation θ and an optimal two-rectangle covering of S whose two rectangles are disjoint and parallel to θ. Then, we have a separating line ℓ between the two rectangles which is either θ-oriented or $(\theta + \pi/2)$-oriented. Thus, we can focus only on the case when ℓ is θ-oriented for all $\theta \in [0, \pi)$.

For a fixed orientation θ, let ℓ be a θ-oriented directed line which partitions S into two subsets L_ℓ and R_ℓ, where L_ℓ contains points in S lying in the left side of ℓ and R_ℓ contains points in S lying in the right side of ℓ; points lying on ℓ, if any, can belong to any of L_ℓ and R_ℓ so that $L_\ell \cup R_\ell = S$. Then the optimal disjoint two-rectangle covering problem with restriction to θ is to find a θ-oriented line ℓ such that $\max\{|B_\theta(L_\ell)|, |B_\theta(R_\ell)|\}$ is minimized, where $|\cdot|$ returns the area of a given rectangle. We denote by $f(\theta)$ the optimal objective value for fixed orientation θ, that is, $f(\theta) = \min_\ell \max\{|B_\theta(L_\ell)|, |B_\theta(R_\ell)|\}$ for all θ-oriented directed lines ℓ. This value can be computed in $O(n \log n)$ time [3]; once the points in S are sorted in direction $\theta + \pi/2$, we can find an optimal partitioning line in linear time by using a plane sweep algorithm over S in the direction.

Obviously, there can be infinitely many optimal partitioning lines. Our algorithm implicitly maintains an optimal partitioning line $\ell(\theta)$ which is uniquely defined for any $\theta \in [0, \pi)$. For the purpose, we consider a bit larger rectangles. For a θ-oriented directed line ℓ, let $B_L(\ell)$ and $B_R(\ell)$ be the minimum θ-oriented rectangles such that both have

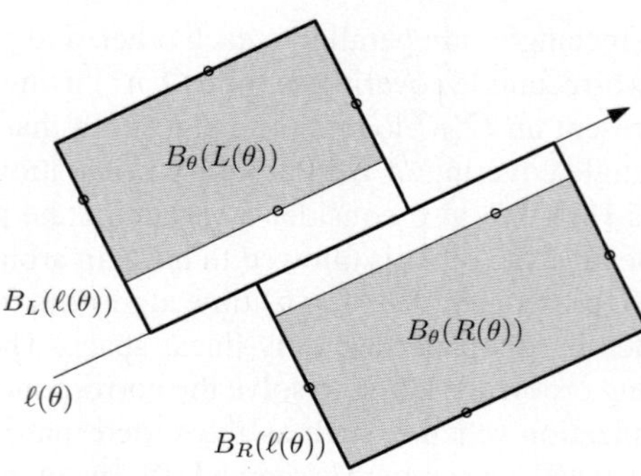

Fig. 1. For an orientation θ, the θ-oriented bisecting line $\ell(\theta)$ and the corresponding bounding boxes. Shaded rectangles are the θ-oriented bounding boxes of $L(\theta)$ and of $R(\theta)$, and rectangles with thick sides are $B_L(\ell(\theta))$ and $B_R(\ell(\theta))$.

one side on ℓ while $B_L(\ell)$ covers L_ℓ and $B_R(\ell)$ covers R_ℓ. If we sweep the plane by ℓ in direction $\theta + \pi/2$, it is easy to see that $|B_L(\ell)|$ is monotonically decreasing and $|B_R(\ell)|$ is monotonically increasing. Note that $|B_L(\ell)|$ and $|B_R(\ell)|$ are discontinuous during the plane sweep but the discontinuity occurs only when a point in S lies on ℓ. We call ℓ a *bisecting line in orientation* θ if $\max\{|B_L(\ell)|, |B_R(\ell)|\}$ is minimized over all such θ-oriented lines, and denote it by $\ell(\theta)$. For each orientation θ, $\ell(\theta)$ is uniquely determined because of the monotonicity of $|B_L(\ell)|$ and $|B_R(\ell)|$. For simplicity of discussion, we let $L(\theta) := L_{\ell(\theta)}$ and $R(\theta) := R_{\ell(\theta)}$. See Figure 1. In the following, we show that the bisecting line $\ell(\theta)$ is indeed an optimal partitioning line. The proof can be found in the full version of the paper.

Lemma 1. *The bisecting line in orientation θ is an optimal partitioning line in θ.*

Consider now that we are allowed to change the orientation θ. Then the optimal disjoint two-rectangle covering problem is to minimize $f(\theta)$ over $\theta \in [0, \pi)$. Before we continue further, we need the following lemma.

Lemma 2 (Saha and Das [11] and Bae et al. [2]). *Let P be a finite set of points in the plane and $(\alpha, \beta) \subset [0, 2\pi)$ be an orientation interval where the sequence of the points touching the sides of $B_\theta(P)$ remains the same for any $\theta \in (\alpha, \beta)$. Then the area of $B_\theta(P)$ can be expressed as a sinusoidal function of θ with angular frequency 2. That is, $|B_\theta(P)|$ is of the form $c_1 \sin(2\theta + c_2) + c_3$, where c_1, c_2, and c_3 are constants depending only on the points on the sides of $B_\theta(P)$.*

On the other hand, we can describe the θ-oriented bounding box $B_\theta(P)$ of a point set P by the sequence of the four touching points, one for each side. Therefore, the optimal two-rectangle covering can be described by a sequence of eight touching points, four from the θ-oriented bounding box of $L(\theta)$ followed by the other four from the other bounding box. We denote by D_θ the sequence of eight touching points at orientation θ and call the points the *determinators* of the two bounding boxes.

During the rotation, we encounter a number of changes in D_θ, which are captured by *events* of following two types:

1. a point crosses over the bisecting line $\ell(\theta)$, that is, a point in $L(\theta)$ moves into $R(\theta)$ or a point in $R(\theta)$ moves into $L(\theta)$, or
2. a side of the bounding boxes touches two points.

We call an event of the first type a *crossing* event, and an event of the second type a *non-crossing* event. Figure 2 shows how a crossing event occurs:

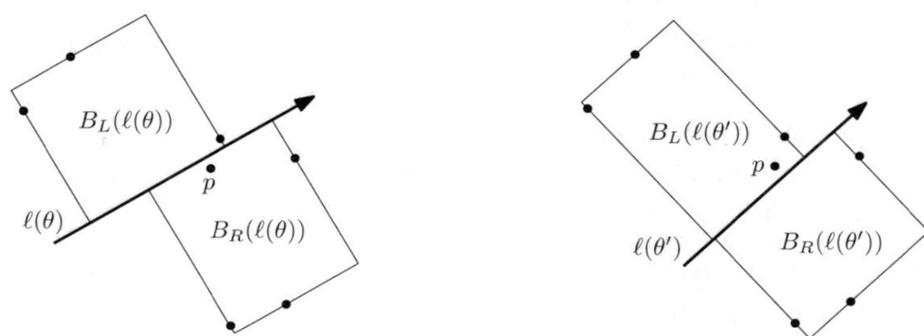

Fig. 2. A crossing event occurs during the rotation from θ to $\theta'(> \theta)$: the point p lies in the right side of the bisecting line $\ell(\theta)$ at θ but it lies in the left side of $\ell(\theta')$

2.1   Algorithm

Our algorithm works by maintaining $L(\theta)$ and $R(\theta)$ as θ increases continuously from 0 to π and minimizing the objective function in orientation intervals where no event occurs.

Algorithm. *ParallelTwoRectangleCover(S)*
(∗ computes the optimal parallel two-rectangle covering over orientations in $[0, \pi)$ ∗)
1. $\theta \leftarrow 0$ and compute $L(0)$, $R(0)$, and D_0
2.   **while** $\theta < \pi$
3.   **do** compute the next non-crossing event at $\theta_n (> \theta)$, assuming no crossing event
4.   compute the next crossing event at θ_c in $[\theta, \theta_n)$, if any
5.   if θ_c is determined, then $\theta' \leftarrow \theta_c$; otherwise, $\theta' \leftarrow \theta_n$
6.   minimize $f(\vartheta)$ in the interval $[\theta, \theta')$
7. $\theta \leftarrow \theta'$
8.   update D_θ, $L(\theta)$ and $R(\theta)$
9.   **return** the minimum objective value with its orientation.

Non-crossing events. A non-crossing event corresponds to an event when two points of $L(\theta)$ (or $R(\theta)$) lie on a side of $B_\theta(L(\theta))$ ($B_\theta(R(\theta))$, respectively). Hence, assuming no further crossing event, the next non-crossing event after θ can be computed in constant time once we know the convex hulls of $L(\theta)$ and of $R(\theta)$, and the determinators D_θ as in rotating caliper [13]. For efficient handling of non-crossing events, we make use of the dynamic convex hull structure by Brodal and Jacob [4]. It supports $O(\log n)$ worst-case time update for insertion/deletion using $O(n)$ space in our case. When updating the invariants, we also update the convex hull of each of $L(\theta)$ and $R(\theta)$ in $O(\log n)$ time. Also, we can easily bound the number of non-crossing events during the algorithm:

when a non-crossing event occurs at θ, two points in S lie on a line which is either θ-oriented or $(\theta + \pi/2)$-oriented. Since we increase θ from 0 to π, the number of non-crossing events to be handled in the algorithm is $O(n^2)$.

Lemma 3. *The number of non-crossing events is at most $O(n^2)$.*

Minimizing the objective function. It can be done in constant time to minimize $f(\vartheta) = \max\{|B_\vartheta(L(\vartheta))|, |B_\vartheta(R(\vartheta))|\}$ in domain $[\theta, \theta')$ where no event occurs due to the nice property of the area functions. The following lemma is based on analysis on sinusoidal functions of the certain form.

Lemma 4. *The sinusoidal functions of the form $c_1 \sin(2\theta + c_2) + c_3$ in the domain $[0, \pi)$ have at most two local minima. For two sinusoidal functions a and b, the equation $a(\vartheta) = b(\vartheta)$ has at most two zeros in the domain.*

As observed in Lemma 2, both $|B_\vartheta(L(\vartheta))|$ and $|B_\vartheta(R(\vartheta))|$ are of the form in Lemma 4. Thus, $f(\vartheta)$ can be expressed by at most three such pieces of functions, and hence $f(\vartheta)$ can be minimized in constant time in domain $[\theta, \theta')$.

2.2   Crossing Events

What remains is to show how to compute the crossing events, and to bound the number of the events. Before we proceed, we need the following lemma. Let $t(\vartheta) := \max\{|B_L(\ell(\vartheta))|, |B_R(\ell(\vartheta))|\}$.

Lemma 5. *The function $t(\vartheta)$ is continuous in the interval $[0, \pi)$.*

Proof. Assume to the contrary that the function is not continuous, that is, there is a discontinuity at some orientation θ in the interval. This means that $\lim_{\vartheta \to \theta+} t(\vartheta) \neq t(\theta)$ or $\lim_{\vartheta \to \theta-} t(\vartheta) \neq t(\theta)$. We consider the latter case, and assume that $t(\theta)$ is larger than $\lim_{\vartheta \to \theta-} t(\vartheta)$ that for every ε in $0 < \varepsilon < \gamma$, $t(\theta) - t(\theta - \varepsilon) > \delta$ for some fixed $\gamma, \delta > 0$. All the other cases can be handled symmetrically. Then there always exists a positive $\varepsilon' < \gamma$ such that

$$t(\theta - \varepsilon') - \max\{|B_L(\ell')|, |B_R(\ell')|\} < \delta,$$

where ℓ' is the θ-oriented through the upper right corner of $B_L(\ell(\theta - \varepsilon'))$. This implies that $t(\theta) > \max\{|B_L(\ell')|, |B_R(\ell')|\}$, which contradicts to the optimality of $\ell(\theta)$.   ☐

The following corollary comes directly from the previous lemma and the uniqueness of the bisecting line.

Corollary 1. *The bisecting line moves continuously during the rotation in the interval $[0, \pi)$.*

Now assume that a point is about to cross the bisecting line at orientation θ. Let $p_L \in L(\theta)$ be the last point lying on the right side of $B_\theta(L(\theta))$ in direction θ. Similarly, we let $p_R \in R(\theta)$ be the first point on the left side of $B_\theta(R(\theta))$ in direction θ. Since the bisecting line moves continuously, the crossing event occurs by p_L or p_R.

   We consider the case that p_L crosses $\ell(\theta)$. The other case is symmetric. Consider the moment of the crossing of p_L: p_L lies on $\ell(\theta)$ and $B_L(\ell(\theta)) = B_\theta(L(\theta))$. Here, we have two possibilities; either another point of S also lies on $\ell(\theta)$ or not. If p_L is the only point on $\ell(\theta)$, the crossing event occurs because $|B_\theta(L(\theta))| \geq |B_\theta(R(\theta))|$ and $|B_{\theta+\varepsilon}(L(\theta))| > |B_{\theta+\varepsilon}(R(\theta) \cup \{p_L\})|$ for some arbitrarily small positive ε.

We now characterize crossing events as follows. Let $f_L(\theta) := |B_\theta(L(\theta))|$ and $f_R(\theta) := |B_\theta(R(\theta))|$. Then, $f(\theta) = \max\{f_L(\theta), f_R(\theta)\}$. We also let $g_L(\theta)$ and $g_R(\theta)$ denote the areas of $B_\theta(L(\theta) \setminus \{p_L\})$ and $B_\theta(R(\theta) \cup \{p_L\})$, respectively, and let $g(\theta) := \max\{g_L(\theta), g_R(\theta)\}$.

Lemma 6. *If a crossing event occurs at θ, either (1) two points of S, including p_L or p_R, lie on $\ell(\theta)$ or (2) $f(\theta) = g(\theta)$.*

Proof. Since no three or more points are co-linear, we consider only case (2). Without loss of generality, we assume that p_L is about to cross the bisecting line. The crossing of p_R can be also handled by symmetry.

Assume to the contrary that $f(\theta) \neq g(\theta)$. If $f(\theta) > g(\theta)$, then there always exists a small positive δ such that $f(\theta - \varepsilon) > g(\theta - \varepsilon)$ for any $0 < \varepsilon < \delta$. This contradicts the optimality of $\ell(\theta - \varepsilon)$. Now assume that $f(\theta) < g(\theta)$. Then again there always exists a small positive δ such that $\max\{|B_{\theta+\varepsilon}(L(\theta))|, |B_{\theta+\varepsilon}(R(\theta))|\} < g(\theta + \varepsilon)$ for any $0 < \varepsilon < \delta$. This contradicts the optimality of $\ell(\theta + \varepsilon)$. $\qquad\square$

The next crossing event of case (1) from the current θ can be predicted easily: we check if the line through p_L and p_R is parallel to some $\theta' \in [\theta, \theta_n)$ for the case where two points lying on the bisecting line are p_L and p_R. Otherwise, it occurs simultaneously with a non-crossing event so that we check if p_L should cross the bisecting line or not at each non-crossing event. Thus, the candidates of the next crossing event of Case (1) can be computed in constant time at each loop of the algorithm.

Hence, in the algorithm, we monitor not only $f(\theta)$ but also $g(\theta)$ and check when they have the the same value after the current orientation θ. This can be done by solving $f(\vartheta) = g(\vartheta)$ in the interval $[\theta, \theta_n)$ where no non-crossing event occurs and then taking the smallest zero, at which we would have the next crossing event. By Lemma 4, each of $f(\vartheta)$ and $g(\vartheta)$ has at most two breakpoints in domain $[\theta, \theta_n)$ (where $f_L(\vartheta) = f_R(\vartheta)$ or $g_L(\vartheta) = g_R(\vartheta)$ holds) and the equation $f(\vartheta) = g(\vartheta)$ has at most constant number of zeros (roughly at most 24.) The case when p_R crosses $\ell(\theta)$ can be handled with functions h_L and h_R defined symmetrically to be $|B_\theta(L(\theta) \cup \{p_R\})|$ and $|B_\theta(R(\theta) \setminus \{p_R\})|$. Therefore, at each step, once we are given D_θ, the functions f, g, and h are determined and then we can find the candidates of the next crossing event of Case (2) in constant time. Note that it is not true that we have a crossing event whenever $f(\theta) = g(\theta)$ or $f(\theta) = h(\theta)$. Thus, when we compute the next crossing event, we should test if it is a "real" crossing event; this can be done simply by checking the local behavior of the functions also in constant time.

To bound the total number of crossing events, we count the number of possible crossing events that occur on a certain point $p \in S$ during the rotation. For this, we need the following lemma. Let $\ell_p(\theta)$ denote the θ-oriented directed line through p.

Lemma 7. *The sequence of the four determinators of the θ-oriented bounding box of the points lying strictly in the left side of $\ell_p(\theta)$ changes at most $O(n)$ times while θ increases continuously from 0 to π.*

Now we are ready to bound the number of crossing events.

Lemma 8. *The number of solutions to $f(\vartheta) = g(\vartheta)$ or $f(\vartheta) = h(\vartheta)$ for $\vartheta \in [0, \pi)$ is at most $O(n^2)$. Also, there are at most $O(n^2)$ crossing events.*

Proof. We discuss about the case where $f(\theta) = g(\theta)$ only since the other case is symmetric. Note that $f(\theta) = g(\theta)$ if and only if $f_L(\theta) = g_R(\theta)$.

Let $f_L^p(\theta) := |B_\theta(L_{\ell_p(\theta)})|$ and $g_R^p(\theta) := |B_\theta(R_{\ell_p(\theta)})|$, where p belongs to $L_{\ell_p(\theta)}$. In orientation θ, if p is the rightmost point of $L(\theta)$, then $f_L^p(\theta) = f_L(\theta)$ and $g_R^p(\theta) = g_R(\theta)$. Hence, in this proof, we rotate $\ell_p(\theta)$ by increasing θ and count the number of times when $f_L^p(\theta) = g_R^p(\theta)$ for all $p \in S$, which upper-bounds the possible number of θ such that $f(\theta) = g(\theta)$.

Consider an orientation interval (α, β) where we have no change in the 8 determinators on the sides of the two bounding boxes. In such an interval, $f_L^p(\theta)$ and $g_R^p(\theta)$ are expressed by sinusoidal functions by Lemma 2, and their graphs intersect at most twice by Lemma 4. Therefore, in (α, β), there are at most two such θ that $f_L^p(\theta) = g_R^p(\theta)$.

Thus, the only thing left is to bound the number of such intervals (α, β). Lemma 7 tells us that the sequence of the four determinators of $B_\theta(R_{\ell_p(\theta)})$ changes at most $O(n)$ times. For $B_\theta(L_{\ell_p(\theta)})$, we have one fixed determinator p; thus the number of changes in its four determinators is also bounded by $O(n)$. For each point $p \in S$, we have $O(n)$ such intervals (α, β), and thus the first statement is shown.

Recall Lemma 6. The number of crossing events falling in the first case that two points of S lie on a line is simply bounded by $O(n^2)$. Also, whenever a crossing event of the second case occurs, a point in S lies on the bisecting line $\ell(\theta)$ and we have $f(\theta) = g(\theta)$ or $f(\theta) = h(\theta)$ by Lemma 6. Thus, the number of crossing events is bounded by $O(n^2)$.   ⊡

Consequently, we spend at most $O(n^2)$ time to compute all events in total, and thus we repeat the main loop $O(n^2)$ times while each run of the main loop takes $O(\log n)$ worst-case time. Finally, we conclude the following.

Theorem 1. *Given a set S of n points, an optimal pair of two disjoint and parallel rectangles containing all points in S can be computed in $O(n^2 \log n)$ worst-case time and $O(n)$ space.*

Tight example construction. Here, we describe how to construct a problem instance which yields at least $\Omega(n^2)$ number of events; thus, our upper bound is asymptotically tight. For any positive integer $n > 3$, let $m = 2\lfloor \frac{n}{4} \rfloor + 1$. Then, m is an odd number, $m = \Theta(n)$, and $n - m = \Theta(n)$. Now, place m points at the corners of a regular m-gon bounded by a unit circle centered at the origin, with one point placed at coordinate $(0, 1)$. The remaining $n - m$ points are placed in a disk U centered at the origin with radius $\varepsilon > 0$, where ε is sufficiently small positive number. Let S_n be this constructed set of points.

The optimal two-rectangle coverings of $S = S_n$ in orientation θ are as follows: If ε is small enough, one of $L(0)$ and $R(0)$ includes all the points in $S \cap U$. Without loss of generality, suppose that $L(0)$ includes $S \cap U$ as shown in Figure 3(a). On the other hand, at orientation π/m, $R(\pi/m)$ consists of the points near the origin; this is easy to see since our point set is symmetric except for points near the origin (see Figure 3(b)). Indeed, $L(2i\pi/m)$ includes the points near the origin $S \cap U$ but $L((2i + 1)\pi/m)$ does

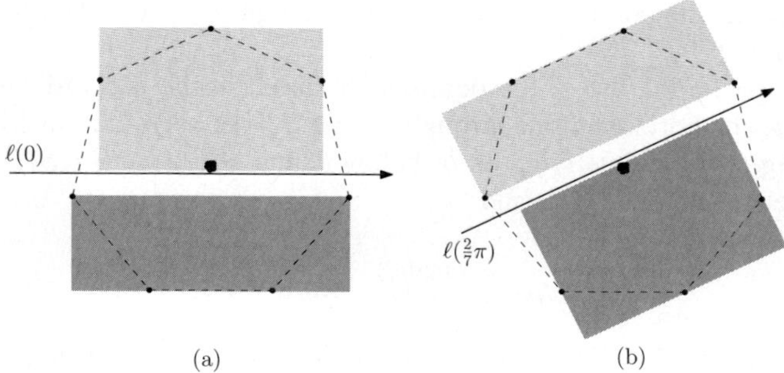

Fig. 3. Tight example construction when $n = 13$ and $m = 7$. The optimal two-rectangle covering in orientation (a) 0 and (b) $2\pi/m = \frac{2}{7}\pi$. The light gray rectangle is $B_\theta(L(\theta))$ and the dark gray rectangle is $B_\theta(R(\theta))$.

not, for $i = 0, \ldots, \lfloor m/2 \rfloor$. Thus, we have at least $n - m$ crossing events per each π/m rotation of orientation, which implies that we have at least $\Omega(n^2)$ number of events for the instance S_n.

3 Covering Points by Two Disjoint Non-parallel Rectangles

In this section, we consider the two-rectangle covering problem where two bounding boxes are not necessarily parallel. More specifically, we find an optimal pair of two disjoint rectangles containing the given set S of points such that one rectangle is axis-parallel and the other rectangle is not necessarily axis-parallel. Informally speaking, one rectangle is free to rotate and we call it the *free* rectangle. We further consider the variant of the problem where both rectangles are free.

Consider an optimal solution for S, consisting of an axis-parallel rectangle B_1 and a θ-oriented rectangle B_2 for $0 \leq \theta < \pi/2$. We observe that there always exists a separating line ℓ between B_1 and B_2 which supports one side of B_1 or of B_2. We have two possibilities:

1. ℓ supports a side of B_1, therefore, is either horizontal or vertical, or
2. ℓ supports a side of B_2.

Our algorithm, to be described below, simply seeks an optimal two-rectangle covering for S in each case above, from which we find the optimal one. Through arguments below, we prove the following.

Theorem 2. *Let S be a set of n points in the plane. One can compute in $O(n^2 \log n)$ time with $O(n)$ space an optimal pair of two disjoint rectangles covering all points in S such that one is axis-parallel and the other can have arbitrary orientation.*

Case 1. We assume without loss of generality that ℓ is horizontal and the axis-parallel rectangle B_1 lies below ℓ. The other cases can be handled in a symmetric way. Since we can assume a determinator on each side of the bounding box as discussed above, a point in S lies on ℓ and we have only n candidates for such ℓ. Thus, in this case, we are done by computing the possible minimum free rectangle B_2 above the horizontal line ℓ through each $p \in S$.

Let $p_1, \ldots, p_n$ be the list of the points in S sorted in the y-coordinate increasing order, and ℓ_i be the horizontal line through p_i. Let $S_1(i) := \{p_1, \ldots, p_i\}$ be the subset of S consisting of the points lying on or below ℓ_i and $S_2(i) := S \setminus S_1(i)$. We define $f_1(i)$ to be the area of $B_0(S_1(i))$ and

$$f_2(i) := \min_{\phi:B_\phi(S_2(i)) \text{ is disjoint from } \ell_i} |B_\phi(S_2(i))|.$$

We then seek a point $p_i \in S$ such that $\max\{f_1(i), f_2(i)\}$ is minimized for all $1 \le i \le n$. Hence, the most difficult part of the algorithm is evaluating $f_2(i)$ in this case.

To evaluate $f_2(i)$, we compute the range Φ_i of ϕ where $B_\phi(S_2(i))$ is disjoint from ℓ_i. Since $S_2(i)$ is fixed, we can compute the description of $|B_\varphi(S_2(i))|$ as a function of $\varphi \in [0, \pi/2)$ in $O(n \log n)$ time by the rotating caliper technique [13]. If $B_\phi(S_2(i))$ and ℓ_i are not disjoint, then one corner of the rectangle lies below ℓ_i. Thus, we compute the locus of the corners of $B_\varphi(S_2(i))$ as φ increases from 0 to $\pi/2$. This locus is a simple closed curve consisting of $O(n)$ circular arcs; this curve is known as the *angle hull* $\mathcal{AH} := \mathcal{AH}(\mathcal{CH}(S_2(i)))$ of the convex hull $\mathcal{CH}(S_2(i))$ of $S_2(i)$, defined to be the locus of points x such that the two tangent lines to $\mathcal{CH}(S_2(i))$ through x make the right angle [6].

Each point x on $\mathcal{AH}$ is mapped to an orientation $\phi \in [0, \pi/2)$ of the bounding box of $S_2(i)$ one of whose corners lies at point x. Moreover, each endpoint of an arc of $\mathcal{AH}$ corresponds to a breakpoint of the function $|B_\varphi(S_2(i))|$, that is, an orientation ϕ where two points of $S_2(i)$ lie on a side of $B_\phi(S_2(i))$. The following lemma follows directly from earlier results [13].

Lemma 9. *The value of $f_2(i)$ is realized as $|B_\phi(S_2(i))|$ such that $\phi \in \Phi_i$, and either (1) two points of $S_2(i)$ lie on one side of $B_\phi(S_2(i))$ or (2) one corner of $B_\phi(S_2(i))$ lies on ℓ_i.*

The first case in the above lemma can be handled by checking each breakpoint of $|B_\varphi(S_2(i))|$; construct $B_\phi(S_2(i))$ for each such breakpoint ϕ and check whether it intersects ℓ_i or not. This takes $O(1)$ time per each breakpoint. For the second case, we compute the intersection of ℓ_i and $\mathcal{AH}$ in $O(n)$ time and take the minimum value among $|B_\phi(S_2(i))|$ for ϕ corresponding to each intersection point between ℓ_i and $\mathcal{AH}$. Then, $f_2(i)$ is the minimum value among those computed as above.

All this process for evaluating $f_2(i)$ takes only $O(n)$ time once the convex hull of $S_2(i)$ is computed. Hence, for each i, we can compute in $O(n \log n)$ time the convex hull of $S_2(i)$, the description of function $|B_\phi(S_2(i))|$, the angle hull $\mathcal{AH}$, the intersection of ℓ_i and $\mathcal{AH}$, and the value of $f_2(i)$.

Lemma 10. *An optimal solution falling in Case 1 can be found in $O(n^2 \log n)$ time with $O(n)$ space.*

Remarks: One might be curious about whether $f_2(i)$ is monotone or not; if it were monotone, one could apply a binary search on i to get a better performance. However, it is not necessarily true by a simple example. See Figure 4.

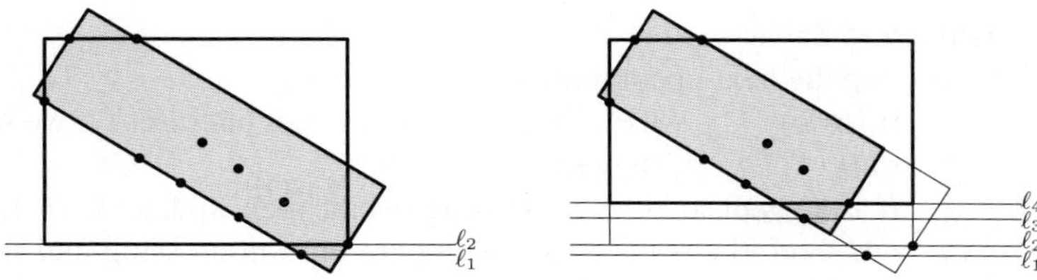

Fig. 4. An example of points S where $f_2(i)$ is not monotone. Observe that $f_2(3) < f_2(1) < f_2(4) < f_2(2)$.

The running time of the algorithm can be improved to $O(n^2)$ by using the dynamic convex hull structure by Brodal and Jacob [4]. But this is not very meaningful since our algorithm for the other case spends $O(n^2 \log n)$ time anyway.

Case 2. In this case, we use a bit different approach. Consider a bipartition L and R of $S = L \cup R$ by a partitioning line ℓ, where R is the set of points lying to the right of ℓ and $L = S \setminus R$. Then, the set of valid orientations for ℓ to get the bipartition is expressed as an orientation interval $(\alpha, \beta) \subset [0, 2\pi)$ (after a proper rotation of the whole S). An optimal solution with an axis-parallel rectangle $B_0(R)$ and a free rectangle $B_\theta(L)$ for $\theta \in (\alpha, \beta)$ can be found by computing $B_0(R)$ and finding the minimum $B_\theta(L)$ for $\theta \in (\alpha, \beta)$ such that both are disjoint from each other.

More formally, we let $\ell_p(\theta)$ be the θ-oriented directed line passing through $p \in S$ and $L_p(\theta)$ and $R_p(\theta)$ be the bipartition of S by $\ell_p(\theta)$ where $p \in L_p(\theta)$. Let $f_1^p(\theta) := |B_0(R_p(\theta))|$ and $f_2^p(\theta) := |B_\theta(L_p(\theta))|$. We rotate $\ell_p(\theta)$ by increasing θ from 0 to 2π and gather local minima of $\max\{f_1^p(\theta), f_2^p(\theta)\}$. By Lemmas 2, 4, and 7, for fixed $p \in S$, $f_2^p(\theta)$ has $O(n)$ breakpoints and is divided into the same asymptotic number of sinusoidal pieces, in each of which we have no change of the determinators of $B_\theta(L_p(\theta))$ and further those of $B_0(R_p(\theta))$. Here, we redefine three types of events to be handled.

- *Non-crossing event*: When the determinators of $B_\theta(L_p(\theta))$ changes while $L_p(\theta)$ remains the same.
- *Crossing event*: When $L_p(\theta)$ changes, that is, another point $q \in S$ lies on $\ell_p(\theta)$.
- *Rectangle-touching event*: When $B_0(R_p(\theta))$ touches $\ell_p(\theta)$, that is, $\ell_p(\theta)$ is tangent to $B_0(R_p(\theta))$.

Our algorithm for this case is described as follows:

Algorithm. *OneAxisParallelAndOneFreeRectangleCoverCase2(S)*
(∗ computes the optimal two rectangles in Case 2 ∗)
1.   **for** each $p \in S$
2.   **do** Initialize a dynamic convex hull $\mathcal{CH}$ and an event queue $\mathcal{Q}$
3.   Compute $L_p(0)$ and add the points of $L_p(0)$ into $\mathcal{CH}$, and compute $R_p(0)$ and $B_0(R_p(0))$
4. $\theta \leftarrow 0$
5.   Compute all crossing events for p and put them into $\mathcal{Q}$
6.   Compute all non-crossing and rectangle-touching events before the first crossing event and put them into $\mathcal{Q}$

7.   **while** $\theta < 2\pi$

8.   **do** Pop the next upcoming event at θ' from $\mathcal{Q}$

9.   If for any $\vartheta \in (\theta, \theta')$, $B_0(R_p(\vartheta))$ dose not intersect ℓ_p, we minimize $\max\{f_1^p(\vartheta), f_2^p(\vartheta)\}$ over $\vartheta \in [\theta, \theta')$

10.   If the event at θ' is a crossing event, then update $L_p(\theta')$, $R_p(\theta')$, $\mathcal{CH}$, and $B_0(R_p(\theta'))$, and compute all non-crossing and rectangle-touching events between θ' and the next crossing event, and put them into $\mathcal{Q}$

11. $\theta \leftarrow \theta'$

12.   **return** the minimum objective value with its orientation

Crossing events occur when $q \in S$ with $q \neq p$ lies on $\ell_p(\theta)$; thus, they can be computed before the **while** loop. Since $L_p(\theta)$ and $R_p(\theta)$ do not change between two consecutive crossing events, we can compute all non-crossing and rectangle-touching events in time proportional to the number of the events with $\mathcal{CH}$ and $B_0(R_p(\theta))$. By Lemma 7, we know that the number of possible non-crossing events is $O(n)$ for each $p \in S$. The number of rectangle-touching events can be bound by the number of crossing events; the number of lines through p which are tangent to a rectangle is at most two unless p is a corner of the rectangle.

Lemma 11. *The total number of events while θ increases from 0 to 2π is at most $O(n)$ for each $p \in S$.*

Finally, we conclude the following. The proof can be found in the full version of the paper.

Lemma 12. *For each $p \in S$, the algorithm above computes all the local minima of $\max\{f_1^p(\theta), f_2^p(\theta)\}$ in $O(n \log n)$ time with $O(n)$ space for $\theta \in [0, 2\pi)$. Thus, an optimal solution of Case 2 can be found in $O(n^2 \log n)$ time and $O(n)$ space.*

3.1 Some Remarks about Two Free Rectangles

We conclude this paper with discussion about the case of two disjoint free rectangles. To express the orientations of two such rectangles, we use two symbols θ and ϕ for their orientations. As observed above, for any optimal disjoint two free rectangles (B_1, B_2) for S, there exists a line ℓ that separates B_1 and B_2 and supports one side of the two rectangles. We thus can search only those rectangles such that B_1 is θ-oriented, B_2 is ϕ-oriented, B_1 and B_2 are disjoint, and a θ-oriented line ℓ supporting a side of B_1 separates B_1 and B_2 for each $\theta, \phi \in [0, 2\pi)$.

For each $p \in S$, we let $\ell_p(\theta)$ be the directed line through p in direction θ and $L_p(\theta)$ and $R_p(\theta)$ be defined as above. Also, let $f_1^p(\theta) := |B_\theta(L_p(\theta))|$ and $f_2^p(\theta) := \min_{\phi \in \Phi_p(\theta)} |B_\phi(R_p(\theta))|$, where $\Phi_p(\theta)$ is the set of orientations ϕ such that $B_\phi(R_p(\theta))$ is disjoint from $\ell_p(\theta)$. As in Case 1 of the above, there always exists $\phi \in \Phi_p(\theta)$ such that $f_2^p(\theta) = |B_\phi(R_p(\theta))|$ and either (1) a corner of $B_\phi(R_p(\theta))$ lies on $\ell_p(\theta)$ or (2) two points in $R_p(\theta)$ lies on a side of $B_\phi(R_p(\theta))$. Note that for fixed θ we can compute each orientation ϕ where a corner of $B_\phi(R_p(\theta))$ lies on $\ell_p(\theta)$ by intersecting the angle hull $\mathcal{AH}$ and $\ell_p(\theta)$. The difficulty here is that we should search every $\theta \in [0, 2\pi)$, and thus each such ϕ is indeed expressed as a function $\phi(\theta)$ of θ. Unfortunately, such a function

$\phi(\theta)$ is very complicated; it appears as the inverse of cosine of a function of $\sin\theta$ and $\cos\theta$ containing a square root term:

$$\arccos\left(\sin\theta\,(c_1\sin\theta)\pm\cos\theta\sqrt{1-(c_1\sin\theta)^2}\right)+c_2,$$

where c_1 and c_2 are constants. It seems very difficult to analyze the functions of this form and to take the same approach as in the previous variations of the problem.

References

1. Agarwal, P., Sharir, M.: Efficient algorithms for geometric optimization. ACM Comput. Surveys 30, 412–458 (1998)
2. Bae, S.W., Lee, C., Ahn, H.-K., Choi, S., Chwa, K.-Y.: Maintaining extremal points and its applications to deciding optimal orientations. In: Proc. 18th Int. Sympos. Alg. Comput. (ISAAC), pp. 788–799 (2007)
3. Bespamyatnikh, S., Segal, M.: Covering a set of points by two axis-parallel boxes. Inform. Process. Lett. 75, 95–100 (2000)
4. Brodal, G.S., Jacob, R.: Dynamic planar convex hull. In: Proc. 43rd Annu. Found. Comput. Sci. (FOCS), pp. 617–626 (2002)
5. Frederickson, G., Johnson, D.: Generalized selection and ranking: sorted matrices. SIAM J. on Comput. 13, 14–30 (1984)
6. Hoffmann, F., Icking, C., Klein, R., Kriegel, K.: The polygon exploration problem. SIAM J. Comput. 31(2), 577–600 (2001)
7. Jaromczyk, J.W., Kowaluk, M.: Orientation indenpendent covering of point sets in R^2 with pairs of rectangles or optimal squares. In: Proc. 12th Euro. Workshop Comput. Geom. (EuroCG), pp. 54–61 (1996)
8. Kats, M., Kedem, K., Segal, M.: Discrete rectilinear 2-center problem. Comput. Geom.: Theory and Applications 15, 203–214 (2000)
9. Katz, M., Sharir, M.: An expander-based approach to geometric optimization. SIAM J. on Comput. 26, 1384–1408 (1997)
10. Megiddo, N.: Applying parallel computation algorithms in the design of serial algorithms. J. ACM 30, 852–865 (1983)
11. Saha, C., Das, S.: Covering a set of points in a plane using two parallel rectangles. In: Proc. 17th Int. Conf. Comput.: Theory and Appl (ICCTA), pp. 214–218 (2007)
12. Segal, M.: Covering point sets and accompanying problems. PhD thesis, Ben-Gurion University, Israel (1999)
13. Toussaint, G.: Solving geometric problems with the rotating calipers. In: Proc. IEEE MELECON 1983, Athens, Greece (1983)

Computing the Maximum Detour of a Plane Graph in Subquadratic Time

Christian Wulff-Nilsen

Department of Computer Science, University of Copenhagen,
Universitetsparken 1, DK-2100 Copenhagen O, Denmark
koolooz@diku.dk
http://www.diku.dk/hjemmesider/ansatte/koolooz/

Abstract. Let G be a plane graph where each edge is a line segment. We consider the problem of computing the maximum detour of G, defined as the maximum over all pairs of distinct points p and q of G of the ratio between the distance between p and q in G and the distance $|pq|$. The fastest known algorithm for this problem has $\Theta(n^2)$ running time where n is the number of vertices. We show how to obtain $O(n^{3/2}\log^3 n)$ expected running time. We also show that if G has bounded treewidth, its maximum detour can be computed in $O(n\log^3 n)$ expected time.

1   Introduction

Given a geometric graph G, its stretch factor (or dilation) is the maximum over all pairs of distinct vertices u and v of the ratio between the distance between u and v in G and the Euclidean distance $|uv|$ between u and v.

A spanner is a network with small stretch factor. Spanners that keep other cost measures low, such as size, weight, degree, and diameter, are important structures in areas such as VLSI design, distributed computing, and robotics. For more on spanners, see [5,9,10].

An interesting dual problem is that of computing the stretch factor of a given geometric graph. In this paper, we consider a related problem, namely that of computing the maximum detour of a plane graph where edges are line segments. Maximum detour is defined like stretch factor except that the maximum is taken over all pairs of distinct *points* of the graph, i.e., interior points of edges as well as the vertices.

If the graph is planar then its stretch factor can be computed in $\Theta(n^2)$ time, where n is the number of vertices, by applying the APSP algorithm in [6]. This bound also holds for the problem of computing the maximum detour of a plane graph [1]. It is an open problem whether subquadratic time algorithms exist.

For more special types of graphs such as paths, trees, cycles, and graphs having bounded treewidth, faster algorithms are known for the stretch factor and maximum detour problem [1,2].

In this paper, we show how to compute the maximum detour of a plane graph with n vertices in $O(n^{3/2}\log^3 n)$ expected time, thereby solving the open problem

S.-H. Hong, H. Nagamochi, and T. Fukunaga (Eds.): ISAAC 2008, LNCS 5369, pp. 740–751, 2008.

of whether a subquadratic time algorithm exists for this problem. We also show that if the graph has bounded treewidth, its maximum detour can be computed in $O(n \log^3 n)$ expected time.

The organization of the paper is as follows. In Section 2, we give various definitions and introduce some notation. In Section 3, we make use of the separator theorem by Lipton and Tarjan which enables us to apply the divide-and-conquer paradigm to the input graph. We define colourings of points of a face of the graph in Section 4, show some properties of these colourings and how to efficiently compute them. In Section 5, we show how the colourings give an efficient way of computing the maximum detour between points on a face of the graph and this in turn gives an efficient algorithm for computing the maximum detour of the entire graph. Finally, we make some concluding remarks in Section 6.

2   Definitions and Notation

Let $G = (V, E)$ be a plane graph where each edge is a line segment and let P_G be the set of points of G (vertices as well as interior points of edges). Given two points $p, q \in P_G$, we define $d_G(p, q)$ as the length of a shortest path in G between p and q, where the length of a path is measured as the sum of the Euclidean lengths of the (parts of) edges on this path. If there is no such path, we define $d_G(p, q) = \infty$. If $p \neq q$ then the *detour* $\delta_G(p, q)$ between p and q (in G) is defined as the ratio $d_G(p, q)/|pq|$. The *maximum detour* δ_G of G is the maximum of this ratio over all pairs of distinct points of P_G.

Where appropriate, we will regard a plane graph as the set of points belonging to the graph. So for instance, if G and H are plane graphs then $G \cap H$ is the set of points belonging to both G and H. If well-defined, we will regard the resulting point set as a graph.

For a graph G, we let $|G|$ denote its size, i.e. the number of vertices plus the number of edges in G. Given two subsets P_1 and P_2 of the set P_G of points of G, we define

$$\delta_G(P_1, P_2) = \max_{p \in P_1, q \in P_2, p \neq q} \delta_G(p, q).$$

If p is a point of G and $P \subseteq P_G$, we write $\delta_G(p, P) = \delta_G(P, p)$ instead of $\delta_G(\{p\}, P)$ and we write $\delta_G(P)$ as a shorthand for $\delta_G(P, P)$. We extend these definitions to subgraphs, edges, and vertices by regarding them as sets of points.

Given paths $P = p_1 \to p_2 \to \ldots \to p_r$ and $Q = q_1 \to q_2 \to \ldots \to q_s$, where $p_r = q_1$, we let PQ denote the combined path $p_1 \to p_2 \to \ldots \to p_{r-1} \to q_1 \to q_2 \to \ldots \to q_s$. For a vertex v, we let $v \to P$ denote the path $v \to p_1 \to p_2 \to \ldots \to p_r$.

3   Separating the Problem

In all the following, let $G = (V, E)$ be an n-vertex plane graph in which edges are line segments. We seek to compute δ_G in $O(n^{3/2} \log^3 n)$ expected time. We will assume that G is connected since otherwise, the problem is trivial.

To compute δ_G, we apply the divide-and-conquer paradigm. Our strategy is in some ways similar to that in [2], the main difference being that in the merge step we use face colourings (Section 4) whereas range searching is used in [2].

In this section, we separate G into two smaller graphs, G_A and G_B, of roughly the same size. We recursively compute $\delta_G(G_A)$ and $\delta_G(G_B)$ and in Section 5, we describe an algorithm for efficiently obtaining the maximum detour $\delta_G(G_A, G_B)$.

To separate our problem, we use the separator theorem of Lipton and Tarjan [8]. This gives us, in $O(n)$ time, a partition of V into three sets, A, B, and P, such that the following three properties hold

1. no edge joins a vertex in A with a vertex in B,
2. neither A nor B contains more than $n/2$ vertices, and
3. P contains no more than $\frac{2\sqrt{2}}{1-\sqrt{2/3}}\sqrt{n}$ vertices.

We refer to vertices of P as *portals*. We must have $P \neq \emptyset$ since otherwise, one of the sets A and B would be empty by property 1, implying that $|A| = n$ or $|B| = n$, contradicting property 2. In the following, let $k \geq 1$ denote the number of portals and let $p_1, \ldots, p_k$ denote the portals.

Having found this partition, we compute and store shortest path lengths from each portal to each vertex of V. Using Dijkstra's SSSP algorithm with, say, binary heaps, this can be done in $O(n \log n)$ time for each portal, giving a total running time of $O(n^{3/2} \log n)$.

Let G_A be the subgraph of G induced by $A \cup P$ and let G_B be the subgraph of G induced by $B \cup P$. We construct G_A and G_B and recursively compute $\delta_G(G_A)$ and $\delta_G(G_B)$. Clearly,

$$\delta_G = \max\{\delta_G(G_A), \delta_G(G_B), \delta_G(G_A, G_B)\}.$$

In the following, we deal with the problem of computing $\delta_G(G_A, G_B)$.

We will need the following lemma which is a generalization of a result in [4] (we omit the proof since it is virtually identical to that in [4]).

Lemma 1. *Maximum detour δ_G is achieved by a pair of co-visible points.*

This result allows us to consider only detours between pairs of points of the same face of G (here we include the external face). In all the following, let f be a face of G, let $f_A = f \cap G_A$, and let $f_B = f \cap G_B$. We will show how to compute $\delta_G(f_A, f_B)$ in $O(|f|k \log^2 n)$ expected time. From this it will follow that $\delta_G(G_A, G_B)$ can be found in $O(n^{3/2} \log^2 n)$ expected time.

We will assume that f is an internal face of G. The external face is dealt with in a similar way. We also assume that f_A and f_B do not share any edges since any edge e shared by them must have portals as both endpoints and the detours from points in e to all points in G will be considered in the two recursive calls that compute $\delta_G(G_A)$ and $\delta_G(G_B)$, respectively. Hence, we may disregard e when computing $\delta_G(f_A, f_B)$.

We assume that f is simple. The case where f is non-simple is handled in a similar way.

4   Colouring Points

In this section, we define colourings of points of f. As we shall see in Section 5, these colourings will prove helpful when computing $\delta_G(f_A, f_B)$. More specifically, they will speed up shortest path computations between pairs of points in f by indicating, for each point pair, which portal is on some shortest path between those two points.

Face f is defined by a simple cycle $v_1 \to v_2 \to \ldots \to v_{n_f} \to v_1$ such that the interior of f is to the left as we walk in the direction specified by the vertices.

For a point $p \in f$, we let $d_f(p)$ denote the Euclidean length of the path from v_1 to p that visits vertices in the order specified above. We define an order $<^f_{v_1}$ of the set of points of f as follows. For two points p and q of f, $p <^f_{v_1} q$ if and only if $d_f(p) < d_f(q)$. Order $p >^f_{v_1} q$ is defined in a similar way. In the following, we assume that the sum of lengths of all edges in f and $d_f(v_i)$ for all $v_i \in f$ are precomputed.

By starting the walk in any other vertex v_i of f, we can similarly define orders $<^f_{v_i}$ and $>^f_{v_i}$. It is easy to see that determining whether e.g. $p <^f_{v_i} q$ holds for any points $p, q \in f$ can be done in constant time using the above precomputed values. Where appropriate, we will regard the points of f (or f_A or f_B) occuring between two points w.r.t. order $<^f_{v_i}$ as an interval of points.

In the following, let a be a vertex in f_A and let $a' \in f \cap V$ be its successor w.r.t. $<^f_a$. We now consider associating with a a colouring of points in f_B using colours $c_1, \ldots, c_k$. A point $p \in f_B$ is given colour c_i if portal p_i is on a shortest path in G from a to p. In case of ties, pick the colour such that the corresponding portal has minimum distance to a in G. In case of further ties, pick the colour with the smaller index. We let $c_a(p)$ denote the colour assigned to p.

We will show that colours occur in intervals as we walk around f_B with each colour assigned to at most one interval. Furthermore, we will show that the order of these intervals is induced by an order of the portals which we define next.

Let u_0 and u_1 be distinct vertices of G connected by an edge and consider a portal p_i. Choose edge $(u_1, u_2) \in E$ such that u_2 is on a shortest path from u_1 to p_i and such that $u_0 \to u_1 \to u_2$ makes the sharpest possible left turn at u_1 (if a left turn is not possible we regard the least possible right turn as a sharpest possible left turn and we regard a turn of angle π as a left turn of angle π).

Repeat this procedure by picking, for $j = 3, \ldots, r$, an edge (u_{j-1}, u_j) such that u_j is on a shortest path from u_{j-1} to p_i and such that $u_{j-2} \to u_{j-1} \to u_j$ makes the sharpest possible left turn at u_{j-1}; here, r is the smallest index such that $u_j = p_i$. The resulting path $u_1 \to u_2 \to \ldots \to u_r$ is uniquely defined and is a shortest path from u_1 to p_i. We denote it by $\overleftarrow{P_i}(u_0, u_1)$. We define $\overleftarrow{P_i}'(u_0, u_1) = u_0 \to \overleftarrow{P_i}(u_0, u_1)$. In the following, we will write $\overleftarrow{P_i}$ resp. $\overleftarrow{P_i}'$ as a shorthand for $\overleftarrow{P_i}(a', a)$ resp. $\overleftarrow{P_i}'(a', a)$, where a and a' are defined as above.

For two distinct portals p_i and p_j we write $p_i \prec^f_a p_j$ if $p_i \in \overleftarrow{P_j}$ or if $\overleftarrow{P_i}'$ makes a sharper left turn than $\overleftarrow{P_j}'$ at some shared interior vertex. As an example, in Figure 1, $p_i \prec^f_a p_j$ since $\overleftarrow{P_i}'$ makes a sharper left turn than $\overleftarrow{P_j}'$ at vertex v.

Lemma 2. *If paths $\overleftarrow{P_i}$ and $\overleftarrow{P_j}$ split at a vertex v they cannot meet after v.*

Lemma 3. *The relation $\prec_a^f$ above is a strict total order of the portals.*

We now show the relation between the order $\prec_a^f$ of portals and the order $<_a^f$ of vertices in f_B.

Theorem 1. *Let p and q be two distinct points of f_B and assume that $c_a(p) = c_i$ and $c_a(q) = c_j$, $i \neq j$. Then $p_i \prec_a^f p_j$ if and only if $p <_a^f q$.*

Proof. By symmetry, it is enough to show that if $p_i \prec_a^f p_j$ then $p <_a^f q$.

So assume that $p_i \prec_a^f p_j$. Since $c_a(q) = c_j$, we have $p_i \notin \overleftarrow{P_j}$. It follows that there is a vertex v at which $\overleftarrow{P_i}$ and $\overleftarrow{P_j}$ split and P_i' makes a sharper left turn at v than $\overleftarrow{P_j}'$. By Lemma 2, the two paths do not meet again after v. In particular, $\overleftarrow{P_i}$ does not cross $\overleftarrow{P_j}$, see Figure 1.

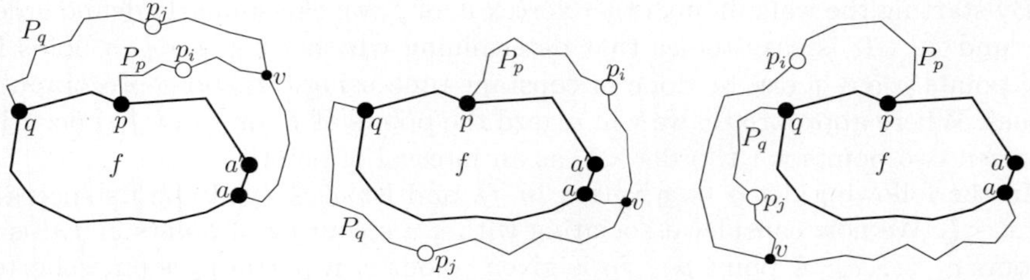

Fig. 1. The possible situations in the proof of Theorem 1 when $p_i \prec_a^f p_j$

Let P_p be a shortest path from p_i to p. Then P_p cannot intersect $\overleftarrow{P_j}$ since then p_i and p_j would both be on a shortest path from a to q with $d_G(a, p_i) < d_G(a, p_j)$, contradicting the assumption that $c_a(q) = c_j$.

Let P_q be a shortest path from p_j to q. Then $\overleftarrow{P_i}$ cannot intersect P_q since otherwise, p_i and p_j would both be on a shortest path from a to p with $d_G(a, p_j) < d_G(a, p_i)$, contradicting the assumption that $c_a(p) = c_i$.

Furthermore, P_q cannot intersect P_p. For assume it did. Then there would be a shortest path from a to p through p_j and a shortest path from a to q through p_i. If $d_G(a, p_i) < d_G(a, p_j)$ then $c_a(q) \neq c_j$ and if $d_G(a, p_j) < d_G(a, p_i)$ then $c_a(p) \neq c_i$. Hence, $d_G(a, p_i) = d_G(a, p_j)$. But then $i < j$ would imply $c_a(q) \neq c_j$ and $i > j$ would imply $c_a(p) \neq c_i$, contradicting the colours assigned to p and q.

It follows from the above that paths $\overleftarrow{P_i}P_p$ and $\overleftarrow{P_j}P_q$ do not intersect except in the vertices they share until reaching v, see Figure 1. This implies that $p <_a^f q$, showing the theorem. $\qquad\square$

Corollary 1. *Interval f_B can be split up into $O(k)$ sub-intervals such that points in the same sub-interval are assigned the same colour w.r.t. vertex $a \in f_A$.*

If the sub-intervals of Corollary 1 are picked such that they have maximal size, we refer to them as *colour intervals of a*.

We now show how the colour intervals of a can be computed efficiently.

Lemma 4. *The order of colour intervals of a can be computed in $O(k \log^2 n)$ time assuming $O(kn \log n)$ time for preprocessing. The preprocessing step is independent of a and f.*

Proof. We will prove the lemma by presenting a data structure that, given distinct portals p_i and p_j, determines whether $p_i \prec_a^f p_j$ and does so for any a and f. We will show how to construct this data structure in $O(kn \log n)$ time such that it can determine whether $p_i \prec_a^f p_j$ in $O(\log n)$ time. This will allow us to sort the portals according to $\prec_a^f$ in time $O(k \log^2 n)$ time using a sorting algorithm like merge or heap sort since such an algorithm performs $O(k \log k)$ comparisons. This result together with Theorem 1 will show the lemma.

We first show how to construct the data structure. In the following, let E' be the set of directed edges obtained by regarding each edge of E as two oppositely directed edges.

With each edge $e = (u, v) \in E'$ and each portal p_i, we associate pointers $\pi_j(e, i)$, $j = 0, \ldots, j_e$. Pointer $\pi_j(e, i)$ points to the edge e_j of $\overleftarrow{P}_i(e)$ such that the number of edges between e and e_j in $\overleftarrow{P}_i(e)$ is $2^j - 1$, see Figure 2. Here, j_e is the largest j such that e_j exists. Note that $j_e = O(\log n)$ so the total number of pointers over all edges e and all portals p_i is $O(kn \log n)$.

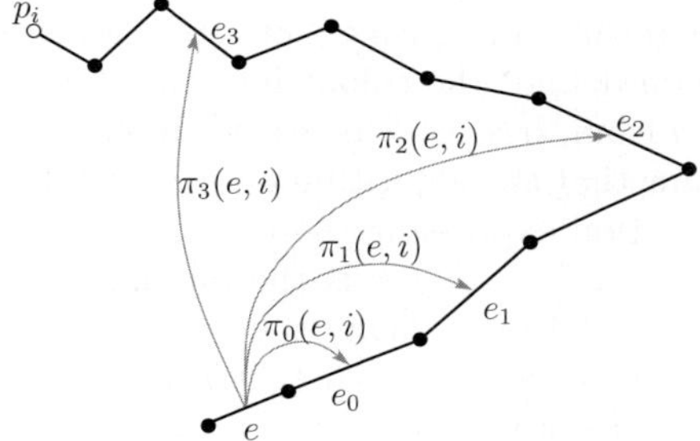

Fig. 2. Example with π-pointers from edge e to edges $e_0 = \pi_0(e, i)$, $e_1 = \pi_1(e, i)$, $e_2 = \pi_2(e, i)$, and $e_3 = \pi_3(e, i)$ on $\overleftarrow{P}_i(e)$. Here, $j_e = 3$.

It is easy to show that all π-pointers can be computed in $O(kn \log n)$ time. Now, given these pointers, for any edge e and any portals p_i and p_j, binary search allows us to determine, in $O(\log n)$ time, the vertex at which $\overleftarrow{P}_i(e)$ and $\overleftarrow{P}_j(e)$ split. If they do not split, binary search will also detect this and determine whether $p_i \in \overleftarrow{P}_j(e)$ or $p_j \in \overleftarrow{P}_i(e)$. From this it follows that we can determine whether $p_i \prec_a^f p_j$ in $O(\log n)$ time, given these pointers.   □

Theorem 2. *Endpoints of colour intervals of a can be computed in $O(k \log^2 n)$ time assuming $O(kn \log n)$ preprocessing time. The preprocessing step is independent of a and f.*

Proof. We start by computing the order of colour intervals of a. By Lemma 4, this can be done in $O(k \log^2 n)$ time with $O(kn \log n)$ preprocessing time. Let π be the permutation of $\{1, \ldots, k\}$ such that $p_{\pi(1)} \prec_a^f \ldots \prec_a^f p_{\pi(k)}$.

We then compute the colour of the first point $b_{\min}$ and the last point $b_{\max}$ of f_B w.r.t. the order $<_a^f$. This can be done in time proportional to the number k of colours since SSSP lengths for each portal have been precomputed.

If $c_a(b_{\min}) = c_a(b_{\max})$ then by Theorem 1, all points between $b_{\min}$ and $b_{\max}$ have this colour. In this case, the algorithm associates this colour with the sub-interval between the two vertices and returns (the sub-interval is not stored explicitly, only its end vertices $b_{\min}$ and $b_{\max}$).

Otherwise, a vertex $b \in f_B$ is picked, such that the number of edges in f_B before resp. after b w.r.t. the order $<_a^f$ is (approximately) the same, and its colour $c_a(b)$ is computed. Let i be the index such that $c_{\pi(i)} = c_a(b)$. The algorithm calls itself recursively on vertices between $b_{\min}$ and b with colours $c_{\pi(1)}, \ldots, c_{\pi(i)}$. And it calls itself recursively on vertices between b and $b_{\max}$ with colours $c_{\pi(i)}, \ldots, c_{\pi(k)}$.

The recursion stops when $b_{\min}$ and $b_{\max}$ are the endpoints of a single edge e of f_B. If $c_a(b_{\min}) = c_a(b_{\max})$, the algorithm associates this colour with e and returns. Otherwise, we need the following simple observation: there is a point p on e such that all points on $b_{\min}p$ have colour $c_a(b_{\min})$ and all points on $pb_{\max}$ have colour $c_a(b_{\max})$. Furthermore, p can be computed in $O(1)$ time, given these two colours. The algorithm associates the two colours with their respective segments of e and returns.

When the algorithm terminates, each colour c_i is associated with $O(\log n)$ sub-intervals and their union defines the colour interval of a with colour c_i. Finding the colour intervals of a from these $O(k \log n)$ sub-intervals takes $O(k \log n)$ time. What remains is to show that the algorithm above has $O(k \log^2 n)$ running time.

Let $T(m, k)$ be a function expressing the time for the above algorithm where m is the number of edges of f_B and k is the number of colours. If we assume that vertices of f_B are stored in an array then the point b that splits points of f_B into two equal halves can be found in $O(1)$ time. Thus, there is a constant $c' > 0$ such that the algorithm uses at most $c'k$ time steps excluding time spent in recursive calls. There is also a constant c'' such that $T(m, k) \le c''k$ when $m \le 2$. Let $c = \max\{c', c''\}$. Then (ignoring floors and ceilings)

$$T(m, k) \le ck + T(m/2, k_1) + T(m/2, k_2).$$

where $m > 2$ and $k_1, k_2 \in \{1, \ldots, k\}$, $k_1 + k_2 = k + 1$.

Let $\tilde{T}(m, k)$ be the running time $T(m, k)$ minus a value of $c \log n$ charged to each split vertex b encountered in the current and in recursive calls. Then it can be shown by induction on $m \ge 2$ that $\tilde{T}(m, k) < ck \log m$, implying that $T(m, k) \le ck \log m + xc \log n$, where x is the total number of split vertices. This number is proportional to the number of sub-intervals returned by the algorithm which is $O(k \log m)$. $\qquad\square$

Note that a colour interval I need not be closed, i.e. one or both of the endpoints need not belong to I. We conclude this section with the following simple result which will prove useful when we compute detours between points that may be interior points of edges.

Lemma 5. *Let p a point of edge $e = (u, v)$ of G and let q be a point in G. Suppose that p_i is on a shortest path from u to q and that p_j is on a shortest path from v to q. Then either p_i or p_j is on a shortest path from p to q.*

Proof. A shortest path from p to q goes through either u or v. $\qquad\qquad$ $\square$

5   The Detour of Points in a Face

In this section, we show how to compute $\delta_G(f_A, f_B)$ in $O(|f|k \log^2 n)$ expected time.

We start by computing, for each edge $e = (u, v) \in f_A$, $O(k)$ colour intervals of u and of v using $O(k \log^2 n)$ time (with $O(kn \log n)$ preprocessing) and take the union of the endpoints of these colour intervals. This gives $O(k)$ smaller sub-intervals which we associate with e. The total running time for this over all edges is $O(|f|k \log^2 n)$.

Now, let P be one of the sub-intervals associated with edge $e = (u, v)$. Then there are $i, j \in \{1, \ldots, k\}$ such that $c_u(p) = c_i$ and $c_v(p) = c_j$ for all $p \in P$. Hence, for any point $q \in P$, p_i is on a shortest path from u to q and p_j is on a shortest path from v to q. Lemma 5 implies that for any point $p \in e$ and any point $q \in P$, either p_i or p_j is on a shortest path from p to q.

We refer to P as a *type 1-interval* (of e) if $c_i \neq c_j$ and a *type 2-interval* (of e) if $c_i = c_j$.

For any edge e of f_A and any type i-interval P of e, $i = 1, 2$, we may assume that P is a closed interval having endpoints in vertices of f_B. For otherwise, we could compute the maximum detour between e and the first resp. last edge e' of P (all other edges of P have endpoints in vertices of f_B). Computing $\delta_G(e, e')$ is a constant-size problem (when SSSP lengths for each portal have been precomputed) since we know that for each point $p \in e$ and each point $p' \in e$, there is a shortest path from p to p' through either of two portals p_i and p_j. Thus, it takes $O(1)$ time to compute $\delta_G(e, e')$ (see also [1]). Over all e and P, this amounts to $O(|f|k)$ time.

The value $\delta_G(f_A, f_B)$ is computed in two phases. In phase i, the maximum detour between points in edges of f_A and points in associated type i-intervals is computed, $i = 1, 2$.

5.1   Phase 1

We will show that phase 1 takes $O(|f|k)$ time when shortest path lengths from portals and colour intervals have been computed.

The algorithm for this phase is straightforward. For each edge e of f_A it considers all edges e' of each type 1-interval of e and computes $\delta_G(e, e')$ in constant time.

To show that phase 1 takes $O(f|k|)$ time, we need to show that the number of edge pairs (e, e') is $O(f|k|)$. To this end, we introduce a so called *dual colouring* of vertices of f_A for each vertex of f_B. Let b be a vertex of f_B. Then vertex $a \in f_A$ is given *dual colour* $c_b(a)$, defined as the colour $c_a(b)$.

Assigning dual colours to all vertices of f_A partitions this set into maximal sub-intervals with vertices in each sub-interval having the same dual colour. We call these sub-intervals the *dual colour intervals of b*. These dual colour intervals will help us bound the number of edge pairs (e, e'). First, we need two lemmas.

Lemma 6. *Let b be a vertex of f_B and let a, a_1, a_2 be vertices of f_A such that $a_1 <_b^f a <_b^f a_2$, $c_b(a_1) = c_b(a_2) = c_i$, and $c_b(a) = c_j$, $i \neq j$. Then for any vertex a' of f_A, $c_b(a') \neq c_j$ if either $a' <_b^f a_1$ or $a' >_b^f a_2$.*

Lemma 7. *The number of dual colour intervals of a vertex $b \in f_B$ is $O(k)$.*

Proof. Let $N(k)$ be the maximum number of dual colour intervals of b when the number of distinct colours in these intervals is exactly k. We will show that $N(k) \leq 2k - 1$. The proof is by induction on $k \geq 1$. If $k = 1$ then there is only one dual colour interval and we have $N(k) = 1 = 2k - 1$.

Now, suppose that $k > 1$ and that $N(k') \leq 2k' - 1$ for all k' less than k. Let c_i be the colour of the first dual colour interval w.r.t. the order $<_b^f$. By Lemma 6, there is a finite number r of dual colour intervals with colour c_i (in fact at most k). Let $I_1, \ldots, I_s$ be the intervals between each consecutive pair of these dual colour intervals and let $k_1, \ldots, k_s$ be the number of colours in each of them. Note that $s \geq r - 1$. Also note that by the choice of c_i and by Lemma 6, two points in two different I-intervals cannot have the same colour since for at least one of the two intervals, the colour of the dual colour interval preceding and succeeding it is c_i. From this and from the fact that the I-intervals do not contain colour c_i, $\sum_{j=1}^s k_j = k - 1$. Applying the induction hypothesis, this gives

$$N(k) \leq r + \sum_{j=1}^s N(k_j) \leq r + \sum_{j=1}^s 2k_j - 1 = r - s + 2(k-1) \leq 1 + 2(k-1) = 2k - 1.$$

$\square$

We are now ready to bound the running time for the phase 1 algorithm.

Theorem 3. *Phase 1 runs in $O(|f|k)$ time.*

Proof. Consider edges $e = (u, v) \in f_A$ and $e' = (u', v') \in f_B$ such that e' is an edge of a type 1-interval of e. Let $c_i = c_u(u') = c_u(v')$ and let $c_j = c_v(u') = c_v(v')$. Since u' is a vertex of f_B, dual colours $c_{u'}(u) = c_i$ and $c_{u'}(v) = c_j$ are well-defined and since $i \neq j$ by assumption, these two dual colours are distinct, implying that two dual colour intervals of b meet at e.

It follows by the above and by Lemma 7 that for each e', $O(k)$ edges e are picked. Thus, the total number of edge pairs considered by the algorithm is $O(|f|k)$. By our earlier discussion, this suffices to show the theorem. $\square$

5.2   Phase 2

We now consider the problem of computing the maximum detour between points of edges of f_A and points of type 2-intervals.

For any pair (e, P), where e is an edge of f_A and P is a type 2-interval of e, there is a portal p_i such that for any point $p \in e$ and any point $q \in P$, p_i is on a shortest path from p to q. In the following, we consider all such pairs for a fixed p_i. We will show that computing the maximum detour δ_i over all these pairs can be done in $O(|f| \log^2 |f|)$ expected time. From this, it will follow that phase 2 takes $O(|f| k \log^2 |f|)$ expected time.

Before showing how to compute δ_i, we need the idea of a canonical decomposition of f_B, defined next.

Canonical Decomposition. Define $b_1, \ldots, b_m$ as the interval of vertices of f_B ordered according to $<_a^f$ for some arbitrary vertex $a \in f_A$. Consider splitting this interval at vertex $b_j = b_{\lceil m/2 \rceil}$ and repeat this process recursively on the two sub-intervals, stopping when an interval containing only two vertices is reached. This gives us $O(m) = O(|f|)$ intervals of total size $O(|f| \log |f|)$ which we refer to as *canonical intervals*. The subgraphs of f_B induced by these canonical intervals are referred to as *canonical subgraphs*. The set of these subgraphs, which we denote by $\mathcal{C}$, can be found in $O(|f| \log |f|)$ time.

Every sub-interval of $b_1, \ldots, b_m$ can be decomposed into $O(\log |f|)$ canonical intervals in $O(\log |f|)$ time. This is easily seen by applying a greedy algorithm that picks canonical intervals as large as possible. We refer to such a decomposition as a *canonical decomposition*.

Let e be an edge of f_A. By the assumption earlier that type 2-intervals end in vertices and by Theorem 1, the union of all type 2-intervals of e of colour c_i is exactly the set of points of f_B between two vertices of $b_1, \ldots, b_m$. We compute a canonical decomposition of the sub-interval between these two vertices and add a pointer to e from each canonical subgraph corresponding to canonical intervals in this decomposition. This is done for all e in f_A.

When finished we have a total of $O(|f| \log |f|)$ pointers from canonical subgraphs in $\mathcal{C}$ to edges of f_A. Note that some canonical subgraphs may contain pointers to several edges. The total time spent on constructing $\mathcal{C}$ and on finding pointers is $O(|f| \log |f|)$.

Observe that δ_i is the maximum of $\delta_G(e, C)$ over all pairs consisting of an edge $e \in f_A$ and a canonical subgraph $C \in \mathcal{C}$ with a pointer to e. We now show how to compute this maximum in $O(|f| \log^2 |f|)$ expected time.

Sweep-plane Algorithm. To efficiently compute the maximum over pairs of edges and canonical subgraphs, we will use the idea of lifting and lowering points followed by a sweep-plane algorithm as described in [7]. In order to do this, we consider the following decision problem below: given $\delta \in \mathbb{R}$, is $\delta_i \geq \delta$? If we can answer this quickly we can compute δ_i in low expected time using a randomized algorithm by Chan [3] as described in [7].

For each canonical subgraph $C \in \mathcal{C}$, we lift each point $p \in C$ to height $d_G(p_i, p)$. And for each edge $e \in f_A$, we lower each point $p \in e$ to height $-d_G(p, p_i)$. Since we have precomputed SSSP lengths for portal p_i, this lifting/lowering can be done in $O(|f| \log |f|)$ time since the total size of all canonical subgraphs and edges is $O(|f| \log |f|)$.

Let e be a vertex in f_A and let C be a canonical subgraph with a pointer to e. Then it is clear that the height difference between a point $p \in e$ and a point $q \in C$ equals $d_G(p, q)$.

For each lifted and lowered point p, we associate a cone extending downwards from p and spanning an angle of $\alpha = 2 \arctan(1/\delta)$. Then as shown in [7], $\delta_i \geq \delta$ if and only if a cone of a lowered point of some edge is contained in a cone of a lifted point of some canonical subgraph with a pointer to that edge.

Now, we sweep a plane over the cones. The sweep-plane is parallel to the x-axis and forms an angle of $(\pi - \alpha)/2$ with the xy-plane. During the sweep, we maintain, for each canonical subgraph C, the intersection between the sweep-plane and the upper envelope of lifted points of C together with lowered points in edges that C points to. If it is detected that a cone of a lowered point is contained in a cone of a lifted point, the algorithm reports that $\delta_i \geq \delta$ and if no such event occurs, the algorithm reports that $\delta_i < \delta$.

It follows from the results of [7] that maintaining intersections between the sweep-plane and upper envelopes takes a total of $O(|f| \log^2 |f|)$ time since the number of sweep-plane event points is $O(|f| \log |f|)$ and each event point takes $O(\log |f|)$ time to handle. However, this is under the assumption that no cone of a lifted resp. lowered point is contained in the interior of another cone of a lifted resp. lowered point (see [7] for details). It can be shown that this assumption is satisfied if we only consider values $\delta \geq \max\{\delta_G(G_A), \delta_G(G_B)\}$.

So by recursively computing $\delta_G(G_B)$ and $\delta_G(G_A)$ before computing $\delta_G(G_A, G_B)$ it follows that the above decision problem can be solved in $O(|f| \log^2 |f|)$ time and Chan's algorithm gives us the following result.

Theorem 4. *Phase 2 runs in $O(|f| k \log^2 |f|)$ expected time.*

We have shown that $\delta_G(G_A, G_B)$ can be computed in $O(n^{3/2} \log^2 n)$ expected time. Due to space constraints, we leave the details for recursively computing $\delta_G(G_A)$ and $\delta_G(G_B)$. But it can be shown that the time spent in a level of the recursion tree is $O(n^{3/2} \log^2 n)$ and we get the main result of our paper.

Theorem 5. *The maximum detour of a plane graph with n vertices can be computed in $O(n^{3/2} \log^3 n)$ expected time.*

The second main result of this paper shows that if the graph has bounded *treewidth* (see definition in [2]), faster running time can be obtained.

Theorem 6. *The maximum detour of a plane graph with n vertices and bounded treewidth can be computed in $O(n \log^3 n)$ expected time.*

The proof of Theorem 6 follows easily from our results above and from [2].

6   Concluding Remarks

In this paper, we showed how to compute the maximum detour of a plane graph in $O(n^{3/2} \log^3 n)$ expected time. This is an improvement over the best known algorithm with $\Theta(n^2)$ running time. We also showed that if the graph has bounded treewidth, its maximum detour can be computed in $O(n \log^3 n)$ expected time.

We believe that by using parametric search as described in [1], we can obtain an algorithm computing the maximum detour of a plane graph in $O(n^{3/2} \operatorname{polylog} n)$ *worst-case* time and in $O(n \operatorname{polylog} n)$ *worst-case* time when the graph has bounded treewidth.

It would be interesting to try to beat the quadratic time bound also for the problem of computing the stretch factor of a plane geometric graph. This problem appears harder since pairs of vertices achieving the maximum detour need not be co-visible.

References

1. Agarwal, P.K., Klein, R., Knauer, C., Langerman, S., Morin, P., Sharir, M., Soss, M.: Computing the Detour and Spanning Ratio of Paths, Trees and Cycles in 2D and 3D. Discrete and Computational Geometry 39(1), 17–37 (2008)
2. Caballo, S., Knauer, C.: Algorithms for Graphs With Bounded Treewidth Via Orthogonal Range Searching, Berlin (manuscript, 2007)
3. Chan, T.M.: Geometric applications of a randomized optimization technique. Discrete Comput. Geom. 22(4), 547–567 (1999)
4. Ebbers-Baumann, A., Klein, R., Langetepe, E., Lingas, A.: A Fast Algorithm for Approximating the Detour of a Polygonal Chain. Comput. Geom. Theory Appl. 27, 123–134 (2004)
5. Eppstein, D.: Spanning trees and spanners. In: Sack, J.-R., Urrutia, J. (eds.) Handbook of Computational Geometry, pp. 425–461. Elsevier Science Publishers, Amsterdam (2000)
6. Frederickson, G.N.: Fast algorithms for shortest paths in planar graphs, with applications. SIAM J. Comput. 16, 1004–1022 (1987)
7. Langerman, S., Morin, P., Soss, M.: Computing the Maximum Detour and Spanning Ratio of Planar Paths, Trees and Cycles. In: Alt, H., Ferreira, A. (eds.) STACS 2002. LNCS, vol. 2285, pp. 250–261. Springer, Heidelberg (2002)
8. Lipton, R.J., Tarjan, R.E.: A Separator Theorem for Planar Graphs. STAN-CS-77-627 (October 1977)
9. Narasimhan, G., Smid, M.: Geometric Spanner Networks. Cambridge University Press, Cambridge (2007)
10. Smid, M.: Closest point problems in computational geometry. In: Sack, J.-R., Urrutia, J. (eds.) Handbook of Computational Geometry, pp. 877–935. Elsevier Science Publishers, Amsterdam (2000)

Finding Long Paths, Cycles and Circuits

Harold N. Gabow and Shuxin Nie

Department of Computer Science, University of Colorado, Boulder, CO 80309-0430
{hal,nies}@cs.colorado.edu

Abstract. We present a polynomial-time algorithm to find a cycle of length $\exp(\Omega(\sqrt{\log \ell}))$ in an undirected graph having a cycle of length $\geq \ell$. This is a slight improvement over previously known bounds. In addition the algorithm is more general, since it can similarly approximate the longest circuit, as well as the longest S-circuit (i.e., for S an arbitrary subset of vertices, a circuit that can visit each vertex in S at most once). We also show that any algorithm for approximating the longest cycle can approximate the longest circuit, with a square root reduction in length. For digraphs, we show that the long cycle and long circuit problems have the same approximation ratio up to a constant factor. We also give an algorithm to find a vw-path of length $\geq \log n / \log \log n$ if one exists; this is within a $\log \log n$ factor of a hardness result.

1   Introduction

Finding a long path or cycle is a basic problem in graph theory – it obviously includes the Hamiltonian path and cycle problems. There is a growing body of work on approximation algorithms for these problems. But for undirected graphs there remains a huge gap between known performance guarantees and known hardness results. This paper improves the performance guarantee for undirected graphs, generalizes it to the edge-simple case of finding long trails and circuits, and presents an algorithm for long directed vw-paths. We now survey previous work and then summarize our results. Throughout this paper, n and m denote the number of vertices and edges of the given graph, respectively; ℓ denotes the length of a longest cycle or circuit, depending on context.

Previous Work. Early work by Monien [18] showed that finding paths and cycles of length exactly k is fixed parameter tractable, as is finding an undirected cycle of length $\geq k$. Monien's work implies that a path of length $\log n / \log \log n$ can be found in polynomial time. See also [2,8].

Alon, Yuster and Zwick [1] introduced the technique of color coding, which finds a path of length $\log n$ in polynomial time if one exists. For digraphs, this result is still the best known. See also [11,16].

Progress for undirected graphs was achieved by Vishwanathan [20] and also Björklund and Husfeldt [3], the latter showing how to find an undirected path of length $\Omega(\log^2 \ell / \log \log \ell)$, for ℓ the length of a longest path (see also [13]). Gabow [12] extended Björklund and Husfeldt's approach to find undirected paths and cycles of length $\exp(\Omega(\sqrt{\log \ell / \log \log \ell}))$.

S.-H. Hong, H. Nagamochi, and T. Fukunaga (Eds.): ISAAC 2008, LNCS 5369, pp. 752–763, 2008.

Much better results are known for special classes of undirected graphs [6,10]. For instance Feder and Motwani [9] find a cycle of length $\exp(\Omega(\log n/\log\log n))$ in Hamiltonian graphs. This extends to graphs that are close to Hamiltonian.

We turn to digraphs. Gabow and Nie [13] find a directed cycle of length $\Omega(\log n/\log\log n)$ in polynomial time. The best known hardness results are by Björklund, Husfeldt and Khanna [4]: Assuming $P \neq NP$, for any $\epsilon > 0$, no polynomial-time algorithm can find a path or cycle of length $\geq n^\epsilon$ in a Hamiltonian digraph. But assuming the Exponential Time Hypothesis [15], the lower bound improves to essentially $\Omega(\log^2 n)$ for paths and $\Omega(\log n)$ for cycles. So for instance the results of [13] are within a $\log\log n$ factor of optimal under this hypothesis.

Hardness results for undirected graphs are much weaker. Karger, Motwani and Ramkumar [16] showed that getting any constant factor approximation to the longest undirected path is NP-hard. Furthermore for any $\epsilon > 0$, approximating to within a factor $\exp(O(\log^{1-\epsilon} n))$ is quasi-NP-hard. These two results extend to approximating the longest path in cubic Hamiltonian graphs, as shown by Bazgan, Santha and Tuza [5].

Our Contribution. Our main concern is undirected graphs. We refine the approach of [12] to find long cycles. Our algorithm is simpler and more practical (we give objective facts supporting this claim). We improve the performance guarantee for finding paths or cycles from $\exp(\Omega(\sqrt{\log \ell/\log\log \ell}))$ to $\exp(\Omega(\sqrt{\log \ell}))$. In addition our algorithm is more general: We allow the requirement of vertex-simplicity to be discarded at an arbitrarily chosen subset of vertices. More precisely let S be an arbitrary set of vertices. An *S-trail* is a trail that visits each vertex of S at most once. (So an S-trail is an edge-simple path, but it needn't be vertex-simple unless S contains every vertex.) An *S-circuit* is defined similarly. For example taking S empty shows that our algorithm can approximate the longest trail or circuit.

To the best of our knowledge the long trail and circuit problems have not been investigated seriously. Of course the problem can be solved exactly if some circuit contains every edge, i.e, the graph is Eulerian. But the general problem is more difficult, and again we achieve approximation ratio $\exp(\Omega(\sqrt{\log \ell}))$.

Regarding hardness observe that in graphs with maximum degree 3 a cycle (path) is the same as a circuit (trail). Hence the previously mentioned results of Bazgan et.al. show that the long circuit/trail problem is hard. Thus like 2-SAT, the long trail/circuit problem is easy when there is a perfect solution but hard in general.

We also give transformations between the long circuit and long cycle problems. For digraphs it is easy to see the problems are essentially equivalent. For undirected graphs we give an efficient reduction from circuits to cycles. More precisely, for any function $\mathrm{CYC}(\cdot)$, an algorithm that finds a cycle of length $\geq \mathrm{CYC}(\ell)$ on graphs having a cycle of length $\geq \ell$ yields an algorithm that finds a circuit of length $\geq \sqrt{\mathrm{CYC}(\ell)}$ on graphs having a circuit of length $\geq \ell$. However our reduction is not valid for S-circuits that have $S \neq \emptyset, V$.

Our last result concerns digraphs. It generalizes the algorithm of [13] for finding a long cycle: We show for any two given vertices v, w, a vw-path of length $\geq \log n / \log \log n$ can be found in polynomial time, if one exists[1]. The main structural tool of [13], the Gap Theorem, fails for vw-paths. We achieve our results using algorithmic techniques.

This section concludes by reviewing our terminology. Section 2 presents the transformations between circuit and cycle problems. Section 3 sketches our algorithm for undirected S-circuits (the detailed algorithm is given in the complete version of this paper). Section 4 gives the algorithm for directed vw-paths.

Terminology. Our graph terminology is consistent with [21] whenever possible. An *element* of a graph is a vertex or an edge. All graphs are assumed simple. If H is a subgraph of G and S is a set of elements of G then $S(H)$ denotes the set of elements of H that belong to S. (For instance if $G = (V, E)$, the notation $V(H)$ and $E(H)$ have their usual meaning.) For X and Y disjoint vertex sets, $E[X, Y]$ consists of all edges joining X and Y. Furthermore by convention writing $xy \in E[X, Y]$ means $x \in X$ and $y \in Y$.

A *path* has no repeated vertices. A *trail* has no repeated edges. A *circuit* is a closed trail. If T is a trail we use T to denote a set of vertices or edges, as is convenient; $|T|$ always denotes the number of edges in T. If e is an element of a graph an *e-trail* is a trail that includes e. Note that an algorithm that finds a long e-circuit, for a given edge e, applies to all other variants of the problem – e.g., it can find a long trail, circuit, vw-trail, v-trail or v-circuit, all with the same length guarantee and time increased by only a polynomial factor.

For a trail T containing elements x,y with x preceding y, $T[x, y]$ denotes the subtrail from x to y. For a circuit C containing elements x, y and e, $C_e[x, y]$ denotes the subtrail of C from x to y that contains e. We extend this notation to allow open-ended intervals, e.g., $T(x, y]$ excludes x.

2   Long Cycles Versus Long Circuits

This section relates the approximability of the long cycle and long circuit problems: These problems, and even the S-circuit problem, are essentially the same in digraphs. In undirected graphs we show that an algorithm for long cycles can be used to find long circuits, with only moderate degradation of performance guarantee. But our reduction does not generalize to S-circuits.

Let $\mathrm{CYC}(\cdot)$ denote a function such that some polynomial-time algorithm finds a cycle of length $\geq \mathrm{CYC}(\ell)$ on any graph that has a cycle of length $\geq \ell$. Let $\mathrm{CIRC}(\cdot)$ denote the analogous function for circuits.

Consider undirected graphs. Throughout this discussion Δ denotes the maximum degree. Given a graph $G = (V, E)$ in which we seek a long circuit, construct a new graph H by replacing every vertex by a clique. More precisely replace each edge $e = (v, w) \in E$ by an edge (v_e, w_e) that joins two new vertices v_e, w_e, and

---

[1] Going from long cycles to long vw-paths is easy in undirected graphs, by Menger's Theorem. But it is nontrivial in digraphs, because the analogous tool is the 2-Vertex-Disjoint Directed Paths Problem which is NP-hard [4].

for each vertex $v \in V$ make all vertices v_e adjacent. The relation between circuits in G and cycles in H is simple:

Lemma 1. *A circuit of length ℓ in G gives a cycle of length 2ℓ in H. A cycle of length $\ell > \Delta$ in H gives a circuit of length $\geq \ell/\Delta$ in G.*

The lemma enables us to find a circuit of length $\geq \text{CYC}(2\ell)/\Delta$ if $2\ell > \Delta$. We supplement this with a good method for finding circuits when Δ is large. Recall [19] that in a connected graph $G = (V, E)$ with $T \subseteq V$, a *T-join* is a subgraph that has T equal to the set of all vertices of odd degree. A T-join exists precisely when $|T|$ is even. A T-join can be found in linear time by a bottom-up traversal of any spanning tree of G.

Let G be a bridgeless graph, with x a vertex of degree Δ.

Lemma 2. *An x-circuit of length $\geq \Delta$ can be found in linear time.*

Proof. Form the graph $G - x$. For each connected component C, apply the following procedure. Let N be the set of all neighbors of x that belong to C.

Suppose $|N|$ is odd. Take any $v \in N$. Form an $(N - v)$-join, and enlarge it to an x-circuit by adding all the edges from x to $N - v$. Note that $|N| \geq 3$ since G is bridgeless. So the x-circuit is nontrivial and contains $> |N| - 1$ edges. The case $|N|$ even is similar.

Since $2\ell > \ell \geq \Delta$ and $\max\{\text{CYC}(2\ell)/\Delta, \Delta\} \geq \sqrt{\text{CYC}(2\ell)}$ we get the following:

Theorem 1. $CIRC(\ell) \geq \sqrt{CYC(2\ell)}$.

Our approach does not appear to extend to S-circuits since Lemma 2 can fail. For example take a graph consisting of two vertices v, w joined by many paths of length two. $\Delta = n - 2$ yet for any set S containing v but not w, a longest S-circuit has only four edges.

For digraphs our functions are essentially identical, even for S-circuits and cycles: Transformations similar to Lemma 1 show $\text{CYC}(2\ell)/2 \leq \text{CIRC}(\ell) \leq 2\text{CYC}(\lceil \ell/2 \rceil)$.

3   Algorithm for Undirected Graphs

For any $S \subseteq V$ and $e \in E$, an *(S, e)-circuit* is both an S-circuit and an e-circuit. This section presents the approximation algorithm for long (S, e)-circuits on undirected graphs. We start with a high-level overview. To simplify terminology we describe the case of finding a long e-cycle in a given graph $G = (V, E)$ (i.e., we assume $S = V$) and we interject modifications needed for (S, e)-circuits.

As in [12] our algorithm is organized as a family of recursive algorithms $\mathcal{A}_p$, where $\mathcal{A}_p$ invokes both itself and $\mathcal{A}_{p-1}$. An execution is always working on a current e-cycle C and attempting to enlarge it by means of recursive invocations. To improve the approximation ratio of [12], our goal is to increase the length of C by a multiplicative factor rather than by an additive constant as in [12].

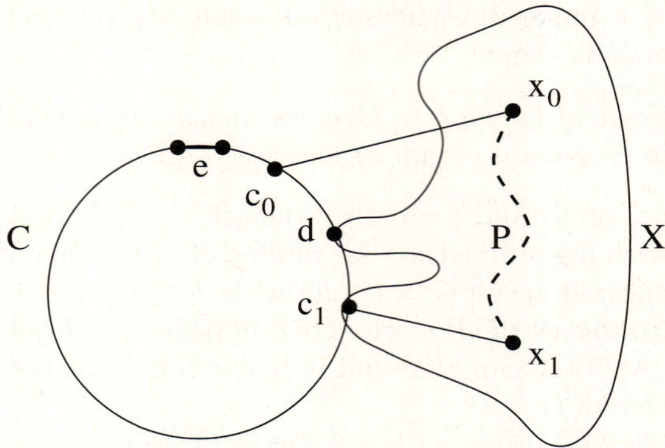

Fig. 1. A circuit C and connected component X of $G_{C,S}$. c_1 and d belong to $\overline{S}$. The Extend Step combines the recursively found trail P with $C_e[c_0, c_1]$.

In more detail $\mathcal{A}_p$ will return an e-cycle of length $\geq (3/2)^p$ if G contains a sufficiently long e-cycle, say a cycle C^* of length $\geq L_p$, for L_p a parameter of the algorithm. $\mathcal{A}_p$ begins by invoking $\mathcal{A}_{p-1}$ to find an e-cycle C of length $\geq (3/2)^{p-1}$. Assume $|C| < (3/2)^p$ since otherwise C is the desired cycle.

$C^* - V(C)$ consists of $\leq |C|$ segments, so at least one of those segments has length $\geq L_p/(3/2)^p$. We will require this quantity to be $\geq L_{p-1}$, i.e., we define

$$L_p = (3/2)^p L_{p-1}. \tag{1}$$

Each segment of $C^* - V(C)$ is contained in a connected component of $G - V(C)$. The algorithm invokes $\mathcal{A}_{p-1}$ on each connected component X of $G - V(C)$ to find a long path P. (Strictly speaking $\mathcal{A}_{p-1}$ finds P by finding an f-cycle in X, for an appropriate dummy edge f.) Condition (1) with the corresponding long segment of $C^* - V(C)$ guarantees that some invocation finds a path P of length $\geq (3/2)^{p-1}$. We will combine all of P with at least half of C to get an e-cycle of length $\geq (3/2)(3/2)^{p-1} = (3/2)^p$ as promised.

(1) implies that $L_p = (3/2)^p (3/2)^{p-1} \cdots (3/2) = (3/2)^{\Theta(p^2)}$. So if $|C^*| = \ell$, taking $p = \Theta(\sqrt{\log \ell})$ shows that $\mathcal{A}_p$ finds a cycle of length $\exp(\Omega(\sqrt{\log \ell}))$ as desired.

How do we combine C and P? We use the *Extend Step* illustrated in Fig.1, where the combined cycle consists of P, $C_e[c_0, c_1]$, and edges $c_0 x_0$, $c_1 x_1$. (In Fig.1 X contains vertices of C, specifically c_1 and d – this can occur when we are looking for a circuit rather than a cycle.) If

$$|C_e[c_0, c_1]| \geq |C|/2 \tag{2}$$

then the combined cycle has the desired length.

Suppose (2) fails. A simple example is when X has only two neighbors c_0, c_1 in C, and both neighbors are close to e but on opposite sides of it. Note that in this case c_0 and c_1 separate the long segment of C^* from e. Surprisingly [12] shows that such a separating pair always exists whenever the Extend Step fails.

We prove this holds for S-circuits as well, if we allow each c_i to be an element of G. More precisely, if a component X contains a long segment of C^* and the Extend Step does not enlarge C as desired then there is a separation pair r_0, r_1 such that some connected component T of $G - \{r_0, r_1\}$ excludes e and contains a long segment of C^* that joins elements r_0 and r_1.

Given this theorem, the rest of our algorithm is similar to [12]: The algorithm finds every possible candidate for the above separator r_0, r_1. It calls itself recursively on the corresponding component T of $G - \{r_0, r_1\}$ to find a long $r_0 r_1$-path Q. (This is efficient since the various components T are disjoint.) Then the desired e-cycle is constructed according to one of two cases:

The first case is when C^* has long segments passing through two of the above components T, say T_i, $i = 1, 2$. The recursive call on T_i will find a path Q_i in T_i of length $\geq (3/2)^{p-1}$, for $i = 1, 2$. G has an e-cycle that contains both Q_1 and Q_2. (To see this simply replace the portions of C^* in T_i by Q_i.) Our algorithm can find such an e-cycle, for two reasons: First, although the algorithm doesn't know which of its components T are actually T_1 and T_2, it can try all possibilities. Second, we find the desired e-cycle containing Q_1, Q_2 and e by using the algorithm of [17] that finds a cycle through 3 given vertices. (This algorithm easily generalizes from cycles to S-circuits.) Note that in this first case $\mathcal{A}_p$ returns a cycle of length $\geq 2(3/2)^{p-1}$, even better than what we require.

The second case is when C^* has only one long segment passing through a component T. There appears to be no useful structure in this case. The algorithm proceeds by brute force: Each recursive call[2] adds ≥ 1 edge to the cycle that will eventually be returned.. So roughly speaking we may need $(3/2)^p$ recursive calls to find the desired e-cycle. In other words the time for our algorithm obeys the recurrence (1) if we take L_p to be the time for $\mathcal{A}_p$. Since S-circuits have $\ell \leq m$, our overall algorithm can choose $p = \Theta(\sqrt{\log m})$. So the total time is at most $(3/2)^{\Theta(p^2)} = (3/2)^{\Theta(\log m)} = m^{\Theta(1)}$ as desired.

This concludes the overview. Now we return to the general case of finding (S, e)-circuits. The next two subsections present the crucial ideas and supporting lemmas for our algorithm. A detailed statement of the algorithm and its analysis is deferred to the complete paper.

3.1   The Extend Step

For the given graph $G = (V, E)$, define $\overline{S} = V - S$ and $ES = E \cup S$. After constructing a circuit C, we are interested in the graph

$$G_{C,S} = G - ES(C).$$

The Extend Step enlarges C by combining it with a recursively found path in a connected component of $G_{C,S}$ (recall Fig.1). This subsection culminates in a more detailed description of this step. We start with the concept of ES-block.

---

[2] In order to program these recursive calls correctly, algorithm $\mathcal{A}_p$ actually calls $\mathcal{A}_p$, not $\mathcal{A}_{p-1}$, on each component T. This does not apply to the Extend Step, which as we said calls $\mathcal{A}_{p-1}$ on each connected component X. But this entire issue can be safely ignored in this overview.

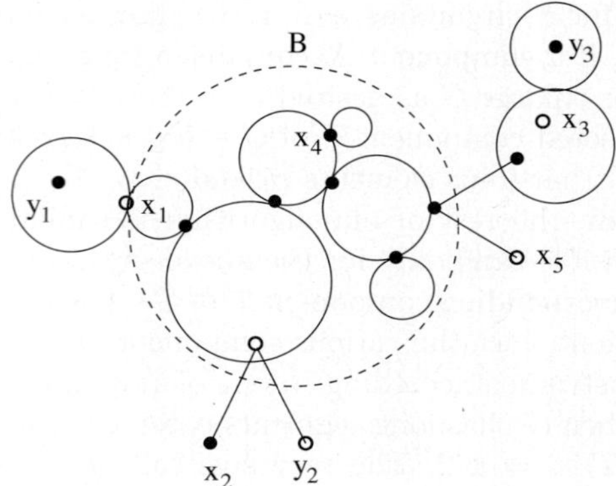

Fig. 2. An ES-block B, drawn dashed. Each solid circle represents a biconnected component that is not a bridge. S vertices are drawn hollow, $\overline{S}$ solid. Any two labelled vertices are $E\overline{S}$-separated by B except for the pairs x_i, y_i, $i = 1, 2, 3$.

An ES-*separator* is a set of one or more elements of ES whose removal disconnects the graph. Recall that a *block* of a graph is a biconnected component [21]. Analogously we define an ES-*block* to be a maximal subgraph that has no singleton ES-separator (see Fig.2). Thus an ES-block B is a union of (ordinary) blocks of G that is connected, and each of its cutpoints belongs to $\overline{S}$. Furthermore for any such cutpoint x, B contains every block of G that contains x, except for bridges of G.

Clearly an ES-block is bridgeless, i.e., it is 2-edge connected. More precisely Menger's Theorem shows that any two vertices v, w in the same ES-block are joined by two paths that do not share an edge or a vertex of $S - \{v, w\}$. Finally it is easy to see that any S-circuit is entirely contained within one ES-block. This explains our interest in ES-blocks.

ES-blocks can be used to enlarge trails and circuits. This statement will be made precise in the next two lemmas, which involve two more concepts. The first is another variant of separation: A subgraph B is a *(weak)* $E\overline{S}$-*separator* for vertices x and y if every xy-path contains an element of $E\overline{S}(B)$ (this notation is defined in Section 1). Fig.2 illustrates this concept. Note two special cases of the definition: B is an $E\overline{S}$-separator for any two of its vertices. Also if $x \in \overline{S}(B)$, B is a weak $E\overline{S}$-separator for x and any other vertex, including itself. (Such self-separation is important.) We usually drop the modifier "weak" for convenience, but we keep it when it is significant, as in the case of self-separation. Here are some useful properties of this notion; the straightforward proofs are omitted.

Lemma 3. *Let B be an ES-block, and let x, y be two arbitrary distinct vertices of G. Then the following conditions are equivalent.*

(i) B is an $E\overline{S}$-separator for x and y.

(ii) Some xy-path contains an element of $E\overline{S}(B)$.

(iii) Some xy-path contains an edge of B or a bridge of G that is incident to a vertex of $\overline{S}(B)$.

(iv) Some xy S-trail contains an element of $E\overline{S}(B)$.
(v) Every xy S-trail contains an element of $E\overline{S}(B)$.

For any real value $b > 2$, an ES-block is *b-round* if it contains an S-circuit of $\geq b$ edges. The following lemma shows this notion captures the situation of when an ES-block can be used to enlarge a trail. The lemma and its proof are similar to Lemma 2.4 of [12]. For two vertices x, y, let $d(x, y)$ $(D^S(x, y))$ denote the length of a shortest xy-path (longest xy S-trail), respectively.

Lemma 4. *Consider a connected graph and two distinct vertices x, y.*

(i) If x and y are $E\overline{S}$-separated by a b-round ES-block then $D^S(x, y) \geq b/2$. This holds even if $x = y$ (which implies $x \in \overline{S}$).
(ii) If $b = (D^S(x, y) - d(x, y))/(d(x, y) + 1) > 2$ then x and y are $E\overline{S}$-separated by a b-round ES-block.

Proof.

(i) We just discuss the key case. Let B be an ES-block containing an S-circuit C of length $\geq b$. We prove the lemma assuming that x and y are vertices of B (we allow $x = y \in \overline{S}$).

It is easy to see that the previously mentioned Menger's Theorem for B implies B contains two paths, P_x going from x to C and P_y going from y to C, that do not share an edge or a vertex of $S - \{x, y\}$. Let P_x and P_y end at vertices r and s on C. Let $C[r, s]$ be a subtrail of C having length $\geq b/2$. We get the desired xy S-trail by combining P_x, P_y and $C[r, s]$.

(ii) Let T (P) be a longest (shortest) xy S-trail. The proof analyzes $P \oplus T$, which is a collection of circuits. The bound of (ii) is tight, e.g., form T from P by attaching a long circuit of c edges at each vertex.

An S-circuit is *maximal* if it cannot be enlarged by adding a cycle, i.e., no cycle of $G_{C,S}$ contains a vertex of $\overline{S}(C)$. The algorithm always makes its current circuit C maximal.

We conclude this section with a more precise statement of the Extend Step. For this statement, recall the notation $E[\cdot, \cdot]$ from Section 1. Define

$$b = 2L_{p-1}.$$

Consider a maximal S-circuit C and a connected component X of $G_{C,S}$. Suppose there are two edges $c_i x_i \in E[C, X - C]$, $i = 0, 1$ such that c_0 and c_1 satisfy condition (2) and X contains a b-round ES-block B that is a weak $E\overline{S}$-separator for c_0 and c_1 in the graph $\overline{X} = X \cup \{c_0 x_0, c_1 x_1\}$.[3] The Extend Step invokes $\mathcal{A}_{p-1}$ to find an $(S, c_0 c_1)$-circuit in the graph $\overline{X} \cup \{c_0 c_1\}$. Then it enlarges the current circuit C, as illustrated in Fig.1.

---

[3] This long hypothesis subsumes two different cases: If a c_i is in $\overline{S}$ then $c_i x_i$ is already an edge of X, but if $c_i \in S$ we add edge $c_i x_i$ to X. In that case $c_i x_i$ is a bridge of $\overline{X}$. ($E\overline{S}$-separation and $c_i \in S$ implies $c_0 \neq c_1$.) For that reason B is an ES-block of $\overline{X}$ iff it is an ES-block of X. In particular this allows us to apply Lemma 4(i) in the next paragraph.

Note that the hypothesis of the Extend Step makes Lemma 4(i) applicable. The lemma implies $D^S(c_0, c_1) \geq L_{p-1}$. Hence the call to $\mathcal{A}_{p-1}$ returns a circuit of the length $\geq (3/2)^{p-1}$ as desired (recall the overview).

It may not be clear how to implement this Extend Step efficiently – the algorithm must decide whether or not an ES-block B is b-round, and this seems to be a circular task. However the algorithm can simply try every ES-block B. If B is in fact b-round, $\mathcal{A}_{p-1}$ is guaranteed to find the desired circuit.

Note that the Extend Step is guaranteed to find its desired circuit of length $\geq (3/2)^{p-1}$ whenever X contains an appropriately long segment of the longest (S, e)-circuit C^*. More precisely assume the following setting:

($*$) X is a connected component of $G_{C,S}$. C^* contains a subtrail $C^*[c_0, c_1] = c_0, x_0, \ldots, x_1, c_1$ with $c_0 x_0, c_1 x_1 \in E[C, X - C]$, $C^*(c_0, c_1) \subseteq X$, and $|C^*[c_0, c_1]| \geq L$ for some lower bound value L.

The graph $\overline{X} = X \cup \{c_0 x_0, c_1 x_1\}$ has $D^S(c_0, c_1) \geq L$. So an appropriately large value for L ensures the hypothesis of Lemma 4(ii) is satisfied for $x, y = c_0, c_1$. (Note $c_0 \neq c_1$: Otherwise $c_0 \in \overline{S}$ since $e \in C^*$. But then the circuit $C^*[c_0, c_1]$ contradicts maximality of C.) This lemma implies c_0 and c_1 are $E\overline{S}$-separated (in $\overline{X}$) by a b-round ES-block B of X. B is the ES-block needed by the Extend Step. So if c_0 and c_1 satisfy condition (2) all hypotheses of the Extend Step are satisfied, and it finds a circuit of the desired length.

3.2   Long Circuits Produce Separating Pairs

To show how the algorithm finds the separating pairs r_0, r_1 of the overview we first define the notion of b-closeness. As in [12] it models the situation when the Extend Step fails.

Definition 1. *Consider a maximal S-circuit C and a connected component X of $G_{C,S}$. Let $b > 2$ be a real value. A subset F of $E[C, X - C]$ is b-close if for every pair of its edges, $c_i x_i \in F$, $i = 0, 1$ no b-round ES-block of X is a weak $E\overline{S}$-separator for c_0 and c_1 in the graph $X \cup \{c_0 x_0, c_1 x_1\}$.*

Let us summarize the situation when the Extend Step fails. Specifically assume this setting: ($*$) holds. So as in the previous subsection c_0 and c_1 are $E\overline{S}$-separated (in $\overline{X} = X \cup \{c_0 x_0, c_1 x_1\}$) by a b-round ES-block B of X. Assume the Extend Step for X fails, i.e., no choice of edges $c_i x_i$ satisfies the hypothesis of the Extend Step.

First observe that $E[C, X - C]$ can be partitioned into two nonempty subsets F_i, $i = 0, 1$, with each set F_i being b-close. Here's why: Clearly the two vertices c_0 and c_1 of X do not satisfy condition (2). So $C_e[c_0, c_1]$ is a short trail, with c_0 on one side of e and c_1 on the opposite side. For $i = 0, 1$ let F_i be the subset of edges $cx \in E[C, X - C]$ with c on the same side of e as c_i. Any two edges in the same set F_i satisfy condition (2). So each set F_i is b-close – if it weren't, the hypothesis of the Extend Step would hold.

We now show how, as promised in the overview, we find a pair r_0, r_1 that separates a long subtrail of C^* from e. In fact r_0, r_1 will separate B from e. (The portion of C^* in B is easily seen to be long by Lemma 4(i).) Also r_0, r_1 will be an ES-separating pair, i.e., $r_0, r_1 \in ES$. This is needed to properly handle the structureless case, discussed in the last paragraph of the overview. The r_i come from the following lemma.

Lemma 5. *Suppose F_i is b-close. There is an element $r_i \in ES$ such that for every edge $cx \in F_i$, every path from c to B in $X \cup \{cx\}$ contains r_i.*

Proof. First observe that we never have $c \in B$. This follows since $c \in X$ implies $c \in \overline{S}$. In that case the maximality of C shows $c \notin B$.

Take any edge $cx \in F_i$. B is an ES-block of $X \cup \{cx\}$. The definition of ES-block, with $c \notin B$, implies c is separated from B (in graph $X \cup \{cx\}$) by an element of ES. As illustrated in Fig.2, the element is either a cutpoint of $S(B)$ or a bridge with an end in $\overline{S}(B)$. Define $\rho(cx)$ to be that separating element. To prove the lemma we need only show that any two edges $c_j x_j \in F_i$, $j = 0, 1$ have the same value $\rho(c_j x_j)$. Suppose $\rho(c_0 x_0) \neq \rho(c_1 x_1)$. We will assume $c_0 \neq c_1$. (That case makes the case $c_0 = c_1$ straightforward, so we omit it.)

Let P_j be a path from c_j to $\rho(c_j x_j)$. P_0 and P_1 are edge-disjoint. (If either $\rho(c_j x_j)$ is a vertex of S then P_0 and P_1 are vertex-disjoint. In the opposite case, P_0 and P_1 end at two distinct bridges.) Join P_0 and P_1 by a path in B. The resulting $c_0 c_1$-path shows that in the graph $X \cup \{c_0 x_0, c_1 x_1\}$, c_0 and c_1 are $E\overline{S}$-separated by B. (Here we use $c_0 \neq c_1$ along with Lemma 3(ii).) But this contradicts the b-closeness of F_i.

Note that $V(B) \subseteq V(X - C)$, because C is maximal. Since $E[C, X - C] = F_0 \cup F_1$, any path P from e to B goes through an edge of F_i. The lemma shows P contains r_0 or r_1. Thus r_0, r_1 separates e from B and is the desired ES-separating pair.

These are the crucial ideas for the algorithm. Our algorithm improves the approximation ratio of [12] and is also simpler and more practical: [12] uses a brute-force base case for $\ell \leq 2^{256}$, while our algorithm uses the natural base case based on ordinary breadth-first search; we do not use color coding or the Gap Theorem of [13]; we use fewer numeric parameters (2 versus 6); our Extend Step does not divide into different cases, and furthermore, it is not iterated, so our algorithm makes fewer recursive calls.

4   Algorithm for Digraphs

Let G be a digraph with two given vertices v, w (possibly $v = w$) and a given integer k. A vw-path of length $\geq k$ is called *long*. We prove the following result:

Theorem 2. *A long directed vw-path can be found in time $k^{O(k)} nm \log n$ if one exists. In particular a directed vw-path of length $\geq \log n / \log \log n$ can be found in polynomial time if one exists.*

We present the algorithm as randomized and then give a trivial derandomization. Call the last k edges of the desired path the *tail*. The idea is to guess the vertices in the tail (not necessarily in correct order). The algorithm starts by guessing

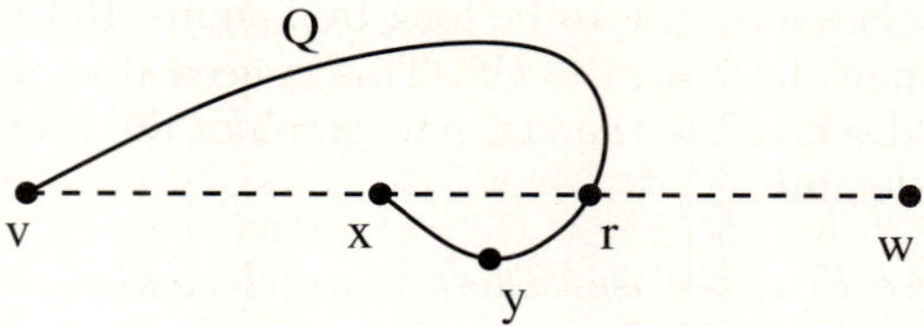

Fig. 3. The dashed path is a long vw-path with x the first vertex of the tail. Q is a vx-path found by the algorithm. If P doesn't exist, Q intersects the tail, as illustrated by vertex r. The algorithm attempts to guess r.

the first vertex of the tail. Each iteration guesses a new vertex in the tail. As long as the algorithm hasn't yet found the desired path, the probability of guessing correctly is high.

The algorithm is stated below and is illustrated in Fig.3. It uses color coding [1] to find an ab-path of length exactly k, for specified vertices a, b.

It is obvious that any path returned by the algorithm is a long vw-path. So we need only show that if the desired path exists, the probability of guessing correctly is high. Let L be an arbitrary long vw-path, and let x be the first vertex of its tail.

Lemma 6. *Consider an iteration that starts with T contained in the tail of L. If the iteration fails to find a long vw-path then it chooses r as a new vertex in the tail of L with probability $\geq 1/k$.*

Proof. The hypothesis implies the path Q exists. The algorithm defines y so that $|Q[v, y]| \leq k$. So the algorithm's choice of r shows we need only prove $Q(v, y)$ contains a vertex t that is in the tail of L but not in T. In fact since Q is chosen disjoint from $T - x$, it suffices to show t is in the tail of L and $t \neq x$.

We are assuming the path P does not exist. Since we search for P in $G - V(Q) + x$, it must be that Q contains a vertex $u \neq x$ in the tail of L. If $|Q| \leq k$ then we can take $t = u$. The case $|Q| > k$ is similar.

Applying the lemma with $|T| = k + 1$ shows in that case the iteration must find the desired path. So it is easy to see that the algorithm finds the desired path with probability $\geq 1/nk^k$. The algorithm is simple to derandomize: The sample space has $< nk^k$ points, so we try each one.

Since the derandomized version of color coding finds an ab-path of length k in time $2^{O(k)}m \log n$, the total time is $nk^k \times 2^{O(k)}m \log n = k^{O(k)}nm \log n$.

References

1. Alon, N., Yuster, R., Zwick, U.: Color-coding. J. ACM 42, 844–856 (1995)
2. Bodlaender, H.L.: Minor tests with depth-first search. J. Algorithms 14, 1–23 (1993)
3. Björklund, A., Husfeldt, T.: Finding a path of superlogarithmic length. SIAM J. Comput. 32, 1395–1402 (2003)

4. Björklund, A., Husfeldt, T., Khanna, S.: Approximating longest directed paths and cycles. In: D(i) az, J., Karhumäki, J., Lepistö, A., Sannella, D. (eds.) ICALP 2004. LNCS, vol. 3142, pp. 222–233. Springer, Heidelberg (2004)
5. Bazgan, C., Santha, M., Tuza, Z.: On the approximability of finding a(nother) Hamiltonian cycle. In: Proceedings of STACS 1998 (1998)
6. Chen, G., Xu, J., Yu, X.: Circumference of graphs with bounded degree. SIAM J. Comput. 33, 1136–1170 (2004)
7. Fellows, M.R., Langston, M.A.: Nonconstructive tools for proving polynomial-time decidability. J. ACM 35, 727–739 (1988)
8. Fellows, M.R., Langston, M.A.: On search, decision and he efficiency of polynomial-time decidability. In: Proceedings of the 21st Annual ACM Symposium on Theory of Computing, pp. 501–512 (1989)
9. Feder, T., Motwani, R.: Finding large cycles in Hamiltonian graphs. In: Proceedings of the 16th Annual ACM-SIAM Symposium on Discrete Algorithms, Vancouver, BC, pp. 166–175 (2005)
10. Feder, T., Motwani, R., Subi, C.: Approximating the longest cycle problem in sparse graphs. SIAM J. Comput. 31, 1596–1607 (2002)
11. Fürer, M., Raghavachari, B.: Approximating the minimum degree spanning tree to within one from the optimal degree. In: Proceedings of the 3rd Annual ACM-SIAM Symposium on Discrete Algorithms, pp. 317–324 (1992)
12. Gabow, H.N.: Finding paths and cycles of superpolylogarithmic length. SIAM J. Comput. 36(6), 1648–1671 (2007)
13. Gabow, H.N., Nie, S.: Finding a long directed cycle. ACM Trans. on Algorithms 4(1) Article 7, 21 pages (2008)
14. Hopcroft, J.E., Tarjan, R.E.: Dividing a graph into triconnected components. SIAM J. Comput. 2, 135–158 (1973)
15. Impagliazzo, R., Paturi, R.: On the complexity of k-SAT. J. Comp. Sys. Sci. 62(2), 367–375 (2001)
16. Karger, D., Motwani, R., Ramkumar, G.D.S.: On approximating the longest path in a graph. Algorithmica 18, 82–98 (1997)
17. LaPaugh, A.S., Rivest, R.L.: The subgraph homeomorphism problem. J. Comput. Sys. Sci. 20, 133–149 (1980)
18. Monien, B.: How to find long paths efficiently. Annals Disc. Math. 25, 239–254 (1985)
19. Schrijver, A.: Combinatorial Optimization: Polyhedra and Efficiency, vol. A. Springer, NY (2003)
20. Vishwanathan, S.: An approximation algorithm for finding a long path in Hamiltonian graphs. In: Proceedings of the 11th Annual ACM-SIAM Symposium on Discrete Algorithms, pp. 680–685 (2000)
21. West, D.B.: Introduction to Graph Theory, 2nd edn. Prentice Hall, Upper Saddle River (2001)

Computing Best and Worst Shortcuts of Graphs Embedded in Metric Spaces

Jun Luo[1,*] and Christian Wulff-Nilsen[2]

[1] Department of Information and Computing Sciences, Universiteit Utrecht,
Centrumgebouw Noord, office A210, Padualaan 14, De Uithof, 3584CH Utrecht,
The Netherlands
ljroger@cs.uu.nl
http://www.cs.uu.nl/staff/ljroger.html
[2] Department of Computer Science, University of Copenhagen,
Universitetsparken 1, DK-2100 Copenhagen O, Denmark
koolooz@diku.dk
http://www.diku.dk/hjemmesider/ansatte/koolooz/

Abstract. Given a graph embedded in a metric space, its dilation is the maximum over all distinct pairs of vertices of the ratio between their distance in the graph and the metric distance between them. Given such a graph G with n vertices and m edges and consisting of at most two connected components, we consider the problem of augmenting G with an edge such that the resulting graph has minimum dilation. We show that we can find such an edge in $O((n^4 \log n)/\sqrt{m})$ time using linear space which solves an open problem of whether a linear-space algorithm with $o(n^4)$ running time exists. We show that $O(n^2 \log n)$ time is achievable if G is a simple path or the union of two vertex-disjoint simple paths. Finally, we show how to find an edge that maximizes the dilation of the resulting graph in $O(n^3)$ time with $O(n^2)$ space and in $O(n^3 \log n)$ time with linear space.

1 Introduction

Given a set of cities, represented by points in the plane, consider the problem of finding a road network that interconnects these cities. We seek a network with low cost in which no large detours are required to get from any city to any other city, that is, the road distance between the two cities should be at most a constant factor larger than the Euclidean distance between them. Typical cost measures of the network include size, length, diameter, and maximum degree.

Spanners are networks in which the largest detour is small, and the problem of finding a low-cost spanner for a given point set has received a lot of attention in recent years [3,5,6].

Typically however, one is not interested in constructing a network from scratch but rather to modify a given network. Consider the following problem. Suppose

---

* This research has been partially funded by the Netherlands Organisation for Scientific Research (NWO) under FOCUS/BRICKS grant number 642.065.503.

S.-H. Hong, H. Nagamochi, and T. Fukunaga (Eds.): ISAAC 2008, LNCS 5369, pp. 764–775, 2008.

we are given a network and we would like to extend it with new edge connections to reduce the largest detour. A cost is involved in adding a connection and we can only afford to add, say, a constant $k \geq 1$ number of connections. The problem is to pick these connections such that the largest detour in the resulting network is minimized.

When $k = 1$, an optimal edge to add is called a best shortcut. Farshi et al. [4] considered the problem of finding a best shortcut in a graph embedded in an arbitrary metric space and showed how to solve it in $O(n^4)$ time with $O(n^2)$ space and in $O(n^3 m + n^4 \log n)$ time with linear space (linear in the size of the input graph), where n is the number of vertices and m is the number of edges of the given network. Various approximation algorithms with faster running times were also presented. The authors posed the following open problem: is there an (exact) algorithm with running time $o(n^4)$ using linear space?

An algorithm with $O(n^3 \log n)$ running time was presented in [7]. It has $O(n^2)$ space requirement however, leaving the problem in [4] open.

In this paper, we present a linear-space algorithm with $O((n^4 \log n)/\sqrt{m})$ running time. Since it may be assumed that the input graph consists of at most two connected components (otherwise, the problem is trivial), $m = \Omega(n)$. Hence, we solve the open problem in [4].

For more special types of graphs, we give faster algorithms. We show how to obtain $O(n^2 \log n)$ running time when the graph is a simple path or the union of two vertex-disjoint simple paths.

Finally, we consider the problem of finding a worst shortcut, i.e. an edge that *maximizes* the largest detour of the resulting graph. This relates to a problem in [1] where edge-deletions were considered. We show how to solve this for general graphs in $O(n^3)$ time with $O(n^2)$ space and in $O(n^3 \log n)$ time with linear space.

The organization of the paper is as follows. In Section 2, we give some definitions and introduce some notation. In Section 3, we present our algorithm for computing a best shortcut in a graph embedded in a metric space together with the algorithms for special types of graphs. In Section 4, we present the algorithm for finding a worst shortcut in a graph. Finally, we make some concluding remarks in Section 5.

2 Definitions and Notation

Given a non-empty set M, a *metric* on M is a function $d : M \times M \to \mathbb{R}$ such that for all $x, y, z \in M$,

$$d(x, y) \geq 0$$
$$d(x, y) = 0 \Leftrightarrow x = y$$
$$d(x, y) = d(y, x)$$
$$d(x, y) \leq d(x, z) + d(z, y).$$

The pair (M, d) is called a *metric space*.

Let $G = (V, E)$ be a graph embedded in metric space (V, d). We regard G as a weighted graph where each edge $e \in E$ is assigned cost $d(e)$. For any two vertices $u, v \in V$, we define $d_G(u, v)$ as the length of a shortest path between u and v in G. If no such path exists, we define $d_G(u, v) = \infty$.

If $u \neq v$, the *dilation* (or detour) between u and v is defined as $d_G(u, v)/d(u, v)$ and is denoted by $\delta_G(u, v)$. The *dilation* δ_G of G is defined as

$$\delta_G = \max_{u,v \in V, u \neq v} \delta_G(u, v).$$

We define a *shortcut* (in G) as a vertex pair $(u, v) \in V \times V$ and call it a *best shortcut* resp. *worst shortcut* if it minimizes resp. maximizes $\delta_{G \cup \{(u,v)\}}$.

3   Finding a Best Shortcut

In the following, let $G = (V, E)$ be a graph embedded in metric space (V, d). In this section, we present our algorithm for finding a best shortcut in G. Due to space constraints, we only consider the case where G is connected. The disconnected case may be handled in a way similar to that in [7].

3.1   Staircase Functions

We start by introducing so called staircase functions as in [7]. Let $u, v, w_1, w_2 \in V$ be given with $u \neq v$. Define G' as the graph obtained by augmenting G with shortcut $e = (w_1, w_2)$.

A shortest path from u to v in G' either avoids e, visits e in direction $w_1 \rightarrow w_2$, or visits e in direction $w_2 \rightarrow w_1$. Hence,

$$\begin{aligned}
\delta_{G'}(u, v) = \min\{&d_G(u, w_1) + d(w_1, w_2) + d_G(w_2, v), \\
&d_G(u, w_2) + d(w_2, w_1) + d_G(w_1, v), \\
&d_G(u, v)\}/d(u, v).
\end{aligned}$$

Let us assume that $d_G(u, w_2) < d_G(u, w_1) + d(w_1, w_2)$. Then no shortest path from u in G' visits e in direction $w_1 \rightarrow w_2$. Thus, letting

$$\begin{aligned}
x &= d_G(u, w_2) + d(w_2, w_1) \\
a &= 1/d(u, v) \\
b &= d_G(w_1, v)/d(u, v) \\
c &= \delta_G(u, v),
\end{aligned}$$

we have

$$\delta_{G'}(u, v) = \min\{ax + b, c\} = \begin{cases} ax + b & \text{for } x \leq (c - b)/a \\ c & \text{for } x \geq (c - b)/a \end{cases}.$$

If we keep u, v, and w_1 fixed, we see that a, b, and c are constants and that the dilation between u and v in G' may be expressed as a piecewise linear function $\delta_{(u,v,w_1)}(x)$ of $x = d_G(u, w_2) + d(w_2, w_1)$.

We define *staircase function* $s_{(u,w_1)} : [0, \infty) \rightarrow [0, \infty)$ by

$$s_{(u,w_1)}(x) = \max\{\delta_{(u,v,w_1)}(x) | v \in V \setminus \{u\}\}.$$

Note that this function is piecewise linear and non-decreasing.

3.2   The Algorithm

In this section, we present our $O((n^4 \log n)/\sqrt{m})$ time algorithm with linear space for finding a best shortcut in G. In fact, we will show a more general result. Letting M be a positive function of n and m with $M = O(n^2)$, we present an algorithm with $O(M + n)$ space requirement (in addition to the space for storing G and a representation of the metric space) that finds a best shortcut in G in time $O((mn^4 + n^5 \log n)/(M\sqrt{M}) + (n^4 \log n)/\sqrt{M})$.

In the following, let $v_1, \ldots, v_n$ be an arbitrary ordering of the vertices of G. Pseudo-code of the algorithm is shown in Figure 1. To simplify the code and the following analysis, we have made the following assumptions: $\sqrt{M}$ is an integer that divides n and M/n is an integer that divides $\sqrt{M}$. It should be clear how to modify the algorithm to handle the case where these assumptions are not satisfied without affecting the asymptotic time and space bounds.

Let us give a high-level overview of the algorithm before proving its correctness and bounding its time and space requirement. The main ideas are similar to those in [7]: build a table T with n^2 entries, one entry for each vertex pair, and fill

1. set $\delta_{\min} := \infty$ and let $u_{\min}, v_{\min}$ be uninitialized vertex variables
2. **for** $i := 1$ **to** $n - \sqrt{M}$ **step** $\sqrt{M}$
3.   **for** $j := 1$ **to** $n - \sqrt{M}$ **step** $\sqrt{M}$
4.   initialize a table $T[i, \ldots, i + \sqrt{M} - 1][j, \ldots, j + \sqrt{M} - 1]$
5.   with all entries set to zero
6.   **for** $r := i$ **to** $i + \sqrt{M} - M/n$ **step** M/n
7.   **for** $s := 1$ **to** $n - M/n$ **step** M/n
8.   **for** $a := r$ **to** $r + M/n - 1$
9.   compute and store SSSP distances in G with source v_a
10.   **for** $b := s$ **to** $s + M/n - 1$
11.   compute and store SSSP distances in G with source v_b
12.   **for** $a := r$ **to** $r + M/n - 1$
13.   **for** $b := s$ **to** $s + M/n - 1$
14.   let $(w_1, u) = (v_a, v_b)$
15.   compute staircase function $s_{(u,w_1)}$
16.   **for** $c := j$ **to** $j + \sqrt{M} - 1$
17.   let $w_2 = v_c$
18.   **if** $d_G(u, w_2) < d_G(u, w_1) + d(w_1, w_2)$
19.   set $T[a][c] := \max\{T[a][c], s_{(u,w_1)}(d_G(u, w_2) + d(w_2, w_1))\}$
20.   free memory used to store SSSP distances
21.   repeat lines 4 to 19 with the values of i and j swapped and
22.   store values in another table $T'[j, \ldots, j + \sqrt{M} - 1][i, \ldots, i + \sqrt{M} - 1]$
23.   **for** $r := i$ **to** $i + \sqrt{M} - 1$
24.   **for** $s := j$ **to** $j + \sqrt{M} - 1$
25.   **if** $r \neq s$ and $\max\{T[r][s], T'[s][r]\} < \delta_{\min}$
26.   set $\delta_{\min} := \max\{T[r][s], T'[s][r]\}$
27.   set $(u_{\min}, v_{\min}) := (v_r, v_s)$
28. **return** $(u_{\min}, v_{\min})$ as a best shortcut

Fig. 1. Pseudo-code for algorithm to find a best shortcut in G

in entries such that, at termination, $\max\{T[u][v], T[v][u]\}$ equals $\delta_{G\cup\{(u,v)\}}$ for each pair of distinct vertices (u, v).

However, now we only have $O(M)$ space available so we subdivide T into n^2/M sub-tables each of width and height $\sqrt{M}$, see Figure 2.

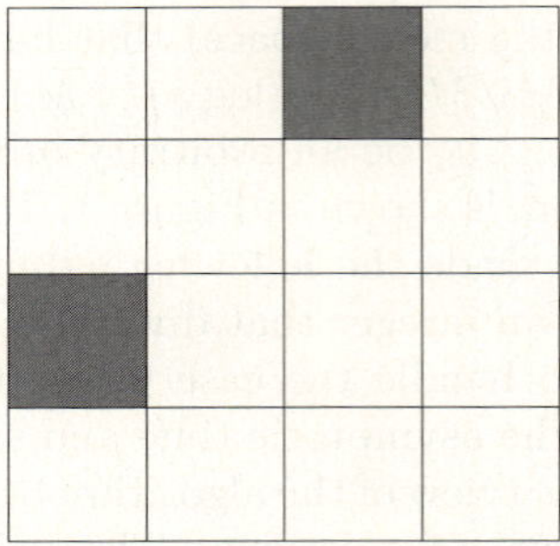

Fig. 2. T is decomposed into n^2/M sub-tables. Here, $n/\sqrt{M} = 4$. If the entries of T in the two marked sub-tables are computed then $\max\{T[u][v], T[v][u]\} = \delta_{G\cup\{(u,v)\}}$ can be obtained for all M vertex pairs (u, v) belonging to those two subtables.

We only keep two sub-tables in memory at a time. More precisely, for $i = 1, \ldots, n/\sqrt{M}$ and $j = 1, \ldots, n/\sqrt{M}$, we fill in entries of the sub-table in row i and column j and entries of the sub-table in row j and column i (this is the reason why lines 4 to 19 in the pseudo-code are repeated with indices i and j swapped) and from these entries we obtain $\max\{T[u][v], T[v][u]\}$ for vertex pairs in those two sub-tables.

In the following, we show the correctness of the algorithm and then bound its running time and space requirement.

Correctness. In order to prove the correctness of the algorithm it is enough to show that each value $\max\{T[r][s], T'[s][r]\}$ computed in line 25 equals $\delta_{G\cup\{(v_r,v_s)\}}$ for $r \neq s$ since (r, s) covers all distinct pairs in $\{1, \ldots, n\}^2$ throughout the course of the algorithm.

So consider any i-iteration and any j-iteration of the algorithm. At the end of each iteration of the for-loop in lines 7 to 20, we have (with $w_1 = w_a$ and $w_2 = w_c$),

$$T[a][c] = \max\{s_{(u,w_1)}(d_G(u, w_2) + d(w_2, w_1)) | u \in \{v_s, \ldots, v_{s+M/n-1}\},$$
$$d_G(u, w_2) < d_G(u, w_1) + d(w_1, w_2)\}.$$

or $T[a][c] = 0$ if there is no $u \in \{v_s, \ldots, v_{s+M/n-1}\}$ such that $d_G(u, w_2) < d_G(u, w_1) + d(w_1, w_2)$, for $a = r, \ldots, r + M/n - 1$ and $c = j, \ldots, j + \sqrt{M} - 1$, $a \neq c$. Hence, at the end of each iteration of the for-loop in lines 6 to 20,

$$T[a][c] = \max\{s_{(u,w_1)}(d_G(u, w_2) + d(w_2, w_1)) | u \in V,$$
$$d_G(u, w_2) < d_G(u, w_1) + d(w_1, w_2)\}$$
$$= \max\{\delta_{G\cup\{(w_1,w_2)\}}(u, v) | u, v \in V, u \neq v, d_G(u, w_2) < d_G(u, w_1) + d(w_1, w_2)\}.$$

for $a = i, \ldots, i + \sqrt{M} - 1$ and $c = j, \ldots, j + \sqrt{M} - 1$, $a \neq c$ (this is well-defined since $u = w_2$ satisfies $d_G(u, w_2) < d_G(u, w_1) + d(w_1, w_2)$.

Similarly, after lines 21 to 22,

$$T'[c][a] = \max\{\delta_{G \cup \{(w_1, w_2)\}}(u, v) \mid u, v \in V, u \neq v, d_G(u, w_1) < d_G(u, w_2) + d(w_2, w_1)\}.$$

for $a = i, \ldots, i + \sqrt{M} - 1$ and $c = j, \ldots, j + \sqrt{M} - 1$, $a \neq c$.

For any $u \in V$, any $w_1 \in \{v_i, \ldots, v_{i+\sqrt{M}-1}\}$, and any $w_2 \in \{v_j, \ldots, v_{j+\sqrt{M}-1}\}$ with $w_1 \neq w_2$, either $d_G(u, w_2) < d_G(u, w_1) + d(w_1, w_2)$ or $d_G(u, w_1) < d_G(u, w_2) + d(w_2, w_1)$ for otherwise we get the contradiction

$$d_G(u, w_2) + d(w_2, w_1) \leq d_G(u, w_1)$$
$$\leq d_G(u, w_2) - d(w_1, w_2)$$
$$< d_G(u, w_2) + d(w_2, w_1).$$

Hence, after lines 6 to 22,

$$\max\{T[a][c], T'[c, a]\} = \max_{u, v \in V, u \neq v} \delta_{G \cup \{(w_1, w_2)\}}(u, v) = \delta_{G \cup \{(w_1, w_2)\}}.$$

for $a = i, \ldots, i + \sqrt{M} - 1$ and $c = j, \ldots, j + \sqrt{M} - 1$, $a \neq c$. This shows the correctness of the algorithm.

Space Requirement. As for space requirement, we observe that each table takes up $\sqrt{M}^2 = M$ space. In lines 8 to 11, we store shortest path distances from $O(M/n)$ vertices to all vertices in G. This takes up a total of $O(M)$ space. Staircase function $s_{(u, w_1)}$ can be represented using $O(n)$ space since it consists of $O(n)$ line segments. Hence, the algorithm uses $O(M + n)$ space.

Running Time. What remains is to bound the running time of the algorithm. The total time spent in lines 4 to 5 is $O(n^2)$. The total number of iterations of the for-loop in lines 7 to 19 is $(n/\sqrt{M})^2(\sqrt{M}/(M/n))(n/(M/n)) = n^5/(M^2\sqrt{M})$.

If we use Dijkstra's SSSP algorithm in lines 9 and 11, the total time spent in lines 8 to 11 is

$$O((n^5/(M^2\sqrt{M}))(M/n)(m + n \log n)) = O((n^4/(M\sqrt{M}))(m + n \log n)).$$

Computing staircase function $s_{(u, w_1)}$ in line 15 can be done in $O(n \log n)$ time since it is the upper envelope of $O(n)$ line segments (or halflines) and each line segment can be found in constant time using the precomputed shortest path distances. Since the total number of iterations of the for-loop in lines 13 to 19 is $(n^5/(M^2\sqrt{M}))(M/n)^2 = n^3/\sqrt{M}$, it follows that the time spent in line 15 throughout the course of the algorithm is $O((n^4 \log n)/\sqrt{M})$.

The number of iterations of the for-loop in lines 16 to 19 is $(n^3/\sqrt{M})\sqrt{M} = n^3$. The test in line 18 can be performed in constant time using the precomputed shortest path distances. Computing value $s_{(u, w_1)}(d_G(u, w_2) + d(w_2, w_1))$ can be done in $O(\log n)$ time with binary search since $s_{(u, w_1)}$ is a non-decreasing piecewise linear function. Hence, the time spent in lines 16 to 19 throughout the course of the algorithm is $O(n^3 \log n)$.

Adding up, it follows that the total time spent in lines 4 to 20 is

$$O((n^4/(M\sqrt{M}))(m + n \log n) + (n^4 \log n)/\sqrt{M} + n^3 \log n)$$

which is

$$O((mn^4 + n^5 \log n)/(M\sqrt{M}) + (n^4 \log n)/\sqrt{M})$$

since $M = O(n^2)$. A similar argument shows that the total time spent in lines 21 to 22 is

$$O((mn^4 + n^5 \log n)/(M\sqrt{M}) + (n^4 \log n)/\sqrt{M}).$$

Finally, the total time spent in lines 23 to 27 is $O(n^2)$.

We have now shown the following.

Theorem 1. *A best shortcut in G can be found in $O((mn^4 + n^5 \log n)/(M\sqrt{M}) + (n^4 \log n)/\sqrt{M})$ time using $O(M + n)$ space.*

Note the tradeoff between time and space. Setting $M = m + n$ solves the open problem in [4].

Corollary 1. *A best shortcut in G can be found in $O((n^4 \log n)/\sqrt{m})$ time using linear space.*

Also note that by setting $M = n^2$, it follows that a best shortcut can be found in $O(n^3 \log n)$ time using $O(n^2)$ space which is the result in [7].

Next, we consider special types of graphs and show that we can achieve faster running times for these.

3.3   Best Shortcut of Two Vertex-Disjoint Simple Paths

In this section, we assume that G is the union of two simple vertex disjoint simple paths L_1 and L_2. We will show that a best shortcut in G can be found in $O(n^2 \log n)$ time.

Clearly, we may restrict our attention to finding shortcuts that connect L_1 and L_2. In the following, let (w_1, w_2) denote a shortcut and assume w.l.o.g. that $w_1 \in L_1$ and $w_2 \in L_2$. Let G' denote the graph $G \cup \{(w_1, w_2)\}$. Denote the endpoints of L_1 by v_1 and v_3 and the endpoints of L_2 by v_2 and v_4. We do not yet know how to pick w_1 and w_2 so let us regard them as variables and introduce real parameters x_1, x_2, x_3, and x_4, defined by

$$x_1 = d_G(v_1, w_1) + d(w_1, w_2) + d_G(w_2, v_2),$$
$$x_2 = d_G(v_3, w_1) + d(w_1, w_2) + d_G(w_2, v_4),$$
$$x_3 = d_G(v_3, w_1) + d(w_1, w_2) + d_G(w_2, v_2),$$
$$x_4 = d_G(v_1, w_1) + d(w_1, w_2) + d_G(w_2, v_4),$$

see Figure 3.

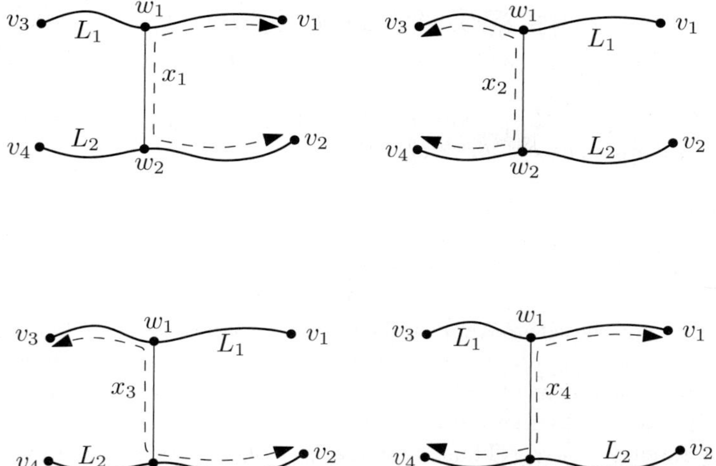

Fig. 3. Definition of four parameters for two vertex-disjoint simple paths

Now, let us express the dilation between vertices $u \in L_1$ and $v \in L_2$ by these parameters. First define

$$\delta^1_{G'}(u,v) = \frac{x_1 - d_G(u,v_1) - d_G(v,v_2)}{d(u,v)},$$

$$\delta^2_{G'}(u,v) = \frac{x_2 - d_G(u,v_3) - d_G(v,v_4)}{d(u,v)},$$

$$\delta^3_{G'}(u,v) = \frac{x_3 - d_G(u,v_3) - d_G(v,v_2)}{d(u,v)},$$

$$\delta^4_{G'}(u,v) = \frac{x_4 - d_G(u,v_1) - d_G(v,v_4)}{d(u,v)}.$$

The following lemma shows how to express $\delta_{G'}(u,v)$ as a function of x_1, x_2, x_3, and x_4.

Lemma 1. $\delta_{G'}(u,v) = \max\{\delta^1_{G'}(u,v), \delta^2_{G'}(u,v), \delta^3_{G'}(u,v), \delta^4_{G'}(u,v)\}.$

Proof. Define $L_{w_1 v_i}$ as the subpath of L_1 from w_1 to v_i, $i = 1, 3$. Similarly, define $L_{w_2 v_i}$ as the subpath of L_2 from w_2 to v_i, $i = 2, 4$. Then $\delta_{G'}(u,v) \geq \delta^1_{G'}(u,v)$, $\delta_{G'}(u,v) \geq \delta^2_{G'}(u,v)$, $\delta_{G'}(u,v) \geq \delta^3_{G'}(u,v)$, and $\delta_{G'}(u,v) \geq \delta^4_{G'}(u,v)$.
Furthermore,

$$\delta_{G'}(u,v) = \delta^1_{G'}(u,v) \Leftrightarrow u \in L_{w_1 v_1} \text{ and } v \in L_{w_2 v_2},$$

$$\delta_{G'}(u,v) = \delta^2_{G'}(u,v) \Leftrightarrow u \in L_{w_1 v_3} \text{ and } v \in L_{w_2 v_4},$$

$$\delta_{G'}(u,v) = \delta_{G'}^3(u,v) \Leftrightarrow u \in L_{w_1 v_3} \text{ and } v \in L_{w_2 v_2},$$
$$\delta_{G'}(u,v) = \delta_{G'}^4(u,v) \Leftrightarrow u \in L_{w_1 v_1} \text{ and } v \in L_{w_2 v_4}.$$

Since $u \in L_1 = L_{w_1 v_i} \cup L_{w_1 v_3}$ and $v \in L_2 = L_{w_2 v_2} \cup L_{w_2 v_4}$, we have $\delta_{G'}(u,v) = \max\{\delta_{G'}^1(u,v), \delta_{G'}^2(u,v), \delta_{G'}^3(u,v), \delta_{G'}^4(u,v)\}$. $\qquad\square$

From Lemma 1, it follows that

$$\delta_{G'} = \max\{\delta_{L_1}, \delta_{L_2}, \max_{i=1,2,3,4}\{\max_{u \in L_1, v \in L_2} \delta_{G'}^i(u,v)\}\}.$$

We now give an algorithm that finds a vertex pair (w_1, w_2) that maximizes the value $\max_{u \in L_1, v \in L_2} \delta_{G'}^1(u,v)$. By symmetry, it is enough to show that this problem can be solved in $O(n^2 \log n)$ time in order to prove our claim.

For each pair of distinct vertices u and v, $\delta_{G'}^1(u,v)$ is a linear and non-decreasing function of x_1. Thus, $\max_{u \in L_1, v \in L_2} \delta_{G'}^1(u,v)$ is a piecewise linear, non-decreasing function of x_1 consisting of $O(n^2)$ line segments (and one halfline) and it can be found in $O(n^2 \log n)$ time.

Having found this upper envelope function, we determine an x_1-value for each pair of vertices w_1 and w_2 and compute the corresponding value of the upper envelope function. Using binary search, this takes $O(\log n)$ time for each vertex pair. The pair (w_1, w_2) with the largest value is the pair maximizing $\max_{u \in L_1, v \in L_2} \delta_{G'}^1(u,v)$.

We have now obtained the following result.

Theorem 2. *If G is the union of two vertex-disjoint simple paths, a best shortcut in G can be found in $O(n^2 \log n)$ time.*

3.4   Best Shortcut of a Simple Path

We now show how to find a best shortcut in G when G is a simple path L.

Let v_1 and v_2 be the end vertices of L. By symmetry, we may restrict our attention to shortcuts (w_1, w_2) where $d_G(w_1, v_1) < d_G(w_2, v_1)$. And when finding a pair (u,v) achieving the dilation of $\delta_{G'}$, $G' = G \cup \{(w_1, w_2)\}$, we only need to consider pairs where $d_G(u, v_1) < d_G(u, v_2)$.

We will present an algorithm for the above problem with $O(n^2 \log n)$ running time. The basic idea is the same as in Section 3.3. We introduce real parameters x_1, x_2, x_3, and x_4, defined by

$$x_1 = d(w_1, w_2) + d_G(w_1, w_2),$$
$$x_2 = d_G(v_1, w_1) + d(w_1, w_2) + d_G(w_2, v_2),$$
$$x_3 = d_G(v_1, w_1) + d(w_1, w_2) + d_G(w_2, v_1),$$
$$x_4 = d_G(v_2, w_2) + d(w_1, w_2) + d_G(w_1, v_2),$$

see Figure 4. And we define

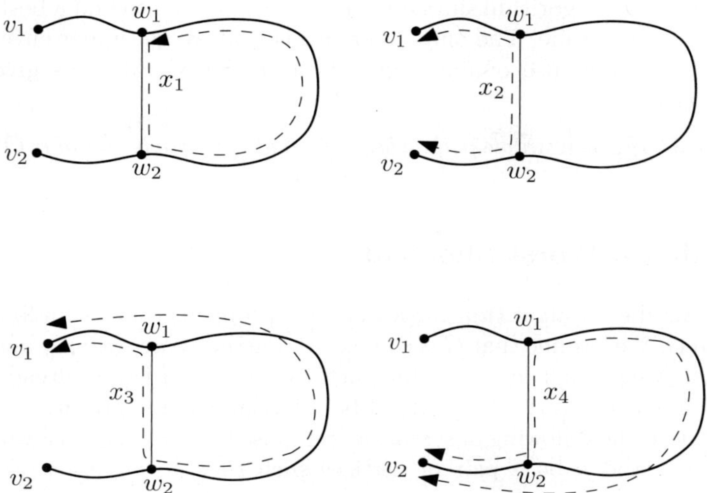

Fig. 4. Definition of four parameters for simple path

$$\delta^1_{G'}(u, v) = \min\{\frac{x_1 - d_G(u, v)}{d(u, v)}, \delta_G(u, v)\},$$

$$\delta^2_{G'}(u, v) = \min\{\frac{x_2 - d_G(u, v_1) - d_G(v, v_2)}{d(u, v)}, \delta_G(u, v)\},$$

$$\delta^3_{G'}(u, v) = \min\{\frac{x_3 - d_G(u, v_1) - d_G(v, v_1)}{d(u, v)}, \delta_G(u, v)\},$$

$$\delta^4_{G'}(u, v) = \min\{\frac{x_4 - d_G(u, v_2) - d_G(v, v_2)}{d(u, v)}, \delta_G(u, v)\}.$$

We have the following result.

Lemma 2. $\delta_{G'}(u, v) = \max\{\delta^1_{G'}(u, v), \delta^2_{G'}(u, v), \delta^3_{G'}(u, v), \delta^4_{G'}(u, v)\}$

Proof. For each pair of vertices $u', v' \in L$, define $L_{u'v'}$ as the subpath of L from u' to v'. As in the proof of Lemma 1, we have $\delta_{G'}(u, v) \geq \delta^1_{G'}(u, v)$, $\delta_{G'}(u, v) \geq \delta^2_{G'}(u, v)$, $\delta_{G'}(u, v) \geq \delta^3_{G'}(u, v)$, and $\delta_{G'}(u, v) \geq \delta^4_{G'}(u, v)$.

Furthermore,

$$u, v \in L_{w_1 w_2} \Rightarrow \delta_{G'}(u, v) = \delta^1_{G'},$$

$$u \in L_{v_1 w_1}, v \in L_{w_2 v_2} \Rightarrow \delta_{G'}(u, v) = \delta^2_{G'},$$

$$u \in L_{v_1 w_1}, v \in L_{v_1 w_2} \Rightarrow \delta_{G'}(u, v) = \delta^3_{G'},$$

$$u \in L_{w_1 v_2}, v \in L_{w_2 v_2} \Rightarrow \delta_{G'}(u, v) = \delta^4_{G'}.$$

Since either $u, v \in L_{w_1 w_2}$, or $u \in L_{v_1 w_1}, v \in L_{w_2 v_2}$, or $u \in L_{v_1 w_1}, v \in L_{v_1 w_2}$, or $u \in L_{w_1 v_2}, v \in L_{w_2 v_2}$, the lemma follows. $\square$

We can now use an algorithm similar to that in Section 3.3 to find a best shortcut in G in $O(n^2 \log n)$ time. The only difference is that we get upper envelopes not just of halflines but also of line segments as in Section 3. This gives us the following result.

Theorem 3. *If G is a simple path, a best shortcut in G can be found in $O(n^2 \log n)$ time.*

4   Finding a Worst Shortcut

We now describe an algorithm for a different problem, that of finding a worst shortcut in G. We assume that G is connected. Furthermore, we only allow vertex pairs (w_1, w_2) such that $w_1 \neq w_2$ and such that (w_1, w_2) is not already an edge in G since any other vertex pair would be a trivial worst shortcut.

We will need the following observation. Suppose that (w_1, w_2) is a worst shortcut and let u and v be a pair of vertices such that $\delta_{G'} = \delta_{G'}(u, v)$. Then if $d_G(u, w_2) < d_G(u, w_1) + d(w_1, w_2)$, we may assume that w_2 is a vertex in $\{w_2' \in V | d_G(u, w_2') < d_G(u, w_1) + d(w_1, w_2')\}$ that maximizes $d_G(u, w_2') + d(w_2', w_1)$.

The following algorithm finds a worst shortcut in G. A main loop iterates over all vertices w_1. In the following, consider one of these iterations.

For each u, a vertex w_2 is picked (if any) such that $d_G(u, w_2) < d_G(u, w_1) + d(w_1, w_2)$, $w_1 \neq w_2$, $(w_1, w_2) \notin E$, and such that $d_G(u, w_2) + d(w_2, w_1)$ is maximized and the maximum of $\delta_{G'}(u, v)$ over all $v \neq u$ is computed. Over all u, this gives at most n dilation values and we keep the largest of them together with a w_2 giving this dilation.

This is done for all w_1, again giving at most n dilation values and we pick the largest of them together with the (w_1, w_2)-pair giving this dilation.

From the observation above and the observation from Section 3.2 that either $d_G(u, w_2) < d_G(u, w_1) + d(w_1, w_2)$ or $d_G(u, w_1) < d_G(u, w_2) + d(w_2, w_1)$ for all u and $w_1 \neq w_2$, it follows that the algorithm picks a worst shortcut in G.

If we precompute APSP distances using, say, the algorithm in [2], we obtain an $O(n^3)$ time algorithm with $O(n^2)$ space requirement. We can also obtain linear space requirement but then in each iteration of the main loop, we have to compute SSSP distances with source w_1 and with source u for each u. With Dijkstra's algorithm, this gives $O(n^3 \log n)$ running time.

Theorem 4. *A worst shortcut in G can be found in $O(n^3)$ time with $O(n^2)$ space requirement and in $O(n^3 \log n)$ time with linear space requirement.*

5   Concluding Remarks

In this paper, we considered the problem of finding a best shortcut in a graph G with at most two connected components embedded in a metric space. We presented an algorithm with $O((n^4 \log n)/\sqrt{m})$ running time and linear space requirement, where n is the number of vertices and m is the number of edges in G. This solves an open problem of whether an $o(n^4)$ time algorithm with linear space requirement exists for this problem. We showed that if G is a simple path or the union of two simple vertex-disjoint paths, a best shortcut in G can be found in $O(n^2 \log n)$ time. Finally, we considered the problem of finding a worst shortcut in G. We gave an $O(n^3)$ time algorithm with $O(n^2)$ space requirement and an $O(n^3 \log n)$ time algorithm with linear space requirement for this problem.

It would be interesting to consider the more general case where a fixed number of edges are inserted and to consider edge-removals. Another direction for future research is to consider other special types of graphs like cycles and trees. Finally, not much is known about lower bounds on running time. It is quite easy to show that to find a best shortcut it is sometimes necessary to look at all entries of the $n \times n$-matrix defining the metric. Can this $\Omega(n^2)$ bound be improved?

References

1. Ahn, H.-K., Farshi, M., Knauer, C., Smid, M., Wang, Y.: Dilation-optimal edge deletion in polygonal cycles. In: Tokuyama, T. (ed.) ISAAC 2007. LNCS, vol. 4835, pp. 88–99. Springer, Heidelberg (2007)
2. Chan, T.M.: More algorithms for all-pairs shortest paths in weighted graphs. In: Proceedings of the thirty-ninth annual ACM symposium on Theory of computing, pp. 590–598 (2007)
3. Eppstein, D.: Spanning trees and spanners. In: Sack, J.-R., Urrutia, J. (eds.) Handbook of Computational Geometry, pp. 425–461. Elsevier Science Publishers, Amsterdam (2000)
4. Farshi, M., Giannopoulos, P., Gudmundsson, J.: Finding the Best Shortcut in a Geometric Network. In: 21st Ann. ACM Symp. Comput. Geom., pp. 327–335 (2005)
5. Narasimhan, G., Smid, M.: Geometric Spanner Networks. Cambridge University Press, Cambridge (2007)
6. Smid, M.: Closest point problems in computational geometry. In: Sack, J.-R., Urrutia, J. (eds.) Handbook of Computational Geometry, pp. 877–935. Elsevier Science Publishers, Amsterdam (2000)
7. Wulff-Nilsen, C.: Computing the Dilation of Edge-Augmented Graphs in Metric Spaces. In: 24th European Workshop on Computational Geometry, Nancy, pp. 123–126 (2008)

On Labeled Traveling Salesman Problems

Basile Couëtoux[1], Laurent Gourvès[1], Jérôme Monnot[1],
and Orestis A. Telelis[2,⋆]

[1] CNRS UMR 7024 LAMSADE, Université de Paris-Dauphine, France
`basile.couetoux@dauphine.fr,`
`{laurent.gourves,monnot}@lamsade.dauphine.fr`
[2] Department of Computer Science, University of Aarhus, Denmark
`telelis@daimi.au.dk`

Abstract. We consider labeled Traveling Salesman Problems, defined upon a complete graph of n vertices with colored edges. The objective is to find a tour of maximum (or minimum) number of colors. We derive results regarding hardness of approximation, and analyze approximation algorithms for both versions of the problem. For the maximization version we give a $\frac{1}{2}$-approximation algorithm and show that it is **APX**-hard. For the minimization version, we show that it is not approximable within $n^{1-\epsilon}$ for every $\epsilon > 0$. When every color appears in the graph at most r times and r is an increasing function of n the problem is not $O(r^{1-\epsilon})$-approximable. For fixed constant r we analyze a polynomial-time $(r+H_r)/2$-approximation algorithm (H_r is the r-th harmonic number), and prove **APX**-hardness. Analysis of the studied algorithms is shown to be tight.

1 Introduction

We consider labeled versions of the Traveling Salesman Problem (TSP), defined upon a complete graph K_n of n vertices along with an edge-labeling (or coloring) function $\mathcal{L} : E(K_n) \rightarrow \{c_1, \ldots, c_q\}$. The objective is to find a hamiltonian tour T of K_n optimizing (either maximizing or minimizing) $|\mathcal{L}(T)|$, where $\mathcal{L}(T) = \{\mathcal{L}(e) : e \in T\}$. We refer to the corresponding problems with MaxLTSP and MinLTSP respectively. The *color frequency* of a MinLTSP instance is the maximum number of equi-colored edges. We use MinLTSP$_{(r)}$ to refer to the class of MinLTSP instances with fixed color frequency r.

Labeled network optimization over colored graphs has seen extended study [17,18,3,5,12,2,4,14,10,11,15]. Minimization of used colors models naturally the need for using links with common properties, whereas the maximization case can be viewed as a maximum covering problem with a certain network structure (in our case such a structure is a hamiltonian cycle). If for example every color represents a technology consulted by a different vendor, then we wish to use as few colors as possible, so as to diminish incompatibilities among different technologies. For the maximization case, consider the situation of designing a

---

⋆ Center for Algorithmic Game Theory, funded by the Carlsberg Foundation, Denmark.

S.-H. Hong, H. Nagamochi, and T. Fukunaga (Eds.): ISAAC 2008, LNCS 5369, pp. 776–787, 2008.

metropolitan peripheral ring road, where every color represents a different suburban area that a certain link would traverse. In order to maximize the number of suburban areas that such a peripheral ring covers, we seek a tour of a maximum number of colors. It was shown in [4] that both MAXLTSP and MINLTSP are **NP**-hard.

Contribution. We present approximation algorithms and hardness results for MAXLTSP and MINLTSP. In section 2 we provide a $\frac{1}{2}$-approximation local improvement algorithm for the MAXLTSP problem and show that the problem is **APX**-hard. In section 3 we show that MINLTSP is not approximable within a factor $n^{1-\epsilon}$ for every $\epsilon > 0$ or within a factor $O(r^{1-\epsilon})$ when color frequency r is an increasing function of n (paragraph 3.1). For the case of *fixed constant* r, we analyze a simple greedy algorithm with approximation ratio $(r + H_r)/2$, where $H_r = \sum_{i=1}^{r} \frac{1}{i}$ is the r-th harmonic number (paragraph 3.2). For $r = 2$ MINLTSP$_{(2)}$ is shown to be **APX**-hard. We conclude with open problems.

Related Work. Identification of conditions for the existence of single-colored or multi-colored cycles on colored graphs was first treated in [6]. A great amount of work that followed concerned identification of such conditions and bounds on the number of colors [4,1,7,9]. The optimization problems that we consider here were shown to be **NP**-hard in [4]. To the best of our knowledge no further theoretical development prior our work exists with respect to MAXLTSP and MINLTSP. An experimental study of MINLTSP appeared in [19]. *TSP under categorization* [17,18] generalizes several TSP problems, and is also a weighted generalization of MINLTSP. For metric edge weights and at most q colors appearing in the graph a $2q$ approximation is achieved in [17,18].

The recent literature on labeled network optimization problems includes several interesting results from both perspectives of hardness and approximation algorithms. In [10] the authors investigate weighted generalizations of labeled minimum spanning tree and shortest paths problems, where each label is also associated with a positive weight and the objective generalizes to minimization of the weighted sum of different labels used. They analyze approximation algorithms and prove inapproximability results for both problems. **NP**-hardness of finding paths with the fewest different colors was shown in [4]. The labeled minimum spanning tree problem was introduced in [5]. In [12] a greedy approximation algorithm is analyzed, and in [2] bounded color frequency is considered. The labeled perfect matching problems were studied in [14,15], while Maffioli *et al.* worked on a labeled matroid problem [13]. Complexity of approximation of bottleneck labeled problems is studied in [11].

2   MaxLTSP: Constant Factor Approximation

A simple greedy algorithm yields a 1/3 approximation of MAXLTSP (see full version).We analyze a $\frac{1}{2}$-approximation algorithm based on local search. The algorithm grows iteratively by local improvements a subset $S \subseteq E$ of edges, such that *(i)* each label of $\mathcal{L}(S)$ appears at most once in S and *(ii)* S does not induce vertices of degree three or more, or a cycle of length less than n. We call S a *labeled valid* subset of edges. Finding a labeled valid subset S of maximum size is clearly equivalent to MAXLTSP.

Given a labeled valid subset S of $(K_n, \mathcal{L})$, a *1-improvement* of S is a labeled valid subset $S \cup \{e_1\}$ where $e_1 \notin S$, whereas a *2-improvement* of S is a labeled valid subset $(S \setminus \{e\}) \cup \{e_1, e_2\}$ where $e \in S$ and $e_1, e_2 \notin S \setminus \{e\}$. An 1- or 2-improvement of S is a labeled valid subset S' such that $|S'| = |S| + 1$. An 1-improvement can be viewed as a particular 2-improvement but we separate the two cases for ease of presentation. The local improvement algorithm - denoted by LOCIM - initializes $S = \emptyset$ and performs iteratively either an 1- or a 2-improvement on the current S as long as such an improvement exists. This algorithm works clearly in polynomial-time. We denote by S the solution returned by LOCIM and by S^* an optimal solution.

We introduce further notations. Given $e \in S$, let $\ell(e)$ be the edge of S^* with the same label if such an edge exists. Formally, $\ell : S \to S^* \cup \{\bot\}$ is defined by:

$$\ell(e) = \begin{cases} \bot & \text{if } \mathcal{L}(e) \notin \mathcal{L}(S^*) \\ e^* \in S^* & \text{such that } \mathcal{L}(e^*) = \mathcal{L}(e) \text{ otherwise.} \end{cases}$$

For $e = [i, j] \in S$, let $N(e)$ be the edges of S^* incident to i or j.

$$N(e) = \{[k, l] \in S^* \mid \{k, l\} \cap \{i, j\} \neq \emptyset\}$$

$N(e)$ is partitionned into $N_1(e)$ and $N_0(e)$ as follows: $e^* \in N_1(e)$ iff $(S \setminus \{e\}) \cup \{e^*\}$ is a labeled valid subset, and $N_0(e) = N(e) \setminus N_1(e)$. In particular, $N_0(e)$ contains the edges $e^* \in S^*$ of $N(e)$ such that $(S \setminus \{e\}) \cup \{e^*\}$ is not labeled valid subset. Finally, for $e^* = [k, l] \in S^*$, let $N^{-1}(e^*)$ be the edges of S incident to k or l.

$$N^{-1}(e^*) = \{[i, j] \in S \mid \{k, l\} \cap \{i, j\} \neq \emptyset\}$$

Property 1. Let $e = [i, j] \in S$ and $e^* = [i, k] \in N_1(e)$ with $k \neq j$, $e^* \neq \ell(e)$. Either S has two edges incident to i, or $S \cup \{e^*\}$ contains a cycle through e, e^*.

Property 1 holds at the end of the algorithm since otherwise $S \cup \{e^*\}$ would be an 1-improvement of S.

Property 2. Let $e = [i, j] \in S$ and $e_1^*, e_2^* \in N_1(e)$. Either both e_1^* and e_2^* are adjacent to i (or to j) or there is a cycle in $S \cup \{e_1^*, e_2^*\}$ passing through e_1^*, e_2^*.

Property 2 holds at the end of the algorithm since otherwise $(S \setminus \{e\}) \cup \{e_1^*, e_2^*\}$ would be a 2-improvement of S. In order to prove the $\frac{1}{2}$ approximation factor for LOCIM we use charging/discharging arguments based on the following function $g : S \to \mathbb{R}$:

$$g(e) = \begin{cases} |N_0(e)|/4 + |N_1(e)|/2 + 1 - |N^{-1}(\ell(e))|/4 & \text{if } \ell(e) \neq \bot \\ |N_0(e)|/4 + |N_1(e)|/2 & \text{otherwise} \end{cases}$$

For simplicity the proof of the 1/2-approximation is cut into two propositions.

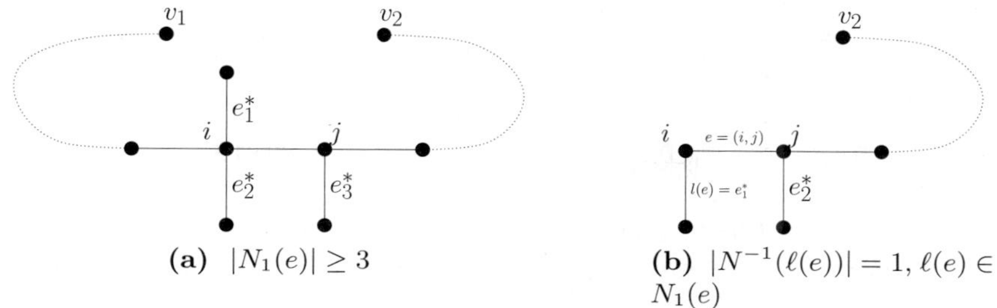

(a) $|N_1(e)| \geq 3$ **(b)** $|N^{-1}(\ell(e))| = 1$, $\ell(e) \in N_1(e)$

Fig. 1. Cases studied in proof of proposition 1

Proposition 1. $\forall e \in S$, $g(e) \leq 2$.

Proof. Let $e = [i, j]$ be an edge of S. We study two cases, when $e \in S \cap S^*$ and when $e \in S \setminus S^*$. If $e \in S \cap S^*$ then $\ell(e) = e$. Observe that $|N^{-1}(e)| \geq |N_1(e)|$, since otherwise an 1- or 2-improvement would be possible. Since $|N(e)| = |N_0(e)| + |N_1(e)| \leq 4$ we obtain $g(e) \leq (|N_0(e)| + |N_1(e)|)/4 + 1 \leq 2$.

Suppose now that $e \in S \setminus S^*$. Let us first show that $|N_1(e)| \leq 2$. By contradiction, suppose that $\{e_1^*, e_2^*, e_3^*\} \subseteq N_1(e)$ and w.l.o.g., assume that e_1^* and e_2^* are incident to i (see Fig. 1a for an illustration).

The pairs e_1^*, e_3^* and e_2^*, e_3^* cannot be simultaneously adjacent since otherwise $\{e_1^*, e_2^*, e_3^*\}$ would form a triangle. Then e_1^*, e_3^* is a matching. Property 2 implies that $(S \setminus \{e\}) \cup \{e_1^*, e_3^*\}$ contains a cycle. This cycle must be $(P_e \setminus \{e\}) \cup \{e_1^*, e_3^*\}$ where P_e is the path containing e in S (see Fig. 1a: $e_1^* = [i, v_2]$ and $e_3^* = [j, v_1]$. Note that $e_2^* \neq [i, v_1]$ since $e_2^* \in N_1(e)$). Then $(S \setminus \{e\}) \cup \{e_2^*, e_3^*\}$ would be a 2-improvement of S, a contradiction.

- If $\ell(e) = \perp$ or $|N^{-1}(\ell(e))| \geq 2$, we deduce from $|N_1(e)| \leq 2$ that $g(e) \leq 2$.
- If $\ell(e) \neq \perp$ and $|N^{-1}(\ell(e))| = 1$, then $|N_1(e)| \leq 1$. Otherwise, let $\{e_1^*, e_2^*\} \subseteq N_1(e)$. We have $\ell(e) \neq e_1^*$ and $\ell(e) \neq e_2^*$ since otherwise $(S \setminus \{e\}) \cup \{e_1^*, e_2^*\}$ is a 2-improvement of S, see Fig. 1b for an illustration.
  In this case, we deduce that $(S \setminus \{e\}) \cup \{\ell(e), e_2^*\}$ or $(S \setminus \{e\}) \cup \{\ell(e), e_1^*\}$ is a 2-improvement of S, a contradiction. Hence, $|N_1(e)| \leq 1$ and $g(e) \leq 2$.
- If $\ell(e) \neq \perp$ and $|N^{-1}(\ell(e))| = 0$, then $|N_1(e)| = 0$. Hence, $g(e) \leq 2$. $\square$

We apply a discharging method to establish a relationship between g and $|S^*|$.

Proposition 2. $\sum_{e \in S} g(e) \geq |S^*|$.

Proof. Let $f : S \times S^* \to \mathbb{R}$ be defined as:

$$f(e, e^*) = \begin{cases} 1/4 & \text{if } e^* \in N_0(e) \text{ and } \ell(e) \neq e^* \\ 1/2 & \text{if } e^* \in N_1(e) \text{ and } \ell(e) \neq e^* \\ 1 - |N^{-1}(e^*)|/4 & \text{if } e^* \notin N(e) \text{ and } \ell(e) = e^* \\ 5/4 - |N^{-1}(e^*)|/4 & \text{if } e^* \in N_0(e) \text{ and } \ell(e) = e^* \\ 3/2 - |N^{-1}(e^*)|/4 & \text{if } e^* \in N_1(e) \text{ and } \ell(e) = e^* \\ 0 & \text{otherwise} \end{cases}$$

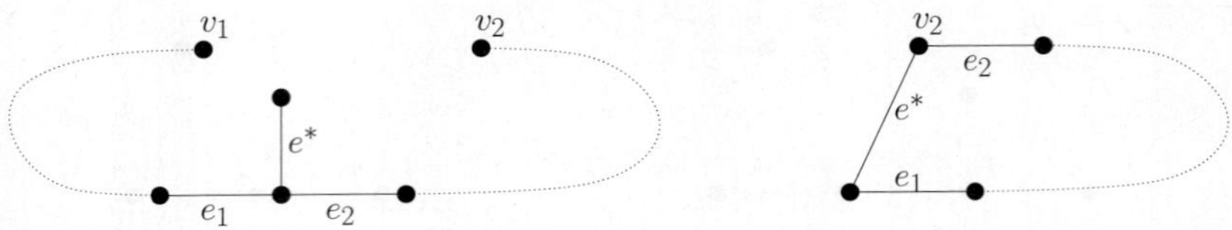

Fig. 2. The case where $N^{-1}(e^*) = \{e_1, e_2\}$

For all $e \in S$ it is $\sum_{\{e^* \in S^*\}} f(e, e^*) = g(e)$. Because of the following:

$$\sum_{\{e \in S\}} g(e) = \sum_{\{e^* \in S^*\}} \sum_{\{e \in S\}} f(e, e^*)$$

it is enough to show that $\sum_{\{e \in S\}} f(e, e^*) \geq 1$ for all $e^* \in S^*$. For an edge $e^* \in S^*$, we study two cases: $\mathcal{L}(e^*) \in \mathcal{L}(S)$ and $\mathcal{L}(e^*) \notin \mathcal{L}(S)$. If $\mathcal{L}(e^*) \in \mathcal{L}(S)$ then there is $e_0 \in S$ such that $\ell(e_0) = e^*$. We distinguish two possibilities:

- $e^* \in N(e_0)$: it is possible that $e_0 = e^*$ if $e^* \in N_1(e_0)$. Then $\sum_{\{e \in S\}} f(e, e^*) \geq$
  $f(e_0, e^*) + \sum_{\{e \in (N^{-1}(e^*)) \setminus \{e_0\}\}} f(e, e^*) \geq \frac{5}{4} - \frac{|N^{-1}(e^*)|}{4} + \frac{|N^{-1}(e^*)|-1}{4} = 1$
- $e^* \notin N(e_0)$: then $\sum_{\{e \in S\}} f(e, e^*) \geq f(e_0, e^*) + \sum_{\{e \in N^{-1}(e^*)\}} f(e, e^*) \geq 1 -$
  $\frac{|N^{-1}(e^*)|}{4} + \frac{|N^{-1}(e^*)|}{4} = 1$.

Now consider $\mathcal{L}(e^*) \notin \mathcal{L}(S)$. Then $|N^{-1}(e^*)| \geq 2$, otherwise $S \cup \{e^*\}$ would be an 1-improvement. We examine the following situations:

- $N^{-1}(e^*) = \{e_1, e_2\}$: By Property 1 e_1 and e_2 are adjacent, or there is a cycle passing through e^*, e_1 and e_2. In this case $e^* \in N_1(e_1)$ and $e^* \in N_1(e_2)$ (see Fig. 2). Thus $\sum_{\{e \in S\}} f(e, e^*) \geq f(e_1, e^*) + f(e_2, e^*) = \frac{1}{2} + \frac{1}{2} = 1$.
- $N^{-1}(e^*) = \{e_1, e_2, e_3\}$: Then, $e^* \in N_1(e_1) \cup N_1(e_2)$ where e_1 and e_2 are assumed adjacent. In the worst case e_3 is the ending edge of a path in S containing both e_1 and e_2. Assuming that e_2 is between e_1 and e_3 in this path we obtain $e^* \in N_1(e_2)$. In conclusion, we deduce $\sum_{\{e \in S\}} f(e, e^*) \geq$
  $\sum_{i=1}^{3} f(e_i, e^*) \geq \frac{1}{2} + 2\frac{1}{4} = 1$.
- $N^{-1}(e^*) = \{e_1, e_2, e_3, e_4\}$: then $\sum_{\{e \in S\}} f(e, e^*) \geq \sum_{i=1}^{4} f(e_i, e^*) \geq 4\frac{1}{4} = 1$. $\square$

Theorem 1. LOCIM *is a 1/2-approximation algorithm and this ratio is tight.*

Proof. By propositions 1 and 2, we have $2|S| \geq \sum_{e \in S} g(e) \geq |S^*|$. Fig. 3 gives an example with approximation ratio $\frac{6}{10}$ achieved by LOCIM. This example can be generalized to asymptotic $\frac{1}{2}$ (to appear in the full version). $\square$

Theorem 2. MAXLTSP *is* **APX**-*hard.*

Proof. (Sketch) We construct an L-reduction from the *maximum hamiltonian path problem on graphs with distances 1 and 2* (complete proof appears in the full version). $\square$

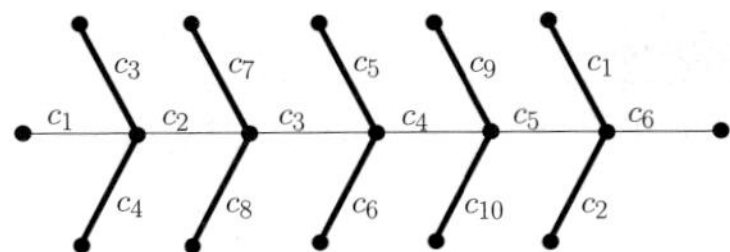

Fig. 3. A critical instance: undrawn edges have label c_1. LOCIM returns the horizontal path (colors c_1 to c_6). An optimum contains the other edges, using colors c_1 to c_{10}.

3   MinLTSP: Hardness and Approximation

We show that the MINLTSP is generally inapproximable, unless $\mathbf{P} = \mathbf{NP}$: MINLTSP$_{(r)}$ where r is any increasing function of n is not $r^{1-\epsilon}$ approximable for any $\epsilon > 0$. We focus subsequently on fixed color frequency r, and show that a simple greedy algorithm exhibits a tight non-trivial approximation ratio equal to $(r + H_r)/2$, where H_r is the harmonic number of order r. Finally we consider the simple case of $r = 2$, for which the algorithm's approximation ratio becomes $\frac{7}{4}$, and show that MINLTSP$_{(2)}$ is **APX**-hard.

3.1   Hardness of MinLTSP

Without restrictions on color frequency, any algorithm for MINLTSP will trivially achieve an approximation factor of n. We show that this ratio is optimal, unless **P=NP**, by reduction from the hamiltonian $s - t$-path problem which is defined as follows: given a graph $G = (V, E)$ with two specified vertices $s, t \in V$, decide whether G has a hamiltonian path from s to t. See [8] (problem [GT39]) for this problem's **NP**-completeness. The restriction of the hamiltonian $s - t$-path problem on graphs where vertices s, t are of degree 1 remains **NP**-complete. In the following let $OPT(\cdot)$ be the optimum solution value to some problem instance.

Theorem 3. *For all $\varepsilon > 0$, MINLTSP is not $n^{1-\varepsilon}$-approximable unless $\boldsymbol{P}$=$\boldsymbol{NP}$, where n is the number of vertices.*

Proof. Let $\varepsilon > 0$ and let $G = (V, E)$ be an instance of the hamiltonian $s - t$-path problem on a graph with two specified vertices $s, t \in V$ having degree 1 in G. Let $p = \lceil \frac{1}{\varepsilon} \rceil - 1$. We construct the following instance I of MINLTSP: take a graph consisting of n^p copies of G, where the i-th copy is denoted by $G_i = (V_i, E_i)$ and s_i, t_i are the corresponding copies of vertices s, t. Set $\mathcal{L}(e) = c_0$ for every $e \in \cup_{i=1}^{n^p} E_i$, $\mathcal{L}([t_i, s_{i+1}]) = c_0$ for all $i = 1, \ldots, n^p - 1$, and $\mathcal{L}([t_{n^p}, s_1]) = c_0$. Complete this graph by taking a new color per remaining edge. This construction can obviously be done in polynomial time, and the resulting graph has n^{p+1} vertices.

If G has a hamiltonian $s - t$-path, then $OPT(I) = 1$. Otherwise, G has no hamiltonian path for any pair of vertices since vertices $s, t \in V$ have a degree 1 in G. Hence $OPT(I) \geq n^p + 1$, because for each copy G_i either the restriction of an optimal tour T^* (with value $OPT(I)$) in copy G_i is a hamiltonian path, and

---

Algorithm 1. Greedy Tour

> Let $T \leftarrow \emptyset$;
> Let $K \leftarrow \{c_1, \ldots, c_q\}$;
> **while** T *is not a tour* **do**
>   Consider $c_j \in K$ maximizing $|E'|$ such that $E' \subseteq \mathcal{L}^{-1}(c_j)$ and $T \cup E'$ is valid;
> $T \leftarrow T \cup E'$;
> $K \leftarrow K \setminus \{c_j\}$;
> **end**
> return T ;

T^* uses a new color (distinct of c_0) or T^* uses at least two new colors linking G_i to the other copies. Since $|V(K_{n^{p+1}})| = n^{p+1}$, we deduce that it is **NP**-complete to distinguish between $OPT(I) = 1$ and $OPT(I) \geq |V(K_{n^{p+1}})|^{1-\frac{1}{p+1}} + 1 > |V(K_{n^{p+1}})|^{1-\varepsilon}$. $\square$

The hamiltonian $s - t$-path problem is also **NP**-complete in graphs of maximum degree 3 (problem [GT39] in [8]). Thus, applying the reduction given in Theorem 3 to this restriction, we deduce that the color frequency r of I is upper bounded by $(\frac{3n+2}{2})n^p = O(n^{p+1})$. Thus, when r grows with n we obtain:

Corollary 1. *There exists $c > 0$ such that for all $\varepsilon > 0$, MINLTSP is not $cr^{1-\varepsilon}$-approximable where r is the color frequency, unless **P=NP**.*

3.2   The Case of Fixed Color Frequency

We describe and analyze a greedy approximation algorithm (referred to as *Greedy Tour* - algorithm 1) for the $\text{MINLTSP}_{(r)}$, for fixed $r = O(1)$. In the description of the algorithm *Greedy Tour* we use the notion of a *valid* subset of edges which do not induce vertices of degree three or more and also do not induce a cycle of length less than n. The algorithm augments iteratively a valid subset of edges by a chosen subset E', until a feasible tour of the input graph is formed. It initializes the set of colors K and iteratively identifies the color that offers the largest set of edges that is valid with respect to the current (partial) tour T and adds it to the tour, while also eliminating the selected color from the current set of colors. For constant $r \geq 1$ Greedy Tour is of polynomially bounded complexity proportional to $O(n^{r+1})$. We introduce some definitions and notations that we use in the analysis of Greedy Tour. Let T^* denote an optimum tour and T be a tour produced by Greedy Tour.

Definition 1. *(Blocks) For $j = 1, \ldots, r$, the j-block with respect to the execution of Greedy Tour is the subset of iterations during which it was $|E'| \geq j$. Let T_j be the subset of edges selected by Greedy Tour during the j-block and $V_j = V(T_j)$ be the set of vertices that are endpoints of edges in T_j.*

Definition 2. *(Color Degree) For a color $c \in \mathcal{L}(T^*)$ define its color degree $f_j(c)$ in V_j to be $f_j(c) = \sum_{v \in V_j} d_{G_c}(v)$, where $G_c = (V, \mathcal{L}^{-1}(c) \cap T^*)$ and $d_{G_c}(v)$ is the degree of v in graph G_c.*

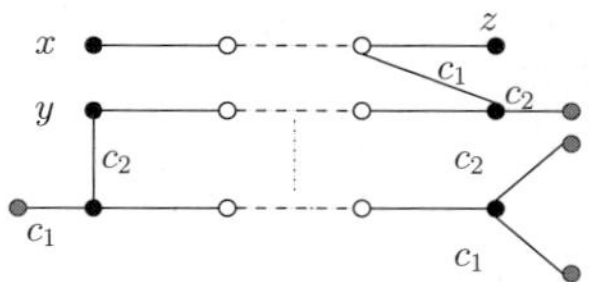

Fig. 4. Graphical illustration of definitions: if $c_1, c_2 \in \mathcal{L}_j(T^*)$, apart from vertices x, y, z, the remaining endpoints of paths are *black terminals*. Inner vertices are *white terminals* (drawn white), while vertices outside the paths are *optional vertices*.

For $j \in \{2, \ldots, r\}$ let $\mathcal{L}_j(T^*)$ be the set of colors that appear at least j times in T^*: $\mathcal{L}_j(T^*) = \{c \in \mathcal{L}(T^*) : |\mathcal{L}^{-1}(c) \cap T^*| \geq j\}$. In general T_j contains $k \geq 0$ paths (in case $k = 0$, T_j is a tour). We consider p vertices $\{v_1, \ldots, v_p\} \subseteq V_j$ of degree 1 in T_j (i.e. they are endpoints of paths), such that each such vertex is adjacent to two edges of T^* that have colors in $\mathcal{L}_j(T^*)$. We refer to vertices of $\{v_1, \ldots v_p\}$ as *black terminals*. We refer to vertices in $V_j \setminus \{v_1, \ldots, v_p\}$ as *white terminals* and to vertices in $V \setminus V_j$ as *optional* (see Fig. 4 for an illustration). We also assume the existence of $q \geq 0$ path endpoints of T_j adjacent to one edge of T^* with color in $\mathcal{L}_j(T^*)$. Clearly $p + q \leq 2k$. We consider a partition of $\mathcal{L}_j(T^*)$: $\mathcal{L}^*_{j,in}$ and $\mathcal{L}^*_{j,out}$. A color $c \in \mathcal{L}_j(T^*)$ belongs in $\mathcal{L}^*_{j,out}$ if there is an edge with this color incident to a black terminal of V_j. Then $\mathcal{L}^*_{j,in} = \mathcal{L}_j(T^*) \setminus \mathcal{L}^*_{j,out}$.

Lemma 1 (Color Degree Lemma). *For any $j = 2, \ldots, r$ the following hold:*

(i) *If $c \in \mathcal{L}^*_{j,in}$, then $f_j(c) \geq |\mathcal{L}^{-1}(c) \cap T^*| + 1 - j$.*

(ii) $\displaystyle \sum_{c \in \mathcal{L}^*_{j,out}} f_j(c) \geq \sum_{c \in \mathcal{L}^*_{j,out}} (|\mathcal{L}^{-1}(c) \cap T^*| + 1 - j) + p.$

Proof.

(i): Except of the $|\mathcal{L}^{-1}(c) \cap T^*| \geq j$ edges of color c in T^*, at most $j - 1$ valid ones (with respect to T_j) may be missing from T_j (and possibly collected in T_{j-1}): if there are more than $j - 1$, then they should have been collected by Greedy Tour in T_j. Then at least $|\mathcal{L}^{-1}(c) \cap T^*| - (j - 1)$ edges of color c must have one endpoint in V_j, and the result follows.

(ii): First we note an important fact for each color $c \in \mathcal{L}^*_{j,out}$: exactly one of the two edges incident to a black terminal (suppose one with color c) belongs to the set of at most $j - 1$ valid c-colored edges, that were not collected in T_j. Using the same argument as in statement **(i)**, we have that at least $|\mathcal{L}^{-1}(c) \cap T^*| - (j-1)$ c-colored edges that are incident to at least one vertex of V_j.

The fact that we mentioned can help us tighten this bound even further, by counting to the color degree the contribution of one edge belonging to the set of at most $j - 1$ valid ones: an edge incident to a black terminal is also incident to either an optional vertex, or a terminal (black or white). Take one black terminal v_i of the two edges $[x, v_i]$, $[v_i, y]$ of T^* incident to it, and consider the cases:

- If x is a white or black terminal: then the color degree must be increased by one, because this edge can be counted twice in the color degree. The same fact also holds for y.

- If x and y are optional vertices: then the color degree must be increased by at least one, because each edge set $\{[x, v_i]\} \cup T_j$ or $\{[v_i, y]\} \cup T_j$ is valid (and was subtracted from $|\mathcal{L}^{-1}(c) \cap T^*|$ with the at most $j-1$ valid ones). However, if the both edges have the same color, the color degree only increases by one unit since the set $\{[x, v_i], [v_i, y]\} \cup T_j$ is not valid.

Therefore we have an increase of one in the color degree of some colors in $\mathcal{L}^*_{j,out}$ and, in fact, of p of them at least. Thus statement **(ii)** follows. $\qquad\square$

Let y_i^* and y_i be the number of colors appearing exactly i times in T^* and T respectively. Then we show that:

Lemma 2. *For* $j = 2, \ldots, r$: $\displaystyle\sum_{i=j}^{r}(i+1-j)y_i^* \leq \sum_{i=j}^{r} 2i\, y_i$

Proof. We prove the inequality by upper and lower bounding the quantity $F_j^* = \sum_{c \in \mathcal{L}_j(T^*)} f_j(c)$. A lower bound stems from Lemma 1:

$$F_j^* \geq \sum_{i=j}^{r}(i+1-j)y_i^* + p \tag{1}$$

Assume now that T_j consists of k disjoint paths. Then $|V_j| = \sum_{i=j}^{r} iy_i + k$ and the number of internal vertices on all k paths of T_j is: $\sum_{i=j}^{r} iy_i - k$. Each internal vertex of V_j may contribute at most twice to F_j^*. Furthermore, each black terminal of T_j, i.e. each vertex of $\{v_1, \ldots, v_p\}$, also contributes twice by definition. Assume that there are q endpoints of paths in T_j, each contributing once to F_j^*. Clearly $p + q \leq 2k$. Then:

$$F_j^* \leq 2\Big(\sum_{i=j}^{r} iy_i - k\Big) + 2p + q \leq \sum_{i=j}^{r} i2y_i + p \tag{2}$$

The result follows by combination of (1) and (2). $\qquad\square$

We prove the approximation ratio of Greedy Tour by using Lemma 2:

Theorem 4. *For any $r \geq 1$ fixed, Greedy tour gives a $\frac{r+H_r}{2} - approximation$ for* MINLTSP$_{(r)}$ *and the analysis is tight.*

Proof. By summing up inequality of Lemma 2 with coefficient $\frac{1}{2(j-1)j}$ for $j = 2, \ldots, r$, we obtain:

$$\sum_{j=2}^{r}\sum_{i=j}^{r} \frac{i+1-j}{2j(j-1)}\, y_i^* \leq \sum_{j=2}^{r}\sum_{i=j}^{r} \frac{i}{j(j-1)}\, y_i \tag{3}$$

For the right-hand part of inequality (3) we have:

$$\sum_{j=2}^{r}\sum_{i=j}^{r} \frac{i}{j(j-1)}\, y_i = \sum_{i=2}^{r} i\, y_i \sum_{j=2}^{i} \frac{1}{j(j-1)} = \sum_{i=2}^{r} i\, y_i \sum_{j=2}^{i}\Big(\frac{1}{j-1} - \frac{1}{j}\Big)$$

$$= \sum_{i=2}^{r} i\, y_i\Big(1 - \frac{1}{i}\Big) = \sum_{i=2}^{r}(i-1)y_i$$

For the left-hand part of inequality (3) we have:

$$\sum_{j=2}^{r}\sum_{i=j}^{r}\frac{i+1-j}{2j(j-1)}\,y_i^* = \sum_{i=2}^{r}\frac{y_i^*}{2}\sum_{j=2}^{i}\frac{i+1-j}{j(j-1)} \tag{4}$$

But we also have:

$$\sum_{j=2}^{i}\frac{i+1-j}{j(j-1)} = \sum_{j=2}^{i}\left(\frac{i-(j-1)}{j-1}-\frac{i-j}{j}\right) - (H_i - 1) = i - H_i \tag{5}$$

where $H_i = \sum_{k=1}^{i}\frac{1}{k}$. Therefore relation (4) becomes by (5):

$$\sum_{j=2}^{r}\sum_{i=j}^{r}\frac{i+1-j}{2j(j-1)}\,y_i^* = \sum_{i=2}^{r}\frac{i-H_i}{2}y_i^* \tag{6}$$

By plugging the right-hand equality and (6) into inequality (3), we obtain:

$$\sum_{i=2}^{r}\frac{i-H_i}{2}y_i^* \le \sum_{i=2}^{r}(i-1)y_i \tag{7}$$

Denote by APX and OPT the number of colors used by Greedy Tour and by the optimum solution respectively. Then

$$OPT = \sum_{i=1}^{r}y_i^*, \quad APX = \sum_{i=1}^{r}y_i, \quad \text{and} \quad \sum_{i=1}^{r}iy_i = \sum_{i=1}^{r}iy_i^* = n \tag{8}$$

where $n = |T| = |T^*|$ is the number of vertices of the graph. By (8) we can write $APX = n - \sum_{i=2}^{r}(i-1)y_i$, and using inequality (7), we deduce:

$$APX \le \sum_{i=1}^{r}iy_i^* - \sum_{i=2}^{r}\frac{i-H_i}{2}y_i^* = \sum_{i=1}^{r}\frac{i+H_i}{2}y_i^*$$

Finally, since $i + H_i \le r + H_r$ when $i \le r$, we obtain:

$$APX \le \frac{r+H_r}{2}\sum_{i=1}^{r}y_i^* = \frac{r+H_r}{2}OPT$$

Fig. 5 illustrates tightness for $r = 2$. Only colors appearing twice are drawn. The optimal tour uses colors c_1 to c_4, whereas Greedy Tour takes c_5 and completes the tour with 6 new colors appearing once. This yields factor $\frac{7}{4} = \frac{2+H_2}{2}$ approximation. A detailed example for $r \ge 3$ is given in the full version. □

MINLTSP$_{(2)}$ proves as hard to approximate as the min-cost hamiltonian path on a complete graph with edge costs 1 and 2 (MINHPP$_{1,2}$ - [ND22] in [8]).

Theorem 5. *A ρ-approximation for MINLTSP$_{(2)}$ can be polynomially transformed into a $(\rho + \varepsilon)$-approximation for MINHPP$_{1,2}$, for all $\varepsilon > 0$.*

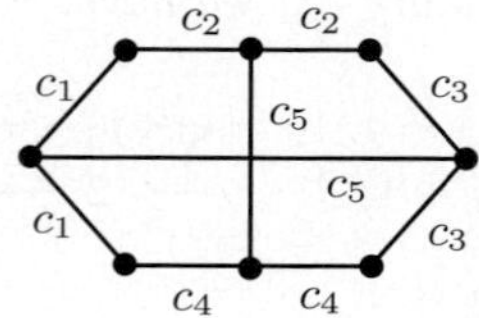

Fig. 5. Only colors appearing twice are represented. The others appears once.

Proof. Let I be an instance of $\text{MINHPP}_{1,2}$, with $V(K_n) = \{v_1, \ldots, v_n\}$, and $d : E(K_n) \to \{1, 2\}$. We construct an instance I' of $\text{MINLTSP}_{(2)}$ on K_{2n} as follows. The vertex set of K_{2n} is $V(K_{2n}) = \{v_1, \ldots, v_n\} \cup \{v'_1, \ldots, v'_n\}$. For every edge $e = [x, y] \in E(K_n)$ with $d(x, y) = 1$ we define two edges $[x, y], [x', y'] \in E(K_{2n})$ with the same color $\mathcal{L}([x, y]) = \mathcal{L}([x', y']) = c_e$. We complete the coloring of K_{2n} by adding a new color for each of the rest of the edges K_{2n}.

Let P^* be an optimum hamiltonian path (with endpoints s and t) of K_n with cost $OPT(I)$. Build a tour T' of K_{2n} by taking P^*, the edges $[x, x'], [y, y']$ and a copy of P^* on vertices $\{v'_1, \ldots, v'_n\}$. Then $|\mathcal{L}(T')| = OPT(I) + 2$, and:

$$OPT(I') \leq OPT(I) + 2 \tag{9}$$

Now let T' be a feasible solution of I'. Assume that n_2 colors appear twice in T' (thus $2n - 2n_2$ colors appear once in T'). In K_n, the set of edges with these colors corresponds to a collection of disjoint paths $P_1, \ldots, P_k$ with edges of distance 1. Then, by adding exactly $n - 1 - n_2$ edges we obtain a hamiltonian path P of K_n with cost at most:

$$d(P) \leq |\mathcal{L}(T')| - 2 \tag{10}$$

where $d(P) = \sum_{e \in P} d(e)$. Using inequalities (9) and (10), we deduce $OPT(I') = OPT(I) + 2$. Now, if T is a ρ-approximation for $\text{MINLTSP}_{(2)}$, we deduce $d(P) \leq \rho OPT(I) + 2(\rho - 1) \leq (\rho + \varepsilon)OPT(I)$ when n is large enough. $\qquad\square$

Since the traveling salesman problem with distances 1 and 2 ($\text{MINTSP}_{1,2}$) is **APX**-hard, [16] (then, $\text{MINHPP}_{1,2}$ is also **APX**-hard), we conclude by Theorem 5 that $\text{MINLTSP}_{(2)}$ is **APX**-hard. Moreover, $\text{MINLTSP}_{(2)}$ belongs to **APX** because any feasible tour is 2-approximate.

Corollary 2. $\text{MINLTSP}_{(2)}$ *is **APX**-complete.*

4   Open Questions

Is there a better approximation algorithm for $\text{MINLTSP}_{(r)}$, when r is a fixed small constant (e.g. $r = 2$)? For MAXLTSP, using $k-$improvements for fixed $k \geq 3$ could yield better performance but analysis appears quite non-trivial. It is also interesting to study $\text{MAXLTSP}_{(r)}$ with bounded color frequency r.

References

1. Albert, M., Frieze, A., Reed, B.: Multicoloured Hamilton Cycles. The Electronic Journal of Combinatorics, 2, R10 (1995)
2. Brueggemann, T., Monnot, J., Woeginger, G.J.: Local search for the minimum label spanning tree problem with bounded color classes. Oper. Res. Lett. 31(3), 195–201 (2003)
3. Broersma, H., Li, X.: Spanning Trees with Many or Few Colors in Edge-Colored Graphs. Discussiones Mathematicae Graph Theory 17(2), 259–269 (1997)
4. Broersma, H., Li, X., Woeginger, G.J., Zhang, S.: Paths and Cycles in Colored Graphs. Australasian Journal on Combinatorics 31, 299–311 (2005)
5. Chang, R.-S., Leu, S.-J.: The Minimum Labeling Spanning Trees. Inf. Process. Lett. 63(5), 277–282 (1997)
6. Erdős, P., Nesetril, J., Roedl, V.: Some problems related to partitions of edges of a graph. In: Graphs and Other Combinatorial Topics, Teubner, Leipzig, pp. 54–63 (1983)
7. Frieze, A.M., Reed, B.A.: Polychromatic Hamilton cycles. Discrete Mathematics 118, 69–74 (1993)
8. Garey, M.R., Johnson, D.S.: Computers and Intractability: A Guide to the Theory of NP-Completeness. W. H. Freeman and Company, New York (1979)
9. Hahn, G., Thomassen, C.: Path and cycle sub-Ramsey numbers and an edge-coloring conjecture. Discrete Mathematics 62(1), 29–33 (1986)
10. Hassin, R., Monnot, J., Segev, D.: Approximation Algorithms and Hardness Results for Labeled Connectivity Problems. J. Comb. Opt. 14(4), 437–453 (2007)
11. Hassin, R., Monnot, J., Segev, D.: The Complexity of Bottleneck Labeled Graph Problems. In: Brandstädt, A., Kratsch, D., Müller, H. (eds.) WG 2007. LNCS, vol. 4769. Springer, Heidelberg (2007)
12. Krumke, S.O., Wirth, H.-C.: On the Minimum Label Spanning Tree Problem. Inf. Process. Lett. 66(2), 81–85 (1998)
13. Maffioli, F., Rizzi, R., Benati, S.: Least and most colored bases. Discrete Applied Mathematics 155(15), 1958–1970 (2007)
14. Monnot, J.: The labeled perfect matching in bipartite graphs. Inf. Process. Lett. 96(3), 81–88 (2005)
15. Monnot, J.: A note on the hardness results for the labeled perfect matching problems in bipartite graphs. RAIRO-Operations Research (to appear, 2007)
16. Papadimitriou, C.H., Yannakakis, M.: The traveling salesman problem with distances one and two. Math. of Operations Research 18(1), 1–11 (1993)
17. Punnen, A.P.: Traveling Salesman Problem under Categorization. Oper. Res. Lett. 12(2), 89–95 (1992)
18. Punnen, A.P.: Erratum: Traveling Salesman Problem under Categorization. Oper. Res. Lett. 14(2), 121 (1993)
19. Xiong, Y., Golden, B., Wasil, E.: The Colorful Traveling Salesman Problem. In: Extending the Horizons: Advances in Computing, Optimization, and Decision Technologies, pp. 115–123. Springer, Heidelberg (2007)

Navigating in a Graph by Aid of Its Spanning Tree[*]

Feodor F. Dragan[1] and Martin Matamala[2]

[1] Algorithmic Research Laboratory, Department of Computer Science
Kent State University, Kent, OH 44242, USA[**]
dragan@cs.kent.edu
[2] Departmento de Ingeniería Matemática, Universidad de Chile
Centro de Modelamiento Matemático UMR 2071CNRS, Santiago, Chile
mmatamal@dim.uchile.cl

Abstract. Let $G = (V, E)$ be a graph and T be a spanning tree of G. We consider the following strategy in advancing in G from a vertex x towards a target vertex y: from a current vertex z (initially, $z = x$), unless $z = y$, go to a neighbor of z in G that is closest to y in T (breaking ties arbitrarily). In this strategy, each vertex has full knowledge of its neighborhood in G and can use the distances in T to navigate in G. Thus, additionally to standard local information (the neighborhood $N_G(v)$), the only global information that is available to each vertex v is the topology of the spanning tree T (in fact, v can know only a very small piece of information about T and still be able to infer from it the necessary tree-distances). For each source vertex x and target vertex y, this way, a path, called a greedy routing path, is produced. Denote by $g_{G,T}(x, y)$ the length of a longest greedy routing path that can be produced for x and y using this strategy and T. We say that a spanning tree T of a graph G is an *additive r-carcass* for G if $g_{G,T}(x, y) \leq d_G(x, y) + r$ for each ordered pair $x, y \in V$. In this paper, we investigate the problem, given a graph family $\mathcal{F}$, whether a small integer r exists such that any graph $G \in \mathcal{F}$ admits an additive r-carcass. We show that rectilinear $p \times q$ grids, hypercubes, distance-hereditary graphs, dually chordal graphs (and, therefore, strongly chordal graphs and interval graphs), all admit additive 0-carcasses. Furthermore, every chordal graph G admits an additive $(\omega(G) + 1)$-carcass (where $\omega(G)$ is the size of a maximum clique of G), each 3-sun-free chordal graph admits an additive 2-carcass, each chordal bipartite graph admits an additive 4-carcass. In particular, any k-tree admits an additive $(k+2)$-carcass. All those carcasses are easy to construct.

1 Introduction

As part of the recent surge of interest in different kind of networks, there has been active research exploring strategies for navigating synthetic and real-world

---

[*] This work was partially supported by CONICYT through grants Anillo en Redes ACT08 and Fondap.

[**] These results were obtained while the first author was visiting the Universidad de Chile, Santiago.

S.-H. Hong, H. Nagamochi, and T. Fukunaga (Eds.): ISAAC 2008, LNCS 5369, pp. 788–799, 2008.

networks (modeled usually as graphs). These strategies specify some rules to be used to advance in a graph (a network) from a given vertex towards a target vertex along a path that is close to shortest. Current strategies include (but not limited to): routing using full-tables, interval routing, routing labeling schemes, greedy routing, geographic routing, compass routing, etc. in wired or wireless communication networks and in transportation networks (see [19,20,27,33,25,40] and papers cited therein); routing through common membership in groups, popularity, and geographic proximity in social networks and e-mail networks (see [2,3,27,32] and literature cited therein).

In this paper we use terminology used mostly for communication networks. Thus, navigation is performed using a routing scheme, i.e., a mechanism that can deliver packets of information from any vertex of a network to any other vertex. In most strategies, each vertex v of a graph has full knowledge of its neighborhood and uses a piece of global information available to it about the graph topology – some "sense of direction" to each destination, stored locally at v. Based only on this information and the address of a destination vertex, vertex v needs to decide whether the packet has reached its destination, and if not, to which neighbor of v to forward the packet.

1.1   Some Known Strategies

In *routing using full-tables*, each vertex v of G knows for each destination u the first edge along some shortest path from v to u (so-called *complete routing table*). When v needs to send a message to u, it just sends the message along the edge stored for destination u. While this approach guarantees routing along a shortest path, it is too expensive for large systems since it requires to store locally $O(n \log \delta)$ bits of global information for an n-vertex graph with maximum degree δ.

Unfortunately, if one insists on a routing via shortest paths, $\Omega(n \log \delta)$ bits is the lower bound on the memory requirements per vertex [24] (this much each vertex needs to know at least). To obtain routing schemes for general graphs that use $o(n)$ of memory at each vertex, one has to abandon the requirement that packets are always delivered via shortest paths, and settle instead for the requirement that packets are routed on paths that are relatively close to shortest. The efficiency of a routing scheme is measured in terms of its additive stretch, called *deviation* (or multiplicative stretch, called *delay*), namely, the maximum surplus (or ratio) between the length of a route, produced by the scheme for a pair of vertices, and the shortest route. There is a tradeoff between the memory requirements of a routing scheme (how much of global information is available locally at a vertex) and the worst case stretch factor it guarantees. Any multiplicative t-stretched routing scheme must use $\Omega(n)$ bits for some vertices in some graphs for $t < 3$ [21] (see also [17]), and $\Omega(n \log n)$ bits for $t < 1.4$ [24]. These lower bounds show that it is not possible to lower memory requirements of a routing scheme for an arbitrary network if it is desirable to route messages along paths close to optimal. Therefore, it is interesting, both from a theoretical and a practical view point, to look for specific routing strategies on graph families with certain topological properties.

One specific way of routing, called *interval routing*, has been introduced in [36] and later generalized in [31]. In this method, the complete routing tables are compressed by grouping the destination addresses which correspond to the same output edge. Then each group is encoded as an interval, so that it is easy to check whether a destination address belongs to the group. This approach requires $O(\delta \log n)$ bits of memory per vertex, where δ is the maximum degree of a vertex of the graph. A graph must satisfy some topological properties in order to support interval routing, especially if one insists on paths close to optimal. Routing schemes for many graph classes were obtained by using interval routing techniques. The classical and most recent results in this field are presented in [19,20].

Recently, so-called *routing labeling schemes* [33] become very popular. A number of interesting results for general graphs and particular classes of graphs were obtained. These are schemes that label the vertices of a graph with short labels (describing some global topology information) in such a way that given the label of a source vertex and the label of a destination, it is possible to compute efficiently the edge from the source that heads in the direction of the destination. In [18,40], a shortest path routing scheme for trees with $O(\log^2 n/\log\log n)$-bit labels is described. For general graphs, the most general result to date is a multiplicative $(4k-5)$-stretched routing labeling scheme that uses labels of size $\tilde{O}(kn^{1/k})$ bits[1] is obtained in [40] for every $k \geq 2$. For planar graphs, a shortest path routing labeling scheme which uses $8n+o(n)$ bits per vertex is developed in [22], and a multiplicative $(1+\epsilon)$-stretched routing labeling scheme for every $\epsilon > 0$ which uses $O(\epsilon^{-1}\log^3 n)$ bits per vertex is developed in [39]. Routing in graphs with doubling dimension α has been considered in [1,10,37,38]. It was shown that any graph with doubling dimension α admits a multiplicative $(1+\epsilon)$-stretched routing labeling scheme with labels of size $\epsilon^{-O(\alpha)}\log^2 n$ bits. Recently, the routing result for trees of [18,40] was used in designing additive $O(1)$-stretched routing labeling schemes with $O(\log^{O(1)} n)$ bit labels for several families of graphs, including chordal graphs, chordal bipartite graphs, circular-arc graphs, AT-free graphs and their generalizations, the graphs with bounded longest induced cycle, the graphs of bounded tree–length, the bounded clique-width graphs, etc. (see [12,13,14,15] and papers cited therein).

In wireless networks, the most popular strategy is the *geographic routing* (sometimes called also the *greedy geographic routing*), were each vertex forwards the packet to the neighbor geographically closest to the destination (see survey [25]). Each vertex of the network knows its position (e.g., Euclidean coordinates) in the underlying physical space and forwards messages according to the coordinates of the destination and the coordinates of neighbors. Although this greedy method is effective in many cases, packets may get routed to where no neighbor is closer to the destination than the current vertex. Many recovery schemes have been proposed to route around such voids for guaranteed packet delivery as long as a path exists [4,26,30]. These techniques typically exploit planar subgraphs (e.g., Gabriel graph, Relative Neighborhood graph), and packets traverse faces on such graphs using the well-known right-hand rule.

---

[1] Here, $\tilde{O}(f)$ means $O(f \text{ polylog } n)$.

All earlier papers assumed that vertices are aware of their physical location, an assumption which is often violated in practice for various of reasons (see [16,28,35]). In addition, implementations of recovery schemes are either based on non-rigorous heuristics or on complicated planarization procedures. To overcome these shortcomings, recent papers [16,28,35] propose routing algorithms which assign virtual coordinates to vertices in a metric space X and forward messages using geographic routing in X. In [35], the metric space is the Euclidean plane, and virtual coordinates are assigned using a distributed version of Tutte's "rubber band" algorithm for finding convex embeddings of graphs. In [16], the graph is embedded in R^d for some value of d much smaller than the network size, by identifying d beacon vertices and representing each vertex by the vector of distances to those beacons. The distance function on R^d used in [16] is a modification of the ℓ_1 norm. Both [16] and [35] provide substantial experimental support for the efficacy of their proposed embedding techniques – both algorithms are successful in finding a route from the source to the destination more than 95% of the time – but neither of them has a provable guarantee. Unlike embeddings of [16] and [35], the embedding of [28] guarantees that the geographic routing will always be successful in finding a route to the destination, if such a route exists. Algorithm of [28] assigns to each vertex of the network a virtual coordinate in the hyperbolic plane, and performs greedy geographic routing with respect to these virtual coordinates. More precisely, [28] gets virtual coordinates for vertices of a graph G by embedding in the hyperbolic plane a spanning tree of G. The proof that this method guaranties delivery is relied only on the fact that the hyperbolic greedy route is no longer than the spanning tree route between two vertices; even more, it could be much shorter as greedy routes take enough short cuts (edges which are not in the spanning tree) to achieve significant saving in stretch. However, although the experimental results of [28] confirm that the greedy hyperbolic embedding yields routes with low stretch when applied to typical unit-disk graphs, the worst-case stretch is still linear in the network size.

1.2 Our Approach

Motivated by the work of Robert Kleinberg [28], in this paper, we initiate exploration of the following strategy in advancing in a graph from a source vertex towards a target vertex. Let $G = (V, E)$ be a graph and T be a spanning tree of G. To route/move in G from a vertex x towards a target vertex y, use the following rule:

> *from a current vertex z (initially, $z = x$), unless $z = y$,*
> *go to a neighbor of z in G that is closest to y in T*
> *(break ties arbitrarily).*

In this strategy, each vertex has full knowledge of its neighborhood in G and can use the distances in T to navigate in G. Thus, additionally to standard local information (the neighborhood $N_G(v)$), the only global information that is available to each vertex v is the topology of the spanning tree T. In fact, v

can know only a very small piece of information about T and still be able to infer from it the necessary tree-distances. It is known [23,34] that the vertices of an n-vertex tree T can be labeled in $O(n \log n)$ total time with labels of up to $O(\log^2 n)$ bits such that given the labels of two vertices v, u of T, it is possible to compute in constant time the distance $d_T(v, u)$, by merely inspecting the labels of u and v. Hence, one may assume that each vertex v of G knows, additionally to its neighborhood in G, only its $O(\log^2 n)$ bit distance label. This distance label can be viewed as a virtual coordinate of v.

For each source vertex x and target vertex y, by this routing strategy, a path, called a *greedy routing path*, is produced (clearly, this routing strategy will always be successful in finding a route to the destination). Denote by $g_{G,T}(x, y)$ the length of a longest greedy routing path that can be produced for x and y using this strategy and T. We say that a spanning tree T of a graph G is an *additive r-carcass* for G if $g_{G,T}(x, y) \leq d_G(x, y) + r$ for each ordered pair $x, y \in V$ (in a similar way one can define also a *multiplicative t-carcass* of G).

In this paper, we start investigating the problem, given a graph family $\mathcal{F}$, whether a small integer r exists such that any graph $G \in \mathcal{F}$ admits an additive r-carcass, and give our preliminary results. We show that rectilinear $p \times q$ grids, hypercubes, distance-hereditary graphs, dually chordal graphs (and, therefore, strongly chordal graphs and interval graphs), all admit additive 0-carcasses. Furthermore, every chordal graph G admits an additive $(\omega(G)+1)$-carcass (where $\omega(G)$ is the size of a maximum clique of G), each 3-sun-free chordal graph admits an additive 2-carcass, each chordal bipartite graph admits an additive 4-carcass. In particular, any k-tree admits an additive $(k + 2)$-carcass. All those carcasses are easy to construct.

2   Preliminaries

All graphs occurring in this paper are connected, finite, undirected, unweighted, loopless and without multiple edges. In a graph $G = (V, E)$ $(n = |V|, m = |E|)$ the *length* of a path from a vertex v to a vertex u is the number of edges in the path. The *distance* $d_G(u, v)$ between the vertices u and v is the length of a shortest path connecting u and v. The *neighborhood* of a vertex v of G is the set $N_G(v) = \{u \in V : uv \in E\}$ and the *closed neighborhood* of v is $N_G[v] = N_G(v) \cup \{v\}$. The *disk* of radius k centered at v is the set of all vertices at distance at most k to v, i.e., $D_k(v) = \{u \in V : d_G(u, v) \leq k\}$. A set $S \subseteq V$ is a *clique* (an *independent set*) of G if all vertices of S are pairwise adjacent (respectively, nonadjacent) in G. A clique of G is *maximal* if it is not contained in any other clique of G.

Next we recall the definitions of special graph classes mentioned in this paper (see survey [8]). A graph is *chordal* if it does not have any induced cycle of length greater than 3. A *p-sun* $(p \geq 3)$ is a chordal graph on $2p$ vertices whose vertex set can be partitioned into two sets, $U = \{u_0, \ldots, u_{p-1}\}$ and $W = \{w_0, \ldots, w_{p-1}\}$, such that W is an independent set, U is a clique, and every w_i is adjacent only to u_i and u_{i+1} (mod p). A chordal graph having no induced subgraphs isomorphic to p-suns (for any $p \geq 3$) is called a *strongly chordal graph*. A chordal graph having no induced subgraphs isomorphic to 3-sun is called a *3-sun-free chordal graph*. A graph is *chordal bipartite* if it is bipartite and has no induced cycles of length greater than 4. A *dually chordal graph* is the intersection graph of the maximal cliques of a chordal graph (see [8,7] for many equivalent definitions of dually chordal graphs and strongly chordal graphs). A graph is *interval* if it is the intersection graph of intervals of a line. It is known that interval graphs are strongly chordal and strongly chordal graphs are dually chordal (see [8,7]). A graph G is *distance-hereditary* if every induced path of G is shortest (see [8] for many equivalent definitions of distance-hereditary graphs). The *k-trees* are defined recursively: a clique of size k (denoted by K_k) is a k-tree; if G is a k-tree, then a graph obtained from G by adding a new vertex v adjacent to all vertices of some clique K_k of G is a k-tree. It is known (see [8]) that all k-trees are chordal graphs and that maximal cliques of a k-tree have size at most $k + 1$.

Let now $G = (V, E)$ be a graph and T be a spanning tree of G. In what follows, we will use the following notations. For vertices v and u from V, denote by vTu the (unique) path of T connecting vertices v and u. For a source vertex x and a target vertex y in G, denote by $R_{G,T}(x,y)$ a greedy routing path obtained for x and y by using tree T and the strategy described in Subsection 1.2. Clearly, for the same pair of vertices x and y, breaking ties differently, different greedy routing paths $R_{G,T}(x,y)$ can be produced. Denote, as before, by $g_{G,T}(x,y)$, the length of a longest greedy routing path that can be produced for x and y. If no confusion can arise, we will omit indexes G and T, i.e., use $R(x,y)$ and $g(x,y)$ instead of $R_{G,T}(x,y)$ and $g_{G,T}(x,y)$.

For $r \geq 0$ and $t \geq 1$, a spanning tree T of a graph G is called an *additive r-carcass* (a *multiplicative t-carcass*) for G if $g_{G,T}(x,y) \leq d_G(x,y)+r$ (respectively, $g_{G,T}(x,y) \leq t\, d_G(x,y)$) for each ordered pair $x,y \in V$.

Let x^* be the neighbor of x in $R_{G,T}(x,y)$ and x' be the neighbor of x in xTy. Since both x^* and x' are in $N_G(x)$ and $d_T(x',y) = d_T(x,y) - 1$, according to our strategy $d_T(x^*,y) \leq d_T(x',y) = d_T(x,y) - 1$ must hold. Furthermore, any subpath of a greedy routing path $R_{G,T}(x,y)$ containing y is a greedy routing path to y as well. Hence, one can conclude, by induction, that the length of any greedy routing path $R_{G,T}(x,y)$ never exceeds $d_T(x,y)$. It is clear also, that a greedy routing path $R_{G,T}(x,y) := (x := x_0, x_1, x_2, \ldots, y := x_\ell)$ cannot have a chord $x_i x_j \in E$ with $j > i + 1$ (since $d_T(x_{i+1},y) > d_T(x_j,y)$), i.e., any greedy routing path is an induced path. Thus, we have the following.

Observation 1. *Let G be an arbitrary graph and T be its arbitrary spanning tree. Then, for any vertices x, y of G,*

(a) $g_{G,T}(x,y) \leq d_T(x,y)$,
(b) any greedy routing path $R_{G,T}(x,y)$ is an induced path of G,
(c) a tale of any greedy routing path is a greedy routing path.

Since in distance-hereditary graphs each induced path is a shortest path, by Observation 1(b), we conclude.

Corollary 1. *Any spanning tree of a distance-hereditary graph G is a 0-carcass of G.*

There are well-known notions of additive tree r-spanners and multiplicative tree t-spanners. For $r \geq 0$ and $t \geq 1$, a spanning tree T of a graph G is called an *additive tree r-spanner* (a *multiplicative tree t-spanner*) of G if $d_T(x,y) \leq d_G(x,y) + r$ (respectively, $d_T(x,y) \leq t\, d_G(x,y)$) for each pair $x, y \in V$ [9]. By Observation 1(a), we obtain.

Corollary 2. *Any additive tree r-spanner (multiplicative tree t-spanner) of a graph G is an additive r-carcass (multiplicative t-carcass) of G.*

Note that the converse of Corollary 2 is not generally true. As we will see in next sections, there are many families of graphs which do not admit any tree r-spanners (additive as well as multiplicative) for any constant r, yet they admit very good carcasses. For example, there is no constant r such that any 2-tree or any chordal bipartite graph has a tree r-spanner (additive or multiplicative), but both these families of graphs admit additive 4-carcasses (see Section 5 for details).

In what follows, in a rooted tree T, by $f(v)$ we will denote the father of a vertex v.

3   Rectilinear Grids and Hypercubes

In this section we show that the rectilinear grids and the hypercubes admit additive 0-carcasses.

Consider a rectilinear $p \times q$ grid G and assume that it is naturally embedded into the plane such that all inner faces of G are squares (see Fig. 1). First we notice that G does not admit any good tree spanner. For this, consider an arbitrary spanning tree T of G, and assume that p and q are odd integers and $p \leq q$. Since T is a planar graph with only the outer face, we can connect by a Jordan curve $\mathcal{C}$ a point of the plane inside the central square of G with a point

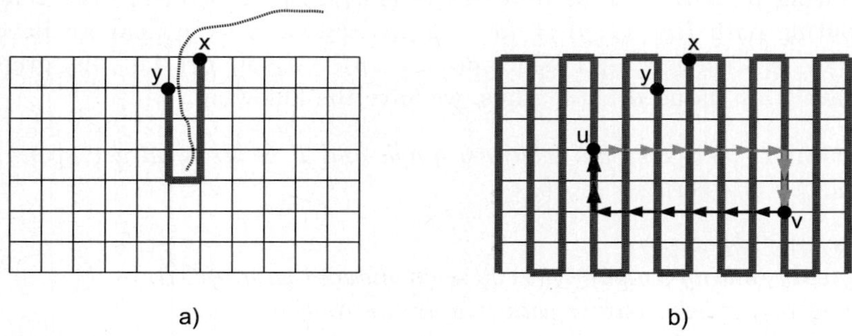

a)   b)

Fig. 1. Rectilinear grids do not admit any (additive or multiplicative) tree r-spanners with a constant r, but have additive 0-carcasses

in the outer face of G without intersecting the tree T. Let R be the first square of G crossed by $\mathcal{C}$ and x and y be two opposite vertices of R (see Fig. 1(a) for an illustration). Clearly, for x and y, $d_T(x, y) \geq p + 1$ holds, while $d_G(x, y) = 2$. Here, we considered nonadjacent vertices of G since adjacent vertices are of no interest in our greedy routing. Thus, there are no good tree spanners for rectilinear grids. On the other hand, G admits an additive 0-carcass. Consider a Hamiltonian path of G depicted on Fig. 1(b), called *column-wise Hamiltonian path*. This path is an additive 0-carcass of G. We leave verification of this fact to the reader.

Now we turn to the hypercubes. Let $H_q = (V, E)$ be the q-dimensional hypercube whose vertices are binary words of length q and two vertices are adjacent if they differ in exactly one letter. Let $a \in \{0, 1\}$ and $i \in \{1, \ldots, q\}$. Let $H_q^{a,i}$ be a subgraph of H_q induced by vertices having letter a in the position i. Then, $H' := H_q^{a,i}$ is isomorphic to the $(q - 1)$-dimensional hypercube and $d_{H_q}(x, y) = d_{H'}(x, y)$, whenever the letter in the position i of x and y is a.

Let T be the *Gray-Hamiltonian* path of H_q defined recursively as follows. If $x_1, \ldots, x_{2^q-1}$ is the Gray-Hamiltonian path for H_{q-1}, then T is given by $T_0 T_1$, where $T_0 = x_1 0, \ldots, x_{2^q-1} 0$ and $T_1 = x_{2^q-1} 1, \ldots, x_1 1$. By applying two steps of previous recursion, it is clear that T can be decomposed into four consecutive subpaths $T = T_{00} T_{10} T_{11} T_{01}$, where the subpath T_w contains all the vertices of H_q ending with w. Notice that T_0 is the Gray-Hamiltonian path for $H_q^{0,q}$, T_1 is the reverse of the Gray-Hamiltonian path for $H_q^{1,q}$ and $T_{10} T_{11}$ is the Gray-Hamiltonian path for the hypercube $H_q^{1,q-1}$.

By using induction on q, we prove that $g(x, y) := g_{G,T}(x, y) = d_G(x, y)$, where $G = H_q$ and T is the Gray-Hamiltonian path of H_q. When x and y belong to T_0 (resp. T_1), conclusion is obtained by applying induction hypothesis to vertices x and y in the hypercube $H_q^{0,q}$ (resp. $H_q^{1,q}$) with the Gray-Hamiltonian path T_0 (resp. T_1). Similarly, when x belongs to T_{10} and y belongs to T_{11}, we can apply induction in the hypercube $H_q^{1,q-1}$ with the Gray-Hamiltonian path $T_{10} T_{11}$.

For the remaining cases, let x^* denote the vertex next to x in a greedy routing path $R_{G,T}(x, y)$. If x^* and y belong to T_1, and x belongs to T_0, then $d_G(x^*, y) = d_G(x, y) - 1$, since x^* and y agree in their last letters. By applying induction to x^* and y using T_1, we get that $g(x^*, y) = d_{H_q^{1,q}}(x^*, y) = d_G(x^*, y)$. Hence, $g(x, y) = 1 + g(x^*, y) = 1 + d_G(x^*, y) = d_G(x, y)$. Let us now consider the case when x and x^* belong to T_0, and y belongs to T_1. As x has a neighbor in T_1 and we are assuming that x^* belongs to T_0, vertex y must belong to T_{11}. We have already considered the case when x belongs to T_{10}. Hence, let us assume that x belongs to T_{00}. As x has a neighbor in T_{10}, vertex x^* must belong to T_{10}. Since in this case $d_G(x^*, y) = d_G(x, y) - 1$, we can conclude as before, by applying induction to x^* and y in $H_q^{1,q-1}$ with the Gray-Hamiltonian path $T_{10} T_{11}$.

So, we can state the following theorem.

Theorem 1. *Every rectilinear grid and every hypercube admits an additive 0-carcass (which is a Hamiltonian path) constructible in linear time.*

4   Locally Connected Spanning Trees Are Additive 0-Carcasses: Dually Chordal Graphs

In this section, we show that every dually chordal graph $G = (V, E)$ admits an additive 0-carcass constructible in linear time. Recall that every dually chordal graph has an additive tree 3-spanner and there are dually chordal graphs without any additive tree 2-spanners (see [6]). Clearly, those additive tree 3-spanners are additive 3-carcasses, but it is not hard to see that they are not necessarily additive 0-carcasses (see, e.g., Fig. 2).

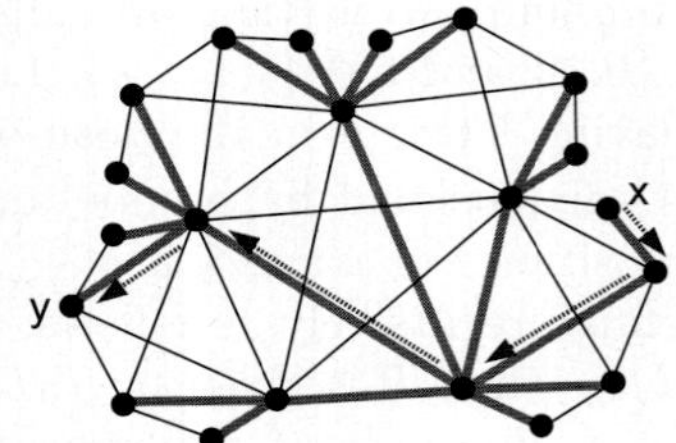

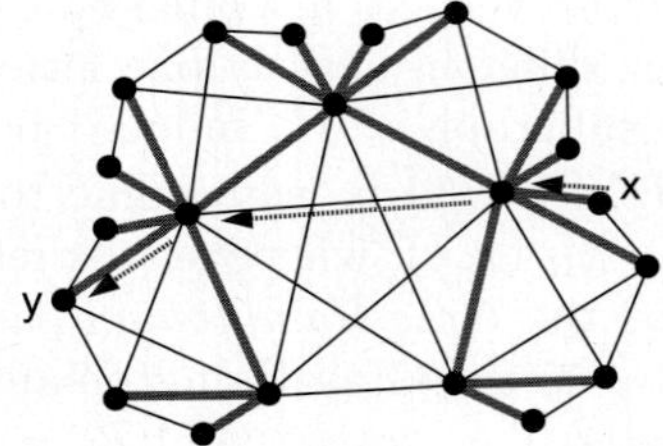

Fig. 2. A dually chordal graph with an additive tree 3-spanner (on the left) and an additive 0-carcass (on the right). This dually chordal graph does not have any additive tree 2-spanner. A greedy routing path from x to y with respect to the corresponding tree is shown on both pictures.

Let G be a graph. We say that a spanning tree T of G is *locally connected* if the closed neighborhood $N_G[v]$ of any vertex v of G induces a subtree in T (i.e., $T \cap N_G[v]$ is a connected subgraph of T). See the right picture on Fig. 2 for an example of a locally connected spanning tree.

Theorem 2. *If T is a locally connected spanning tree of a graph G, then T is an additive 0-carcass of G.*

Proof. Assume that we want to route from a source vertex x to a target vertex y in G. Let v be an arbitrary vertex of G and v^* be a vertex from $N_G[v]$ closest to y in T. Since $T \cap N_G[v]$ is a connected subgraph of T, for each vertex $v \in V$ such a neighbor v^* is unique (any subtree of a tree has only one vertex closest in T to a given vertex y). Moreover, $v^* \neq v$, unless $v = y$. In what follows, we will assume that the tree T is rooted at vertex y.

Claim 1. *For any vertex $v \in V$, the vertex v^* belongs to a shortest path of G connecting v and y.*

Proof. We prove by induction on $d_G(v, y)$. If $d_G(v, y) \leq 1$, then $v^* = y$ and therefore v^* belongs to any shortest path between v and y. So, assume that $d_G(v, y) \geq 2$. Consider a shortest path $P(v, y) := (v, a, b, \ldots, y)$ in G connecting v and y, where a and b are the first two (after v) vertices of this path. They exist since $d_G(v, y) \geq 2$.

To obtain the conclusion, we prove that v^* and b are adjacent. For sake of contradiction, let us assume that they are not adjacent. Since $a, b^* \in N_G(b)$ and $T \cap N_G[b]$ is a connected subgraph of T, we get that v^* is not on path aTb^*. We also know that $a \in N_G[v] \cap N_G[b]$ and therefore both v^* and b^* are ancestors in T of a (recall that we have rooted T at y). Hence, b^* is on the path aTv^*. As $a, v^* \in N_G(v)$ and $T \cap N_G[v]$ is a connected subgraph of T, we get that v and b^* are adjacent. By induction, b^* belongs to a shortest path of G between b and y, which leads to the following contradiction: $d_G(v, y) \leq 1 + d_G(b^*, y) = d_G(b, y) < d_G(v, y)$. $\Box$(of Claim)

Now we prove by induction on $d_T(x, y)$ that $g(x, y) = d_G(x, y)$. Indeed, $g(x, y) = 1 + g(x^*, y)$ and, by induction, $g(x^*, y) = d_G(x^*, y)$ as $d_T(x, y) > d_T(x^*, y)$. By Claim 1, we conclude $g(x, y) = 1 + d_G(x^*, y) = d_G(x, y)$. $\Box$

It has been shown in [7] that the graphs admitting locally connected spanning trees are precisely the dually chordal graphs. Furthermore, [7] showed that the class of dually chordal graphs contains such known families of graphs as strongly chordal graphs, interval graphs and others. Thus, we have the following corollary.

Corollary 3. *Every dually chordal graph admits an additive 0-carcass constructible in linear time. In particular, any strongly chordal graph (any interval graph) admits an additive 0-carcass constructible in linear time.*

Note that, in [5,7], it was shown that dually chordal graphs can be recognized in linear time, and if a graph G is dually chordal, then a locally connected spanning tree of G can be efficiently constructed.

5 Additive Carcasses for Chordal Graphs and Chordal Bipartite Graphs

In this section, we just list our results for chordal graphs and chordal bipartite graph. The proofs can be found in the journal version of this paper.

Theorem 3. *Every chordal bipartite graph admits an additive 4-carcass constructible in linear time.*

Recall that chordal bipartite graphs do not have any tree r-spanners (additive or multiplicative) with a constant r (see, e.g., [11]).

Theorem 4. *Any shortest path tree of a chordal graph G is an additive $(\omega(G) + 1)$-carcass of G. Here $\omega(G)$ is the size of a maximum clique of G.*

Since k-trees are chordal graphs with the size of a maximum clique at most $k+1$, we conclude.

Corollary 4. *Every k-tree admits an additive $(k + 2)$-carcass constructible in linear time. In particular, any 2-tree admits an additive 4-carcass constructible in linear time.*

Recall that 2-trees do not have any tree r-spanners (additive or multiplicative) with a constant r (see, e.g., [29]). We also have the following result.

Corollary 5. *Every 3-sun-free chordal graph admits an additive 2-carcass.*

References

1. Abraham, I., Gavoille, C., Goldberg, A.V., Malkhi, D.: Routing in Networks with Low Doubling Dimension. In: ICDCS 2006, p. 75 (2006)
2. Adamic, L.A., Lukose, R.M., Huberman, B.A.: Local Search in Unstructured Networks. Willey, New-York (2002)
3. Adamic, L.A., Lucose, R.M., Puniyani, A.R., Huberman, B.A.: Search in power-law networks. Physical Review E 64, 046135, 1–8 (2001)
4. Bose, P., Morin, P., Stojmenovic, I., Urrutia, J.: Routing with guaranteed delivery in ad hoc wireless networks. In: Proceedings of the 3rd International Workshop on Discrete algorithms and Methods for Mobile Computing and Communications, pp. 48–55. ACM Press, New York (1999)
5. Brandstädt, A., Chepoi, V.D., Dragan, F.F.: The algorithmic use of hypertree structure and maximum neighbourhood orderings. Discrete Appl. Math. 82, 43–77 (1998)
6. Brandstädt, A., Chepoi, V.D., Dragan, F.F.: Distance approximating trees for chordal and dually chordal graphs. Journal of Algorithms 30, 166–184 (1999)
7. Brandstädt, A., Dragan, F.F., Chepoi, V.D., Voloshin, V.I.: Dually chordal graphs. SIAM J. Discrete Math. 11, 437–455 (1998)
8. Brandstädt, A., Le, V., Bang, S.J.P.: Graph Classes: A Survey. SIAM Monographs on Discrete Mathematics and Applications, Philadelphia (1999)
9. Cai, L., Corneil, D.G.: Tree spanners. SIAM J. Disc. Math. 8, 359–387 (1995)
10. Chan, H.T.-H., Gupta, A., Maggs, B.M., Zhou, S.: On hierarchical routing in doubling metrics. In: Proceedings of the Sixteenth Annual ACM-SIAM Symposium on Discrete Algorithms (SODA 2005), pp. 762–771. SIAM, Philadelphia (2005)
11. Chepoi, V.D., Dragan, F.F., Yan, C.: Additive Sparse Spanners for Graphs with Bounded Length of Largest Induced Cycle. Theoretical Computer Science 347, 54–75 (2005)
12. Dourisboure, Y.: Compact Routing Schemes for Bounded Tree-Length Graphs and for k-Chordal Graphs. In: Guerraoui, R. (ed.) DISC 2004. LNCS, vol. 3274, pp. 365–378. Springer, Heidelberg (2004)
13. Dragan, F.F., Yan, C.: Collective Tree Spanners in Graphs with Bounded Genus, Chordality, Tree-width, or Clique-width. In: Deng, X., Du, D.-Z. (eds.) ISAAC 2005. LNCS, vol. 3827, pp. 583–592. Springer, Heidelberg (2005)
14. Dragan, F.F., Yan, C., Corneil, D.G.: Collective Tree Spanners and Routing in AT-free Related Graphs. Journal of Graph Algorithms and Applications 10, 97–122 (2006)
15. Dragan, F.F., Yan, C., Lomonosov, I.: Collective tree spanners of graphs. SIAM J. Discrete Math. 20, 241–260 (2006)
16. Fonseca, R., Ratnasamy, S., Zhao, J., Ee, C.T., Culler, D., Shenker, S., Stoica, I.: Beacon vector routing: Scalable point-to-point routing in wireless sensornets. In: Proceedings of the Second USENIX/ACM Syposium on Networked Systems Design and Implementation (NSDI 2005) (2005)
17. Fraigniaud, P., Gavoille, C.: Memory requirements for univesal routing schemes. In: Proceedings of the 14th Annual ACM Symposium on Principles of Distributed Computing, Ontario, Canada, pp. 223–230 (1995)
18. Fraigniaud, P., Gavoille, C.: Routing in trees. In: Orejas, F., Spirakis, P.G., van Leeuwen, J. (eds.) ICALP 2001. LNCS, vol. 2076, pp. 757–772. Springer, Heidelberg (2001)

19. Gavoille, C.: A survey on interval routing schemes. Theoretical Computer Science 245, 217–253 (1999)
20. Gavoille, C.: Routing in distributed networks: Overview and open problems. ACM SIGACT News - Distributed Computing Column 32 (2001)
21. Gavoille, C., Gengler, M.: Space-efficiency of routing schemes of stretch factor three. Journal of Parallel and Distributed Computing 61, 679–687 (2001)
22. Gavoille, C., Hanusse, N.: Compact Routing Tables for Graphs of Bounded Genus. In: Wiedermann, J., Van Emde Boas, P., Nielsen, M. (eds.) ICALP 1999. LNCS, vol. 1644, pp. 351–360. Springer, Heidelberg (1999)
23. Gavoille, C., Peleg, D., Pérennès, S., Raz, R.: Distance labeling in graphs. J. Algorithms 53, 85–112 (2004)
24. Gavoille, C., Pérennès, S.: Memory requirements for routing in distributed networks. In: Proceedings of the 15th Annual ACM Symposium on Principles of Distributed Computing, Philadelphia, Pennsylvania, pp. 125–133 (1996)
25. Giordano, S., Stojmenovic, I.: Position based routing algorithms for ad hoc networks: A taxonomy. In: Cheng, X., Huang, X., Du, D. (eds.) Ad Hoc Wireless Networking, pp. 103–136. Kluwer, Dordrecht (2004)
26. Karp, B., Kung, H.T.: GPSR: greedy perimeter stateless routing for wireless networks. In: Proceedings of the 6th ACM/IEEE MobiCom, pp. 243–254. ACM, New York (2000)
27. Kleinberg, J.M.: The small-world phenomenon: an algorithm perspective. In: Proceedings of the Thirty-Second Annual ACM Symposium on Theory of Computing (STOC 2000), Portland, OR, USA, pp. 163–170. ACM, New York (2000)
28. Kleinberg, R.: Geographic routing using hyperbolic space. In: INFOCOM 2007, pp. 1902–1909 (2007)
29. Kratsch, D., Le, H.-O., Müller, H., Prisner, E., Wagner, D.: Additive tree spanners. SIAM J. Discrete Math. 17, 332–340 (2003)
30. Kuhn, F., Wattenhofer, R., Zhang, Y., Zollinger, A.: Geometric ad-hoc routing: of theory and practice. In: Proceedings of the 22nd Annual Symposium on Principles of Distributed Computing, pp. 63–72. ACM Press, New York (2003)
31. van Leeuwen, J., Tan, R.B.: Interval routing. The Computer Journal 30, 298–307 (1987)
32. Liben-Nowell, D., Novak, J., Kumar, R., Raghavan, P., Tomkins, A.: Geographic routing in social networks. PNAS 102, 11623–11628 (2005)
33. Peleg, D.: Distributed Computing: A Locality-Sensitive Approach. In: SIAM Monographs on Discrete Math. Appl. SIAM, Philadelphia (2000)
34. Peleg, D.: Proximity-Preserving Labeling Schemes and Their Applications. J. of Graph Theory 33, 167–176 (2000)
35. Rao, A., Papadimitriou, C., Shenker, S., Stoica, I.: Geographical routing without location information. In: Proceedings of MobiCom 2003, pp. 96–108 (2003)
36. Santoro, N., Khatib, R.: Labeling and implicit routing in networks. The Computer Journal 28, 5–8 (1985)
37. Slivkins, A.: Distance estimation and object location via rings of neighbors. In: PODC 2005, pp. 41–50 (2005)
38. Talwar, K.: Bypassing the embedding: Algorithms for low dimensional metrics. In: STOC 2004, pp. 281–290 (2004)
39. Thorup, M.: Compact oracles for reachability and approximate distances in planar digraphs. Journal of the ACM 51, 993–1024 (2004)
40. Thorup, M., Zwick, U.: Compact routing schemes. In: 13th Ann. ACM Symp. on Par. Alg. and Arch., pp. 1–10 (2001)

Single Vehicle Scheduling Problems on Path/Tree/Cycle Networks with Release and Handling Times

Binay Bhattacharya[1,*], Paz Carmi[2], Yuzhuang Hu[1], and Qiaosheng Shi[1]

[1] School of Computing Science, Simon Fraser University, Burnaby B.C.,
Canada. V5A 1S6
{binay,yhu1,qshi1}@cs.sfu.ca
[2] School of Computer Science, Carleton University,
Ottawa, Ontario, Canada. K1S 5B6
paz@cg.scs.carleton.ca

Abstract. In this paper, we consider the single vehicle scheduling problem (SVSP) on networks. Each job, located at some node, has a release time and a handling time. The vehicle starts from a node (depot), processes all the jobs, and then returns back to the depot. The processing of a job cannot be started before its release time, and its handling time indicates the time needed to process the job. The objective is to find a routing schedule of the vehicle that minimizes the completion time. When the underlying network is a path, we provide a simple 3/2-approximation algorithm for SVSP where the depot is arbitrarily located on the path, and a 5/3-approximation algorithm for SVSP where the vehicle's starting depot and the ending depot are not the same. For the case when the network is a tree network, we show that SVSP is polynomially approximable within 11/6 of optimal. All these results are improvements of the previous results [2,4]. The approximation ratio is improved when the tree network has constant number of leaf nodes. For cycle networks, we propose a 9/5-approximation algorithm and show that SVSP without handling times can be solved exactly in polynomial time. No such results on cycle networks were previously known.

1   Introduction

In this paper, we study the single vehicle scheduling problem (SVSP) on a network $G = (V, E)$ with node set V and edge set E. Each edge $e \in E$ is associated with a positive travel time $l(e)$. Let $d(u, v)$ denote the total travel time cost of a shortest path between two nodes u, v in G. A job is located at each node $v \in V$. Each job v becomes available for processing at a time $r(v)$ known as its *release time*. Job v requires a specific amount of time $h(v)$ for its completion known as its *handling time*. Job handling is *non-preemptive* in the sense that if the vehicle starts processing a job, it is required to complete the processing without

---

* Research was partially supported by MITACS and NSERC.

S.-H. Hong, H. Nagamochi, and T. Fukunaga (Eds.): ISAAC 2008, LNCS 5369, pp. 800–811, 2008.

interruption. The vehicle starts from the starting node called *starting depot* and after processing all the jobs the vehicle returns to the ending node called *ending depot*. Unless mentioned otherwise, we assume that the starting depot and the ending depot are the same. Without any loss of generality, we assume a unit vehicle speed, so that the travel time between any two nodes is equal to the corresponding distances. Our goal is to determine the processing schedule of the jobs such that the completion time is minimized.

It is known that SVSP is NP-hard even when G is a path [9]. The majority of the research efforts on SVSP are focused on finding approximate solutions with bounded cost (relative to the optimum) [1,2,3,4,6,7].

Psaraftis et al. [8] showed that SVSP on paths is 2-approximable, and that if all handling times are zero, the problem is easily solvable in linear time when the vehicle is required to return to the starting depot. When the vehicle is not required to return back to the depot, a quadratic time solution exists. Tsitsiklis [9] showed that if the jobs have non-zero release and handling times, SVSP on paths is NP-complete. Karuno et al. [3] gave a 3/2-approximation algorithm for SVSP on paths with non-zero release and handling times with some restriction on the depot location, and in [2] Gaur et al. presented a $\frac{5}{3}$-approximation for the problem with the same setting except that the depot is located arbitrarily on the path network.

Nagamochi et al. showed that SVSP on trees, with release times only, is NP-hard [7]. For SVSP on trees with release and handling times, Karuno et al. [4] proposed a 2-approximation algorithm. PTAS algorithms are known [1,5] for SVSP on trees with a fixed number of leaves. These schemes can also be applied to the case when more than one vehicle is available for processing the jobs.

Not many results are known for SVSP in general settings e.g., when the underlying graph is an undirected complete graph and the edge costs satisfy the triangle inequality. Nagamochi et. al. [7] proposed a 2.5-approximation algorithm for SVSP on general networks satisfying the triangle inequality. Tsitsiklis [9] considered the complexity of SVSP with release and (or) handling times, and proposed polynomial time solutions for some special cases of the problem, e.g., when the number of vertices in the graph is bounded by some constant (the number of jobs may be arbitrary).

In this paper we focus on SVSP in the cases where the underlying network is a path network, a tree network, or a cycle network.

2   SVSP on Path Networks

In this section we first present a 3/2-approximation algorithm for SVSP on a path network P where the depot is located at some arbitrary location on the path. This improves the previous best $\frac{5}{3}$-approximation result [2]. We also consider a more generalized SVSP on P where the starting and the ending depots are different.

The path P has node set $V = \{v_1, \ldots, v_n\}$ and edge set $E = \{v_i v_{i+1} : i = 1, \ldots, n-1\}$ with nodes v_1 and v_n designating the beginning and the end of the path respectively. The depot o is assumed to be located at $v_s, 1 \leq s \leq n$ i.e., $v_s = o$. We use C_P^* to denote the optimal completion time of the problem.

Given a positive value t, let $V_{\geq t}$ be the set of jobs whose release times are at least t, and let $V_{>t}$ be the set of jobs whose release times are larger than t. We also define the following notations. $r_{max} = \max_{v \in V} r(v); D = 2d(v_1, v_n); H_{\geq t} = \sum_{v \in V_{\geq t}} h(v), H = H_{\geq 0}; H_{>t} = \sum_{v \in V_{>t}} h(v)$.

2.1   Related Works

In [3], Karuno et al. showed that the optimal completion time C_P^* is at least $\max\{H + D, r(v) + h(v) + d(v, o), t + H_{\geq t}\}$ for any node $v \in V$ and any t with $0 \leq t \leq r_{max}$. Based on this observation, Karuno et al. [3] proposed a 3/2-approximation algorithm for SVSP where the depot is located either at v_1 or at v_n. Gaur et al. [2] provided a $\frac{5}{3}$-approximation algorithm for SVSP where the depot is at some arbitrary location of the path.

In the following, we give a brief overview of the algorithm of Karuno et al. [3]. The algorithm contains two phases, the forward and the backward phases. These two phases are delineated by a suitably chosen t where $H_{>t} \leq t \leq H_{\geq t}$. Clearly $H_{>t} \leq t \leq 0.5C_P^*$ since $t + H_{\geq t} \leq C_P^*$. In the forward phase, the vehicle starts from depot v_1 after waiting t time units, and then goes to the other end v_n of the path, processing all the nodes whose release times are at most t. This may involve waiting at some nodes for their jobs to be released. During the backward phase, the vehicle processes all the remaining jobs on its way back to v_1.

In the event when there is no time spent on waiting in the backward phase, Karuno et al. [3] showed that the completion time of this schedule is no more than $t + H + D$ which is at most $\frac{3}{2}C_P^*$. Otherwise, let v_k be the last job in the schedule where the vehicle is forced to wait before it is released during the backward phase. In this case the completion time of this schedule is no more than $r(v_k) + H_{>t} + d(v_k, o)$ which is at most $\frac{3}{2}C_P^*$ [3].

2.2   Our Approach

In this section we study SVSP on P where the depot is located at $v_s, 1 \leq s \leq n$. Our approach is based on a new lower bound of the optimal completion time C_P^* (shown in Lemma 1), where, for any time $t(\leq r_{max})$, the optimal completion time is at least the sum of the time t, handling time $H_{\geq t}$, and the minimum traveling time needed to visit all nodes in $V_{\geq t}$ starting from the depot $o = v_s$.

We describe below a few more notations.

$$d_l(t) = \max\{0, \max\{d(v_i, o)|1 \leq i < s, v_i \in V_{\geq t}\}\};$$

$$d_r(t) = \max\{0, \max\{d(v_i, o)|s \leq i \leq n, v_i \in V_{\geq t}\}\}.$$

Here $v_l(t)$ is the leftmost node of v_s in $V_{\geq t}$, and if there is no such node, then $v_l(t) = v_s$. Similarly, $v_r(t)$ is the rightmost node of v_s in $V_{\geq t}$ and if there is no such node then $v_r(t) = v_s$. Note that $d(v_l(t), o) = d_l(t)$ and $d(v_r(t), o) = d_r(t)$.

Lemma 1. *The optimal completion time C_P^* is at least $t + H_{\geq t} + \max\{d_l(t), d_r(t)\} + 2\min\{d_l(t), d_r(t)\}$, for any $t \leq r_{max}$.*

Proof. In an optimal schedule $S :< v_{i_1}, \cdots, v_{i_n} >$, let v' be the first job in this schedule that is in $V_{\geq t}$. Firstly, the starting time t' in S to process job v' must be at least $r(v')$. All the jobs in $V_{\geq t}$ are processed after time t'. Hence, in the sechdule S, the processing time needed after time t' is at least $H_{\geq t}$. In order to process all the jobs in $V_{\geq t}$ and return to the depot v_s, the traveling time required is at least $\max\{d_l(t), d_r(t)\} + 2\min\{d_l(t), d_r(t)\}$.

Therefore, the completion time of schedule S is at least $r(v') + H_{\geq t} + \max\{d_l(t), d_r(t)\} + 2\min\{d_l(t), d_r(t)\}$, which is at least $t + H_{\geq t} + \max\{d_l(t), d_r(t)\} + 2\min\{d_l(t), d_r(t)\}$ since $v' \in V_{\geq t}$. $\square$

We compute t_P^* such that $t_P^* = H_{>t_P^*} + \min\{d_l(t_P^*), d_r(t_P^*)\}$ or $H_{>t_P^*} + \min\{d_l(t_P^*), d_r(t_P^*)\} < t_P^* \leq H_{\geq t_P^*} + \min\{d_l(t_P^*), d_r(t_P^*)\}$. Such $t = t_P^*$ always exists, since $H_{>t}$, $H_{\geq t}$, and $\min\{d_l(t), d_r(t)\}$ monotonically decrease with t. Without any loss of generality we assume that $d_l(t_P^*) \leq d_r(t_P^*)$. Therefore, $H_{>t_P^*} + 2d_l(t_P^*) \leq t_P^* + d_l(t_P^*) \leq 0.5C_P^*$, since $d_l(t_P^*) \leq d_r(t_P^*)$ and $t_P^* + H_{\geq t_P^*} + 2d_l(t_P^*) + d_r(t_P^*) \leq C_P^*$ (Lemma 1).

Our routing schedule of the vehicle to process jobs consists of four phases, which are described as follows (Fig. 1)

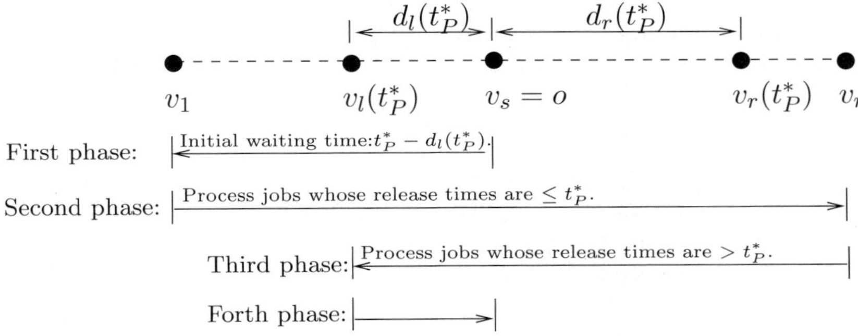

Fig. 1. Four phases in our routing schedule of the vehicle

First phase: The vehicle waits for $t_P^* - d_l(t_P^*)$ time (note that $t_P^* \geq d_l(t_P^*)$) at the depot o and then travels from o to v_1 without processing any job. The time spent in the first phase is exactly $t_P^* - d_l(t_P^*) + d(v_1, o)$, which is at least t_P^*.

Second phase: The vehicle processes jobs on its way from v_1 to v_n whose release times are no more than t_P^*. Clearly, the vehicle does not wait for a job to be released during this phase.

Third phase: The vehicle processes the remaining unprocessed jobs on its way from v_n to $v_l(t_P^*)$.

Fourth phase: The vehicle travels back to the depot node o from $v_l(t_P^*)$.

We have two cases in the third phase. If there is no waiting in the third phase, then the completion time of the routing schedule is equal to

$$t_P^* - d_l(t_P^*) + d(v_1, o) + d(v_1, v_n) + d(v_n, v_l(t_P^*)) + d(v_l(t_P^*), o) + H$$

$$= t_P^* + d(v_1, o) + d(v_1, v_n) + d(v_n, o) + d(v_l(t_P^*), o) + H$$
$$= H + D + t_P^* + d_l(t_P^*) \leq 1.5 C_P^*,$$

since $d(v_l(t_P^*), o) = d_l(t_P^*)$, $H + D \leq C_P^*$, and $t_P^* + d_l(t_P^*) \leq 0.5 C_P^*$. Recall that $D = 2d(v_1, v_n)$.

If there is some waiting in the third phase, let v_k be the last job in the schedule where the vehicle waited for the release of the job at v_k. The completion time of the schedule in this case is no more than $r(v_k) + d(v_k, v_l(t_P^*)) + d(v_l(t_P^*), o) + H_{>t_P^*} \leq r(v_k) + d(v_k, o) + d(o, v_l(t_P^*)) + d(v_l(t_P^*), o) + H_{>t_P^*} = r(v_k) + d(v_k, o) + H_{>t_P^*} + 2d_l(t_P^*) \leq 1.5 C_P^*$, since $r(v_k) + d(v_k, o) \leq C_P^*$, $d(v_l(t_P^*), o) = d_l(t_P^*)$, and $H_{>t_P^*} + 2d_l(t_P^*) \leq 0.5 C_P^*$. Therefore,

Theorem 1. *The approximation ratio of the algorithm for the SVSP problem on a path network is at most 1.5.*

2.3 SVSP with Different Starting and Ending Depots

Here we only consider the case where the starting depot is located at v_1 and the ending depot is located at v_n. The algorithm generalizes easily for different pairs of starting and ending depots.

The optimal completion time, denoted by $\overline{C}_P^*$, is lower bounded by $\max \{H + d(v_1, v_n), r(v_k) + h(v_k) + d(v_k, v_n), \max_{t \leq r_{max}} \{t + H_{\geq t} + d(t)\}\}$, where $d(t) = \max_{v \in V_{>t}} d(v, v_n)$.

We now consider the following two cases.

Case 1 $(H \leq \frac{2}{3}\overline{C}_P^*)$: The vehicle processes jobs one by one on its way from v_1 to v_n. If there is no waiting, then the completion time is equal to $\overline{C}_P^*$. Otherwise, if v_k is the last job where the vehicle waited before processing, then the completion time is no more than $r(v_k) + d(v_k, v_n) + H \leq \frac{5}{3}\overline{C}_P^*$.

Case 2 $(H > \frac{2}{3}\overline{C}_P^*)$: In this case $d(v_1, v_n) < \frac{1}{3}\overline{C}_P^*$ since $H + d(v_1, v_n) \leq \overline{C}_P^*$. We compute $\overline{t}^*$ such that $\overline{t}^* + d(\overline{t}^*) = 0.5 H_{>\overline{t}^*}$ or $0.5 H_{>\overline{t}^*} < \overline{t}^* + d(\overline{t}^*) \leq 0.5 H_{\geq \overline{t}^*}$. Such $\overline{t}^*$ always exists since $H_{>t}$ and $H_{\geq t}$ monotonically decrease with t and $d(0) = d(v_1, v_n) < 0.5 H = 0.5 H_{\geq 0}$. Therefore, $0.5 H_{>\overline{t}^*} \leq \overline{t}^* + d(\overline{t}^*) \leq \frac{1}{3}\overline{C}_P^*$ since $\overline{t}^* + d(\overline{t}^*) + H_{\geq \overline{t}_P^*} \leq \overline{C}_P^*$.

The schedule of the vehicle in this case consists of three phases. In phase one the vehicle waits for $\overline{t}^*$ time at the starting depot v_1 and then processes jobs whose release times are no more than $\overline{t}^*$ on its way from v_1 to v_n. In phase two the vehicle travels to the farthest node of $V_{>\overline{t}^*}$ to v_n without processing any job. In the last phase the vehicle processes the remaining unprocessed jobs on its way back to v_n.

If there is no waiting in the last phase, then the completion time is no more than $\overline{t}^* + H + d(v_1, v_n) + 2d(\overline{t}^*) < \frac{5}{3}\overline{C}_P^*$ since $d(\overline{t}^*) \leq d(v_1, v_n) < \frac{1}{3}\overline{C}_P^*$. Otherwise, if v_k is the last job where the vehicle waited before processing, then the completion time is no more than $r(v_k) + d(v_k, v_n) + H_{>\overline{t}^*} \leq \frac{5}{3}\overline{C}_P^*$.

Combining the results for the cases $H \leq \frac{2}{3}\overline{C}_P^*$ and $H > \frac{2}{3}\overline{C}_P^*$, we have the following theorem.

Theorem 2. *The approximation ratio of the algorithm for SVSP where the starting and ending depots are located at the two ends of the path is $\leq 5/3$.*

3   SVSP on a Tree Network

In this section, we present our approximation algorithm for the SVSP on a tree network $T = (V(T), E(T))$ where the depot is located at the root node. Let C_T^* denote the optimal completion time of the SVSP on T.

Let $T_{\geq t}$ (resp. $T_{>t}$) denote the spanning subtree of the node set $V_{\geq t} \cup \{o\}$ (resp. $V_{>t} \cup \{o\}$). For a subtree T' of T, let $E(T')$ denote the edge set of T' and $L(T')$ denote the total travel time of T', i.e., $L(T') = \sum_{e \in E(T')} l(e)$. Let $v_{\geq t}$ (resp. $v_{>t}$) be the farthest node of T to o in $V_{\geq t}$ (resp. $V_{>t}$). Let $d_{max} = \max_{v \in V(T)} d(o, v)$.

We have the following bounds for the optimum C_T^* of SVSP on T.

Lemma 2. *The completion time C_T^* of an optimal routing schedule for SVSP on a tree T is at least $t + H_{\geq t} + 2L(T_{\geq t}) - d(o, v_{\geq t})$, for any $t \leq \max \{r_{max}, H + 2L(T)\}$.*

Proof. The proof is similar to the proof of the lower bound in Lemma 1.   □

The following lemma establishes a useful monotonicity property which will be used later.

Lemma 3. $L(T_{\geq t}) - d(o, v_{\geq t}) \geq L(T_{>t}) - d(o, v_{>t}) \geq L(T_{\geq t'}) - d(o, v_{\geq t'}) \geq L(T_{>t'}) - d(o, v_{>t'})$, *for any t and t' with $0 \leq t < t' \leq r_{max}$.*

Proof. Observe that $T_{>t}$ (resp. $T_{>t'}$) is a subtree of $T_{\geq t}$ (resp. $T_{\geq t'}$) and $T_{\geq t'}$ is a subtree of $T_{>t}$ for any times t and t' with $0 \leq t < t' \leq r_{max}$.   □

3.1   First Routing Schedule of the Vehicle

Our first schedule on T is described as follows. We first find a time t_1^* such that $t_1^* = H_{>t_1^*}$ or $H_{>t_1^*} < t_1^* \leq H_{\geq t_1^*}$. In phase one the vehicle starts from the depot after waiting for t_1^* time units and then visits all the nodes in T and then stops at $v_{>t_1^*}$ (the farthest node of $V_{>t_1^*}$). The vehicle will process a node if its release time is $\leq t_1^*$. Obviously, there is no waiting time in phase one. Phase two starts from node $v_{>t_1^*}$ and travels the subtree $T_{>t_1^*}$. The vehicle processes all the remaining unprocessed jobs on its return trip back to o. Note that in phase two, the vehicle might have to wait at nodes for their jobs to be released.

Initial analysis on the first routing schedule. If there is no waiting time during phase two of the schedule, then the completion time is no more than $t_1^* + 2L(T) - d(o, v_{>t_1^*}) + 2L(T_{>t_1^*}) - d(o, v_{>t_1^*}) + H \leq C_T^* + t_1^* + 2L(T_{>t_1^*}) - 2d(o, v_{>t_1^*})$, since $H + 2L(T) \leq C_T^*$. Otherwise, if v_k is the last job in the schedule where the vehicle has to wait, we observe that the remaining traveling time after visiting node v_k in phase two is no more than $2L(T_{>t_1^*}) - 2d(o, v_{>t_1^*}) + d(v_k, o)$. Hence, the completion time of the first routing schedule in this case is no more than $r(v_k) + H_{>t_1^*} + 2L(T_{>t_1^*}) - 2d(o, v_{>t_1^*}) + d(v_k, o) \leq C_T^* + H_{>t_1^*} + 2L(T_{>t_1^*}) - 2d(o, v_{>t_1^*})$.

3.2   Second Routing Schedule of the Vehicle

Let t_2^* be $H + L(T)$. The second schedule also consists of two phases. In phase one, the vehicle starts from the depot o after waiting for t_2^* time units, and then the vehicle traverses the tree in some order using $L(T)$ time units and processes all the jobs whose release times are at most t_2^*. Obviously, there is no waiting time in phase one. In phase two, the vehicle processes all the remaining unprocessed jobs on its way back to o. Note that the vehicle starts its phase two from the location where it finishes phase one. The path of visiting nodes in two phases is described below. It is similar to the depth-first search order.

Let $\mathcal{P}_T$ be a depth-first path. Clearly, the length of $\mathcal{P}_T$ is $2L(T)$. Let p_{mid} denote the middle point of the path $\mathcal{P}_T$, which partitions $\mathcal{P}_T$ into two sub-paths of same length denoted by $\mathcal{P}_T^1$ and $\mathcal{P}_T^2$. We consider two cases: the traveling path of the vehicle in phase one is (a) $\mathcal{P}_T^1$ or (b) the reverse of $\mathcal{P}_T^2$. In both cases, the vehicle stops at p_{mid} at the end of phase one and the vehicle processes all the remaining unprocessed jobs on its way back to o from p_{mid} in phase two. The case that has smaller total travel time is chosen to be the ordering of visiting nodes.

Lemma 4. *The total travel time of the above routing schedule is no more than* $2L(T) + L(T_{>t_2^*})$.

Proof. Let T_1 and T_2 be the subtrees visited in phase one and phase two, respectively. It is easy to see that the total travel time in the case one and in the case two described above are $2L(T)+2L((T_1\setminus T_2)\cap T_{>t_2^*})$ and $2L(T)+2L((T_2\setminus T_1)\cap T_{>t_2^*})$, respectively. Since $2L((T_1\setminus T_2)\cap T_{>t_2^*}) + 2L((T_2\setminus T_1)\cap T_{>t_2^*}) \leq 2L(T_{>t_2^*})$, the smaller total travel time is no more than $2L(T) + L(T_{>t_2^*})$, as desired. $\qquad\square$

Initial analysis on the second routing schedule. If the vehicle did not spend any time waiting in phase two, then the completion time is no more than $t_2^* + L(T) + L(T) + L(T_{>t_2^*}) + H = H + 2L(T) + t_2^* + L(T_{>t_2^*}) \leq C_T^* + H + L(T) + L(T_{>t_2^*})$. Otherwise, if v_k is the last job in the schedule where the vehicle has to wait in phase two, then the completion time of this schedule is no more than $r(v_k) + H + L(T) + L(T_{>t_2^*}) \leq C_T^* + H + L(T) + L(T_{>t_2^*})$.

3.3   Final Analysis

In the first routing schedule on T, the selected time t_1^* satisfies that $H_{>t_1^*} \leq t_1^* \leq H_{\geq t_1^*}$. Suppose that $L(T) - d_{max} \leq \alpha \cdot C_T^*$, where $0 < \alpha < 1$. We can see that $C_T^* \geq t_1^* + H_{\geq t_1^*} + 2L(T_{\geq t_1^*}) - d(o, v_{\geq t_1^*})$ (Lemma 2) $\Rightarrow t_1^* + L(T_{\geq t_1^*}) - \frac{1}{2}d(o, v_{\geq t_1^*}) \leq \frac{C_T^*}{2}$ (since $t_1^* \leq H_{\geq t_1^*}$). Therefore, the solution of the first schedule has a cost no more than

$$C_T^* + H_{>t_1^*} + 2L(T_{>t_1^*}) - 2d(o, v_{>t_1^*})$$
$$\leq C_T^* + t_1^* + 2L(T_{>t_1^*}) - 2d(o, v_{>t_1^*}) \text{ (since } H_{>t_1^*} \leq t_1^*)$$
$$\leq C_T^* + \frac{C_T^*}{2} + L(T_{>t_1^*}) - \frac{3d(o, v_{>t_1^*})}{2}$$

$$\leq C_T^* + \frac{C_T^*}{2} + L(T_{\geq t_1^*}) - d(o, v_{>t_1^*})$$
$$\leq 1.5C_T^* + L(T) - d_{max} \text{ (Lemma 3)}$$
$$\leq (1.5 + \alpha)C_T^* \text{ (by assumption)}.$$

We consider the other case $L(T) - d_{max} > \alpha \cdot C_T^*$ in the following. Clearly, $H + L(T) + d_{max} < (1 - \alpha)C_T^*$ since $H + 2L(T) \leq C_T^*$.

According to Lemmas 2 and 3, $2L(T_{t_2^*}) \leq 2L(T_{\geq t_2^*}) \leq C_T^* - t_2^* + d(o, v_{\geq t_2^*})$. Therefore, the solution of the second schedule has a cost no more than $C_T^* + H + L(T) + L(T_{>t_2^*}) \leq C_T^* + H + L(T) + 0.5(C_T^* - t_2^* + d(o, v_{\geq t_2^*})) = 1.5C_T^* + 0.5(H + L(T) + d(o, v_{\geq t_2^*})) \leq (2 - 0.5\alpha)C_T^*$.

Thus the minimum completion time of the above routing schedules is bounded within $\frac{11}{6}C_T^*$ (assuming $\alpha = \frac{1}{3}$).

Theorem 3. *The approximation ratio of the algorithm for SVSP on a tree network is at most $11/6$.*

3.4 T Has b Leaf Nodes (Fixed b)

In the following, we consider the case where T has b (a fixed number) leaf nodes.

Let S' be the routing schedule of the vehicle in which the nodes are visited in the order given by the optimal algorithm due to Karuno et al. [7]. This algorithm does not take into account the handling times, and its time complexity is $O(n^b)$. Let $C_{S'}$ be the completion time of the schedule S' assuming all handling times are 0. We now consider a schedule in which the vertices are serviced (and handled) in the same order as specified by S'. The completion time of such schedule is no more than $C_{S'} + H$.

Let C_T^* denote the completion time of an optimal schedule on T with b leaf nodes. It is easy to see that $C_T^* \geq C_{S'}$. If $H \leq \frac{2}{3}C_T^*$, then $C_{S'} + H \leq \frac{5}{3}C_T^*$. We assume that $H > \frac{2}{3}C_T^*$. Obviously, $2L(T) < \frac{1}{3}C_T^*$.

In the first routing schedule on T described above, $C_T^* + H_{>t_1^*} + 2L(T_{\geq t_1^*}) - 2d(o, v_{\geq t_1^*}) \leq C_T^* + t_1^* + 2L(T_{\geq t_1^*}) - 2d(o, v_{\geq t_1^*}) = C_T^* + (t_1^* + L(T_{\geq t_1^*}) - \frac{1}{2}d(o, v_{\geq t_1^*})) + L(T_{\geq t_1^*}) - \frac{3}{2}d(o, v_{\geq t_1^*}) \leq C_T^* + \frac{1}{2}C_T^* + L(T_{\geq t_1^*}) \leq \frac{5}{3}C_T^*$.

Therefore,

Theorem 4. *The approximation ratio of the algorithm for SVSP on a tree network with a fixed number of leaf nodes is at most $\frac{5}{3}$.*

4   SVSP on a Cycle Network

In this section we study the SVSP problem on a cycle network $A = (V, E)$. The nodes on A are indexed as $v_1, v_2, \ldots, v_n$ in the counterclockwise order. It is assumed here that the depot is located at v_1.

Let δ denote the travel time needed to visit all the edges of A. In the following we assume that there is no edge whose travel cost is at least $\delta/2$, otherwise, the problem is reducible to SVSP on a path network. It is not hard to see that the

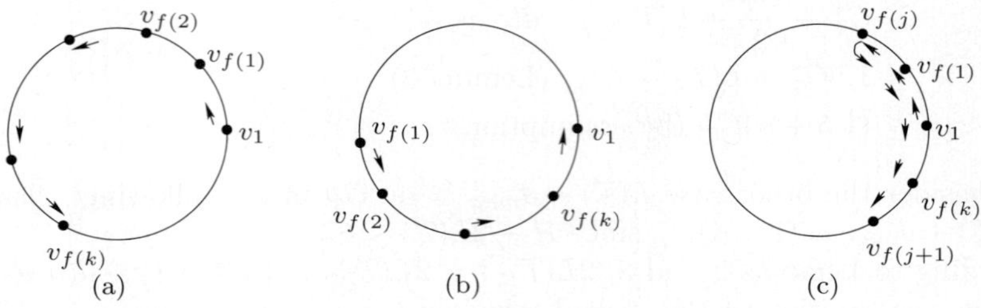

Fig. 2. Lemma 5

optimal completion time, denoted by C_A^*, is at least $\max\{H+\delta, r(v_k)+h(v_k)+ d(v_k,v_1)$ for any v_k, $\max\{t+H_{\geq t}+\delta_{\geq t}\}$, for any t $\leq r_{max}\}$. Here $\delta_{\geq t}$ denotes the shortest travel time needed to visit all the nodes in $V_{\geq t}$ starting from the depot v_1, and $d(v_k,v_1)$ denote the shortest travel time between v_1 and v_k.

We denote by $\pi_{\geq t}$ a path, starting from v_1, of travel time $\delta_{\geq t}$ that goes through all the nodes in $V_{\geq t}$ at least once. Let $w(\pi_{\geq t})$ be the other endpoint of $\pi_{\geq t}$. We use $d^{cc}(v_i, v_j)$ to denote the travel time of the counterclockwise arcs from v_i to v_j.

Lemma 5. *For any $t \leq r_{max}$, there exists a path $\pi_{\geq t}$ of one of the following types (Let $V_{\geq t} = \{v_{f(1)}, v_{f(2)}, \ldots, v_{f(k)}\}$, $f(1) < f(2) < \cdots < f(k)$):*

- *the counterclockwise path from v_1 to $v_{f(k)}$ (Fig. 2(a));*
- *the counterclockwise path from $v_{f(1)}$ to v_1 if $v_{f(1)} \neq v_1$ (Fig. 2(b)) and the counterclockwise path from $v_{f(2)}$ to v_1 otherwise;*
- *for some $j, 1 \leq j \leq k-1$, (assuming that $d^{cc}(v_1, v_{f(j)}) \leq d^{cc}(v_{f(j+1)}, v_1)$), $\pi_{\geq t}$ is composed of the counterclockwise path from v_1 to $v_{f(j)}$ and the clockwise path from $v_{f(j)}$ to $v_{f(j+1)}$ (Fig. 2(c)).*

From Lemma 5, it is not hard to see that $\delta_{\geq t}+d(w(\pi_{\geq t}), v_1) \leq \delta$ for any $t \leq r_{max}$. We consider the following two cases for a fixed constant $0 \leq \beta \leq 0.5$.

Case 1. $\delta \leq (1-\beta)C_A^*$: We compute $t^* \in [0, r_{max}]$ such that $t^*+\delta = H_{>t^*}+\delta_{>t^*}$ or $H_{>t^*} + \delta_{>t^*} < t^* + \delta \leq H_{\geq t^*} + \delta_{>t^*}$. Such a t^* always exists since $H_{>t}$, $H_{\geq t}$, and $\delta_{>t}$ monotonically decrease with t and $\delta < H_{\geq 0} + \delta_{>0}$ & $r_{max} + \delta > H_{>r_{max}} + \delta_{>r_{max}}$.

We thus see that $H_{>t^*}+\delta_{>t^*} \leq t^*+\delta \leq (1-\frac{\beta}{2})C_A^*$, since $t^*+H_{>t^*}+\delta_{>t^*} \leq C_A^*$ and $\delta \leq (1-\beta)C_A^*$.

The routing schedule in this case consists of three phases. Refer to Fig. 3(a). Each phase is represented by a set of arrows with the phase number attached. In phase one, the vehicle waits for t^* time at v_1 and then processes jobs whose release time is no more than t^* along the cycle $< v_1, v_2, \ldots, v_n, v_1 >$. In phase two, the vehicle travels to node $w(\pi_{\geq t^*})$ without processing any job. In phase three, the vehicle processes the remaining jobs along the path $\pi_{\geq t^*}$.

Analysis. Clearly, only the last phase might encounter some waiting time. If there is no waiting time in the last phase, then the completion time is no more than

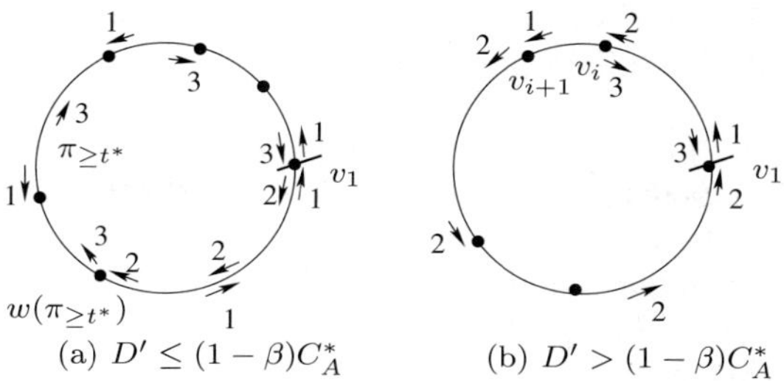

Fig. 3. Routing schedules for the two cases

$t^* + H + \delta + \delta_{>t^*} + d(w(\pi_{\geq t^*}), v_1)) \leq (2 - \frac{\beta}{2})C_A^*$. Otherwise, if v_k is the last job in the schedule in phase three where the vehicle is forced to wait before its release time, then the completion time is no more than $r(v_k) + \delta_{>t^*} + H_{>t^*} \leq (2 - \frac{\beta}{2})C_A^*$ since $r(v_k) > t^*$.

Thus, the completion time of the algorithm for the case when $\delta \leq (1 - \beta)C_A^*$ is no more than $(2 - \frac{\beta}{2})C_A^*$.

Case 2. $\delta > (1 - \beta)C_A^*$: In this case some edges are traveled at most once in an optimal routing schedule, otherwise, if the vehicle travels each edge at least twice in an optimal routing schedule, then the completion time is at least $2\delta > (2 - 2\beta)C_A^* \geq C_A^*$. Moreover, if there is an edge that is not traveled in an optimal routing schedule, then our 1.5-approximation algorithm for path networks in Section 2.2 will provide a routing schedule with completion time $\leq 1.5C_A^*$ by considering all n candidate cut edges. Thus, we assume that each edge is traveled at least once and that $\overline{v_i v_{i+1}}$ is the edge that was traveled once, and all the edges from v_1 to v_i are traveled at least twice in an optimal schedule. Without any loss of generality, we assume that the edge $\overline{v_i v_{i+1}}$ is traveled from v_i to v_{i+1}. Clearly, $H + \delta + d^{cc}(v_1, v_i) \leq C_A^*$.

The routing schedule in this case is as follows. Refer to Fig. 3(b). Firstly the vehicle travels counterclockwise from v_1 to v_{i+1} without processing any job, then it processes jobs one by one along the path $< v_{i+1}, \ldots, v_n, v_1, \ldots, v_i >$. In the last phase, the vehicle travels (clockwise) back to v_1 from v_i.

Analysis. If there is no waiting time in the above routing schedule then its completion time is equal to $H + \delta + 2d^{cc}(v_1, v_i) \leq (1 + \beta)C_A^*$, since $H + d^{cc}(v_1, v_i) < \beta C_A^*$. Otherwise, if v_k is the last job in the schedule where the vehicle is forced to wait before its release in phase two, then the completion time is no more than $r(v_k) + H + d^{cc}(v_k, v_i) + d^{cc}(v_1, v_i) \leq (1 + 2\beta)C_A^*$ since $r(v_k) + d^{cc}(v_k, v_1) \leq C_A^*$ if $i < k \leq n$ (the edge $\overline{v_i v_{i+1}}$ is visited once in the optimal routing schedule) and $d^{cc}(v_k, v_i) \leq d^{cc}(v_1, v_i)$ if $1 \leq k \leq i$. Thus, the completion time of the algorithm for the case when $D' > (1 - \beta)C_A^*$ is no more than $(1 + 2\beta)C_A^*$.

In summary, the completion times of the above schedules is bounded within $\frac{9}{5}C_A^*$ (when $\beta = \frac{2}{5}$).

Theorem 5. *The approximation ratio of the algorithm for SVSP on a cycle network is at most $\frac{9}{5}$.*

When the handling times of all the jobs in cycle networks are zero, it is possible to design an optimal schedule in polynomial time.

No handling time for the jobs. Here we show that SVSP on A can be solved optimally when $h(v_i) = 0 (i = 1, \ldots, n)$. This is based on a simple observation that if the vehicle visits a node v_j more than once in a routing schedule, then the completion time remains unchanged when the vehicle only processes the job (v_j) during the last visit. Similar observation was also used in [8].

It is known that this problem (zero handling time) on a path where the vehicle returns to the starting depot can be solved exactly in linear time [8]. Hence, if there is one edge not used in an optimal schedule, then the problem on A can be solved in $O(n^2)$ time. We therefore assume that all the edges of A are used at least once in an optimal schedule.

We call the routing path $v_{j+1} \rightarrow v_j \rightarrow v_{j+1}$ a *counterclockwise turn* at node v_j and the routing path $v_{j-1} \rightarrow v_j \rightarrow v_{j-1}$ a *clockwise turn* at node v_j.

Lemma 6. *There exists an optimal routing schedule such that the routing path between any two consecutive turn nodes goes through the depot v_1.*

From Lemma 6, we can see that in some optimal schedule there exist two nodes x and y such that all edges in the path between them, which does not contain v_1, are visited only once and the other edges are visited at least twice. By combining the result of Lemma 6 and the fact that every edge is visited at least once, we can conclude that there exists one routing subpath without turn in an optimal schedule that goes through all nodes on A. Therefore, there exists an optimal schedule of the type $< A, \hat{P} >$ where $\hat{P}$ is the optimal schedule of the path from y to x that contains v_1. Using the algorithm in [8] we can determine $\hat{P}$ in linear time.

Therefore, we can compute an optimal schedule for A by considering all possible pairs of nodes x and y. For each such pair, the vehicle first travels to x without processing any job, and then processes all jobs along the path from x to y. After traveling back to v_1 from y, we use the algorithm in [8] to find an optimal schedule for all remaining jobs on the path between x and y that contains v_1. Hence we have the following theorem.

Theorem 6. *SVSP on a cycle network without handling times can be solved in $O(n^3)$ time.*

5   Conclusion

We considered the single vehicle scheduling problems on paths, trees and cycles, all of which are known to be NP-hard. In this paper, we improved the approximation ratio of the optimal schedule from $\frac{5}{3}$ to 1.5 when the underlying network is a path and the depot is situated at some arbitrary node. The main idea used in the improved algorithms is a new lower bound of the optimal completion time.

For the case where the underlying network is a tree, we proposed a 11/6-approximation algorithm, which improved on the previous best result [4]. When the underlying tree network only contains a fixed number of leaf nodes, a better approximation ratio of 5/3 can be obtained.

We then showed that for cycle networks, the ideas developed for paths and trees can be extended, and as a result we obtained a 9/5-approximation algorithm. When all the jobs have zero handling times, SVSP on cycle networks can be optimally solved.

There are many issues that are still unresolved. We would like to know whether the approximation bounds obtained in this paper are tight. It is straightforward to design a 2.5-approximation algorithm for SVSP on general networks satisfying the triangle inequality [7]. A better approximation bound for the general network is desirable. There are many variants of this problem class. Psaraftis et al [8] in their paper discussed these variants in greater detail. Most of these variants are still open.

References

1. Augustine, J.E., Seiden, S.: Linear time approximation schemes for vehicle scheduling problems. Theoretical Comput. Sci. 324(2-3), 147–160 (2004)
2. Gaur, D.R., Gupta, A., Krishnamurti, R.: A 5/3-approximation algorithm for scheduling vehicles on a path with release and handling times. Inform. Proc. Let. 86(2), 87–91 (2003)
3. Karuno, Y., Nagamochi, H., Ibaraki, T.: A 1.5-approximation for single vehicle scheduling problem on a line with release and handling times. In: Proc. ISCIE/ASME, vol. 3, pp. 1363–1368 (1998)
4. Karuno, Y., Nagamochi, H., Ibaraki, T.: Vehicle scheduling on a tree with release and handling time. Ann. Oper. Res. 69, 193–207 (1997)
5. Karuno, Y., Nagamochi, H.: An approximability result of the multi-vehicle scheduling problem on a path with release and handling times. Theoretical Comput. Sci. 312, 267–280 (2004)
6. Karuno, Y., Nagamochi, H.: 2-approximation algorithms for the multi-vehicle scheduling problem on a path with release and handling times. Discrete Appl. Math. 129, 433–447 (2003)
7. Nagamochi, H., Mochizuki, K., Ibaraki, T.: Complexity of the single vehicle scheduling problem on graphs. Inform. Systems Oper. Res. 35(4), 256–276 (1997)
8. Psaraftis, H., Solomon, M., Magnanti, T., Kim, T.-U.: Routing and scheduling on a shoreline with release times. Management Sci. 36, 212–223 (1990)
9. Tsitsiklis, J.: Special cases of traveling salesman and repairman problems with time windows. Networks 22(3), 263–282 (1992)

Bidirectional Core-Based Routing in Dynamic Time-Dependent Road Networks[*]

Daniel Delling[1] and Giacomo Nannicini[2,3]

[1] Universität Karlsruhe (TH), 76128 Karlsruhe, Germany
delling@ira.uka.de
[2] LIX, École Polytechnique, F-91128 Palaiseau, France
giacomon@lix.polytechnique.fr
[3] Mediamobile, 27 bd Hyppolite Marques, 94200 Ivry Sur Seine, France
giacomo.nannicini@v-trafic.com

Abstract. In a recent work [1], we proposed a point-to-point shortest paths algorithm which applies bidirectional search on time-dependent road networks. The algorithm is based on A^* and runs a backward search in order to bound the set of nodes that have to be explored by the forward search. In this paper we extend the bidirectional time-dependent search algorithm in order to allow core routing, which is a very effective technique introduced for static graphs that consists in carrying out most of the search on a subset of the original node set. Moreover, we tackle the dynamic scenario where the piecewise linear time-dependent arc cost functions are not fixed, but can have their coefficients updated. We provide extensive computational results to show that our approach is a significant speed-up with respect to the original algorithm, and it is able to deal with the dynamic scenario requiring only a small computational effort to update the cost functions and related data structures.

1   Introduction

The Shortest Path Problem (SPP) on static graphs has received a great deal of attention in recent years, because it has interesting practical applications (e.g. route planners for GPS devices, web services) and provides an algorithmic challenge. Several works propose efficient algorithms for the SPP: see [2] for a review, and [3] for an interesting analysis of possible combinations of speed-up techniques.

Much of the focus is now moving to the Time-Dependent Shortest Path Problem (TDSPP), which can be formally stated as follows: given a directed graph $G = (V, A)$, a source node $s \in V$, a destination node $t \in V$, an interval of time instants T, a departure time $\tau_0 \in T$ and a time-dependent arc cost function $c : A \times T \to \mathbb{R}_+$, find a path $p = (s = v_1, \ldots, v_k = t)$ in G such that its *time-dependent cost* $\gamma_{\tau_0}(p)$, defined recursively as follows:

$$\gamma_{\tau_0}(v_1, v_2) = c(v_1, v_2, \tau_0) \tag{1}$$

$$\gamma_{\tau_0}(v_1, \ldots, v_i) = \gamma_{\tau_0}(v_1, \ldots, v_{i-1}) + c(v_{i-1}, v_i, \tau_0 + \gamma_{\tau_0}(v_1, \ldots, v_{i-1})) \tag{2}$$

for all $2 \leq i \leq k$, is minimum.

[*] Partially supported by the DFG (project WA 654/16-1) and the EU under grant ARRIVAL (contract no. FP6-021235-2).

S.-H. Hong, H. Nagamochi, and T. Fukunaga (Eds.): ISAAC 2008, LNCS 5369, pp. 812–823, 2008.

The TDSPP has been first addressed by [4] with a recursion formula; Dijkstra's algorithm [5] is then extended to the dynamic case in [6], and the FIFO property, which is necessary to guarantee correctness, is implicitly assumed. The FIFO property is also called the *non-overtaking property*, because it states that if T_1 leaves u at time τ_0 and T_2 at time $\tau_1 > \tau_0$, T_2 cannot arrive at v before T_1 using the arc (u, v). The TDSPP in FIFO networks is polynomially solvable [7], while it is NP-hard in non-FIFO networks [8]. We focus on the FIFO variant. The A^* algorithm [9] has been adapted to efficiently compute shortest paths on static road networks in [10,11]. Those ideas have been used in [12] on dynamic graphs as well, while the time-dependent case on graphs with the FIFO property has been addressed in [13,12,1]. The SHARC-algorithm [14], which employs a hierarchical approach combined with goal directed search via arc flags [15], allows fast unidirectional shortest path calculations in large scale networks; it has been recently extended in [16] to compute optimal paths even on time-dependent graphs, and represents the fastest known algorithm so far for time-dependent shortests path computations.

Bidirectional search cannot be directly applied on time-dependent graphs, the optimal arrival time at the destination being unknown. In [1], we tackled this problem running a forward search on the time-dependent graph, and a backward search on a time-independent graph with the purpose of bounding the set of nodes explored by the forward search. To the best of our knowledge, it was the first method allowing practical shortest path computations (i.e., in less than 300 msec) on large scale time-dependent road networks. In this paper we extend those concepts in order to include *core routing* on the time-dependent graph, and we analyze a dynamic scenario as well, in order to take into account updates of the cost function. Core routing is a well known technique for shortest path algorithms on static graphs [3], whose main idea is to shrink the original graph in order to get a new graph (*core*) with a smaller number of vertices. Most of the search is then carried out on the core, yielding a reduced search space.

Throughout the rest of this paper we will consider a lower bounding function $\lambda : A \to \mathbb{R}_+$ such that $\forall (u, v) \in A, \tau \in T$ we have $\lambda(u, v) \leq c(u, v, \tau)$. In practice, λ can easily be computed, given an arc length and the maximum allowed speed on that arc. In the experimental evaluation we will consider piecewise linear time-dependent arc cost functions.

The rest of this paper is organized as follows. In Section 2 we briefly review A^* and the bidirectional A^* algorithm applied on a time-dependent graph described in [1]. In Section 3 we describe core routing on static graphs and generalize it to the time-dependent case. In Section 4 we discuss the dynamic scenario. In Section 5 we provide a detailed experimental evaluation of our method, and analyze the results.

2 A^* with Landmarks

A^* is an algorithm for goal-directed search which is very similar to Dijkstra's algorithm. The difference between the two algorithms lies in the priority key.

For A^*, the priority key of a node v is made up of two parts: the length of the tentative shortest path from the source to v (as in Dijkstra's algorithm), and an underestimation of the distance to reach the target from v. The function which estimates the distance between a node and the target is called potential function π; the use of π has the effect of giving priority to nodes that are (supposedly) closer to target node t. If the potential function is such that $\pi(v) \leq d(v, t) \, \forall v \in V$, where $d(v, t)$ is the distance from v to t, then A^* always finds shortest paths [9]; otherwise, it becomes a heuristic. A^* is guaranteed to explore no more nodes than Dijkstra's algorithm.

On a road network, Euclidean distances can be used to compute the potential function, possibly dividing by the maximum allowed speed if arc costs are travelling times instead of distances. A significant improvement over Euclidean potentials can be achieved using *landmarks* [10]. The main idea is to select a small set of nodes in the graph, sufficiently spread over the whole network (several heuristic selection strategies have been proposed — see [17]), and pre-compute all distances between landmarks and any node of the vertex set. Then, by triangle inequalities, it is possible to derive lower bounds to the distance between any two nodes. Suppose we have selected a set $L \subset V$ of landmarks, and we have stored all distances $d(v, \ell), d(\ell, v) \, \forall v \in V, \ell \in L$; the following triangle inequalities hold: $d(u, t) + d(t, \ell) \geq d(u, \ell)$ and $d(\ell, u) + d(u, t) \geq d(\ell, t)$. Therefore $\pi_f(u) = \max_{\ell \in L} \{ d(u, \ell) - d(t, \ell), d(\ell, t) - d(\ell, u) \}$ is a lower bound for the distance $d(u, t)$, and it can be used as a valid potential function for the forward search [10]. Bidirectional search can be applied, but the potential function must be consistent for the forward and backward search [11]. Bidirectional A^* with the potential function described above is called ALT; an experimental evaluation on static graphs can be found in [11]. It is straightforward to observe that, if arc costs can only increase with respect to their original value, the potential function associated with landmarks yields valid lower bound, even on a time-dependent graph; in [12] this idea is applied to a real road network in order to analyse the algorithm's performance both in the case of arc cost updates and of time-dependent cost functions, but in the latter scenario the ALT algorithm is applied in an unidirectional way.

In a recent work [1], a bidirectional ALT algorithm on time-dependent road networks was proposed. The algorithm is based on restricting the scope of a time-dependent A^* search from the source using a set of nodes defined by a time-*independent* A^* search from the destination. The backward search is a reverse search on the graph G weighted by the lower bounding function λ.

Given a graph $G = (V, A)$, source and destination vertices $s, t \in V$, and a departure time $\tau_0 \in T$, let p^* be the shortest path from s to t leaving node s at τ_0. The algorithm for computing p^* works in three phases.

1. A bidirectional A^* search occurs on G, where the forward search is run on the graph weighted by c with the path cost defined by (1)-(2), and the backward search is run on the graph weighted by the lower bounding function λ. All nodes settled by the backward search are included in a set M. Phase 1 terminates as soon as the two search scopes meet.

2. Suppose that $v \in V$ is the first vertex in the intersection of the heaps of the forward and backward search; then the time dependent cost $\mu = \gamma_{\tau_0}(p_v)$ of the path p_v going from s to t passing through v is an upper bound to $\gamma_{\tau_0}(p^*)$. Let β be the key of the minimum element of the backward search queue; phase 2 terminates as soon as $\beta > \mu$. Again, all nodes settled by the backward search are included in M.

3. Only the forward search continues, with the additional constraint that only nodes in M can be explored. The forward search terminates when t is settled.

We call this algorithm TIME-DEPENDENT ALT (TDALT). Given a constant $K > 1$, K-approximated solutions can be computed switching from phase 2 to phase 3 as soon as $\beta > K\mu$; as the search stops sooner, the number of explored nodes decreases. We use the backward potential function $\pi_b^*(w) = \max\{\pi_b(w), d(s, v, \tau_0) + \pi_f(v) - \pi_f(w)\}$ described in [1], where π_f and π_b are the landmark potential functions for, respectively, the forward and the backward search, and v is a node already settled by the forward search. To guarantee correctness of this approach (see [1]), we do the following: we set up 10 checkpoints during the query; when a checkpoint is reached, the node v is updated, and the backward search queue is flushed and filled again using the updated π_b^*. We always pick v as the last node settled by the forward search before the checkpoint. The checkpoints are calculated comparing the initial lower bound $\pi_f(t)$ and the current distance from the source node, both for the forward search.

3 Time-Dependent Core-Based Routing

Core-based routing is a powerful approach which has been widely used for shortest paths algorithms on static graphs [3]. The main idea is to use contraction [18]: a routine iteratively removes nodes and adds edges to preserve correct distances between the remaining nodes, so that we have a smaller network where most of the search can be carried out. Note that in principle we can use any contraction routine which removes nodes from the graph and inserts edges to preserve distances. When the contracted graph $G_C = (V_C, A_C)$ has been computed, it is merged with the original graph to obtain $G_F = (V, A \cup A_C)$.

Suppose that we have a contraction routine which works on a time-dependent graph: that is, $\forall u, v \in V_C$, for each departure time $\tau_0 \in T$ there is a shortest path between u and v in G_C with the same cost as the shortest path between u and v in G with the same departure time. We propose the following query algorithm.

1. Initialization phase: start a Dijkstra search from both the source and the destination node on G_F, using the time-dependent costs for the forward search and the time-independent costs λ for the backward search, pruning the search (i.e. not relaxing outgoing arcs) at nodes $\in V_C$. Add each node settled by the forward search to a set S, and each node settled by the backward search to a set T. Iterate between the two searches until: (i) $S \cap T \neq \emptyset$ or (ii) the priority queues are empty.

2. Main phase: (i) If $S \cap T \neq \emptyset$, then start an unidirectional Dijkstra search from the source on G_F until the target is settled. (ii) If the priority queues are empty and we still have $S \cap T = \emptyset$, then start TDALT on the graph G_C, initializing the forward search queue with all leaves of S and the backward search queue with all leaves of T, using the distance labels computed during the initialization phase. The forward search is also allowed to explore any node $v \in T$, throughout the 3 phases of the algorithm. Stop when t is settled by the foward search.

In other words, the forward search "hops on" the core when it reaches a node $u \in S \cap V_C$, and "hops off" at all nodes $v \in T \cap V_C$. Note that landmark distances need be computed and stored only for vertices in V_C (see [3]). This means that the landmark potential function cannot be used to apply the forward A^* search on the nodes in T. However, we can use the backward distance labels computed with Dijkstra's algorithm during the initialization phase, which are valid distances on G_λ. We call this algorithm TIME-DEPENDENT CORE-BASED ALT (TDCALT).

Proposition 3.1. TDCALT *is correct.*

Since landmark distances are available only for nodes in V_C, the ALT potential function cannot be used "as is" whenever the source or the destination node do not belong to the core. In order to compute valid lower bounds to the distances from s or to t, proxy nodes have been introduced in [19] and used for the CALT algorithm (i.e. core-based ALT on a static graph) in [3]. We briefly report here the main idea: on the graph G weighted by λ, let $t' = \arg\min_{v \in V_C}\{d(t, v)\}$ be the core node closest to t. By triangle inequalities it is easy to derive a valid potential function for the forward search which uses landmark distances for t' as a proxy for t: $\pi_f(u) = \max_{\ell \in L}\{d(u, \ell) - d(t', \ell) - d(t, t'), d(\ell, t') - d(\ell, u) - d(t, t')\}$. The same calculations yield the potential function for the backward search π_b using a proxy node s' for the source s and the distance $d(s', s)$.

Contraction. For the contraction phase, i.e., the routine which selects which nodes have to be bypassed and then adds shortcuts to preserve shortest paths, we use the same algorithm proposed in [16]. We define the *expansion* [19] of a node u as the quotient between the number of added shortcuts and the number of edges removed if u is bypassed, and the *hop-number* of a shortcut as the number of edges that the shortcut represents. We iterate the contraction routine until the expansion of all remaining nodes exceeds a limit c or the hop-number exceeds a limit h. At the end of contraction, we perform an edge-reduction step which removes unnecessary shortcuts from the graph (cf. [16] for details).

Outputting Shortest Paths. TDCALT adds shortcuts to the graph in order to accelerate queries. Hence, if we want to retrieve the complete shortest path (and not only the distance) we must expand those shortcuts. In [20], an efficient unpacking routine based on storing all the edges a shortcut represents is introduced. However, in the static case a shortcut represents exactly one path, whereas in the time-dependent case a shortcut may represent a different path for each different traversal times. We solve this problem by allowing multi-edges: whenever a node is bypassed, a shortcut is inserted to represent each pair of incoming and outgoing edges, even if another edge between the two endpoints already exists. With this modification each shortcut represents exactly one path, so we can directly apply the unpacking routine from [20].

4 Dynamic Time-Dependent Costs

Up to now, time-dependent routing algorithms assumed complete knowledge of the time-dependent cost functions on arcs. However, since the speed profiles on which these functions are based are generated using historical data gathered from sensors (or cams), it is reasonable to assume that also real-time traffic information is available through these sensors. Moreover, other technologies exist to be aware of traffic jams even without having access to real-time speed information (e.g., TMC[1]). In the end, a procedure to update the time-dependent cost functions depending on real-time traffic information would be desirable for practical applications. Since we use piecewise linear functions stored as a list of breakpoints, we will consider modifications in the value of these.

Update procedure. Let (V_C, A_C) be the core of G. Suppose that the cost function of one arc $a \in A$ is modified; the set of core nodes V_C need not change, as long as A_C is updated in order to preserve distances with respect to the uncontracted graph $G = (V, A)$ with the new cost function. There are two possible cases: either the new values of the modified breakpoints are smaller than the previous ones, or they are larger. In the first case, then all arcs on the core A_C must be recomputed by running a label-correcting algorithm between the endpoints of each shortcut, as we do not know which shortcuts the updated arc may contribute to. In the second case, then the cost function for core arcs (i.e. shortcuts) may change for all those arcs $a' \in A_C$ such that a' contains a in its decomposition for at least one time instant τ. In other words, if a contributed to a shortcut a', then the cost of a' has to be recomputed. As the cost of a has increased, then a cannot possibly contribute to other shortcuts, thus we can restrict the update only to the shortcuts that contain the arc. To do so, we store for each $a \in A$ the set $S(a)$ of all shortcuts that a contributes to. Then, if one or more breakpoints of a have their value changed, we do the following.

Let $[\tau_1, \tau_{n-1}]$ be the smallest time interval that contains all modified breakpoints of arc a. If the breakpoints preceding and following $[\tau_1, \tau_{n-1}]$ are, respectively, at times τ_0 and τ_n the cost function of a changes only in the interval $[\tau_0, \tau_n]$. For each shortcut $a' \in S(a)$, let $a'_0, \ldots, a'_d$, with $a'_i \in A \, \forall i$, be its decomposition in terms of the original arcs, let $\lambda_j = \sum_{i=0}^{j-1} \lambda(a'_i)$ and $\mu_j = \sum_{i=0}^{j-1} \mu(a'_i)$, where $\forall a \in A$ we define $\mu(a) = \max_{\tau \in T} c(a, \tau)$, i.e., $\mu(a)$ is an upper bound on the cost of arc a. If a is the arc with index j in the decomposition of a', then a' may be affected by the change in the cost function of a only if the departure

---

[1] http://www.tmcforum.com/

time from the starting point of a' is in the interval $[\tau_0 - \mu_j, \tau_n - \lambda_j]$. This is because a can be reached from the starting node of a' no sooner than λ_j, and no later than μ_j. Thus, in order to update the shortcut a', we need to run a label-correcting algorithm between its two endpoints only in the time interval $[\tau_0 - \mu_j, \tau_n - \lambda_j]$, as the rest of the cost function is not affected by the change. In practice, if the length of the time interval $[\tau_0, \tau_n]$ is larger than a given threshold we run a label-correcting algorithm between the shortcut's endpoints over the whole time period, as the gain obtained by running the algorithm over a smaller time interval does not offset the overhead due to updating only a part of the profile with respect to computing from scratch.

The procedure described above is valid only when the value of breakpoints increases. In a typical realistic scenario, this is often the case: the initial cost profiles are used to model normal traffic conditions, and cost updates occur only to add temporary slowdowns due to unexpected traffic jams. When the temporary slowdowns are no longer valid we would like to restore the initial cost profiles, i.e. lower breakpoints to their initial values, without recomputing the whole core. If we want to allow fast updates as long as the new breakpoint values are larger than the ones used for the initial core construction, without requiring that the values can only increase, then we have to manage the sets $S(a) \,\forall a \in A$ accordingly. We provide an example that shows how problems could arise.

Example 4.1. Given $a \in A$, suppose that the cost of its breakpoint at time $\tau \in T$ increases, and all shortcuts $\in S(a)$ are updated. Suppose that, for a shortcut $a' \in S(a)$, a does not contibute to a' anymore due to the increased breakpoint value. If a' is removed from $S(a)$ and at a later time the value of the breakpoint at τ is restored to the original value, then a' would not be updated because $a' \notin S(a)$, thus a' would not be optimal.

Our approach to tackle this problem is the following: for each arc $a \in A$, we update the sets $S(a)$ whenever a breakpoint value changes, with the additional constraint that elements of $S(a)$ after the initial core construction phase cannot be removed from the set. Thus, $S(a)$ contains all shortcuts that a contributes to with the current cost function, plus all shortcuts that a contributed to during the initial core construction. As a consequence we may update a shortcut $a' \in S(a)$ unnecessarily, if a contributed to a' during the initial core construction but ceased contributing after an update step; however, this guarantees correctness for all changes in the breakpoint values, as long as the new values are not strictly smaller than the values used during the initial graph contraction. From a practical point of view, this is a reasonable assumption.

Since the sets $S(a) \,\forall a \in A$ are stored in memory, the computational time required by the core update is largely dominated by the time required to run the label-correcting algorithm between the endpoints of shortcuts. Thus, we have a trade-off between query speed and update speed: if we allow the contraction routine to build long shortcuts (in terms of number of bypassed nodes, i.e. "hops", as well as travelling time) then we obtain a faster query algorithm, because we are able to skip more nodes during the shortest path computations. On the other hand, if we allow only limited-length shortcuts, then the query search space is larger, but the core update is significantly faster as the label-correcting algorithm takes less time. In Section 5 we provide an experimental evaluation for different scenarios.

5   Experiments

In this section, we present an extensive experimental evaluation of our time-dependent ALT algorithm. Our implementation is written in C++ using solely the STL. As priority queue we use a binary heap. Our tests were executed on one core of an AMD Opteron 2218 running SUSE Linux 10.3. The machine is clocked at 2.6 GHz, has 16 GB of RAM and 2 x 1 MB of L2 cache. The program was compiled with GCC 4.2, using optimization level 3.

We use 32 *avoid* landmarks [10], computed on the core of the input graph using the lower bounding function λ to weight edges, and we use the tightened potential function π_b^* described in Section 2 as potential function for the backward search, with 10 checkpoints. When performing random s-t queries, the source s, target t, and the starting time τ_0 are picked uniformly at random and results are based on 10 000 queries. In the following, we restrict ourselves to the scenario where only distances — not the complete paths — are required. However, our shortcut expansion routine for TDCALT needs less than 1 ms to output the whole path; the additional space overhead is $\approx$ 4 bytes per node.

Input. We tested our algorithm on the road network of Western Europe provided by PTV AG for scientific use, which has approximately 18 million vertices and 42.6 million arcs. A travelling time in uncongested traffic situation was assigned to each arc using that arc's category (13 different categories) to determine the travel speed. Since we are not aware of a *large* publicly available real-world road network with time-dependent arc costs we used artificially generated costs. In order to model the time-dependent costs on each arc, we developed a heuristic algorithm, based on statistics gathered using real-world data on a limited-size road network, which is described in [1] and ensures spatial coherency for traffic jams.

Contraction Rates. Table 1 shows the performance of TDCALT for different contraction parameters (cf. Section 3). In this setup, we fix the approximation constant K to 1.15, which was found to be a good compromise between speed and quality of computed paths (see [1]). As the performed TDCALT queries may compute approximated results instead of optimal solutions when $K > 1$, we record three different statistics to characterize the solution quality: error rate, average relative error, maximum relative error. By *error rate* we denote the percentage of computed suboptimal paths over the total number of queries. By *relative error* on a particular query we denote the relative percentage increase of the approximated solution over the optimum, computed as $\omega/\omega^* - 1$, where ω is the cost of the approximated solution computed by our algorithm and ω^* is the cost of the optimum computed by Dijkstra's algorithm. We report *average* and *maximum* values of this quantity over the set of all queries. Note that contraction parameters of $c = 0.0$ and $h = 0$ yield a pure TDALT setup.

Table 1. Performance of TDCALT for different contraction rates. c denotes the maximum expansion of a bypassed node, h the hop-limit of added shortcuts. The third column records how many nodes have *not* been bypassed applying the corresponding contraction parameters. Preprocessing effort is given in time and *additional* space in bytes per node. Moreover, we report the increase in number of edges and interpolation points of the merged graph compared to the original input.

CORE			PREPROCESSING				ERROR			QUERY	
param.		core	time	space	increase in			relative		#settled	time
c	h	nodes	[min]	[B/n]	#edges	#points	rate	avg.	max	nodes	[ms]
0.0	0	100.0%	28	256	0.0%	0.0%	40.1%	0.303%	10.95%	250 248	188.2
0.5	10	35.6%	15	99	9.8%	21.1%	38.7%	0.302%	11.14%	99 622	78.2
1.0	20	6.9%	18	41	12.6%	69.6%	34.7%	0.288%	10.52%	19 719	21.7
2.0	30	3.2%	30	45	9.9%	114.1%	34.9%	0.287%	10.52%	9 974	13.2
2.5	40	2.5%	39	50	9.1%	138.0%	34.1%	0.275%	8.74%	8 093	11.4
3.0	50	2.0%	50	56	8.7%	161.2%	32.8%	0.267%	9.58%	7 090	10.3
3.5	60	1.8%	60	61	8.5%	181.1%	33.8%	0.280%	8.69%	6 227	9.2
4.0	70	1.5%	88	74	8.5%	223.1%	32.8%	0.265%	8.69%	5 896	8.8
5.0	90	1.2%	134	89	8.6%	273.5%	32.6%	0.266%	8.69%	5 812	8.4

As expected, increasing the contraction parameters has a positive effect on query performance. Interestingly, the space overhead first decreases from 256 bytes per node to 41 ($c = 1.0$, $h = 20$), and then increases again. The reason for this is that the core shrinks very quickly, hence we store landmark distances only for 6.9% of the nodes. On the other hand, the number of interpolation points for shortcuts increases by up to a factor ≈ 4 with respect to the original graph. Storing these additional points is expensive and explains the increase in space consumption.

It is also interesting to note that the maximum error rate decreases when we allow more and longer shortcuts to be built. We believe that this is due to the fact that long shortcuts decrease the number of settled nodes and have large costs, so at each iteration of TDCALT the key of the backward search priority queue β increases by a large amount. As the algorithm switches from phase 2 to phase 3 when $\mu/\beta < K$, and β increases by large steps, phase 3 starts with a smaller maximum approximation value for the current query μ/β. This is especially true for short distance queries, where the value of μ is small.

Query speed. Table 2 reports the results of TDCALT for different approximation values K using the European road network as input. In this experiment we used contraction parameters $c = 3.5$ and $h = 60$, i.e. we allow long shortcuts to be built to favour query speed. For comparison, we also report the results on the same road network for the time-dependent versions of Dijkstra, unidirectional ALT, TDALT and the time-dependent SHARC algorithm [16].

Table 2 shows that TDCALT yields a significant improvement over TDALT with respect to error rates, preprocessing space, size of the search space and query times. The latter two figures are improved by one order of magnitude. For exact queries, TDCALT is faster than unidirectional ALT by one order of

Table 2. Performance of time-dependent Dijkstra, unidirectional ALT, SHARC, TDALT and TDCALT with different approximation values K

technique	K	PREPROC. time [min]	PREPROC. space [B/n]	ERROR rate	ERROR relative av.	ERROR relative max	QUERY # settled nodes	QUERY time [ms]
Dijkstra	-	0	0	0.0%	0.000%	0.00%	8 877 158	5 757.4
uni-ALT	-	28	256	0.0%	0.000%	0.00%	2 056 190	1 865.4
SHARC	-	511	112	0.0%	0.000%	0.00%	84 234	75.3
TDALT	1.00	28	256	0.0%	0.000%	0.00%	2 931 080	2 939.3
	1.15	28	256	40.1%	0.303%	10.95%	250 248	188.2
	1.50	28	256	52.8%	0.734%	21.64%	113 040	71.2
TDCALT	1.00	60	61	0.0%	0.000%	0.00%	60 961	121.4
	1.05	60	61	2.7%	0.010%	3.94%	32 405	62.5
	1.10	60	61	16.6%	0.093%	7.88%	12 777	21.9
	1.15	60	61	33.0%	0.259%	8.69%	6 365	9.2
	1.20	60	61	39.8%	0.435%	12.37%	4 707	6.4
	1.30	60	61	43.0%	0.611%	16.97%	3 943	5.0
	1.50	60	61	43.7%	0.679%	20.73%	3 786	4.8
	2.00	60	61	43.7%	0.682%	27.61%	3 781	4.8

Table 3. CPU time required to update the core in case of traffic jams for different contraction parameters. The length of shortcuts is limited to 20 minutes of travel time (10 minutes for the values in parentheses).

cont. c h	space [B/n]	single traffic jam av.[ms]	single traffic jam max[ms]	batch update (1 000 jams) av.[ms]	batch update (1 000 jams) max[ms]	query time [ms]
0.0 0	256 (256)	0 (0)	0 (0)	0 (0)	0 (0)	188.2 (188.2)
0.5 10	100 (103)	1 (1)	49 (49)	820 (619)	1200 (799)	76.8 (85.2)
1.0 20	45 (50)	37 (21)	2231 (778)	30787 (20329)	39470 (22734)	22.8 (27.1)
2.0 30	51 (56)	220 (90)	5073 (3868)	187595 (79092)	206569 (85259)	16.4 (22.8)

magnitude, and the improvement over Dijkstra's algorithm is of a factor ≈ 50. Comparing TDCALT to SHARC, we see that for exact queries SHARC yields better query times by a factor ≈ 1.6, although preprocessing time and space for SHARC are larger. However, SHARC cannot efficiently deal with dynamic scenarios. If we can accept a maximum approximation factor $K \geq 1.05$ then TDCALT is faster than SHARC, by one order of magnitude for $K \geq 1.20$. The size of the search space decreases by even larger factors, but in terms of time spent per node SHARC is faster than TDCALT, as we observed in [1].

Dynamic Updates. In order to evaluate the performance of the core update procedure (see Section 4) we generated several traffic jams as follows: for each traffic jam, we select a path in the network covering 4 minutes of uncongested travel time on motorways. Then we randomly select a breakpoint between 6AM and 9 PM, and for all edges on the path we multiply the corresponding breakpoint value by a factor 5. As also observed in [12], updates on motorway edges are the most difficult to deal with, since those edges contribute to a large number of shortcuts. In Table 3 we report average and maximum required time over 1 000 runs to update the core in case of a single traffic jam, applying different contraction parameters. Moreover, we report the corresponding figures for a batch-update of 1000 traffic jams (100 runs), in order to reduce the fluctuations and give a clearer indication of required CPU time when performing multiple updates. Note that for this experiment we limit the length of shortcuts to 20 minutes (10 for the values in parentheses) of uncongested travel time. This is because in the dynamic scenario the length of shortcuts plays the most important role when determining the required CPU effort for an update operation, and if we allow the shortcuts length to grow indefinitely we may have unpractical update times. Hence, we also report query times with $K = 1.15$.

As expected, the effort to update the core becomes more expensive with increasing contraction parameters. However, for $c = 1.0$, $h = 20$ with maximum shortcut length of 20 minutes, we have reasonable update times together with query times of 22.8 ms: an update of 1 000 traffic jams can be done in less than 40 seconds, which should be sufficient in most applications. In most cases, the required time to update the core for a single traffic jam is of a few milliseconds, and query times are fast even with limited length shortcuts. We observe a clear trade off between query times and update times depending on the contraction parameters, so that for those applications which require frequent updates we can minimize update costs while keeping query times < 100 ms, and for applications which require very few or no updates we can minimize query times. If most of the graph's edges have their cost changed we can rerun the core edges computation, which takes less than 15 minutes.

6   Conclusion

We have proposed a bidirectional ALT algorithm for time-dependent graphs which uses a hierarchical approach: the bidirectional search starts on the full graph, but is soon restricted to a smaller network in order to reduce the number of explored nodes. This algorithm is flexible and allows us to deal with the dynamic scenario, where the piecewise linear time-dependent cost functions on arcs are not fixed, but can have their coefficients updated. Extensive computational experiments show a significant improvement over existing time-dependent algorithms, with query times reduced by at least an order of magnitude in almost all scenarios, and a faster and less space consuming preprocessing phase. Updates in the cost functions are dealt with in a practically efficient way, so that traffic jams can be added in a few milliseconds, and we can parameterize the preprocessing phase in order to balance the trade off between query speed and update speed.

References

1. Nannicini, G., Delling, D., Liberti, L., Schultes, D.: Bidirectional A* search for time-dependent fast paths. In: [22], pp. 334–346
2. Wagner, D., Willhalm, T.: Speed-up techniques for shortest-path computations. In: Thomas, W., Weil, P. (eds.) STACS 2007. LNCS, vol. 4393, pp. 23–36. Springer, Heidelberg (2007)

3. Bauer, R., Delling, D., Sanders, P., Schieferdecker, D., Schultes, D., Wagner, D.: Combining hierarchical and goal-directed speed-up techniques for dijkstra's algorithm. In: [22], pp. 303–318

4. Cooke, K., Halsey, E.: The shortest route through a network with time-dependent internodal transit times. J. of Math. Analysis and Applications 14, 493–498 (1966)

5. Dijkstra, E.: A note on two problems in connexion with graphs. Numerische Mathematik 1, 269–271 (1959)

6. Dreyfus, S.: An appraisal of some shortest-path algorithms. Operations Research 17(3), 395–412 (1969)

7. Kaufman, D.E., Smith, R.L.: Fastest paths in time-dependent networks for intelligent vehicle-highway systems application. Journal of Intelligent Transportation Systems 1(1), 1–11 (1993)

8. Orda, A., Rom, R.: Shortest-path and minimum delay algorithms in networks with time-dependent edge-length. Journal of the ACM 37(3), 607–625 (1990)

9. Hart, E., Nilsson, N., Raphael, B.: A formal basis for the heuristic determination of minimum cost paths. IEEE Transactions on Systems, Science and Cybernetics SSC 4(2), 100–107 (1968)

10. Goldberg, A., Harrelson, C.: Computing the shortest path: A^* meets graph theory. In: Proceedings of SODA 2005, pp. 156–165. SIAM, Philadelphia (2005)

11. Goldberg, A., Kaplan, H., Werneck, R.: Reach for A^*: Efficient point-to-point shortest path algorithms. In: Proceedings of ALENEX 2006. LNCS, pp. 129–143. Springer, Heidelberg (2006)

12. Delling, D., Wagner, D.: Landmark-based routing in dynamic graphs. In: [21], pp. 52–65

13. Chabini, I., Lan, S.: Adaptations of the A^* algorithm for the computation of fastest paths in deterministic discrete-time dynamic networks. IEEE Transactions on Intelligent Transportation Systems 3(1), 60–74 (2002)

14. Bauer, R., Delling, D.: SHARC: Fast and Robust Unidirectional Routing. In: Proceedings of the ALENEX 2008, pp. 13–26. SIAM, Philadelphia (2008)

15. Möhring, R.H., Schilling, H., Schütz, B., Wagner, D., Willhalm, T.: Partitioning graphs to speed up dijkstra's algorithm. In: Nikoletseas, S.E. (ed.) WEA 2005. LNCS, vol. 3503, pp. 189–202. Springer, Heidelberg (2005)

16. Delling, D.: Time-Dependent SHARC-Routing. In: Halperin, D., Mehlhorn, K. (eds.) Esa 2008. LNCS, vol. 5193, pp. 332–343. Springer, Heidelberg (2008)

17. Goldberg, A., Werneck, R.: Computing point-to-point shortest paths from external memory. In: Demetrescu, C., Sedgewick, R., Tamassia, R. (eds.) Proceedings of ALENEX 2005, pp. 26–40. SIAM, Philadelphia (2005)

18. Geisberger, R., Sanders, P., Schultes, D., Delling, D.: Contraction hierarchies: Faster and simpler hierarchical routing in road networks. In: [22], pp. 319–333

19. Goldberg, A., Kaplan, H., Werneck, R.: Better landmarks within reach. In: [21], pp. 38–51

20. Delling, D., Sanders, P., Schultes, D., Wagner, D.: Highway Hierarchies Star. In: Shortest Paths: Ninth DIMACS Implementation Challenge. DIMACS Book. American Mathematical Society (to appear, 2008)

21. Demetrescu, C. (ed.): WEA 2007. LNCS, vol. 4525. Springer, Heidelberg (2007)

22. McGeoch, C.C. (ed.): WEA 2008. LNCS, vol. 5038. Springer, Heidelberg (2008)

Bandwidth of Bipartite Permutation Graphs

(Extended Abstract)

Ryuhei Uehara

School of Information Science, JAIST, Asahidai 1-1, Nomi, Ishikawa 923-1292, Japan
uehara@jaist.ac.jp

Abstract. The bandwidth problem is finding a linear layout of vertices in a graph in such a way that minimizes the maximum distance between two vertices joined by an edge. The bandwidth problem is one of the classic $\mathcal{NP}$-complete problems. Especially, the problem is $\mathcal{NP}$-complete even for trees. The bandwidth problem can be solved in polynomial time for a few graph classes. Efficient algorithms for computing the bandwidth for three graph classes are presented. The first one is a linear time algorithm for a threshold graph, and the second one is a linear time algorithm for a chain graph. The last algorithm solves the bandwidth problem for a bipartite permutation graph in $O(n^2)$ time. The former two algorithms improve the previously known upper bounds to optimal, and the last one improves recent result, and they give positive answers to some open problems.

Keywords: Bandwidth, bipartite permutation graphs, chain graphs, interval graphs, threshold graphs.

1   Introduction

A layout of a graph $G = (V, E)$ is a bijection π between the vertices in V and the set $\{1, 2, \ldots, |V|\}$. The bandwidth of a layout π equals $\max\{|\pi(u) - \pi(v)| \mid \{u, v\} \in E\}$. The bandwidth of G is the minimum bandwidth of all layouts of G. The bandwidth has been studied since the 1950s; it has applications in sparse matrix computations (see [5,17] for survey). From the graph theoretical viewpoint, the bandwidth of a graph is strongly related to the proper interval completion problem [10], which is motivated by problems in molecular biology, and hence it attracts much attention (see, e.g., [11]). However, computing the bandwidth of a graph is one of basic and classic $\mathcal{NP}$-complete problems [21] (see also [7, GT40]). Especially, it is $\mathcal{NP}$-complete even if G is restricted to a caterpillar with hair length 3 [20]; that is, it is $\mathcal{NP}$-complete for trees and chordal graphs. Moreover, it is also $\mathcal{NP}$-complete for split graphs [13]. Few graph classes have been known for which the bandwidth problem can be solved in polynomial time; chain graphs [14], cographs [15], and interval graphs [12,19,23] (see [15] for a survey).

One of the interesting graph classes above is the class of interval graphs. The class was introduced in the 1950's by Hajös and Benzer independently (see [8]). In 1987, Kratsch proposed a polynomial time algorithm of the bandwidth problem for the class [16]. Unfortunately, it has a flaw, which was fixed by Mahesh et al. [19]. Kleitman and Vohra also showed a polynomial time algorithm [12], and Sprague improved the time complexity to $O(n \log n)$ by a smart implementation of the algorithm [23]. All

S.-H. Hong, H. Nagamochi, and T. Fukunaga (Eds.): ISAAC 2008, LNCS 5369, pp. 824–835, 2008.

the algorithms above solve the decision problem that asks if an interval graph G has bandwidth at most k for given G and k. Thus, using binary search for k, we can compute the bandwidth $\mathsf{bw}(G)$ of an interval graph G in $O(M(n) \cdot \log \mathsf{bw}(G))$ time, where $M(n) = O(n \log n)$ is the time complexity to solve the decision problem [23]. There are two unsolved problems for interval graphs mentioned in the literature. The first one is direct computation of the bandwidth of an interval graph. All the known algorithms are strongly depending on the given bound k to construct a desired layout. The second one is to improve the time complexity to linear time. Interval graphs have so simple structure that many $\mathcal{NP}$-hard problems can be solved in linear time on an interval graph (see, e.g., [2,18]).

Another interesting class is the class of chain graphs that plays an important role as a compact subclass of bipartite permutation graphs (the intersection of bipartite graphs and permutation graphs; see [4,25]). Kloks, Kratsch, and Müller gave an $O(n^2 \log n)$ time algorithm for a chain graph [14]. It uses the algorithm for an interval graph as a subroutine, and the factor $O(n \log n)$ comes from the time complexity of the subroutine.

We propose three algorithms of the bandwidth problem for three graph classes.

The first algorithm computes the bandwidth of a threshold graph G in $O(n)$ time and space. We note that threshold graphs form a proper subclass of interval graphs, and each graph can be represented in $O(n)$ space. The algorithm directly constructs an optimal layout, that is, we give a partial answer to the open problem for interval graphs, and improve the previously known upper bound $O(n \log n \log \mathsf{bw}(G))$ to optimal.

Extending the first algorithm, we next show an algorithm that computes the bandwidth of a chain graph in $O(n)$ time and space. It also directly constructs an optimal layout, and improves the previously known bound $O(n^2 \log n)$ in [14] to optimal.

The last algorithm solves the decision problem for bandwidth of a bipartite permutation graph. That is, for any given bipartite permutation graph G of n vertices and any integer k, the algorithm determines if G has a layout of bandwidth at most k in $O(n^2)$ time and $O(n)$ space. Thus the bandwidth of a bipartite permutation graph G of n vertices can be computed in $O(n^2 \log \mathsf{bw}(G))$ time and $O(n)$ space.

Recently, Heggernes, Kratsch, and Meister also propose a polynomial time algorithm of the bandwidth problem for bipartite permutation graphs [9]. They take a new approach to the bandwidth problem; they do not use the previously known techniques and algorithms for the other graph classes, and obtain an $O(n^4 \log n)$ time algorithm.

Here we compare our last result with the recent result in [9]. Essentially, Heggernes et al. only use a characterization of a bipartite permutation graph based on a vertex ordering called strong ordering. They do not rely on the previously known techniques and results for the bandwidth problem in literature, and give a new approach to the bandwidth problem for the class. On the other hand, our result is an extension of the known results for interval graphs and chain graphs with a new characterization of bipartite permutation graphs based on a graph decomposition in [4,25]. We sophisticate and adapt the known results to threshold graphs and chain graphs that admit us to extend and apply these results on bipartite permutation graphs.

Due to space limitation, some proofs are outlined.

2 Preliminaries

The *neighborhood* of a vertex v in a graph $G = (V, E)$ is the set $N_G(v) = \{u \in V \mid \{u, v\} \in E\}$, and the *degree* of a vertex v is $|N_G(v)|$ denoted by $d_G(v)$. If no confusion arise we will omit the index G. For a subset U of V, the subgraph of G induced by U is denoted by $G[U]$. Given a graph $G = (V, E)$, its *complement* $\bar{G} = (V, \bar{E})$ is defined by $\bar{E} = \{\{u, v\} \mid \{u, v\} \notin E\}$. A vertex set I is an *independent set* iff $G[I]$ contains no edges, and then the graph $\bar{G}[I]$ is said to be a *clique*. For a graph $G = (V, E)$, a sequence of distinct vertices $v_0, v_1, \ldots, v_l$ is a *path*, denoted by $(v_0, v_1, \ldots, v_l)$, if $\{v_j, v_{j+1}\} \in E$ for each $0 \le j < l$. The *length* of a path is the number of edges on the path. For two vertices u and v, the *distance* of the vertices, denoted by $dist(u, v)$, is the minimum length of the paths joining u and v. A graph $G = (V, E)$ is *bipartite* iff V can be partitioned into two sets X and Y such that every edge joins a vertex in X and the other vertex in Y.

A graph $G = (V, E)$ is called a *threshold graph* when there exist nonnegative weights $w(v)$ for $v \in V$ and t such that $\{u, v\} \in E$ iff $w(u) + w(v) \ge t$. A graph (V, E) with $V = \{v_1, v_2, \ldots, v_n\}$ is an *interval graph* if there is a finite set of intervals $\mathcal{I} = \{I_{v_1}, I_{v_2}, \ldots, I_{v_n}\}$ on the real line such that $\{v_i, v_j\} \in E$ iff $I_{v_i} \cap I_{v_j} \ne \emptyset$ for each i and j with $0 < i, j \le n$. We call the set $\mathcal{I}$ of intervals an *interval representation* of the graph. For each interval I, we denote by $R(I)$ and $L(I)$ the right and left endpoints of the interval, respectively (therefore we have $L(I) \le R(I)$ and $I = [L(I), R(I)]$). For any interval representation $\mathcal{I}$ and a point p, $N[p]$ denotes the set of intervals that contain the point p. An interval representation is called *proper* iff $L(I) \le L(J)$ and $R(I) \le R(J)$ for every pair of intervals I and J or vice versa. An interval graph is *proper* iff it has a proper interval representation. It is known that the class of proper interval graphs coincide with the class of unit interval graphs [22]. That is, any proper interval graph has a proper interval representation that consists of intervals of unit length (explicit and simple construction is given in [1]). Moreover, each connected proper interval graph has essentially unique proper (or unit) interval representation up to reversal in the following sense (see, e.g., [6, Corollary 2.5]):

Proposition 1. *For any proper interval graph $G = (V, E)$, there is a unique ordering (up to reversal) $v_1, v_2, \ldots, v_n$ of n vertices such that G has a unique proper interval representation $\mathcal{I}(G)$ such that $L(I_{v_1}) < L(I_{v_2}) < \cdots < L(I_{v_n})$.*

A bipartite graph (X, Y, E) with $X = \{x_1, x_2, \ldots, x_n\}$ and $Y = \{y_1, y_2, \ldots, y_{n'}\}$ is an *interval bigraph* if there are families of intervals $\mathcal{I}_X = \{I_{x_1}, I_{x_2}, \ldots, I_{x_n}\}$ and $\mathcal{I}_Y = \{I_{y_1}, I_{y_2}, \ldots, I_{y_{n'}}\}$ such that $\{x_i, y_j\} \in E$ iff $I_{x_i} \cap I_{y_j} \ne \emptyset$ for each i and j with $1 \le i \le n$ and $1 \le j \le n'$. Let $G = (X, Y, E)$ be a bipartite graph with $X = \{x_1, x_2, \ldots, x_n\}$ and $Y = \{y_1, y_2, \ldots, y_{n'}\}$. The ordering of X has the *adjacency property* iff, for each vertex $y \in Y$, $N(y)$ consists of vertices that are consecutive in the ordering of X. A bipartite graph $G = (X, Y, E)$ is *biconvex* iff there are orderings of X and Y that fulfill the adjacency property, and that is *convex* iff there is an ordering of X or Y that fulfills the adjacency property. A biconvex graph $G = (X, Y, E)$ is said to be a *chain graph* iff it has a vertex ordering of X such that $N(x_n) \subseteq N(x_{n-1}) \subseteq \cdots \subseteq N(x_1)$. (This property implies that we can sort Y as $N(y_1) \subseteq N(y_2) \subseteq \cdots \subseteq N(y_{n'})$; see, e.g., [24].)

A graph $G = (V, E)$ with $V = \{v_1, v_2, \ldots, v_n\}$ is said to be a *permutation graph* iff there is a permutation σ over V such that $\{v_i, v_j\} \in E$ iff $(i - j)(\sigma(v_i) - \sigma(v_j)) < 0$.

Intuitively, each vertex v in a permutation graph corresponds to a line segment ℓ_v joining two points on two parallel lines L_1 and L_2. Then two vertices v and u are adjacent iff the corresponding line segments ℓ_v and ℓ_u intersect. The ordering of vertices gives the ordering of the points on L_1, and the permutation of the ordering gives the ordering of the points on L_2. We call the intersection model a *line representation* of the permutation graph. When a permutation graph is bipartite, it is said to be a *bipartite permutation graph*. The following proper inclusions are known (see, e.g., [3,24]):

Lemma 1. *(1) Threshold graphs $\subset$ interval graphs, (2) chain graphs $\subset$ bipartite permutation graphs.*

A *layout* of a graph $G = (V, E)$ on n vertices is a bijection π between the vertices in V and the set $\{1, 2, \ldots, n\}$. The *bandwidth of a layout* π equals $\max\{|\pi(u) - \pi(v)| \mid \{u, v\} \in E\}$. The *bandwidth of G*, denoted by $\mathsf{bw}(G)$, is the minimum bandwidth of all layouts of G. A layout achieving $\mathsf{bw}(G)$ is called an *optimal layout*.

For given graph $G = (V, E)$, a *proper interval completion of G* is a superset E' of E such that $G' = (V, E')$ is a proper interval graph. Hereafter, we will omit the "proper interval" since we always consider proper interval completions. We say a completion E' is *minimum* iff $|C'| \leq |C''|$ for maximum cliques C' in $G' = (V, E')$ and C'' in $G'' = (V, E'')$ for any other completion E''.

For any graph G and its minimum completion E', it is known that $\mathsf{bw}(G) = |C'| - 1$, where C' is a maximum clique in $G' = (V, E')$ [10]. Let $G = (V, E)$ be an interval graph with interval representation $\mathcal{I} = \{I_{v_1}, I_{v_2}, \ldots, I_{v_n}\}$. For each maximal clique C, there is a point p such that $N[p]$ induces the clique C by Helly property. Thus we can compute $\mathsf{bw}(G)$ by the following algorithm for *any* given graph G;

   Input : Graph $G = (V, E)$
   Output: $\mathsf{bw}(G)$
 1 generate a proper interval graph $G' = (V, E')$ that gives a minimum completion of G;
 2 make a unique interval representation $\mathcal{I}(G')$ of G';
 3 find a point p such that $|N[p]| \geq |N[p']|$ for any other point p' on $\mathcal{I}(G')$;
 4 **return** $(|N[p]| - 1)$.

The following observation can be derived from the results in [10]:

Observation 1. *For a minimum completion $G' = (V, E')$ of $G = (V, E)$, let $\mathcal{I}(G') = (I_{v_1}, I_{v_2}, \ldots, I_{v_n})$ be the unique proper interval representation of G' given in Proposition 1. Then the ordering $v_1, v_2, \ldots, v_n$ gives an optimal layout of G, and vice versa.*

Here we show a technical lemma for proper interval subgraphs of an interval graph that will play an important role of our results.

Lemma 2. *Let $G = (V, E)$ be an interval graph with $V = \{v_1, v_2, \ldots, v_n\}$, and $\mathcal{I} = \{I_{v_1}, I_{v_2}, \ldots, I_{v_n}\}$ an interval representation of G. Let $\mathcal{J} = \{J_{u_1}, J_{u_2}, \ldots, J_{u_k}\}$ be a subset of $\mathcal{I}$ such that $\mathcal{J}$ forms a proper interval representation. That is, we have $U = \{u_1, \ldots, u_k\} \subseteq V$, and we can order $\mathcal{J}$ as $L(J_{u_i}) \leq L(J_{u_{i+1}})$ and $R(J_{u_i}) \leq R(J_{u_{i+1}})$ for each $1 \leq i < k$. Let ρ be the injection from $\mathcal{J}$ to $\mathcal{I}$ with $J_{v_i} = I_{v_{\rho(i)}}$ for each $1 \leq i \leq k \leq n$. Then, G has an optimal layout π such that each interval J_{u_i} appears according to the ordering in $\mathcal{J}$. More precisely, for each i with $1 \leq i < k$, we have $\pi(I_{v_{\rho(i)}}) < \pi(I_{v_{\rho(i+1)}})$.*

828   R. Uehara

Proof (Outline). The main idea is based on the algorithm by Kleitman and Vohra in [12]. We assume that we give the interval representation $\mathcal{I}$ and $\mathsf{bw}(G)$ as an input to the algorithm in [12]. It constructs an optimal layout π that achieves $\mathsf{bw}(G)$. By careful analysis of the algorithm, we can show that it does not change the ordering in $\mathcal{J}$ in the optimal layout π. Thus it labels all intervals in $\mathcal{J}$ from left to right. $\qquad\square$

3   Polynomial Time Algorithms

3.1   Linear Time Algorithm for Threshold Graphs

We first show a linear time algorithm for computing $\mathsf{bw}(G)$ of a threshold graph G. For a threshold graph $G = (V, E)$, there exist nonnegative weights $w(v)$ for $v \in V$ and t such that $\{u, v\} \in E$ if and only if $w(u) + w(v) \geq t$. We assume that G is connected and V is already ordered as $\{v_1, v_2, \ldots, v_n\}$ with $w(v_i) \leq w(v_{i+1})$ for $1 \leq i < n$ (this sort can be done in $O(n)$ time by bucket sort using the degrees of vertices). We can find ℓ such that $w(v_{\ell-1}) + w(v_\ell) < t$ and $w(v_\ell) + w(v_{\ell+1}) \geq t$ in $O(n)$ time. Then G has the following interval representation $\mathcal{I}(G)$:

· For $1 \leq i \leq \ell$, v_i corresponds to the point i, that is, $I_{v_i} = [i, i]$.
· For $\ell < i \leq n$, v_i corresponds to the interval $[j, \ell]$, where j is the minimum index with $w(v_i) + w(v_j) \geq t$.

For example, Fig. 1(a) is a threshold graph; each number in a circle is its weight, and threshold value is 5. We have $\ell = 5$ and its interval representation is given in Fig. 1(b).

Theorem 2. *Assume that a connected threshold graph $G = (V, E)$ is given in the interval representation $\mathcal{I}(G)$ stated above. We can compute* $\mathsf{bw}(G)$ *in $O(n)$ time and space.*

Proof (Outline). First observe that $L(I_{v_i}) < L(I_{v_{i+1}})$ and $R(I_{v_i}) < R(I_{v_{i+1}})$ for each i with $1 \leq i < \ell$, and $L(I_{v_i}) \geq L(I_{v_{i+1}})$ and $R(I_{v_i}) = R(I_{v_{i+1}}) = \ell$ for each i with $\ell < i < n$. That is, G consists of two proper interval graphs induced by $\{v_1, v_2, \ldots, v_\ell\}$ and $\{v_\ell, v_{\ell+1}, \ldots, v_n\}$ (v_ℓ is shared). Their proper interval representations also appear in $\mathcal{I}(G)$. Hence, by Lemma 2, there exists an optimal layout π of $V = \{v_1, \ldots, v_n\}$ such that $\pi(v_1) < \pi(v_2) < \cdots < \pi(v_\ell)$ and $\pi(v_\ell) > \pi(v_{\ell+1}) > \pi(v_{\ell+2}) > \cdots > \pi(v_n)$. Thus we can obtain an optimal layout by merging two sequences of vertices.

To obtain an optimal layout, by Observation 1, we construct a minimum completion of G from two sequences. Since G is connected, $[L(I_{v_n}), R(I_{v_n})] = [1, \ell]$ is the longest interval. Hence we extend all intervals (except I_{v_n}) to length $\ell - 1$ and construct a minimum completion. We denote the extended interval I_{v_i} by I'_{v_i}. The extension of intervals

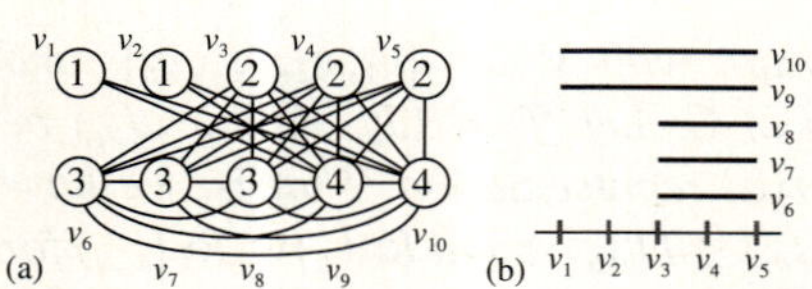

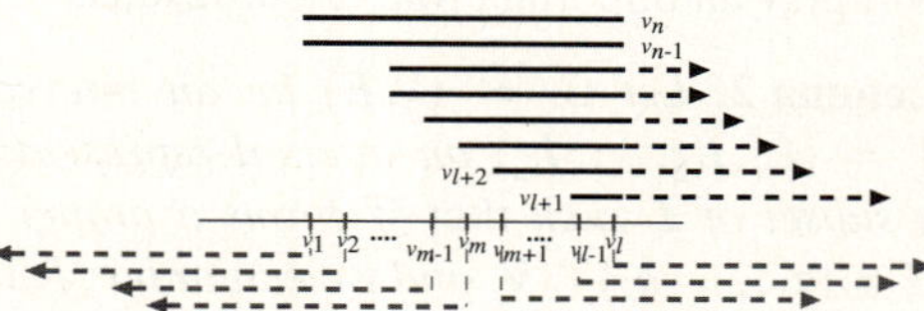

Fig. 1. (a) Threshold graph and (b) its interval representation

Fig. 2. Construction of a minimum completion

Algorithm 1. Bandwidth of a threshold graph

Input : Threshold graph $G = (V, E)$ with $w(v_1) \leq w(v_2) \leq \cdots \leq w(v_n)$ and t
Output: $\mathrm{bw}(G)$

1   let ℓ be the minimum index with $w(v_\ell) + w(v_{\ell+1}) \geq t$;
2   set $bw := \infty$;
3   **for** $m = 1, 2, \ldots, \ell - 1$ **do**
4   set $lc := 0$; // size of a maximum clique at points in $[1..m]$
5   **for** $i = 1, 2, \ldots, m$ **do**
6   let j be the minimum index with $w(v_i) + w(v_j) \geq t$;
7   **if** $lc < (m - i + 1) + (n - j + 1)$ **then** set $lc := (m - i + 1) + (n - j + 1)$;
8   **end**
9   **if** $\max\{lc, n - m\} < bw$ **then** $bw := \max\{lc, n - m\}$;
10   **end**
11   **return** $(bw - 1)$.

I_{v_i} for $i > \ell$ is straightforward (Fig. 2); just extend them to right, which does not increase the size of a maximum clique. Thus we focus on the points $I_{v_i} = [i, i]$ with $i \leq \ell$, which are extended to I'_{v_i} with length $\ell - 1$. We can observe that I'_{v_i} contains either 1 or ℓ. In the former case we can set $R(I'_{v_i}) = i$, and otherwise $L(I'_{v_i}) = i$ with loss of generality.

Thus, a minimum completion is given by the following proper interval representation of n intervals of length $\ell - 1$ for some m with $1 \leq m < \ell$: (0) for each $i > \ell$, $L[I_{v_i}] = j$, where j is the minimum index with $w(v_i) + w(v_j) \geq t$; (1) for each $1 \leq i \leq m$, $R[I_{v_i}] = i$, and (2) for each $m < i \leq \ell$, $L[I_{v_i}] = i$ (Fig. 2). Thus, to construct a minimum completion, we search the index m that minimizes a maximum clique in the proper interval graph represented by above proper interval representation determined by m.

On the minimum completion, there are ℓ distinct cliques C_i induced at each point i with $1 \leq i \leq \ell$. Now we consider a maximum clique of the corresponding proper interval graph for a fixed $m \in [1..\ell]$.

At points in $[m + 1, \ell]$, $N[m + 1] \subseteq \cdots \subseteq N[\ell]$ and hence $N[\ell]$ induces a maximum clique of size $n - m$. At each point i in $[1, m]$, $N[i]$ induces a clique that consists of $\{v_i, v_{i+1}, \ldots, v_m\}$ and $\{v_j, v_{j+1}, \ldots, v_n\}$, where j is the minimum index with $w(v_i)+w(v_j) \geq t$. Hence we have a clique of size $(m - i + 1) + (n - j + 1)$ for each point i in $[1, m]$.

Thus, for a fixed m, we compute these two candidates of a maximum clique from $[1..m]$ and $[m + 1..\ell]$, compare them, and obtain a maximum one. We compute the minimum size of the maximum cliques for all m, which gives $\mathrm{bw}(G) + 1$. Therefore, we can compute $\mathrm{bw}(G)$ by Algorithm 1. The correctness of Algorithm 1 follows from Observation 1, Lemma 2 and above discussions.

Algorithm 1 runs in $O(n^2)$ time and $O(n)$ space by a straightforward implementation. However, careful implementation achieves linear time, which is omitted here.   □

3.2   Linear Time Algorithm for Chain Graphs

We next show a linear time algorithm for computing $\mathrm{bw}(G)$ of a connected chain graph $G = (X, Y, E)$. We assume that $X = \{x_1, x_2, \ldots, x_n\}$ and $Y = \{y_1, y_2, \ldots, y_{n'}\}$ are already ordered by inclusion of neighbors; $N(x_n) \subseteq N(x_{n-1}) \subseteq \cdots \subseteq N(x_1) = Y$ and $N(y_1) \subseteq N(y_2) \subseteq \cdots \subseteq N(y_{n'}) = X$. We assume that a chain graph $G = (X, Y, E)$ with

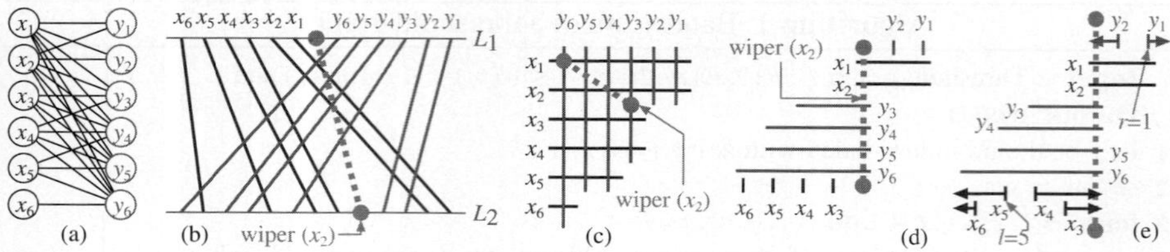

Fig. 3. (a) Chain graph and (b)-(e) its corresponding representations

$|X| = n$ and $|Y| = n'$ is given in $O(n + n')$ space; each vertex $y \in Y$ stores two endpoints 1 and $d(y)$ such that $N(y) = \{x_1, x_2, \ldots, x_{d(y)}\}$, and each vertex $x \in X$ stores two endpoints n' and $n' - d(x) + 1$ such that $N(x) = \{y_{n'}, y_{n'-1}, \ldots, y_{n'-d(x)+1}\}$. (We abuse the degree $d(\cdot)$ as a maximum index of the neighbors.) A chain graph has an intersection model of horizontal and vertical line segments (Fig. 3(a)(c)); X corresponds to horizontal line segments, and all left endpoints have the same coordinates, and Y corresponds to vertical line segments, and all top endpoints have the same coordinates. By the property of the inclusions of neighbors, vertices in X can be put from top to bottom, and vertices in Y can be put from right to left. It also can be transformed to the line representation of a bipartite permutation graph in a natural way (Fig. 3(b)). The endpoints on L_1 are sorted as $x_n, \ldots, x_1, y_{n'}, \ldots, y_1$ from left to right.

For a chain graph $G = (X, Y, E)$, a supergraph $H_i = (X \cup Y, E_i)$ is defined as follows [14]: We first define $H_0 = (X \cup Y, E_0)$ by a graph obtained from G by making a clique of X. For $1 \le i \le n - 1$, let C_i be a set $\{x_1, x_2, \ldots, x_i\} \cup N(x_{i+1})$. Then the graph H_i is obtained from G by making a clique of C_i. More precisely, $E_i = E \cup \{\{x_{i'}, x_{i''}\} \mid 1 \le i', i'' \le i\} \cup \{\{y_{j'}, y_{j''}\} \mid (n' - d(x_{i+1}) + 1) \le j', j'' \le n'\}$. The following lemma plays an important role in the algorithm in [14].

Lemma 3 ([14]). *(1) H_i is an interval graph for each i. (2) $\mathsf{bw}(G) = \min_i \mathsf{bw}(H_i)$.*

We first observe that H_0 is a threshold graph that has an interval representation in the form shown in Fig. 3(c) by regarding each $y \in Y$ as a point and each $x \in X$ as an interval. Thus, by Theorem 2, $\mathsf{bw}(H_0)$ can be computed in $O(n + n')$ time and space. Hereafter, we construct all minimum completions of H_i directly for $1 \le i \le n - 1$.

We introduce a $\mathsf{wiper}(x_i)$ which is a line segment joining two points p_1 on L_1 and p_2 on L_2 of the line representation of a chain graph $G = (X, Y, E)$ as follows (Fig. 3(b)); p_1 is fixed on L_1 between x_1 and $y_{n'}$, and p_2 is a point on L_2 between x_{i+1} and q, where q is the right neighbor point of x_{i+1} on L_2. More precisely, q is either (1) x_i if $N(x_i) = N(x_{i+1})$, or (2) the maximum vertex y_j in $N(x_i) \setminus N(x_{i+1})$ if $N(x_i) \setminus N(x_{i+1}) \neq \emptyset$. Using $\mathsf{wiper}(x_i)$, H_i is obtained from G by making a clique C_i which consists of the vertices corresponding to line segments intersecting $\mathsf{wiper}(x_i)$ on the line representation.

Intuitively, the interval representation of H_i can be obtained as follows; first, we construct a line representation of G and put the $\mathsf{wiper}(x_i)$ (Fig. 3(b)), second, we modify it to the intersection model of horizontal and vertical line segments with $\mathsf{wiper}(x_i)$, which is placed between x_i and x_{i+1} on y_j, where y_j is the minimum vertex in $N(x_{i+1})$ (Fig. 3(c)), and finally, we stretch the $\mathsf{wiper}(x_i)$ to vertical, or regard as a point at 0 on an interval representation, and arrange the line segments corresponding to the vertices

---

Algorithm 2. Bandwidth of a chain graph

Input : Chain graph $G = (X, Y, E)$ with $N(x_n) \subseteq \cdots \subseteq N(x_1)$ and $N(y_1) \subseteq \cdots \subseteq N(y_{n'})$

Output: bw(G)

1   $bw := $ bw(H_0) // by Algorithm 1

2   **for** $i = 1, 2, \ldots, n - 1$ **do**

3   construct the interval representation $\mathcal{I}(H_i)$ of the graph H_i with wiper(x_i);

4   **for** $\ell = n, n - 1, \ldots, i + 1$ **do**

5   **for** $r = 1, 2, \ldots, n' - d(x_{i+1})$ **do**

6   **if** $\max\{|\mathsf{RC}_i(\ell, r)|, |\mathsf{CC}_i(\ell, r)|, |\mathsf{LC}_i(\ell, r)|\} < bw$ **then**

          $bw = \max\{|\mathsf{RC}_i(\ell, r)|, |\mathsf{CC}_i(\ell, r)|, |\mathsf{LC}_i(\ell, r)|\}$;

7   **end**

8   **end**

9   **end**

10 **return** $(bw - 1)$.

---

in X and Y (Fig. 3(d)). We note that the interval representation of H_i is a combination of two interval representations of two threshold graphs (Fig. 1(b)). Precisely, formal construction of the interval representation of H_i is as follows. By Helly property, the intervals in the clique C_i share a common point 0, corresponding to wiper(x_i). For the point, we can construct a symmetric interval representation as follows (Fig. 3(d)); (1) each $x_{i'} \in X$ with $i' \leq i$ corresponds to an interval $[0, (d(x_{i'}) - d(x_i))]$, (2) each $x_{i'} \in X$ with $i' > i$ corresponds to the point $i - i' (< 0)$, (3) each $y_j \in Y$ with $j > n' - d(x_{i+1})$ corresponds to an interval $[(i - d(y_j)), 0]$, and (4) each $y_j \in Y$ with $j \leq n' - d(x_{i+1})$ corresponds to the point $i - j + 1$. Let $X_i^R = \{x_{i'} \in X \mid i' \leq i\}$, $X_i^L = \{x_{i'} \in X \mid i' > i\}$, $Y_i^L = \{y_j \in Y \mid j > n' - d(x_{i+1})\}$, and $Y_i^R = \{y_j \in Y \mid j \leq n' - d(x_{i+1})\}$. Two induced subgraphs $H_i[X_i^L \cup Y_i^L]$ and $H_i[X_i^R \cup Y_i^R]$ of H_i are threshold graphs, which allow us to use the algorithm in Section 3.1. We are ready to show the main theorem in this section.

Theorem 3. *We assume that a chain graph $G = (X, Y, E)$ is given in $O(n + n')$ space stated above. Then we can compute* bw(G) *in $O(n + n')$ time and space.*

Proof (Outline). By Lemma 3, we can compute bw(G) by computing the minimum bw(H_i) for $i = 0, 1, 2, \ldots, n - 1$. Since H_0 is a threshold graph, we can compute bw(H_0) in linear time and space by Theorem 2. We assume that $1 \leq i \leq n - 1$.

We here fix the index i. The basic idea is the combination of the algorithm for a threshold graph. We directly construct a minimum completion of H_i. When G is a threshold graph, we put a midpoint m such that each point i less than or equal to m is extended to an interval with $R[i] = i$, and each point i greater than m is extended to an interval with $L[i] = i$. Similarly, we put two midpoints ℓ in $H_i[X_i^L \cup Y_i^L]$ and r in $H_i[X_i^R \cup Y_i^R]$. Now we make a proper interval representation. For two midpoints ℓ and r, we make a proper interval representation as follows; (1) for each $x_{i'} \in X_i^L$ with $i' \geq \ell$, $I'_{x_{i'}} = [i - n, i - i']$, (2) for each $x_{i'} \in X_i^L$ with $\ell < i'$, $I'_{x_{i'}} = [i - i', 0]$, (3) for each $y_j \in Y_i^R$ with $r < j (\leq n' - d(x_{i+1}))$, $I'_{y_j} = [0, i - j + 1]$, and (4) for each $y_j \in Y_i^R$ with $j \leq r$, $I'_{y_j} = [i - j + 1, i]$. In Fig. 3(e), we give an example with $\ell = 5$ and $r = 1$. For each possible pair of (ℓ, r) with $i + 2 \leq \ell \leq n$ and $1 \leq r \leq n' - d(x_{i+1}) - 1$, we compute the size of a maximum clique in the proper interval representation. In this time, we have three candidates of a maximum clique at the left, center, and right parts of the proper interval representation. Precisely, for fixed i, ℓ, and r, we define three maximum cliques $\mathsf{RC}_i(\ell, r)$, $\mathsf{CC}_i(\ell, r)$, and $\mathsf{LC}_i(\ell, r)$ in three proper interval graphs induced by $\{x_\ell, x_{\ell+1}, \ldots, x_n\} \cup \{y_j, y_{j+1}, \ldots, y_{n'}\}$ where y_j is the minimum vertex in $N(x_\ell)$, $\{x_1, x_2, \ldots, x_{\ell-1}\} \cup \{y_{r+1}, y_{r+2}, \ldots, y_{n'}\}$, and $\{x_1, x_2, \ldots, x_{i'}\} \cup \{y_1, y_2, \ldots, y_r\}$ where $x_{i'}$ is the maximum vertex in $N(y_r)$, respectively. For each pair (ℓ, r), we compute $\max\{|\mathsf{RC}_i(\ell, r)|, |\mathsf{CC}_i(\ell, r)|, |\mathsf{LC}_i(\ell, r)|\}$, and we take the minimum value of $\max\{|\mathsf{RC}_i(\ell, r)|, |\mathsf{CC}_i(\ell, r)|, |\mathsf{LC}_i(\ell, r)|\}$ for all pairs, which is equal to $\mathsf{bw}(H_i) + 1$ for the fixed i. We next compute the minimum one for all i, which gives $\mathsf{bw}(G) + 1$. Summarizing up, we have Algorithm 2. Correctness and linear time implementation are omitted here.   □

3.3 Polynomial Time Algorithm for Bipartite Permutation Graphs

Now we turn to a bipartite permutation graph. Let $G = (X, Y, E)$ be a connected bipartite permutation graph. We assume that G is given in the line representation (like Fig. 3(b)), and the vertices in $X = \{x_1, x_2, \ldots, x_n\}$ and $Y = \{y_1, y_2, \ldots, y_{n'}\}$ are ordered from left to right on the representation. The main theorem in this section is the following.

Theorem 4. *For a bipartite permutation graph $G = (X, Y, E)$ and a positive integer k, we can find a layout that achieves $\mathsf{bw}(G) \leq k$ if it exists in $O((|X| + |Y|)^2)$ time and space.*

We note that we aim at solving the decision problem for given k.

We first observe that the vertex orderings $x_1, x_2, \ldots, x_n$ and $y_1, y_2, \ldots, y_{n'}$ correspond to the strong ordering for G in [9]. Therefore we have the following lemma:

Lemma 4 ([9, Observation 4]). *Let $G = (X, Y, E)$ be a bipartite permutation graph with the line representation, where $X = \{x_1, x_2, \ldots, x_n\}$ and $Y = \{y_1, y_2, \ldots, y_{n'}\}$ are ordered from left to right in this orderings. Then, there is an optimal layout π such that $\pi(x_i) < \pi(x_{i+1})$ for each $1 \leq i < n$ and $\pi(y_{i'}) < \pi(y_{i'+1})$ for each $1 \leq i' < n'$.*

Thus, intuitively, we can obtain an optimal layout by merging these two sequences of vertices. We here define a partition of the vertex sets X and Y by $V_0 = \{x_1\}$ and $V_j = \{v \mid dist(x_1, v) = j\}$. Thus, there is no edge between V_j and $V_{j'}$ if $|j - j'| > 1$, $V_0 \cup V_2 \cup V_4 \cup \cdots = X$ and $V_1 \cup V_3 \cup \cdots = Y$, and each V_i is an independent set. Let m denote the index with $V_m \neq \emptyset$ and $V_{m+1} = \emptyset$. The partition can be computed in $O(n + n')$ time and space. Hereafter, we denote the induced subgraph $G[V_j \cup V_{j+1}]$ for each $j = 0, 1, \ldots, m - 1$ by $G_j = (V_j \cup V_{j+1}, E_j)$. Then we have the following lemma:

Lemma 5 ([4,25]). *For a bipartite permutation graph $G = (X, Y, E)$ with the partition $V_0, V_1, V_2, \ldots, V_m$ of $X \cup Y$, each induced subgraph $G_j = (V_j \cup V_{j+1}, E_j)$ is a connected chain graph with $0 \leq j < m$.*

The algorithm is based on dynamic programming technique for the sequence of the chain graphs. For each $j = 0, 1, \ldots, m - 1$, we compute an optimal layout of $G_0 \cup G_1 \cup \cdots \cup G_j$ with constraint $\mathsf{bw}(G_0 \cup \cdots \cup G_j) \leq k$.

In the construction, we fix each chain graph G_j, and use the algorithm in Theorem 3. Then we have several H_i^j, each of which is a combination of two threshold graphs. We consider two midpoints ℓ^j and r^j for H_i^j, and obtain three candidates of maximum

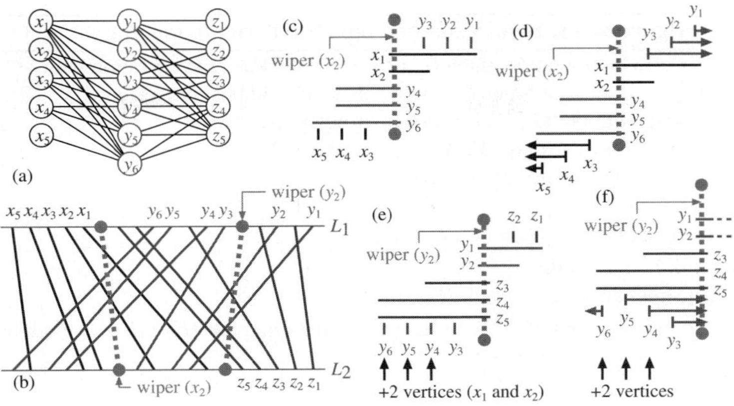

Fig. 4. A part of a bipartite permutation graph

cliques $\mathsf{LC}_i^j(\ell^j, r^j)$, $\mathsf{CC}_i^j(\ell^j, r^j)$, and $\mathsf{RC}_i^j(\ell^j, r^j)$. The difficulty occurs when we switch to $G_{j+1} = (V_{j+1}, V_{j+2}, E_{j+1})$. The vertices in V_{j+1} were already used in the current graph $G_j = (V_j, V_{j+1}, E_j)$, and some vertices in V_j are placed among them on the current layout. More precisely, when we deal with G_{j+1}, since V_{j+1} is already used in G_j, we have some vertices in V_j carried over from G_j to G_{j+1} since they have been put among V_{j+1} in the layout already. One example is shown in Fig. 4. Here we define the *carry set* S_i^j to deal with the vertices in V_j carried over G_{j+1}; S_i^j contains the vertices in V_j that have influences to the layout of the vertices in G_{j+1}. We have the following observation.

Observation 5. *The carry set S_i^j is equal to $\mathsf{CC}_i^j \cap V_j$.*

By Lemma 5, we can compute the partition $V_0, V_1, \ldots, V_m$ in $O(n + n')$ time and space. For each j with $0 \le j < m$, we compute a layout up to $\mathsf{bw}(G_0 \cup \cdots \cup G_j) \le k$ in this order. The modification of Algorithm 2 to deal with a carry set is tedious, and hence omitted here. Summarizing up, we have Algorithm 3. Now we are ready to prove the main theorem.

Proof (Sketch). If Algorithm 3 outputs a layout, it achieves that $\mathsf{bw}(G) \le k$. Hence we show that if $\mathsf{bw}(G) \le k$, a layout should be found by Algorithm 3. By Lemma 4, it is sufficient to consider the orderings that satisfy $\pi(x_1) < \pi(x_2) < \cdots < \pi(x_n)$ and $\pi(y_1) < \pi(y_2) < \cdots < \pi(y_{n'})$. Each chain graph G_i is a subgraph of G. Hence we have $\mathsf{bw}(G_i) \le \mathsf{bw}(G)$. When all carry sets are empty, the correctness follows from the proof of Theorem 3. Hereafter, we assume that there exists a nonempty carry set S_i^j. During the computation from left to right, each carry set S_i^j achieves the best possible layout for each pair of i and j; LC_i^j and CC_i^j are saturated for given k in general, and the vertices in $S_i^j = \mathsf{CC}_i^j \cap V_j$ (Observation 5) are put at as left-side as possible. Intuitively, the "margin" in RC_i^j is maximized for each i and j, and we can move no vertex in S_i^j to left any more. Therefore, there exists no better layout than the output of Algorithm 3 under the constraint of k.

For each i, time complexity for performing the modified algorithm of Algorithm 2 is linear. In steps 6 and 7, when H_i^j is constructed, we have to deal with the table for V_j

---

Algorithm 3. Bandwidth of a bipartite permutation graph

Input : Bipartite permutation graph $G = (X, Y, E)$ in a line representation with
$\quad\quad X = \{x_1, x_2, \ldots, x_n\}$ and $Y = \{y_1, y_2, \ldots, y_{n'}\}$, and positive integer k
Output: Layout with $\mathsf{bw}(G) \le k$ if it exists, or otherwise "No"

1   compute the partition $V_0 = \{x_1\}, V_1, V_2, \ldots, V_m$ of $X \cup Y$;
2   let $S_i^{-1} = \emptyset$ for each i // `no carry for` G_0
3   **for** $j = 0, 1, \ldots, m-1$ **do**
4   construct chain graph $G_j = (V_j, V_{j+1}, E_j)$;
5   **for** $i = 0, 1, \ldots, |V_j|$ **do**
6   construct an interval graph H_i^j;
7   compute a layout of H_i^j with carry S_i^{j-1} satisfying $\mathsf{bw}(H_i^j) \le k$ by a modified
          algorithm of Algorithm 2;
8   **end**
9   **if** *there is no layout that achieves* $\mathsf{bw}(H_i^j) \le k$ **then**
10   **return** ("No")
11   **end**
12 **end**
13 **return** (obtained layout).

---

which is fixed in H_i^{j-1}. This step requires $O(|V_{j-1}||V_j|)$. Hence, the total running time can be bounded above by $O(\sum_{j=0}^{m-1} |V_j||V_{j+1}|) = O((|X| + |Y|)^2)$. $\qquad\square$

Corollary 1. *For a bipartite permutation graph* $G = (X, Y, E)$, *we can compute* $\mathsf{bw}(G)$ *and find an optimal layout in* $O((|X| + |Y|)^2 \log \mathsf{bw}(G))$ *time and* $O(|X| + |Y|)$ *space.*

4   Concluding Remarks

In the proof of Theorem 4, we implement the algorithm straightforwardly. Using a subtle implementation based on the maintenance of "differences" of V_j with a carry set, which is a similar idea in the proof of Theorem 3, we would be able to improve the time complexity to linear. However, the algorithm is pretty complicated, and hence the straightforward implementation seems to be better from the practical viewpoint.

A straightforward extension of our algorithm to permutation graphs yields an exponential time algorithm. Thus the complexity for permutation graphs is still open.

Acknowledgment

The author is grateful to Pinar Heggernes for informing their quite recent work about bandwidth of bipartite permutation graphs [9].

References

1. Bogart, K.P., West, D.B.: A Short Proof That 'Proper=Unit'. Discrete Mathematics 201, 21–23 (1999)
2. Booth, K.S., Lueker, G.S.: Testing for the Consecutive Ones Property, Interval Graphs, and Graph Planarity Using PQ-Tree Algorithms. JCSS 13, 335–379 (1976)

3. Brandstädt, A., Le, V.B., Spinrad, J.P.: Graph Classes: A Survey. SIAM, Philadelphia (1999)
4. Brandstädt, A., Lozin, V.V.: On the Linear Structure and Clique-Width of Bipartite Permutation Graphs. Ars Combinatoria 67, 273–281 (2003)
5. Chinn, P.Z., Chvátalová, J., Dewdney, A.K., Gibbs, N.E.: The Bandwidth Problem for Graphs and Matrices — A Survey. Journal of Graph Theory 6, 223–254 (1982)
6. Deng, X., Hell, P., Huang, J.: Linear-time Representation Algorithms for Proper Circular-arc Graphs and Proper Interval Graphs. SIAM J. on Comp. 25(2), 390–403 (1996)
7. Garey, M.R., Johnson, D.S.: Computers and Intractability — A Guide to the Theory of NP-Completeness. W.H. Freeman, New York (1979)
8. Golumbic, M.C.: Algorithmic Graph Theory and Perfect Graphs, 2nd edn. Annals of Discrete Mathematics 57. Elsevier, Amsterdam (2004)
9. Heggernes, P., Kratsch, D., Meister, D.: Bandwidth of bipartite permutation graphs in polynomial time. In: Laber, E.S., Bornstein, C., Nogueira, L.T., Faria, L. (eds.) LATIN 2008. LNCS, vol. 4957, pp. 216–227. Springer, Heidelberg (2008)
10. Kaplan, H., Shamir, R.: Pathwidth, Bandwidth, and Completion Problems to Proper Interval Graphs with Small Cliques. SIAM J. on Comp. 25(3), 540–561 (1996)
11. Kaplan, H., Shamir, R., Tarjan, R.E.: Tractability of Parameterized Completion Problems on Chordal, Strongly Chordal, and Proper Interval Graphs. SIAM J. on Comp. 28(5), 1906–1922 (1999)
12. Kleitman, D.J., Vohra, R.V.: Computing the Bandwidth of Interval Graphs. SIAM J. Disc. Math. 3(3), 373–375 (1990)
13. Kloks, T., Kratsch, D., Borgne, Y.L., Müller, H.: Bandwidth of split and circular permutation graphs. In: Brandes, U., Wagner, D. (eds.) WG 2000. LNCS, vol. 1928, pp. 243–254. Springer, Heidelberg (2000)
14. Kloks, T., Kratsch, D., Müller, H.: Bandwidth of Chain Graphs. IPL 68, 313–315 (1998)
15. Kloks, T., Tan, R.B.: Bandwidth and Topological Bandwidth of Graphs with Few P_4's. Discrete Applied Mathematics 115, 117–133 (2001)
16. Kratsch, D.: Finding the Minimum Bandwidth of an Interval Graph. Inf. and Comp. 74, 140–158 (1987)
17. Lai, Y.L., Williams, K.: A Survey of Solved Problems and Applications on Bandwidth, Edgesum, and Profile of Graphs. Journal of Graph Theory 31(2), 75–94 (1999)
18. Lueker, G.S., Booth, K.S.: A Linear Time Algorithm for Deciding Interval Graph Isomorphism. Journal of the ACM 26(2), 183–195 (1979)
19. Mahesh, R., Rangan, C.P., Srinivasan, A.: On Finding the Minimum Bandwidth of Interval Graphs. Information and Computation 95, 218–224 (1991)
20. Monien, B.: The Bandwidth Minimization Problem for Caterpillars with Hair Length 3 is NP-Complete. SIAM J. Alg. Disc. Meth. 7(4), 505–512 (1986)
21. Papadimitriou, C.H.: The NP-Completeness of the Bandwidth Minimization Problem. Computing 16, 263–270 (1976)
22. Roberts, F.S.: Indifference Graphs. In: Harary, F. (ed.) Proof Techniques in Graph Theory, pp. 139–146. Academic Press, London (1969)
23. Sprague, A.P.: An $O(n \log n)$ Algorithm for Bandwidth of Interval Graphs. SIAM Journal on Discrete Mathematics 7, 213–220 (1994)
24. Uehara, R., Uno, Y.: On Computing Longest Paths in Small Graph Classes. International Journal of Foundations of Computer Science 18(5), 911–930 (2007)
25. Uehara, R., Valiente, G.: Linear Structure of Bipartite Permutation Graphs with an Application. Information Processing Letters 103(2), 71–77 (2007)

König Deletion Sets and Vertex Covers above the Matching Size

Sounaka Mishra[1], Venkatesh Raman[2], Saket Saurabh[3], and Somnath Sikdar[2]

[1] Indian Institute of Technology,
Chennai 600 036, India
`sounak@iitm.ac.in`
[2] The Institute of Mathematical Sciences,
Chennai 600 113, India
`{vraman,somnath}@imsc.res.in`
[3] The University of Bergen, Norway
`saket.saurabh@ii.uib.no`

Abstract. A graph is König-Egerváry if the size of a minimum vertex cover equals the size of a maximum matching in the graph. We show that the problem of deleting at most k vertices to make a given graph König-Egerváry is fixed-parameter tractable with respect to k. This is proved using interesting structural theorems on matchings and vertex covers which could be useful in other contexts.

We also show an interesting parameter-preserving reduction from the vertex-deletion version of red/blue-split graphs [4,9] to a version of VERTEX COVER and as a by-product obtain

1. the best-known exact algorithm for the optimization version of ODD CYCLE TRANSVERSAL [15];
2. fixed-parameter algorithms for several vertex-deletion problems including the following: deleting k vertices to make a given graph (a) bipartite [17], (b) split [5], and (c) red/blue-split [7].

1   Introduction

The classical notions of *matchings* and *vertex covers* have been at the center of serious study for several decades in the area of Combinatorial Optimization [11]. In 1931, König and Egerváry independently proved a result of fundamental importance: in a bipartite graph the size of a maximum matching equals that of a minimum vertex cover [11]. This led to a polynomial-time algorithm for finding a minimum vertex cover in bipartite graphs. In fact, a maximum matching can be used to obtain a 2-approximation algorithm for the MINIMUM VERTEX COVER problem in general graphs, which is still the best-known approximation algorithm for this problem. Interestingly, this min-max relationship holds for a larger class of graphs known as König-Egerváry graphs and it includes bipartite graphs as a proper subclass. König-Egerváry graphs will henceforth be called König graphs.

König graphs have been studied for a fairly long time from a structural point of view [1,3,9,10,18,15]. Both Deming [3] and Sterboul [18] gave independent characterizations of König graphs and showed that König graphs can be recognized in polynomial

S.-H. Hong, H. Nagamochi, and T. Fukunaga (Eds.): ISAAC 2008, LNCS 5369, pp. 836–847, 2008.

time. Lovász [10] used the theory of matching-covered graphs to give an excluded-subgraph characterization of König graphs that contain a perfect matching. Korach et al. [9] generalized this and gave an excluded-subgraph characterization for the class of all König graphs.

A natural optimization problem associated with a graph class $\mathcal{G}$ is the following: given a graph G, what is the minimum number of vertices to be deleted from G to obtain a graph in $\mathcal{G}$? For example, when $\mathcal{G}$ is the class of empty graphs, forests or bipartite graphs, the corresponding problems are VERTEX COVER, FEEDBACK VERTEX SET and ODD CYCLE TRANSVERSAL, respectively. We call the vertex-deletion problem corresponding to class of König graphs the KÖNIG VERTEX DELETION problem. A set of vertices whose deletion makes a given graph König is called a König vertex deletion set. In the parameterized setting, the parameter for vertex-deletion problems is the solution size, that is, the number of vertices to be deleted so that the resulting graph belongs to the given graph class.

An algorithmic study of the KÖNIG VERTEX DELETION problem was initiated in [12], where it was shown that when restricted to the class of graphs with a perfect matching, KÖNIG VERTEX DELETION fixed-parameter reduces to a problem known as MIN 2-CNF DELETION. This latter problem was shown to be fixed-parameter tractable by Razgon and O'Sullivan [16]. This immediately implies that KÖNIG VERTEX DELETION is fixed-parameter tractable for graphs with a perfect matching. But the parameterized complexity of the problem in general graphs remained open.

In this paper, we first establish interesting structural connections between minimum vertex covers and minimum König vertex deletion sets. Using these, we show that

1. the parameterized KÖNIG VERTEX DELETION problem is fixed-parameter tractable, and
2. there exists an $O^*(1.221^n)$ algorithm[1] for the optimization version of KÖNIG VERTEX DELETION problem, where n denotes the number of vertices in the graph.

Note that König graphs are not hereditary, that is, not closed under taking induced subgraphs. For instance, a 3-cycle is not König but attaching an edge to one of the vertices of the 3-cycle results in a König graph. In fact, KÖNIG VERTEX DELETION is one of the few vertex-deletion problems associated with a non-hereditary graph class whose parameterized complexity has been resolved. Another such example can be found in [13].

Our second result is an interesting parameter-preserving reduction from the vertex-deletion version of red/blue-split graphs [4,9] to a version of VERTEX COVER called ABOVE GUARANTEE VERTEX COVER. A red/blue graph [7] is a tuple $(G = (V, E), c)$, where $G = (V, E)$ is a simple undirected graph and $c : E \rightarrow 2^{\{r,b\}} \setminus \emptyset$ is an assignment of "colors" red and blue to the edges of the graph. An edge may be assigned both red and blue simultaneously and we require that R, the set of red edges, and B, the set of blue edges, both be nonempty. A red/blue graph $G = (V, R \cup B)$ is red/blue-split if its vertex set can be partitioned into a red independent set V_R and a blue independent set V_B. A red (resp. blue) independent set is an independent set in the red graph $G_R = (V, R)$ (resp. blue graph $G_B = (V, B)$). A graph G is split if its vertex set can be partitioned into an

---

[1] The $O^*(\cdot)$ notation suppresses polynomial terms. We write $O^*(T(n))$ for a time complexity of the form $O(T(n) \cdot \text{poly}(n))$, where $T(n)$ grows exponentially with n.

independent set and a clique. A 2-clique graph is a graph whose vertex set can be partitioned into two cliques. Note that a graph is 2-clique if and only if it is the complement of a bipartite graph. We will see that red/blue-split graphs are a generalization of König (and hence bipartite) graphs, split and 2-clique graphs [7].

As a by-product of the reduction from RED/BLUE-SPLIT VERTEX DELETION to ABOVE GUARANTEE VERTEX COVER we obtain:

1. an $O^*(1.49^n)$ algorithm for the optimization version of RED/BLUE-SPLIT VERTEX DELETION, ODD CYCLE TRANSVERSAL, SPLIT VERTEX DELETION and 2-CLIQUE VERTEX DELETION.[2]
2. fixed parameter algorithms for all the above problems.

For ODD CYCLE TRANSVERSAL, this gives the best-known exact algorithm for the optimization version improving over the previous best of $O^*(1.62^n)$ [15].

This paper is organized as follows. In Section 2 we give a brief outline of parameterized complexity, the notations and known results that we use in the rest of the paper. In Section 3 we show the KÖNIG VERTEX DELETION problem to be fixed-parameter tractable. In Section 4 we show that a number of vertex-deletion problems fixed-parameter reduce to RED/BLUE-SPLIT VERTEX DELETION which fixed-parameter reduces to ABOVE GUARANTEE VERTEX COVER. Finally in Section 5 we end with some concluding remarks and directions for further research.

2 Preliminaries

In this section we summarize the necessary concepts concerning parameterized complexity, fix our notation and outline some results that we make use of in the paper.

2.1 Parameterized Complexity

A parameterized problem is a subset of $\Sigma^* \times \mathbb{Z}^{\geq 0}$, where Σ is a finite alphabet and $\mathbb{Z}^{\geq 0}$ is the set of nonnegative numbers. An instance of a parameterized problem is therefore a pair (I, k), where k is the parameter. In the framework of parameterized complexity, the running time of an algorithm is viewed as a function of two quantities: the size of the problem instance *and* the parameter. A parameterized problem is said to be *fixed-parameter tractable (FPT)* if there exists an algorithm that takes as input (I, k) and decides whether it is a YES or NO-instance in time $O(f(k) \cdot |I|^{O(1)})$, where f is a function depending only on k. The class FPT consists of all fixed parameter tractable problems.

A parameterized problem π_1 is *fixed-parameter reducible* to a parameterized problem π_2 if there exist functions $f, g : \mathbb{Z}^{\geq 0} \to \mathbb{Z}^{\geq 0}$, $\Phi : \Sigma^* \times \mathbb{Z}^{\geq 0} \to \Sigma^*$ and a polynomial $p(\cdot)$ such that for any instance (I, k) of π_1, $(\Phi(I, k), g(k))$ is an instance of π_2 computable in time $f(k) \cdot p(|I|)$ and $(I, k) \in \pi_1$ if and only if $(\Phi(I, k), g(k)) \in \pi_2$. Two parameterized problems are *fixed-parameter equivalent* if they are fixed-parameter reducible to each other. The basic complexity class for fixed-parameter intractability is $W[1]$ as

---

[2] Since a 2-clique graph is the complement of a bipartite graph, the 2-CLIQUE VERTEX DELETION problem is NP-complete [17].

there is strong evidence that $W[1]$-hard problems are not fixed-parameter tractable. To show that a problem is $W[1]$-hard, one needs to exhibit a fixed-parameter reduction from a known $W[1]$-hard problem to the problem at hand. For more on parameterized complexity see [14].

2.2 Notation

Given a graph G, we use $\mu(G)$, $\beta(G)$ and $\kappa(G)$ to denote, respectively, the size of a maximum matching, a minimum vertex cover and a minimum König vertex deletion set of G. When the graph being referred to is clear from the context, we simply use μ, β and κ. Given a graph $G = (V, E)$ and two disjoint vertex subsets V_1, V_2 of V, we let (V_1, V_2) denote the bipartite graph with vertex set $V_1 \cup V_2$ and edge set $\{\{u, v\} : \{u, v\} \in E$ and $u \in V_1, v \in V_2\}$. If B is a bipartite graph with vertex partition $L \uplus R$ then we let $\mu(L, R)$ denote the size of the maximum matching of B. If M is matching and $\{u, v\} \in M$ then we say that *u is the partner of v in M*. If the matching being referred to is clear from the context we simply say *u is a partner of v*. The vertices of G that are the endpoints of edges in the matching M are said to be *saturated by M*; all other vertices are *unsaturated by M*.

2.3 Related Results

We next mention some known results about König graphs and the ABOVE GUARANTEE VERTEX COVER problem.

Fact 1 *(See for instance [12]). A graph $G = (V, E)$ is König if and only if there exists a polynomial-time algorithm that partitions $V(G)$ into V_1 and V_2 such that V_1 is a minimum vertex cover of G and there exists a matching across the cut (V_1, V_2) saturating every vertex of V_1.*

Given a graph G it is clear that $\beta(G) \geq \mu(G)$. The ABOVE GUARANTEE VERTEX COVER problem is this: given a graph G and an integer parameter k decide whether $\beta(G) \leq \mu(G) + k$. As was shown in [12], for this problem we may assume that the input graph $G = (V, E)$ has a perfect matching.

Theorem 1 [12]. *Let $G = (V, E)$ be a graph with a maximum matching M and let $I :=$ $V \setminus V(M)$. Construct G' by replacing every vertex $u \in I$ by a vertex pair u, u' and adding the edges $\{u, u'\}$ and $\{u', v\}$ for all $\{u, v\} \in E$. Then G has a vertex cover of size $\mu(G) + k$ if and only if G' has a vertex cover of size $\mu(G') + k$.*

In [12], we also showed that the ABOVE GUARANTEE VERTEX COVER problem fixed-parameter reduces to MIN 2-SAT DEL which is the problem of deciding whether k clauses can be deleted from a given 2-CNF SAT formula to make it satisfiable. Since MIN 2-SAT DEL is fixed-parameter tractable [16], so is ABOVE GUARANTEE VERTEX COVER. The algorithm for ABOVE GUARANTEE VERTEX COVER actually outputs a vertex cover of size $\mu(G) + k$ if there exists one.

Corollary 1. *Given a graph $G = (V, E)$ and an integer k, one can decide whether G has a vertex cover of size at most $\mu(G) + k$ in time $O(15^k \cdot k \cdot |E|^3)$. If G has a vertex cover of size $\mu(G) + k$ then the algorithm actually outputs one such vertex cover.*

Note that Theorem 1 says that for the Above Guarantee Vertex Cover problem it is sufficient to consider graphs with a *perfect matching*. In [12], we showed that the König Vertex Deletion problem on graphs with a perfect matching fixed-parameter reduces to Above Guarantee Vertex Cover. This shows that König Vertex Deletion on graphs with a perfect matching is fixed-parameter tractable. However this does not seem to resolve the parameterized complexity status of König Vertex Deletion in general graphs. We do not know of a fixed-parameter reduction from the general case to the case with a perfect matching as in the case of Above Guarantee Vertex Cover. However, in the next section, we show the general problem to be fixed-parameter tractable using some new structural results between maximum matchings and vertex covers.

3   The König Vertex Deletion Problem

We now consider the König Vertex Deletion Problem in general graphs and show it fixed-parameter tractable.

Suppose Y is a vertex cover in a graph $G = (V, E)$. Consider a maximum matching M between Y and $V \setminus Y$. If M saturates every vertex of Y then the graph is König. If not, then $Y \setminus V(M)$, the set of vertices of Y unsaturated by M, is a König deletion set by Fact 1. What we prove in this section is that if Y is a minimum vertex cover, then $Y \setminus V(M)$ is a minimum König vertex deletion set.

Our first observation is that any minimum König vertex deletion set is contained in some minimum vertex cover.

Theorem 2. *Let G be an undirected graph with a minimum König vertex deletion set K. Let $V(G \setminus K) = V_1 \uplus V_2$ where V_2 is independent and there is a matching M from V_1 to V_2 saturating V_1. Then $V_1 \cup K$ is a minimum vertex cover for G.*

Proof. Suppose S is a vertex cover of G such that $|S| < |V_1| + |K|$. We will show that there exists a König vertex deletion set of size smaller than $|K|$, contradicting our hypothesis. Define $V_1' = V_1 \cap S$, $V_2' = V_2 \cap S$ and $K' = K \cap S$. Let A_1 be the vertices of V_1' whose partner in M is in V_2' and let A_2 be the vertices of V_1' whose partner in M is not in V_2'. See Figure 1. We claim that $A_1 \cup K'$ is a König vertex deletion set of G and $|A_1 \cup K'| < |K|$, which will produce the required contradiction and prove the theorem. This claim is proved using the following three claims:

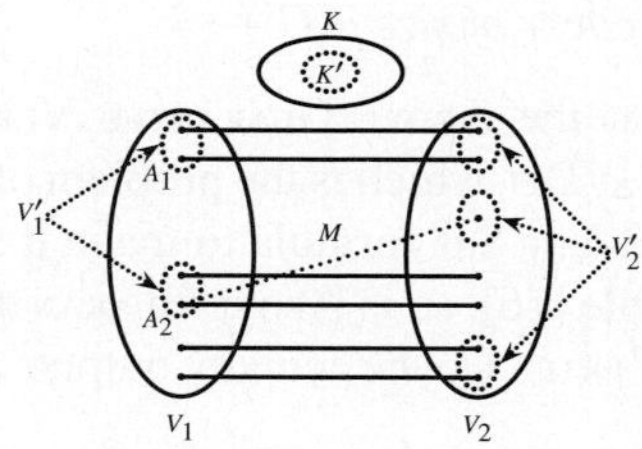

Fig. 1. The sets that appear in the proof of Theorem 2. The matching M consists of the solid edges across V_1 and V_2.

Claim 1. $|A_1 \cup K'| < |K|$.

Claim 2. $A_2 \cup V_2'$ *is a vertex cover in* $G \setminus (A_1 \cup K')$.

Claim 3. *There exists a matching between* $A_2 \cup V_2'$ *and* $V \setminus (V_1' \cup K' \cup V_2')$ *saturating every vertex of* $A_2 \cup V_2'$.

Proof of Claim 1. Clearly $|S| = |V_1'| + |V_2'| + |K'|$. Note that S intersects $|A_1|$ of the edges of M in both end points and $|M| - |A_1|$ edges of M in one end point (in either V_1' or V_2'). Furthermore V_2' has $|V_2' \setminus V(M)|$ vertices of S that do not intersect any edge of M. Hence $|M| + |A_1| + |V_2' \setminus V(M)| = |V_1'| + |V_2'|$. That is, $|V_1'| + |V_2'| = |V_1| + |A_1| + |V_2' \setminus V(M)|$ (as $|M| = |V_1|$). Hence $|S| < |V_1| + |K|$ implies that $|A_1| + |V_2' \setminus V(M)| + |K'| < |K|$ which implies that $|A_1| + |K'| < |K|$ proving the claim.

Proof of Claim 2. Since $S = A_1 \cup A_2 \cup V_2' \cup K'$ is a vertex cover of G, clearly $A_2 \cup V_2'$ covers all edges in $G \setminus (A_1 \cup K')$.

Proof of Claim 3. Since the partner of a vertex in A_2 in M is in $V \setminus (V_1' \cup K' \cup V_2')$, we can use the edges of M to saturate vertices in A_2. To complete the proof, we show that in the bipartite graph $(V_2', (V_1 \setminus V_1') \cup (K \setminus K'))$ there is a matching saturating V_2'. To see this, note that any subset $D \subseteq V_2'$ has at least $|D|$ neighbors in $(V_1 \setminus V_1') \cup (K \setminus K')$. For otherwise, let D' be the set of neighbors of D in $(V_1 \setminus V_1') \cup (K \setminus K')$ where we assume $|D| > |D'|$. Then $(S \setminus D) \cup D'$ is a vertex cover of G of size strictly less than $|S|$, contradicting the fact that S is a minimum vertex cover. To see that $(S \setminus D) \cup D'$ is indeed a vertex cover of G, note that $S \setminus V_2'$ covers all edges of G except those in the graph $(V_2', (V_1 \setminus V_1') \cup (K \setminus K'))$ and all these edges are covered by $(V_2' \setminus D) \cup D'$. Hence by Hall's theorem, there exists a matching saturating all vertices of V_2' in the bipartite graph $(V_2', (V_1 \setminus V_1') \cup (K \setminus K'))$, proving the claim.

    This completes the proof of the theorem. $\qquad\qquad\square$

Theorem 2 has interesting consequences.

Corollary 2. *For any two minimum König vertex deletion sets (KVDSs)* K_1 *and* K_2, $\mu(G \setminus K_1) = \mu(G \setminus K_2)$.

Proof. Since K_1 is a minimum KVDS of G, $\beta(G \setminus K_1) = \mu(G \setminus K_1)$. By Theorem 2, $\beta(G \setminus K_1) + |K_1| = \beta(G)$ and $\beta(G \setminus K_2) + |K_2| = \beta(G)$. Since $|K_1| = |K_2|$, it follows that $\beta(G \setminus K_1) = \beta(G \setminus K_2)$ and hence $\mu(G \setminus K_1) = \mu(G \setminus K_2)$. $\qquad\square$

From Theorem 2 and Fact 1, we get

Corollary 3. *Given a graph* $G = (V, E)$ *and a minimum König vertex deletion set for* G, *one can construct a minimum vertex cover for* G *in polynomial time.*

Our goal now is to prove the "converse" of Corollary 3. In particular, we would like to construct a minimum König vertex deletion set from a minimum vertex cover. Our first step is to show that if we know that a given minimum vertex cover contains a minimum König vertex deletion set then we can find the König vertex deletion set in polynomial time. Recall that given a graph $G = (V, E)$ and $A, B \subseteq V$ such that $A \cap B = \emptyset$, we use $\mu(A, B)$ to denote a maximum matching in the bipartite graph comprising of the vertices in $A \cup B$ and the edges in $\{\{u, v\} \in E : u \in A, v \in B\}$. We denote this graph by (A, B).

Lemma 1. *Let K be a minimum KVDS and Y a minimum vertex cover of a graph $G = (V, E)$ such that $K \subseteq Y$. Then $\mu(G \setminus K) = \mu(Y, V \setminus Y)$ and $|K| = |Y| - \mu(Y, V \setminus Y)$.*

Proof. If G is König then the theorem clearly holds. Therefore assume that $K \neq \emptyset$. Note that $Y \setminus K$ is a minimum vertex cover of the König graph $G \setminus K$. Thus $\mu(G \setminus K) = \mu(Y \setminus K, V \setminus Y)$. We claim that $\mu(Y \setminus K, V \setminus Y) = \mu(Y, V \setminus Y)$. For if not, we must have $\mu(Y \setminus K, V \setminus Y) < \mu(Y, V \setminus Y)$. Then let M be a maximum matching in the bipartite graph $(Y, V \setminus Y)$ and $K' \subseteq Y$ be the set of vertices unsaturated by M. Note that $K' \neq \emptyset$ is a KVDS for G. Since $\mu(Y, V \setminus Y) = |Y| - |K'|$ and $\mu(Y \setminus K, V \setminus Y) = |Y| - |K|$ we have $|K'| < |K|$, a contradiction, since by hypothesis K is a smallest KVDS for G. Therefore we must have $\mu(G \setminus K) = \mu(Y, V \setminus Y)$ and $|K| = |Y| - \mu(Y, V \setminus Y)$. $\square$

The next lemma says that $\mu(Y, V \setminus Y)$ is the same for all minimum vertex covers Y of a graph G. Together with Lemma 1, this implies that if K is a minimum König vertex deletion set and Y is a minimum vertex cover of a graph $G = (V, E)$, then $\mu(G \setminus K) = \mu(Y, V \setminus Y)$. This result is crucial to our FPT-algorithm for the KÖNIG VERTEX DELETION problem.

Lemma 2. *For any two minimum vertex covers Y_1 and Y_2 of G, $\mu(Y_1, V \setminus Y_1) = \mu(Y_2, V \setminus Y_2)$.*

Proof. Suppose without loss of generality that $\mu(Y_1, V \setminus Y_1) > \mu(Y_2, V \setminus Y_2)$. Let M_1 be a maximum matching in the bipartite graph $(Y_1, V \setminus Y_1)$. To arrive at a contradiction, we study how Y_2 intersects the sets Y_1 and $V \setminus Y_1$ with respect to the matching M_1. To this end, we define the following sets (see Figure 2):

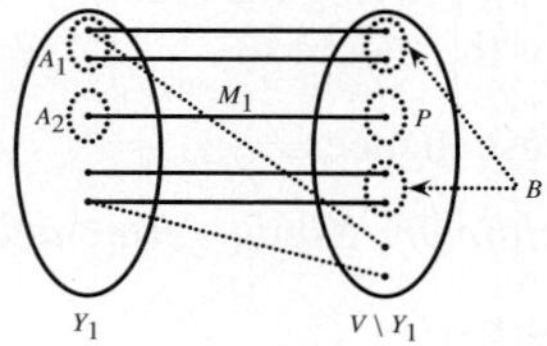

Fig. 2. The sets that appear in the proof of Lemma 2. The solid edges across Y_1 and $V \setminus Y_1$ constitute the matching M_1.

- $A = Y_2 \cap Y_1 \cap V(M_1)$.
- $B = Y_2 \cap (V \setminus Y_1) \cap V(M_1)$.
- A_1 is the set of vertices in A whose partners in M_1 are also in Y_2.
- A_2 is the set of vertices in A whose partners in M_1 are not in Y_2.

We first show that

Claim. In the bipartite graph $(Y_2, V \setminus Y_2)$ there is a matching saturating each vertex in $A_2 \cup B$.

It will follow from the claim that $\mu(Y_2, V \setminus Y_2) \geq |A_2| + |B|$. However, note that Y_2 intersects every edge of M_1 at least once (as Y_2 is a vertex cover). More specifically, Y_2

intersects $|A_1|$ edges of M_1 twice and $|M_1| - |A_1|$ edges once (either in Y_1 or in $V \setminus Y_1$). Hence, $|A|+|B| = |M_1|+|A_1|$ and so $|A_2|+|B| = |M_1|$ and so $\mu(Y_2, V \setminus Y_2) \geq |A_2|+|B| = |M_1|$ a contradiction to our assumption at the beginning of the proof. Thus it suffices to prove the claim.

Proof of Claim. Let P denote the partners of the vertices of A_2 in M_1. Since $P \subseteq V \setminus Y_2$, we use the edges of M_1 to saturate vertices of A_2. Hence it is enough to show that the bipartite graph $\mathcal{B} = (B, (V \setminus Y_2) \setminus P)$ contains a matching saturating the vertices in B. Suppose not. By Hall's Theorem there exists a set $D \subseteq B$ such that $|N_{\mathcal{B}}(D)| < |D|$. We claim that the set $Y_2' := Y_2 \setminus D + N_{\mathcal{B}}(D)$ is a vertex cover of G. To see this, note that the vertices in $Y_2 \setminus D$ cover all the edges of G except those in the bipartite graph $(D, Y_1 \cap (V \setminus Y_2))$ and these are covered by $N_{\mathcal{B}}(D)$. Therefore Y_2' is a vertex cover of size strictly smaller than Y_2, a contradiction. This proves that there exists a matching in $(Y_2, V \setminus Y_2)$ saturating each vertex in $A_2 \cup B$.

This completes the proof of the lemma. $\qquad\square$

The next theorem shows how we can obtain a minimum König vertex deletion set from a minimum vertex cover in polynomial time.

Theorem 3. *Given a graph $G = (V, E)$, let Y be any minimum vertex cover of G and M a maximum matching in the bipartite graph $(Y, V \setminus Y)$. Then $K := Y \setminus V(M)$ is a minimum König vertex deletion set of G.*

Proof. Clearly K is a KVDS. Let K_1 be a minimum KVDS of G. By Theorem 2, there exists a minimum vertex cover Y_1 such that $K_1 \subseteq Y_1$ and

$$
\begin{aligned}
|K_1| &= |Y_1| - \mu(Y_1, V \setminus Y_1) \text{ (By Lemma 1.)} \\
&= |Y| \ - \mu(Y_1, V \setminus Y_1) \text{ (Since } Y_1 \text{ and } Y \text{ are minimum vertex covers.)} \\
&= |Y| \ - \mu(Y, V \setminus Y) \quad \text{(By Lemma 2.)} \\
&= |K|
\end{aligned}
$$

This proves that K is a minimum KVDS. $\qquad\square$

Corollary 4. *Given a graph $G = (V, E)$ and a minimum vertex cover for G, one can construct a minimum König vertex deletion set for G in polynomial time.*

Note that although both these problems–Vertex Cover and König Vertex Deletion Set–are NP-complete, we know of very few pairs of such parameters where we can obtain one from the other in polynomial time (e.g. edge dominating set and minimum maximal matching, see [8]). In fact, there are parameter-pairs such as dominating set and vertex cover where such a polynomial-time transformation is not possible unless $P = NP$. This follows since in bipartite graphs, for instance, a minimum vertex cover is computable in polynomial time whereas computing a minimum dominating set is NP-complete.

To show that the König Vertex Deletion problem is fixed-parameter tractable we make use of the following

Lemma 3. *[12] If G is a graph such that $\beta(G) = \mu(G) + k$, then $k \leq \kappa(G) \leq 2k$.*

We are now ready to prove that the KÖNIG VERTEX DELETION problem is fixed-parameter tractable in general graphs.

Theorem 4. *Given a graph $G = (V, E)$ and an integer parameter k, the problem of whether G has a subset of at most k vertices whose deletion makes the resulting graph König can be decided in time $O(15^k \cdot k^2 \cdot |E|^3)$.*

Proof. Use the FPT algorithm from Corollary 1 to test whether G has a vertex cover of size at most $\mu(G) + k$. If not, by Lemma 3, we know that the size of a minimum König vertex deletion set is strictly more than k. Therefore return NO. If yes, then find the size of a minimum vertex cover by applying Corollary 1 with every integer value between 0 and k for the excess above $\mu(G)$. Note that for YES-instances of the ABOVE GUARANTEE VERTEX COVER problem, the FPT algorithm actually outputs a vertex cover of size $\mu(G) + k$. We therefore obtain a minimum vertex cover of G. Use Theorem 3 to get a minimum König vertex deletion set in polynomial time and depending on its size answer the question. It is easy to see that all this can be done in time $O(15^k \cdot k^2 \cdot |E|^3)$. $\square$

We know that computing a maximum independent set (or equivalently a minimum vertex cover) in an n-vertex graph can be done in time $O^*(2^{0.288n})$ [6]. By Corollary 4, we can compute a minimum König vertex deletion set in the same exponential time. Given a graph G together with a tree-decomposition for it of width w, one can obtain a minimum vertex cover in time $O^*(2^w)$ [14]. For the definitions of treewidth and tree-decomposition, refer [14]. In general, algorithms on graphs of bounded treewidth are based on dynamic programming over the tree-decomposition of the graph. It is not obvious how to find such a dynamic programming algorithm for the KÖNIG VERTEX DELETION problem. By applying Corollary 4, we can find a minimum König deletion set in time $O^*(2^w)$ in graphs of treewidth w. The above discussion results in the following corollary.

Corollary 5

1. *Given a graph $G = (V, E)$ on n vertices we can find a minimum König vertex deletion set in time $O^*(2^{0.288n}) = O^*(1.221^n)$.*
2. *If a tree-decomposition for G of width w is given, we can find a minimum König vertex deletion set in time $O^*(2^w)$.*

4 Red/Blue-Split Graphs and above Guarantee Vertex Cover

In this section we introduce the RED/BLUE-SPLIT VERTEX DELETION problem and show that a number of vertex-deletion problems fixed-parameter reduce to it. Recall that a red/blue-graph is one in which the edges are colored red or blue and where an edge may receive multiple colors. A red/blue-graph $G = (V, R \cup B)$ is red/blue-split if its vertex set can be partitioned into a red independent set and a blue independent set, where a red (blue) independent set is an independent set in the red graph $G = (V, R)$ (blue graph $G = (V, B)$). In what follows we use r/b as an abbreviation for red/blue and E_c to denote the set of edges assigned color c.

The R/B-Split Vertex Deletion problem is the following: given an r/b-graph $G = (V, R \cup B)$ and an integer k, are there k vertices whose deletion makes G r/b-split? We first show that R/B-Split Vertex Deletion fixed-parameter reduces to the Above Guarantee Vertex Cover problem. Since Above Guarantee Vertex Cover is fixed-parameter tractable, this will show that R/B-Split Vertex Deletion is fixed-parameter tractable too.

At this point, we note that r/b-split graphs can be viewed as a generalization of König graphs as follows. A graph $G = (V, E)$ with a maximum matching M is König if and only if the 2-colored graph $G' = (V, R \cup B)$, where $R = E$ and $B = M$, is r/b-split. It is important to realize that while this gives a recognition algorithm for König graphs using one for r/b-split graphs, it does not seem to give any relationship between the corresponding vertex-deletion problems. In fact, we do not know of any parameter-preserving reduction from König Vertex Deletion to the R/B-Split Vertex Deletion problem for general graphs.

For graphs with a perfect matching, we show by an independent argument based on the structure of a minimum König vertex deletion set that the König Vertex Deletion problem does indeed fixed-parameter reduce to Above Guarantee Vertex Cover (and also to R/B-Split Vertex Deletion) [12]. But this structural characterization of minimum König vertex deletion sets does not hold in general graphs. Therefore the fixed-parameter tractability result for König Vertex Deletion Set cannot be obtained from that of R/B-Split Vertex Deletion.

Theorem 5. *Let $G = (V, E = E_r \cup E_b)$ be an r/b graph. Construct $G' = (V', E')$ as follows: the vertex set V' consists of two copies V_1, V_2 of V and for all $u \in V$, $u_1 \in V_1$ and $u_2 \in V_2$ the edge set $E' = \{\{u_1, u_2\} : u \in V\} \cup \{\{u_1, v_1\} : \{u, v\} \in E_r\} \cup \{\{u_2, v_2\} : \{u, v\} \in E_b\}$. Then there exists k vertices whose deletion makes G r/b-split if and only if G' has a vertex cover of size $\mu(G') + k$.*

Proof. Clearly G' has $2|V|$ vertices and a perfect matching of size $|V|$. It suffices to show that G has an r/b-split subgraph on t vertices if and only if G' has an independent set of size t. This would prove that there exists $|V| - t$ vertices whose deletion makes G r/b-split if and only if G' has a vertex cover of size $2|V| - t$. Finally, plugging in $k = |V| - t$ will prove the theorem.

Therefore let H be an r/b-split subgraph of G on t vertices with a red independent set V_r and a blue independent set V_b. Then the copy V_r^1 of V_r in V_1 and the copy V_b^2 of V_b in V_2 are independent sets in G'. Since $V_r \cap V_b = \emptyset$, $V_r^1 \cup V_b^2$ is an independent set in G' on t vertices. Conversely if H' is an independent set in G' of size t, then for $i = 1, 2$ let $V(H') \cap V_i = W_i$ and $|W_i| = t_i$ so that $t_1 + t_2 = t$. For $i = 1, 2$, let $\widetilde{W}_i$ be the vertices in $V(G)$ corresponding to the vertices in W_i. Then $\widetilde{W}_1$ is an independent set of size t_1 in the red graph $G_r = (V(G), E_r)$ and $\widetilde{W}_2$ is an independent set of size t_2 in the blue graph $G_b = (V(G), E_b)$. Since W_1 and W_2 do not both contain copies of the same vertex of $V(G)$, as $W_1 \cup W_2$ is independent, we have $\widetilde{W}_1 \cap \widetilde{W}_2 = \emptyset$. Thus $G[\widetilde{W}_1 \cup \widetilde{W}_2]$ is an r/b-split graph of size t in G. $\qquad\square$

Since a maximum independent set in an n-vertex graph can be obtained in time $O^*(2^{0.288n})$ [6], we immediately have the following

Corollary 6. *The optimization version of the* R/B-SPLIT VERTEX DELETION *problem can be solved in time* $O^*(2^{0.576n}) = O^*(1.49^n)$ *on input graphs on n vertices.*

Since ABOVE GUARANTEE VERTEX COVER is fixed-parameter tractable (Corollary 1) we have

Corollary 7. *The parameterized version of the* R/B-SPLIT VERTEX DELETION *problem is fixed-parameter tractable and can be solved in time* $O(15^k \cdot k^2 \cdot m^3)$, *where m is the number of edges in the input graph.*

As mentioned before, König (and hence bipartite) and split graphs can be viewed as r/b-split graphs. Since 2-clique graphs are complements of bipartite graphs it follows that they can also be viewed as r/b-split graphs. The vertex-deletion problems ODD CYCLE TRANSVERSAL, SPLIT VERTEX DELETION and 2-CLIQUE VERTEX DELETION fixed-parameter reduce to R/B-SPLIT VERTEX DELETION. We show this reduction for ODD CYCLE TRANSVERSAL as the proofs in the other cases are quite similar.

Theorem 6. *Given a simple undirected graph* $G = (V, E)$, *construct an r/b-graph* $G' = (V', E')$ *as follows: define* $V' = V$ *and* $E' = E$; $E_r(G') = E$ *and* $E_b(G') = E$. *Then there exists k vertices whose deletion makes G bipartite if and only if there exist k vertices whose deletion makes* G' *r/b-split.*

Proof. Suppose deleting k vertices from G makes it bipartite with vertex bipartition $V_1 \cup V_2$. Then V_1 and V_2 are independent in both the red graph $G'_r = (V', E_r)$ and in the blue graph $G'_b = (V', E_b)$. Thus $G'[V_1 \cup V_2]$ is r/b-split. Conversely if k vertices can be deleted from G' to make it r/b-split, let V_r and V_b be the red and blue independent sets respectively. Then both these sets must be independent in G and therefore the subgraph of G induced on $V_r \cup V_b$ is bipartite. $\qquad\square$

From Theorem 6 and Corollaries 6 and 7 the following result follows immediately.

Corollary 8. *The parameterized version of* ODD CYCLE TRANSVERSAL, SPLIT VERTEX DELETION *and* 2-CLIQUE VERTEX DELETION *are fixed-parameter tractable and their optimization versions can be solved in time* $O^*(2^{0.576n}) = O^*(1.49^n)$ *on input graphs on n vertices.*

5   Conclusion

We showed that the KÖNIG VERTEX DELETION problem is fixed-parameter tractable in general graphs. To prove this, we made use of a number of structural results involving minimum vertex covers, minimum König vertex deletion sets and maximum matchings. We also showed that a number of vertex-deletion problems, in particular, R/B-SPLIT VERTEX DELETION, and ODD CYCLE TRANSVERSAL fixed-parameter reduce to ABOVE GUARANTEE VERTEX COVER. Since the latter problem is FPT, all these vertex-deletion problems are also FPT.

An interesting open problem is the parameterized complexity of the KÖNIG EDGE DELETION problem: given $G = (V, E)$ and an integer parameter k, does there exist at most k edges whose deletion makes the resulting graph König? Deriving a problem kernel for KÖNIG VERTEX DELETION SET is an interesting open problem. Another natural open problem to design better FPT algorithms for ABOVE GUARANTEE VERTEX COVER perhaps without using the reduction to MIN 2-SAT DELETION.

Acknowledgements. We thank anonymous referees of ISAAC 2008 for suggestions that helped improve the presentation.

References

1. Bourjolly, J.M., Pulleyblank, W.R.: König-Egerváry Graphs, 2-Bicritical Graphs and Fractional Matchings. Disc. Appl. Math. 24, 63–82 (1989)
2. Cai, L.: Fixed-Parameter Tractability of Graph Modification Problems for Hereditary Properties. Inform. Proc. Lett. 58(4), 171–176 (1996)
3. Deming, R.W.: Independence Numbers of Graphs – An Extension of the König-EgerváryTheorem. Disc. Math. 27, 23–33 (1979)
4. Faigle, U., Fuchs, B., Wienand, B.: Covering Graphs by Colored Stable Sets. Electronic Notes in Disc. Math. 17, 145–149 (2004)
5. Földes, S., Hammer, P.L.: Split graphs. In: Proc. of the 8th Southeastern Conference on Combinatorics, Graph Theory and Computing (Louisiana State Univ., Baton Rouge, La, pp. 311-315, Congressus Numerantium, Winnipeg, vol. XIX, Utilitas Math. (1977)
6. Fomin, F.V., Grandoni, F., Kratsch, D.: Measure and Conquer: A Simple $O(2^{0.288n})$ Independent Set Algorithm. In: Proc. of SODA 2006, pp. 18–25 (2006)
7. Gavril, F.: An Efficiently Solvable Graph Partition Problem to Which Many Problems are Reducible. Inform. Proc. Lett. 45, 285–290 (1993)
8. Gavril, F., Yannakakis, M.: Edge Dominating Sets in Graphs. SIAM Jour. Appl. Math. 38(3), 364–372 (1980)
9. Korach, E., Nguyen, T., Peis, B.: Subgraph Characterization of Red/Blue-Split Graphs and König-Egerváry Graphs. In: Proc. of SODA 2006, pp. 842–850 (2006)
10. Lovász, L.: Ear-Decompositions of Matching-covered Graphs. Combinatorica 3, 105–118 (1983)
11. Lovász, L., Plummer, M.D.: Matching Theory. North Holland, Amsterdam (1986)
12. Mishra, S., Raman, V., Saurabh, S., Sikdar, S., Subramanian, C.R.: The Complexity of Finding Subgraphs Whose Matching Number Equals the Vertex Cover Number. In: Tokuyama, T. (ed.) ISAAC 2007. LNCS, vol. 4835, pp. 268–279. Springer, Heidelberg (2007)
13. Moser, H., Thilikos, D.M.: Parameterized Complexity of Finding Regular Induced Subgraphs. In: Proc. of ACiD 2006, pp. 107–118 (2006)
14. Niedermeier, R.: An Invitation to Fixed-Parameter Algorithms. Oxford University Press, Oxford (2006)
15. Raman, V., Saurabh, S., Sikdar, S.: Efficient Exact Algorithms Through Enumerating Maximal Independent Sets and Other Techniques. Theory Comput. Systems 41, 563–587 (2007)
16. Razgon, I., O'Sullivan, B.: Almost 2-SAT is Fixed-Parameter Tractable. In: Aceto, L., Damgård, I., Goldberg, L.A., Halldórsson, M.M., Ingólfsdóttir, A., Walukiewicz, I. (eds.) ICALP 2008, Part I. LNCS, vol. 5125, pp. 551–562. Springer, Heidelberg (2008)
17. Reed, B., Smith, K., Vetta, A.: Finding Odd Cycle Transversals. Operations Research Letters 32, 299–301 (2004)
18. Sterboul, F.: A Characterization of Graphs in which the Transversal Number Equals the Matching Number. Jour. of Comb. Theory, Ser. B 27, 228–229 (1979)

Independent Sets of Maximum Weight in Apple-Free Graphs

Andreas Brandstädt[1], Tilo Klembt[1], Vadim V. Lozin[2], and Raffaele Mosca[3]

[1] Institut für Informatik, Universität Rostock, D-18051 Rostock, Germany
ab,tilo.klembt@informatik.uni-rostock.de
[2] DIMAP and Mathematics Institute, University of Warwick,
Coventry, CV4 7AL, UK
V.Lozin@warwick.ac.uk
[3] Dipartimento di Scienze, Universitá degli Studi "G. d'Annunzio",
Pescara 65121, Italy
r.mosca@unich.it

Abstract. We present the first polynomial-time algorithm to solve the Maximum Weight Independent Set problem for apple-free graphs, which is a common generalization of several important classes where the problem can be solved efficiently, such as claw-free graphs, chordal graphs and cographs. Our solution is based on a combination of two algorithmic techniques (modular decomposition and decomposition by clique separators) and a deep combinatorial analysis of the structure of apple-free graphs. Our algorithm is robust in the sense that it does not require the input graph G to be apple-free; the algorithm either finds an independent set of maximum weight in G or reports that G is not apple-free.

Keywords: Maximum independent set, clique separators, modular decomposition, polynomial-time algorithm, claw-free graphs, apple-free graphs.

1 Introduction

In 1965, Edmonds solved the MAXIMUM MATCHING problem [15] by implementing the idea of augmenting chains due to Berge [1]. Moreover, in [16] Edmonds showed how to solve the problem in case of weighted graphs. Lovász and Plummer observed in their book "Matching Theory" [22] that Edmonds' solution is "among the most involved of combinatorial algorithms." This algorithm also witnesses that the MAXIMUM WEIGHT INDEPENDENT SET (MWIS) problem, being NP-hard in general, admits a polynomial-time solution when restricted to the class of line graphs. In 1980, Minty [24] and Sbihi [27] independently of each other generalized the idea of Edmonds and extended his solution from line graphs to claw-free graphs. With a small repair from Nakamura and Tamura [25], the Minty's algorithm also works for weighted graphs. In the present paper, we further develop this fundamental line of research and extend polynomial-time solvability of the MWIS problem from claw-free graphs to apple-free graphs.

S.-H. Hong, H. Nagamochi, and T. Fukunaga (Eds.): ISAAC 2008, LNCS 5369, pp. 848–858, 2008.

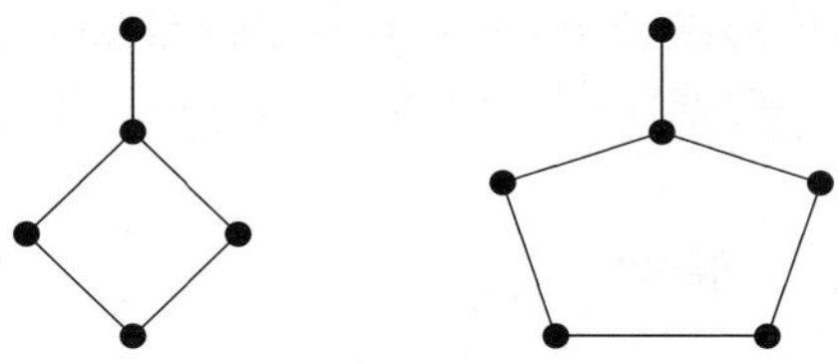

Fig. 1. Two smallest apples A_4 and A_5

An *apple* A_k is a graph obtained from a chordless cycle C_k of length $k \geq 4$ by adding a vertex that has exactly one neighbor on the cycle (see Figure 1). A graph G is *apple-free* if it contains no A_k, $k \geq 4$, as an induced subgraph. Odd apples were introduced by De Simone in [14], and Olariu in [26] called the apple-free graphs *pan-free*. The fact that the apple-free graphs include all claw-free graphs follows from the observation that every apple contains an induced claw centered at the unique vertex of degree 3 (see Figure 1). Along with maximum independent sets in claw-free graphs, our solution extends several other key results in algorithmic graph theory.

In particular, the class of apple-free graphs generalizes that of chordal graphs, since each apple contains a chordless cycle of length at least 4. Chordal graphs enjoy many attractive properties, one of which is that any non-complete graph in this class has a separating clique, i.e., a clique deletion of which increases the number of connected components (see e.g. [21]). This decomposability property finds applications in many algorithmic graph problems, including the problem of our interest. An efficient procedure to detect a separating clique in a graph was proposed by Tarjan [29] in 1985. Recently [2], an interest to this technique was revived by combining it with another important decomposition scheme, known as *modular decomposition*. The graphs that are completely decomposable with respect to modular decomposition are called *complement reducible graphs* [13], or *cographs*, and this is one more important class covered by our solution.

Our approach is based on a combination of the two decomposition techniques mentioned above and a deep combinatorial analysis of the structure of apple-free graphs. An important feature of our solution is that it does not require the input graph G to be apple-free; it either finds an independent set of maximum weight in G or reports that G is not apple-free. Such algorithms are called *robust* in [28].

In an obvious way, our algorithm can be used to find a minimum weight vertex cover in G or a maximum weight clique in the complement of G. We also believe that the structural analysis given in the paper can be used to extend many of the combinatorial properties of claw-free graphs established in the recent line of research by Chudnovsky and Seymour [5,6,7,8,9,10,11,12] to apple-free graphs.

This paper is organized as follows. In Section 2, we introduce basic notations and definitions, describe algorithmic techniques used in our approach and present preliminary information about the structure of apple-free graphs. In Section 3, we give a robust polynomial time algorithm for a subclass of apple-free graphs. In Section 4, we develop more tools for proving the main result, and in Section 5, we present the main result and a robust polynomial-time algorithm for the MWIS problem on apple-free graphs. Due to space limitation, most of the proofs are given in an appendix.

2   Preliminaries

Throughout this paper, $G = (V, E)$ is a finite undirected simple graph with $|V| = n$ and $|E| = m$. We also denote the vertex set of G as $V(G)$. For a vertex $v \in V$, let $N(v) = \{w \in V \mid vw \in E\}$ denote the *neighborhood* of v and let $\overline{N}(v) = \{w \in V \mid w \neq v$ and $vw \notin E\}$ denote the *antineighborhood* of v. If $vw \in E$ then v *sees* w and vice versa, and if $vw \notin E$ then v *misses* w and vice versa. For a subset $U \subseteq V$ of vertices let $G[U]$ denote the *induced subgraph* of G, i.e., the subgraph of G with vertex set U and two vertices of U being adjacent in $G[U]$ if and only if they are adjacent in G. We say that G is a *vertex-weighted* graph if each vertex of G is assigned a positive integer, the *weight* of the vertex. Let P_k, $k \geq 2$, denote a chordless path with k vertices and $k - 1$ edges, and let C_k, $k \geq 4$, denote a chordless cycle with k vertices, say $1, 2, \ldots, k$ and k edges, say $(i, i + 1)$ for $i \in \{1, 2, \ldots, k\}$ (index arithmetic modulo k). A chordless cycle with at least 6 vertices will be called a *long cycle*. A graph is *chordal* if it contains no induced subgraph C_k, $k \geq 4$. A graph is a *cograph* if it contains no induced P_4. See [4] for various properties of chordal graphs and of cographs. A *claw K* consists of four vertices, say a, b, c, d with edges ab, ac, ad; then a is the *midpoint of K* and b, c, d are the *endpoints of K*, also denoted as $K = (a; b, c, d)$ to emphasize the difference between midpoint and endpoints. A graph is *claw-free* if it contains no induced claw.

An *independent set* in G is a subset of pairwise nonadjacent vertices. A *clique* is a set of pairwise adjacent vertices. For disjoint vertex sets $X, Y \subset V$, we say that there is a *join between X and Y* if each vertex in X sees each vertex in Y. A $K_{2,3}$ has 5 vertices, say a_1, a_2 and b_1, b_2, b_3 such that $A = \{a_1, a_2\}$ and $B = \{b_1, b_2, b_3\}$ are independent vertex sets and there is a join between A and B. In an undirected graph G with vertex weight function w, the maximum total weight of an independent set in G is called the *weighted independence number* of G and is denoted by $\alpha_w(G)$. The following identity obviously holds.

$$\alpha_w(G) = \max_{x \in V(G)} \{w(x) + \alpha_w(G[\overline{N}(x)])\}. \tag{1}$$

An immediate consequence of (1) is the following.

Proposition 1. *If for every vertex $x \in V(G)$, the MWIS problem can be solved for $G[\overline{N}(x)]$ in time T, then it can be solved for G in time $n{\cdot}T$, where $n = |V(G)|$.*

If, for instance, for every vertex $x \in V(G)$, $G[\overline{N}(x)]$ is chordal then by the linear-time algorithm for the MWIS problem given in [18], these problem can be solved for G in time $\mathcal{O}(n \cdot m)$; we call such graphs *nearly chordal*. More generally, for a graph class $\mathcal{C}$, a graph G is *nearly $\mathcal{C}$* if for all $x \in V(G)$, $G[\overline{N}(x)]$ is in $\mathcal{C}$.

Now we recall two decomposition techniques used in our algorithm. A *clique separator* in a connected graph G is a subset Q of vertices of G which induces a complete graph, such that the graph $G[V \setminus Q]$ is disconnected. Tarjan showed in [29] that the MWIS problem can be reduced in polynomial time to graphs without clique separators (also called *atoms*), and a clique separator decomposition of a given graph can be determined in polynomial time (see also [30]).

Let $G = (V, E)$ be a graph, $U \subset V$ and $x \in V \setminus U$. We say that x *distinguishes* U if x has both a neighbor and a non-neighbor in U. A subset $U \subset V$ is a *module* in G if no vertex outside U distinguishes U. A module U is *nontrivial* if $1 < |U| < |V|$, otherwise it is *trivial*. A graph is *prime* if all its modules are trivial.

It is well known that the MWIS problem can be reduced in polynomial time from any hereditary (i.e., closed under taking induced subgraphs) class $\mathcal{C}$ to prime graphs in $\mathcal{C}$. Recently, in [2], decomposition by clique separators was combined with modular decomposition in a more general decomposition scheme:

Theorem 1 ([2]). *Let $\mathcal{C}$ be a hereditary class of graphs. If the MWIS problem can be solved in time T for those induced subgraphs of graphs in $\mathcal{C}$ which are prime and have no clique separators, then MWIS is solvable in time $\mathcal{O}(n^2 \cdot T)$ for graphs in $\mathcal{C}$.*

In [2,3], some examples are given where this technique can be applied. The aim of this paper is to show that this approach leads to a polynomial-time solution for MWIS on apple-free graphs. By Theorem 1, we can assume that throughout this paper, we deal with prime apple-free atoms. The following lemma will also allow us to assume that our graphs are $K_{2,3}$-free.

Lemma 1. *Prime A_4-free graphs are $K_{2,3}$-free.*

Proof. Suppose to the contrary that a prime A_4-free graph G contains an induced $K_{2,3}$, say with vertices a_1, a_2 in one color class and b_1, b_2, b_3 in the other, i.e., the edges are $a_i b_j$ for $i \in \{1, 2\}$ and $j \in \{1, 2, 3\}$. Then the co-connected component Q in $G[N(b_1) \cap N(b_2) \cap N(b_3)]$ containing a_1 and a_2 is no module. Let a_1', a_2' be two vertices in Q such that $a_1' a_2' \notin E$ which are distinguished by a vertex $z \notin Q$, say $z a_1' \in E$ and $z a_2' \notin E$. If $z b_1 \notin E$ then z is adjacent to at most one of the vertices b_2 and b_3 else b_2, z, b_3, a_2', b_1 is an A_4. Suppose that $z b_2 \notin E$; then b_2, a_2', b_1, a_1', z is an A_4 - contradiction. Thus $z b_1 \in E$, and by an analogous argument, also $z b_2 \in E$ and $z b_3 \in E$, but now z is in Q, a contradiction. $\square$

Let G be a prime apple-free atom, C a chordless cycle C_k with $k \geq 4$ in G, and v a vertex of G outside C. Denote by $N_C(v)$ the set of neighbors of v in C. We say that v *is universal for* C if v is adjacent to every vertex of C. Let F_u denote the set of all universal vertices for C. Moreover, for $i \geq 0$, we say that v is a *vertex of type i* or *an i-vertex* for C if it has exactly i neighbors in C, and we denote by F_i the set of all vertices of type i (for C). The following facts are easy to see:

Fact 1. *Every nonuniversal vertex for C is of type 0, 2, 3 or 4.*

Fact 2. *Every vertex of type 2 has two consecutive neighbors in C, every vertex of type 3 has three consecutive neighbors in C and every vertex of type 4 has two pairs of consecutive neighbors in C.*

Directly from Fact 2 we conclude that

Fact 3. *Any vertex v outside C adjacent to vertex i in C is also adjacent to $i-1$ or $i+1$ (i.e., v has no isolated neighbor in C).*

Fact 4. *If two distinct nonadjacent vertices $x, y \in F_2$ see vertices of the same connected component in $G[F_0]$ then $N_C(x) \neq N_C(y)$.*

Now we analyze the adjacencies between vertices of different types. Observe that in the next five facts we assume that the cycle C has length at least 6.

Fact 5. *If C is long, then no vertex in F_0 can see a vertex in $F_3 \cup F_4$.*

Fact 6. *If C is long, then every vertex in F_u sees every vertex in $F_3 \cup F_4$.*

Fact 7. *If C is long, then the set F_u of C-universal vertices is a clique.*

Fact 8. *Let C be a chordless cycle of length at least 6 and let $v \in F_2$ and $w \in F_3$. If $N_C(v) \subset N_C(w)$ then $vw \in E$. If $N_C(v) \cap N_C(w) = \emptyset$, then $vw \notin E$.*

Fact 9. *Let C be a chordless cycle of length at least 6 and let $v \in F_2$ and $w \in F_4$. If $|N_C(v) \cap N_C(w)| \leq 1$ then $vw \notin E$. If $N_C(v) \subset N_C(w)$ then $vw \in E$ unless $N_C(w) = \{i-1, i, i+1, i+2\}$ and $N_C(v) = \{i, i+1\}$ in which case $vw \notin E$.*

Before we proceed to detailed analysis of the structure of apple-free graphs and algorithms in the next sections, let us introduce more notations. Let $F_2(i, i+1)$ denote the set of vertices in F_2 which see exactly i and $i+1$ on C. Similarly, we denote by $F_3(i, i+1, i+2)$ the set of vertices in F_3 which see exactly $i, i+1$ and $i+2$, and we denote by $F_4(i, i+1, j, j+1)$ the set of vertices in F_4 which see exactly $i, i+1$ and $j, j+1$. We will distinguish between vertices of type 4 with *consecutive neighbors* and vertices of type 4 with *opposite neighbors* (if the neighbors are not consecutive in C). Also, for a vertex $v \in F_0$, let

- $F_0(v)$ denote the connected component in $G[F_0]$ containing v and let
- $S(v) := \{x \mid x \text{ sees } F_0(v) \text{ and } C\}$ (we call $S(v)$ the set of *contact vertices of C and $F_0(v)$*).

Obviously, $S(v)$ is a separator between v and C, and since G is an atom, $S(v)$ cannot be a clique.

We complete this section with an outline of our proof of polynomial-time solvability of the MWIS problem in the class of apple-free graphs. As one of the main results (Theorem 3 (i)), we show that a prime apple-free atom containing a chordless cycle C_k with $k \geq 7$ is claw-free. This reduces the problem to $(A_4, A_5, A_6, C_7, C_8, \ldots)$-free graphs.

To solve the problem for a $(A_4, A_5, A_6, C_7, C_8, \ldots)$-free graph G, we apply the decomposition scheme of Theorem 1 twice: first to the graph G and then to the subgraph $G[\overline{N}(v)]$ for each vertex $v \in V(G)$. We show that if the resulting graph (obtained by double application) contains a C_6, then it is claw-free (Theorem 3 (ii)), otherwise it is nearly chordal (Section 3).

3   MWIS on $(A_4, A_5, C_6, C_7, \ldots)$-Free Graphs

Lemma 2. *Prime $(A_4, C_5, C_6, C_7, \ldots)$-free atoms are nearly chordal.*

Proof. Suppose that a prime $(A_4, C_5, C_6, C_7, \ldots)$-free atom G is not nearly chordal. Then there is a vertex v in G such that $G[\overline{N}(v)]$ contains an induced C_4, say C. Since G is A_4-free, the set $S(v)$ of contact vertices contains only 2-vertices and universal vertices. We claim that $S(v)$ is a clique: By Lemma 1, and since G is A_4-free, $S(v) \cap F_u$ is a clique. Moreover, by Lemma 1, 2-vertices have consecutive neighbors in C and since G is $(A_4, C_5, C_6, C_7, \ldots)$-free, every set $S(v) \cap F_2(i, i+1)$ of 2-vertices in $S(v)$ is a clique. Finally, since G is $(A_4, C_5, C_6, C_7, \ldots)$-free, there is a join between $S(v) \cap F_u$ and every $S(v) \cap F_2(i, i+1)$, $i \in \{1, \ldots, 4\}$, and there is a join between $S(v) \cap F_2(i, i+1)$ and $S(v) \cap F_2(j, j+1)$ for $i \neq j$ (note that if $S(v) \cap F_2(i, i+1) \neq \emptyset$ then $S(v) \cap F_2(i+1, i+2) = \emptyset$). Now $S(v)$ is a clique cutset between v and C, a contradiction. Thus, G is nearly chordal. $\square$

Lemma 3. *Prime $(A_4, A_5, C_6, C_7, \ldots)$-free atoms are nearly C_5-free.*

Proof. Suppose that a prime $(A_4, A_5, C_6, C_7, \ldots)$-free atom G is not nearly C_5-free. Then there is a vertex v such that $G[\overline{N}(v)]$ contains an induced C_5, say C. Clearly, C has no 1-vertex. We first claim that $S(v)$ contains only 2-vertices and universal vertices: If $x \in S(v)$ is a 3-vertex for C, it must have consecutive neighbors $i, i+1, i+2$ in C, but then a neighbor y of x in $F_0(v)$, x and $i, i+2, i+3, i-1$ would induce an A_5 in G. A similar argument holds for 4-vertices of C.

Next, we claim that $S(v)$ is a clique: By Lemma 1, and since G is A_4-free, $S(v) \cap F_u$ is a clique. Moreover, 2-vertices have consecutive neighbors in C, and every set $S(v) \cap F_2(i, i+1)$ is a clique since G is $(A_4, A_5, C_6, C_7, \ldots)$-free. Finally, since G is $(A_4, A_5, C_6, C_7, \ldots)$-free, there is a join between $S(v) \cap F_u$ and every $S(v) \cap F_2(i, i+1)$, $i \in \{1, 2, \ldots, 5\}$, and there is a join between $S(v) \cap F_2(i, i+1)$ and $S(v) \cap F_2(j, j+1)$ for $i \neq j$ (note that if $S(v) \cap F_2(i, i+1) \neq \emptyset$ then $S(v) \cap F_2(i+1, i+2) = \emptyset$). Now $S(v)$ is a clique cutset between v and C, a contradiction. Thus, G is nearly C_5-free. $\square$

Corollary 1. *In a prime $(A_4, A_5, C_6, C_7, \ldots)$-free atom G, for every vertex $v \in V(G)$, the prime atoms of $G[\overline{N}(v)]$ are nearly chordal.*

Together with Theorem 1, Corollary 1 implies polynomial-time solvability of the MWIS problem in the class of $(A_4, A_5, C_6, C_7, \ldots)$-free graphs. We formally state this conclusion in Theorem 2 below and describe the solution in the following Algorithm 1. To simplify the description, we assume, by Theorem 1, that the input graph is a prime atom.

Algorithm 1

Input: A vertex-weighted prime atom G.
Output: A maximum weight independent set in G or the output
'G is not $(A_4, A_5, C_6, C_7, \ldots)$-free'.

(a) Check whether for every vertex $v \in V(G)$, the prime atoms of $G[\overline{N}(v)]$ are nearly chordal.
(b) If yes, apply a polynomial time algorithm for the MWIS problem on chordal graphs [18,19] and combine the partial results according to equation (1) and Theorem 1.
(c) Otherwise, G is not $(A_4, A_5, C_6, C_7, \ldots)$-free.

Theorem 2. *Algorithm 1 solves the MWIS problem on $(A_4, A_5, C_6, C_7, \ldots)$-free graphs (and an even larger class) in polynomial time in a robust way.*

4   More Tools

According to Theorem 2, we may restrict ourselves to apple-free graphs containing long cycles, i.e., chordless cycles of length at least 6. Throughout this section, C will denote a chordless cycle C_k, $k \geq 6$, with vertices $\{1, \ldots, k\}$ and edges $\{i, i+1\}$ (index arithmetic modulo k). A fundamental separator property which will be frequently used, is the following:

Lemma 4. *Let G be a prime apple-free atom and let C be an induced cycle C_k with $k \geq 6$ in G, and let v miss C, i.e., $v \in F_0$. Then for the set $S(v)$ of contact vertices, the following properties hold:*

(i) $S(v) \subseteq F_2 \cup F_u$;
(ii) $S(v)$ *is no clique;*
(iii) *if x and y are two distinct nonadjacent vertices in $S(v)$, then $x, y \in F_2$ and $N_C(x) \neq N_C(y)$.*

Proof. Condition (i) follows from Fact 5. For (ii), note that $S(v)$ is a separator between v and C and G is an atom. For (iii), let $x, y \in S(v)$ with $x \neq y$ and $xy \notin E$. Recall that F_u is a clique (Fact 7). Thus, at least one of x, y is not in F_u. Moreover, if $x \in F_u$ and $y \in F_2$, say $y \in F_2(1, 2)$ then let P_{xy} denote a shortest path in $F_0(v)$ connecting x and y. Now, $3, x, P_{xy}, y, 1$ is an apple. Thus, $x, y \in F_2$, and by Fact 4, $N_C(x) \neq N_C(y)$. $\quad\square$

Theorem 3, as our main result, says that a prime apple-free atom G containing a chordless cycle of length at least 7 is claw-free. If G is $(A_4, A_5, A_6, C_7, C_8, \ldots)$-free but contains a C_6 then the situation is slightly more complicated: We need to exclude two other graphs, the D_6 and E_6 (see Figure 2).

Lemma 5. *Prime $(A_4, A_5, A_6, C_7, C_8, \ldots)$-free atoms are nearly D_6- and E_6-free.*

Proof. Let G be a prime $(A_4, A_5, A_6, C_7, C_8, \ldots)$-free atom. Suppose that v is a vertex with a cycle C of length 6 in $\overline{N}(v)$. By Lemma 4, there are $w_1, w_2 \in S(v) \cap F_2$ with $w_1 w_2 \notin E$. Since G is $(C_7, C_8, \ldots)$-free, w_1 and w_2 see opposite edges of C, i.e., if w_1 sees i and $i+1$ then w_2 sees $i+3$ and $i+4$, and we can assume that w_1 and w_2 have a common neighbor v' in $F_0(v)$ (otherwise, there is a C_k with $k \geq 7$, in G).

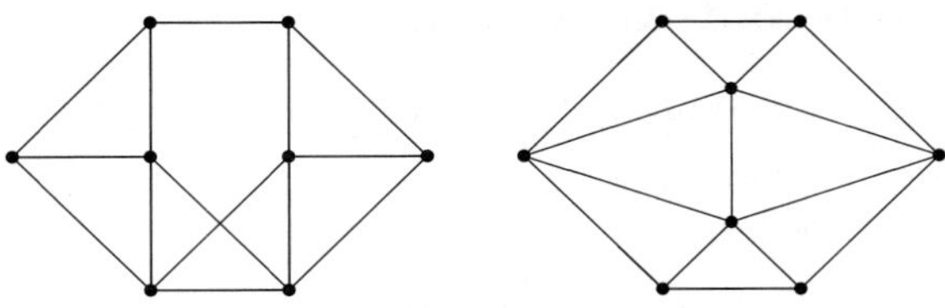

Fig. 2. The graphs D_6 (left) and E_6 (right)

Now assume that x is a vertex of type 4 with $x \in F_4(1, 2, 3, 4)$. Then if w_1 sees 2 and 3, and w_2 sees 5 and 6 then by Fact 3, $w_1 x \notin E$ since otherwise w_1 has the isolated neighbor x in the $C_5 = (1, x, 4, 5, 6)$, and very similarly, $w_2 x \notin E$, but then $(v', w_1, 2, x, 4, 5, w_2)$ is a C_7 in G, a contradiction.

Thus, up to symmetry, the only possibility for contact vertices to a C_6 with a 4-vertex $x \in F_4(1, 2, 3, 4)$ is $w_1 \in F_2(3, 4)$ and $w_2 \in F_2(6, 1)$ in which case $w_1 x \in E$ by Fact 3 with respect to the $C_5 = (1, x, 4, 5, 6)$ and w_1, and $w_2 x \notin E$ by Fact 3 with respect to the $C_4 = (v', w_1, x, w_2)$ and vertex 2.

For a D_6 in $\overline{N}(v)$ with 4-vertices $x \in F_4(1, 2, 3, 4)$ and $y \in F_4(3, 4, 5, 6)$, $xy \notin E$, the above discussion implies that $w_1 \in F_2(2, 3)$ and $w_2 \in F_2(5, 6)$ is impossible, and also $w_1 \in F_2(1, 2)$ and $w_2 \in F_2(4, 5)$ is impossible. Thus, $w_1 \in F_2(3, 4)$ and $w_2 \in F_2(6, 1)$. Then w_1 sees x and y but now $v', w_1, y, 6, 1, x$ is an A_5, a contradiction.

For an E_6 in $\overline{N}(v)$ with 4-vertices $x \in F_4(1, 2, 3, 4)$ and $y \in F_4(4, 5, 6, 1)$ with $xy \in E$, the above discussion implies, without loss of generality, that $w_1 \in F_2(3, 4)$ and $w_2 \in F_2(6, 1)$. Then w_1 sees x but not y while w_2 sees y but not x but now $v', w_1, x, y, w_2, 5$ is an A_5, a contradiction. $\qquad\square$

The proof of our main result, namely Theorem 3, will be prepared by various other lemmas.

Lemma 6. *If C is either*

(i) *a chordless cycle C_k with $k \geq 7$ in a prime apple-free atom or*
(ii) *a chordless cycle C_6 in a prime $(A_4, A_5, A_6, D_6, C_7, C_8, \ldots)$-free atom,*

then C has no universal vertex.

Lemma 7. *If C is either*

(i) *a chordless cycle C_k with $k \geq 7$ in a prime apple-free atom or*
(ii) *a chordless cycle C_6 in a prime $(A_4, A_5, A_6, D_6, C_7, C_8, \ldots)$-free atom,*

then every set $F_2(i, i + 1)$, $i \in \{1, 2, \ldots, k\}$, is a clique.

Lemma 8. *Let G be an apple-free atom containing a claw K and a long cycle C. If K and C are chosen so that the distance between them is as small as possible and $K \cap C = \emptyset$, then C sees the midpoint of K.*

5   Main Result and Algorithm

Theorem 3 is the main result of this paper; the algorithmic consequences are based on it.

Theorem 3. *Let G be a prime apple-free atom.*

(i) *If G contains a chordless cycle C_k with $k \geq 7$, then G is claw-free.*
(ii) *If G is $(D_6, E_6, C_7, C_8, \ldots)$-free and contains C_6 then G is claw-free.*

In order to solve the MWIS problem for apple-free graphs, we apply twice the decomposition scheme of Theorem 1 that reduces the problem from general graphs to prime atoms. We outline the procedure solving the problem in Algorithm 2 below. To simplify the description, we assume that the input graph is a prime atom.

Algorithm 2

Input: A vertex-weighted prime atom G.
Output: A maximum weight independent set in G or 'G is not apple-free'.

(a) If G is claw-free apply a polynomial time algorithm for MWIS on claw-free graphs [24,25], and output the solution.
(b) For every vertex $v \in V(G)$, apply the decomposition scheme of Theorem 1 to $G[\overline{N}(v)]$, and let $G_1, \ldots, G_k$ be the list of all prime atoms obtained in this way.
(c) If for each $i = 1, \ldots, k$,
   - either G_i is claw-free
   - or for each vertex $u \in V(G_i)$, the prime atoms of $G_i[\overline{N}(u)]$ are nearly chordal,

   then solve the problem for G_i and use the obtained solutions to compose a solution S for G, and output S.
(d) Otherwise, output 'G is not apple-free'.

Theorem 4. *Algorithm 2 is correct and solves the MWIS problem in polynomial time for apple-free graphs.*

Proof. Let G be an input graph (not necessarily apple-free). If Algorithm 2 outputs an independent set S, then obviously S is a solution for the MWIS problem in G. Therefore, to prove the correctness, we have to show that if G is apple-free, then the output of the algorithm is an independent set.

Let G be a prime apple-free atom. If the algorithm does not return an independent set after step (a), then G contains a claw, and hence by Theorem 3 (i), G is $(A_4, A_5, A_6, C_7, C_8, \ldots)$-free. In step (b), we apply the decomposition scheme of Theorem 1 to $G[\overline{N}(v)]$ for each vertex $v \in V(G)$, and reduce the problem to prime atoms $G_1, \ldots, G_k$. Let G_i be an arbitrary graph in this list. If G_i is claw-free, we can solve the problem for it. Otherwise, by Theorem 3 (ii), G_i is C_6-free and hence the problem can be solved for G_i by Corollary 1. This completes the proof of the correctness.

A polynomial-time bound follows from polynomial-time solvability of the problem on claw-free graphs and chordal graphs, and polynomial-time recognition algorithms for claw-free and chordal graphs. $\qquad\square$

6 Conclusion

The class of apple-free graphs is a natural generalization of claw-free graphs, chordal graphs, cographs, and various other classes (such as (A_4, P_5)-free graphs [23,3], $(A_4, C_5, C_6, \ldots)$-free graphs [20]) which have been extensively studied in the literature. For each of these classes, the Maximum Weight Independent Set problem is efficiently solvable in completely different ways; for cographs, it uses the cotree in a bottom-up way, for claw-free graphs, it is based on matching and for chordal graphs, it uses perfect elimination orderings and perfection or clique separator decomposition.

In this paper, we have shown that the Maximum Weight Independent Set problem can be solved in polynomial time on apple-free graphs. Our approach is based on a combination of clique separator decomposition and modular decomposition, and our algorithm does not require to recognize whether the input graph is apple-free; it solves the MWIS problem on a larger class (which is recognizable in polynomial time) given by the conditions in Algorithm 2. Some other important problems such as Maximum Clique and Chromatic Number are known to be NP-complete on claw-free graphs (see e.g. [17]) and thus remain hard on apple-free graphs.

Open Problem. What is the complexity of recognizing apple-free graphs?

References

1. Berge, C.: Two theorems in graph theory. Proc. Nat. Acad. Sci. USA 43, 842–844 (1957)
2. Brandstädt, A., Hoàng, C.T.: On Clique Separators, Nearly Chordal Graphs, and the Maximum Weight Stable Set Problem. In: Jünger, M., Kaibel, V. (eds.) IPCO 2005. LNCS, vol. 3509, pp. 265–275. Springer, Heidelberg (2005); Theoretical Computer Science 389, 295–306 (2007)
3. Brandstädt, A., Le, V.B., Mahfud, S.: New applications of clique separator decomposition for the Maximum Weight Stable Set Problem. In: Liśkiewicz, M., Reischuk, R. (eds.) FCT 2005. LNCS, vol. 3623, pp. 505–516. Springer, Heidelberg (2005); Theoretical Computer Science 370, 229–239 (2007)
4. Brandstädt, A., Le, V.B., Spinrad, J.P.: Graph Classes: A Survey. SIAM Monographs on Discrete Math. Appl., vol. 3. SIAM, Philadelphia (1999)
5. Chudnovsky, M., Seymour, P.: The roots of the independence polynomial of a clawfree graph. J. Combin. Theory Ser. B 97(3), 350–357 (2007)
6. Chudnovsky, M., Seymour, P.: Clawfree graphs I – Orientable prismatic graphs. J. Combin. Theory Ser. B 97, 867–901 (2007)
7. Chudnovsky, M., Seymour, P.: Clawfree graphs II – Nonorientable prismatic graphs. J. Combin. Theory Ser. B 98, 249–290 (2008)
8. Chudnovsky, M., Seymour, P.: Clawfree graphs III – Circular interval graphs. J. Combin. Theory Ser. B (to appear)
9. Chudnovsky, M., Seymour, P.: Clawfree graphs IV – Decomposition theorem. J. Combin. Theory Ser. B (to appear)
10. Chudnovsky, M., Seymour, P.: Clawfree graphs V – Global structure. J. Combin. Theory Ser. B (to appear)

11. Chudnovsky, M., Seymour, P.: Clawfree graphs VI – The structure of quasi-line graphs (submitted)
12. Chudnovsky, M., Seymour, P.: Clawfree graphs VII – Coloring claw-free graphs (submitted)
13. Corneil, D.G., Lerchs, H., Stewart, L.: Complement reducible graphs. Discrete Appl. Math. 3, 163–174 (1981)
14. De Simone, C.: On the vertex packing problem. Graphs and Combinatorics 9, 19–30 (1993)
15. Edmonds, J.: Paths, trees and flowers. Canad. J. of Mathematics 17, 449–467 (1965)
16. Edmonds, J.: Maximum matching and a polyhedron with 0, 1-vertices. J. Res. Nat. Bur. Standards Sect. B 69B, 125–130 (1965)
17. Faudree, R., Flandrin, E., Ryjáček, Z.: Claw-free graphs – a survey. Discrete Math. 164, 87–147 (1997)
18. Frank, A.: Some polynomial algorithms for certain graphs and hypergraphs. In: Proceedings of the Fifth British Combinatorial Conference (Univ. Aberdeen, Aberdeen 1975), pp. 211–226 (1975); Congressus Numerantium No. XV, Utilitas Math., Winnipeg, Man. (1976)
19. Gavril, F.: Algorithms for minimum coloring, maximum clique, minimum covering by cliques, and maximum independent set of a chordal graph. SIAM J. Computing 1, 180–187 (1972)
20. Gerber, M.U., Lozin, V.V.: Robust algorithms for the stable set problem. Graphs and Combinatorics 19, 347–356 (2003)
21. Golumbic, M.C.: Algorithmic graph theory and perfect graphs, 2nd edn. Annals of Discrete Mathematics, vol. 57. Elsevier Science B.V, Amsterdam (2004)
22. Lovász, L., Plummer, M.D.: Matching theory. Annals of Discrete Mathematics 29 (1986)
23. Lozin, V.V.: Stability in P_5- and banner-free graphs. European Journal of Operational Research 125, 292–297 (2000)
24. Minty, G.J.: On maximal independent sets of vertices in claw-free graphs. J. Combinatorial Theory, Ser. B 28, 284–304 (1980)
25. Nakamura, D., Tamura, A.: A revision of Minty's algorithm for finding a maximum weight stable set of a claw-free graph. J. Oper. Res. Soc. Japan 44, 194–204 (2001)
26. Olariu, S.: The strong perfect graph conjecture for pan-free graphs. J. Combin. Th. (B) 47, 187–191 (1989)
27. Sbihi, N.: Algorithme de recherche d'un stable de cardinalité maximum dans un graphe sans étoile. Discrete Math. 29, 53–76 (1980)
28. Spinrad, J.P.: Efficient Graph Representations, Fields Institute Monographs 19. American Mathematical Society, Providence (2003)
29. Tarjan, R.E.: Decomposition by clique separators. Discrete Math. 55, 221–232 (1985)
30. Whitesides, S.H.: A method for solving certain graph recognition and optimization problems, with applications to perfect graphs. In: Berge, C., Chvátal, V. (eds.) Topics on perfect graphs. North-Holland, Amsterdam (1984)

Enumeration of Perfect Sequences of Chordal Graph

Yasuko Matsui[1], Ryuhei Uehara[2], and Takeaki Uno[3]

[1] Department of Mathematical Sciences, School of Science, Tokai University,
Kitakaname 1117, Hiratsuka, Kanagawa 259-1292, Japan
`yasuko@ss.u-tokai.ac.jp`
[2] School of Information Science, JAIST, Asahidai 1-1,
Nomi, Ishikawa 923-1292, Japan
`uehara@jaist.ac.jp`
[3] National Institute of Informatics, Hitotsubashi 2-1-2,
Chiyoda-ku, Tokyo 101-8430, Japan
`uno@nii.jp`

Abstract. A graph is chordal if and only if it has no chordless cycle of
length more than three. The set of maximal cliques in a chordal graph
admits special tree structures called clique trees. A perfect sequence is
a sequence of maximal cliques obtained by using the reverse order of
repeatedly removing the leaves of a clique tree. This paper addresses the
problem of enumerating all the perfect sequences. Although this problem
has statistical applications, no efficient algorithm has been proposed.
There are two difficulties with developing this type of algorithms. First,
a chordal graph does not generally have a unique clique tree. Second,
a perfect sequence can normally be generated by two or more distinct
clique trees. Thus it is hard using a straightforward way to generate
perfect sequences from each possible clique tree. In this paper, we propose
a method to enumerate perfect sequences without constructing clique
trees. As a result, we have developed the first polynomial delay algorithm
for dealing with this problem. In particular, the time complexity of the
algorithm on average is $O(1)$ for each perfect sequence.

Keywords: Chordal graph, clique tree, enumeration, perfect sequence.

1   Introduction

A graph is said to be chordal if every cycle of length at least 4 has a chord.
Chordal graphs have been investigated for a long time for many areas. From
the viewpoint of graph theory, this class has a simple characterization; a graph
is chordal if and only if it is an intersection graph of the subtrees of a tree.
That is, there is a set of subtrees T_v of a tree $\mathcal{T}$ that correspond to the vertices
v of a chordal graph G such that u and v are adjacent in G if and only if
the corresponding subtrees T_u and T_v have non-empty intersection. From an
algorithmic point of view, a chordal graph is characterized by a simple vertex
ordering called a perfect elimination ordering (PEO). Its geometrical property

S.-H. Hong, H. Nagamochi, and T. Fukunaga (Eds.): ISAAC 2008, LNCS 5369, pp. 859–870, 2008.

of rigidity plays an important role in many practical areas, and the property of its adjacency matrix is useful in matrix manipulations. Since the class appears in different contexts in so many areas they also are called "rigid circuit graphs" or "triangulated graphs" (see e.g., [12,5]).

Recently, the probabilistic relationships between random variables are represented by a graph called "graphical model" in the science of statistics. On a graphical model, the random variables are represented by the vertices and the conditional dependencies are represented by the edges. In particular, if a graphical model is chordal, we can easily compute its maximum likelihood estimator. In the computation, a chordal graph is decomposed into its subgraphs by removing a separator which induces a clique. Therefore, chordal graphs are also called "decomposable graphs" or "decomposable models" in this area (see, e.g., [9]). Even if a given graphical model is not decomposable, we sometimes add edges to make it decomposable to obtain a good approximation value of the maximum likelihood estimator of the graphical model (see, e.g., [4]). This corresponds to the notion of "chordal completion" in the area of graph algorithms [8]. Hence chordal graphs play an important role in graphical modeling in statistics.

Perfect sequences are the sequences of maximal cliques in a given chordal graph that satisfy certain properties. The notion arises from the decomposable models, and all perfect sequences are required and must not have repetitions (see, e.g., [7]) From the viewpoint of graph theory, perfect sequences can be seen in the following way. As mentioned, a chordal graph G can be represented by the intersection graph of the subtrees of a tree $\mathcal{T}$. That is, each vertex v of $G = (V, E)$ corresponds to a subtree T_v of $\mathcal{T}$, and $\{u, v\} \in E$ if and only if T_v and T_u intersect. We can make each node c_i of tree $\mathcal{T}$ correspond to a maximal clique C_i of G; C_i consists of all the vertices v in G such that T_v contains the node c_i. Therefore, the tree $\mathcal{T}$ is called a clique tree of G. From the clique tree $\mathcal{T}$, we make an ordering π over the set of maximal cliques $\{C_1, C_2, \ldots, C_k\}$ of G such that $C_{\pi(i)}$ is a leaf of tree $\mathcal{T}_i$ which is a subgraph of $\mathcal{T}$ induced by $C_{\pi(1)}$, $C_{\pi(2)}$, $\ldots$, $C_{\pi(i)}$ for each i. Intuitively, we can construct such a sequence from $\mathcal{T}$ by repeatedly pruning leaves and put them on the top of the sequence until $\mathcal{T}$ is empty. Then the sequence of maximal cliques in a chordal graph is called a perfect sequence, and it is known that a graph is chordal if and only if it has a perfect sequence.

In 2006, Hara and Takemura proposed a sampling algorithm for the perfect sequences of a given chordal graph [6] that used the Lauritzen's method [9]. However their algorithm does not generate each perfect sequence uniformly at random, and to our knowledge, no enumeration algorithm of perfect sequences exists. There are two major reasons for the difficulty in enumerating perfect sequences. First, the clique tree is not generally unique for a chordal graph. That is, a chordal graph has many distinct (non-isomorphic) clique trees in general. For a clique tree, we can define a set of perfect sequences consistent to the clique tree. Then, secondly, the sets of perfect sequences consistent with the distinct clique trees are not disjoint. That is, we can obtain one perfect sequence from possibly many distinct clique trees. Therefore, a straightforward algorithm based on a simple idea (generate all clique trees, and generate all perfect sequences for each clique tree) cannot avoid redundancy.

In this paper, we propose an algorithm enumerating all the perfect sequences of a chordal graph. The algorithm enumerates all the perfect sequences on an average of $O(1)$ time per sequence. In order to avoid redundancy, our algorithm makes a weighted intersection graph of maximal cliques first instead of explicitly constructing the clique trees. The intersection graph is uniquely constructed, and each maximum weighted spanning tree of the intersection graph gives a clique tree of a chordal graph. Then the algorithm generates each perfect sequence from the union of maximum weighted spanning trees without any repetitions. The algorithm is based on a new idea that characterizes the union of maximum weighted spanning trees, and that also gives us insight into the properties of the set of clique trees of a chordal graph.

We note that the set of perfect sequences is strongly related to the set of PEOs. The PEO is a standard characterization of a chordal graph in the area of graph algorithms. Any PEO can be obtained by repeatedly removing a simplicial vertex, and any sequence of removals of simplicial vertices is a PEO. This property admits us to enumerate all PEOs by recursively removing simplicial vertices. The enumeration of all PEOs is investigated by Chandran et al. [3]. Intuitively, any perfect sequence can be obtained by removing a set of "equivalent" simplicial vertices together repeatedly. However, each removal has to follow (a kind of) lexicographical ordering. A removal of a set of simplicial vertices may produce a new set of simplicial vertices or change another set of simplicial vertices, and the new set cannot be chosen before the old sets. Thus the family of sets of equivalent simplicial vertices has to follow a partial ordering defined by the appearance in the graph to obtain a perfect sequence. This fact also implies that while some PEOs can be obtained from a perfect sequence easily, but some PEOs cannot be obtained from any perfect sequence straightforwardly by the constraint of the partial ordering. Therefore, the correspondence between the set of perfect sequences and the set of PEOs is not straightforward and hence we have to analyze some special cases. Indeed, using this approach, we can obtain another enumeration algorithm of all perfect sequences from the enumeration algorithm of all PEOs. However this approach does not allow us to enumerate efficiently; the algorithm takes $O(|V| + |E|)$ time for each perfect sequence. This is the reason why we take completely different approach based on a maximum weighted spanning tree of a weighted clique graph, which admits us to improve the time to $O(1)$ on average.

2   Preliminaries

The *neighborhood* of a vertex v in a graph $G = (V, E)$ is the set $N_G(v) = \{u \in V \mid \{u, v\} \in E\}$, and the *degree* of a vertex v is $|N_G(v)|$ and is denoted by $\deg_G(v)$. For a vertex subset U of V we denote by $N_G(U)$ the set $\{v \in V \mid v \in N_G(u) \text{ for some } u \in U\}$. If no confusion arises we will omit the subscript G. Given a graph $G = (V, E)$ and a subset $U \subseteq V$, the *subgraph of G induced* *by* U is the graph (U, F), where $F = \{\{u, v\} \in E \mid u, v \in U\}$, and denoted by $G[U]$. A vertex set I is an *independent set* of G if $G[I]$ contains no edge, and a vertex set C is a *clique* in G if any pair of vertices in C is connected by an edge in G. An edge e in a connected graph $G = (V, E)$ is called a *bridge* if its removal partitions G into two connected components.

For a given graph $G = (V, E)$, consider a sequence $(v_0, v_1, \ldots, v_\ell)$ of vertices in V such that $\{v_{j-1}, v_j\} \in E$ for each $0 < j \leq \ell$. Such a sequence is a *path* if the vertices $v_0, \ldots, v_\ell$ are all distinct, and it is a *cycle* if the vertices $v_0, \ldots, v_{\ell-1}$ are distinct and $v_0 = v_\ell$. The *length* of such a path and a cycle is the number ℓ. An edge that joins the two vertices of a cycle, but is not itself an edge of the cycle, is a *chord* of the cycle. A graph is *chordal* if each cycle of length at least 4 has a chord.

Given a graph $G = (V, E)$, a vertex $v \in V$ is *simplicial* in G if $N(v)$ is a clique in G. An ordering $v_1, \ldots, v_n$ of the vertices of V is a *perfect elimination ordering* of G if the vertex v_i is simplicial in $G[\{v_i, v_{i+1}, \ldots, v_n\}]$ for all $i = 1, \ldots, n$. It is known that a graph is chordal if and only if it has a perfect elimination ordering [2, Section 1.2]. Given a chordal graph, a perfect elimination ordering of the graph can be found in linear time [11,13].

For a chordal graph $G = (V, E)$, we can associate a tree $\mathcal{T}$, called a *clique tree* of G, as satisfying the following three properties. (A) The nodes[1] of $\mathcal{T}$ are the maximal cliques of G. (B) Two nodes of $\mathcal{T}$ are adjacent only if the intersection of the corresponding maximal cliques is non-empty. (C) For every vertex v of G, the subgraph T_v of $\mathcal{T}$ induced by the maximal cliques containing v is connected. (Here, condition (A) is sometimes weakened as each node is not necessarily maximal.) It is well known that a graph is chordal if and only if it has a clique tree, and in such a case a clique tree can be constructed in linear time. On the tree, each vertex v in V corresponds to a subtree T_v of $\mathcal{T}$. That is, T_v consists of maximal cliques that contain v. Then, the graph G is an intersection graph of subtrees T_v of a tree $\mathcal{T}$. Some of these details are explained in books [2,12].

For a given chordal graph $G = (V, E)$, we denote the set of all maximal cliques of G by $\mathcal{C}(G)$. (It is known that $|\mathcal{C}(G)| \leq |V|$.) Let $k = |\mathcal{C}(G)|$, $\mathcal{C}(G) = \{C_1, C_2, \ldots, C_k\}$, and π be a permutation of k elements. Then, the ordering $C_{\pi(1)}, C_{\pi(2)}, \ldots, C_{\pi(k)}$ on $\mathcal{C}(G)$ is said to be a *perfect sequence* if there is a clique tree $\mathcal{T}$ such that each $C_{\pi(i)}$ is a leaf of the subtree $\mathcal{T}[\{C_{\pi(1)}, C_{\pi(2)}, \ldots, C_{\pi(i)}\}]$ which is the subgraph of $\mathcal{T}$ induced by $\{C_{\pi(1)}, C_{\pi(2)}, \ldots, C_{\pi(i)}\}$, for each i with $1 \leq i \leq k$. Intuitively, we have two explanations for this. One is that we can prune all the leaves off a clique tree in the reverse order $C_{\pi(k)}, \ldots, C_{\pi(2)}, C_{\pi(1)}$ of any perfect sequence. On the other hand, according to the perfect sequence $C_{\pi(1)}, C_{\pi(2)}, \ldots, C_{\pi(k)}$, we can construct a clique tree $\mathcal{T}$ by repeatedly attaching $C_{\pi(i)}$ as a leaf.

We here note that, in general, a clique tree for a chordal graph G is not uniquely determined up to isomorphism. For example, for the chordal graph $G = (V, E)$ in Fig. 1(a), there are four distinct clique trees given in Fig. 1(b).

---

[1] In this paper, "vertex" is in a chordal graph G, and "node" corresponds to a clique in G to distinguish them.

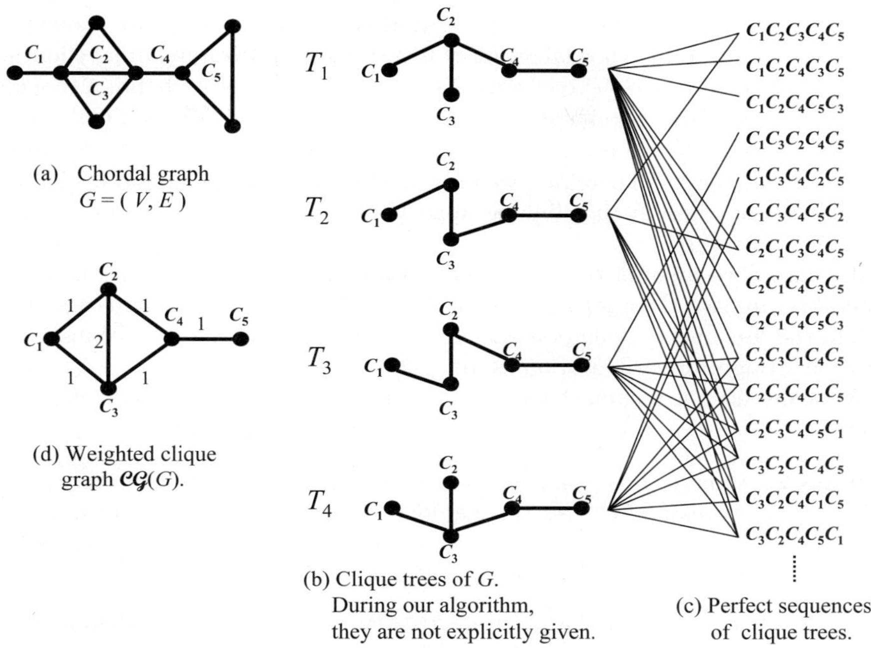

(a) Chordal graph
$G = (V, E)$

(d) Weighted clique
graph $\mathcal{CG}(G)$.

(b) Clique trees of G.
During our algorithm,
they are not explicitly given.

(c) Perfect sequences
of clique trees.

Fig. 1. (a) Chordal graph G, (b) its clique trees, (c) corresponding perfect sequences, and (d) its weighted clique graph $\mathcal{CG}(G) = (\mathcal{C}(G), \mathcal{E})$

Moreover, two or more distinct clique trees of the same chordal graph G can generate the same perfect sequence. For example, for the chordal graph $G = (V, E)$ in Fig. 1(a), a perfect sequence $C_2 C_1 C_3 C_4 C_5$ can be generated from two trees, T_1 and T_2, as depicted in Fig. 1(b)(c).

For a chordal graph $G = (V, E)$ and the set $\mathcal{C}(G)$ of all maximal cliques, we define the *weighted clique graph* $\mathcal{CG}(G) = (\mathcal{C}(G), \mathcal{E})$ with a weight function $w : \mathcal{E} \to \mathbb{Z}$ as follows. For two maximal cliques C_1 and C_2 in $\mathcal{C}(G)$, $\mathcal{E}$ contains the edge $\{C_1, C_2\}$ if and only if $C_1 \cap C_2 \neq \emptyset$. For each edge $\{C_1, C_2\}$ in $\mathcal{E}$, $w(\{C_1, C_2\})$ is defined by $|C_1 \cap C_2|$ (therefore every edge in $\mathcal{E}$ has a positive integer weight less than $|V|$). The $\mathcal{CG}(G) = (\mathcal{C}(G), \mathcal{E})$ of a chordal graph $G = (V, E)$ in Fig. 1(a) is given in Fig. 1(d). The weights of the edges are all 1 except for $\{C_2, C_3\}$ which has a weight 2.

We recall that each edge in a clique tree $\mathcal{T}$ of G corresponds to a nonempty intersection of two maximal cliques. Thus, $\mathcal{T}$ is a spanning tree of $\mathcal{CG}(G)$. However, some spanning trees of $\mathcal{CG}(G)$ may not be clique trees of G. The characterization of a clique tree is given as follows.

Lemma 1 (e.g., [1,10]). *Let $G = (V, E)$ be a chordal graph and $\mathcal{CG}(G) = (\mathcal{C}(G), \mathcal{E})$ be the weighted clique graph with a weight function w. A spanning tree $\mathcal{T}$ of $\mathcal{CG}(G)$ is a clique tree of G if and only if it has the maximum weight.*

For the $\mathcal{CG}(G) = (\mathcal{C}(G), \mathcal{E})$ in Fig. 1(d) of the chordal graph $G = (V, E)$ in Fig. 1(a), the only spanning trees that contain the edge $\{C_2, C_3\}$ are clique trees of G. We note that any chordal graph of n vertices contains n maximal cliques at the most. Therefore, $\mathcal{CG}(G)$ contains $O(n)$ nodes. On the other hand, although a star S_n of n vertices contains n vertices, $n-1$ edges, and $n-1$ maximal cliques, the clique graph $\mathcal{CG}(S_n)$ is a complete graph K_{n-1} with $n-1$ nodes that contains $\binom{n-1}{2} = O(n^2)$ edges. Therefore, we only have a trivial upper bound $O(|V|^2)$ for the number of edges in the clique graph $\mathcal{CG}(G)$ of a chordal graph $G = (V, E)$, even if $|E| = O(|V|)$.

Hereafter, we assume that the input graph G is connected without loss of generality; in case G is not connected, we allow the use of h weight-zero edges to join the h connected components. The modification of the algorithms in this paper is straightforward, and omitted.

We show here a technical lemma for a clique tree that will be referred to later:

Lemma 2. *Suppose $\mathcal{T}$ is a clique tree of a chordal graph $G = (V, E)$ that consists of at least two maximal cliques (in the other words, G is not a complete graph). Let C be a leaf in $\mathcal{T}$ and C' be the unique neighbor of C. Then for each vertex v in C, v is simplicial in G if and only if v is in $C \setminus C'$.*

Proof. If v is in $C \setminus C'$, it is easy to see that $N(v) = C \setminus \{v\}$, and therefore, v is simplicial. Now suppose a simplicial vertex v is in $C \cap C'$ to derive a contradiction. Since v is in C, $N(v)$ contains all the vertices in C, except v. On the other hand, if v is also in C', $N(v)$ contains all the vertices in C', except v. However, C and C' are distinct maximal cliques. Therefore, there are two vertices $u \in C$ and $w \in C'$ with $\{u, w\} \notin E$, which contradicts that v is simplicial. Therefore, v is in $C \setminus C'$. $\qquad\square$

3   Enumeration Algorithm

The idea for enumerating perfect sequences is simple. We construct a graph representing the adjacency of maximal cliques, and recursively remove the maximal cliques that can be leaves of a clique tree. Since the clique tree is a spanning tree of the graph and the removed maximal clique is a leaf of the clique tree, after removing the maximal cliques, we still have a clique (sub)tree that is a spanning tree of the resultant graph. Since any tree has at least two leaves, we always have at least two maximal cliques that correspond to the leaves of the spanning tree. Therefore, we invariably get a perfect sequence by repeating the removal process. During the algorithm, the spanning tree is not explicitly given, and we have to deal with all the potential spanning trees that can generate perfect sequences. The following outlines a description of the algorithm.

The graph $\mathcal{CG}(G)^*$ is a subgraph of $\mathcal{CG}(G)$ that excludes unnecessary edges described later.

Algorithm 1. Outline of Enumeration

Input : Chordal graph $G = (V, E)$;
Output: All perfect sequences of G;

1 construct weighted clique graph $\mathcal{CG}(G)$;
2 compute arbitrary maximum weighted spanning tree $\mathcal{T}^*$ of $\mathcal{CG}(G)$;
3 construct graph $\mathcal{CG}(G)^*$ composed of edges that can be included in clique trees from $\mathcal{CG}(G)$ and $\mathcal{T}^*$;
4 enumerate all sequences of maximal cliques obtained by repeatedly removing maximal cliques that can be leaves of some clique trees.

To efficiently find the maximal cliques that can be leaves, we first compute any maximum weighted spanning tree $\mathcal{T}^*$. Then, we use a (unweighted) graph $\mathcal{CG}(G)^*$ obtained from $\mathcal{CG}(G)$ and $\mathcal{T}^*$. We say an edge in $\mathcal{CG}(G)$ is *unnecessary* if it cannot be included in any maximum weighted spanning tree of $\mathcal{CG}(G)$. On the other hand, an edge is *indispensable* if it appears in every maximum weighted spanning tree of $\mathcal{CG}(G)$. The other edges are called *dispensable*, which means they appear in some (but not all) maximum weighted spanning trees. Let e be an edge not in $\mathcal{T}^*$. Since $\mathcal{T}^*$ is a spanning tree of $\mathcal{CG}(G)$, the addition of e to $\mathcal{T}^*$ produces a unique cycle C_e which consists of e and the other edges in $\mathcal{T}^*$. We call C_e an *elementary cycle of* e (the terminology comes from the matroid theory). The unnecessary, dispensable, and indispensable edges are characterized by the following lemmas.

Lemma 3. *For an edge e not in $\mathcal{T}^*$, $w(e) \leq w(e')$ holds for any $e' \in C_e \setminus \{e\}$. Moreover, e is unnecessary if and only if $w(e) < w(e')$ holds for any $e' \in C_e \setminus \{e\}$. On the other hand, e is dispensable if and only if $w(e) = w(e')$ holds for some $e' \in C_e \setminus \{e\}$.*

Proof. If we have $w(e) > w(e_i)$ for some $1 \leq i < k$, by swapping e and e_i, we can obtain a heavier spanning tree, which contradicts the fact that $\mathcal{T}^*$ is a maximum weighted spanning tree. Therefore, $w(e) \leq w(e_i)$ for each $1 \leq i < k$. When $w(e) = w(e_i)$ for some $1 \leq i < k$, we can obtain a maximum weight spanning tree $\mathcal{T}'$ by removing e' and adding e to $\mathcal{T}^*$. $\mathcal{T}^*$ does not include e while $\mathcal{T}'$ includes e, which implies e is dispensable. $\square$

Lemma 4. *An edge e in $\mathcal{T}^*$ is an indispensable edge if $w(e) > w(e')$ for all edges e' such that e' is not on $\mathcal{T}^*$ and $C_{e'}$ contains e.*

Proof is analogous to proof of Lemma 3, and hence omitted.

We note that any bridge is indispensable by Lemma 4; there is no edge e' not in $\mathcal{T}^*$ such that $C_{e'}$ contains e and $w(e') \geq w(e)$.

We denote the sets of unnecessary, indispensable, and dispensable edges by $\mathcal{E}_u$, $\mathcal{E}_i$, and $\mathcal{E}_d$, respectively. They partition the edge set $\mathcal{E}$ of $\mathcal{CG}(G)$ into three disjoint sets. The sets can be computed by the following algorithm in $O(|\mathcal{C}(G)|^3) = O(|V|^3)$ time. We note that by using a dynamic programming technique starting from the bottom of the tree, the run time can be reduced to $O(|\mathcal{C}(G)|^2)$, which is omitted here since it is too complex and tedious.

Algorithm 2. Search for Unnecessary, Indispensable, and Dispensable Edges

---

Input : The weighted clique graph $\mathcal{CG}(G) = (\mathcal{C}(G), \mathcal{E})$ and an arbitrary maximum weighted spanning tree $\mathcal{T}^*$ of $\mathcal{CG}(G)$;

Output: Sets $\mathcal{E}_u$, $\mathcal{E}_i$, and $\mathcal{E}_d$ of the unnecessary, indispensable, and dispensable edges;

1 set $\mathcal{E}_u := \emptyset; \mathcal{E}_d := \emptyset; \mathcal{E}_i := \emptyset$;
2 **foreach** e *not in* $\mathcal{T}^*$ **do**
3   **if** $w(e) < w(e')$ *for all* $e' \in C_e$ **then**
4          $\mathcal{E}_u := \mathcal{E}_u \cup \{e\}$;
5   **else**
6          $\mathcal{E}_d := \mathcal{E}_d \cup \{e\}$;
7   **foreach** $e' \in C_e$ *satisfying* $w(e) = w(e')$ **do** $\mathcal{E}_d := \mathcal{E}_d \cup \{e'\}$;
8   **end**
9 **end**
10 $\mathcal{E}_i := E \setminus (\mathcal{E}_u \cup \mathcal{E}_d)$;
11 **return** $(\mathcal{E}_i, \mathcal{E}_u, \mathcal{E}_d)$;

We now define an unweighted graph $\mathcal{CG}(G)^*$ by $(\mathcal{C}(G), \mathcal{E}_i \cup \mathcal{E}_d)$.

Observation 1. *Any spanning tree of $\mathcal{CG}(G)^*$ that contains all the edges in $\mathcal{E}_i$ gives a maximum weighted spanning tree of $\mathcal{CG}(G)$.*

Now, we have characterized clique trees by using spanning trees in $\mathcal{CG}(G)$. Next, we take the characterization of maximal cliques that can be leaves of some clique trees into consideration.

Lemma 5. *A maximal clique C can be a leaf of a clique tree if and only if C satisfies (1) C is incident to at most one edge in $\mathcal{E}_i$, and (2) C is not a cut vertex in $\mathcal{CG}(G)^*$.*

Proof. First, we suppose that C is a leaf of a clique tree $\mathcal{T}$. Since $\mathcal{T}$ is a clique tree of G, $\mathcal{T}$ is a spanning tree in $\mathcal{CG}(G)^*$ that includes all the edges in $\mathcal{E}_i$. Since C is a leaf of $\mathcal{T}$, C is incident to at most one edge of $\mathcal{E}_i$, and C is not a cut vertex of $\mathcal{CG}(G)^*$. Thus, C satisfies the conditions.

We next suppose that C satisfies the conditions. We assume that $\mathcal{CG}(G)$ contains two or more nodes. We choose any edge e from $\mathcal{E}_i \cup \mathcal{E}_d$ that is incident to C. We always can choose e since $\mathcal{CG}(G)$ is connected. Then, we remove C from $\mathcal{CG}(G)^*$. Since C is not a cut vertex, the resultant graph $\mathcal{CG}(G)'$ is still connected. Therefore, $\mathcal{CG}(G)'$ has a spanning tree $\mathcal{T}'$ which contains all the edges in $\mathcal{E}_i \setminus \{e\}$. Then, by adding e to $\mathcal{T}'$, we have a spanning tree $\mathcal{T}$ that contains all the edges in $\mathcal{E}_i$, and C is a leaf of $\mathcal{T}$. This concludes the proof. $\square$

Hereafter, the pair of two conditions in Lemma 5 is said to be a *leaf condition*. A perfect sequence is obtained by removing a leaf of a clique tree $\mathcal{T}$. Thus, any perfect sequence is obtained by iteratively removing the maximal cliques satisfying a leaf condition. The converse is shown by the following lemma.

Lemma 6. *Any maximal clique sequence $\mathcal{S} = (C_1, \ldots, C_k)$ obtained by iteratively removing a maximal clique satisfying the leaf condition is a perfect sequence.*

Proof. Let $\mathcal{T}_1$ be a tree in $\mathcal{CG}(G)^*$ that consists of one vertex C_1, and $\mathcal{T}_i$ be a tree in $\mathcal{CG}(G)^*$ obtained by adding C_i to $\mathcal{T}_{i-1}$ and an edge e of $\mathcal{CG}(G)^*$ connecting C_i and a vertex in $\mathcal{T}_{i-1}$. If there is an edge of $\mathcal{E}_i$ connecting C_i and a vertex in $\mathcal{T}_{i-1}$, we choose the edge as e. Note that there is at most one such edge by the leaf condition. We observe that any C_i is a leaf of $\mathcal{T}_i$, and $\mathcal{T}_{i-1}$ is obtained by removing a leaf C_i from $\mathcal{T}_i$. Thus, $\mathcal{S} = (C_1, \ldots, C_k)$ is a perfect sequence. $\square$

This lemma ensures that by repeatedly removing maximal cliques satisfying the leaf condition, we can obtain any perfect sequence. This yields the following algorithm to enumerate all perfect sequences.

Algorithm 3. All Perfect Sequences

    Input   : Chordal graph $G = (V, E)$;
    Output: All perfect sequences of G;
1 construct $\mathcal{CG}(G)$;
2 find maximum weighted spanning tree $\mathcal{T}^*$ of $\mathcal{CG}(G)$;
3 by using $\mathcal{T}^*$, compute sets $\mathcal{E}_u$, $\mathcal{E}_i$, $\mathcal{E}_d$ of unnecessary, indispensable, and
    dispensable edges, respectively;
4 set P to empty sequence;   `// keep current perfect sequence`
5 let $\mathcal{CG}(G)^* := (\mathcal{C}(G), \mathcal{E}_i \cup \mathcal{E}_d)$;
6 call Enumerate$(\mathcal{CG}(G)^*, P)$;

Procedure Enumerate$(\mathcal{CG}(G)^* = (\mathcal{C}(G), \mathcal{E}_i \cup \mathcal{E}_d), P)$

    [H]
    Output: A perfect sequence at the last node;
7 **if** $\mathcal{C}(G)$ *contains one node C* **then**
8   **output** $(C + P)$;   `// ` $C + P$ ` denotes concatenation of node ` C ` and`
        `sequence ` P
9 **else**
10   compute $S := \{ C \in \mathcal{C}(G) \mid C$ satisfies the leaf condition$\}$;
11   **foreach** $C \in S$ **do**
12   call Enumerate$(\mathcal{CG}(G) \setminus C, C + P)$;
13   **end**
14 **end**

A part of the computation tree for the chordal graph in Fig. 1(a) is given in Fig. 2. The unweighted graph $\mathcal{CG}(G)^*$ contains two indispensable edges, $\{C_2, C_3\}$ and $\{C_4, C_5\}$. Now we are ready to show the main theorem in this paper.

Theorem 1. *For any chordal graph $G = (V, E)$, with $O(|V|^3)$ time and $O(|V|^2)$ space pre-computation, all perfect sequences can be enumerated in $O(1)$ time per sequence on average and $O(|V|^2)$ space.*

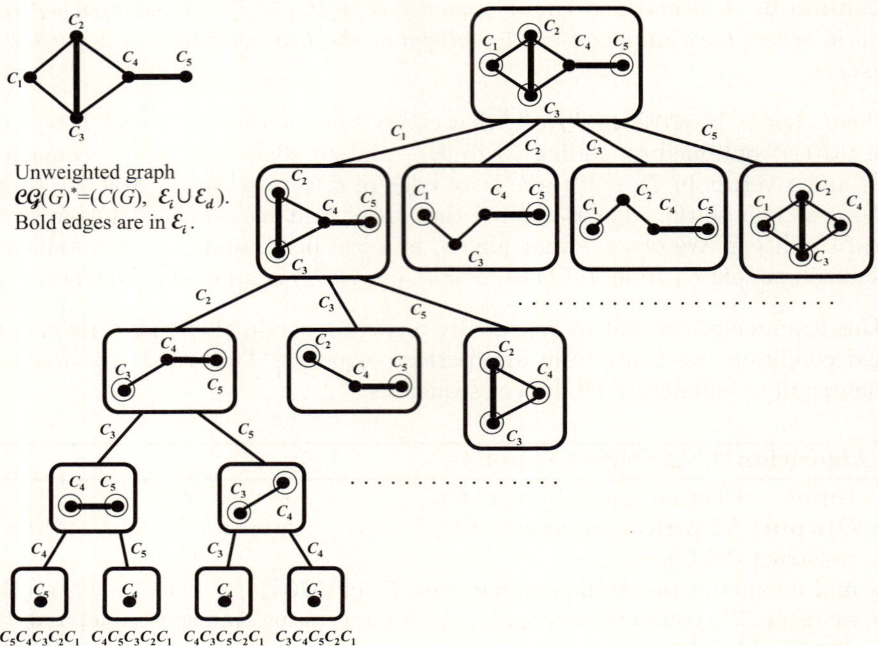

Fig. 2. Part of computation tree that enumerates all perfect sequences

Proof. From Lemmas 5 and 6, we can see that Algorithm 3 generates all perfect sequences. Since each iteration adds a maximal clique on the top of the sequence when it generates a recursive call, no two iterations can output the same perfect sequence. Thereby any perfect sequence is generated exactly once. This shows the correctness of the algorithm.

Therefore, we concentrate on the analysis of the time complexity. The space complexity is easy to see. We first observe that the computation of set S in step 10 takes $O(n^2)$ time, where $n = |\mathcal{C}(G)|$ in the procedure. Therefore, the time complexity of each procedure call of **Enumerate** can be bounded above by cn^2 time for a positive constant c except for the computation time spent for the recursive calls generated by the procedure.

Now, a procedure call of **Enumerate** where $\mathcal{CG}(G) = (\mathcal{C}(G), \mathcal{E}_i \cup \mathcal{E}_d)$ is called a *k-level call* if $|\mathcal{C}(G)| = k$. Let $t(k)$ be the total computation time for k-level calls. When $k = 1$, we have a perfect sequence for each call. Therefore, we have $t(1) \leq cN$, where N is the number of perfect sequences.

When $k > 1$, there are at least two maximal cliques in $\mathcal{C}(G)$ satisfying the leaf condition. Therefore, there are at least two cliques in S, and the number of k-level calls is at most half of the number of $(k - 1)$-level calls. Thus, we have $t(k) \leq \frac{k^2}{2^{k-1}} cN$ if $k > 1$. We can see that $t(2)+t(3)+t(4) \leq (2+\frac{9}{4}+2)cN = \frac{25}{4}cN$.

For any $k > 4$, we have $\frac{k^2}{2^{k-1}} \leq \frac{8}{9}\frac{(k-1)^2}{2^{k-2}}$. Thus, we have $\sum_{k=5}^{\infty} t(k) \leq (\frac{8}{9} + (\frac{8}{9})^2 + \cdots)t(4) < 8t(4) \leq 16cN$. Therefore, each perfect sequence can be obtained on average in $O(1)$ time. $\square$

We note here that we can include the time to output in $O(1)$ time for each perfect sequence on average. The first idea is to output the difference of the previous output. The second idea is to output at each step when the clique of the sequence is found. More precisely, we replace step 8 of **Enumerate** by the following three steps

    output $+C$;
    output "end of reverse of a perfect sequence";
    output $-C$;

and replace step 12 of **Enumerate** by the following three steps

    output $+C$;
    call **Enumerate**$(\mathcal{CG}(G) \setminus C, C + P)$;
    output $-C$;

Then we will incrementally have all perfect sequences, and the time complexity is still $O(1)$ time for each perfect sequence on average.

We also note that the *delay* in Algorithm 3, which is the maximum time between two consecutive perfect sequences, is $O(|V|^3)$ with a straightforward implementation. This time complexity comes from (1) the distance between two perfect sequences in the computation tree is $O(n)$ and (2) each computation of the set S takes $O(n^2)$ time in step 10. However, we do not need to compute S completely every time. We can update S incrementally for each removal and addition with a suitable data structure, and the computation time can be reduced to $O(n)$ time. Hence it is not difficult to reduce the delay to $O(n^2)$, but the details are omitted here.

4 Conclusion

We propose an algorithm to enumerate all the perfect sequences of a given chordal graph. The time complexity for each perfect sequence is $O(1)$, which is the optimal time complexity. From the proof of the main theorem, we can see that the number of perfect sequences might be exponential in the size of the graph in general, and thus for large graphs the algorithm is impractical. Therefore, one of our future works is to construct an efficient random sampling algorithm. Our approach does not use clique trees, and thus, with polynomial time convergence, there is the possibility for efficient sampling.

References

1. Blair, J.R.S., Peyton, B.: An Introduction to Chordal Graphs and Clique Trees. In: George, A., Gilbert, J.R., Liu, J.W.H. (eds.) Graph Theory and Sparse Matrix Computation. IMA, vol. 56, pp. 1–29. Springer, Heidelberg (1993)
2. Brandstädt, A., Le, V.B., Spinrad, J.P.: Graph Classes: A Survey. SIAM, Philadelphia (1999)

3. Chandran, L.S., Ibarra, L., Ruskey, F., Sawada, J.: Generating and Characterizing the Perfect Elimination Orderings of a Chordal Graph. Theoretical Computer Science 307, 303–317 (2003)
4. Dahl, J., Vandenberghe, L., Roychowdhury, V.: Covariance Selection for Non-chordal Graphs via Chordal Embedding. Optimization Methods and Software 23(4), 501–520 (2008)
5. Golumbic, M.C.: Algorithmic Graph Theory and Perfect Graphs, 2nd edn. Annals of Discrete Mathematics 57. Elsevier, Amsterdam (2004)
6. Hara, H., Takemura, A.: Boundary Cliques, Clique Trees and Perfect Sequences of Maximal Cliques of a Chordal Graph. METR 2006-41, Department of Mathematical Informatics, University of Tokyo (2006)
7. Hara, H., Takemura, A.: Bayes Admissible Estimation of the Means in Poisson Decomposable Graphical Models. Journal of Statistical Planning and Inference (in press, 2008)
8. Kaplan, H., Shamir, R., Tarjan, R.E.: Tractability of Parameterized Completion Problems on Chordal, Strongly Chordal, and Proper Interval Graphs. SIAM Journal on Computing 28(5), 1906–1922 (1999)
9. Lauritzen, S.L.: Graphical Models. Clarendon Press, Oxford (1996)
10. McKee, T.A., McMorris, F.R.: Topics in Intersection Graph Theory. SIAM, Philadelphia (1999)
11. Rose, D.J., Tarjan, R.E., Lueker, G.S.: Algorithmic Aspects of Vertex Elimination on Graphs. SIAM Journal on Computing 5(2), 266–283 (1976)
12. Spinrad, J.P.: Efficient Graph Representations. American Mathematical Society (2003)
13. Tarjan, R.E., Yannakakis, M.: Simple Linear-Time Algorithms to Test Chordality of Graphs, Test Acyclicity of Hypergraphs, and Selectively Reduce Acyclic Hypergraphs. SIAM Journal on Computing 13(3), 566–579 (1984)

From Tree-Width to Clique-Width:
Excluding a Unit Interval Graph

Vadim V. Lozin

DIMAP and Mathematics Institute, University of Warwick, Coventry, CV4 7AL, UK
V.Lozin@warwick.ac.uk

Abstract. From the theory of graph minors we know that the class of planar graphs is the only critical class with respect to tree-width. In the present paper, we reveal a critical class with respect to clique-width, a notion generalizing tree-width. This class is known in the literature under different names, such as unit interval, proper interval or indifference graphs, and has important applications in various fields, including molecular biology. We prove that the unit interval graphs constitute a minimal hereditary class of unbounded clique-width. As an application, we show that LIST COLORING is fixed parameter tractable in the class of unit interval graphs.

Keywords: Tree-width, Clique-width, Unit interval graphs, Fixed parameter tractability.

1 Introduction

The tree- and clique-width are two graph parameters which are of primary importance in algorithmic graph theory due to the fact that many problems, being NP-hard in general, admit polynomial-time solutions when restricted to graphs of bounded tree- or clique-width. Determining whether one of these parameters is bounded in a class of graphs X is not a trivial question. For tree-width, an answer to this question can be found with the help of the following two results:

(1) in the study of tree-width one can be restricted to minor-closed graph classes, as the tree-width of a graph cannot be less than the tree-width of any of its minors. In other words, if X is not a minor-closed graph class, it can be extended, without loss of generality, to a minor-closed class by adding to X all minors of graphs in X. Observe that such an extension can be characterized by finitely many forbidden minors [24].
(2) in the family of minor-closed graph classes there is a unique minimal class of unbounded tree-width, namely, the class of planar graphs [23]. In other words, the tree-width of graphs in a minor-closed class X is bounded if and only if X excludes at least one planar graph.

In the present paper we study clique-width of graphs, a notion generalizing that of tree-width in the sense that graphs of bounded tree-width have bounded

S.-H. Hong, H. Nagamochi, and T. Fukunaga (Eds.): ISAAC 2008, LNCS 5369, pp. 871–882, 2008.

clique-width. In the study of clique-width the restriction to minor-closed graph classes is not valid anymore. Instead, we restrict ourselves to hereditary classes, i.e., those containing with each graph G all induced subgraphs of G. This restriction can be made without loss of generality, because the clique-width of a graph cannot be less than the clique-width of any of its induced subgraphs [6]. The family of hereditary classes generalizes that of minor-closed classes, and this generalization provides a natural analog of (1). However, (2) does not admit such a simple generalization. Moreover, apparently the family of hereditary classes contains many (if not infinitely many) classes which are critical with respect to clique-width. In the present paper, we reveal the first class of this type, the *unit interval graphs*, also known in the literature as proper interval graphs [3] and indifference graphs [22]. Graphs in this class enjoy many attractive properties and find important applications in various fields, including molecular biology [17]. The structure of unit interval graphs is relatively simple, allowing efficient algorithms for recognizing and representing these graphs [16], as well as for many other computational problems (see e.g. [2,4,5]). Nonetheless, some algorithmic problems remain NP-hard when restricted to the class of unit interval graphs [20], the complexity of some others is unknown [7], and most width parameters (including clique-width [14]) are unbounded in this class. In the present paper, we show that the unit interval graphs constitute a minimal hereditary class of unbounded clique-width. As an application, we show that the LIST COLORING problem is fixed parameter tractable in this class.

The organization of the paper is as follows. In the rest of this section, we introduce basic notations and terminology. In section 2, we describe some structural properties of unit interval graphs. Section 3 is devoted to clique-width of unit interval graphs and Section 4 discusses algorithmic applications.

All graphs in this paper are undirected, without loops and multiple edges. For a graph G, we denote by $V(G)$ and $E(G)$ the vertex set and the edge set of G, respectively. The neighborhood of a vertex $v \in V(G)$, denote $N_G(v)$, is the set of vertices adjacent to v. If there is no confusion about G we simply write $N(v)$. We say that G is an H-free graph if no induced subgraph of G is isomorphic to H. The subgraph of G induced by a subset $U \subseteq V(G)$ is denoted $G[U]$. Two vertices of $U \subseteq V(G)$ will be called U-similar if they have the same neighborhood outside U. Clearly, the similarity is an equivalence relation. The number of equivalence classes of U in G will be denoted $\mu_G(U)$ (or $\mu(U)$ if no confusion arises). A graph G is said to be *prime* if $\mu_G(U) = 1$ implies $|U| = 1$ or $U = V(G)$. When determining the clique-width of graphs in a hereditary class X one can be restricted to prime graphs in X, because the clique-width of a graph G equals the clique-width of a maximal prime induced subgraph of G [6].

2   Unit Interval Graphs

A graph G is an interval graph if it is the intersection graph of intervals on the real line. G is a *unit* interval graph if all the intervals in the intersection model are of the same length. In this section, we describe some structural properties of unit interval graphs. We start by introducing unit interval graphs of a special form that will play an important role in our considerations. Denote by $H_{m,n}$ the graph with mn vertices which can be partitioned into m cliques

$$V_1 = \{v_{1,1}, \ldots, v_{1,n}\}$$

$$\cdots$$

$$V_m = \{v_{m,1}, \ldots, v_{m,n}\}$$

so that for each $i = 1, \ldots, m-1$ and for each $j = 1, \ldots, n$, vertex $v_{i,j}$ is adjacent to vertices $v_{i+1,1}, v_{i+1,2}, \ldots, v_{i+1,j}$ and there are no other edges in the graph. An example of the graph $H_{5,5}$ is given in Figure 1 (for clarity of the picture, each clique V_i is represented by an oval without inside edges). We will call the vertices of V_i the i-th row of $H_{m,n}$, and the vertices $v_{1,j}, \ldots, v_{m,j}$ the j-th column of $H_{m,n}$.

It is not difficult to see (and will follow from Theorem 1) that $H_{m,n}$ is a unit interval graph. Moreover, below we will prove that $H_{n,n}$ contains every unit interval graph on n vertices as an induced subgraph. That's why we call the graph $H_{m,n}$ a *canonical* unit interval graph.

Now consider the special case of $H_{m,n}$ when $m = 2$. The complement of this graph is bipartite and is known in the literature under various names such as a *difference graph* [15] or a *chain graph* [18]. The latter name is due to the fact that the neighborhoods of vertices in each part of the graph form a chain, i.e., the vertices can be ordered under inclusion of their neighborhoods. We will call

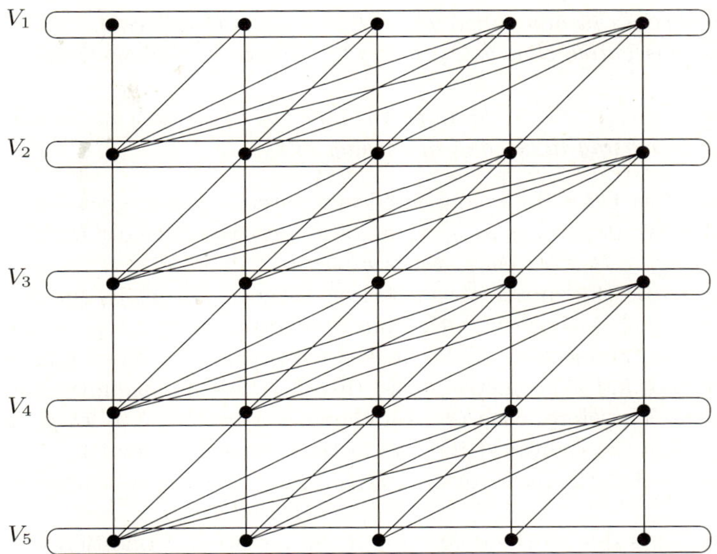

Fig. 1. Canonical graph $H_{5,5}$

an ordering $x_1, \ldots, x_k$ *increasing* if $i < j$ implies $N(x_i) \subseteq N(x_j)$ and *decreasing* if $i < j$ implies $N(x_j) \subseteq N(x_i)$. The class of all bipartite chain graphs can be characterized in terms of forbidden induced subgraphs as $2K_2$-free bipartite graphs ($2K_2$ is the complement of the chordless cycle on 4 vertices). In general, the two parts of a bipartite chain graph can be of different size. But a prime graph in this class has equally many vertices in both parts, i.e., it is of the form $H_{2,n}$ with V_1 and V_2 being independent sets (see e.g. [12]).

In what follows, we call the complements of bipartite chain graphs *co-chain graphs*. Let G be a co-chain graph with a given bipartition into two cliques V_1 and V_2, and let n be a maximum number such that G contains the graph $H_{2,n}$ as an induced subgraph. Consider two vertices $w_1 \in V_1$ and $w_2 \in V_2$ in the same column of $H_{2,n}$ and let $W_1 = \{v \in V_1 \mid N(v) \cap V_2 = N(w_1) \cap V_2\}$ and $W_2 = \{v \in V_2 \mid N(v) \cap V_1 = N(w_2) \cap V_1\}$. It is not difficult to see that $W_1 \cup W_2$ is a clique and we will call this clique a *cluster* of G. The vertices of V_1 that have no neighbors in V_2 will be called a *trivial cluster* of G. Similarly, we define a trivial cluster which is a subset of V_2. Clearly the set of all clusters of G defines a partition of $V(G)$. For instance, the graph $H_{2,n}$ consists of n clusters, each containing two vertices of the same column.

To derive a structural characterization of unit interval graphs, we use the ordinary intersection model: with each vertex v we associate an interval $I(v)$ on the real line with endpoints $l(v)$ and $r(v)$ such that $r(v) = l(v) + 1$. We will write $I(u) \leq I(v)$ to indicate that $l(u) \leq l(v)$.

Theorem 1. *A connected graph G is a unit interval graph if and only if the vertex set of G can be partitioned into cliques $Q_0, \ldots, Q_t$ in such a way that*

(a) any two vertices in non-consecutive cliques are non-adjacent,
(b) any two consecutive cliques Q_{j-1} and Q_j induce a co-chain graph, denoted G_j,
(c) for each $j = 1, \ldots, t-1$, there is an ordering of vertices in the clique Q_j, which is decreasing in G_j and increasing in G_{j+1}.

Proof. **Necessity.** Let G be a connected unit interval graph given by an intersection model. We denote by p_0 a vertex of G with the leftmost interval in the model, i.e., $I(p_0) \leq I(v)$ for each vertex v.

Define Q_j to be the subset of vertices of distance j from p_0 (in the graph-theoretic sense, i.e., a shortest path from any vertex of Q_j to p_0 consists of j edges). From the intersection model, it is obvious that if u is not adjacent to v and is closer to p_0 in the geometric sense, then it is closer to p_0 in the the graph-theoretic sense. Therefore, each Q_j is a clique. For each $j > 0$, let p_j denote a vertex of Q_j with the rightmost interval in the intersection model.

We will prove that the partition $Q_0 \cup Q_1 \cup \ldots \cup Q_t$ satisfies all three conditions of the theorem.

Condition (a) is due to the definition of the partition. Condition (b) will be proved by induction. Moreover, we will show by induction on j that

(1) G_j is a co-chain graph,
(2) p_{j-1} is adjacent to each vertex in Q_j,
(3) for every $v \in Q_i$ with $i \geq j$, $I(p_{j-1}) \leq I(v)$.

For $j = 1$, statements (1), (2), (3) are obvious. To make the inductive step, assume by contradiction that vertices $x_1, x_2 \in Q_{j-1}$ and $y_1, y_2 \in Q_j$ induce a chordless cycle with edges $x_1 y_1$ and $x_2 y_2$ (i.e., these vertices induce a $2K_2$ in the complement of G_j). By the induction hypothesis, both $I(x_1)$ and $I(x_2)$ intersect $I(p_{j-2})$, and also $I(p_{j-2}) \leq I(y_1), I(y_2)$. Assuming without loss of generality that $I(x_1) \leq I(x_2)$, we must conclude that $I(y_1)$ intersects both $I(x_1)$ and $I(x_2)$, which contradicts the assumption. Hence, (1) is correct. To prove (2) and (3), consider a vertex $v \in Q_i$, $i \geq j$, non-adjacent to p_{j-1}. By the induction hypothesis, $I(p_{j-2})$ intersects $I(p_{j-1})$, and also $I(p_{j-2}) \leq I(v)$, therefore $I(p_{j-1}) \leq I(v)$, which proves (3). Moreover, by the choice of p_{j-1}, this also implies that v does not have neighbors in Q_{j-1}. Therefore, $v \notin Q_j$ and hence (2) is valid.

To prove (c), we will show that for every pair of vertices u and v in Q_j, $N_{G_j}(u) \subset N_{G_j}(v)$ implies $N_{G_{j+1}}(v) \subseteq N_{G_{j+1}}(u)$. Assume the contrary: $s \in N_{G_j}(v) - N_{G_j}(u)$ and $t \in N_{G_{j+1}}(v) - N_{G_{j+1}}(u)$. From (2) we conclude that $s \neq p_{j-1}$. Therefore, $j > 1$. Due to the choice of p_{j-1} we have $I(s) \leq I(p_{j-1})$, and from (3) we have $I(p_{j-1}) \leq I(u)$ and $I(p_{j-1}) \leq I(v)$. Therefore, $I(v) \leq I(u)$ by geometric considerations. But now, geometric arguments lead us to the conclusion that $tv \in E(G)$ implies $tu \in E(G)$. This contradiction proves (c).

Sufficiency. Consider a graph G with a partition of the vertex set into cliques $Q_0, Q_1, \ldots, Q_t$ satisfying conditions (a), (b), (c). We assume that the vertices of $Q_j = \{v_{j,1}, v_{j,2}, \ldots, v_{j,k_j}\}$ are listed in the order that agrees with (c). Let us construct an intersection model for G as follows. Each clique Q_j will be represented in the model by a set of intervals in such a way that $l(v_{j,i}) < l(v_{j,k}) < r(v_{j,i})$ whenever $i < k$. For $j = 0$, there are no other restrictions. For $j > 0$, we proceed inductively: for every vertex $u \in Q_j$ with neighbors $v_{j-1,s}, v_{j-1,s+1}, \ldots, v_{j-1,k_{j-1}}$ in Q_{j-1}, we place $l(u)$ between $l(v_{j-1,s})$ and $l(v_{j-1,s+1})$ (or simply to the right of $l(v_{j-1,s})$ if $v_{j-1,s+1}$ does not exist). It is not difficult to see that the constructed model represents G. $\qquad\square$

From this theorem it follows in particular that $H_{m,n}$ is a unit interval graph. Any partition of a connected unit interval graph G agreeing with (a), (b) and (c) will be called a *canonical partition* of G and the cliques $Q_0, \ldots, Q_t$ the *layers* of the partition; cliques Q_0 and Q_t will be called marginal layers. Any cluster of any co-chain graph G_j in a canonical partition of G will be also called a cluster of G. Thus, $H_{m,n}$ consists of $(m-1)n$ clusters.

Now we turn to showing that any connected unit interval graph with n vertices that admits a canonical partition into m layers is contained in the graph $H_{m,n}$ as an induced subgraph. The proof will be given by induction on the number of layers and we start with the basis of the induction.

Lemma 1. *The graph $H_{2,n}$ contains every co-chain graph with n vertices as an induced subgraph.*

Proof. Let G be an n-vertex co-chain graph with a bipartition into cliques V_1 and V_2. We will assume that the vertices of V_1 are ordered increasingly according to their neighborhoods in V_2, while the vertices of V_2 are ordered decreasingly. The graph $H_{2,n}$ containing G will be created by adding to G some new vertices and edges. Let $W^1, \ldots, W^p$ be the clusters of G and $W_i^j = V_i \cap W^j$.

For each W_1^j we add to G a set U_2^j of new vertices of size $k = |W_1^j|$ and create on $W_1^j \cup U_2^j$ the graph $H_{2,k}$. Also, create a clique on the set $V_2' = U_2^1 \cup W_2^1 \cup \ldots \cup U_2^p \cup W_2^p$, and for each $i < j$ connect every vertex of W_1^j to every vertex of U_2^i. Symmetrically, for each W_2^j we add to G a set U_1^j of new vertices of size $k = |W_2^j|$ and create on $W_2^j \cup U_1^j$ the graph $H_{2,k}$. Also, create a clique on the set $V_1' = W_1^1 \cup U_1^1 \cup \ldots \cup W_1^p \cup U_1^p$, and for each $i < j$ connect every vertex of U_1^j to every vertex of $W_2^i \cup U_2^i$. It is not difficult to see that the set $V_1' \cup V_2'$ induces the graph $H_{2,n}$ and this graph contains G as an induced subgraph. $\square$

Now we proceed to the general case.

Lemma 2. *Let G be a connected unit interval graph with n vertices that admits a canonical partition into m layers $Q_1, \ldots, Q_m$. Then G is an induced subgraph of $H_{m,n}$.*

Proof. We will show by induction on m that $H_{m,n}$ contains G as an induced subgraph, moreover, the i-th layer Q_i of G belongs to the i-th row V_i of $H_{m,n}$. Lemma 1 provides the basis of the induction. Now assume that the statement is valid for any connected unit interval graph with less than m layers, and let G contain m layers. For $j = 1, \ldots, m$, denote $n_j = |Q_j|$ and $p = n_1 + \ldots + n_{m-1}$.

Let $H_{m-1,p}$ be the canonical graph containing the first $m-1$ layers of G as an induced subgraph. Now we create an auxiliary graph H' out of $H_{m-1,p}$ by

(1) adding to $H_{m-1,p}$ the clique Q_m,
(2) connecting the vertices of $Q_{m-1} \subseteq V_{m-1}$ to the vertices of Q_m as in G,
(3) connecting the vertices of $V_{m-1} \setminus Q_{m-1}$ to the vertices of Q_m so as to make the existing order of vertices in V_{m-1} decreasing in the subgraph induced by V_{m-1} and Q_m. More formally, whenever vertex $w_{m-1,i} \in V_{m-1} \setminus Q_{m-1}$ is connected to a vertex $v \in Q_m$, every vertex $w_{m-1,j}$ with $j < i$ must be connected to v too.

According to (2) and (3) the subgraph of H' induced by V_{m-1} and Q_m is a co-chain graph. We denote this subgraph by G'. Clearly H' contains G as an induced subgraph. To extend H' to a canonical graph containing G we apply the induction hypothesis twice. First, we extend G' to a canonical co-chain graph as described in Lemma 1. This will add p new vertices to the m-th row and n_m new vertices to $(m-1)$-th row of the graph. Then we use the induction once more to extend the first $m-1$ rows to a canonical form. $\square$

We summarize the above two lemmas as follows.

Theorem 2. *$H_{n,n}$ is an n-universal unit interval graph, i.e., every unit interval graph with n vertices is an induced subgraph of $H_{n,n}$.*

Proof. We use induction on the number of connected components of a unit interval graph G. If G is connected, the result follows from Lemma 2. Now assume that G is disconnected. Denote by G_1 a connected component of G and by G_2 the rest of the graph. Also let $k_1 = |V(G_1)|$ and $k_2 = |V(G_2)|$. The intersection of the first k_1 columns and the first k_1 rows of $H_{n,n}$ induces the graph H_{k_1,k_1}, which, according to Lemma 2, contains G_1 as an induced subgraph. The remaining k_2 columns and k_2 rows of $H_{n,n}$ induce the graph H_{k_2,k_2}, which contains G_2 according to the induction hypothesis. Notice that no vertex of the H_{k_1,k_1} is adjacent to a vertex of the H_{k_2,k_2}. Therefore, $H_{n,n}$ contains G as an induced subgraph. $\square$

3   Clique-Width of Unit Interval Graphs

The *clique-width* of a graph G is the minimum number of labels needed to construct G by means of the following four operations:

- Creation of a new vertex v with label i (denoted $i(v)$).
- Disjoint union of two labeled graphs G and H (denoted $G \oplus H$).
- Joining by an edge each vertex with label i to each vertex with label j (denoted $\eta_{i,j}$).
- Renaming label i to j (denoted $\rho_{i \to j}$).

Finding the exact value of the clique-width of a graph is known to be an NP-hard problem [11]. In general, this value can be arbitrarily large. Moreover, it is unbounded in many restricted graph families, including the unit interval graphs [14]. On the other hand, in some specific classes of graphs the clique-width is bounded by a constant. Consider, for instance, a chordless path P_5 on five consecutive vertices a, b, c, d, e. By means of the four operations described above this graph can be constructed as follows:

$$\eta_{3,2}(3(e) \oplus \rho_{3 \to 2}(\rho_{2 \to 1}(\eta_{3,2}(3(d) \oplus \rho_{3 \to 2}(\rho_{2 \to 1}(\eta_{3,2}(3(c) \oplus \eta_{2,1}(2(b) \oplus 1(a))))))))).$$

This expression uses only three different labels. Therefore, the clique-width of P_5 is at most 3. Obviously, in a similar way we can construct any chordless path with the help of at most three labels. This simple example suggests the main idea for the construction of $H_{k,k}$-free unit interval graphs with bounded number of labels. We describe this idea in the following lemma.

Lemma 3. *If the vertices of a graph G can be partitioned into subsets $V_1, V_2, \ldots$ in such a way that for every i, the clique-width of $G[V_i]$ is at most $k \geq 2$, $\mu(V_i) \leq l$ and $\mu(V_1 \cup \ldots \cup V_i) \leq l$, then the clique-width of G is at most kl.*

Proof. If $G[V_1]$ can be constructed with at most k labels and $\mu(V_1) \leq l$, then $G[V_1]$ can be constructed with at most kl different labels in such a way that in the process of constructing, any two vertices in different equivalence classes of V_1 have different labels, and by the end of the process, any two vertices in the same equivalence class of V_1 have the same label. So, the construction of $G[V_1]$ finishes with at most l different labels corresponding to the equivalence classes of V_1.

Now assume we have constructed the graph $G_i = G[V_1 \cup \ldots \cup V_i]$ with the help of kl different labels making sure that the construction finishes with a set A of at most l different labels corresponding to the equivalence classes of $V_1 \cup \ldots \cup V_i$. Separately, we construct $G[V_{i+1}]$ with the help of kl different labels and complete the construction with a set B of at most l different labels corresponding to the equivalence classes of V_{i+1}. We choose the labels so that the sets A and B are disjoint. Now we use operations $\oplus$ and η to build the graph $G_{i+1} = G[V_1 \cup \ldots \cup V_i \cup V_{i+1}]$ out of G_i and $G[V_{i+1}]$. Notice that any two vertices in the same equivalence class of $V_1 \cup \ldots \cup V_i$ or V_{i+1} belong to the same equivalence class of $V_1 \cup \ldots \cup V_i \cup V_{i+1}$. Therefore, the construction of G_{i+1} can be completed with a set of at most l different labels corresponding to the equivalence classes of the graph.

The conclusion now follows by induction. $\qquad\square$

This lemma implies in particular that

Corollary 1. *The clique-width of $H_{m,n}$ is at most $3m$.*

Proof. To build $H_{m,n}$ we partition it into subsets $V_1, V_2, \ldots, V_n$ by including in V_i the vertices of the i-th column of $H_{m,n}$. Then the clique-width of $G[V_i]$ is at most 3. Trivially, $\mu(V_i) = m$. Also, it is not difficult to see that $\mu(V_1 \cup \ldots \cup V_i) = m$. Therefore, the conclusion follows by Lemma 3. $\qquad\square$

Now we prove the key lemma of the paper.

Lemma 4. *For every natural k, there is a constant $c(k)$ such that the clique-width of any $H_{k,k}$-free unit interval graph G is at most $c(k)$.*

Proof. Without loss of generality, we will assume that G is prime. In particular, G is connected. To better understand the global structure of G, let us associate with it another graph which will be denoted $B(G)$. To define $B(G)$ we first partition the vertices of G into layers $Q_0, \ldots, Q_t$ as described in Theorem 1 and then partition each co-chain graph G_j induced by two consecutive cliques Q_{j-1}, Q_j into clusters as described in the last but one paragraph before Theorem 1. Without loss of generality we may assume that no G_j contains a trivial cluster. Indeed, if such a cluster exists, it contains at most one vertex due to primality of G. Each G_j contains at most two trivial clusters. Therefore, by adding at most two vertices to each layer of G, we can extend it to a unit interval graph G' such that G' has no trivial clusters, G' contains G as an induced subgraph and G' is $H_{k,k+2}$-free.

With each cluster of G we associate a vertex of the graph $B(G)$ and connect two vertices of $B(G)$ if and only if the respective clusters have a non-empty intersection. For instance, $B(H_{m,n})$ is the set of n disjoint paths of length $m-2$ each. Clearly the vertices of $B(G)$ representing clusters of the same co-chain graph G_j in the partition of G form an independent set and we will call this set a level of $B(G)$. In the proof we will use a graphical representation of $B(G)$ obtained by arranging the vertices of the same level on the same horizontal line (different lines for different levels) according to the order of the respective clusters in the canonical partition of G. From this representation it is obvious that $B(G)$ is a plane graph.

Since G is prime, any two clusters of G have at most one vertex in common. Therefore, each edge of $B(G)$ corresponds to a vertex of G (this correspondence can be made one-to-one by adding to the two marginal levels of $B(G)$ pendant edges representing the vertices of the two marginal layers of G).

Now let us consider any k consecutive layers in the canonical partition of G and denote the subgraph of G induced by these layers G^*. The respective graph $B(G^*)$ will be denoted B^*; it has $k-1$ levels denoted $B_1, \ldots, B_{k-1}$. Since G (and G^*) is $H_{k,k}$-free, the two marginal levels of B^* are connected to each other by a set $\mathcal{P}$ of at most $k-1$ disjoint paths. Denote $s = |\mathcal{P}|$. Without loss of generality we may assume that the first path in $\mathcal{P}$ is formed by the leftmost vertices of B^*, while the last one by the rightmost vertices of B^*. The s paths of $\mathcal{P}$ cut B^* into $s-1$ stripes, i.e., subgraphs induced by two consecutive paths and all the vertices between them.

Since s is the maximum number of disjoint paths connecting B_1 to B_{k-1}, by Menger's Theorem (see e.g. [9]), these two levels can be separated from each other by a set S of $s \leq k-1$ vertices, containing exactly one vertex in each of the paths. To visualize this situation, let us draw a curve Ω that separates B_1 from B_{k-1} and crosses B^* at precisely s points (the vertices of S; no edge of B^* is crossed by or belongs to Ω). We claim that without loss of generality we may assume that this curve traverses each stripe of B^* "monotonically", meaning that its "y-coordinate" changes within a stripe either non-increasingly or non-decreasingly. Indeed, assume Ω has a "local maximum" within a stripe, and let v be a vertex (below the curve) that gives rise to this maximum. Obviously, v does not belong to B_1 (since otherwise B_1 is not separated from B_{k-1}), and v must have a neighbor at a higher level within the stripe (since there are no trivial clusters in G). But then the edge connecting v to that neighbor would cross Ω, which is impossible according to the definition of Ω.

The above discussion allows us to conclude that whenever Ω separates vertices of the same level within a stripe, the two resulting sets form "intervals", i.e., their vertices appear in the representation of B^* consecutively.

Now let us translate the above discussion in terms of the graph G^*. The partition of the edges of B^* defined by Ω results in a respective partition of the vertices of G^* into two parts, say X and Y. Let Q_i be a layer of G^*. As we mentioned before, the vertices of Q_i correspond to the edges between two consecutive levels of B^*. We partition these edges and the respective vertices of Q_i into at most $4s-1$ subsets $Q_{i,1}, \ldots, Q_{i,4s-1}$ of three types as follows. The first type consists of s 1-element subsets corresponding to the edges of the s paths of $\mathcal{P}$. For each such an edge e, we form at most two subsets of the second type, each consisting of the edges that have a common vertex with e and belong to the same stripe. The remaining edges form the third group consisting of at most $s - 1$ subsets, each representing the edges of the same stripe. Observe that the vertices of each $Q_{i,j}$ form an "interval", i.e., they are consecutive in Q_i. The curve Ω partitions each $Q_{i,j}$ into at most two "sub-intervals" corresponding to X and Y, respectively. We claim that no vertex of Y can distinguish the vertices of $Q_{i,j} \cap X$. Assume the contrary: a vertex $y \in Y$ is not adjacent to $x_1 \in Q_{i,j} \cap X$ but is adjacent to $x_2 \in Q_{i,j} \cap X$. Then $y \in Q_{i+1}$, x_2 and y belong to the same cluster U of G_{i+1}, while x_1 does not belong to U. Let u denote the vertex of B^* representing U. Also, let e_{x_1}, e_{x_2}, e_y be the edges of B^* corresponding to vertices x_1, x_2, and y, respectively. Since e_{x_2} and e_y are incident to u but separated by Ω, vertex u belongs to Ω and hence to the separator S. Therefore, u belongs to a path from $\mathcal{P}$. But then $Q_{i,j}$ is of the second type and therefore e_{x_1} must also be incident to u. This contradicts the fact that x_1 does not belong U and shows that any two vertices of the same $Q_{i,j} \cap X$ have the same neighborhood in Y. Therefore, $\mu_{G^*}(X)$ is at most the number of different $Q_{i,j}$s, which is at most $k(4s - 1) \leq 4k^2 - 5k$. Symmetrically, $\mu_{G^*}(Y) \leq 4k^2 - 5k$.

To complete the proof, we partition G into subsets $V_1, V_2 \ldots$ according to the following procedure. Set $i := 1$. If the canonical partition of G consists of less than k layers, then define $V_i := V(G)$ and stop. Otherwise consider the first k layers of G and partition the subgraph induced by these layers into sets X and Y as described above. Denote $V_i := X$ and repeat the procedure with $G := G - V_i$ and $i := i + 1$. By Lemma 2 and Corollary 1, each V_i induces a graph of clique-width at most $3k$, and from the above discussion we know that $\mu(V_i) \leq 4k^2 - 5k$ and $\mu(V_1 \cup \ldots \cup V_i) \leq 4k^2 - 5k$. Therefore, by Lemma 3 the clique-width of G is at most $12k^3 - 15k^2$. With the correction on the possible existence of trivial clusters, we conclude that the clique-width of G is at most $12k^3 + 72k^2 - 36k + 96$. $\qquad\square$

Theorem 3. *Let X be a proper hereditary subclass of unit interval graphs. Then the clique-width of graphs in X is bounded by a constant.*

Proof. Since X is hereditary, it admits a characterization in terms of forbidden induced subgraphs. Since X is a proper subclass of unit interval graphs, it must exclude at least one unit interval graph. Let G be such a graph with minimum number of vertices. If $|V(G)| = k$, then G is an induced subgraph of $H_{k,k}$ by Theorem 2. Therefore, X is a subclass of $H_{k,k}$-free unit interval graphs. But then the clique-width of graphs in X is bounded by a constant by Lemma 4. $\qquad\square$

4   Application and Open Problem

One of the consequences of Theorem 3 is that the *local* clique-width of unit interval graphs is bounded. The local variant of width parameters was first introduced with respect to tree-width: we say that the local tree-width of a graph G is bounded if the tree-width of G is bounded by a function of its diameter. The local tree-width has been shown to be bounded in the class of planar graphs and, more generally, in any minor-closed graph class that does not contain all apex graphs [8,10]. Graphs of bounded local tree-width allow polynomial-time approximation schemes for some combinatorial optimization problems that are NP-hard in general [1,13].

In the class of unit interval graphs, the local tree-width is not bounded (as the tree-width is not bounded on the cliques, i.e., graphs of diameter 1), but the local clique-width is bounded. Indeed, a unit interval graph G of diameter d has a canonical partition with at most $d + 1$ layers and hence the clique-width of G is at most $3d + 3$. We now apply this result to the following problem:

LIST k-COLORING

Input: A graph G, a set C of k colors and a list $l(v) \subseteq C$ associated with each vertex v of G.

Question: Is there a proper coloring $\psi : V(G) \to C$ such that $\psi(v) \in l(v)$ for each $v \in V(G)$?

This problem is NP-complete on unit interval graphs, because PRECOLORING EXTENSION (a restricted variant of LIST COLORING) is NP-complete in this class [20]. However, parameterized complexity of this problem is unknown. We propose the following parameterization of LIST k-COLORING on unit interval graphs.

It is known that LIST k-COLORING on n-vertex graphs of clique-width at most c can be solved in $O(2^{2kc}kc^3n)$ time [19] provided that a c-expression of the input graph is given. If an expression defining the input graph G is not given, we can apply a procedure from [21] that either concludes that the clique-width of G is strictly greater than c or outputs a $(2^{3c+2} - 1)$-expression defining G in $O(n^4)$ time. This discussion can be summarized as follows.

Theorem 4. *Let G be a unit interval graph with n vertices and diameter d. Then the LIST k-COLORING problem can be solved for G in time $O(f(k,d)n^4)$.*

Theorem 3 provides many more ways to parameterize the problem, for instance, with respect to the maximum vertex degree of G, the size of a maximum clique in G, etc.

Among various open problems, let us distinguish the following one. Motivated by molecular biology, Kaplan and Shamir studied the problem of completion of a given graph to a unit interval graph. In general, this is an NP-hard problem. They showed in [17] that completion to a unit interval graph with clique size at most k is polynomial-time solvable (though not fixed parameter tractable). We believe that Theorem 3 can be used to generalize this result.

References

1. Baker, B.S.: Approximation algorithms for NP-complete problems on planar graphs. Journal of the ACM 41, 153–180 (1994)
2. Bodlaender, H.L., Kloks, T., Niedermeier, R.: Simple MAX-CUT for unit interval graphs and graphs with few P_4s. Electron. Notes Discrete Math. 3, 1–8 (1999)
3. Bogart, K.P., West, D.B.: A short proof that proper = unit. Discrete Math. 201, 21–23 (1999)

4. Chang, R.-S.: Jump number maximization for proper interval graphs and series-parallel graphs. Inform. Sci. 115, 103–122 (1999)
5. Chen, C., Chang, C.-C., Chang, G.J.: Proper interval graphs and the guard problem. Discrete Math. 170, 223–230 (1997)
6. Courcelle, B., Olariu, S.: Upper bounds to the clique-width of a graph. Discrete Applied Math. 101, 77–114 (2000)
7. de Figueiredo, C.M.H., Meidanis, J., de Mello, C.P., Ortiz, C.: Decompositions for the edge colouring of reduced indifference graphs. Theoret. Comput. Sci. 297, 145–155 (2003)
8. Demaine, E.D., Hajiaghayi, M.T.: Diameter and treewidth in minor-closed graph families, revisited. Algorithmica 40, 211–215 (2004)
9. Diestel, R.: Graph theory, 3rd edn. Graduate Texts in Mathematics, vol. 173, p. 411. Springer, Berlin (2005)
10. Eppstein, D.: Diameter and treewidth in minor-closed graph families. Algorithmica 27, 275–291 (2000)
11. Fellows, M.R., Rosamond, F.A., Rotics, U., Szeider, S.: Clique-width Minimization is NP-hard. In: Proceedings of STOC 2006; 38th ACM Symposium on Theory of Computing, pp. 354–362 (2006)
12. Fouquet, J.-L., Giakoumakis, V.: On semi-P_4-sparse graphs. Discrete Math. 165/166, 277–300 (1997)
13. Grohe, M.: Local tree-width, excluded minors, and approximation algorithms. Combinatorica 23, 613–632 (2003)
14. Golumbic, M.C., Rotics, U.: On the clique-width of some perfect graph classes. International J. Foundations of Computer Science 11, 423–443 (2000)
15. Hammer, P.L., Peled, U.N., Sun, X.: Difference graphs. Discrete Appl. Math. 28, 35–44 (1990)
16. Hell, P., Shamir, R., Sharan, R.: A fully dynamic algorithm for recognizing and representing proper interval graphs. SIAM J. Comput. 31, 289–305 (2001)
17. Kaplan, H., Shamir, R.: Pathwidth, bandwidth, and completion problems to proper interval graphs with small cliques. SIAM J. Comput. 25, 540–561 (1996)
18. Kloks, T., Kratsch, D., Müller, H.: Bandwidth of chain graphs. Inform. Process. Lett. 68, 313–315 (1998)
19. Kobler, D., Rotics, U.: Edge dominating set and colorings on graphs with fixed clique-width. Discrete Applied Mathematics 126, 197–221 (2003)
20. Marx, D.: Precoloring extension on unit interval graphs. Discrete Appl. Math. 154, 995–1002 (2006)
21. Oum, S.-i.: Approximating rank-width and clique-width quickly. In: Kratsch, D. (ed.) WG 2005. LNCS, vol. 3787, pp. 49–58. Springer, Heidelberg (2005)
22. Roberts, F.: Indifference graphs. Proof Techniques in Graph Theory, 139–146 (1969); (Proc. Second Ann Arbor Graph Theory Conf., Ann Arbor, Mich. (1968)
23. Robertson, N., Seymour, P.D.: Graph minors. V. Excluding a planar graph. J. Combinatorial Theory Ser. B 41, 92–114 (1986)
24. Robertson, N., Seymour, P.D.: Graph Minors. XX. Wagner's conjecture. Journal of Combinatorial Theory Ser. B 92, 325–357 (2004)

New Results on the Most Significant Bit of Integer Multiplication

Beate Bollig and Jochen Klump

LS2 Informatik, TU Dortmund, 44221 Dortmund, Germany
`beate.bollig@uni-dortmund.de, jochen.klump@uni-dortmund.de`

Abstract. Integer multiplication as one of the basic arithmetic functions has been in the focus of several complexity theoretical investigations and ordered binary decision diagrams (OBDDs) are one of the most common dynamic data structures for Boolean functions. Analyzing the limits of symbolic graph algorithms for the reachability problem Sawitzki (2006) has presented the first exponential lower bound on the π-OBDD size for the most significant bit of integer multiplication according to one predefined variable order π. Since the choice of the variable order is a main issue to obtain OBDDs of small size the investigation is continued. As a result a new upper bound method and the first non-trivial upper bound on the size of OBDDs according to an arbitrary variable order is presented. Furthermore, Sawitzki's lower bound is improved.

1 Introduction and Results

Integer multiplication is certainly one of the most important functions in computer science and a lot of effort has been spent in designing good algorithms and small circuits and in determining its complexity. For one of the latest results see, e.g., [7]. When working with Boolean functions as in circuit verification, synthesis, and model checking, ordered binary decision diagrams, denoted OBDDs, introduced by Bryant (1986), are one of the most often used data structures supporting all fundamental operations on Boolean functions. Furthermore, in the last years a research branch has emerged which is concerned with the theoretical design and analysis of so-called symbolic algorithms which solve graph problems on OBDD-represented graph instances (see, e.g., [8], [11]). Although many exponential lower bounds on the OBDD size of Boolean functions are known and the lower bound methods are simple, it is often a more difficult task to prove large lower bounds for some predefined and interesting functions. Despite the well-known lower bounds on the OBDD size of the so-called middle bit of multiplication ([6], [14]), only recently it has been shown that the OBDD complexity of the most significant bit of integer multiplication is also exponential answering an open question posed by Wegener [13].

Definition 1. *Let $X_n = \{x_1, \ldots, x_n\}$ be a set of Boolean variables. A variable order π on X_n is a permutation on $\{1, \ldots, n\}$ leading to the ordered list $x_{\pi(1)}, \ldots, x_{\pi(n)}$ of the variables.*

S.-H. Hong, H. Nagamochi, and T. Fukunaga (Eds.): ISAAC 2008, LNCS 5369, pp. 883–894, 2008.
© Springer-Verlag Berlin Heidelberg 2008

In the following a variable order π is sometimes identified with the corresponding order $x_{\pi(1)}, \ldots, x_{\pi(n)}$ of the variables if the meaning in clear from the context.

Definition 2. *A π-OBDD on X_n is a directed acyclic graph $G = (V, E)$ whose sinks are labeled by Boolean constants and whose non sink (or inner) nodes are labeled by Boolean variables from X_n. Each inner node has two outgoing edges one labeled by 0 and the other by 1. The edges between inner nodes have to respect the variable order π, i.e., if an edge leads from an x_i-node to an x_j-node, $\pi^{-1}(i) \leq \pi^{-1}(j)$ (x_i precedes x_j in $x_{\pi(1)}, \ldots, x_{\pi(n)}$). Each node v represents a Boolean function $f_v : \{0,1\}^n \to \{0,1\}$ defined in the following way. In order to evaluate $f_v(b)$, $b \in \{0,1\}^n$, start at v. After reaching an x_i-node choose the outgoing edge with label b_i until a sink is reached. The label of this sink defines $f_v(b)$. The size of the π-OBDD G is equal to the number of its nodes and the π-OBDD size of a function f, denoted by π-OBDD(f), is the size of the minimal π-OBDD representing f.*

The size of the minimal π-OBDD representing a Boolean function f on n variables, i.e., $f \in B_n$, is described by the following structure theorem [12].

Theorem 1. *The number of $x_{\pi(i)}$-nodes of the π-OBDD for f is the number s_i of different subfunctions $f_{|x_{\pi(1)}=a_1, \ldots, x_{\pi(i-1)}=a_{i-1}}$, $a_1, \ldots, a_{i-1} \in \{0,1\}$, essentially depending on $x_{\pi(i)}$ (a function g depends essentially on a variable z if $g_{|z=0} \neq g_{|z=1}$).*

It is well known that the size of an OBDD representing a function f depends on the chosen variable order. Since in applications the variable order is not given in advance we have the freedom (and the problem) to choose a good or even an optimal order for the representation of f.

Definition 3. *The OBDD size or OBDD complexity of f is the minimum of all π-OBDD(f).*

Lower and upper bounds for integer multiplication are motivated by the general interest in the complexity of important arithmetic functions.

Definition 4. *The Boolean function $\mathrm{MUL}_{i,n} \in B_{2n}$ maps two n-bit integers $x = x_{n-1} \ldots x_0$ and $y = y_{n-1} \ldots y_0$ to the ith bit of their product, i.e., $\mathrm{MUL}_{i,n}(x, y) = z_i$, where $x \cdot y = z_{2n-1} \ldots z_0$ and x_0, y_0, z_0 denote the least significant bits.*

The bit z_{2n-1} is the most important bit of integer multiplication in the following sense. Since it has the highest value, for the approximation of the value of the product of two n-bit numbers x and y it is the most interesting one. On the other hand for space bounded models of computation z_{2n-1} is easy to compute in the sense that if it cannot be computed with size $s(n)$, then any other bit z_i, $2n - 1 > i \geq 0$, cannot be computed with size $s(i/4)$.

The first exponential lower bounds have been proved for the middle bit of integer multiplication $\mathrm{MUL}_{n-1,n}$. For OBDDs Bryant [6] has presented a lower bound of $2^{n/8}$ and Woelfel [14] has improved this lower bound to $\Omega(2^{n/2})$ by an approach using universal hashing. Exponential lower bounds for the middle bit of multiplication have also been proved for more general binary decision diagram models (see, e.g., [9], [3], [4], and [10]).

Despite these well-known lower bounds for the middle bit, only recently it has been shown that the OBDD complexity of the most significant bit of multiplication is exponential [2]. Since it is a monotone function it seems to be easier to compute than the middle bit. The known upper bounds on the OBDD size confirms this intuition. Amano and Maruoka [1] have presented an upper bound of $O(2^n)$ on the OBDD size of the most significant bit of multiplication according to the variable order $\pi = (x_{n-1}, y_{n-1}, x_{n-2}, y_{n-2}, \ldots, x_0, y_0)$, whereas the best known upper bound for the middle bit is $O(2^{(6/5)n})$ for the variable order $\pi = (x_0, y_0, x_1, y_1, \ldots, x_{n-1}, y_{n-1})$. In [14] an upper bound of $O(2^{(4/3)n})$ has been presented for the representation of the middle bit and the variable order $\pi = (x_0, \ldots, x_{n-1}, y_0, \ldots, y_{n-1})$. Furthermore, in the lower bound proofs on the OBDD size for $\mathrm{MUL}_{n-1,n}$ it has been shown that for an arbitrary variable order π there exists an assignment b to one of the input vectors such that the π-OBDD size for the resulting subfunction is exponential. In contrast it is not difficult to see that the π-OBDD size for any subfunction of $\mathrm{MUL}_{2n-1,n}$ where one of the input vectors is a constant is $O(n^2)$.

Computing the set of nodes that are reachable from some source $s \in V$ in a digraph $G = (V, E)$ is an important problem in computer-aided design, hardware verification, and model checking. Proving exponential lower bounds on the space complexity of a common class of OBDD-based algorithms for the reachability problem, Sawitzki [11] has presented the first exponential lower bound on the size of π-OBDDs representing the most significant bit of multiplication for the variable order π where the variables are tested according to increasing significance, i.e. $\pi = (x_0, y_0, x_1, y_1, \ldots, x_{n-1}, y_{n-1})$. For the lower bounds on the space complexity of the OBDD-based algorithms he has used the assumption that the output OBDDs use the same variable order as the input OBDDs. But in contrast, practical algorithms usually run variable reordering heuristics on intermediate OBDD results in order to minimize their size. Therefore, the OBDD complexity of $\mathrm{MUL}_{2n-1,n}$ is further investigated.

In this paper we present the following results:

- Let π be an arbitrary variable order, where all x-variables are before the y-variables or vice versa (denoted by $\pi = (x, y)$ or $\pi = (y, x)$). The π-OBDD size for the representation of $\mathrm{MUL}_{2n-1,n}$ is $\Theta(2^n)$.
- Let π be an arbitrary variable order. The π-OBDD size for the representation of $\mathrm{MUL}_{2n-1,n}$ is $O(2^{(4/3)n})$.
- Let $\pi = (x_{n-1}, y_{n-1}, x_{n-2}, y_{n-2}, \ldots, x_0, y_0)$. The π-OBDD size for the representation of $\mathrm{MUL}_{2n-1,n}$ is $O(2^{(4/5)n})$ and $\Omega(2^{n/4})$.

For the last result we construct a so-called fooling set and because of the symmetric definition of fooling sets we also improve Sawitzki's lower bound of $\Omega(2^{n/6})$ on the π-OBDD size of $\mathrm{MUL}_{2n-1,n}$ for the variable order $\pi = (x_0, y_0, x_1, y_1, \ldots, x_{n-1}, y_{n-1})$ [11] up to $\Omega(2^{n/4})$ using a much simpler proof.

In Section 2 the lower and upper bound on the size of π-OBDDs representing the most significant bit of integer multiplication according to a variable order $\pi = (x, y)$ are shown. The main result of the paper, the first non-trivial upper bound on the OBDD size of the most significant bit of integer multiplication, is presented in Section 3. For this result a new upper bound method is applied. Furthermore, in Section 4 a lower bound on the size of π-OBDDs according to the assumed best variable order $\pi = (x_{n-1}, y_{n-1}, x_{n-2}, y_{n-2}, \ldots, x_0, y_0)$ is given which is the best known lower bound on the size of OBDDs according to so-called interleaved variable orders, i.e., variable orders where the x- and y-variables are alternately ordered.

2 The π-OBDD Size for the Most Significant Bit of Integer Multiplication with Respect to $\pi = (x, y)$

In this section we prove that the π-OBDD size for $\mathrm{MUL}_{2n-1,n}$ is $\Theta(2^n)$, where $\pi = (x, y)$ (obviously the same can be shown for $\pi = (y, x)$).

Lemma 1. *Let $\pi = (x, y)$. The π-OBDD size for the representation of* $\mathrm{MUL}_{2n-1,n}$ *is $O(2^n)$.*

Proof. First, the x-variables are tested and the upper part of the OBDD is a complete binary tree of size 2^n. For the upper bound on the lower part of the OBDD we use the following fact. If one factor is given we only have to know the smallest value of the other one such that their product is at least 2^{2n-1}. Since there are only 2^i possible assignments to the remaining y-variables if $n - i$ y-variables have already been tested, we can conclude that the size of the lower part of the OBDD is at most 2^{n+1}. Altogether, we have shown that the OBDD size is $O(2^n)$. $\qquad\qquad\square$

Lemma 2. *Let $\pi = (x, y)$. The π-OBDD size for the representation of* $\mathrm{MUL}_{2n-1,n}$ *is $\Omega(2^n)$.*

Proof. We prove that there are at least $\Omega(2^n)$ different subfunctions of $\mathrm{MUL}_{2n-1,n}$ if all x-variables are replaced by constants. For this reason we use the following observations. An assignment to the x-variables can be seen as the binary representation of an integer. Let $a \in \{2^{n-1}+1, \ldots, 2^n - 1\}$. It is easy to see that the smallest integer b_a such that the product $a \cdot b_a$ is at least 2^{2n-1} is at least $a/2$. Otherwise $a \cdot b_a < a^2/2 < (2^n)^2/2 = 2^{2n-1}$. It is also easy to see that b_a is at most $2^n - 1$. If we can prove that for two arbitrary integers $a, a' \in \{2^{n-1}+1, \ldots, 2^n - 1\}$, where $a \geq a' + 2$, the corresponding smallest integers b_a and $b_{a'}$ such that the products $a \cdot b_a$ and $a' \cdot b_{a'}$ are at least 2^{2n-1} have to be different, we are done. Because of the definition of b_a we know that $(b_a - 1) \cdot a$ is less than 2^{2n-1}. Using the following inequations we can conclude that the product $a' \cdot b_a$ is less than 2^{2n-1}:

$$a \leq 2b_a \Leftrightarrow a \cdot b_a - 2b_a \leq a \cdot b_a - a \Leftrightarrow (a-2) \cdot b_a \leq a \cdot (b_a - 1).$$

Therefore, $a' \cdot b_a < 2^{2n-1}$ and $b_{a'} > b_a$. $\qquad\qquad\square$

3   Upper Bounds on the Size of OBDDs Representing the Most Significant Bit of Integer Multiplication

In this section we prove the main result of the paper. First, an upper bound on the size of OBDDs according to an arbitrary variable order representing the most significant bit of integer multiplication is presented. Afterwards, we prove the best known upper bound on the size of OBDDs representing $\mathrm{MUL}_{2n-1,n}$ according to the variable order $\pi = (x_{n-1}, y_{n-1}, x_{n-2}, y_{n-2}, \ldots, x_0, y_0)$.

We start our proof of the general upper bound with some simple but useful observations. Let $n \in \mathbb{N}$ be arbitrary but fixed in the rest of the paper.

Lemma 3. *Let $f : \mathbb{R}^+ \rightarrow \mathbb{R}^+$ be defined as $f(x) := \frac{2^{2n-1}}{x}$. For arbitrary $\Delta x, \Delta y > 0$ there exists exactly one value $x \in \mathbb{R}^+$ with $f(x) - f(x + \Delta x) = \Delta y$.*

Definition 5. *For $c, d \in \mathbb{R}$ and $n \in \mathbb{N}$ we define the function $f_{c,d} : \mathbb{R} \rightarrow \mathbb{R}$ in the following way:*

$$f_{c,d}(x) := \frac{2^{2n-1}}{c + x} - d.$$

A tuple (x, y) belongs to the functions $f_{c,d}$ or the function $f_{c,d}$ contains the tuple (x, y) iff $f_{c,d}(x) = y$.

Obviously the function value $f_{c,d}(x)$ is y, iff $c \neq -x$ and d satisfies

$$d = \frac{2^{2n-1}}{c + x} - y.$$

Since $f_{c,d}$ is considered throughout the whole proof we start our investigations with some of its properties.

Lemma 4. *Increasing the parameter c by Δc (parameter d by Δd) shifts the graph of the function $f_{c,d}$ Δc units to the left (Δd units downwards).*

Lemma 5. *Let (x_0, y_0) with $x_0, y_0 > 0$ be given and (c, d) and (c', d') with c and c' both positive, $c' < c$, and d respectively d' the corresponding parameters, for which (x_0, y_0) belongs to the functions $f_{c,d}$ and $f_{c',d'}$. Then the function value $f_{c',d'}(x)$ is greater than $f_{c,d}(x)$ for $0 \leq x < x_0$, and smaller for $x > x_0$.*

If we replace (c, d) to (c', d'), the curve of the function $f_{c,d}$ seems visually to rotate to the graph of the function $f_{c',d'}$, because point (x_0, y_0) stays on the graph, whereas all points left of x_0 are shifted upwards and the other ones downwards. Nevertheless the graph's shape does not change since the rotation can be decomposed to a vertical and a horizontal movement (see Figure 1).

Our proof idea of the upper bound on the size of OBDDs representing $\mathrm{MUL}_{2n-1,n}$ is to use the functions $f_{c,d}$ in order to analyze the number of different subfunctions of $\mathrm{MUL}_{2n-1,n}$ that can be obtained by replacing i x- and j y-variables to constants. For this reason we have to relate the functions $f_{c,d}$ to subfunctions of $\mathrm{MUL}_{2n-1,n}$.

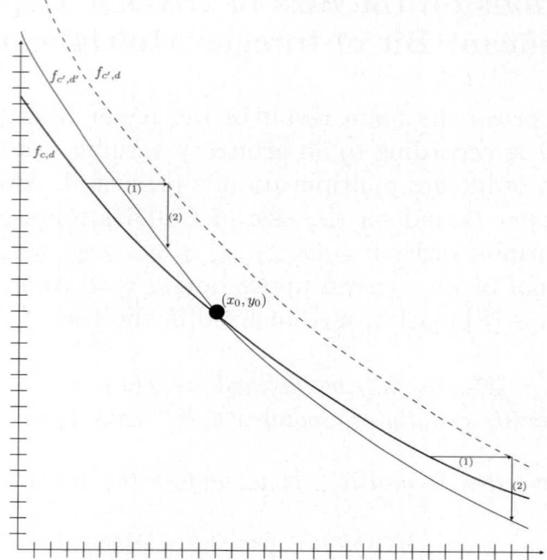

Fig. 1. Rotation of the graph of the function $f_{c,d}$

Definition 6. *For a given function $f_{c,d}$ and two arbitrary finite sets $A, B \subseteq \mathbb{N}$ and $A, B \neq \emptyset$ the corresponding step function $f_{c,d}^{A,B} : A \rightarrow B$ is defined as*

$$f_{c,d}^{A,B}(x) := \min\{y \in B | y \geq f_{c,d}(x)\}.$$

$\mathrm{MUL}_{2n-1,n}$ can be described by $f_{0,0}^{A,B}$, where $A, B = \{0, \ldots, 2^n - 1\}$. The tuples (x, y), $x \in A$ and $f_{0,0}^{A,B}(x) = y$, are significant points of $\mathrm{MUL}_{2n-1,n}$. Obviously there can be several functions $f_{c,d}$ that lead to the same step function. Our aim is to show that a not too large number of bits is sufficient to represent functions $f_{c,d}$ whose corresponding step functions identify the subfunctions of the most significant bit of integer multiplication. It is easy to see that each function $f_{c,d}$ can be characterized by two tuples (x_1, y_1) and (x_2, y_2), where $f_{c,d}(x_i) = y_i$ and $x_i, y_i \in \mathbb{R}$ for $i \in \{1, 2\}$, but the length of the numbers could be large. In order to find a small representation for $f_{c,d}$ our idea is to modify $f_{c,d}$ without changing essentially the corresponding step function.

We start to analyze the effect of moderate modifications of the parameters c and d.

Lemma 6. *Let $c, d \in \mathbb{R}^+$ and A, B be two arbitrary finite, nonempty subsets of $\mathbb{N}$. Let y be the largest element in B that is smaller than $f_{c,d}(x)$ and $\epsilon_x := f_{c,d}^{-1}(y) - x$, if y is defined, otherwise $\epsilon_x := \infty$. We define $\epsilon_{min} := \min\{\epsilon_x | x \in A\}$. Then $f_{c,d}^{A,B} = f_{c+\epsilon_{min}/2, d}^{A,B}$.*

Lemma 7. *Let $c, d \in \mathbb{R}^+$ and A, B be two arbitrary finite, nonempty subsets of $\mathbb{N}$. For $x \in A$ let $\epsilon_x := f_{c,d}^{A,B}(x) - f_{c,d}(x)$ if $f_{c,d}^{A,B}(x)$ is defined and ∞ otherwise. We define $\epsilon_{min} := \min\{\epsilon_x | x \in A\}$. Then $f_{c,d-\epsilon_{min}}^{A,B} = f_{c,d}^{A,B}$.*

Lemma 7 tells us that we are allowed to move the graph of the function $f_{c,d}$ upwards, right until it hits its corresponding step function for the first time, without changing the step function.

Lemma 8. *Let $c, d \in \mathbb{R}^+$ and A, B be two arbitrary finite, nonempty subsets of $\mathbb{N}$, such that there exists at least one element $x_0 \in A$ where $f_{c,d}(x_0) = f_{c,d}^{A,B}(x_0)$, and there are at least two elements $x_1, x_2 \in A$ where $\min\{y|y \in B\} < f_{c,d}(x_i) \leq \max\{y|y \in B\}$, $i \in \{1,2\}$. We define the following rotation operation for $f_{c,d}$ with respect to (x_0, y_0): decrease c continuously to c' and adjust d to d' at the same time such that $f_{c,d}(x_0) = f_{c',d'}(x_0)$ is always fulfilled until there exists another element $x' \in A$ with $f_{c',d'}(x') \in B$.*

1. The rotation operation is finite.

2. The function $f_{c,d}^{A,B}$ can be reconstructed from $f_{c',d'}$ in the following way:

$$f_{c,d}^{A,B}(x) = \begin{cases} \min\{y \in B|y \geq f_{c',d'}(x)\}, & \text{if } x \leq x_0, \\ \min\{y \in B|y > f_{c',d'}(x)\}, & \text{if } x > x_0. \end{cases}$$

Now applying Lemma 4 - 8 we are able to prove our general upper bound on the π-OBDD size for $\text{MUL}_{2n-1,n}$.

Theorem 2. *Let π be an arbitrary variable order. The π-OBDD size for the representation of $\text{MUL}_{2n-1,n}$ is $O(2^{(4/3)n})$.*

Proof. Let π be an arbitrary variable order. Our aim is to prove an upper bound of $2^{2(n-i)+2(n-j)+2}$ on the number of subfunctions of $\text{MUL}_{2n-1,n}$ which can be obtained by replacing i x- and j y-variables by constants.

If $i = 0$ or $j = 0$, we are done. Therefore, we assume that $i \neq 0$ and $j \neq 0$.

In the following let X_S be the set of i arbitrary x-variables and Y_S be the set of j arbitrary y-variables, $X_T := \{x_0, \ldots x_{n-1}\}\backslash X_S$, and $Y_T := \{y_0, \ldots, y_{n-1}\}\backslash Y_S$. Let a_S be an assignment to the X_S-variables, $a_S(x_k) \in \{0,1\}$ the assignment to $x_k \in X_S$, and $\|a_S\| := \sum_{x_k \in X_S} a_S(x_k) \cdot 2^k$. Let a_T, $\|a_T\|$, b_S, $\|b_S\|$, b_T, $\|b_T\|$ be defined in the same way.

$\text{MUL}_{2n-1,n}$ answers the question, whether for a given assignment (a, b) of the variables, the product $\|a\| \cdot \|b\|$ is at least 2^{2n-1}. Therefore, the function $\text{MUL}_{2n-1,n}$ can be described by specifying for every possible assignment a of the x-variables, the assignment b of the y-variables with $\|b\| = \left\lceil \frac{2^{2n-1}}{\|a\|} \right\rceil$. Figure 2 shows $\text{MUL}_{2n-1,n}$, where for a value $\|a\|$ the smallest corresponding value $\|b\|$ that fulfills $\text{MUL}_{2n-1,n}$ is dotted. Such pairs of assignments are called significant points. (For sake of simplicity the possible values are at least 2^{n-1} because for smaller numbers the product cannot be at least 2^{2n-1}.)

Let $c := \|a_S\|$ and $d := \|b_S\|$. We define A_T as the set of possible values $\|a_T\|$ that can be expressed by the variables from X_T. Let B_T be defined in the same way. These sets A_T and B_T are independent of the choice of c and d. A grid can be defined for the $\|a_T\|$- and $\|b_T\|$-values, which has the same appearance for all possible assignments c and d. A subfunction of $\text{MUL}_{2n-1,n}$

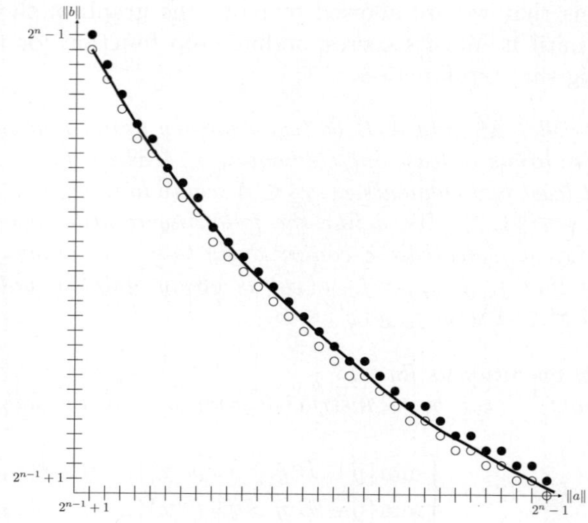

Fig. 2. Significant points for the evaluation of $MUL_{2n-1,n}$

obtained by replacing the variables in X_S to a_S and Y_S to b_S can be described by the pairs of A_T- and B_T-values $(\|a_T\|, \|b_T\|)$, so that $\|b_T\|$ is the minimal value that fulfills $\|b_T\| \geq \frac{2^{2n-1}}{c+\|a_T\|} - d$. Therefore, the subfunction of $MUL_{2n-1,n}$ can be characterized by the step function $f_{c,d}^{A_T,B_T}$ (see Definition 6) for the underlying function $f_{c,d}$. Figure 3 shows an example for two different step functions that result from two different assignments to the variables in $X_S \cup Y_S$.

Since the subfunctions obtained by replacing the variables in X_S and Y_S by constants can uniquely be described by their step functions, our aim is to prove the existence of a small representation such that the corresponding step function and therefore the corresponding subfunction of $MUL_{2n-1,n}$ can be reconstructed later on. Since each representation implicates at most one possible step function, the number of different representations is an upper bound on the number of different subfunctions. If the length of such a representation is not too large, there cannot be too many different representations.

The idea is to transform the function $f_{c,d}$ in a moderate way into a function $f_{c',d'}$, such that $f_{c',d'}$ contains at least two points from $A_T \times B_T$ and the step function $f_{c,d}^{A_T,B_T}$ can easily be obtained from $f_{c',d'}$. In the following we assume that for at least two A_T-values, the function $f_{c,d}$ is greater than 0 and smaller or equal to the greatest value in B_T. The other cases will be considered later on. If c equals 0, we have to make some extra considerations. Since the function $f_{c,d}$ is not defined for the value $\|a_T\| = 0$, we use Lemma 6 to move the graph a tiny distance to the left. As a result we obtain the function $f_{c',d}$ and $f_{c',d}^{A_T,B_T} = f_{c,d}^{A_T,B_T}$.

According to Lemma 7 we now move the graph upwards by decreasing the parameter d, right until the graph cuts the graph of its step function. Let $f_{c',d'}$ be the resulting function and $f_{c',d'}^{A_T,B_T}$ its step function. Obviously $f_{c',d'}^{A_T,B_T} = f_{c',d}^{A_T,B_T}$.

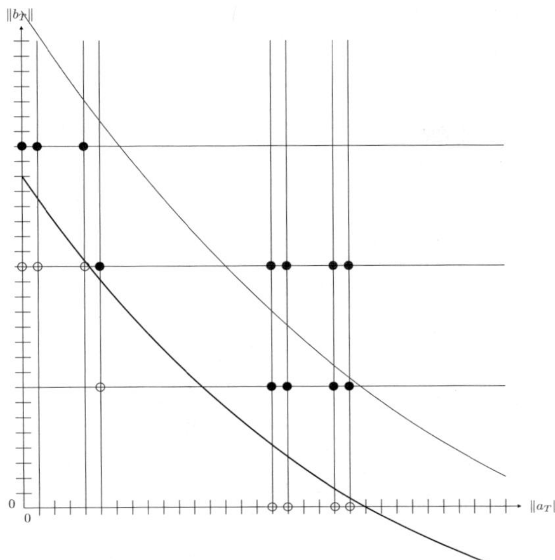

Fig. 3. Two different step functions

We now have at least one position $p_1 \in A_T$, so that $f_{c',d'}(p_1) = f^{A_T,B_T}_{c',d'}(p_1) = q_1$. If $f_{c',d'}$ contains another point $(p_2, q_2) \in A_T \times B_T$, we can be sure that q_2 is not equal to q_1 because the function is strictly monotonic. In this case we stop the transformation and encode the step function $f^{A,B}_{c',d'}$ by the triple $((p_1, q_1), (p_2, q_2), z)$ and $z = 1$, where z indicates that we stopped at this point.

In all other cases we modify the function $f_{c',d'}$ again to hit a second point of $A_T \times B_T$. Using Lemma 8 we rotate the graph clockwise by decreasing c' and adjusting d', so that the point (p_1, q_1) stays on the graph. We get a new function $f_{c'',d''}$ and another point $(p_2, q_2) \in A_T \times B_T$ with $f_{c'',d''}(p_2) = q_2$.

Now we have achieved that the function $f_{c'',d''}$ contains two points (p_1, q_1) and (p_2, q_2) that can be addressed by the variables in $X_T \cup Y_T$. In order to use Lemma 3, we have to be sure that the distance between these points can be specified without knowledge of the assignment to the variables in $X_S \cup Y_S$. Therefore let (c^*, d^*) be an arbitrary assignment to these variables. In composition with the assignments to the variables in $X_T \cup Y_T$ we can calculate the decimal value of the investigated points:

$$\|p_1^*\| := \|c^*\| + \|p_1\|, \qquad \|q_1^*\| := \|d^*\| + \|q_1\|,$$
$$\|p_2^*\| := \|c^*\| + \|p_2\|, \qquad \|q_2^*\| := \|d^*\| + \|q_2\|.$$

Obviously the distances $\Delta p = |\,\|p_1^*\| - \|p_2^*\|\,|$ and $\Delta q = |\,\|q_1^*\| - \|q_2^*\|\,|$, are independent of (c^*, d^*). Moreover, we have to be sure, that (p_1, q_1) and (p_2, q_2) can be used to identify a so-called shifted cutting of the initial graph $\frac{2^{2n-1}}{x}$, i.e., $\frac{2^{2n-1}}{x} \to \frac{2^{2n-1}}{c''+x} - d''$, with positive numbers in the denominator. The modification

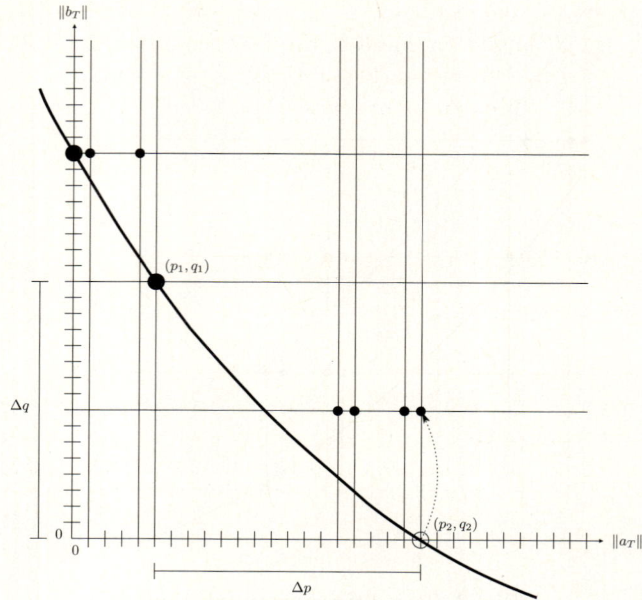

Fig. 4. Reconstruction of the step function

of d is not critical, because it does not have any influence on the denominator. For the values c we assure at the beginning that c is greater that 0 (either because $c = \|a_S\|$ is greater than 0 or by using $\epsilon_{min}/2$). Just the rotation operation decreases c. But as we continuously check, whether the value of $f_{c'',d''}$ hits a point in $A_T \times B_T$, it is impossible that the function's pole will be translated across any point of the grid. Therefore, Lemma 3 can be used to identify the underlying function $f_{c'',d''}$ with (p_1, q_1) and (p_2, q_2).

Our last step is now the reconstruction of the original step function $f_{c,d}^{A_T,B_T}$. If we have just moved the graph upwards without rotating it, then for every $x \in A_T$ the corresponding value of the step function $f_{c,d}^{A_T,B_T}$ is the smallest value of B_T that is at least $f_{c',d'}(x)$. In the other case we can use the second statement of Lemma 8 to reconstruct the original step function. Figure 4 illustrates the reconstruction of the step function $f_{c,d}^{A_T,B_T}$.

As we have seen a triple that consists of two points and an additional bit z can encode any possible step function that itself represents a subfunction of $MUL_{2n-1,n}$ obtained by replacing i x- and j y-variables by constants. As this subfunction can by uniquely reconstructed by this representative, there cannot be two different subfunctions with the same representations. The maximal number of these representations is

$$\underbrace{2^{n-i} \cdot 2^{n-j}}_{(p_1,q_1)} \cdot \underbrace{2^{n-i} \cdot 2^{n-j}}_{(p_2,q_2)} \cdot \underbrace{2}_{\text{bit } z} = 2^{2(n-i)+2(n-j)+1}.$$

Up to now we have assumed that for at least two A_T-values the function $f_{c,d}$ is greater than 0 and smaller or equal to the greatest value in B_T. A subfunction that is not of this type can be characterized by only one point (p, q) of the step function $f_{c,d}^{A_T, B_T}$. So there can be at most

$$\underbrace{2^{n-i}}_{p} \cdot \underbrace{2^{n-j}}_{q} < 2^{2(n-i)+2(n-j)+1}$$

of these functions. All in all there are less than

$$2^{2(n-i)+2(n-j)+1} + 2^{2(n-i)+2(n-j)+1} = 2^{2(n-i)+2(n-j)+2}$$

different subfunctions.

Obviously there are at most 2^{i+j} different subfunctions obtained by the replacement of $i + j$ variables by constants. Using the minimum of the two upper bounds for each layer we obtain the result that the π-OBDD size for $\mathrm{MUL}_{2n-1,n}$ is $O(2^{(4/3)n})$ for any variable order π. $\qquad\square$

Corollary 1. *Let* $\pi = (x_{n-1}, y_{n-1}, x_{n-2}, y_{n-2}, \ldots, x_0, y_0)$. *The π-OBDD size for the representation of* $\mathrm{MUL}_{2n-1,n}$ *is* $O(2^{(4/5)n})$.

Proof. In [1] it has been proved that the number of subfunctions obtained by replacing the variables $x_{n-1}, \ldots, x_{n-i}$ and $y_{n-1}, \ldots, y_{n-i}$ by constants is at most 2^{i+6} and therefore the number of subfunctions obtained by replacing the variables $x_{n-1}, \ldots, x_{n-(i+1)}$ and $y_{n-1}, \ldots, y_{n-i}$ can be at most 2^{i+7}. From the proof of Theorem 2 we know that there are at most $2^{4n-4i+2}$ different subfunctions after replacing these variables by constants. Using the minimum of the two upper bounds for each layer we can prove that the π-OBDD size for $\mathrm{MUL}_{2n-1,n}$ is $O(2^{(4/5)n})$. $\qquad\square$

4　A Lower Bound on the Size of π-OBDDs Representing $\mathrm{MUL}_{2n-1,n}$ with Respect to $\pi = (x_{n-1}, y_{n-1}, x_{n-2}, y_{n-2}, \ldots, x_0, y_0)$

Using techniques from analytical number theory Sawitzki [11] has presented a lower bound of $2^{n/6}$ on the size of π-OBDDs representing the most significant bit of integer multiplication for the variable order π where the variables are tested according to increasing significance, i.e. $\pi = (x_0, y_0, x_1, y_1, \ldots, x_{n-1}, y_{n-1})$. A larger lower bound can be proved in an easier way and without analytical number theory using the following two observations. For a number $2^{n-1} + \ell 2^{n/2}$, $\ell \leq 2^{n/4-1}$, the corresponding smallest number such that the product of the two numbers is at least 2^{2n-1} is $2^n - \ell 2^{n/2+1} + 4\ell^2 - \left\lfloor \frac{4\ell^3}{2^{n/2-1}+\ell} \right\rfloor$. Furthermore,

$$2^{n/2} > 4\ell^2 - \left\lfloor \frac{4\ell^3}{2^{n/2-1} + \ell} \right\rfloor > 4(\ell - 1)^2$$

for $0 < \ell \leq 2^{n/4-1}$. Using these two facts it is not difficult to construct a so-called fooling set of size $2^{n/4-1}$ and therefore to prove a lower bound of $\Omega(2^{n/4})$ on the π-OBDD size. The same facts can be used for the proof of a lower bound of $2^{n/4-1}$ on the size of π'-OBDDs for the most significant bit, where $\pi' = (x_{n-1}, y_{n-1}, x_{n-2}, y_{n-2}, \ldots, x_0, y_0)$.

The next challenge is to improve the general lower bound on the OBDD complexity of $MUL_{2n-1,n}$ which is up to now $\Omega(2^{n/288})$. Furthermore, the complexity of $MUL_{2n-1,n}$ for more general models than OBDDs is open.

References

1. Amano, K., Maruoka, A.: Better upper bounds on the QOBDD size of integer multiplication. Discrete Applied Mathematics 155, 1224–1232 (2007)
2. Bollig, B.: On the OBDD complexity of the most significant bit of integer multiplication. In: Agrawal, M., Du, D.-Z., Duan, Z., Li, A. (eds.) TAMC 2008. LNCS, vol. 4978, pp. 306–317. Springer, Heidelberg (2008)
3. Bollig, B., Woelfel, P.: A read-once branching program lower bound of $\Omega(2^{n/4})$ for integer multiplication using universal hashing. In: Proc. of 33rd STOC, pp. 419–424 (2001)
4. Bollig, B., Waack, St., Woelfel, P.: Parity graph-driven read-once branching programs and an exponential lower bound for integer multiplication. Theoretical Computer Science 362, 86–99 (2006)
5. Bryant, R.E.: Graph-based algorithms for boolean function manipulation. IEEE Trans. on Computers 35, 677–691 (1986)
6. Bryant, R.E.: On the complexity of VLSI implementations and graph representations of boolean functions with application to integer multiplication. IEEE Trans. on Computers 40, 205–213 (1991)
7. Führer, M.: Faster integer multiplication. In: Proc. of 39th STOC, pp. 57–66 (2007)
8. Gentilini, R., Piazza, C., Policriti, A.: Computing strongly connected components in a linear number of symbolic steps. In: Proc. of SODA, pp. 573–582. ACM Press, New York (2003)
9. Ponzio, S.: A lower bound for integer multiplication with read-once branching programs. SIAM Journal on Computing 28, 798–815 (1998)
10. Sauerhoff, M., Woelfel, P.: Time-space trade-off lower bounds for integer multiplication and graphs of arithmetic functions. In: Proc. of 35th STOC, pp. 186–195 (2003)
11. Sawitzki, D.: Exponential lower bounds on the space complexity of OBDD-based graph algorithms. In: Correa, J.R., Hevia, A., Kiwi, M. (eds.) LATIN 2006. LNCS, vol. 3887, pp. 781–792. Springer, Heidelberg (2006)
12. Sieling, D., Wegener, I.: NC-algorithms for operations on binary decision diagrams. Parallel Processing Letters 48, 139–144 (1993)
13. Wegener, I.: Branching Programs and Binary Decision Diagrams - Theory and Applications. SIAM Monographs on Discrete Mathematics and Applications (2000)
14. Woelfel, P.: New bounds on the OBDD-size of integer multiplication via universal hashing. Journal of Computer and System Science 71(4), 520–534 (2005)

Sorting with Complete Networks of Stacks

Felix G. König and Marco E. Lübbecke

Technische Universität Berlin, Institut für Mathematik, MA 5-1
Straße des 17. Juni 136, 10623 Berlin, Germany
{fkoenig,m.luebbecke}@math.tu-berlin.de

Abstract. Knuth introduced the problem of sorting numbers using a sequence of stacks. Tarjan extended this idea to sorting with acyclic networks of stacks (and queues), where items to be sorted move from a source through the network to a sink while they may be stored temporarily at nodes (the stacks). Both characterized which permutations are *sortable* this way; but they ignored the associated optimization problem (minimize the number of moves) and its complexity.

Given a complete, thus cyclic, network of $k \geq 2$ stacks, any permutation is obviously sortable. The important question now is how to actually sort with a minimum number of *shuffles*, i.e., moves in between stacks. This is a natural algorithmic complement to the structural questions asked by Knuth, Tarjan, and others. It is the first time shuffles are considered in stack sorting—despite of the great practical importance of this optimization version.

We show that it is NP-hard to approximate the minimum number of shuffles within $\mathcal{O}(n^{1-\varepsilon})$ for networks of $k \geq 4$ stacks, even when the problem is restricted to complete networks, by relating stack sorting to MIN k-PARTITION on circle graphs (for which we prove a stronger inapproximability result of independent interest). For complete networks, a simple merge sort algorithm achieves an approximation ratio of $\mathcal{O}(n \log n)$ for $k \geq 3$; however, closing the logarithmic gap to our lower bound appears to be an intriguing open question. Yet, on the positive side, we present a tight approximation algorithm which computes a solution with a linear approximation guarantee, using a resource augmentation to $\alpha k + 1$ stacks, given an α-approximation algorithm for coloring circle graphs.

When there are constraints as to which items may be placed on top of each other, deciding about sortability becomes non-trivial again. We show that this problem is PSPACE-complete, for every fixed $k \geq 3$.

1   Introduction

Stacks, as a fundamental data structure, play an important role in theoretical computer science, artificial intelligence, and combinatorial optimization. At the same time, stacks also model a wide range of applications in logistics and motion planning, where the access to items is restricted in a last-in first-out fashion.

We investigate a problem, in which k stacks are used for sorting. The k stacks form a directed network which we assume to be complete for most of the paper. A permutation of items to be sorted is given at a source node, and all items must arrive at a sink in correct order. Items may move along arcs in the network and may be stored temporarily on a stack at each node. When the network contains a cycle, any permutation can be

S.-H. Hong, H. Nagamochi, and T. Fukunaga (Eds.): ISAAC 2008, LNCS 5369, pp. 895–906, 2008.

sorted. Popping an item from one stack, moving it along an arc and then pushing it to the next stack is called a *shuffle*, and our goal is to minimize the number of shuffles needed to sort the given permutation.

Our Contribution. Our paper is the first primarily algorithmic view on stack sorting; it explicitly captures the essence of shuffles in sorting with a network of stacks. We prove that it is NP-hard to approximate the minimum number of shuffles within $\mathcal{O}(n^{1-\varepsilon})$ for $k \geq 4$ and any fixed $\varepsilon > 0$, even when the network of stacks is restricted to be complete, by relating STACK SORTING to the MIN k-PARTITION problem on circle graphs. For the latter problem we prove inapproximability within $\mathcal{O}(n^{2-\varepsilon})$ as an intermediate result which is of interest in its own. For the case of complete networks and $k \geq 3$, a simple merge sort algorithm computes an $\mathcal{O}(n \log n)$-approximation, but closing the gap to our lower bound appears to require significantly new insight into the problem (or into graph coloring, a we discuss). Still, we present an $\mathcal{O}(n)$-approximation algorithm which needs a resource augmentation to $\alpha k + 1$ stacks instead of only k, using an α-approximation algorithm for coloring circle graphs. We discuss that this is best possible in a certain sense. Furthermore, we prove that it is PSPACE-complete to decide whether a given permutation is sortable using a complete network of $k \geq 3$ stacks, when there are constraints as to which items may be placed on top of one another. We conclude with several challenging open problems.

Our results have direct consequences for various practical stacking problems from the operations research literature, e.g., [2,3,6,9,16], as well as for blocks world models in artificial intelligence [12,17], among others.

Related Work

Stack Sorting. Knuth introduced the idea of stack sorting using the language of railway sidings [15]; he characterized permutations which can be sorted using k stacks in series. Tarjan extended these ideas to sorting with acyclic networks of stacks (and queues) [18]. Even and Itai considered the sortability of permutations using k parallel stacks [7]. They related this question to the problem of deciding k-colorability of a circle graph, which was proven to be NP-complete for $k \geq 4$ by Unger [19].

In all these papers (and those which followed), an item, once popped from a stack, may never be pushed back on it again. The point of interest has always been a characterization of which permutations can be sorted using a particular configuration of stacks. This (mathematically beautiful) point of view is surveyed by Bóna in [4]; he states, that "virtually nothing can be proved" for general networks of stacks.

Inspired by personal discussions about our work on this paper, Felsner and Pergel recently considered stack sorting from the perspective of extremal combinatorics [8]. Assuming that all items have to be moved to stacks before the first item may move to the sink, they consider instances in which an optimal solution has a maximum number of shuffles. They give good bounds on this number for different magnitudes of k.

Applications. In recent years, the operations research literature dealt with a number of practical applications involving stack sorting: Assigning incoming trains [6] or trams [3] to tracks of a switching yard or depot; parking buses in parking lots [9]; and stowage planning for container ships [2], to mention only a few. All of these ask whether it is possible to assign items to stacks such that items can be retrieved in a desired order without blocking each other. Even though shuffles are a natural part of real life stacking, researchers asked for sortability with parallel stacks, where shuffles are not an issue, instead of asking for sorting with few shuffles using a complete network of stacks, which would model many of the above applications more accurately: shunting rail cars with fewest moves, stowing containers with minimal number of re-stacking operations, etc. Our paper is a first step towards investigating these optimization problems.

König et al. have recently introduced a mathematical model and a heuristic for a particularly rich stacking problem which has the minimization of shuffles as objective. They model in great detail stacking problems occurring in the logistics of integrated steel production and in container terminal operation [16], which among other things include limited stack heights. PSPACE-completeness of deciding sortability is shown.

2   Problem Formulation and Relations to Graph Coloring

A formal definition of STACK SORTING is as follows. We are given a directed graph $G = (V \cup \{s,t\}, E)$ where s has no in- and t has no out-edges. To avoid trivialities, we require that any $v \in V$ is on an s-t-path in G, $|V| = k \geq 2$, and that G contains a cycle. A permutation π of items $1, \ldots, n$ (the input sequence) is given at s, and has to be sorted, i.e., all items have to arrive at t in ascending order. Items may only move along arcs in G. When an item arrives at node $v \neq t$, it is stored on a stack S_v. A stack may be accessed on one end only, its *top*, so items may only leave in the reverse order they arrived. Naturally, items may only leave s in the order prescribed by π. Whenever an item moves along an arc (v, w) where $v \neq s$ and $w \neq t$, we say it is *shuffled*. The question is how to move items such that all items arrive at t in the correct order, using the smallest number of shuffles?

When $(s,v), (v,t) \in E$ for all $v \in V$, i.e., items can be moved from the source to any stack and from any stack to the sink, there is an interesting relationship between STACK SORTING and graph coloring. A graph coloring is an assignment of colors to the nodes of a graph. We call a coloring *proper*, if nodes which share an edge receive different colors; a k-coloring uses at most k colors. A *circle graph* is a graph the nodes of which can be drawn as chords of a circle such that two chords intersect iff the corresponding nodes share an edge. Even and Itai noted that deciding if a permutation π is sortable with k parallel stacks, i.e., $E = \{(s,v) : v \in V\} \cup \{(v,t) : v \in V\}$, is equivalent to deciding k-colorability of a circle graph the nodes of which are the items in π [7], which is hard for $k \geq 4$ [19]. Obviously, shuffles are impossible in the case of parallel stacks.

The class of circle graphs is equivalent to the so called *overlap graphs* [10]: Their nodes can be represented as intervals such that two nodes share an edge iff their corresponding intervals intersect but none of the two contains the other. With the latter representation, the correspondence of k-colorings to sorting with k parallel stacks becomes clear immediately: We define n intervals with unique start and end points in a discrete set of $2n$ points in time. The start points of the intervals are ordered corresponding to π. Immediately after the start of an interval, we insert the endpoints of all intervals corresponding to items i such that all intervals corresponding to items $1, \ldots, i$ have already started, in ascending order. When two nodes do not share an edge, either their intervals

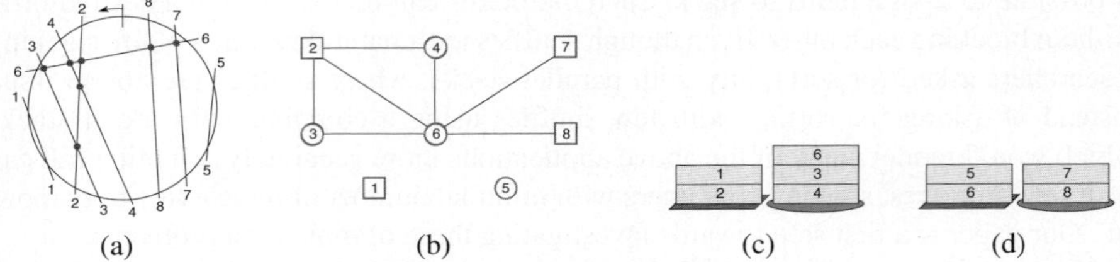

Fig. 1. Optimally sorting $\pi = 24361875$ with one shuffle using $k = 2$ stacks; (a) shows the circle representation of the permutation, (b) the circle graph with a shuffle-optimal coloring of the nodes; (c) depicts the assignment of items to stacks before the shuffle, (d) the assignment after the shuffle and after moving the first four items out and all remaining items to the stacks

do not intersect, which means the latter of the corresponding items arrives after the first has already left the buffer, or one of their intervals contains the other, which means that the corresponding items arrive at a stack in the reverse order they need to leave it. In both cases, the items can be put on the same stack. In any other case, the two items would block each other leaving the stack, which could be a reason for not putting them on the same stack.

In a proper k-coloring of the nodes of the circle graph, the color of a node determines to which of the k stacks the corresponding item has to be moved to from s in order for all items to be able to arrive at t in correct order without shuffles. We now make the important observation that, if a proper k-coloring is impossible, *monochromatic* edges are unavoidable (edges with both endpoints in the same color), and this relates STACK SORTING to another coloring problem, the MIN k-PARTITION problem. The following example, cf. Fig. 1, illustrates the connection between circle graphs and STACK SORTING assuming G to be complete, i.e., $E = \{(s,v) : v \in V\} \cup \{(v,w) : v, w \in V\} \cup \{(v,t) : v \in V\}$. It also demonstrates that, quite counterintuitively, while a proper k-coloring of a circle graph does correspond to a solution to STACK SORTING without shuffles, asking for a coloring with fewest monochromatic edges is not the same as asking for few shuffles.

The circle graph in Fig. 1(b) is clearly not 2-colorable. So obviously, at least one shuffle is necessary. The improper 2-coloring of the nodes represents the sorting of the permutation on two stacks shown in 1(c) and 1(d), which has exactly one shuffle and thus is optimal. The coloring has two monochromatic edges and is suboptimal in this sense: Changing the color of node 6 from green (circle) to red (square) would yield a coloring with just one monochromatic edge. However, the sorting of the permutation implied by this new coloring would have at least two shuffles.

At the core of the relationship between monochromatic edges in a coloring of a circle graph and the number of shuffles in sorting with complete networks of stacks lies the following observation: Suppose in a k-coloring of a circle graph, there is a color class with c nodes and many, say $c - 1$, monochromatic edges. Consider a stack containing the corresponding c items. On one hand, all monochromatic edges could be incident to the item on top of the stack. In this case, shuffling this one item would, so to speak, resolve all $c - 1$ monochromatic edges. On the other hand, each of the $c - 1$ monochromatic edges could connect two neighboring items on the stack—in this case, at least $\frac{c}{2}$ shuffles would be necessary. The point is, that in a coloring, the number of monochromatic edges for one color class does not depend on a certain ordering of its nodes, while on a stack, the order of its items is the single most important factor determining the number of shuffles necessary.

Yet, as we will see, the circle graph representation for sorting with complete networks of stacks is useful for proving hardness of approximation for STACK SORTING.

3 Hardness of Approximation

We show that it is NP-hard to approximate the minimum number of shuffles in STACK SORTING within a factor better than $\mathcal{O}(n)$. We start with the MIN k-PARTITION problem on graphs: one asks for deleting a minimum number of edges such that the remaining graph is k-colorable. We prove a strong inapproximability for this problem on circle graphs using ideas from [14] where the same result was shown for dense graphs.

Theorem 1. *Let G be a circle graph. For any $k \geq 4$, it is NP-hard to approximate the minimum number of monochromatic edges $\gamma(G,k)$ in a k-coloring of G within $\mathcal{O}(n^{2-\varepsilon})$.*

Proof. Let $I = (H,k)$ an instance of the k-COLORING problem for a circle graph $H = (V',E')$ with n nodes. Deciding if I is a yes-instance (which is equivalent to deciding if the minimum number γ^* of monochromatic edges in a k-coloring of H is 0) is NP-complete [19]. We will construct an instance $J = (G,k)$ of MIN k-PARTITION where G is a circle graph with N nodes, such that approximating the minimum number $\gamma(G,k)$ of monochromatic edges in a k-coloring of G within $\mathcal{O}(N^{2-\varepsilon})$ is equivalent to deciding $\gamma^* = 0$, and thus to deciding I. In other words, our construction will create a quadratic gap in the possible optimal values of J, thus amplifying the hardness of deciding I to the hardness of approximating the optimal value of J.

Let $s := n^{\frac{2}{\varepsilon}-1}$. We construct $G = (V,E)$ as follows:

$$V := \{v_1, v_2, \ldots, v_s : v \in V'\},$$
$$E := \{(v_i, w_j) : (v,w) \in E'; \ i,j = 1, \ldots, s; \ i \neq j\}.$$

In terms of the chord diagram of the circle graph, we obtain G from H by replacing each chord in H by s parallel chords in G. So G is a circle graph with $N := sn$ nodes and $s^2 m$ edges.

Now every k-coloring c of G can be transformed to a coloring c' in which all copies $v_i \in V$ of a node $v \in V'$ have the same color without increasing the number of monochromatic edges: For each $v \in V'$, pick the $v_i \in V$ with the fewest incident monochromatic edges and color all of v's copies in v_i's color. By this, it follows immediately that every optimal k-coloring c of G either has no or at least s^2 monochromatic edges: When there is at least one monochromatic edge, compute c' from c as described above. Let (v_i, w_j) be a monochromatic edge in c'. Since all s copies of w_j are neighbors of all s copies of v_i, there are at least s^2 monochromatic edges. Furthermore, we have

$$s^2 = n^{\frac{4}{\varepsilon}-2} = n^{\frac{2}{\varepsilon}(2-\varepsilon)} = N^{2-\varepsilon}.$$

Now suppose there was an algorithm computing an $\mathcal{O}(N^{2-\varepsilon})$-approximate solution to J with γ monochromatic edges. Then,

- $\gamma \in o(N^{2-\varepsilon}) \Rightarrow \gamma^* = 0 \Rightarrow H$ is k-colorable
- $\gamma \in \Omega(N^{2-\varepsilon}) \Rightarrow \gamma^* \geq 1 \Rightarrow H$ is not k-colorable.

Thus, such an algorithm would decide I. $\qquad\square$

We now establish some bounds relating monochromatic edges in the coloring of a circle graph and shuffles in solutions to the corresponding instance of STACK SORTING.

Lemma 1. *Let ℓ be a solution to an instance I of STACK SORTING with k stacks requiring L shuffles. Let $c_\ell : V \to \{1,\ldots,k\}$ be the coloring of the corresponding circle graph $G = (V,E)$ obtained by assigning each node the color corresponding to the stack its item was first placed on, and let γ_ℓ denote the number of monochromatic edges in c_ℓ. Then,*

$$\gamma_\ell \leq (n-1) \cdot L.$$

Proof. From the correspondence between G and I it is clear, that for each monochromatic edge in c_ℓ, at least one of the items corresponding with its end points must be shuffled from its original stack in order to move both items to the output sequence. On the other hand, each item can only be incident to at most $n-1$ other items in G, so each shuffle can only "pay" for the need to shuffle items of at most $n-1$ monochromatic edges. $\qquad\square$

Lemma 2. *Let $c : V \to \{1,\ldots,k\}$ be a coloring of a circle graph $G = (V,E)$ with γ monochromatic edges. Let I be the instance of STACK SORTING with a complete network of stacks corresponding to G. One can easily obtain a solution ℓ_c to I with $L_c = 2 \cdot \gamma$ shuffles from c.*

Proof. The construction of ℓ_c happens in phases. A phase p ends whenever a prefix a_p of the identity permutation of all items not moved to the sink yet has been removed from the source, e.g., phase one ends immediately after item 1 has been removed from the source. During each phase, items are moved from the source to stacks as prescribed by c. When phase p ends, we move all items in a_p to the sink before continuing with the next phase: If the item needed next, say i, is not on top of its stack, we shuffle away all items above it to some arbitrary stack, move i to the sink, and then reverse all shuffles.

Note that for any item j which has to be shuffled in order to access i, we have $i < j$. On the other hand, when j was removed from the source, not all items m with $m < i$ had been removed from the source yet (otherwise i would have been moved to the sink in a previous phase). So from the correspondence between G and I, it is clear that $(i,j) \in E$, and this edge is monochromatic in c since i and j are on the same stack. So for each two shuffles in ℓ_c, there is a distinct monochromatic edge in c. $\qquad\square$

We are now ready to proof the main result of this section.

Theorem 2. *Approximating the minimum number L^* of shuffles in STACK SORTING within $\mathcal{O}(n^{1-\varepsilon})$ is NP-hard, even when restricted to complete graphs with a fixed number of $k \geq 4$ stacks.*

Proof. For an arbitrary circle graph G, let I denote the corresponding instance of STACK SORTING with a complete network of $k \geq 4$ stacks. Now suppose there was an $n^{1-\varepsilon}$-approximation algorithm for STACK SORTING. Then, for any instance, we can compute a solution ℓ with $L \leq n^{1-\varepsilon}L^*$ shuffles, where L^* denotes the number of shuffles in an optimal solution. As before, let c_ℓ be the coloring with γ_ℓ monochromatic edges obtained from assigning each node in G the color corresponding to the stack which the corresponding item was first placed on in ℓ. By Lem. 1, we have

$$\gamma_\ell \ \leq \ (n-1) \cdot L \ \leq \ n^{2-\varepsilon}L^* \ \leq \ n^{2-\varepsilon}L_{c^*} \ \leq \ 2 \cdot n^{2-\varepsilon} \cdot \gamma^*$$

where c^* denotes a coloring of G with the minimum number γ^* of monochromatic edges. The last inequalities follow from the fact that L^* is the minimum number of shuffles possible and Lem. 2.

Thus, an $\mathcal{O}(n^{1-\varepsilon})$-approximation algorithm for STACK SORTING immediately implies an $\mathcal{O}(n^{2-\varepsilon})$-approximation algorithms for MIN k-PARTITION on circle graphs. $\square$

Due to the generality of STACK SORTING, Thm. 2 has numerous consequences: The hardness of approximation immediately carries over to most applications involving sorting with stacks, and also to many blocks world planning models in artificial intelligence where table capacity, i.e., the number of stacks, is limited.

4   Approximation Algorithms for Complete Networks

We will now state our positive results. Even though their tightness w.r.t. our hardness result is unsatisfactory, we will discuss that they are the best we may currently hope for.

Fact 1. STACK SORTING *with* $k \geq 3$ *stacks can be done with* $2 \cdot (n \log n)$ *shuffles.*

This fact follows immediately from the application of a merge sort algorithm on stacks. An elaborate proof of this can be found in [8], where the authors also argue that there are instances for which $\Omega(n \log n)$ shuffles are needed when k is constant.

Remark 1. It follows from Fact 1, that in order to close the gap to the lower bound from Thm. 2, it would suffice to obtain an algorithm A which has a linear approximation guarantee only for instances, for which an optimal solution requires $o(\log n)$ shuffles. Returning the better solution of algorithm A and the mentioned merge sort would result in a linear approximation algorithm.

Closing the gap between hardness of sublinear approximation and the straight-forward $\mathcal{O}(n \log n)$ approximation appears to be very intriguing. One way to achieve this is at the expense of a resource augmentation.

Theorem 3. *There is an efficient algorithm which computes a solution with* $3n \cdot L^*$ *shuffles using* $\alpha k + 1$ *stacks where* L^* *denotes the minimum number of shuffles using k stacks, given an* α-*approximation algorithm for coloring circle graphs.*

Proof. Our algorithm proceeds in phases, the lengths of which are determined by repeatedly employing an α-approximation algorithm A for coloring circle graphs. Iteratively considering longer prefixes of π, we apply A to the circle graph defined by the current prefix, to determine the longest prefix which is still αk-colorable by A. We move the corresponding set P of items of the current phase to the first αk stacks according to the computed αk-coloring, meanwhile moving items to the output whenever possible. Since there may be items, which the coloring of the subgraph of this phase assumes to be movable to the output, but which we cannot actually move out yet, we use stack $S_{\alpha k+1}$ to store these items temporarily.

As a result, the items in P on each stack are now ordered from top to bottom on the first αk stacks and from bottom to top on $S_{\alpha k+1}$. We will now perform shuffles in order to have all these items on one stack, linearly ordered from bottom to top.

Let S_1 be the stack containing the set Q of all items from previous phases. We pick another arbitrary stack S_2 and merge all items in $X := P \cap (S_1 \cup S_2)$ onto a third stack S_3. We then merge all items in $Q \cup X \cup S_{\alpha k+1}$ onto S_2, now having all stacks ordered from top to bottom and S_1 empty. Finally, we merge all items from all stacks onto S_1 while moving items to the output whenever possible and, quite importantly, inserting the next item in π at the correct position, thus having sorted a part of π for which A could not find an αk-coloring. We have obtained a stack containing all items up to the current phase ordered from bottom to top.

Note that in one phase, we have shuffled each item at most three times. Hence, our algorithm needs $L \leq 3pn$ shuffles where p denotes the number of phases.

On the other hand, any solution to STACK SORTING requires at least one shuffle for each phase of our algorithm: Whenever the number of colors needed by A exceeds αk, the chromatic number of the circle graph exceeds k, thus at least one shuffle is necessary. Also, at the beginning of the next phase, our algorithm has moved the maximum number of items possible to the output and it may use all stacks available without restrictions. This is naturally the best possible situation for an optimal solution as well. Thus, the number of shuffles needed by an optimal solution is $L^* \geq p$, and $L \leq 3n \cdot L^*$.

The runtime of our algorithm is obviously dominated by the n calls to A. $\square$

Remark 2. As the chromatic number of circle graphs cannot be determined exactly in polynomial time, a resource augmentation of at least one stack is unavoidable with our algorithm. This is rooted in the structure of circle graphs itself: In the stack assignment obtained from a proper coloring of the corresponding circle graph, the stacks do only remain sorted as needed by the algorithm as long as items are moved to the output immediately whenever possible. Thus, it is impossible to assign more than one color class to one stack: It may always happen, that items in an additional color class on the same stack keep items in the first color class from being moved to the output in time, thus destroying the linear order of items within one color class in the stack and possibly causing a non-constant number of additional shuffles in each phase of the algorithm. In that sense, our resource augmentation is best possible.

The relation to (improper) coloring circle graphs is our only algorithmic handle to approximating STACK SORTING (despite considerable efforts). Lower bounds for approximate colorings, in turn, classically rely on (a) maximum clique size ω, or (b) the number of vertices divided by the cardinality of a largest independent set. While for

(b) a simple example shows that the bound is trivial (linear gap) for circle graphs, it is known for (a) that no better factor than $\log \omega$ can be obtained (this is mentioned in [1], citing a Russian paper by Kostochka; note that this falsifies Unger's claim of having obtained a 2-approximation [19]). In fact, the best known approximation factor for coloring circle graphs is $\alpha = \log n$ [5], and no improvement to a constant factor is possible without a new lower bound on the chromatic number of graphs.

Alternatively, abandoning colorings, one is tempted to characterize instances which need "few" shuffles (in the sense of Remark 1), yet, even deciding whether *no* shuffles are needed is NP-hard. On the other hand, if the permutation π avoids the pattern 1-2-3, no shuffles are needed if $k \geq 5$ (this is the result that every triangle free circle graph is 5-colorable, see again [1]). It becomes clear once more why circle graphs "frustrated mathematicians for some years" [11], and still continue to do so.

5   Stacking Constraints

We finally consider the generalization in which items may not be placed arbitrarily on top of others. An instance of STACK SORTING WITH CONFLICTS is an instance of STACK SORTING plus such constraints, which can be modeled as a directed graph D with node set $\{1,\dots,n\}$ such that an edge (i, j) in D signifies that item i may not be put directly on top of item j. This is a practically relevant extension [16], and sortability becomes a justified question again.

Deciding sortability in a more general setting is PSPACE-complete: In [16], initially, some items need to be placed on stacks in a well-defined configuration; in addition, there are height bounds on the stacks. Also, the number of stacks k is part of the input and this fact is crucial in the reduction proving hardness there.

We give a more elaborate construction, eliminating all these additional assumptions and prove the following significantly stronger result.

Theorem 4. *For any fixed $k \geq 3$, deciding whether there exists a feasible solution to an instance of* STACK SORTING WITH CONFLICTS *is PSPACE-complete.*

Proof. We give a reduction from a special case of CONFIGURATION TO EDGE shown to be PSPACE-complete in [13]: An instance $NCL = (G, C, e^*)$ of this problem is based on a 3-regular undirected graph $G = (V, E)$, node weights $c : V \to \mathbb{R}$ with $c \equiv 2$ and edge weights $w : E \to \mathbb{R}$ with $w(e) \in \{1, 2\} \; \forall e \in E$. C denotes a *feasible configuration*, given by an orientation of the edges in E such that the sum of the weights of edges pointing into a node is at least the node weight, i.e., $\sum_{e \in \delta^-(v)} w(e) \geq c(v) \; \forall v \in V$, where $\delta^-(v)$ and $\delta^+(v)$ denote the sets of incoming and outgoing edges of a node $v \in V$ in C, respectively. $e^* \in E$ denotes a certain edge of the graph and the question is: Is there a sequence of feasible edge reversals such that the orientation of e^* is finally reversed?

We call edges with weight two *heavy*, all other edges *light*. G may only contain two types of nodes: Nodes with three heavy incident edges, called OR-nodes, and nodes with one heavy and two light incident edges, called AND-nodes.

The construction of an instance J of STACK SORTING WITH CONFLICTS for fixed $k \geq 3$ from an instance NCL of this special case works in three steps: First, we define basic gadgets consisting of items on two stacks for OR- and AND-nodes, respectively;

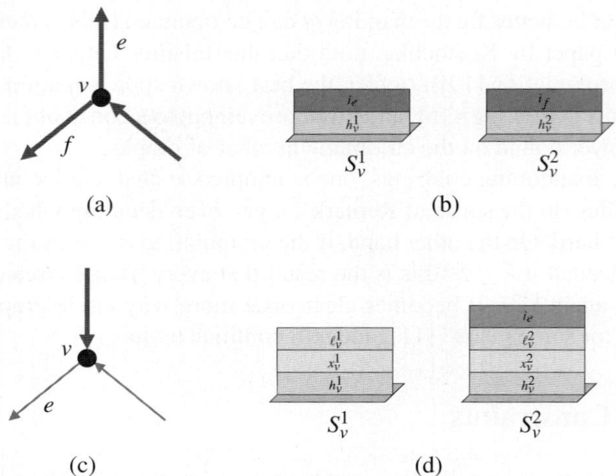

Fig. 2. Basic gadgets: Figs. (b) and (d) show stack representations of feasible configurations (a) and (c) at an OR-node and an AND-node respectively

then, we prove that the items of all gadgets can be put on any fixed number of $k \geq 3$ stacks without losing their properties crucial to the reduction; finally, we prove that we can number all items in the construction such that there exist a permutation π which forces the items into a configuration on stacks corresponding to the initial configuration C in *NCL*, and that a subsequent feasible reversal of e^* in *NCL* corresponds exactly to sortability of π. A complete proof for Thm. 4 is given in the full version of this paper; here, we only state its main ideas.

Fig. 2 shows the basic gadget for each OR- and AND-node. Each node $v \in V$ is associated with two stacks S_v^1 and S_v^2, and for each edge $e \in E$, we introduce an item i_e. A feasible configuration C in *NCL* corresponds to a configuration of items on stacks as follows: Item i_e is on stack S_v^1 or S_v^2 iff $e \in \delta^+(v)$ in C. We specify stacking constraints, such that only items corresponding to edges incident to a node v can be placed on the stacks corresponding to v. Also, we introduce some additional items, which may only be placed on the stacks of one single basic gadget (cf. Fig 2). Items i_e corresponding to heavy edges may only be placed on items h_v, items corresponding to light items only on items ℓ_v, and in any case it is required that $e \in \delta(v)$, i.e., e is incident to v. It is fairly easy to check, that feasible stackings on S_v^1 and S_v^2 correspond exactly with configurations in *NCL* which are feasible at v.

Fig. 3 shows how we can now assign all items from all basic gadgets to only three stacks while preserving their properties essential for the reduction. We separate the items of each stack of a basic gadget by items t with proper stacking constraints. Then we have one stack S_3 holding the items of stacks in the basic gadgets in reverse order. In order to change the orientation of one edge, we shuffle items from S_3 to the other two stacks S_1, S_2, such that the stacks of two basic gadgets in between which we would like to shuffle an item i_e are exposed on top of S_1 and S_2 (cf. Fig. 3). Due to the stacking constraints specified, items from different basic gadgets never mix in the process.

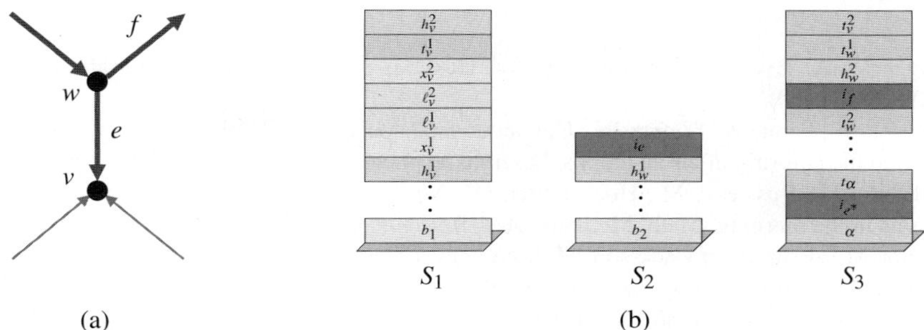

$$S_1 \qquad\qquad S_2 \qquad\qquad S_3$$

(a)   (b)

Fig. 3. Only using three stacks: Fig. (b) shows the stacking in which item i_e can be shuffled from S^1_w to S^2_v. This corresponds to changing the orientation of e in Fig. (a).

The stack of the basic gadget containing i_{e^*} is at the bottom of S_3, in correct order. Directly underneath i_{e^*} is an item α, which clearly may only be moved, if the direction of e^* is feasibly changed before. With the help of some more additional items b, we can now define a numbering of the items and a permutation π in which α comes before all other items of basic gadgets, such that the constructed instance of STACK SORTING WITH CONFLICTS is sortable iff the orientation of e^* can be feasibly changed.   □

6   Conclusions

Sorting with stacks is not a surprising connection between a fundamental data structure and a classic algorithmic theme. It is surprising however, that theory avoided the apparent need for shuffles—only sortability, not sorting itself has been considered so far. Our hardness results partially explain this lack of elegant and efficient algorithms.

Open Problems

Our work spawns some challenging open complexity issues. Hardness of approximation, i.e., Thm. 2, only holds for $k \geq 4$. Indeed, it is not even known whether polynomial time algorithms exist for $k = 2$ and $k = 3$.

There is still an annoying logarithmic gap between our inapproximability result and the best known approximation algorithm, which we only manage to close by resource augmentation. As pointed out in Sec. 4, the lower bound we use—the minimum number of monochromatic edges in a k-coloring of a circle graph—is not suited for a better result. Since (improper) coloring circle graphs is the only known handle to approximate STACK SORTING, what is an alternative lower bound on the number of shuffles?

Finally, also queues could be considered for intermediate storage of items (as e.g., Tarjan did). This is closely related to the so-called "midnight constraint" present in some applications, where all items have to be removed from the source, before the first item may be moved to the sink. The circle graphs then become permutation graphs which can be properly colored in polynomial time; thus, our hardness of approximation does not apply. But is this case really significantly easier?

References

1. Ageev, A.A.: Every circle graph of girth at least 5 is 3-colourable. Discrete Math. 195(1–3), 229–233 (1999)
2. Avriel, M., Penn, M., Shpirer, N.: Container ship stowage problem: complexity and connection to the coloring of circle graphs. Discrete Appl. Math. 103(1-3), 271–279 (2000)
3. Blasum, U., Bussieck, M., Hochstättler, W., Moll, C., Scheel, H., Winter, T.: Scheduling trams in the morning. Math. Methods Oper. Res. 49(1), 137–148 (1999)
4. Bóna, M.: A survey of stack-sorting disciplines. Electron. J. Comb. 9(2) (2002)
5. Černý, J.: Coloring circle graphs. Electr. Notes Discrete Math. 29, 457–461 (2007)
6. Cornelsen, S., Di Stefano, G.: Track assignment. J. Discrete Algorithms 5(2), 250–261 (2007)
7. Even, S., Itai, A.: Queues, stacks, and graphs. In: Proceedings of the International Symposium on the Theory of Machines and Computations, pp. 71–86. Academic Press, New York (1971)
8. Felsner, S., Pergel, M.: Sorting using networks of stacks and queues. In: Halperin, D., Mehlhorn, K. (eds.) ESA 2008. LNCS, vol. 5193, pp. 417–429. Springer, Heidelberg (2008)
9. Gallo, G., Di Miele, F.: Dispatching buses in parking depots. Transp. Sci. 35(3), 322–330 (2001)
10. Gavril, F.: Algorithms for a maximum clique and a maximum independent set of a circle graph. Networks 3(3), 261–273 (1973)
11. Golumbic, M.C.: Algorithmic Graph Theory and Perfect Graphs, 2nd edn. Elsevier B.V., Amsterdam (2004)
12. Gupta, N., Nau, D.S.: On the complexity of blocks-world planning. Artif. Intell. 56(2-3), 223–254 (1992)
13. Hearn, R.A., Demaine, E.D.: PSPACE-completeness of sliding-block puzzles and other problems through the nondeterministic constraint logic model of computation. Theor. Comput. Sci. 343(1–2), 72–96 (2005)
14. Kann, V., Khanna, S., Lagergren, J., Panconesi, A.: On the hardness of approximating max k-cut and its dual. Chic. J. Theor. Comput. Sci. (1997)
15. Knuth, D.E.: The Art of Computer Programming, Volume 3: Sorting and Searching, 2nd edn. Addison Wesley Longman, Amsterdam (1998)
16. König, F.G., Lübbecke, M.E., Möhring, R.H., Schäfer, G., Spenke, I.: Practical solutions to PSPACE-complete stacking. In: Proceedings of the 15th European Symposium on Algorithms. Lect. Notes Comput. Sci, vol. 4698, pp. 729–740. Springer, Heidelberg (2007)
17. Nilsson, N.J.: Principles of Artificial Intelligence. Morgan Kaufmann, San Francisco (1980)
18. Tarjan, R.: Sorting using networks of queues and stacks. J. ACM 19(2), 341–346 (1972)
19. Unger, W.: On the k-colouring of circle graphs. In: Cori, R., Wirsing, M. (eds.) STACS 1988. LNCS, vol. 294, pp. 61–72. Springer, Heidelberg (1988)

Quantum Query Complexity of Boolean Functions with Small On-Sets

Andris Ambainis[1], Kazuo Iwama[2], Masaki Nakanishi[3], Harumichi Nishimura[4], Rudy Raymond[5], Seiichiro Tani[6,7], and Shigeru Yamashita[3]

[1] Institute of Mathematics and Computer Science, University of Latvia, Riga, Latvia
ambainis@math.uwaterloo.ca
[2] School of Informatics, Kyoto University, Kyoto, Japan
iwama@kuis.kyoto-u.ac.jp
[3] Graduate School of Information Science, NAIST, Nara, Japan
{m-naka,ger}@is.naist.jp
[4] School of Science, Osaka Prefecture University, Osaka, Japan
hnishimura@mi.s.osakafu-u.ac.jp
[5] Tokyo Research Laboratory, IBM Japan, Kanagawa, Japan
raymond@jp.ibm.com
[6] NTT Communication Science Laboratories, NTT Corporation, Kyoto, Japan
[7] JST ERATO-SORST QCI Project, Tokyo, Japan
tani@theory.brl.ntt.co.jp

Abstract. The main objective of this paper is to show that the quantum query complexity $Q(f)$ of an N-bit Boolean function f is bounded by a function of a simple and natural parameter, i.e., $M = |\{x \mid f(x) = 1\}|$ or the size of f's *on-set*. We prove that: (i) For $poly(N) \leq M \leq 2^{N^d}$ for some constant $0 < d < 1$, the upper bound of $Q(f)$ is $O(\sqrt{N \log M / \log N})$. This bound is tight, namely there is a Boolean function f such that $Q(f) = \Omega(\sqrt{N \log M / \log N})$. (ii) For the same range of M, the (also tight) lower bound of $Q(f)$ is $\Omega(\sqrt{N})$. (iii) The average value of $Q(f)$ is bounded from above and below by $Q(f) = O(\log M + \sqrt{N})$ and $Q(f) = \Omega(\log M / \log N + \sqrt{N})$, respectively. The first bound gives a simple way of bounding the quantum query complexity of testing some graph properties. In particular, it is proved that the quantum query complexity of planarity testing for a graph with n vertices is $\Theta(N^{3/4})$ where $N = \frac{n(n-1)}{2}$.

1   Introduction

Query complexities for Boolean functions are one of the most fundamental and popular topics in quantum computation. It is well known that a quadratic speed-up, i.e., $\Omega(N)$ classically to $O(\sqrt{N})$ quantumly, is possible for several Boolean functions including OR, AND, AND-OR trees [18,19,17,6]. On the other hand, we can obtain only a constant-factor speed-up (i.e., $\Omega(N)$ are needed both classically and quantumly) for other Boolean functions such as PARITY [10], and this is also the case for almost all Boolean functions with N variables [3,13]. Thus

S.-H. Hong, H. Nagamochi, and T. Fukunaga (Eds.): ISAAC 2008, LNCS 5369, pp. 907–918, 2008.

our knowledge about the quantum query complexity for Boolean functions is relatively good for these typical cases, but much less is known for the others, especially for quantitative properties based on nontrivial parameters. In this paper, we show that the *size of the on-set* of a Boolean function f plays a key role for this purpose, i.e., to non-trivially bound the f's quantum query complexity.

Obviously this line of research started with the Grover's quantum search algorithm [18], which can be directly used to compute the Boolean OR function of N variables in the same $O(\sqrt{N})$ queries. Since then, a sequence of results have extensively appeared in the literature, showing that similar speed-ups are possible for many other, more general Boolean functions. For example, if a Boolean function is given by a constant-depth balanced AND-OR trees (OR is by a single-depth tree), it can be computed in $O(\sqrt{N})$ queries [19]. This was recently extended to any AND-OR tree with $O(N^{\frac{1}{2}+o(1)})$ queries by using the quantum walk technique [6]. Another famous example is monotone Boolean functions describing monotone graph properties (if a graph has n vertices then the corresponding Boolean function has $N = n(n-1)/2$ variables, one for each possible edge). In the quantum setting, it is known that $O(N^{13/20})$ queries suffice to decide if the given graph G includes a triangle [22,21], $O(N^{3/4})$ queries if G includes a star [11] (with zero error), and $O(N^{3/4})$ queries if G is connected [16]. Classical query complexities for those functions are all $\Omega(N)$.

Note that each of these Boolean functions, for which quantum query complexities are significantly smaller than classical query complexities, has a certain kind of "structure". In other words, researchers have been working on the question of what kind of structures help for efficient quantum computation. Our question in this paper is quite different; namely we ask if there is a "non-structural" parameter that greatly affects the quantum complexity of Boolean functions.

Our contribution: Let $\mathcal{F}_M$ be a family of N-variable Boolean functions f whose on-set is of size M, namely, f has output 1 (true) for M 0/1 assignments among the total 2^N ones. Then we can show that for *any* Boolean function f in $\mathcal{F}_{poly(N)}$, its query complexity is $\Theta(\sqrt{N})$ and the complexity gradually increases as M grows up to 2^{N^d} for some constant d ($0 < d < 1$). More in detail, let $Q(f)$ be the (true) query complexity of f. Then we investigate the upper bound $C(\mathcal{F}_M)$, the lower bound $c(\mathcal{F}_M)$, and the average value, $\tilde{C}(\mathcal{F}_M)$, of $Q(f)$ over all functions f in $\mathcal{F}_M$. Our results are as follows: (i) For $poly(N) \leq M \leq 2^{N^d}$ for some constant $0 < d < 1$, $C(\mathcal{F}_M) = \Theta(\sqrt{N \log M / \log N})$. This means that for any function in $\mathcal{F}_M$, its query complexity is $O(\sqrt{N \log M / \log N})$ and there exists a function in $\mathcal{F}_M$ such that its complexity is $\Omega(\sqrt{N \log M / \log N})$. (ii) For the same range of M, $c(\mathcal{F}_M) = \Theta(\sqrt{N})$, meaning that for any function in $\mathcal{F}_M$ its complexity is $\Omega(\sqrt{N})$ and there exists a function such that its complexity is $O(\sqrt{N})$. Thus our results are tight for both $C(\mathcal{F}_M)$ and $c(\mathcal{F}_M)$. Unfortunately, there is a $\log N$ factor gap in the evaluation of $\tilde{C}(\mathcal{F}_M)$, namely (iii) $\tilde{C}(\mathcal{F}_M) = O(\log M + \sqrt{N})$ and $\tilde{C}(\mathcal{F}_M) = \Omega(\log M / \log N + \sqrt{N})$.

A direct application of our upper bound result reduces bounding the query complexity of graph property testing to counting all graphs with a given property: For the family $\mathcal{F}$ of all graphs with a given property, $O(\sqrt{N \log |\mathcal{F}| / \log N})$ queries suffice to test if a given graph has the property. An interesting special case is that $O(N^{3/4})$ queries can decide if a given graph G is isomorphic to an arbitrary fixed graph G'. The bound is optimal in the worst case over all G'. Another interesting case is to test planarity, one of the most fundamental graph properties. We show the tight bound of $\Theta(N^{3/4})$ for the quantum query complexity of planarity testing. The upper bound is by just bounding the number of planar graphs with the fact that they are sparse. The interesting part is its lower bound. Our proof is based on the quantum adversary method [4], which requires us to find carefully two graphs which are almost the same but have different answers. We also prove that the lower bound of the classical query complexity is $\Omega(N)$, thus adding the new nontrivial property into the class of graph properties for which there is a significant gap between the quantum and classical query complexities.

Related works: A large literature exists for the quantum query complexity of Boolean functions. Other than OR and AND-OR trees, the complexity of the threshold function [10] was tightly characterized in the early stages. Element distinctness was also tightly shown to be $\Theta(N^{2/3})$ while the upper bound [5] and lower bound [2] of its complexity were obtained after exhaustive work of many researchers, which gave many technical contributions in (not for only quantum) complexity theory. For the monotone graph properties, Dürr et al. [16] showed the tight complexity $\Theta(N^{3/4})$ of connectivity. The quantum query complexities of total functions are polynomially related to the classical equivalents [10], and the maximum gap is conjectured to be quadratic.

The two major lower bound methods of quantum query complexity are polynomial methods [10] and adversary methods [4] (see [20] for its excellent survey.) Our lower bound of planarity testing is inspired by the application of adversary methods to connectivity in [16], which uses one-cycle vs. two-cycles as the two graphs with different answers. A similar choice of graphs is also used in [24] to get the lower bounds of several graph problems such as bipartiteness.

There have been few studies on the complexity of "non-structural" Boolean functions. All Boolean functions have quantum query complexity at most $N/2 + O(\sqrt{N})$ [13] while almost all functions have quantum query complexity at least $N/4 + \Omega(\sqrt{N})$ [3,15].

2   Preliminaries

We assume the oracle (or black-box) model in the quantum setting (e.g., [10]). In this model, an input (i.e., a problem instance) is given as an oracle. For any input $x = (x_1, \ldots, x_N) \in \{0, 1\}^N$, a unitary operator O, corresponding to a single query to an oracle, maps $|i\rangle|b\rangle|w\rangle$ to $|i\rangle|b \oplus x_i\rangle|w\rangle$ for each $i \in [N] = \{1, 2, \ldots, N\}$ and $b \in \{0, 1\}$, where w denotes workspace. A *quantum computation* of the oracle model is a sequence of unitary transformations $U_0 \to O \to U_1 \to O \to \cdots \to O \to U_t$, where U_j may be any unitary transformation that does not depend on the input. The above computation sequence involves t oracle calls, which is our measure of the complexity: The *quantum query complexity $Q(P)$* of a problem P whose input is given as an N-bit string is defined to be the number of quantum queries needed to solve P with bounded-error, i.e., with success probability at least $1/2 + c$ where c is some constant.

In this paper, our problem P is to evaluate the value (0 or 1) of a Boolean function $f(x_1, \ldots, x_N)$ over N variables, assuming that the truth table of f is known. The *on-set* of f is the set of assignments $(x_1, \ldots, x_N)$ satisfying $f(x_1, \ldots, x_N) = 1$. We denote the family of all functions whose on-set is of size M by $\mathcal{F}_M$.

Our algorithms in this paper use the algorithm in [8] for the oracle identification problem defined as follows. Notice that there are 2^N different oracles with length N.

Definition 1 (Oracle Identification Problem (OIP) [7,8]). *Given an oracle x and a set S of M oracle candidates out of 2^N ones, determine which oracle in S is identical to x with the promise that x is a member of S.*

Improving the previous result in [7], Ambainis et al. [8] showed the following upper bound for the quantum query complexity of OIP when M is not so large, which is asymptotically optimal.

Theorem 1 (Optimal bound of OIP [8]). *OIP can be quantumly solved with a constant success probability by making $O(\sqrt{N \frac{\log M}{\log N}})$ queries to the given oracle if $poly(N) \le M \le 2^{N^d}$ for some constant d $(0 < d < 1)$.*

3   Worst-Case Analysis

In this section, we study both upper and lower bounds for the quantum query complexity of Boolean functions in $\mathcal{F}_M$. First, we show the upper bound.

Theorem 2 (Upper Bound). *Any function $f \in \mathcal{F}_M$ has quantum query complexity $O(\sqrt{N \frac{\log M}{\log N}})$ if $poly(N) \le M \le 2^{N^d}$ for some constant d $(0 < d < 1)$.*

Proof. Recall that OIP is the problem that we are requested to find a hidden oracle, with the promise that it is a member of oracle candidate set S. To use this for evaluation of the Boolean function f, let S be the on-set of f, which can be constructed from the known truth table of f. Note that $|S| = M$ since $f \in \mathcal{F}_M$. We then invoke the OIP algorithm of Theorem 1 to find the hidden oracle with $O(\sqrt{N \frac{\log M}{\log N}})$ queries, assuming the promise that the current oracle x is in S (actually, the promise does not hold if $f(x) = 0$). Let $z \in \{0,1\}^N$ be the string obtained by the OIP algorithm.

If $f(x) = 1$, the promise of the above OIP is indeed satisfied; z is equal to x with high probability.

If $f(x) = 0$, the promise does not hold; the OIP algorithm outputs some answer $z \in S$ such that $z \neq x$. To recognize this case, it suffices to check whether z is equal to x by using Grover search [18] with $O(\sqrt{N})(\in O(\sqrt{N\frac{\log M}{\log N}}))$ queries. This completes the proof. $\square$

In fact, this upper bound is optimal for the threshold function with threshold $\Theta(\log M / \log N)$, which has a query complexity of $\Omega(\sqrt{N\frac{\log M}{\log N}})$ due to the results in [10]. The following corollary is immediate.

Corollary 1. *For $M \in poly(N)$, any function $f \in \mathcal{F}_M$ has quantum query complexity $O(\sqrt{N})$.*

This corollary together with the next lower bound theorem implies that if M is in $poly(N)$, then any function $f \in \mathcal{F}_M$ has essentially the same complexity up to a constant factor as the OR function.

Theorem 3 (Lower Bound). *If $M \leq 2^{\frac{N}{2+\epsilon}}$ for any positive constant ϵ, any $f \in \mathcal{F}_M$ has quantum query complexity $\Omega(\sqrt{N})$.*

Proof. We use the sensitivity argument. Recall that the sensitivity $s_x(f)$ of a Boolean function f on $x \in \{0,1\}^N$ is the number of variables x_i such that $f(x) \neq f(x^i)$, where x^i is the string obtained from x by flipping the value of x_i. The sensitivity $s(f)$ of f is the maximum of $s_x(f)$ over all x. The results of Beals et al. [10] implies $Q(f) = \Omega(\sqrt{s(f)})$. By the definition of $s(f)$ and the result by Beals et al., we can see that $Q(f) = \Omega(\sqrt{|Z|})$, where Z is the set of 0-*points*, elements whose values of f is 0, "around" an arbitrarily chosen element in the on-set (1-*point*). Here, "around" means the Hamming distance is 1. Therefore, if there is a 1-point around which there are $\Omega(N)$ 0-points, $Q(f) = \Omega(\sqrt{N})$.

To prove by contradiction, we assume that, around every 1-point, there are $o(N)$ 0-points, i.e., there are $(N - o(N))$ 1-points. Suppose that $(0,0,\ldots,0)$ is a 1-point (otherwise, we can give a similar argument using some 1-point). Set $S_0 = \{(0,0,\ldots,0)\}$. Define S_k inductively to be the set of all 1 points around all points in S_{k-1}, whose Hamming weight is k. By assumption, the number of 1-points around every point in S_{k-1} is $N - o(N) = N(1-\alpha)$ for any small $\alpha = o(1)$. For each point x in S_{k-1}, there exist at most $(k-1)$ 1-points around x in S_{k-2}. Thus, for each point x in S_{k-1}, there exist at least $(N(1-\alpha) - (k-1))$ 1-points around x in S_k. Similarly, for each point x in S_k, there exist at most k 1-points around x in S_{k-1}. Thus, $|S_k| \geq |S_{k-1}|(N(1-\alpha) - (k-1))/k$. From this inductive inequality and $|S_0| = 1$, we have $|S_k| \geq (N(1-\alpha))(N(1-\alpha) - 1)(N(1-\alpha) - 2) \cdots (N(1-\alpha) - (k-1))/k!$. The number of inputs x such that $f(x) = 1$ and the Hamming weight of x is at most k is $T(k) = |S_0| + \cdots + |S_k|$. We will show $T(k) > M$ for some $k \leq N/2$, a contradiction, as follows. $T(k) > |S_k| \geq (N(1-\alpha))(N(1-\alpha) - 1) \cdots (N(1-\alpha) - (k-1))/k! > \left(\frac{N(1-\alpha)}{k}\right)^k$. For $k = \frac{N}{2+\epsilon}$, we obtain $T(k) > 2^{\frac{N}{2+\epsilon}} \geq M$. $\square$

The above lower bound is tight, since it is easy to construct a Boolean function for any $M \leq 2^{\frac{N}{2+\epsilon}}$ such that its query complexity is $O(\sqrt{N})$. Thus we have shown that there are Boolean functions, f_1 and f_2, which are "easiest" and "hardest" in class $\mathcal{F}_M$, such that $Q(f_1) = \Theta(\sqrt{N})$ and $Q(f_2) = \Theta\left(\sqrt{N \frac{\log M}{\log N}}\right)$.

4   Average-Case Analysis

This section considers upper and lower bounds for the quantum query complexity for almost all functions in $\mathcal{F}_M$. They are essentially $O(\log M)$ and $\Omega(\log M / \log N)$, thus having a $\log N$ factor gap. Note that the bounds hold for the entire range of M.

Theorem 4 (Upper Bound). *Almost all Boolean functions in $\mathcal{F}_M$ have quantum query complexity $O(\log M + \sqrt{N})$.*

Proof. It suffices to show the statement for $poly(N) < M < 2^{\frac{N}{3}}$ since the case of $M \in poly(N)$ is obtained by Corollary 1, and the case of $M \geq 2^{\frac{N}{3}}$ leads to the trivial bound. We can make the following claim (proof will be given later):

Claim. *If we generate a random Boolean function f whose on-set S_f has size M, then, for almost all cases, any two of M elements in S_f differ from each other in the first k bits, where $k = 3 \log M$ (which is smaller than N).*

Now the following algorithm works by using the claim. Below we will use k as $k = 3 \log M$. First, we identify the first k bits of the current input x by making k classical queries. Then we can decide that $f(x) = 0$ regardless of the remaining bits, if the k-bit string is different from the first-k-bit string of any element in the on-set. Otherwise, $f(x)$ can have value 1, depending on the remaining $N - k$ bits. For the latter case, the claim implies that, for almost all functions, there is only one possible way of assigning $0/1$ to the remaining $N - k$ bits that determines $f(x) = 1$. Thus, we just check whether the remaining bits are subject to such one possibility or not by using Grover search. In total, the query complexity is $O(k + \sqrt{N}) = O(\log M + \sqrt{N})$.

What remains is to show the above claim. The number of all functions whose on-set has size M is $\binom{2^N}{M}$. Among such functions, we count the number of our desired functions, i.e., the functions such that any two inputs in the on-set differ from each other in the first k bits. We first consider the number of possible assignments to the first k bits of M inputs. The number of possibilities is $\binom{2^k}{M}$ since we need to choose different M k-bit strings among the 2^k possibilities. We then choose the remaining $(N - k)$ bits arbitrarily, i.e., 2^{N-k} possibilities for each of the M assignments to the first k bits. Thus, the number of possibilities of assigning the remaining $(N - k)$ bits for all M inputs is $(2^{(N-k)})^M = 2^{M(N-k)}$. In conclusion, the number of our desired functions is $\binom{2^k}{M} 2^{M(N-k)}$. The ratio of our desired functions is $\dfrac{\binom{2^k}{M} 2^{M(N-k)}}{\binom{2^N}{M}}$. We can show that, by using $k = 3 \log M$, this ratio is larger than $\exp(-\frac{M^2-M}{2(M^3-M+1)})$. Hence, the ratio approaches 1 as N (and hence M) goes to infinity. $\qquad\square$

Theorem 5 (Lower Bound). *Almost all Boolean functions in $\mathcal{F}_M$ have quantum query complexity $\Omega(\log M/\log N + \sqrt{N})$.*

Proof. For almost all $f \in \mathcal{F}_M$, we prove that $Q(f) = \Omega(\log M/\log N)$ when $M > N^{\sqrt{N}}$ since by Theorem 3 the lower bound $\Omega(\sqrt{N})$ holds for $M \leq N^{\sqrt{N}}$. Moreover, we assume that $M \leq 2^{N-1}$ without loss of generality. We shall show the lower bound for the *unbounded-error* setting, where the success probability suffices to be at least $1/2 + \epsilon$, for any positive ϵ (which does not need to be a constant as in the bounded-error setting). Obviously, any lower bound in this setting also holds in the bounded-error setting.

The lower bound of unbounded-error query complexity of a Boolean function f is characterized by the minimum degree of its *sign-representing polynomial p*, a real-valued polynomial with properties that $p(x)$ is positive whenever $f(x) = 0$ and $p(x)$ is negative whenever $f(x) = 1$, for all N-bit strings x. There are two nice properties of sign-representing polynomials, which are useful for our proof. First, as shown in [23] (also, implicitly in [12]), the unbounded-error quantum query complexity of f is exactly half of the minimum degree of its sign-representing polynomial. Secondly, the number of Boolean functions of N variables whose minimum degrees of sign-representing polynomials are at most d, denoted as $T(N,d)$, is also known to be at most $2\sum_{k=0}^{D-1}\binom{2^N-1}{k}$, for $D = \sum_{i=0}^{d}\binom{N}{i}$ as proved in [9].

Hence to show that almost all Boolean functions in $\mathcal{F}_M$ have unbounded-error quantum query complexity $\Omega(\log M/\log N)$, it suffices to show that $T(N, \frac{\log M}{2\log N})$ is small compared to $\binom{2^N}{M}$, i.e., the size of $\mathcal{F}_M$. Notice that in this case $D = \sum_{i=0}^{\log M/(2\log N)}\binom{N}{i} \leq N^{\frac{\log M}{2\log N}} = \sqrt{M}$, and therefore,

$$\frac{T\left(N, \log M/(2\log N)\right)}{\binom{2^N}{M}} \leq \frac{2\sum_{k=0}^{D-1}\binom{2^N-1}{k}}{\binom{2^N}{M}} \leq \frac{2\sum_{k=0}^{\sqrt{M}-1}\binom{2^N}{k}}{\binom{2^N}{M}} \leq \frac{2\sqrt{M}\binom{2^N}{\sqrt{M}}}{\binom{2^N}{M}}.$$

$$(1)$$

Moreover, the right-hand side of (1) is bounded by

$$\frac{2\sqrt{M}\binom{2^N}{\sqrt{M}}}{\binom{2^N}{M}} = 2\sqrt{M}\frac{M\cdots(\sqrt{M}+1)}{(2^N - \sqrt{M})\cdots(2^N - M + 1)} \leq \frac{2\sqrt{M}}{2^N - M + 1}. \qquad (2)$$

By the assumption that $M \leq 2^{N-1}$, the right-hand side of (2) goes to 0 as N goes to infinity, which completes the proof. $\qquad\square$

5   Applications

As application of Theorem 2, we consider the problem of graph property testing, i.e., the problem of testing, for a given graph G as an oracle, if G has a certain property. More precisely, an n-vertex graph is given as $n(n-1)/2$ Boolean variables, x_i for $i \in \{1, \ldots, n(n-1)/2\}$, representing the existence of the ith possible edge e_i, i.e., $x_i = 1$ if and only if e_i exists. In this setting, graph property testing is just to evaluate a Boolean function f depending on the $n(n-1)/2$ variables such that $f(x_1, \ldots, x_{n(n-1)/2}) = 1$ if and only if the graph has a certain property. An interpretation of graph property testing according to Theorem 2 is to decide if G is a member of $\mathcal{F}$ for the family $\mathcal{F}$ of all graphs with certain properties. Thus, Theorem 2 directly gives the next theorem with $M = |\mathcal{F}|$ and $N = n(n-1)/2$.

Theorem 6. *For graph family $\mathcal{F}$ defined as in the above, graph property testing can be solved with $O(\sqrt{n^2 \log |\mathcal{F}| / \log n})$ quantum queries, if $poly(n) \leq |\mathcal{F}| \leq 2^{n^{2d}}$ for some constant d $(0 < d < 1)$.*

An interesting special case of the problem is graph isomorphism testing: the problem of deciding if a given graph G is isomorphic to an arbitrary fixed graph G'.

Corollary 2. *Graph isomorphism testing can be solved with $O(n^{1.5})$ quantum queries.*

Proof. The number of graphs isomorphic to G' is at most the number of permutations over the vertex set, which is $n! = 2^{O(n \log n)}$. $\qquad\square$

This upper bound is optimal in the worst case, since the lower bound $\Omega(n^{1.5})$ of connectivity testing problem in [16] is essentially that of deciding whether a given graph is isomorphic to one cycle or two cycles. Another interesting special case is *planarity testing*, the problem of testing if a given graph is planar.

Corollary 3. *Planarity testing can be solved with $O(n^{1.5})$ quantum queries.*

Proof. Since the number of edges of any planar graph is at most $3n-6 < 3n$ [14], the number of planar graphs is at most $\binom{n(n-1)/2}{3n} < n^{2 \cdot 3n} = 2^{6n \log n}$. $\qquad\square$

For verifying optimality, we prove a lower bound of $\Omega(n^{1.5})$ for this problem. We also prove a tight classical lower bound of $\Omega(n^2)$.

To prove the lower bounds, we use the next two theorems.

Theorem 7 (Quantum adversary method [4], reformulated by [1]). *Let $\mathcal{A} \subseteq F^{-1}(0)$ and $\mathcal{B} \subseteq F^{-1}(1)$ be sets of inputs to function F. Let $R(A, B) \geq 0$ be a real-valued function, and for $A \in \mathcal{A}$, $B \in \mathcal{B}$, and location i, let*

$$\theta(A, i) = \frac{\sum_{B^* \in \mathcal{B} \,:\, A(i) \neq B^*(i)} R(A, B^*)}{\sum_{B^* \in \mathcal{B}} R(A, B^*)}, \theta(B, i) = \frac{\sum_{A^* \in \mathcal{A} \,:\, A^*(i) \neq B(i)} R(A^*, B)}{\sum_{A^* \in \mathcal{A}} R(A^*, B)},$$

where $A(i)$ and $B(i)$ denotes the value of the ith variable for A and B, respectively, the denominators are all nonzero. Then the number of quantum queries needed to evaluate F with probability at least $9/10$ is $\Omega(1/v_{\text{geom}})$, where

$$v_{\text{geom}} = \max_{\substack{A \in \mathcal{A}, \ B \in \mathcal{B}, \ i \,: \\ R(A,B) > 0, \ A(i) \neq B(i)}} \sqrt{\theta(A, i) \theta(B, i)}.$$

Theorem 8 (Classical adversary method [1]). *Let $\mathcal{A}, \mathcal{B}, R, \theta$ be the same as in Theorem 7. Then the number of randomized queries needed to evaluate F with probability at least $9/10$ is $\Omega\left(1/\upsilon_{\min}\right)$, where*

$$\upsilon_{\min} = \max_{\substack{A \in \mathcal{A},\ B \in \mathcal{B},\ i\ :\\ R(A,B)>0,\ A(i)\neq B(i)}} \min\left\{\theta\left(A, i\right), \theta\left(B, i\right)\right\}.$$

Now we are ready to present our lower bound.

Theorem 9. *Solving planarity testing needs $\Omega(n^{1.5})$ queries in the quantum setting, and $\Omega(n^2)$ queries in the classical setting.*

Proof. Before giving our formal proof, we briefly describe our proof idea. To apply the adversary method, we need to take two sets, $\mathcal{A}$ and $\mathcal{B}$, of non-planar graphs and planar graphs, respectively, such that for "many" pairs $(A, B) \in \mathcal{A} \times \mathcal{B}$, A and B are hard to distinguish. We define $\mathcal{A}$ as a set of subdivisions of the complete graph K_5. Focusing on one cycle included in each $A \in \mathcal{A}$, we define $\mathcal{B}$ be a set of planar graphs B obtained by dividing the cycle to two cycles. We then give a relation between graphs $A \in \mathcal{A}$ and graphs $B \in \mathcal{B}$. The relation is seemingly similar to that defined in [16] for proving the quantum lower bound of connectivity. However, their relation is not enough to keep "many" pairs in our case. Different from the case of [16], we place a careful restriction on the way of transforming A to B as well as B to A.

Our formal proof is as follows. To get lower bounds by using Theorems 7 and 8, we will define sets $\mathcal{A}$ and $\mathcal{B}$, and function $R : \mathcal{A} \times \mathcal{B} \to \{0, 1\}$. Let a, b, c, d be any four vertices of complete graph K_5 with five vertices. Let $\mathcal{A}$ be the set of graphs obtained by replacing edges (a, c) and (b, d) of the K_5 with path P_{ac} between a and c and path P_{bd} between b and d, respectively, on which there are $n - 5$ vertices except a, b, c, d, and each one of which is at most three times longer than the other. Every graph in $\mathcal{A}$ is not planar and has n vertices, since it is a subdivision of K_5, i.e., it becomes K_5 by contracting all but one edges on each of P_{ac} and P_{bd}. An example of an instance in $\mathcal{A}$ is shown at the left side in Fig. 1. Let $\mathcal{B}$ be the set of graphs obtained by replacing (a, c) and (b, d) of the K_5 with path P_{ad} between a and d and path P_{bc} between b and c, respectively, on which there are $n - 5$ vertices except a, b, c, d, and one of which is at most three times longer than the other. It is easy to see that every graph in $\mathcal{B}$ is planar and has n vertices. An example of an instance in $\mathcal{B}$ is shown at the right side in Fig. 1.

Now, we define $B \in \mathcal{B}$ such that $R(A, B) = 1$ for every $A \in \mathcal{A}$. For every graph $A \in \mathcal{A}$, let (e_a, e_c) be any edge on P_{ac}, where e_a is assumed to be closer to a on the path, and let (e_b, e_d) be any edge on P_{bd}, where e_b is assumed to be closer to b on the path (see the left graph in Fig. 1). If we replace (e_a, e_c) and (e_b, e_d) with (e_a, e_d) and (e_b, e_c), the resulting graph has paths P_{ad} and P_{bc} instead of P_{ac} and P_{bd} (see the right graph in Fig. 1). We can guarantee that each of P_{ad} and P_{bc} is at most three times longer than the other by imposing some restriction on the choice of (e_b, e_d) for each (e_a, e_c), which will be proved later; the resulting graph is a member of $\mathcal{B}$. Similarly, we define $A \in \mathcal{A}$ such that $R(A, B) = 1$ for every $B \in \mathcal{B}$. For every graph $B \in \mathcal{B}$, let (e_a, e_d) be any edge

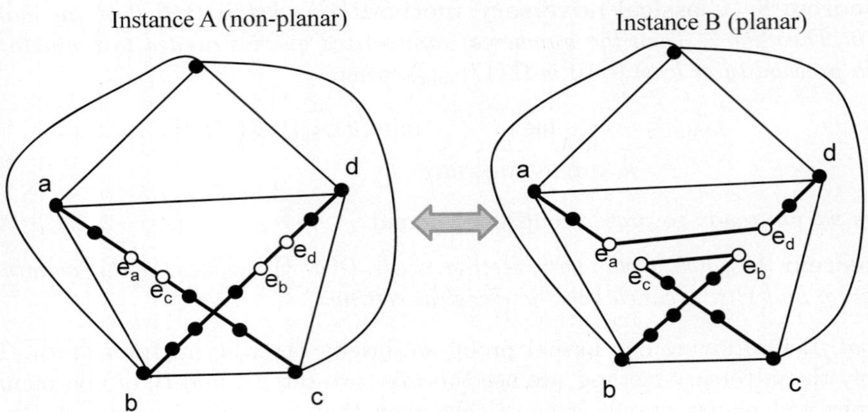

Fig. 1. Instance $A \in \mathcal{A}$ and instance $B \in \mathcal{B}$

on P_{ad}, where e_a is assumed to be closer to a on the path, and let (e_b, e_c) be any edge on P_{bc}, where e_b is assumed to be closer to b on the path. If we replace (e_a, e_d) and (e_b, e_c) with (e_a, e_c) and (e_b, e_d), the resulting graph has paths P_{ac} and P_{bd} instead of P_{ad} and P_{bc}. Since we can guarantee that each of P_{ac} and P_{bd} is at most three times longer than the other by imposing a similar restriction (shown later) on the choice of (e_b, e_c) for each (e_a, e_d), the resulting graph is a member of $\mathcal{A}$.

We here show the restriction on the choice of (e_b, e_d) for each (e_a, e_c) when relating A to B (a similar restriction works when relating B to A). Without loss of generality, P_{ac} is shorter than or equal to P_{bd} (otherwise, we just switch P_{ac} and P_{bd}). Let the length of paths P_{ac} and P_{bd} be cL and $(1-c)L$, respectively, for $1/4 \le c \le 1/2$, where $L = n - 3$. Notice that the sum of the lengths of P_{ac} and P_{bd} is always $(n-3)$. If the subpath between a and e_a of P_{ac} is of length $k \in \{0, \ldots, cL-1\}$, we choose (e_b, e_d) such that the subpath of P_{bd} between e_b and d has the length of at least $\max\{L/4 - k, 1\}$ and at most $\min\{3L/4 - k, (1-c)L\}$. Then after replacing (e_a, e_c) and (e_b, e_d) with (e_a, e_d) and (e_b, e_c), the lengths of P_{ad} and P_{bc} are each at least $L/4$ and at most $3L/4$. This edge replacement is always possible in many ways, since there are many edges (e_b, e_d) that satisfies the condition, as proved below. More precisely, there are at least $L/4$ choices of (e_b, e_d): $\Delta \equiv \min\{3L/4 - k, (1-c)L\} - \max\{L/4 - k, 1\} \ge L/4$.

If $\min\{3L/4 - k, (1-c)L\} = 3/4L - k$,

$$\Delta = 3/4L - k - \max\{L/4 - k, 1\} = \min\{L/2, 3L/4 - k - 1\},$$

which is at least $\min\{L/2, 3L/4 - cL\} \ge L/4$.

If $\min\{3L/4 - k, (1-c)L\} = (1-c)L$,

$$\Delta = (1-c)L - \max\{L/4 - k, 1\} = \min\{(3/4 - c)L + k, (1-c)L - 1\},$$

which is at least $\min\{(3/4 - c)L, (1-c)L - 1\} \ge L/4$.

This means that, for each (e_a, e_c), there are $\Omega(L)$ choices of (e_b, e_d). Since there are cL choices of (e_a, e_c), $\sum_{B^* \in \mathcal{B}} R(A, B^*) = \Omega(cL \cdot L) = \Omega(n^2)$, implying $\Theta(n^2)$. Similarly, $\sum_{A^* \in \mathcal{A}} R(A^*, B) = \Theta(n^2)$.

For any fixed $A \in \mathcal{A}$, $\sum_{B^* \in \mathcal{B} \ : \ A(i)=1, B^*(i)=0} R(A, B^*)$ attains the maximum value of $\Theta(n)$, when i is the index of an edge on the shorter path of P_{ac} and P_{bd}. For any fixed $B \in \mathcal{B}$, $\sum_{A^* \in \mathcal{A} \ : \ A^*(i)=1, B(i)=0} R(A^*, B)$ is a positive constant $\Theta(1)$ for all i such that there exists at least one A^* satisfying that $A^*(i) = 1, B(i) = 0$ and $R(A^*, B) = 1$. This is because, to flip x_i, we need to pick up a pair of edges which are adjacent to the ith possible edge, and replace the edge pair with another pair of edges including the ith possible edge. Thus, for every A and B such that $R(A, B) = 1$, $\max_{i: A(i)=1, B(i)=0} \sqrt{\theta(A, i)\theta(B, i)} = \Theta(\sqrt{(n/n^2)(1/n^2)}) = \Theta(1/n^{1.5})$. Similarly, $\max_{i: A(i)=0, B(i)=1} \sqrt{\theta(A, i)\theta(B, i)} = \Theta(1/n^{1.5})$. The quantum lower bound follows from Theorem 7.

The classical lower bound can be obtained by a similar argument. For any A and B such that $R(A, B) = 1$, $\max_{i: A(i)=1, B(i)=0} \min\{\theta(A, i), \theta(B, i)\} = \Theta(1/n^2)$, and $\max_{i: A(i)=0, B(i)=1} \min\{\theta(A, i), \theta(B, i)\} = \Theta(1/n^2)$, since the smallest value of $\theta(A, i)$ and $\theta(B, i)$ is $\Theta(1/n^2)$. The classical lower bound follows from Theorem 8. $\qquad\square$

References

1. Aaronson, S.: Lower bounds for local search by quantum arguments. SIAM J. Comput. 35(4), 804–824 (2006)
2. Aaronson, S., Shi, Y.: Quantum lower bounds for the collision and the element distinctness problems. J. ACM 51(4), 595–605 (2004)
3. Ambainis, A.: A note on quantum black-box complexity of almost all Boolean functions. Inf. Process. Lett. 71(1), 5–7 (1999)
4. Ambainis, A.: Quantum lower bounds by quantum arguments. J. Comput. Sys. Sci. 64, 750–767 (2002)
5. Ambainis, A.: Quantum walk algorithm for element distinctness. SIAM J. Comput. 37(1), 210–239 (2007)
6. Ambainis, A., Childs, A.M., Reichardt, B.W., Špalek, R., Zhang, S.: Any AND-OR formula of size N can be evaluated in time $N^{1/2+o(1)}$ on a quantum computer. In: Proc. 48th FOCS, pp. 363–372 (2007)
7. Ambainis, A., Iwama, K., Kawachi, A., Masuda, H., Putra, R.H., Yamashita, S.: Quantum identification of Boolean oracles. In: Diekert, V., Habib, M. (eds.) STACS 2004. LNCS, vol. 2996, pp. 105–116. Springer, Heidelberg (2004)
8. Ambainis, A., Iwama, K., Kawachi, A., Raymond, R., Yamashita, S.: Improved algorithms for quantum identification of Boolean oracles. Theor. Comput. Sci. 378(1), 41–53 (2007)
9. Anthony, M.: Classification by polynomial surfaces. Discrete Applied Mathematics 61, 91–103 (1995)
10. Beals, R., Buhrman, H., Cleve, R., Mosca, M., de Wolf, R.: Quantum lower bounds by polynomials. J. ACM 48(4), 778–797 (2001)
11. Buhrman, H., Cleve, R., de Wolf, R., Zalka, C.: Bounds for small-error and zero-error quantum algorithms. In: Proc. 40th FOCS, pp. 358–368 (1999)
12. Buhrman, H., Vereschagin, N., de Wolf, R.: On computation and communication with small bias. In: Proc. 22nd CCC, pp. 24–32 (2007)

13. van Dam, W.: Quantum oracle interrogation: getting all information for almost half the price. In: Proc. 39th FOCS, pp. 362–367 (1998)
14. Diestel, R.: Graph Theory, 2nd edn. Graduate Texts in Mathematics. Springer, Heidelberg (2000)
15. O'Donnell, R., Servedio, R.A.: Extremal properties of polynomial threshold functions. In: Proc. 18th CCC, pp. 3–12 (2003)
16. Dürr, C., Heiligman, M., Høyer, P., Mhalla, M.: Quantum query complexity of some graph problems. SIAM J. Comput. 35(6), 1310–1328 (2006)
17. Farhi, E., Goldstone, J., Gutmann, S.: A quantum algorithm for the Hamiltonian NAND tree. quant-ph/0702144 (2007)
18. Grover, L.K.: A fast quantum mechanical algorithm for database search. In: Proc. 28th STOC, pp. 212–219 (1996)
19. Høyer, P., Mosca, M., de Wolf, R.: Quantum search on bounded-error inputs. In: Baeten, J.C.M., Lenstra, J.K., Parrow, J., Woeginger, G.J. (eds.) ICALP 2003. LNCS, vol. 2719, pp. 291–299. Springer, Heidelberg (2003)
20. Høyer, P., Špalek, R.: Lower bounds on quantum query complexity. Bulletin of the EATCS 87, 78–103 (2005)
21. Magniez, F., Nayak, A., Roland, J., Santha, M.: Search via quantum walk. In: Proc. 39th STOC, pp. 575–584 (2007)
22. Magniez, F., Santha, M., Szegedy, M.: Quantum algorithms for the triangle problem. In: Proc. ACM-SIAM SODA, pp. 1109–1117 (2005)
23. Montanaro, A., Nishimura, H., Raymond, R.: Unbounded-error quantum query complexity. In: Proc. 19th ISAAC (to appear, 2008)
24. Zhang, S.: On the power of Ambainis lower bounds. Theor. Comput. Sci. 339(2-3), 241–256 (2005)

Unbounded-Error Quantum Query Complexity

Ashley Montanaro[1], Harumichi Nishimura[2], and Rudy Raymond[3]

[1] Department of Computer Science, University of Bristol, UK
`montanar@cs.bris.ac.uk`
[2] School of Science, Osaka Prefecture University, Japan
`hnishimura@mi.s.osakafu-u.ac.jp`
[3] Tokyo Research Laboratory, IBM Japan, Japan
`raymond@jp.ibm.com`

Abstract. This work studies the quantum query complexity of Boolean functions in an *unbounded-error* scenario where it is only required that the query algorithm succeeds with a probability strictly greater than $1/2$. We first show that, just as in the communication complexity model, the unbounded-error quantum query complexity is exactly half of its classical counterpart for any (partial or total) Boolean function. Next, connecting the query and communication complexity results, we show that the "black-box" approach to convert quantum query algorithms into communication protocols by Buhrman-Cleve-Wigderson [STOC'98] is optimal even in the unbounded-error setting. We also study a related setting, called the *weakly* unbounded-error setting. In contrast to the case of communication complexity, we show a tight multiplicative $\Theta(\log n)$ separation between quantum and classical query complexity in this setting for a *partial* Boolean function.

1   Introduction

Many models in computational complexity have several settings where different restrictions are placed on the success probability to evaluate a Boolean function f. The most basic one is the *exact* setting: it requires that the computation of f is always correct. In the polynomial-time complexity model, this corresponds to the complexity class P. If we require the success probability to be only "high" (say, $2/3$), such a setting is called *bounded error*. The corresponding polynomial-time complexity class is known as BPP. The *unbounded-error* setting is also standard. In this setting, it suffices to have a "positive hint", even infinitesimal, towards the right answer. That is, the unbounded-error setting requires that the success probability to compute Boolean functions is strictly larger than $1/2$. The most famous model for this setting is also polynomial-time complexity, and PP is the corresponding complexity class. This setting also has connections with important concepts such as *polynomial threshold functions* in computational learning theory.

There are two major computing models which have been introduced to develop the lower bound method in complexity theory. The first one is the *communication complexity (CC)* model. The CC model measures the amount of communication

S.-H. Hong, H. Nagamochi, and T. Fukunaga (Eds.): ISAAC 2008, LNCS 5369, pp. 919–930, 2008.

for several parties, which have distributed inputs, to compute Boolean functions. The second one is the *query complexity (QC)* model. The QC model measures the amount of queries required for a machine with no input to compute a Boolean function by querying the input given in a black-box. The CC and QC models are also studied in the *quantum* setting, and there are many results on the performance gaps between classical and quantum computation [25].

So far, the unbounded-error setting has also been studied in the CC and QC models. In the classical CC model, a large literature has developed since its introduction by Paturi and Simon [22]. In the quantum case, Iwama et al. [18] showed that the quantum CC of *any* Boolean function is almost half of its classical CC. Furthermore, a variant of the unbounded-error setting was studied, which is often called the *weakly unbounded-error* setting. Here the cost of a protocol is defined by $q + \log(1/2(p - 1/2))$, where q is the number of communication (qu)bits and $p > 1/2$ is the success probability. This concept appeared in [5,16], and was later studied in [19]. Halstenberg and Reischuck [16] showed that weakly unbounded-error protocols correspond to so-called "majority nondeterministic" protocols, while Klauck [19] showed a close connection between this setting and the discrepancy method in communication complexity. Recently, Buhrman et al. [10] and Sherstov [24] independently showed that there is a Boolean function that exponentially separates the classical weakly unbounded-error CC and unbounded-error CC. On the other hand, there can only be a constant gap between quantum and classical CCs for Boolean functions in the weakly unbounded-error setting [18,19].

The study of the QC model in the unbounded-error setting has been developed implicitly as the study of the sign-representing polynomial (say, [4,7]) since Beals et al. [6] gave the nice characterization of the (quantum) QC by polynomials. In fact, Buhrman et al. [10] mentioned the close relation between sign-representing polynomials and QCs of Boolean functions. However, there is no explicit literature on unbounded-error quantum QC.

Our Results. In this paper we deal with the unbounded-error quantum QC and study its relations to the other unbounded-error concepts. In Section 3, we show that, as in the case of CC, the unbounded-error quantum QC of some (total/partial) Boolean function is always exactly half of its classical counterpart. In Section 4, we discuss the relation between the unbounded-error quantum QC and CC. A powerful result by Buhrman, Cleve and Wigderson [9] is often used to "reduce" quantum CC to quantum QC, which is a "black-box" approach to convert quantum query algorithms into communication protocols with $O(\log n)$ overhead. It is a natural question whether their black-box approach is *optimal*, that is, $\Omega(\log n)$ overhead is inevitable. We show that the overhead of the black-box approach of [9] is optimal in the unbounded-error setting. Moreover, we show that this bound on overhead factor also holds under nondeterministic and exact settings. In Section 5, we develop the weakly unbounded-error QC, which is a natural measure to trade off queries and success probability, as the correspondence of the weakly unbounded-error CC. We show a multiplicative separation, $T(n)$ vs. $\Omega(T(n)\log(n/T(n)))$ for any monotone increasing function satisfying $T(n) \leq n$, between the weakly unbounded-error quantum and classical QCs of some *partial* function. This result contrasts with the only constant quantum-classical gaps of the weakly unbounded-error CC [18,19] as well as the unbounded-error QC.

Related Work. In a similar direction, de Wolf [26] characterized the nondeterministic (or *one-sided unbounded-error*) quantum QC and CC by, respectively, *nondeterministic degree* of approximating polynomials and *nondeterministic rank* of communication matrices. When comparing classical and quantum complexities under these models, de Wolf showed strong separations; an unbounded gap for QC and an exponential gap for CC (the first unbounded gap for CC was shown before in [20]). Under a different (certificate based) type of nondeterminism, a quadratic separation between quantum and classical CC is known for some total function [15].

2   Definition and Models

We first list some useful definitions, starting with unbounded-error polynomials.

Definition 1. *Let $f : \{0,1\}^n \rightarrow \{0,1\}$ be a Boolean function of n variables, and $q : \{0,1\}^n \rightarrow [0,1]$ be a real multilinear polynomial. We say that q is an unbounded-error polynomial for f if for any $x \in \{0,1\}^n$, $q(x) > 1/2$ if $f(x) = 1$ and $q(x) < 1/2$ if $f(x) = 0$. We denote the lowest degree among all unbounded-error polynomials for f as $udeg(f)$.*

Note that this definition is given in terms of total Boolean functions, but we can naturally extend it to partial functions. Throughout this paper, we use the term "Boolean functions" for results that hold for both partial and total functions; if not, we mention it explicitly.

There have been many studies and extensive results in the literature on polynomials that *sign-represent* Boolean functions [4,7,21]. A polynomial $p :$ $\{0,1\}^n \rightarrow \mathcal{R}$ is said to sign-represent f if $p(x) > 0$ whenever $f(x) = 1$, and $p(x) < 0$ whenever $f(x) = 0$. If $|p(x)| \leq 1$ for all x, we say that p is *normalized*. The *bias* of a normalized polynomial p is defined as $\beta = \min_x |p(x)|$. Denoting the minimum degree of polynomials that sign-represent f as $sdeg(f)$, it is easy to see that $udeg(f) = sdeg(f)$. In the following, we will use some results about polynomials that sign-represent f to characterize the unbounded-error QC.

It is folklore that every real multilinear polynomial q of degree at most d can be represented in the so-called *Fourier basis*. Namely,

$$q(x) = \sum_{S \in S_d} \hat{q}(S)(-1)^{x_S}, \tag{1}$$

where S_d denotes the set of all index sets $S \subseteq [n]$ of at most d variables, and x_S denotes the XOR (or PARITY) of the bits of x on index set S, namely, $x_S = \oplus_{i \in S} x_i$.

Next, we give the definitions of unbounded-error QC and CC, as well as their weak counterparts.

Definition 2. *Let $UQ(f)$ and $UC(f)$ be the unbounded-error quantum and classical, respectively, QCs of a Boolean function f. Namely, $UQ(f)$ (resp. $UC(f)$) is the minimum number of quantum (resp. classical) queries to a black box that holds the input $x \in \{0,1\}^n$ to f such that f can be computed with a success probability greater than $1/2$. Let $UQ_{cc}(g)$ and $UC_{cc}(g)$ be the unbounded-error quantum and classical, respectively, CCs of a distributed Boolean function $g : A \times B \rightarrow \{0,1\}$, where $A \subseteq \{0,1\}^{n_1}$ and $B \subseteq \{0,1\}^{n_2}$ denote the sets of inputs each given to Alice and Bob, respectively. Namely, $UQ_{cc}(g)$ (resp. $UC_{cc}(g)$) is the minimum number of quantum (resp. classical) bits exchanged between Alice and Bob to compute g with success probability greater than $1/2$.*

Define the bias β of a quantum or classical query algorithm (resp. communication protocol) which succeeds with probability $p > 1/2$ as $p - 1/2$. Then the weakly unbounded-error cost of such an algorithm (resp. protocol) is equal to the number of queries (resp. communicated bits or qubits) plus $\log 1/2\beta$. Let $WUQ(f)$, $WUC(f)$, $WUQ_{cc}(g)$ and $WUC_{cc}(g)$ be the weakly unbounded-error counterparts of the previous measures, given by the minimum weakly unbounded cost over all quantum or classical query algorithms and communication protocols, respectively.

Notice that some previous work defines the weakly unbounded-error cost as the number of queries plus $\log 1/\beta$ [10,18]. However, we prefer the present definition, as it ensures that weakly unbounded QC or CC is never greater than its exact counterpart.

3　Unbounded-Error Quantum and Classical QCs

In [18], it was shown that $UQ_{cc}(f)$ is always almost half of $UC_{cc}(f)$ for any (partial or total) Boolean function f. We will show that in the unbounded-error QC model, the equivalent (and rather tighter) result – that quantum query complexity is always *exactly* half of its classical counterpart – also holds for any Boolean function. For this purpose, we need the following three lemmas. The first lemma, shown by Beals et al. [6], gives a lower bound on the number of queries in terms of the minimum degree of representing polynomials.

Lemma 1 ([6]). *The amplitude of the final basis states of a quantum algorithm using T queries can be written as a multilinear polynomial of degree at most T.*

The second lemma, shown by Beals et al. [6] and Farhi et al. [13], gives an exact quantum algorithm for computing the parity of n variables with just $n/2$ queries.

Lemma 2 ([6,13]). *Let $S \subseteq [n]$ be a set of indices of variables. There exists a quantum algorithm for computing x_S with $\lceil |S|/2 \rceil$ queries. That is, there exists a unitary transformation U_f which needs exactly $\lceil |S|/2 \rceil$ queries: for any $b \in \{0,1\}$,*

$$U_f|S\rangle|0^m\rangle|b\rangle = |S\rangle|\psi_S\rangle|b \oplus x_S\rangle,$$

where $|0^m\rangle$ and $|\psi_S\rangle$ are the workspace quantum registers before and after the unitary transformation, respectively.

The third lemma was shown recently by Buhrman et al. [10]. It turns out to be very useful in characterizing the unbounded-error QC of Boolean functions.

Lemma 3 ([10]). *Suppose that there exists a multilinear polynomial p of d-degree that sign-represents $f : \{0,1\}^n \to \{0,1\}$ with bias β. Define $N = \sum_{i=0}^{d} \binom{n}{i}$. Then there also exists a multilinear polynomial $q(x) = \sum_{S \in S_d} \hat{q}(S)(-1)^{x_S}$ of the same degree and bias $\beta/\sqrt{N}$ that sign-represents f such that $\sum_{S \in S_d} |\hat{q}(S)| = 1$.*

Now, we are ready to prove the exact relation between UQ and UC.

Theorem 1. *For any Boolean function $f : X \to \{0,1\}$ such that $X \subseteq \{0,1\}^n$, it holds that: $UQ(f) = \left\lceil \frac{UC(f)}{2} \right\rceil = \left\lceil \frac{udeg(f)}{2} \right\rceil$.*

Proof. $[UC(f) = udeg(f)]$ This follows from a result in Buhrman et al. [10]: an unbounded-error randomized algorithm for f using d queries is equivalent to a d-degree polynomial p that sign-represents f.

$[UQ(f) \geq udeg(f)/2]$ By Lemma 1, the acceptance probability of a quantum algorithm using $UQ(f)$ queries can be written as a multilinear polynomial of degree at most $2UQ(f)$. Hence, $udeg(f) \leq 2UQ(f)$.

$[UQ(f) \leq \lceil udeg(f)/2 \rceil]$ This follows from Lemmas 3 and 2. First, let $\delta(y) = 1$ if $y > 0$, and $\delta(y) = 0$ otherwise. With regard to the Fourier representation of polynomial p that sign-represents f as in Eq. (1), and for a fixed $x \in \{0,1\}^n$, we can write $p(x) = \sum_{S \in S_{udeg(f)}} \hat{p}(S)(-1)^{x_S} = \sum_{S \in S_x^+} |\hat{p}(S)| - \sum_{S \in S_x^-} |\hat{p}(S)|$, such that $S_x^+ = \{S | x_S \oplus \delta(\hat{p}(S)) = 1\}$, and $S_x^- = \{S | x_S \oplus \delta(\hat{p}(S)) = 0\}$. By Lemma 3, we can assume that $\sum_{S \in S_{udeg(f)}} |\hat{p}(S)| = 1$.

Then, we have $\sum_{S \in S_x^+} |\hat{p}(S)| > 1/2$ if $f(x) = 1$, and $\sum_{S \in S_x^+} |\hat{p}(S)| < 1/2$ otherwise. Thus, the unbounded-error quantum algorithm for f can be obtained by computing x_S (by Lemma 2 with $\lceil |S|/2 \rceil \leq \lceil udeg(f)/2 \rceil$ queries), and its XOR with $\delta(\hat{p}(S))$, and measure its result. $\qquad \square$

From Theorem 1 and classical results in [12], we immediately obtain the following corollary, which implies that almost every function has unbounded-error quantum QC $n(1/4 + o(1))$. By contrast, there remains a gap in the bounded-error setting: Almost every function has bounded-error quantum QC between $n/4 + \Omega(\sqrt{n})$ [2,12] and $n/2 + O(\sqrt{n})$ [11].

Corollary 1. *Almost every function $f : \{0,1\}^n \to \{0,1\}$ has unbounded-error quantum QC bounded by $n/4 \leq UQ(f) \leq n/4 + O(\sqrt{n \log n})$.*

4   Tightness of Reducing CC to QC

Buhrman, Cleve and Wigderson [9] gave a method for reducing a quantum communication protocol to a quantum query algorithm with $O(\log n)$ overhead, which we call the *BCW reduction*, as follows.

Theorem 2 ([9]). *Let $F : \{0,1\}^n \to \{0,1\}$, and $F^L : \{0,1\}^n \times \{0,1\}^n \to \{0,1\}$ denote the distributed function of F induced by the bitwise function $L : \{0,1\} \times \{0,1\} \to \{0,1\}$, that is defined by $F^L(x,y) = F(z)$ such that each bit of z is $z_i = L(x_i, y_i)$. If there is a quantum algorithm that computes F using T queries with some success probability, then there is a $T(2\log n + 4)$-qubit communication protocol for F^L, where Alice has input x and Bob has input y, with the same success probability.*

In the reverse direction, this implies that any lower bound C in the CC side is translated into a lower bound $\Omega(C/\log n)$ in the QC side. The BCW reduction is exact: the success probability of the communication protocol is the same as that of the query computation. In fact, [9] proved some interesting results using this reduction, such as the first non-trivial quantum protocol for the disjointness problem, which used $O(\sqrt{n}\log n)$ communication by a reduction to Grover's quantum search algorithm. This upper bound was eventually improved to $O(\sqrt{n})$, which matches the lower bound shown in [23], by Aaronson and Ambainis using ingenious techniques [1]. However, unlike the results of [9], those techniques seem to be limited only to specific functions such as disjointness.

Thus, it is of interest to know whether there exists a universal reduction similar to the BCW reduction but with $o(\log n)$ overhead and preserving the success probability. This might be achieved by designing new reduction methods. In addition, smaller overhead might be achieved by relaxing the success probability condition, that is, allowing the success probability of the resulting protocol to be significantly lower than that of the original algorithm. To look for the possibility of such a universal reduction, one can consider relations between quantum QC and CC by using such a reduction as a "black-box" under various settings for the required success probability.

Our result in this section is the optimality of the BCW reduction in the exact, nondeterministic, and unbounded-error settings, as showed in the following theorem. Here, for a Boolean function $F : \{0,1\}^n \to \{0,1\}$, the distributed function of F induced by the bitwise XOR (resp. AND), denoted by $F^\oplus$ (resp. $F^\wedge$), is defined by $F^\oplus(x,y) = F(x \oplus y)$ (resp. $F^\wedge(x,y) = F(x \wedge y)$).

Theorem 3. *Let $T(n)$ be a monotone increasing function satisfying $T(n) \le n$. The following hold:*

(1) Assume that there is a procedure $\mathcal{A}$ that, for any function $F : \{0,1\}^n \to \{0,1\}$, converts a nondeterministic (exact, resp.) quantum algorithm for F using $T(n)$ queries into a nondeterministic (exact, resp.) quantum communication protocol for $F^\oplus$ using $O(T(n)D(n))$ qubits. Then, $D(n) = \Omega(\log(n/T(n)))$.

(2) Assume that there is a procedure $\mathcal{A}$ that, for any function $F : \{0,1\}^n \to \{0,1\}$, converts an unbounded-error quantum algorithm for F using $T(n)$ queries into an unbounded-error quantum communication protocol for $F^\wedge$ which uses $O(T(n)D(n))$ qubits. Then, $D(n) = \Omega(\log(n/T(n)))$.

The proof is based on the existence of Boolean functions whose quantum QCs are $T(n)$, while the CCs of their distributed counterparts are $\Omega(T(n)\log(n/T(n)))$, where n is the input length. We should remark that for bounded-error case, the same function for showing the optimality of BCW reduction in nondeterministic (exact) case can also be used to show that $\log\log n$ factor is necessary while we do not know if $\log n$ factor is necessary (i.e., BCW reduction is optimal).

4.1   Nondeterministic and Exact Cases

The following partial Boolean function, which is a variant of the Fourier Sampling problem of Bernstein and Vazirani [8], will be the base of the proof of the first part of Theorem 3.

Definition 3. *For $x, r \in \{0,1\}^m$, let F^r be a bit string of length $n = 2^m$ whose x-th bit is $F_x^r = \sum_i x_i \cdot r_i \bmod 2$. Let also g be another bit string of length n. The Fourier Sampling (FS) of F^r and g is defined by $\mathrm{FS}(F^r, g) = g_r$. When Alice and Bob are given (F^a, g) and (F^b, h), respectively, as their inputs where $a, b \in \{0,1\}^m$ and $g, h \in \{0,1\}^n$, the Distributed Fourier Sampling (DFS) on their inputs is $\mathrm{DFS}((F^a, g), (F^b, h)) = \mathrm{FS}(F^a \oplus F^b, g \oplus h)$.*

Now, let us consider the AND of $T = T(N)$ instances of FS, namely, $\mathrm{FS}(F^{r_1}, g^1) \wedge \ldots \wedge \mathrm{FS}(F^{r_T}, g^T)$ where $N = 2Tn$ is the length of the input string (note that n, the length of F^{r_i} and g^i, is a function of N). The proof of the first part of Theorem 3 is given in the following lemma.

Lemma 4. *The exact quantum QC of the AND of T instances of FS is $O(T)$, while the nondeterministic quantum CC of its distributed function induced by the bitwise XOR is $\Omega(T \log(N/T))$.*

Proof. For the QC part, note that for each instance of FS we can construct the following two-query quantum algorithm: Given input (F^r, g) (in a black-box), (i) Determine r with one query to F^r with certainty by the quantum algorithm of [8]. (ii) Output $\mathrm{FS}(F^r, g) = g_r$ with one query to g. Thus, the exact quantum QC of the AND of T instances of FS is $O(T)$.

For the CC part, we first prove that the nondeterministic (and hence exact) CC of DFS is $\Omega(\log n)$, and use this result for showing the lower bound of the AND of T instances of DFS. For this purpose, let us consider the set of inputs $((F^a, g), (F^b, h)) \in (\{0,1\}^n)^2 \times (\{0,1\}^n)^2$ such that $g \oplus h = 10^{n-1}$. For such inputs, $\mathrm{DFS}((F^a, g), (F^b, h)) = 1$ (resp. 0) implies $a = b$ (resp. $a \neq b$) since $\mathrm{DFS}((F^a, g), (F^b, h)) = \mathrm{FS}(F^a \oplus F^b, g \oplus h) = \mathrm{FS}(F^{a \oplus b}, 10^{n-1}) = (10^{n-1})_{a \oplus b}$ (the $(a \oplus b)$-th bit of 10^{n-1}). This means that if the nondeterministic CC of DFS is $o(\log n)$, then that of $\mathrm{EQ}_{\log n}$ (the equality on two $\log n$ bits a and b) is also $o(\log n)$, which contradicts the fact that the nondeterministic quantum CC of $\mathrm{EQ}_{\log n}$ is $\Omega(\log n)$ [26]. Next, notice that any protocol for the AND of T instances of DFS problem can also be used for $\mathrm{EQ}_{T \log n}$. Thus, its nondeterministic quantum CC should be $\Omega(T \log(N/T))$, as claimed. $\square$

4.2   Unbounded-Error Case

Here we show the proof of the second part of Theorem 3, that is, the impossibility of converting a quantum query algorithm into the corresponding communication protocol with $o(\log n)$ overhead, even with the success probability of the resulting protocol becomes very close to half. The base function is ODD-MAX-BIT function, a total Boolean function introduced in [7]. First, we give the definition of ODD-MAX-BIT$_n$ (or OMB$_n$ for short) and its distributed variant DOMB$_n$.

Definition 4. *For any $x \in \{0,1\}^n$, let us define* OMB$_n(x) = k \mod 2$, *where k is the largest index of x such that $x_k = 1$ ($k = 0$ for $x = 0^n$). For any $a, b \in \{0,1\}^n$, let us also define* DOMB$_n(a,b) =$ OMB$_n(a \wedge b)$, *where $a \wedge b$ is the bitwise AND of a and b.*

The proof of the second part of Theorem 3 follows from the complexities of the XOR of T instances of OMB and DOMB, as given in the following lemma.

Lemma 5. *The unbounded-error quantum QC of the XOR of T instances of OMB is $O(T)$, while the unbounded-error quantum CC of the distributed function induced by the bitwise AND is $\Omega(T \log(N/T))$ where $N = Tn$ is the input length.*

Proof. The QC part is easy since one instance of OMB$_n$ can be solved with only one query by the following classical algorithm: Query x_i with probability $p_i = \frac{2^i}{2^{n+1}-2}$. Then, output $i \mod 2$ if $x_i = 1$, and the result of a random coin flip if $x_i = 0$. It can be seen that the success probability is always bigger than $1/2$ for all positive integers n. It is not difficult to see that if each instance can be solved with probability more than half, so can the XOR of T instances.

For the CC part, we first show $UQ_{cc}(\mathrm{DOMB}_n) \geq (\log n - 3)/2$, and use this result for proving the lower bound of the XOR of T instances of DOMB. The bound for $UQ_{cc}(\mathrm{DOMB}_n) \geq (\log n - 3)/2$ follows from the lower bound on quantum random access coding (which is also known as the INDEX function) shown in [17]. For $a \in \{0,1\}^n$ and $b \in \{0,1\}^{\log n}$, INDEX$_n(a,b)$ is defined as the value of the b-th bit of a, or a_b. Then, we consider the case when Alice uses $x = a_1 0 a_2 0 \ldots a_n 0$, and Bob uses $y = y_1 y_2 y_3 \ldots y_{2n}$ such that $y_j = 1$ iff $j = 2b - 1$, as inputs to the protocol for DOMB$_{2n}$. Clearly, DOMB$_{2n}(x,y) =$ INDEX$_n(a,b)$. However, according to [17], $UQ_{cc}(\mathrm{INDEX}_n) \geq \frac{1}{2}\log(n+1) - 1$ (Here, -1 comes from the difference between two-way and one-way CC [18]). Therefore, $UQ_{cc}(\mathrm{DOMB}_n) \geq \frac{1}{2}\log(n/2+1) - 1 \geq (\log n - 3)/2$.

Next, we show that the XOR of T instances of DOMB$_n$ can be used to compute the inner product of two distributed $T\log(n/2)$ bits. Let Alice and Bob's inputs for the inner product be x and y, respectively. To compute the answer $\sum_{i=1}^{T\log(n/2)} x_i y_i \mod 2$, they divide their input strings into T parts, each of length $\log(n/2)$ bits, and for each part they compute the inner product, which is done by the reduction to INDEX$_{n/2}$ (and hence DOMB$_n$) as follows: Alice writes the inner product of her part with all possible Bob's part, which results in a bit string of length $n/2$ as her input to INDEX$_{n/2}$. By setting the corresponding Bob's part as the other input to INDEX$_{n/2}$, they can compute the inner product for each part by applying the protocol for INDEX$_{n/2}$. Since the quantum CC lower bound of the inner product on distributed inputs of $T\log(n/2)$ bits is $\Omega(T\log(n/2)) = \Omega(T\log(N/T))$ [14,18], we obtain the desired result. $\square$

5   Weakly Unbounded-Error Quantum and Classical QCs

In this section, we study the weakly unbounded-error QC (WUQ). There are two reasons why this model, which at first sight seems somewhat contrived, may be of interest: (i) The separation between quantum and classical QCs appears to be different for different success probability settings. For example, the best known separation for a total function is quadratic in the bounded-error setting, but we showed earlier that only a factor of two is possible in the unbounded-error setting. The WUQ model gives a natural way to trade off queries and success probability. (ii) Weakly unbounded-error CC is closely related to the well-studied notion of *discrepancy* [19]. WUQ is thus a QC analogue of a natural CC quantity.

5.1   Tight Gaps between *WUQ* and *WUC* for Partial Functions

In the CC model, Klauck [19] showed that weakly unbounded-error quantum and classical CCs are within some constant factor (see also [18]). It turns out that the gap is a bit different in the QC model: there exists a Boolean function f such that its classical weakly unbounded-error QC is $\Omega(\log n)$-times worse than its quantum correspondence. To show this, we will use a probabilistic method requiring the following Chernoff bound lemma from Appendix A of [3].

Lemma 6. *Let* $S = \{X_i\}$ *be a set of* N *independent random variables with* $\Pr[X_i = 1] = \Pr[X_i = -1] = \frac{1}{2}$. *Then* $\Pr\left[\left|\sum_{i=1}^{N} X_i\right| > a\right] < 2e^{-a^2/2N}$.

Lemma 7. *There exists a partial Boolean function* f *such that* $WUC(f) = \Omega(\log n)$ *and* $WUQ(f) = 2$.

Proof. We will again consider the Fourier Sampling problem $\mathrm{FS}(F^a, g) = g_a$, which (as shown in Section 4) can be solved exactly with two quantum queries for any choice of g. For the classical lower bound, we fix a string g (to be determined shortly), and assume that g is already known, so the algorithm need only make queries to F^a.

We use the Yao minimax principle that the minimum number of queries required in the worst case for a randomized algorithm to compute some function f with success probability at least p for any input is equal to the maximum, over all distributions on the inputs, of the minimum number of queries required for a *deterministic* algorithm to compute f correctly on a p fraction of the inputs. Thus, in order to show a lower bound on the number of queries used by any randomized algorithm that succeeds with probability $1/2 + \beta$, it suffices to show a lower bound on the number of queries required for a deterministic algorithm to successfully output g_a for a $1/2 + \beta$ fraction of the functions F^a (under some distribution). We will use the uniform distribution over all strings F^a – recall that $F_x^a = \sum_i x_i \cdot a_i \mod 2$ – which are also known as *Hadamard codewords*.

Now consider a fixed deterministic algorithm which makes an arbitrary sequence of $\frac{1}{3} \log n$ distinct queries to F^a, and then guesses the bit g_a. Assume without loss of generality that the algorithm makes exactly $\frac{1}{3} \log n$ queries on all inputs and that $n^{1/3}$ is an integer. There are at most $n^{1/3}$ possible answers to the queries, dividing the set S of n inner product functions into at most $k \leq n^{1/3}$ non-empty subsets $\{S_i\}$, where $|S_i| \geq n^{2/3}$ for all i (this is because each query will either split the set of remaining functions exactly in half, or will do nothing). Each subset will contain between 0 and $|S_i|$ functions F^a such that $g_a = 0$, with the remainder of the functions having $g_a = 1$. For any i, define m_0^i (resp. m_1^i) as the number of remaining functions $F^a \in S_i$ such that $g_a = 0$ (resp. $g_a = 1$). To succeed on the largest possible fraction of the inputs, the deterministic algorithm should guess the value with which the majority of the remaining bit strings in the subset S_i picked out by the answers to the queries are associated. It is thus easy to see that this deterministic algorithm can succeed on at most a p fraction of the inputs, where $p = \frac{1}{2} + \frac{1}{2n} \sum_{i=1}^{k} |m_0^i - m_1^i|$. We now turn to finding a g such that this expression is close to $1/2$ for all possible deterministic algorithms.

Our string g will be picked uniformly at random from the set of all n-bit strings. This implies that, for an arbitrary *fixed* deterministic algorithm and for any i, m_0^i and m_1^i are random variables. Lemma 6 can thus be used to upper bound the fraction of the inputs on which this algorithm succeeds:

$$\Pr[p > 1/2 + \beta] \leq \Pr[\frac{1}{2n^{2/3}}|m_0^1 - m_1^1| > \beta] < 2e^{-2\beta^2 n^{2/3}},$$

where it is sufficient for the bound to consider a fixed i with $|S_i| = n^{2/3}$, w.l.o.g. assuming that this is true for $i = 1$. The remainder of the proof is a simple counting argument. We find a rough upper bound on the number of deterministic algorithms using exactly q queries on every input by noting that such an algorithm is a complete binary tree with $q + 1$ levels, where each leaf is labelled with 0 or 1 (corresponding to the output of the algorithm) and each internal node is labelled with a number from $[n]$ (corresponding to the input variable to query). There are thus fewer than $n^{2^{q+1}}$ deterministic algorithms using exactly q queries. For $q = \frac{1}{3}\log n$, there are fewer than $2^{2n^{1/3}\log n}$ algorithms. We can now use a union bound to determine an upper bound on the probability p', taken over random strings g, that *any* of these algorithms succeeds on a $1/2 + \beta$ fraction of the inputs.

$$\Pr[p' > 1/2 + \beta] < 2^{2n^{1/3}\log n + 1} e^{-2\beta^2 n^{2/3}} < 2e^{2n^{1/3}(\log n - \beta^2 n^{1/3})}.$$

Let us pick $\beta = n^{-1/7}$. It can easily be verified that $\Pr[p' > 1/2 + \beta] < 1$ for sufficiently large n, so there exists *some* g such that no classical algorithm that uses at most $\frac{1}{3}\log n$ queries can succeed on more than $1/2 + n^{-1/7}$ of the inputs. By Yao's principle, this implies that for this g, no randomized algorithm that uses at most $\frac{1}{3}\log n$ queries can solve $\mathrm{FS}(F^a, g)$ with a bias greater than $n^{-1/7}$. Therefore, we have the desired separation: $WUQ(\mathrm{FS}) = 2$ (by the proof of Lemma 4) while $WUC(\mathrm{FS}) = \Omega(\log n)$. $\qquad\square$

Indeed, we can use this function to obtain a more general multiplicative separation between WUC and WUQ.

Theorem 4. *Let $T(N)$ be any monotone increasing function satisfying $T(N) \leq N$. Then there exists a partial Boolean function g on N variables such that $WUC(g) = \Omega(T(N)\log(N/T(N)))$ and $WUQ(g) \leq T(N) + 2$.*

Proof. Let $k = T(N)$ and $n = N/T(N)$. For an arbitrary (partial/total) function $f(x_1, \ldots, x_n)$ on n bits, define a new function f^k on $N = nk$ bits by encoding each input bit x_i by the parity of k bits $(y_{i1}, \ldots, y_{ik})$, i.e. $f^k(y_{11}, \ldots, y_{nk}) = f(y_{11} \oplus \cdots \oplus y_{1k}, \ldots, y_{n1} \oplus \cdots \oplus y_{nk})$. By Lemma 2, $WUQ(f^k) \leq \lceil k/2 \rceil WUQ(f) \leq (k/2+1)WUQ(f)$. On the contrary, it is essentially immediate that $WUC(f^k) = k\,WUC(f)$ since no sequence of queries to fewer than k of the bits $(y_{i1}, \ldots, y_{ik})$ can guess the parity $(y_{i1} \oplus \cdots \oplus y_{ik})$ with probability $> 1/2$. Taking f to be the FS function, the theorem follows from Lemma 7. $\qquad\square$

This gap is asymptotically almost optimal, as we show with the following lemma.

Lemma 8. *For any function $f : \{0,1\}^n \to \{0,1\}$, $WUC(f) \leq 2WUQ(f)\log n$.*

Proof. Let $\mathcal{A}$ be an algorithm achieving $WUQ(f)$, i.e., $\mathcal{A}$ uses d queries and has the success probability $1/2 + \beta$ such that $WUQ(f) = d + \log(1/2\beta)$. By the result of [6], we know that there exists a polynomial that sign-represents f such that its degree is $2d$, and its bias is β. Now we can use Lemma 3, which says that given such a polynomial, we can produce a randomized algorithm using at most $2d$ queries with success probability at least $1/2 + \beta/\sqrt{n^d}$. This implies that $WUC(f) \leq 2d + \log(1/2\beta) + d\log n \leq 2WUQ(f)\log n$. $\qquad\square$

6   Concluding Remarks

We have given the tight quantum-classical gap between weakly unbounded-error QC for partial functions. We conjecture that for all total functions f, it holds that $WUC(f) = O(WUQ(f))$. Indeed, we have the following theorem (where $|x|$ denotes the Hamming weight of x) that proves the conjecture for threshold functions (a class of total Boolean functions that includes AND and OR). The proof is omitted due to space restriction.

Theorem 5. *For the threshold function defined by $TH_k(x) = 1$ iff $|x| > k$, $UC(TH_k) = UQ(TH_k) = 1$ and $WUC(TH_k) = WUQ(TH_k) = \Theta(\log n)$.*

Acknowledgements. AM was supported by the EC-FP6-STREP network QICS, and would like to thank Richard Low for helpful discussions. HN was supported in part by Scientific Research Grant, Ministry of Japan, 19700011.

References

1. Aaronson, S., Ambainis, A.: Quantum search of spatial regions. Theory of Computing 1, 47–79 (2005)
2. Ambainis, A.: A note on quantum black-box complexity of almost all Boolean functions. Inform. Process. Lett. 71, 5–7 (1999)
3. Alon, N., Spencer, J.: The probabilistic method. Discrete Mathematics and Optimization. Wiley Interscience, Hoboken (2000)

4. Aspnes, J., Beigel, R., Furst, M., Rudich, S.: The expressive power of voting polynomials. Combinatorica 14, 1–14 (1994)
5. Babai, L., Frankl, P., Simon, J.: Complexity classes in communication complexity. In: Proc. 27th FOCS, pp. 303–312 (1986)
6. Beals, R., Buhrman, H., Cleve, R., Mosca, M., de Wolf, R.: Quantum lower bounds by polynomials. J. ACM 48, 778–797 (2001)
7. Beigel, R.: Perceptrons, PP, and the polynomial hierarchy. Comput. Complexity 4, 339–349 (1994)
8. Bernstein, E., Vazirani, U.: Quantum complexity theory. SIAM J. Comput. 26, 1411–1473 (1997)
9. Buhrman, H., Cleve, R., Wigderson, A.: Quantum vs. classical communication and computation. In: Proc. 30th STOC, pp. 63–68 (1998)
10. Buhrman, H., Vereshchagin, N., de Wolf, R.: On computation and communication with small bias. In: Proc. 22nd CCC, pp. 24–32 (2007)
11. van Dam, W.: Quantum oracle interrogation: getting all information for almost half the price. In: Proc. 39th FOCS, pp. 362–367 (1998)
12. O'Donnell, R., Servedio, R.A.: Extremal properties of polynomial threshold functions. In: Proc. 18th CCC, pp. 3–12 (2003)
13. Farhi, E., Goldstone, J., Gutmann, S., Sipser, M.: A limit on the speed of quantum computation in determining parity. Phys. Rev. Lett. 81, 5442–5444 (1998)
14. Forster, J.: A linear lower bound on the unbounded error probabilistic communication complexity. J. Comput. Syst. Sci. 65, 612–625 (2002)
15. Le Gall, F.: Quantum weakly nondeterministic communication complexity. In: Královič, R., Urzyczyn, P. (eds.) MFCS 2006. LNCS, vol. 4162, pp. 658–669. Springer, Heidelberg (2006)
16. Halstenberg, B., Reischuk, R.: Relations between communication complexity classes. J. Comput. Syst. Sci. 41, 402–429 (1990)
17. Iwama, K., Nishimura, H., Raymond, R., Yamashita, S.: Unbounded-error oneway classical and quantum communication complexity. In: Arge, L., Cachin, C., Jurdziński, T., Tarlecki, A. (eds.) ICALP 2007. LNCS, vol. 4596, pp. 110–121. Springer, Heidelberg (2007)
18. Iwama, K., Nishimura, H., Raymond, R., Yamashita, S.: Unbounded-error classical and quantum communication complexity. In: Tokuyama, T. (ed.) ISAAC 2007. LNCS, vol. 4835, pp. 100–111. Springer, Heidelberg (2007)
19. Klauck, H.: Lower bounds for quantum communication complexity. SIAM J. Comput. 37, 20–46 (2007)
20. Massar, S., Bacon, D., Cerf, N., Cleve, R.: Classical simulation of quantum entanglement without local hidden variables. Phys. Rev. A 63, 052305 (2001)
21. Minsky, M.L., Papert, S.A.: Perceptrons. MIT Press, Cambridge (1988)
22. Paturi, R., Simon, J.: Probabilistic communication complexity. J. Comput. Syst. Sci. 33, 106–123 (1986)
23. Razborov, A.A.: Quantum communication complexity of symmetric predicates. Izvestiya Math. (English version) 67, 145–149 (2003)
24. Sherstov, A.: Halfspace matrices. In: Proc. 22nd CCC, pp. 83–95 (2007)
25. de Wolf, R.: Quantum Computing and Communication Complexity, University of Amsterdam (2001)
26. de Wolf, R.: Nondeterministic quantum query and communication complexities. SIAM J. Comput. 32, 681–699 (2003)

Super-Exponential Size Advantage of Quantum Finite Automata with Mixed States*

Rūsiņš Freivalds

Department of Computer Science, University of Latvia,
Raiņa bulvāris 29, Rīga, Latvia

Abstract. Quantum finite automata with mixed states are proved to be super-exponentially more concise rather than quantum finite automata with pure states. It was proved earlier by A.Ambainis and R.Freivalds that quantum finite automata with pure states can have exponentially smaller number of states than deterministic finite automata recognizing the same language. There was a never published "folk theorem" proving that quantum finite automata with mixed states are no more than super-exponentially more concise than deterministic finite automata. It was not known whether the super-exponential advantage of quantum automata is really achievable.

We use a novel proof technique based on Kolmogorov complexity to prove that there is an infinite sequence of distinct integers n such that there are languages L_n in a 4-letter alphabet such that there are quantum finite automata with mixed states with $2n + 1$ states recognizing the language L_n with probability $\frac{3}{4}$ while any deterministic finite automaton recognizing L_n needs to have at least $e^{O(nlnn)}$ states.

1 Introduction

A.Ambainis and R.Freivalds proved in [4] that for recognition of some languages the quantum finite automata can have smaller number of the states than deterministic ones, and this difference can even be exponential. The proof contained a slight non-constructiveness, and the exponent was not shown explicitly. For probabilistic finite automata exponentiality of such a distinction was not yet proved. The best (smaller) gap was proved by Ambainis [2]. The languages recognized by automata in [4] were presented explicitly but the exponent was not. In a very recent paper by R.Freivalds [13] the non-constructiveness is modified, and an explicit (and seemingly much better) exponent is obtained at the expense of having only non-constructive description of the languages used. Moreover, the best estimate proved in this paper is proved under assumption of the well-known Artin's Conjecture (1927) in Number Theory. [13] contains also a theorem that does not depend on any open conjectures but the estimate is worse, and the description of the languages used is even less constructive. This seems to be the first result in finite automata depending on open conjectures in Number Theory.

---

* This research is supported by Grant No.05.1528 from the Latvian Council of Science.

S.-H. Hong, H. Nagamochi, and T. Fukunaga (Eds.): ISAAC 2008, LNCS 5369, pp. 931–942, 2008.
© Springer-Verlag Berlin Heidelberg 2008

The following two theorems are proved in [13]:

Theorem 1. *Assume Artin's Conjecture. There exists an infinite sequence of regular languages $L_1, L_2, L_3, \ldots$ in a 2-letter alphabet and an infinite sequence of positive integers $z(1), z(2), z(3), \ldots$ such that for arbitrary j:*

1. *there is a probabilistic reversible automaton with ($z(j)$ states recognizing L_j with the probability $\frac{19}{36}$,*
2. *any deterministic finite automaton recognizing L_j has at least $(2^{1/4})^{z(j)} = = (1.1892071115\ldots)^{z(j)}$ states,*

Theorem 2. *There exists an infinite sequence of regular languages $L_1, L_2, L_3, \ldots$ in a 2-letter alphabet and an infinite sequence of positive integers $z(1), z(2), z(3), \ldots$ such that for arbitrary j:*

1. *there is a probabilistic reversible automaton with $z(j)$ states recognizing L_j with the probability $\frac{68}{135}$,*
2. *any deterministic finite automaton recognizing L_j has at least $(7^{\frac{1}{14}})^{z(j)} = = (1.1149116725\ldots)^{z(j)}$ states,*

The paper [13] concluded a long research on relative size of probabilistic and deterministic automata [8,9,10,11,16,2,15,13,14]. The two theorems above are formulated in [13] as assertions about reversible probabilistic automata. For probabilistic automata (reversible or not) it was unknown before the paper [13] whether the gap between the size of probabilistic and deterministic automata can be exponential. It is easy to re-write the proofs in order to prove counterparts of Theorems 1 and 2 for quantum finite automata with pure states. The aim of this paper is to prove a counterpart of these theorems for quantum finite automata with mixed states.

Quantum algorithms with mixed states were first considered by D.Aharonov, A.Kitaev, N.Nisan [1]. More detailed description of quantum finite automata with mixed states can be found in A.Ambainis, et al.[3].

The automaton is defined by the initial density matrix ρ_0. Every symbol a_i in the input alphabet is associated with a unitary matrix A_i. When the automaton reads the symbol a_i the current density matrix ρ is transformed into $A_i^* \rho A_i$. When the reading of the input word is finished and the end-marker \$ is read, the current density matrix ρ is transformed into $A_{end}^* \rho A_{end}$ and separate measurements of all states are performed. After that the probabilities of all the accepting states are totaled, and the probabilities of all the rejecting states are totaled.

Like quantum finite automata with pure states described by A.Kondacs and J.Watrous [19] we allow measurement of the accepting states and rejecting states after every step of the computation.

The main result in our paper is:

Theorem 3. *There is an infinite sequence of distinct integers n such that there are languages L'_n in a 4-letter alphabet such that there are quantum finite automata with mixed states with $2n + 1$ states recognizing the language L'_n with probability $\frac{3}{4}$ while any deterministic finite automaton recognizing L'_n needs to have at least $e^{O(n \ln n)}$ states.*

Proof is delayed till Section 5.

Since the numbers of the states for deterministic automata and quantum automata with pure states differ no more than exponentially, we have

Theorem 4. *There is an infinite sequence of distinct integers n such that there are languages L_n in a 4-letter alphabet such that there are quantum finite automata with mixed states with $2n+1$ states recognizing the language L_n with probability $\frac{3}{4}$ while any quantum finite automaton with pure states recognizing L_n with bounded error needs to have at least $e^{O(n \ln n)}$ states.*

2   Permutations and Hamming Distance

Permutation of the set N_n is a 1-1 correspondence from N_n onto itself. Let f be such a permutation. The fact that it is *onto* means that for any $k \in N_n$ there exists $i \in N_n$ such that $f(i) = k$.

We need a notion similar to Hamming codes for permutations. Since Hamming distance between permutations is already considered in several well-known textbooks (e.g. [6]), it seemed natural that the corresponding theory might be already published. Very far from truth!

Hamming distance between two objects is the number of changes one needs to perform to obtain one object from the. The Hamming distance between two binary words (of the same length) is defined to be the number of positions at which they differ. For instance, we consider a set of three binary words $\{0011, 0110, 1100\}$. The first word is at Hamming distance 3 from the other two. Additionally, every word in this set is at Hamming distance at least 2 from any other. Such systems of words are called codes. They are important because they allow us to eliminate accidental errors when transmitting the words through noisy information channels.

We consider Hamming distance between permutations. Hamming distance between the permutation s of the set $\{1, 2, 3, \cdots n\}$ and the permutation r of the same set is the number of distinct numbers i such that $s(i) \neq r(i)$. For instance, let s be a permutation of the set $\{1, 2, 3, \cdots n\}$ and the number of it's fixed points be p. Then the Hamming distance between the permutation s and the identity permutation is the number $n - p$.

Lemma 1. *Let d be an arbitrary real number such that $0 \leq d \leq 1$. No more than $2^{dn \ln n}$ permutations can be on Hamming distance less or equal than dn from the identity permutation.*

Proof. By Stirling formula, $n! = e^{n.\ln n - o(n \ln n)}$. Let π be an arbitrary n-permutation. How many there are distinct n-permutations differing from the permutation π in no more than dn positions? The differing positions can be chosen in

$$\leq \binom{n}{d} < 2^n$$

ways and these $\leq dn$ positions are permuted. Hence there are no more than $2^n.2^{dn \ln n - o(n \ln n)} \leq 2^{dn \ln n}$ permutations of this type.

Theorem 5. *For arbitrary constant $c < 1$ such that for arbitrary n there is a set G_n of n-permutations containing $e^{\Omega(n \log n)}$ permutations such that the Hamming distance of any two permutations is at least $c \cdot n$.*

Proof. Immediately from Lemma 1.

3   Permutations and Automata

Definition 1. *The* Hamming distance *or simply* distance *$d(r, s)$ between two n-permutations r and s on the set S is the number of elements $x \in$ S such that $r(x) \neq s(x)$. The* similarity *$e(r, s)$ is the number of $x \in$ S such that $r(x) = s(x)$. Note that $d(r, s) + e(r, s) = |\text{S}| = n$.*

Theorem 6. *Let c be a fixed constant and let there be an infinite sequence of distinct integers n such that for each n there exists a group G_n of permutations of the set $\{1, 2, \ldots, n\}$, the group has order(G_n) elements and k generating elements, and the Hamming distance of any two permutations is at least $c \cdot n$. Then there is an infinite sequence of distinct integers n such that for each n there is a language L_n in a k-letter alphabet that can be recognized with probability $\frac{c}{2}$ by a quantum finite automata with mixed states that has $2n$ states, while any deterministic finite automaton recognizing L_n must have at least order(G_n) states.*

Proof. For each permutation group G_n we define the language L_n as follows:

The letters of L_n are the k generators of the group G_n and it consists of words $s_1 s_2 s_3 \ldots s_m$ such that the product $s_1 \circ s_2 \circ s_3 \circ \cdots \circ s_m$ differs from the identity permutation.

(A) Any deterministic automaton recognizing L_n is to remember the first input letter by a specific state.

(B) We will construct a quantum automaton with mixed states. It has $4n$ states and the initial density matrix ρ_0 is a diagonal block-matrix that consists of n blocks $\widetilde{\rho}_0$:

$$\widetilde{\rho}_0 = \frac{1}{2n} \begin{pmatrix} 1 & 1 & 0 & 0 \\ 1 & 1 & 0 & 0 \\ 0 & 0 & 0 & 0 \\ 0 & 0 & 0 & 0 \end{pmatrix}$$

For each of k generators $g_i \in G_n$ we will construct the corresponding unitary matrix U_i as follows – it is a $2n \times 2n$ permutation matrix, that permutes the elements in the even positions according to permutation g_i, but leaves the odd positions unpermuted.

For example, $g = 3241$ can be expressed as the following permutation matrix that acts on a column vector:

$$g = \begin{pmatrix} 0 & 0 & 1 & 0 \\ 0 & 1 & 0 & 0 \\ 0 & 0 & 0 & 1 \\ 1 & 0 & 0 & 0 \end{pmatrix}$$

The initial density matrix ρ_0 for $n = 4$ and the unitary matrix U that corresponds to the permutation matrix (3) of permutation g are as follows:

$$\rho_0 = \frac{1}{8}\begin{pmatrix} 1&1&0&0&0&0&0&0 \\ 1&1&0&0&0&0&0&0 \\ 0&0&1&1&0&0&0&0 \\ 0&0&1&1&0&0&0&0 \\ 0&0&0&0&1&1&0&0 \\ 0&0&0&0&1&1&0&0 \\ 0&0&0&0&0&0&1&1 \\ 0&0&0&0&0&0&1&1 \end{pmatrix}, U = \begin{pmatrix} 1&0&0&0&0&0&0&0 \\ 0&0&0&0&0&1&0&0 \\ 0&0&1&0&0&0&0&0 \\ 0&0&0&1&0&0&0&0 \\ 0&0&0&0&1&0&0&0 \\ 0&0&0&0&0&0&0&1 \\ 0&0&0&0&0&0&1&0 \\ 0&1&0&0&0&0&0&0 \end{pmatrix}$$

The unitary matrix $U_\$$ for the end-marker is also a diagonal block-matrix. It consists of n blocks that are the *Hadamard matrices*

$$\widetilde{H} = \frac{1}{\sqrt{2}}\begin{pmatrix} 1 & 1 \\ 1 & -1 \end{pmatrix}$$

Notice how the Hadamard matrix $\widetilde{H}$ acts on two specific 2×2 density matrices:

$$\text{if } \rho = \frac{1}{2n}\begin{pmatrix} 1&1 \\ 1&1 \end{pmatrix}, \text{ then } \widetilde{H}\rho\widetilde{H}^\dagger = \frac{1}{2n}\begin{pmatrix} 2&0 \\ 0&0 \end{pmatrix},$$

$$\text{if } \rho = \frac{1}{2n}\begin{pmatrix} 1&0 \\ 0&1 \end{pmatrix}, \text{ then } \widetilde{H}\rho\widetilde{H}^\dagger = \frac{1}{2n}\begin{pmatrix} 1&0 \\ 0&1 \end{pmatrix}.$$

For example, when the letter g is read, the unitary matrix U is applied to the density matrix ρ_0 (both are given in equation (3)) and the density matrix $\rho_1 = U\rho_0 U^\dagger$ is obtained. When the end-marker "$\$$" is read, the density matrix becomes $\rho_\$ = U_\$\rho_1 U_\$^\dagger$. Matrices ρ_1 and $\rho_\$$ are as follows:

$$\rho_1 = \frac{1}{8}\begin{pmatrix} 1&0&0&0&0&0&0&1 \\ 0&1&0&0&1&0&0&0 \\ 0&0&1&1&0&0&0&0 \\ 0&0&1&1&0&0&0&0 \\ 0&1&0&0&1&0&0&0 \\ 0&0&0&0&0&1&1&0 \\ 0&0&0&0&0&1&1&0 \\ 1&0&0&0&0&0&0&1 \end{pmatrix}, \rho_\$ = \frac{1}{8}\begin{pmatrix} 1 & 0 & 0 & 0 & \frac{1}{2} & \frac{1}{2} & \frac{1}{2} & -\frac{1}{2} \\ 0 & 1 & 0 & 0 & -\frac{1}{2} & -\frac{1}{2} & \frac{1}{2} & -\frac{1}{2} \\ 0 & 0 & 2 & 0 & 0 & 0 & 0 & 0 \\ 0 & 0 & 0 & 0 & 0 & 0 & 0 & 0 \\ \frac{1}{2} & -\frac{1}{2} & 0 & 0 & 1 & 0 & \frac{1}{2} & \frac{1}{2} \\ \frac{1}{2} & -\frac{1}{2} & 0 & 0 & 0 & 1 & -\frac{1}{2} & -\frac{1}{2} \\ \frac{1}{2} & \frac{1}{2} & 0 & 0 & \frac{1}{2} & -\frac{1}{2} & 1 & 0 \\ -\frac{1}{2} & -\frac{1}{2} & 0 & 0 & \frac{1}{2} & -\frac{1}{2} & 0 & 1 \end{pmatrix}.$$

Finally, we declare the states in the even positions to be accepting, but the states in the odd positions to be rejecting. Therefore one must sum up the diagonal entries that are in the even positions of the final density matrix to find the probability that a given word is accepted.

In our example the final density matrix $\rho_\$$ is given in (3). It corresponds to the input word "$g\$$", which is accepted with probability $\frac{1}{8}(1 + 0 + 1 + 1) = \frac{3}{8}$ and rejected with probability $\frac{1}{8}(1 + 2 + 1 + 1) = \frac{5}{8}$. Note that the accepting and rejecting probabilities sum up to 1.

It is easy to see, that the words that do not belong to the language L_n are rejected with certainty, because the matrix $U_\$ \rho_0 U_\$^\dagger$ has all zeros in the even positions on the main diagonal. However, the words that belong to L_n are accepted with the probability at least $\frac{d}{2n} = \frac{cn}{2n} = \frac{c}{2}$, because all permutations are at least at the distance d from the identity permutation.

It is also easy to see that any deterministic automaton that recognizes the language L_n must have at least $order(G_n)$ states. If the number of states is less than $order(G_n)$, then there are two distinct words u and v such that the deterministic automaton ends up in the same state no matter which one of the two words it reads. Since G_n is a group, for each word we can find an inverse, that returns the automaton in the initial state (the only rejecting state). Since u and v are different, they have different inverses and $u \circ u^{-1}$ is the identity permutation and must be rejected, but $v \circ u^{-1}$ is not the identity permutation and must be accepted – a contradiction. $\square$

4   Super-Exponential Size Advantage

Now we wish to prove a theorem differing from Theorem 3 only in one way. In this section we allow the alphabets of the languages L_n to grow with n.

Consider the following infinite sequence of languages. For every n take the set G_n considered in Theorem 5. The language L'_n consists of all the words aa (of the length 2) where a is a symbol for an arbitrary element from G_n. Hence there are $e^{\Omega(n \log n)}$ letters in the alphabet of the language L'_n and equally many words in L'_n.

Theorem 7. *There is an infinite sequence of distinct integers n such that there are languages L'_n such that there are quantum finite automata with mixed states with $2n+1$ states recognizing the language L'_n with probability $\frac{3}{4}$ while any deterministic finite automaton recognizing L'_n needs to have at least $e^{O(nlnn)}$ states.*

Proof is similar to the proof of Theorem 6.

It would be nice to improve our theorem in the natural way by finding algebraic groups G_n of n-permutations such that they are generated by a small (constant) number of generating elements and at the same time the order of G_n (the number of elements) would be superexponentially large with the respect to n. This would give us a natural proof for Theorem 3. Unfortunately, there exists a paper [17] by J.Kempe et al. where it is proven that large Hamming distance between all the n-permutations in a group implies that the order of the group is no more than exponential with the respect to n.

5   4-Letter Input Alphabet

We consider 2 algebraic groups in this Section. The set of all permutations of the set $1, ..., n$ with algebraic operation "product of permutations" can be considered as the group G_1. This group has two generating elements, the permutations

$(123\cdots n)$ and $(12)(3)(4)\cdots(n)$. The other group G_2 is defined as follows. Two binary strings α_1 and α_2 , each one of the length $nlog_2n$ are taken and two real-valued unitary matrices are constructed. The corresponding unitary operators are rotations around axes depending, correspondingly, on α_1 and α_2. The angle of rotation is a multiple of $\frac{2\pi}{p}$ where p is a prime slightly exceeding n^3. The inverses of these operators again are unitary, and their matrices are real-valued. The group direct product $G_1 \times G_2$ is again a group G with 4 generating elements and the elements of this group can be represented by n-dimensional unitary operators. G has the following properties.

Property 1. For arbitrary fixed pair (α_1, α_2) distinct n-permutations generate distinct unitary operators.

Property 2. For arbitrary fixed n-permutation distinct pairs (α_1, α_2) generate distinct unitary operators.

Property 3. The n-permutation and the pair (α_1, α_2) is transformed into the unitary operator described above by an algorithmic process.

Property 4. The n-permutation and the unitary operator can be transformed into the pair (α_1, α_2) by an algorithmic process.

These properties will be used below to prove the main result, to prove Theorem 3. At first we construct a quantum finite automaton with mixed states having 2n states. This automaton has the following initial density matrix (we use an 8×8-matrix instead of formal description of $n \times n$-matrix for the sake of easier readability).

$$\rho_0 = \begin{pmatrix} \frac{1}{8} & 0 & 0 & 0 & \frac{1}{8} & 0 & 0 & 0 \\ 0 & \frac{1}{8} & 0 & 0 & 0 & \frac{1}{8} & 0 & 0 \\ 0 & 0 & \frac{1}{8} & 0 & 0 & 0 & \frac{1}{8} & 0 \\ 0 & 0 & 0 & \frac{1}{8} & 0 & 0 & 0 & \frac{1}{8} \\ \frac{1}{8} & 0 & 0 & 0 & \frac{1}{8} & 0 & 0 & 0 \\ 0 & \frac{1}{8} & 0 & 0 & 0 & \frac{1}{8} & 0 & 0 \\ 0 & 0 & \frac{1}{8} & 0 & 0 & 0 & \frac{1}{8} & 0 \\ 0 & 0 & 0 & \frac{1}{8} & 0 & 0 & 0 & \frac{1}{8} \end{pmatrix}$$

At every step an input letter a_i (one of four possible input letters a_1, a_2, a_3 or a_4) is read from the input. Hence the density matrix ρ_x is changed into $U_i^* \rho_x U_i$. Hence if $x = x_1 x_2 x_3 \cdots x_t$ then the current U_x equals $U_t^* U_{t-1}^* \cdots U_2^* U_1^* \rho_0 U_1 U_2 \cdots U_{t-1} U_t$.

Let the matrix

$$U_x = \begin{pmatrix} u_{11} & u_{12} & \cdots & u_{1n} \\ u_{21} & u_{22} & \cdots & u_{2n} \\ u_{31} & u_{32} & \cdots & u_{3n} \\ \cdots & & & \\ u_{n1} & u_{n2} & \cdots & u_{nn} \end{pmatrix}$$

be the $n \times n$-matrix corresponding to the input word x. We associate a $2n \times 2n$-matrix M_x

$$M_x = \begin{pmatrix} 1 & 0 & \cdots & 0 & 0 & 0 & \cdots & 0 \\ 0 & 1 & \cdots & 0 & 0 & 0 & \cdots & 0 \\ 0 & 0 & \cdots & 0 & 0 & 0 & \cdots & 0 \\ & & \cdots & & & & & \\ 0 & 0 & \cdots & 1 & 0 & 0 & \cdots & 0 \\ 0 & 0 & \cdots & 0 & u_{11} & u_{12} & \cdots & u_{1n} \\ 0 & 0 & \cdots & 0 & u_{21} & u_{22} & \cdots & u_{2n} \\ 0 & 0 & \cdots & 0 & u_{31} & u_{32} & \cdots & u_{3n} \\ & & \cdots & & & & & \\ 0 & 0 & \cdots & 0 & u_{n1} & u_{n2} & u & \cdots & u_{nn} \end{pmatrix}$$

to it. We use the initial density matrix and the following property

$$\begin{pmatrix} 1 & 0 & 0 & 0 & 0 & 0 & 0 & 0 \\ 0 & 1 & 0 & 0 & 0 & 0 & 0 & 0 \\ 0 & 0 & 1 & 0 & 0 & 0 & 0 & 0 \\ 0 & 0 & 0 & 1 & 0 & 0 & 0 & 0 \\ 0 & 0 & 0 & 0 & u_{11} & u_{12} & u_{13} & u_{14} \\ 0 & 0 & 0 & 0 & u_{21} & u_{22} & u_{23} & u_{24} \\ 0 & 0 & 0 & 0 & u_{31} & u_{32} & u_{33} & u_{34} \\ 0 & 0 & 0 & 0 & u_{41} & u_{42} & u_{43} & u_{44} \end{pmatrix} \begin{pmatrix} \frac{1}{8} & 0 & 0 & 0 & \frac{1}{8} & 0 & 0 & 0 \\ 0 & \frac{1}{8} & 0 & 0 & 0 & \frac{1}{8} & 0 & 0 \\ 0 & 0 & \frac{1}{8} & 0 & 0 & 0 & \frac{1}{8} & 0 \\ 0 & 0 & 0 & \frac{1}{8} & 0 & 0 & 0 & \frac{1}{8} \\ \frac{1}{8} & 0 & 0 & 0 & \frac{1}{8} & 0 & 0 & 0 \\ 0 & \frac{1}{8} & 0 & 0 & 0 & \frac{1}{8} & 0 & 0 \\ 0 & 0 & \frac{1}{8} & 0 & 0 & 0 & \frac{1}{8} & 0 \\ 0 & 0 & 0 & \frac{1}{8} & 0 & 0 & 0 & \frac{1}{8} \end{pmatrix} \begin{pmatrix} 1 & 0 & 0 & 0 & 0 & 0 & 0 & 0 \\ 0 & 1 & 0 & 0 & 0 & 0 & 0 & 0 \\ 0 & 0 & 1 & 0 & 0 & 0 & 0 & 0 \\ 0 & 0 & 0 & 1 & 0 & 0 & 0 & 0 \\ 0 & 0 & 0 & 0 & u_{11} & u_{21} & u_{31} & u_{41} \\ 0 & 0 & 0 & 0 & u_{12} & u_{22} & u_{32} & u_{42} \\ 0 & 0 & 0 & 0 & u_{13} & u_{23} & u_{33} & u_{43} \\ 0 & 0 & 0 & 0 & u_{14} & u_{24} & u_{34} & u_{44} \end{pmatrix} =$$

$$= \begin{pmatrix} \frac{1}{8} & 0 & 0 & 0 & \frac{u_{11}}{8} & \frac{u_{21}}{8} & \frac{u_{31}}{8} & \frac{u_{41}}{8} \\ 0 & \frac{1}{8} & 0 & 0 & \frac{u_{12}}{8} & \frac{u_{22}}{8} & \frac{u_{32}}{8} & \frac{u_{42}}{8} \\ 0 & 0 & \frac{1}{8} & 0 & \frac{u_{13}}{8} & \frac{u_{23}}{8} & \frac{u_{33}}{8} & \frac{u_{43}}{8} \\ 0 & 0 & 0 & \frac{1}{8} & \frac{u_{14}}{8} & \frac{u_{24}}{8} & \frac{u_{34}}{8} & \frac{u_{44}}{8} \\ \frac{u_{11}}{8} & \frac{u_{12}}{8} & \frac{u_{13}}{8} & \frac{u_{14}}{8} & \frac{1}{8} & 0 & 0 & 0 \\ \frac{u_{21}}{8} & \frac{u_{22}}{8} & \frac{u_{23}}{8} & \frac{u_{24}}{8} & 0 & \frac{1}{8} & 0 & 0 \\ \frac{u_{31}}{8} & \frac{u_{32}}{8} & \frac{u_{33}}{8} & \frac{u_{34}}{8} & 0 & 0 & \frac{1}{8} & 0 \\ \frac{u_{41}}{8} & \frac{u_{42}}{8} & \frac{u_{43}}{8} & \frac{u_{44}}{8} & 0 & 0 & 0 & \frac{1}{8} \end{pmatrix}$$

to justify the correspondence between the $n \times n$-matrices U_x and the $2n \times 2n$-matrices M_x at every moment. When all the input word is read the end-marker $\$$ comes in, the unitary matrix M_x corresponding to the end-marker $\$$ is

$$M_\$ = \begin{pmatrix} \frac{1}{\sqrt{2}} & 0 & 0 & 0 & \frac{1}{\sqrt{2}} & 0 & 0 & 0 \\ 0 & \frac{1}{\sqrt{2}} & 0 & 0 & 0 & \frac{1}{\sqrt{2}} & 0 & 0 \\ 0 & 0 & \frac{1}{\sqrt{2}} & 0 & 0 & 0 & \frac{1}{\sqrt{2}} & 0 \\ 0 & 0 & 0 & \frac{1}{\sqrt{2}} & 0 & 0 & 0 & \frac{1}{\sqrt{2}} \\ \frac{1}{\sqrt{2}} & 0 & 0 & 0 & -\frac{1}{\sqrt{2}} & 0 & 0 & 0 \\ 0 & \frac{1}{\sqrt{2}} & 0 & 0 & 0 & -\frac{1}{\sqrt{2}} & 0 & 0 \\ 0 & 0 & \frac{1}{\sqrt{2}} & 0 & 0 & 0 & -\frac{1}{\sqrt{2}} & 0 \\ 0 & 0 & 0 & \frac{1}{\sqrt{2}} & 0 & 0 & 0 & -\frac{1}{\sqrt{2}} \end{pmatrix}$$

This matrix can be regarded as a block matrix with blocks of size 2×2. It is worth to remember that

$$\begin{pmatrix} \frac{1}{\sqrt{2}} & \frac{1}{\sqrt{2}} \\ \frac{1}{\sqrt{2}} & -\frac{1}{\sqrt{2}} \end{pmatrix} \begin{pmatrix} a_{11} & a_{12} \\ a_{21} & a_{22} \end{pmatrix} \begin{pmatrix} \frac{1}{\sqrt{2}} & \frac{1}{\sqrt{2}} \\ \frac{1}{\sqrt{2}} & -\frac{1}{\sqrt{2}} \end{pmatrix} = \begin{pmatrix} \frac{a_{11}+a_{21}+a_{12}+a_{22}}{2} & \frac{a_{11}+a_{21}-a_{12}-a_{22}}{2} \\ \frac{a_{11}-a_{21}+a_{12}-a_{22}}{2} & \frac{a_{11}-a_{21}-a_{12}+a_{22}}{2} \end{pmatrix}.$$

If the input word is such that the corresponding member of the free group G_1 equals the empty word then the final density matrix becomes

$$\rho_{final} = \begin{pmatrix} \frac{1}{4} & 0 & 0 & 0 & 0 & 0 & 0 & 0 \\ 0 & \frac{1}{4} & 0 & 0 & 0 & 0 & 0 & 0 \\ 0 & 0 & \frac{1}{4} & 0 & 0 & 0 & 0 & 0 \\ 0 & 0 & 0 & \frac{1}{4} & 0 & 0 & 0 & 0 \\ 0 & 0 & 0 & 0 & 0 & 0 & 0 & 0 \\ 0 & 0 & 0 & 0 & 0 & 0 & 0 & 0 \\ 0 & 0 & 0 & 0 & 0 & 0 & 0 & 0 \\ 0 & 0 & 0 & 0 & 0 & 0 & 0 & 0 \end{pmatrix}.$$

Otherwise the total of the first n elements on the main diagonal is less than 1.

At the end separate measurements of all states are performed. After that the probabilities of all the accepting states are totaled, and the probabilities of all the rejecting states are totaled. What is the probability of the acceptance in the case when the member of the free group G_1 does not equal the empty word? It depends on the values of (α_1, α_2). We will prove in the subsequent sections that if the Kolmogorov complexity of (α_1, α_2) is maximal then this probability is MUCH less than 1.

6   Kolmogorov Complexity

The theorems in this section are well-known results in spite of the fact that it is not easy to find exact references for all of them.

Definition 2. *We say that the numbering* $\Psi = \{\Psi_0(x), \Psi_1(x), \Psi_2(x), \ldots\}$ *of 1-argument partial recursive functions is* **computable** *if the 2-argument function* $U(n,x) = \Psi_n(x)$ *is partial recursive.*

Definition 3. *We say that a numbering* Ψ *is reducible to the numbering* η *if there exists a total recursive function* $f(n)$ *such that, for all n and x,* $\Psi_n(x) = \eta_{f(n)}(x)$.

Definition 4. *We say that a computable numbering* φ *of all 1-argument partial recursive functions is a* **Gödel numbering** *if every computable numbering (of any class of 1-argument partial recursive functions) is reducible to* φ.

Theorem. [7] There exists a Gödel numbering.

Definition 5. *We say that a Gödel numbering ϑ is a* **Kolmogorov numbering** *if for arbitrary computable numbering Ψ (of any class of 1-argument partial recursive functions) there exist constants $c > 0, d > 0$, and a total recursive function $f(n)$ such that:*

1. *for all n and x, $\Psi_n(x) = \vartheta_{f(n)}(x)$,*
2. *for all n, $f(n) \leq c \cdot n + d$.*

Kolmogorov Theorem. [18] There exists a Kolmogorov numbering.

7   Back to Automata

We denote by $U(\tau, \alpha)$ the pair of unitary matrices generated from the n-permutation τ and the binary string α of the length $2|\tau|log_2|\tau|$ (later divided into the pair (α_1, α_2)) each one of the length $|\tau|log_2|\tau|$.

There exist many distinct Kolmogorov numberings. We now fix one of them and denote it by η. Since Kolmogorov numberings give indices for all partial recursive functions, for arbitrary τ and n there is an i such that $\eta_i(\tau, n) = x$. Let $i(\tau, n)$ be the minimal i such that $\eta_i(\tau, n) = \alpha$. It is easy to see that if $x_1 \neq x_2$, then $i(\alpha_1, n) \neq i(\alpha_2, n)$. We consider all binary words α corresponding to n and denote by $\alpha(n)$ the word α such $i(\alpha, n)$ exceeds $i(\beta, n)$ for all binary words β corresponding to n different from α. It is obvious that $i \geq 2^{2n(log_2 n)-1}$.

Until now we considered generating matrices $U(\tau, \alpha)$ for independently chosen τ and α. From now on we consider only α corresponding to $i(\tau, n)$. We wish to prove that if n is sufficiently large, then Hamming distances between the vectors $\frac{1}{n}, \frac{1}{n}, \frac{1}{n}, \cdots, \frac{1}{n}$ and $a_{11}, a_{22}, \cdots, a_{nn}$ in the matrix ρ_{final} exceeds $\frac{1}{18}$.

We introduce a partial recursive function $\mu(z, n)$ defined as follows. Above when defining $U(\tau, \alpha)$ we considered auxiliary function $U(\tau, \alpha)$. To define $\mu(z, n)$ we consider all binary words α of the length $2|\tau|log_2|\tau|$. If z is not a binary word of such a length, then $\mu(z, n)$ is not defined. If z is a binary word of length $2|\tau|log_2|\tau|$, then we consider all $\alpha \in \{0, 1\}^{2|\tau|log_2|\tau|}$ such that $U(\tau, \alpha) = z$. If there are no such α, then $\mu(z, n)$ is not defined. If there is only one such α, then $\mu(z, n) = \alpha$.

If there are more than two such x, then $\mu(z, n)$ is not defined.

Now we introduce a computable numbering of some partial recursive functions. This numbering is independent of n.

For each n (independently from other values of n) we order the set of all the $2^{2|\tau|log_2|\tau|}$ binary words z of the length $2|\tau|log_2|\tau|$: $z_0, z_1, z_2, \ldots, z_{2^{2|\tau|log_2|\tau|}-1}$. We define z_0 as the word $000\ldots0$. The words $z_1, z_2, \ldots, z_{2^{2p}}$ are words with exactly one symbol 1. We strictly follow a rule "if the word z_i contains less symbols 1 than the word z_j, then $i < j$". Words with equal number of the symbol 1 are ordered lexicographically. Hence $z_{2^{2p}-1} = 111\ldots1$.

For each n, we define $w = |\tau|log_2|\tau|$ and

$$\Psi_0(n) = \mu(z_0, n)$$
$$\Psi_1(n) = \mu(z_0, n)$$
$$\Psi_2(n) = \mu(z_1, n)$$

$$\cdots$$

$$\Psi_{2^{2w+1}-2}(p) = \mu(z_{2^{2w}-1}, n)$$
$$\Psi_{2^{2w+1}-1}(p) = \mu(z_{2^{2w}-1}, n)$$

For $j \geq 2^{2w+1}$, $\Psi_j(w)$ is undefined.

We have fixed a Kolmogorov numbering η and we have just constructed a computable numbering Ψ of some partial recursive functions.

Lemma 2. *There exist constants $c > 0$ and $d > 0$ (independent of p) such that for arbitrary i there is a j such that*

1. $\Psi_i(t) = \eta_j(t)$ for all t, and
2. $j \leq ci + d$.

Proof. Immediately from Kolmogorov Theorem.

The rest of the proof of Theorem 3 is from the contrary. Suppose that n is sufficiently large (because Kolmogorov numbering is optimal only for large lengths of the words) and for $\alpha(n)$ the value of $a_{11} + a_{22} + \cdots + a_{nn}$ had exceeded $\frac{17}{18}$ in the unitary matrix $U(\tau, \alpha)$. Then our α could be uniquely and algorithmically be described in more concise way than our obvious assertion $i(\tau) \geq 2nlog_2n - 1$ allows. Contradiction.

This way we have proved that the languages L_n can be recognized by 2n-state quantum finite automata with mixed states so that every word in the language is accepted with the probability 1 and every word not in L_n is accepted with probability less than $\frac{17}{18}$. If we add one more state to the automata we can get automata with a cut-point $\frac{1}{2}$ and a uniformly bounded error.

It remains only to find the complexity (the number of states) of deterministic finite automata recognizing these languages. Theorem 5 above shows that there are $e^{O(nlnn)}$ n-permutations τ_j with pairwise Hamming distances no less than $const.n$. These n-permutations can be constructed using no more than n^3 generating elements $(123\cdots n)$ and $(12)(3)(4)\cdots(n)$. Input words corresponding to these $e^{O(nlnn)}$ n-permutations τ_j should be remembered by distinct states of any deterministic finite automaton recognizing the language.

Acknowledgments

I would like to thank Andris Ambainis for pointing at a possibility to use random unitary operators instead of permutations. In the final version of my proof the unitary operators are far from being random but this idea helped to move my proof off a deadlock.

References

1. Aharonov, D., Kitaev, A., Nisan, N.: Quantum circuits with mixed states. In: Proc. STOC 1998, pp. 20–30 (1998)
2. Ambainis, A.: The complexity of probabilistic versus deterministic finite automata. LNCS, vol. 1178, pp. 233–237. Springer, Heidelberg (1996)
3. Ambainis, A., Beaudry, M., Golovkins, M., Ķikusts, A., Mercer, M., Thérien, D.: Algebraic Results on Quantum Automata. Theory Comput. Syst. 39(1), 165–188 (2006)
4. Ambainis, A., Freivalds, R.: 1-way quantum finite automata: strengths, weaknesses and generalizations. In: Proc. IEEE FOCS 1998, pp. 332–341 (1998)
5. Artin, E.: Beweis des allgemeinen Reziprozitätsgesetzes. Mat. Sem. Univ. Hamburg B.5, 353–363 (1927)
6. Cameron, P.: Permutation groups. London Mathematical Society Student Texts series. Cambridge University Press, Cambridge (1999)
7. Ershov, Y.L.: Theory of numberings. In: Griffor, E.R. (ed.) Handbook of computability theory, pp. 473–503. North-Holland, Amsterdam (1999)
8. Freivalds, R.: Recognition of languages with high probability on different classes of automata. Doklady Akademii Nauk SSSR 239(1), 60–62 (1978) (Russian)
9. Freivalds, R.: Projections of languages recognizable by probabilistic and alternating finite multi-tape automata. Information Processing Letters 13(4–5), 195–198 (1981)
10. Freivalds, R.: On the growth of the number of states in result of the determinization of probabilistic finite automata. Avtomatika i Vichislitel'naya Tekhnika 3, 39–42 (1982) (Russian)
11. Freivalds, R.: Complexity of probabilistic versus deterministic automata. In: Barzdins, J., Bjorner, D. (eds.) Baltic Computer Science. LNCS, vol. 502, pp. 565–613. Springer, Heidelberg (1991)
12. Freivalds, R.: Languages Recognizable by Quantum Finite Automata. In: Farré, J., Litovsky, I., Schmitz, S. (eds.) CIAA 2005. LNCS, vol. 3845, pp. 1–14. Springer, Heidelberg (2006)
13. Freivalds, R.: Non-constructive methods for finite probabilistic automata. In: Harju, T., Karhumäki, J., Lepistö, A. (eds.) DLT 2007. LNCS, vol. 4588, pp. 169–180. Springer, Heidelberg (2007)
14. Freivalds, R.: Hamming, Permutations and Automata. In: Hromkovič, J., Královič, R., Nunkesser, M., Widmayer, P. (eds.) SAGA 2007. LNCS, vol. 4665, pp. 18–29. Springer, Heidelberg (2007)
15. Freivalds, R., Ozols, M., Mančinska, L.: Permutation Groups and the Strength of Quantum Finite Automata with Mixed States. In: Proceedings of the Workshop Probabilistic and Quantum Algorithms colocated with the 11th International Conference Developments in Language Theory DLT 2007, Turku, Finland, July 3-5, 2007, vol. 45, pp. 13–27. TUCS General Publication, No 45 (June 2007)
16. Kaņeps, J., Freivalds, R.: Running time to recognize nonregular languages by 2-way probabilistic automata. In: Leach Albert, J., Monien, B., Rodríguez-Artalejo, M. (eds.) ICALP 1991. LNCS, vol. 510, pp. 174–185. Springer, Heidelberg (1991)
17. Kempe, J., Pyber, L., Shalev, A.: Permutation groups, minimal degrees and quantum computing. Groups, Geometry, and Dynamics 1(4), 553–584 (2007)
18. Kolmogorov, A.N.: Three approaches to the quantitative definition of information. Problems in Information Transmission 1, 1–7 (1965)
19. Kondacs, A., Watrous, J.: On the power of quantum finite state automata. In: Proc. IEEE FOCS 1997, pp. 66–75 (1997)

Author Index